D0152694

PHYSIOLOGY OF SPORT AND EXERCISE

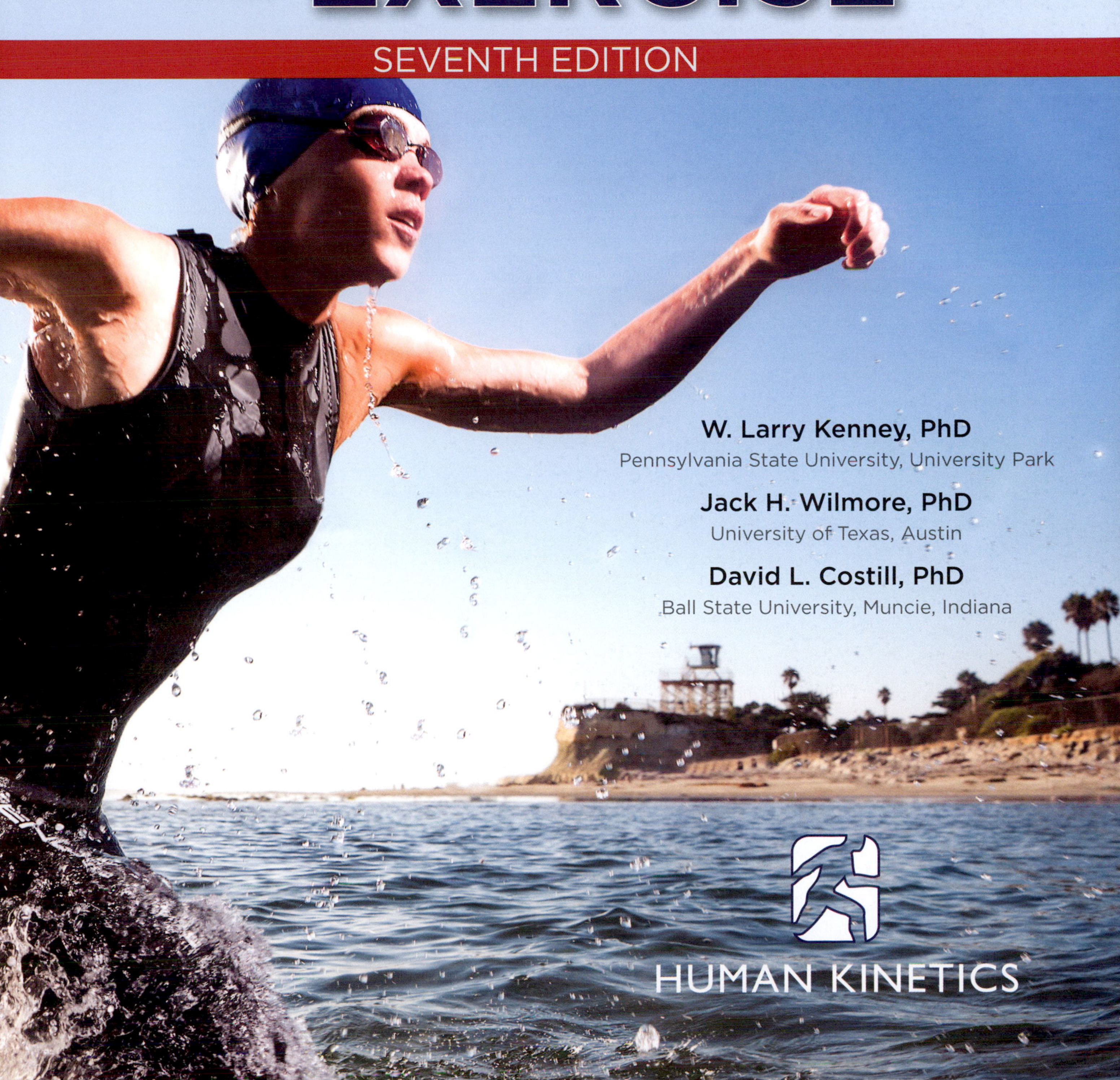

Library of Congress Cataloging-in-Publication Data

Names: Kenney, W. Larry, author. | Wilmore, Jack H., 1938-2014, author. | Costill, David L., author.
Title: Physiology of sport and exercise / W. Larry Kenney, Jack H. Wilmore, David L. Costill.
Description: Seventh edition. | Champaign, IL : Human Kinetics, [2020] | Includes bibliographical references and index.
Identifiers: LCCN 2018040753 (print) | LCCN 2018041421 (ebook) | ISBN 9781492574859 (epub) | ISBN 9781492589198 (PDF) | ISBN 9781492572299 (hardback)
Subjects: | MESH: Sports--physiology | Exercise--physiology | Physical Fitness--physiology | Physical Endurance--physiology
Classification: LCC QP301 (ebook) | LCC QP301 (print) | NLM QT 260 | DDC 612/.044--dc23
LC record available at https://lccn.loc.gov/2018040753

ISBN: 978-1-4925-7229-9 (hardback)
ISBN: 978-1-4925-7486-6 (loose-leaf)

Permission notices for material reprinted in this book from other sources can be found on page xix.

The web addresses cited in this text were current as of January 2019, unless otherwise noted.

Senior Acquisitions Editor: Amy N. Tocco; **Developmental Editor:** Judy Park; **Managing Editor:** Anna Lan Seaman; **Copyeditor:** Joy Hoppenot; **Indexer:** Alisha Jeddeloh; **Permissions Manager:** Dalene Reeder; **Senior Graphic Designer:** Nancy Rasmus; **Cover Designer:** Keri Evans; **Cover Design Associate:** Susan Rothermel Allen; **Photograph (cover):** AMR Image/Getty Images; **Photo Asset Manager:** Laura Fitch; **Visual Production Assistant:** Joyce Brumfield; **Photo Production Manager:** Jason Allen; **Senior Art Manager:** Kelly Hendren; **Illustrations:** © Human Kinetics, unless otherwise noted; **Printer:** Walsworth

Printed in the United States of America 10 9 8 7 6 5 4 3 2 1

The paper in this book was manufactured using responsible forestry methods.

Human Kinetics
P.O. Box 5076
Champaign, IL 61825-5076
Website: www.HumanKinetics.com

In the United States, email info@hkusa.com or call 800-747-4457.
In Canada, email info@hkcanada.com.
In the United Kingdom/Europe, email hk@hkeurope.com.

For information about Human Kinetics' coverage in other areas of the world, please visit our website: **www.HumanKinetics.com**

E7426 (hardback)/E7449 (loose-leaf)

Tell us what you think!
Human Kinetics would love to hear what we can do to improve the customer experience. Use this QR code to take our brief survey.

Jack H. Wilmore was an exceptional teacher, researcher, writer, and lecturer. His ability to communicate the complexities of exercise physiology to students, health professionals, and the general public is evident in this textbook. As the lead author of the first four editions of *Physiology of Sport and Exercise*, Jack took great pride in the clarity and accuracy of its contents. This book was his brainchild.

Jack began his career in exercise physiology at Ithaca College in New York. He then held professorships at the University of California at Berkley, University of California at Davis, University of Arizona, University of Texas, and Texas A&M. He published more than 300 scientific and lay articles, 15 books, and 55 chapters in other texts. In addition to serving as president of the American College of Sports Medicine (ACSM) and the American Academy of Kinesiology and Physical Education, Jack was active in many other professional organizations. His star status in sports medicine was rewarded with a long list of honors including the Citation and Honor Awards from ACSM. The achievements in his 50-year career are the basis for our current knowledge of the critical importance of regular physical activity in health, disease, and aging. His impact on students and the general public was the envy of all his colleagues. *Physiology of Sport and Exercise* lives on as a legacy of an exceptional scientist in sport and exercise and friend to many. He is missed by family, friends, and colleagues alike, and this book remains an enduring part of his legacy.

Jack H. Wilmore
April 23, 1938 - November 15, 2014

Contents

PART III Exercise Training

PART IV Environmental Influences on Performance

PART V Optimizing Performance in Sport

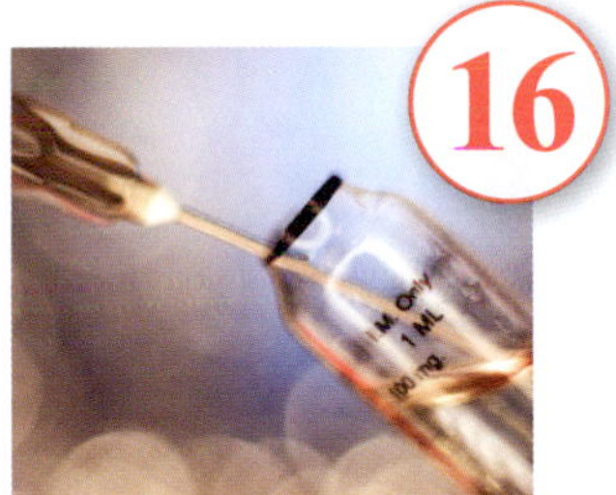

PART VI Age and Sex Considerations in Sport and Exercise

Research Perspectives Finder

Preface

Physiology is the study of how the human body functions. Cells, tissues, organs, and systems intricately and precisely communicate and integrate to coordinate the body's myriad physiological functions. Even at rest, the body is physiologically quite active. Imagine, then, how much more active all of these body systems become when you engage in exercise. During exercise, nerves excite muscles to contract. Exercising muscles are metabolically active and require more nutrients, more oxygen, and efficient clearance of waste products. The autonomic nervous system and endocrine glands combine to fine-tune these processes. How does the whole body respond to the increased physiological demands of physical activity in all its forms?

That is the key question when you study the physiology of sport and exercise. *Physiology of Sport and Exercise, Seventh Edition,* introduces you to the fields of sport and exercise physiology. Our goal is to build on the knowledge that you developed during basic coursework in human anatomy and physiology and to apply those principles in studying how the body (1) performs and responds to the added demands of an acute bout of exercise and (2) adapts to repeated bouts of exercise (i.e., exercise training).

What's New in the Seventh Edition

The seventh edition of *Physiology of Sport and Exercise* maintains the previous edition's high standard for illustrations, photos, and medical artwork. This visual detail, clarity, and realism allow both a greater insight into the physiological responses to exercise and a better understanding of the underlying research. In addition, the text is now augmented with animations, audio clips, and video clips, provided online in the student web study guide and separately for instructors. Throughout the text, you will find icons to identify pieces of artwork that are the basis for an animation or that have an accompanying audio clip. Accessing these resources will further aid understanding of the illustrations and the physiological processes they represent. In addition, video clips feature experts in the field discussing exciting current topics of research.

The new edition also brings back the Research Perspective elements introduced in the last edition that highlight interesting current research. These inserts discuss a wide range of important new or developing topics in sport and exercise physiology, providing interested students with additional insight into the state of research in the field. We have also revised the introductory chapter to include information on new frontiers in exercise physiology in the 21st century such as genomics and epigenetics, chapter 5 to include more detailed coverage of mechanisms associated with fatigue and muscle cramps, chapter 11 to expand coverage of high-intensity interval training, and chapter 17 to focus more on health benefits of physical activity in children and adolescents. In addition, we have extensively updated the text to include the latest research on important topics in the field, including the following:

- New information on length–tension and force–velocity relations in muscle (chapter 1) and individual variability in appetite hormones (chapter 4)
- Newly added sections on the crossover concept (chapter 2), critical power (chapter 5), functional sympatholysis (chapter 6), the oxygen cascade (chapters 7 and 13), group exercise (chapter 9), and exercise and mobility in aging (chapter 18)
- Updated information on the role of maximal stroke volume in determining maximal aerobic capacity (chapter 8)
- Expanded mechanistic discussion of protein synthesis in muscle hypertrophy (chapter 10)
- Revision of information to reflect published guidelines by professional organizations on nutrition and athletic performance (chapter 15), exercise and pregnancy (chapter 19), and health-related screening, fitness testing, and exercise prescription (chapter 20)

All of these changes were made while retaining our emphasis on the ease of reading and understanding that have made this book the leading text for introducing students to this exciting field. The overall structure and progression of the text have been retained from the sixth edition. Our first focus is on muscle and how its needs are altered as an individual goes from a resting to an active state and how these needs are supported by—and interact with—other body systems. In later chapters we address principles of exercise training; considerations of environmental factors of heat, cold, and altitude; sport performance; and exercise for disease prevention.

Organization of the Seventh Edition

We begin in the introduction with a historical overview of sport and exercise physiology as they have emerged from the parent disciplines of anatomy and physiology, and we explain basic concepts that are used throughout the text. In parts I and II, we review the major physiological systems, focusing on their responses to acute bouts of exercise. In part I, we examine how the muscular, metabolic, nervous, and endocrine systems interact to produce body movement. In part II, we look at how the cardiovascular and respiratory systems continue to deliver nutrients and oxygen to the active muscles and transport waste products away during physical activity. In part III, we consider how these systems adapt to chronic exposure to exercise (i.e., training).

We change perspective in part IV to examine the impact of the external environment on physical performance. We consider the body's response to heat and cold, and then we examine the impact of low atmospheric pressure experienced at altitude. In part V, we shift attention to how athletes can optimize physical performance. We evaluate the effects of different types and volumes of training. We consider the importance of appropriate body composition for optimal performance and examine athletes' special dietary needs, as well as how nutrition can be used to enhance performance. Finally, we explore the use of ergogenic aids—substances purported to improve athletic ability.

In part VI, we examine unique considerations for specific populations. We look first at the processes of growth and development and how they affect the performance capabilities of young athletes. We evaluate changes that occur in physical performance as people age and explore the ways in which physical activity can help maintain health and independence. Finally, we examine issues and special physiological concerns of female athletes.

In the final part of the book, part VII, we turn our attention to the application of sport and exercise physiology to prevent and treat various diseases and the use of exercise for rehabilitation. We look at prescribing exercise for maintaining health and fitness, and we then close the book with a discussion of cardiovascular disease, obesity, and diabetes.

Exercise at Altitude

13

In this chapter and in the web study guide

331

Special Features in the Seventh Edition

This seventh edition of *Physiology of Sport and Exercise* is designed with the goal of making learning easy and enjoyable. This text is comprehensive, but the many special features included will help you progress through the book without being overwhelmed by its scope. In addition to these features, the fully updated web study guide that accompanies this text provides opportunities for interactive learning and review, along with animations, audio clips, and video clips to enhance your understanding of the text.

Each chapter in the book begins with a chapter outline with page numbers to help you locate material. Also noted in the chapter outline are the web study guide activities relating to each section of the chapter. Each chapter begins with a brief story that explores a real-world application of the concepts presented.

Within each chapter, the Research Perspective elements introduce you to important topics in current exercise physiology research. You will also find icons to alert you to animations and audio clips that will help you understand important figures and to video clips that provide expanded discussion on current topics in the field:

- **Animation** icons identify figures that are also provided as animations.
- **Audio** icons identify figures that are further explained in an accompanying audio clip.
- **Video** icons let you know when a video clip on a topic is available.

As you read through, you will also find In Review elements that summarize the major points presented in the previous sections. And at the end of the chapter, the In Closing wraps things up and notes how what you have learned sets the stage for the topics to come.

Key terms are in bold in the text, listed at the end of each chapter, and defined in the glossary at the end of the book. At each chapter's end, you will also find study questions to test your knowledge of

Research Perspective

RESEARCH PERSPECTIVE 7.1

Sprint Interval Training for Respiratory Muscles

Typical respiratory muscle endurance training (RMET) improves exercise capacity and thus performance; such improvements are largely attributed to reductions in the development of respiratory muscle fatigue. However, can a shorter version of RMET based on the principle of high-intensity interval training (respiratory muscle sprint-interval training, or RMSIT) elicit similar improvements in respiratory muscle function?

A team of investigators recently sought to compare the effects of traditional RMET versus RMSIT on respiratory muscle function.[7] Mechanical airway properties and respiratory muscle testing (e.g., respiratory muscle strength) were measured before and after experimental sessions of RMET and RMSIT. RMET consisted of continuous volitional hyperpnea (increased depth and rate of breathing) performed for 30 min using a commercially available training device. The RMSIT was a novel respiratory muscle training regimen developed by the researchers. Using the same training device as with RMET, the RMSIT regimen consisted of six short maximal respiratory sprints with additional airway resistance to maximize respiratory muscle work. In this fashion, RMSIT is characterized by higher respiratory muscle power output and tension-time indices, but considerably lower total work compared to RMET. The standard RMET and the novel RMSIT regimens reduced respiratory muscle contractility to the same extent, triggering similar muscular adaptations in response to training. Neither protocol altered mechanical airway properties. Therefore, RMSIT appears to be a safe and time-saving alternative to RMET.

RMET can improve overall function for individuals who have undergone a median sternotomy (splitting of the sternum to access underlying organs) during cardiac surgery. Clinical exercise physiologists are interested in exercise training adaptations that occur with structured cardiac rehabilitation programs. The results of a 2013 study suggest that it would be beneficial to include exercises that improve the strength of the inspiratory muscles as part of a cardiac rehabilitation program.[4] This type of training would reduce inspiratory muscle effort and further improve ventilatory efficiency in patients after open-heart surgery.

alveoli into the blood in the pulmonary capillaries, and carbon dioxide diffuses from the blood into the alveoli in the lungs. The **alveoli** are grapelike clusters, or air sacs, at the ends of the terminal bronchioles.

Blood from the body (except for that returning from the lungs) returns through the vena cava to the right side of the heart. From the right ventricle, this blood is pumped through the pulmonary artery to the lungs, ultimately working its way into the pulmonary capillaries. These capillaries form a dense network around the alveolar sacs and are so small that the red blood cells must pass through them in single file, such that the maximal surface area of each cell is exposed to the surrounding lung tissue. This is where pulmonary diffusion occurs.

Blood Flow to the Lungs at Rest

At rest the lungs receive approximately 4 to 6 L/min of blood flow, depending on body size. Because cardiac output from the right side of the heart approximates cardiac output from the left side of the heart, blood flow to the lungs matches blood flow to the systemic circulation. However, pressure and vascular resistance in the blood vessels in the lungs are different than in the system circulation. The mean pressure in the pulmonary artery is ~15 mmHg (systolic pressure is ~25 mmHg and diastolic pressure is ~8 mmHg) compared to the mean pressure in the aorta of ~95 mmHg. The pressure in the left atrium, where blood is returning to the heart from the lungs, is ~5 mmHg; thus, there is not a great pressure difference across the pulmonary circulation (15 − 5 mmHg). Figure 7.4 illustrates the differences in pressures between the pulmonary and systemic circulation.

Recalling the discussion of blood flow in the cardiovascular system from chapter 6, pressure = flow × resistance. Since blood flow to the lungs is equal to that of the systemic circulation, and there is a substantially lower change in pressure across the pulmonary vascular system, resistance is proportionally lower compared to that in the systemic circulation. This is reflected in differences in the anatomy of the vessels in the pulmonary versus systemic circulation: The pulmonary blood vessels are thin walled, with relatively little smooth muscle.

Respiratory Membrane

Gas exchange between the air in the alveoli and the blood in the pulmonary capillaries occurs across the **respiratory membrane** (also called the alveolar-capillary membrane). This membrane, depicted in figure 7.5, is composed of

- the alveolar wall,
- the capillary wall, and
- their respective basement membranes.

182

The Respiratory System and Its Regulation 183

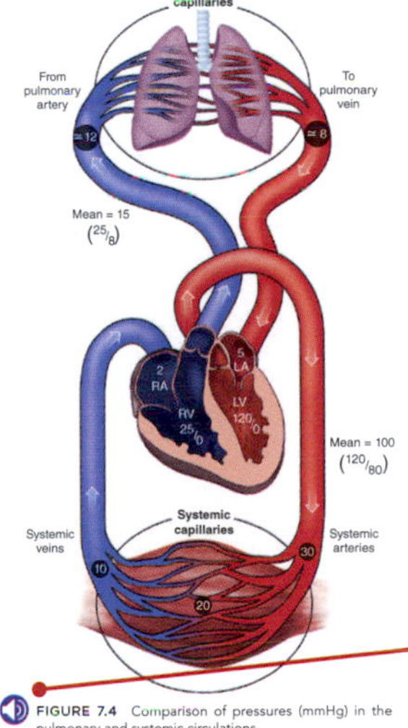

FIGURE 7.4 Comparison of pressures (mmHg) in the pulmonary and systemic circulations.

Icons

The primary function of these membranous surfaces is for gas exchange. The respiratory membrane is very thin, measuring only 0.5 to 4 µm. As a result, the gases in the nearly 300 million alveoli are in close proximity to the blood circulating through the capillaries.

Partial Pressures of Gases

The air we breathe is a mixture of gases. Each exerts a pressure in proportion to its concentration in the gas mixture. The individual pressures from each gas in a mixture are referred to as **partial pressures**. According to **Dalton's law**, the total pressure of a mixture of gases equals the sum of the partial pressures of the individual gases in that mixture.

Consider the air we breathe. It is composed of 79.04% nitrogen (N_2), 20.93% oxygen (O_2), and 0.03% carbon dioxide (CO_2). These percentages remain constant regardless of altitude. At sea level, the atmospheric (or barometric) pressure is approximately 760 mmHg, which is also referred to as standard atmospheric pressure. Thus, if the total atmospheric pressure is 760 mmHg, then the partial pressure of nitrogen (PN_2) in air is 600.7 mmHg (79.04% of the total 760 mmHg pressure). Oxygen's partial pressure (PO_2) is 159.1 mmHg (20.93% of 760 mmHg), and carbon dioxide's partial pressure (PCO_2) is 0.2 mmHg (0.03% of 760 mmHg).

In the human body, gases are usually dissolved in fluids, such as blood plasma. According to **Henry's law**, gases dissolve in liquids in proportion to their partial pressures, depending also on their solubilities in the specific fluids and on the temperature. A gas's solubility in blood is a constant, and blood temperature also remains relatively constant at rest. Thus, the most critical factor for gas exchange between the alveoli and the blood is the pressure gradient between the gases in the two areas.

Key terms

Gas Exchange in the Alveoli

Differences in the partial pressures of the gases in the alveoli and the gases in the blood create a pressure gradient across the respiratory membrane. This forms the basis of gas exchange during pulmonary diffusion. If the pressures on each side of the membrane were equal, the gases would be at equilibrium and would not move. But the pressures are not equal, so gases move according to partial pressure gradients.

Oxygen Exchange

The PO_2 of air outside the body at standard atmospheric pressure is 159 mmHg. But this pressure decreases to about 105 mmHg when air is inhaled and enters the alveoli, where it is moistened and mixes with the air in the alveoli. The alveolar air is saturated with water vapor (which has its own partial pressure) and contains more carbon dioxide than the inspired air. Both the increased water vapor pressure and increased partial pressure of carbon dioxide contribute to the total pressure in the alveoli. Fresh air that ventilates the lungs is constantly mixed with the air in the alveoli while some of the alveolar gases are exhaled to the environment. As a result, alveolar gas concentrations remain relatively stable.

The blood, stripped of much of its oxygen by the metabolic needs of the tissues, typically enters the pulmonary capillaries with a PO_2 of about 40 mmHg (see figure 7.6). This is about 60 to 65 mmHg less than the

the chapter's contents and a reminder of the study guide activities that are available, along with the web address of the online study guide.

At the end of the book is a comprehensive glossary that includes definitions of all key terms, a listing of numbered references for the sources cited in each chapter, and a thorough index. Finally, printed on the inside front and back covers for easy reference are lists of common abbreviations and conversions.

Instructors using this text in their courses will find a wealth of updated ancillary materials available at www.HumanKinetics.com/PhysiologyOfSportAndExercise, including an instructor guide, a presentation package, a test package, chapter quizzes, and an image bank. The instructor ancillaries also include convenient access to the animations, video clips, and audio clips.

You might read this book only because it is a required text for a course. But we hope that the information will entice you to continue to study this relatively new and exciting area. We hope at the very least to further your interest in and understanding of your body's marvelous abilities to perform various types and intensities of exercise and sports, to adapt to stressful situations, and to improve its physiological capacities. This book is useful not only for anyone who pursues a career in exercise or sport science but also for anyone who wants to be active, healthy, and fit.

This is the case even when elderly subjects are exposed to what we would consider a mild cold stress.[7] However, people can easily offset these decrements by wearing clothing that is appropriate for the environmental conditions and activity level. By performing so-called behavioral thermoregulation, older athletes can offset the decrements in physiological thermoregulation and continue to exercise safely in cold environments.

During exposure to high altitude, there is little reason to expect older athletes to respond differently than their younger counterparts. Unfortunately, data are lacking regarding aging and the rate and magnitude of acclimatization to altitude. Likewise, it is unclear whether aging per se increases the incidence of any of the altitude illnesses. We can expect the performance of an older athlete at altitude to be similar to that of a younger athlete of comparable physical fitness.

Longevity

Regular physical activity is an important contributor to good health. Does training throughout adulthood affect longevity? Because the aging rate in rats is more rapid than in humans, they have been used as subjects in studies conducted to determine the influence of chronic exercise (training) on **longevity** (the length of one's life). A study by Goodrick[11] demonstrated that rats that exercised freely lived about 15% longer than sedentary rats. But an investigation at Washington University in St. Louis showed no significant increase in the life span of rats that voluntarily ran on an exercise wheel.[17] More of the active rats lived to old age, but on average, they still died at the same age as their sedentary counterparts. It is interesting that rats that had restricted food intake and maintained a lower body weight lived 10% longer than the freely eating, sedentary rats. Although exercise training is a key component to energy balance, the only known way to increase longevity is through caloric restriction.

Of course, we cannot directly apply these findings to humans, but these results raise some interesting questions that might be relevant to our health and longevity. Although it is true that an endurance exercise program can reduce a number of the risk factors associated with cardiovascular disease, only limited information supports the contention that people will live longer if they exercise regularly. Data collected from the alumni at Harvard University and the University of Pennsylvania and from participants at the Aerobic Center in Dallas suggest that there is a decrease in mortality rate and a small increase in longevity (about 2 years) among people who remain physically active throughout life. At a minimum, regular physical activity can increase the **health span**—the number of years of generally healthy living, free from serious disease or chronic disability.

In Review

- The impaired ability of older individuals to tolerate exercise in the heat is due to reduced $\dot{V}O_{2max}$ and impaired cardiovascular adaptations rather than a direct effect of aging on thermoregulatory control or sweating.
- Regular exercise training can increase skin blood flow and sweating rate and improve the redistribution of cardiac output in older individuals as well as young men and women.
- Older people generally have an impaired ability to tolerate cold, but they can compensate by wearing appropriate clothing.
- Adaptation to altitude appears to be independent of age.
- An active lifestyle appears to be associated with a small increase in longevity. Just as important, an active lifestyle leads to a higher quality of life!
- There is an increased risk of injury from exercise as people age, and injuries tend to be slower to heal.
- The risk of death during exercise is not increased in those who are regularly active but is increased in those who seldom exercise.

In Review

IN CLOSING

In this chapter, we examined the effects of aging on physical performance. We evaluated changes in cardiorespiratory endurance and strength with age. We considered the effect of aging on body composition, which we know can affect performance. And yet, in the course of our discussion, it became clear that much of the change that occurs with aging is attributable to the inactivity that often accompanies aging. When older people participate in training, most of the changes associated with aging are lessened, and the resulting degree of change is similar to that seen in young and middle-aged adults. Thus, we have dispelled many of the myths about the capacity for physical activity of older people.

In the next chapter, we turn our attention to women, who as a group are often considered less capable of physical activity than men. We consider the physiology of girls and women, the impact of this physiology on athletic ability, how performances of female athletes compare with those of male athletes, and special issues associated with being female.

KEY TERMS

cardiovascular deconditioning
endothelial dysfunction
forced expiratory volume in 1 s ($FEV_{1.0}$)
health span
longevity
maximal expiratory ventilation ($\dot{V}O_{2max}$)
osteopenia
osteoporosis
peripheral blood flow
primary aging

STUDY QUESTIONS

1. What changes occur in height, weight, and body composition with aging? What accounts for these changes? How do these changes affect maximal oxygen uptake?
2. What changes occur in muscle with aging? How do they affect strength and athletic performance?
3. Describe the changes in HR_{max} with age. How does training alter this relationship?
4. How does aging affect maximal stroke volume and maximal cardiac output? What mechanisms can potentially explain these changes?
5. How does the respiratory system change with aging? What happens to vital capacity, $FEV_{1.0}$, residual volume, total lung capacity, and the ratio RV/TLC?
6. $\dot{V}O_{2max}$ declines with age across the entire population. Describe the physiological mechanisms that account for this decline. How do trained older individuals maintain a relatively high $\dot{V}O_{2max}$?
7. How do aging and habitual exercise affect blood vessel function?
8. How does age affect anaerobic function?
9. Differentiate between biological aging and physical inactivity.
10. What influence do aging and training have on body composition?
11. Describe the trainability of the older individual for both strength and aerobic endurance.
12. What can be done to minimize age-related losses in motor function and mobility?
13. Describe the changes in strength and endurance performance records with aging.
14. What concerns should we have about older people exercising in hot and cold environments or at altitude?
15. What is the difference between life span and health span?

STUDY GUIDE ACTIVITIES

In addition to the activities listed in the chapter opening outline, two other activities are available in the web study guide, located at

www.HumanKinetics.com/PhysiologyOfSportAndExercise

The **KEY TERMS** activity reviews important terms, and the end-of-chapter **QUIZ** tests your understanding of the material covered in the chapter.

In Closing

Student and Instructor Resources

Student Resources

Students, visit the free web study guide at www.HumanKinetics.com/PhysiologyOfSportAndExercise for interactive learning activities—all of which can be conducted outside the lab or classroom—as well as animations, video clips, and audio clips to aid your learning. You'll be able to apply key concepts by conducting experiments and recording your own physiological responses to exercise. The guide includes activities and quizzes that test your knowledge of the material as you prepare for classroom quizzes or tests. You'll also have access to links to professional journals and information on organizations and careers in the field.

Updated for the seventh edition, the web study guide includes the following multimedia content:

 26 animated versions of artwork from the text that will help you to understand physiological processes

 27 video discussions with experts in the field of exercise physiology

 66 audio clips that describe the processes shown in figures

Look for the icons in the text to know when this additional content is available. As you work to understand a concept illustrated in a figure, refer to the audio or animation for an explanation and to build your understanding. In combination with the web study guide activities, the animations, video, and audio allow you to practice, review, and develop knowledge and skills about the physiology of sport and exercise.

Instructor Resources

Instructor Guide

Specifically developed for instructors who have adopted *Physiology of Sport and Exercise, Seventh Edition,* the instructor guide includes sample lecture outlines, key points, and student assignments for every chapter in the text, along with sample laboratory exercises and direct links to a range of detailed sources on the internet.

Test Package

The test package includes a bank of 1,609 questions created especially for *Physiology of Sport and Exercise, Seventh Edition.* Various types of questions are included:

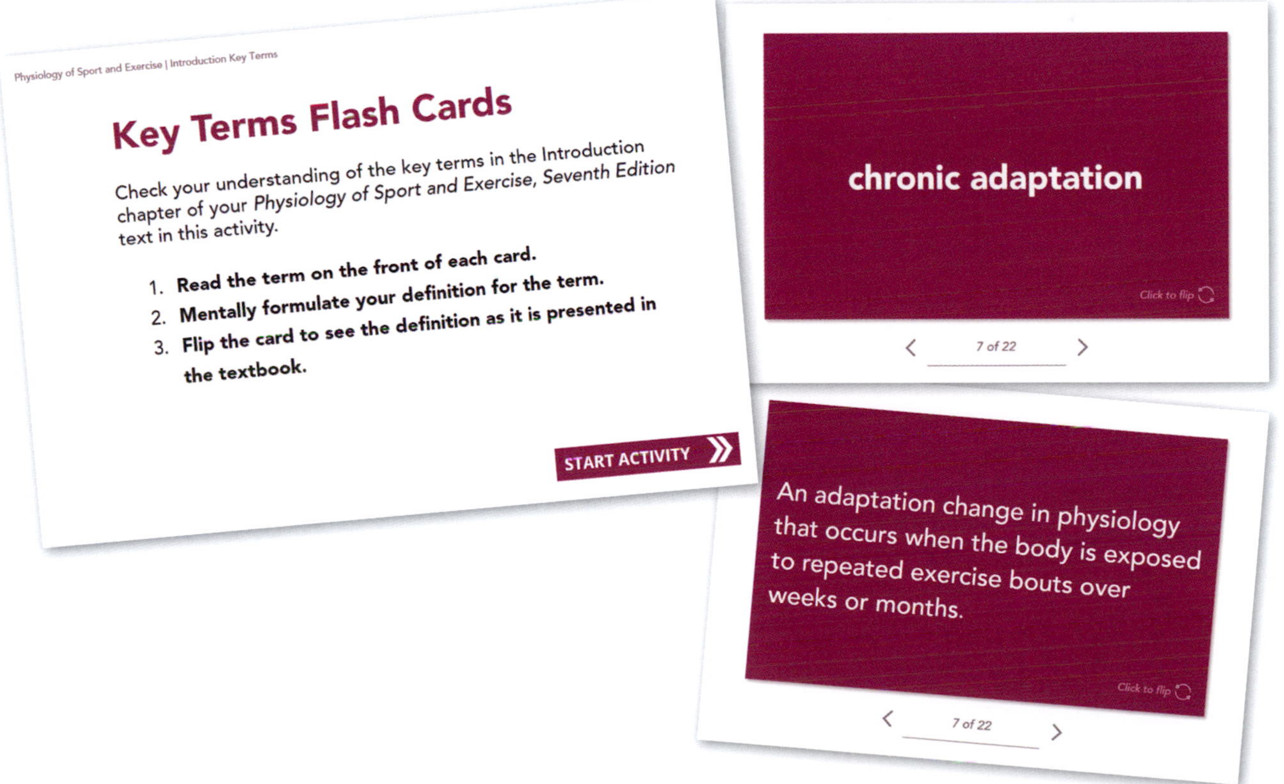

true or false, fill in the blank, essay, and multiple choice. The test package is available for use through multiple formats, including a learning management system, Respondus, and rich text.

Chapter Quizzes

Updated for the seventh edition, these ready-to-use quizzes test students' understanding of the most important concepts in each chapter. Chapter quizzes can be imported into learning management systems or printed for use as written quizzes.

Presentation Package

The presentation package includes a comprehensive series of PowerPoint slides for each chapter. Slides of learning objectives present the major topics covered in each chapter, text slides list key points, and illustration and photo slides contain graphics found in the text. The presentation package has more than 1,000 slides that can be used directly with PowerPoint and for printing transparencies or slides or making copies for distribution to students. Instructors can easily add to, modify, or rearrange the order of the slides as well as search for slides based on key words. You may access the presentation package by visiting www.HumanKinetics.com/PhysiologyOfSportAndExercise.

Image Bank

The image bank includes most of the illustrations, artwork, and tables from the text, sorted by chapter. These are provided as separate files for easy insertion into tests, quizzes, handouts, and other course materials, which provides instructors with greater flexibility when creating customized resources.

Instructor Video, Animations, and Audio

Within the instructor resources, the multimedia content in the web study guide is compiled for convenient access and inclusion in lectures and classroom presentations.

The instructor guide, test package, chapter quizzes, presentation package, image bank, video clips, animations, and audio clips are free to course adopters.

Acknowledgments

We would like to thank the staff at Human Kinetics for their continued support of the seventh edition of *Physiology of Sport and Exercise* and their dedication to publishing a high-quality product that meets the needs of instructors and students alike. Recognition goes to Amy Tocco (acquisitions editor) as well as our capable developmental editors: Lori Garrett (first edition), Julie Rhoda (second and third editions), Maggie Schwarzentraub (fourth edition), and Kate Maurer (fifth and sixth editions); Judy Park took over the reins as developmental editor for this seventh edition and has worked tirelessly and expertly to keep all phases of the project on schedule while continuing to demand the high quality for which our book is known. They have all been a true pleasure to work with, and their competence and skill are evident throughout the book. Special thanks go to Joanne Brummett for her artistic expertise and contributions to continuously improving the artwork.

For the seventh edition, special thanks also go to a handful of colleagues who provided their valued expertise and time. In particular, direct feedback and input from Drs. Gustavo Nader, Jinger Gottschall, Lacy Alexander, and Jim Pawelczyk at Penn State were invaluable in making substantive changes that not only updated and enhance the content but also provided high quality feedback from an instructor's viewpoint. Special recognition goes to the "postdoc dream team" of Drs. Jody Greaney and Anna Stanhewicz for all of their hard work in helping update all of the Research Perspective elements. In addition to Larry Kenney's Penn State colleagues, thanks also go to Dr. Bob Murray, who once again contributed his vast knowledge about ergogenic aids to chapter 16.

Finally, we thank our families for their constant love, support, and patience while we were writing, rewriting, editing, and proofing this book across all seven editions.

W. Larry Kenney
David L. Costill
Jack H. Wilmore (posthumously)

Photo Credits

Chapter and part opener photos

Introduction: Echo/Juice Images/Getty Images; **Part I:** David Davies/Press Association Images; **Chapter 1:** BSIP/Medical Images; **Chapter 2:** Hero Images/DigitalVision/Getty Images; **Chapter 3:** Carolina Biological/Medical Images; **Chapter 4:** Hank Grebe/Getty Images; **Chapter 5:** Buda Mendes/Getty Images; **Part II:** Press Association Images; **Chapter 6:** Biophoto Associates/Science Source; **Chapter 7:** 3D4Medical /Medical Images; **Chapter 8:** Sam Edwards/Caiaimage/Getty Images; **Part III:** © Human Kinetics; **Chapter 9:** Alexander Hassenstein/Getty Images; **Chapter 10:** Grady Reese/E+/Getty Images; **Chapter 11:** Alex Goodlett - International Skating Union (ISU)/ISU via Getty Images; **Part IV:** © E Simanor/Robert Harding Picture Library/age fotostock; **Chapter 12:** Technotr/E+/Getty Images; **Chapter 13:** FRANCK FIFE/AFP/Getty Images; **Part V:** Joshua Sarner/Icon Sportswire; **Chapter 14:** Hero Images/Getty Images; **Chapter 15:** Sanjeri/E+/Getty Images; **Chapter 16:** Simon Hausberger/Getty Images; **Part VI:** © Human Kinetics; **Chapter 17:** Hero Images/Getty Images; **Chapter 18:** Westend61/Getty Images; **Chapter 19:** AMR Image/E+/Getty Images; **Part VII:** © Human Kinetics; **Chapter 20:** FatCamera/E+/Getty Images; **Chapter 21:** ISM / SOVEREIGN/Medical Images; **Chapter 22:** Science Photo Library/Getty Images

Photos courtesy of the authors

Figures 0.2, 0.3, 0.4, 0.6*b*, 0.6*c*, 0.7, 0.9, 1.1, 1.11, 1.12, 1.13*a*, 1.13*b*, 5.9, 18.6, 22.7*a*; photos on pp. 2 (*a* and *b*)

Additional photos

Photo *c* on p. 2: Photo courtesy of Dr. Larry Golding, University of Nevada, Las Vegas. Photographer Dr. Moh Youself; **figures 0.1, 0.5*a*, 0.5*b*, and 0.6*a*:** Photos courtesy of American College of Sports Medicine Archives. All rights reserved; **figure 0.5*c*:** Courtesy of Noll Laboratory, The Pennsylvania State University; **figure 0.12:** Andy Cross/The Denver Post via Getty Images; **figure 0.13:** © Human Kinetics; **photo on p. 21:** © Human Kinetics; **photo in figure 1.2:** ISM/Medical Images; **figure 1.4:** BSIP/Medical Images; **photo on p. 35:** © Human Kinetics; **figure 1.17*b*:** Reprinted from J.C. Bruusgaard et al., "Myonuclei Acquired by Overload Exercise Precede Hypertrophy and are Not Lost on Detraining," Proceedings of the National Academy of Sciences 107 (2010): 15111-15116. By permission of J.C. Bruusgaard; **photo on p. 71:** © Human Kinetics; **photo in figure 3.2:** Carolina Biological/Medical Images; **photos on pp. 93 and 110:** © Human Kinetics; **photo on p. 112:** Photo courtesy of Larry Kenney; **figure 5.2:** © Human Kinetics; **photo on p. 142:** © Human Kinetics; **figure 6.16*b*:** Westend61/Getty Images; **photo in figure 7.3:** © Human Kinetics; **photos on pp. 190, 194, 212, 217, and 222:** © Human Kinetics; **figures 9.1, 9.3, and 9.5:** © Human Kinetics; **photos on pp. 240, 242, 244, and 252:** © Human Kinetics; **figure 10.2:** Photos courtesy of Dr. Michael Deschene's laboratory; **photos on pp. 259 and 262:** © Human Kinetics; **photo on p. 292:** Dylan Buell/Getty Images**; photo in figure 12.2:** Carolina Biological/Medical Images; **figure 12.3:** From Department of Health and Human Performance, Auburn University, Alabama. Courtesy of John Eric Smith, Joe Molloy, and David D. Pascoe. By permission of David Pascoe; **photos on pp. 307 and 325:** © Human Kinetics; **photo on p. 326:** ©Wojciech Gajda/fotolia.com; **photo on p. 369:** Photo courtesy of Larry Kenney; **figure 15.2:** © Human Kinetics; **figure 15.3:** Photos courtesy of Hologic, Inc.; **figure 15.4:** David Cooper/Toronto Star via Getty Images; **figure 15.5:** © Human Kinetics; **figure 15.6:** Courtesy of Rice Lake Weighing Systems; **photos on pp. 454, 488, and 493:** © Human Kinetics; **figure 19.9:** Dee Breger/Science Source; **photo on p. 505:** © Human Kinetics; **figure 20.1:** © Human Kinetics; **figure 22.7*b*:** ISM/ Pr Jean-Denis LAREDO/Medical Images; **photo on p. 571:** © Human Kinetics

568
8
231.7

INTRODUCTION

An Introduction to Exercise and Sport Physiology

In this chapter and in the web study guide

Much of the history of exercise physiology in the United States can be traced to the effort of a Kansas farm boy, David Bruce (D.B.) Dill, whose interest in physiology first led him to study the composition of crocodile blood. Fortunately for what would eventually grow into the discipline of exercise physiology, this young scientist redirected his research to humans when he became the first research director of the Harvard Fatigue Laboratory in 1927. Throughout his life he was intrigued by the physiology and adaptability of many animals that survive extreme exercise and environmental conditions, but he is best remembered for his research on *human* responses to exercise, heat, high altitude, and other environmental factors. Dr. Dill always served as one of the human guinea pigs in his own studies. During the Harvard Fatigue Laboratory's 20-year existence, he and his coworkers produced approximately 350 scientific papers along with a classic book titled *Life, Heat, and Altitude.*[10]

After the Harvard Fatigue Laboratory closed its doors in 1947, Dr. Dill began a second career as deputy director of medical research for the Army Chemical Corps, a position he held until his retirement from that post in 1961. Dr. Dill was then 70 years old—an age he considered too young for retirement—so he moved his research to Indiana University, where he served as a senior physiologist until 1966. In 1967 he obtained funding to establish the Desert Research Laboratory at the University of Nevada at Las Vegas. Dr. Dill used this laboratory as a base for his studies on human tolerance to exercise in the desert and at high altitude. He continued his research and writing until his final retirement at age 93, the same year he produced his last publication, titled *The Hot Life of Man and Beast.*[11]

Dr. David Bruce (D.B.) Dill *(a)* at the beginning of his career, *(b)* as director of the Harvard Fatigue Laboratory at age 42, and *(c)* at age 92 just before his fourth retirement.

The human body is an amazing machine. As you sit reading this introduction, countless perfectly coordinated and integrated events are occurring simultaneously in your body. These events allow complex functions, such as hearing, seeing, breathing, and information processing, to continue without any conscious effort. If you stand up, walk out the door, and jog around the block, almost all of your body's systems will be activated, enabling you to successfully shift from rest to exercise. If you continue this routine regularly for weeks or months and gradually increase the duration and intensity of your jogging, your body will adapt so that you can perform better. Therein lie the two basic components of the study of exercise physiology: the acute responses of the body to exercise in all its forms and the adaptation of the body's systems to repeated or chronic exercise, often called exercise training.

For example, as a point guard directs her team down the basketball court on a fast break, her body makes many adjustments that require a series of complex interactions involving many body systems. Adjustments occur even at the cellular and molecular levels. To enable the coordinated leg muscle actions as she moves rapidly down court, nerve cells from the brain, referred to as motor neurons, conduct electrical impulses down the spinal cord to the legs. On reaching the muscles, these neurons release chemical messengers that cross the gap between the nerve and muscle, each neuron exciting a number of individual muscle cells or fibers. Once the nerve impulses cross this gap, they spread along the length of each muscle fiber and attach to specialized receptors. Binding of the messenger to its receptor sets into motion a series of steps that activate the muscle fiber's contraction processes, which involve specific protein molecules—actin and myosin—and an elaborate energy system to provide the fuel necessary to sustain a single contraction and subsequent contractions. It is at this level that other molecules, such as adenosine triphosphate (ATP) and phosphocreatine (PCr), become critical for providing the energy necessary to fuel contraction.

In support of this sustained and rhythmic muscular contraction and relaxation, multiple additional systems are called into action, including the following:

- The skeletal system provides the basic framework around which muscles act.
- The cardiovascular system delivers fuel to working muscle and to all of the cells of the body and removes waste products.
- The cardiovascular and respiratory systems work together to provide oxygen to the cells and remove carbon dioxide.
- The integumentary system (skin) helps maintain body temperature by allowing the exchange of heat between the body and its surroundings.

- The nervous and endocrine systems coordinate this activity, while helping to maintain fluid and electrolyte balance and assisting in the regulation of blood pressure.

For centuries, scientists have studied how the human body functions at rest in health and disease. During the past 100 years or so, a specialized group of physiologists have focused their studies on how the body functions during physical activity and sport. This introduction presents a historical overview of exercise and sport physiology and then explains some basic concepts that form the foundation for the chapters that follow.

Focus of Exercise and Sport Physiology

Exercise and sport physiology have evolved from the fundamental disciplines of anatomy and physiology. Anatomy is the study of an organism's structure, or morphology. While anatomy focuses on the basic *structure* of various body parts and their interrelationships, **physiology** is the study of body *function*. Physiologists study how the body's organ systems, tissues, cells, and the molecules within cells work and how their functions are integrated to regulate the body's internal environment, a process called **homeostasis**. Because physiology focuses on the functions of body structures, understanding anatomy is essential to learning physiology. Furthermore, both anatomy and physiology rely on a working knowledge of biology, chemistry, physics, and other basic sciences.

Exercise physiology is the study of how the body's functions are altered when we are physically active, since exercise presents a challenge to homeostasis. Because the environment in which one performs exercise has a large impact, **environmental physiology** has emerged as a subdiscipline of exercise physiology. **Sport physiology** further applies the concepts of exercise physiology to enhancing sport performance and optimally training athletes. Thus, sport physiology derives its principles from exercise physiology. Because exercise physiology and sport physiology are so closely related and integrated, it is often hard to clearly distinguish between them. Because the same underlying scientific principles apply, exercise and sport physiology are often considered together, as they are in this text.

Acute and Chronic Responses to Exercise

The study of exercise and sport physiology involves learning the concepts associated with two distinct exercise patterns. First, exercise physiologists are concerned with how the body responds to an individual bout of exercise, such as running on a treadmill for an hour or lifting weights. An individual bout of exercise is called **acute exercise**, and the responses to that exercise bout are referred to as acute responses. When examining the acute response to exercise, we are concerned with the body's immediate response to, and sometimes its recovery from, a single exercise bout.

The other major area of interest in exercise and sport physiology is how the body responds over time to the stress of repeated bouts of exercise, sometimes referred to as **chronic adaptation** or **training effects**. When one performs regular exercise over a period of days and weeks, the body adapts. The physiological adaptations that occur with chronic exposure to exercise or training improve both exercise capacity and efficiency. With resistance training, the muscles become stronger. With aerobic training, the heart and lungs become more efficient and endurance capacity of the muscles increases. As discussed later in this introductory chapter and in more detail in chapters 10 and 11, these adaptations are highly specific to the type of training the person does.

In Review

- Exercise physiology evolved from its parent discipline, physiology. The two primary concerns of exercise physiology are
  - how the body responds to the acute stress of a single bout of exercise or physical activity; and
  - how the body adapts to the chronic stress of repeated bouts of exercise—that is, exercise training.
- Some exercise physiologists use exercise or environmental conditions (heat, cold, altitude, and so on) to stress the body in ways that uncover basic physiological mechanisms. Others examine exercise training's effects on health, disease, and well-being. Sport physiologists apply these concepts to athletes and sport performance.

The Evolution of Exercise Physiology

To students, contemporary exercise physiology may seem like a vast collection of new ideas never before subjected to rigorous scientific scrutiny. On the contrary, our current understanding of exercise physiology is based on the lifelong efforts of hundreds of outstanding scientists. The theories and hypotheses of modern physiologists have been shaped by the efforts of scientists who may be long forgotten. What we consider original or new is most often an assimilation of previous findings or the application of basic science to problems in exercise physiology. As with every discipline, there are, of course, a number of key scientists and pivotal scientific contributions that brought about significant advances in our knowledge of the physiological responses to exercise. The following section reflects on the history of the field of exercise physiology and on a few of the people who shaped it. It is impossible in this short section to do justice to the hundreds of pioneering scientists who paved the way and laid the foundation for modern exercise physiology.

Beginnings of Anatomy and Physiology

One of the earliest descriptions of human anatomy and physiology was Claudius Galen's Greek text *De fascius*, published in the first century. As a physician to the gladiators, Galen had ample opportunity to study and experiment on human anatomy and was a great proponent of science based on observation and experimentation. He was aware of the dire consequences of sedentary living and linked regular exercise to overall health and well-being by including regular exercise as one of his laws of health:

- Breathe fresh air.
- Eat the proper foods.
- Drink the right drinks.
- Exercise.
- Get adequate sleep.
- Have a daily bowel movement.
- Control your emotions.

Galen's theories of anatomy and physiology were so widely accepted that they remained unchallenged for nearly 1,400 years. Not until the 1500s were any truly significant contributions made to the understanding of both the structure and function of the human body. A landmark text by Andreas Vesalius, titled *Fabrica Humani Corporis [Structure of the Human Body]*, presented his findings on human anatomy in 1543. Although Vesalius' book focused primarily on anatomical descriptions of various organs, he occasionally attempted to explain their functions as well. British historian Sir Michael Foster said, "This book is the beginning, not only of modern anatomy, but of modern physiology. It ended, for all time, the long reign of fourteen centuries of precedent and began in a true sense the renaissance of medicine" (p. 354).[14]

Most early attempts at explaining physiology were either incorrect or so vague that they could be considered no more than speculation. Attempts to explain how a muscle generates force, for example, were usually limited to a description of its change in size and shape during action because observations were limited to what could be seen with the naked eye. From such observations, Hieronymus Fabricius (ca. 1574) suggested that a muscle's contractile power resided in its fibrous tendons, not in its "flesh." Anatomists did not discover the existence of individual muscle fibers until Dutch scientist Antonie van Leeuwenhoek introduced the microscope (ca. 1660). How these fibers shorten and create force would remain a mystery until the middle of the 20th century, when the intricate workings of muscle proteins could be studied by electron microscopy.

Early History of Exercise Physiology

Although exercise physiology is a relative newcomer to the world of science, one of its first publications appeared in 1793, when a paper by Séguin and Lavoisier described the oxygen consumption of a young man measured both in the resting state and while he repeatedly lifted a 7.3 kg (16 lb) weight for 15 min.[26] At rest the man used 24 L of oxygen per hour (L/h), which increased to 63 L/h during exercise. Lavoisier believed that the site of oxygen utilization and carbon dioxide production was in the lungs. Even though this concept was doubted by other physiologists of the time, it remained accepted doctrine until the middle of the 1800s, when several German physiologists demonstrated that combustion of oxygen occurred in cells throughout the entire body.

Although many advances in the understanding of circulation and respiration occurred during the 1800s, few efforts were made to focus on the physiology of physical activity. However, in 1888, an apparatus was described that enabled scientists to study subjects during mountain climbing, even though the subjects had to carry a 7 kg (15.4 lb) "gasometer" on their backs.[31]

Arguably the first published textbook on exercise physiology, *Physiology of Bodily Exercise*, was written in French by Fernand LaGrange in 1889.[19] Considering the small amount of research on exercise that had been conducted up to that time, it is intriguing to read the author's accounts of such topics as "Muscular Work," "Fatigue," "Habituation to Work," and "The Office of the Brain in Exercise." This early attempt to explain the response to exercise was, in many ways, limited to speculation and theory. Although some basic concepts of exercise biochemistry were emerging at that time, LaGrange was quick to admit that many details were still in the formative stages. For example, he stated that

> Vital combustion [energy metabolism] has become very complicated of late; we may say that it is somewhat perplexed, and that it is difficult to give in a few words

> a clear and concise summary of it. It is a chapter of physiology which is being rewritten, and we cannot at this moment formulate our conclusions. (p. 395)[19]

Because the early text by LaGrange offered only limited physiological insights regarding bodily functions during physical activity, some argue that the third edition of a text by F.A. Bainbridge titled *The Physiology of Muscular Exercise*, published in 1931, should be considered the earliest scientific text on this subject.[2] Interestingly, that third edition was written by A.V. Bock and D.B. Dill, at the request of A.V. Hill, three key pioneers of exercise physiology discussed in this introductory chapter.

Archibald V. (A.V.) Hill was a significant figure in the history of exercise physiology. In his inaugural address as Joddrell Professor of Physiology at University College London, Hill stated the principles that subsequently shaped the field of exercise physiology:

> It is strange how often a physiological truth discovered on an animal may be developed and amplified, and its bearings more truly found, by attempting to work it out on man. Man has proved, for example, far the best subject for experiments on respiration and on the carriage of gases by the blood, and an excellent subject for the study of kidney, muscular, cardiac and metabolic function. . . . Experiment on man is a special craft requiring a special understanding and skill, and "human physiology," as it may be called, deserves an equal place in the list of those main roads which are leading to the physiology of the future. The methods, of course, are those of biochemistry, of biophysics, of experimental physiology; but there is a special kind of art and knowledge required of those who wish to make experiments on themselves and their friends, the kind of skill that the athlete and the mountaineer must possess in realizing the limits to which it is wise and expedient to go.

During the late 1800s, many theories were proposed to explain the source of energy for muscle contraction. Muscles were known to generate much heat during exercise, so some theories suggested that this heat was used directly or indirectly to cause muscle fibers to shorten. After the turn of the century, Walter Fletcher and Sir Frederick Gowland Hopkins observed a close relation between muscle action and lactate formation.[12] This observation led to the realization that energy for muscle action is derived from the breakdown of muscle glycogen to lactic acid (see chapter 2), although the details of this reaction remained obscure. Because of the high energy demands of exercising muscle, this tissue served as an ideal model to help unravel the mysteries of cellular metabolism. In 1921, A.V. Hill (figure 0.1) was awarded the Nobel Prize for his findings on energy metabolism. At that time, biochemistry was in its infancy, although it was rapidly gaining recognition

FIGURE 0.1 1921 Nobel Prize winner Archibald V. Hill (1927).

through the research efforts of such other Nobel laureates as Albert Szent-Györgyi, Otto Meyerhof, August Krogh, and Hans Krebs, all of whom were actively studying how living cells generate and use energy.

Although much of Hill's research was conducted with isolated frog muscle, he also conducted some of the first physiological studies of runners. Such studies were made possible by the technical contributions of John S. Haldane, who developed the methods and equipment needed to measure oxygen use during exercise. These and other investigators provided the basic framework for our understanding of whole-body energy production, which became the focus of considerable research during the middle of the 20th century and is incorporated into the manual and computer-based systems that are used to measure oxygen uptake in exercise physiology laboratories throughout the world today. In his address, A.V. Hill went on to acknowledge Haldane's contributions and discuss the wide range of applications he saw for his work in exercise physiology:

> Quite apart from direct physiological research on man, the study of instruments and methods applicable to man, their standardization, their description, their reduction to routine, together with the setting up of standards of normality in man are bound to prove of great advantage to medicine; and not only to medicine but to all those activities and arts where normal man is the object of study. Athletics, physical training, flying, working, submarines, or coal mines, all require a knowledge of the physiology of man, as does also the study of conditions in factories. The observation of sick men in hospitals is not the best training for the study of normal man at work. It is necessary to build up a sound body of trained scientific opinion versed in the study of normal man, for such trained opinion is likely to prove of the greatest service, not merely to medicine, but in our ordinary social and industrial life.

> Haldane's unsurpassed knowledge of the human physiology of respiration has often rendered immeasurable service to the nation in such activities as coal mining or diving; and what is true of the human physiology of respiration is likely also to be true of many other normal human functions.

Era of Scientific Exchange and Interaction

From the early 1900s through the 1930s, the medical and scientific environment in the United States was changing. This was an era of revolution in the education of medical students, led by changes at Johns Hopkins. More medical and graduate programs based their educational endeavors on the European model of experimentation and development of scientific insights. There were important advances in physiology in areas such as bioenergetics, gas exchange, and blood chemistry that served as the basis for advances in the physiology of exercise. Building on collaborations forged in the late 1800s, interactions of laboratories and scientists were promoted, and international meetings of organizations such as the International Union of Physiological Sciences created an atmosphere for free scientific exchange, discussion, and debate. Research laboratories and collaborations created during this period would go on to do some of the most important exercise physiology research of the 20th century.

Research on Athletes

For more than 100 years, athletes have served as subjects for study of the upper limits of human endurance. Perhaps the first physiological studies on athletes occurred in 1871. Austin Flint studied one of the most celebrated athletes of that era, Edward Payson Weston, an endurance runner-walker. Flint's investigation involved measuring Weston's energy balance (i.e., food intake versus energy expenditure) during Weston's attempt to walk 400 mi (644 km) in 5 days. Although the study resolved few questions about muscle metabolism during exercise, it did demonstrate that some protein is lost from the body during prolonged heavy exercise.[13]

Throughout the 20th century, athletes were used repeatedly to assess the physiological capabilities of human strength and endurance and ascertain characteristics needed for record-setting performances. Some attempts have been made to use the technology and knowledge derived from exercise physiology to predict performance, prescribe training, or identify athletes with exceptional potential. In most cases, however, these applications of physiological testing are of little more than academic interest because few laboratory or field tests can accurately assess all the qualities required for someone to become a champion.

The Harvard Fatigue Laboratory

Perhaps no university has had more influence on the field of exercise physiology than Harvard. From 1891 to 1898, Harvard offered a degree in anatomy, physiology, and physical training under the direction of Dr. George Wells Fitz to "provide necessary knowledge about the science of exercise." While that department changed its focus with Fitz's departure in 1899, many other U.S. universities developed programs over the next 25 years that coupled basic science coursework with physical education.

A visit by A.V. Hill to Harvard University in 1926 had a significant impact on the founding and early activities of the Harvard Fatigue Laboratory (HFL), which was established a year later in 1927. Interestingly, the early home of the HFL was the basement of Harvard's Business School, and its stated early mission was to conduct research on fatigue and other hazards in industry. Creation of this laboratory was due to the insightful planning of world-famous biochemist Lawrence J. (L.J.) Henderson. A young biochemist from Stanford University, David Bruce (D.B.) Dill, was appointed as the first director of research, a title Dill held until the HFL closed in 1947.

As noted earlier, Dill had aided Arlen "Arlie" Bock in writing the third edition of Bainbridge's text on exercise physiology. Later in his career Dill credited the writing of that textbook with shaping the program of the HFL. Although he had little experience in applied human physiology, Dill's creative thinking and ability to surround himself with young, talented scientists created an environment that would lay the foundation for modern exercise and environmental physiology. For example, HFL personnel examined the physiology of endurance exercise and described the physical requirements for success in events such as distance running. Some of the most outstanding HFL investigations were conducted not in the laboratory but in the Nevada desert, on the Mississippi Delta, and in the White Mountains in California (with an altitude of 3,962 m, or 13,000 ft). These and other studies provided the foundation for future investigations on the effects of the environment on physical performance and in exercise and sport physiology.

In its early years, the HFL focused primarily on general problems of exercise, nutrition, and health. For example, the first studies on exercise and aging were conducted in 1939 by Sid Robinson (see figure 0.2), a student at the HFL. On the basis of his studies of subjects ranging in age from 6 to 91 years, Robinson described the effect of aging on maximal heart rate and oxygen uptake.[24] But with the onset of World War II, Henderson and Dill realized the HFL's potential contribution to the war effort, and research at the

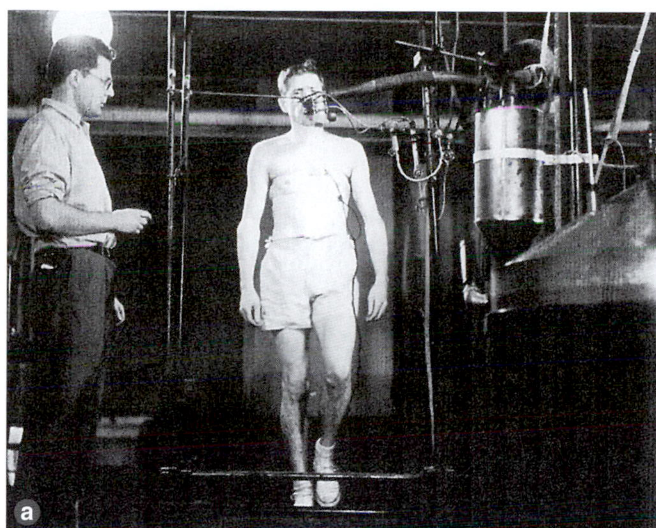

FIGURE 0.2 Sid Robinson *(a)* being tested by R.E. Johnson on the treadmill in the Harvard Fatigue Laboratory and *(b)* as a Harvard student and athlete in 1938.

HFL took a different direction. Harvard Fatigue Lab scientists and support personnel were instrumental in forming new laboratories for the Army, Navy, and Army Air Corps (now the Air Force). They also published the methodologies necessary for relevant military research, methods that are still in use throughout the world.

Today's exercise physiology students would be amazed at the methods and devices used in the early days of the HFL and at the time it took to conduct research projects in those days. What is now accomplished in milliseconds with the aid of computers and automatic analyzers literally demanded days of effort by HFL scientists. Measurements of oxygen uptake during exercise, for example, required collecting expired air in Douglas bags and analyzing it for oxygen and carbon dioxide by using a manually operated chemical analyzer, without the help of a computer, of course (see figure 0.3). The analysis of a single 1 min sample of expired air required 20 to 30 min of effort by one or more laboratory workers. Today, scientists make such measurements almost instantaneously and with little physical effort. One must marvel at the dedication, diligence, and hard work of the HFL's exercise physiology pioneers. Using the equipment and methods available at the time, HFL scientists published approximately 350 research papers over a 20-year period.

The HFL was an intellectual environment that attracted young physiologists and physiology doctoral students from all over the globe. Scholars from 15 countries worked in the HFL between 1927 and its closure in 1947. Most went on to develop their own laboratories and become noteworthy figures in exercise physiology

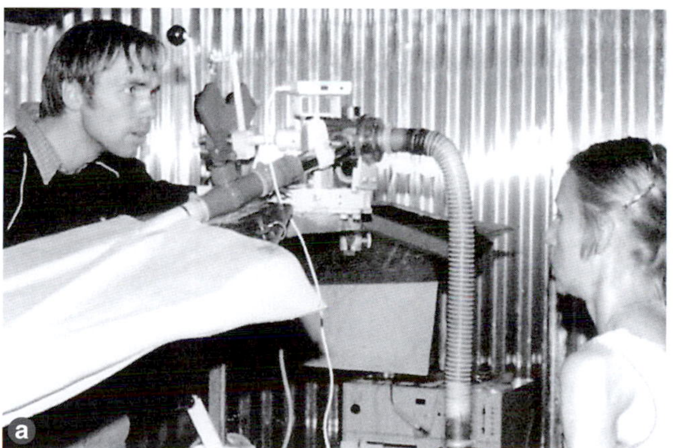

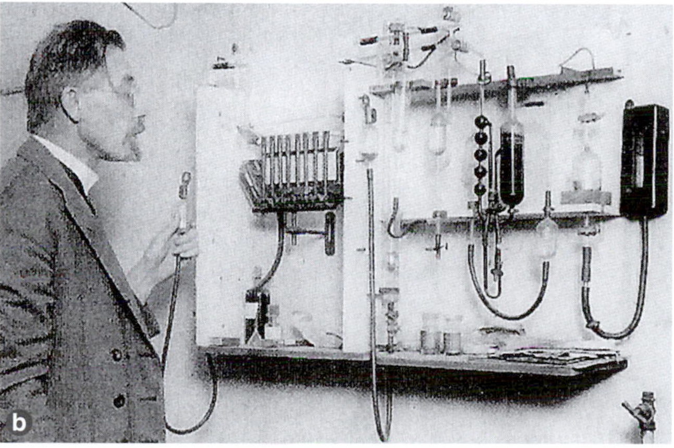

FIGURE 0.3 *(a)* Early measurements of metabolic responses to exercise required the collection of expired air in a sealed bag known as a Douglas bag. *(b)* A sample of that gas then was measured for oxygen and carbon dioxide using a chemical gas analyzer, as illustrated by this photo of Nobel laureate August Krogh.

in the United States, including Sid Robinson, Henry Longstreet Taylor, Lawrence Morehouse, Robert E. Johnson, Ancel Keys, Steven Horvath, C. Frank Consolazio, and William H. Forbes. Notable international scientists who spent time at the HFL included August Krogh, Lucien Brouha, Edward Adolph, Walter B. Cannon, Peter Scholander, and Rodolfo Margaria, along with several other notable Scandinavian scientists discussed later. Thus, the HFL planted seeds of intellect at home and around the world that resulted in an explosion of knowledge and interest in this new field. Most contemporary exercise physiologists can trace the roots of their research training back to the HFL.

Scandinavian Influence

In 1909, Johannes Lindberg established a laboratory that became a fertile breeding ground for scientific contributions at the University of Copenhagen in Denmark. Lindberg and 1920 Nobel Prize winner August Krogh teamed up to conduct many classic experiments and published seminal papers on topics ranging from the metabolic fuels for muscle to gas exchange in the lungs. This work was continued from the 1930s into the 1970s by Erik Hohwü-Christensen, Erling Asmussen, and Marius Nielsen.

As a result of contacts between D.B. Dill and August Krogh, these three Danish physiologists came to the HFL in the 1930s, where they studied exercise in hot environments and at high altitude. After returning to Europe, each man established a separate line of research. Asmussen and Nielsen became professors at the University of Copenhagen, where Asmussen studied the mechanical properties of muscle and Nielsen conducted studies on control of body temperature. Both remained active at the University of Copenhagen's August Krogh Institute until their retirements.

In 1941, Hohwü-Christensen (see figure 0.4*a*) moved to Stockholm, Sweden, to become the first physiology professor at the College of Physical Education at Gymnastik-och Idrottshögskolan (GIH). In the late 1930s, he teamed with Ole Hansen to conduct and publish a series of five studies of carbohydrate and fat metabolism during exercise. These studies are still cited frequently and are considered to be among the first and most important sport nutrition studies. Hohwü-Christensen introduced Per-Olof Åstrand to the field of exercise physiology. Åstrand, who conducted numerous studies related to physical fitness and endurance capacity during the 1950s and 1960s, became the director of GIH after Hohwü-Christensen retired in 1960. During his tenure at GIH, Hohwü-Christensen mentored a

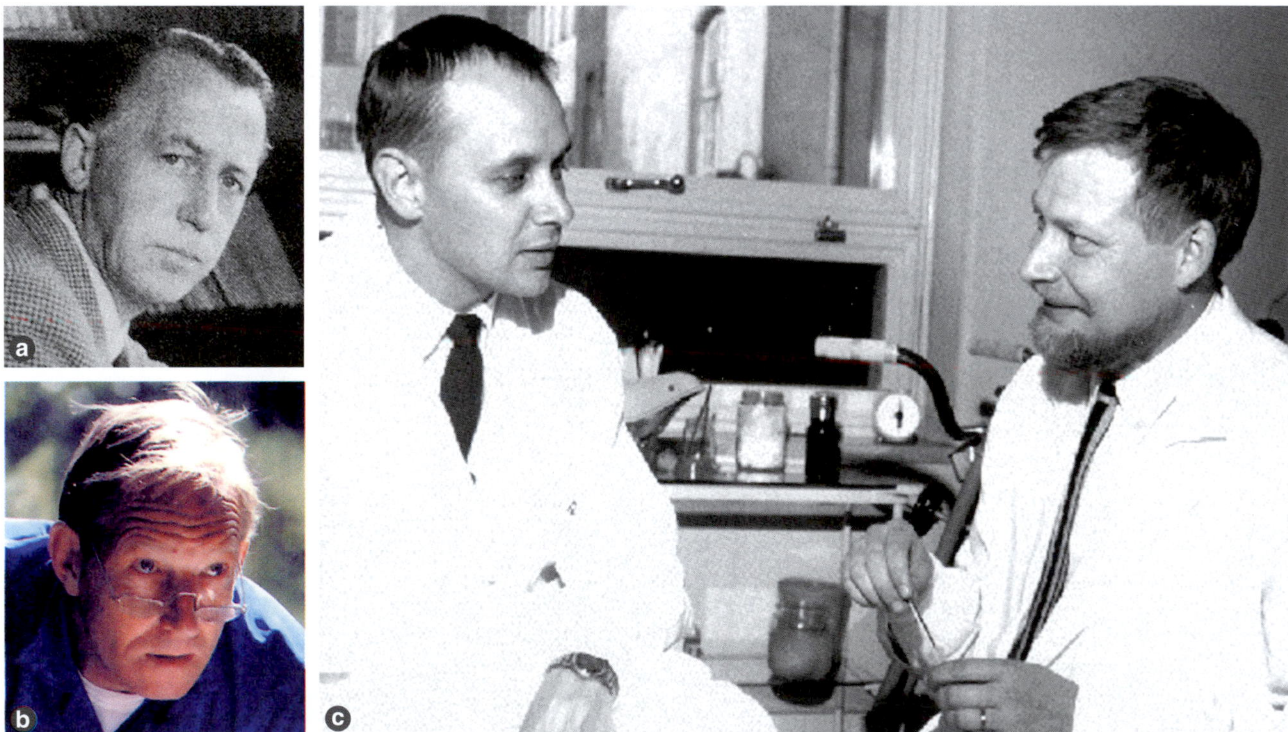

FIGURE 0.4 *(a)* Erik Hohwü-Christensen was the first physiology professor at the College of Physical Education at Gymnastik-och Idrottshögskolan in Stockholm, Sweden. *(b)* Bengt Saltin, winner of the 2002 Olympic Prize. *(c)* Jonas Bergstrom (left) and Eric Hultman (right) were the first to use muscle biopsy to study muscle glycogen use and restoration before, during, and after exercise.

number of outstanding scientists, including Bengt Saltin, who was the 2002 Olympic Prize winner for his many contributions to the field of exercise and clinical physiology (see figure 0.4*b*).

In addition to their work at GIH, both Hohwü-Christensen and Åstrand interacted with physiologists at the Karolinska Institute in Stockholm, Sweden, who studied clinical applications of exercise. It is hard to single out the most exceptional contributions from this institute, but Jonas Bergstrom's (figure 0.4*c*) reintroduction of the biopsy needle (ca. 1966) to sample muscle tissue was a pivotal point in the study of human muscle biochemistry and muscle nutrition. This technique, which involves withdrawing a tiny sample of muscle tissue with a needle inserted into the muscle through a small incision, was originally introduced in the early 1900s to study muscular dystrophy. The needle biopsy enabled physiologists to conduct histological and biochemical studies of human muscle before, during, and after exercise.

Other invasive studies of blood circulation were subsequently conducted by physiologists at GIH and at the Karolinska Institute. Just as the HFL had been the mecca of exercise physiology research between 1927 and 1947, the Scandinavian laboratories were equally noteworthy beginning in the late 1940s. Many leading investigations over the following 35 years were collaborations between American and Scandinavian exercise physiologists. Norwegian Per Scholander introduced a gas analyzer in 1947. Finn Martti Karvonen published a formula for calculating exercise heart rate that is still widely used today.

Other Research Milestones

Physiology has always been the basis for clinical medicine. In the same way, exercise physiology has provided essential knowledge for many other areas, such as physical education, physical fitness, physical therapy, and health promotion. In the late 1800s and early 1900s, physicians such as Amherst College's Edward Hitchcock, Jr. and Harvard's Dudley Sargent studied body proportions (anthropometry) and the effects of physical training on strength and endurance. Although a number of physical educators introduced biological science to the undergraduate physical education curriculum, Peter Karpovich, a Russian immigrant who had been briefly associated with the HFL (figure 0.5*a*), played a major role in introducing physiology to physical education. Karpovich established his own research facility and taught physiology at Springfield College (Massachusetts) from 1927 until his death in 1968.

Although he made numerous contributions to physical education and exercise physiology research, he is best remembered for the outstanding students he advised, including Charles Tipton and Loring Rowell, both recipients of the American College of Sports Medicine Honor and Citation Awards.

Another Springfield faculty member, swim coach Thomas K. (T.K.) Cureton (figure 0.5*b*), created an exercise physiology laboratory at the University of Illinois at Urbana-Champaign in 1941. He continued his research and taught many of today's leaders in physical fitness and exercise physiology until his retirement in 1971. Physical fitness programs developed by Cureton and his students, as well as Kenneth Cooper's 1968 book, *Aerobics*, established a physiological rationale for using exercise to promote a healthy lifestyle.[9]

Another contributor to the establishment of exercise physiology as an academic endeavor was Elsworth R. "Buz" Buskirk (figure 0.5*c*). After holding positions as chief of the environmental physiology section at the Quartermaster Research and Development Center in Natick, Massachusetts (1954-1957), and research physiologist at the National Institutes of Health (1957-

FIGURE 0.5 *(a)* Peter Karpovich introduced the field of exercise physiology during his tenure at Springfield College. *(b)* Thomas K. Cureton directed the exercise physiology laboratory at the University of Illinois at Urbana-Champaign from 1941 to 1971. *(c)* At Penn State, Elsworth Buskirk founded an intercollege graduate program focusing on applied physiology (1966) and constructed The Laboratory for Human Performance Research (1974).

1963), Buskirk moved to Pennsylvania State University, where he stayed for the remainder of his career. At Penn State, Buz founded the Intercollege Graduate Program in Physiology (1966) and constructed The Laboratory for Human Performance Research (1974), the nation's first freestanding research institute devoted to the study of human adaptation to exercise and environmental stress. He remained an active scholar until his death in April of 2010.

Although there was some awareness as early as the mid-1800s of a need for regular physical activity to maintain optimal health, this idea did not gain popular acceptance until the late 1960s. Subsequent research has continued to support the importance of exercise in slowing the physical decline associated with aging, preventing or mitigating the problems associated with chronic diseases, and rehabilitating injuries.

Development of Contemporary Approaches

Much advancement in exercise physiology must be credited to improvements in technology. In the late 1950s, Henry L. Taylor and Elsworth R. Buskirk published two seminal papers[6,28] describing the criteria for measuring maximal oxygen uptake and establishing that measure as the gold standard for cardiorespiratory fitness. In the 1960s, development of electronic analyzers to measure respiratory gases made studying energy metabolism much easier and more productive than before. This technology and radio telemetry (which uses radio-transmitted signals), used to monitor heart rate and body temperature during exercise, were developed as a result of the U.S. space program. Although such instruments took much of the labor out of research, they did not alter the direction of scientific inquiry. Until the late 1960s, most exercise physiology studies focused on the whole body's response to exercise. The majority of investigations involved measurements of such variables as oxygen uptake, heart rate, body temperature, and sweat rate. Cellular responses to exercise received little attention.

Biochemical Approaches

In the mid-1960s, three biochemists emerged who were to have a major impact on the field of exercise physiology. John Holloszy (figure 0.6*a*) at Washington University in St. Louis, Missouri, Charles "Tip" Tipton (figure 0.6*b*) at the University of Iowa, and Phil Gollnick (figure 0.6*c*) at Washington State University first used rats and mice to study muscle metabolism and examine factors related to fatigue. Their publications and their training of graduate and postdoctoral students resulted in a more biochemical approach to exercise physiology research. Holloszy was ultimately awarded the 2000 Olympic Prize for his contributions to exercise physiology and health.

Before the 1960s, there were few biochemical studies on the adaptations of muscle to training. Although the field of biochemistry can be traced to the early part of the 20th century, this special area of chemistry was not applied to human muscle until Bergstrom and Hultman reintroduced and popularized the needle biopsy procedure in 1966. Initially, this procedure was used to examine glycogen depletion during exhaustive exercise and its resynthesis during recovery. In the early 1970s, as noted earlier, a number of exercise physiologists used the muscle biopsy method, histological staining, and the light microscope to determine human muscle fiber types.

Around the time Bergstrom reintroduced the needle biopsy procedure, exercise physiologists who were well trained as biochemists emerged. In Stockholm, Bengt Saltin realized the value of this procedure for studying

FIGURE 0.6 *(a)* John Holloszy was the winner of the 2000 Olympic Prize for scientific contributions in the field of exercise science. *(b)* Charles Tipton was a professor at the University of Iowa and the University of Arizona and a mentor to many students who have become the leaders in molecular biology and genomics. *(c)* Phil Gollnick conducted muscle and biochemical research at Washington State University.

human muscle structure and biochemistry. He first collaborated with Bergstrom in the late 1960s to study the effects of diet on muscle endurance and muscle nutrition. About the same time, Reggie Edgerton (University of California at Los Angeles) and Phil Gollnick were using rats to study the characteristics of individual muscle fibers and their responses to training. Saltin subsequently combined his knowledge of the biopsy procedure with Gollnick's biochemical talents. These researchers were responsible for many early studies on human muscle fiber's characteristics and use during exercise. Although many biochemists have used exercise to study metabolism, few have had more impact on the current direction of human exercise physiology than Bergstrom, Saltin, Tipton, Holloszy, and Gollnick.

Other Tools and Techniques

The history of exercise physiology has, in some ways, been driven by advancements in technologies adapted from basic sciences. The early studies of energy metabolism during exercise were made possible by the invention of gas-collecting equipment and chemical analysis of oxygen and carbon dioxide. Chemical determination of blood lactic acid seemed to provide some insights regarding the aerobic and anaerobic aspects of muscular activity, but these data told us little regarding the production and removal of this by-product of exercise. Likewise, blood glucose measurements taken before, during, and after exhaustive exercise proved to be interesting data but were of limited value for understanding the energy exchange at the cellular level.

Over the last 30 years, muscle physiologists have used various chemical procedures to understand how muscles generate energy and adapt to training. Test tube experiments (in vitro) with muscle biopsy samples have been used to measure muscle proteins (enzymes) and determine the muscle fiber's capacity to use oxygen. Although these studies provided a snapshot of the fiber's potential to generate energy, they often left more questions than answers. It was natural, therefore, for the sciences of cell biology to move to an even deeper level. It was apparent that the answers to those questions must lie within the fiber's molecular makeup.

Although not a new science, molecular biology has become a useful tool for exercise physiologists who wish to delve more deeply into the cellular regulation of metabolism and adaptations to the stress of exercise. Physiologists like Frank Booth and Ken Baldwin (figure 0.7) have dedicated their careers to understanding the molecular regulation of muscle fiber characteristics and function and have laid the groundwork for our current understanding of the genetic controls of muscle growth and atrophy. The use of molecular biological techniques to study the contractile characteristics of single muscle fibers is discussed in chapter 1.

FIGURE 0.7 *(a)* Frank Booth and *(b)* Ken Baldwin.

Well before James Watson and Francis Crick unraveled the structure of deoxyribonucleic acid (DNA) in 1953, scientists appreciated the importance of genetics in predetermining the structure and function of all living organisms. The newest frontier in exercise physiology combines the study of molecular biology and genetics. Since the early 1990s, scientists have attempted to explain how exercise causes signals that affect the expression of genes within skeletal muscle.

In retrospect, it is apparent that since the beginning of the 20th century, the field of exercise physiology has evolved from measuring whole-body function (i.e., oxygen consumption, respiration, and heart rate) to molecular studies of muscle fiber genetic expression. There is little doubt that exercise physiologists of the future will need to be well grounded in biochemistry, molecular biology, and genetics.

Integrative Physiology

VIDEO 0.1 Presents Jim Pawelczyk discussing the integration of cellular-level processes with a view of the entire organism.

With the announcement of the sequencing of the human genome in 2001, it was hoped that one day, scientists could simply analyze cheek cells from a mouth swab and, by looking at your gene sequence, predict whether you were at risk for developing diabetes or cardiovascular disease.[7,8] More promising was the idea that detection of these predictive genetic variations could aid in developing more effective treatments for these debilitating diseases.

These advances in biotechnology have produced huge volumes of data over the past several years, but the initial optimism regarding the prediction and treatment of human disease has not been fulfilled.[17] While there are a few specific gene mutations that have reliable predictive power, such as the breast cancer gene BRCA1, the translation of genetic technologies into predictive diagnostics or therapies has largely not occurred. In fact, analysis of traditional risk factors still has much more predictive power for evaluating risk for type 2 diabetes than the evaluation of genetic risk scores based on 20 different gene variants associated with this disease.[27]

In the era of mega-genomic data, where does the study of physiology fit in? And is it still relevant to human health and disease? One outspoken advocate for the field of **integrative physiology** is Dr. Michael J. Joyner. Dr. Joyner is an award-winning, distinguished investigator at the Mayo Clinic who has critically questioned the functional value of so-called reductionist thinking in molecular biology. In contrast to examining biological processes at the lowest common level (for example, how genes code for proteins in cells), integrative physiology examines how whole organisms function and adapt to internal and external stresses (including exercise). This approach is informed by the concepts of homeostasis, regulated organ systems, and redundancy in physiological systems. Moreover, integrative physiologists strive to ask hypothesis-driven research questions and design defensible experiments to test those hypotheses.

The importance of seeking to study biological questions from an integrative, regulated approach is highlighted by the influences of culture, environment, and behavior on disease pathology. The challenge for integrative physiologists is to incorporate key findings from genetics and molecular biology and to examine how behavioral patterns, including physical activity, diet, and stress, interplay with this genetic variation to affect health and disease.

Translational Physiology

Exercise physiologists, by the nature of the topics we study and the variety of approaches we use in those studies, make valuable contributions to what has become known as **translational physiology**. *Translational physiology* is a term that was originally used in the early 1990s to refer to the research process needed to link cancer risk with its predisposing genetic factors.[25] The field of translational physiology has broadened substantially from that time to include the processes by which basic research findings are extended to the clinical research setting, then to the realm of clinical practice, and finally to health policy (figure 0.8). This translational research continuum, however, works best in a bidirectional manner, such that population-based problems, like obesity, also drive the basic research questions that exercise physiologists ask. In turn, these basic research findings eventually drive changes in clinical practice and overall community health.

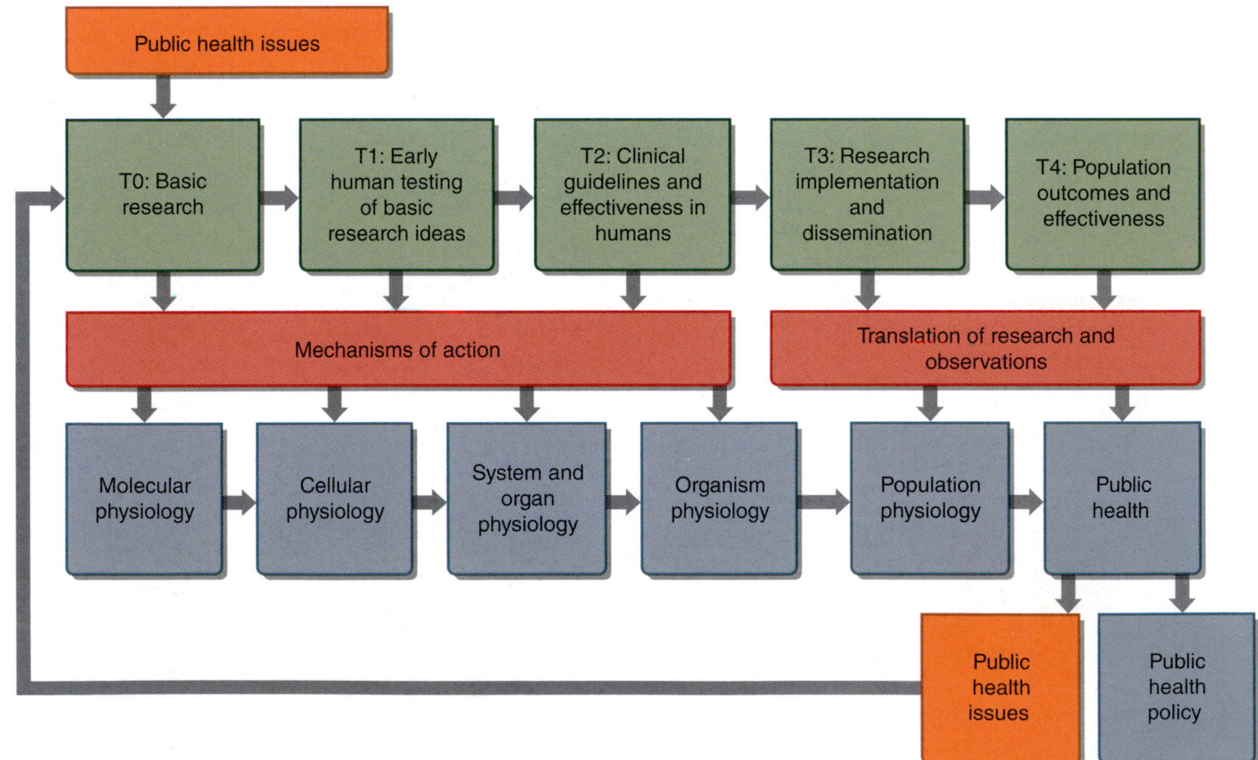

FIGURE 0.8 Flow chart for translational physiology.
Adapted from Seals (2013).

A good example of opportunities in translational physiology is in the field of aging. Advancing age by itself is a risk factor for many chronic diseases and presents a significant challenge to our health care system and to society in general. In order to fully understand the underlying physiology of aging and be able to engage in appropriate interventions to keep the aging population healthy, we must understand the aging process from the molecular level all the way to the community and population levels. Being able to successfully contribute to the translational physiology process requires a broad skill set to critically examine data and approach scientific problems with new goals in mind, from bench to bedside and from bedside to community.

In Review

- In an era that seems to stress a reductionist (genes, molecules) approach to science, there is an acute need for exercise physiologists to continue to study biological questions from an integrative, hypothesis-driven approach.
- The field of translational physiology addresses the processes by which basic research findings are extended to the clinical research setting, then to the realm of clinical practice, and finally to health policy.

Pioneering Women in Exercise Physiology

While outstanding female exercise physiologists are now commonplace, as in many areas of science, the contributions of women to exercise physiology were slow to gain recognition. In 1954, Irma Rhyming collaborated with her future husband, P.-O. Åstrand, to publish a classic study that provided a means to predict aerobic capacity from submaximal heart rate.[1] Although this indirect method of assessing physical fitness has been challenged over the years, its basic concept is still in use today.

In the 1970s, two Swedish women, Birgitta Essén and Karen Piehl (figure 0.9), gained international attention for their research on human muscle fiber composition and function. Essen, who collaborated with Bengt Saltin, was instrumental in adapting microbiochemical methods to study the small amounts of tissue obtained with the needle biopsy procedure. Her efforts enabled others to conduct studies on the muscle's use of carbohydrates and fats and to identify different muscle fiber types. Piehl published a number of studies that illustrated which muscle fiber types were activated during both aerobic and anaerobic exercise.

In the 1970s and 1980s, a third Scandinavian female physiologist, Bodil Nielsen, daughter of Marius Nielsen, actively conducted studies on human responses to environmental heat stress and dehydration. Her studies even encompassed measurements of body temperature during immersion in water. At about the same time, an American exercise physiologist, Barbara Drinkwater (figure 0.9*c*), was doing similar work at the University of California at Santa Barbara. Her studies were often conducted in collaboration with Steven Horvath, D.B. Dill's son-in-law and director of the UCSB's environmental physiology laboratory. Drinkwater's contributions to environmental physiology and study of the physiological problems confronting the female athlete gained international recognition. In addition to their scientific contributions, the legacy of these and other women in physiology includes the credibility they earned and the roles they played in attracting other

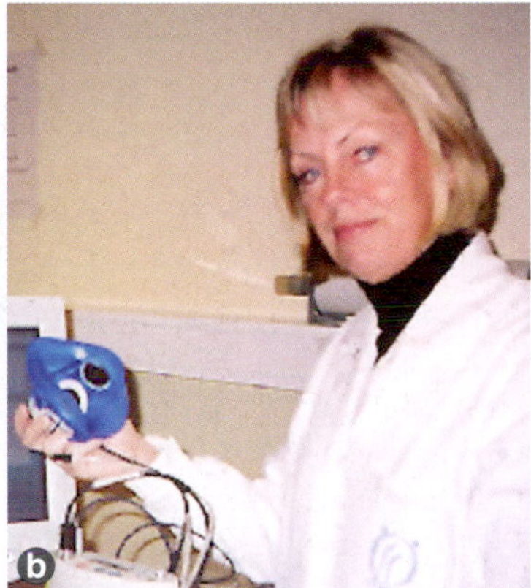

FIGURE 0.9 *(a)* Birgitta Essén collaborated with Bengt Saltin and Phil Gollnick in publishing the earliest studies on muscle fiber types in human muscle. *(b)* Karen Piehl was among the first physiologists to demonstrate that the nervous system selectively recruits type I (slow-twitch) and type II (fast-twitch) fibers during exercise of differing intensities. *(c)* Barbara Drinkwater was among the first to conduct studies on female athletes and to address issues specifically related to the female athlete.

young women to the fields of exercise physiology and medicine.

The intent of this section has been to provide readers with an overview of the personalities and technologies that have helped to shape the field of exercise physiology. Naturally, a comprehensive review of all the scientists and research associated with this field is not possible in a text intended as an introduction to exercise physiology, but for those students who wish to take an in-depth look at the historical background in exercise physiology, there are several good sources. Now that we understand the historical basis for the discipline of exercise physiology, from which sport physiology emerged, we can explore some basic principles of, and tools used in, exercise and sport physiology.

Exercise Physiology in the 21st Century

The field of exercise physiology is rapidly evolving. Ever-expanding technological developments and new approaches to science have substantial implications for health, medicine, and biomedical research. Exercise physiology and our understanding of the physiological processes that underpin physical activity are often at the forefront of this new age of science.

Exercise in Personalized Medicine

VIDEO 0.2 Presents Jim Pawelczyk discussing the four P's of medicine and the important role of exercise in individualized health strategies.

In 2007, the United States Congress passed the Genomics and Personalized Medicine Act. The intent of this legislation was to implement and support research related to formulating a personalized prescription to fit each patient's unique genetic and environmental characteristics in order to optimize health care strategies.[15,16] This personalized medicine concept first emerged in the field known as pharmacogenomics, which provided scientific insights into why some individuals respond favorably to certain drugs while others do not respond (or may even respond adversely). For example, studies have identified two different genes that influence an individual's ability to metabolize the blood thinner warfarin and make it possible for a physician to prescribe an appropriate dosage to optimize the drug's therapeutic effectiveness for each individual patient.[29]

Similarly, there has been a recent push to personalize each individual's exercise prescription.[5] Exercise is a powerful intervention for the treatment of many different medical conditions—cardiovascular disease, diabetes mellitus, osteoporosis, metabolic diseases, and many more. However, there is significant heterogeneity or variability in people's abilities to perform exercise and adapt to the effects of exercise training,[21] especially in individuals with different clinical disease manifestations. Moreover, researchers are just beginning to understand and formulate optimal training programs or personalized dosages of exercise to produce beneficial responses in these patients.

Researchers are designing experimental paradigms to determine (1) the mechanisms through which exercise produces effects (either positive or negative) on a cellular and systems level, (2) the optimal dosage of exercise to produce results in different clinical populations, (3) the best way to evaluate a person's responses to exercise on an individual and a group level, and (4) the benefit of adding exercise therapy to existing disease treatment strategies. Part of the challenge in personalizing exercise medicine is to understand, on a genomic and systems level, the mechanisms responsible for the huge variability in individuals' responses to exercise training. Eventually, the long-term outcomes from large randomized clinical trials examining intraindividual variability in responses to exercise in humans will make it possible to develop *personalized* strategies to be implemented in preventive health care interventions,[5] including the health benefits of regular exercise.

The "-Omics" Revolution

As a part of the Human Genome Project, scientists sequenced all 3.2 billion nucleotides that compose the human genome. This was mostly completed in 2003 (the last chromosome was sequenced in 2006) at an estimated cost of $2.7 billion. Today, the entire human genome can be sequenced for less than $1,000. This has opened up new fields of science, often called *-omics*. This new research area in turn has fostered the development of new technologies aimed at the universal detection of gene sequences and variants (genomics), the expression of genes at the messenger RNA (mRNA) level (transcriptomics), the proteins that are produced (proteomics), and other products of metabolic reactions (i.e., metabolites, studied in metabolomics)[30] involved in all aspects of physiological function (see figure 0.10). One purported appeal of -omics research is that a highly complex system (e.g., the exercising human) can be more fully understood if it is examined at each of the most basic levels of inquiry. As it is applied to exercise physiology, the primary goal of -omics research is to illuminate exercise physiology and behavior in order to better understand the preventive and therapeutic values of exercise.[4]

Exercise **genomics** research examines the role of individual (or groups of) genes in modifying the impact

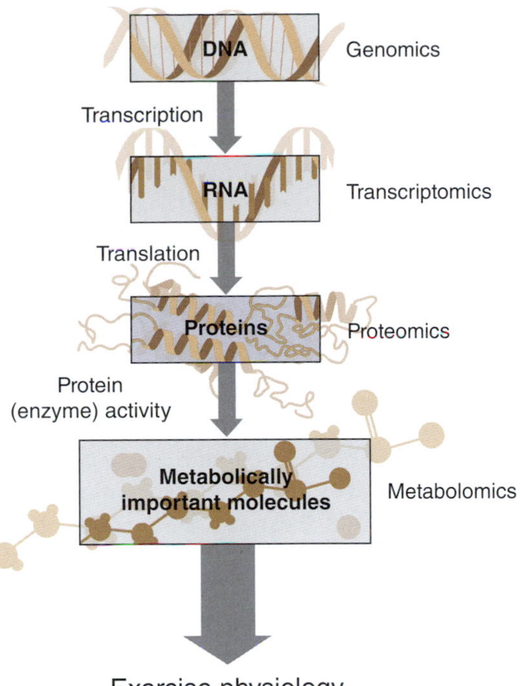

FIGURE 0.10 The link between genomics, transcriptomics, proteomics, and metabolomics in the context of exercise physiology.

of exercise training and physical activity on performance and health- and fitness-related traits. This is based on accumulating evidence that variations within the DNA sequences (called *single nucleotide polymorphisms*, or SNPs) of one or more genes may contribute to differences in exercise behavior, cardiorespiratory and muscular fitness, cardiovascular and metabolic function during acute exercise, and adaptations to exercise training.[3]

Using genomics approaches, researchers are also attempting to examine the genetic basis of these highly complex traits by examining tissue-specific mRNA levels. The technologies used to confirm a gene target and define its biological function are increasingly sophisticated and now include DNA and RNA sequencing, *in vitro* cell-based investigations, genetic modifications in animal models, and selective breeding of animals for extreme performance traits to identify target genes and their variants.

More recently, researchers have started to combine genomics and transcriptomics. That is, by examining the RNA strands produced during transcription (transcript abundance) in relevant tissues, researchers can predict a certain trait and identify gene targets for subsequent genomics research. These new gene targets can then be probed for their DNA sequence variants and their relation with other traits of interest. This integrated strategy within the -omics world has the potential to expand our understanding of exercise physiology at a level of detail that was not possible in the past. For example, understanding the exercise training–induced alterations in gene expression may provide novel candidates for genomics and genetics research aimed at further understanding the physiology of exercise.[3]

Exercise proteomics aims to study the entire protein content of a biological tissue in a particular situation (e.g., immediately after a resistance training session) or over a predetermined period (e.g., before and after months of endurance training), enabling investigators to examine the molecular mechanisms that underlie physiological adaptations to exercise.[22] The original tool for proteomic analysis was a procedure called two-dimensional polyacrylamide gel electrophoresis. During the current genomic research era, these gel-based methodologies have continued to improve and are now being coupled with newer techniques based on protein labeling, peptide fragmentation, and high-throughput mass spectrometry to improve proteomic analyses. Research combining proteomic data with the genomic approaches described previously will continue to expand our understanding of exercise physiology by providing a big picture of how exercise affects the various organs and systems in the body to improve physiological function, exercise performance, and health in general.

Epigenetics

It is now apparent that exercise alters gene expression, expression of transcription factors, and other regulatory proteins. These exercise-induced alterations have functional consequences at many levels, including metabolism, cardiovascular regulation, and fitness in general. However, more mechanisms and more tissues are likely involved in the integrative response to habitual exercise. **Epigenetics** is the study of changes in gene expression that occur without changing the genetic code itself. For example, inherited factors clearly influence an individual's response to exercise. However, additional environmental factors can alter those genes by epigenetic modifications, which are changes in how genes function that do not change the nucleotide sequence of the genes themselves. Environmental stimuli can alter the epigenome in a stable and inheritable fashion. Epigenetic modifications include DNA methylation, histone modification, and noncoding RNAs.[20] Although this area of research is relatively new, recent studies have demonstrated that epigenetic modifications contribute to altered gene expression in response to regular exercise; these findings have implications for improving our understanding of exercise-induced health benefits. The field of exercise epigenetics is still in its infancy but will increasingly provide new insights into human adaptations to exercise.

Bioinformatics

The techniques described in the previous sections generate an enormous amount of complex data. Sophisticated technologies, computer software, and statistical methods are therefore critical to analyzing the vast amount of genetic and molecular data generated from a single study, not to mention the integration of information from tens, hundreds, or thousands of experiments. **Bioinformatics** is essentially the management information system for molecular biology, serving as an intersection between molecular data and advanced mathematical and statistical approaches.[18]

Bioinformatics techniques allow us to address physiological questions that are otherwise unattainable using conventional methods. Using robotics, software for data processing and control, liquid handling devices, and sensitive detectors, high-throughput biology allows researchers to quickly conduct millions of chemical, genetic, or pharmacological tests by automating experiments on a large scale. By doing this, it becomes feasible to repeat experiments thousands of times. Using high-throughput methods, we can rapidly identify active compounds, antibodies, or genes that control or alter a particular physiological pathway.

Over the past decade, much has been learned from applying -omics approaches to the field of exercise physiology, and as this area of research continues to advance, bioinformatics will continue to play a role. The development of software-based analyses that would consider the genetic profile of an individual and then predict his or her response to aerobic exercise training is one example of a potential application of bioinformatics and functional -omics in exercise physiology. As more and more research laboratories begin to incorporate -omics approaches in exercise physiology, the need for the bioinformatics tools to analyze and interpret the data will only increase.

One of the main goals of exercise physiology in the 21st century is to map function from **genotype** (the genetic makeup of an individual) to **phenotype** (observable characteristics of an individual resulting from the interaction of its genotype with the environment). In essence, exercise is a powerful stimulus that influences gene transcription across multiple tissues with implications for multiple phenotypes. It is tempting to speculate that, in the future, perhaps a person's genotype will be fed into an algorithm that can make predictions about exercise-related attributes, such as endurance, speed, strength, or adaptability. From there, an individualized and optimized training program could be developed.

However, it is important to remember that, while these reductionist methods and -omics approaches have provided important new information about the genes and pathways that underlie the physiological responses to exercise, a much more comprehensive understanding of the complex interaction among various genetic and epigenetic factors is required to fully optimize the use of exercise for disease prevention and treatment.

Exercise Physiology Beyond Earth's Boundaries

An important segment of exercise physiology concerns the response and adaptation of people to extremes of heat, cold, and altitude. Understanding and controlling the physiological stresses and adaptations that occur at these environmental limits have contributed directly to notable societal achievements such as construction of the Brooklyn Bridge, the Hoover Dam, pressurized aircraft, and underwater habitats for the commercial diving industry.

The next generation of environmental challenges will also require such physiological expertise. Commercial space vehicles now travel routinely to low Earth orbit. NASA recently announced a set of new initiatives that will place humans in deep orbits near the moon in the late 2020s, followed by regular trips to Martian orbit in the 2030s. Indeed, we are on the verge of becoming an interplanetary civilization!

There are tremendous physiological and psychological challenges imposed on humans living in space and on planetary bodies for extended periods of time. The continuous pull of gravity contributes to the growth and adaptation of postural skeletal muscles, loads bones, which increases their size and density, and requires the cardiovascular system to maintain blood pressure and brain blood flow. In a microgravity environment (free fall around the Earth or constant-velocity conditions in deep space), the reduction in loading leads to dramatic losses in muscle mass and strength, osteoporosis, and exercise intolerance at rates that mimic those seen in spinal cord–injured patients.

Beginning in the 1980s, experiments done aboard a series of dedicated space shuttle flights investigated these problems in detail. The National Aeronautics and Space Administration (NASA) began flying the European Space Agency–developed Spacelab module, ushering in a new era of internationally sponsored scientific research into low Earth orbit. The Spacelab Life Sciences (SLS-1, SLS-2) missions (STS-40 and STS-58) emphasized the study of cardiorespiratory, vestibular, and musculoskeletal adaptations to microgravity, and the Life and Microgravity Sciences mission (STS-78) concentrated on neuromuscular adaptation. The 1998

Neurolab mission (STS-90), with an exclusive neuroscience theme, concluded flights of the Spacelab module. Dr. James A. Pawelczyk, a Penn State exercise physiologist and payload specialist on that flight, cotaught the first exercise physiology class from space!

With the end of the space shuttle program, work continues today aboard the International Space Station, which has provided a continuous human presence in space for nearly 20 years. The tools of modern molecular biology are helping to elucidate how loading, radiation, and stress interact to affect all physiological systems.

For the exercise physiologist, the question is what combination of resistance and aerobic exercise training can prevent or diminish the changes that occur during space exploration. At this time, the answer is not complete. Furthermore, if physical conditioning is required before, during, and after space missions that could last up to 30 months, how should exercise prescriptions be individualized, evaluated, and updated? Without doubt, further research in exercise and environmental physiology will be essential to complete what is destined to be the largest exploration feat of the 21st century.

Research: The Foundation for Understanding

Exercise and sport scientists actively engage in research to better understand the mechanisms that regulate the body's physiological responses to acute bouts of exercise, as well as its adaptations to training and detraining. Most of this research is conducted at major research universities, medical centers, and specialized institutes using standardized research approaches and select tools of the exercise physiologist.

The Research Process

Science and research (the process by which science is developed) involve a process designed to pose and answer appropriate questions, develop testable hypotheses, test those hypotheses appropriately, generate usable data, interpret those data, and either accept or refute the original hypotheses. The research process is illustrated in figure 0.11. Scientists are constantly challenged to make careful observations either from nature or the scientific literature, then ask focused questions that can be examined using a well-designed and well-controlled experimental process. The usual result of this overall process is submitting a research manuscript to an appropriate scientific journal, where it is peer reviewed, revised, and (hopefully) published. As other scientists read the research paper, they may in turn craft their own follow-up questions, and the process continues.

Research Settings

Research can be conducted either in the laboratory or the field. Laboratory tests are usually more accurate because more specialized and sophisticated equipment can be used and conditions can be carefully controlled. As an example, the direct laboratory measurement of maximal oxygen uptake ($\dot{V}O_{2max}$) is considered the most accurate estimate of cardiorespiratory endurance capacity. However, some field tests, such as the 1.5 mi (2.4 km) run, are also used to estimate $\dot{V}O_{2max}$. These field tests, which measure the time it takes to run a set distance or the distance that can be covered in a fixed time, are not totally accurate, but they provide a reasonable estimate of $\dot{V}O_{2max}$, are inexpensive to conduct, and allow many people to be tested in a short time. Field tests can be conducted in the workplace, on a running

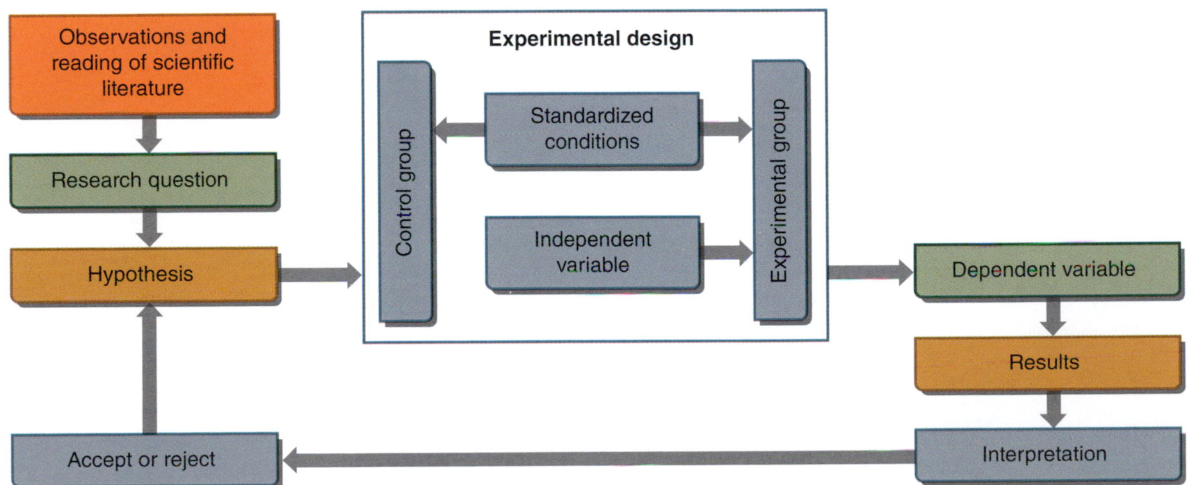

FIGURE 0.11 A simplified diagram of the typical process involved in scientific research.

track or in a swimming pool, or during athletic competitions. To measure $\dot{V}O_{2max}$ directly and accurately, one would need to go to a university or clinical laboratory.

Research Tools: Ergometers

When physiological responses to exercise are assessed in a laboratory setting, the participant's physical effort must be controlled to provide a measurable exercise intensity. This is generally accomplished through use of ergometers. An **ergometer** (*ergo* = work; *meter* = measure) is an exercise device that allows the intensity of exercise to be controlled (standardized) and measured.

Treadmills

Treadmills are the ergometers of choice for most researchers and clinicians, particularly in the United States. With these devices, a motor drives a large belt on which a subject can either walk or run; thus, these ergometers are often called motor-driven treadmills (see figure 0.12). Belt length and width must accommodate the individual's body size and stride length. For example, it is nearly impossible to test elite athletes on treadmills that are too short, or obese subjects on treadmills that are too narrow or not sturdy enough.

Treadmills offer a number of advantages. Walking is a natural activity for almost everyone, so individuals normally adjust to the skill required for walking on a treadmill within a few minutes. Also, most people can achieve their peak values for most physiological variables (heart rate, ventilation, oxygen uptake) on the treadmill, although some athletes (e.g., competitive cyclists) achieve higher values on ergometers that more closely match their mode of training or competition.

Treadmills do have some disadvantages. They are generally more expensive than simpler ergometers, like the cycle ergometers discussed next. They are also bulky, require electrical power, and are not very portable. Accurate measurement of blood pressure during treadmill exercise can be difficult because both the noise associated with normal treadmill operation and subject movement can make hearing through a stethoscope difficult.

Cycle Ergometers

For many years, the **cycle ergometer** was the primary testing device in use, and it is still used extensively in both research and clinical settings. Cycle ergometers can be designed to allow subjects to pedal either in the normal upright position (see figure 0.13) or in reclining or semireclining positions.

Cycle ergometers in a research setting generally use either mechanical friction or electrical resistance. With mechanical friction devices, a belt encompassing a flywheel is tightened or loosened to adjust the resistance against which the cyclist pedals. The power output depends on the combination of the resistance and the pedaling rate—the faster one pedals, the greater the power output. To maintain the same power output throughout the test, one must maintain the same pedaling rate, so pedaling rate must be constantly monitored.

With electrically braked cycle ergometers, the resistance to pedaling is provided by an electrical conductor that moves through a magnetic or electromagnetic field. The strength of the magnetic field determines the resistance to pedaling. These ergometers can be controlled so that the resistance increases automatically

FIGURE 0.12 A motor-driven treadmill.

FIGURE 0.13 A cycle ergometer.

as pedal rate decreases, and decreases as pedal rate increases, to provide a constant power output.

Similar to treadmills, cycle ergometers offer some advantages and disadvantages compared to other ergometers. Exercise intensity on a cycle ergometer does not depend on the subject's body weight. This is important when one is investigating physiological responses to a standard rate of work (power output). As an example, if someone lost 5 kg (11 lb), data derived from treadmill testing could not be compared with data obtained before the weight loss because physiological responses to a set speed and grade on the treadmill vary with body weight. After the weight loss, the rate of work at the same speed and grade would be less than before. With the cycle ergometer, weight loss does not have as great an effect on physiological response to a standardized power output. Thus, walking or running is often referred to as weight-dependent exercise, while cycling is weight independent.

Cycle ergometers also have disadvantages. If the subject does not regularly engage in that form of exercise, the leg muscles will likely fatigue early in the exercise bout. This may prevent a subject from attaining a true maximal intensity. When exercise is limited in this way, responses are often referred to as peak exercise intensity rather than maximal exercise intensity. This limitation may be attributable to local leg fatigue, blood pooling in the legs (less blood returns to the heart), or the use of a smaller muscle mass during cycling than during treadmill exercise. Trained cyclists, however, tend to achieve their highest peak values on the cycle ergometer.

Other Ergometers

Other ergometers allow athletes who compete in specific sports or events to be tested in a manner that more closely approximates their training and competition. For example, an arm ergometer may be used to test athletes or nonathletes who use primarily their arms and shoulders in physical activity. Arm ergometry has also been used extensively to test and train athletes paralyzed below arm level. The rowing ergometer was devised to test competitive rowers.

Valuable research data have been obtained by instrumenting swimmers and monitoring them during swimming in a pool. However, the problems associated with turns and constant movement led to the use of two devices—tethered swimming and swimming flumes. In tethered swimming, the swimmer is attached to a harness connected to a rope, a series of pulleys, and counterbalancing weights and must swim against the pull of the apparatus to maintain a constant position in the pool. A swimming flume allows swimmers to more closely simulate their natural swimming strokes. The swimming flume operates by pumps that circulate water past the swimmer, who attempts to maintain body position in the flume. The pump circulation can be increased or decreased to vary the speed at which the swimmer must swim. The swimming flume, which unfortunately is very expensive, has at least partially resolved the problems with tethered swimming and has created new opportunities to investigate the sport of swimming.

When one is choosing an ergometer, the concept of specificity is particularly important with highly trained athletes. The more specific the ergometer is to the actual pattern of movement used by the athlete in his or her sport, the more meaningful will be the test results.

In Review

- Treadmills generally produce higher peak values than other ergometers for almost all assessed physiological variable, such as heart rate, ventilation, and oxygen uptake.
- Cycle ergometers are the most appropriate devices for evaluating changes in submaximal physiological function before and after training in people whose weights have changed. Unlike treadmill exercise, cycle ergometer intensity is largely independent of body weight.

Research Designs

In exercise physiology research, there are two basic types of research design: cross-sectional and longitudinal. With a **cross-sectional research design**, a cross section of the population of interest (that is, a representative sample) is tested at one specific time, and the differences between subgroups from that sample are compared. With a **longitudinal research design**, the same research subjects are retested periodically after initial testing to measure changes over time in variables of interest.

The differences between these two approaches are best understood through an example. The objective of a research study is to determine whether a regular program of distance running increases the concentration of cardioprotective high-density lipoprotein cholesterol (HDL-C) in the blood. High-density lipoprotein cholesterol is the desirable form of cholesterol; increased concentrations are associated with reduced risk for heart disease. Using the cross-sectional approach, one could, for example, test a large number of people who fall into the following categories:

- A group of subjects who do no training (the control group)

- A group of subjects who run 24 km (15 mi) per week
- A group of subjects who run 48 km (30 mi) per week
- A group of subjects who run 72 km (45 mi) per week
- A group of subjects who run 96 km (60 mi) per week

One would then compare the results from all the groups, basing one's conclusions on how much running was done. Using this approach, exercise scientists found that weekly running results in elevated HDL-C levels, suggesting a positive health benefit related to running distance. Furthermore, as illustrated in figure 0.14, there was a **dose–response relation** between these variables—the higher the dose of exercise training, the higher the resulting concentration of HDL-C. It is important to remember, however, that with a cross-sectional design, these are different groups of runners, not the same runners at different training volumes.

Using the longitudinal approach to test the same question, one could design a study in which untrained people would be recruited to participate in a 12-month distance-running program. One could, for example, recruit 40 people willing to begin running and then randomly assign 20 to a training group and the remaining 20 to a **control group**. Both groups would be followed for 12 months. Blood samples would be tested at the beginning of the study and then at 3-month intervals, concluding at 12 months when the program ended. With this design, both the running group and the control group would be followed over the entire period of the study, and changes in their HDL-C levels could be determined across each period. Studies have been conducted using this longitudinal design to examine changes in HDL-C with training, but their results have not been as clear as the results of the cross-sectional studies. See figure 0.15 as an example. Note that in this figure, in contrast to figure 0.14, there is only a small increase in HDL-C in the subjects who are training. The control group stays relatively stable, with only minor fluctuations in their HDL-C from one 3-month period to the next.

A longitudinal research design is usually best suited to studying changes in variables over time. Too many factors that may taint results can influence cross-sectional designs. For example, genetic factors might interact so that those who perform well in long-distance running are also those who have high HDL-C levels. Also, different populations might follow different diets. In a longitudinal study, diet and other variables can be more easily controlled. However, longitudinal studies are time consuming and expensive to conduct, and are not always possible; cross-sectional studies do provide some insight into the questions at hand.

Research Controls

When we conduct research, it is important to be as careful as possible in designing the study and collecting the data. We see from figure 0.15 that changes in a variable over time resulting from an intervention such as exercise can be very small. Yet, even small changes

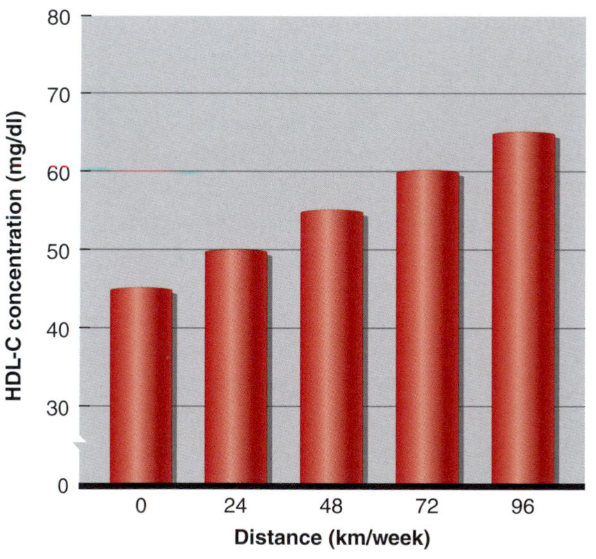

FIGURE 0.14 The relation between distance run per week and average high-density lipoprotein cholesterol (HDL-C) concentrations across five groups: nontraining control (0 km/week), 24 km/week, 48 km/week, 72 km/week, and 96 km/week. This illustrates a cross-sectional study design.

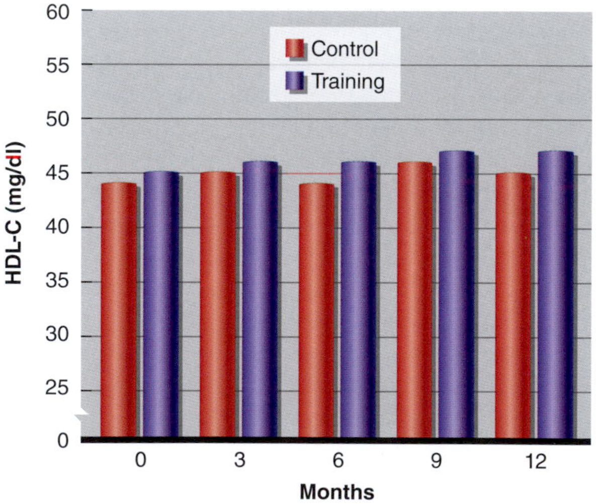

FIGURE 0.15 The relation between months of distance-running training and average high-density lipoprotein cholesterol (HDL-C) concentrations in an experimental group (20 subjects, distance training) and a sedentary (20 subjects) control group. This illustrates a longitudinal study design.

in a variable such as HDL-C can mean a substantial reduction in risk for heart disease. Recognizing this, scientists design studies aimed at providing results that are both accurate and reproducible. This requires that studies be carefully controlled.

Research controls are applied at various levels. Starting with the design of the research project, the scientist must determine how to control for variation in the subjects used in the study. The scientist must determine if it is important to control for the subjects' sex, age, or body size. To use age as an example, for certain variables, the response to an exercise training program might be different for a child or an aged person compared with a young or middle-aged adult. Is it important to control for the subject's smoking or dietary status? Considerable thought and discussion are needed to make sure that the subjects used in a study are appropriate for the specific research question being asked.

For almost all studies, it is critical to have a control group. In the longitudinal research design for the cholesterol study described earlier, the control group acts as a comparison group to make certain that any changes observed in the running group are attributable solely to the training program and not to any other factors, such as the time of the year or aging of the subjects during the course of the study. Experimental designs often employ a **placebo group**. Thus, in a study in which a subject might expect to have a benefit from the proposed intervention, such as the use of a specific food or drug, a scientist might decide to use three groups of subjects: an intervention group that receives the actual food or drug, a placebo group that receives an inert substance that looks exactly like the actual food or drug, and a control group that receives nothing. (The last group often serves as a time control, accounting for nonexperimentally induced changes that may occur over the course of the study period.) If the intervention and placebo groups improve their performance to the same level and the control group does not improve performance, then the improvement is likely the result of the placebo effect, or the expectation that the substance will improve performance. If the intervention group improves performance, and the placebo and control groups do not, then we can conclude that the intervention does improve performance.

One other way of controlling for the placebo effect is to conduct a study that uses a **crossover design**. In this case, each group undergoes both treatment and control trials at different times. For example, one group is administered the intervention for the first half of the study (e.g., 6 months of a 12-month study) and serves as a control during the last half of the study. The second group serves as a control during the first half of the study and receives the intervention during the second half. In some cases, a placebo can be used in the control phase of the study. Chapter 16, Ergogenic Aids in Sport, provides further discussion of placebo groups.

It is equally important to control data collection. The equipment must be calibrated so the researcher knows that the values generated by a given piece of equipment are accurate, and the procedures used in collecting data must be standardized. For example, when using a scale to measure the weight of subjects, researchers need to calibrate that scale by using a set of calibrated weights (e.g., 10 kg, 20 kg, 30 kg, and 40 kg) that have been measured on a precision scale. These weights are placed on the scale to be used in the study, individually and in combination, at least once a week to provide certainty that the scale is measuring the weights accurately. As another example, electronic analyzers used to measure

respiratory gases need to be calibrated frequently with gases of known concentration to ensure the accuracy of these analyses.

Finally, it is important to know that all test results are reproducible. In the example illustrated in figure 0.15, the HDL-C of an individual is measured every 3 months. If that person is tested 5 days in a row before he or she starts the training program, one would expect the HDL-C results to be similar across all 5 days, provided that diet, exercise, sleep, and time of day for testing remained the same. In figure 0.15, the values for the control group across 12 months varied from about 44 to 45 mg/dl, whereas the exercise group values increased from 45 to 47 mg/dl. Over five consecutive days, the measurements should not vary by more than 1 mg/dl for any one person if the researcher is going to pick up this small change over time. To control for reproducibility of results, scientists generally take several measurements, sometimes on different days, and then average the results before, during, and at the end of an intervention.

Confounding Factors in Exercise Research

Many factors can alter the body's acute response to a bout of exercise. For example, environmental conditions such as the temperature and humidity of the laboratory and the amount of light and noise in the test area can markedly affect physiological responses, both at rest and during exercise. Even the timing, volume, and content of the last meal and the quantity and quality of sleep the night before must be carefully controlled in research studies.

To illustrate this, table 0.1 shows how varying environmental and behavioral factors can alter heart rate at rest and during running on a treadmill at 14 km/h (9 mph). The subject's heart rate response during exercise differed by 25 beats/min when the air temperature was increased from 21 °C (70 °F) to 35 °C (95 °F). Most physiological variables that are normally measured during exercise are similarly influenced by environmental fluctuations. Whether one is comparing a person's exercise results from one day to another or comparing the responses of two different subjects, all of these factors must be controlled as carefully as possible.

Physiological responses, both at rest and during exercise, also vary throughout the day. The term **diurnal variation** refers to fluctuations that occur during a 24 h day. Because such variables as body temperature and heart rate vary naturally during a 24 h period, testing the same person in the morning on one day and in the afternoon on the next will produce different results. Test times must be standardized to control for this diurnal effect.

TABLE 0.1 **Heart Rate Responses to Running Differ with Variations in Environmental and Behavioral Conditions**

	Heart rate (beats/min)	
Environmental and behavioral factors	Rest	Exercise
Temperature (at 50% humidity)		
21 °C (70 °F)	60	165
35 °C (95 °F)	70	190
Humidity (at 21 °C)		
50%	60	165
90%	65	175
Noise level (at 21 °C, 50% humidity)		
Low	60	165
High	70	165
Food intake (at 21 °C, 50% humidity)		
Small meal 3 h before exercising	60	165
Large meal 30 min before exercising	70	175
Sleep (at 21 °C, 50% humidity)		
8 h or more	60	165
6 h or less	65	175

At least one other physiological cycle must also be considered. The normal 28-day menstrual cycle often involves considerable variations in

- body weight,
- total body water and blood volume,
- body temperature,
- metabolic rate, and
- heart rate and stroke volume (the amount of blood leaving the heart with each contraction).

Exercise scientists must control for menstrual cycle phase or the use of oral contraceptives (which similarly alter hormonal status), or both, when testing women. When older women are being tested, testing strategies must take into account menopause and hormone replacement therapies.

In summary, the conditions under which research participants are monitored, at rest and during exercise, must be carefully controlled. Environmental factors, such as temperature, humidity, altitude, and noise, can affect the magnitude of response of all basic physiological systems, as can behavioral factors such as eating patterns and sleep. Likewise, physiological measurements must be well controlled for diurnal and menstrual cycle variations.

Units and Scientific Notation

A set of international standards for units and abbreviations (SI, Le Système International d'Unités) serves as the preferred units of measurement in exercise and sport physiology. In this text, alternate units in common use (such as weight in pounds) are often provided as well. Many of these units are provided on the inside front cover of this text, and conversions for units in common use are found on the inside back cover.

In common writing and even in mathematics, the ratio between two numbers is typically written using a slash (/). For example, in dry air at 20 °C, the speed of sound is 343 m/s. That notation works well for simple fractions or ratios, and we have maintained it in this text. However, the notation gets confusing for relations of several—that is, more than two—variables. Take, for instance, one of the cornerstone measurements in exercise physiology, an individual's maximal oxygen uptake or maximal aerobic capacity, abbreviated $\dot{V}O_{2max}$. This important physiological measurement is the maximal volume of oxygen that an individual can use during exhaustive aerobic exercise, and can be measured in liters per minute, or L/min. However, because a large person can use more oxygen yet not be more aerobically fit, we often standardize this value to body weight in kilograms, that is, milliliters per kilogram per minute. Now the notation becomes a bit more complex and potentially more confusing. We could write the units as ml/kg/min, but what is being divided by what in this notation? Recall that L/min can also be written as $L \cdot min^{-1}$, just as the fraction $1/4 = 1 \cdot 4^{-1}$. To avoid errors and ambiguity, in exercise physiology, we use the exponent notation any time more than two variables are involved. Therefore, milliliters per kilogram per minute is written as $ml \cdot kg^{-1} \cdot min^{-1}$ rather than ml/kg/min.

Reading and Interpreting Tables and Graphs

This book contains references to specific research studies that have had a major impact on our understanding of exercise and sport physiology. Once scientists complete a research project, they submit the results of their research to one of the many research journals in sport and exercise physiology.

As in other areas of science, most of the quantitative research in exercise physiology is presented in the form of tables and graphs. Tables and graphs provide an efficient way for researchers to communicate the results of their studies to other scientists. For the student in exercise and sport physiology, a working knowledge of how to read and interpret tables and graphs is critical.

Tables are usually used to convey a large number of data points or complex data that are affected by several factors. Take table 0.1 as an example. It is important to first look at the title of the table, which identifies what information is being presented. In this case, the table is designed to illustrate how various conditions affect heart rate, at rest and during exercise. The left-hand column, along with the horizontal subheadings (like "Humidity (at 21 °C)"), specify the conditions under which the heart rate was measured. Columns 2 and 3 provide the mean heart rate values that correspond to each condition, with the middle column giving the resting value and the right-most column the exercise value. In every good table and graph, the units for each variable are clearly presented; in this table, heart rate is expressed in beats/min, or beats per minute. Pay careful attention to the units of measure used when interpreting a table or graph. From this table—a relatively simple one—we see that both resting and exercise heart rate were increased by increased ambient temperature and humidity, while noise level affected only resting heart rate. Similarly, consuming a large meal or getting less than 6 h of sleep also raises heart rate. These data could not easily have been shown in graphical form.

Graphs can provide a better view of trends in data, response patterns, and comparisons of data collected from two or more groups of subjects. For some students, graphs can be more difficult to read and interpret, but graphs are, and will remain, a critical tool in the understanding of exercise physiology. First, every graph has a horizontal axis, or *x*-axis, for the **independent variable** and a vertical axis, or *y*-axis, (or sometimes two) for the **dependent variable** or variables. Independent variables are those factors that are manipulated or controlled by the researcher, while dependent variables are those that change with—that is, depend on—the independent variables.

In figure 0.16, time of day is the independent variable and is therefore placed along the *x*-axis of the graph, while heart rate is the dependent variable (since heart rate *depends on* the time of day) and is therefore plotted on the *y*-axis. The units of measure for each variable are clearly displayed on the graph. Figure 0.16 is in the form of a line graph. Line graphs are useful in illustrating patterns or trends in data but should be used only to compare two variables that change in a continuous manner (for example, across time) and only if both the dependent and independent variables are numbers.

In a line graph, if the dependent variable goes up or down at a constant rate with the independent variable, the result will be a straight line. However, in physiology, the response pattern between variables is often not a straight line but a curve of one shape or another. In such cases, pay close attention to the slope of various parts of the curve as it changes across the graph. For

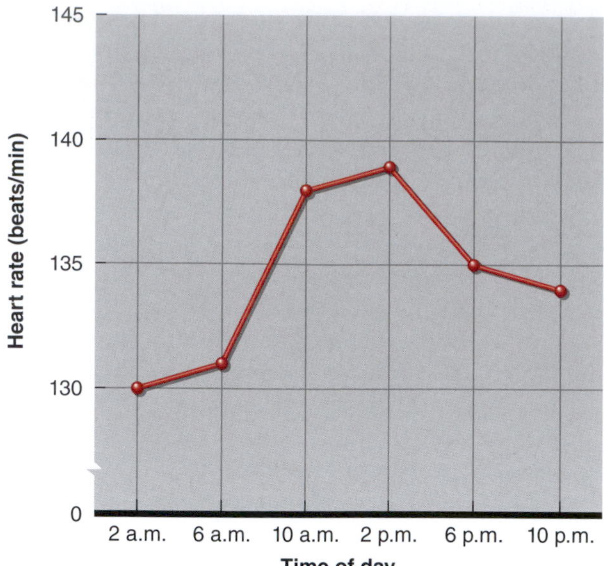

FIGURE 0.16 This line graph depicts the relation between the time of day (on the *x*-axis, independent variable) and heart rate during low-intensity exercise (on the *y*-axis, dependent variable) that was measured at that time of day with no change in the exercise intensity.

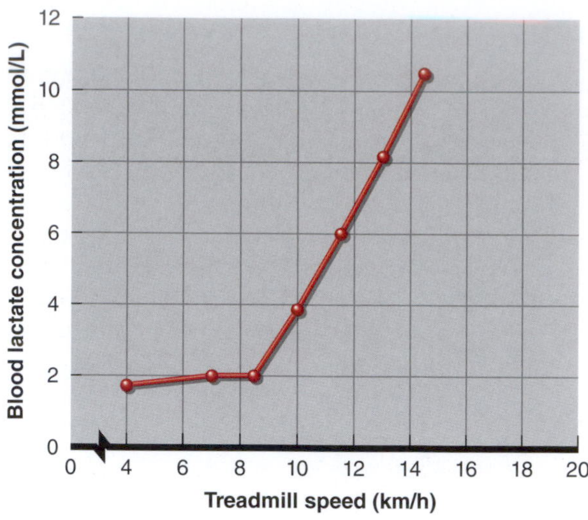

FIGURE 0.17 A line graph showing the nonlinear nature of many physiological responses. This graph shows that above a threshold (onset of response) of about 8.5 km/h, the slope of the blood lactate response increases sharply.

instance, figure 0.17 shows the concentration of lactate in the blood as subjects walk-run on a treadmill at various increasing speeds. At low treadmill speeds of 4 to 8 km/h, lactate increases very little. However, at about 8.5 km/h, a threshold is reached beyond which lactate increases more dramatically. In many physiological responses, both the threshold (onset of response) and the slope of the response beyond that threshold are important.

Data can also be plotted in the format of a bar graph. Bar graphs are commonly used when only the dependent variable is a number and the independent variable is a category. Bar graphs often show treatment effects, as in figure 0.14, which was previously discussed. Figure 0.14 shows the effect of distance run per week (a category) on HDL-C (a numerical response) in the bar graph format.

In Review

- Exercise physiologists make use of both cross-sectional (finding differences between groups at one point in time) and longitudinal (retesting the same subjects at different points in time) research designs.
- For all sound research studies, it is critical to have a control group as well as an experimental group. The control group often involves a placebo treatment rather than no treatment at all.
- Exercise physiologists use the SI units of measurement and abbreviations.

IN CLOSING

In this introduction, we highlighted the historical roots and scientific underpinnings of exercise and sport physiology. We learned that the current state of knowledge in these fields builds on the past and is a bridge to the future—many questions remain unanswered. New and exciting approaches and techniques are being developed continually. While reductionist (e.g., genomics) approaches are growing in popularity, being able to integrate those findings into a systems and whole-body perspective will never go out of style. Exercise and sport physiology is an important part of integrative and translational physiology. We briefly defined the acute responses to exercise bouts and chronic adaptations to long-term training. We concluded with an overview of the principles used in sport and exercise physiology research as well as an introduction to interpreting graphs, some important terminology, and SI units and their notation.

In part I, we begin examining physical activity the way exercise physiologists do as we explore the essentials of movement. In the next chapter, we examine the structure and function of skeletal muscle, how it produces movement, and how it responds during exercise.

KEY TERMS

acute exercise
bioinformatics
chronic adaptation
control group
crossover design
cross-sectional research design
cycle ergometer
dependent variable
diurnal variation
dose–response relation
environmental physiology
epigenetics
ergometer
exercise physiology
genomics
genotype
homeostasis
independent variable
integrative physiology
longitudinal research design
phenotype
physiology
placebo group
sport physiology
training effect
translational physiology
treadmill

STUDY QUESTIONS

1. What is exercise physiology? How does sport physiology differ?
2. Provide an example of "studying acute responses to a single bout of exercise."
3. Describe what is meant by "studying chronic adaptations to exercise training."
4. Describe the evolution of exercise physiology from the early studies of anatomy. Who were some of the key figures in the development of this field?
5. Describe the founding and the key areas of research emphasized by the Harvard Fatigue Laboratory. Who was the first research director of this laboratory?
6. Name the three Scandinavian physiologists who conducted research in the Harvard Fatigue Laboratory.
7. What is an ergometer? Name the two most commonly used ergometers and explain their advantages and disadvantages.
8. What factors must researchers consider when designing a research study to ensure that they get accurate and reproducible results?
9. What is translational physiology?
10. Define the following terms and discuss their relevance to exercise physiology: genomics, epigenetics, bioinformatics, genotype, and phenotype.
11. List several environmental conditions that could affect one's response to an acute bout of exercise.
12. What are the advantages and disadvantages of a cross-sectional versus a longitudinal study design?
13. When should data be depicted as a bar graph as opposed to a line graph? What purpose does a line graph serve?

STUDY GUIDE ACTIVITIES

In addition to the activities listed in the chapter opening outline, two other activities are available in the web study guide, located at

www.HumanKinetics.com/PhysiologyOfSportAndExercise

The **KEY TERMS** activity reviews important terms, and the end-of-chapter **QUIZ** tests your understanding of the material covered in the chapter.

PART I

Exercising Muscle

In the introduction, we explored the foundations of exercise and sport physiology. We defined these fields of study, gained a historical perspective of their development, looked at present trends as well as the future of exercise physiology, and established some basic concepts that we will follow through the remainder of this book. We also examined the tools and research methods used by exercise physiologists along with some tips about interpreting graphs and scientific notation. With this foundation, we can begin our main objective—understanding how the human body performs, and adapts to, exercise and physical activity. Because muscle is the foundation of all movement, we start with chapter 1, Structure and Function of Exercising Muscle, where we focus on skeletal muscle, examining the structure and function of skeletal muscles and muscle fibers and how they contract. We learn how muscle fiber types differ and why these differences are important to specific types of activity. Because movement requires energy, in chapter 2, Fuel for Exercise: Bioenergetics and Muscle Metabolism, we study the principles of metabolism, focusing on the primary form of usable energy, adenosine triphosphate (ATP), and how it is provided from the foods that we eat through three energy systems. In chapter 3, Neural Control of Exercising Muscle, we discuss how the nervous system initiates and controls muscle actions. Chapter 4, Hormonal Control During Exercise, presents an overview of the complex endocrine system, with a focus on hormonal control of energy metabolism, body fluid and electrolyte balance during exercise, and caloric intake. Finally, chapter 5, Energy Expenditure, Fatigue, and Muscle Soreness, discusses the measurement of energy expenditure, how energy expenditure changes from rest to varying types and intensities of exercise, the various causes of fatigue that limits exercise performance, and the causes of muscle cramping and feelings of soreness.

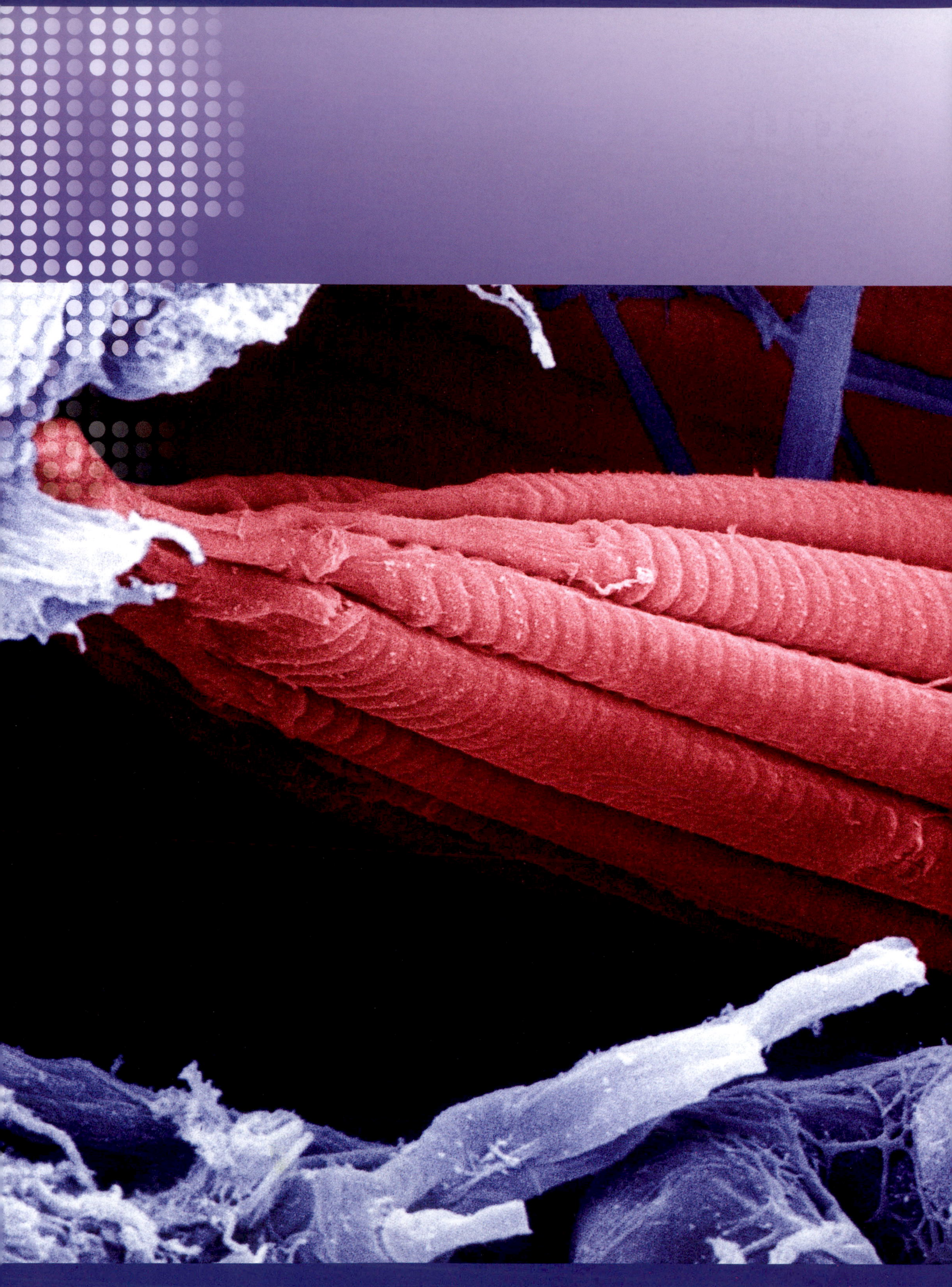

Structure and Function of Exercising Muscle

In this chapter and in the web study guide

Liam Hoekstra possesses a physique and physical attributes like many professional athletes: ripped abdominal muscles, enough strength to perform feats like an iron cross and inverted sit-ups, and amazing speed and agility. Not bad when you consider the fact that Liam could do all this when he was just 19 months old and weighed 10 kg (22 lb)! Liam has a rare genetic condition called myostatin-related muscle hypertrophy, a condition that was first described in an abnormally muscular breed of beef cattle in the late 1990s. Myostatin is a protein that inhibits the growth of skeletal muscles; myostatin-related muscle hypertrophy is a genetic mutation that blocks production of this inhibitory growth factor and thus promotes the rapid growth and development of skeletal muscles.

Liam's condition is extremely rare in humans, with fewer than 100 cases documented worldwide. However, studying this genetic phenomenon has helped scientists unlock secrets of how skeletal muscles grow and deteriorate. Research on Liam's condition could lead to new treatments for debilitating muscular conditions such as muscular dystrophy. On the darker side, it could open up a whole new realm of abuse by athletes who are looking for shortcuts to develop muscle size and strength, not unlike the illicit and dangerous use of anabolic steroids.

When the heart beats, when partially digested food moves through the intestines, and when the body moves in any way, muscle is involved. These many and varied functions of the muscular system are performed by three distinct types of muscle (see figure 1.1): smooth muscle, cardiac muscle, and skeletal muscle.

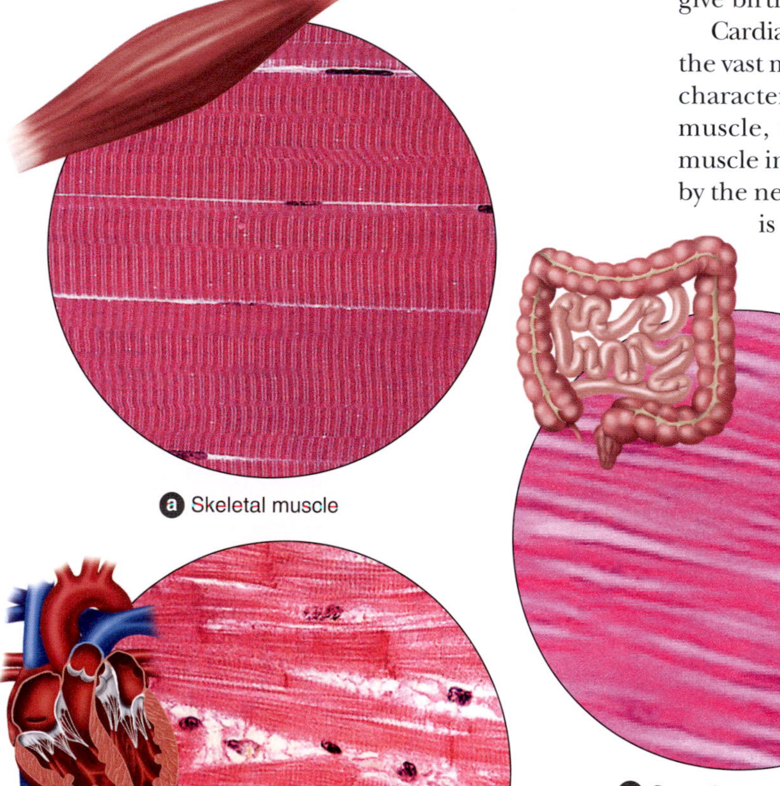

FIGURE 1.1 Microscopic photographs of the three types of muscle: *(a)* skeletal, *(b)* cardiac, and *(c)* smooth.

Smooth muscle is sometimes called involuntary muscle because it is not under direct conscious control. It is found in the walls of most blood vessels, where its contraction or relaxation leads to vessel constriction or dilation, respectively, to regulate blood flow. It is also found in the walls of most internal organs, allowing them to contract and relax, for example, to move food through the digestive tract, to expel urine, or to give birth.

Cardiac muscle is found only in the heart, composing the vast majority of the heart's structure. It shares some characteristics with skeletal muscle, but like smooth muscle, it is not under conscious control. Cardiac muscle in essence controls itself, with some fine-tuning by the nervous and endocrine systems. Cardiac muscle is discussed more fully in chapter 6.

Skeletal muscles are under conscious control and are so named because most attach to and move the skeleton. Together with the bones of the skeleton, they make up the **musculoskeletal system**. The names of many of these muscles have found their way into our everyday vocabulary—such as deltoids, pectorals (or "pecs"), and biceps—but the human body contains more than 600 skeletal muscles. The thumb alone is controlled by nine separate muscles!

Exercise requires movement of the body, which is accomplished through the action of skeletal muscles. Because exercise and sport physiology depend on human movement, the primary focus of this chapter is on the structure and function of skeletal muscle. Although the anatomical structures and control of smooth, cardiac, and skeletal muscle differ in many respects, their principles of action—for example, creating tension, shortening, and lengthening—are similar.

Anatomy of Skeletal Muscle

When we think of muscles, we visualize each muscle as a single unit. This is natural because a skeletal muscle most often acts as a single entity. But skeletal muscles are far more complex than that.

If a person were to dissect a muscle, he or she would first cut through an outer connective tissue covering known as the **epimysium** (see figure 1.2). It surrounds the entire muscle and functions to hold it together and give it shape. Once through the epimysium, one would see small bundles of fibers wrapped in a connective tissue sheath. These bundles are called fascicles (or fasciculi), and the connective tissue sheath surrounding each **fascicle** is the **perimysium**.

Finally, by cutting through the perimysium and using a microscope, one would see the individual **muscle fibers**, each of which is a muscle cell. Unlike most cells in the body, which have a single nucleus, muscle cells are multinucleated. A sheath of connective tissue, called the **endomysium**, also covers each muscle fiber. It is generally thought that muscle fibers extend from one end of the muscle to the other, but under the microscope, muscle bellies (the thick middle parts of muscles) often divide into compartments or more transverse fibrous bands (inscriptions).

Because of this compartmentalization, the longest human muscle fibers are about 12 cm (4.7 in.), which corresponds to about 500,000 sarcomeres, the basic functional unit of the myofibril. The number of fibers in different muscles ranges from several hundred (e.g., in the tensor tympani, attached to the eardrum) to more than a million (e.g., in the medial gastrocnemius muscle).[12]

Muscle Fibers

Muscle fibers range in diameter from 10 to 120 μm, so they are nearly invisible to the naked eye. The following sections describe the structure of the individual muscle fiber.

Plasmalemma

Looking closely at an individual muscle fiber, it is surrounded by a plasma membrane, called the **plasmalemma** (figure 1.3). The plasmalemma is part of a larger unit referred to as the **sarcolemma**. The sarcolemma is composed of the plasmalemma and the basement membrane. (Some textbooks use the term *sarcolemma* to refer to just the plasmalemma.[12]) At the end of each

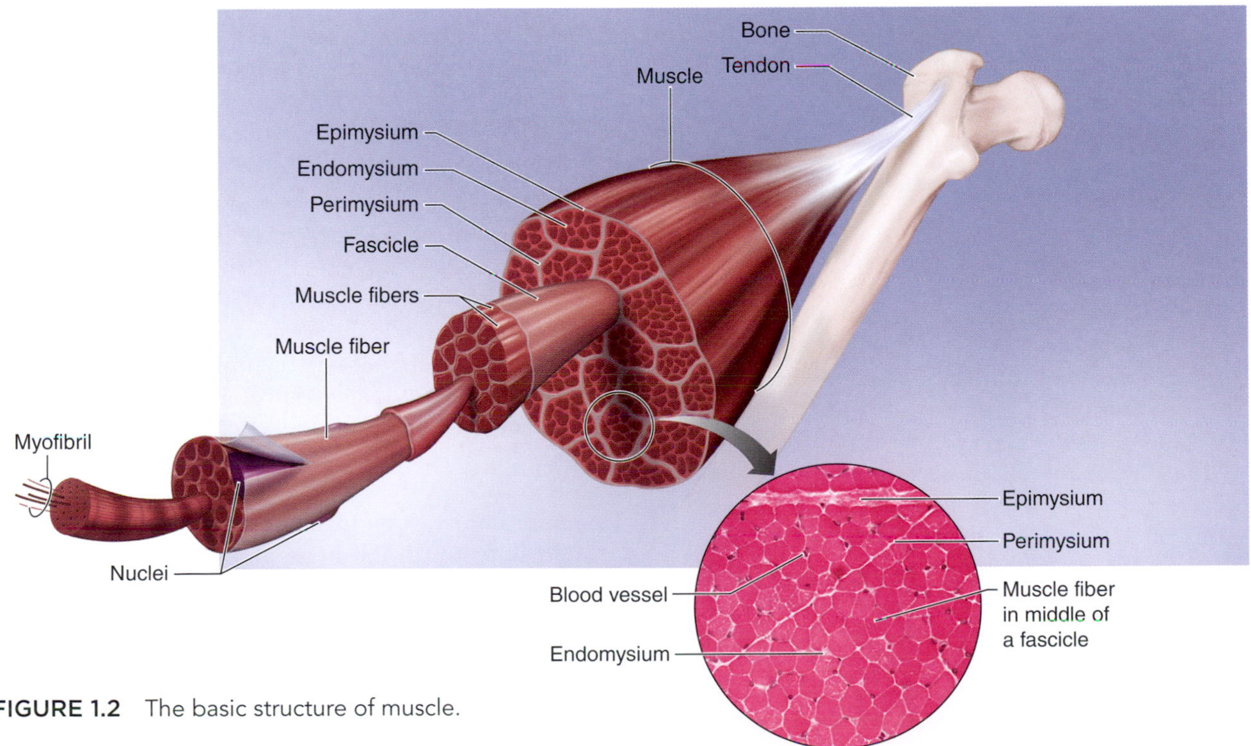

FIGURE 1.2 The basic structure of muscle.

RESEARCH PERSPECTIVE 1.1

Muscle Changes After Only 6 Weeks of Training

Architectural characteristics of a muscle, such as its thickness, pennation angle (the angle at which the fibers are oriented within the muscle), and fascicle length, all contribute to its ability to produce force. Changes in structural characteristics have been demonstrated in many muscles in response to mechanical stimuli, including long-term exercise training. Understanding how—and how quickly—muscle architecture adapts to exercise training is important for recreational exercisers beginning a training program and athletes getting ready for competition.

A group of investigators recently used ultrasound imaging to examine the architectural adaptations of the biceps femoris in a group of young men before and after 6 weeks of either concentric or eccentric strength training.[17] Eccentric strength training increased muscle fascicle length and reduced pennation angle (the fascicles aligned better with the direction of the muscle). In contrast, concentric strength training reduced fascicle length and increased pennation angle (fascicles were angled more away from the full muscle's direction). After 4 weeks of detraining, eccentric training-induced alterations were reversed, but the adaptations in response to concentric training were maintained. Thus, short-term resistance training can cause structural adaptations in the biceps that are highly specific to the mode of training. Understanding architectural alterations that occur in response to training is important for injury prevention and development of proper rehabilitation programs.

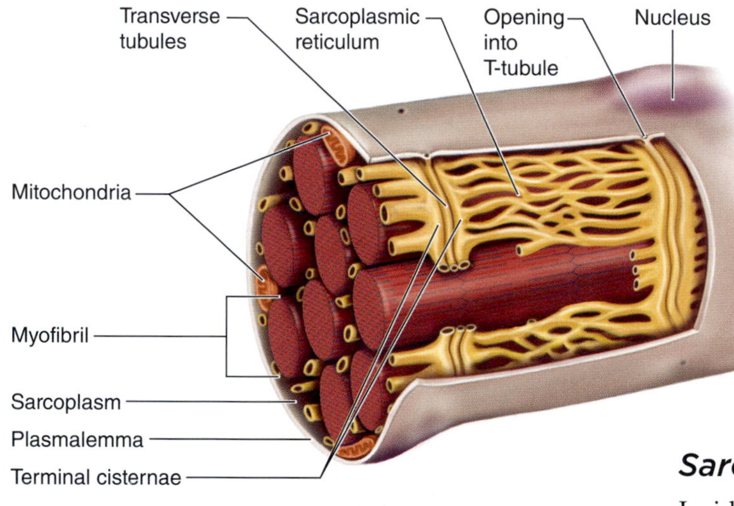

FIGURE 1.3 The structure of a single muscle fiber.

muscle fiber, its plasmalemma fuses with the tendon, which inserts into the bone. Tendons are made of fibrous cords of connective tissue that transmit the force generated by muscle fibers to the bones, thereby creating motion. So typically, individual muscle fibers are ultimately attached to bone via the tendon.

The plasmalemma has several unique features that are critical to muscle fiber function. It appears as a series of shallow folds along the surface of the fiber when the fiber is contracted or in a resting state, but these folds disappear when the fiber is stretched. This folding allows stretching of the muscle fiber without disrupting the plasmalemma. The plasmalemma also has junctional folds in the innervation zone at the motor end plate, which assists in the transmission of the action potential from the motor neuron to the muscle fiber, as discussed later in this chapter. Finally, the plasmalemma helps to maintain acid–base balance and transport of metabolites from the capillary blood into the muscle fiber.[12]

Satellite cells are located between the plasmalemma and the basement membrane. These cells are involved in the growth and development of skeletal muscle and in muscle's adaptation to injury, immobilization, and training. This is discussed in greater detail in subsequent chapters.

Sarcoplasm

Inside the plasmalemma, a muscle fiber contains successively smaller subunits, as shown in figure 1.3. The largest of these are myofibrils, the contractile element of the muscle, which are described later. A gelatin-like substance fills the spaces within and between the myofibrils. This is the **sarcoplasm**. It is the fluid part of the muscle fiber—its cytoplasm. The sarcoplasm contains mainly dissolved proteins, minerals, glycogen, fats, and necessary organelles. It differs from the cytoplasm of most cells because it contains a large quantity of stored glycogen as well as the oxygen-binding compound myoglobin, which is similar in structure and function to the hemoglobin found in red blood cells.

Transverse Tubules The sarcoplasm also houses an extensive network of **transverse tubules (T-tubules)**, which are extensions of the plasmalemma that pass laterally through the muscle fiber. These tubules are interconnected as they pass among the myofibrils, allowing nerve impulses received by the plasmalemma to be transmitted rapidly to individual myofibrils. The tubules also provide pathways from outside the fiber to its interior, enabling substances to enter the cell and waste products to leave the fibers.

Sarcoplasmic Reticulum A longitudinal network of tubules, known as the **sarcoplasmic reticulum (SR)**, is also found within the muscle fiber. These membranous channels parallel the myofibrils and loop around them. The SR serves as a storage site for calcium, which is essential for muscle contraction. Figure 1.3 depicts the T-tubules and the SR. Their functions are discussed in more detail later in this chapter when we consider the process of muscle contraction.

Myofibrils

Each muscle fiber contains several hundred to several thousand **myofibrils**. These small fibers are made up of the basic contractile elements of skeletal muscle—the sarcomeres. Under the electron microscope, myofibrils appear as long strands of sarcomeres.

Sarcomeres

Under a light microscope, skeletal muscle fibers have a distinctive striped appearance. Because of these markings, or striations, skeletal muscle is also called striated muscle. This striation pattern is also seen in cardiac muscle, so it too can be considered striated muscle.

Refer to figure 1.4, showing myofibrils within a single muscle fiber, and note the striations. Note that dark regions, known as A-bands, alternate with light regions, known as I-bands. Each dark A-band has a lighter region in its center, the H-zone, which is visible only when the myofibril is relaxed. There is a dark line in the middle of the H-zone called the M-line. The light I-bands are interrupted by a dark stripe referred to as the Z-disk, also known as the Z-line.

A **sarcomere** is the basic functional unit of a myofibril and the basic contractile unit of muscle. Each myofibril is composed of numerous sarcomeres joined end to end at the Z-disks. Each sarcomere includes several elements found between each pair of Z-disks, in this sequence:

- An I-band (light zone)
- An A-band (dark zone)
- An H-zone (in the middle of the A-band)
- An M-line in the middle of the H-zone
- The rest of the A-band
- A second I-band

Looking at individual myofibrils through an electron microscope, one can differentiate two types of small protein filaments that are responsible for muscle contraction. The thinner filaments are composed primarily of **actin**, and the thicker filaments are primarily **myosin**. The striations seen in muscle fibers result from the alignment of these filaments, as illustrated in figure 1.4. The light I-band indicates the region of the sarcomere where there are only thin filaments. The dark A-band represents the regions that contain both thick and thin filaments. The H-zone is the central portion of the A-band and contains only thick filaments. The absence of thin filaments causes the H-zone to appear lighter than the adjacent A-band. In the center of the H-zone is the M-line, which is composed of proteins that serve as the attachment site for the thick filaments and assist in stabilizing the structure of the sarcomere.

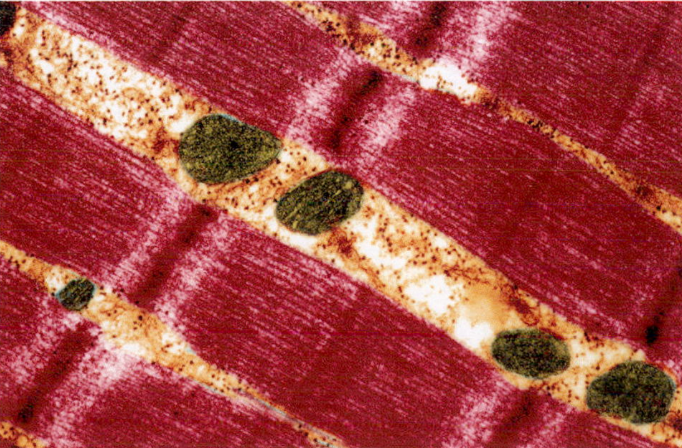

FIGURE 1.4 An electron micrograph of myofibrils within a muscle fiber showing mitochondria (green) between the myofibrils.

In Review

- An individual muscle cell is called a muscle fiber.
- Muscle fibers have a cell membrane and the same organelles—mitochondria, lysosomes, and so on—as other cell types but are uniquely multinucleated.
- A muscle fiber is enclosed by a plasma membrane called the plasmalemma.
- The cytoplasm of a muscle fiber is called the sarcoplasm.
- The extensive tubule network found in the sarcoplasm includes T-tubules, which allow communication and transport of substances throughout the muscle fiber, and the SR, which stores calcium.
- The sarcomere is the smallest functional unit of a muscle.

Z-disks, composed of proteins, appear at each end of the sarcomere. Along with two additional proteins, titin and nebulin, they provide points of attachment and stability for the thin filaments.

Thick Filaments

About two-thirds of all skeletal muscle protein is myosin, the principal protein of the thick filament. Each myosin filament typically contains about 200 myosin molecules.

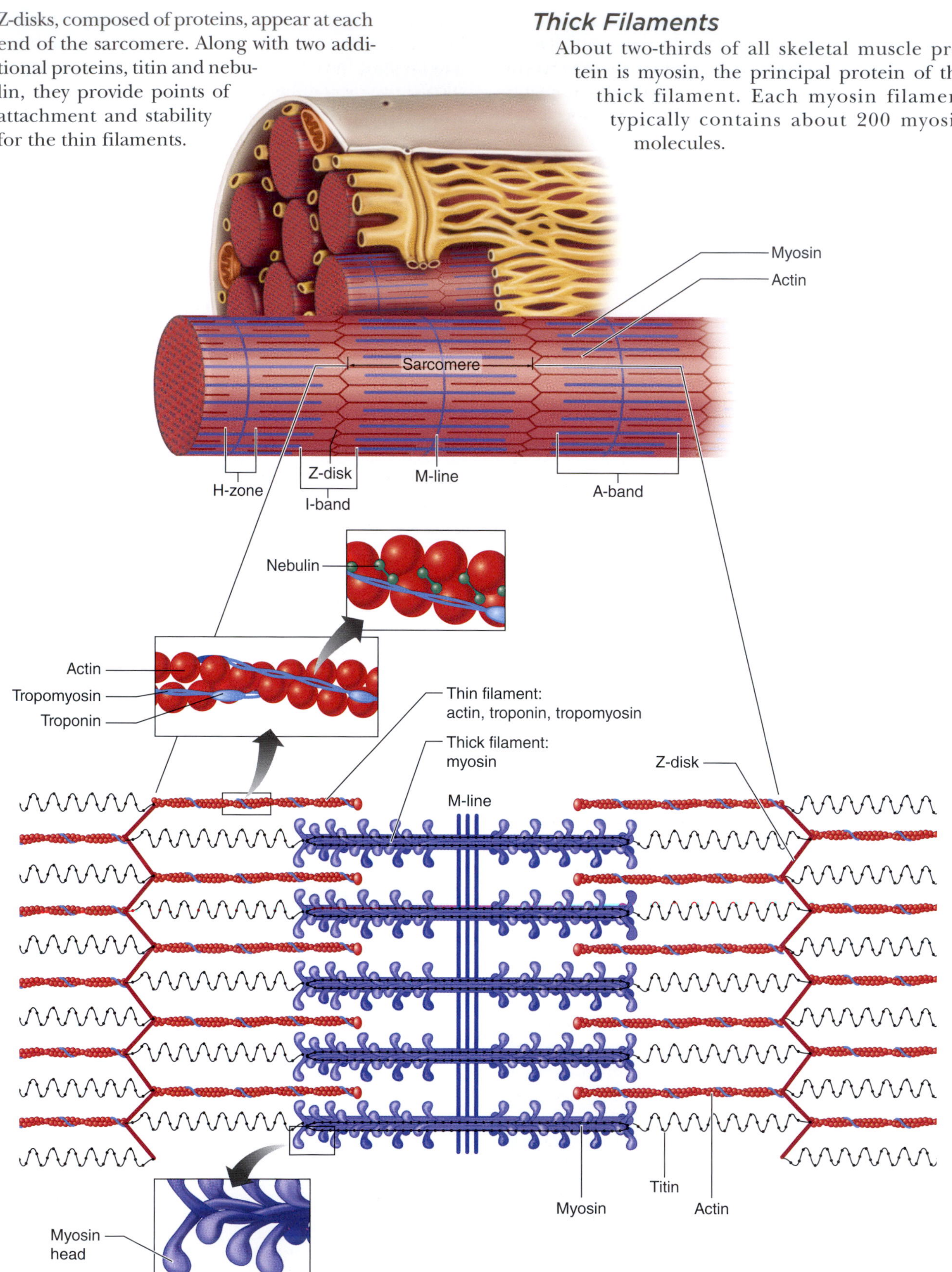

FIGURE 1.5 The sarcomere contains a specialized arrangement of actin (thin) and myosin (thick) filaments. The role of titin is to position the myosin filament to maintain equal spacing between the actin filaments. Nebulin is often referred to as an anchoring protein because it provides a framework that helps stabilize the position of actin.

Each myosin molecule is composed of two protein strands twisted together (see figure 1.5). One end of each strand is folded into a globular head, called the myosin head. Each thick filament contains many such heads, which protrude from the thick filament to form cross-bridges that interact during muscle contraction with specialized active sites on the thin filaments. There is an array of fine filaments, composed of **titin**, that stabilize the myosin filaments along their longitudinal axis (see figure 1.5). Titin filaments extend from the Z-disk to the M-line.

Thin Filaments

Each thin filament, although often referred to simply as an actin filament, is actually composed of three different protein molecules—actin, **tropomyosin**, and **troponin**. Each thin filament has one end inserted into a Z-disk, with the opposite end extending toward the center of the sarcomere, lying in the space between the thick filaments. **Nebulin**, an anchoring protein for actin, coextends with actin and appears to play a regulatory role in mediating actin and myosin interactions (figure 1.5). Each thin filament contains active sites to which myosin heads can bind.

Actin forms the backbone of the filament. Individual actin molecules are globular proteins (G-actin) and join together to form strands of actin molecules. Two strands then twist into a helical pattern, much like two strands of pearls twisted together.

Tropomyosin is a tube-shaped protein that twists around the actin strands. Troponin is a more complex protein that is attached at regular intervals to both the actin strands and the tropomyosin. This arrangement is depicted in figure 1.5. Tropomyosin and troponin work together in an intricate manner along with calcium ions to maintain relaxation or initiate contraction of the myofibril, which we discuss later in this chapter.

Titin: The Third Myofilament

Titin was not discovered until the late 1970s, well after the sliding filament theory of muscle contraction was proposed. The sliding filament theory adequately describes most functions of muscle during shortening (concentric) and constant-length (isometric) contractions. However, traditional cross-bridge theory does not explain why muscles behave as if they have an internal spring—that is, they produce greater force when stretched (eccentric contractions), a mechanism sometimes called passive force enhancement.[9] Recent research has determined that titin's stiffness increases with muscle activation and force development, acting like a spring in active muscle.[9,14,18]

Titin extends from the Z-disk to the M-band in the sarcomere (figure 1.5). It is attached to the myosin filament in the A-band region, but extends freely in the I-band region, where it functions as a spring. Titin has been known for decades to have structural functions, like keeping myosin aligned during contraction and stabilizing adjacent sarcomeres (figure 1.6). However, it is now known that when skeletal muscles are activated by the release of calcium ions (Ca^{2+}), some calcium binds to titin, changing its stiffness. This helps explain why a muscle's ability to generate more force when stretched is not accounted for by traditional actin–myosin cross-bridge theory.

Further, more recently, when titin is included in three-dimensional models of the sarcomere as a third filament, it becomes clear that filaments do not simply slide but actually twist with each cross-bridge interaction. This has led to a new theory called the winding filament theory that better explains how titin contributes to the force produced by muscle sarcomeres at different lengths.[15] In this updated theory, titin is activated by the calcium ion influx and then winds around the thin filaments, rotating them in the process.

The role of titin in regulating skeletal muscle contractile force helps explain the large increase in force that is observed when muscles are actively stretched. That is, titin is increasingly recognized as a third myofilament that is actively involved in the regulation of skeletal muscle force generation. Among its roles are (1) stabilizing sarcomeres and centering myosin filaments

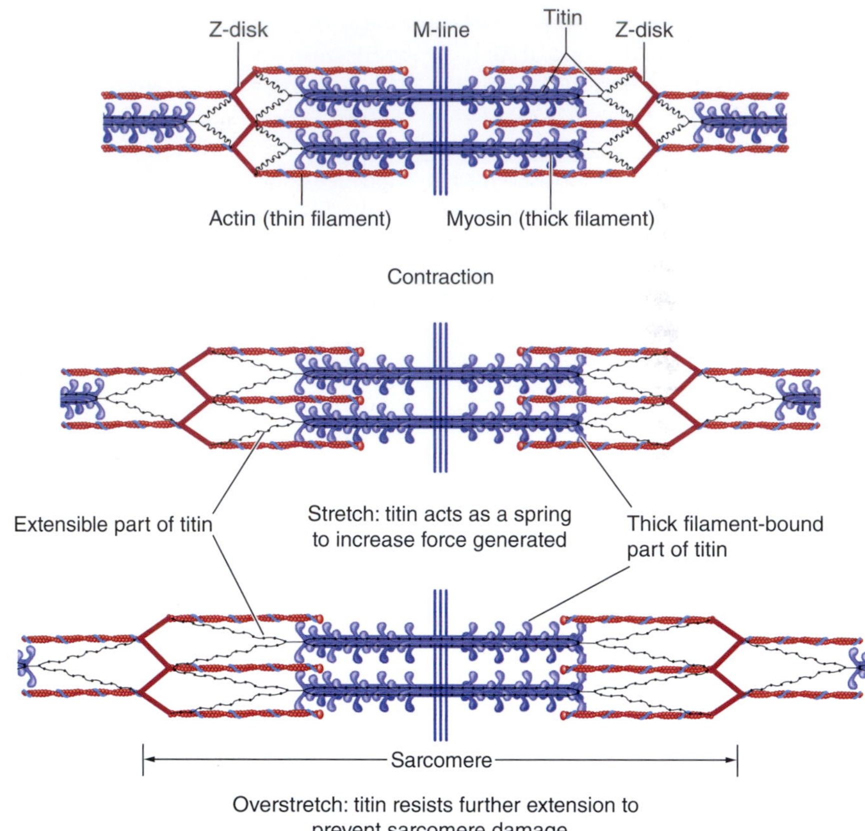

FIGURE 1.6 The mechanism through which the molecule titin acts during muscle contraction. Titin acts as a spring element to increase the force generated and resists overstretch to prevent sarcomere damage.

in the middle of the sarcomere, (2) providing increased force when muscles are stretched, and (3) preventing overstretching and damage to the sarcomere by resisting active stretching.[9]

In Review

- Myofibrils are composed of sarcomeres, the basic contractile units of a muscle.
- A sarcomere is composed of two different-sized filaments, thick and thin filaments, which are responsible for muscle contraction.
- Myosin, the primary protein of the thick filament, is composed of two protein strands, each folded into a globular head at one end.
- The thin filament is composed of actin, tropomyosin, and troponin. One end of each thin filament is attached to a Z-disk.
- A third microfilament, titin, helps stabilize sarcomeres, provides increased force when muscles are stretched, and prevents overstretching and damage to the sarcomere.

Muscle Fiber Contraction

The initiation of contraction of a skeletal muscle occurs in response to a signal from the nervous system. An **α-motor neuron** is a nerve cell that connects with and innervates many muscle fibers. A single α-motor neuron and all the muscle fibers it directly signals are collectively termed a **motor unit** (see figure 1.7). The synapse or gap between the α-motor neuron and a muscle fiber is referred to as a neuromuscular junction. This is where communication between the nervous and muscular systems occurs.

Excitation–Contraction Coupling

The complex sequence of events that triggers a muscle fiber to contract is termed **excitation–contraction coupling** because it begins with the excitation of a motor nerve and results in contraction of the muscle fibers. The process, depicted in figure 1.8, is initiated by a nerve impulse, or **action potential**, from the brain or spinal cord to an α-motor neuron. The action potential arrives at the α-motor neuron's dendrites, specialized

RESEARCH PERSPECTIVE 1.2

Curving Muscle Fascicles

Muscle fascicles are often drawn in a straight line for ease of illustration. Past experimental measures of muscle fascicle characteristics were based on the idea that the muscle fascicles were straight. However, within the muscle, they are actually curved structures, and the curvature of the fascicles is now recognized as an important characteristic relative to muscle function. Two-dimensional (2D) modeling studies demonstrate that muscle fascicles take on a curved path to provide mechanical stability within the muscle, particularly during contraction. Muscle fascicles curve around regions of the muscle that generate high pressures; thus, they curve more where the largest contractions occur. These 2D models also hint that the curvature could extend into three dimensions (3D), but, until recently, this possibility had not been examined during active muscle contraction.

Using sophisticated imaging techniques, researchers have recently quantified 3D fascicle curvature in triceps surae muscles during contractions at different muscle lengths and torques.[16] Fascicle curvatures increased as the muscle contracted more, indicating an increase in intramuscular pressure at greater levels of contraction. Because this study utilized new 3D imaging approaches, the researchers were able to identify details about the fascicle curvature that were not detectable in 2D. This more detailed and precise interpretation of the noted 3D fascicle curvature parameters aids our understanding of how pressure is developed in the contracting muscle and of overall muscle function.

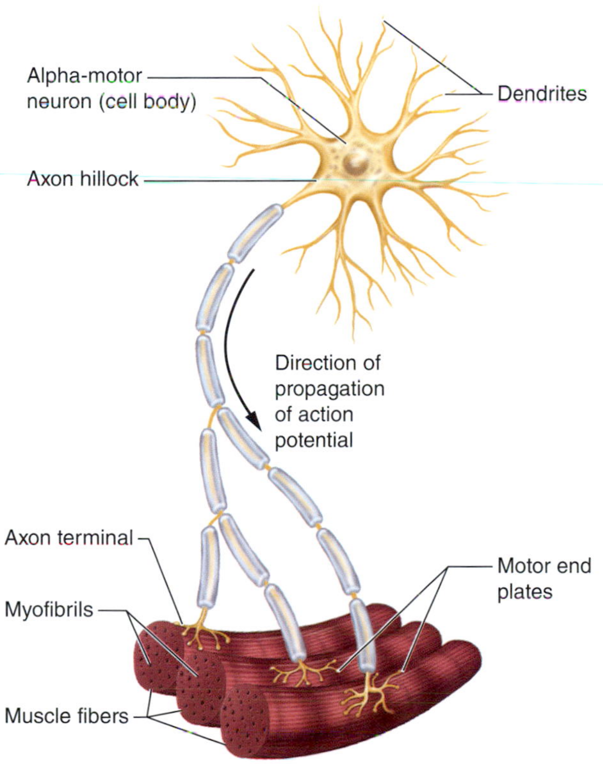

FIGURE 1.7 A motor unit includes one α-motor neuron and all of the muscle fibers it innervates.

receptors on the neuron's cell body. From here, the action potential travels down the axon to the axon terminals, which are located very close to the plasmalemma.

When the action potential arrives at the axon terminals, these nerve endings release a signaling molecule or neurotransmitter called acetylcholine (ACh), which crosses the synaptic cleft and binds to receptors on the plasmalemma (see figure 1.8*a*). If enough ACh binds to the receptors, the action potential will be transmitted the full length of the muscle fiber as ion gates open in the muscle cell membrane and allow sodium to enter. This process is referred to as depolarization. An action potential must be generated in the muscle cell before the muscle cell can act. These neural events are discussed more fully in chapter 3.

Role of Calcium in the Muscle Fiber

In addition to depolarizing the fiber membrane, the action potential travels over the fiber's network of tubules (T-tubules) to the interior of the cell. The arrival of an electrical charge causes the adjacent SR to release large quantities of stored calcium ions (Ca^{2+}) into the sarcoplasm (see figure 1.8*b*).

In the resting state, tropomyosin molecules cover the myosin-binding sites on the actin molecules, preventing the binding of the myosin heads. Once calcium ions are released from the SR, they bind to the troponin on the actin molecules. Troponin, with its strong affinity for calcium ions, is believed to then initiate the contraction process by moving the tropomyosin molecules off the myosin-binding sites on the actin molecules. This is shown in figure 1.8*c*. Because tropomyosin normally

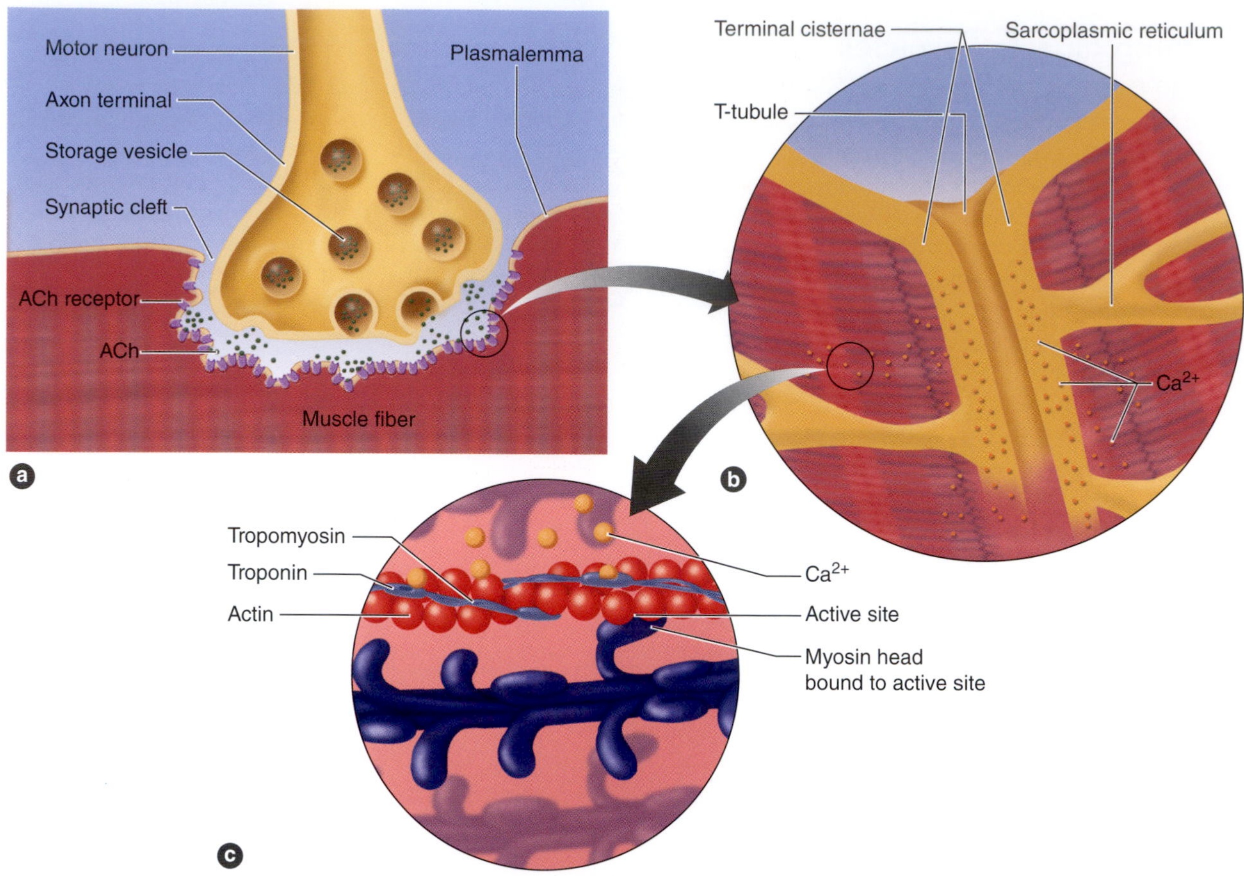

FIGURE 1.8 The sequence of events leading to muscle action, known as excitation–contraction coupling. *(a)* In response to an action potential, a motor neuron releases acetylcholine (ACh), which crosses the synaptic cleft and binds to receptors on the plasmalemma. If enough ACh binds, an action potential is generated in the muscle fiber. *(b)* The action potential triggers the release of calcium ions (Ca^{2+}) from the terminal cisternae of the sarcoplasmic reticulum into the sarcoplasm. *(c)* The Ca^{2+} binds to troponin on the actin filament, and the troponin pulls tropomyosin off the active sites, allowing myosin heads to attach to the actin filament.

covers the myosin-binding sites, it blocks the attraction between the **myosin cross-bridges** and actin molecules. However, once the tropomyosin has been lifted off the binding sites by troponin and calcium, the myosin heads can attach to the binding sites on the actin molecules.

The Sliding Filament Theory: How Muscles Create Movement

When muscle contracts, muscle fibers shorten. How do they shorten? The explanation for this phenomenon is termed the **sliding filament theory**. When the myosin cross-bridges are activated, they bind with actin, resulting in a conformational change in the cross-bridge, which causes the myosin head to tilt and to drag the thin filament toward the center of the sarcomere (see figures 1.9 and 1.10). This tilting of the head is referred to as the **power stroke**. The pulling of the thin filament past the thick filament shortens the sarcomere and generates force. When the fibers are not contracting, the myosin head remains in contact with the actin molecule, but the molecular bonding at the site is weakened or blocked by tropomyosin.

Immediately after the myosin head tilts, it breaks away from the active site, rotates back to its original position, and attaches to a new active site farther along the actin filament. Repeated attachments and power strokes cause the filaments to slide past one another, giving rise to the term *sliding filament theory*. This process continues until the ends of the myosin filaments reach the Z-disks, or until the Ca^{2+} is pumped back into the SR. During this sliding (contraction), the thin filaments move toward the center of the sarcomere and protrude into the H-zone, ultimately overlapping. When this occurs, the H-zone is no longer visible.

Recall that the sarcomeres are joined end to end within a myofibril. Because of this anatomical arrangement, as sarcomeres shorten, the myofibril shortens, causing muscle fibers within a fascicle to shorten. The end result of many such fibers shortening is an organized muscle contraction.

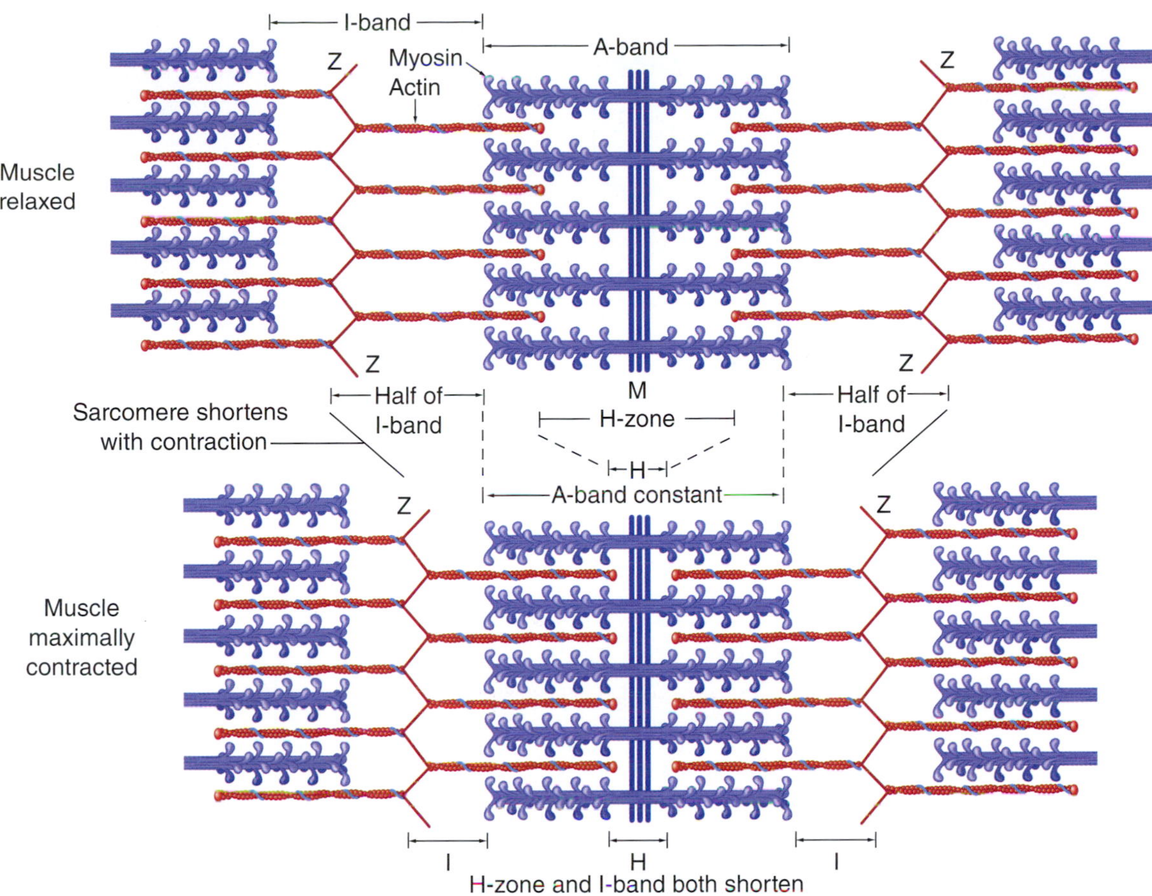

FIGURE 1.9 A sarcomere in its relaxed (top) and contracted (bottom) state, illustrating the sliding of the actin and myosin filaments with contraction.

In Review

- The sequence of events that starts with a motor nerve impulse and results in muscle contraction is called excitation–contraction coupling.
- Muscle contraction is initiated by an α-motor neuron impulse or action potential. The motor neuron releases ACh, which opens up ion gates in the muscle cell membrane, allowing sodium to enter the muscle cell (depolarization). If the cell is sufficiently depolarized, an action potential is generated and muscle contraction occurs.
- When an α-motor neuron is activated, all of the muscle fibers in its motor unit are stimulated to contract.
- The action potential travels along the plasmalemma, then moves through the T-tubule system, causing stored calcium ions to be released from the SR.
- Calcium ions bind with troponin. Then troponin moves the tropomyosin molecules off of the myosin-binding sites on the actin molecules, opening these sites to allow the myosin heads to bind to them.
- Once a strong binding state is established with actin, the myosin head tilts, pulling the thin filament past the thick filament. The tilting of the myosin head is the power stroke.
- Energy is required for muscle contraction to occur. The myosin head binds to the high-energy molecule ATP, and ATPase on the head splits ATP into ADP and P_i, releasing energy to fuel the contraction.
- The end of muscle contraction is signaled when neural activity ceases at the neuromuscular junction. Calcium is actively pumped out of the sarcoplasm and back into the SR for storage. Tropomyosin moves to cover active sites on actin molecules, leading to relaxation between the myosin heads and the binding sites.
- Like muscle contraction, muscle relaxation requires energy supplied by ATP.

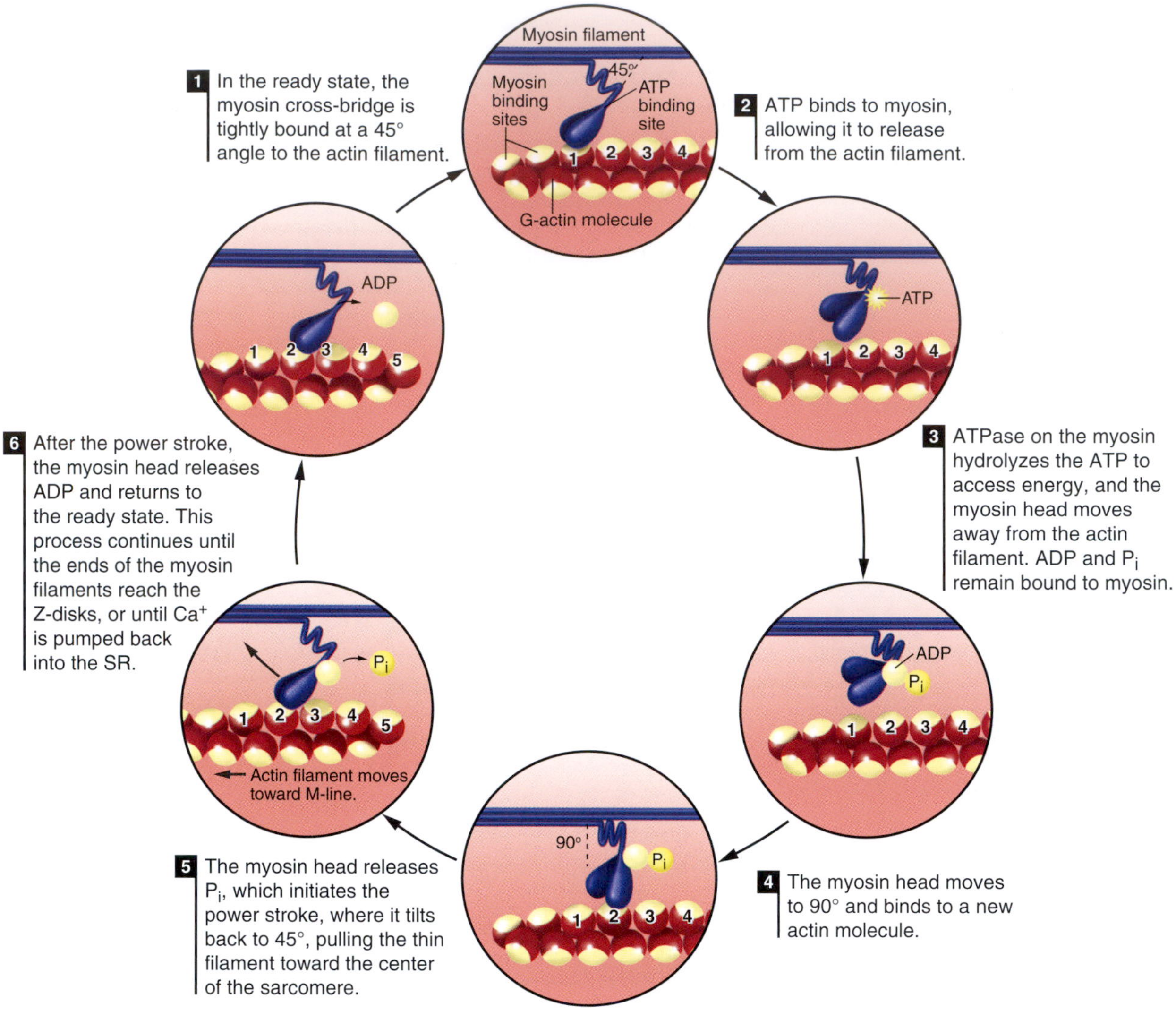

Myosin filament

Myosin binding sites

45°

ATP binding site

G-actin molecule

1 In the ready state, the myosin cross-bridge is tightly bound at a 45° angle to the actin filament.

2 ATP binds to myosin, allowing it to release from the actin filament.

ATP

3 ATPase on the myosin hydrolyzes the ATP to access energy, and the myosin head moves away from the actin filament. ADP and P_i remain bound to myosin.

ADP

P_i

4 The myosin head moves to 90° and binds to a new actin molecule.

90°

P_i

5 The myosin head releases P_i, which initiates the power stroke, where it tilts back to 45°, pulling the thin filament toward the center of the sarcomere.

P_i

Actin filament moves toward M-line.

6 After the power stroke, the myosin head releases ADP and returns to the ready state. This process continues until the ends of the myosin filaments reach the Z-disks, or until Ca^+ is pumped back into the SR.

ADP

FIGURE 1.10 The molecular events of a contractile cycle illustrating the changes in the myosin head during various phases of the power stroke.

Energy for Muscle Contraction

Muscle contraction is an active process, meaning that it requires energy. In addition to the binding site for actin, a myosin head contains a binding site for the molecule **adenosine triphosphate (ATP)**. The myosin molecule must bind with ATP for muscle contraction to occur because ATP supplies the needed energy.

The enzyme **adenosine triphosphatase (ATPase)**, which is located on the myosin head, splits the ATP to yield adenosine diphosphate (ADP), inorganic phosphate (P_i), and energy. The energy released from this breakdown of ATP is used to power the tilting of the myosin head. Thus, ATP is the chemical source of energy for muscle contraction. This process is discussed in much more detail in chapter 2.

Muscle Relaxation

Muscle contraction continues as long as calcium is available in the sarcoplasm. At the end of a muscle contraction, calcium is pumped back into the SR, where it is stored until a new action potential arrives at the muscle fiber membrane. Calcium is returned to the SR by an active calcium-pumping system. This is another energy-demanding process that also relies on ATP. Thus, energy is required for both the contraction and relaxation phases.

When the calcium is pumped back into the SR, troponin and tropomyosin return to the resting conformation. This blocks the linking of the myosin cross-bridges and actin molecules and stops the use of ATP. As a result, the thick and thin filaments return to their original relaxed state.

Muscle Fiber Types

Not all muscle fibers are alike. A single skeletal muscle contains fibers having different speeds of shortening and ability to generate maximal force: type I (also called slow or slow-twitch) fibers and type II (also called fast or fast-twitch) fibers. **Type I fibers** take approximately 110 ms to reach peak tension when stimulated. **Type II fibers**, on the other hand, can reach peak tension in about 50 ms. While the terms *slow twitch* and *fast twitch* continue to be used, scientists now prefer to use the terminology *type I* and *type II*, as is the case in this textbook.

Although only one form of type I fiber has been identified, type II fibers can be further classified. In humans, the two major forms of type II fibers are fast-twitch type a (type IIa) and fast-twitch type x (type IIx). Figure 1.11 is a micrograph of human muscle in which thinly sliced (10 μm) cross sections of a muscle sample have been chemically stained to differentiate the fiber types. The type I fibers are stained black; type IIa fibers are unstained and appear white; and type IIx fibers appear gray. Although not apparent in this figure, a third subtype of fast-twitch fibers has also been identified: type IIc.

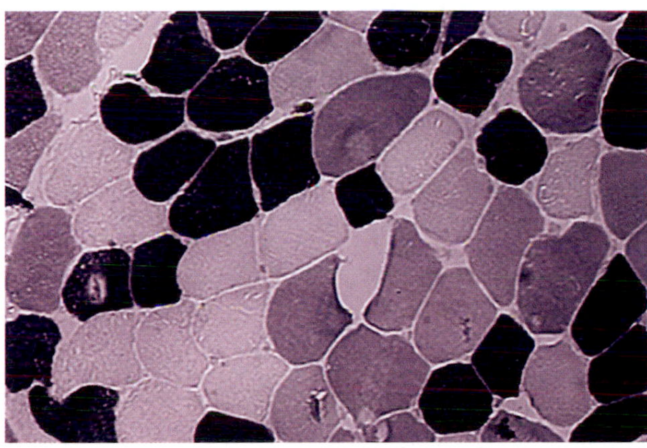

FIGURE 1.11 A photomicrograph showing type I (black), type IIa (white), and type IIx (gray) muscle fibers.

The differences in the type IIa, type IIx, and type IIc fibers are not fully understood, but type IIa fibers are believed to be the most frequently recruited. Only type I fibers are recruited more frequently than type IIa fibers. Type IIc fibers are the least often used. On average, most muscles are composed of roughly 50% type I fibers and 25% type IIa fibers. The remaining 25% are mostly type IIx, with type IIc fibers making up only 1% to 3% of the muscle. Because knowledge about type IIc fibers is limited, we will not discuss them further. The exact percentage of each of these fiber types varies greatly in various muscles and among individuals, so the numbers listed here are only averages. This extreme variation is most evident in athletes, as we will see later in this chapter when we compare fiber types in athletes across sports and events within sports.

In the early 1900s, a needle biopsy procedure was developed to study muscular dystrophy. In the 1960s, this technique was adapted to sample muscle for studies in exercise physiology, specifically to help determine muscle fiber types.

A muscle biopsy (figure 1.12) involves removing a very small piece of muscle tissue from the muscle belly for analysis. The area from which the sample is taken is first numbed with a local anesthetic, and then a small incision (approximately 1 cm, or 0.4 in.) is made with a scalpel through the skin, subcutaneous tissue, and connective tissue. A hollow needle is then inserted to the appropriate depth into the belly of the muscle. A small plunger is pushed through the center of the needle to snip off a very small sample of muscle. The biopsy needle is withdrawn, and the sample, weighing

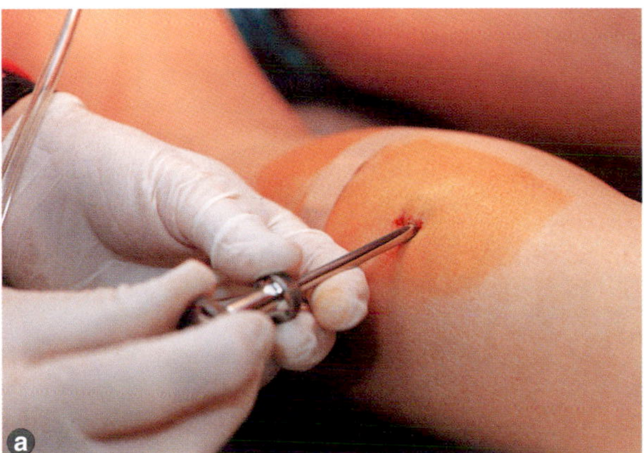

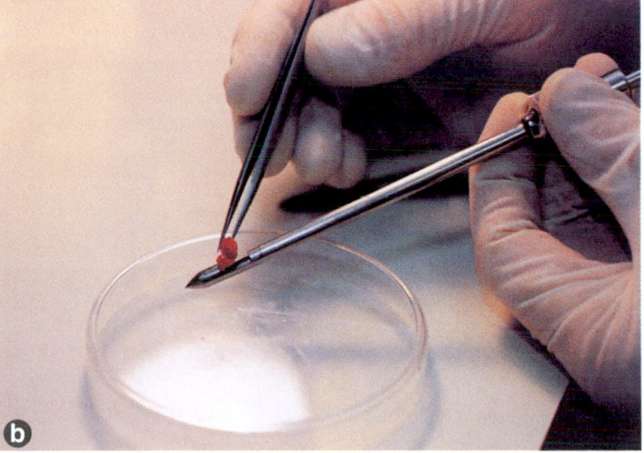

FIGURE 1.12 *(a)* The use of a biopsy needle to obtain a sample from the leg muscle of an elite female runner. *(b)* A close-up view of a muscle biopsy needle and a small piece of muscle tissue.

10 to 100 mg, is removed, cleaned of blood, mounted, and quickly frozen. It is then thinly sliced, stained, and examined under a microscope.

This method allows us to study muscle fibers and gauge the effects of acute exercise and chronic training on fiber composition. Microscopic and biochemical analyses of the samples aid our understanding of the muscles' ability to produce energy for contraction.

Characteristics of Type I and Type II Fibers

Different muscle fiber types play different roles in exercise and sport. This is largely due to differences in their inherent characteristics.

ATPase

Type I and type II fibers differ in their speed of contraction. This difference results primarily from different forms of myosin ATPase. Recall that myosin ATPase is the enzyme that splits ATP to release energy to drive contraction. Type I fibers have a slow form of myosin ATPase, whereas type II fibers have a fast form. In response to neural stimulation, ATP is split more rapidly in type II fibers than in type I fibers. As a result, cross-bridges cycle more rapidly in type II fibers.

One of the methods used to classify muscle fibers is a chemical staining procedure applied to a thin slice of tissue. This staining technique measures the ATPase activity in the fibers. Thus, the type I, type IIa, and type IIx fibers stain differently, as depicted in figure 1.11. This technique makes it appear that each muscle fiber has only one type of ATPase, but fibers can have a mixture of ATPase types. Some have a predominance of type I ATPase, but others have mostly type II ATPase. Their appearance in a stained slide preparation should be viewed as a continuum rather than as absolutely distinct types.

A newer method for identifying fiber types is to chemically separate the different types of myosin molecules (isoforms) by using a process called gel electrophoresis. In electrophoresis, the isoforms are separated by weight in an electric field to show the bands of protein (i.e., myosin) that characterize type I, type IIa, and type IIx fibers. Although our discussion here categorizes fiber types simply as slow twitch (type I) and fast twitch (type IIa and type IIx), scientists have further subdivided these fiber types. The use of electrophoresis has led to the detection of myosin hybrids or fibers that possess two or more forms of myosin. With this method of analysis, the fibers are classified as I, Ic (I/IIa), IIc (IIa/I), IIa, IIax, IIxa, and IIx.[12] In this book, we will use the histochemical method of identifying fibers by their primary isoforms, types I, IIa, and IIx.

Table 1.1 summarizes the characteristics of the different muscle fiber types. The table also includes alternative names that are used in other classification systems to refer to the various muscle fiber types.

Sarcoplasmic Reticulum

Type II fibers have a more highly developed SR than do type I fibers. Thus, type II fibers are more adept at delivering calcium into the muscle cell when stimulated. This ability is thought to contribute to the faster speed of contraction (V_o) of type II fibers. On average, human type II fibers have a V_o that is five to six times faster than that of type I fibers. Although the amount of force (P_o) generated by type II and type I fibers with the same diameter is about the same, the calculated power ($\mu N \cdot \text{fiber length}^{-1} \cdot s^{-1}$) of a type II fiber is three to five times greater than that of a type I fiber because of a faster shortening velocity. This may explain in part why individuals who have a predominance of type II fibers in their leg muscles tend to be better sprinters than individuals who have a high percentage of type I fibers, all other things being equal.

TABLE 1.1 **Classification of Muscle Fiber Types**

	Fiber classification		
System 1 (preferred)	Type I	Type IIa	Type IIx
System 2	Slow twitch (ST)	Fast-twitch a (FTa)	Fast-twitch x (FTx)
System 3	Slow oxidative (SO)	Fast oxidative/glycolytic (FOG)	Fast glycolytic (FG)
	Characteristics of fiber types		
Oxidative capacity	High	Moderately high	Low
Glycolytic capacity	Low	High	Highest
Contractile speed	Slow	Fast	Fast
Fatigue resistance	High	Moderate	Low
Motor unit strength	Low	High	High

RESEARCH PERSPECTIVE 1.3

More About Titin

Eccentric contractions are those during which active muscles are stretched or elongated, such as during the lowering of a weight by the biceps or walking down stairs. Unlike for concentric contractions, the classic theories of muscle contraction—the sliding filament and cross-bridge theories—do not agree well with some aspects of eccentrically contracting muscles. In the cross-bridge model, greater force is developed during an eccentric contraction than for a corresponding isometric or concentric contraction because the attached cross-bridges are more strained. However, skeletal muscles are known to also have history-dependent properties; that is, muscles behave differently depending on preceding contractions. For example, when isometric contractions follow an eccentric contraction, they often demonstrate a longer-lasting, steady-state isometric force compared with the force produced during an isometric contraction not preceded by an eccentric contraction. The classic cross-bridge model involving actin and myosin does not explain such history-dependent properties of skeletal muscle, necessitating a careful reassessment of the currently accepted theory of muscle contraction.

A group of researchers recently proposed a new mechanism that better explains the history-dependent properties of muscle and eccentric contractions by adding a small component to the classic cross-bridge theory.[10] This tweak includes a new key role for the structural protein titin, whose stiffness and force are adjusted upon activation and force production. In this new model, muscle can be stretched passively against little resistance, but upon activation, titin-based force becomes dominant and contributes to eccentric force.

Titin's traditional role (discussed in this chapter) has been associated with centering the thick filaments in the sarcomere and preventing sarcomeres from becoming overstretched by putting the brakes on active stretching. Yet another function of titin is to serve as the third sarcomere myofilament, contributing to active force production during and following eccentric contractions. In this role, titin appears to be critical in the residual force enhancement in skeletal muscle after an eccentric contraction. This theory is based on research evidence that experimentally eliminating titin in single myofibrils abolishes all force transmission across sarcomeres and all residual and passive force enhancement. Thus, titin acts as a kind of molecular spring that increases the muscle's stiffness, and thus its force, in active compared with passive muscle contraction. Researchers now speculate that titin contributes to active force production by changing its stiffness (i.e., when titin becomes stiffer, it can produce more force). The researchers theorize that titin's stiffness increases (1) by binding to calcium upon activation and (2) by binding to actin and decreasing its length.

Although preliminary results and theoretical models provide support for this new three-filament model of force production, the molecular details have not been worked out. If correct, it would provide a substantial update to the classic cross-bridge theory. If proven, the new three-filament model of muscle contraction would simultaneously add to our understanding of eccentric contractions and explain the history-dependent properties of muscle, information not previously explainable using the classical two-filament cross-bridge theory.

Motor Units

Recall that a motor unit is composed of a single α-motor neuron and the muscle fibers it innervates. The α-motor neuron appears to determine whether the fibers are type I or type II. The α-motor neuron in a type I motor unit has a smaller cell body and typically innervates a cluster of ≤300 muscle fibers. In contrast, the α-motor neuron in a type II motor unit has a larger cell body and innervates ≥300 muscle fibers. This difference in the size of motor units means that when a single type I α-motor neuron stimulates its fibers, far fewer muscle fibers contract than when a single type II α-motor neuron stimulates its fibers. Consequently, type II muscle fibers reach peak tension faster and collectively generate more force than type I fibers. The difference in maximal isometric force development between type II and type I motor units is attributable to two characteristics: the number of muscle fibers per individual motor unit and the difference in the size of type II and type I fibers. Type I and type II fibers of the same diameter generate about the same force. On average, however, type II fibers tend to be larger than type I fibers, and type II motor units tend to have more muscle fibers than do the type I motor units.

Distribution of Fiber Types

As mentioned earlier, the percentages of type I and type II fibers are not the same in all the muscles of the body. Generally, arm and leg muscles have similar fiber compositions within an individual. An endurance athlete with a predominance of type I fibers in his or her leg muscles will likely have a high percentage of type I fibers in the arm muscles as well. A similar relationship exists for type II fibers. There are some exceptions, however. The soleus muscle (beneath the gastrocnemius in the calf), for example, is composed of a very high percentage of type I fibers in everyone.

Fiber Type and Exercise

Because of these differences in type I and type II fibers, one might expect that these fiber types would also have different functions when people are physically active. Indeed, this is the case.

Type I Fibers

In general, type I muscle fibers have a high level of aerobic endurance. Aerobic means "in the presence of oxygen," so oxidation is an aerobic process. Type I fibers are very efficient at producing ATP from the oxidation of carbohydrate and fat, which is discussed in chapter 2.

Recall that ATP is required to provide the energy needed for muscle fiber contraction and relaxation. As long as oxidation occurs, type I fibers continue producing ATP, allowing the fibers to remain active. The ability to maintain muscular activity for a prolonged period is known as muscular endurance, so type I fibers have high aerobic endurance. Because of this, they are recruited most often during low-intensity endurance events (e.g., marathon running) and during most daily activities for which the muscle force requirements are low (e.g., walking).

Type II Fibers

Type II muscle fibers, on the other hand, have relatively poor aerobic endurance when compared to type I fibers. They are better suited to perform anaerobically (without oxygen). This means that in the absence of adequate oxygen, ATP is formed through anaerobic pathways, not oxidative pathways. (We discuss these pathways in detail in chapter 2.)

Type IIa motor units generate considerably more force than do type I motor units, but type IIa motor units also fatigue more easily because of their limited endurance. Thus, type IIa fibers appear to be the primary fiber type used during shorter, higher-intensity endurance events, such as the mile run or the 400 m swim.

Although the significance of type IIx fibers is not fully understood, they apparently are not easily activated by the nervous system. Thus, they are used rather infrequently in normal, low-intensity activity but are predominantly used in highly explosive events such as the 100 m dash and the 50 m sprint swim. Characteristics of the various fiber types are summarized in table 1.2.

One of the most advanced methods for the study of human muscle fibers is to dissect fibers out of a muscle biopsy sample, suspend a single fiber between force transducers, and measure its strength and **single-fiber contractile velocity (V_o)**. From figure 1.13, one can see that all of the single fibers tend to reach their peak power when the fibers are generating only about 20% of their peak force. However, it is quite clear that the peak power of the type II fibers is considerably higher than that of the type I fibers.

Determination of Fiber Type

The characteristics of muscle fibers appear to be determined early in life, perhaps within the first few years. Studies with identical twins have shown that muscle fiber type, for the most part, is genetically determined, changing little from childhood to middle age. These studies reveal that identical twins have nearly identical proportions of fiber types, whereas fraternal twins differ in their fiber type profiles. The genes we inherit from our parents likely determine which α-motor neurons innervate our individual muscle fibers. After innervation is established, muscle fibers differentiate (become specialized) according to the type of α-motor neuron that stimulates them. Some recent evidence, however, suggests that endurance training, strength training,

TABLE 1.2 **Structural and Functional Characteristics of Muscle Fiber Types**

	Fiber type		
Characteristic	Type I	Type IIa	Type IIx
Fibers per motor neuron	≤300	≥300	≥300
Motor neuron size	Smaller	Larger	Larger
Motor neuron conduction velocity	Slower	Faster	Faster
Contraction speed (ms)	110	50	50
Type of myosin ATPase	Slow	Fast	Fast
Sarcoplasmic reticulum development	Low	High	High

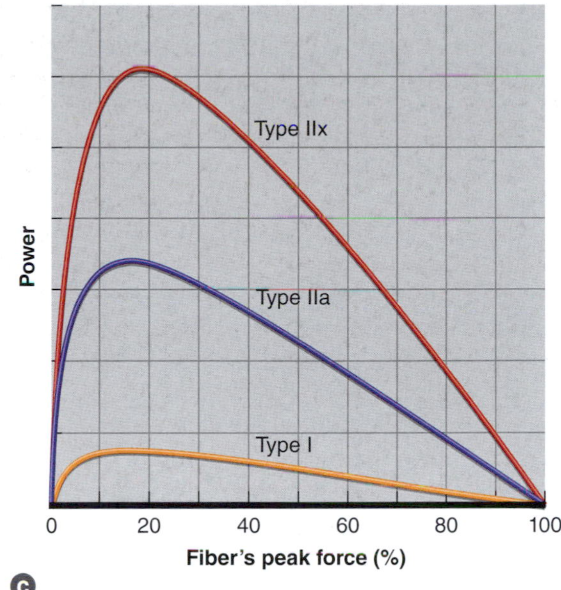

FIGURE 1.13 *(a)* The dissection and *(b)* suspension of a single muscle fiber to study the physiology of different fiber types. *(c)* Differences in peak power generated by each fiber type at various percentages of maximal force.

and muscular inactivity may cause a shift in the myosin isoforms. Consequently, training may induce a small change, perhaps less than 10%, in the percentage of type I and type II fibers. Further, both endurance and resistance training have been shown to reduce the percentage of type IIx fibers while increasing the fraction of type IIa fibers.

Studies of older men and women have shown that aging may alter the distribution of type I and type II fibers. As we grow older, muscles tend to lose type II motor units, which increases the percentage of type I fibers.

In Review

- Most skeletal muscles contain both type I and type II fibers.
- Different fiber types have different myosin ATPase activities. The ATPase in the type II fibers acts faster than the ATPase in type I fibers.
- Type II fibers have a more highly developed SR, enhancing the delivery of calcium needed for muscle contraction.
- α-Motor neurons innervating type II motor units are larger and innervate more fibers than do α-motor neurons for type I motor units. Thus, type II motor units have more (and larger) fibers to contract and can produce more force than type I motor units.
- The proportions of type I and type II fibers in a person's arm and leg muscles are usually similar.
- Type I fibers have higher aerobic endurance and are well suited to low-intensity endurance activities.
- Type II fibers are better suited for anaerobic activity. Type IIa fibers play a major role in high-intensity exercise. Type IIx fibers are activated when the force demanded of the muscle is high.

Skeletal Muscle and Exercise

Having reviewed the overall structure of muscle, the process by which it develops force, and the types of muscle fibers, we now look more specifically at how muscle functions during exercise. Strength, endurance, and speed depend largely on the muscle's ability to produce energy and force. This section examines how muscle accomplishes this task.

Muscle Fiber Recruitment

When an α-motor neuron carries an action potential to the muscle fibers in the motor unit, all fibers in the unit develop force. Activating more motor units is the way muscles produce more force. When little force is needed, only a few motor units are recruited. Recall from our earlier discussion that type IIa and type IIx motor units contain more muscle fibers than type I motor units do. Skeletal muscle contraction involves a progressive recruitment of type I, followed by type II motor units, depending on the requirements of the activity being performed. As the intensity of the activity increases, the number of fibers recruited increases in the following order, in an additive manner: type I → type IIa → type IIx.

Motor units are generally activated on the basis of a fixed order of fiber recruitment. This is known as the **principle of orderly recruitment**, in which the motor units within a given muscle appear to be ranked. Let's use the biceps brachii as an example: Assume a total of 200 motor units, which are ranked on a scale from 1 to 200. For an extremely fine muscle contraction requiring very little force production, the motor unit ranked number 1 would be recruited. As the requirements for force production increase, numbers 2, 3, 4, and so on would be recruited, up to a maximal muscle contraction that would activate most, if not all, of the motor units. For the production of a given force, the same motor units are usually recruited each time and in the same order.

A mechanism that may partially explain the principle of orderly recruitment is the **size principle**, which states that the order of recruitment of motor units is directly related to the size of their motor neuron. Motor units with smaller motor neurons will be recruited first. Because the type I motor units have smaller motor neurons, they are the first units recruited in graded movement (going from very low to very high rates of force production). The type II motor units then are recruited as the force needed to perform the movement increases. It is unclear at this time how the size principle relates to complex athletic movements.

During events that last several hours, exercise is performed at a submaximal pace, and the tension in the muscles is relatively low. As a result, the nervous system tends to recruit those muscle fibers best adapted to endurance activity: the type I and some type IIa fibers. As the exercise continues, these fibers become depleted of their primary fuel supply (glycogen), and the nervous system must recruit more type IIa fibers to maintain muscle tension. Finally, when the type I and type IIa fibers become exhausted, the type IIx fibers may be recruited to continue exercising.

This may explain why fatigue seems to come in stages during events such as the marathon, a 42 km (26.1 mi) run. It also may explain why it takes great conscious effort to maintain a given pace near the finish of the event. This conscious effort results in the activation of muscle fibers that are not easily recruited. Such information is of practical importance to our understanding of the specific requirements of training and performance.

In Review

- Motor units give all-or-none responses. Activating more motor units produces more force.
- In low-intensity activity, most muscle force is generated by type I fibers. As the intensity increases, type IIa fibers are recruited, and at even higher intensities, the type IIx fibers are activated. The same pattern of recruitment is followed during events of long duration.

Fiber Type and Athletic Success

From what we have just discussed, it appears that athletes who have a high percentage of type I fibers might have an advantage in prolonged endurance events, whereas those with a predominance of type II fibers could be better suited for high-intensity, short-term, and explosive activities. But does the relative proportion of an athlete's muscle fiber types determine athletic success?

The muscle fiber makeup of successful athletes from a variety of athletic events and of nonathletes is shown in table 1.3. As anticipated, the leg muscles of distance runners, who rely on endurance, have a predominance of type I fibers.[4] Studies of elite male and female distance runners revealed that many of these athletes' gastrocnemius (calf) muscles may contain more than 90% type I fibers. World champions in the marathon are reported to possess 93% to 99% type I fibers in their gastrocnemius muscles. In contrast, the gastrocnemius muscles are composed principally of type II fibers in sprint runners, who rely on speed and strength. World-class sprinters have only about 25% type I fibers in this muscle. Also, although muscle fiber cross-sectional area varies markedly among elite distance runners, type I fibers in their leg muscles average about 22% more cross-sectional area than type II fibers.[5,6] Swimmers tend to have higher percentages of type I fibers (60%-65%) in their arm muscles than untrained subjects (45%-55%).

The fiber composition of muscles in distance runners and sprinters is markedly different. However, it may be a bit risky to think we can select champion distance runners and sprinters solely on the basis of predominant muscle fiber type. Other factors, such as cardiovascular function, motivation, training, and muscle size, also contribute to success in such events of endurance, speed, and strength. Thus, fiber composition alone is not a reliable predictor of athletic success.

Muscle Contraction

We have examined the different muscle fiber types. We understand that all fibers in a motor unit, when stimu-

TABLE 1.3 **Percentages and Cross-Sectional Areas of Type I and Type II Fibers in Selected Muscles of Male and Female Athletes**

					Cross-sectional area (mm^2)	
	Sex	Muscle	% type I	% type II	Type I	Type II
Sprint runners	M	Gastrocnemius	24	76	5,878	6,034
	F	Gastrocnemius	27	73	3,752	3,930
Distance runners	M	Gastrocnemius	79	21	8,342	6,485
	F	Gastrocnemius	69	31	4,441	4,128
Cyclists	M	Vastus lateralis	57	43	6,333	6,116
	F	Vastus lateralis	51	49	5,487	5,216
Swimmers	M	Posterior deltoid	67	33		
Weightlifters	M	Gastrocnemius	44	56	5,060	8,910
	M	Deltoid	53	47	5,010	8,450
Triathletes	M	Posterior deltoid	60	40		
	M	Vastus lateralis	63	37		
	M	Gastrocnemius	59	41		
Canoeists	M	Posterior deltoid	71	29	4,920	7,040
Shot-putters	M	Gastrocnemius	38	62	6,367	6,441
Nonathletes	M	Vastus lateralis	47	53	4,722	4,709
	F	Gastrocnemius	52	48	3,501	3,141

lated, act at the same time and that different fiber types are recruited in stages, depending on the force required to perform an activity. Now we can turn our attention to how whole muscles work to produce movement.

Types of Muscle Contraction

Muscle movement generally can be categorized into three types of contractions—concentric, static, and eccentric. In many activities, such as running and jumping, all three types of contraction may occur in the execution of a smooth, coordinated movement. For the sake of clarity, though, we will examine each type separately.

A muscle's principal action, shortening, is referred to as a **concentric contraction**, the most familiar type of contraction. To understand muscle shortening, recall our earlier discussion of how the thin and thick filaments slide across each other. In a concentric contraction, the thin filaments are pulled toward the center of the sarcomere. Because joint movement is produced, concentric contractions are considered **dynamic contractions**.

Muscles can also act without moving. When this happens, the muscle generates force, but its length remains static (unchanged). This is called a **static (isometric) muscle contraction** because the joint angle does not change. A static contraction occurs, for example, when one tries to lift an object that is heavier than the force generated by the muscle, or when one supports the weight of an object by holding it steady with the elbow flexed. In both cases, the person feels the muscles tense, but there is no joint movement. In a static contraction, the myosin cross-bridges form and are recycled, producing force, but the external force is too great for the thin filaments to be moved. They remain in their normal position, so shortening can't occur. If enough motor units can be recruited to produce sufficient force to overcome the resistance, a static contraction can become a dynamic one.

Muscles can exert force even while lengthening. This movement is an **eccentric contraction**. Because joint movement occurs, this is also a dynamic contraction. An example of an eccentric contraction is the action of the biceps brachii when one extends the elbow slowly to lower a heavy weight. In this case, the thin filaments are pulled farther away from the center of the sarcomere, essentially stretching it.

Generation of Force

Whenever muscles contract, whether the contraction is concentric, static, or eccentric, the force developed must be graded to meet the needs of the task or activity. Using golf as an example, the force needed to tap in a 1 m (~39 in.) putt is far less than that needed to drive the ball 250 m (273 yd) from the tee to the middle of the fairway. The amount of muscle force developed is dependent on the number and type of motor units activated, the frequency of stimulation of each motor unit, the size of the muscle, the muscle fiber and sarcomere length, and the muscle's speed of contraction.

Motor Units and Muscle Size More force can be generated when more motor units are activated. Type II motor units generate more force than type I motor units because a type II motor unit contains more muscle fibers than a type I motor unit. In a similar manner, larger muscles, having more muscle fibers, can produce more force than smaller muscles.

Frequency of Stimulation of the Motor Units: Rate Coding A single motor unit can exert varying levels of force dependent on the frequency at which it is stimulated. This is illustrated in figure 1.14.[1] The smallest contractile response of a muscle fiber or a motor unit to a single electrical stimulus is termed a **twitch**. A series of three stimuli in rapid sequence, before complete relaxation from the first stimulus, can elicit an even greater increase in force or tension. This is termed **summation**. Continued stimulation at higher frequencies can lead to the state of **tetanus**, resulting in the peak force or tension of the muscle fiber or motor unit. **Rate coding** is the term used to refer to the process by which the tension of a given motor unit can vary from that of a twitch to that of tetanus by increasing the frequency of stimulation of that motor unit.

Length–Tension Relation Each muscle fiber has an optimal length for generating force. Recall that each muscle fiber is composed of sarcomeres connected end to end and that these sarcomeres are made up of both thick and thin filaments. The optimal sarcomere length is defined as that length where there is optimal overlap between the thick and thin filaments. This maximizes the potential for cross-bridge interaction, as illustrated in figure 1.15.[12] When a sarcomere is overly stretched

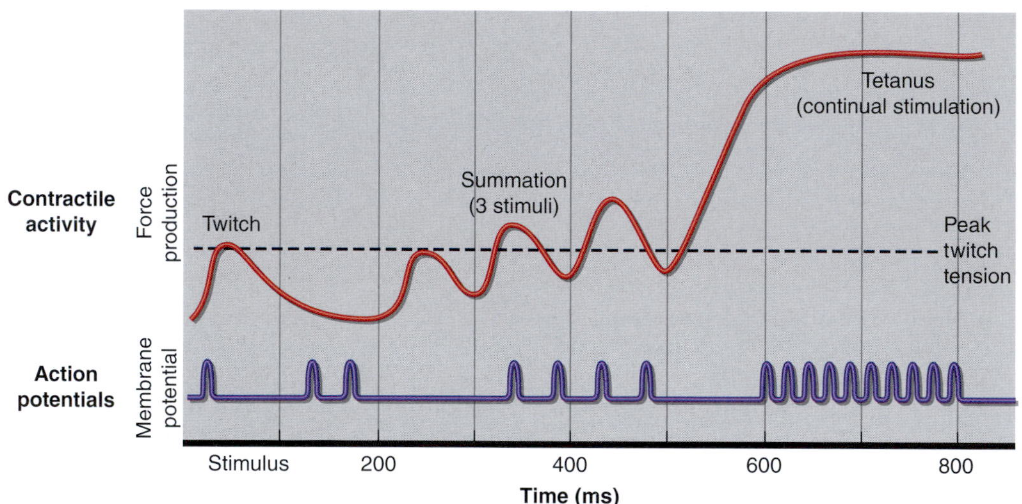

FIGURE 1.14 Variation in force or tension produced based on electrical stimulation frequency, illustrating the concepts of a twitch, summation, and tetanus.

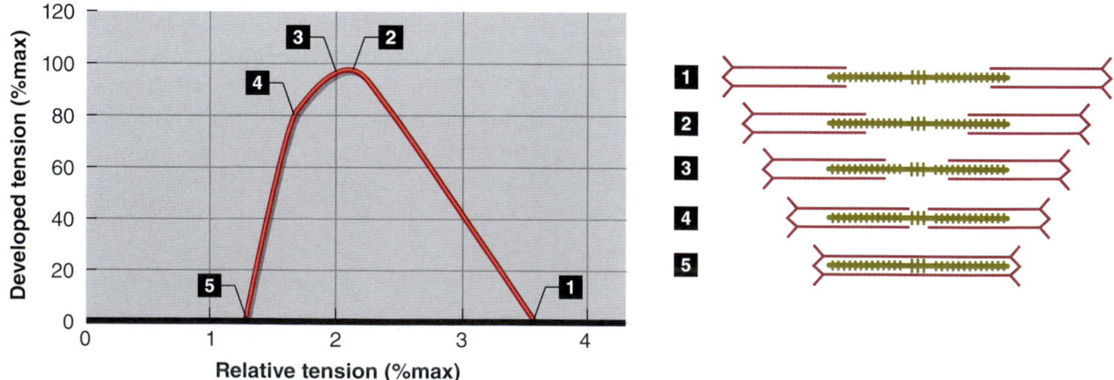

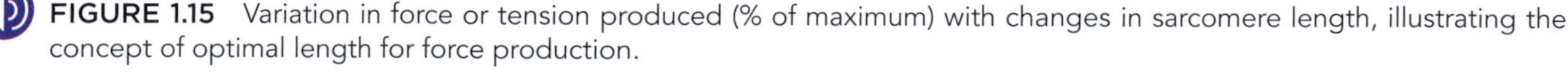
FIGURE 1.15 Variation in force or tension produced (% of maximum) with changes in sarcomere length, illustrating the concept of optimal length for force production.

Adapted by permission from B.R. MacIntosh, P.F. Gardiner, and A.J. McComas, *Skeletal Muscle: Form and Function*, 2nd ed. (Champaign, IL: Human Kinetics, 2006), 156.

(1) or shortened (5), little or no force can be developed since there is little cross-bridge interaction. The muscle's resting length is determined by the tendons that attach muscles to the bones on either end. As it turns out, this natural resting length maximizes the muscle's ability to generate force, referred to as the **length–tension relation**.

The implication of this relation is that muscle length, and therefore joint angle, will provide a mechanical advantage for force generation of a particular muscle or muscle group. The length–tension curve shown in figure 1.15 illustrates this phenomenon. Maximal tension can be achieved at sarcomere lengths between 2.0 and 2.25 μm, where the overlap between myosin and actin is optimal (i.e., the highest number of cross-bridges can be formed). As the sarcomere becomes elongated (>2.25 μm), the number of possible cross-bridges decreases, and therefore tension development (the descending limb of the curve) decreases accordingly. When sarcomeres are shortened to lengths <2.0 μm, the ability of myosin to interact with actin decreases because there are fewer myosin heads available to interact with actin (actin moves close to the M-line where there are few myosin heads; see figure 1.5). Another possible explanation for the reduced ability to produce force at lengths <2.0 μm is the physical constraint imposed by myosin reaching the Z-line of the sarcomere.

Force–Velocity Relation The ability of the muscle to develop force also depends on the speed of contraction. When people try to lift a very heavy object, they tend to do it slowly, maximizing the force they can apply to it. If they grab it and quickly try to lift it, they will likely fail, if not injure themselves. The **force–velocity relation** of a muscle illustrates muscle force as a function of the speed of contraction. During concentric (muscle-shortening) contractions, maximal force development decreases progressively as the speed of contraction increases. However, with eccentric (muscle-lengthening) contractions, the opposite is true.

This relation between force development and speed of contraction can be explained by the number of total cross-bridges attached at various speeds of contraction. When a muscle is contracting slowly, there is more time for cross-bridge formation than when contractions occur at higher speeds. In other words, when cross-bridges are formed at higher velocities, the ability of the muscle to produce force is reduced.

The force–velocity relation applies to both shortening and lengthening contractions. As depicted in figure 1.16, increasing the velocity of contraction while shortening (moving rightward along the *x*-axis) reduces force. Another way to think of the force–velocity relation is in terms of applying an external force to

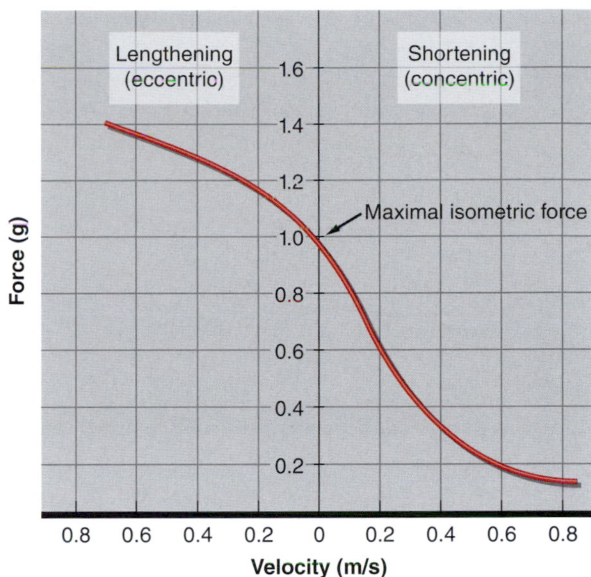

FIGURE 1.16 The relation between muscle lengthening and shortening velocity and force production. Note that the capacity for the muscle to generate force is greater during eccentric (muscle-lengthening) actions than during concentric (muscle-shortening) actions.

the muscle, such as performing a biceps curl. As the load gets heavier, the speed of contraction gets slower. When the load applied equals the maximal isometric force of the muscle, contraction velocity equals zero (by definition, an isometric contraction involves no movement). Now, let's explore what will happen when the load applied to the muscle is higher than the maximal isometric force and the muscle lengthens. In this case, the ability of the muscle to produce force will increase as a function of speed (moving leftward along the *x*-axis in figure 1.16) because as the load increases beyond maximal isometric, the speed of contraction will also increase.

Muscle Memory

Muscle force production depends on muscle mass. Because muscle fibers are large, they need evenly distributed multiple nuclei along their length in order to support all of the protein synthesis that occurs within the vast intracellular volume. Muscle fibers constantly change size, getting smaller with disuse (atrophy) and larger with training (hypertrophy). Standard thinking has been that muscle precursor satellite cells, small mononuclear stem cells, multiply during hypertrophy, fuse with existing muscle fibers, and supply additional nuclei as the fibers grow in size. Alternatively, during atrophy, unnecessary nuclei are cleared by a process called apoptosis, or programmed death.

A newer model has emerged that better explains the mechanisms that underlie changes in muscle fiber size and muscle mass.[3] In a study by Bruusgaard and colleagues, rat hindlimb muscles were hypertrophied by overloading, and their nuclei were measured by injecting labeled nucleotides. Beginning on day 6, the number of nuclei began to increase, increasing by 54% over 21 days. Fiber cross-sectional area did not begin to increase until day 9. In another group of rats, motor nerves were severed, causing muscle atrophy. The cross-sectional area decreased by 60% of the highest value of the hypertrophied group, but the number of nuclei was unchanged.

In trained individuals, retraining after a period of disuse occurs more quickly than in novice exercisers, and such muscle memory has typically been attributed to neural control of the muscle. It now appears that the nuclei may be the site of such memory. However, as pointed out by Lee and Burd,[11] a role for satellite cells in muscle hypertrophy cannot be excluded, and satellite cells undoubtedly contribute to overall skeletal muscle mass. It is possible that satellite cells may be necessary to sustain the mass and may be integral in maintaining muscle quality and function (figure 1.17).

In Review

- Among elite athletes, muscle fiber type composition differs by sport and event, with speed and strength events characterized by higher percentages of type II fibers and endurance events by higher percentages of type I fibers.
- The three main types of muscle contraction are concentric, in which the muscle shortens; static or isometric, in which the muscle acts but the joint angle is unchanged; and eccentric, in which the muscle lengthens.
- Force production can be increased both through the recruitment of more motor units and through an increase in the frequency of stimulation (rate coding) of the motor units.
- Force production is maximized at the muscle's optimal length. At this length, the amount of energy stored and the number of linked actin–myosin cross-bridges are optimal.
- Speed of contraction also affects the amount of force produced. For concentric contraction, maximal force is achieved with slower contractions. The closer to zero the velocity (isometric), the more force can be generated. With eccentric contractions, however, faster movement allows more force production.
- In addition to satellite cells, preserving the number of muscle fiber nuclei may help explain why previously trained muscles adapt more quickly to retraining after a period of disuse.

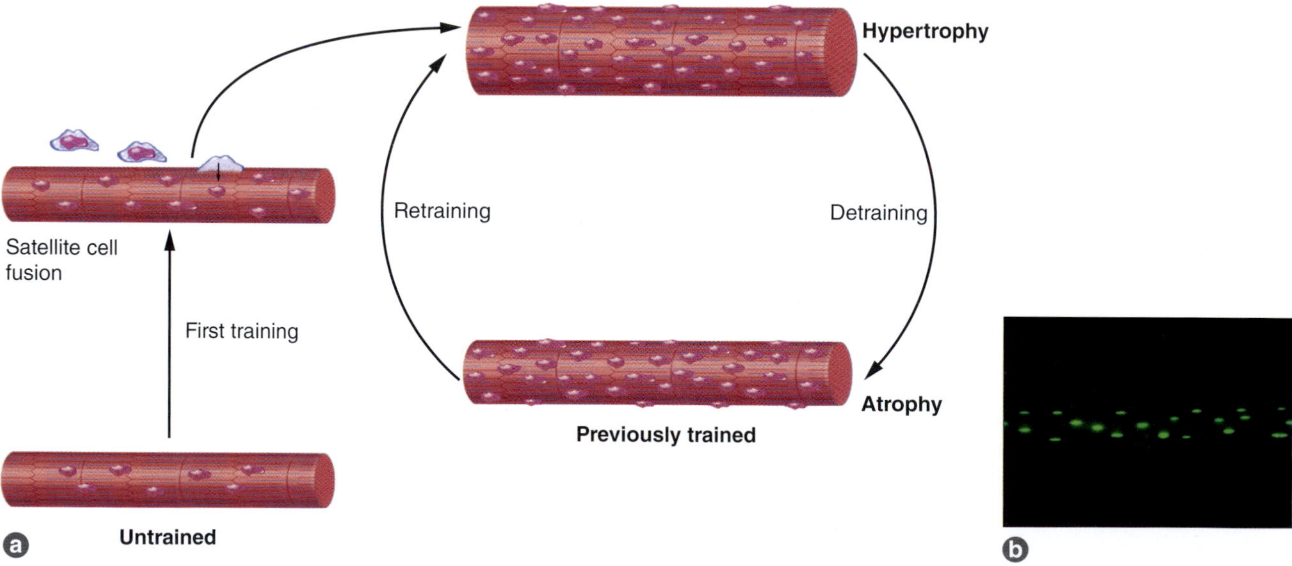

FIGURE 1.17 *(a)* A model to explain how the nuclei of muscle fibers may be the site of muscle memory. This theory explains why previously trained muscles adapt more quickly to retraining after a period of disuse. *(b)* Photomicrograph showing peripheral distribution of nuclei within a muscle fiber.

(a) Reprinted from J.C. Bruusgaard et al., "Myonuclei Acquired by Overload Exercise Precede Hypertrophy and are Not Lost on Detraining," *Proceedings of the National Academy of Sciences* 107 (2010): 15111-15116. By permission of J.C. Bruusgaard

IN CLOSING

In this chapter, we reviewed the components of skeletal muscle. We considered the differences in fiber types and their impact on physical performance. We learned how muscles generate force and produce movement. Now that we understand how movement is produced, we turn our attention to how movement is fueled. In the next chapter, we focus on metabolism and energy production.

KEY TERMS

actin
action potential
adenosine triphosphatase (ATPase)
adenosine triphosphate (ATP)
α-motor neuron
concentric contraction
dynamic contraction
eccentric contraction
endomysium
epimysium
excitation–contraction coupling
fascicle
force–velocity relation
length–tension relation
motor unit
muscle fiber
musculoskeletal system
myofibril
myosin
myosin cross-bridge
nebulin
perimysium
plasmalemma
power stroke
principle of orderly recruitment
rate coding
sarcolemma
sarcomere
sarcoplasm
sarcoplasmic reticulum (SR)
satellite cells
single-fiber contractile velocity (V_o)
size principle
sliding filament theory
static (isometric) muscle contraction
summation
tetanus
titin
transverse tubules (T-tubules)
tropomyosin
troponin
twitch
type I fiber
type II fiber

STUDY QUESTIONS

1. List and describe the anatomical components that make up a muscle fiber.
2. List the components of a motor unit.
3. What are the steps in excitation–contraction coupling?
4. What is the role of calcium in muscle contraction?
5. Describe the sliding filament theory. How do muscle fibers shorten?
6. What are the basic characteristics that differ between type I and type II muscle fibers?
7. What is the role of genetics in determining the proportions of muscle fiber types and the potential for success in selected activities?
8. Describe the relation between muscle force development and the recruitment of type I and type II motor units.
9. Explain, and give examples of, how concentric, static, and eccentric contractions differ.
10. What two mechanisms are used by the body to increase force production in a single muscle?
11. What is the optimal length of a muscle for maximal force development?
12. What is the relation between maximal force development and the speed of shortening (concentric) and lengthening (eccentric) contractions?
13. In muscle contraction, what roles are played by the protein titin?
14. Why do previously trained muscles adapt more quickly to retraining after a period of disuse?

STUDY GUIDE ACTIVITIES

In addition to the activities listed in the chapter opening outline, two other activities are available in the web study guide, located at

www.HumanKinetics.com/PhysiologyOfSportAndExercise

The **KEY TERMS** activity reviews important terms, and the end-of-chapter **QUIZ** tests your understanding of the material covered in the chapter.

Fuel for Exercise: Bioenergetics and Muscle Metabolism

2

In this chapter and in the web study guide

"Hitting the wall" is a common expression heard among marathon runners, and more than half of all nonelite marathon runners report having hit the wall during a marathon regardless of how hard they trained. This phenomenon usually happens around mile 20 to 22. The runner's pace slows considerably and the legs feel like lead. Tingling and numbness are often felt in the legs and arms, and thinking often becomes fuzzy and confused. Hitting the wall is basically running out of available energy.

The runner's primary fuel sources during prolonged exercise are carbohydrates and fats. Fats might seem to be the logical first choice of fuel for endurance events—they are ideally designed to be energy dense, and stores are virtually unlimited. Unfortunately, fat metabolism requires a constant supply of oxygen, and delivery of energy is slower than that provided by carbohydrate metabolism.

Most runners are able to store 2,000 to 2,200 calories of glycogen in their liver and muscles, which is enough to provide energy for about 20 mi (32 km) of moderate-pace running. Since the body is much less efficient at converting fat to energy, running pace slows and the runner suffers from fatigue. Furthermore, carbohydrates are the sole fuel source for brain function. Physiology, not coincidence, dictates why so many marathon runners hit the wall at around the 20 mi mark.

Chemical reactions in plants (photosynthesis) convert light from the sun into stored chemical energy. In turn, humans obtain energy by eating either plants or animals that feed on plants. Nutrients from ingested foods are provided in the form of carbohydrates, fats, and proteins. These three basic fuels, or energy **substrates**, can ultimately be broken down to release the stored energy. Each cell contains chemical pathways that convert these substrates to energy that can then be used by that cell and other cells of the body, a process called **bioenergetics**. All of the chemical reactions in the body are collectively called **metabolism**.

Because all energy eventually degrades to heat, the amount of energy released in a biological reaction can be measured from the amount of heat produced. Energy in biological systems is measured in calories. By definition, 1 calorie (cal) equals the amount of heat energy needed to raise 1 g of water 1 °C, from 14.5 °C to 15.5 °C. In humans, energy is expressed in **kilocalories (kcal)**, where 1 kcal is the equivalent of 1,000 cal. Sometimes the term *Calorie* (with a capital C) is used synonymously with kilocalorie, but *kilocalorie* is more technically and scientifically correct. Thus, when one reads that someone eats or expends 3,000 Cal per day, it really means the person is ingesting or expending 3,000 *kcal* per day.

Some free energy in the cells is used for growth and repair throughout the body. Such processes build muscle mass during training and repair muscle damage after exercise or injury. Energy also is needed for active transport of many substances, such as sodium, potassium, and calcium ions, across cell membranes to maintain homeostasis. Myofibrils use energy to cause sliding of the actin and myosin filaments, resulting in muscle action and force generation, as described in chapter 1.

Energy Substrates

Energy is released when chemical bonds—the bonds that hold elements together to form molecules—are broken. Substrates are composed primarily of carbon, hydrogen, oxygen, and (in the case of protein) nitrogen. The molecular bonds that hold these elements together are relatively weak and therefore provide little energy when broken. Consequently, food is not used directly for cellular operations. Rather, the energy in food's molecular bonds is chemically released within our cells and then stored in the form of the high-energy compound introduced in chapter 1, adenosine triphosphate (ATP), which is discussed in detail later in this chapter.

At rest, the energy that the body needs is derived almost equally from the breakdown of carbohydrates and fats. Proteins serve important functions as enzymes that aid chemical reactions and as structural building blocks but usually provide little energy for metabolism. During intense, short-duration muscular effort, more carbohydrate is used, with less reliance on fat to generate ATP. Longer, less intense exercise uses both carbohydrate and fat for sustained energy production.

Carbohydrate

The amount of **carbohydrate** used during exercise is related to both the carbohydrate availability and the muscles' well-developed system for carbohydrate metabolism. All carbohydrates are ultimately converted to the simple six-carbon sugar, **glucose** (figure 2.1), a monosaccharide (one-unit sugar) that is transported through the blood to all body tissues. Under resting conditions, ingested carbohydrate is stored in muscles and liver in the form of a more complex polysaccharide (multiple linked sugar molecules), **glycogen**. Glycogen is stored in the cytoplasm of muscle cells until those cells use it to form ATP. Additional glycogen stored in the liver is converted back to glucose as needed and then transported by the blood to active tissues, where it is metabolized.

Muscle and liver glycogen stores are limited, especially if the diet contains an insufficient amount of

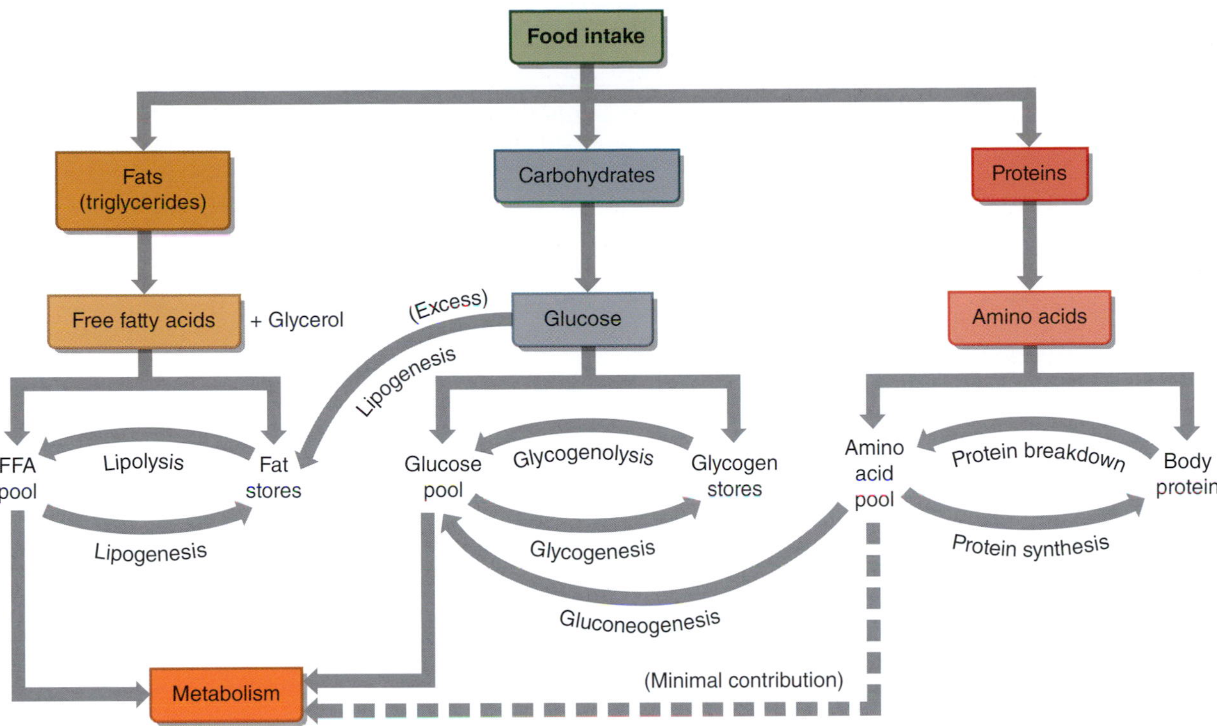

FIGURE 2.1 Cellular metabolism results from the breakdown of three fuel substrates provided by the diet. Once each is converted to its usable form, it either circulates in the blood as an available "pool" to be used for metabolism or is stored in the body.

carbohydrate, and can be depleted during prolonged, intense exercise. Thus, we rely heavily on dietary sources of starches and sugars to continually replenish our carbohydrate reserves. Without adequate carbohydrate intake, muscles can be deprived of their primary energy source. Furthermore, carbohydrates are the only energy source used by brain tissue; therefore, severe carbohydrate depletion results in negative cognitive effects.

Fat

Fats provide a large portion of the energy used during prolonged, less intense exercise. Body stores of potential energy in the form of fat are substantially larger than the reserves of carbohydrate, in terms of both weight and energy availability. Table 2.1 provides an indication of the total body stores of these two energy sources in a lean person (12% body fat). For the average middle-aged adult with more body fat (adipose tissue), the fat stores would be approximately twice as large, whereas the carbohydrate stores would be about the same. But fat is less readily available for cellular metabolism because it must first be reduced from its complex form, **triglyceride**, to its basic components, glycerol and **free fatty acids (FFAs)**. Only FFAs are used to form ATP (figure 2.1).

TABLE 2.1 **Body Stores of Fuels and Associated Energy Availability**

Location	g	kcal
Carbohydrate		
Liver glycogen	110	451
Muscle glycogen	500	2,050
Glucose in body fluids	15	62
Fat		
Subcutaneous and visceral	7,800	73,320
Intramuscular	161	1,513
Total	**7,961**	**74,833**

Note. These estimates are based on a body weight of 65 kg (143 lb) with 12% body fat.

Substantially more energy is derived from breaking down a gram of fat (9.4 kcal/g) than from the same amount of carbohydrate (4.1 kcal/g). Nonetheless, the rate of energy release from fat is too slow to meet all of the energy demands of intense muscular activity.

Other types of fats found in the body serve non-energy-producing functions. Phospholipids are a key structural component of all cell membranes and form protective sheaths around some large nerves. Steroids are found in cell membranes and also function as hormones or as building blocks of hormones, such as estrogen and testosterone.

Protein

Protein can be used as a minor energy source under some circumstances, but it must first be converted into glucose (figure 2.1). In the case of severe energy depletion or starvation, protein may even be used to generate FFAs for energy. The process by which protein or fat is converted into glucose is called **gluconeogenesis**. The process of converting protein into fatty acids is termed **lipogenesis**. Protein can supply up to 10% of the energy needed to sustain prolonged exercise. Only the most basic units of protein—the amino acids—can be used for energy. A gram of protein yields about 4.1 kcal.

Controlling the Rate of Energy Production

To be useful, free energy must be released from chemical compounds at a controlled rate. This rate is determined primarily by two things, the availability of the primary substrate and enzyme activity. The availability of large amounts of a substrate increases the activity of that particular pathway. An abundance of one particular fuel (e.g., carbohydrate) can cause cells to rely more on that source than on alternatives. This influence of substrate availability on the rate of metabolism is termed the *mass action effect*.

Specific protein molecules called **enzymes** also control the rate of free-energy release. Many of these enzymes speed up the breakdown (**catabolism**) of chemical compounds. Chemical reactions occur only when the reacting molecules have sufficient initial energy to start the reaction or chain of reactions. Enzymes do not cause a chemical reaction to occur and do not determine the amount of usable energy that is produced by these reactions. Rather, they speed up reactions by lowering the **activation energy** that is required to begin the reaction (figure 2.2).

Although the enzyme names are quite complex, most end with the suffix *-ase*. For example, an important enzyme that breaks down ATP and releases stored energy is adenosine triphosphatase, better known as ATPase.

Biochemical pathways that result in the production of a product from a substrate almost always involve multiple steps. Each individual step is typically catalyzed by a specific enzyme. Therefore, increasing the amount of enzyme present or the activity of that enzyme (for example, by changing the temperature or pH) results in an increased rate of product formation through that metabolic pathway. Additionally, many enzymes require other molecules called cofactors to function, so cofactor availability may also affect enzyme function and therefore the rate of metabolic reactions.

As illustrated in figure 2.3, metabolic pathways typically have one enzyme that is of particular importance in controlling the reaction's overall rate. This enzyme, usually located in an early step in the pathway, is known as the **rate-limiting enzyme**. The activity of a rate-limiting enzyme is determined by the accumulation of substances farther down the pathway that decrease enzyme activity through **negative feedback**.

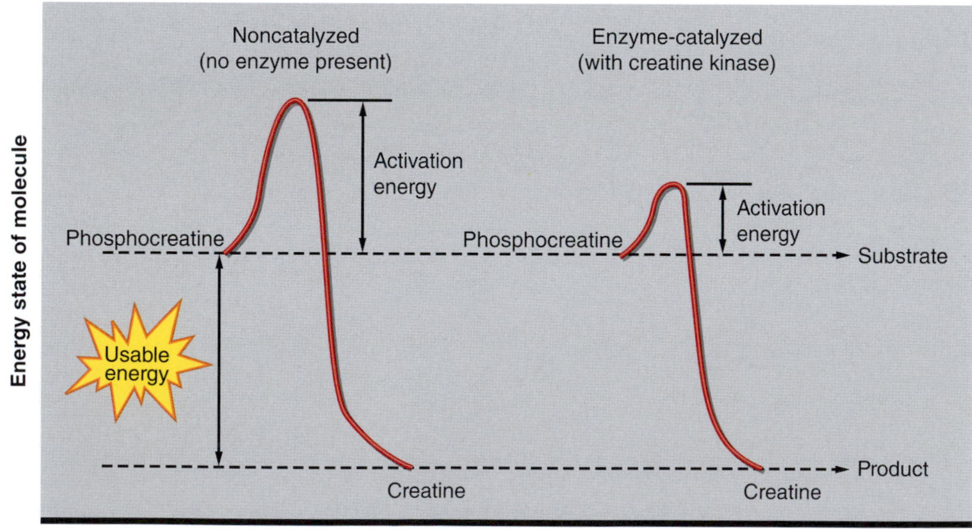

FIGURE 2.2 Enzymes control the rate of chemical reactions by lowering the activation energy required to initiate the reaction. In this example, the enzyme creatine kinase binds to its substrate phosphocreatine to increase the rate of production of creatine.

Adapted from original figure provided by Dr. Martin Gibala, McMaster University, Hamilton, Ontario, Canada.

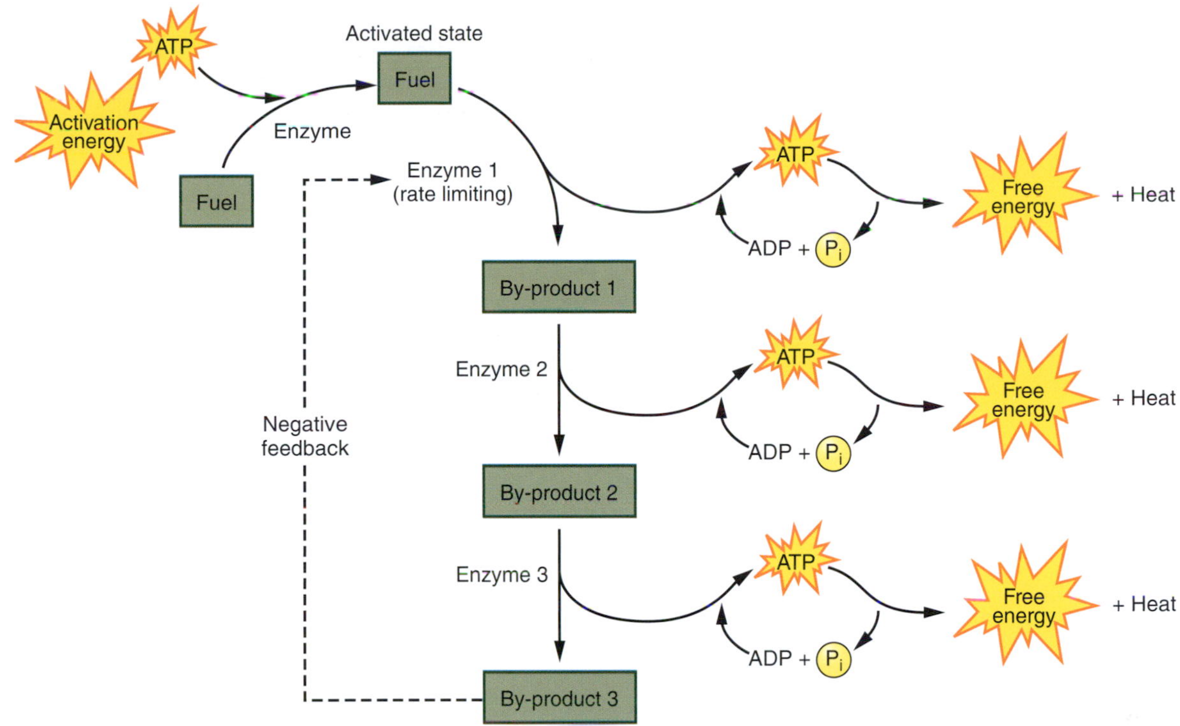

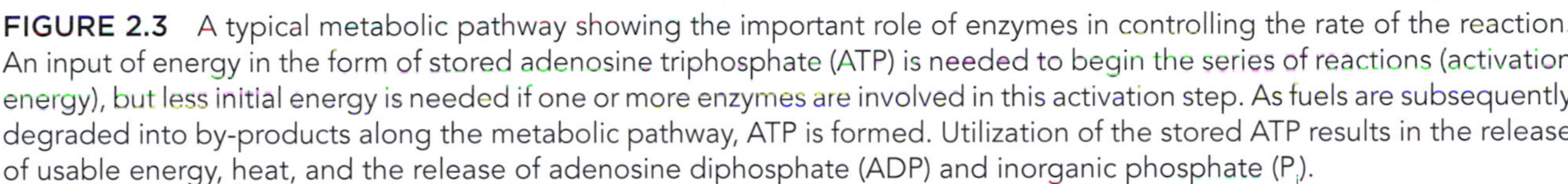

FIGURE 2.3 A typical metabolic pathway showing the important role of enzymes in controlling the rate of the reaction. An input of energy in the form of stored adenosine triphosphate (ATP) is needed to begin the series of reactions (activation energy), but less initial energy is needed if one or more enzymes are involved in this activation step. As fuels are subsequently degraded into by-products along the metabolic pathway, ATP is formed. Utilization of the stored ATP results in the release of usable energy, heat, and the release of adenosine diphosphate (ADP) and inorganic phosphate (P_i).

In Review

- Energy for cell metabolism is derived from three substrates in foods: carbohydrate, fat, and protein. Proteins provide little of the energy used for metabolism under normal conditions.
- Within cells, the usable storage form of the energy we derive from food is the high-energy compound adenosine triphosphate, or ATP.
- Carbohydrate and protein each provide about 4.1 kcal energy per gram, compared with about 9.4 kcal/g for fat.
- Carbohydrate, stored as glycogen in muscle and the liver, is more quickly accessible as an energy source than either protein or fat. Glucose, directly from food or broken down from stored glycogen, is the usable form of carbohydrate.
- Fat, stored as triglycerides in adipose tissue, is an ideal storage form of energy. Free fatty acids from the breakdown of triglycerides are converted to energy.
- Carbohydrate stores in the liver and skeletal muscle are limited to about 2,500 to 2,600 kcal of energy, or the equivalent of the energy needed for about 40 km (25 mi) of running. Fat stores can provide more than 70,000 kcal of energy.
- Enzymes control the rate of metabolism and energy production. Enzymes can speed up the overall reaction by lowering the initial activation energy and by catalyzing various steps along the pathway.
- Enzymes can be inhibited through negative feedback of subsequent pathway by-products (or often ATP), slowing the overall rate of the reaction. This usually involves a particular enzyme located early in the pathway called the rate-limiting enzyme.

One example of a substance that may accumulate and feed back to decrease enzyme activity would be the end product of the pathway; another would be ATP and its breakdown products, ADP and inorganic phosphate. If the goals of a metabolic pathway are to form a chemical product and release free energy in the form of ATP, it makes sense that an abundance of either that end product or ATP would feed back to slow further production and release, respectively.

VIDEO 2.1 Presents Mark Hargreaves discussing the sensitivity of ATP production to muscle activity and control of ATP production during exercise.

Storing Energy: High-Energy Phosphates

The immediately available source of energy for almost all bodily functions, including muscle contraction, is ATP. An ATP molecule (figure 2.4*a*) is composed of adenosine (a molecule of adenine joined to a molecule of ribose) combined with three inorganic phosphate (P_i) groups. Adenine is a nitrogen-containing base, and ribose is a five-carbon sugar. When an ATP molecule is combined with water (hydrolysis) and acted on by the enzyme ATPase, the last phosphate group splits away, rapidly releasing a large amount of free energy (approximately 7.3 kcal per mole of ATP under standard conditions, but possibly up to 10 kcal per mole of ATP or greater within the cell). This reduces the ATP to **adenosine diphosphate (ADP)** and P_i (figure 2.4*b*).

To generate ATP, a phosphate group is added to the relatively low-energy compound, ADP, in a process called **phosphorylation**. This process requires a considerable amount of energy. Some ATP is generated independent of oxygen availability, and such metabolism is called substrate-level phosphorylation. Other ATP-producing reactions (discussed later in the chapter) occur without oxygen, while still others occur with the aid of oxygen, a process called **oxidative phosphorylation**.

As shown in figure 2.3, ATP is formed from ADP and P_i via phosphorylation as fuels are broken down into fuel by-products at various steps along a metabolic pathway. The storage form of energy, ATP, can subsequently release free or usable energy when needed as it is once again broken down into ADP and P_i.

The Basic Energy Systems

Cells can store only very limited amounts of ATP and must constantly generate new ATP to provide needed energy for all cellular metabolism, including muscle contraction. Cells generate ATP through any one of (or a combination of) three metabolic pathways:

1. The ATP-PCr system
2. The glycolytic system (glycolysis)

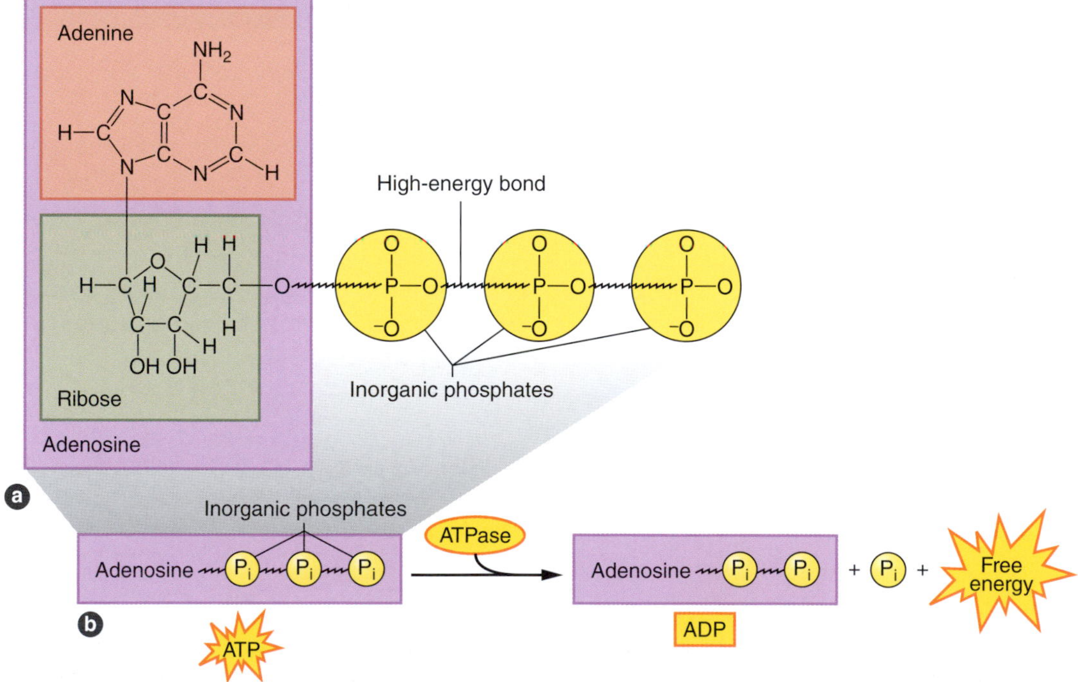

FIGURE 2.4 *(a)* The structure of an adenosine triphosphate (ATP) molecule, showing the high-energy phosphate bonds. *(b)* When the third phosphate on the ATP molecule is separated from adenosine by the action of adenosine triphosphatase (ATPase), energy is released.

3. The oxidative system (oxidative phosphorylation)

The first two systems can act in the absence of oxygen and are jointly termed **anaerobic metabolism**. The third system requires oxygen and therefore comprises **aerobic metabolism**.

ATP-PCr System

The simplest of the energy systems is the **ATP-PCr system**, shown in figure 2.5. In addition to storing a very small amount of adenosine triphosphate (ATP) directly, cells contain another high-energy phosphate molecule that stores energy called **phosphocreatine** (**PCr**; sometimes called creatine phosphate). This simple pathway involves donation of a P_i from PCr to ADP to form ATP. Unlike what occurs with the limited freely available ATP in the cell, energy released by the breakdown of PCr is not directly used for cellular work. Instead, it regenerates ATP to maintain a relatively constant supply under resting conditions.

The release of energy from PCr is catalyzed by the enzyme **creatine kinase**, which acts on PCr to separate P_i from creatine. The energy released can then be used to add a P_i molecule to an ADP molecule, forming ATP. As energy is released from ATP by the splitting of a phosphate group, cells can prevent ATP depletion by breaking down PCr, providing energy and P_i to re-form ATP from ADP.

Following the principles of negative feedback and rate-limiting enzymes discussed earlier, creatine kinase activity is enhanced when concentrations of ADP or P_i increase, and is inhibited when ATP concentrations increase. When intense exercise is initiated, the small amount of available ATP in muscle cells is broken down for immediate energy, yielding ADP and P_i. The increased ADP concentration enhances creatine kinase activity, and PCr is catabolized to form additional ATP. As exercise progresses and additional ATP is generated by the other two energy systems—the glycolytic and oxidative systems—creatine kinase activity is inhibited.

This process of breaking down PCr to allow formation of ATP is rapid and can be accomplished without any special structures within the cell. The ATP-PCr system is classified as substrate-level metabolism. Although it can act in the presence of oxygen, the process does not require oxygen.

During the first few seconds of intense muscular activity, such as sprinting, ATP is maintained at a relatively constant level, but PCr declines steadily as it is used to replenish the depleted ATP (see figure 2.6). At exhaustion, however, both ATP and PCr levels are low and are unable to provide energy for further muscle contraction and relaxation. Thus, the capacity to maintain ATP levels with the energy from PCr is limited. The combination of ATP and PCr stores can sustain the muscles' energy needs for only 3 to 15 s during an all-out sprint. Beyond that time, muscles must rely on other processes for ATP formation: glycolytic and oxidative pathways.

Glycolytic System

The ATP-PCr system has a limited capacity to generate ATP for energy, lasting only a few seconds. The second method of ATP production involves the liberation of energy through the breakdown ("lysis") of glucose. This system is called the glycolytic system because it entails **glycolysis**, the breakdown of glucose through a pathway that involves a sequence of glycolytic enzymes. Glycolysis

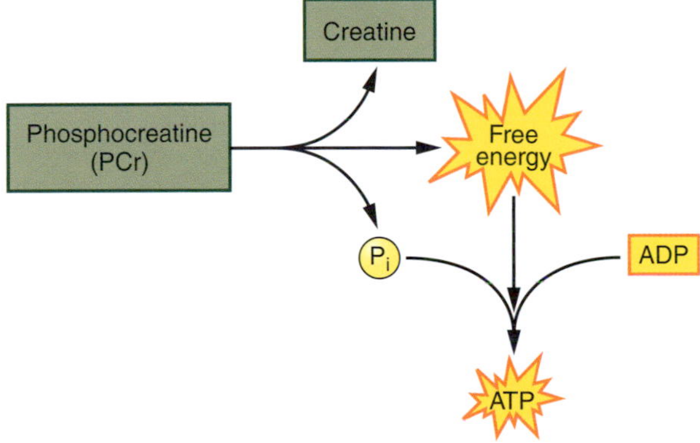

FIGURE 2.5 In the ATP-PCr system, adenosine triphosphate (ATP) can be recreated via the binding of an inorganic phosphate (P_i) to adenosine diphosphate (ADP) with the energy derived from the breakdown of phosphocreatine (PCr).

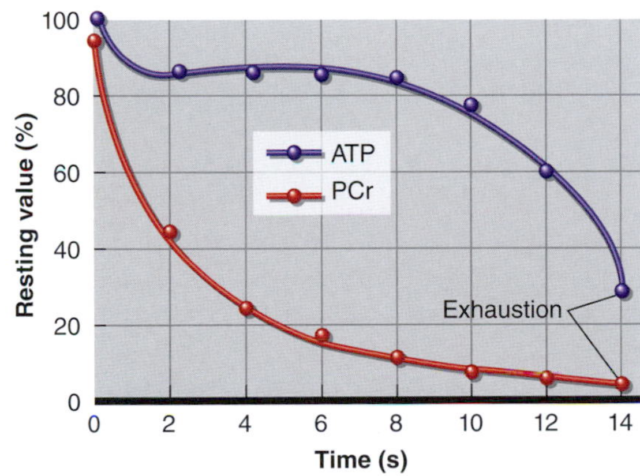

FIGURE 2.6 Changes in type II (fast-twitch) skeletal muscle adenosine triphosphate (ATP) and phosphocreatine (PCr) during 14 s of maximal muscular effort (sprinting). Although ATP is being used at a very high rate, the energy from PCr is used to synthesize ATP, initially preventing the ATP level from decreasing. However, at exhaustion, both ATP and PCr levels are low.

is a more complex pathway than the ATP-PCr system, and the sequence of steps involved in this process is presented in figure 2.7.

Glucose accounts for about 99% of all sugars circulating in the blood. Blood glucose comes from the digestion of carbohydrate and the breakdown of liver

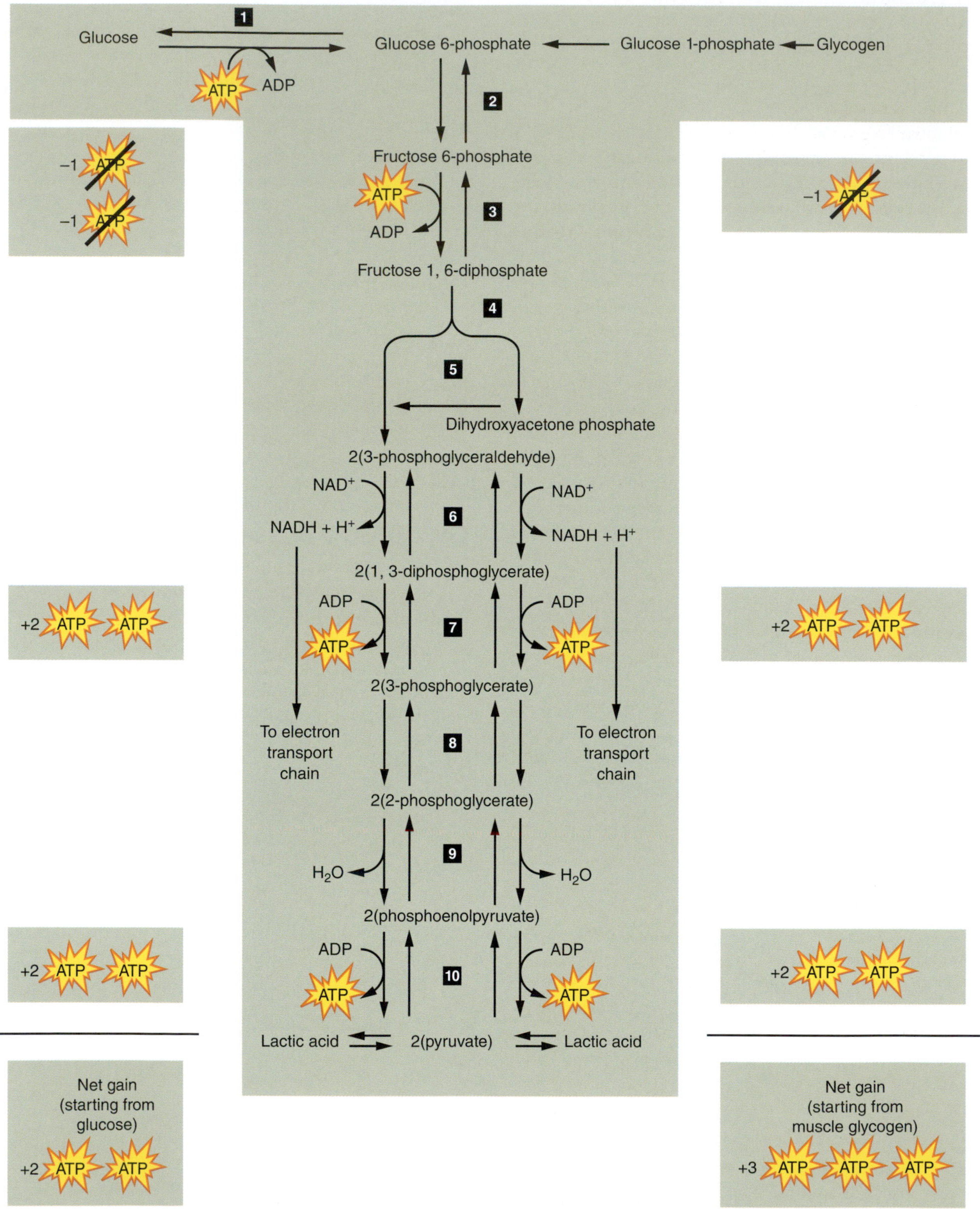

FIGURE 2.7 The derivation of energy (ATP) by glycolysis. Glycolysis involves the breakdown of one glucose (six-carbon) molecule to two three-carbon molecules of pyruvic acid. The process can begin with either glucose circulating in the blood or glycogen (a chain of glucose molecules, the storage form of glucose in muscle and liver). Note that there are roughly 10 separate steps in this anaerobic process, and the net result is the generation of either two or three ATP molecules, depending on whether glucose or glycogen is the initial substrate.

glycogen. Glycogen is synthesized from glucose by a process called glycogenesis and is stored in the liver or in muscle until needed. At that time, the glycogen is broken down to glucose-1-phosphate, which enters the glycolysis pathway, a process termed **glycogenolysis**.

Before either glucose or glycogen can be used to generate energy, it must be converted to a compound called glucose-6-phosphate. Even though the goal of glycolysis is to release ATP, the conversion of a molecule of glucose to glucose-6-phosphate requires the expenditure or input of one ATP molecule. In the conversion of glycogen, glucose-6-phosphate is formed from glucose-1-phosphate without this energy expenditure. Glycolysis technically begins once the glucose-6-phosphate is formed.

Glycolysis requires 10 to 12 enzymatic reactions for the breakdown of glycogen to pyruvic acid, which is then converted to lactic acid. All steps in the pathway and all of the enzymes involved operate within the cell cytoplasm. The net gain from this process is 3 moles (mol) of ATP formed for each mole of glycogen broken down. If glucose is used instead of glycogen, the gain is only 2 mol of ATP because 1 mol was used for the conversion of glucose to glucose-6-phosphate.

This energy system obviously does not produce large amounts of ATP. Despite this limitation, the combined actions of the ATP-PCr and glycolytic systems allow the muscles to generate force even when the oxygen supply is limited. These two systems predominate during the early minutes of high-intensity exercise.

Another major limitation of anaerobic glycolysis is that it causes an accumulation of lactic acid in the muscles and body fluids. Glycolysis produces pyruvic acid. This process does not require oxygen, but the presence of oxygen determines the fate of the pyruvic acid. Without oxygen present, the pyruvic acid is converted directly to lactic acid, an acid with the chemical formula $C_3H_6O_3$ that quickly dissociates, forming lactate. The terms *pyruvic acid* and *pyruvate*, and *lactic acid* and *lactate*, are often used interchangeably in exercise physiology. In each case, the acid form of the molecule is relatively unstable at normal body pH and rapidly loses a hydrogen ion. The remaining molecule is more correctly called pyruvate or lactate. Lactate can itself be a source of energy as discussed later in this chapter.

In all-out sprint events lasting 1 or 2 min, the demands on the glycolytic system are high, and muscle lactic acid concentrations can increase from a resting value of about 1 mmol/kg of muscle to more than 25 mmol/kg. This acidification of muscle fibers inhibits further glycogen breakdown because it impairs glycolytic enzyme function. In addition, the acid decreases the fibers' calcium-binding capacity and thus may impede muscle contraction.

The rate-limiting enzyme in the glycolytic pathway is **phosphofructokinase (PFK)**. Like almost all rate-limiting enzymes, PFK catalyzes an early step in the pathway, the conversion of fructose-6-phosphate to fructose-1,6-diphosphate. Increasing ADP and P_i concentrations enhance PFK activity and therefore speed up glycolysis, while elevated ATP concentrations slow glycolysis by inhibiting PFK. Additionally, because the glycolytic pathway feeds into the Krebs cycle for additional energy production when oxygen is present (discussed later), products of the Krebs cycle, especially citrate and hydrogen ions, likewise feedback to inhibit PFK.

A muscle fiber's rate of energy use during exercise can be 200 times greater than at rest. The ATP-PCr and glycolytic systems alone cannot supply all the needed energy. Furthermore, these two systems are not capable of supplying all of the energy needs for all-out activity lasting more than 2 min or so. Prolonged exercise relies on the third energy system, the oxidative system.

In Review

- The formation of ATP provides cells with a high-energy compound for storing and, when broken down, releasing energy. It serves as the immediate source of energy for most body functions, including muscle contraction.
- Adenosine triphosphate is generated through three primary energy systems:
  1. The ATP-PCr system
  2. The glycolytic system
  3. The oxidative system
- In the ATP-PCr system, P_i is separated from PCr through the action of creatine kinase. The P_i can then combine with ADP to form ATP using the energy released from the breakdown of PCr. This system is anaerobic, and its main function is to maintain ATP levels early in exercise. The energy yield is 1 mol of ATP per 1 mol of PCr.
- The glycolytic system involves the process of glycolysis, through which glucose or glycogen is broken down to pyruvic acid. When glycolysis occurs without oxygen, the pyruvic acid is converted to lactic acid. One mole of glucose yields 2 mol ATP, but 1 mol glycogen yields 3 mol ATP.
- The ATP-PCr and glycolytic systems are major contributors of energy during short-burst activities lasting up to 2 min and during the early minutes of longer high-intensity exercise.

Oxidative System

The final system of cellular energy production is the **oxidative system**. This is the most complex of the three energy systems, and only a brief overview is provided here. The process by which the body breaks down substrates with the aid of oxygen to generate energy is called cellular respiration. Because oxygen is required, this is an aerobic process. Unlike the anaerobic production of ATP that occurs in the cytoplasm of the cell, the oxidative production of ATP occurs within special cell organelles called **mitochondria**. In muscles, these are adjacent to the myofibrils and are also scattered throughout the sarcoplasm (see figure 1.3).

The total number, or density, of mitochondria within a muscle fiber is determined by its demand for ATP production, but the precise location of these mitochondria within the cell is determined by oxygen diffusion. Each individual muscle fiber has an optimal distribution of mitochondria within the cell that allows for a near maximal rate of ATP production while exposing the mitochondria to as little excess oxygen as possible. Excess oxygen exposure in mitochondria creates reactive oxygen species (ROS), which are toxic to the cell at high concentrations.[5,7]

Within a muscle cell, mitochondria tend to be localized along the periphery of the fiber, with higher densities near capillaries. This arrangement functions to create gradients in the oxygen concentration from the capillary to the mitochondria to facilitate the flow of oxygen into the mitochondria. Having mitochondria localized toward the periphery of the cell benefits the muscle fiber by optimizing oxygen delivery to sustain high metabolic rates.[6] However, having the mitochondria located around the periphery of the cell also increases ROS production because of their exposure to oxygen. Thus, mitochondria tend to be distributed nonuniformly around the outside of the cell, depending on capillary location, rather than being evenly spaced. This location is optimal for maintaining high metabolic rates while minimizing risk for increasing ROS production, which can negatively affect the cell.

Muscles need a steady supply of energy to continuously produce the force needed during long-term activity. Unlike what happens with anaerobic ATP production, the oxidative system is slow to turn on, but it has a much larger energy-producing capacity, so aerobic metabolism is the primary method of energy production during endurance activities. This places considerable demands on the cardiovascular and respiratory systems to deliver oxygen to the active muscles. Oxidative energy production can come from carbohydrates (starting with glycolysis) or fats.

Oxidation of Carbohydrate

As shown in figure 2.8, oxidative production of ATP from carbohydrates involves three processes:

- Glycolysis (figure 2.8*a*)
- The Krebs cycle (figure 2.8*b*)
- The electron transport chain (figure 2.8*c*)

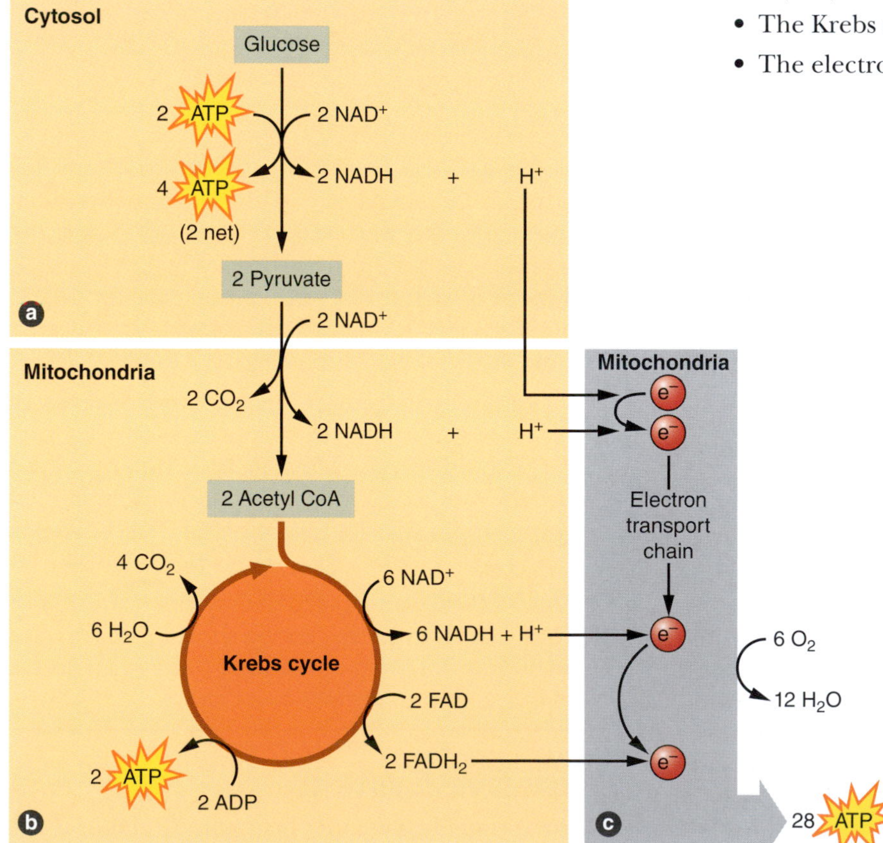

FIGURE 2.8 In the presence of oxygen, after glucose (or glycogen) has been reduced to pyruvate, *(a)* the pyruvate is first catalyzed to acetyl coenzyme A (acetyl CoA), which can enter *(b)* the Krebs cycle, where oxidative phosphorylation occurs. Hydrogen ions released during the Krebs cycle then combine with coenzymes that carry the hydrogen ions to *(c)* the electron transport chain.

Glycolysis In carbohydrate metabolism, glycolysis plays a role in *both* anaerobic and aerobic ATP production. The process of glycolysis is the same regardless of whether oxygen is present. The presence of oxygen determines only the fate of the end product, pyruvic acid. Recall that anaerobic glycolysis produces lactic acid and only three net moles of ATP per mole of glycogen, or two net moles of ATP per mole of glucose. In the presence of oxygen, however, the pyruvic acid is converted into a compound called **acetyl coenzyme A (acetyl CoA)**.

Krebs Cycle Once formed, acetyl CoA enters the **Krebs cycle** (also called the citric acid cycle or tricyclic acid cycle), a complex series of chemical reactions that permit the complete oxidation of acetyl CoA (see figure 2.9). Recall that for every glucose molecule that enters the glycolytic pathway, two molecules of pyruvate are formed. Therefore, each glucose molecule that begins the energy-producing process in the presence of oxygen results in two complete Krebs cycles.

As depicted in 2.8*b* (and shown in more detail in figure 2.9), the conversion of succinyl CoA to succinate in the Krebs cycle results in the generation of guanosine triphosphate, or GTP, a high-energy compound similar to ATP. Guanosine triphosphate then transfers a P_i to ADP to form ATP. These two ATPs (per molecule of glucose) are formed by substrate-level phosphorylation. So at the end of the Krebs cycle, two additional moles of ATP have been formed directly, and the original carbohydrate has been broken down into carbon dioxide and hydrogen.

As in the other pathways involved in energy metabolism, Krebs cycle enzymes are regulated by negative feedback at several steps in the cycle. The rate-limiting enzyme in the Krebs cycle is isocitrate dehydrogenase, which, like PFK, is inhibited by ATP and activated by ADP and P_i as is the electron transport chain. Because muscle contraction relies on the availability of calcium in the cell, excess calcium also stimulates the rate-limiting enzyme isocitrate dehydrogenase.

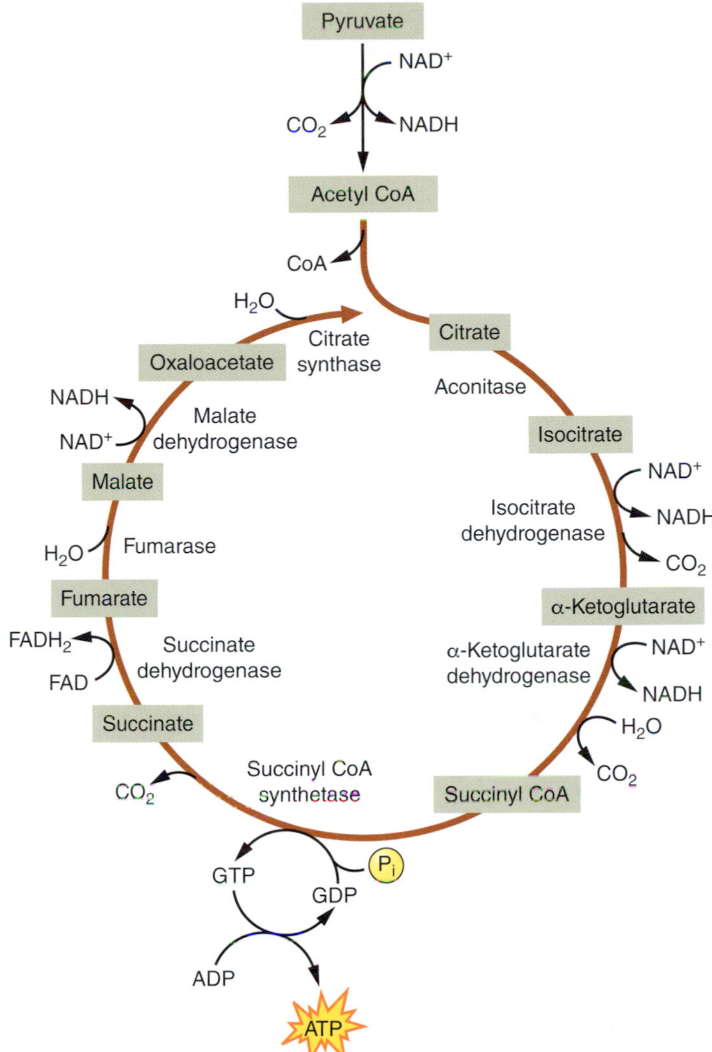

FIGURE 2.9 The series of reactions that take place during the Krebs cycle, showing the compounds formed and enzymes involved.

Electron Transport Chain During glycolysis, hydrogen ions are released when glucose is metabolized to pyruvic acid. Additional hydrogen ions are released in the conversion of pyruvate to acetyl CoA and at several steps in the Krebs cycle. If these hydrogen ions remained in the system, the inside of the cell would become too acidic. What happens to this hydrogen?

The Krebs cycle is coupled to a series of reactions known as the **electron transport chain** (figure 2.8*c*). The hydrogen ions released during the processes of glycolysis, the conversion of pyruvic acid to acetyl CoA, and the Krebs cycle combine with two coenzymes: nicotinamide adenine dinucleotide (NAD) and flavin adenine dinucleotide (FAD), converting each to its reduced form (NADH and $FADH_2$, respectively). During each Krebs cycle, three molecules of NADH and one molecule of $FADH_2$ are produced. These carry the hydrogen atoms (electrons) to the electron transport chain, a group of mitochondrial protein complexes located in the inner mitochondrial membrane (figure 2.10).

These protein complexes contain a series of enzymes and iron-containing proteins known as **cytochromes**. These proteins serve as electron magnets that transfer electrons, where the first complex, flavin mononucleotide (FMN), is a stronger magnet for electrons than NADH, the second complex is a stronger magnet than the first, and so on down the chain. As high-energy electrons are passed from complex to complex along this

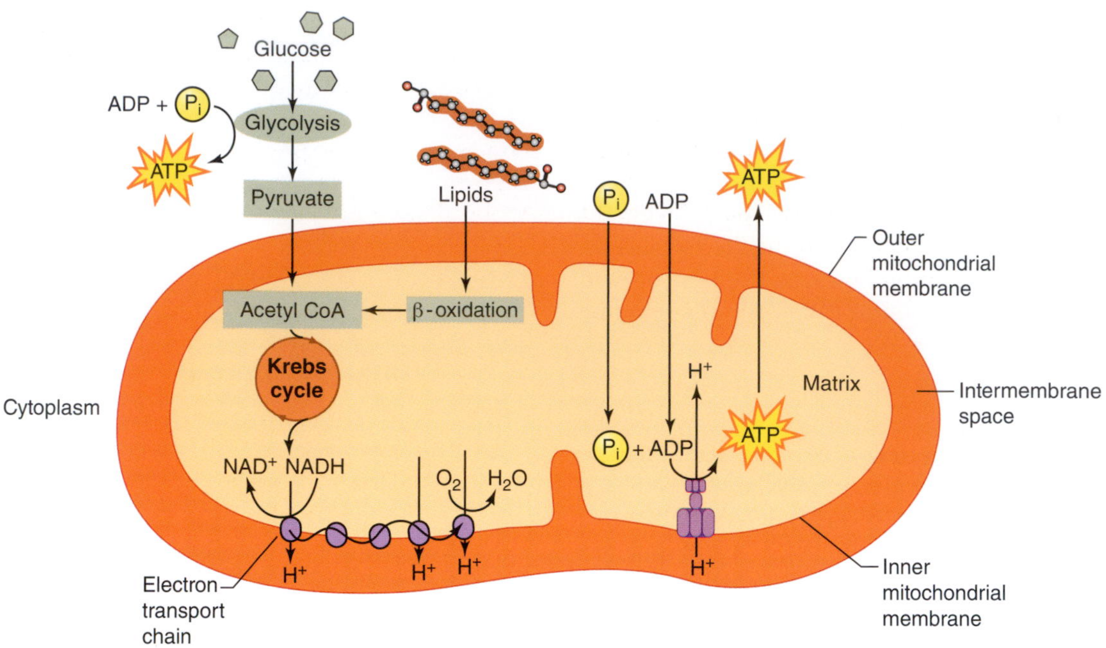

FIGURE 2.10 Locations of the processes of glycolysis (cytoplasm), the Krebs cycle (mitochondria), and the electron transport chain (inner mitochondrial membrane).

chain, some of the energy released by those reactions is used to pump H^+ from the mitochondrial matrix into the outer mitochondrial compartment. As these hydrogen ions move back across the membrane down their concentration gradient, energy is transferred to ADP, and ATP is formed. This final step requires an enzyme known as ATP synthase. At the end of the chain, the H^+ combines with oxygen to form water, thus preventing acidification of the cell. This is illustrated in figure 2.11. Because this overall process relies on oxygen as

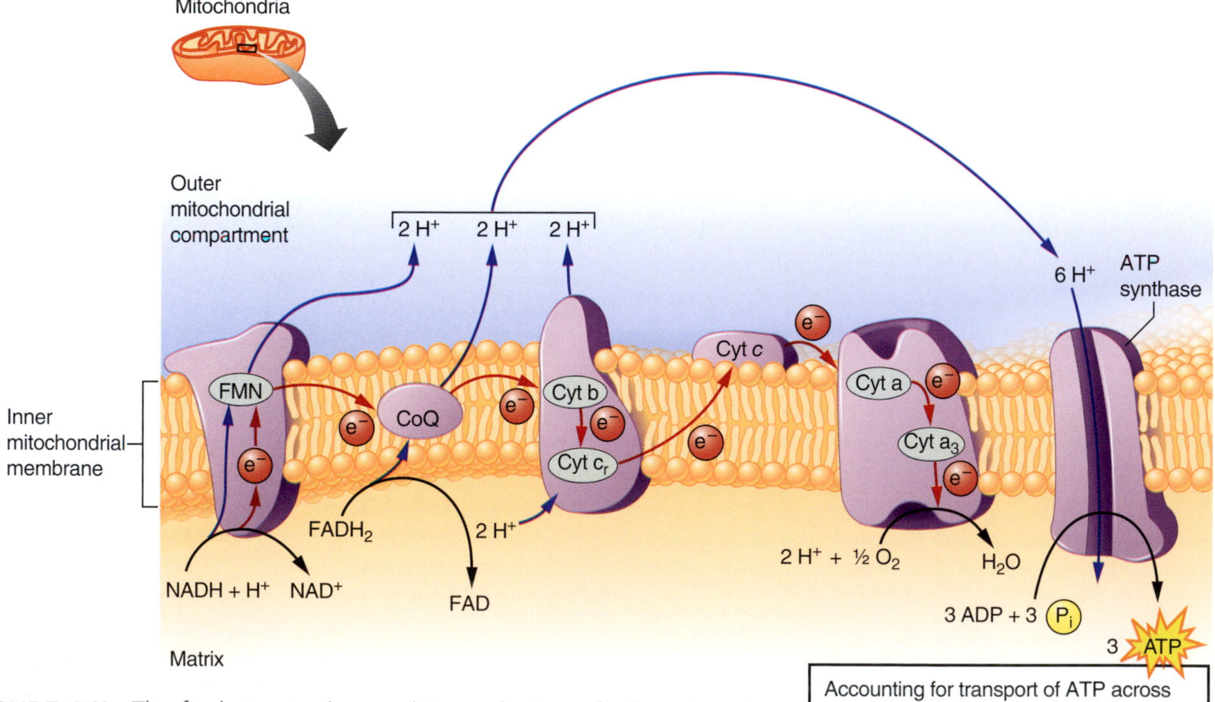

FIGURE 2.11 The final step in the aerobic production of adenosine triphosphate (ATP) is the transfer of energy from the high-energy electrons of reduced nicotinamide adenine dinucleotide (NADH) and reduced flavin adenine dinucleotide ($FADH_2$) within the mitochondria, following a series of steps known as the electron transport chain.

the final acceptor of electrons and H^+, it is referred to as oxidative phosphorylation.

For every pair of electrons transported to the electron transport chain by NADH, three molecules of ATP are formed, while the electrons passed through the electron transport chain by $FADH_2$ yield only two molecules of ATP. However, because the NADH and $FADH_2$ are outside the membrane of the mitochondria, the H^+ must be shuttled through the membrane, which requires energy to be used. So, in reality, the net yields are only 2.5 ATPs per NADH and 1.5 ATPs per $FADH_2$.

Energy Yield From Oxidation of Carbohydrate The complete oxidation of glucose can generate 32 molecules of ATP, while 33 ATPs are produced from one molecule of muscle glycogen. The sites of ATP production are summarized in figure 2.12. The net production of ATP from substrate-level phosphorylation in the glycolytic pathway leading into the Krebs cycle results in a net gain of two ATPs (or three from glycogen). A total of 10 NADH molecules leading into the electron transport chain—two in glycolysis, two in the conversion of pyruvic acid to acetyl CoA, and six in the Krebs cycle—yields 25 net ATP molecules. Remember that while 30 ATPs are actually produced, the energy cost of transporting ATP across membranes uses five of those ATPs. The two FAD molecules in the Krebs cycle that are involved in electron transport result in three additional net ATPs. And finally, substrate-level phosphorylation within the Krebs cycle involving the molecule GTP adds another two ATP molecules.

Accounting for the energy cost of shuttling electrons across the mitochondrial membrane is a relatively new concept in exercise physiology, and many textbooks still refer to net energy productions of 36 to 39 ATPs per molecule of glucose.

Oxidation of Fat

As noted earlier, fat also contributes importantly to the muscle's energy needs. Muscle and liver glycogen stores can provide only ~2,500 kcal of energy, but the fat stored inside muscle fibers and in fat cells can supply at least 70,000 to 75,000 kcal, even in a lean adult. Although many chemical compounds (such as triglycerides, phospholipids, and cholesterol) are classified as fats, only triglycerides are major energy sources. Triglycerides are stored in fat cells and between and within skeletal muscle fibers. To be used for energy, a triglyceride must be broken down to its basic units: one molecule of glycerol and three FFA molecules. This process is called **lipolysis** and is controlled by enzymes known as lipases.

Free fatty acids are the primary energy source for fat metabolism. Once liberated from glycerol, FFAs can enter the blood and be transported throughout the body, entering muscle fibers by either simple diffusion or transporter-mediated (facilitated) diffusion. Their rate of entry into the muscle fibers depends on the concentration gradient. Increasing the concentration

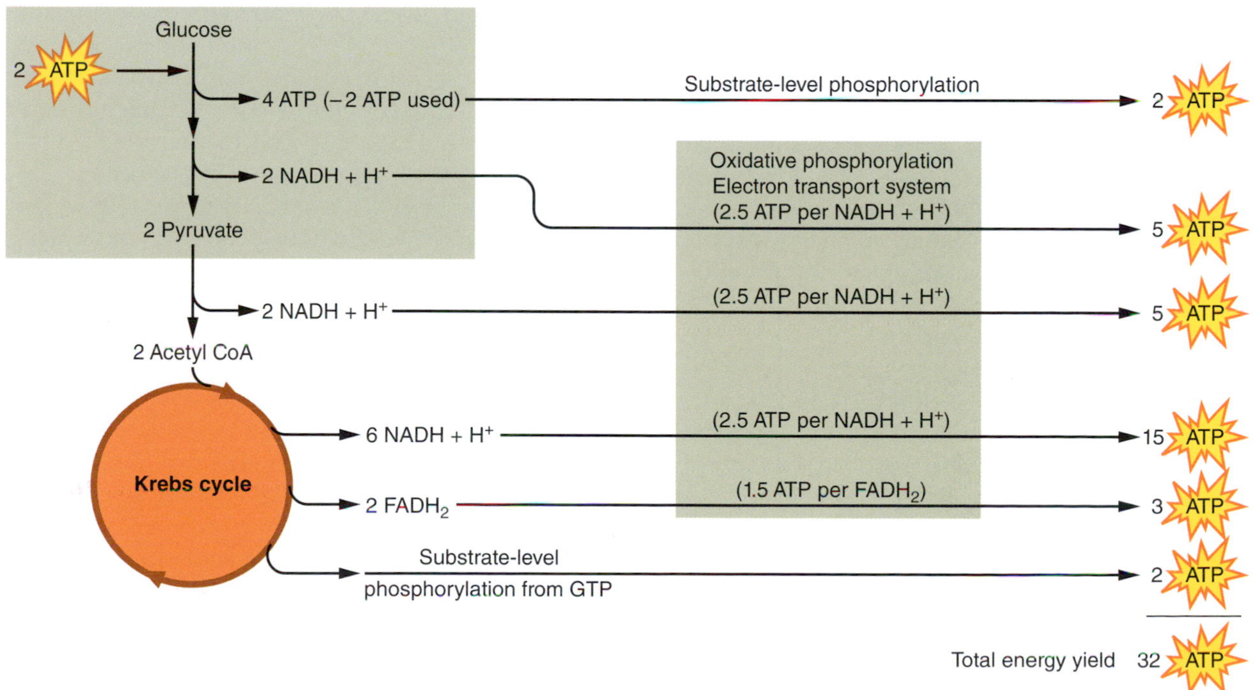

FIGURE 2.12 The net energy production from the oxidation of one molecule of glucose is 32 molecules of adenosine triphosphate (ATP). Oxidation of glycogen as the original substrate would yield one additional ATP.

of FFAs in the blood increases the rate of their transport into muscle fibers.

β-Oxidation Recall that fats are stored in the body in two places, within muscle fibers and in adipose tissue cells called adipocytes. The storage form of fats is triglyceride, which is broken down into FFAs and glycerol for energy metabolism. Before FFAs can be used for energy production, they must be converted to acetyl CoA in the mitochondria, a process called **β-oxidation**. Acetyl CoA is the common intermediate through which all substrates enter the Krebs cycle for oxidative metabolism.

β-Oxidation is a series of steps in which two-carbon acyl units are chopped off of the carbon chain of the FFA. The acyl units become acetyl CoA, which then enters the Krebs cycle for the formation of ATP. The number of steps depends on the number of carbons in the FFA, usually between 14 and 24 carbons. For example, if an FFA originally has a 16-carbon chain, β-oxidation yields eight molecules of acetyl CoA.

On entering the muscle fiber, FFAs must be enzymatically activated with energy from ATP, preparing them for catabolism (breakdown) within the mitochondria. Like glycolysis, β-oxidation requires an input energy of two ATPs for activation but, unlike glycolysis, produces no ATPs directly.

Krebs Cycle and the Electron Transport Chain After β-oxidation, fat metabolism follows the same path as oxidative carbohydrate metabolism. Acetyl CoA formed by β-oxidation enters the Krebs cycle. The Krebs cycle generates hydrogen, which is transported to the electron transport chain along with the hydrogen generated during β-oxidation to undergo oxidative phosphorylation. As in glucose metabolism, the by-products of FFA oxidation are ATP, H_2O, and carbon dioxide (CO_2). However, the complete combustion of an FFA molecule requires more oxygen because an FFA molecule contains considerably more carbon molecules than a glucose molecule.

The advantage of having more carbon molecules in FFAs than in glucose is that more acetyl CoA is formed from the metabolism of a given amount of fat, so more acetyl CoA enters the Krebs cycle and more electrons are sent to the electron transport chain. This is why fat metabolism can generate much more energy than glucose metabolism. Unlike glucose or glycogen, fats are heterogeneous, and the amount of ATP produced depends on the specific fat oxidized.

RESEARCH PERSPECTIVE 2.1

White, Brown, and (Perhaps) Beige Fat in Humans

Brown adipose tissue (BAT), often called brown fat, is found in almost every species of mammal, especially in those that hibernate. Unlike white adipose tissue, which is specialized for lipid storage and breakdown (lipolysis) to meet long-duration metabolic demands, the function of brown adipose is to transfer energy from food directly into heat. Brown adipose cells contain many small lipid droplets and many mitochondria, which give the tissue its brown appearance. BAT cells also have more blood vessels than white adipose cells to supply the tissue with oxygen and nutrients and distribute the heat produced in the cells to the rest of the body. The inner membrane of the mitochondria of BAT cells has a specialized protein called uncoupling protein that uncouples the electron transport chain from the creation of ATP (phosphorylation). While white fat generates ATP for energy, brown fat's primary role is to produce heat and increase metabolism, especially at rest.

Brown adipose is abundant in newborn babies and young children. However, for a long time, it was believed that brown adipose stores were absent in adult humans. In 2009, a study published in the *New England Journal of Medicine* showed that adult humans have functionally active brown adipose tissue.[3] Using positron-emission tomography and computed tomography (PET-CT) scans, researchers found that the most common location for this brown adipose tissue in adults was near the clavicles, that brown adipose was more frequently found in women than in men, and that individuals with a higher body mass index have less brown fat. Because BAT promotes energy dissipation rather than energy storage (the role of white adipose tissue), its discovery in humans sparked great interest in the possibility of increasing the activity of BAT to target diseases like obesity and type 2 diabetes.

Several studies have recently reported that chronic endurance exercise may promote the expression of similar thermogenic (heat-producing) genes in white adipose tissue, resulting in the browning of white fat. In one animal study, training-induced changes in fat type resulted in increases in resting energy expenditure of up to 17% in trained rats.[2] It is still unclear whether exercise training increases BAT mass or promotes the browning of white adipose tissue in humans, but studies are now underway to use the metabolic potential of BAT to increase whole-body energy expenditure. The ultimate goal of these studies is to treat obesity and other metabolic diseases.

Consider the example of palmitic acid, a rather abundant 16-carbon FFA. The combined reactions of oxidation, the Krebs cycle, and the electron transport chain produce 106 molecules of ATP from one molecule of palmitic acid (see table 2.2), compared with only 32 molecules of ATP from glucose or 33 from glycogen.

Oxidation of Protein

As noted earlier, carbohydrates and fatty acids are the preferred fuel substrates. But proteins, or rather the amino acids that compose proteins, are also used for energy under some circumstances. Some amino acids can be converted into glucose, a process called gluconeogenesis (see figure 2.1). Alternatively, some can be converted into various intermediates of oxidative metabolism (such as pyruvate or acetyl CoA) to enter the oxidative process.

Protein's energy yield is not as easily determined as that of carbohydrate or fat because protein also contains nitrogen. When amino acids are catabolized, some of the released nitrogen is used to form new amino acids, but the remaining nitrogen cannot be oxidized by the body. Instead it is converted into urea and then excreted, primarily in the urine. This conversion requires the use of ATP, so some energy is spent in this process.

When protein is broken down through combustion in the laboratory, the energy yield is 5.65 kcal/g. However, because of the energy expended in converting nitrogen to urea when protein is metabolized in the body, the energy yield is only about 4.1 kcal/g.

To accurately assess the rate of protein metabolism, the amount of nitrogen being eliminated from the body must be determined. This requires urine collection for 12 to 24 h periods, a time-consuming process. Because the healthy body uses little protein during rest and exercise (usually not more than 10% of total energy expended), estimates of total energy expenditure generally ignore protein metabolism.

TABLE 2.2 **ATP Produced From One Molecule of Palmitic Acid**

Stage of process	Direct (substrate-level oxidation)	By oxidative phosphorylation
Fatty acid activation	0	−2
β-oxidation (occurs 7 times)	0	28
Krebs cycle (occurs 8 times)	8	72
Subtotal	8	98
Total	106	

Lactic Acid as a Source of Energy During Exercise

Lactic acid is in a state of constant turnover within cells, being produced by glycolysis and removed from the cell, primarily through oxidation. Thus, despite its reputation as a cause of fatigue, lactic acid can be, and is, used as an actual fuel source during exercise. This occurs through several mechanisms.

First, we now know that lactate produced by glycolysis in the cytoplasm of a muscle fiber can be taken up by the mitochondria within that same fiber and directly oxidized. This occurs mostly in cells with a high density of mitochondria like type I (high oxidative) muscle fibers, cardiac muscle, and liver cells.

Second, lactate produced in a muscle fiber can be transported away from its site of production and used elsewhere by a process called the lactate shuttle, first described by Dr. George Brooks. Lactate is produced primarily by type II muscle fibers but can be transported to adjacent type I fibers by diffusion or active transport. In that regard, most of the lactate produced in a muscle never leaves that muscle. It can also be transported through the circulation to sites where it can be directly oxidized. The lactate shuttle allows for glycolysis in one cell to supply fuel for use by another cell. Special transporters called monocarboxylate transport (MCT) proteins facilitate the movement of lactate between cells and tissues and likely within cells. During exercise, approximately 80% to 90% of lactate is transferred across the sarcolemma either by passive diffusion or by facilitated transport through MCTs. These transporters can be expressed in differing numbers, depending on the properties of the cells and tissues helping to move lactate in the cells that are the most metabolically active. Using lactate as a metabolic fuel accounts for approximately 70% to 75% of lactate removal during exercise.

Finally, some of the lactic acid produced in the muscle is transported by the blood to the liver, where it is reconverted to pyruvic acid and back to glucose (gluconeogenesis) and transported back to the working muscle. This is called the Cori cycle. Without this recycling of lactate into glucose for use as an energy source, prolonged exercise would be severely limited. On a more integrative level, lactate produced in exercising skeletal muscle is taken up and oxidized in the brain. Thus, lactate not only is integrally involved as a metabolic fuel but also responds to changes in nutrient sensing as different metabolic fuels are used during exercise.

Summary of Substrate Metabolism

As shown in figure 2.13, the ability to produce muscle contraction for exercise is a matter of energy supply

RESEARCH PERSPECTIVE 2.2

Lifelong Training Can Lead to More Efficient Fuel Utilization

While aging is associated with a decrease in exercise capacity and sport performance, it is clear that the skeletal muscle of older adults can be trained to meet the demands of exercise. Remaining physically active throughout a person's life protects against some of the age-related decrements in muscle-fiber size, fiber type, mitochondrial number, and oxidative capacity when compared to older sedentary people. These age-related changes and adaptations are discussed in greater detail in chapter 18.

A recent study performed at the University of Pittsburgh sought to determine whether the skeletal muscles of older masters athletes had the same substrate storage and capacity for oxidation of those fuels as those of younger athletes who trained similarly (i.e., used the same mode of exercise and frequency of training).[4] That study found that lifelong masters athletes have greater triglyceride stores in their muscle fibers and a greater proportion of oxidative fibers compared to the young athletes. These differences resulted in better metabolic efficiency—a lower reliance on carbohydrate oxidation—during exercise at high intensities in the older athletes (see figure). Lifelong endurance exercise protects against some of the age-associated decreases in oxidative potential and provides older athletes with an increased capacity for fat oxidation to produce ATP during exercise.

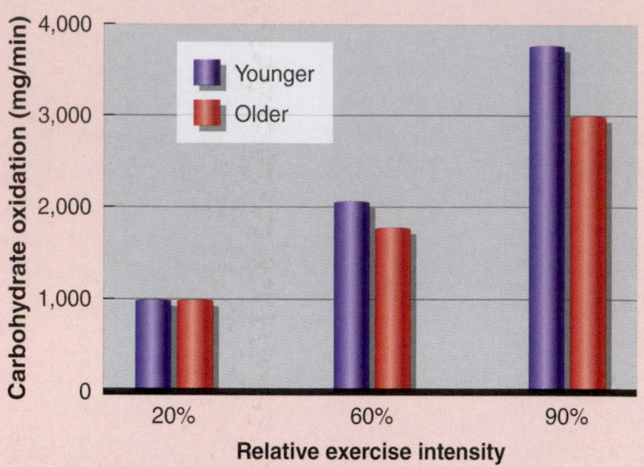

A simplified illustration of carbohydrate oxidation at different relative exercise intensities in younger and older adult athletes. Lifelong masters endurance athletes rely less on carbohydrate oxidation at higher exercise intensities compared to similarly trained younger athletes.

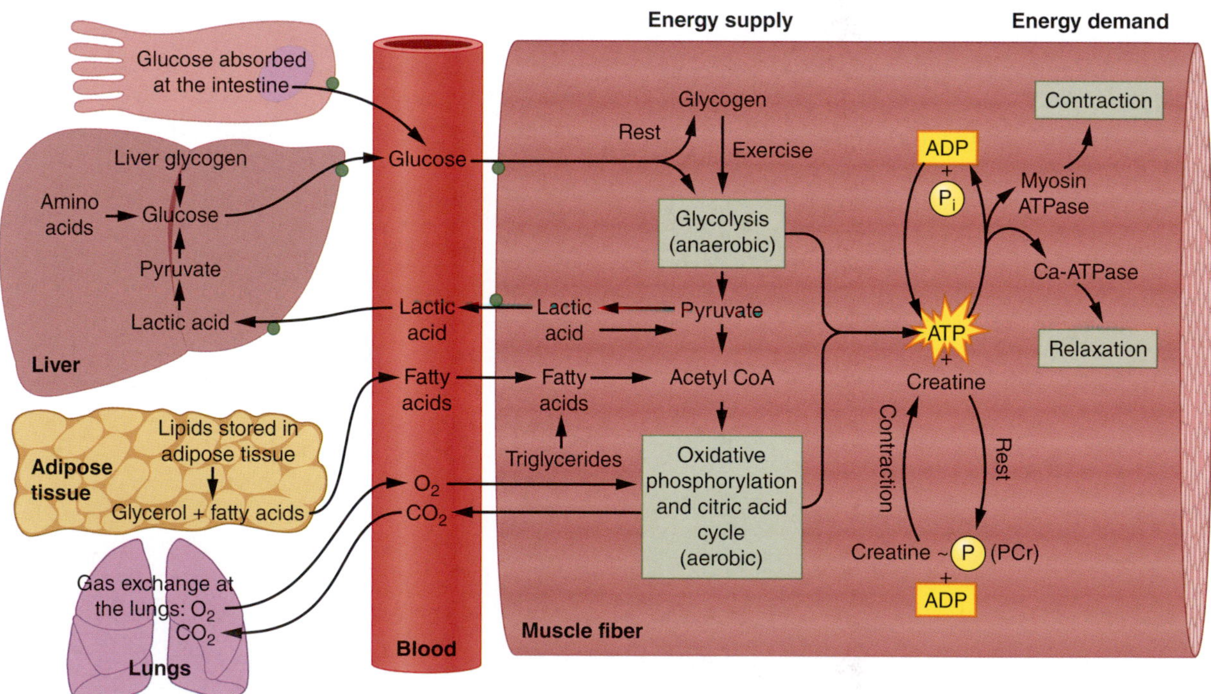

FIGURE 2.13 The metabolism of carbohydrate, fat, and (to a lesser extent) protein shares some common pathways within the muscle fiber. The adenosine triphosphate (ATP) molecules generated by oxidative and nonoxidative metabolism are used by those steps in muscle contraction and relaxation that demand energy.

and energy demand. Both the contraction of skeletal muscle fibers and their relaxation require energy. That energy comes from foodstuffs in the diet and stored energy in the body. The ATP-PCr system operates within the cytosol of the cell, as does glycolysis, and neither requires oxygen for ATP production. Oxidative phosphorylation takes place within the mitochondria. Note that under aerobic conditions, both major substrates—carbohydrates and fats—are reduced to the common intermediate acetyl CoA that enters the Krebs cycle.

In Review

- The oxidative system involves the breakdown of substrates in the presence of oxygen. This system yields more energy than the ATP-PCr or the glycolytic system.
- Oxidation of carbohydrate involves glycolysis, the Krebs cycle, and the electron transport chain. The end result is H_2O, CO_2, and 32 or 33 ATP molecules per carbohydrate molecule.
- Fat oxidation begins with β-oxidation of FFAs and then follows the same path as carbohydrate oxidation: acetyl CoA moving into the Krebs cycle and the electron transport chain. The energy yield for fat oxidation is much higher than for carbohydrate oxidation, and it varies with the FFA being oxidized. However, the maximum rate of high-energy phosphate formation from lipid oxidation is too low to match the rate of utilization of high-energy phosphate during higher-intensity exercise, and the energy yield of fat per oxygen molecule used is much less than that for carbohydrate.
- Although fat provides more kilocalories of energy per gram than carbohydrate, fat oxidation requires more oxygen than carbohydrate oxidation. The energy yield from fat is 5.6 ATP molecules per oxygen molecule used, compared with carbohydrate's yield of 6.3 ATP per oxygen molecule. Oxygen delivery is limited by the oxygen transport system, so carbohydrate is the preferred fuel during high-intensity exercise.
- The maximum rate of ATP production from lipid oxidation is too low to match the rate of utilization of ATP during high-intensity exercise. This explains the reduction in an athlete's race pace when carbohydrate stores are depleted and fat, by default, becomes the predominant fuel source.
- Measurement of protein oxidation is more complex because amino acids contain nitrogen, which cannot be oxidized. Protein contributes relatively little to energy production, generally less than 10%, so its metabolism is often considered negligible.
- Despite its reputation as a potential factor in causing fatigue, lactic acid can be, and is, used as an important fuel source during exercise.

Interaction of the Energy Systems

The three energy systems do not work independently of one another, and no activity is 100% supported by any single energy system. When a person exercises at the highest intensity possible, from the shortest sprints (less than 10 s) to endurance events (greater than 30 min), each of the energy systems is contributing to the total energy needs of the body. Generally, one energy system dominates energy production, except when there is a transition from the predominance of one energy system to another. As an example, in a 10 s, 100 m sprint, the ATP-PCr system is the predominant energy system, but both the anaerobic glycolytic and the oxidative systems provide a small portion of the energy needed. At the other extreme, in a 30 min, 10,000 m (10,936 yd) run, the oxidative system is predominant, but both the ATP-PCr and anaerobic glycolytic systems contribute some energy as well.

Figure 2.14 shows the reciprocal relation among the energy systems with respect to power and capacity. The ATP-PCr energy system can provide energy at a fast rate but has a very low capacity for energy production. Thus

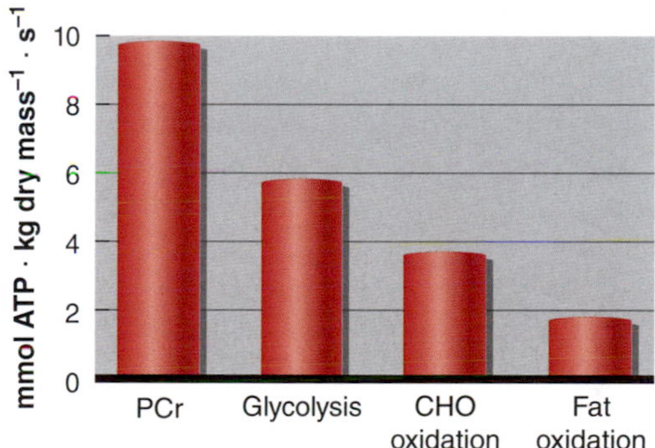

a Maximal rate of ATP generation

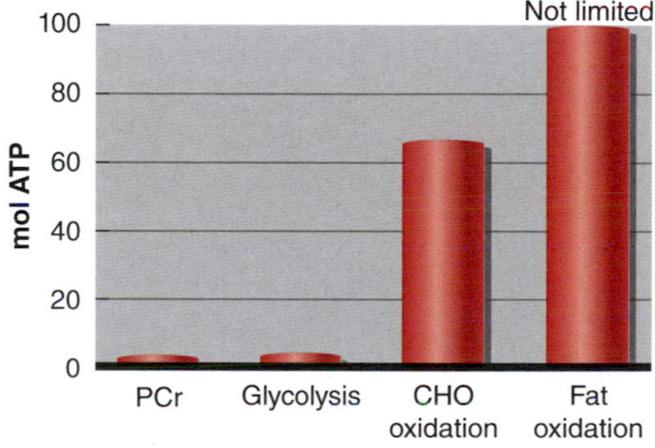

b Maximal available energy

FIGURE 2.14 The various energy systems have a reciprocal relation with respect to *(a)* the maximal rate at which energy can be produced and *(b)* the capacity to produce that energy.

it supports exercise that is intense but of very short duration. By contrast, fat oxidation takes longer to gear up and produces energy at a slower rate; however, the amount of energy it can produce is unlimited. The characteristics of the muscle fiber's energy systems are listed in table 2.3.

The Crossover Concept

The **crossover concept** was first described by Brooks and Mercier[1] to describe the relative balance between carbohydrate (CHO) and fat metabolism during sustained exercise. At rest and during exercise at low to moderate intensities (below 60% of maximal oxygen uptake), lipids serve as the main substrate for generating ATP. During high-intensity exercise (above 75% of maximal oxygen uptake), increases in muscle glycogenolysis and the recruitment of more type II muscle fibers promote a shift to CHO as the predominant substrate for generating ATP. The crossover point is the intensity where fat and carbohydrate utilization intersect (figure 2.15) as the energy from fat decreases and the energy from carbohydrate increases. Beyond this crossover point, further increases in power are met with further increments in CHO utilization and decrements in fat oxidation.

The crossover point is affected by both the exercise intensity and endurance training status. Endurance training results in biochemical adaptations within the muscle fibers that promote and support oxidation of FFAs, including an increase in the number of mitochondria, increased oxidative enzymes, and changes in β-oxidation and the electron transport chain—all important determinants of fat metabolism. The result of training is to allow the body to spare muscle glycogen, since carbohydrate stores within the body are limited. These training-induced adaptations shift the crossover point toward higher exercise intensities. Diet (energy supply and stores) and prior exercise play secondary roles in determining the balance of substrates utilization during submaximal exercise.

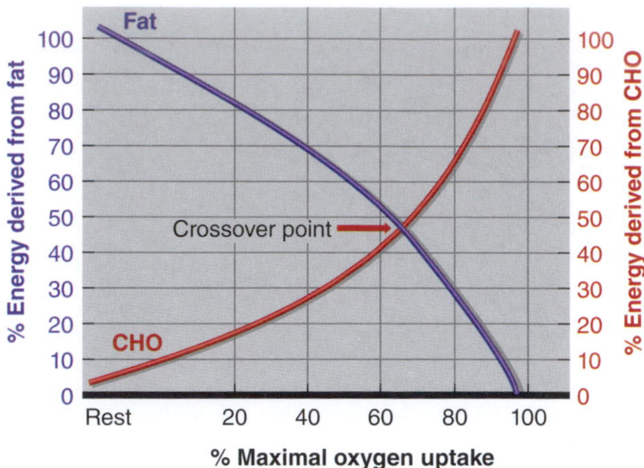

FIGURE 2.15 The relation between the relative contributions of fat and carbohydrate (CHO) utilization to overall energy expenditure as a function of exercise intensity. The point at which the two lines intersect illustrates the classic crossover concept.

The Oxidative Capacity of Muscle

We have seen that the processes of oxidative metabolism have the highest energy yields. It would be ideal if these processes always functioned at peak capacity. But, as with all physiological systems, they operate within certain constraints. The oxidative capacity of muscle ($\dot{Q}O_2$) is a measure of its maximal capacity to use oxygen. This measurement is made in the laboratory, where a small amount of muscle tissue can be tested to determine its capacity to consume oxygen when chemically stimulated to generate ATP. A muscle's

TABLE 2.3 **Characteristics of the Various Energy Supply Systems**

Energy system	Oxygen necessary?	Overall chemical reaction	Relative rate of ATP formed per second	ATP formed per molecule of substrate	Available capacity
ATP-PCr	No	PCr to Cr	10	1	<15 s
Glycolysis	No	Glucose or glycogen to lactate	5	2-3	~1 min
Oxidative (from carbohydrate)	Yes	Glucose or glycogen to CO_2 and H_2O	2.5	36-39*	~90 min
Oxidative (from fat)	Yes	FFA or triglycerides to CO_2 and H_2O	1.5	>100	Days

*Production of 36 to 39 ATP per molecule of carbohydrate excludes energy cost of transport through membranes. The net production is slightly lower (see text).

Courtesy of Dr. Martin Gibala, McMaster University, Hamilton, Ontario, Canada.

oxidative capacity ultimately depends on its oxidative enzyme concentrations, fiber type composition, and oxygen availability.

Enzyme Activity

The capacity of muscle fibers to oxidize carbohydrate and fat is difficult to determine. Numerous studies have shown a close relation between a muscle's ability to perform prolonged aerobic exercise and the activity of its oxidative enzymes. Because many different enzymes are required for oxidation, the enzyme activity of the muscle fibers provides a reasonable indication of their oxidative potential.

Measuring all the enzymes in muscles is impossible, so a few representative enzymes have been selected to reflect the aerobic capacity of the fibers. The enzymes most frequently measured are succinate dehydrogenase and citrate synthase, mitochondrial enzymes involved in the Krebs cycle (see figure 2.9). Figure 2.16 illustrates the close correlation between succinate dehydrogenase activity in the vastus lateralis muscle and that muscle's oxidative capacity. Endurance athletes' muscles have oxidative enzyme activities two to four times greater than those of untrained men and women.

Fiber Type Composition and Endurance Training

A muscle's fiber type composition primarily determines its oxidative capacity. As noted in chapter 1, type I (slow-twitch) fibers have a greater capacity for aerobic activity than type II (fast-twitch) fibers because type I fibers have more mitochondria and higher concentrations of oxidative enzymes. Type II fibers are better suited for glycolytic energy production. Thus, in general, the more type I fibers in one's muscles, the greater the oxidative capacity of those muscles. Elite distance runners, for example, possess more type I fibers, more mitochondria, and higher muscle oxidative enzyme activities than do untrained individuals.

Endurance training enhances the oxidative capacity of all fibers, especially type II fibers. Training that places demands on oxidative phosphorylation stimulates the muscle fibers to develop more mitochondria, larger mitochondria, and more oxidative enzymes per mitochondrion. By increasing the fibers' enzymes for β-oxidation, this training also enables the muscle to rely more on fat for aerobic ATP production. Thus, with endurance training, even people with large percentages of type II fibers can increase their muscles' aerobic capacities. But it is generally agreed that an endurance-trained type II fiber will not develop the same high endurance capacity as a similarly trained type I fiber.

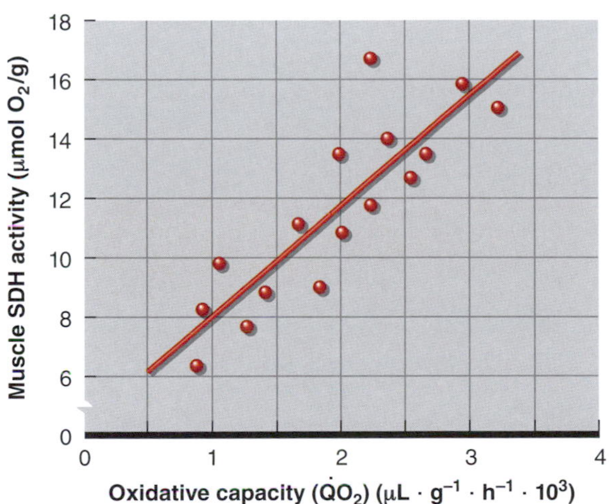

FIGURE 2.16 The relation between muscle succinate dehydrogenase (SDH) activity and its oxidative capacity ($\dot{Q}O_2$), measured in a muscle biopsy sample taken from the vastus lateralis.

Oxygen Needs

Although the oxidative capacity of a muscle is determined by the number of mitochondria and the amount of oxidative enzymes present, oxidative metabolism ultimately depends on an adequate supply of oxygen. At rest, the need for ATP is relatively small, requiring minimal oxygen delivery. As exercise intensity increases, so do energy demands. To meet them, the rate of oxidative ATP production increases. In an effort to meet the muscles' need for oxygen, the rate and depth of respiration increase, improving gas exchange in the lungs, and the heart beats faster and more forcefully, pumping more oxygenated blood to the muscles. Arterioles dilate to facilitate delivery of arterial blood to muscle capillaries.

The human body stores little oxygen. Therefore, the amount of oxygen entering the blood as it passes through the lungs is directly proportional to the amount used by the tissues for oxidative metabolism. Consequently, a reasonably accurate estimate of aerobic energy production can be made by measuring the amount of oxygen consumed at the lungs (see chapter 5).

RESEARCH PERSPECTIVE 2.3

Does the Muscle Fiber's Oxidative Capacity Determine Fitness Level?

Maximal oxygen uptake ($\dot{V}O_{2max}$; discussed in detail in chapter 5) is a measurement of cardiorespiratory fitness, so it is not surprising that well-trained endurance athletes have a high $\dot{V}O_{2max}$. The ability to take in and use oxygen during aerobic exercise may be limited by any number of factors along the pathway of the O_2 molecule as it moves from the atmosphere to the mitochondria to be used for energy: pulmonary ventilation, the oxygen-carrying capacity of the blood, and blood flow to exercising muscle, to name a few. Maximal oxygen uptake is also an important predictor of health, and reductions in $\dot{V}O_{2max}$ are associated with a loss of mobility and independence in the elderly and an increase in mortality in many chronic diseases. Because of its critical role, exercise physiologists are keenly interested in the factors that limit $\dot{V}O_{2max}$ in all people, from chronic heart failure patients to professional endurance athletes.

Since the early development of measurement techniques to quantify $\dot{V}O_{2max}$ in humans, researchers have designed studies to systematically examine each point along the oxygen delivery pathway from inspired air to the mitochondria within muscle fibers. Because it is well accepted that increasing oxygen *supply* to the working muscle improves $\dot{V}O_{2max}$ and exercise capacity, many scientists believed that the ability of the mitochondria themselves to *use* oxygen—mitochondrial oxidative capacity—was not a limiting factor for maximal oxygen uptake. However, a recent study examined how well the mitochondrial oxidative capacity alone was associated with $\dot{V}O_{2max}$ across people of vastly different fitness levels.[8]

In that study, researchers measured $\dot{V}O_{2max}$ during cycling exercise in chronic heart failure patients, healthy subjects, and elite cyclists. They then took muscle biopsy samples from the quadriceps of each subject to measure mitochondrial oxidative capacity. To quantify the muscle fibers' capacity to utilize oxygen, they measured the activity of an important enzyme in the Krebs cycle, succinate dehydrogenase. Interestingly, they found that this measure of mitochondrial oxidative capacity was related to $\dot{V}O_{2max}$ across all subjects, regardless of fitness or health status (see figure). Their results indicated that while limitations in oxygen supply certainly limit $\dot{V}O_{2max}$, maximal oxygen uptake during whole-body exercise is partially determined at the level of the muscle fiber itself.

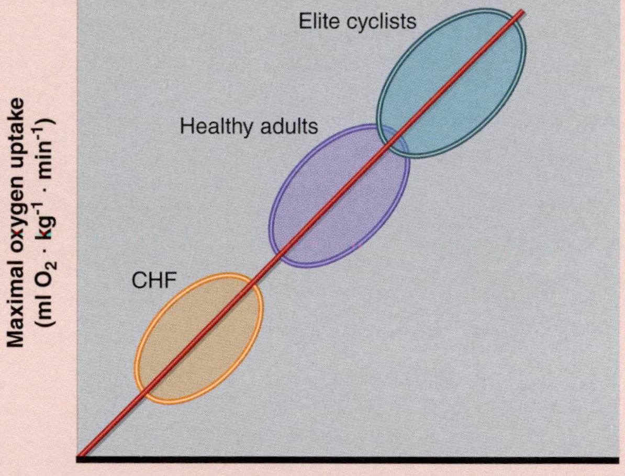

Simplified figure showing the relation between mitochondrial oxidative capacity, measured as succinate dehydrogenase activity in skeletal muscle biopsy samples obtained from the quadriceps after cycle exercise, and maximal oxygen uptake in chronic heart failure patients (CHF), healthy adults, and elite cyclists. Maximal oxygen uptake is closely related to mitochondrial oxidative capacity across all three subject groups.

IN CLOSING

In this chapter, we focused on energy metabolism and the synthesis of the storage form of energy in the body, ATP. We described in some detail the three basic energy systems used to generate ATP and their regulation and interaction. Finally, we highlighted the important role that oxygen plays in the sustained generation of ATP for continued muscle contraction and the three fiber types found in human skeletal muscle. We next look at the neural control of exercising muscle.

KEY TERMS

acetyl coenzyme A (acetyl CoA)
activation energy
adenosine diphosphate (ADP)
aerobic metabolism
anaerobic metabolism
ATP-PCr system
β-oxidation
bioenergetics
carbohydrate
catabolism
creatine kinase
crossover concept
cytochrome
electron transport chain
enzyme
free fatty acids (FFAs)
gluconeogenesis
glucose
glycogen
glycogenolysis
glycolysis
kilocalories (kcal)
Krebs cycle
lipogenesis
lipolysis
metabolism
mitochondria
negative feedback
oxidative phosphorylation
oxidative system
phosphocreatine (PCr)
phosphofructokinase (PFK)
phosphorylation
rate-limiting enzyme
substrate
triglycerides

STUDY QUESTIONS

1. What is ATP, how is it formed, and how does it provide energy during metabolism?
2. What is the primary substrate used to provide energy at rest? During high-intensity exercise?
3. What is the role of PCr in energy production, and what are its limitations? Describe the relationship between muscle ATP and PCr during sprint exercise.
4. Describe the essential characteristics of the three energy systems.
5. Why are the ATP-PCr and glycolytic energy systems considered anaerobic?
6. What role does oxygen play in the process of aerobic metabolism?
7. Describe the by-products of energy production from ATP-PCr, glycolysis, and oxidation.
8. What is lactic acid, and why is it important?
9. Discuss the interaction among the three energy systems with respect to the rate at which energy can be produced and the sustained capacity to produce that energy.
10. What is meant by the crossover concept, and how does it change with endurance exercise training?
11. How do type I muscle fibers differ from type II fibers in their respective oxidative capacities? What accounts for those differences?

STUDY GUIDE ACTIVITIES

In addition to the activities listed in the chapter opening outline, two other activities are available in the web study guide, located at

www.HumanKinetics.com/PhysiologyOfSportAndExercise

The **KEY TERMS** activity reviews important terms, and the end-of-chapter **QUIZ** tests your understanding of the material covered in the chapter.

Neural Control of Exercising Muscle

In this chapter and in the web study guide

Josh Harding retired from the National Hockey League (NHL) in 2015 after an 8-year career as a goalie, posting 60 NHL wins. While warming up for a game, Harding felt a tweak in his neck, followed by dizziness, black spots in front of his eyes, and numbness in his right leg. In December of 2012, just before the 2012-2013 NHL season began, Harding learned he had multiple sclerosis (MS), a disease that attacks the central nervous system and causes a loss of balance and coordination, blurred vision, dehydration, muscle spasms, and weakness. His team, the Minnesota Wild, was made aware of the diagnosis, and Harding eventually made the news public. Despite the challenges that this diagnosis imposes on an NHL goaltender, Harding was determined to continue to play, and play well he did for a while. However, after playing two periods for the minor-league Iowa Wild in 2014, Harding was taken by ambulance to the hospital suffering from severe dehydration, a common effect of MS. In his first full season after being diagnosed, he played 29 games with an 18-7-3 record, a 1.65 goals against average, and a 0.933 save percentage. Harding received the Bill Masterson Memorial Trophy in recognition of his perseverance and dedication to the game. Having found the correct combination of medication and a sleep schedule that works, he now serves as a high school goaltender coach while raising three young children.

All functions within the human body are influenced in some way by the nervous system. Nerves are the wiring through which electrical impulses are sent to and received from virtually all tissues of the body. The brain acts as a central computer, integrating incoming information, selecting an appropriate response, and then signaling the involved organs and tissues to take appropriate action. Thus, the nervous system forms a vital network, allowing communication, coordination, and interaction of the various tissues and systems in the body as well as between the body and the external environment.

The nervous system is one of the body's most complex systems. Because this book is primarily concerned with neural control of muscle contraction and voluntary movement, we will limit our coverage of this complex system. We first review the structure and function of the nervous system and then focus on specific topics relevant to sport and exercise.

Before we examine the intricate details of the nervous system, it is important to look at how the nervous system is organized and how that organization functions to integrate and control movement. The nervous system is commonly divided into two parts: the **central nervous system (CNS)** and the **peripheral nervous system (PNS)**. The CNS is composed of the brain and spinal cord, while the PNS is further divided into **sensory (afferent) nerves** and **effector (efferent) nerves**. Sensory nerves are responsible for informing the CNS about what is going on within and outside the body. Efferent nerves are responsible for sending information from the CNS to the various tissues, organs, and systems of the body in response to the signals coming in from the sensory division. The term **motor neuron (motor nerve)** classically applies to neurons that project their axons outside the CNS to directly or indirectly control muscles. The efferent nervous system is composed of two parts, the autonomic nervous system and the somatic nervous system. Figure 3.1 provides a schematic of these relationships. More detail concerning each of these individual units of the nervous system is presented later in this chapter.

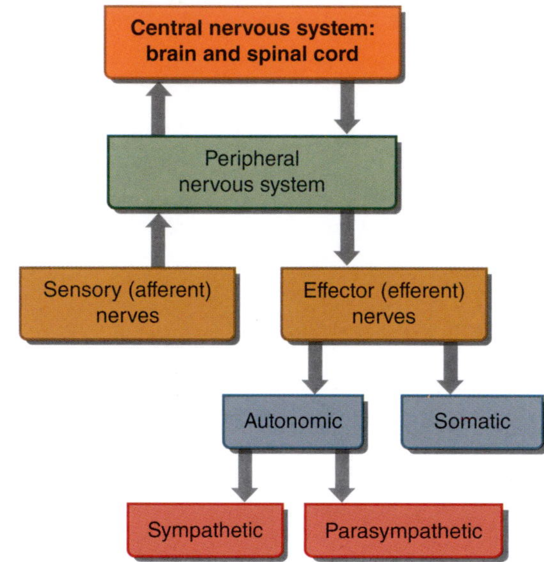

FIGURE 3.1 Organization of the nervous system.

Structure and Function of the Nervous System

The **neuron** is the basic structural unit of the nervous system. We first review the anatomy of the neuron and then look at how it functions—allowing electrical impulses to be transmitted throughout the body.

Neuron

Individual nerve fibers (nerve cells), depicted in figure 3.2, are called neurons. A typical neuron is composed of three regions:

- The cell body, or soma
- The dendrites
- The axon

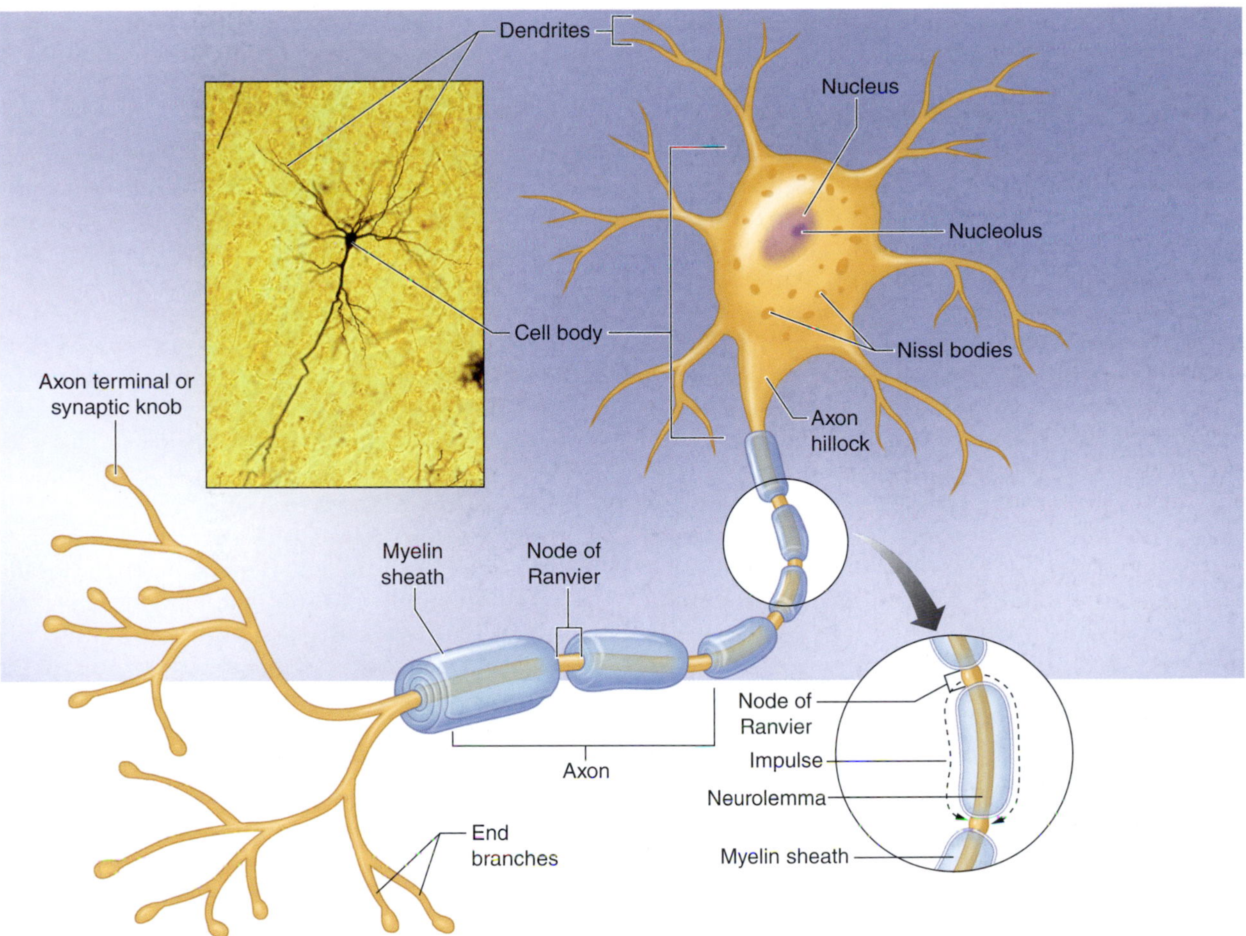

FIGURE 3.2 A drawing and photomicrograph (inset) of a neuron and its structure.

The cell body contains the nucleus. Radiating out from the cell body are the two cell processes: dendrites and the axon. On the side toward the axon, the cell body tapers into a cone-shaped region known as the **axon hillock**. The axon hillock has an important role in impulse conduction, as discussed later.

Most neurons contain only one axon but many dendrites. Dendrites are the neuron's receivers. Most impulses, or action potentials, that enter the neuron from sensory stimuli or from adjacent neurons typically enter the neuron via the dendrites. These processes then carry the impulses toward the cell body.

The axon is the neuron's transmitter and conducts impulses away from the cell body. Near its end, an axon splits into numerous **end branches**. The tips of these branches are dilated into tiny bulbs known as **axon terminals** or synaptic knobs. These terminals or knobs house numerous vesicles (sacs) filled with chemicals known as **neurotransmitters** that are used for communication between a neuron and another cell. (This is discussed in more detail later in the chapter.) The structure of the neuron allows nerve impulses to enter the neuron through the dendrites, and to a lesser extent through the cell body, and to travel through the cell body and axon hillock, down the axon, and out through the end branches to the axon terminals. We next look in more detail at how this happens, including how these impulses travel from one neuron to another and from a somatic motor neuron to muscle fibers.

Nerve Impulse

Neurons are referred to as *excitable tissue* because they can respond to various types of stimuli and convert those messages to an electrical signal called a nerve impulse. A **nerve impulse** arises when a stimulus is strong enough to substantially change the normal electrical charge of the neuron. That signal then moves along the neuron down the axon and toward an end organ, such as another neuron or a group of muscle fibers. A useful analogy is between the nerve impulse traveling through a neuron and electricity traveling through the electrical wires in a home. This section describes how the electrical impulse is generated and how it travels through a neuron.

Resting Membrane Potential

The cell membrane of a typical neuron at rest has a negative electrical potential of about −70 mV. This means that if one were to insert a voltmeter probe inside the cell, the electrical charges found there and the charges found outside the cell would differ by 70 mV, and the inside would be negative relative to the outside. This electrical potential difference is known as the **resting membrane potential (RMP)**. It is caused by an uneven separation of charged ions across the membrane. When the charges across the membrane differ, the membrane is said to be polarized.

The neuron has a high concentration of potassium ions (K^+) on the inside of the membrane and a high concentration of sodium ions (Na^+) on the outside. The imbalance in the number of ions inside and outside the cell causes the RMP. This imbalance is maintained in two ways. First, the cell membrane is much more permeable to K^+ than to Na^+, so the K^+ can move more freely. Because ions tend to move to establish equilibrium, some of the K^+ will move to the area where they are less concentrated, outside the cell. The Na^+ cannot move to the inside as easily. Second, **sodium–potassium pumps** in the neuron membrane, which contain Na^+-K^+ adenosine triphosphatase (Na^+-K^+-ATPase), maintain the imbalance on each side of the membrane by actively transporting potassium ions in and sodium ions out. The sodium–potassium pump moves three Na^+ out of the cell for each two K^+ it brings in. The end result is that more positively charged ions are outside the cell than inside, creating the potential difference across the membrane. Maintenance of a constant RMP of about −70 mV is primarily a function of the sodium–potassium pump.

Depolarization and Hyperpolarization

If the inside of the cell becomes less negative relative to the outside, the potential difference across the membrane decreases. The membrane will be less polarized. When this happens, the membrane is said to be depolarized. Thus, **depolarization** occurs any time the charge difference becomes more positive than the RMP of −70 mV, moving closer to zero. This typically results from a change in the membrane's Na^+ permeability.

The opposite can also occur. If the charge difference across the membrane increases, moving from the RMP to an even more negative value, then the membrane becomes more polarized. This is known as **hyperpolarization**. Changes in the membrane potential control the signals used to receive, transmit, and integrate information within and between cells. These signals are of two types, graded potentials and action potentials. Both are electrical currents created by the movement of ions.

Graded Potentials

Graded potentials are localized changes in the membrane potential, either depolarization or hyperpolarization. The membrane contains ion channels with gates that act as doorways into and out of the neuron. These gates are usually closed, preventing a large number of ions from flowing into and out of the membrane—that is, above and beyond the constant movement of Na^+ and K^+ that maintains the RMP. However, with potent enough stimulation, the gates open, allowing more ions to move from the outside to the inside or vice versa. This ion flow alters the charge separation, changing the polarization of the membrane.

Graded potentials are triggered by a change in the neuron's local environment. Depending on the location and type of neuron involved, the ion gates may open in response to the transmission of an impulse from another neuron or in response to sensory stimuli such as changes in chemical concentrations, temperature, or pressure.

Recall that most neuron receptors are located on the dendrites (although some are on the cell body), yet the impulse is always transmitted from the axon terminals at the opposite end of the cell. For a neuron to transmit an impulse, the impulse must travel almost the entire length of the neuron. Although a graded potential may result in depolarization of the entire cell membrane, it is usually just a local event such that the depolarization does not spread very far along the neuron. To travel the full distance, an impulse must be sufficiently strong to generate an action potential.

Action Potentials

An action potential is a rapid and substantial depolarization of the neuron's membrane. It usually lasts only about 1 ms. Typically, the membrane potential changes from the RMP of about −70 mV to a value of about +30 mV and then rapidly returns to its resting value. This is illustrated in figure 3.3. How does this marked change in membrane potential occur?

All action potentials begin as graded potentials. When enough stimulation occurs to cause a depolarization of at least 15 to 20 mV, an action potential results. In other words, if the membrane depolarizes from the RMP of −70 mV to a value of −50 to −55 mV, an action potential will occur. The membrane voltage at which a graded potential becomes an action potential is called the depolarization **threshold**. Any depolarization that does not attain the threshold will not result in an action potential. For example, if the membrane potential changes from the RMP of −70 mV to −60 mV, the change

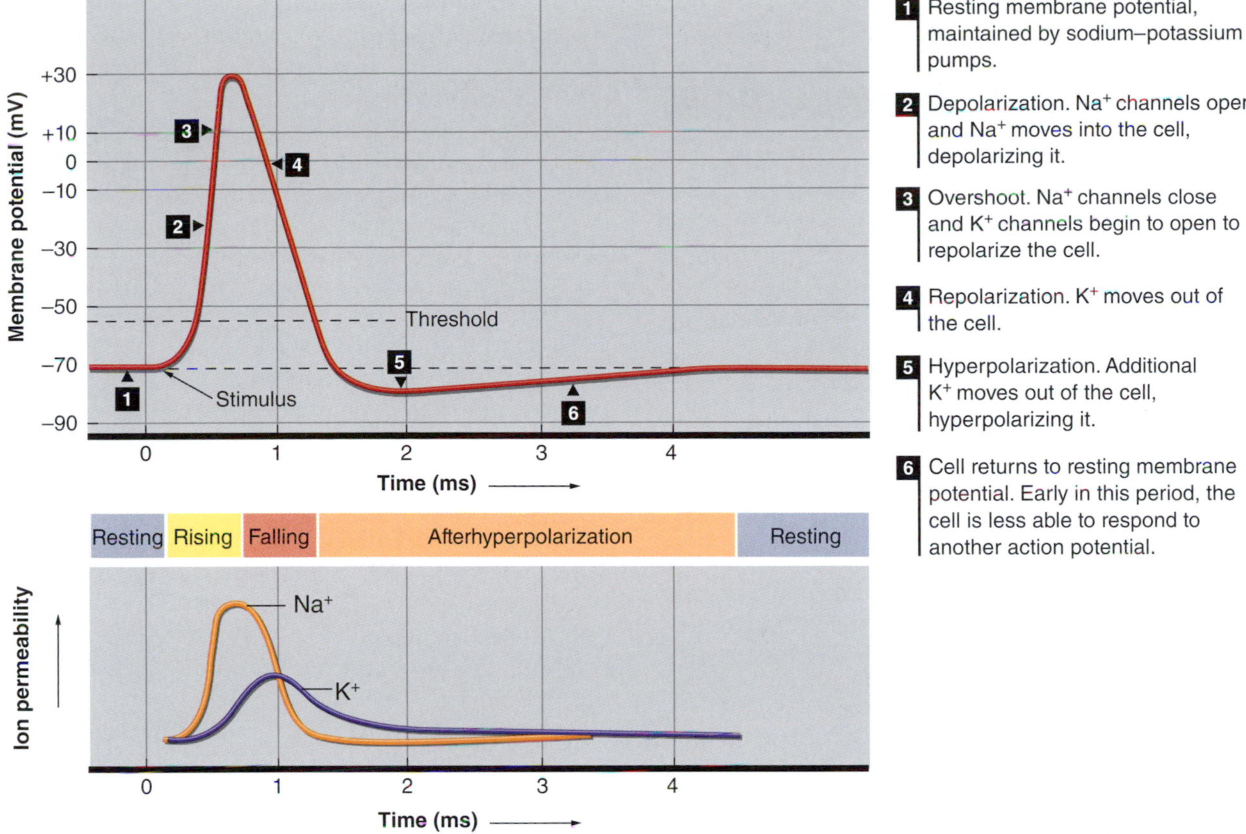

FIGURE 3.3 Voltage and ion permeability changes during an action potential.

is only 10 mV and does not reach the threshold; thus, no action potential occurs. But any time depolarization reaches or exceeds the threshold, an action potential will result. This is commonly referred to as the *all-or-none principle.*

When a segment of an axon's sodium gates is open and it is in the process of generating an action potential, it is unable to respond to another stimulus. This is referred to as the *absolute refractory period.* When the sodium gates are closed, the potassium gates are open, and repolarization is occurring, that segment of the axon can potentially respond to a new stimulus, but the stimulus must be of substantially greater magnitude to evoke an action potential. This is referred to as the *relative refractory period.*

Propagation of the Action Potential

Now that we understand how a neural impulse, in the form of an action potential, is generated, we can look at how the impulse is propagated—that is, how it travels through the neuron. Two characteristics of the neuron determine how quickly an impulse can pass along the axon: myelination and diameter.

Myelination The axons of many neurons, especially large neurons, are myelinated, meaning that they are covered with a sheath formed by myelin, a fatty substance that insulates the cell membrane. This **myelin sheath** (see figure 3.2) is formed by specialized cells called Schwann cells.

The myelin sheath is not continuous. As it spans the length of the axon, the myelin sheath has gaps between adjacent Schwann cells, leaving the axon uninsulated at those points. These gaps are referred to as *nodes of Ranvier* (see figure 3.2). The action potential appears to jump from one node to the next as it traverses a myelinated fiber. This is referred to as **saltatory conduction**, a much faster type of conduction than occurs in unmyelinated fibers.

Myelination of peripheral motor neurons occurs over the first several years of life, partly explaining why children need time to develop coordinated movement. Individuals affected by certain neurological diseases (such as multiple sclerosis, as discussed in our chapter opening) experience degeneration of the myelin sheath and a subsequent loss of coordination.

Diameter of the Neuron The velocity of nerve impulse transmission is also determined by the neuron's size. Neurons of larger diameter conduct nerve impulses faster than neurons of smaller diameter because larger neurons present less resistance to local current flow.

In Review

- Neurons are excitable tissues because they have the ability to respond to various types of stimuli and convert them to an electrical signal or nerve impulse.
- A neuron's RMP of about −70 mV results from the unequal separation of positively charged sodium and potassium ions, with more potassium inside the membrane and more sodium on the outside.
- The RMP is maintained by actions of the sodium–potassium pump, coupled with low sodium permeability and high potassium permeability of the neuron membrane.
- Any change that makes the membrane potential less negative results in depolarization. Any change making this potential more negative is a hyperpolarization. These changes occur when ion gates in the membrane open, permitting more ions to move across the membrane.
- If the membrane is depolarized by 15 to 20 mV, the depolarization threshold is reached and an action potential results. Action potentials are not generated if the threshold is not met.
- In myelinated neurons, the impulse travels through the axon by jumping between nodes of Ranvier (gaps between the cells that form the myelin sheath). This process, saltatory conduction, results in nerve transmission rates 5 to 50 times faster than in unmyelinated fibers of the same size. Impulses also travel faster in neurons of larger diameter.

Synapse

For a neuron to communicate with another neuron, an action potential must occur and travel along the first neuron, ultimately reaching its axon terminals. How does the action potential then move from the neuron in which it starts to another neuron to continue transmitting the electrical signal?

Neurons communicate with each other across junctions called synapses. A **synapse** is the site of action potential transmission from the axon terminals of one neuron to the dendrites or soma of another. There are both chemical and mechanical synapses, but the most common type is the chemical synapse, which is our focus. It is important to note that the signal that is transmitted from one neuron to another changes from electrical to chemical, then back to electrical.

As seen in figure 3.4, a synapse between two neurons includes

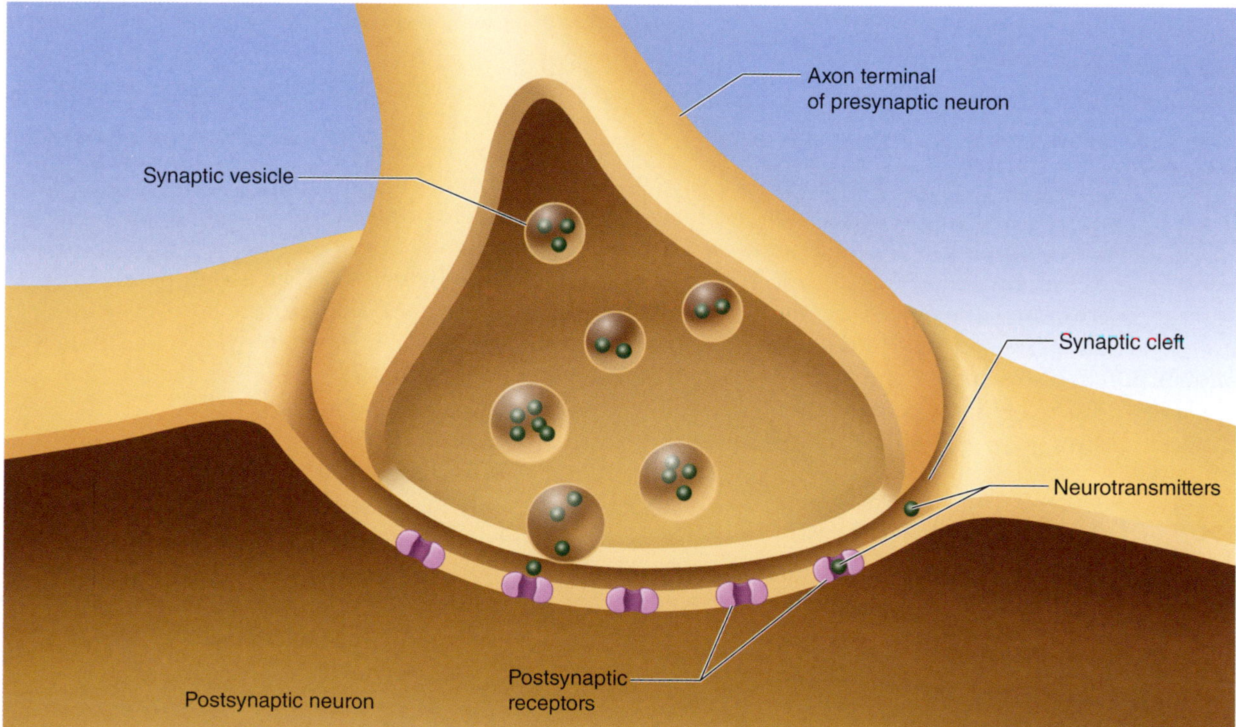

FIGURE 3.4 A chemical synapse between two neurons, showing the synaptic vesicles containing neurotransmitter molecules.

- the axon terminals of the neuron sending the action potential,
- receptors on the neuron receiving the action potential, and
- the space between these structures.

The neuron sending the action potential across the synapse is called the presynaptic neuron, so axon terminals are presynaptic terminals. Similarly, the neuron receiving the action potential on the opposite side of the synapse is called the postsynaptic neuron, and it has postsynaptic receptors. The axon terminals and postsynaptic receptors are not physically in contact with each other. A narrow gap, the synaptic cleft, separates them.

The action potential can be transmitted across a synapse in only one direction: from the axon terminal of the presynaptic neuron to the postsynaptic receptors, about 80% to 95% of which are on the dendrites of the postsynaptic neuron. The remaining 5% to 20% of the postsynaptic receptors are adjacent to the cell body instead of being located on the dendrites. Why can the action potential go in only one direction?

The presynaptic terminals of the axon contain a large number of saclike structures called synaptic (or storage) vesicles. These vesicles contain a variety of chemical compounds called neurotransmitters because they function to transmit the neural signal to the next neuron. When the impulse reaches the presynaptic axon terminals, the synaptic vesicles respond by releasing the neurotransmitters into the synaptic cleft. These neurotransmitters then diffuse across the synaptic cleft to the postsynaptic neuron's receptors. Each neurotransmitter then binds to its specialized postsynaptic receptors. When sufficient binding occurs, a series of graded depolarizations occurs. If the depolarization reaches the threshold, an action potential occurs, and the impulse has been transmitted successfully to the next neuron. Depolarization of the second nerve depends on both the amount of neurotransmitter released and the number of available receptor binding sites on the postsynaptic neuron.

Neuromuscular Junction

Recall from chapter 1 that a single α-motor neuron and all of the muscle fibers it innervates is called a motor unit. Whereas neurons communicate with other neurons at synapses, an α-motor neuron communicates with its muscle fibers at a site known as a **neuromuscular junction**, which functions in essentially the same manner as a synapse. In fact, the proximal part of the neuromuscular junction is the same: It starts with the axon terminals of the motor neuron, which release neurotransmitters into the space between the motor nerve and the muscle fiber in response to an action potential. However, in the neuromuscular junction, the axon terminals protrude into motor end plates, which are invaginated (folded to form cavities) segments on the plasmalemma of the muscle fiber (see figure 3.5).

Neurotransmitters—primarily acetylcholine (ACh)—released from the α-motor neuron axon terminals diffuse across the synaptic cleft and bind to receptors on the muscle fiber's plasmalemma. This binding typically causes depolarization by opening sodium ion channels, allowing more sodium to enter the muscle fiber. Again, if the depolarization reaches the threshold, an action potential is formed. It spreads across the plasmalemma into the T-tubules, initiating muscle fiber contraction. As in the neuron, the plasmalemma, once depolarized, must undergo repolarization. During the period of repolarization, the sodium gates are closed and the potassium gates are open; thus, like the neuron, the muscle fiber is unable to respond to any further stimulation during this refractory period. Once the RMP

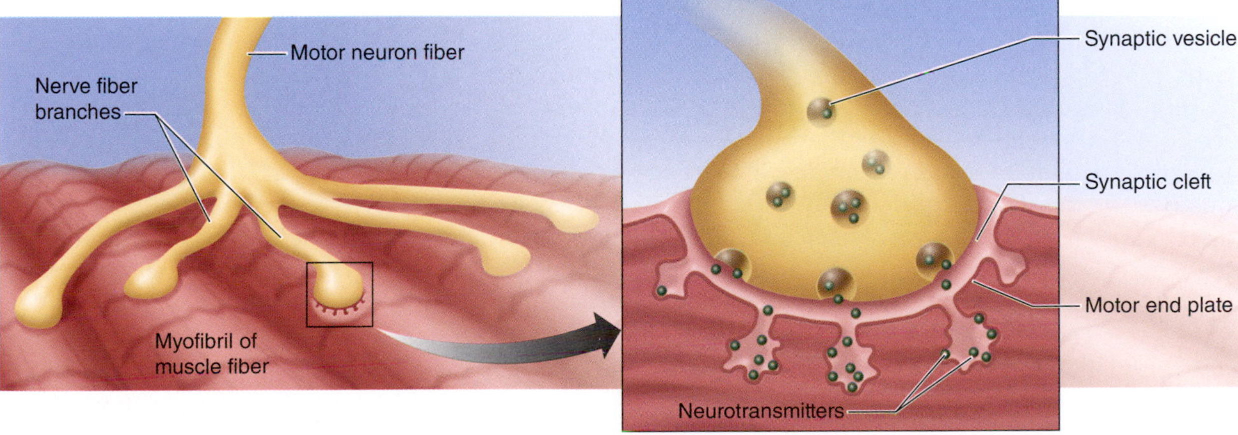

FIGURE 3.5 The neuromuscular junction, illustrating the interaction between the α-motor neuron and the plasmalemma of a single muscle fiber.

of the muscle fiber is restored, the fiber can respond to another stimulus. Thus, the refractory period limits the motor unit's firing frequency.

Exercise training induces changes not only in skeletal muscle, but also at the neuromuscular junction (NMJ) to increase presynaptic release of, and sensitivity of the muscle cell to, acetylcholine. These changes occur through a number of different cellular signaling mechanisms; however, many of the changes induced by training share a common signaling molecule, the *peroxisome proliferator–activated receptor-γ coactivator 1α (PGC-1α)*. PGC-1α contributes to the remodeling of the NMJ in several ways. First, PGC-1α induces adaptations in the motor neuron itself by increasing branching of the presynaptic terminal motor neuron and increasing the number of presynaptic vesicles containing acetylcholine. Second, PGC-1α increases the number of acetylcholine receptors on the cell membrane, thus amplifying the effects for a given amount of acetylcholine released from the motor neuron.[4] Finally, PGC-1α is involved in decreasing the size of the motor end plate (i.e., fewer fibers per motor unit) on glycolytic fibers, making them similar to more oxidative fibers.

Muscular fatigue (discussed in detail in chapter 5) is a complex phenomenon, with many possible contributing factors. One mechanism that may contribute to muscle fatigue is a decline in signal transmission through the NMJ. Prior exercise can decrease motor nerve outflow and neuromuscular transmission rates,[3] which leads to decreased force production.

Now we know how the impulse is transmitted from nerve to nerve or nerve to muscle. But to understand what happens once the impulse is transmitted, we must first examine the chemical signaling molecules, the neurotransmitters, that accomplish this signal transmission.

Neurotransmitters

More than 50 neurotransmitters have been positively identified or are suspected to be potential candidates. These can be categorized as either (a) small-molecule, rapid-acting neurotransmitters or (b) neuropeptide, slow-acting neurotransmitters. The small-molecule, rapid-acting transmitters, which are responsible for most neural transmissions, are our main focus.

Acetylcholine and norepinephrine are the two major neurotransmitters involved in regulating multiple physiological responses to exercise. **Acetylcholine** is the primary neurotransmitter for the motor neurons that innervate skeletal muscle as well as for most parasympathetic autonomic neurons. It is generally an excitatory neurotransmitter in the somatic nervous system but can have inhibitory effects at some parasympathetic nerve endings, such as in the heart. **Norepinephrine** is the neurotransmitter for most sympathetic autonomic neurons, and it too can be either excitatory or inhibitory, depending on the receptors involved. Nerves that primarily release norepinephrine are called **adrenergic**, and those that have acetylcholine as their primary

RESEARCH PERSPECTIVE 3.1

Motor Units Adapt to High-Intensity Interval Training

High-intensity interval training (HIIT, discussed in chapter 11) is a mode of physical activity that involves brief, intermittent bursts of vigorous activity interspersed with periods of low-intensity exercise or rest. An individual can reap the same cardiovascular and musculoskeletal benefits from exercise training using HIIT in far less time compared to traditional long-duration endurance training (END). Because HIIT is now a common alternative to END and exercise training changes the neural control of muscle function, it is important to systematically evaluate HIIT-induced neuromuscular adaptations.

High-density surface electromyography (EMG) is a relatively new technological advancement that allows for both the simultaneous assessment of several motor units over a wide range of forces and the ability to track the same motor units during different sessions over a long period of time (like during exercise training). Recording motor unit activity and function allows investigators to assess the way the nervous system controls muscle force. Researchers recently evaluated differences in the neuromuscular adaptations to HIIT and END using this technique.[6] Two weeks of HIIT and END elicited similar improvements in cardiorespiratory fitness, but there were distinct adjustments in motor unit behavior with the two types of training. HIIT increased both maximum force production and motor unit discharge. In contrast, END did not influence motor unit firing. These findings suggest that HIIT and END have very different effects on motor unit function and provide important new information regarding exercise training–induced neuromuscular adaptations. This study was also the first to demonstrate training-induced changes in motor unit discharge rate by tracking the *same* individual motor units before and after training. This innovative methodology will likely continue to broaden our understanding of neural adaptations to exercise training.

neurotransmitter are termed **cholinergic**. Two major subtypes of cholinergic receptors are muscarinic and nicotinic, with the former involved in motor nerve transmission. The sympathetic and parasympathetic branches of the autonomic nervous systems are discussed later in this chapter.

Once the neurotransmitter binds to the postsynaptic receptor, the nerve impulse has been successfully transmitted. The neurotransmitter then (1) is degraded by enzymes, (2) is actively transported back into the presynaptic terminals for reuse, or (3) diffuses away from the synapse.

In Review

- Neurons communicate with each other across synapses composed of the axon terminals of the presynaptic neuron, the postsynaptic receptors on the dendrite or cell body of the postsynaptic neuron, and the synaptic cleft between the two neurons.
- A nerve impulse causes neurotransmitters to be released from the presynaptic axon terminal into the synaptic cleft.
- Neurotransmitters diffuse across the cleft and bind to the postsynaptic receptors.
- Once sufficient neurotransmitters are bound, the impulse is successfully transmitted and the neurotransmitter is then destroyed by enzymes, is removed by reuptake into the presynaptic terminal for future use, or diffuses away from the synapse.
- Neurotransmitter binding at the postsynaptic receptors opens ion gates in the given membrane and can cause depolarization (excitation) or hyperpolarization (inhibition), depending on the specific neurotransmitter and the receptors to which it binds.
- Neurons communicate with muscle fibers at neuromuscular junctions. A neuromuscular junction involves presynaptic axon terminals, the synaptic cleft, and motor end-plate receptors on the plasmalemma of the muscle fiber and functions much like a neural synapse.
- The neurotransmitters most important in regulating exercise responses are acetylcholine in the somatic nervous system and norepinephrine in the autonomic nervous system.
- Receptors on the motor end plates of the neuromuscular junction are a special subtype of cholinergic receptors called muscarinic receptors. They bind the primary neurotransmitter involved in excitation of muscle fibers, acetylcholine.

Postsynaptic Response

Once the neurotransmitter binds to the receptors, the chemical signal that traversed the synaptic cleft once again becomes an electrical signal. The binding causes a graded potential in the postsynaptic membrane. An incoming impulse may be either excitatory or inhibitory. An excitatory impulse causes depolarization, known as an **excitatory postsynaptic potential (EPSP)**. An inhibitory impulse causes a hyperpolarization, known as an **inhibitory postsynaptic potential (IPSP)**.

The discharge of a single presynaptic terminal generally changes the postsynaptic potential less than 1 mV. Clearly this is not sufficient to generate an action potential, because reaching the threshold requires a change of at least 15 mV. But when a neuron transmits an impulse, several presynaptic terminals typically release their neurotransmitters so that they can diffuse to the postsynaptic receptors. Also, presynaptic terminals from numerous axons can converge on the dendrites and cell body of a single neuron. When multiple presynaptic terminals discharge at the same time, or when only a few fire in rapid succession, more neurotransmitter is released. With an excitatory neurotransmitter, the more that is bound, the greater the EPSP and the more likely it is that an action potential will result.

Triggering an action potential at the postsynaptic neuron depends on the combined effects of all incoming impulses from these various presynaptic terminals. A number of impulses are needed to cause sufficient depolarization to generate an action potential. Specifically, the sum of all changes in the membrane potential

In Review

- Excitatory postsynaptic potentials are graded depolarizations of the postsynaptic membrane; IPSPs are hyperpolarizations of that membrane.
- A single presynaptic terminal cannot generate enough of a depolarization to fire an action potential. Multiple signals are needed. These may come from numerous neurons or from a single neuron when numerous axon terminals release neurotransmitters repeatedly and rapidly.
- The axon hillock keeps a running total of all EPSPs and IPSPs. When their sum meets or exceeds the threshold for depolarization, an action potential occurs. This process of accumulating incoming signals is known as summation.
- *Summation* refers to the cumulative effect of all individual graded potentials as processed by the axon hillock. Once the sum of all individual graded potentials meets or exceeds the depolarization threshold, an action potential occurs.

RESEARCH PERSPECTIVE 3.2

Aging Reduces Rapid Strength

By 2030, it is anticipated that older adults (>65 years) will make up 20% of the total population. Unfortunately, a large percentage of older adults experience functional limitations in their activities of daily living, and one out of three older adults experiences a fall each year. Accidental falls often cause an accelerated deterioration in overall health and impart a significant economic burden on society. Alterations in neuromuscular function have been suggested to contribute to the increased fall risk in older adults.

Although reduced maximal muscle strength is a well-understood characteristic of aging, recent studies have demonstrated that *rapid strength* (the rate of torque development, or RTD) actually decreases at a greater rate than maximal strength. Furthermore, RTD measured within the initial 200 ms from the onset of muscle contraction is more functionally relevant than the peak torque that can be produced by the muscle. Despite this knowledge, until recently, no research has specifically targeted the neural and muscle-specific factors that contribute to the reductions in RTD with aging. This is a clinically relevant topic, since this information may help identify strategies to slow age-related reductions in function, thus reducing the risk of fall-related injuries.

A group of researchers recently sought to determine the mechanisms of age-related reductions in RTD.[2] Young (20 years old) and older men (70 years old) participated in a study that involved ultrasound assessments of muscle properties and measurements of muscle strength during early (the first 50 ms) and late (100 to 200 ms) intervals following the onset of muscle contraction. RTD was reduced in the older men during the late interval of contraction, but surprisingly there were no differences in RTD between young and older men during the early interval of contraction. This suggests that older men have similar initial muscle activation but are unable to sustain the same rates of muscle activation during later muscle contraction. Poorer muscle quality and reductions in pennation angle also contribute to age-related reductions in RTD, likely because they affect muscle fiber shortening and fiber rotation. These age-related alterations in neuromuscular function combine to reduce rapid muscle strength, significantly decreasing neuromuscular function and contributing to falls in older adults.

must equal or exceed the threshold. This accumulation of the individual impulses' effects is called summation.

For summation, the postsynaptic neuron must keep a running total of the neuron's responses, both EPSPs and IPSPs, to all incoming impulses. This task is done at the axon hillock, which lies on the axon just past the cell body. Only when the sum of all individual graded potentials meets or exceeds the threshold can an action potential occur.

Individual neurons are grouped together into bundles. In the CNS (brain and spinal cord), these bundles are referred to as tracts, or pathways. Neuron bundles in the PNS are referred to simply as nerves.

Central Nervous System

To comprehend how even the most basic stimulus can cause muscle activity, we next consider the complexity of the CNS. The CNS comprises more than 100 billion neurons. In this section, we present an overview of the components of the CNS and their functions.

Brain

The brain is a highly complex organ composed of numerous specialized areas. For our purposes, we subdivide it into the four major regions illustrated in figure 3.6: the cerebrum, diencephalon, cerebellum, and brain stem.

Cerebrum

The cerebrum is composed of the right and left cerebral hemispheres. These are connected to each other by the corpus callosum, fiber bundles (tracts) that allow the two hemispheres to communicate with each other. The cerebral cortex forms the outer portion of the cerebral hemispheres and has been referred to as the site of the mind and intellect. It is also called the gray matter, which simply reflects its distinctive color resulting from lack of myelin on the neurons located in this area. The cerebral cortex is the conscious brain. It allows people to think, be aware of sensory stimuli, and voluntarily control their movements.

The cerebrum consists of five lobes—four outer lobes and the central insular lobe—having the following general functions (see figure 3.6):

- Frontal lobe: general intellect and motor control
- Temporal lobe: auditory input and interpretation
- Parietal lobe: general sensory input and interpretation
- Occipital lobe: visual input and interpretation

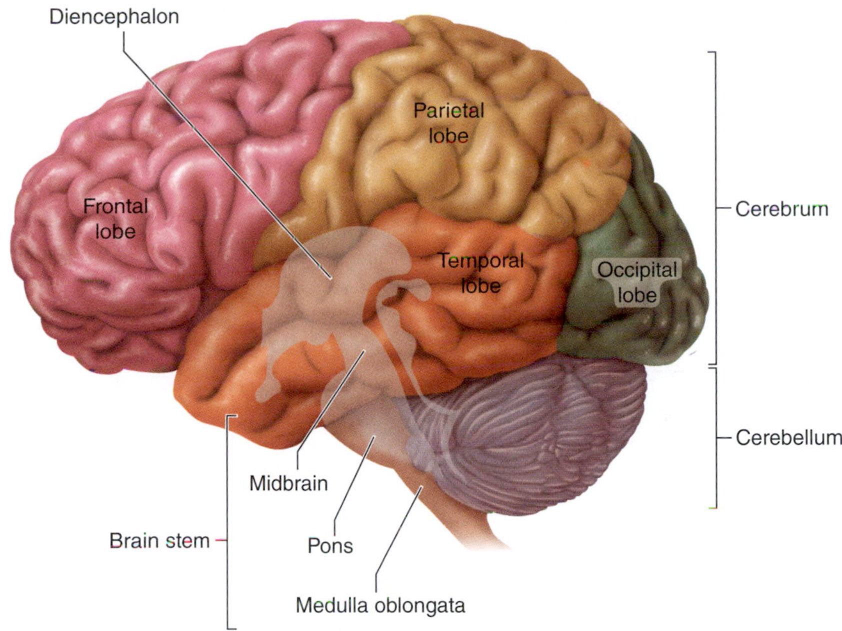

FIGURE 3.6 Four major regions of the brain and four outer lobes of the cerebrum. (Note that the insular lobe is not shown because it is folded deep within the cerebrum between the temporal lobe and the frontal lobe.)

- Insular lobe: diverse functions usually linked to emotion and self-perception

The three areas in the cerebrum that are of primary concern to exercise physiology are the primary motor cortex, located in the frontal lobe; the basal ganglia, located in the white matter below the cerebral cortex; and the primary sensory cortex, located in the parietal lobe. In this section, the focus is on the primary motor cortex and basal ganglia, which work to control and coordinate movement.

Primary Motor Cortex The primary motor cortex is responsible for the control of fine and discrete muscle movements. It is located in the frontal lobe, specifically within the precentral gyrus. Neurons here, known as *pyramidal cells*, let us consciously control movement of skeletal muscles. Think of the primary motor cortex as the part of the brain where decisions are made about what movement one wants to make. For example, in baseball, if a player is in the batter's box waiting for the next pitch, the decision to swing the bat is made in the primary motor cortex, where the entire body is carefully mapped out. The areas that require the finest motor control have a greater representation in the motor cortex; thus, more neural control is provided to them.

The cell bodies of the pyramidal cells are housed in the primary motor cortex, and their axons form the extrapyramidal tracts. These are also known as the corticospinal tracts because the nerve processes extend from the cerebral cortex down to the spinal cord. These tracts provide the major voluntary control of skeletal muscles.

In addition to the primary motor cortex, there is a premotor cortex just anterior to the precentral gyrus in the frontal lobe. Learned motor skills of a repetitious or patterned nature are stored here. This region can be thought of as the memory bank for skilled motor activities.[5]

Basal Ganglia The basal ganglia (nuclei) are not part of the cerebral cortex. Rather, they are in the cerebral white matter, deep in the cortex. These ganglia are clusters of nerve cell bodies. The complex functions of the basal ganglia are not well understood, but the ganglia are known to be important in initiating movements of a sustained and repetitive nature (such as arm swinging during walking), and thus they control complex movements such as walking and running. These cells also are involved in maintaining posture and muscle tone.

Diencephalon

The region of the brain known as the diencephalon (see figure 3.6) contains the thalamus and the hypothalamus. The thalamus is an important sensory integration center. All sensory input (except smell) enters the thalamus and is relayed to the appropriate area of the cortex. The thalamus regulates what sensory input reaches the conscious brain and thus is very important for motor control.

The hypothalamus, directly below the thalamus, is responsible for maintaining homeostasis by regulating almost all processes that affect the body's internal environment. Neural centers here assist in the regulation of most physiological systems, including

- blood pressure, heart rate, and contractility;
- respiration;
- digestion;
- body temperature;
- thirst and fluid balance;
- neuroendocrine control;
- appetite and food intake; and
- sleep–wake cycles.

Cerebellum

The cerebellum is located behind the brain stem. It is connected to numerous parts of the brain and has a crucial role in *coordinating* movement.

The cerebellum is crucial to the control of all rapid and complex muscular activities. It helps coordinate the timing of motor activities and the rapid progression from one movement to the next by monitoring and making corrective adjustments in the motor activities that are elicited by other parts of the brain. The cerebellum assists the functions of both the primary motor cortex and the basal ganglia. It facilitates movement patterns by smoothing out the movement, which would otherwise be jerky and uncontrolled.

The cerebellum acts as an integration system, comparing the programmed or intended activity with the actual changes occurring in the body and then initiating corrective adjustments through the motor system. It receives information from the cerebrum and other parts of the brain and also from sensory receptors (proprioceptors) in the muscles and joints that keep the cerebellum informed about the body's current position. The cerebellum also receives visual and equilibrium input. Thus, it notes all incoming information about the exact tension and position of all muscles, joints, and tendons and the body's current position relative to its surroundings, then it determines the best plan of action to produce the desired movement.

After the primary motor cortex makes the decision to perform a movement, this decision is relayed to the cerebellum. The cerebellum notes the desired action and then compares the intended movement with the actual movement based on sensory feedback from the muscles and joints. If the action is different than planned, the cerebellum informs the higher centers of the discrepancy so corrective action can be initiated.

Brain Stem

The brain stem, composed of the midbrain, the pons, and the medulla oblongata (see figure 3.6), connects the brain and the spinal cord. Sensory and motor neurons pass through the brain stem as they relay information in both directions between the brain and the spinal cord. This is the site of origin for 10 of the 12 pairs of cranial nerves. The brain stem also contains the major autonomic centers that control the respiratory and cardiovascular systems.

A specialized collection of neurons in the brain stem, known as the *reticular formation*, is influenced by, and has an influence on, nearly all areas of the CNS. These neurons help

- coordinate skeletal muscle function,
- maintain muscle tone,
- control cardiovascular and respiratory functions, and
- determine state of consciousness (arousal and sleep).

The brain has a pain control system located in the reticular formation, a group of nerve fibers in the brain stem. Opioid substances such as enkephalins and β-endorphin act on the opiate receptors in this region to help modulate pain. Research has demonstrated that exercise of long duration increases the concentrations of these substances. While this has been interpreted as the mechanism causing the "endorphin calm" or "runner's high" experienced by some exercisers, the cause–effect association between endogenous opioids and these sensations has not been substantiated.

Spinal Cord

The lowest part of the brain stem, the medulla oblongata, is continuous with the spinal cord below it. The spinal cord is composed of tracts of nerve fibers that allow two-way conduction of nerve impulses. The sensory (afferent) fibers carry neural signals from sensory receptors, such as those in the skin, muscles, and joints, to the upper levels of the CNS. Motor (efferent) fibers from the brain and upper spinal cord transmit action potentials to end organs (e.g., muscles, glands).

In Review

- The CNS includes the brain and the spinal cord.
- The four major divisions of the brain are the cerebrum, the diencephalon, the cerebellum, and the brain stem.

- The cerebral cortex is the conscious brain. The primary motor cortex, located in the frontal lobe, is the center of conscious motor control.
- The basal ganglia, in the cerebral white matter, help initiate some movements (sustained and repetitive ones) and help control posture and muscle tone.
- The diencephalon includes the thalamus, which receives all sensory input entering the brain, and the hypothalamus, which is a major control center for homeostasis.
- The cerebellum, which is connected to numerous parts of the brain, is critical for coordinating movement. It is an integration center that decides how to best execute the desired movement, given the body's current position and the muscles' current status.
- The brain stem is composed of the midbrain, the pons, and the medulla oblongata.
- The spinal cord contains both sensory and motor fibers that transmit action potentials between the brain and the periphery.

Peripheral Nervous System

The PNS contains 43 pairs of nerves: 12 pairs of cranial nerves that connect with the brain and 31 pairs of spinal nerves that connect with the spinal cord. Cranial and spinal nerves directly supply the skeletal muscles. Functionally, the PNS has two major divisions: the sensory division and the motor division.

Sensory Division

The sensory division of the PNS carries sensory information toward the CNS. Sensory (afferent) neurons originate in such areas as blood vessels, internal organs, muscles and tendons, the skin, and sensory organs (taste, touch, smell, hearing, vision).

Sensory neurons in the PNS end in either the spinal cord or the brain and continuously convey information to the CNS concerning the body's constantly changing status, position, and internal and external environment. Sensory neurons within the CNS carry the sensory input to appropriate areas of the brain, where the information can be processed and integrated with other incoming information.

The sensory division receives information from five primary types of receptors:

1. *Mechanoreceptors* that respond to mechanical forces such as pressure, touch, vibrations, or stretch
2. *Thermoreceptors* that respond to changes in temperature
3. *Nociceptors* that respond to painful stimuli
4. *Photoreceptors* that respond to electromagnetic radiation (light) to allow vision
5. *Chemoreceptors* that respond to chemical stimuli, such as from foods, odors, or changes in blood or tissue concentrations of substances like oxygen, carbon dioxide, glucose, and electrolytes

Virtually all of these receptors are important in exercise and sport. Special muscle and joint nerve endings are of many types and functions, and each type is sensitive to a specific stimulus. Here are some important examples:

- Free nerve endings detect crude touch, pressure, pain, heat, and cold. Thus, they function as mechanoreceptors, nociceptors, and thermoreceptors. These nerve endings are important for preventing injury during athletic performance.
- Joint kinesthetic receptors located in the joint capsules are sensitive to joint angles and rates of change in these angles. Thus, they sense the position and any movement of the joints.
- Muscle spindles sense muscle length and rate of change in length.
- Golgi tendon organs detect the tension applied by a muscle to its tendon, providing information about the strength of muscle contraction.

Muscle spindles and Golgi tendon organs are discussed in more detail later in this chapter.

Motor Division

The CNS transmits information to various parts of the body through the motor, or efferent, division of the PNS. Once the CNS has processed the information it receives from the sensory division, it determines how the body should respond to that input. From the brain and spinal cord, intricate networks of neurons go out to all parts of the body, providing detailed instructions to the target areas including—and central to exercise and sport physiology—muscles.

Autonomic Nervous System

The autonomic nervous system, often considered part of the motor division of the PNS, controls the body's involuntary internal functions. Some of these functions that are important to sport and activity are heart rate, blood pressure, blood distribution, and lung function.

The autonomic nervous system has two major divisions: the sympathetic nervous system and the parasympathetic nervous system. These originate from different sections of the spinal cord and from the base of the brain. The effects of the two systems are often antagonistic, but the systems always function together.

Sympathetic Nervous System

The sympathetic nervous system is sometimes called the fight-or-flight system: It prepares the body to face a crisis and sustains its function during the crisis. When fully engaged, the sympathetic nervous system can produce a massive discharge throughout the body, preparing it for action. A sudden loud noise, a life-threatening situation, or those last few seconds before the start of an athletic competition are examples of circumstances in which this massive sympathetic excitation may occur. The effects of sympathetic stimulation are important during exercise. To give a few examples:

- Heart rate and strength of cardiac contraction increase.
- Coronary vessels dilate, increasing the blood supply to the heart muscle to meet its increased demands.
- Peripheral vasodilation increases blood flow to active skeletal muscles.
- Vasoconstriction in most other tissues diverts blood away from them and to the active muscles.
- Blood pressure increases, allowing better perfusion of the muscles and improving the return of venous blood to the heart.
- Bronchodilation improves ventilation and effective gas exchange.
- Metabolic rate increases, reflecting the body's effort to meet the increased demands of physical activity.
- Mental activity increases, allowing better perception of sensory stimuli and more concentration on performance.
- Glucose is released from the liver into the blood as an energy source.
- Functions not directly needed at that time are slowed (e.g., renal function, digestion).

These basic alterations in bodily function facilitate motor responses, demonstrating the importance of the autonomic nervous system in preparing the body for and sustaining it during acute stress or physical activity.

Parasympathetic Nervous System

The parasympathetic nervous system can be thought of as the body's housekeeping system. It has a major role in carrying out such processes as digestion, urination, glandular secretion, and conservation of energy. This system is more active when one is calm and at rest. Its effects tend to oppose those of the sympathetic system. The parasympathetic division causes decreased heart rate, constriction of coronary vessels, and bronchoconstriction.

The various effects of the sympathetic and parasympathetic divisions of the autonomic nervous system are summarized in table 3.1.

In Review

- The PNS contains 43 pairs of nerves: 12 cranial and 31 spinal.
- The PNS can be subdivided into the sensory and motor divisions. The motor division also includes the autonomic nervous system.
- The sensory division carries information from sensory receptors to the CNS. The motor division carries motor impulses from the CNS to the muscles and other organs.
- The autonomic nervous system includes the sympathetic nervous system and the parasympathetic system. Although these systems often oppose each other, they always function together to create an appropriately balanced response.

Sensory-Motor Integration

Having discussed the components and divisions of the nervous system, we now discuss how a sensory stimulus gives rise to a motor response. How, for example, do the muscles in the hand know to pull one's finger away from a hot stove? When someone decides to run, how do the muscles in the legs coordinate while supporting weight and propelling the person forward? To accomplish these tasks, the sensory and motor systems must communicate with each other.

This process is called **sensory-motor integration**, and it is depicted in figure 3.7. For the body to respond to sensory stimuli, the sensory and motor divisions of the nervous system must function together in the following sequence of events:

1. A sensory stimulus is received by sensory receptors (e.g., pinprick).
2. The sensory action potential is transmitted along sensory neurons to the CNS.
3. The CNS interprets the incoming sensory information and determines which response is most appropriate, or reflexively initiates a motor response.

TABLE 3.1 **Effects of the Sympathetic and Parasympathetic Nervous Systems on Various Organs**

Target organ or system	Sympathetic effects	Parasympathetic effects
Heart muscle	Increases rate and force of contraction	Decreases rate of contraction
Heart: coronary blood vessels	Causes vasodilation	Causes vasoconstriction
Lungs	Causes bronchodilation; mildly constricts blood vessels	Causes bronchoconstriction
Blood vessels	Increases blood pressure; causes vasoconstriction in abdominal viscera and skin to divert blood when necessary; causes vasodilation in the skeletal muscles and heart during exercise	Has little or no effect
Liver	Stimulates glucose release	Has no effect
Cellular metabolism	Increases metabolic rate	Has no effect
Adipose tissue	Stimulates lipolysis[a]	Has no effect
Sweat glands	Increases sweating	Has no effect
Adrenal glands	Stimulates secretion of epinephrine and norepinephrine	Has no effect
Digestive system	Decreases activity of glands and muscles; constricts sphincters	Increases peristalsis and glandular secretion; relaxes sphincters
Kidney	Causes vasoconstriction; decreases urine formation	Has no effect

[a]Lipolysis is the process of breaking down triglyceride into its basic units to be used for energy.

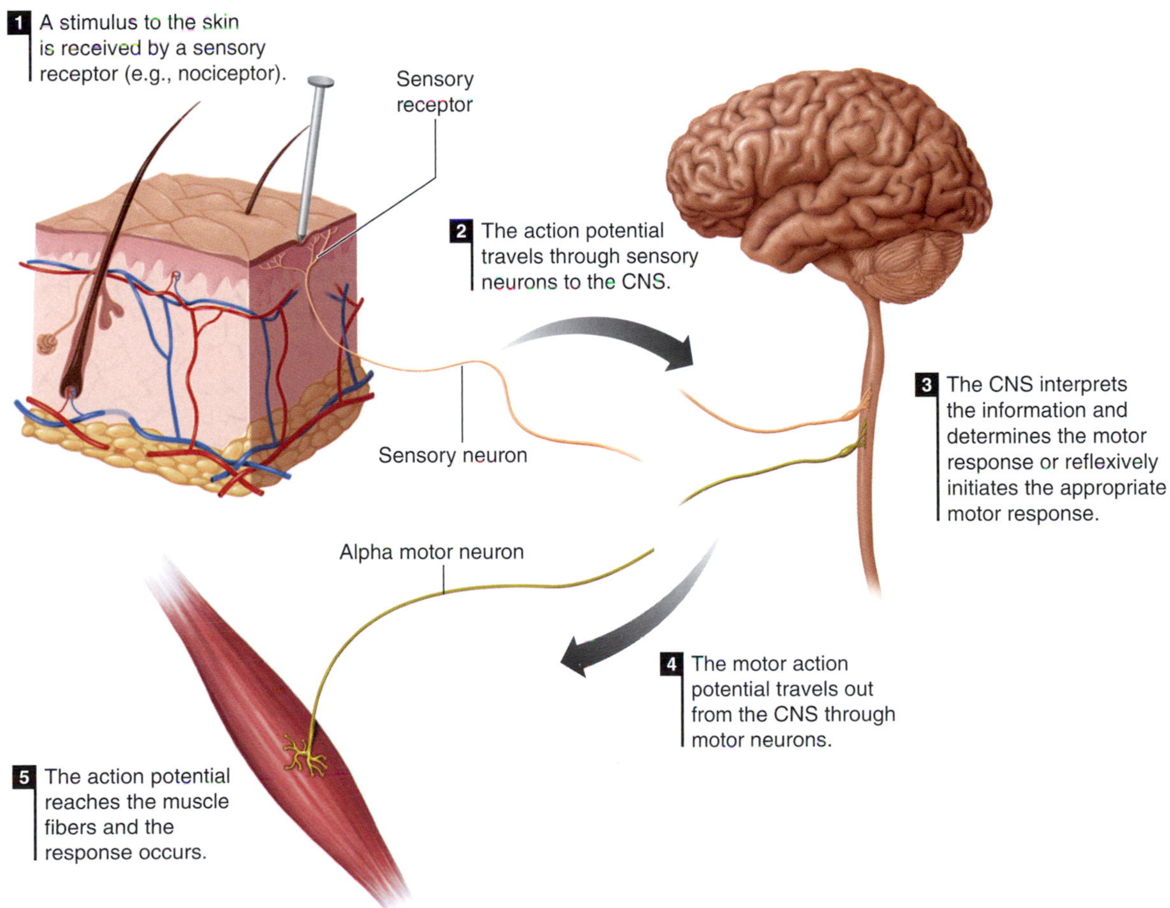

FIGURE 3.7 The sequence of events in sensory-motor integration.

4. The action potentials for the response are transmitted from the CNS along α-motor neurons.
5. The motor action potential is transmitted to a muscle, and the response occurs.

Sensory Input

Recall that sensations and physiological status are detected by sensory receptors throughout the body. The action potentials resulting from sensory stimulation are transmitted via the sensory nerves to the spinal cord. When they reach the spinal cord, they can either trigger a local reflex at that level or travel to the upper regions of the spinal cord or to the brain. Sensory pathways to the brain can terminate in sensory areas of the brain stem, the cerebellum, the thalamus, or the cerebral cortex. An area in which the sensory impulses terminate is referred to as an integration center. This is where the sensory input is interpreted and linked to the motor system. Figure 3.8 illustrates various sensory receptors and their nerve pathways back to the spinal cord and up into various areas of the brain. The integration centers vary in function:

- Sensory impulses that terminate in the spinal cord are integrated there. The response is typically a simple motor reflex, which is the simplest type of integration.
- Sensory signals that terminate in the lower brain stem result in subconscious motor reactions of a higher and more complex nature than simple spinal cord reflexes. Postural control during sitting, standing, or moving is an example of this level of sensory input.
- Sensory signals that terminate in the cerebellum also result in subconscious control of movement. The cerebellum appears to be the center of coordination, smoothing out movements by coordinating the actions of the various contracting muscle groups to perform the desired movement. Both fine and gross motor movements appear to be coordinated by the cerebellum in concert with the basal ganglia. Without the control exerted by the cerebellum, all movement would be uncontrolled and uncoordinated.

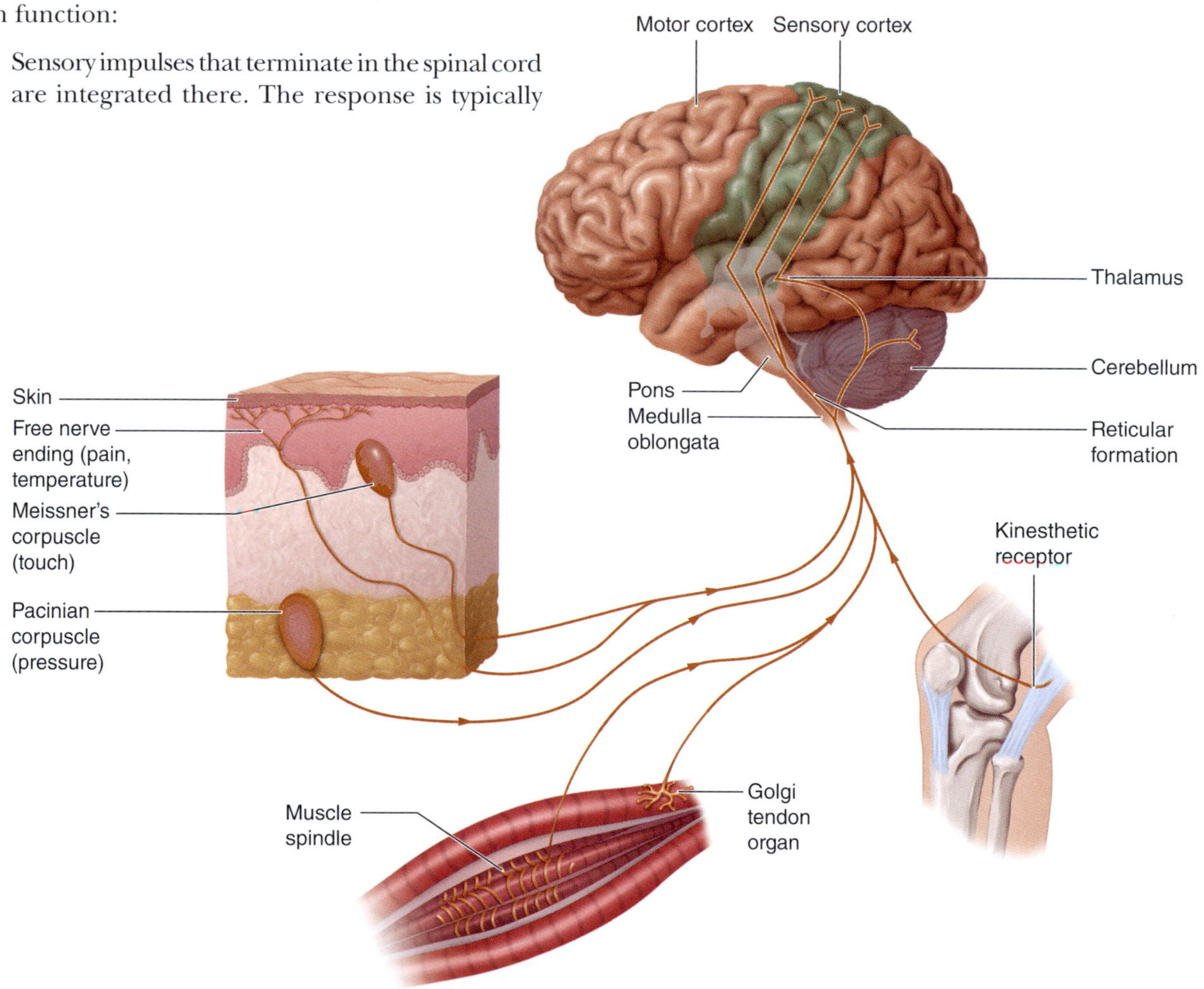

FIGURE 3.8 The sensory receptors and their afferent pathways back to the spinal cord and brain.

- Sensory signals that terminate at the thalamus begin to enter the level of consciousness, and the person begins to distinguish various sensations.
- Only when sensory signals enter the cerebral cortex can one discretely localize the signal. The primary sensory cortex, located in the postcentral gyrus (in the parietal lobe), receives general sensory input from receptors in the skin and from proprioceptors in the muscles, tendons, and joints. This area has a map of the body. Stimulation in a specific area of the body is recognized, and its exact location is known instantly. Thus, this part of the conscious brain allows us to be constantly aware of our surroundings and our relationship to them.

Once a sensory impulse is received, it may evoke a motor response, regardless of the level at which the sensory impulse stops. This response can originate from any of three levels:

- The spinal cord
- The lower regions of the brain
- The motor area of the cerebral cortex

As the level of control moves from the spinal cord to the motor cortex, the degree of movement complexity increases from simple reflex control to complicated movements requiring basic thought processes. Motor responses for more complex movement patterns typically originate in the motor cortex of the brain, and the basal ganglia and cerebellum help to coordinate repetitive movements and to smooth out overall movement patterns. Sensory-motor integration may also involve reflex pathways for quick responses and specialized sensory organs within muscles.

Reflex Activity

What happens when one unknowingly puts one's hand on a hot stove? First, the stimuli of heat and pain are received by the thermoreceptors and nociceptors in the hand, and then sensory action potentials travel to the spinal cord, terminating at the level of entry. Once in the spinal cord, these action potentials are integrated instantly by interneurons that connect the sensory and motor neurons. The action potentials move to the motor neurons and travel to the effectors, the muscles controlling the withdrawal of the hand. The result is that the person reflexively withdraws the hand from the hot stove without giving the action any thought.

A **motor reflex** is a preprogrammed response; any time the sensory nerves transmit certain action potentials, the body responds instantly and identically. In examples like the one just used, whether one touches something that is too hot or too cold, thermoreceptors will elicit a reflex to withdraw the hand. Whether the pain arises from heat or from a sharp object, the nociceptors will also cause a withdrawal reflex. By the time one is consciously aware of the specific stimulus (after sensory action potentials also have been transmitted to the primary sensory cortex), the reflex activity is well under way, if not completed. All neural activity occurs extremely rapidly, but a reflex is the fastest mode of response because the impulse is not transmitted up the spinal cord to the brain before an action occurs. Only one response is possible; no options need to be considered.

RESEARCH PERSPECTIVE 3.3

Sex Differences in Skeletal Muscle Fiber Types

As discussed in this chapter, and in chapter 1, skeletal muscle is made up of different types of fibers that vary in terms of their structure, biochemistry, and function. The fiber-type composition of different skeletal muscles depends in part on the anatomical location and function of the muscle. However, relatively little is known about whether the proportion of the different fiber types within a skeletal muscle differs between men and women. To date, the few studies that have assessed differential fiber-type composition between sexes have been conducted in rats and mice. In studies that examined sex differences in humans, the fibers measured in men had significantly larger cross-sectional areas, which is not surprising because men have an overall greater muscle mass. However, it appears that women have more type I fibers and fewer type II fibers than their male counterparts on average. When fiber-type composition was examined in the vastus lateralis muscle of a group of men, the average fiber type percentages were 34% type I, 46% type IIa, and 20% type IIx. In women, the percentages were 41% type I, 36% type IIa, and 23% type IIx. This greater prevalence of slow-twitch fibers in women corresponds with a lower contractile velocity in women compared to men but allows for increased endurance and recovery in women.[8] These data highlight sex differences in muscle fiber-type composition beyond that associated with muscle size alone. This has important implications. Future studies examining skeletal muscle composition, function, and adaptive responses to different forms of exercise training, as well as in pathophysiological conditions, should consider potential sex differences.

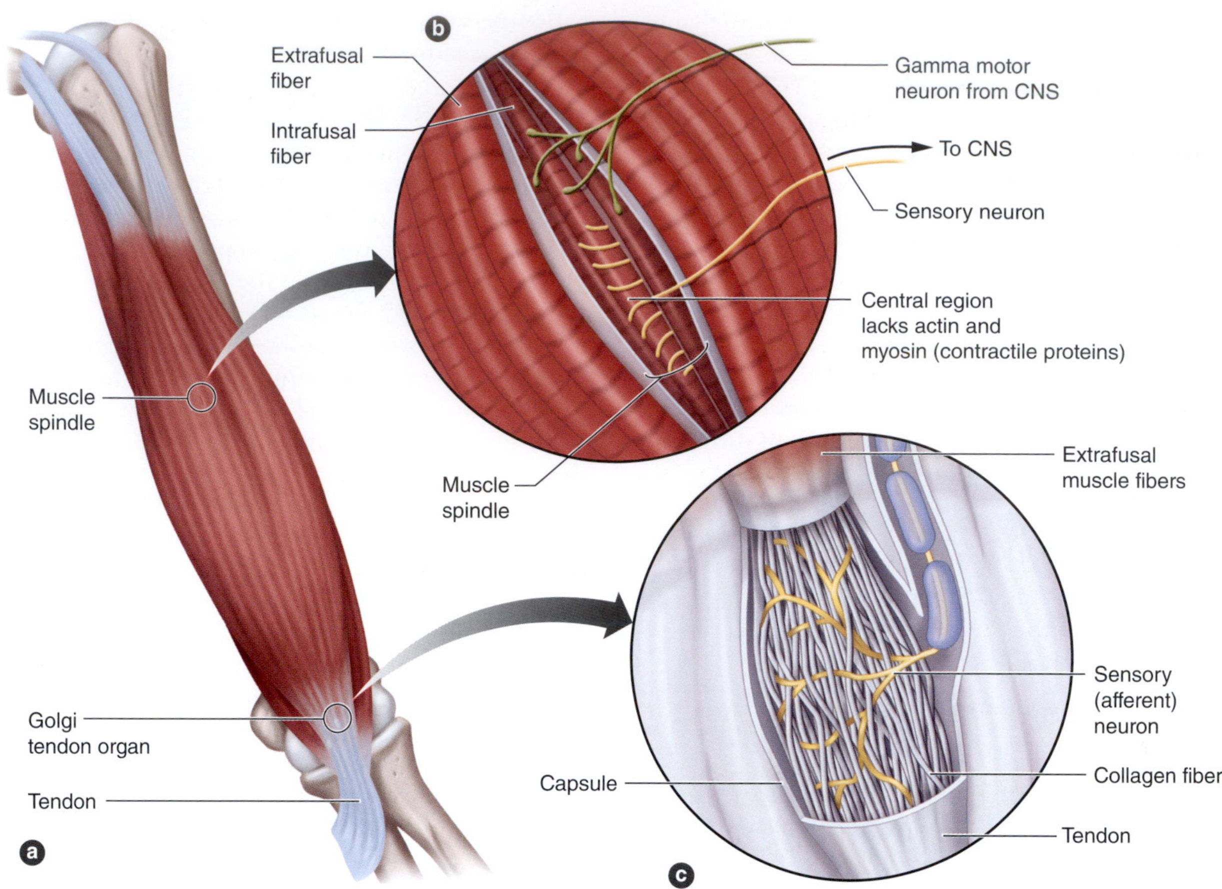

FIGURE 3.9 *(a)* A muscle belly showing *(b)* a muscle spindle and *(c)* a Golgi tendon organ.

Muscle Spindles

Now that we have covered the basics of reflex activity, we can look more closely at two specific reflexes that help control muscle function. The first involves a special structure: the muscle spindle (see figure 3.9).

The **muscle spindle** is a group of specialized muscle fibers found between regular skeletal muscle fibers, referred to as *extrafusal* (outside the spindle) fibers. A muscle spindle consists of 4 to 20 small, specialized *intrafusal* (inside the spindle) fibers and the nerve endings, sensory and motor, associated with these fibers. A connective tissue sheath surrounds the muscle spindle and attaches to the endomysium of the extrafusal fibers. The intrafusal fibers are controlled by specialized motor neurons, referred to as *γ-motor neurons* (or *gamma motor neurons*). In contrast, extrafusal fibers (the regular fibers) are controlled by *α-motor neurons*.

The central region of an intrafusal fiber cannot contract because it contains no or only a few actin and myosin filaments. This central region can only stretch. Because the muscle spindle is attached to the extrafusal fibers, any time those fibers are stretched, the central region of the muscle spindle is also stretched.

Sensory nerve endings wrapped around this central region of the muscle spindle transmit information to the spinal cord when this region is stretched, transmitting a signal to the CNS about the muscle's length. In the spinal cord, the sensory neuron synapses with an α-motor neuron, which triggers reflexive muscle contraction (in the extrafusal fibers) to resist further stretching.

Let's illustrate this action with an example. A person's arm is bent at the elbow, and the hand is extended, palm up. Suddenly someone places a heavy weight in the palm. The forearm starts to drop, which stretches the muscle fibers in the elbow flexors (e.g., biceps brachii), which in turn stretch the muscle spindles. In response to that stretch, the sensory neurons send action potentials to the spinal cord, which then activates the α-motor neurons of motor units in the same muscles. This activation causes the muscles to increase their force production, overcoming the stretch.

γ-Motor neurons excite the intrafusal fibers, prestretching them slightly. Although the midsection of the intrafusal fibers cannot contract, the ends can. The γ-motor neurons cause slight contraction of the ends of these fibers, which stretches the central region slightly. This prestretch makes the muscle spindle highly sensitive to even small degrees of stretch.

The muscle spindle also assists normal muscle action. It appears that when the α-motor neurons are stimulated to contract the extrafusal muscle fibers, the γ-motor neurons are also activated, contracting the ends of the intrafusal fibers. This stretches the central region of the muscle spindle, giving rise to sensory impulses that travel to the spinal cord and then to the α-motor neurons. In response, the muscle increases its force production. Thus, muscle force production is enhanced through this function of the muscle spindles.

Information brought into the spinal cord from the sensory neurons associated with muscle spindles does not merely end at that level. Impulses are also sent up to higher parts of the CNS, supplying the brain with continuous feedback on the exact length of the muscle and the rate at which that length is changing. This information is essential for maintaining muscle tone and posture and for executing movements. The muscle spindle functions as a servomechanism to continuously correct movements that do not proceed as planned. The brain is informed of errors in the intended movement at the same time that the error is being corrected at the spinal cord level.

Golgi Tendon Organs

Golgi tendon organs are encapsulated sensory receptors through which a small bundle of muscle tendon fibers pass. These organs are located just proximal to the tendon fibers' attachment to the muscle fibers, as shown in figure 3.9. Approximately 5 to 25 muscle fibers are usually connected with each Golgi tendon organ. Whereas muscle spindles monitor the length of a muscle, Golgi tendon organs are sensitive to tension in the muscle–tendon complex and operate like a strain gauge, a device that senses changes in tension. Their sensitivity is so great that they can respond to the contraction of a single muscle fiber. These sensory receptors are inhibitory in nature, performing a protective function by reducing the potential for injury. When stimulated, these receptors inhibit the contracting (agonist) muscles and excite the antagonist muscles.

In Review

- Sensory-motor integration is the process by which the PNS relays sensory input to the CNS and the CNS interprets this information and then sends out the appropriate motor signal to elicit the desired motor response.
- The level of nervous system response to sensory input varies according to the complexity of movement necessary. Most simple reflexes are handled by the spinal cord, whereas complex reactions and movements require activation of higher centers in the brain.
- Sensory input can terminate at various levels of the CNS. Not all of this information reaches the brain.
- Reflexes are the simplest form of motor control. These are not conscious responses. For a given sensory stimulus, the motor response is always identical and instantaneous.
- Muscle spindles trigger reflexive muscle action when stretched.
- Golgi tendon organs trigger a reflex that inhibits contraction if the tendon fibers are stretched from high muscle tension.

RESEARCH PERSPECTIVE 3.4

Nontraditional Factors That Impair Neuromuscular Control

Lower-extremity musculoskeletal injuries that occur during sport and physical activity, such as anterior cruciate ligament (ACL) tears, are far too common and extremely costly. Furthermore, these injuries are associated with serious long-term consequences beyond the injury itself, including the accelerated development of osteoarthritis. The first step toward effectively preventing lower-extremity injuries is the appropriate identification of the important risk factors for injury.

Perhaps the most commonly considered primary injury risk factors are measures of neuromuscular control, such as balance and movement technique. However, beyond these traditional risk factors, it is important to consider nontraditional factors that may predispose the athlete to injury, such as alterations in hydration, increases in body temperature, and fatigue. Importantly, hypohydration (below-optimal body fluid balance), hyperthermia (increased body core temperature), and fatigue, which are likely to be encountered during physical activity, all impair neuromuscular control. A 2012 study has substantiated this finding.[3] In particular, hypohydration combined with hyperthermia negatively affected movement technique and, to a lesser extent, balance. These findings emphasize the need for adequate hydration during exercise, especially when performed in hot environments, not only to optimize performance and prevent heat-related complications (see chapter 12), but also to reduce the risk of lower-extremity injury.[1]

Golgi tendon organs are important in resistance exercise. They function as safety devices, helping to prevent the muscle from developing excessive force during a contraction that may ultimately damage the muscle. Additionally, some researchers speculate that reducing the influence of Golgi tendon organs disinhibits the active muscles, allowing a more forceful muscle action. This mechanism may explain at least part of the gains in muscular strength that accompany strength training.

Motor Response

Now that we have discussed how sensory input is integrated to determine the appropriate motor response, the last step in the process is how muscles respond to motor action potentials once they reach the muscle fibers.

Once an action potential reaches an α-motor neuron, it travels the length of the neuron to the NMJ. From there, the action potential spreads to all muscle fibers innervated by that particular α-motor neuron. Recall that the α-motor neuron and all muscle fibers it innervates form a single motor unit. Each muscle fiber is innervated by only one α-motor neuron, but each α-motor neuron innervates up to several thousand muscle fibers, depending on the function of the muscle. Muscles controlling fine movements have only a small number of muscle fibers per α-motor neuron. The muscles that control eye movements (the extraocular muscles) have an innervation ratio of 1:15, meaning that one α-motor neuron controls only 15 muscle fibers. Muscles with more general functions have many fibers per α-motor neuron. For example, the gastrocnemius and tibialis anterior muscles of the lower leg have innervation ratios of almost 1:2,000.

The muscle fibers in a specific motor unit are homogeneous with respect to fiber type. Thus, one will not find a motor unit that has both type II and type I fibers. In fact, as mentioned in chapter 1, it is generally believed that the characteristics of the α-motor neuron actually determine the fiber type in the given motor unit.[7]

IN CLOSING

In this chapter, we examined how the nervous system is organized and how that organization functions to control movement. We covered the central nervous system as it relates to movement and the sensory and effector arms of the peripheral nervous system. We have seen how muscles respond to neural stimulation, whether through reflexes or under complex control of the higher brain centers, and the role of individual motor units in determining this response. Thus, we have learned how the body functions to allow people to move. In the next chapter, we examine the role of hormones in the body's response to exercise.

KEY TERMS

acetylcholine
adrenergic
axon hillock
axon terminal
central nervous system (CNS)
cholinergic
depolarization
effector (efferent) nerves
end branches
excitatory postsynaptic potential (EPSP)
Golgi tendon organ
graded potential
hyperpolarization
inhibitory postsynaptic potential (IPSP)
motor neurons (motor nerves)
motor reflex
muscle spindle
myelin sheath
nerve impulse
neuromuscular junction
neuron
neurotransmitter
norepinephrine
peripheral nervous system (PNS)
resting membrane potential (RMP)
saltatory conduction
sensory (afferent) nerves
sensory-motor integration
sodium–potassium pump
synapse
threshold

STUDY QUESTIONS

1. What are the major divisions of the nervous system? What are their major functions?
2. Name the different anatomical parts of a neuron, and discuss their function.
3. Explain the resting membrane potential. What causes it? How is it maintained?
4. Describe an action potential. What is required before an action potential is activated?
5. Explain how an action potential is transmitted from a presynaptic neuron to a postsynaptic neuron. Describe a synapse and a neuromuscular junction.
6. What brain centers have major roles in controlling movement, and what are these roles?
7. How do the sympathetic and parasympathetic systems differ? What is their significance in performing physical activity?
8. Explain how reflex movement occurs in response to touching a hot object.
9. Describe the role of the muscle spindle in controlling muscle contraction.
10. Describe the role of the Golgi tendon organ in controlling muscle contraction.

STUDY GUIDE ACTIVITIES

In addition to the activities listed in the chapter opening outline, two other activities are available in the web study guide, located at

www.HumanKinetics.com/PhysiologyOfSportAndExercise

The **KEY TERMS** activity reviews important terms, and the end-of-chapter **QUIZ** tests your understanding of the material covered in the chapter.

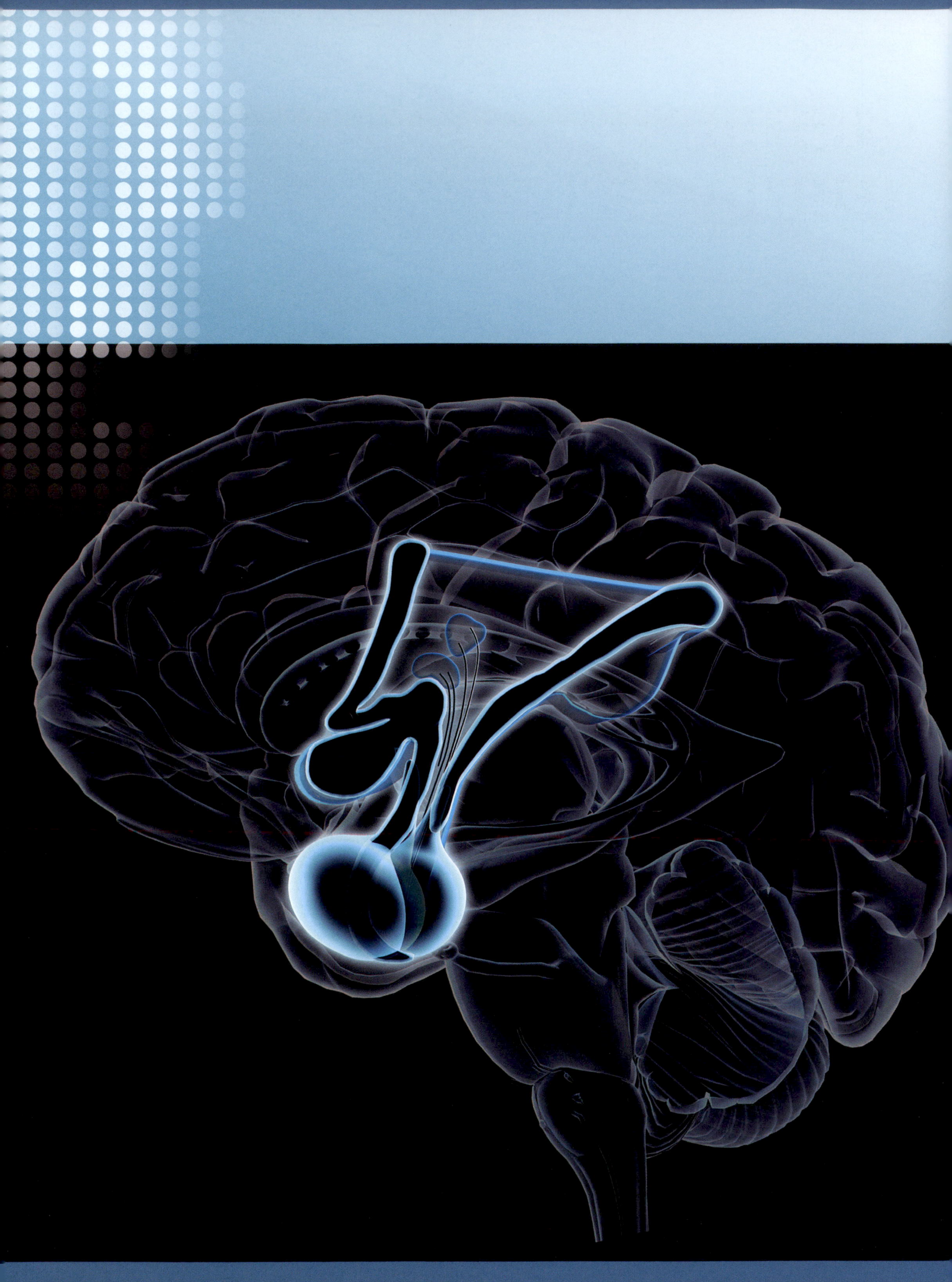

Hormonal Control During Exercise

In this chapter and in the web study guide

On May 22, 2010, a 13-year-old American boy became the youngest climber to reach the top of Mount Everest, a grueling trek to an altitude 29,035 ft (8,850 m) above sea level. The climb was extremely controversial because of the boy's age. In fact, because the Nepalese government would not give the family permission to climb Everest from Nepal, the climbing team ascended from the more difficult Chinese side where there was no age restriction. To prepare for the climb, the boy and his father (and climbing partner) slept for months in a hypoxic tent to prepare their bodies for ascent to high altitude. One goal of high-altitude acclimation is to increase the concentration of oxygen-carrying red blood cells in the blood. Two important hormones facilitated this goal. An increase in the hormone erythropoietin signaled the bone marrow to produce more red blood cells, and a decrease in vasopressin (also called antidiuretic hormone) caused the kidneys to produce excess urine to better concentrate the red blood cells. Because of these adaptations, the climbers were able to summit Mount Everest with less time spent in the various base camps along the way.

During exercise and exposure to extreme environments, the body must make a multitude of physiological adjustments. Energy production must increase, and metabolic by-products must be cleared. Cardiovascular and respiratory function must be constantly adjusted to match the demands placed upon these and other body systems, such as those regulating temperature. While the body's internal environment is in a constant state of flux even at rest, during exercise these well-orchestrated changes must occur rapidly and in a well-coordinated manner.

While much of the physiological regulation and integration required during exercise is accomplished by the nervous system (discussed in chapter 3), another physiological system—the endocrine system—affects virtually every cell, tissue, and organ in the body. It constantly monitors the body's internal environment, noting all changes that occur and rapidly releasing hormones to ensure that homeostasis is not dramatically disrupted. In this chapter, we focus on the importance of hormones in maintaining homeostasis and aiding all the internal processes that support physical activity. Because we cannot cover all aspects of endocrine control during exercise, the focus is on hormonal control of metabolism and body fluid balance during exercise. Because diet plays an important role in exercise metabolism, hormonal regulation of food intake is also covered. Additional hormones—including those that regulate growth and development, muscle mass, and reproductive function—are covered in other chapters of this book.

The Endocrine System

As the body transitions from a resting to an active state, the rate of metabolism must increase to provide necessary energy. This requires the coordinated integration and communication of many physiological and biochemical systems. Although the nervous system is responsible for much of this communication, fine-tuning the physiological responses to any disturbance in homeostasis is primarily the responsibility of the endocrine system. The endocrine and nervous systems, often collectively called the neuroendocrine system, work in concert to control all of the physiological processes that support exercise. The nervous system functions quickly, having short-lived, localized effects, whereas the endocrine system responds more slowly but has longer-lasting effects.

The endocrine system is defined as all tissues or glands that secrete **hormones**. The major endocrine glands and tissues are illustrated in figure 4.1. Endocrine glands typically secrete their hormones directly into the blood where they act as chemical signals throughout the body. When secreted by the specialized endocrine cells, hormones are transported via the blood to specific **target cells**—cells that possess specific hormone receptors. On reaching their destinations, hormones can control the activity of the target tissue. Historically, hormones were defined as chemicals made by a gland that traveled to a remote tissue in the body to exert their action. Now hormones are more broadly defined as any chemical that controls and regulates the activity of certain cells or organs. Some hormones affect many body tissues, including the brain, whereas others target specific cells within a tissue.

Hormones are involved in most physiological processes, so their actions are relevant to many aspects of exercise and physical activity. Because hormones play key roles in almost every system of the body, total coverage of that topic is well beyond the scope of this book. In the following sections, the chemical nature of hormones and the general mechanisms through which they act are discussed. An overview of the major endocrine glands and their hormones is presented for completeness. With respect to exercise, the focus is on two major aspects of hormonal control, the control of exercise metabolism and the regulation of body fluids and electrolytes during exercise. Finally, new information about hormonal regulation of food intake is presented, since caloric intake and specific nutrients consumed have a profound influence on exercise metabolism.

Chemical Classification of Hormones

Hormones are traditionally categorized as steroid hormones and nonsteroid hormones. **Steroid hormones**

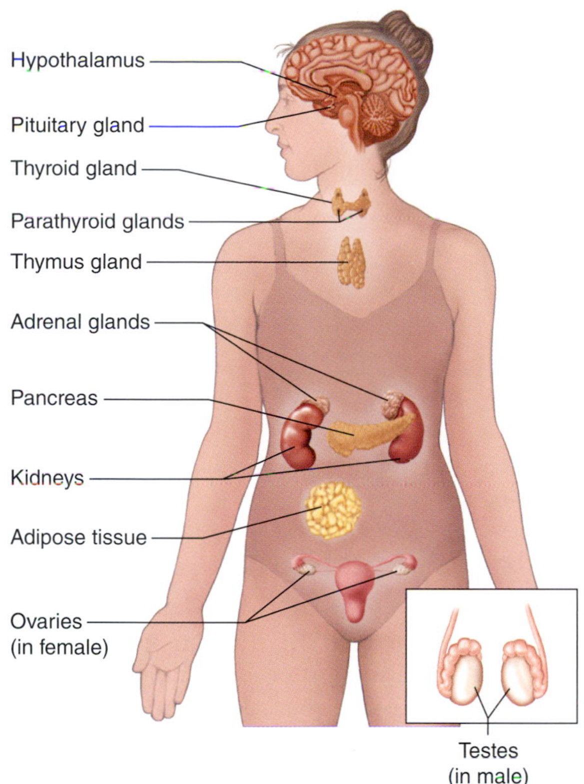

FIGURE 4.1 Location of the major endocrine organs of the body.

have a chemical structure similar to cholesterol, since most are derived from cholesterol. For this reason, they are soluble in lipids so they diffuse rather easily through cell membranes. This group includes the reproductive hormones testosterone (secreted by the testes) and estrogen and progesterone (secreted by the ovaries and placenta), as well as cortisol and aldosterone (secreted by the adrenal cortex).

Nonsteroid hormones are not lipid soluble, so they cannot easily cross cell membranes. The nonsteroid hormone group can be subdivided into two groups: protein or peptide hormones and amino acid–derived hormones. The two hormones produced by the thyroid gland (thyroxine and triiodothyronine) and the two from the adrenal medulla (epinephrine and norepinephrine) are amino acid–derived hormones. All other nonsteroid hormones are protein or peptide hormones. The chemical structure of a hormone determines its mechanism of action on target cells and tissues.

Hormone Secretion and Plasma Concentration

Control of hormone secretion must be rapid in order to meet the demands of changing bodily functions. Hormones are not secreted constantly or uniformly, but often in a pulsatile manner, that is, in irregularly timed brief bursts. Therefore, plasma concentrations of specific hormones fluctuate over short periods of an hour or less. But plasma concentrations of many hormones also fluctuate over longer periods of time, showing daily or even monthly cycles (such as monthly menstrual cycles). How do endocrine glands know when to release their hormones and how much to release?

Negative feedback is the primary mechanism through which the endocrine system maintains homeostasis. Secretion of a hormone causes some change in the body, and this change in turn inhibits further hormone secretion. Consider how a home thermostat works. When the room temperature decreases below some preset level, the thermostat signals the furnace to produce heat. When the room temperature increases to the preset level, the thermostat's signal ends, and the furnace stops producing heat. In the body, secretion of a specific hormone is similarly turned on or off (or up or down) by specific physiological changes.

Using the example of plasma glucose concentrations and the hormone insulin, when the plasma glucose concentration is high, the pancreas releases insulin. Insulin increases cellular uptake of glucose, lowering plasma glucose concentration. When plasma glucose concentration returns to normal, insulin release is inhibited until the plasma glucose level increases again. Because the endocrine system works in concert with the nervous system, the central nervous system is also involved in maintenance of appropriate hormonal balance.

The plasma concentration of a specific hormone is not always the best indicator of that hormone's activity because hormones must bind to specific cellular receptors to exert an effect. Accordingly, the number of receptors on target cells can be altered to increase or decrease that cell's sensitivity to the hormone. With fewer receptors, fewer hormone molecules can bind, and the cell becomes less sensitive to the given hormone. This is referred to as **downregulation**, or desensitization. In people with **insulin resistance**, for example, the number of insulin receptors on their cells appears to be reduced. Their bodies respond by increasing insulin secretion from the pancreas, so their plasma insulin concentrations increase. To obtain the same degree of plasma glucose control as normal, healthy people, these individuals must release much more insulin.

In a few instances, a cell may respond to the prolonged presence of large amounts of a hormone by increasing its number of available receptors. When this happens, the cell becomes more sensitive to that hormone because more can be bound at one time. This is referred to as **upregulation**. For example, individuals with a high **insulin sensitivity**, the opposite of insulin resistance, need relatively normal or low levels of insulin to process a given concentration of blood glucose.

Hormone Actions

Because hormones travel in the blood, they contact virtually all body tissues. How, then, do they limit their effects to specific targets? This ability is attributable to the specific hormone receptors on target tissues that can bind only specific hormones. Each cell typically has from 2,000 to 10,000 receptors. The combination of a hormone and its bound receptor is referred to as a hormone–receptor complex.

Recall that steroid hormones are lipid soluble and can therefore pass through cell membranes whereas nonsteroid hormones cannot. Receptors for nonsteroid hormones are located on the cell membrane, while those for steroid hormones are found either in the cytoplasm or in the nucleus of the cell. Each hormone is usually highly specific for a single type of receptor and binds only with its specific receptors, thus affecting only tissues that contain those specific receptors. Once hormones are bound to a receptor, numerous mechanisms allow them to control the actions of those cells.

Steroid Hormones

The general mechanism of action of steroid hormones is illustrated in figure 4.2. Once through the cell membrane and inside the cell, a steroid hormone binds to its specific receptors. The hormone–receptor complex then enters the nucleus, binds to part of the cell's DNA (deoxyribonucleic acid), and activates certain genes. This process is referred to as **direct gene activation**. In response to this activation, mRNA (messenger ribonucleic acid) is synthesized within the nucleus. The mRNA then enters the cytoplasm and promotes protein synthesis. These proteins may be

- enzymes that can have numerous effects on cellular processes,
- structural proteins for tissue growth and repair, or
- regulatory proteins that can alter enzyme function.

Nonsteroid Hormones

Because nonsteroid hormones cannot cross the cell membrane, they bind with specific receptors on the cell membrane. A nonsteroid hormone molecule binds to its membrane receptor and triggers a series of reactions that lead to the formation of an intracellular **second messenger**. In addition to relaying signals, second mes-

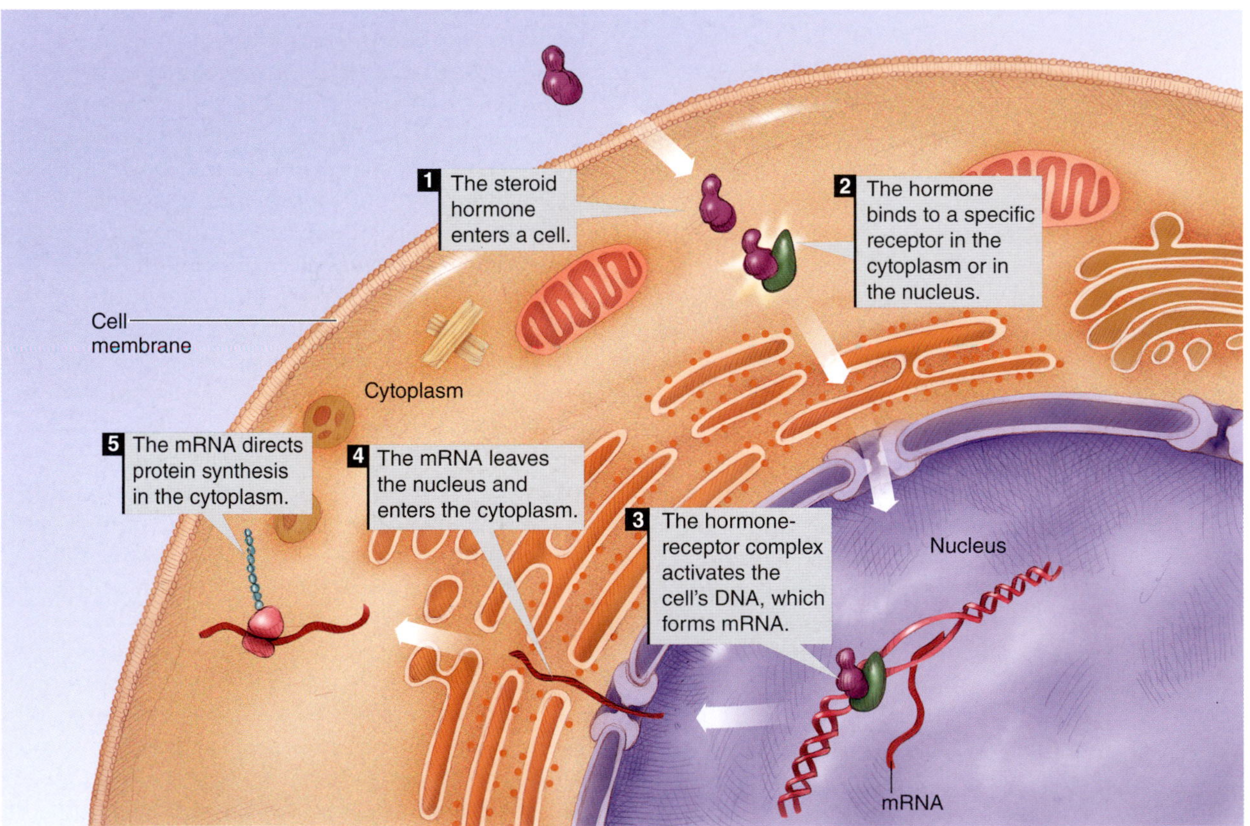

FIGURE 4.2 The general mechanism of action of a typical steroid hormone, leading to direct gene activation and protein synthesis.

sengers can also help intensify the strength of the signal. While there are many second messenger molecules, one important second messenger that mediates multiple hormone–receptor responses is **cyclic adenosine monophosphate** (**cAMP**, or cyclic AMP); its mechanism of action is depicted in figure 4.3. In this case, attachment of the hormone to the appropriate membrane receptor activates an enzyme, adenylate cyclase, situated within the cell membrane. This enzyme regulates the formation of cAMP from cellular adenosine triphosphate (ATP). Cyclic AMP then controls specific physiological responses that can include

- activation of cellular enzymes,
- change in membrane permeability,
- promotion of protein synthesis,
- change in cellular metabolism, or
- stimulation of cellular secretions.

Some of the hormones that employ cAMP as a second messenger are epinephrine, glucagon, and luteinizing hormone. In addition to cAMP, other important second messengers include cyclic guanosine monophosphate (cGMP), inositol trisphosphate (IP3), diacylglycerol (DAG), and calcium ions (Ca^{2+}).

Although by strict definition not hormones, **prostaglandins** are often considered to be a third class of hormones. These substances are derived from a fatty acid, arachidonic acid, and they are associated with the plasma membranes of almost all body cells. Prostaglandins typically act as *local* hormones or **autocrines**, exerting their effects in the immediate area where they are produced. But some also survive long enough to circulate through the blood to affect distant tissues. Prostaglandin release can be triggered by many stimuli, such as other hormones or a local injury. Their functions are quite numerous because there are several different types of prostaglandins. They often mediate the effects of other hormones. They are also known to act directly on blood vessels, increasing vascular permeability (which promotes swelling) and vasodilation. In this capacity, they are important mediators of the inflammatory response. They also sensitize the nerve endings of pain fibers; thus, they mediate both inflammation and pain.

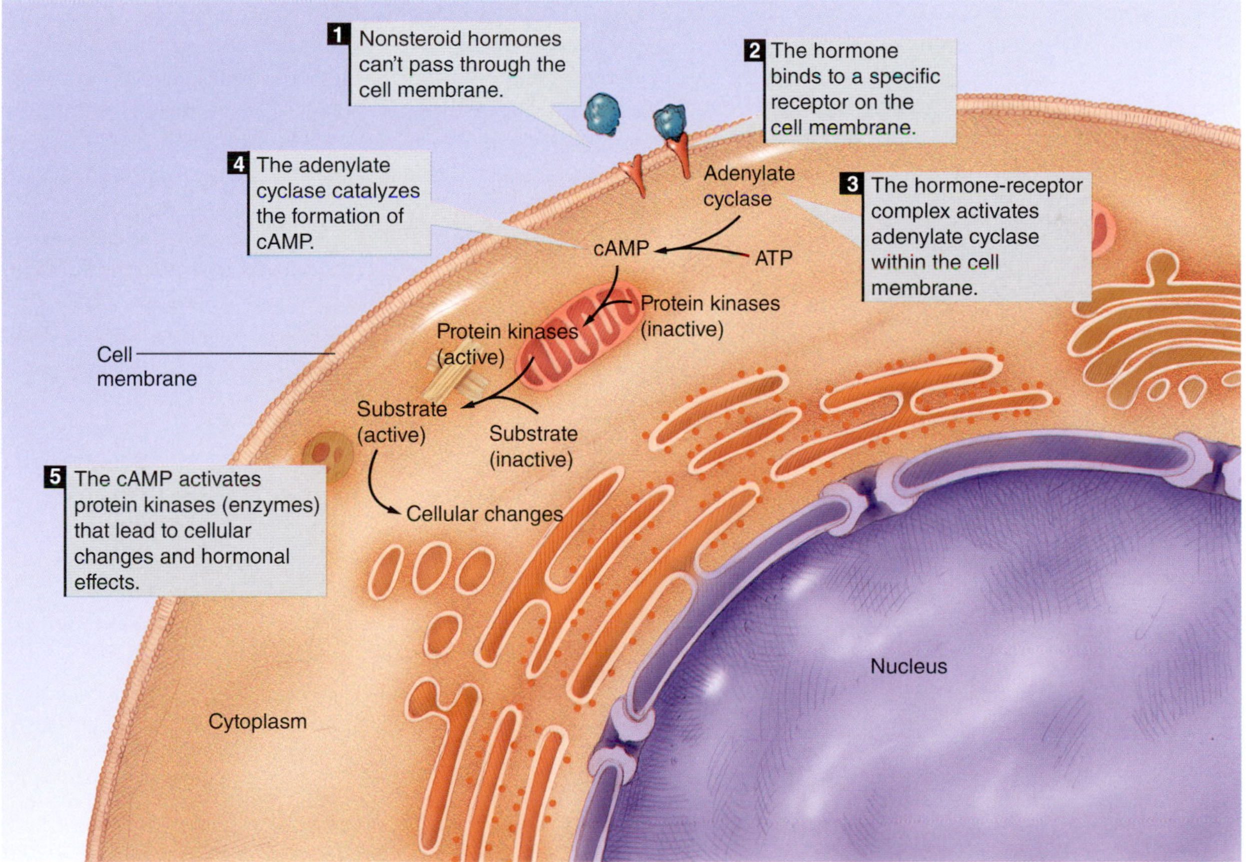

FIGURE 4.3 The mechanism of action of a nonsteroid hormone, in this case activating a second messenger (cyclic adenosine monophosphate) within the cell to activate cellular functions.

Endocrine Glands and Their Hormones: An Overview

The major endocrine glands and their respective hormones are listed in table 4.1. This table also lists each hormone's primary target and actions. Because the endocrine system is extremely complex, the presentation here has been greatly simplified to focus on those endocrine glands and hormones of greatest importance to exercise and physical activity.

Because hormones play such an important role in regulation of many physiological variables during exercise, it is not surprising that hormone release changes during acute bouts of activity. The hormonal responses to an acute bout of exercise and to exercise training are summarized in table 4.2. This table is limited to those hormones that play major roles in sport and physical activity. Further details of these exercise-induced hormonal responses are provided in the following discussion of specific endocrine glands and their hormones.

TABLE 4.1 **The Major Endocrine Glands and Their Hormones, Target Organs, Controlling Factors, and Functions**

Endocrine gland	Hormone	Target organ	Controlling factor	Major functions
Anterior pituitary	Growth hormone (GH)	All cells in the body	Hypothalamic GH-releasing hormone; GH-inhibiting hormone (somatostatin)	Promotes development and enlargement of all body tissues until maturation; increases rate of protein synthesis; increases mobilization of fats and use of fat as an energy source; decreases rate of carbohydrate use
	Thyrotropin (TSH)	Thyroid gland	Hypothalamic TSH-releasing hormone	Controls the amount of thyroxin and triiodothyronine produced and released by the thyroid gland
	Adrenocorticotropin (ACTH)	Adrenal cortex	Hypothalamic ACTH-releasing hormone	Controls the secretion of hormones from the adrenal cortex
	Prolactin	Breasts	Prolactin-releasing and -inhibiting hormones	Stimulates milk production by the breasts
	Follicle-stimulating hormone (FSH)	Ovaries and testes	Hypothalamic FSH-releasing hormone	Initiates growth of follicles in the ovaries and promotes secretion of estrogen from the ovaries; promotes development of the sperm in the testes
	Luteinizing hormone (LH)	Ovaries and testes	Hypothalamic FSH-releasing hormone	Promotes secretion of estrogen and progesterone and causes the follicle to rupture, releasing the ovum; causes testes to secrete testosterone
Posterior pituitary	Antidiuretic hormone (ADH or vasopressin)	Kidneys	Hypothalamic secretory neurons	Assists in controlling water excretion by the kidneys; elevates blood pressure by constricting blood vessels
	Oxytocin	Uterus and breasts	Hypothalamic secretory neurons	Controls contraction of uterus; milk secretion
Thyroid	Thyroxine (T_4) and triiodothyronine (T_3)	All cells in the body	TSH, T_3, and T_4 concentrations	Increase the rate of cellular metabolism; increase rate and contractility of the heart
	Calcitonin	Bones	Plasma calcium concentrations	Controls calcium ion concentration in the blood

Endocrine gland	Hormone	Target organ	Controlling factor	Major functions
Parathyroid	Parathyroid hormone (PTH, or parathormone)	Bones, intestines, and kidneys	Plasma calcium concentrations	Controls calcium ion concentration in the extracellular fluid through its influence on bones, intestines, and kidneys
Adrenal medulla	Epinephrine	Most cells in the body	Baroreceptors, glucose receptors, and brain and spinal centers	Stimulates breakdown of glycogen in liver and muscle and lipolysis in adipose tissue and muscle; increases skeletal muscle blood flow; increases heart rate and contractility; increases oxygen consumption
	Norepinephrine	Most cells in the body	Baroreceptors, glucose receptors, and brain and spinal centers	Stimulates lipolysis in adipose tissue and in muscle to a lesser extent; constricts arterioles and venules, thereby elevating blood pressure
Adrenal cortex	Mineralocorticoids (aldosterone)	Kidneys	Angiotensin and plasma potassium concentrations; renin	Increase sodium retention and potassium excretion through the kidneys
	Glucocorticoids (cortisol)	Most cells in the body	ACTH	Control metabolism of carbohydrates, fats, and proteins; exert an anti-inflammatory action
	Androgens and estrogens	Ovaries, breasts, and testes	ACTH	Assist in the development of female and male sex characteristics
Pancreas	Insulin	All cells in the body	Plasma glucose and amino acid concentrations	Controls blood glucose levels by lowering glucose levels; increases use of glucose and synthesis of fat
	Glucagon	All cells in the body	Plasma glucose and amino acid concentrations	Increases blood glucose; stimulates the breakdown of protein and fat
	Somatostatin	Islets of Langerhans and intestines	Plasma glucose, insulin, and glucagon concentrations	Depresses the secretion of both insulin and glucagon
Kidney	Renin	Adrenal cortex	Plasma sodium concentrations	Assists in blood pressure control
	Erythropoietin (EPO)	Bone marrow	Low tissue oxygen concentrations	Stimulates erythrocyte production
Testes	Testosterone	Sex organs and muscle	FSH and LH	Promotes development of male sex characteristics, including growth of testes, scrotum, and penis, facial hair, and change in voice; also promotes muscle growth
Ovaries	Estrogens and progesterone	Sex organs and adipose tissue	FSH and LH	Promote development of female sex organs and characteristics; increase storage of fat; assist in regulating the menstrual cycle

TABLE 4.2 **Hormone Responses to Acute Exercise and Change in Response With Exercise Training**

Endocrine gland	Hormone	Response to acute exercise (untrained)	Effect of exercise training
Anterior pituitary	Growth hormone (GH)	Increases with increasing rates of work	Attenuated response at same rate of work
	Thyrotropin (TSH)	Increases with increasing rates of work	No known effect
	Adrenocorticotropin (ACTH)	Increases with increasing rates of work and duration	Attenuated response at same rate of work
	Prolactin	Increases with exercise	No known effect
	Follicle-stimulating hormone (FSH)	Small or no change	No known effect
	Luteinizing hormone (LH)	Small or no change	No known effect
Posterior pituitary	Antidiuretic hormone (ADH or vasopressin)	Increases with increasing rates of work	Attenuated response at same rate of work
	Oxytocin	Unknown	Unknown
Thyroid	Thyroxine (T_4) and triiodothyronine (T_3)	Free T_3 and T_4 increase with increasing rates of work	Increased turnover of T_3 and T_4 at same rate of work
	Calcitonin	Unknown	Unknown
Parathyroid	Parathyroid hormone (PTH or parathormone)	Increases with prolonged exercise	Unknown
Adrenal medulla	Epinephrine	Increases with increasing rates of work, starting at about 75% of $\dot{V}O_{2max}$	Attenuated response at same rate of work
	Norepinephrine	Increases with increasing rates of work, starting at about 50% of $\dot{V}O_{2max}$	Attenuated response at same rate of work
Adrenal cortex	Aldosterone	Increases with increasing rates of work	Unchanged
	Cortisol	Increases only at high rates of work	Slightly higher values
Pancreas	Insulin	Decreases with increasing rates of work	Attenuated response at same rate of work
	Glucagon	Increases with increasing rates of work	Attenuated response at same rate of work
Kidney	Renin	Increases with increasing rates of work	Unchanged
	Erythropoietin (EPO)	Unknown	Unchanged
Testes	Testosterone	Small increases with exercise	Resting levels decreased in male runners
Ovaries	Estrogens and progesterone	Small increases with exercise	Resting levels might be decreased in highly trained women

As mentioned earlier, a comprehensive description of neuroendocrine control is well beyond the scope of this textbook. Two important exercise-related functions of the endocrine glands and their hormones are the regulation of metabolism during exercise and the regulation of body fluids and electrolytes. The endocrine system also plays an important role in regulating appetite and food intake. The sections that follow detail these three important functions. Each section provides a description of the primary endocrine glands involved, the hormones produced, and how those hormones serve the given regulatory role.

VIDEO 4.1 Presents Katarina Borer on the contributing role of sex hormones to ACL tears in women.

In Review

- The nervous system functions quickly, having short-lived, localized effects, whereas the endocrine system typically responds more slowly but has longer-lasting effects.
- Hormones are classified chemically as either steroid or nonsteroid. Steroid hormones are lipid soluble, and most are formed from cholesterol. Nonsteroid hormones are formed from proteins, peptides, or amino acids.
- Hormones influence specific target tissues or cells through a unique interaction between the hormone and the specific receptors for that hormone on the cell membrane (nonsteroid hormones) or within the cytoplasm or nucleus of the cell (steroid hormones).
- Hormones generally are secreted nonuniformly, often in brief pulsatile bursts, into the blood and then circulate to target cells.
- A negative feedback system regulates secretion of most hormones.
- The number of receptors for a specific hormone can be altered to meet the body's demands. *Upregulation* refers to an increase in available receptors, and *downregulation* refers to a decrease. These two processes change a cell's sensitivity to a given hormone.
- Steroid hormones pass through cell membranes and bind to receptors in the cytoplasm or nucleus of the cell. At the nucleus, they use a mechanism called direct gene activation to cause protein synthesis.
- Nonsteroid hormones cannot easily enter cells, so they bind to receptors on the cell membrane. This activates a second messenger within the cell, often cAMP, which in turn can trigger numerous cellular processes.
- Prostaglandins are not hormones by strict definition but act as local hormones, exerting their effect in the immediate area where they are produced.

Hormonal Regulation of Metabolism During Exercise

As noted in chapter 2, carbohydrate and fat metabolism are responsible for maintaining muscle ATP during prolonged exercise. Various hormones work to ensure adequate glucose and free fatty acid (FFA) availability for muscle energy metabolism. In the next sections, we examine (1) the major endocrine glands and hormones responsible for metabolic regulation and (2) how the metabolism of glucose and fat is regulated by these hormones during exercise.

Endocrine Glands Involved in Metabolic Regulation

While many complex systems interact to regulate metabolism at rest and during exercise, the major endocrine glands responsible are the anterior pituitary gland, the thyroid gland, the adrenal glands, and the pancreas.

Anterior Pituitary

The pituitary gland is a marble-sized gland attached to the hypothalamus at the base of the brain. It has three lobes: anterior, intermediate, and posterior. The intermediate lobe is very small and is thought to play little or no role in humans, but both the anterior and posterior lobes serve major endocrine functions. Hormonal release from the anterior pituitary is controlled by hormones secreted by the hypothalamus, while the posterior pituitary releases hormones in response to direct nerve signals from the hypothalamus. Therefore, the pituitary gland can be thought of as the relay between CNS control centers and peripheral endocrine glands. The posterior pituitary is discussed later in the chapter.

The anterior pituitary, also called the adenohypophysis, secretes six hormones in response to **releasing factors** or **inhibiting factors** (which are also categorized as hormones) secreted by the hypothalamus. Hormonal communication between the hypothalamus and the anterior lobe of the pituitary occurs through a specialized circulatory system. The major functions of each of the anterior pituitary hormones, along with their releasing and inhibiting factors, are listed in table 4.1. Exercise is a strong stimulus to the hypothalamus because exercise increases the release of most anterior pituitary hormones (see table 4.2).

Of the six anterior pituitary hormones, four are tropic hormones, meaning they affect the functioning

RESEARCH PERSPECTIVE 4.1

Does Having More Testosterone Give You a Competitive Advantage?

Androgens (testosterone and its chemical derivatives) stimulate the development and maintenance of primary and secondary male sex characteristics. Although androgens are typically described as male sex hormones, they are found naturally in both men and women and can improve sport performance in both male and female athletes, particularly in strength-dependent events. Because of their ergogenic effects (enhanced physical performance, stamina, and recovery), androgens have been widely abused by athletes despite advances in tests to detect their abuse. In fact, androgens are the most common ergogenic aid used by female athletes. However, some women have naturally higher circulating androgens, and a great deal of controversy has surrounded the debate about whether these women should be allowed to compete with this natural ergogenic advantage. Because of this controversy, regulatory committees are keenly interested in scientific evidence that may link circulating natural androgens and athletic performance.

A recent study of 2,127 elite track and field athletes competing in the 2011 and 2013 International Association of Athletics Federations World Championships provided more scientific data on this controversial topic.[1] Researchers measured blood androgens, particularly testosterone concentrations, in male and female athletes and compared these concentrations to each athlete's best performances at the World Championships. Male sprinters showed higher testosterone concentrations, and men involved in throwing events had lower testosterone concentrations than male athletes in other events. The type of event had no association with testosterone concentration in women. However, women (but not men) with the highest testosterone concentrations performed better in the 400 m, 400 m hurdles, 800 m, hammer throw, and pole vault when compared to women with the lowest testosterone. The study concluded that female athletes with high natural testosterone concentrations may have a competitive advantage over those with low testosterone competing in these specific track-and-field events. Thus, the quantitative relation between elevated testosterone and improved athletic performance should be considered when regulatory and governing bodies discuss the eligibility of women with hyperandrogenism in competitive events.

of other endocrine glands. The exceptions are growth hormone and prolactin. **Growth hormone (GH)** is a potent anabolic agent (a substance that builds up organs and tissues, producing growth and cell differentiation and an increase in size of tissues). It promotes muscle growth and hypertrophy by facilitating amino acid transport into the cells. In addition, GH directly stimulates fat metabolism (lipolysis) by increasing the synthesis of lipolytic enzymes. Growth hormone concentrations are elevated during both aerobic and resistance exercise in proportion to the exercise intensity and typically remain elevated for some time after exercise.

Thyroid Gland

The thyroid gland is located along the midline of the neck, immediately below the larynx. It secretes two important nonsteroid hormones, **triiodothyronine (T_3)** and **thyroxine (T_4)**, which regulate metabolism in general, and an additional hormone, calcitonin, which assists in regulating calcium metabolism.

The two metabolic thyroid hormones share similar functions. Triiodothyronine and thyroxine increase the metabolic rate of almost all tissues and can increase the body's basal metabolic rate by as much as 100%. These hormones also

- increase protein synthesis (including enzymes),
- increase the size and number of mitochondria in most cells,
- promote rapid cellular uptake of glucose,
- enhance glycolysis and gluconeogenesis, and
- enhance lipid mobilization, increasing FFA availability for oxidation.

Acute exercise causes the release of **thyrotropin** (**TSH**, or thyroid-stimulating hormone) from the anterior pituitary. Thyroid-stimulating hormone controls the release of triiodothyronine and thyroxine, so the exercise-induced increase in TSH would be expected to stimulate the thyroid gland. Exercise increases plasma thyroxine concentrations, but a delay occurs between the increase in TSH concentrations during exercise and the increase in plasma thyroxine concentration. Furthermore, during prolonged submaximal exercise, thyroxine concentration increases sharply, then remains relatively constant while triiodothyronine concentrations tend to decrease over time.

Adrenal Glands

The adrenal glands are situated directly atop each kidney and are composed of the inner adrenal medulla and the outer adrenal cortex. The hormones secreted by these two areas are quite distinct. The adrenal medulla produces and releases two hormones, **epinephrine** and norepinephrine, which are collectively referred to as **catecholamines**. Because of its origin in the adrenal gland, a synonym for epinephrine is **adrenaline**. When the adrenal medulla is stimulated by the sympathetic nervous system, approximately 80% of its secretion is epinephrine and 20% is norepinephrine, although these percentages vary with different physiological conditions. Circulating catecholamines have powerful effects similar to those of the sympathetic nervous system. Recall that these same catecholamines function as neurotransmitters in the sympathetic nervous system; however, the hormones' effects last longer because these substances are removed from the blood relatively slowly compared to the quick reuptake and degradation of the neurotransmitters. These two hormones prepare a person for immediate action, often called the fight-or-flight response.

Although some of the specific actions of these two hormones differ, the two work together. Their combined effects include

- increased heart rate and force of contraction,
- increased metabolic rate,
- increased glycogenolysis (breakdown of glycogen to glucose) in the liver and muscle,
- increased release of glucose and FFAs into the blood,
- redistribution of blood to the skeletal muscles,
- increased blood pressure, and
- increased respiration.

Release of epinephrine and norepinephrine is affected by a wide variety of factors, including psychological stress and exercise. Plasma concentrations of these hormones increase as individuals increase their exercise intensity. Plasma norepinephrine concentrations increase markedly at intensities above 50% of $\dot{V}O_{2max}$, but epinephrine concentrations do not increase significantly until the exercise intensity exceeds 60% to 70% of $\dot{V}O_{2max}$. During long-duration steady-state exercise at a moderate intensity, blood concentrations of both hormones increase. When the exercise bout ends, epinephrine returns to resting concentrations within only a few minutes of recovery, but norepinephrine can remain elevated for several hours.

The adrenal cortex secretes more than 30 different steroid hormones, referred to as corticosteroids. These generally are classified into three major types: mineralocorticoids (discussed later in the chapter), glucocorticoids, and gonadocorticoids (sex hormones).

The **glucocorticoids** are essential to the ability to adapt to exercise and other forms of stress. They also help maintain fairly consistent plasma glucose concentrations even during long periods without ingestion of food. **Cortisol**, also known as hydrocortisone, is the major corticosteroid. It is responsible for about 95% of all glucocorticoid activity in the body. Cortisol

- stimulates gluconeogenesis to ensure an adequate fuel supply;
- increases mobilization of FFAs, making them more available as an energy source;
- decreases glucose utilization, sparing it for the brain;
- stimulates protein catabolism to release amino acids for use in repair, enzyme synthesis, and energy production;
- acts as an anti-inflammatory agent;
- depresses immune reactions; and
- increases the vasoconstriction caused by epinephrine.

We discuss cortisol's important role in exercise later in this chapter when we consider the regulation of glucose and fat metabolism.

Pancreas

The pancreas is located behind and slightly below the stomach. Its two major hormones are insulin and glucagon. The balance of these two opposing hormones provides the major control of plasma glucose concentration. When plasma glucose is elevated (**hyperglycemia**), as occurs after a meal, the pancreas releases **insulin** into the blood. Among its actions, insulin

- facilitates glucose transport into the cells, especially muscle fibers;
- promotes glycogenesis; and
- inhibits gluconeogenesis.

Insulin's main function is to reduce the amount of glucose circulating in the blood. But it is also involved in protein and fat metabolism, promoting cellular uptake of amino acids and enhancing synthesis of protein and fat.

The pancreas secretes **glucagon** when the plasma glucose concentration falls below normal concentrations (**hypoglycemia**). The effects of glucagon generally oppose those of insulin. Glucagon promotes increased breakdown of liver glycogen to glucose (glycogenolysis) and increased gluconeogenesis, both of which increase plasma glucose levels.

During exercise lasting 30 min or longer, the body attempts to maintain plasma glucose concentrations;

however, insulin concentrations tend to decline. The ability of insulin to bind to its receptors on muscle cells increases during exercise, due in large part to increased blood flow to muscle. This increases the body's sensitivity to insulin and reduces the need to maintain high plasma insulin concentrations for transporting glucose into the muscle cells. Plasma glucagon, on the other hand, shows a gradual increase throughout exercise. Glucagon primarily maintains plasma glucose concentrations by stimulating liver glycogenolysis. This increases glucose availability to the cells, maintaining adequate plasma glucose concentrations to meet increased metabolic demands. The responses of these hormones are usually blunted in trained individuals, and those who are well trained are better able to maintain plasma glucose concentrations.

Regulation of Carbohydrate Metabolism During Exercise

As we learned in chapter 2, the heightened energy demands of exercise require that more glucose be made available to the muscles. Because glucose is stored in the body as glycogen, primarily in the muscles and the liver, glycogenolysis must increase to free the glucose from this storage form. Glucose freed from the liver enters the blood to circulate throughout the body, allowing it access to active tissues. Plasma glucose concentration also can be increased through gluconeogenesis, the production of new glucose from noncarbohydrate sources like lactate, amino acids, and glycerol.

Regulation of Plasma Glucose Concentration

The plasma glucose concentration during exercise depends on a balance between glucose uptake by exercising muscles and its release by the liver. Four hormones work to increase the circulating plasma glucose:

- Glucagon
- Epinephrine
- Norepinephrine
- Cortisol

At rest, glucose release from the liver is facilitated by glucagon, which promotes both liver glycogen breakdown and glucose formation from amino acids. During exercise, glucagon secretion increases, as does the rate of catecholamine release from the adrenal medulla; these three hormones (glucagon, epinephrine, and norepinephrine) work in concert to further increase glycogenolysis. After a slight initial drop, cortisol concentration increases during the first 30 to 45 min of exercise. Cortisol increases protein catabolism, freeing amino acids to be used within the liver for gluconeogenesis. Thus, all four of these hormones can increase plasma glucose by enhancing the processes of glycogenolysis (breakdown of glycogen) and gluconeogenesis (making glucose from other substrates). In addition to the effects of the four major glucose-controlling hormones, GH increases mobilization of FFAs and decreases cellular uptake of glucose, so less glucose is used by the cells and more remains in circulation. The thyroid hormones promote glucose catabolism and fat metabolism.

The amount of glucose released by the liver depends on both exercise intensity and duration. As intensity increases, so does the rate of catecholamine release. This can cause the liver to release more glucose than is being taken up by the active muscles. Consequently, during or shortly after an explosive, short-term sprint, blood glucose concentrations may be 40% to 50% above the resting value, since glucose is released by the liver at a greater rate than the rate of uptake by the muscles.

The greater the exercise intensity, the greater the catecholamine release, and thus the rate of glycogenolysis is significantly increased. This process occurs not only in the liver but also in the muscle. Glucose released from the liver enters the blood, where it becomes available to the muscle fibers. But the muscle has a more readily available source of glucose: its own glycogen stores. The muscle uses its own glycogen stores before using the plasma glucose during explosive, short-term exercise. Glucose released from the liver is not used as readily, so it remains in the circulation, elevating the plasma glucose. Following exercise, plasma glucose concentration decreases as the glucose enters the muscle to replenish the depleted muscle glycogen stores (glycogenolysis).

During exercise bouts that last for several hours, however, the rate of liver glucose release more closely matches the muscles' needs, keeping plasma glucose at or only slightly above the resting concentrations. As muscle uptake of glucose increases, the liver's rate of glucose release also increases. In most cases, plasma glucose does not begin to decline until late in the activity as liver glycogen stores become depleted, at which time the glucagon concentration increases significantly. Glucagon and cortisol together enhance gluconeogenesis, providing more fuel.

Figure 4.4 illustrates the changes in plasma concentrations of epinephrine, norepinephrine, glucagon, cortisol, and glucose during 3 h of cycling. Although the hormonal regulation of glucose remains intact throughout such long-term activities, the liver's glycogen supply may become limiting and the liver's rate of glucose release may be unable to keep pace with the muscles' rate of glucose uptake. Under this condition, plasma glucose may decline despite strong hormonal

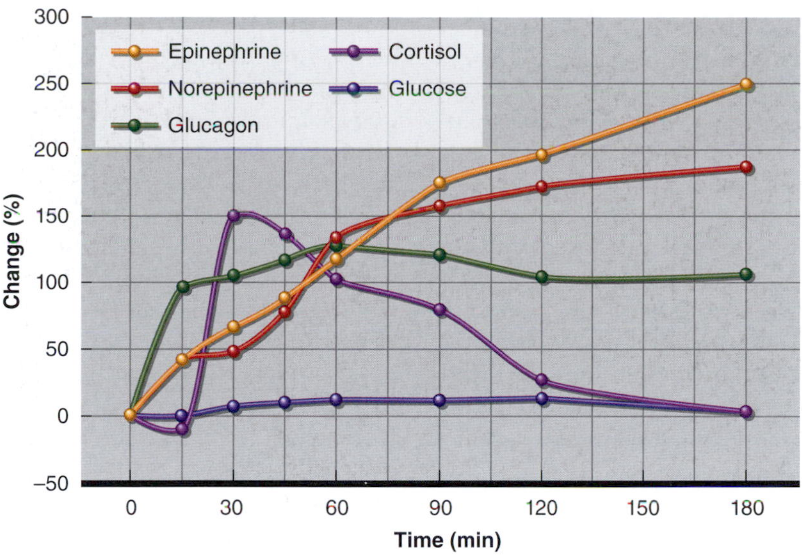

FIGURE 4.4 Changes (as a percentage of preexercise values) in plasma concentrations of epinephrine, norepinephrine, glucagon, cortisol, and glucose during 3 h of cycling at 65% $\dot{V}O_2$.

stimulation. Glucose ingestion during the activity can play a major role in maintaining plasma glucose concentrations.

Glucose Uptake by Muscle

Merely releasing sufficient amounts of glucose into the blood does not ensure that the muscle cells will have enough glucose to meet their energy demands. Not only must the glucose be released and delivered to these cells, it also must be taken up by the cells. Transport of glucose through the cell membranes and into muscle cells is controlled by insulin. Once glucose is delivered to the muscle, insulin facilitates its transport into the fibers.

Surprisingly, as seen in figure 4.5, plasma insulin concentration tends to decrease during prolonged exercise, despite a slight increase in plasma glucose concentration and glucose uptake by muscle. This apparent contradiction between the plasma insulin concentrations and the muscles' need for glucose serves as a reminder that a hormone's activity is determined not only by its concentration in the blood but also by a cell's sensitivity to that hormone. In this case, the cell's sensitivity to insulin is at least as important as the concentration of circulating hormone. Exercise may enhance insulin's binding to receptors on the muscle fiber, thereby reducing the need for high concentrations of plasma insulin to transport glucose across the muscle cell membrane into the cell. This is important, because during exercise, four hormones are working to release glucose from its storage sites and create new glucose. High insulin concentrations would oppose their action, preventing this needed increase in plasma glucose supply.

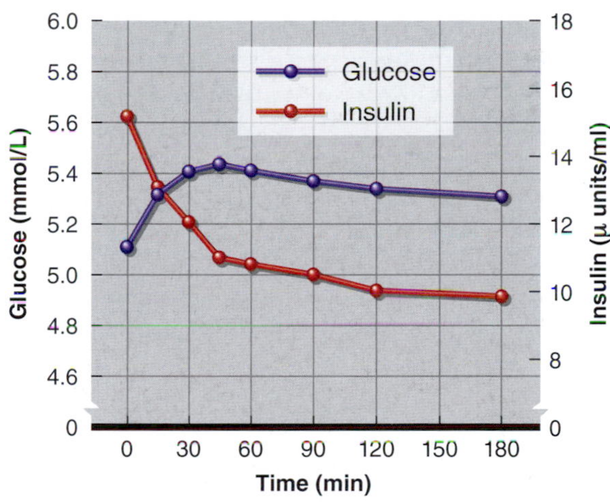

FIGURE 4.5 Changes in plasma concentrations of glucose and insulin during prolonged cycling at 65% to 70% of $\dot{V}O_2$. Note the gradual decline in insulin throughout the exercise, suggesting an increased sensitivity to insulin during prolonged effort.

CNS–Endocrine System Interaction

The central nervous system (CNS) integrates the activities of the nervous and endocrine systems. Therefore, it is not surprising that the CNS is involved in the regulation of carbohydrate metabolism through the sensing of hormones (especially insulin) and nutrients (including glucose, fatty acids, and amino acids).

The actions of insulin on the CNS were clarified through studies using a mouse model of insulin resistance, a condition commonly associated with obesity.[4] In this model, insulin signaling directly to neurons in the brain regulated how the tissues elsewhere around the body regulated glucose metabolism. Other studies have similarly demonstrated the important regulatory actions of the CNS on insulin's control of carbohydrate metabolism throughout the body. In these studies, the researchers directly injected glucose into areas of the brain to specifically stimulate receptors sensitive to glucose. Then they measured the hormones that regulate glucose metabolism as well as how much glucose was taken up and stored by the liver and muscle, respectively.[7] By injecting glucose into the brain to examine central signaling and measuring glucose uptake throughout the body, they were able to show that the brain itself is indeed sensitive to glucose and helps to control the hormones released throughout the body in the regulation of carbohydrate metabolism.

Other hormones have a similar integration with the CNS. *Leptin* is a hormone that is released by adipose tissue in response to feeding, suppressing food intake. It also acts through specific CNS neurons called pro-opiomelanocortin (POMC) neurons to decrease glucose production in the liver since more glucose is not required after feeding. *Glucagon-like peptide 1* (GLP-1), a hormone released in the gut that signals β-cells in the pancreas to release insulin, also works through CNS POMC cells to decrease liver glucose production through both a decrease in gluconeogenesis and increased glycogenolysis. The integration of these hormonal effects through the CNS and subsequent peripheral actions is illustrated in figure 4.6.

Within the brain itself, glucose regulation is particularly important because glucose is the only substrate that can be used for the brain's metabolism. Neuronal activity is tightly coupled with glucose utilization, and neurons preferentially use glucose derived from lactate (see chapter 2) as an oxidative fuel source.[13] As in exercising muscle, lactate can be shuttled between cells in the brain to support oxidative metabolism.[8] Together these findings illustrate the important role of the CNS in regulating hormones associated with carbohydrate metabolism and glucose homeostasis, both within the CNS and throughout the body.

Regulation of Fat Metabolism During Exercise

Free fatty acids are a primary source of energy at rest and during prolonged endurance exercise. They are derived from triglycerides through the action of the enzyme lipase, which breaks down triglycerides into FFA and glycerol. Although fat generally contributes less than carbohydrate does to muscles' energy needs during most bouts of exercise, mobilization and oxidation of FFAs are critical to performance in endurance exercise. During such prolonged activity, carbohydrate reserves become depleted, and muscle must rely more heavily on the oxidation of fat for energy production. When carbohydrate reserves are low (low plasma glucose and low muscle glycogen), the endocrine system can accelerate the oxidation of fats (lipolysis), thus ensuring that muscles' energy needs can be met.

Free fatty acids are stored as triglycerides in adipose tissue and within muscle fibers. Adipose tissue

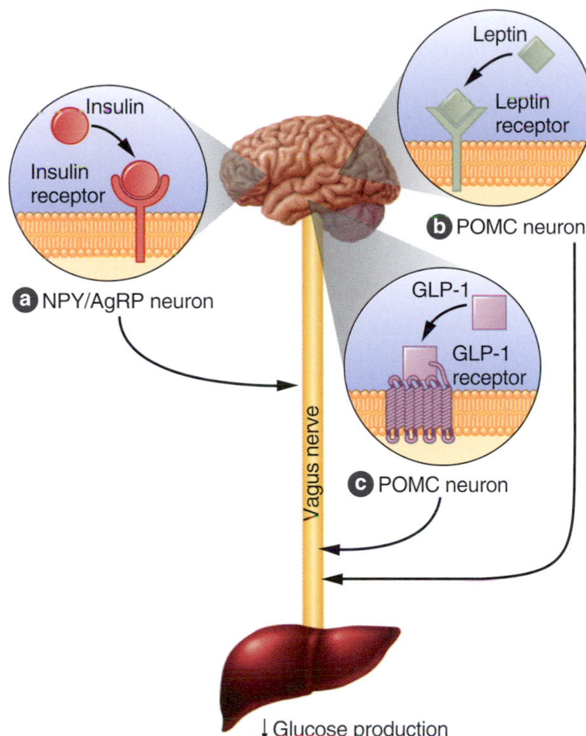

FIGURE 4.6 Hormones secreted throughout the peripheral tissues in the body, including the gastrointestinal tract and the pancreas, stimulate specific receptors in the hypothalamus to control glucose metabolism in the liver. *(a)* Insulin released by the β-cells in the pancreas acts through appetite-stimulating (NPY/AgRP) neurons in the arcuate nucleus of the hypothalamus. These neurons are stimulated by the peptide neurotransmitter neuropeptide Y (NPY) and release agouti-related peptide; insulin receptors are also present on these specialized neurons. *(b and c)* The pro-opiomelanocortin (POMC) neurons are stimulated by both leptin and glucagon-like peptide 1 (GLP-1). Together these hormones act on neurons in the brain, signaling through the vagus nerve to the liver to decrease glucose production.

Based on Lam et al. (2005).

triglycerides, once broken down to release the FFAs, must be transported to the muscle fibers. The rate of FFA uptake by active muscle correlates highly with the plasma FFA concentration. Increasing this concentration would increase cellular uptake of the FFA. Therefore, the rate of triglyceride breakdown may determine, in part, the rate at which muscles use fat as a fuel source during exercise.

The rate of lipolysis is controlled by at least five hormones:

- (Decreased) insulin
- Epinephrine
- Norepinephrine
- Cortisol
- Growth hormone

The major factor responsible for adipose tissue lipolysis during exercise is a fall in circulating insulin. Lipolysis is also enhanced through the elevation of epinephrine and norepinephrine. In addition to having a role in gluconeogenesis, cortisol accelerates the mobilization and use of FFAs for energy during exercise. Plasma cortisol concentration peaks after 30 to 45 min of exercise and then decreases to near-normal levels. But the plasma FFA concentration continues to increase throughout the activity, meaning that lipase continues to be activated by other hormones. The hormones that continue this process are the catecholamines and GH. The thyroid hormones also contribute to the mobilization and metabolism of FFAs but to a much lesser degree.

Thus, the endocrine system plays a critical role in regulating ATP production during exercise as well as controlling the balance between carbohydrate and fat metabolism.

In Review

- Plasma glucose concentration is increased by the combined actions of glucagon, epinephrine, norepinephrine, and cortisol. These hormones promote glycogenolysis and gluconeogenesis, thus increasing the amount of glucose available for use as a fuel source. This is important during exercise, particularly long-duration or high-intensity exercise, when blood glucose concentrations might otherwise decline.
- Insulin allows circulating glucose to enter the cells, where it can be used for energy production. But insulin concentrations decline during prolonged exercise, indicating that exercise increases cell sensitivity to insulin so that less of the hormone is required during exercise than at rest.
- When carbohydrate reserves are low, the body turns more to fat oxidation for energy and lipolysis increases. This process is facilitated by a decreased insulin concentration and increased concentrations of epinephrine, norepinephrine, cortisol, and GH.

Hormonal Regulation of Fluid and Electrolytes During Exercise

Fluid balance during exercise is critical for optimal metabolic, cardiovascular, and thermoregulatory function. At the onset of exercise, water shifts from the plasma volume to the interstitial and intracellular spaces. This water shift is specific to the amount of muscle that is active and the intensity of effort. Metabolic by-products begin to accumulate in and around the muscle fibers, increasing the osmotic pressure there. Water then moves passively into these areas by diffusion. Also, increased muscular activity increases blood pressure, which in turn drives water out of the blood (hydrostatic forces). In addition, sweating increases during exercise. The combined effect of these actions is that plasma volume decreases. For example, prolonged running at approximately 75% of $\dot{V}O_{2max}$ decreases plasma volume by 5% to 10%. Reduced plasma volume can decrease blood pressure and increase the strain on the heart to pump blood to the working muscles. Both of these effects can impede athletic performance.

Endocrine Glands Involved in Fluid and Electrolyte Homeostasis

The endocrine system plays a major role in monitoring fluid levels and electrolyte balance, especially that of sodium. The two major endocrine glands involved in these processes are the posterior pituitary and the adrenal cortex. Additionally, the kidneys not only serve as the primary target organ for hormones released by these glands but also function as endocrine glands themselves.

Posterior Pituitary

The pituitary's posterior lobe is an outgrowth of neural tissue from the hypothalamus. For this reason, it is also referred to as the neurohypophysis. It secretes two hormones: **antidiuretic hormone (ADH)**—also called vasopressin or arginine vasopressin—and oxytocin. Both of these hormones are actually produced in the hypothalamus, travel through nerves, and are stored in vesicles within nerve endings in the posterior pituitary. These hormones are released into capillaries as needed in response to neural impulses from the hypothalamus.

Of the two posterior pituitary hormones, only ADH is known to play an important role during exercise. Antidiuretic hormone promotes water conservation by increasing water reabsorption by the kidneys. As a result, less water is excreted in the urine, creating an "antidiuresis."

Muscular activity and sweating cause electrolytes to become concentrated in the blood plasma as more fluid, compared to electrolytes, leaves the plasma. This is called **hemoconcentration**, and it increases the plasma **osmolality**. Osmolality refers to the ionic concentration of dissolved substances in the plasma. The presence of dissolved molecules and minerals in various body fluid compartments (i.e., intracellular, plasma, and interstitial spaces) generates an osmotic pressure or attraction to retain water within a compartment. The amount of osmotic pressure exerted by a body fluid is proportional to the number of molecular particles (osmoles, or Osm) in solution. A solution that has 1 Osm of solute dissolved in each kilogram (the weight of a liter) of water is said to have an osmolality of 1 osmole per kilogram (1 Osm/kg), whereas a solution that has 0.001 Osm/kg has an osmolality of 1 milliosmole per kilogram (1 mOsm/kg). Normally, body fluids have an osmolality of 300 mOsm/kg. Increasing the osmolality of the solutions in one body compartment generally causes water to be drawn away from adjacent compartments that have a lower osmolality (i.e., more water).

An increased plasma osmolality is the primary physiological stimulus for ADH release. The increased osmolality is sensed by osmoreceptors in the hypothalamus. A second and related stimulus for ADH release is a low plasma volume sensed by baroreceptors in the cardiovascular system. In response to either stimulus, the hypothalamus sends neural impulses to the posterior pituitary, stimulating ADH release. The ADH enters the blood, travels to the kidneys, and promotes water retention in an effort to dilute the plasma electrolyte concentration back to normal levels. This hormone's role in conserving body water minimizes the extent of water loss and therefore the risk of severe dehydration during periods of heavy sweating and hard exercise. Figure 4.7 illustrates this process.

Adrenal Cortex

A group of hormones called **mineralocorticoids**, secreted from the adrenal cortex, maintain electrolyte balance, especially that of sodium (Na^+) and potassium (K^+), in the extracellular fluids. **Aldosterone** is the major mineralocorticoid, responsible for at least 95% of all mineralocorticoid activity. It works primarily by promoting renal reabsorption of sodium, thus causing the body to retain sodium. When sodium is retained, so is water, which follows the osmotic gradient; thus, aldosterone, like ADH, results in water retention. Sodium retention also enhances potassium excretion, so aldosterone plays a role in potassium balance as well. For these reasons, aldosterone secretion is stimulated by many factors, including decreased plasma sodium, decreased blood volume, decreased blood pressure, and increased plasma potassium concentration.

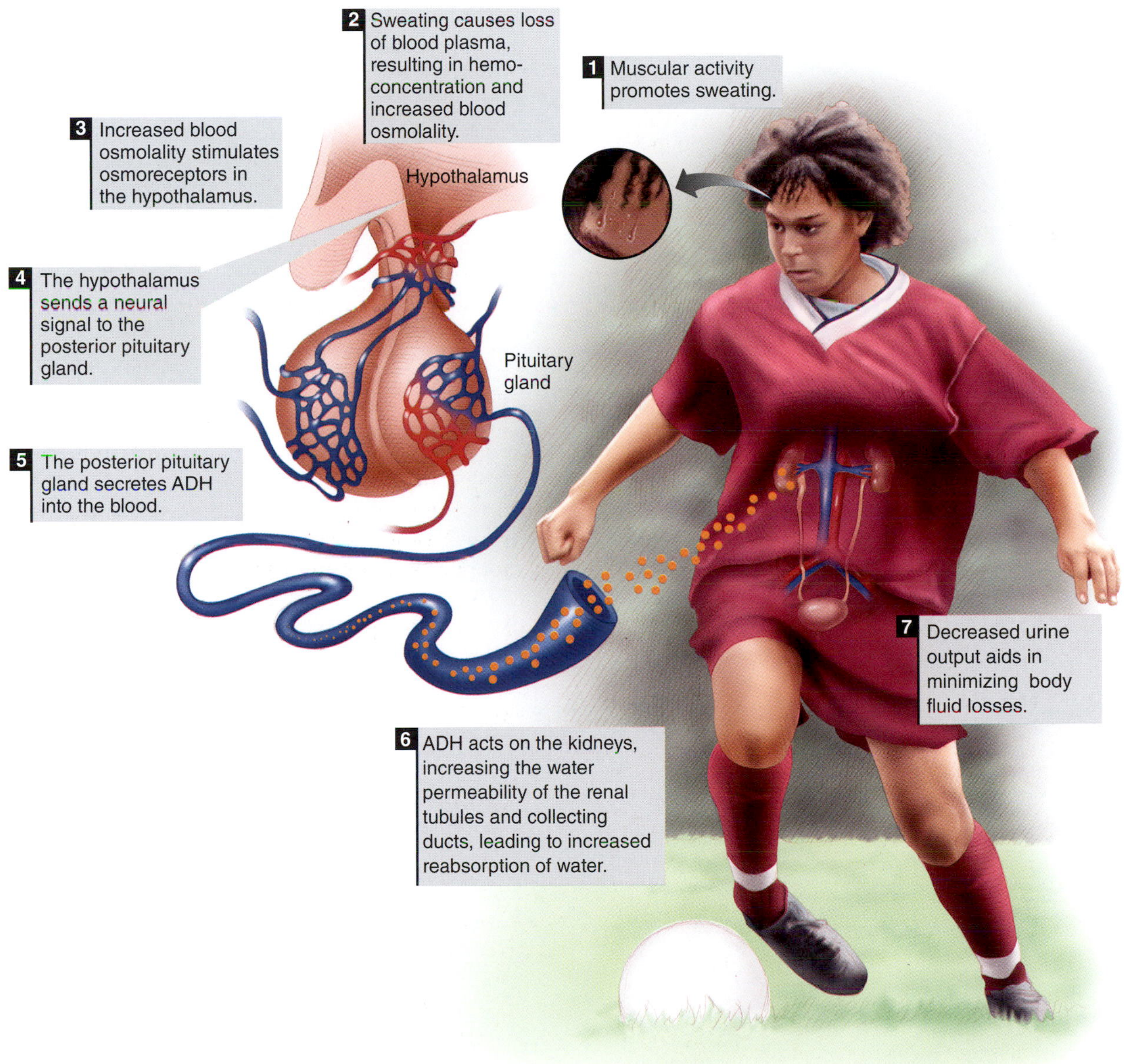

FIGURE 4.7 The mechanism by which antidiuretic hormone (ADH) helps to conserve body water.

The Kidneys as Endocrine Organs

Although the kidneys are not typically considered major endocrine organs, they do release two important hormones. The kidneys play a role in determining the aldosterone concentration in the blood. While the primary regulator of aldosterone release is plasma electrolyte concentration, a second set of hormones also determines aldosterone concentration and thus helps regulate body fluid balance. In response to a fall in blood pressure or plasma volume, blood flow to the kidneys decreases. Stimulated by activation of the sympathetic nervous system, the kidneys release **renin**. Renin is an enzyme that is released into the circulation, where it converts a molecule called angiotensinogen to angiotensin I. Angiotensin I is subsequently converted to its active form, angiotensin II, in the lungs with the aid of an enzyme, **angiotensin-converting enzyme (ACE)**. Angiotensin II stimulates aldosterone release from the adrenal cortex for sodium and water resorption at the kidneys. Figure 4.8 shows the mech-

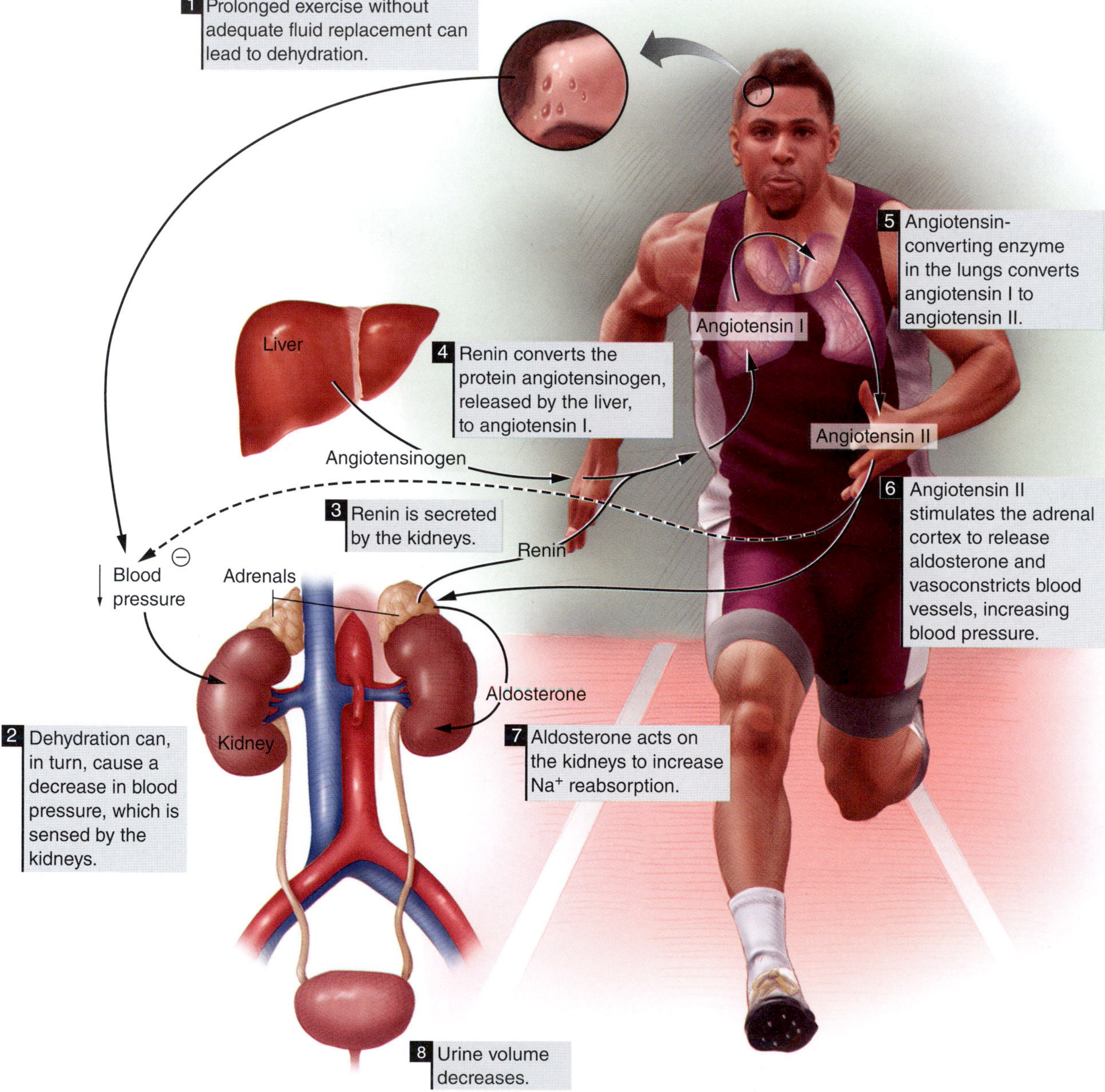

FIGURE 4.8 Water loss from plasma during exercise leads to a sequence of events that promotes sodium (Na^+) and water reabsorption from the renal tubules, thereby reducing urine production. In the hours after exercise when fluids are consumed, the elevated aldosterone concentration causes an increase in the extracellular volume and an expansion of plasma volume.

anism involved in renal control of blood pressure, the **renin-angiotensin-aldosterone mechanism**. In addition to stimulating aldosterone release from the adrenal cortex, angiotensin II causes blood vessels to constrict. Because ACE catalyzes the conversion of angiotensin I to angiotensin II, ACE inhibitors are sometimes prescribed for individuals with hypertension, since relaxation of the blood vessels lowers blood pressure.

Recall that aldosterone's primary action is to promote sodium reabsorption in the kidneys. Because water follows sodium, this renal conservation of sodium causes the kidneys to also retain water. The net effect is to conserve the body's fluid content, thereby minimizing the loss of plasma volume while keeping blood pressure near normal. Figure 4.9 illustrates the changes in plasma volume and aldosterone concentrations during 2 h of exercise. The hormonal influences of ADH and aldosterone persist for up to 48 h after exercise, reducing urine production and protecting the body from further dehydration.

The kidneys also release a hormone called erythropoietin. **Erythropoietin (EPO)** regulates red blood cell (erythrocyte) production by stimulating bone marrow cells. The red blood cells are essential for transporting oxygen to the tissues and removing carbon dioxide, so this hormone is extremely important in our adaptation to training and altitude.

Most athletes involved in heavy training have an expanded plasma volume, which dilutes various blood constituents. As proteins leave working muscle, they reenter the plasma through the lymphatic system, and water follows. This is a relatively short-term phenomenon, and new protein synthesis eventually supports this expanded plasma volume. During the early phases of plasma volume expansion, however, hemoglobin concentration decreases; that is, a **hemodilution** occurs.

The actual amount of hemoglobin has not changed; it is simply diluted. For this reason, some athletes who actually have normal hemoglobin concentrations may appear to be anemic as a consequence of Na^+-induced hemodilution. This condition, not to be confused with true anemia, can be remedied with a few days of rest, allowing time for aldosterone concentrations to return to normal and for the kidneys to unload the extra Na^+ and water.

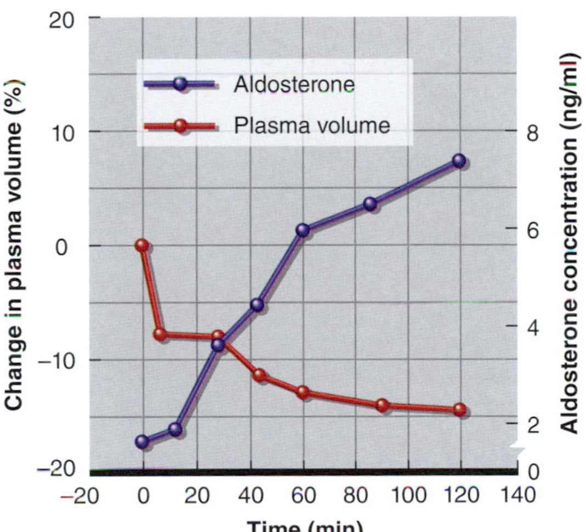

FIGURE 4.9 Changes in plasma volume and aldosterone concentration during 2 h of cycling exercise. Note that plasma volume declines rapidly during the first few minutes of exercise and then shows a smaller rate of decline despite large sweat losses. Plasma aldosterone concentration, on the other hand, increases rather steadily throughout the exercise.

In Review

- Loss of fluid (plasma) from the blood results in a concentration of the constituents of the blood, a phenomenon referred to as hemoconcentration. Conversely, a gain of fluid in the blood results in a dilution of the constituents of the blood, which is referred to as hemodilution.
- The presence of dissolved particles in body fluid compartments generates an osmotic pressure or attraction to retain water. The osmotic pressure is proportional to the number of molecular particles in solution. A solution that has 1 osmole of solute dissolved in each kilogram (the weight of a liter) of water is said to have an osmolality of 1 osmole per kilogram (1 Osm/kg).
- Body fluids normally have an osmolality of 300 mOsm/kg. Increasing the osmolality of the solutions in one body compartment generally causes water to be drawn away from adjacent compartments.
- The two primary hormones involved in the regulation of fluid balance are ADH and aldosterone.
- Antidiuretic hormone is released in response to increased plasma osmolality. When osmoreceptors in the hypothalamus sense this increase, the hypothalamus triggers ADH release from the posterior pituitary. Low blood volume is a secondary stimulus for ADH release.
- Antidiuretic hormone acts on the kidneys, directly promoting water reabsorption and thus fluid conservation. As more fluid is resorbed, plasma volume increases and plasma osmolality decreases.
- When plasma volume or blood pressure decreases, the kidneys release an enzyme called renin that converts angiotensinogen into angiotensin I, which later becomes angiotensin II in the lung

circulation. Angiotensin II is a powerful constrictor of blood vessels and increases peripheral resistance, increasing the blood pressure.

- Angiotensin II also triggers the release of aldosterone from the adrenal cortex. Aldosterone promotes sodium reabsorption in the kidneys, which in turn causes water retention, thus minimizing the loss of plasma volume.

Hormonal Regulation of Caloric Intake

The regulation of appetite, the sensations of hunger and satiety, and the feeling of fullness are part of a complex system that involves hormonal signaling from all over the body, including the gastrointestinal system and fat cells. Food intake is primarily under the control of the hypothalamus with some input from higher brain centers. The satiety area of the brain is located in the ventromedial nucleus, while the hunger center is located in the lateral hypothalamus. The hypothalamus, as it does for many aspects of homeostasis, integrates neural and hormonal signals for both the short- and long-term regulation of eating behavior and calorie intake.

Hormones that influence these brain centers are synthesized in, and released from, peripheral tissues including the gut and fat cells (adipocytes). These hormones can be categorically split into those that are anorexigenic, meaning that they suppress appetite, and those that are orexigenic, meaning that they stimulate appetite. The main hormones that regulate appetite and satiety are cholecystokinin, leptin, peptide YY, GLP-1, insulin, and ghrelin.

Gastrointestinal Tract Hormones

Short-term control of food intake is regulated by plasma concentrations of nutrients including amino acids, glucose, and lipids. However, another significant influence on short-term regulation of food intake involves hormones released in the gastrointestinal (GI) tract. Gastrointestinal distention caused by a full stomach triggers the release of the hormone **cholecystokinin (CCK)**, which stimulates afferent fibers of the vagus nerve to send signals to the brain to suppress hunger. In addition, other hormones including GLP-1 and peptide YY (PYY) are secreted from the large and small intestines during and after eating. These hormones travel through the blood to the brain where they suppress hunger. Peptide YY also acts on the hypothalamus to inhibit

RESEARCH PERSPECTIVE 4.2

Endurance Training for More Red Blood Cells

The direct association between endurance training and increased red blood cell volume was first discovered in 1949.[6] This adaptation to endurance training, called erythropoiesis, increases oxygen delivery to the exercising muscle by increasing the number of red blood cells available to carry oxygen in addition to increasing the blood volume pumped with every beat of the heart. Increasing red blood cell volume is a fundamental component of the increase in aerobic capacity (maximal oxygen uptake) that occurs with regular endurance exercise training (discussed further in chapter 11). Although erythropoiesis is a central mechanism by which endurance training adaptations occur, relatively little is known about how red blood cell volume expansion is regulated during repeated bouts of endurance training.

A recent study conducted in Switzerland examined erythropoiesis and its physiological regulators during an 8-week endurance-training program in a group of healthy young men and women.[12] Researchers measured body composition, heart rate, blood pressure, maximal exercise capacity, total blood volume, red blood cell volume, and erythropoietin (a hormone that stimulates red blood cell production) concentrations in the blood throughout the 8-week training period. At the end of the training period, average red blood cell volume had doubled and maximal exercise capacity had increased by ~10%. Across the 8 weeks of training, total blood volume increased during week 2 and remained high throughout; erythropoietin also increased in week 2 but subsequently returned to baseline values by week 4. Red blood cell volume was increased by week 4 and continued to increase through week 8, coming after the preceding increases in blood volume and circulating erythropoietin concentration. By week 8, exercise-induced increases in erythropoietin concentration had ended but red blood cell volume continued to increase, suggesting that there are still unexplained mechanisms that control erythropoiesis. These findings provided novel insight into the time course of expansion of red blood cell volume as an adaptation to endurance training, while at the same time uncovering new questions for future research.

gastric motility. Insulin released from the pancreas in response to eating also acts as a satiety hormone.

Conversely, the hormone ghrelin is secreted from the stomach and pancreas when the stomach is empty; it can be thought of as a hunger hormone. **Ghrelin** is transmitted through the blood to the brain where it crosses the blood–brain barrier to act on the hunger areas in the lateral hypothalamus. After eating, ghrelin concentrations decrease.

Adipose Tissue as an Endocrine Organ

In addition to hormones secreted by the stomach and intestines to signal hunger or fullness, additional hormones are secreted by adipocytes (fat cells) that likewise act on the hunger and satiety centers in the hypothalamus. Because the level of these hormones depends on the amount of adipose tissue in the body, which changes slowly, these hormones are more involved in the long-term regulation of food intake. The hormone **leptin** is primarily secreted by fat cells and acts on receptors in the hypothalamus to decrease hunger. Leptin is also an indicator of energy balance, as its circulating concentrations are proportional to body fat. A simple schematic of how leptin and ghrelin interact to modify appetite and satiety is presented in figure 4.10.

A great deal has been discovered about what leptin does in terms of energy balance from a mouse model using mice that lack the ability to make leptin in their fat cells. These mice have a voracious appetite and are massively obese. In obese humans, circulating concentrations of leptin are elevated, but many obese humans are leptin resistant. This suggests that despite an elevated signal that they are in an overfed state, the signal is not being transmitted through the hypothalamus to initiate the sensations of satiety. Interestingly, obese humans also appear to have a dampened ghrelin signal. Researchers are only beginning to understand how hormonal appetite signaling changes with weight gain and obesity. This is critical in order to determine how best to treat obesity, as well as how exercise may influence appetite and satiety hormones.

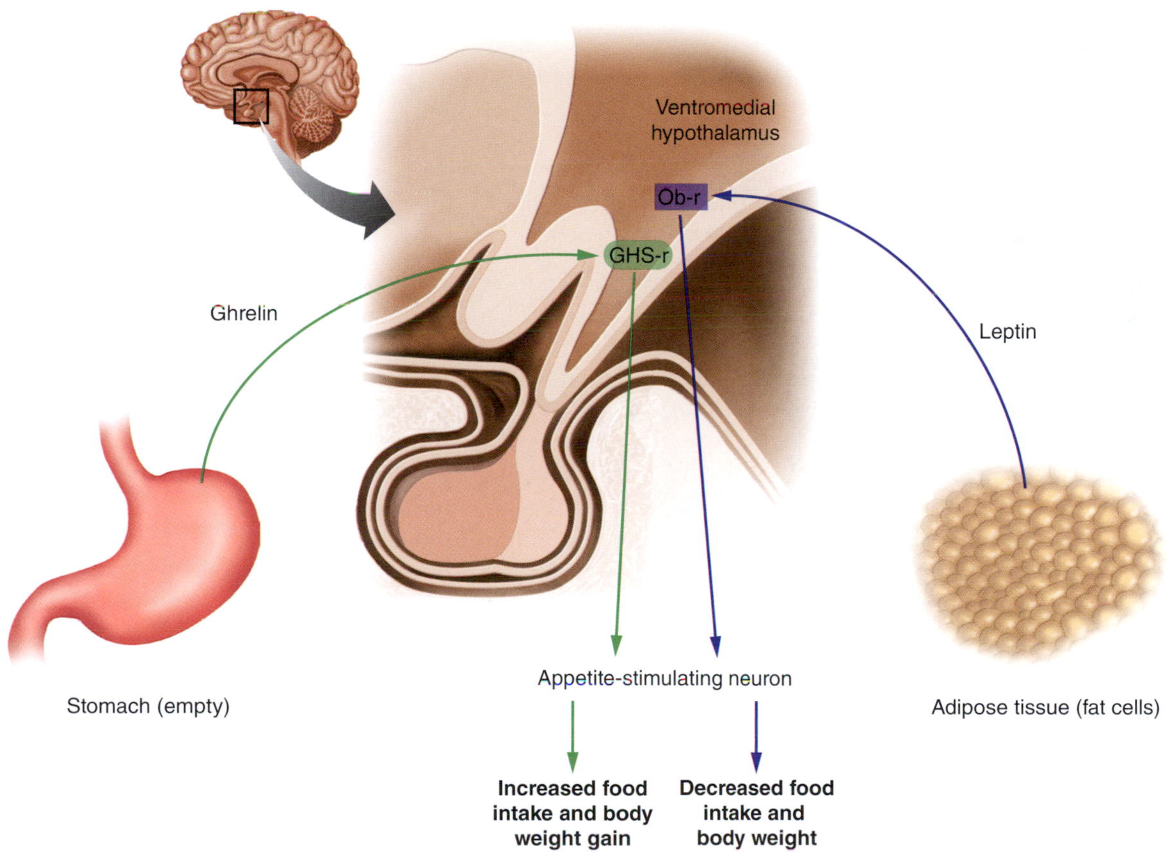

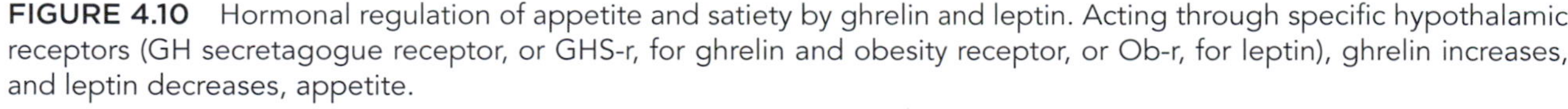
FIGURE 4.10 Hormonal regulation of appetite and satiety by ghrelin and leptin. Acting through specific hypothalamic receptors (GH secretagogue receptor, or GHS-r, for ghrelin and obesity receptor, or Ob-r, for leptin), ghrelin increases, and leptin decreases, appetite.

RESEARCH PERSPECTIVE 4.3

Does Environmental Temperature Alter the Hormones That Control Appetite?

The interactions among exercise, appetite, and energy intake are important for the control and maintenance of energy homeostasis and body weight. Scientifically, these interactions have received widespread attention because they may hold the key to treating excess weight gain and obesity. Leptin and ghrelin are hormones that regulate the perception of hunger and lead to changes in appetite. Leptin (the "satiety hormone") decreases energy intake, while ghrelin (the "hunger hormone") increases energy intake. Both exercise and exposure to extreme temperatures can affect the concentrations of these appetite-regulating hormones. Circulating ghrelin concentration and perception of hunger both decrease immediately after a single bout of moderate- to high-intensity exercise but have no influence on the total energy intake throughout the day. Environmental temperature has an impact on resting metabolic rate. Indigenous populations who live in polar climates have elevated basal metabolic rates, while those who live in tropical climates have decreased basal metabolic rates. Additionally, exercise in a hot environment reduces appetite, while exercise in the cold stimulates appetite; however, it is unknown if these effects involve changes in circulating leptin or ghrelin.

Recently, a group of researchers at the University of Nebraska Omaha conducted an experiment to examine how exercise in different environmental temperatures would affect the leptin and ghrelin responses to exercise.[9] Research subjects completed three separate 1 h bouts of cycling in hot (33 °C [91°F]), neutral (20 °C [68 °F]), and cold (7 °C [45 °F]) air temperatures. The research team measured leptin and ghrelin in blood samples collected preexercise, immediately postexercise, and after a 3 h recovery. Similar to previous studies, circulating leptin was increased immediately after exercise and remained elevated 3 h later. Circulating ghrelin concentrations did not change. Although the researchers hypothesized that there would be larger increases in leptin after exercise in the heat and larger increases in ghrelin after exercise in the cold, there was no effect of air temperature on any hormone measurements. The conclusion from this study was that environmental temperature does not alter the leptin or ghrelin responses to short bouts of aerobic exercise. Future research studies are needed to determine what other variables might affect regulatory hormone and hunger responses following exercise in extreme environments.

Effects of Acute and Chronic Exercise on Satiety Hormones

Acute bouts of moderate- to vigorous-intensity exercise temporarily suppress appetite, likely by decreasing ghrelin and increasing GLP-1 and PYY released from the GI tract.[14] These hormonal changes are most pronounced with aerobic exercise and are not observed after resistance exercise training.[3]

With chronic exercise training comes a shift in energy balance due to the calorie deficit induced by exercise. This is accompanied by a partial compensation to increase hunger and therefore caloric intake through changes in the appetite-regulating hormones. Several studies have observed an increase in plasma PYY concentrations after exercise training, which would be consistent with improved satiety. Counterintuitively, the hunger hormone ghrelin does not change in people who do not lose weight during exercise training but increases significantly in those who do lose weight.[10] In general, appetite and satiety hormones are sensitive to the total energy balance that is modulated by regular exercise. It has been suggested that for elite athletes who need to monitor their energy balance, measures of circulating leptin and ghrelin may help to determine when the athlete is overtraining and may help predict states of energy deficit.[5]

IN CLOSING

In this chapter, we focused on the role of the endocrine system in regulating some of the many physiological processes that accompany exercise. We discussed the role of hormones in regulating the metabolism of glucose and fat for energy metabolism and the role of other hormones in maintaining fluid balance. We touched on some of the relatively new findings about how hormones regulate appetite and calorie consumption. We next look at the related topics of energy expenditure and fatigue during exercise.

KEY TERMS

adrenaline
aldosterone
angiotensin-converting enzyme (ACE)
antidiuretic hormone (ADH)
autocrines
catecholamines
cholecystokinin (CCK)
cortisol
cyclic adenosine monophosphate (cAMP)
direct gene activation
downregulation
epinephrine
erythropoietin (EPO)
ghrelin
glucagon
glucocorticoids
growth hormone (GH)
hemoconcentration
hemodilution
hormone
hyperglycemia
hypoglycemia
inhibiting factors
insulin
insulin resistance
insulin sensitivity
leptin
mineralocorticoids
nonsteroid hormones
osmolality
prostaglandins
releasing factors
renin
renin-angiotensin-aldosterone mechanism
second messenger
steroid hormones
target cells
thyrotropin (TSH)
thyroxine (T_4)
triiodothyronine (T_3)
upregulation

STUDY QUESTIONS

1. What is an endocrine gland, and what are the functions of hormones?
2. Explain the difference between steroid hormones and nonsteroid hormones in terms of their actions at target cells.
3. How can hormones have very specific functions when they reach nearly all parts of the body through the blood?
4. What determines plasma concentrations of specific hormones? What determines their effectiveness on target cells and tissues?
5. Define the terms *upregulation* and *downregulation*. How do target cells become more or less sensitive to hormones?
6. What are second messengers, and what role do they play in hormonal control of cell function?
7. Briefly outline the major endocrine glands, their hormones, and the specific action of these hormones.
8. Which of the hormones outlined in question 7 are of major significance during exercise?
9. What hormones are involved in the regulation of metabolism during exercise? How do they influence the availability of carbohydrates and fats for energy during exercise lasting for several hours?
10. Discuss how the central nervous system helps integrate glucose regulation and the hormones involved in this process.
11. Describe the hormonal regulation of fluid balance during exercise.
12. Discuss the sources and function of the hormones cholecystokinin, leptin, and ghrelin, and explain how they are interrelated.

STUDY GUIDE ACTIVITIES

In addition to the activities listed in the chapter opening outline, two other activities are available in the web study guide, located at

www.HumanKinetics.com/PhysiologyOfSportAndExercise

The **KEY TERMS** activity reviews important terms, and the end-of-chapter **QUIZ** tests your understanding of the material covered in the chapter.

LACERDA
BRA

Energy Expenditure, Fatigue, and Muscle Soreness

In this chapter and in the web study guide

ACTIVITY 5.1 *Evaluating Energy Use* explores six methods of measuring energy use and the advantages and disadvantages of each.

AUDIO FOR FIGURE 5.3 describes the relationship between oxygen uptake and power output during exercise.

AUDIO FOR FIGURE 5.4 describes the relationship between exercise intensity and oxygen uptake in a trained and an untrained subject.

ACTIVITY 5.2 *Energy Expenditure at Rest and During Exercise* reviews the measurement of basal metabolic rate and the common terms used to refer to the best single measurement of cardiorespiratory endurance and aerobic fitness.

ANIMATION FOR FIGURE 5.5 breaks down the concepts of oxygen deficit and EPOC.

AUDIO FOR FIGURE 5.7 describes the effect of running economy.

AUDIO FOR FIGURE 5.8 describes the relationship between subjective fatigue and muscle glycogen concentration.

AUDIO FOR FIGURE 5.10 describes the use of muscle glycogen in various leg muscles during running.

AUDIO FOR FIGURE 5.12 describes changes in muscle pH during sprint exercise and recovery.

ACTIVITY 5.3 *Sources of Fatigue Affecting Athletic Performance* considers various sources of fatigue and how they affect athletic performance.

AUDIO FOR FIGURE 5.13 describes the concept of critical power.

VIDEO 5.1 presents Mike Bergeron on the two types of muscle cramping and the best ways to prevent muscle cramps.

ACTIVITY 5.4 *Muscle Soreness and Cramps* investigates the causes and treatment of muscle soreness and muscle cramps.

The causes and sites of what exercisers and athletes call "fatigue" are as numerous as the sensations that characterize it. Although it is usually considered in terms of how different parts of the body feel—burning lungs, aching legs, unyielding tiredness—researchers have begun to focus on the role the brain plays in fatigue. After all, the brain collects all of the sensory feedback from the body and determines when physical exertion simply cannot continue. Recent research has shown that a fatigued brain can squash successful sport performance as much as tired muscles can. An article by Dr. Samuele Marcora titled "Mental Fatigue Impairs Physical Performance in Humans," published in the *Journal of Applied Physiology*, suggests that perceptions of fatigue cause us to reach our physical limits long before the body does. A group of rugby players exercised to fatigue during an endurance test but were subsequently able to do a 5 s sprint. That is, the brain's perceptions of fatigue stopped the endurance trial before the athletes had reached their physical limits. The brain may be acting as a regulatory brake, slowing down activity before a somatic limit is reached. This does not mean that fatigue is imagined. Rather, its causes are complex (for example, both the muscles and the brain rely on glucose and glycogen for fuel).

One cannot understand exercise physiology without understanding some key concepts about energy expenditure at rest and during exercise. In chapter 2, we discussed the formation of adenosine triphosphate (ATP), the major form of chemical energy stored within, and used by, cells. Adenosine triphosphate is produced from substrates by a series of processes that are known collectively as *metabolism*. In the first half of this chapter we discuss various techniques for measuring the whole-body energy expenditure or metabolic rate, then we describe how energy expenditure varies from basal or resting conditions up to maximal exercise intensities. If exercise is sustained for a prolonged time, eventually muscular contraction cannot be sustained and performance will diminish. This inability to maintain muscle contractions is broadly called fatigue. Fatigue is a complex, multidimensional phenomenon that may or may not result from an inability to maintain metabolism and expend energy. Because fatigue often has a metabolic component, it is discussed in this chapter along with energy expenditure. Muscle soreness and cramping are also discussed as additional factors that can limit exercise.

Measuring Energy Expenditure

The energy used by contracting muscle fibers during exercise cannot be directly measured. But numerous laboratory methods can be used to calculate whole-body energy expenditure at rest and during exercise. Several of these methods have been in use since the early 1900s. Others are new and have only recently been used in exercise physiology.

Direct Calorimetry

Only about 40% of the energy liberated during the metabolism of glucose and fats is used to produce ATP. The remaining 60% is converted to heat, so one way to gauge the rate and quantity of energy production is to measure the body's heat production. This technique is called **direct calorimetry** ("measuring heat"), since the basic unit of heat is the **calorie (cal)**.

This approach was first described by Zuntz and Hagemann in the late 1800s.[29] They developed the **calorimeter**, which consists of an insulated, airtight chamber as illustrated in figure 5.1. The walls of the chamber contain tubing through which water is circulated. In the chamber, the heat produced by the body radiates to the walls and warms the water. The water temperature and temperature changes of the air entering and leaving the chamber vary with the heat the body generates. One's metabolism can be calculated from the resulting values.

Calorimeters are expensive to construct and operate and are slow to generate results, so very few are in actual operation. Their only real advantage is that they measure heat directly, but they have several disadvantages for exercise physiology. Although a calorimeter can provide an accurate measure of total body energy expenditure over time, it cannot follow rapid changes in energy expenditure. Therefore, while direct calorimetry is useful for measuring resting metabolism and energy expended during prolonged, steady-state aerobic exercise, energy metabolism during more typical exercise situations cannot be adequately studied with a direct calorimeter. Second, exercise equipment such as a motor-driven treadmill gives off its own heat that must be accounted for in the calculations. Third, not all heat is liberated from the body; some is stored in the body, causing body temperature to rise. And finally, sweating affects the measurements and the constants used in the calculations of heat produced. Consequently, it is easier and less expensive to quantify energy expenditure by measuring the exchange of oxygen and carbon dioxide that occurs during oxidative phosphorylation.

Indirect Calorimetry

As discussed in chapter 2, oxidative metabolism of glucose and fat—the main substrates for aerobic exercise—uses O_2 and produces CO_2 and water. The rate of

FIGURE 5.1 A direct calorimeter for the measurement of energy expenditure by an exercising human subject. The heat generated by the subject's body is transferred to the air and walls of the chamber (through conduction, convection, and evaporation). This heat produced by the subject—a measure of his or her metabolic rate—is measured by recording the temperature change in the air entering and leaving the calorimeter as well as in the water flowing through its walls.

O_2 and CO_2 exchanged in the lungs normally equals the rate of their usage and release by the body tissues. Based on this principle, energy expenditure can be determined by measuring the respiratory exchange of O_2 and CO_2. This method of estimating total body energy expenditure is called **indirect calorimetry** because heat production is not measured directly.

In order for oxygen consumption to reflect energy metabolism accurately, energy production must be almost completely oxidative. If a large portion of energy is being produced anaerobically, respiratory gas measurements will not reflect all metabolic processes and will underestimate the total energy expenditure. Therefore, this technique is limited to steady-state aerobic activities lasting a few minutes or longer, which fortunately takes into account most daily activities including exercise.

Respiratory gas exchange is determined through measurement of the volume of O_2 and CO_2 that enters and leaves the lungs during a given period of time. Because O_2 is removed from the inspired air in the alveoli and CO_2 is added to the alveolar air, the expired O_2 concentration is less than the inspired, whereas the CO_2 concentration is higher in expired air than in inspired air. Consequently, the differences in the concentrations of these gases between the inspired and the expired air tell us how much O_2 is being taken up and how much CO_2 is being produced by the body. Because the body has only limited O_2 storage, the amount taken up at the lungs accurately reflects the body's use of O_2. Although a number of sophisticated and expensive methods are available for measuring the respiratory exchange of O_2 and CO_2, the simplest and oldest methods (i.e., Douglas bag to collect expired air and chemical analysis of collected gas sample) are probably the most accurate, but they are relatively slow and permit only a few measurements during each session. Modern electronic computer systems for respiratory gas exchange measurements offer the ability to make rapid and repeated measurements.

Notice in figure 5.2 that the gas expired by the subject passes through a hose into a mixing chamber. The subject is wearing a nose clip so that all expired gas is collected from the mouth and none is lost to the air. From the mixing chamber, samples are pumped to electronic oxygen and carbon dioxide analyzers. In this setup, a computer uses the measurements of expired gas (air) volume and the fraction (percentage) of oxygen and carbon dioxide in a sample of that expired air to calculate O_2 uptake and CO_2 production. Sophisticated equipment can do these calculations breath by breath, but calculations are more typically done over discrete time periods lasting from one to several minutes.

Calculating Oxygen Consumption and Carbon Dioxide Production

Using equipment like that shown in figure 5.2, exercise physiologists can measure the three variables needed to calculate the actual volume of oxygen consumed (VO_2) and volume of CO_2 produced (VCO_2). Generally, values are presented as oxygen consumed per minute ($\dot{V}O_2$) and CO_2 produced per minute ($\dot{V}CO_2$). The dot over the V ($\dot{V}$) indicates the rate of O_2 consumption or CO_2 production per minute.

In simplified form, $\dot{V}O_2$ is equal to the volume of O_2 inspired minus the volume of O_2 expired. To calculate the volume of O_2 inspired, we multiply the volume of air inspired by the fraction of that air that is composed of O_2; the volume of O_2 expired is equal to the volume of air expired multiplied by the fraction of the expired air that is composed of O_2. The same holds true for CO_2.

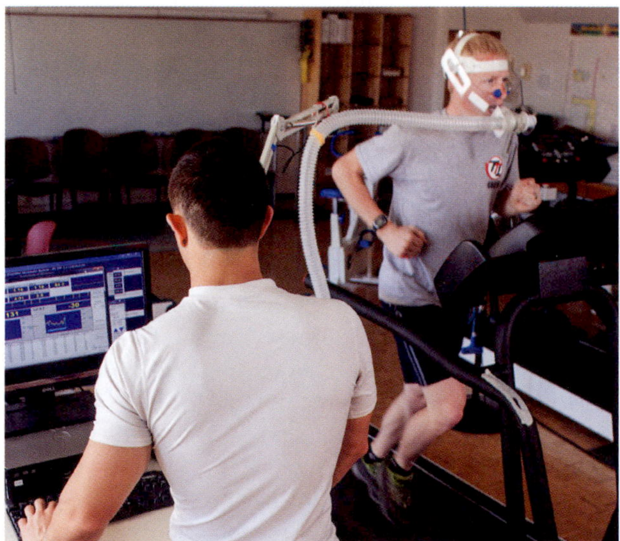

FIGURE 5.2 Typical equipment that is routinely used by exercise physiologists to measure O_2 consumption and CO_2 production. These values can be used to calculate $\dot{V}O_{2max}$ and the respiratory exchange ratio and therefore energy expenditure. Although this equipment is cumbersome and limits movement, smaller versions have recently been adapted for use under a variety of conditions in the laboratory, on the playing field, in industry, and elsewhere.

Thus, calculation of $\dot{V}O_2$ and $\dot{V}CO_2$ requires the following information:

- Volume of air inspired ($\dot{V}_I$)
- Volume of air expired ($\dot{V}_E$)
- Fraction of oxygen in the inspired air (F_IO_2)
- Fraction of CO_2 in the inspired air (F_ICO_2)
- Fraction of oxygen in the expired air (F_EO_2)
- Fraction of CO_2 in the expired air (F_ECO_2)

The oxygen consumption, in liters of oxygen consumed per minute, can then be calculated as follows:

$$\dot{V}O_2 = (\dot{V}_I \times F_IO_2) - (\dot{V}_E \times F_EO_2)$$

The CO_2 production is similarly calculated as follows:

$$\dot{V}CO_2 = (\dot{V}_E \times F_ECO_2) - (\dot{V}_I \times F_ICO_2)$$

These equations provide reasonably good estimates of $\dot{V}O_2$ and $\dot{V}CO_2$. However, the equations are based on the idea that inspired air volume exactly equals expired air volume and there are no changes in gases stored within the body. Since there are differences in gas storage during exercise (discussed next), more accurate equations can be derived from the variables listed.

Haldane Transformation

Over the years, scientists have attempted to simplify the actual calculation of oxygen consumption and CO_2 production. Several of the measurements needed in the preceding equations are known and do not change. The gas concentrations of the three gases that make up inspired air are known: oxygen accounts for 20.93% (or 0.2093), CO_2 accounts for 0.03% (0.0003), and nitrogen accounts for 79.03% (0.7903) of the inspired air. What about the volume of inspired and expired air? Aren't they the same, such that we would need to measure only one of the two?

Inspired air volume equals expired air volume only when the volume of O_2 consumed equals the volume of CO_2 produced. When the volume of oxygen consumed is greater than the volume of CO_2 produced, $\dot{V}_I$ is greater than $\dot{V}_E$. Likewise, $\dot{V}_E$ is greater than $\dot{V}_I$ when the volume of CO_2 produced is greater than the volume of oxygen consumed. However, the one thing that is constant is that the volume of nitrogen inspired ($\dot{V}_IN_2$) is equal to the volume of nitrogen expired ($\dot{V}_EN_2$). Because $\dot{V}_IN_2 = \dot{V}_I \times F_IN_2$ and $\dot{V}_EN_2 = \dot{V}_E \times F_E N_2$, we can calculate $\dot{V}_I$ from $\dot{V}_E$ by using the following equation, which has been referred to as the **Haldane transformation**:

$$(1)\ \dot{V}_I \times F_IN_2 = \dot{V}_E \times F_EN_2,$$

which can be rewritten as

$$(2)\ \dot{V}_I = (\dot{V}_E \times F_E N_2) / F_I N_2.$$

Furthermore, because we are actually measuring the concentrations of O_2 and CO_2 in the expired gases, we can calculate $F_E N_2$ from the sum of $F_E O_2$ and $F_E CO_2$, or

$$(3)\ F_E N_2 = 1 - (F_E O_2 + F_E CO_2).$$

So, in pulling all of this information together, we can rewrite the equation for calculating $\dot{V}O_2$ as follows:

$$\dot{V}O_2 = (\dot{V}_I \times F_I O_2) - (\dot{V}_E \times F_E O_2)$$

By substituting equation 2, we get the following:

$$\dot{V}O_2 = [(\dot{V}_E \times F_E N_2) / (F_I N_2 \times F_I O_2)] - [(\dot{V}_E) \times (F_E O_2)]$$

By substituting known values for $F_I O_2$ of 0.2093 and for $F_I N_2$ of 0.7903, we get the following:

$$\dot{V}O_2 = [(\dot{V}_E \times F_E N_2) / (0.7903 \times 0.2093)] - (\dot{V}_E \times F_E O_2)$$

By substituting equation 3, we get the following:

$$\dot{V}O_2 = \{(\dot{V}_E) \times [1 - (F_E O_2 + F_E CO_2)] \times (0.2093 / 0.7903)\} - (\dot{V}_E \times F_E O_2)$$

or, simplified,

$$\dot{V}O_2 = (\dot{V}_E) \times \{[1 - (F_E O_2 + F_E CO_2)] \times 0.265\} - (\dot{V}_E \times F_E O_2)$$

or, further simplified,

$$\dot{V}O_2 = (\dot{V}_E) \times \{[1 - (F_E O_2 + F_E CO_2)] \times 0.265\} - (F_E O_2).$$

This final equation is the one actually used in practice by exercise physiologists, although computers now do the calculating automatically in most laboratories.

One final correction is necessary. When air is expired, it is at body temperature (BT), is at the prevailing atmospheric or ambient pressure (P), and is saturated (S) with water vapor, or what are referred to as BTPS conditions. Each of these influences would not only add error to the measurement of $\dot{V}O_2$ and $\dot{V}CO_2$ but also would make it difficult to compare measurements made in laboratories at different altitudes, for example. For that reason, every gas volume is routinely converted to its standard temperature (ST: 0 °C or 273 K) and pressure (P: 760 mmHg), dry equivalent (D), or STPD. This is accomplished by a series of correction equations.

Respiratory Exchange Ratio

To estimate the amount of energy used by the body, it is necessary to know the type of food substrate (combination of carbohydrate, fat, protein) being oxidized. The carbon and oxygen contents of glucose, free fatty acids (FFAs), and amino acids differ dramatically. As a result, the amount of oxygen used during metabolism depends on the type of fuel being oxidized. Indirect calorimetry measures the rate of CO_2 release ($\dot{V}CO_2$) and oxygen consumption ($\dot{V}O_2$). The ratio between these two values is termed the **respiratory exchange ratio (RER)**.

$$RER = \dot{V}CO_2 / \dot{V}O_2$$

In general, the amount of oxygen needed to completely oxidize a molecule of carbohydrate or fat is proportional to the amount of carbon in that fuel. For example, glucose ($C_6H_{12}O_6$) contains six carbon atoms. During glucose combustion, six molecules of oxygen are used to produce 6 CO_2 molecules, 6 H_2O molecules, and 32 ATP molecules:

$$6\ O_2 + C_6H_{12}O_6 \rightarrow 6\ CO_2 + 6\ H_2O + 32\ ATP$$

By evaluating how much CO_2 is released compared with the amount of O_2 consumed, we find that the RER is 1.0:

$$RER = \dot{V}CO_2 / \dot{V}O_2 = 6\ CO_2 / 6\ O_2 = 1.0$$

As shown later in the chapter, the RER value varies with the type of fuels being used for energy. Free fatty acids have considerably more carbon and hydrogen but less oxygen than glucose. Consider palmitic acid, $C_{16}H_{32}O_2$. To completely oxidize this molecule to CO_2 and H_2O requires 23 molecules of oxygen:

$$16\ C + 16\ O_2 \rightarrow 16\ CO_2$$

$$32\ H + 8\ O_2 \rightarrow 16\ H_2O$$

Total = $24\ O_2$ needed

$-1\ O_2$ provided by the palmitic acid

$23\ O_2$ must be added

Ultimately, this oxidation results in 16 molecules of CO_2, 16 molecules of H_2O, and 129 molecules of ATP:

$$C_{16}H_{32}O_2 + 23\ O_2 \rightarrow 16\ CO_2 + 16\ H_2O + 129\ ATP$$

Combustion of this fat molecule requires significantly more oxygen than combustion of a carbohydrate molecule. During carbohydrate oxidation, approximately 6.3 molecules of ATP are produced for each molecule of O_2 used (32 ATP per 6 O_2), compared with 5.6 molecules of ATP per molecule of O_2 during palmitic acid metabolism (129 ATP per 23 O_2).

Although fat provides more energy than carbohydrate, more oxygen is needed to oxidize fat than carbohydrate. This means that the RER value for fat is substantially lower than for carbohydrate. For palmitic acid, the RER value is 0.70:

$$RER = \dot{V}CO_2 / \dot{V}O_2 = 16 / 23 = 0.70$$

Once the RER value is determined from the calculated respiratory gas volumes, the value can be compared with a table (table 5.1) to determine the food mixture being oxidized. If, for example, the RER value is 1.0, the cells are using only glucose or glycogen, and each liter of oxygen consumed would generate 5.05 kcal. The oxidation of only fat would yield 4.69 kcal/L of O_2, and the oxidation of protein would yield 4.46 kcal/L of O_2 consumed. Thus, if the muscles were using only glucose and the body were consuming 2 L of O_2/min, then the rate of heat energy production would be 10.1 kcal/min (2 L/min · 5.05 kcal/L).

Limitations of Indirect Calorimetry

While indirect calorimetry is a common and extremely important tool of exercise physiologists, it has some limitations. Calculations of gas exchange assume that the body's O_2 content remains constant and that CO_2 exchange in the lung is proportional to its release from the cells. Arterial blood remains almost completely oxygen saturated (about 98%), even during intense effort, and we can accurately assume that the oxygen being removed from the air we breathe is in proportion to its cellular uptake. Carbon dioxide exchange, however, is less constant. Body CO_2 pools are quite large and can be altered simply by deep breathing or by performance of highly intense exercise. Under these conditions, the amount of CO_2 released in the lung may not represent that being produced in the tissues, so calculations of carbohydrate and fat used based on gas measurements are accurate only at rest or during steady-state exercise.

Use of the RER can also lead to inaccuracies. Recall that protein is not completely oxidized in the body. This makes it impossible to calculate the body's protein use from the RER. As a result, the RER is sometimes referred to as nonprotein RER because it simply ignores any protein oxidation. Traditionally, protein was thought to contribute little to the energy used during exercise, so exercise physiologists felt justified in using the nonprotein RER when making calculations. But more recent evidence suggests that in exercise lasting for several hours, protein may contribute up to 5% of the total energy expended under certain circumstances.

The body normally uses a combination of fuels. Respiratory exchange ratio values vary depending on the specific mixture being oxidized. At rest, the RER value is typically in the range of 0.78 to 0.80. During exercise, though, muscles rely increasingly on carbohydrate for energy, resulting in a higher RER. As exercise intensity increases, the muscles' carbohydrate demand also increases. As more carbohydrate is used, the RER value approaches 1.0.

This increase in the RER value to 1.0 reflects the demands on blood glucose and muscle glycogen, but it also may indicate that more CO_2 is being unloaded from the blood than is being produced by the muscles. At or near exhaustion, lactate accumulates in the blood. The body tries to reverse this acidification by releasing more CO_2. Lactate accumulation increases CO_2 production because excess acid causes carbonic acid in the blood to be converted to CO_2. As a consequence, the excess CO_2 diffuses out of the blood and into the lungs for exhalation, increasing the amount of CO_2 released. For this reason, RER values approaching 1.0 may not accurately estimate the type of fuel being used by the muscles.

Another complication is that glucose production from the catabolism of amino acids and fats in the liver produces an RER below 0.70. Thus, calculations of carbohydrate oxidation from the RER value will be underestimated if energy is derived from this process.

Despite its shortcomings, indirect calorimetry still provides the best estimate of energy expenditure at rest and during aerobic exercise and is widely used in laboratories throughout the world.

TABLE 5.1 **Respiratory Exchange Ratio (RER) as a Function of Energy Derived From Various Fuel Mixtures**

% Kcal from			
Carbohydrates	Fats	RER	Energy (kcal/L O_2)
0	100	0.71	4.69
16	84	0.75	4.74
33	67	0.80	4.80
51	49	0.85	4.86
68	32	0.90	4.92
84	16	0.95	4.99
100	0	1.00	5.05

Isotopic Measurements of Energy Metabolism

In the past, determining an individual's total daily energy expenditure depended on recording food intake over several days and measuring body composition changes during that period. This method, although widely used, is limited by the individual's ability to keep accurate records and by the ability to match the individual's activities to accurate energy costs.

Fortunately, the use of chemical isotopes has expanded our ability to investigate energy metabolism. Isotopes are elements with an atypical atomic weight. They can be either radioactive (radioisotopes) or nonradioactive (stable isotopes). As an example, carbon-12 (^{12}C) has a molecular weight of 12, is the most common natural form of carbon, and is nonradioactive. In contrast, carbon-14 (^{14}C) has two more neutrons than ^{12}C, giving it an atomic weight of 14. ^{14}C is radioactive.

Carbon-13 (^{13}C) constitutes about 1% of the carbon in nature and is used frequently for studying energy metabolism. Because ^{13}C is nonradioactive, it is less easily traced within the body than ^{14}C. But although radioactive isotopes are easily detected in the body, they pose a hazard to body tissues and thus are used infrequently in human research.

^{13}C and other isotopes such as hydrogen-2 (deuterium, or ^2H) are used as tracers, meaning that they can be selectively followed in the body. Tracer techniques involve infusing isotopes into an individual and then following their distribution and movement.

Although the method was first described in the 1940s, studies that used doubly labeled water for monitoring energy expenditure during normal daily living in humans were not conducted until the 1980s. The subject ingests a known amount of water labeled with two isotopes ($^2H_2$ and ^{18}O), hence the term *doubly labeled water*. The deuterium (^2H) diffuses throughout the body's water, and the oxygen-18 (^{18}O) diffuses throughout both the water and the bicarbonate stores (where much of the CO_2 derived from metabolism is stored). The rate at which the two isotopes leave the body can be determined by analysis of their presence in a series of urine, saliva, or blood samples. These turnover rates then can be used to calculate how much CO_2 is produced, and that value can be converted to energy expenditure through the use of calorimetric equations.

Because isotope turnover is relatively slow, energy metabolism must be measured for several weeks. Thus, this method is not well suited for measurements of acute exercise metabolism. However, its accuracy (more than 98%) and low risk make it well suited for determining day-to-day energy expenditure.

In Review

- Direct calorimetry involves a large sophisticated chamber that directly measures heat produced by the body; while it can provide very accurate measures of resting metabolism, it is not a commonly used tool for exercise physiologists.
- Indirect calorimetry involves measuring whole-body O_2 consumption and CO_2 production from expired gases. Since we know the fraction of O_2 and CO_2 in the inspired air, three additional measurements are needed: the volume of air inspired ($\dot{V}_I$) or expired ($\dot{V}_E$), the fraction of oxygen in the expired air (F_EO_2), and the fraction of CO_2 in the expired air (F_ECO_2).
- By calculating the RER (the ratio of CO_2 production to O_2 consumption) and determining the metabolic substrates being oxidized, we can convert $\dot{V}O_2$ into energy expenditure in kilocalories.
- The RER value at rest is usually 0.78 to 0.80. The RER value for the oxidation of fat is 0.70 and is 1.00 for carbohydrates.
- Isotopes can be used to determine metabolic rate over longer periods of time. They are injected or ingested into the body. The rates at which they are cleared can be used to calculate CO_2 production and then caloric expenditure.

Energy Expenditure at Rest and During Exercise

With the techniques described in the previous section, exercise physiologists can measure the amount of energy a person expends in a variety of conditions. This section deals with the body's rates of energy expenditure (metabolic rates) at rest, during submaximal and maximal exercise, and during the period of recovery following an acute exercise bout.

Basal and Resting Metabolic Rates

The rate at which the body uses energy is called the metabolic rate. Estimates of energy expenditure at rest and during exercise are often based on measurement of whole-body oxygen consumption ($\dot{V}O_2$) and its caloric equivalent. At rest, an average person consumes about 0.3 L of O_2/min.

Knowing an individual's $\dot{V}O_2$ allows us to calculate that person's caloric expenditure. Recall that at rest, the body usually burns a mixture of carbohydrate and fat. An RER value of approximately 0.80 is fairly common for most resting individuals eating a mixed diet. The caloric equivalent associated with an RER value of 0.80 is 4.80 kcal per liter of O_2 consumed (see table 5.1). Using these values and an estimate of 0.3 L of O_2/min, we can calculate this individual's caloric expenditure as follows:

$$\begin{aligned}\text{kcal/day} &= \text{liters of } O_2 \text{ consumed per day} \times \text{kcal used per liter of } O_2\\ &= 432\text{ L } O_2\text{/day} \times 4.80\text{ kcal/L } O_2\\ &= 2{,}074\text{ kcal/day}\end{aligned}$$

This value closely agrees with the average resting energy expenditure expected for a 70 kg (154 lb) man. Of course, it does not include the extra energy needed for normal daily activity or any excess energy used for exercise.

One standardized measure of energy expenditure at rest is the **basal metabolic rate (BMR)**. The BMR is the rate of energy expenditure for an individual at rest in a supine position, measured in a thermoneutral environment immediately after at least 8 h of sleep and at least 12 h of fasting. This value reflects the minimum amount of energy required to carry on essential physiological functions.

Because muscle has high metabolic activity, the BMR is directly related to an individual's fat-free mass and is generally reported in kilocalories per kilogram of fat-free mass per minute ($kcal \cdot kg\ FFM^{-1} \cdot min^{-1}$). The higher the fat-free mass, the more total calories expended in a day. Because women tend to have a lower fat-free mass and a greater percent body fat than men, women tend to have a lower BMR than men of a similar weight.

Body surface area also affects BMR. The higher the surface area, the more heat loss occurs from the skin, which raises the BMR because more energy is needed to maintain body temperature. For this reason, the BMR is sometimes reported in kcal per square meter of body surface area per hour ($kcal \cdot m^{-2} \cdot h^{-1}$). Because we are discussing daily energy expenditure, we will use the simpler unit, kcal/day.

Many other factors affect BMR, including these:

- Age: BMR gradually decreases with increasing age, generally because of a decrease in fat-free mass.
- Body temperature: BMR increases with increasing temperature.
- Psychological stress: Stress increases activity of the sympathetic nervous system, which increases the BMR.
- Hormones: For example, increased release of thyroxine from the thyroid gland or epinephrine from the adrenal medulla can both increase the BMR.

Instead of BMR, most researchers measure **resting metabolic rate (RMR)**, which is similar to BMR but does not require the stringent standardized conditions associated with a true BMR. Basal metabolic rate and RMR values are typically within 5% to 10% of each other, with BMR slightly lower, and range from 1,200 to 2,400 kcal/day. But the average total metabolic rate of an individual engaged in normal daily activity ranges from 1,800 to 3,000 kcal. However, the energy expenditure for large athletes engaged in intense training—for example, large football players in two-a-day practice sessions—can exceed 10,000 kcal/day!

Metabolic Rate During Submaximal Exercise

Exercise increases the energy requirement well in excess of RMR. Metabolism increases in direct proportion to the increase in exercise intensity, as shown in figure 5.3*a*. As this subject exercised on a cycle ergometer for 5 min at 50 watts (W), oxygen consumption ($\dot{V}O_2$) increased from its resting value to a steady-state value within 1 min or so. The same subject then cycled for 5 min at 100 W, and again a steady-state $\dot{V}O_2$ was reached in 1 to 2 min. In a similar manner, the subject cycled for 5 min at 150 W, 200 W, 250 W, and 300 W, respectively, and steady-state values were achieved at each power output. The steady-state $\dot{V}O_2$ value represents the energy cost for that specific power output. The steady-state $\dot{V}O_2$ values were plotted against their respective power outputs (right half in figure 5.3*a*), showing clearly that there is a linear increase in the $\dot{V}O_2$ with increases in power output.

From more recent studies, it is clear that the $\dot{V}O_2$ response at higher rates of work does not follow the steady-state response pattern shown in figure 5.3*a* but rather looks more like the graphs presented in figure 5.3*b*. At power outputs above the lactate threshold (the lactate response is indicated by the dashed line in the right half of figure 5.3, *a* and *b*), the oxygen consumption continues to increase beyond the typical 1 to 2 min needed to reach a steady-state value. This increase has been called the slow component of oxygen uptake kinetics.[11] The most likely mechanism for this slow component is an alteration in muscle fiber recruitment patterns, with the recruitment of more type II muscle fibers, which are less efficient (i.e., they require a higher $\dot{V}O_2$ to achieve the same power output).[11]

A similar, but unrelated, phenomenon is referred to as the **$\dot{V}O_2$ drift**, defined as a slow increase in $\dot{V}O_2$ during prolonged, submaximal, constant power output exercise. Unlike the slow component, $\dot{V}O_2$ drift is observed at power outputs well below lactate threshold, and the increase in $\dot{V}O_2$ drift is more gradual. Although not understood completely, $\dot{V}O_2$ drift is likely attributable to an increase in ventilation and effects of increased circulating catecholamines.

Maximal Capacity for Aerobic Exercise

In figure 5.3*a*, it is clear that when the subject cycled at 300 W, the $\dot{V}O_2$ response was not different from that achieved at 250 W. This indicates that the subject had reached the maximal limit of his ability to increase his $\dot{V}O_2$. This value is referred to as aerobic capacity, **maximal oxygen uptake ($\dot{V}O_{2max}$)**. $\dot{V}O_{2max}$ is widely regarded as the best single measurement of **cardiorespiratory endurance** or aerobic fitness. This concept is further illustrated in figure 5.4, which compares the $\dot{V}O_{2max}$ of a trained and an untrained man.

In some exercise settings, as intensity increases, a subject reaches volitional fatigue before a plateau occurs in the $\dot{V}O_2$ response (the criterion for a true $\dot{V}O_{2max}$). In such cases, the highest oxygen uptake achieved is more

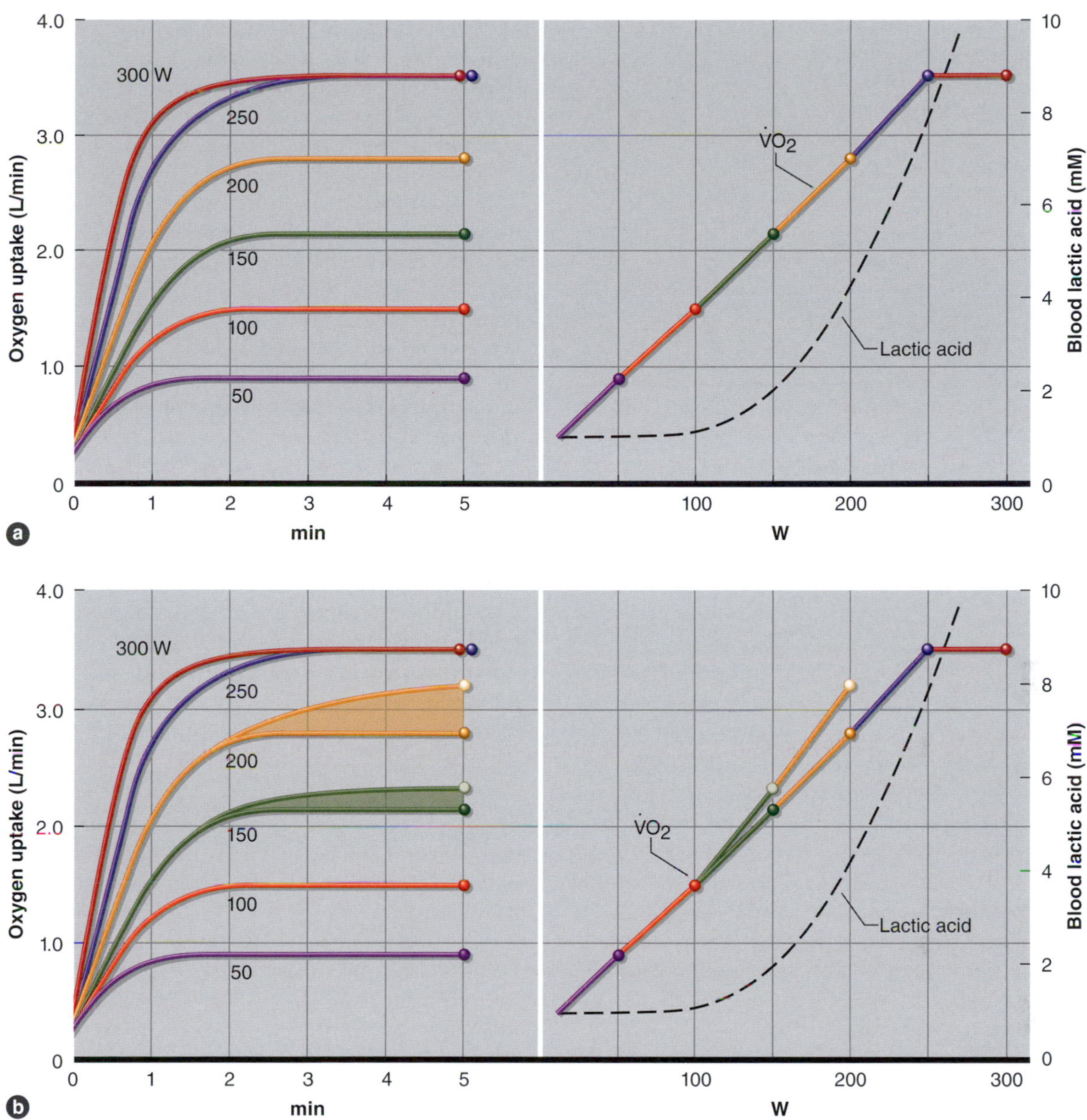

FIGURE 5.3 The increase in oxygen uptake with increasing power output *(a)* as originally proposed by P.-O. Åstrand and K. Rodahl, *Textbook of work physiology: Physiological bases of exercise*, 3rd ed. (New York: McGraw-Hill, 1986), p. 300; and *(b)* as redrawn by Gaesser and Poole (1996, p. 36). See the text for a detailed explanation of this figure.

Reprinted by permission from G.A. Gaesser and D.C. Poole, "The Slow Component of Oxygen Uptake Kinetics in Humans," *Exercise and Sport Sciences Reviews* 24 (1996): 36.

correctly termed the **peak oxygen uptake ($\dot{V}O_{2peak}$)**. For example, a highly trained marathon runner will almost always achieve a higher $\dot{V}O_2$ value ($\dot{V}O_{2max}$) on a treadmill than when he or she is tested to volitional fatigue on a cycle ergometer ($\dot{V}O_{2peak}$). In the latter case, fatigue of the quadriceps muscles is likely to occur before a true maximal oxygen uptake is achieved.

Although $\dot{V}O_{2max}$ is a good measure of aerobic fitness, the winner of a marathon race cannot be predicted from the runner's laboratory-measured $\dot{V}O_{2max}$. This suggests that while a relatively high $\dot{V}O_{2max}$ is a necessary attribute for elite endurance athletes, a stellar endurance athlete requires more than a high $\dot{V}O_{2max}$, a concept discussed in chapter 11.

Also, research has documented that $\dot{V}O_{2max}$ typically increases with physical training for only 8 to 12 weeks and then plateaus despite continued higher-intensity training. Although $\dot{V}O_{2max}$ does not continue to increase, participants continue to improve their endurance performance. It appears that these individuals

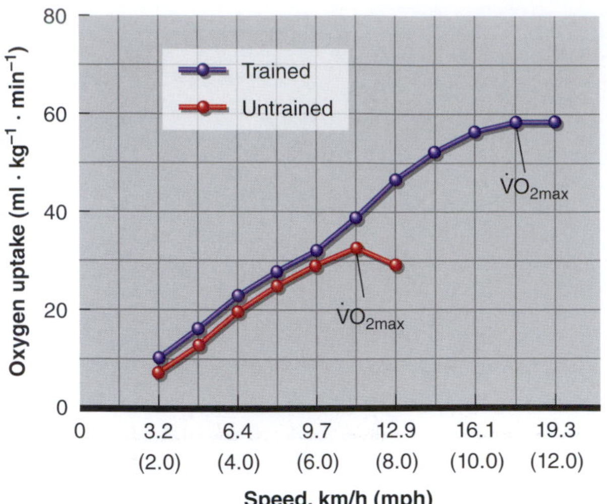

FIGURE 5.4 The relation between exercise intensity (running speed) and oxygen uptake, illustrating $\dot{V}O_{2max}$ in a trained and an untrained man.

develop the ability to perform at a higher percentage of their $\dot{V}O_{2max}$. Well-trained marathon runners, for example, can complete a 42 km (26.1 mi) marathon at an average pace that equals approximately 75% to 80% of their $\dot{V}O_{2max}$ or higher.

Consider the case of Alberto Salazar, arguably the premier marathon runner in the world in the 1980s. His measured $\dot{V}O_{2max}$ was 70 ml · kg^{-1} · min^{-1}. That is below the $\dot{V}O_{2max}$ one might expect based on his best marathon performance of 2 h 8 min. He was, however, able to run at a race pace in the marathon at 86% of his $\dot{V}O_{2max}$, a percentage considerably higher than that of other world-class runners. This may partly explain his world-class running ability.

Because individuals' energy requirements vary with body size, $\dot{V}O_{2max}$ generally is expressed relative to body weight, in milliliters of oxygen consumed per kilogram of body weight per minute (ml · kg^{-1} · min^{-1}). This allows a more accurate comparison of the cardiorespiratory endurance capacity of different-sized individuals who exercise in weight-bearing events, such as running. In nonweight-bearing activities, such as swimming and cycling, endurance is better reflected by $\dot{V}O_{2max}$ measured in liters per minute.

Normally active but untrained 18- to 22-year-old college students have an average $\dot{V}O_{2max}$ of about 38 to 42 ml · kg^{-1} · min^{-1} for women and 44 to 50 ml · kg^{-1} · min^{-1} for men. In contrast, poorly conditioned adults may have values below 20 ml · kg^{-1} · min^{-1}. At the other end of the spectrum, $\dot{V}O_{2max}$ values of 80 to 84 ml · kg^{-1} · min^{-1} have been measured for elite male long-distance runners and cross-country skiers. (The highest $\dot{V}O_{2max}$ value recorded for a man is that of a champion Norwegian cross-country skier who had a $\dot{V}O_{2max}$ of 94 ml · kg^{-1} · min^{-1}! The highest value recorded for a woman is 77 ml · kg^{-1} · min^{-1} for a Russian cross-country skier.)

After the age of 25 to 30 years, the $\dot{V}O_{2max}$ of inactive individuals decreases at a rate of about 1% per year, attributable to the combination of biological aging and sedentary lifestyle. Two physiological reasons why adult women generally have $\dot{V}O_{2max}$ values considerably below those of adult men (discussed further in chapter 19) are sex differences in body composition (women generally have less fat-free mass and more fat mass) and blood hemoglobin content (lower in women, so they have a lower oxygen-carrying capacity).

Anaerobic Effort and Exercise Capacity

No exercise is 100% aerobic or 100% anaerobic. The methods we have discussed thus far ignore the anaerobic processes that accompany aerobic exercise. How can the interaction of the aerobic (oxidative) processes and the anaerobic processes be evaluated? The most common methods for estimating anaerobic contribution to sustained exercise involve examination of either the excess postexercise oxygen consumption (EPOC) or the lactate threshold.

Postexercise Oxygen Consumption

The matching of energy requirements during exercise with oxygen delivery is not perfect. When aerobic exercise begins, the oxygen transport system (respiration and circulation) does not immediately supply the needed quantity of oxygen to the active muscles. Oxygen consumption requires several minutes to reach the required (steady-state) level at which the aerobic processes are fully functional, even though the body's oxygen requirements increase the moment exercise begins.

Because oxygen needs and oxygen supply differ during the transition from rest to exercise, the body incurs an **oxygen deficit**, as shown in figure 5.5. This deficit accrues even at low exercise intensities. The oxygen deficit is calculated simply as the difference between the oxygen required for a given exercise intensity (steady state) and the actual oxygen consumption. Despite the insufficient oxygen delivery at the onset of exercise, the active muscles are able to generate the ATP needed through the anaerobic pathways described in chapter 2.

During the initial minutes of recovery, even though active muscle activity has stopped, oxygen consumption does not immediately decrease to a resting value. Rather, oxygen consumption decreases gradually toward resting values (figure 5.5). This excess oxygen consumption, which exceeds that required at rest, was traditionally referred to as the "oxygen debt." The more common

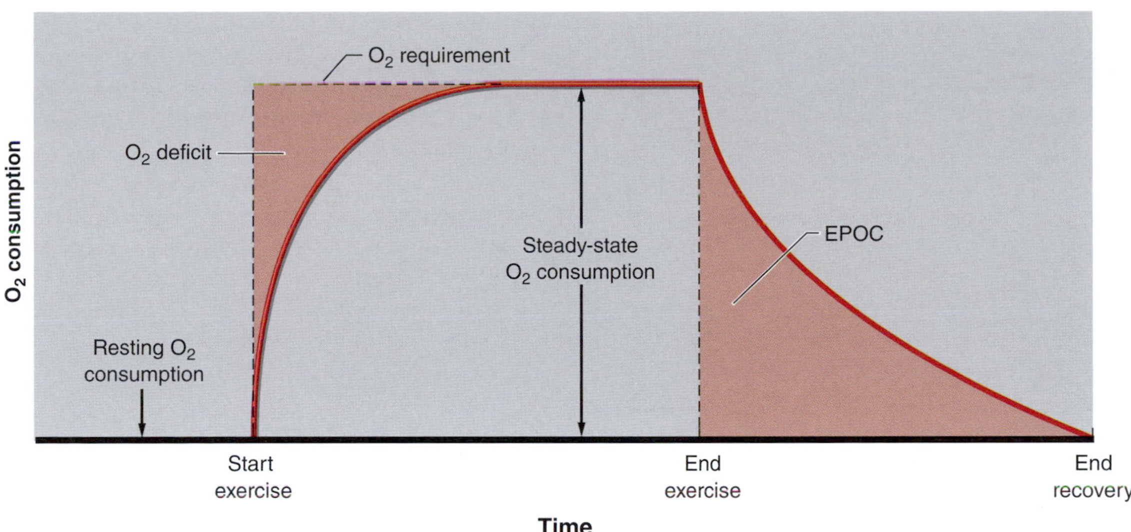

FIGURE 5.5 Oxygen requirement (dashed line) and oxygen consumption (red solid line) during exercise and recovery, illustrating the oxygen deficit and the concept of excess postexercise oxygen consumption (EPOC).

term today is **excess postexercise oxygen consumption (EPOC)**. The EPOC is the volume of oxygen consumed during the minutes immediately after exercise ceases that is above that normally consumed at rest. Everyone has experienced this phenomenon at the end of an intense exercise bout: A fast climb up several flights of stairs leaves one with a rapid pulse and breathing hard, physiological adjustments that serve to support the EPOC. After several minutes of recovery, heart rate and breathing return to resting rates.

For many years, the EPOC curve was described as having two distinct components: an initial fast component and a secondary slow component. According to classical theory, the fast component of the curve represented the oxygen required to rebuild the ATP and phosphocreatine (PCr) used during the initial stages of exercise. Without sufficient oxygen available, the high-energy phosphate bonds in these compounds were broken to supply the required energy. During recovery, these bonds would need to be re-formed, via oxidative processes, to replenish the energy stores or to repay the debt. The slow component of the curve was thought to result from removal of accumulated lactate from the tissues, by either conversion to glycogen or oxidation to CO_2 and H_2O, thus providing the energy needed to restore glycogen stores.

According to this theory, both the fast and slow components of the curve reflected the anaerobic activity that had occurred during exercise. The belief was that by examining the postexercise oxygen consumption, one could estimate the amount of anaerobic activity that had occurred.

However, more recently researchers have concluded that the classical explanation of EPOC is too simplistic. For example, during the initial phase of exercise, some oxygen is borrowed from the oxygen stores (hemoglobin and myoglobin), and that oxygen must be replenished during early recovery as well. Also, respiration remains temporarily elevated following exercise partly in an effort to clear CO_2 that has accumulated in the tissues as a by-product of metabolism. Body temperature also is elevated, which keeps the metabolic and respiratory rates high, thus requiring more oxygen; and elevated concentrations of norepinephrine and epinephrine during exercise have similar effects.

Thus, the EPOC depends on many factors other than merely the replenishing of ATP and PCr and the clearing of lactate produced by anaerobic metabolism.

Lactate Threshold

Many investigators consider the lactate threshold a good indicator of an athlete's potential for endurance exercise. The **lactate threshold** is defined as the point at which blood lactate begins to substantially accumulate above resting concentrations during exercise of increasing intensity. For example, a runner might be required to run on the treadmill at different speeds with a rest between each speed. After each run, a blood sample is taken and blood lactate is measured. Figure 5.6 depicts the relation between blood lactate and running velocity. At low running velocities, blood lactate concentrations remain near resting levels. But as running speed increases, the blood lactate concentration increases rapidly beyond some threshold exercise intensity. The point at which blood lactate first appears to increase disproportionately above resting values is called the lactate threshold.

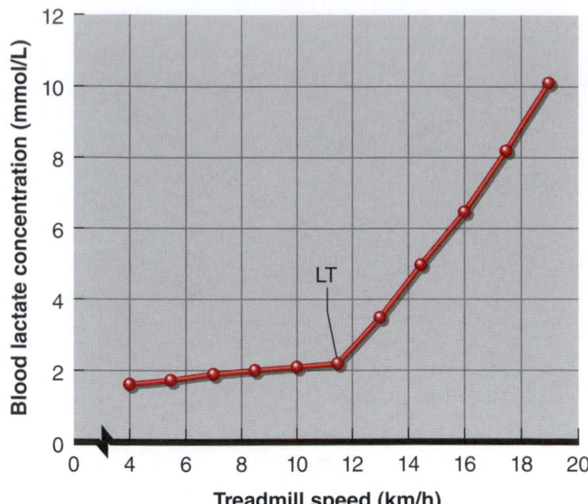

FIGURE 5.6 The relation between exercise intensity (running velocity) and blood lactate concentration. Blood samples were taken from a runner's arm vein and analyzed for lactate after the subject ran at each speed for 5 min. LT = lactate threshold.

The lactate threshold has been thought to reflect the interaction of the aerobic and anaerobic energy systems. Some researchers have suggested that the lactate threshold represents a significant shift toward anaerobic glycolysis, which forms lactate from pyruvic acid. Consequently, the sudden increase in blood lactate with increasing effort has also been referred to as the *anaerobic threshold.* However, blood lactate concentration is determined not only by the production of lactate in skeletal muscle or other tissues but also by the clearance of lactate from the blood by the liver and its use as a fuel source by muscle and other tissues in the body. Thus, lactate threshold is best defined as that point in time during exercise of increasing intensity when the rate of lactate production exceeds the rate of lactate clearance.

The lactate threshold is usually expressed as the percentage of maximal oxygen uptake (% $\dot{V}O_{2max}$) at which it occurs. In untrained people, the lactate threshold typically occurs at approximately 50% to 60% of their $\dot{V}O_{2max}$, while elite endurance athletes may not reach lactate threshold until closer to 70% or 80% of $\dot{V}O_{2max}$.

From the previous section, we learned that in addition to a high $\dot{V}O_{2max}$, the percentage of $\dot{V}O_{2max}$ that an athlete can maintain for a prolonged period is a major determinant of successful endurance performance. The lactate threshold is likely the major determinant of the fastest pace that can be tolerated during a long-term endurance event. So the ability to perform at a higher percentage of $\dot{V}O_{2max}$ probably reflects a higher lactate threshold. Consequently, a lactate threshold at 80% $\dot{V}O_{2max}$ suggests a greater aerobic exercise tolerance than a threshold at 60% $\dot{V}O_{2max}$. Generally, in two individuals with the same maximal oxygen uptake, the person with the highest lactate threshold usually exhibits the best endurance performance, although other factors contribute as well, including economy of movement.

Economy of Effort

As people become more skilled at performing an exercise, the energy demands during exercise at a given pace are reduced. In a sense, people become more economical. (Note that we avoid calling this efficiency, which has a more stringent mechanical definition.) This is illustrated in figure 5.7 by the data from two distance runners. At all running speeds faster than 11.3 km/h (7.0 mph), runner B used significantly less oxygen than runner A. These men had similar $\dot{V}O_{2max}$ values (64-65 ml · kg^{-1} · min^{-1}), so runner B's lower submaximal energy use would be a decided advantage during competition.

These two runners competed on numerous occasions. During marathon races, they ran at paces requiring them to use 85% of their $\dot{V}O_{2max}$. On average, runner B beat runner A by 13 min in their competitions. Because their $\dot{V}O_{2max}$ values were so similar but their energy needs so different during these events, much of runner B's competitive advantage could be attributed to his greater running economy. Unfortunately, there is no single specific explanation for the underlying causes of differences in economy, which are likely due to a variety of complex physiological and biomechanical factors.

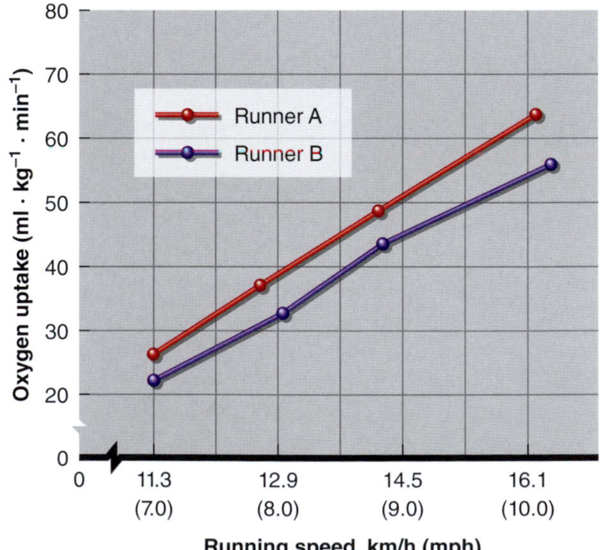

FIGURE 5.7 The oxygen requirements for two distance runners running at various speeds. Although they had similar $\dot{V}O_{2max}$ values (64-65 ml · kg^{-1} · min^{-1}), runner B was more economical and therefore could run at a faster pace for a given oxygen cost.

Various studies with sprint, middle-distance, and distance runners have shown that marathon runners are generally very economical. It is not uncommon for ultra-long-distance runners to use 5% to 10% less energy than middle-distance runners and sprinters at a given pace. However, this economy of effort has been studied at only relatively slow speeds (paces of 10-19 km/h, or 6-12 mph). We can reasonably assume that distance runners are less economical at sprinting than runners who train specifically for short, faster races. It is probable that runners self-select their chosen events in part because they achieve early success, success achieved in part due to better running economy at that distance.

Variations in running form and the specificity of training for sprint and distance running may account for at least part of these differences in running economy. Film analyses reveal that middle-distance runners and sprinters have significantly more vertical body movement when running at 11 to 19 km/h (7-12 mph) than marathoners do. But such speeds are well below those required during middle-distance races and probably do not accurately reflect the running economy of competitors in shorter events of 1,500 m (1 mi) or less.

Performance in other athletic events might be even more affected by economy of movement. Part of the energy expended during swimming, for example, is used to support the body on the surface of the water and to generate enough force to overcome the water's resistance to motion. Although the energy needed for swimming depends on body size and buoyancy, the efficient application of force against the water is the major determinant of swimming economy.

Characteristics of Successful Athletes in Aerobic Endurance Events

From our discussion of the metabolic characteristics of aerobic endurance athletes in this chapter and of their muscle fiber type characteristics in chapter 1, it is clear that to be successful in aerobic endurance activities, one needs some combination of the following:

- High $\dot{V}O_{2max}$
- High lactate threshold when expressed as a percentage of $\dot{V}O_{2max}$
- High economy of effort, or a low $\dot{V}O_2$ for a given absolute exercise intensity
- High percentage of type I muscle fibers

From the limited data available, these four characteristics appear to be properly ranked in their order of importance. As an example, running velocity at lactate threshold and $\dot{V}O_{2max}$ are the best predictors of actual race pace among a group of elite distance runners. However, each of those runners already has a high $\dot{V}O_{2max}$, which, in elite athletes, is supported by having a large heart and an expanded blood volume. Although economy of effort is important, it does not vary much between elite runners. Finally, having a high percentage of type I muscle fibers is helpful but not essential. The bronze medal winner in one of the Olympic marathon races had only 50% type I muscle fibers in his gastrocnemius muscle, one of the primary muscles used in running.

Energy Cost of Various Activities

The amount of energy expended for different activities varies with the intensity and type of exercise. Despite subtle differences in economy, the *average* energy costs of many activities have been determined, usually through the measurement of oxygen consumption during the activity to determine an average oxygen uptake per unit of time. The amount of energy expended per minute (kcal/min) then can be calculated from this value.

These values typically ignore the anaerobic aspects of exercise and the EPOC. This omission is important because an activity that costs a total of 300 kcal during the actual exercise period may cost an additional 100 kcal during the recovery period. Thus, the total cost of that activity would be 400, not 300, kcal. Because of these nuances along with individual variation, the "calories burned" readouts on exercise machines can be highly inaccurate.

The body requires 0.16 to 0.35 L of oxygen per minute to satisfy its resting energy requirements. This would amount to 0.80 to 1.75 kcal/min, 48 to 105 kcal/h, or 1,152 to 2,520 kcal/day. Obviously, any activity above resting levels will add to the projected daily expenditure. The range for total daily caloric expenditure is highly variable and depends on

- physical activity (by far the largest influence),
- age,
- sex,
- body size,
- weight, and
- body composition.

The energy costs of sport activities also differ. Some, such as archery or bowling, require only slightly more energy than rest. Others, such as sprinting, require such a high rate of energy delivery that they can be maintained for only seconds. Clearly, both exercise intensity and duration of the activity must be considered to determine energy expended. For example, approximately 29 kcal/min is expended during running at 25 km/h (15.5 mph), but this pace can be endured for only brief periods. Jogging at 11 km/h (7 mph), on the other hand, expends only 14.5 kcal/min, half that of running at 25 km/h (15.5 mph). But jogging can be maintained for considerably longer, resulting in greater total energy expenditure for an exercise session.

Table 5.2 provides estimates of average energy expenditure during various activities for men and women. Remember that these values are merely averages, and these figures vary considerably with individual differences such as those on the preceding list and with individual skill (economy of movement).

In Review

- The basal metabolic rate (BMR) is the minimum amount of energy required by the body to sustain basic cellular functions and is related to fat-free body mass and, to a lesser extent, body surface area. It typically ranges from 1,100 to 2,500 kcal/day, but when daily activity is added, typical daily caloric expenditure is 1,700 to 3,100 kcal/day.
- $\dot{V}O_2$ increases linearly with increased exercise intensity but eventually reaches a plateau. Its maximal value is called the $\dot{V}O_{2max}$. When volitional fatigue limits exercise before a true maximum is reached, the term $\dot{V}O_{2peak}$ is used.
- Successful aerobic performance is linked to a high $\dot{V}O_{2max}$, to the ability to perform for long periods at a high percentage of $\dot{V}O_{2max}$, to the running velocity at lactate threshold, and to a good economy of movement.
- The EPOC is the elevated metabolic rate above resting levels that occurs during the recovery period immediately after exercise has ceased.
- Lactate threshold is that point at which blood lactate production begins to exceed the body's ability to clear lactate, resulting in a rapid increase in blood lactate concentration during exercise of increasing intensity. Generally, individuals with higher lactate thresholds, expressed as a percentage of their $\dot{V}O_{2max}$, are capable of better endurance performances. Lactate threshold is a strong determinant of an athlete's optimal pace in endurance events such as distance running and cycling.

Fatigue and Its Causes

The term **fatigue** means different things to different people. Sensations that exercising individuals describe as fatigue are markedly different for a 400 m (437 yd) runner (an event lasting 45 to 60 s) than for a marathoner nearing the end of a 42.2 km (26.2 mi) endurance event. Therefore, it is not surprising that

TABLE 5.2 **Average Values for Energy Expenditure During Various Physical Activities**

Activity	Men (kcal/min)	Women (kcal/min)	Relative to body mass ($kcal \cdot kg^{-1} \cdot min^{-1}$)
Basketball	8.6	6.8	0.123
Cycling			
11.3 km/h (7.0 mph)	5.0	3.9	0.071
16.1 km/h (10.0 mph)	7.5	5.9	0.107
Handball	11.0	8.6	0.157
Running			
12.1 km/h (7.5 mph)	14.0	11.0	0.200
16.1 km/h (10.0 mph)	18.2	14.3	0.260
Sitting	1.7	1.3	0.024
Sleeping	1.2	0.9	0.017
Standing	1.8	1.4	0.026
Swimming (crawl), 4.8 km/h (3.0 mph)	20.0	15.7	0.285
Tennis	7.1	5.5	0.101
Walking, 5.6 km/h (3.5 mph)	5.0	3.9	0.071
Weightlifting	8.2	6.4	0.117
Wrestling	13.1	10.3	0.187

Note. The values presented are for a 70 kg (154 b) man and a 55 kg (121 lb) woman. These values will vary depending on individual differences.

RESEARCH PERSPECTIVE 5.1

Energy Expenditure of Walking

Understanding how much metabolic energy is expended during walking has many applications, from clinical rehabilitation programs to fitness and activity tracking and even military maneuvers. The metabolic energy required for walking can be accurately determined by directly measuring oxygen consumption during activity. However, this measurement technique is impractical outside the laboratory. Published equations that predict this energy requirement are frequently used. The two most established and commonly used equations are those developed by the American College of Sports Medicine (ACSM)[1] and by a group of research scientists at the U.S. Army Research Institute of Environmental Medicine (USARIEM).[21] Both of these equations are specific to body mass and divide the person's metabolic rate into resting and nonresting components, where the nonresting component is speed dependent. Although these equations are considered the gold standards for predicting energy expenditure during walking, they were developed based on studies that included only young, healthy, male subjects of a relatively similar body size.

A recent study conducted at Southern Methodist University in Dallas, Texas, derived a new mathematical model to predict the metabolic energy requirements of walking.[15] Data from 10 previous studies were compiled to create a data set of over 400 subjects of both sexes who varied in age, height, body weight, and fitness level. Researchers then developed mathematical models to identify the variables required for accurate predictions of metabolic energy requirements across this heterogeneous subject pool and found that the most accurate predictions accounted for the walker's height (ht), a variable that is absent in the commonly used equations. Like previous equations, that accuracy of the prediction was increased when the walking metabolism was quantified as two separate components, the minimum walking component (different from the resting component) and the velocity-dependent component. The study resulted in a new model for predicting metabolic energy expenditure that predicted over 90% of the actual metabolic cost across all walking speeds:

$$\dot{V}O_2 \text{ total} = (\dot{V}O_2 \text{ rest} + 3.85) + (5.97 \cdot v^2/\text{ht})$$

where walking velocity (v) is measured in m/s, ht in m, and $\dot{V}O_2$ in ml $O_2 \cdot kg^{-1} \cdot min^{-1}$. In this equation, ($\dot{V}O_2$ rest + 3.85) is the minimum walking energy expenditure, ($5.97 \cdot v^2/ht$) is the velocity- and height-dependent energy expenditure, and [$3.85 + (5.97 \cdot v^2/ht)$] quantifies the total walking component.

This model can now be used to inform exercise prescriptions for numerous health and fitness outcomes in wide-ranging populations.

the causes of fatigue are different in those two scenarios as well. In exercise physiology, we typically describe *fatigue* as decrements in muscular performance with continued effort accompanied by general sensations of tiredness. An alternative definition used in research studies to quantify fatigue is the inability to maintain the required power output to continue muscular work at a given intensity. The fact that fatigue is reversible by rest distinguishes it from muscle weakness or damage (discussed later in the chapter).

Ask most exercisers what causes fatigue during exercise, and the most common two-word answer is "lactic acid." Not only is this common misconception an oversimplification but also there is mounting evidence that lactic acid has beneficial effects on exercise performance (see chapter 2). Fatigue is an extremely complex phenomenon, and its causes can range from the molecular level to the entire body. Most efforts to describe the underlying causes and sites of fatigue have focused on

- a decreased rate of energy delivery (ATP-PCr, anaerobic glycolysis, and oxidative metabolism);
- accumulation of metabolic by-products, such as lactate and H^+;
- failure of the muscle fiber's contractile mechanism; and
- alterations in neural control of muscle contraction.

The first three causes occur within the muscle itself; along with alterations in motor nerve control of muscle function, these are often referred to as *peripheral* fatigue. In addition to alterations at the motor unit level, changes in the brain or central nervous system may also cause what has become known as *central* fatigue.

However, none of these alone can explain all aspects or all types of fatigue, and several causes may act synergistically to bring about fatigue. Mechanisms of fatigue depend on the type and intensity of the exercise, the fiber type of the involved muscles, the subject's training status, and even his or her diet. Many questions about fatigue remain unanswered, including the cellular sites of fatigue within the muscle fibers themselves. It is important to remember that while fatigue arises at least in part from failure of cross-bridge cycling within the muscle cells, this machinery depends on the nervous, cardiovascular, and energy systems to support it.[14] Fatigue is rarely caused by a single factor but typically by multiple factors acting synergistically at multiple sites. Some potential sites of fatigue are discussed next.

Energy Systems and Fatigue

The energy systems are an obvious area to explore when one is considering possible causes of fatigue. When we feel fatigued, we often express this by saying, "I have no energy." But this use of the term *energy* is far removed from its physiological meaning. What role does energy availability play in fatigue during exercise, in the true physiological sense of providing ATP from substrates?

PCr Depletion

Recall that PCr is used for short-term high-intensity effort, to rebuild ATP as it is used and thus to maintain ATP stores within the muscle. Biopsy studies of human thigh muscles have shown that during repeated maximal contractions, fatigue coincides with PCr depletion. Although ATP is directly responsible for the energy used during such activities, it is depleted less rapidly than PCr during muscular effort because ATP is being produced by other systems (see figure 2.6). But as PCr is depleted, the ability to quickly replace the spent ATP is hindered. Use of ATP continues, but the ATP-PCr system is less able to replace it. Thus, ATP concentration also decreases. At exhaustion, both ATP and PCr may be depleted.

To delay fatigue, an athlete must control the rate of effort through proper pacing to ensure that PCr and ATP are not prematurely exhausted. This holds true even in endurance-type events. If the beginning pace is too rapid, available ATP and PCr concentrations will quickly decrease, leading to early fatigue and an inability to maintain the pace in the event's later stages. Training and experience allow the athlete to judge the optimal pace that permits the most efficient use of ATP and PCr for the entire event.

Glycogen Depletion

Muscle ATP concentrations are also maintained by the breakdown of muscle glycogen. In events lasting longer than a few seconds, muscle glycogen becomes the primary energy source for ATP synthesis. Unfortunately, glycogen reserves are limited and are depleted quickly. Since the muscle biopsy technique was first established, studies have shown a correlation between muscle glycogen depletion and fatigue during prolonged exercise.

Muscle glycogen is used more rapidly during the first few minutes of exercise than in the later stages, as seen in figure 5.8.[6] The illustration shows the change in muscle glycogen content in the subject's gastrocnemius (calf) muscle during the test. Although the subject ran the test at a steady pace, the rate of muscle glycogen metabolized from the gastrocnemius was greatest during the first 75 min.

The subject also reported his perceived exertion (how difficult his effort seemed to be) at various times during the test. He felt only moderately stressed early

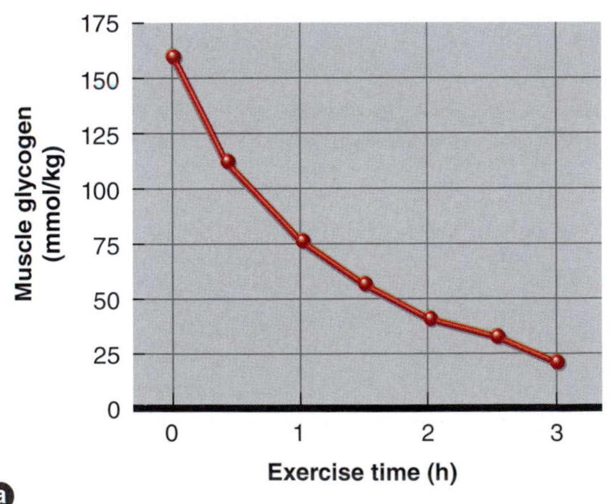

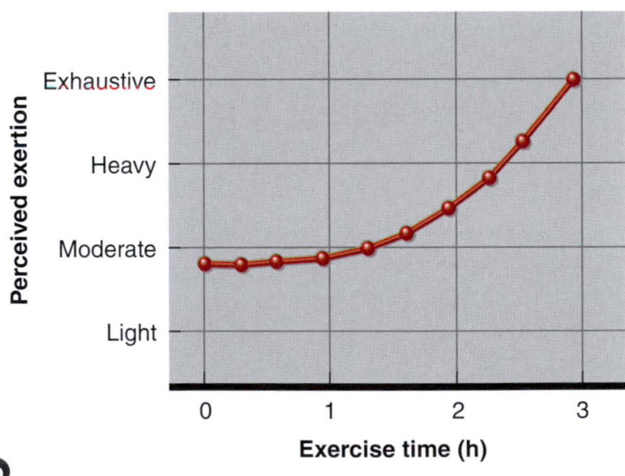

FIGURE 5.8 *(a)* The decline in gastrocnemius (calf) muscle glycogen during 3 h of treadmill running at 70% of $\dot{V}O_{2max}$, and *(b)* the subject's subjective rating of the effort. Note that the effort was rated as moderate for nearly 1.5 h of the run, although glycogen was decreasing steadily. Not until the muscle glycogen became quite low (less than 50 mmol/kg) did the rating of effort increase.

Adapted by permission from D.L. Costill, *Inside Running: Basics of Sports Physiology* (Indianapolis: Benchmark Press, 1986). Copyright 1986 Cooper Publishing Group, Carmel, IN.

in the run, when his glycogen stores were still high, even though he was using glycogen at a high rate. He did not perceive severe fatigue until his muscle glycogen levels were nearly depleted. Thus, the sensation of fatigue in long-term exercise coincides with a decreased concentration of muscle glycogen but not with its rate of depletion. Marathon runners commonly refer to the sudden onset of fatigue that they experience at 29 to 35 km (18-22 mi) as "hitting the wall." At least part of this sensation can be attributed to muscle glycogen depletion.

Glycogen Depletion in Different Fiber Types

Muscle fibers are recruited and deplete their energy reserves in selected patterns. The individual fibers most frequently recruited during exercise may become depleted of glycogen. This reduces the number of fibers capable of producing the muscular force needed for exercise.

This glycogen depletion is illustrated in figure 5.9, which shows a micrograph of muscle fibers taken from a runner after a 30 km (18.6 mi) run. Figure 5.9*a* has been stained to differentiate type I and type II fibers. One of the type II fibers is circled. Figure 5.9*b* shows a second sample from the same muscle, stained to show glycogen. The redder (darker) the stain, the more glycogen is present. Before the run, all fibers were full of glycogen and appeared red (not depicted). In figure 5.9*b* (after the run), the lighter type I fibers are almost completely depleted of glycogen. This suggests that type I fibers are used more heavily during endurance exercise that requires only moderate force development, such as the 30 km run.

The pattern of glycogen depletion from type I and type II fibers depends on the exercise intensity. Recall that type I fibers are the first fibers to be recruited during light exercise. As muscle tension requirements increase, type IIa fibers are added to the workforce. In exercise approaching maximal intensities, the type IIx fibers are added to the pool of recruited fibers.

Depletion in Different Muscle Groups

In addition to selectively depleting glycogen from type I or type II fibers, exercise may place unusually heavy demands on select muscle groups. In one study, subjects ran on a treadmill positioned for uphill, downhill, and level running for 2 h at 70% of $\dot{V}O_{2max}$. Figure 5.10

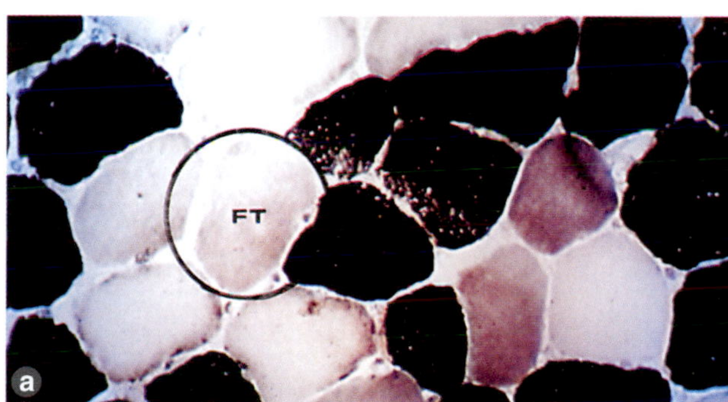

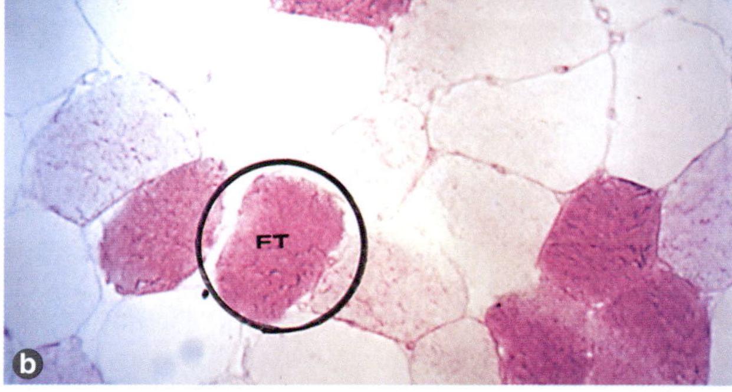

FIGURE 5.9 *(a)* Histochemical staining for fiber type after a 30 km run; a type II (fast-twitch) fiber is circled. *(b)* Histochemical staining for muscle glycogen after the run. Note that a number of type II fibers still have glycogen, as noted by their darker stain, whereas most of the type I (slow-twitch) fibers are depleted of glycogen.

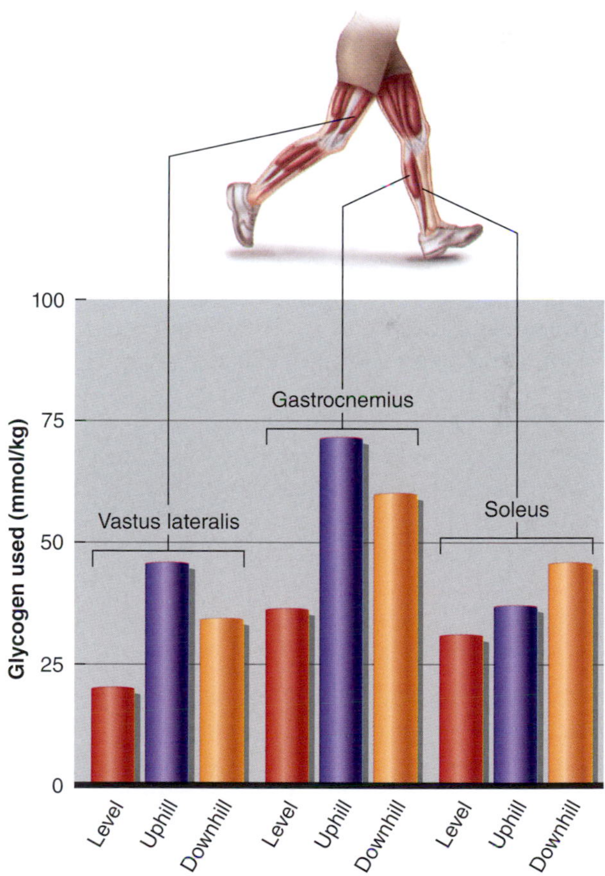

FIGURE 5.10 Muscle glycogen use in the vastus lateralis, gastrocnemius, and soleus muscles during 2 h of level, uphill, and downhill running on a treadmill at 70% of $\dot{V}O_{2max}$. Note that the greatest glycogen use is in the gastrocnemius during uphill and downhill running.

compares the resultant glycogen depletion in three muscles of the lower extremity: the vastus lateralis (knee extensor), the gastrocnemius (ankle extensor), and the soleus (another ankle extensor).

The results show that whether one runs uphill, downhill, or on a level surface, the gastrocnemius uses more glycogen than does the vastus lateralis or the soleus. This suggests that the ankle extensor muscles are more likely to become depleted during distance running than are the thigh muscles, isolating the site of fatigue to the lower leg muscles.

Glycogen Depletion and Blood Glucose

Muscle glycogen alone cannot provide enough carbohydrate for exercise lasting several hours. Glucose delivered by the blood to the muscles contributes a lot of energy during endurance exercise. The liver breaks down its stored glycogen to provide a constant supply of blood glucose. In the early stages of exercise, energy production requires relatively little blood glucose, but in the later stages of an endurance event, blood glucose may make a large contribution. To keep pace with the muscles' glucose uptake, the liver must break down increasingly more glycogen as exercise duration increases.

Liver glycogen stores are limited, and the liver cannot produce glucose rapidly from other substrates. Consequently, blood glucose concentration can decrease when muscle uptake exceeds the liver's glucose output. Unable to obtain sufficient glucose from the blood, the muscles must rely more heavily on their glycogen reserves, accelerating muscle glycogen depletion and leading to earlier exhaustion.

Not surprisingly, endurance performances improve when the muscle glycogen supply is elevated before the start of activity. On the other hand, most studies have shown no effect of carbohydrate ingestion on net muscle glycogen utilization during prolonged, strenuous exercise. The importance of muscle glycogen storage for endurance performance is discussed in chapter 15. For now, note that glycogen depletion and hypoglycemia (low blood sugar) limit performance in activities lasting longer than 60 min.

Mechanisms of Fatigue with Glycogen Depletion It does not appear likely that glycogen depletion *directly* causes fatigue during endurance exercise performance, but it may play an *indirect* role. We cannot explain precisely why muscle function is impaired when muscle glycogen is low, but this is usually explained by a compromised rate of ATP production. Glycogen is more than simply a form for carbohydrate storage; it also acts as a regulator of several cellular functions. To aid in that role, glycogen is not distributed homogeneously throughout the muscle fiber but localized in distinct pools. Evidence suggests that depletion of glycogen granules localized within the myofibrils interferes with excitation–contraction coupling and Ca^{2+} release from the sarcoplasmic reticulum.[20]

Metabolic By-Products and Fatigue

Various by-products of metabolism have been implicated as factors causing, or contributing to, fatigue. The metabolic by-products that have received the most attention in discussions of fatigue are inorganic phosphate, heat, lactate, and hydrogen ions.

Inorganic Phosphate

Inorganic phosphate increases during intense short-term exercise as PCr and ATP are being broken down. It now appears that P_i, which accumulates during intense short-term exercise from the breakdown of ATP, may be the largest contributor to fatigue in this type of exercise.[26] Excess P_i directly impairs contractile function of the myofibrils and can reduce Ca^{2+} release from the sarcoplasmic reticulum. Increases in both P_i and ADP also inhibit ATP breakdown through negative feedback.

Heat and Muscle Temperature

Recall that energy expenditure results in a relatively large heat production, some of which is retained in the body, causing core temperature to rise. Exercise in the heat can increase the rate of carbohydrate utilization and hasten glycogen depletion, effects that may be stimulated by the increased secretion of epinephrine. It is hypothesized that high muscle temperatures impair both skeletal muscle function and muscle metabolism.

The ability to continue moderate- to high-intensity cycle performance is affected by ambient temperature. Galloway and Maughan[12] studied performance time to exhaustion of male cyclists at four different air temperatures: 4 °C (38 °F), 11 °C (51 °F), 21 °C (70 °F), and 31 °C (87 °F). Results of that study are shown in figure 5.11. Time to exhaustion was longest when subjects exercised at an air temperature of 11 °C and was shorter at colder and warmer temperatures. Fatigue set in earliest at 31 °C. Similarly, at a given warm air temperature, increasing relative humidity caused early fatigue.[16] Precooling of muscles similarly prolongs exercise, while preheating causes earlier fatigue. Heat acclimation, discussed in chapter 12, spares glycogen and reduces lactate accumulation.

Lactic Acid

Recall that lactic acid is a by-product of anaerobic glycolysis. Although most lay people believe that lactic acid is responsible for fatigue in all types of exercise, lactic

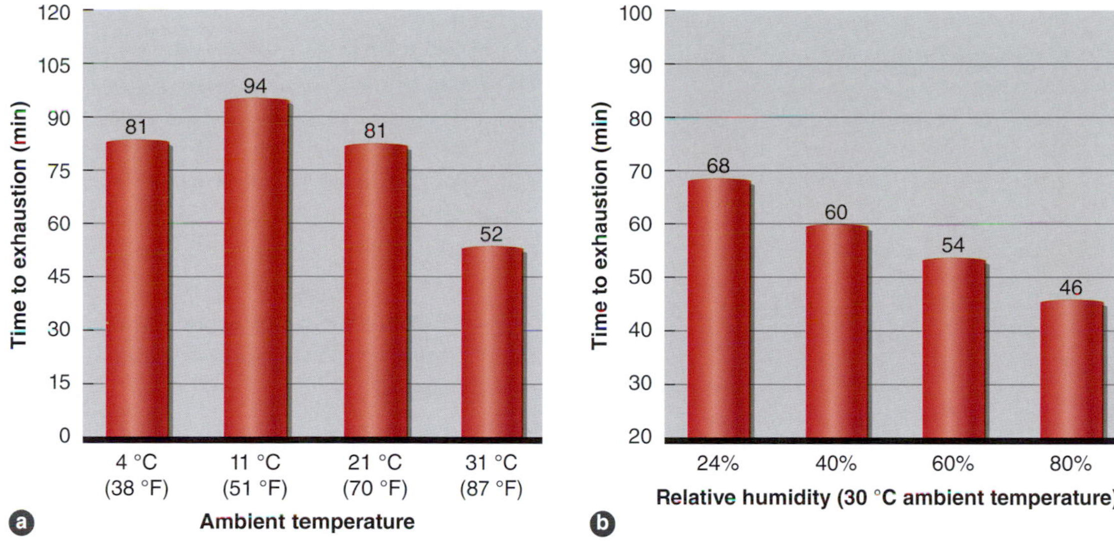

FIGURE 5.11 Time to exhaustion for a group of men performing cycle exercise at about 70% $\dot{V}O_{2max}$. *(a)* The subjects were able to perform longer (delay fatigue longer) in a cool environment of 11 °C. Exercising in colder or warmer conditions hastened fatigue. *(b)* At an ambient temperature of 30 °C, increased relative humidity decreased time to exhaustion.

(a) Adapted by permission from S.D.R. Galloway and R.J. Maughan, "Effects of Ambient Temperature on the Capacity to Perform Prolonged Cycle Exercise in Man," *Medicine and Science in Sports and Exercise* 29 (1997): 1240-1249. *(b)* Reprinted by permission from R.J. Maughan et al., "Influence of Relative Humidity on Prolonged Exercise Capacity in a Warm Environment," *European Journal of Applied Physiology* 112 (2012): 2313-2321.

acid undergoes constant turnover and, as described in chapter 2, is recycled to provide energy. Lactic acid produced within the cytoplasm of a muscle fiber can be taken up by mitochondria within that same muscle fiber and oxidized for ATP formation. Lactic acid can also be shuttled to other sites where it can be oxidized. In fact, lactic acid only accumulates within a muscle fiber during relatively brief, highly intense muscular effort. Marathon runners often have near-baseline lactic acid concentrations at the end of the race, despite notable fatigue. As noted in the previous section, their fatigue is likely caused by inadequate energy supply, not excess lactic acid.

Short sprints in running, cycling, and swimming can all lead to large accumulations of lactic acid. While the presence of lactic acid in itself cannot be blamed for the feeling of fatigue, if it is not cleared, the lactic acid dissociates, converting to lactate and causing an accumulation of hydrogen ions.

Hydrogen Ions

While the lactate ion does not appear to have any major negative effects on the ability to generate force, H^+ accumulation causes muscle acidosis (decreased pH).

Activities of short duration and high intensity, such as sprint running and sprint swimming, depend heavily on anaerobic glycolysis and produce large amounts of lactate and H^+ within the muscles. Fortunately, the cells and body fluids possess buffers, such as bicarbonate (HCO_3^-), that minimize the disrupting influence of the H^+. Without these buffers, H^+ would lower the pH to about 1.5, killing the cells. Because of the body's buffering capacity, the H^+ concentration remains low even during the most severe exercise, allowing muscle pH to decrease from a resting value of 7.1 to no lower than 6.4 at exhaustion after high-intensity activity.

However, pH changes of this magnitude can adversely affect energy production and muscle contraction. An intracellular pH below 6.9 inhibits the action of phosphofructokinase, an important glycolytic enzyme, slowing the rate of glycolysis and ATP production. At a pH of 6.4, the influence of H^+ stops any further glycogen breakdown, causing a rapid decrease in ATP and ultimately exhaustion. In addition, H^+ may lower the amount of calcium released from the sarcoplasmic reticulum, interfering with the coupling of the actin–myosin cross-bridges and decreasing the muscle's contractile force. However, the impact on muscle force production is small. A larger impact comes from H^+ acting to decrease the myofilaments' sensitivity to calcium, causing a loss of contractile force and velocity.[8] Because of those effects, low muscle pH may be a primary cause of fatigue during maximal, all-out exercise lasting 20 to 30 s.

As seen in figure 5.12, reestablishing the preexercise muscle pH after an exhaustive sprint bout requires about 30 to 35 min of recovery. Even when normal pH is restored, blood and muscle lactate levels can remain

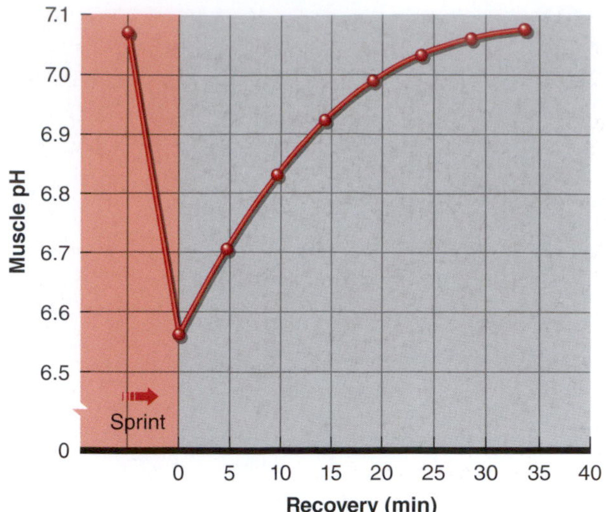

FIGURE 5.12 Changes in muscle pH during sprint exercise and recovery. Note the drastic decrease in muscle pH during the sprint and the gradual recovery to normal after the effort. Note that it took more than 30 min for pH to return to its preexercise level.

quite elevated. However, experience has shown that an athlete can continue to exercise at relatively high intensities even with a muscle pH below 7.0 and a blood lactate level above 6 or 7 mmol/L, four to five times the resting value.

Neuromuscular Fatigue

Thus far we have considered only factors within the muscle that might be responsible for fatigue. Evidence also suggests that under some circumstances, fatigue may result from an inability to activate the muscle fibers, a function of the nervous system. As noted in chapter 3, the nerve impulse is transmitted across the neuromuscular junction to activate the fiber's membrane, and it causes the fiber's sarcoplasmic reticulum to release calcium. The calcium, in turn, binds with troponin to initiate muscle contraction, a process collectively called excitation–contraction coupling. Several possible neural mechanisms could disrupt this process and possibly contribute to fatigue, and two of those—one peripheral and one central—are discussed next.

Neural Transmission

Fatigue may occur at the neuromuscular junction, preventing nerve impulse transmission to the muscle fiber membrane. Studies in the early 1900s clearly established such a failure of nerve impulse transmission in fatigued muscle. This failure may involve one or more of the following processes:

- The release or synthesis of acetylcholine (ACh), the neurotransmitter that relays the nerve impulse from the motor nerve to the muscle membrane, might be reduced.
- Cholinesterase, the enzyme that breaks down ACh once it has relayed the impulse, might become hyperactive, preventing sufficient concentration of ACh to initiate an action potential.
- Cholinesterase activity might become hypoactive (inhibited), allowing ACh to accumulate excessively, inhibiting relaxation.
- The muscle fiber membrane might develop a higher threshold for stimulation by motor neurons.
- Some substance might compete with ACh for the receptors on the muscle membrane without activating the membrane.
- Potassium might leave the intracellular space of the contracting muscle, decreasing the membrane potential to half of its resting value.

Although most of these causes for a neuromuscular block have been associated with neuromuscular diseases (such as myasthenia gravis), they may also cause some forms of neuromuscular fatigue. Some evidence suggests that fatigue also may be attributable to calcium retention within the sarcoplasmic reticulum, which would decrease the calcium available for muscle contraction. In fact, depletion of PCr and lactate buildup might simply increase the rate of calcium accumulation within the sarcoplasmic reticulum. However, these theories of fatigue remain speculative.

Central Nervous System

The discussion to this point suggests that fatigue is due to *peripheral* changes that limit or completely stop further effective muscular actions. The recruitment of muscle depends, in part, on conscious or subconscious control by the brain. An alternate theory to peripheral fatigue, termed the **central governor theory**, proposes that processes occur in the brain that regulate power output by the muscles to maintain homeostasis and prevent unsafe levels of exertion that may damage tissues or cause catastrophic events. The central governor limits exercise by decreasing the recruitment of muscle fibers, which in turn causes fatigue. While this theory has been hotly debated in recent years, the concept of a central "governor" was first proposed by A.V. Hill (see introductory chapter) in 1924.

In a 2012 study, researchers in Switzerland sought to separate out the central and peripheral contributors to muscle fatigue during low-intensity isometric contraction of the knee extensor muscles using an innovative protocol.[18] Subjects performed a sustained isometric muscle contraction at 20% of the maximal voluntary contraction (MVC) until they experienced fatigue, defined as the point at which they could no longer maintain the 20% MVC force output. Then, the muscle was immediately electrically stimulated to maintain the same force output for 1 min, followed by an immediate voluntary effort to maintain that same force once again. In essence, fatigue was induced, and then the external electrical stimulation took over for the brain and the motor neuron to continue to produce force. If the initial fatigue was caused by problems with excitation–contraction coupling (peripheral factors), then electrically stimulated muscle would still be fatigued and not able to produce force. Conversely, if fatigue was due to problems with the motor neuron or central neural factors, electrical stimulation would cause the muscle to generate force once again. The researchers also measured the maximal amount of force that could be voluntarily generated before and after fatigue was induced.

They found that when the electrical stimulation was applied after fatigue, the muscle was again able to maintain 20% MVC, indicating that, given the proper neural stimulation, the muscle itself maintains the ability to contract and generate force. When they recorded the muscle's electrical activity during the MVC after fatigue,

RESEARCH PERSPECTIVE 5.2

Can You Talk Yourself Out of Fatiguing?

As endurance sports become increasingly popular, the number of people participating in competitive endurance events is growing. Successful performance in endurance events requires the ability to sustain aerobic exercise over an extended period of time (that is, to delay the onset of fatigue as long as possible), and performance-enhancing strategies to increase the intensity and duration of the activity are important for athletes in these sports. Because fatigue defines the upper limit of endurance, research efforts have targeted understanding the physiological and psychological causes of fatigue and how to delay their effects.

During long-duration endurance events such as marathons or triathlons, many exercise physiologists believe that fatigue is the result of depleting energy stores within the body, and the physiological causes of fatigue are discussed in detail in this chapter. Alternatively, the *psychobiological model of endurance performance* suggests that fatigue is caused by the conscious decision to terminate a given activity, rather than by physiological limits. According to this model, the ultimate determinant of endurance performance in highly motivated athletes is the conscious perception of how hard, heavy, and strenuous the effort is. Based on this reasoning, strategies to reduce the *perception* of effort may delay the onset of fatigue in endurance athletes.

One strategy that may reduce fatigue according to the psychobiological model of endurance performance is *self-talk*, or self-addressed verbalizations, either aloud or silently. Self-talk can be both instructional and motivational for athletes, and it has been suggested to improve performance on effort-based tasks by motivating athletes to push themselves further even when the perceptual drive to terminate exercise is high. A 2014 study of 24 recreationally active men and women investigated the effect of self-talk on endurance performance during high-intensity cycling exercise.[3] After baseline data were recorded, research subjects were divided into two groups: One group received 2 weeks of coaching and practice in the use of self-talk while the other group (control group) did not. After the 2 weeks, all of the participants returned to the laboratory for retesting, and their performance results were compared to those from the beginning of the study. The group of subjects who received the self-talk coaching had a lower rating of perceived exertion (RPE) and a longer time to exhaustion (by almost 2 min) compared to their first visit (see figure); however, the results for the control group did not change. Motivational self-talk can reduce the perceived effort and increase endurance performance during aerobic activity. Training with psychobiological interventions that reduce the perception of effort may improve endurance performance in endurance athletes by delaying fatigue.

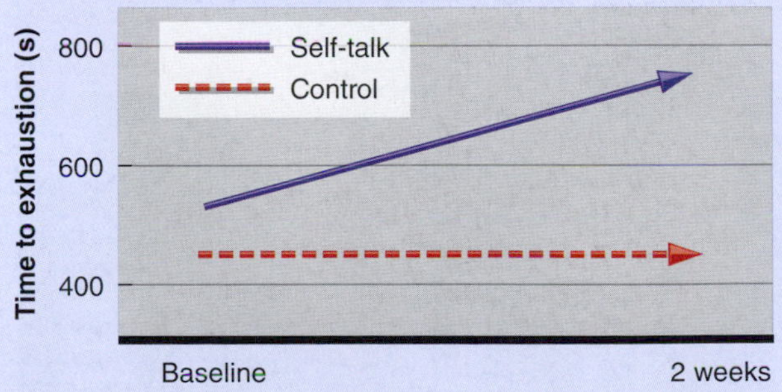

Changes in time to exhaustion after 2 weeks in the control group (dotted line) and the group of subjects who received 2 weeks of self-talk training (solid line). The self-talk group showed improvements in endurance after the self-talk training, while the control group did not change across the 2 weeks.

they found that muscle activation by the nervous system *increased* greatly, suggesting that the reduced force of the maximal contraction was due to impairments of the contractile elements. Overall, this study suggests that the initial fatigue experienced after a bout of submaximal exercise is likely due to a reduction in central neural factors, whereas the impairments in maximal contraction are due to peripheral factors related to changes in excitation–contraction coupling.

Undoubtedly, there is some central nervous system (CNS) involvement in most types of fatigue. When a subject's muscles appear to be nearly exhausted, verbal encouragement, shouting, music, or even direct electrical stimulation of the muscle can increase the strength of muscle contraction. The precise mechanisms underlying the CNS role in causing, sensing, and even overriding fatigue are not fully understood. Unless they are highly motivated, most individuals terminate exercise before their muscles are physiologically exhausted. To achieve peak performance, athletes train to learn proper pacing and tolerance for fatigue.

Other Contributors to Fatigue

As one can appreciate from the previous sections, the underlying causes of fatigue are many and varied and depend largely on the intensity and duration of the exercise being performed. Recent research has also uncovered roles for impaired mitochondrial function and reactive oxygen species in some types of fatigue. A group of French investigators reported that impaired mitochondrial function led to both a slower rate of PCr recovery and a reduction in oxidative ATP production after dynamic submaximal exercise.[10] However, the extent to which those effects can be attributed to cellular acidosis or muscle damage is unclear.

Michael Reid provided a compelling case that reactive oxygen species (ROS) accumulation in working muscle contributes to the loss of function that occurs in muscle fatigue.[25] These molecules, including hydrogen peroxide, superoxide, and hydroxyl radicals, increase during strenuous muscle contractions. They are present in myofibrillar cytoplasm and organelles, in interstitial fluid, and within the vascular space. Two lines of evidence point to a role for these chemically reactive molecules: (1) Directly exposing muscle cells to ROS evokes many of the same changes that occur with fatigue during exercise and (2) pretreating the muscle with antioxidants delays fatigue.

In Review

- Depending on the circumstances, fatigue may result from depletion of PCr or glycogen; both situations impair ATP production. Glycogen depletion may occur in select fiber types or specific muscle groups depending on the exercise.
- Increased metabolic by-products like phosphate ions and heat may contribute to fatigue.
- Lactic acid often has been blamed for fatigue, but it is generally not directly related to fatigue during prolonged endurance exercise, and may serve as a fuel source (see chapter 2).
- In short-duration exercise, like sprinting, the H^+ generated by dissociation of lactic acid may contributes to fatigue. The accumulation of H^+ decreases muscle pH, which impairs the cellular processes that produce energy and muscle contraction.
- Failure of neural transmission may be a cause of some types of fatigue. Many mechanisms can lead to such failure, and further research is needed.
- The CNS plays a role in most types of fatigue, perhaps limiting exercise performance as a pro-

tective mechanism. Perceived fatigue usually precedes physiological fatigue, and athletes who feel exhausted can often be encouraged to continue by various cues that stimulate the CNS, such as music or self-talk.

Critical Power: The Link Between Energy Expenditure and Fatigue

An athlete who can sustain a high level of exercise intensity for a prolonged period without fatiguing will be successful. Exercise physiologists have a name for the link between optimal performance and fatigue: **critical power**. The critical power defines the tolerable duration of high-intensity exercise. If we graph the relation between power output (or exercise intensity, or speed) and the maximal time that intensity can be maintained, the line is curvilinear, as depicted in figure 5.13. At very high power outputs, exercise can be performed only for short durations. But as intensity is progressively decreased, exercise can be performed for longer and longer durations. At some point, this relation levels off and reaches an asymptote, defining the critical power for that activity—the maximal intensity that can be sustained without fatigue limiting performance.

Critical power represents the highest metabolic rate that is maintained entirely by oxidative metabolism. In that regard, it is related to the lactate threshold (discussed earlier in this chapter), but occurs at slightly higher intensities. Not surprisingly, critical power is increased with endurance or high-intensity interval training and decreased with aging and in chronic disease states. Hypoxia, such as that encountered at altitude (discussed in chapter 13), also reduces critical power, while breathing elevated oxygen concentrations elevates it.

Critical power is a useful measure in sport and exercise physiology because it correlates well with performance in running, rowing, swimming, and even team sport activities lasting from a few min to 2 h.[28] However, while exercise at or below the critical power should theoretically be able to be continued indefinitely, in reality exercise at the critical power cannot be sustained beyond 30 min or so. With much attention being paid to breaking the 2 h barrier for the marathon, the critical power concept would dictate that a runner has to sustain a critical speed of only 21.1 km/h (13.1 mph, or slightly under 4.6 min miles); however, sustaining that heavy-intensity pace for 2 h has proven virtually impossible.[28]

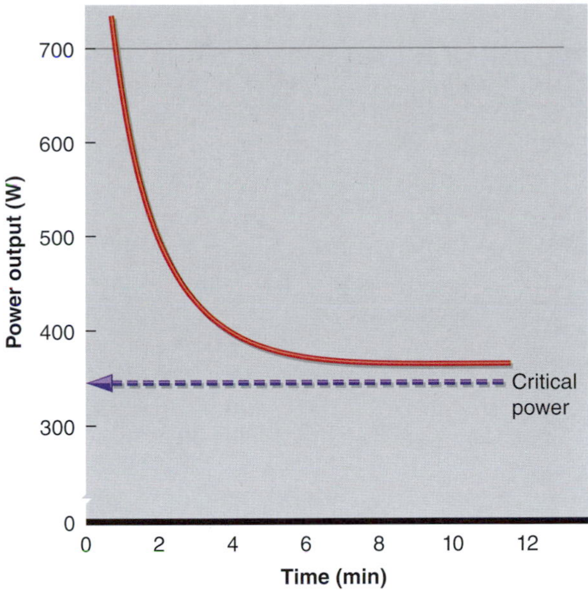

FIGURE 5.13 The relation between power output (in watts [W]) and the time that power output can be maintained. The critical power is defined as the asymptote in the relation, i.e., the maximal power output that can be sustained without fatigue limiting the duration of performance.

Muscle Soreness and Muscle Cramps

Muscle soreness generally results from exercise that is exhaustive or of very high intensity. This is particularly true when people perform a specific exercise for the first time. While muscle soreness can be felt at any time, there is generally a period of mild muscle soreness that can be felt during and immediately after exercise and then a more intense soreness felt a day or two later.

Acute Muscle Soreness

Pain felt during and immediately after exercise is classified as a muscle strain and is perceived as muscle stiffness, aching, or tenderness. It can result from accumulation of the end products of exercise, such as H^+, and from tissue edema that is caused by fluid shifting from the blood plasma into the tissues. Edema is the cause of the acute muscle swelling that people feel after heavy endurance or strength training. The pain and soreness usually disappear within several hours after the exercise. Thus, this soreness is often referred to as **acute muscle soreness**.

Delayed-Onset Muscle Soreness

The precise causes of muscle soreness felt a day or two after a heavy bout of exercise are not totally understood.

RESEARCH PERSPECTIVE 5.3

Are Muscle Fatigue and Exercise Inefficiency the Same Thing?

During whole-body exercise, fatigue and decreased efficiency (the ratio of mechanical energy output, or external work performed, to metabolic energy production) are major causes of exercise intolerance and resulting early termination of an acute exercise bout. Exercise physiologists agree that these two concepts, fatigue and decreased efficiency, contribute to exercise intolerance, but are they linked?

Decreased efficiency typically precedes exercise termination during high-intensity exercise. There is an increased oxygen cost of work during constant power output and incremental exercise above the lactate threshold. This is most clearly seen with $\dot{V}O_2$ drift during steady-state exercise, where $\dot{V}O_2$ slowly increases despite a constant power output; this increased oxygen cost for the same amount of work is evidence of a decreased efficiency of muscle contraction. Similarly, during incremental exercise, $\dot{V}O_2$ increases in excess of predicted need for power outputs above the lactate threshold. In fact, during incremental exercise, efficiency is reduced by approximately 20% for power outputs above the lactate threshold. As a consequence of this inefficiency, the exerciser reaches his or her peak $\dot{V}O_2$ at a lower power output, resulting in muscle fatigue and ultimately failure to continue. Overall, the decline in skeletal muscle efficiency during high-intensity exercise above the lactate threshold dictates a greater oxygen demand to produce the same mechanical power, termed inefficiency. Because the energy produced by the muscle is limited, the rate at which this inefficiency develops is a major determinant of fatigue and task failure.

But what is causing muscle inefficiency at high power outputs? The fact that muscle fatigue during whole-body exercise occurs only at intensities above the lactate threshold (where ATP production relies on contributions from substrate-level phosphorylation) may provide clues, and the factors affecting the ratio between ATP resynthesis and oxygen consumption by the muscle mitochondria may provide the answer. A 2014 study using combined magnetic resonance spectroscopy and pulmonary gas exchange in humans found that the tight relation between ATP production and $\dot{V}O_2$ that is observed at moderate-intensity exercise was lost at exercise intensities above the lactate threshold.[5] This finding suggests that muscle inefficiency is due to impairments in ATP production and turnover.

As discussed in this chapter, many intracellular mechanisms, including changes in oxygen and substrate availability, impaired function of the ATPases, decreased pH, increased temperature, altered Na^+/K^+ pump function, and changes in motor unit recruitment patterns have all been studied and shown to contribute to reduced muscle efficiency and muscle fatigue. Impairments in ATP turnover and production challenge the cellular homeostasis of the muscle fiber. In this scenario, muscle fatigue may be a protective mechanism that prevents muscle fiber damage. Changes in muscle cellular processes during high-intensity aerobic exercise provide a link between this inefficiency and muscle fatigue that ultimately leads to failure.[13] However, no one has shown a clear cause–effect relation between muscle fatigue and decreased efficiency. Future research is necessary to tell if fatigue and inefficiency are in fact the same thing.

Because this pain does not occur immediately, it is referred to as **delayed-onset muscle soreness (DOMS)**. Delayed-onset muscle soreness can vary from slight muscle stiffness to severe, debilitating pain that restricts movement. In the following sections, we discuss some theories that attempt to explain this form of muscle soreness.

Almost all current theories acknowledge that eccentric muscle action is the primary initiator of DOMS. This has been clearly demonstrated in a number of studies examining the relationship of muscle soreness to eccentric, concentric, and static actions. Individuals who train solely with eccentric actions experience extreme muscle soreness, whereas those who train using only static and concentric actions experience little soreness. This idea has been further explored in studies in which subjects ran on a treadmill for 45 min on two separate days, one day on a level grade and the other day on a 10% downhill grade.[26,27] No muscle soreness was associated with the level running. But the downhill running, which required extensive eccentric action, resulted in considerable soreness within 24 to 48 h, even though blood lactate concentrations, previously thought to cause muscle soreness, were much higher with level running.

In the following sections we examine some of the proposed explanations for exercise-induced DOMS. In general, the pathway for developing DOMS begins with structural damage to muscle fibers (microtrauma) and to the surrounding connective tissues. This damage is followed by an inflammatory process that leads to edema as fluid and electrolytes shift into the area. To make matters worse, muscle spasms can occur, prolonging the condition and making the soreness worse.

Structural Damage

The presence of increased concentrations of several specific muscle enzymes in blood after intense exercise suggests that some structural damage may occur in the muscle membranes. These enzyme concentrations in the blood increase from 2 to 10 times following bouts of heavy training. Recent studies support the idea that these changes might indicate various degrees of muscle tissue breakdown. Examination of tissue from the leg muscles of marathon runners has revealed remarkable damage to the muscle fibers after both training and marathon competition. The onset and timing of these muscle changes parallel the degree of muscle soreness experienced by the runners.

Figure 5.14 shows changes in the contractile filaments and Z-disks before and after a marathon race. Recall that Z-disks are the points of contact for the contractile proteins. They provide structural support for the transmission of force when the muscle fibers are activated to shorten. Figure 5.14*b*, after the marathon, shows moderate Z-disk streaming and major disruption of the thick and thin filaments in a parallel group of sarcomeres as a result of the force of eccentric actions or stretching of the tightened muscle fibers.

Although the effects of muscle damage on performance are not fully understood, it is generally agreed that this damage is responsible in part for the localized muscle pain, tenderness, and swelling associated with DOMS. However, blood enzyme concentrations might increase and muscle fibers might be damaged frequently during daily exercise that produces no muscle soreness. Also, remember that muscle damage appears to be a precipitating factor for muscle hypertrophy.

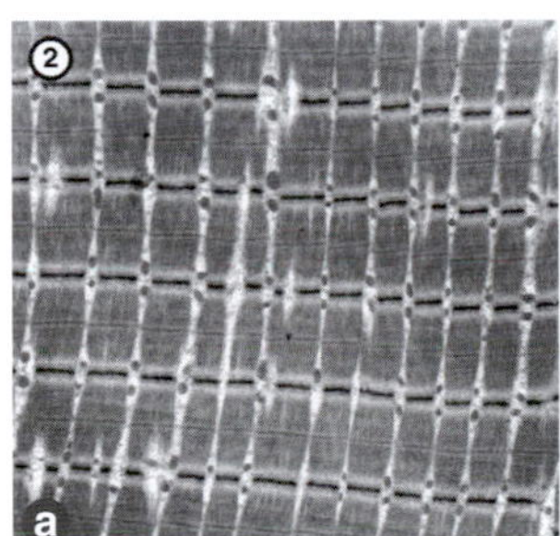

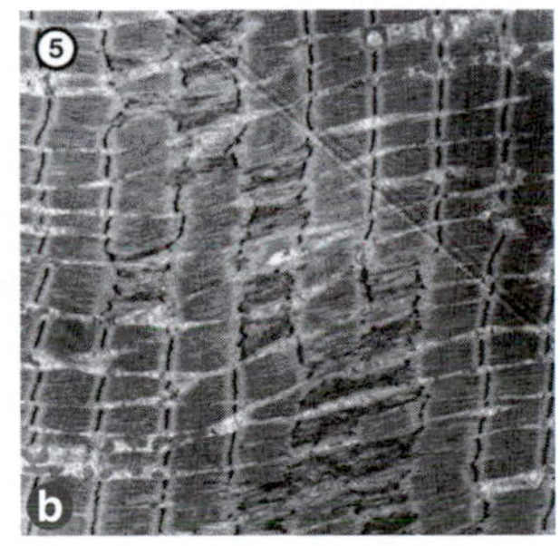

FIGURE 5.14 *(a)* An electron micrograph showing the normal arrangement of the actin and myosin filaments and Z-disk configuration in the muscle of a runner before a marathon race. *(b)* A muscle sample taken immediately after a marathon race shows moderate Z-disk streaming and major disruption of the thick and thin filaments in a parallel group of sarcomeres, caused by the eccentric actions of running.

Reprinted by permission from S.M. Roth et al., "High-Volume, Heavy-Resistance Strength Training and Muscle Damage in Young and Older Women," *Journal of Applied Physiology* 88 (2000): 1112-1118. Image courtesy of Dr. Roth.

Inflammatory Reaction

White blood cells serve as a defense against foreign materials that enter the body and against conditions that threaten the normal function of tissues. The white blood cell count tends to increase following activities that induce muscle soreness, leading some investigators to suggest that soreness results from inflammatory reactions in the muscle. But the link between these reactions and muscle soreness has been difficult to establish.

In early studies, researchers attempted to use drugs to block the inflammatory reaction, but these efforts were unsuccessful in reducing either the amount of muscle soreness or the degree of inflammation. These early results did not support a link between simple inflammatory mediators and DOMS. However, more recent studies have begun to establish a link between muscle soreness and inflammation. It is now recognized that substances released from injured muscle can act as attractants, initiating the inflammatory process. Mononucleated cells in muscle are activated by the injury, providing the chemical signal to circulating inflammatory cells. Neutrophils (a type of white blood cell) invade the injury site and release cytokines (immunoregulatory substances), which then attract and activate additional inflammatory cells. Neutrophils possibly also release oxygen free radicals that can damage cell membranes. The invasion of these inflammatory cells is also associated with the incidence of pain, thought to be caused by a release of substances from the inflammatory cells stimulating the pain-sensitive nerve endings. Macrophages (another type of cell of the immune system) then invade the damaged muscle fibers, removing debris through a process known as phagocytosis. Last, a second phase of macrophage invasion occurs, which is associated with muscle regeneration.

Sequence of Events in DOMS

The general consensus among researchers is that a single theory or hypothesis cannot explain the mechanism causing DOMS. Instead researchers have proposed a sequence of events that may explain the DOMS phenomenon, including the following:

1. High tension in the contractile-elastic system of muscle results in structural damage to the muscle and its cell membrane. This is also accompanied by excessive strain of the connective tissue.
2. The cell membrane damage disturbs calcium homeostasis in the injured fiber, inhibiting cellular respiration. The resulting high calcium concentrations activate enzymes that degrade the Z-lines.

3. Within a few hours there is a significant elevation in circulating neutrophils that participate in the inflammatory response.
4. The products of macrophage activity and intracellular contents (such as histamine, kinins, and K^+) accumulate outside the cells. These substances then stimulate the free nerve endings in the muscle. This process appears to be accentuated in eccentric exercise, in which large forces are distributed over relatively small cross-sectional areas of the muscle.
5. Fluid and electrolytes shift into the area, creating edema, which causes tissue swelling and activates pain receptors. Muscle spasms may also be present.

DOMS and Performance

With DOMS comes a reduction in the force-generating capacity of the affected muscles. Whether the DOMS is the result of injury to the muscle or edema, the affected muscles are not able to exert as much force when the person is asked to apply maximal force, as in the performance of a 1-repetition maximum strength test. Maximal force-generating capacity gradually returns over days or weeks. The loss of strength is the result of

1. the physical disruption of the muscle as illustrated in figure 5.14,
2. failure within the excitation–contraction coupling process, and
3. loss of contractile protein.

Failure in excitation–contraction coupling appears to be the most important, particularly during the first 5 days. This is illustrated in figure 5.15.

Muscle glycogen resynthesis also is impaired when a muscle is damaged. Resynthesis is generally normal for the first 6 to 12 h after exercise, but it slows or stops completely as the muscle undergoes repair, thus limiting the fuel storage capacity of the injured muscle. Figure 5.16 illustrates the time sequence of the various markers of muscle damage associated with intense eccentric exercise of the elbow flexor muscles as compared to concentric exercise. As shown in the figure, changes in function (MVC and range of motion), muscle swelling (circumference), soreness, and molecular indicators of damage (creatine kinase activity and myoglobin concentration) persist for several days.

Minimizing DOMS

Reducing the negative effects of DOMS is important for maximizing training gains. The eccentric component of muscle action could be minimized during early training, but this is not possible for athletes in most sports.

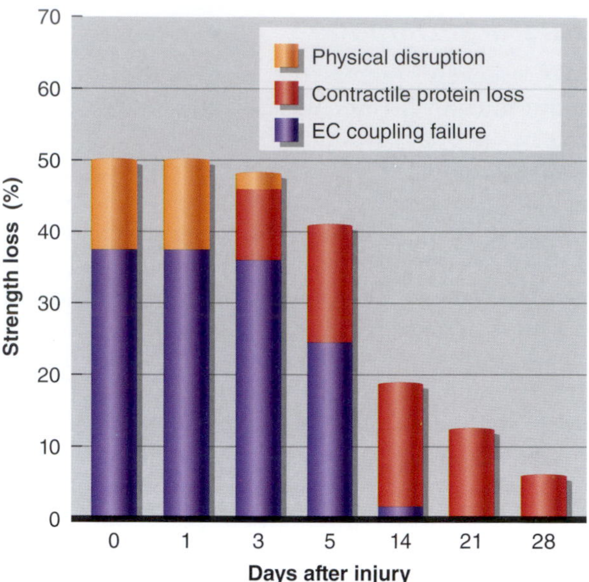

FIGURE 5.15 Estimated contributions of excitation–contraction (EC) coupling failure, decreased contractile protein content, and physical disruption to the decrease in strength following muscle injury.

Reprinted by permission from G. Warren et al., "Excitation-Contraction Uncoupling: Major Role in Contraction-Induced Muscle Injury," *Exercise and Sport Sciences Reviews* 29, no. 2 (2001): 82-87

An alternative approach is to start training at a very low intensity and progress slowly through the first few weeks. Yet another approach is to initiate the training program with a high-intensity, exhaustive training bout. Muscle soreness would be great for the first few days, but evidence suggests that subsequent training bouts would cause considerably less muscle soreness. Because the factors associated with DOMS are also potentially important in stimulating muscle hypertrophy, DOMS is most likely necessary to maximize the training response.

Statins and Skeletal Muscle Soreness

3-Hydroxy-3-methylglutaryl (HMG)-CoA reductase inhibitors, or statins, are the most commonly prescribed cardiovascular drugs in the world. Statins are extremely effective at reducing serum cholesterol concentration and reducing the risk of future cardiovascular events. The most common side effect associated with taking statins is muscle pain, which is reported to occur in up to 25% of patients.[23] Muscle pain from statins can range from mild soreness including cramps and weakness to a life-threatening condition associated with a severe breakdown of muscle tissue called rhabdomyolysis. While the precise mechanism for how statins may contribute to muscle soreness and damage is unclear, it has been linked to excessive production of reactive oxygen molecules by the mitochondria and changes in the way muscle cells get rid of damaged proteins.

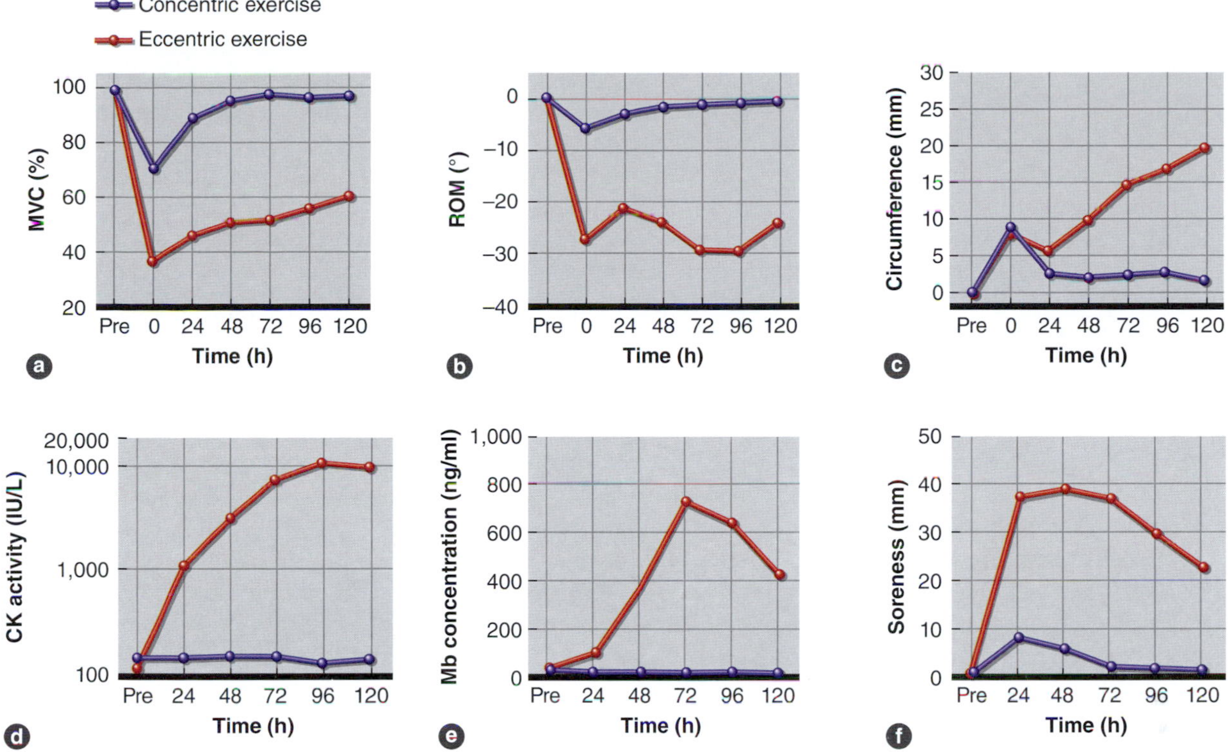

FIGURE 5.16 The responses of various physiological markers of muscle damage after eccentric and concentric exercise by elbow flexors. The changes persist for several days and include *(a)* MVC and *(b)* range of motion (ROM), both indicators of muscle function; *(c)* muscle swelling (circumference); *(d)* creatine kinase (CK) and *(e)* plasma myoglobin (Mb) concentration, both molecular indicators of damage; and *(f)* soreness.

Reprinted by permission from K. Nosaka, Muscle Soreness and Damage and the Repeated-Bout Effect, in *Skeletal Muscle Damage and Repair,* edited by Peter Tiidus (Champaign, IL: Human Kinetics, 2008). Data from A.P. Lavender and K. Nosaka, "Changes in Steadiness of Isometric Force Following Eccentric and Concentric Exercise," *European Journal of Applied Physiology* 96 (2006): 235-240.

Statin use increases creatine kinase concentration after eccentric exercise, a clinical marker of muscle protein breakdown. However, patients who take statins may have muscle pain without an increase in creatine kinase, suggesting that other mechanisms might be causing pain.[22] While trained individuals may be able to tolerate pain during vigorous exercise, in some people, statin-associated muscle pain may limit even leisure-type physical activity.[4] Additionally, recent exercise training studies in older people indicate that statin users do not fully adapt to the training stimulus.[17] Because exercise is a cornerstone therapy for treating and preventing cardiovascular disease, much more research needs to be done to more fully understand the effects of statins on skeletal muscle physiology and how to optimize the beneficial effects of both therapies.

In Review

- Acute muscle soreness occurs during or immediately after an exercise bout.
- Delayed-onset muscle soreness usually peaks a day or two after the exercise bout. Eccentric muscle action seems to be the primary initiator of this type of soreness.
- Proposed causes of DOMS include structural damage to muscle cells and inflammatory reactions within the muscles. The proposed sequence of events includes structural damage, impaired calcium homeostasis, inflammatory response, increased macrophage activity, and edema.
- Reduced muscle strength with DOMS is likely the result of physical disruption of the muscle, failure of the excitation–contraction process, and loss of contractile protein.
- Muscle soreness can be minimized through the use of lower intensity and fewer eccentric contractions early in training. However, muscle soreness may ultimately be an important part of maximizing the resistance training response.

RESEARCH PERSPECTIVE 5.4

Delayed-Onset Muscle Soreness May Be Different in Men and Women

Creatine kinase is the enzyme that catalyzes the exchange of high-energy phosphate bonds between phosphocreatine and ADP to supply ATP to the working muscle during exercise. When creatine kinase appears in the blood, it can indicate metabolic and mechanical disturbances in the muscle cell. Following eccentric exercise in men, creatine kinase activity measured in the blood correlates with muscle soreness and decrements in maximal isometric strength. Men have higher blood creatine kinase activity compared to women, which may be due to the actions of circulating estrogens. The creatine kinase response to exercise may be lower when women are tested during periods of higher circulating estrogen concentration (e.g., the late follicular phase preceding ovulation) compared to periods of low circulating estrogen. Some studies of sex differences in reported muscle soreness suggest that women have a reduced perception of soreness following eccentric exercise compared to men, but other studies have found no differences. Interestingly, the creatine kinase response to exercise is correlated with delayed-onset muscle soreness (DOMS) in men but not in women. Altogether, there is still debate about whether sex differences in the creatine kinase response to exercise, and in DOMS, exist in humans.

A recent study conducted in South Africa set out to determine whether the serum creatine kinase response and the perception of DOMS following a bout of downhill running were influenced by sex and if the magnitudes of those responses depended on the circulating estrogen concentrations in the women.[19] In that study, 21 sedentary subjects (6 men and 15 women) performed 20 min of downhill running on a treadmill in the laboratory. Blood samples were collected before exercise, immediately after exercise, and then 24, 48, and 72 h after exercise. Blood samples were analyzed for creatine kinase activity and for estrogen and progesterone concentrations in the women. The researchers also assessed DOMS at the same time points by having the subjects rate their perception of soreness when standing from a seated position on a visual scale from "no pain" to "the worst pain ever experienced."

The 24 h peak creatine kinase response to this bout of downhill running was the same between men and women; however, circulating creatine kinase activity was restored to preexercise baseline faster in women (by 48 h postexercise) than in men (72 h after exercise). Neither estrogen nor progesterone influenced the creatine kinase response in the women. Interestingly, both men and women still reported muscle soreness at 72 h after the eccentric exercise bout despite the recovery of creatine kinase in women 24 h earlier. While feelings of muscle soreness were prolonged in women who participated during the follicular phase of their menstrual cycle, the researchers were not able to determine if the associated hormones (estrogen or progesterone) were responsible for those findings.

Overall, the study concluded that both creatine kinase and DOMS responses to downhill running are affected by sex. Creatine kinase recovers more quickly in women, regardless of circulating reproductive hormone concentrations, but the recovery of muscle soreness is only correlated with creatine kinase concentrations in men. While the DOMS response in women may be affected by menstrual phase, a direct link to circulating hormones has not been demonstrated.

Exercise-Induced Muscle Cramps

Skeletal muscle cramps are a frustrating problem in sport and physical activity and commonly occur even in highly fit athletes. Skeletal muscle cramps can come during the height of competition, immediately after competition, or at night during deep sleep. Muscle cramps are equally frustrating to scientists, because there are multiple and unknown triggers and causes of muscle cramping, and little is known about the best treatment and prevention strategies. Nocturnal muscle cramps, especially in the calf muscle, have been experienced by 60% of adults. This type of cramp is probably caused by muscle fatigue and nerve dysfunction and may or may not be associated with exercise. Electrolyte imbalances and hydration do not seem to play an important role.

Exercise-associated muscle cramps (EAMCs), on the other hand, are defined as painful, spasmodic, involuntary contractions of skeletal muscles that occur during or immediately after exercise. Two distinct theories have been proposed to explain the causes of EAMCs, termed the *neuromuscular control theory* and the *electrolyte depletion theory*.[2]

The neuromuscular control theory proposes that EAMCs occur when some aspect of control between the motor neuron and the muscle itself becomes altered. As muscle fatigue develops, excitation of the muscle spindle and inhibition of the Golgi tendon organ occur, resulting in abnormal α-motor neuron activity and reduced inhibitory feedback. This abnormal firing of motor neurons initially presents as muscle twitches or prefasciculations. If muscle contraction continues, an EAMC occurs.

Risk factors associated with this type of cramping are age, cramping history, and excessive exercise intensity and duration.

Several lines of evidence support this theory:

1. This type of cramping is generally localized to the overworked muscle.
2. Lack of conditioning, improper training, and depletion of muscle energy stores, which are all associated with muscle fatigue, can lead to the development of EAMCs.
3. Muscle cramping can be induced in the laboratory by electrical stimulation or voluntary muscle contraction (with no changes in electrolytes), which suggests that the mechanism is neuromuscular in origin.
4. Often the most effective treatment for relieving cramps is stretching of the muscle. Stretching increases tension in the muscle and in the Golgi tendon organ that inhibits the α-motor neuron.
5. Changing excitatory properties of the motor neuron—for example, by ingesting transient receptor potential (TRP) channel agonists—has been efficacious in attenuating both electrically induced and voluntarily induced EAMCs.[7]

The second theory, electrolyte depletion, better describes a different type of exercise-associated muscle cramp, often called *heat cramps*. This type of muscle cramp typically occurs in athletes who have been sweating extensively and have significant electrolyte disturbances, mainly of sodium and chloride. These cramps involve large muscle groups, and their occurrence is sometimes described as "locking up." This type of exertional heat cramp usually evolves from small localized visible muscle fasciculations to severe and debilitating muscle spasms. The cramps often begin in the legs but can become widespread.

Because a significant amount of sodium can be lost only along with a large loss of fluid, this theory is typically coupled with dehydration. Progressive dehydration and electrolyte depletion cause fluid to shift from the interstitial compartment to the intravascular compartment. This contracts the extracellular fluid compartment, increasing surrounding neurotransmitter concentrations and causing selected motor nerve terminals to become hyperexcitable, leading to spontaneous discharge, initiation of action potentials in the muscles, and ultimately EAMCs.

Proponents of this theory have espoused the following:

1. Anecdotal evidence has existed for centuries that laborers working in hot and humid conditions suffered from cramping.
2. In those laborers, ingesting small volumes of salt water prevented or alleviated the cramps.
3. Increases in sweat sodium concentration ("salty sweating") is evident in those athletes, specifically tennis and American football players, who are most prone to cramping.[9]

VIDEO 5.1 Presents Mike Bergeron on the two types of muscle cramping and the best ways to prevent muscle cramps. 

Treatment of heat cramps involves the prompt ingestion of a high-salt solution (3 g in 500 ml of a sodium-containing beverage every 5 to 10 min) or intravenous fluid and sodium loading. In addition, massage and ice application may help to calm the affected muscles and relieve pain. Electrolyte-containing fluids should be continued if dehydration and electrolyte loss are suspected.

To prevent EAMCs, the athlete should

- be well conditioned, to reduce the likelihood of muscle fatigue;
- regularly stretch the muscle groups prone to EAMCs;
- maintain fluid and electrolyte balance and carbohydrate stores; and
- reduce exercise intensity and duration if necessary.

In Review

- Muscle fatigue-associated cramps are related to sustained α-motor neuron activity, with increased muscle spindle activity and decreased Golgi tendon organ activity.
- Exercise-associated muscle cramps may be caused by altered neuromuscular control, fluid or electrolyte imbalances, or both.
- Heat-associated cramps, which typically occur in athletes who have been sweating excessively, involve a shift in fluid from the interstitial space to the intravascular space, resulting in a hyperexcitable neuromuscular junction.
- Rest, passive stretching, holding the muscle in the stretched position, and fluid and electrolyte restoration can be effective in treating EAMCs. Proper conditioning, stretching, and nutrition are also possible prevention strategies.

IN CLOSING

In previous chapters, we discussed how muscles and the nervous system function together to produce movement. In this chapter we focused on energy expenditure during exercise and fatigue. We considered the energy needed by the body at rest and during movement. We explored how energy production and availability can limit performance and learned that metabolic needs vary considerably. We discussed the many potential factors involved in fatigue, including those resulting from decreased energy delivery and accumulation of metabolic by-products and those associated with the peripheral and central nervous systems. We also introduced the concept of critical power as a link between energy expenditure and fatigue. We also examined delayed-onset muscle soreness and muscle cramps as additional limiting factors in exercise. In the next chapter, we turn our attention to the cardiovascular system and its control.

KEY TERMS

acute muscle soreness
basal metabolic rate (BMR)
calorie (cal)
calorimeter
cardiorespiratory endurance
central governor theory
critical power
delayed-onset muscle soreness (DOMS)
direct calorimetry
excess postexercise oxygen consumption (EPOC)
exercise-associated muscle cramps (EAMCs)
fatigue
Haldane transformation
indirect calorimetry
lactate threshold
maximal oxygen uptake ($\dot{V}O_{2max}$)
oxygen deficit
peak oxygen uptake ($\dot{V}O_{2peak}$)
respiratory exchange ratio (RER)
resting metabolic rate (RMR)
$\dot{V}O_2$ drift

STUDY QUESTIONS

1. Define direct calorimetry and indirect calorimetry, and describe how they are used to measure energy expenditure.
2. What is the respiratory exchange ratio (RER)? Explain why it is used to determine the relative contributions of carbohydrate and fat to energy expenditure.
3. What are basal metabolic rate and resting metabolic rate, and how do they differ?
4. What is maximal oxygen uptake? How is it measured? What is its relationship to sport performance?
5. Describe two possible markers of anaerobic capacity.
6. What is the lactate threshold? How is it measured? What is its relation to sport performance?
7. What is economy of effort? How is it measured? What is its relationship to sport performance?
8. What is the relationship between oxygen consumption and energy production?
9. Why do athletes with high $\dot{V}O_{2max}$ values perform better in endurance events than those with lower values?
10. Why is oxygen consumption often expressed as milliliters of oxygen per kilogram of body weight per minute ($ml \cdot kg^{-1} \cdot min^{-1}$)?
11. Describe the possible causes of fatigue during exercise bouts lasting 15 to 30 s and those lasting 2 to 4 h.
12. Discuss three mechanisms through which lactate can be used as an energy source.
13. Define critical power and explain its usefulness in sport physiology. What is its relation to sport performance?
14. What is the physiological basis for delayed-onset muscle soreness?
15. What two theories have been proposed to explain the physiological basis for exercise-associated muscle cramps? Provide support for each.

STUDY GUIDE ACTIVITIES

In addition to the activities listed in the chapter opening outline, two other activities are available in the web study guide, located at

www.HumanKinetics.com/PhysiologyOfSportAndExercise

The **KEY TERMS** activity reviews important terms, and the end-of-chapter **QUIZ** tests your understanding of the material covered in the chapter.

PART II

Cardiovascular and Respiratory Function

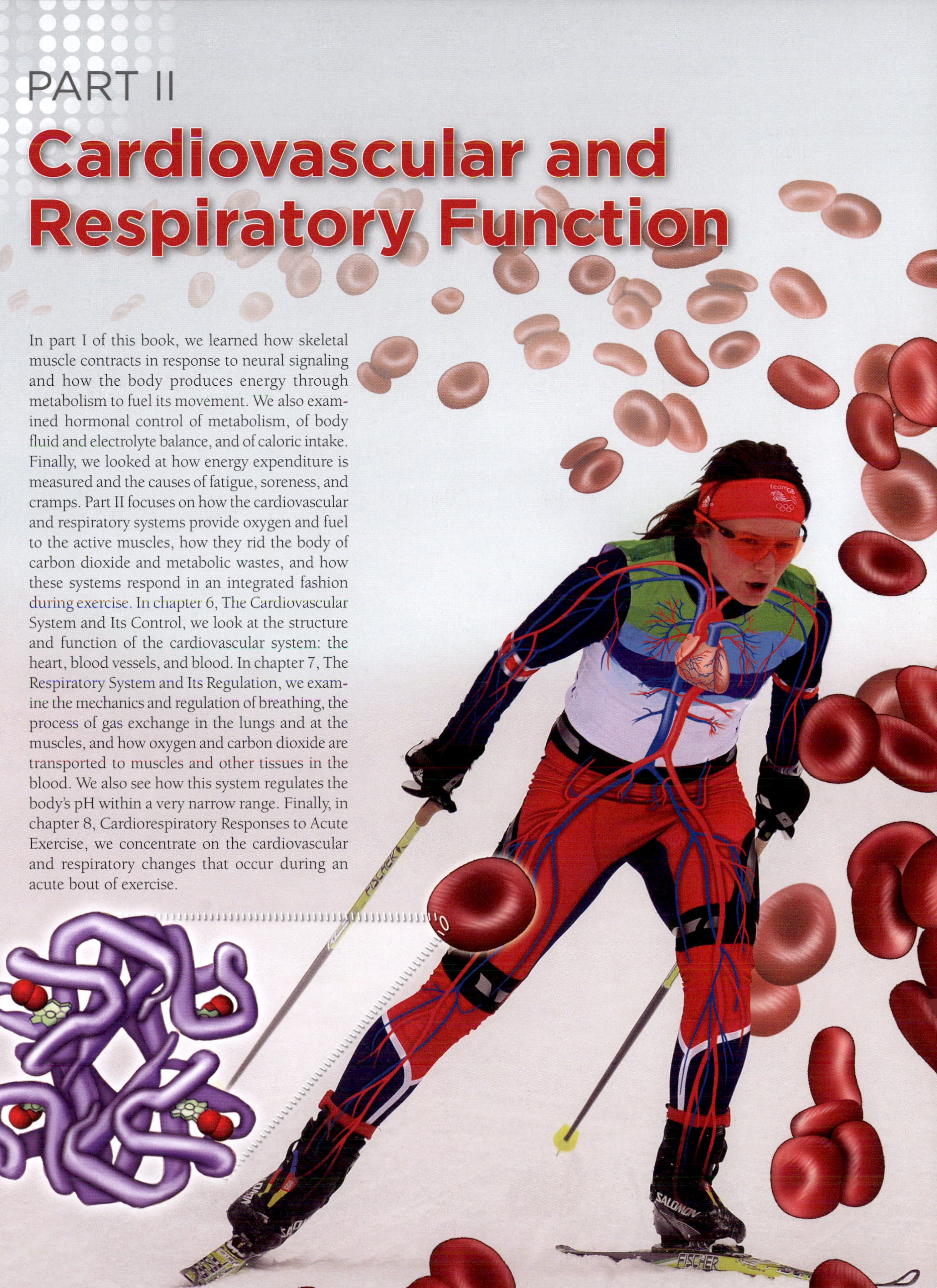

In part I of this book, we learned how skeletal muscle contracts in response to neural signaling and how the body produces energy through metabolism to fuel its movement. We also examined hormonal control of metabolism, of body fluid and electrolyte balance, and of caloric intake. Finally, we looked at how energy expenditure is measured and the causes of fatigue, soreness, and cramps. Part II focuses on how the cardiovascular and respiratory systems provide oxygen and fuel to the active muscles, how they rid the body of carbon dioxide and metabolic wastes, and how these systems respond in an integrated fashion during exercise. In chapter 6, The Cardiovascular System and Its Control, we look at the structure and function of the cardiovascular system: the heart, blood vessels, and blood. In chapter 7, The Respiratory System and Its Regulation, we examine the mechanics and regulation of breathing, the process of gas exchange in the lungs and at the muscles, and how oxygen and carbon dioxide are transported to muscles and other tissues in the blood. We also see how this system regulates the body's pH within a very narrow range. Finally, in chapter 8, Cardiorespiratory Responses to Acute Exercise, we concentrate on the cardiovascular and respiratory changes that occur during an acute bout of exercise.

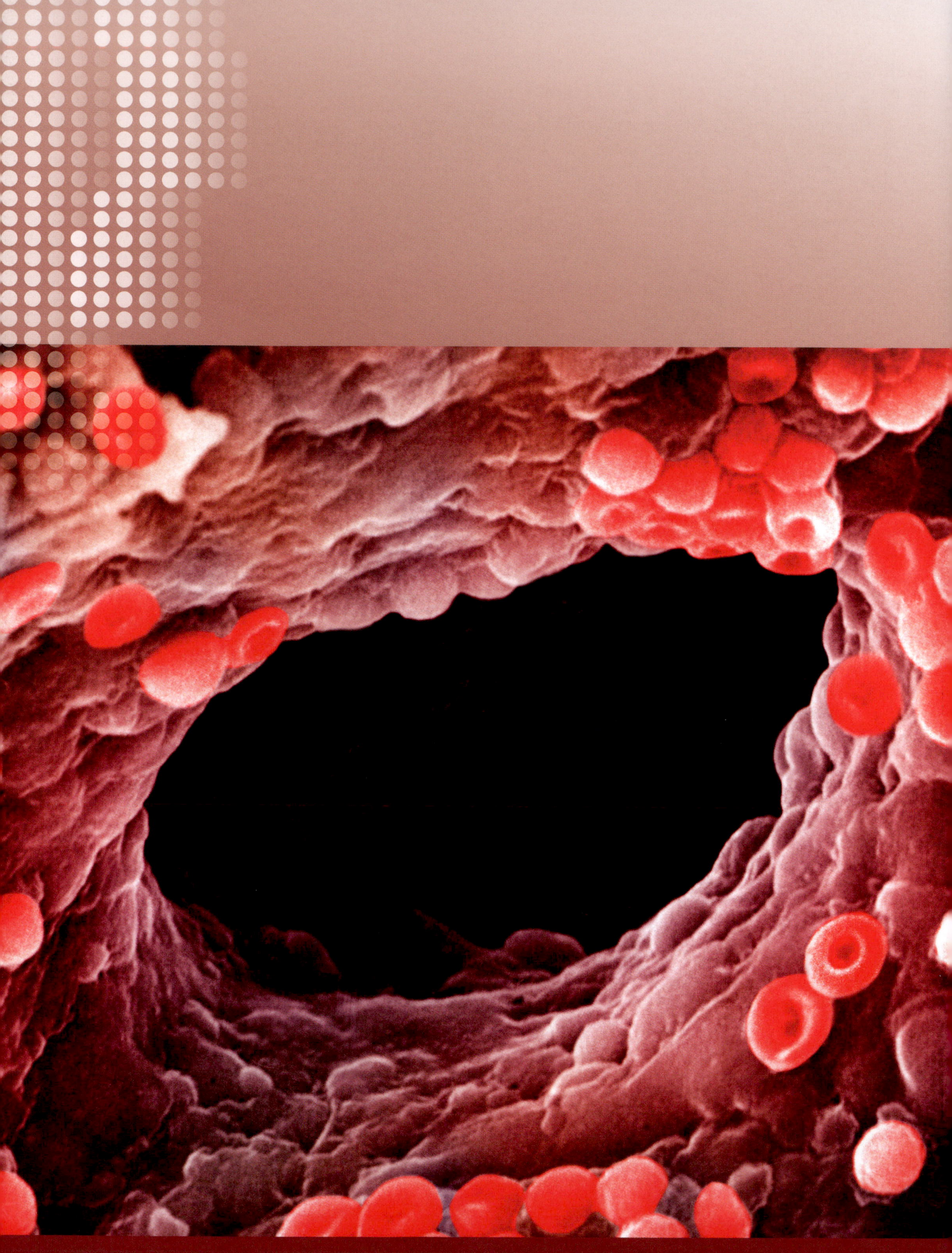

The Cardiovascular System and Its Control

In this chapter and in the web study guide

Rod Williams was a 17-year-old high school junior, an offensive lineman on the football team. On September 22, 2015, Williams collapsed on the football field with no heartbeat or respirations. Despite successful CPR and hospitalization, he died 2 weeks later. Like many tragic, sudden cardiac arrest deaths in young athletes, an autopsy revealed that Williams had a preexisting heart condition that went undetected. In fact, sudden cardiac arrest is the leading cause of death in high school athletes, resulting from underlying heart anomalies that were exposed only during intense physical activity. The most common of these is hypertrophic cardiomyopathy, a genetic disease of the heart muscle, but many other causes exist, such as long QT syndrome, an electrical abnormality. Most young people living with these problems have no symptoms. Sadly, the initial manifestation is sudden cardiac arrest. While routine intensive screening seems like the logical answer, the type of advanced screening that is necessary has significant financial constraints, since well over 1 million high school football players are competing in the United States at any one time. Further, because of the low incidence of these abnormalities in the young, healthy population, the incidences of false positives (the test shows a problem where there is none) and false negatives (the test is normal but the problem actually exists) limit the tests' predictive value. Hopefully, more accurate and cost-effective screening tools are on the horizon.

The cardiovascular system serves a number of important functions in the body and supports every other physiological system. Major cardiovascular functions can be grouped into six categories:

- Delivery of oxygen and energy substrates
- Removal of carbon dioxide and other metabolic waste products
- Transport of hormones and other molecules
- Support of thermoregulation and control of body fluid balance
- Maintenance of acid–base balance to help control the body's pH
- Regulation of immune function

Although this is just an abbreviated list of roles, the cardiovascular functions listed here are important for understanding the physiological basis of exercise and sport. Obviously these roles change and become even more critical with the challenges imposed by exercise.

All physiological functions and virtually every cell in the body depend in some way on the cardiovascular system. Any system of circulation requires three components:

- A pump (the heart)
- A system of channels or tubes (the blood vessels)
- A fluid medium (the blood)

In order to keep blood circulating, the heart must generate sufficient pressure to drive blood through the continuous network of blood vessels in a closed-loop system. Thus, the primary goal of the cardiovascular system is to ensure that there is adequate blood flow throughout the circulation to meet the metabolic demands of the tissues. We look first at the heart.

The Heart

About the size of a fist and located in the center of the thoracic cavity, the heart is the primary pump that circulates blood through the entire cardiovascular system. As shown in figure 6.1, the heart has two atria that act as receiving chambers and two ventricles that serve as the pumping chambers. It is enclosed in a tough membranous sac called the **pericardium**. The thin cavity between the pericardium and the heart is filled with pericardial fluid, which reduces friction between the sac and the beating heart.

Blood Flow Through the Heart

The heart is sometimes considered to be two separate pumps, with the right side of the heart pumping deoxygenated blood to the lungs through the pulmonary circulation and the left side of the heart pumping oxygenated blood to all other tissues in the body through the systemic circulation. Blood that has circulated through the body, delivering oxygen and nutrients and picking up waste products, returns to the heart through the great veins—the superior vena cava and inferior vena cava—to the right atrium. This chamber receives all the deoxygenated blood from the systemic circulation.

From the right atrium, blood passes through the tricuspid valve into the right ventricle. This chamber pumps the blood through the pulmonary valve into the pulmonary artery, which carries the blood to the lungs. Thus, the right side of the heart is known as the pulmonary side, sending the blood that has circulated throughout the body into the lungs for reoxygenation.

After blood is oxygenated in the lungs, it is transported back to the heart through the pulmonary veins. All freshly oxygenated blood is received from the pulmonary veins by the left atrium. From the left atrium, the blood passes through the mitral valve into the left ventricle. Blood leaves the left ventricle by passing through the aortic valve into the aorta and is distributed to the systemic circulation. The left side of the heart is known as the systemic side. It receives the oxygenated blood from the lungs and then sends it out to supply all other body tissues.

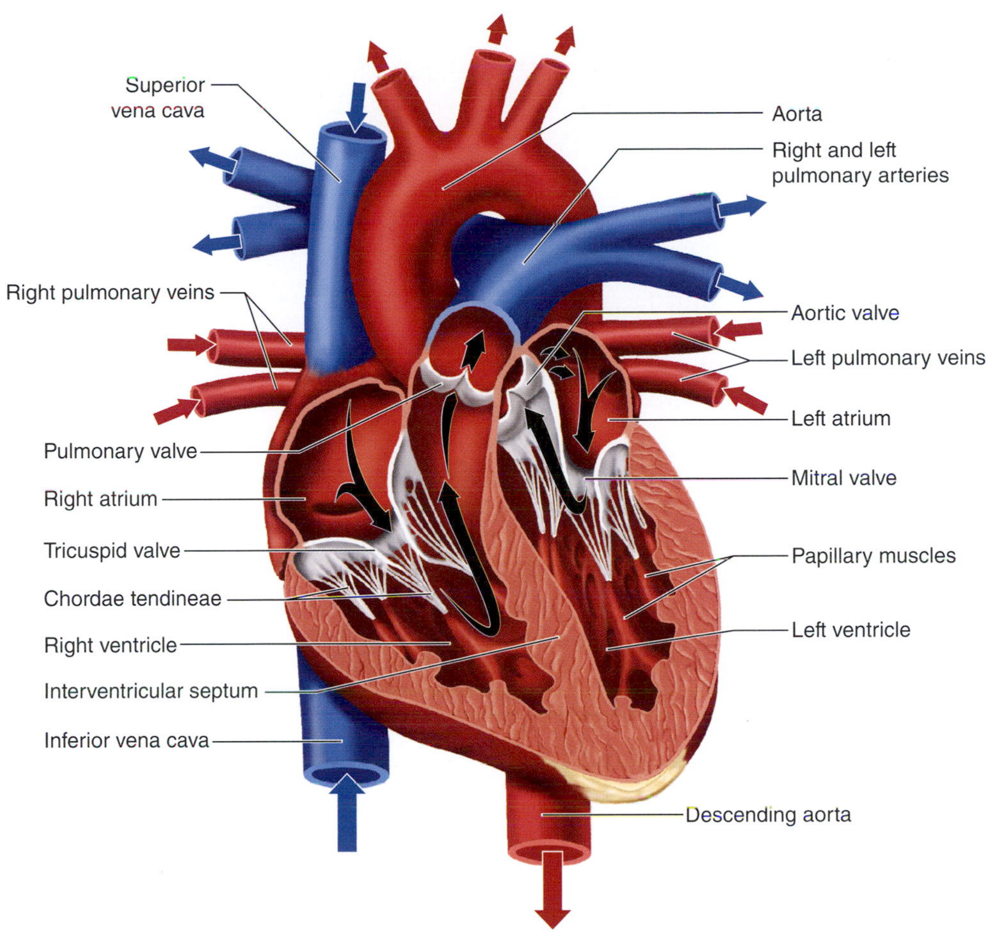

FIGURE 6.1 An anterior (as if the person is facing you) cross-sectional view of the human heart.

The four heart valves prevent backflow of blood, ensuring one-way flow through the heart. These valves maximize the amount of blood pumped out of the heart during contraction. A **heart murmur** is a condition in which abnormal sounds are detected with the aid of a stethoscope. This abnormal sound can indicate the turbulent flow of blood through a narrowed valve (stenosis) or retrograde flow back toward the atria through a leaky valve (prolapse). When valves leak as a result of disease, this condition can require surgical replacement of the valve. With *mitral valve prolapse*, the mitral valve allows some blood to flow back into the left atrium during ventricular contraction. This disorder, relatively common in adults (6%-17% of the population), usually has little clinical significance unless there is significant backflow.

Mild heart murmurs are fairly common in growing children and adolescents. Likewise, most murmurs heard in athletes are benign, affecting neither the heart's pumping nor the athlete's performance. Only when there is a functional consequence, such as light-headedness or dizziness, are murmurs a cause for immediate concern.

The Myocardium

Cardiac or myocardial muscle is collectively called the **myocardium**. Myocardial thickness at various locations in the heart varies according to the amount of stress regularly placed on the myocardium. The left ventricle is the most powerful pump because it must generate sufficient pressure to pump blood through the entire body. When a person is sitting or standing, the left ventricle must contract with enough force to overcome the effect of gravity, which tends to pool blood in the lower extremities.

Because the left ventricle must generate considerable force to pump blood to the systemic circulation, it has the thickest muscular wall compared with the other heart chambers. This hypertrophy is the result of the pressure placed on the left ventricle at rest or under normal conditions of moderate activity. With more vigorous exercise—particularly intense aerobic activity, during which the working muscles' need for blood increases considerably—the demand on the left ventricle to deliver blood to exercising muscles is much higher. In response to both intense aerobic and resistance training, the left ventricle will hypertrophy. In contrast to this positive adaptation that occurs as a result of exercise training, cardiac muscle also hypertrophies as a result of several diseases, such as high blood pressure or valvular heart disease. In response to either training or disease, over time the left ventricle adapts by increasing its size and pumping capacity, similar to the way skeletal muscle adapts to physical training. However, the mechanisms for adaptation and cardiac performance with disease are different from those observed with aerobic training.

Although striated in appearance, the myocardium differs from skeletal muscle in several important ways. First, because the myocardium has to contract as if it were a single unit, individual cardiac muscle fibers are anatomically interconnected end to end by dark-staining regions called **intercalated disks**. These disks have desmosomes, which are structures that anchor the individual cells together so that they do not pull apart during contraction, and gap junctions, which allow rapid transmission of the action potentials that signal the heart to contract as one unit. Secondly, the myocardial fibers are rather homogeneous in contrast to the mosaic of fiber types in skeletal muscle. The myocardium contains only one fiber type, similar to type I fibers in skeletal muscle in that it is highly oxidative, has a high capillary density, and has a large number of mitochondria.

In addition to these differences, the mechanism of muscle contraction also differs between skeletal and cardiac muscle. Cardiac muscle contraction occurs by calcium-induced calcium release (figure 6.2). The action potential spreads rapidly along the myocardial sarcolemma from cell to cell via gap junctions and also to the inside of the cell through the T-tubules. Upon stimulation, calcium enters the cell by the dihydropyridine receptor in the T-tubules. Unlike what happens in skeletal muscle, the amount of calcium that enters the cell is not sufficient to directly cause the cardiac muscle to contract, but it serves as a trigger to another type of receptor, called the ryanodine receptor, to release calcium from the sarcoplasmic reticulum. Figure 6.3 summarizes some of the similarities and differences between cardiac and skeletal muscle.

The myocardium, just like skeletal muscle, must have a blood supply to deliver

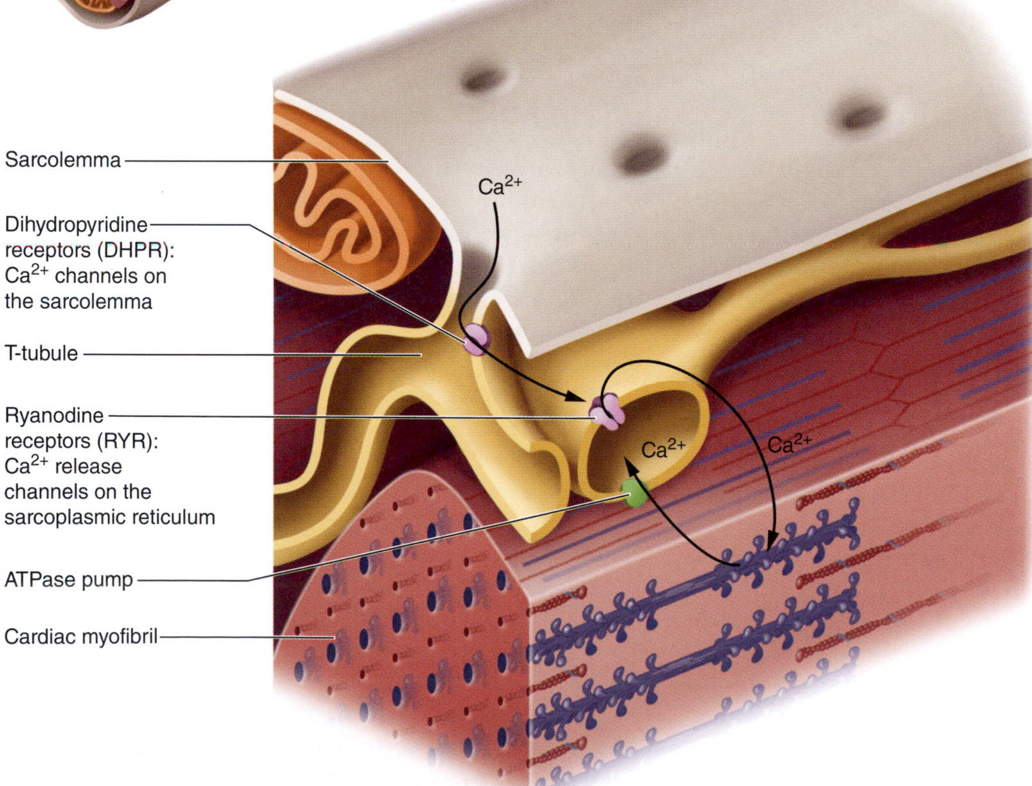

FIGURE 6.2 The mechanism of contraction in a cardiac muscle fiber, termed calcium-induced calcium release.

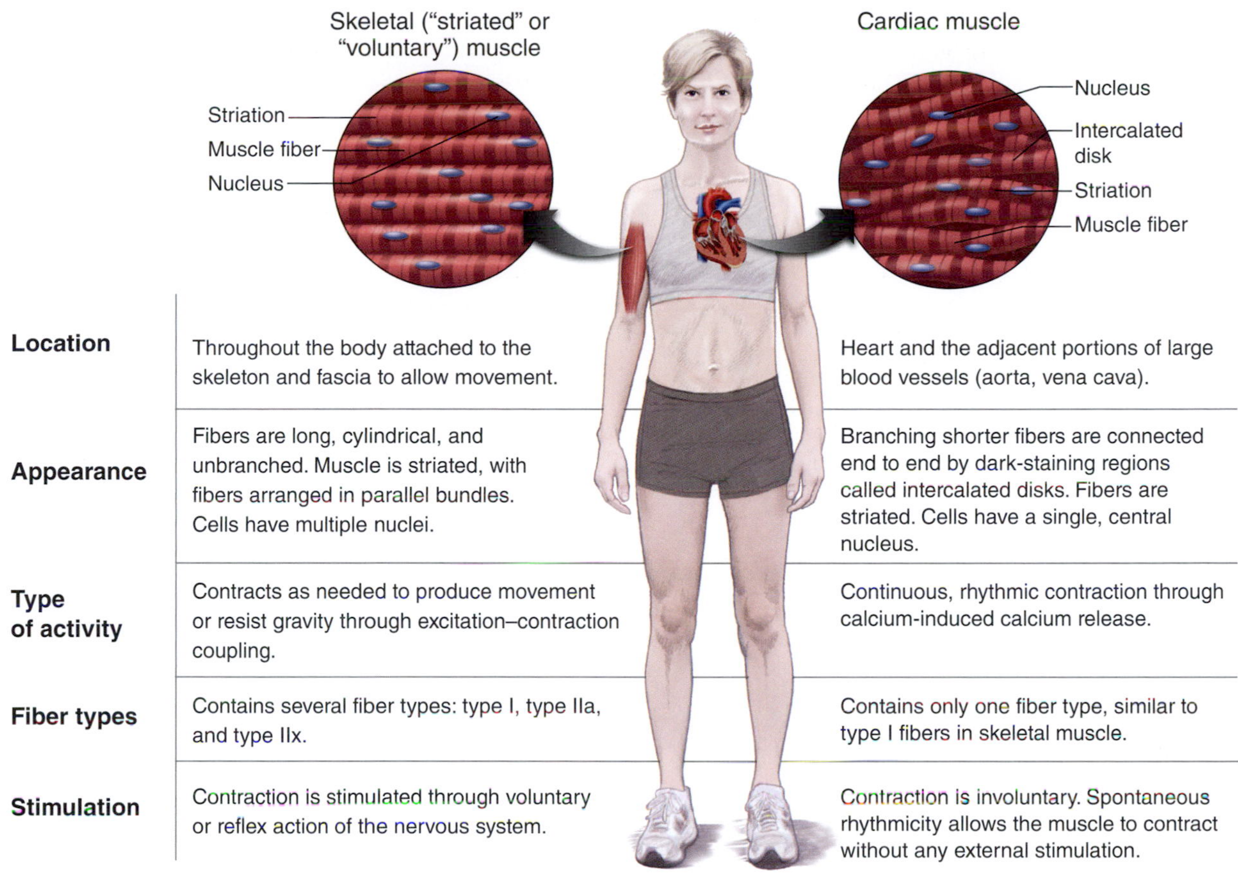

	Skeletal ("striated" or "voluntary") muscle	Cardiac muscle
Location	Throughout the body attached to the skeleton and fascia to allow movement.	Heart and the adjacent portions of large blood vessels (aorta, vena cava).
Appearance	Fibers are long, cylindrical, and unbranched. Muscle is striated, with fibers arranged in parallel bundles. Cells have multiple nuclei.	Branching shorter fibers are connected end to end by dark-staining regions called intercalated disks. Fibers are striated. Cells have a single, central nucleus.
Type of activity	Contracts as needed to produce movement or resist gravity through excitation–contraction coupling.	Continuous, rhythmic contraction through calcium-induced calcium release.
Fiber types	Contains several fiber types: type I, type IIa, and type IIx.	Contains only one fiber type, similar to type I fibers in skeletal muscle.
Stimulation	Contraction is stimulated through voluntary or reflex action of the nervous system.	Contraction is involuntary. Spontaneous rhythmicity allows the muscle to contract without any external stimulation.

FIGURE 6.3 Functional and structural characteristics of skeletal and cardiac muscle.

oxygen and nutrients and remove waste products. Although blood courses through each chamber of the heart, little nourishment comes from the blood within the chambers. The primary blood supply to the heart is provided by the coronary arteries, which arise from the base of the aorta and encircle the outside of the myocardium (figure 6.4). The right coronary artery supplies the right side of the heart, dividing into two primary branches, the marginal artery and the posterior interventricular artery. The left coronary artery, also referred to as the left main coronary artery, also divides into two major branches, the circumflex artery and the anterior descending artery. The posterior interventricular artery and the anterior descending artery merge, or anastomose, in the lower posterior area of the heart, as does the circumflex. Blood flow increases through the coronary arteries when the heart is between contractions (during diastole).

The mechanism of blood flow to and through the coronary arteries is quite different from that of blood flow to the rest of the body. During contraction, when blood is forced out of the left ventricle under high pressure, the aortic valve is forced open. When this valve is open, its flaps block the entrances to the coronary arteries. As the pressure in the aorta decreases, the aortic valve closes, and blood can then enter the coronary arteries. This design ensures that the coronary arteries are spared the very high blood pressure created by contraction of the left ventricle, thus protecting these critical vessels from damage.

The coronary arteries are, however, very susceptible to **atherosclerosis**, or narrowing by the accumulation of plaque and inflammation, leading to coronary artery disease. This disease is discussed in greater detail in chapter 21. Anomalies—vessel shortenings, blockages, or flow misdirections—sometimes occur in the coronary arteries, and such congenital abnormalities are a common cause of sudden death in athletes.

In addition to its unique anatomical structure, the ability of the myocardium to contract as a single unit also depends on initiation and propagation of an electrical signal through the heart, the cardiac conduction system.

The Cardiac Conduction System

Myocardial cells are unique in that they have the ability to spontaneously depolarize and directionally conduct that electrical signal throughout the heart. The rate of depolarization is set by depolarization of a unique type of myocardial cells located in the upper right atrium and also determined by extrinsic influences, including

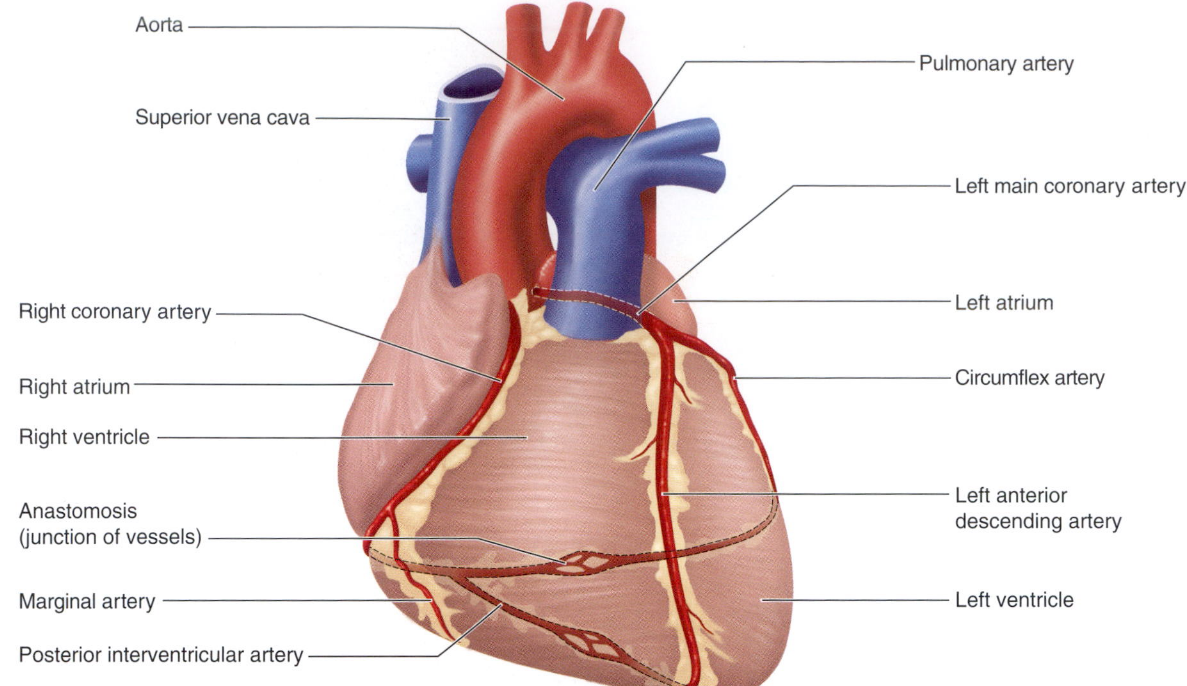

FIGURE 6.4 The coronary circulation, illustrating the right and left coronary arteries and their major branches.

the autonomic nervous system and circulating hormones. The following sections describe the intrinsic and extrinsic mechanisms that combine to determine heart rate and rhythm at rest and during exercise.

Intrinsic Control of Electrical Activity

Cardiac muscle has the unique ability to generate its own electrical signal, called spontaneous rhythmicity, which allows it to contract without any external stimulation. The contraction is rhythmical, in part because of the anatomical coupling of the myocardial cells through gap junctions. Without neural or hormonal stimulation, the intrinsic heart rate averages ~100 beats (contractions) per minute. This resting heart rate of about 100 beats/min can be observed in patients who have undergone cardiac transplant surgery, because their transplanted hearts lack autonomic innervation.

Even though all myocardial fibers have inherent rhythmicity, the heart has a series of specialized myocardial cells that function to coordinate the heart's excitation and contraction and maximize the efficient pumping of blood. These are specialized cardiac muscle fibers, and not nerve tissue, even though they function to generate and transmit a signal. Figure 6.5 illustrates the four main components of the cardiac conduction system:

- Sinoatrial (SA) node
- Atrioventricular (AV) node
- AV bundle (bundle of His)
- Purkinje fibers

The impulse for a normal heart contraction is initiated in the **sinoatrial (SA) node**, a group of specialized fibers located in the upper posterior wall of the right atrium. These specialized cells spontaneously depolarize at a faster rate than other myocardial muscle cells because they are especially leaky to sodium ions. Because this tissue has the fastest intrinsic firing rate, typically at a frequency of about 100 beats/min, the SA node is known as the heart's pacemaker, and the rhythm it establishes is called the *sinus rhythm*. The electrical impulse generated by the SA node spreads through both atria and reaches the **atrioventricular (AV) node**, located in the right atrial wall near the center of the heart. As the electrical impulse spreads through the atria, the atrial myocardium is signaled to contract.

The AV node conducts the electrical impulse from the atria into the ventricles. The impulse is delayed by about 0.13 s as it passes through the AV node, and then it enters the AV bundle. This delay is important because it allows blood from the atria to completely empty into the ventricles to maximize ventricular filling before the ventricles contract. While most blood moves passively from the atria to the ventricles, active contraction of the atria (sometimes called the "atrial kick") completes the process. The AV bundle travels along the ventricular septum and then sends right and left bundle branches into the respective ventricles. These branches send the impulse toward the apex of the heart and then outward. Each bundle branch subdivides into many smaller ones that spread throughout the entire ventricular wall. These terminal branches of the AV bundle are the

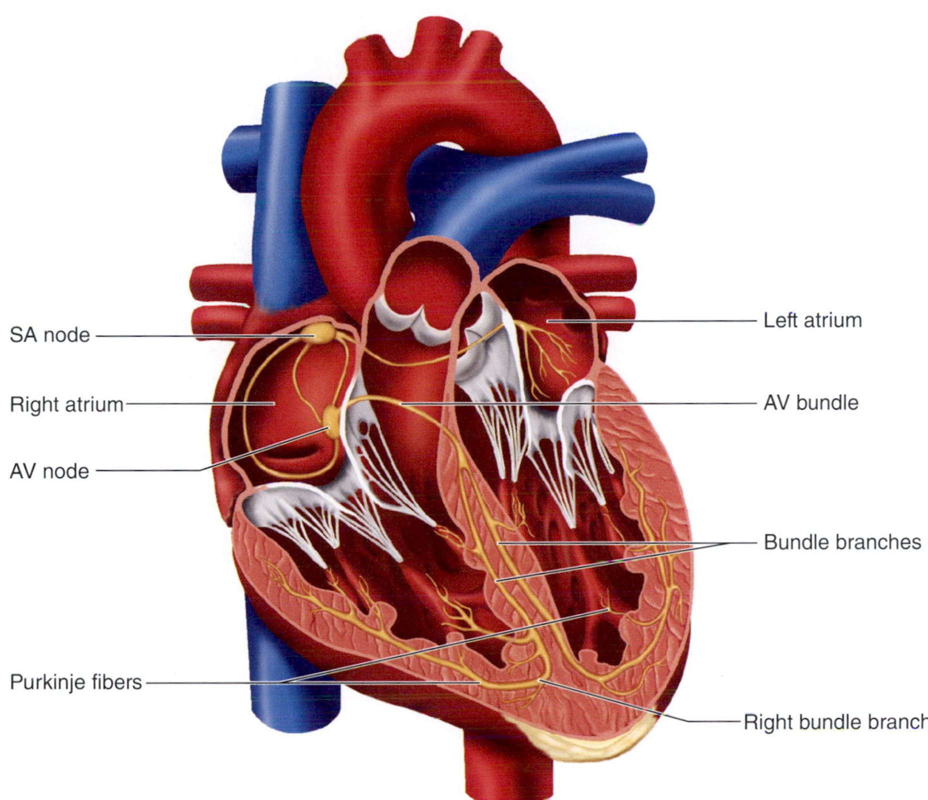

FIGURE 6.5 The specialized conduction system of the heart.

Purkinje fibers. They transmit the impulse through the ventricles approximately six times faster than through the rest of the cardiac conduction system. This rapid conduction allows all parts of the ventricle to contract at virtually the same time.

Occasionally, chronic problems develop within the cardiac conduction system, hampering its ability to maintain appropriate sinus rhythm throughout the heart. In such cases, an artificial pacemaker can be surgically installed. This small, battery-operated electrical stimulator, usually implanted under the skin, has tiny electrodes attached to the right ventricle. For example, in a condition called *AV block,* the SA node creates an impulse, but the impulse is blocked at the AV node and cannot reach the ventricles, resulting in the heart rate's being controlled by the intrinsic firing rate of the pacemaker cells in the ventricles (closer to 40 beats/min). The artificial pacemaker takes over the role of the disabled AV node, supplying the needed impulse and thus controlling ventricular contraction.

Extrinsic Control of Heart Rate and Rhythm

Although the heart initiates its own electrical impulses (intrinsic control), both the heart rate and force of contraction can be altered. Under normal conditions, this is accomplished primarily through three extrinsic systems:

- The parasympathetic nervous system
- The sympathetic nervous system
- The endocrine system (hormones)

Although an overview of these systems' effects is offered here, they are discussed in more detail in chapters 3 and 4.

The parasympathetic system, a branch of the autonomic nervous system, originates centrally in a region of the brain stem called the medulla oblongata and reaches the heart through the vagus nerve (cranial nerve X). The vagus nerve carries impulses to the SA and AV nodes, and when stimulated it releases acetylcholine, which causes hyperpolarization of the conduction cells. The result is a slower spontaneous depolarization and a decrease in heart rate. At rest, parasympathetic system activity predominates and the heart is said to have "vagal tone." Recall that, in the absence of vagal tone, intrinsic heart rate would be approximately 100 beats/min, but the normal resting adult heart rate is typically 60 to 80 beats/min. The vagus nerve has a depressant effect on the heart: It slows impulse generation and conduction and thus decreases the heart rate. Maximal vagal stimulation can decrease the heart rate to as low as 20 beats/min. The vagus nerve also decreases the force of cardiac muscle contraction.

The sympathetic nervous system, the other branch of the autonomic system, has opposite effects. Sympathetic stimulation increases the rate of depolarization of the SA node as well as conduction speed, and thus heart rate. Maximal sympathetic stimulation can increase the heart rate to 250 beats/min. Sympathetic input also increases the force of contraction of the ventricles. Sympathetic control predominates during times of physical or emotional stress when the heart rate is greater than 100 beats/min. The parasympathetic system dominates when heart rate is less than 100. Thus, when exercise begins, or if exercise is at a low intensity, heart rate first increases due to withdrawal of vagal tone, then increases further due to sympathetic activation, as shown in figure 6.6.

The third extrinsic influence, the endocrine system, exerts its effect through two hormones released by the adrenal medulla: norepinephrine and epinephrine (see chapter 4). These hormones are also known as catecholamines. Like the norepinephrine released as a neurotransmitter by the sympathetic nervous system, circulating norepinephrine and epinephrine stimulate the heart, increasing its rate and contractility. In fact, release of these hormones from the adrenal medulla is triggered by sympathetic stimulation during times of stress, and their actions prolong the sympathetic response.

Normal resting heart rate (RHR) is defined as between 60 and 100 beats/min. With extensive endurance training (over months to years), RHR can decrease to 35 beats/min or less. A RHR as low as 28 beats/min has been observed in a world-class long-distance runner. While it has been widely accepted that these lower training-induced RHRs result primarily from increased parasympathetic stimulation (vagal tone), the actual mechanisms responsible for this training-induced sinus bradycardia remain an area of much debate (see Research Perspective 6.1).

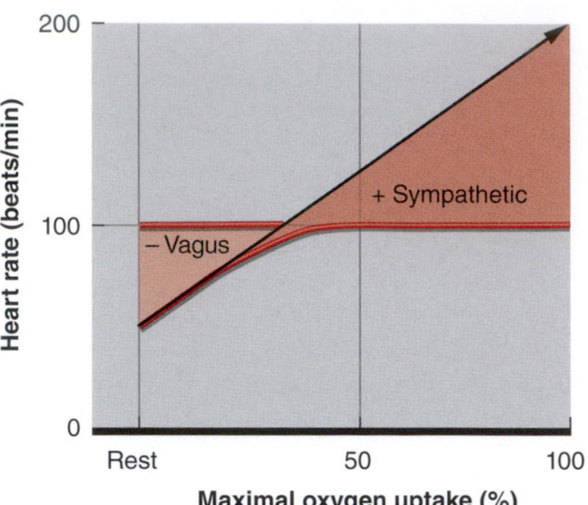

FIGURE 6.6 Relative contribution of sympathetic and parasympathetic nervous systems to the rise in heart rate from rest to exercise of increasing intensity.

Adapted from Rowell (1993).

Electrocardiogram

The electrical activity of the heart can be recorded to monitor cardiac changes or diagnose potential cardiac problems. Because body fluids contain electrolytes, they are good electrical conductors. Electrical impulses generated in the heart are conducted through body fluids to the skin, where they can be amplified, detected, and printed out on an **electrocardiograph**. This printout is called an **electrocardiogram (ECG)**. A standard ECG is recorded from 10 electrodes placed in specific anatomical locations. These 10 electrodes correspond to 12 leads that represent different views of the heart. Three basic components of the ECG represent important aspects of cardiac function (figure 6.7):

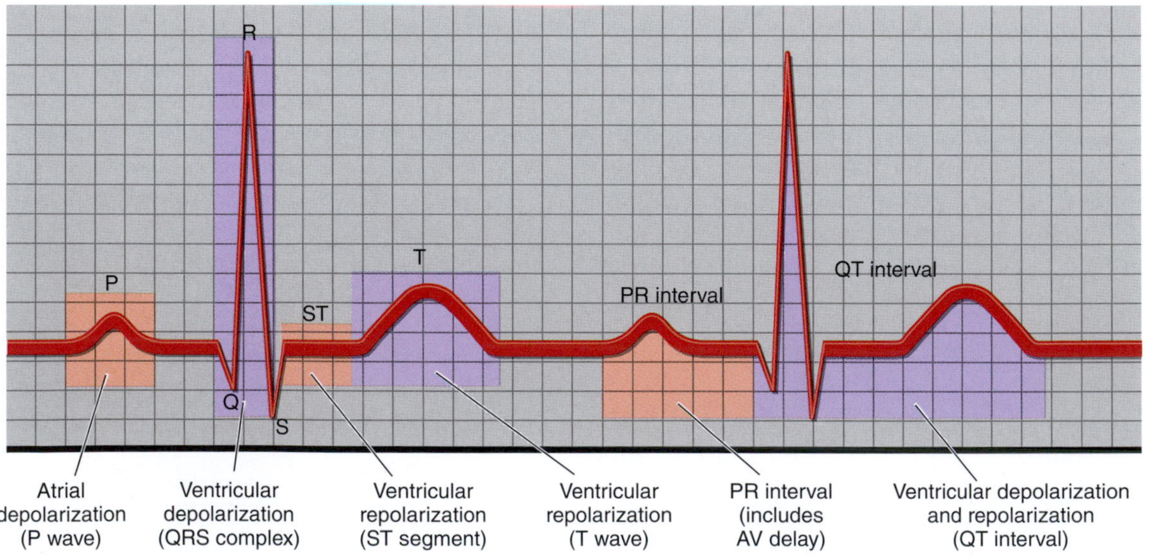

FIGURE 6.7 A graphic illustration of the various phases of the resting electrocardiogram.

- The P wave
- The QRS complex
- The T wave

The P wave represents atrial depolarization and occurs when the electrical impulse travels from the SA node through the atria to the AV node. The QRS complex represents ventricular depolarization and occurs as the impulse spreads from the AV bundle to the **Purkinje fibers** and through the ventricles. The T wave represents ventricular repolarization. Atrial repolarization cannot be seen, because it occurs during ventricular depolarization (QRS complex).

It is important to remember that an ECG measures only the electrical activity of the heart and does not provide any information about its function as a pump. Electrocardiograms are often obtained at rest, then again during exercise as clinical diagnostic tests of cardiac function. As exercise intensity increases, the heart must beat faster and work harder to deliver more blood to active muscles. Indications of coronary artery disease not evident at rest may show up on the ECG as the strain on the heart increases.

In Review

- The atria serve primarily as filling chambers, receiving blood from the veins; the ventricles are the primary pumps that eject blood from the heart.
- Because the left ventricle must produce more force than other chambers to pump blood throughout the systemic circulation, its myocardial wall is thicker.
- Cardiac tissue is capable of spontaneous rhythmicity and has its own specialized conduction system made up of myocardial fibers that serve specialized functions.
- Because it has the fastest inherent rate of depolarization, the SA node is normally the heart's pacemaker.
- Heart rate and force of contraction can be altered by the autonomic nervous system (sympathetic and parasympathetic) and the endocrine system through circulating catecholamines (epinephrine and norepinephrine).
- Electrocardiograms are often obtained during exercise as clinical diagnostic tests of cardiac function. Indications of coronary artery disease not evident at rest may show up on the ECG as the strain on the heart increases.
- The ECG provides no information about the pumping capacity of the heart, only its electrical activity.

Cardiac Arrhythmias

Occasionally, disturbances in the normal sequence of cardiac events can lead to an irregular heart rhythm, called an arrhythmia. These disturbances vary in degree of seriousness. Bradycardia and tachycardia are two types of arrhythmias. **Bradycardia** is defined as a RHR lower than 60 beats/min, whereas **tachycardia** is defined as a resting rate greater than 100 beats/min. With these arrhythmias, the sinus rhythm is normal, but the rate is altered. In extreme cases, bradycardia or tachycardia can affect maintenance of blood pressure. Symptoms of both arrhythmias include fatigue, dizziness, light-headedness, and fainting. Tachycardia can sometimes be sensed as palpitations or a racing pulse. Interestingly, highly trained endurance athletes also develop a resting bradycardia, an advantageous adaptation. This adaptation should not be confused with pathological causes of bradycardia. Nor should the elevated heart rate during exercise be confused with a tachycardia indicative of underlying disease or dysfunction.

Other arrhythmias may also occur. For example, **premature ventricular contractions (PVCs)**, which result in the feeling of skipped or extra beats, are relatively common and result from impulses originating outside the SA node. Atrial flutter, in which the atria depolarize at rates of 200 to 400 beats/min, and atrial fibrillation, in which the atria depolarize in a rapid and uncoordinated manner, are more serious arrhythmias that may cause ventricular filling problems. **Ventricular tachycardia**, defined as three or more consecutive PVCs, is a very serious arrhythmia that compromises the pumping capacity of the heart and can lead to **ventricular fibrillation**, in which depolarization of the ventricular tissue is random and uncoordinated. When this happens, the heart is extremely inefficient, and little or no blood is pumped out of the heart. Under such conditions, the use of a defibrillator to shock the heart back into a normal sinus rhythm must occur within minutes if the victim is to survive.

The Cardiac Cycle

The **cardiac cycle** includes all the mechanical and electrical events that occur during one heartbeat. In mechanical terms, all heart chambers undergo a relaxation phase (diastole) and a contraction phase (systole). During diastole, the chambers fill with blood. During systole, the ventricles contract and expel blood into the aorta and pulmonary arteries. The diastolic phase is approximately twice as long as the systolic phase. Consider an individual with a heart rate of 74 beats/min. At this heart rate, the entire cardiac cycle takes 0.81 s to complete (60 s divided by 74 beats). Of the total cardiac cycle at this rate, diastole accounts for 0.50 s, or 62% of the cycle, and systole accounts for 0.31 s, or

RESEARCH PERSPECTIVE 6.1

The Debate Surrounding Exercise Training–Induced Reductions in Heart Rate

It is well established that endurance exercise training lowers resting heart rate. Sinus bradycardia (a slow but otherwise normal heart rate) is evident in endurance athletes, whose resting heart rates can be half of that of their sedentary age-matched peers. However, despite substantial research, the mechanisms responsible for this training-induced reduction in resting heart rate remain an area of much debate.[1,2]

Two primary hypotheses have been proposed to explain exercise training–induced reductions in heart rate. The first, termed the autonomic neural hypothesis, posits that reductions in heart rate result from a shift in autonomic neural balance (sympathetic versus parasympathetic influences) toward increased parasympathetic activity. The second, referred to as the intrinsic rate hypothesis, suggests that changes in inherent cardiac pacemaker rate (i.e., rate of spontaneous depolarization of the sinoatrial [SA] node cells) govern reductions in heart rate following training.

Evidence put forth in support of the autonomic neural hypothesis is largely derived from the indirect evaluation of cardiac autonomic regulation from changes in heart rate variability or systemic pharmacological interventions that were not selective to cardiac function. For this hypothesis to be correct, the selective surgical elimination of cardiac autonomic innervation should prevent reductions in resting heart rate following exercise training. This has been demonstrated in an experimental dog model as well as in human cardiac transplant patients. That is, surgical elimination of all cardiac innervation completely prevented the exercise training–induced bradycardia. These data provide direct support for the autonomic neural hypothesis because they demonstrate that intact autonomic innervation of the heart (specifically, parasympathetic) is necessary for exercise training to result in reductions in resting heart rate.

However, equally convincing data have been set forth in support of the intrinsic rate hypothesis. With this alternate hypothesis, training bradycardia results not from increased vagal tone but instead from an electrical remodeling of the SA node itself. In support of this hypothesis, exercise-trained rodents display downregulation of cardiac ion channels, directly changing the function of the pacemaker of the heart, the SA node. Furthermore, prevention of ion channel downregulation abolished the difference in heart rate between trained and untrained animals. When considered collectively, these data provide support for the concept that alterations in SA node function mediate training-induced reductions in heart rate.

Regardless of whether the autonomic neural hypothesis or the intrinsic rate hypothesis, or a synthesis of the two, is ultimately proven correct, this ongoing debate highlights the critical importance of the scientific method in reaching these conclusions. That is, experiments must be conducted to test a specific hypothesis, and these experiments must be well controlled and adequately powered in order for the results to advance scientific discussion and discovery.

38%. As the heart rate increases, these time intervals shorten proportionately.

Refer to the normal ECG in figure 6.7. One cardiac cycle spans the time between one systole and the next. Ventricular contraction (systole) begins during the QRS complex and ends in the T wave. Ventricular relaxation (diastole) occurs during the T wave and continues until the next contraction. Although the heart is continually working, it spends slightly more time in diastole (~2/3 of the cardiac cycle) than in systole (~1/3 of the cardiac cycle).

The pressure inside the heart chambers rises and falls during each cardiac cycle. When the atria are relaxed, blood from the venous circulation fills the atria. About 70% of the blood filling the atria during this time passively flows directly through the mitral and tricuspid valves into the ventricles. When the atria contract, the atria push the remaining 30% of their volume into the ventricles.

During ventricular diastole, the pressure inside the ventricles is low, allowing the ventricles to passively fill with blood. As atrial contraction provides the final filling volume of blood, the pressure inside the ventricles increases slightly. As the ventricles contract, pressure inside the ventricles rises sharply. This increase in ventricular pressure forces the atrioventricular valves (i.e., tricuspid and mitral valves) closed, preventing any backflow of blood from the ventricles to the atria. The closing of the atrioventricular valves results in the first heart sound. Then, when ventricular pressure exceeds the pressure in the pulmonary artery and the aorta, the pulmonary and aortic valves open, allowing blood to flow into the pulmonary and systemic circulations, respectively. Following ventricular contraction, pressure inside the ventricles falls and the pulmonary and aortic valves close. The closing of these valves corresponds to the second heart sound. The two sounds together, the result of valves closing, results in the typical "lub, dub" heard through a stethoscope during each heartbeat.

The interactions of the various events that take place during one cardiac cycle are illustrated in figure 6.8, called a Wiggers diagram after the physiologist who

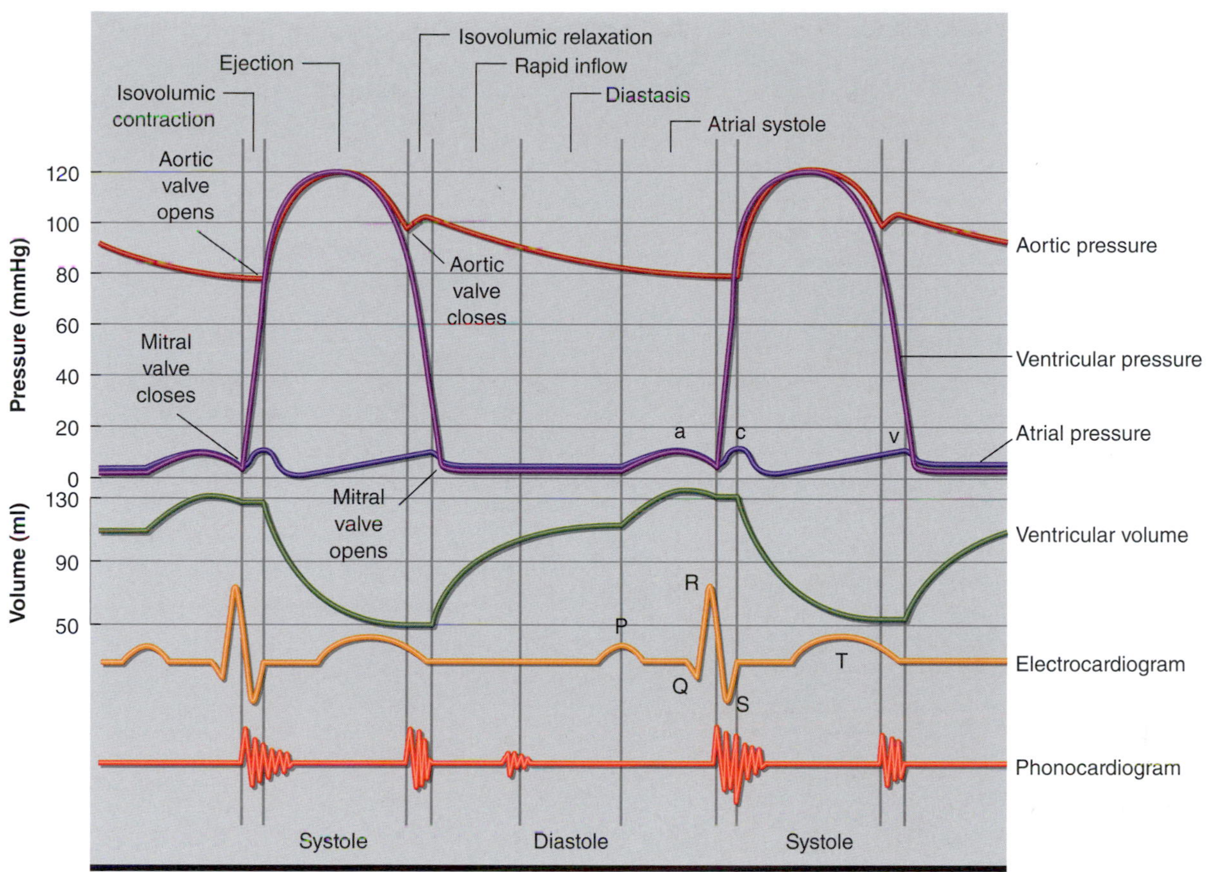

FIGURE 6.8 The Wiggers diagram, illustrating the events of the cardiac cycle for left ventricular function. Integrated into this diagram are the changes in left atrial and ventricular pressure, aortic pressure, ventricular volume, electrical activity (electrocardiogram), and heart sounds.

created it. The diagram integrates information from the electrical conduction signals (ECG), heart sounds from the heart valves, pressure changes within the heart chambers, and left ventricular volume.

Determinants of Cardiac Output

The heart's primary function is as a pump. The volume of blood pumped by the heart each minute governs blood flow to living tissues and, in the case of working muscle, is a key determinant of exercise performance.

Stroke Volume

During systole, most, but not all, of the blood in the ventricles is ejected. This volume of blood pumped during one beat (contraction) is the **stroke volume (SV)**. This is depicted in figure 6.9*a*. To understand SV, consider the amount of blood in the ventricle before and after contraction. At the end of diastole, just before contraction, the ventricle has finished filling. The volume of blood it now contains is called the **end-diastolic volume (EDV)**. At rest in a normal healthy adult, this value is approximately 100 ml. At the end of systole, just after the contraction, the ventricle has completed its ejection phase, but not all the blood is pumped out of the heart. The volume of blood remaining in the ventricle is called the **end-systolic volume (ESV)** and is approximately 40 ml under resting conditions. Stroke volume is the volume of blood that was ejected and is merely the difference between the volume of the filled ventricle and the volume remaining in the ventricle after contraction. So, SV is simply the difference between EDV and ESV; that is, SV = EDV – ESV (example: SV = 100 ml – 40 ml = 60 ml).

Ejection Fraction

The fraction of the blood pumped out of the left ventricle in relation to the amount of blood that was in the ventricle before contraction is called the **ejection fraction (EF)**. Ejection fraction is determined by dividing the SV by EDV (60 ml / 100 ml = 60%), as in figure 6.9*b*. The EF, generally expressed as a percentage, averages about 60% at rest in healthy, active young adults. Thus, 60% of the blood in the ventricle at the end of diastole is ejected with the next contraction, and 40% remains in the ventricle. Ejection fraction is often used clinically as an index of the pumping ability of the heart.

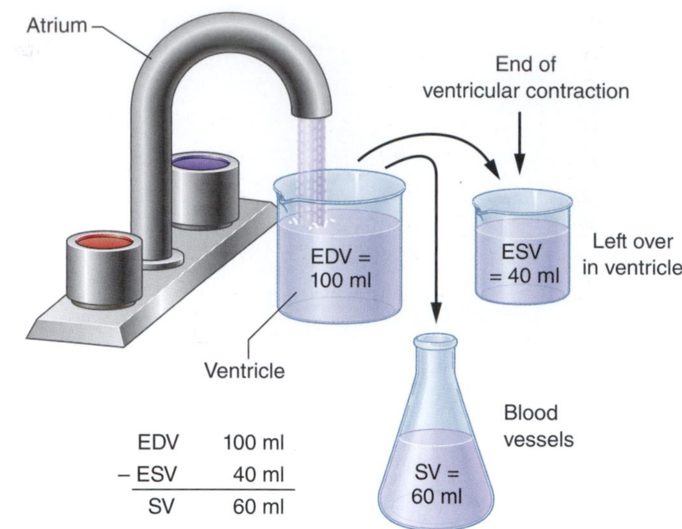

(a) Calculation of stroke volume (SV), the difference between end-diastolic volume (EDV) and end-systolic volume (ESV)

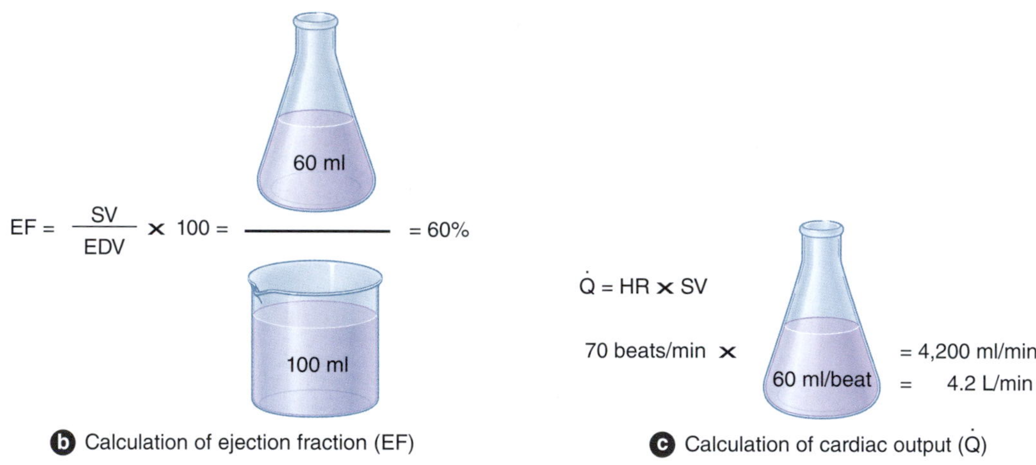

(b) Calculation of ejection fraction (EF)

(c) Calculation of cardiac output ($\dot{Q}$)

FIGURE 6.9 Calculations of stroke volume, ejection fraction, and cardiac output based on volumes of blood flowing into and out of the heart.

Cardiac Output

Cardiac output ($\dot{Q}$), as shown in figure 6.9*c*, is the total volume of blood pumped by the ventricle per minute, the product of heart rate (HR) and SV. The SV at rest in the standing posture averages between 60 and 80 ml of blood in most adults. Thus, at a RHR of 70 beats/min, the resting cardiac output will vary between 4.2 and 5.6 L/min. The average adult body contains about 5 L of blood, so this means that the equivalent of our total blood volume is pumped through our hearts about once every minute.

Pumping Action of the Heart During Exercise

As described earlier in this chapter, the myocardium has to contract as if it were a single unit in order to efficiently pump blood. For that reason, myocardial cells are anatomically interconnected end to end by intercalated disks that anchor the individual cells together so that they do not pull apart during contraction. This better allows the heart to contract as one unit, often called a functional syncytium.

VIDEO 6.1 Presents Ben Levine on torsional contraction of the heart muscle and its role in ventricular filling. 

During intense exercise when heart rates are high, the time available between contractions for diastolic filling is very short. Yet complete filling of the left ventricle must occur in order to appropriately increase cardiac output. The heart actually uses the increased contractility that occurs during exercise to enhance left ventricle filling, a process called *torsional contraction*. As the heart beats, contraction and relaxation of the atria and the ventricles create a twisting and untwisting action, similar to wringing out a towel. During systole

RESEARCH PERSPECTIVE 6.2

Can Too Much Exercise Be Bad for Your Heart?

It is well known that habitual physical activity reduces cardiovascular disease risk and that exercise dose is also important; higher physical activity levels further reduce mortality risk, and the most active individuals demonstrate the highest overall life expectancy. However, few studies have included individuals engaging in lifelong *high-intensity* endurance exercise. This is an important gap in our knowledge, since recent evidence suggests that such intense exercise may paradoxically *increase* cardiovascular risk.

Intense exercise performed regularly elicits structural, functional, and electrical cardiac adaptations, collectively known as the athlete's heart. As elegantly reviewed by Eijsvogels and colleagues,[8] these adaptations may also impart deleterious effects. In response to exercise training, all four chambers of the heart enlarge. Although this adaptation facilitates exercise performance, it may also have adverse cardiac effects. For example, atrial fibrillation becomes more common, potentially resulting from increased vagal tone and left atrial size. Further, right ventricular wall stress is increased, possibly due to exercise-induced increases in pulmonary artery systolic pressure. Together, these changes in heart structure may hasten cardiac disease in susceptible individuals.

In addition to structural adaptations, exercise also acutely increases circulating biomarkers for cardiovascular disease, including creatine kinase, cardiac troponin, and B-type natriuretic peptide. Although the source of these circulating molecules remains unclear, the increases likely result from both skeletal muscle damage and stress-activated cardiac muscle. These increases may be cause for concern because prolonged exercise reduces ventricular function and acutely injures cardiac muscle, resulting in cardiac fatigue. Myocardial fibrosis (a buildup of scar tissue in the cardiac muscle or valves) has also been documented in some lifelong endurance athletes. The interrelation between increases in cardiac biomarkers, reductions in ventricular function, and cardiac fibrosis may impart increased risk, though the precise mechanisms remain incompletely understood.

While the possibility that lifelong intense endurance exercise may increase cardiac risk cannot be ignored, for the majority of the population, the evidence supporting the beneficial cardiovascular outcomes attributed to exercise and physical activity is overwhelming. It is these health benefits that led to the development of strategies to increase physical activity, with very specific guidelines put forth by both the American College of Sports Medicine and the American Heart Association. Unfortunately, the majority of Americans fail to meet these exercise criteria, putting themselves at an ever-increased risk of cardiovascular disease.

(contraction), the heart twists gradually, storing energy and compressing the springlike titin molecules in the sarcomere (see chapter 1 for a description of titin's similar role in skeletal muscle). When the aortic valve closes, the ventricle abruptly untwists. This recoil creates a 1 to 2 mmHg pressure difference between the base (top) and apex (bottom) of the heart, which pulls blood from the atrium, across the mitral valve, and into the ventricle.

The torsion during systole stores energy that is then released during isovolumetric relaxation (the period during the cardiac cycle after contraction when all of the valves are closed and the myocardium is relaxing) to generate the diastolic suction that allows enhanced atrial filling during exercise. This twisting action is enhanced almost threefold during exercise and assists in efficient ventricular filling. This left ventricle twisting and rapid untwisting, inducing diastolic suction in the ventricle, is called *dynamic relaxation.*[10] Cardiac twisting mechanics are improved with exercise training and reduced by detraining.[7]

Understanding the electrical and mechanical activity of the heart provides a basis for understanding the

In Review

- The electrical and mechanical events that occur in the heart during one heartbeat make up one cardiac cycle. A Wiggers diagram depicts the intricate timing of these events.
- Cardiac output, the volume of blood pumped by each ventricle per minute, is the product of HR and SV.
- Not all of the blood in the ventricles is ejected during systole. The ejected volume is the SV, while the percentage of blood pumped with each beat is the EF.
- To calculate SV, EF, and cardiac output:

$$\text{SV (ml/beat)} = \text{EDV} - \text{ESV}$$

$$\text{EF (\%)} = (\text{SV/EDV}) \times 100$$

$$\dot{Q}\ \text{(L/min)} = \text{HR} \times \text{SV}$$

- As the heart beats, contraction and relaxation of the atria and the ventricles create a twisting and untwisting action that allows the ventricles to fill even at high heart rates.

cardiovascular system, but the heart is only one part of this system. In addition to this pump, the cardiovascular system contains an intricate network of tubes that serve as a delivery system carrying the blood to all body tissues.

Vascular System

The vascular system contains a series of vessels that transport blood from the heart to the tissues and back: the arteries, arterioles, capillaries, venules, and veins.

Arteries are large, muscular, elastic conduit vessels for transporting blood away from the heart to the arterioles. The aorta is the largest artery, transporting blood from the left ventricle to all regions of the body as it eventually branches into smaller and smaller arteries, finally branching into arterioles. The **arterioles** are the site of greatest control of the circulation by the sympathetic nervous system, so arterioles are sometimes called *resistance vessels.* Arterioles are heavily innervated by the sympathetic nervous system and are the main site of control of blood flow to specific tissues.

From the arterioles, blood enters the **capillaries**, the narrowest and simplest vessels in terms of their structure, with walls only one cell thick. Virtually all exchange between the blood and the tissues occurs at the capillaries. Blood leaves the capillaries to begin the return trip to the heart in the **venules**, and the venules form larger vessels—the **veins**. The vena cava is the great vein transporting blood back to the right atrium from all regions of the body above (superior vena cava) and below (inferior vena cava) the heart.

Blood Pressure

Blood pressure is the pressure exerted by the blood on the arterial walls. It is expressed by two numbers: the **systolic blood pressure (SBP)** and the **diastolic blood pressure (DBP)**. The higher number is the SBP; it represents the highest pressure in the artery that occurs during ventricular systole. Ventricular contraction pushes the blood through the arteries with tremendous force, and that force exerts high pressure on the arterial walls. The lower number is the DBP and represents the lowest pressure in the artery, corresponding to ventricular diastole when the ventricle is filling.

Mean arterial pressure (MAP) represents the average pressure exerted by the blood as it travels through the arteries. Since diastole takes about twice as long as systole in a normal cardiac cycle, MAP can be estimated from the DBP and SBP as follows:

$$\text{MAP} = 2/3\ \text{DBP} + 1/3\ \text{SBP}$$

Alternatively,

$$\text{MAP} = \text{DBP} + [0.333 \times (\text{SBP} - \text{DBP})]$$

(SBP – DBP) is also called pulse pressure.

To illustrate, with a normal resting blood pressure of 120 mmHg over 80 mmHg, the MAP = 80 + [0.333 × (120 – 80)] = 93 mmHg.

General Hemodynamics

The cardiovascular system is a continuous closed-loop system. Blood flows through this closed loop because of the pressure gradient that exists between the arterial and venous sides of the circulation. To understand regulation of blood flow to the tissues, it is necessary to understand the intricate relationship between pressure, blood flow, and resistance.

In order for blood to flow through a vessel, there must be a pressure difference from one end of the vessel to the other end. Blood will flow from the region of the vessel with high pressure to the region of the vessel with low pressure. Alternatively, if there is no pressure difference across the vessel, there is no driving force and therefore no blood flow. In the circulatory system, the MAP in the aorta is approximately 100 mmHg at rest, and the pressure in the right atrium is very close to 0 mmHg. Therefore, the pressure difference across the entire circulatory system is 100 mmHg – 0 mmHg = 100 mmHg.

The reason for the pressure differential from the arterial to the venous circulation is that the blood vessels themselves provide resistance to blood flow. The resistance that the vessel provides is largely dictated by the properties of the blood vessel and the blood itself. These properties include the length and radius of the blood vessel and the viscosity or thickness of the blood flowing through the vessel. Resistance to flow can be calculated as

$$\text{Resistance} = \eta \times L / r^4$$

where η is the viscosity (thickness) of the blood, L is the length of the vessel, and r is the radius of the vessel, which is raised to the fourth power. Blood flow is proportional to the pressure difference across the system and is inversely proportional to resistance. This relationship can be illustrated by the following equation:

$$\text{Blood flow} = \Delta\text{pressure} / \text{resistance}$$

Notice that blood flow can increase by either an increase in the pressure difference (Δpressure), a decrease in resistance, or a combination of the two. Altering resistance to control blood flow is much more advantageous because very small changes in blood vessel radius equate to large changes in resistance. This is due to the fourth-power mathematical relationship between vascular resistance and vessel radius.

Changes in vascular resistance are largely due to changes in the radius or diameter of the vessels, since the viscosity of the blood and the length of the vessels do not change significantly under normal conditions. Therefore, regulation of blood flow to organs is accomplished by small changes in vessel radius through **vasoconstriction** and **vasodilation**. This allows the cardiovascular system to divert blood flow to the areas where it is needed most.

As mentioned earlier, most resistance to blood flow occurs in the arterioles. Figure 6.10 shows the blood pressure changes across the entire vascular system. The arterioles are responsible for ~70% to 80% of the drop in MAP across the entire cardiovascular system. This is important because small changes in arteriole radius can greatly affect the regulation of mean arteriole pressure and the local control of blood flow. At the capillary level, changes due to systole and diastole are no longer evident, and the flow is smooth (laminar) rather than turbulent.

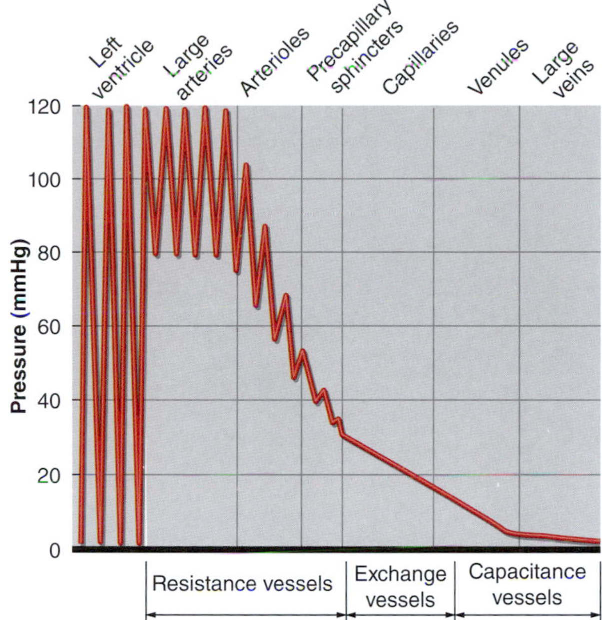

FIGURE 6.10 Pressure changes across the systemic circulation. Notice the large pressure drop that occurs across the arteriole portion of the system.

In Review

- Systolic blood pressure is the highest pressure within the vascular system, whereas DBP is the lowest pressure.
- Mean arterial pressure is the average pressure on the vessel walls during a cardiac cycle; however, it is not the mathematical mean of SBP and DBP because diastole takes about twice as long as systole.
- In terms of the entire cardiovascular system, cardiac output is the blood flow to the entire system, the Δpressure is the difference between aortic pressure when blood leaves the heart and venous pressure when blood returns to the heart, and resistance is the impedance to blood flow from the blood vessels.
- Blood flow is mainly controlled by small changes in blood vessel (arteriole) radius that greatly change resistance.

Distribution of Blood

Distribution of blood to the various body tissues varies considerably depending on the immediate needs of a specific tissue compared with those of other areas of the body. As a general rule, the most metabolically active tissues receive the greatest blood supply. At rest under normal conditions, the liver and kidneys combine to receive approximately half of the cardiac output, while resting skeletal muscles receive only about 15% to 20%.

During exercise, blood is redirected to the areas where it is needed most. During heavy endurance exercise, contracting muscles may receive 80% or more of the blood flow, and flow to the liver and kidneys decreases. This redistribution, along with increases in cardiac output (discussed in chapter 8), allows up to 25 times more blood flow to active muscles (see figure 6.11).

Alternatively, after one eats a big meal, the digestive system receives more of the available cardiac output than when the digestive system is empty. Along the same lines, during increasing environmental heat stress, skin blood flow increases to a greater extent as the body attempts to maintain normal temperature. The cardiovascular system responds accordingly to redistribute blood, whether it is to the exercising muscle to match metabolism, for digestion, or to facilitate thermoregulation. These changes in the distribution of cardiac output are controlled by the sympathetic nervous system, primarily by increasing or decreasing the diameter of the arterioles providing blood flow to the given tissue or organ. Arterioles have a strong muscular wall that can significantly alter vessel diameter, are highly innervated by sympathetic nerves, and have the capacity to respond to local control mechanisms.

Intrinsic Control of Blood Flow

Intrinsic control of blood distribution refers to the ability of the local tissues to dilate or constrict the arterioles that serve them and alter regional blood flow depending on the immediate needs of those tissues. With exercise and the increased metabolic demand of

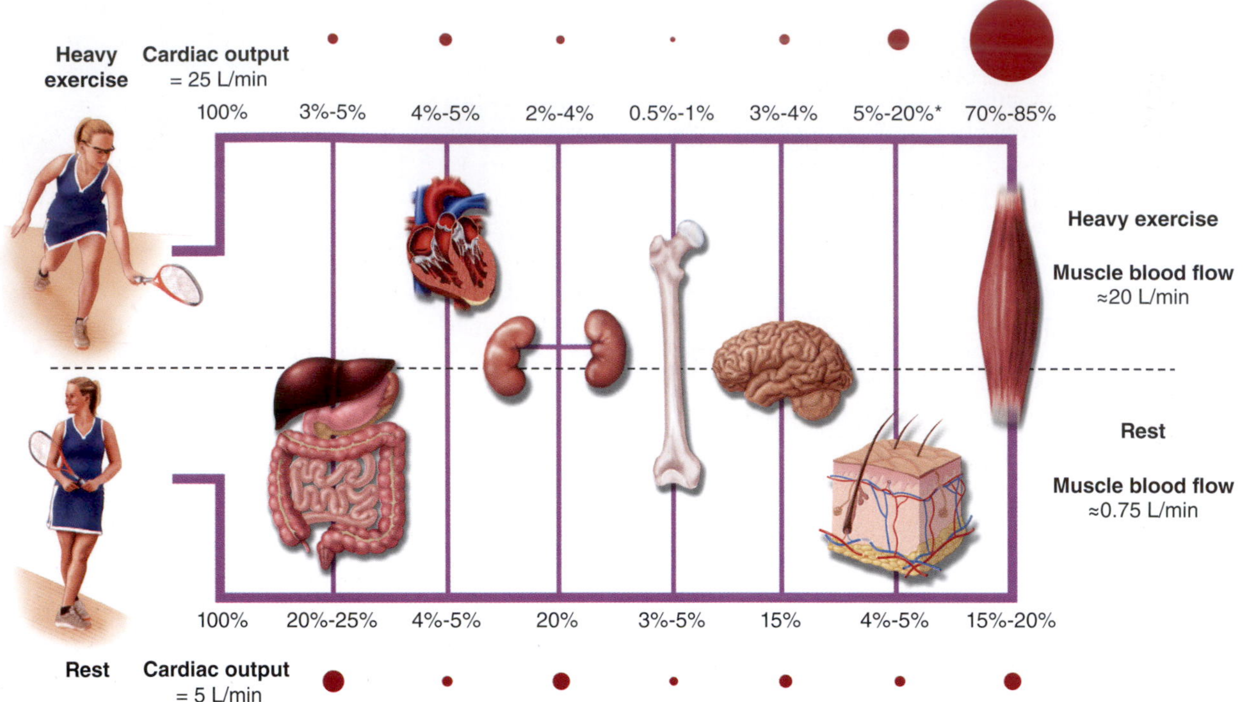

FIGURE 6.11 Distribution of cardiac output at rest and maximal exercise.

*Depends on ambient and body temperatures.

Reprinted by permission from P.O. Åstrand et al., *Textbook of Work Physiology: Physiological Bases of Exercise*, 4th ed. (Champaign, IL: Human Kinetics, 2003), 143.

the exercising skeletal muscles, the arterioles undergo intrinsic vasodilation, opening up to allow more blood to enter the highly active tissue.

There are essentially three types of intrinsic control of blood flow. *Metabolic regulation*, in response to an increased oxygen demand, is the strongest stimulus for the release of local vasodilating chemicals. As oxygen uptake by metabolically active tissues increases, available oxygen is diminished. Local arterioles dilate to allow more blood to perfuse the area, delivering more oxygen. Other chemical changes that can stimulate increased blood flow are decreases in other nutrients and increases in by-products (carbon dioxide, K^+, H^+, lactic acid) or inflammatory molecules.

Second, many dilator substances can be produced within the endothelium (inner lining) of arterioles that can initiate vasodilation in the vascular smooth muscle of those arterioles *(endothelium-mediated vasodilation)*. These substances include nitric oxide (NO), prostaglandins, and endothelium-derived hyperpolarizing factor (EDHF). These endothelium-derived vasodilators are important regulators of blood flow at rest and during exercise in humans. Additionally, acetylcholine and adenosine have been proposed as potential vasodilators for the increase in muscle blood flow during exercise.

Third, pressure changes within the vessels themselves can also cause vasodilation and vasoconstriction. This is referred to as the *myogenic response*. The vascular smooth muscle contracts in response to an increase in pressure across the vessel wall and relaxes in response to a decrease in pressure across the vessel wall. Figure 6.12 illustrates the three types of intrinsic control of vascular tone.

Extrinsic Neural Control

The concept of intrinsic local control explains redistribution of blood within an organ or tissue; however, the cardiovascular system must divert blood flow to where it is needed, beginning at a site upstream of the local environment. Redistribution at the system or organ level is controlled by neural mechanisms. This is known as **extrinsic neural control** of blood flow, because the control comes from outside the specific area (extrinsic) instead of from inside the tissues (intrinsic).

Blood flow to all body parts is regulated largely by the sympathetic nervous system. Sympathetic nerves are abundant in the circular layers of smooth muscle within the artery and arteriole walls. In most vessels, an increase in sympathetic nerve activity causes these circular smooth muscle cells to contract, constricting blood vessels and thereby decreasing blood flow.

Under normal resting conditions, sympathetic nerves transmit impulses continuously to the blood vessels (in particular, the arterioles), keeping the vessels moderately constricted to maintain adequate blood pressure. This state of tonic vasoconstriction is referred

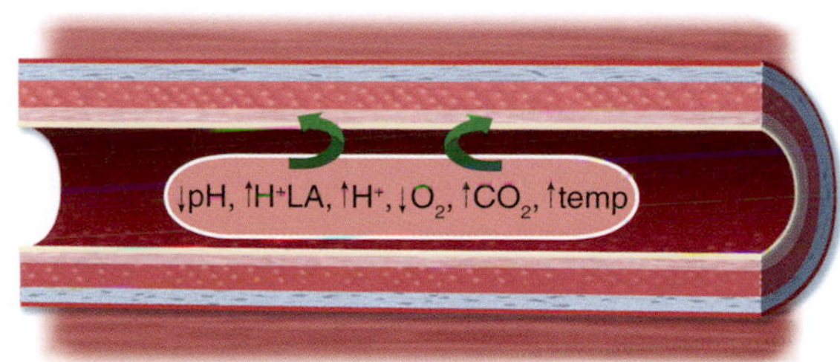

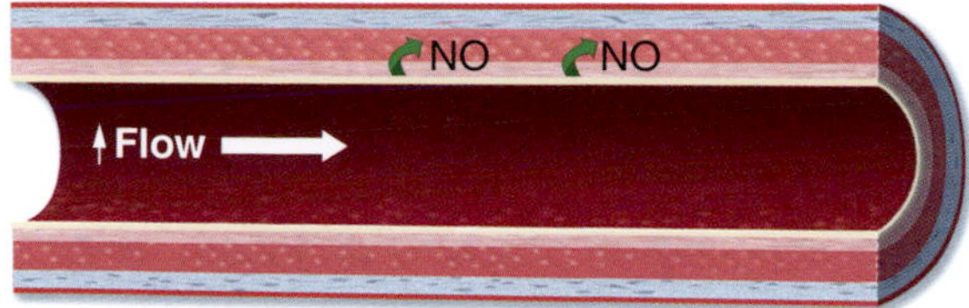

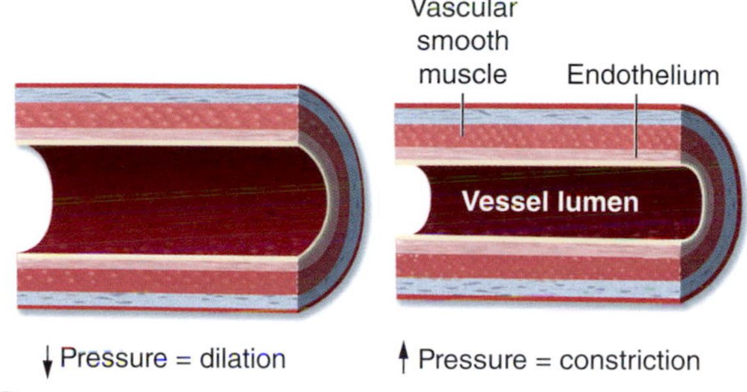

FIGURE 6.12 Intrinsic control of blood flow. Arterioles are signaled to dilate or constrict at the local level *(a)* by changes in the local concentration of oxygen or metabolic products, *(b)* by the effects of local pressure within the arterioles, and *(c)* by endothelium-derived factors.

Figure courtesy of Dr. Donna H. Korzick, Pennsylvania State University.

to as *vasomotor tone.* When sympathetic stimulation increases, further constriction of the blood vessels in a specific area decreases blood flow into that area and allows more blood to be distributed elsewhere. But if sympathetic stimulation decreases below the level needed to maintain tone, constriction of vessels in the area is lessened, so the vessels passively dilate, increasing blood flow into that area. Therefore, sympathetic stimulation will cause vasoconstriction in most vessels. Blood flow can passively be increased through a lowering of the normal tonic level of sympathetic outflow.

Local Control of Muscle Blood Flow

The previous two sections discuss intrinsic and extrinsic control of blood flow, mechanisms that pertain to controlling flow to all tissues of the body. However, muscle blood flow deserves special attention because (1) contracting muscle is the hallmark response of exercise physiology and (2) unique mechanisms exist to support increased muscle blood flow. During aerobic exercise, blood flow to exercising muscle must increase to match the metabolic demand of that muscle. Enhanced oxygen delivery to exercising muscle can occur via a number of different mechanisms, including local alteration of blood flow, improved oxygen extraction at the tissue level, or both.

Exercise is accompanied by a general increase in sympathetic nerve activity, including that directed to muscle, which causes vasoconstriction. How does working muscle overcome systemic vasoconstriction and actually increase blood flow? The primary mechanism is termed **functional sympatholysis**. Vasoactive molecules released from the active skeletal muscle and endothelium have been shown to inhibit sympathetic vasoconstriction by reducing vascular responsiveness to α-adrenergic receptor activation. Endothelial cells release molecules called endothelial-derived hyperpolarizing factors (EDHFs) that make it more difficult for smooth muscle cells to constrict in response to sympathetic stimulation. For example, we now know that ATP released from the endothelium and from red blood cells can cause hyperpolarization of vascular smooth muscle cells that helps override α-adrenergic vasoconstriction. Functional sympatholysis helps optimize muscle blood flow distribution to match tissue perfusion with metabolic demand.

When oxygen availability is limited under conditions of reduced arterial O_2 content (e.g., hypoxia) or decreased perfusion pressure, the skeletal muscle arterioles dilate to compensate for reduced O_2 delivery, allowing for a greater O_2 extraction at the tissue level.[4] This phenomenon is termed compensatory vasodilation.

In order to examine the mechanisms by which the local control of skeletal muscle blood flow is altered in exercising muscle in humans, investigators have used acute systemic hypoxia, usually by having subjects breathe air mixtures with a low O_2 content, to reduce arterial O_2 content during exercise[11] or have temporarily limited blood flow to the exercising muscle by partially occluding flow to the exercising limb. During submaximal hypoxic exercise, blood flow to the exercising muscle is the same as that seen during normoxic exercise due to the individual and combined roles of β-adrenergic receptors, adenosine, and nitric oxide (NO), as shown in figure 6.13.

Interestingly, the contribution of these different dilator mechanisms can change, depending on the exercise intensity and whether blood flow to the exercising muscle is limited. For example, during low-intensity exercise under hypoxic conditions, stimulation of β-adrenergic receptors contributes to vasodilation; however, as exercise intensity increases, NO release from the endothelium contributes to a greater extent in the compensatory vasodilator response.[3] The molecule adenosine can also contribute to compensatory vasodilation, especially under conditions in which blood flow is limited. At higher exercise intensities in which the muscle fibers' oxygen needs are even greater, NO and several other vasodilator molecules, including prostaglandins and adenosine triphosphate (ATP), mediate vasodilation. However, there are redundancies in these vasodilator mechanisms, such that when one is blocked or downregulated, another vasodilator can compensate and cause vasodilation.

RESEARCH PERSPECTIVE 6.3

Vascular Adaptations to Exercise Training in Postmenopausal Women

Despite recent declines in prevalence, cardiovascular disease remains the leading cause of death in the United States. Interestingly, cardiovascular mortality risk differs greatly between the sexes. The establishment of the National Institutes of Health (NIH) Office of Women's Health Research in 1990 and the passage of the NIH Revitalization Act in 1993 mandated the inclusion of women in NIH-funded research. Since the adoption of these requirements, much has been learned about the mechanisms and manifestation of cardiovascular disease in women. Yet despite these advancements, the reasons for the sex disparity in cardiovascular morbidity and mortality remain unclear.

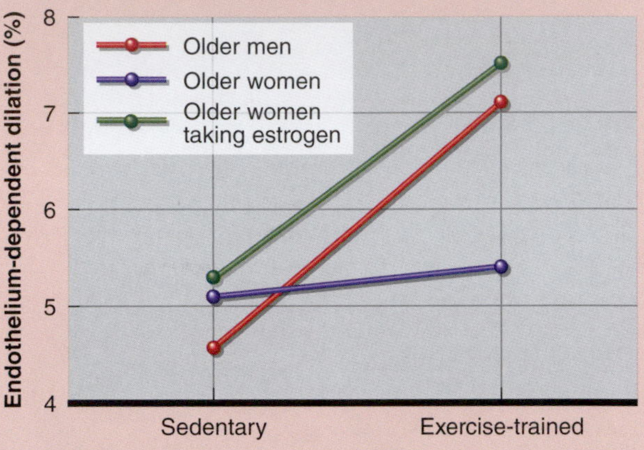

Estrogen is required to elicit exercise training–induced improvements in vascular endothelial function in older postmenopausal women. In older men, moderate-intensity exercise training significantly improves vascular function, as assessed by endothelium-dependent dilation. This beneficial vascular adaptation to exercise did not occur in a group of postmenopausal women of similar age. However, when postmenopausal women were supplemented with estrogen, the expected improvements in vascular endothelial function following exercise training were similar to those observed in age-matched men.

Vascular aging (age-related changes in blood vessels) is considered a primary risk factor for age-associated cardiovascular disease. Endothelial dysfunction, defined as impaired endothelium-dependent dilation, is the first functional manifestation of atherosclerosis and accelerates the progression of cardiovascular disease. Lifestyle modifications, such as habitual physical activity, are commonly recommended as a first-line strategy to mitigate age-associated declines in vascular endothelial function. However, there is now an increasing awareness of potential sex differences in the beneficial effects of exercise training on vascular health in older, postmenopausal women.

In a 2006 review of the prevailing literature, researchers provided support for the notion that sex hormones, specifically estrogen, modulate the exercise training–induced improvements in vascular function in women.[11] That is, the reduction in estrogen that occurs during menopause consequently prevents improvements in vascular function with exercise training, as highlighted in the accompanying figure. As expected, in previously sedentary middle-aged and older men, a moderate-intensity exercise training program significantly improved vascular function. In contrast, this exercise training paradigm had no effect in older postmenopausal women. However, in sedentary postmenopausal women treated with estrogen, a moderate-intensity training program did significantly improve vascular endothelial function. These findings have been corroborated in studies using varied methodology, and provide direct evidence suggesting that estrogen is required for training-induced improvements in vascular function in postmenopausal women.

Given the rapidly aging population, effective preventive strategies to mitigate the untoward consequences of cardiovascular disease are paramount. Habitual physical activity is an important strategy for the primary prevention of cardiovascular disease—in men and women. Yet, it is exceedingly apparent that the vasculature of older men and women responds differently to exercise training, and this differential responsiveness can likely be attributed to sex hormones, or the lack thereof. Further research is necessary to determine if additional pharmacological or nonpharmacological strategies should be coupled with exercise training in order to provide a viable therapeutic intervention that permits improvements in vascular endothelial function during exercise training in estrogen-deficient postmenopausal women.

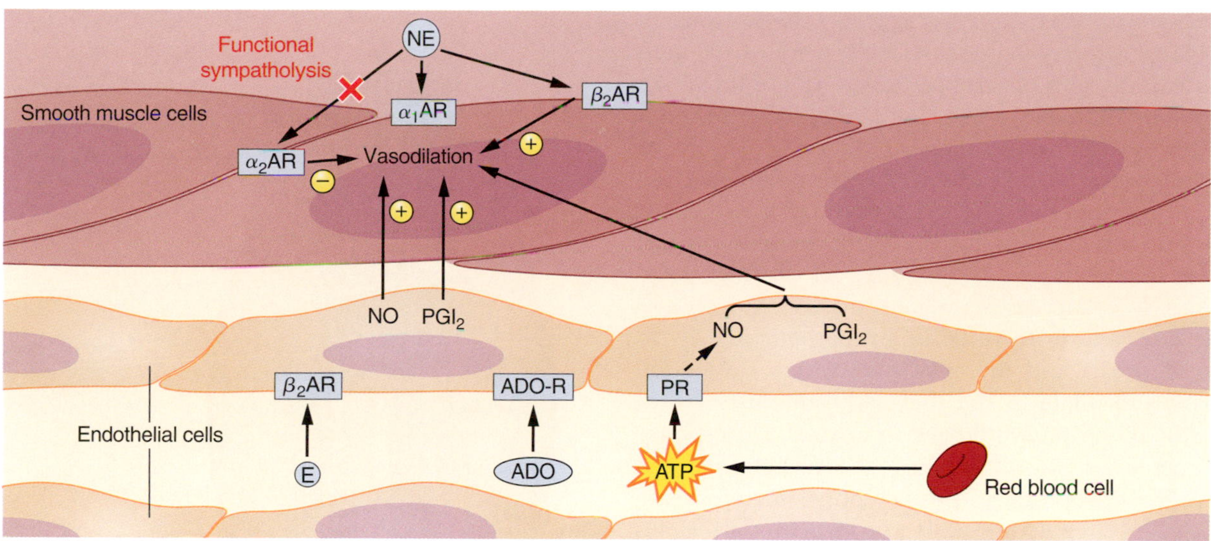

FIGURE 6.13 The proposed mechanisms for functional sympatholysis and hypoxia-induced vasodilation during exercise. During hypoxic exercise, nitric oxide (NO) is the final common pathway for the compensatory dilator response. Systemic epinephrine (E) release, acting via β-adrenergic receptors, contributes to the NO-mediated vasodilation at lower exercise intensities, but this β-adrenergic contribution decreases with increasing exercise intensity. Adenosine triphosphate (ATP) released from the red blood cell (RBC), endothelial-derived prostacyclin (PGI$_2$), or both remain, also stimulating NO during higher-intensity hypoxic exercise. α_1AR, α_2AR, and β_2AR = α_1-, α_2-, and β_2-adrenergic receptors, respectively; NE = norepinephrine; PR = purinergic receptors that are stimulated by ATP; ADO = adenosine.

Because of the biological importance of NO, these mechanisms have significant implications in clinical populations such as older individuals and patients with cardiovascular disease, in whom NO production and availability may be limited.[5,6] For example, as humans age, there is a reduction in NO synthesis and an increase in NO breakdown, and compensatory vasodilation is blunted in healthy older humans.[11]

Distribution of Venous Blood

While flow to tissues is controlled by changes on the arterial side of the system, most of the blood *volume* resides in the venous side of the system. At rest, the blood volume is distributed among the vasculature as shown in figure 6.14. The venous system has a great capacity to store blood volume because veins have little vascular smooth muscle and are very elastic and balloon-like. Thus, the venous system provides a large reservoir of blood available to be rapidly distributed back to the heart (venous return) and from there to the arterial circulation. This is accomplished through sympathetic stimulation of the venules and veins, which causes the vessels to constrict (venoconstriction).

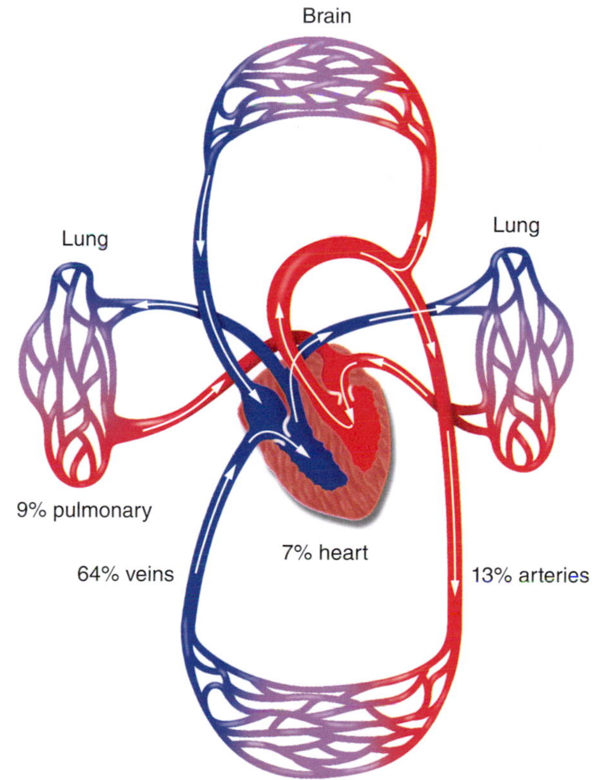

FIGURE 6.14 Blood volume distribution within the vasculature when the body is at rest.

Integrative Control of Blood Pressure

Blood pressure is controlled by reflexes. Specialized pressure sensors located in the aortic arch and the carotid arteries, called **baroreceptors**, are sensitive to changes in arterial pressure. When the pressure in these large arteries changes, afferent signals are sent to the cardiovascular control centers in the brain where autonomic reflexes are initiated, and efferent signals are sent to respond to changes in blood pressure. For example, when blood pressure is elevated, the baroreceptors are stimulated by an increase in stretch. They relay this information to the cardiovascular control center in the brain. In response to the increased pressure is an increase in vagal tone, which decreases heart rate, and a decrease in sympathetic activity to both the heart and the arterioles, which causes the arterioles to dilate. All of these adjustments serve to decrease blood pressure back to normal. In response to a decrease in blood pressure, less stretch is sensed by the baroreceptors, and the response is to increase heart rate (vagal withdrawal) and constrict arterioles (through increased sympathetic nervous activity), thus correcting the low-pressure signal.

Other specialized receptors, called **chemoreceptors** and **mechanoreceptors**, send information about the chemical environment in the muscle and the length and tension of the muscle, respectively, to the cardiovascular control centers. These receptors also modify the blood pressure response and are especially important during exercise.

Return of Blood to the Heart

Because humans spend so much time in an upright position, the cardiovascular system requires mechanical assistance to overcome the force of gravity and assist the return of venous blood from the lower parts of the body to the heart. Three basic mechanisms assist in this process:

- Valves in the veins
- The muscle pump
- The respiratory pump

The veins contain valves that allow blood to flow in only one direction, thus preventing backflow and further pooling of blood in the lower body. These venous valves also complement the action of the skeletal **muscle pump**, the rhythmic mechanical compression of the veins that occurs during the rhythmic skeletal muscle contraction accompanying many types of movement and exercise, for example, during walking and running (figure 6.15). The muscle pump pushes blood volume in the veins back toward the heart. Finally, changes in pressure in the abdominal and thoracic cavities during breathing assist blood return to the heart by creating a pressure gradient between the veins and the chest cavity.

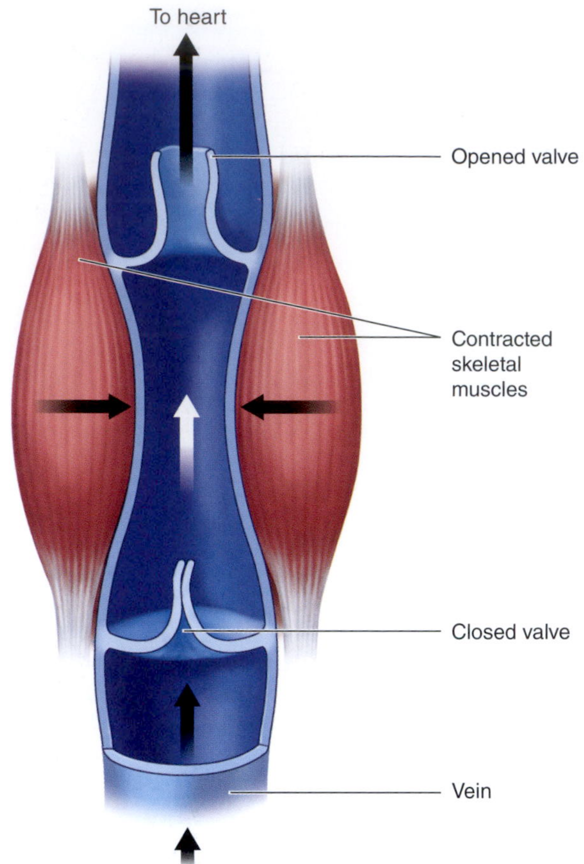

FIGURE 6.15 The muscle pump. As the skeletal muscles contract, they squeeze the veins in the legs and assist in the return of blood to the heart. Valves within the veins ensure the unidirectional flow of blood back to the heart.

In Review

- Blood is distributed throughout the body based primarily on the metabolic needs of the individual tissues. The most active tissues receive the highest blood flow.
- Skeletal muscle normally receives about 15% of the cardiac output at rest. This can increase to 80% or more during heavy endurance exercise.
- Redistribution of blood flow is controlled locally by the release of dilators from either the tissue (metabolic regulation) or the endothelium of the blood vessel (endothelium-mediated dilation). A third type of intrinsic control involves the response of the arteriole to pressure. Decreased arteriolar pressure causes vasodilation, thus increasing blood flow to the area, while increased pressure causes local constriction.

- Extrinsic neural control of blood flow distribution is accomplished by the sympathetic nervous system, primarily through vasoconstriction of small arteries and arterioles.
- During aerobic exercise, blood flow to exercising muscle must increase to match the metabolic demand of that muscle. This is accomplished by (1) functional sympatholysis (which overcomes sympathetic vasoconstriction) and (2) compensatory vasodilation (involving molecules such as adenosine and nitric oxide).
- Blood pressure is maintained under normal conditions by reflexes within the autonomic system.
- Blood returns to the heart through the veins, assisted by valves within the veins, the muscle pump, and changes in respiratory pressure.

Blood

Blood serves many diverse purposes in regulating normal body function. The three functions of primary importance to exercise and sport are

- transportation,
- temperature regulation, and
- acid–base (pH) balance.

We are most familiar with blood's transporting functions, delivering oxygen and fuel substrates and carrying away metabolic by-products. In addition, blood is critical in temperature regulation during physical activity because it picks up heat from the exercising muscle and delivers it to the skin, where it can be dissipated to the environment (see chapter 12). Blood also buffers the acids produced by anaerobic metabolism and maintains proper pH for metabolic processes (see chapters 2 and 7).

Blood Volume and Composition

The total volume of blood in the body varies considerably with an individual's size, body composition, and state of training. Larger blood volumes are associated with greater lean body mass and higher levels of endurance training. The blood volume of people of average body size and normal physical activity generally ranges from 5 to 6 L in men and 4 to 5 L in women.

Blood is composed of plasma and formed elements (see figure 6.16). Plasma normally constitutes about 55% to 60% of total blood volume but can decrease by 10% of its normal amount or more with intense exercise in hot conditions, or can increase by 10% or more with endurance training or acclimation to heat. Approximately 90% of the plasma volume is water;

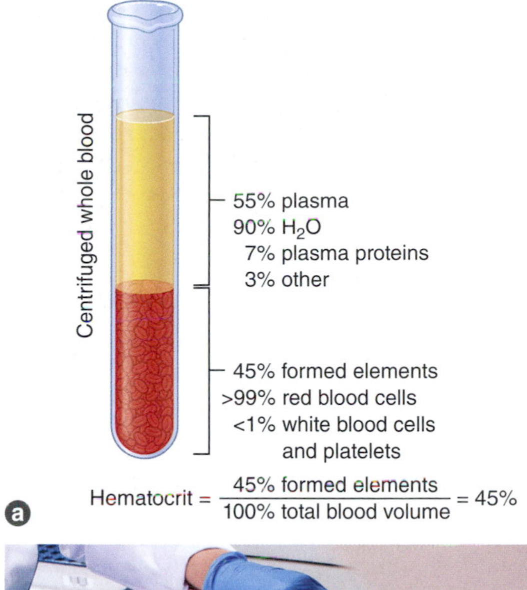

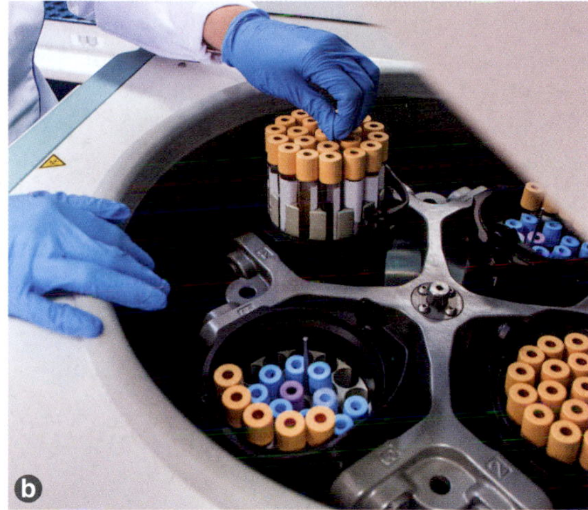

FIGURE 6.16 *(a)* The composition of whole blood, illustrating the plasma volume (fluid portion) and the cellular volume (red cells, white cells, and platelets) after the blood sample has been centrifuged to separate its components. *(b)* A centrifuge.

7% consists of plasma proteins; and the remaining 3% includes cellular nutrients, electrolytes, enzymes, hormones, antibodies, and wastes.

The formed elements, which normally constitute the other 40% to 45% of total blood volume, are the red blood cells (erythrocytes), white blood cells (leukocytes), and platelets (thrombocytes). Red blood cells constitute more than 99% of the formed-element volume; white blood cells and platelets together account for less than 1%. The percentage of the total blood volume composed of cells or formed elements is referred to as the **hematocrit**. Hematocrit varies among individuals, but a normal range is 41% to 50% in adult men and 36% to 44% in adult women.

White blood cells protect the body from infection either by directly destroying the invading agents through phagocytosis (ingestion) or by forming antibodies to destroy them. Adults have about 7,000 white blood cells per cubic millimeter of blood.

The remaining formed elements are the blood platelets. These are cell fragments that are required for blood coagulation (clotting), which prevents excessive blood loss. Exercise physiologists are most concerned with red blood cells.

Red Blood Cells

Mature red blood cells (erythrocytes) have no nucleus, so they cannot reproduce as other cells can. They must be replaced with new cells on a recurring basis, a process called **hematopoiesis**. The normal life span of a red blood cell is about 4 months. Thus, these cells are continuously produced and destroyed at equal rates. This balance is very important, because adequate oxygen delivery to tissues depends on having a sufficient number of red blood cells to transport oxygen. Decreases in their number or function can hinder oxygen delivery and thus affect exercise performance.

When we donate blood, the removal of one unit, or nearly 500 ml, represents approximately an 8% to 10% reduction in both the total blood volume and the number of circulating red blood cells. Donors are advised to drink plenty of fluids. Because plasma is primarily water, simple fluid replacement returns plasma volume to normal within 24 to 48 h. However, it takes at least 6 weeks to reconstitute the red blood cells because they must go through full development before they are functional. Blood loss greatly compromises the performance of endurance athletes by reducing oxygen delivery capacity.

Red blood cells transport oxygen, which is primarily bound to hemoglobin. **Hemoglobin** is composed of a protein (globin) and a pigment (heme). Heme contains iron, which binds oxygen. Each red blood cell contains approximately 250 million hemoglobin molecules, each able to bind four oxygen molecules—so each red blood cell can bind up to a billion molecules of oxygen! There is an average of 15 g of hemoglobin per 100 ml of whole blood. Each gram of hemoglobin can combine with 1.33 ml of oxygen, so as much as 20 ml of oxygen can be bound for each 100 ml of blood. Therefore, when arterial blood is saturated with oxygen, it has an oxygen-carrying capacity of 20 ml of oxygen per 100 ml of blood.

Blood Viscosity

Viscosity refers to the thickness of the blood. Recall from our discussion of vascular resistance that the more viscous a fluid, the more resistant it is to flow. Syrup is more viscous than water and thus flows more slowly when poured. The viscosity of blood is normally about twice that of water and increases as hematocrit increases.

Because of oxygen transport by the red blood cells, an increase in their number would be expected to maximize oxygen transport. But if an increase in red blood cell count is not accompanied by a similar increase in plasma volume, blood viscosity and vascular resistance will increase, which could result in reduced blood flow. This generally is not a problem unless the hematocrit reaches 60% or more.

Conversely, the combination of a low hematocrit with a high plasma volume, which decreases the blood's viscosity, appears to have certain benefits for the blood's transport function because the blood can flow more easily. Unfortunately, a low hematocrit frequently results from a reduced red blood cell count, as in diseases such as anemia. Under these circumstances, the blood can flow easily, but it contains fewer carriers, so oxygen transport is impeded. For optimal physical performance, a low-normal hematocrit with a normal or slightly elevated number of red blood cells is desirable. This combination facilitates oxygen transport. Many endurance athletes achieve this combination as part of their cardiovascular system's normal adaptation to training. This adaptation is discussed in chapter 11.

In Review

- Blood is about 55% to 60% plasma and 40% to 45% formed elements. Red blood cells compose about 99% of the formed elements.
- The hematocrit is the ratio of the formed elements in the blood (red cells, white cells, and platelets) to the total blood volume. An average hematocrit for adult men is 42% and for adult women is 38%.
- Oxygen is transported primarily by binding to the hemoglobin in red blood cells.
- During endurance training, athletes respond with both a higher red cell volume (RCV) and an expanded plasma volume (PV). Since the PV increase is higher than the RCV increase, the hematocrit in these athletes tends to be somewhat lower than that of sedentary individuals.
- As blood viscosity increases, so does resistance to flow. Increasing the number of red blood cells is advantageous to aerobic performance but only up to the point (a hematocrit approaching 60%) where viscosity limits flow.

IN CLOSING

In this chapter, we reviewed the structure and function of the cardiovascular system. We learned how blood flow and blood pressure are regulated to meet the body's needs and explored the role of the cardiovascular system in transporting and delivering oxygen and nutrients to the body's cells while clearing away metabolic wastes, including carbon dioxide. Knowing how substances are moved within the body, we now look more closely at the transport of oxygen and carbon dioxide. In the next chapter, we explore the role of the respiratory system in delivering oxygen to, and removing carbon dioxide from, the cells of the body.

KEY TERMS

arteries
arterioles
atherosclerosis
atrioventricular (AV) node
baroreceptor
bradycardia
capillaries
cardiac cycle
cardiac output ($\dot{Q}$)
chemoreceptor
diastolic blood pressure (DBP)
ejection fraction (EF)
electrocardiogram (ECG)
electrocardiograph
end-diastolic volume (EDV)
end-systolic volume (ESV)
extrinsic neural control
functional sympatholysis
heart murmur
hematocrit
hematopoiesis
hemoglobin
intercalated disks
mean arterial pressure (MAP)
mechanoreceptors
muscle pump
myocardium
pericardium
premature ventricular contraction (PVC)
Purkinje fibers
sinoatrial (SA) node
stroke volume (SV)
systolic blood pressure (SBP)
tachycardia
vasoconstriction
vasodilation
veins
ventricular fibrillation
ventricular tachycardia
venules

STUDY QUESTIONS

1. Describe the structure of the heart, the pattern of blood flow through the valves and chambers of the heart, how the heart as a muscle is supplied with blood, and what happens when the resting heart must suddenly supply an exercising body.
2. What events take place that allow the heart to contract, and how is heart rate controlled?
3. What is torsional contraction of the heart, and why is it important during exercise?
4. What is the difference between systole and diastole, and how do they relate to SBP and DBP?
5. What is the relationship between pressure, flow, and resistance?
6. How is blood flow to the various regions of the body controlled?
7. How does muscle blood flow increase during exercise despite increased sympathetic nerve activity that favors vasoconstriction?
8. Describe the three important mechanisms for returning blood back to the heart when someone is exercising in an upright position.
9. Describe the primary functions of blood.

STUDY GUIDE ACTIVITIES

In addition to the activities listed in the chapter opening outline, two other activities are available in the web study guide, located at

www.HumanKinetics.com/PhysiologyOfSportAndExercise

The **KEY TERMS** activity reviews important terms, and the end-of-chapter **QUIZ** tests your understanding of the material covered in the chapter.

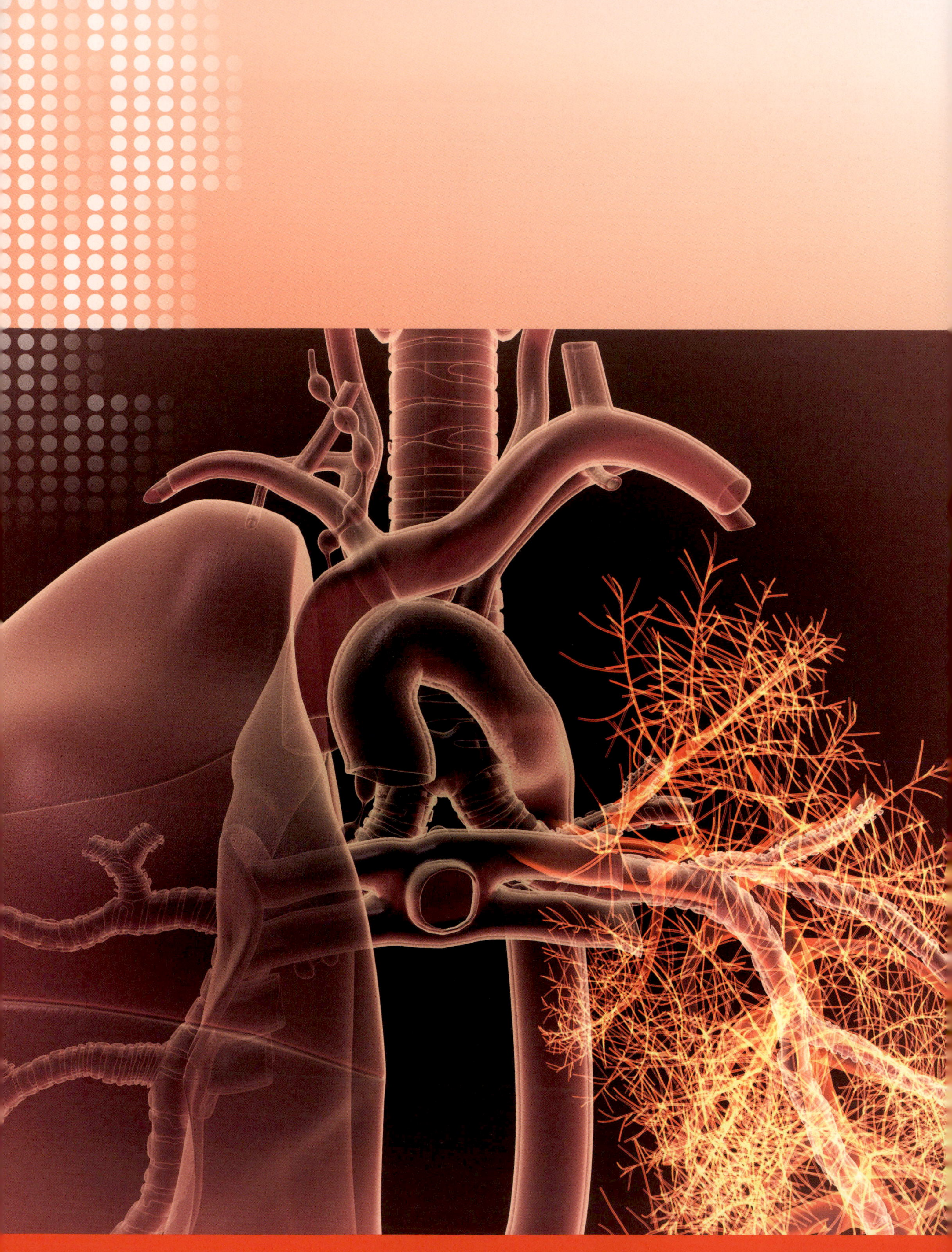

The Respiratory System and Its Regulation

In this chapter and in the web study guide

By any standard, Beijing, China, is one of the most polluted cities on the planet. In preparation for the 2008 Olympic Games, nearly $17 billion was spent in attempts to temporarily improve air quality, including cloud seeding to increase the likelihood of rain showers in the region overnight. Factories were closed, traffic was halted, and construction was put on hold for the duration of the Games. Yet air pollution at the Olympics was still about two to four times higher than that of Los Angeles on an average day, exceeding levels considered safe by the World Health Organization. Several athletes opted out of events because of respiratory problems or concerns, including Ethiopian marathon record holder Haile Gebrselassie and 2004 cycling silver medalist Sérgio Paulinho of Portugal. Athletes previously diagnosed with asthma were allowed to use rescue inhalers. For the first time ever, soccer matches were interrupted to give athletes time to recover from the pollutants, smog, heat, and humidity. Athletes and spectators endured these conditions for a few weeks, and there are no reports of long-term health problems among athletes or spectators from exposure to the Beijing air. However, the residents of Beijing encounter these adverse respiratory conditions on a daily basis.

The respiratory and cardiovascular systems combine to provide an effective delivery system that carries oxygen to, and removes carbon dioxide from, all tissues of the body.

This transportation involves four separate processes:

- Pulmonary ventilation (breathing): movement of air into and out of the lungs
- Pulmonary diffusion: the exchange of oxygen and carbon dioxide between the lungs and the blood
- Transport of oxygen and carbon dioxide via the blood
- Capillary diffusion: the exchange of oxygen and carbon dioxide between the capillary blood and metabolically active tissues

The first two processes are referred to as **external respiration** because they involve moving gases from outside the body into the lungs and then the blood. Once the gases are in the blood, they must be transported to the tissues. When blood arrives at the tissues, the fourth step of respiration occurs. This gas exchange between the blood and the tissues is called **internal respiration**. Thus, external and internal respiration are linked by the circulatory system. The following sections examine all four components of respiration.

Pulmonary Ventilation

Pulmonary ventilation, or breathing, is the process by which we move air into and out of the lungs. The anatomy of the respiratory system is illustrated in figure 7.1. At rest, air is typically drawn into the lungs through the nose, although the mouth must also be used when the demand for air exceeds the amount that can comfortably be brought in through the nose. Nasal breathing is advantageous because the air is warmed and humidified as it swirls through the bony irregular sinus surfaces (turbinates or conchae). Of equal importance, the turbinates churn the inhaled air, causing dust and other particles to contact and adhere to the nasal mucosa. This filters out all but the tiniest particles, minimizing irritation and the threat of respiratory infections. From the nose and mouth, the air travels through the pharynx, larynx, trachea, and bronchial tree.

This transport zone also has physiological significance because it comprises the so-called anatomical **dead space**. Because part of each expired breath stays within this space, air from outside the body mixes with this air with each inspiration, and the resulting mixture reaches the alveoli.

These anatomical structures serve a transport function only, because gas exchange does not occur in these structures. Exchange of oxygen and carbon dioxide occurs when air finally reaches the smallest respiratory units: the respiratory bronchioles and the alveoli. The respiratory bronchioles are primarily transport tubes also but are included in this region because they contain clusters of alveoli. This is known as the respiratory zone because it is the site of gas exchange in the lungs.

The lungs are not directly attached to the ribs. Rather, they are suspended by the pleural sacs. The pleural sacs have a double wall: the parietal pleura, which lines the thoracic wall, and the visceral or pulmonary pleura, which lines the outer aspects of the lung. These pleural walls envelop the lungs and have a thin film of fluid between them that reduces friction during respiratory movements. In addition, these sacs are connected to the lungs and the inner surface of the thoracic cage, causing the lungs to take the shape and size of the rib or thoracic cage as the chest expands and contracts.

The anatomy of the lungs, the pleural sacs, the diaphragm muscle, and the thoracic cage determines airflow into and out of the lungs, that is, inspiration and expiration.

Inspiration

Inspiration is an active process involving the diaphragm and the external intercostal muscles. Figure 7.2*a* shows the resting positions of the diaphragm and the thoracic

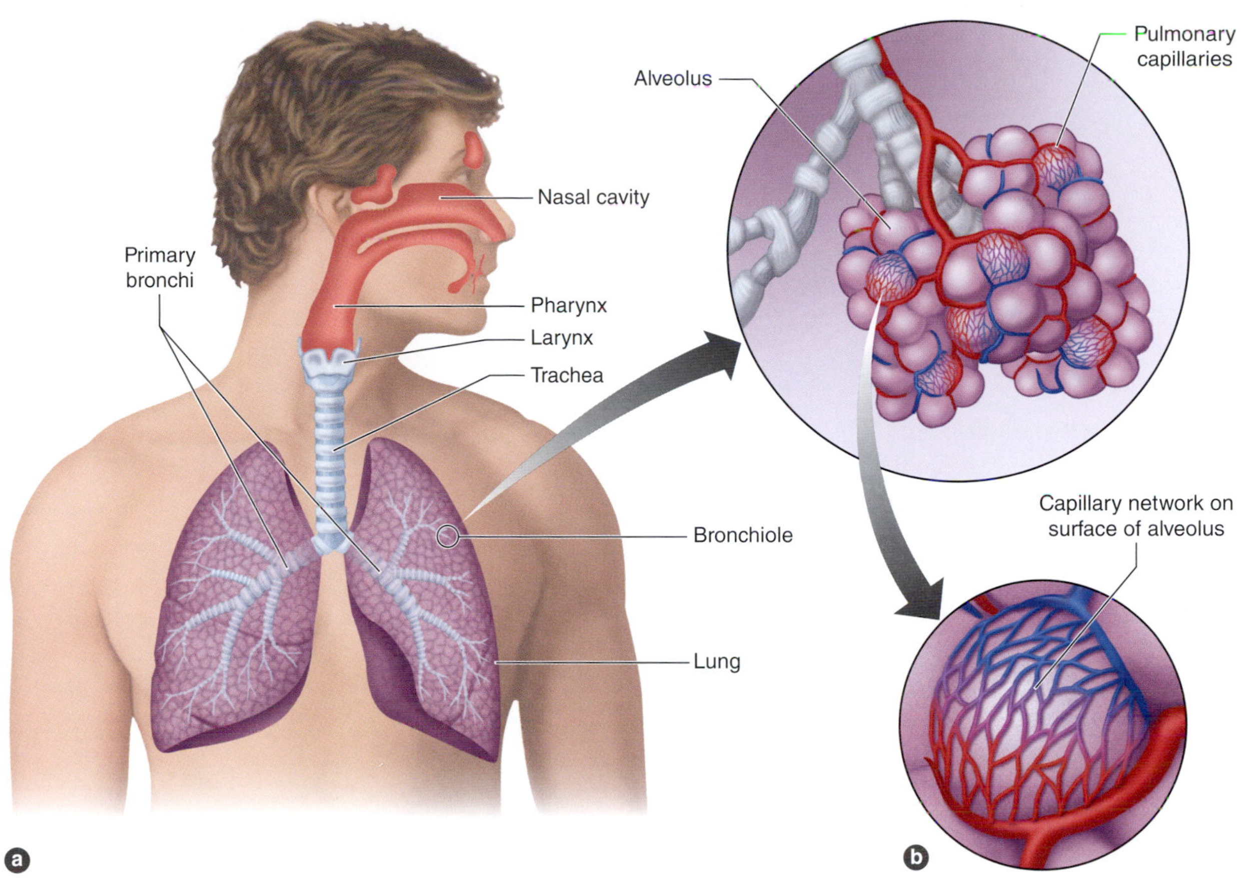

FIGURE 7.1 *(a)* The anatomy of the respiratory system, illustrating the respiratory tract (i.e., nasal cavity, pharynx, trachea, and bronchi). *(b)* The enlarged view of an alveolus shows the regions of gas exchange between the alveolus and pulmonary blood in the capillaries.

cage, or thorax. With inspiration, the ribs and sternum are moved by the external intercostal muscles. The ribs swing up and out and the sternum swings up and forward. At the same time, the diaphragm contracts, flattening down toward the abdomen.

These actions, illustrated in figure 7.2*b*, expand all three dimensions of the thoracic cage, increasing the volume inside the lungs. When the lungs are expanded they have a greater volume, and the air within them has more space to fill. According to **Boyle's gas law**, which states that pressure × volume is constant (at a constant temperature), the pressure within the lungs decreases. As a result, the pressure in the lungs (intrapulmonary pressure) is less than the air pressure outside the body. Because the respiratory tract is open to the outside, air rushes into the lungs to reduce this pressure difference. This is how air moves into the lungs during inspiration.

The pressure changes required for adequate ventilation at rest are really quite small. For example, at the standard atmospheric pressure at sea level (760 mmHg), inspiration may decrease the pressure in the lungs (intrapulmonary pressure) by only about 2 to 3 mmHg. However, during maximal respiratory effort, such as during exhaustive exercise, the intrapulmonary pressure can decrease by 80 to 100 mmHg.

During forced or labored breathing, as during heavy exercise, inspiration is further assisted by the action of other muscles, such as the scalenes (anterior, middle, and posterior) and sternocleidomastoid in the neck and the pectorals in the chest. These muscles help raise the ribs even more than during regular breathing.

Expiration

At rest, **expiration** is a passive process involving relaxation of the inspiratory muscles and elastic recoil of the lung tissue. As the diaphragm relaxes, it returns to its normal upward, arched position. As the external intercostal muscles relax, the ribs and sternum move back into their resting positions (figure 7.2*c*). While this happens, the elastic nature of the lung tissue causes it to recoil to its resting size. This increases the pressure in the lungs and causes a proportional decrease in volume in the thorax, and therefore air is forced out of the lungs.

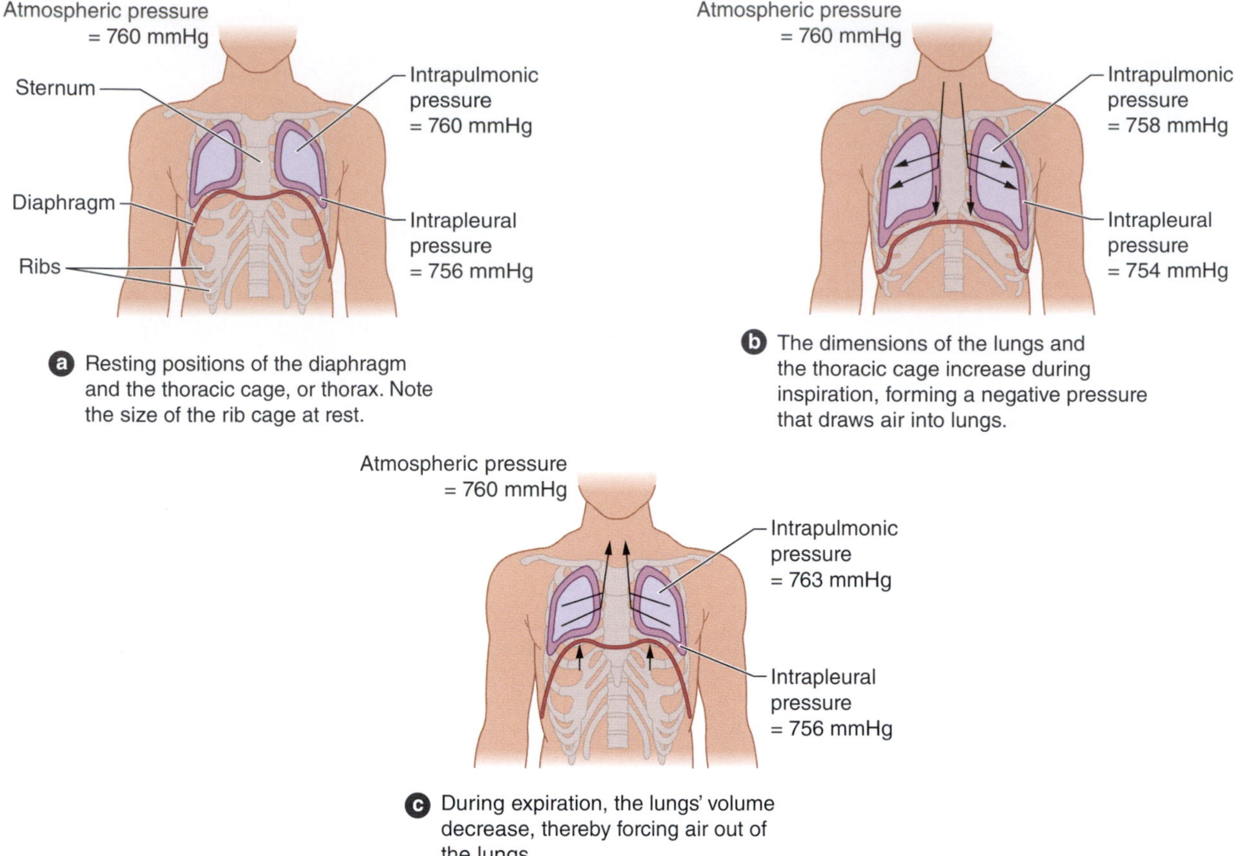

FIGURE 7.2 The process of inspiration and expiration, showing *(a)* the positions of the ribs and thorax at rest, and how movement of the ribs and diaphragm *(b)* increase the size of the thorax during inspiration and *(c)* decrease the size of the thorax during expiration.

During forced breathing, expiration becomes a more active process. The internal intercostal muscles actively pull the ribs down. This action can be assisted by the latissimus dorsi and quadratus lumborum muscles. Contracting the abdominal muscles increases the intra-abdominal pressure, forcing the abdominal viscera upward against the diaphragm and accelerating its return to the domed position. These muscles also pull the rib cage down and inward.

The changes in intra-abdominal and intrathoracic pressure that accompany forced breathing also help return venous blood back to the heart, working together with the muscle pump in the legs to assist the return of venous volume. As intra-abdominal and intrathoracic pressure increases, it is transmitted to the great veins—the pulmonary veins and superior and inferior venae cavae—that transport blood back to the heart. When the pressure decreases, the veins return to their original size and fill with blood. The changing pressures within the abdomen and thorax squeeze the blood in the veins, assisting its return through a milking action. This phenomenon is known as the **respiratory pump** and is essential in maintaining adequate venous return.

Pulmonary Volumes

The volume of air in the lungs can be measured with a technique called **spirometry**. A spirometer measures the volumes of air inspired and expired and therefore changes in lung volume. Although more sophisticated spirometers are used today, a simple spirometer contains a bell filled with air that is partially submerged in water. A tube runs from the subject's mouth under the water and emerges inside the bell, just above the water level. As the person exhales, air flows down the tube and into the bell, causing the bell to rise. The bell is attached to a pen, and its movement is recorded on a simple rotating drum (figure 7.3).

This technique is used clinically to measure lung volumes, capacities, and flow rates as an aid in diagnosing such respiratory diseases as asthma, chronic obstructive pulmonary disease (COPD), and emphysema.

The amount of air entering and leaving the lungs with each breath is known as the **tidal volume**. The **vital capacity (VC)** is the greatest amount of air that can be expired after a maximal inspiration. Even after a full expiration, some air remains in the lungs. The

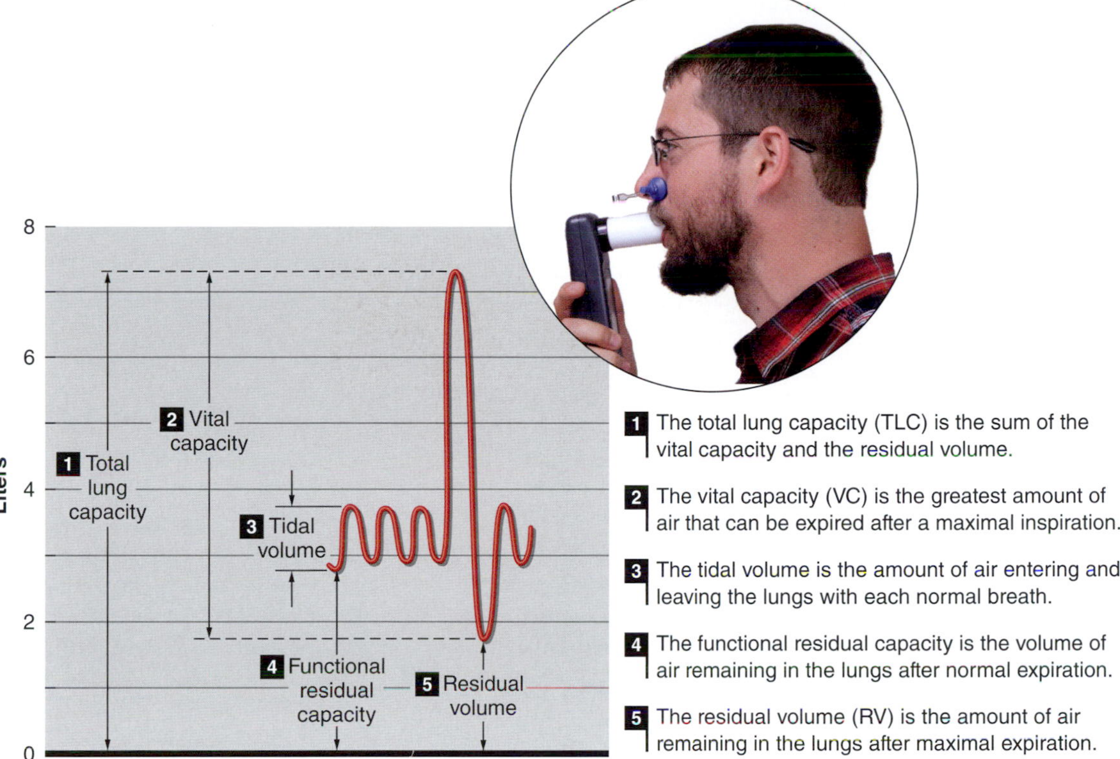

FIGURE 7.3 Lung volumes measured by spirometry.

amount of air remaining in the lungs after a maximal expiration is the **residual volume (RV)**. The RV cannot be measured with spirometry. The **total lung capacity (TLC)** is the sum of the VC and the RV.

In Review

- Pulmonary ventilation (breathing) is the process by which air is moved into and out of the lungs. It has two phases: inspiration and expiration.
- Inspiration is an active process in which the diaphragm and the external intercostal muscles contract, increasing the dimensions, and thus the volume, of the thoracic cage. This decreases the pressure in the lungs, causing air to flow in.
- Expiration at rest is normally a passive process. The inspiratory muscles and diaphragm relax and the elastic tissue of the lungs recoils, returning the thoracic cage to its smaller, normal dimensions. This increases the pressure in the lungs and forces air out.
- The pressure changes required for ventilation at rest are small, as little as 2 mmHg. However, during maximal respiratory effort, the intrapulmonary pressure can decrease by 80 to 100 mmHg.
- Forced or labored inspiration and expiration are active processes and involve accessory muscle actions.
- Breathing through the nose helps humidify and warm the air during inhalation and filters out foreign particles from the air. Mouth breathing dominates at moderate to high exercise intensities.
- Lung volumes and capacities, along with rates of airflow into and out of the lungs, are measured by spirometry.

Pulmonary Diffusion

Gas exchange in the lungs between the alveoli and the capillary blood, called **pulmonary diffusion**, serves two major functions:

- It replenishes the blood's oxygen supply, which is depleted at the tissue level as it is used for oxidative energy production.
- It removes carbon dioxide from venous blood returning from systemic tissues.

Air is brought into the lungs during pulmonary ventilation, enabling gas exchange to occur through pulmonary diffusion. Oxygen from the air diffuses from the

RESEARCH PERSPECTIVE 7.1

Sprint Interval Training for Respiratory Muscles

Typical respiratory muscle endurance training (RMET) improves exercise capacity and thus performance; such improvements are largely attributed to reductions in the development of respiratory muscle fatigue. However, can a shorter version of RMET based on the principle of high-intensity interval training (respiratory muscle sprint-interval training, or RMSIT) elicit similar improvements in respiratory muscle function?

A team of investigators recently sought to compare the effects of traditional RMET versus RMSIT on respiratory muscle function.[7] Mechanical airway properties and respiratory muscle testing (e.g., respiratory muscle strength) were measured before and after experimental sessions of RMET and RMSIT. RMET consisted of continuous volitional hyperpnea (increased depth and rate of breathing) performed for 30 min using a commercially available training device. The RMSIT was a novel respiratory muscle training regimen developed by the researchers. Using the same training device as with RMET, the RMSIT regimen consisted of six short maximal respiratory sprints with additional airway resistance to maximize respiratory muscle work. In this fashion, RMSIT is characterized by higher respiratory muscle power output and tension-time indices, but considerably lower total work compared to RMET. The standard RMET and the novel RMSIT regimens reduced respiratory muscle contractility to the same extent, triggering similar muscular adaptations in response to training. Neither protocol altered mechanical airway properties. Therefore, RMSIT appears to be a safe and time-saving alternative to RMET.

RMET can improve overall function for individuals who have undergone a median sternotomy (splitting of the sternum to access underlying organs) during cardiac surgery. Clinical exercise physiologists are interested in exercise training adaptations that occur with structured cardiac rehabilitation programs. The results of a 2013 study suggest that it would be beneficial to include exercises that improve the strength of the inspiratory muscles as part of a cardiac rehabilitation program.[4] This type of training would reduce inspiratory muscle effort and further improve ventilatory efficiency in patients after open-heart surgery.

alveoli into the blood in the pulmonary capillaries, and carbon dioxide diffuses from the blood into the alveoli in the lungs. The **alveoli** are grapelike clusters, or air sacs, at the ends of the terminal bronchioles.

Blood from the body (except for that returning from the lungs) returns through the vena cava to the right side of the heart. From the right ventricle, this blood is pumped through the pulmonary artery to the lungs, ultimately working its way into the pulmonary capillaries. These capillaries form a dense network around the alveolar sacs and are so small that the red blood cells must pass through them in single file, such that the maximal surface area of each cell is exposed to the surrounding lung tissue. This is where pulmonary diffusion occurs.

Blood Flow to the Lungs at Rest

At rest the lungs receive approximately 4 to 6 L/min of blood flow, depending on body size. Because cardiac output from the right side of the heart approximates cardiac output from the left side of the heart, blood flow to the lungs matches blood flow to the systemic circulation. However, pressure and vascular resistance in the blood vessels in the lungs are different than in the system circulation. The mean pressure in the pulmonary artery is ~15 mmHg (systolic pressure is ~25 mmHg and diastolic pressure is ~8 mmHg) compared to the mean pressure in the aorta of ~95 mmHg. The pressure in the left atrium, where blood is returning to the heart from the lungs, is ~5 mmHg; thus, there is not a great pressure difference across the pulmonary circulation (15 − 5 mmHg). Figure 7.4 illustrates the differences in pressures between the pulmonary and systemic circulation.

Recalling the discussion of blood flow in the cardiovascular system from chapter 6, pressure = flow × resistance. Since blood flow to the lungs is equal to that of the systemic circulation, and there is a substantially lower change in pressure across the pulmonary vascular system, resistance is proportionally lower compared to that in the systemic circulation. This is reflected in differences in the anatomy of the vessels in the pulmonary versus systemic circulation: The pulmonary blood vessels are thin walled, with relatively little smooth muscle.

Respiratory Membrane

Gas exchange between the air in the alveoli and the blood in the pulmonary capillaries occurs across the **respiratory membrane** (also called the alveolar-capillary membrane). This membrane, depicted in figure 7.5, is composed of

- the alveolar wall,
- the capillary wall, and
- their respective basement membranes.

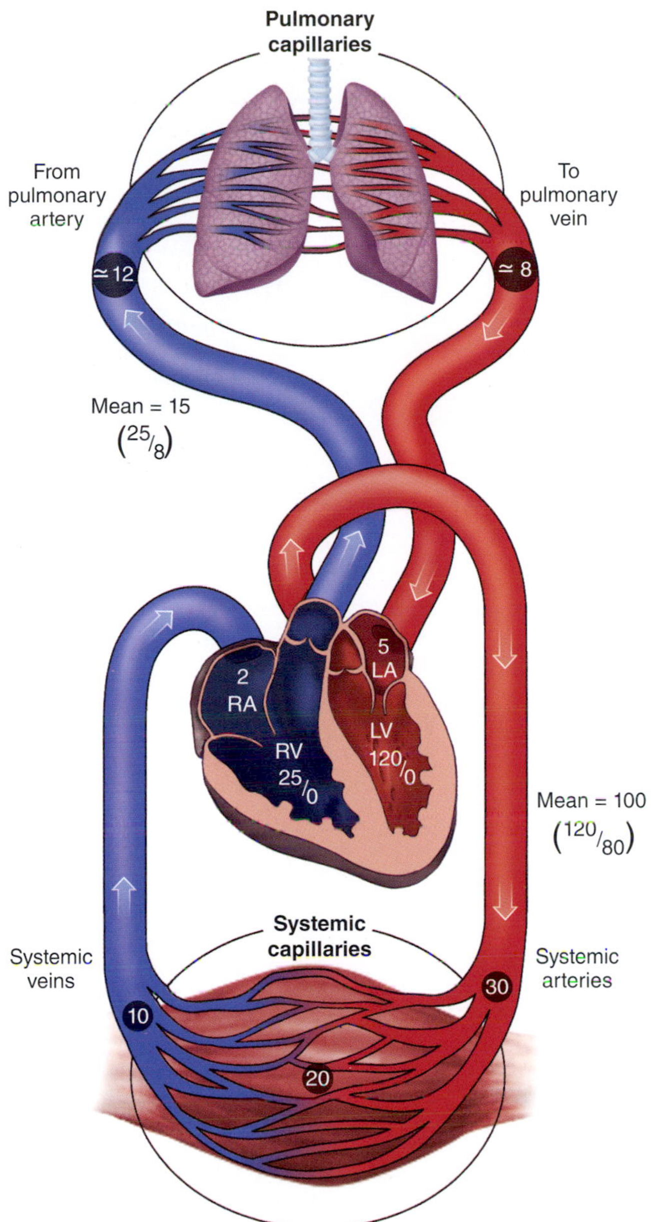

FIGURE 7.4 Comparison of pressures (mmHg) in the pulmonary and systemic circulations.

The primary function of these membranous surfaces is for gas exchange. The respiratory membrane is very thin, measuring only 0.5 to 4 µm. As a result, the gases in the nearly 300 million alveoli are in close proximity to the blood circulating through the capillaries.

Partial Pressures of Gases

The air we breathe is a mixture of gases. Each exerts a pressure in proportion to its concentration in the gas mixture. The individual pressures from each gas in a mixture are referred to as **partial pressures**. According to **Dalton's law**, the total pressure of a mixture of gases equals the sum of the partial pressures of the individual gases in that mixture.

Consider the air we breathe. It is composed of 79.04% nitrogen (N_2), 20.93% oxygen (O_2), and 0.03% carbon dioxide (CO_2). These percentages remain constant regardless of altitude. At sea level, the atmospheric (or barometric) pressure is approximately 760 mmHg, which is also referred to as standard atmospheric pressure. Thus, if the total atmospheric pressure is 760 mmHg, then the partial pressure of nitrogen (PN_2) in air is 600.7 mmHg (79.04% of the total 760 mmHg pressure). Oxygen's partial pressure (PO_2) is 159.1 mmHg (20.93% of 760 mmHg), and carbon dioxide's partial pressure (PCO_2) is 0.2 mmHg (0.03% of 760 mmHg).

In the human body, gases are usually dissolved in fluids, such as blood plasma. According to **Henry's law**, gases dissolve in liquids in proportion to their partial pressures, depending also on their solubilities in the specific fluids and on the temperature. A gas's solubility in blood is a constant, and blood temperature also remains relatively constant at rest. Thus, the most critical factor for gas exchange between the alveoli and the blood is the pressure gradient between the gases in the two areas.

Gas Exchange in the Alveoli

Differences in the partial pressures of the gases in the alveoli and the gases in the blood create a pressure gradient across the respiratory membrane. This forms the basis of gas exchange during pulmonary diffusion. If the pressures on each side of the membrane were equal, the gases would be at equilibrium and would not move. But the pressures are not equal, so gases move according to partial pressure gradients.

Oxygen Exchange

The PO_2 of air outside the body at standard atmospheric pressure is 159 mmHg. But this pressure decreases to about 105 mmHg when air is inhaled and enters the alveoli, where it is moistened and mixes with the air in the alveoli. The alveolar air is saturated with water vapor (which has its own partial pressure) and contains more carbon dioxide than the inspired air. Both the increased water vapor pressure and increased partial pressure of carbon dioxide contribute to the total pressure in the alveoli. Fresh air that ventilates the lungs is constantly mixed with the air in the alveoli while some of the alveolar gases are exhaled to the environment. As a result, alveolar gas concentrations remain relatively stable.

The blood, stripped of much of its oxygen by the metabolic needs of the tissues, typically enters the pulmonary capillaries with a PO_2 of about 40 mmHg (see figure 7.6). This is about 60 to 65 mmHg less than the

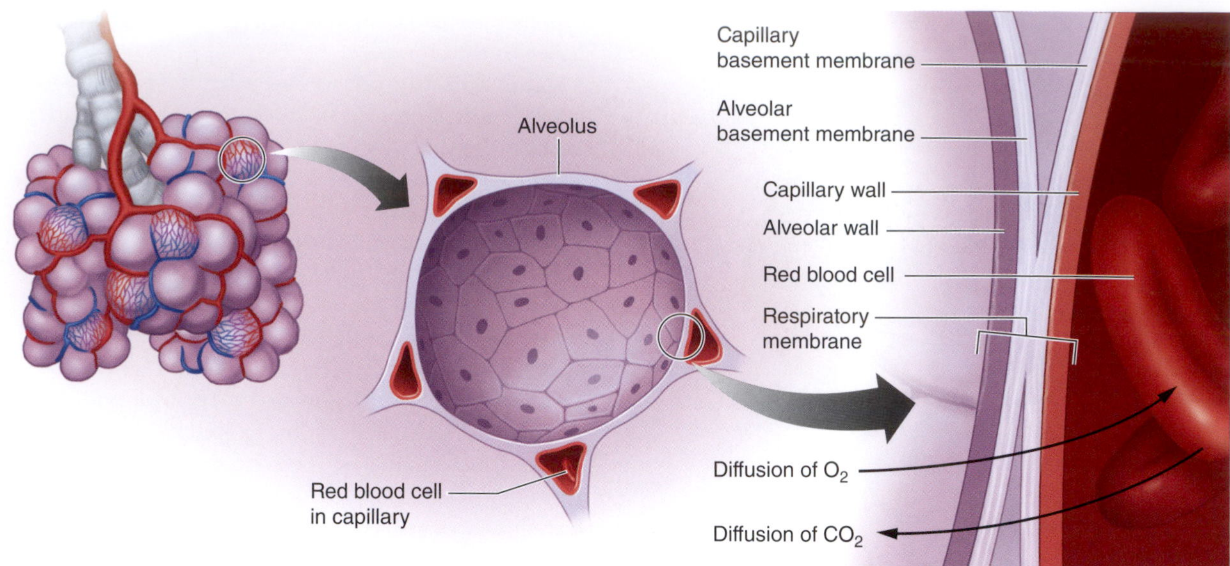

FIGURE 7.5 The anatomy of the respiratory membrane, showing the exchange of oxygen and carbon dioxide between an alveolus and pulmonary capillary blood.

PO_2 in the alveoli. In other words, the pressure gradient for oxygen across the respiratory membrane is typically about 65 mmHg. As noted earlier, this pressure gradient drives the oxygen from the alveoli into the blood to equilibrate the pressure of the oxygen on each side of the membrane.

RESEARCH PERSPECTIVE 7.2

Exercise Training Offsets Decreases in Lung Diffusing Capacity with Aging

The structure and function of the pulmonary vasculature contributes to maximal aerobic capacity ($\dot{V}O_{2max}$), such that a larger, more distensible vascular network in the lungs is associated with greater aerobic exercise capacity. During exercise, increased cardiac output and pulmonary perfusion pressure cause an expansion of the highly compliant pulmonary capillary network, resulting in increased lung diffusing capacity, alveolar-capillary membrane conductance, and pulmonary capillary blood volume.

As we age, the structure and function of the pulmonary circulation changes, resulting in increased pulmonary vascular stiffness, pulmonary vascular pressures, and pulmonary vascular resistance, all of which impair recruitment and distension of pulmonary capillaries during exercise. However, these age-related alterations do not appear to limit the expansion of pulmonary capillaries during exercise in healthy older adults. The pulmonary vascular response to exercise in endurance-trained, highly fit older adults is not well defined. It is plausible that a higher $\dot{V}O_{2max}$ may cause the demand for cardiac output and pulmonary blood flow during exercise to remain elevated in older athletes, thus predisposing highly fit older adults to impairments in pulmonary vascular expansion and pulmonary gas exchange relative to the metabolic demands of exercise.

This concept was recently tested by a group of investigators who characterized lung diffusing capacity, alveolar-capillary membrane conductance, and pulmonary capillary blood volume in response to incremental exhaustive exercise in aerobically trained older adults.[3] The authors hypothesized that older athletes would be limited in their ability to expand the pulmonary vascular network during high-intensity exercise. Their findings confirmed the negative age-related reductions in lung diffusing capacity, alveolar-capillary membrane conductance, and pulmonary capillary blood volume during exercise; however, these variables were increased in exercise-trained older adults during exercise relative to age-matched, nontrained individuals. In contrast to the original hypothesis, there was a progressive increase in lung diffusing capacity throughout exercise in exercise-trained adults, suggesting that the expansion of the pulmonary capillary network during exercise is not limited during exercise in highly fit older adults. Follow-up studies should include measures of pulmonary vascular pressures to more specifically determine the relation between increases in lung diffusing capacity and the pulmonary vascular response to exercise.

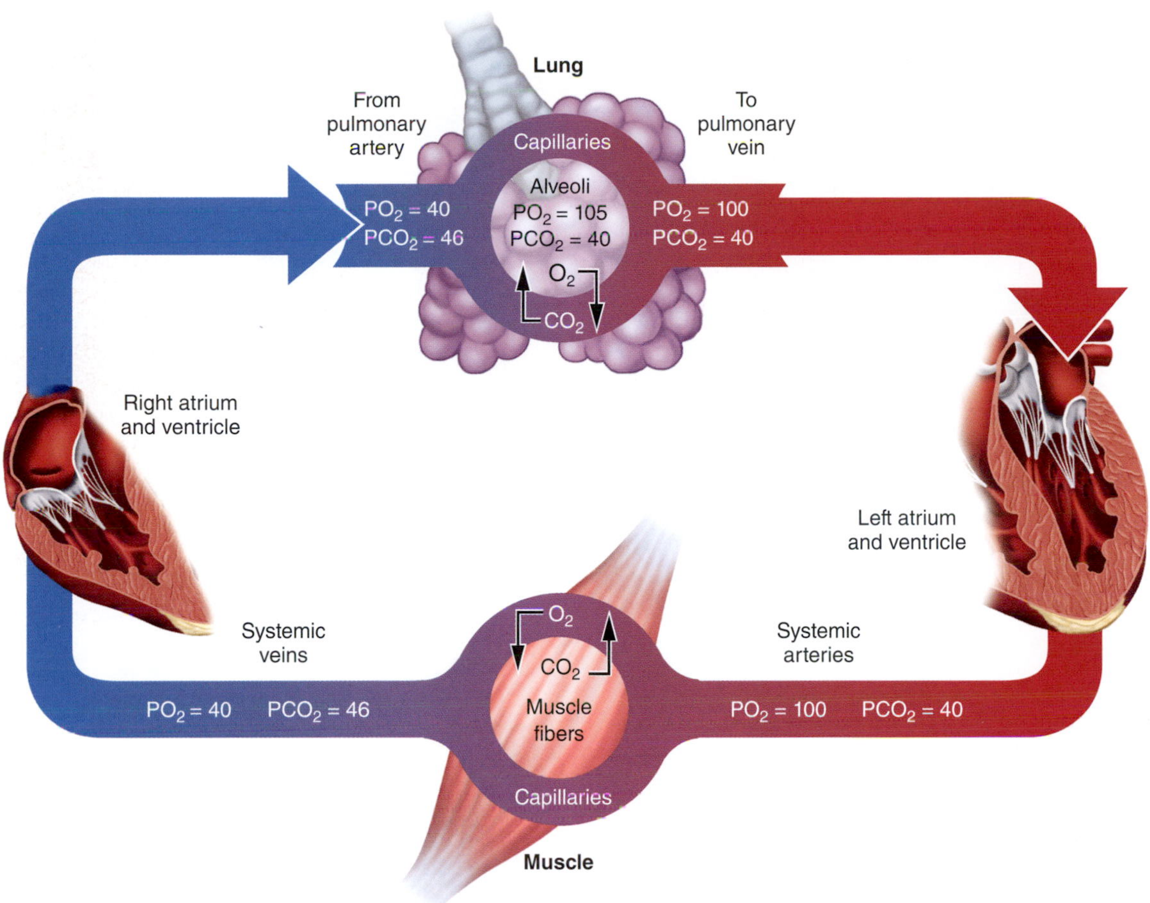

FIGURE 7.6 Partial pressure of oxygen (PO_2) and carbon dioxide (PCO_2) in blood as a result of gas exchange in the lungs and gas exchange between the capillary blood and tissues.

The PO_2 in the alveoli stays relatively constant at about 105 mmHg. As the deoxygenated blood enters the pulmonary artery, the PO_2 in the blood is only about 40 mmHg. But as the blood moves along the pulmonary capillaries, gas exchange occurs. By the time the pulmonary blood reaches the venous end of these capillaries, the PO_2 in the blood equals that in the alveoli (approximately 105 mmHg), and the blood is now considered to be saturated with oxygen at its full carrying capacity. The blood leaving the lungs through the pulmonary veins and subsequently returning to the systemic (left) side of the heart has a rich supply of oxygen to deliver to the tissues. Notice, however, that the PO_2 in the pulmonary vein is 100 mmHg, not the 105 mmHg found in the alveolar air and pulmonary capillaries. This difference is attributable to the fact that about 2% of the blood is shunted from the aorta directly to the lung to meet the oxygen needs of the lung itself. This blood has a lower PO_2 and reenters the pulmonary vein along with fully saturated blood returning to the left atrium that has just completed gas exchange. This blood mixes and thus decreases the PO_2 of the blood returning to the heart.

Diffusion through tissues is described by **Fick's law** (figure 7.7). Fick's law states that the rate of diffusion through a tissue such as the respiratory membrane is proportional to the surface area and the difference in the partial pressure of gas between the two sides of the tissue. For example, the greater the pressure gradient for oxygen is across the respiratory membrane, the more rapidly oxygen diffuses across it. The rate of diffusion is also inversely proportional to the thickness of the tissue in which the gas must diffuse. Additionally, the diffusion constant, which is unique to each gas, influences the rate of diffusion across the tissue. Carbon dioxide has a much lower diffusion constant than oxygen; therefore, even though there is not as great a difference between alveolar and capillary partial pressure of carbon dioxide as there is for oxygen, carbon dioxide still diffuses easily.

The rate at which oxygen diffuses from the alveoli into the blood is referred to as the **oxygen diffusion capacity** and is expressed as the volume of oxygen that diffuses through the membrane each minute for a pressure difference of 1 mmHg. At rest, the oxygen diffusion capacity is about 21 ml of oxygen per minute

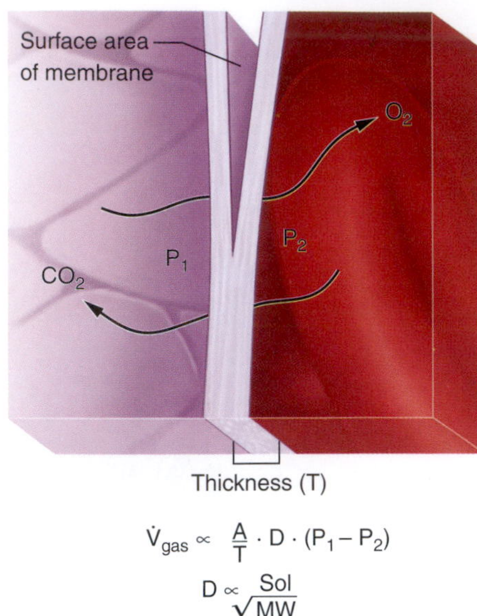

$$\dot{V}_{gas} \propto \frac{A}{T} \cdot D \cdot (P_1 - P_2)$$

$$D \propto \frac{Sol}{\sqrt{MW}}$$

FIGURE 7.7 Diffusion through a sheet of tissue. The amount of gas ($\dot{V}_{gas}$) transferred is proportional to the area (A), a diffusion constant (D), and the difference in partial pressure ($P_1 - P_2$) and is inversely proportional to the thickness (T). The constant is proportional to the gas solubility (Sol) but inversely proportional to the square root of its molecular weight (MW).

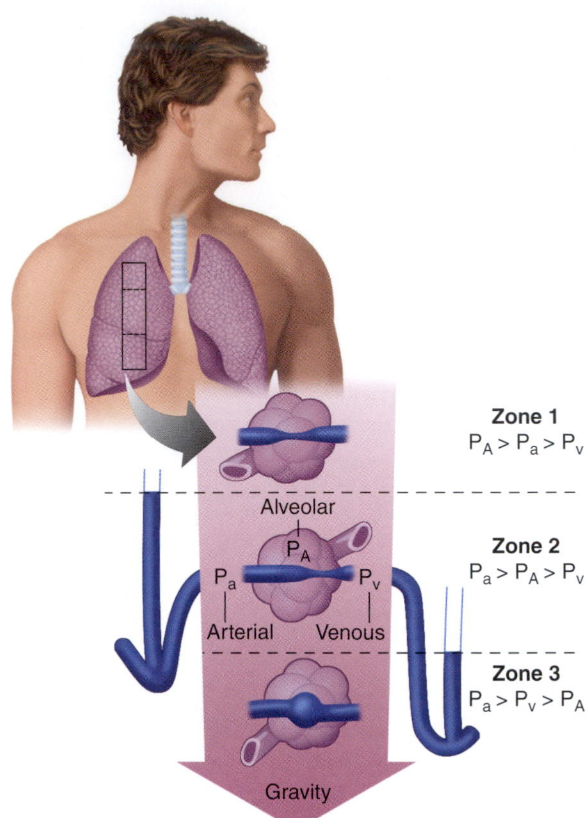

FIGURE 7.8 Explanation of the uneven distribution of blood flow in the lung.

per 1 mmHg of pressure difference between the alveoli and the pulmonary capillary blood. Although the partial pressure gradient between venous blood coming into the lung and the alveolar air is about 65 mmHg (105 mmHg – 40 mmHg), the oxygen diffusion capacity is calculated on the basis of the mean pressure in the pulmonary capillary, which has a substantially higher PO_2. The gradient between the mean partial pressure of the pulmonary capillary and the alveolar air is approximately 11 mmHg, which would provide a diffusion of 231 ml of oxygen per minute through the respiratory membrane. During maximal exercise, the oxygen diffusion capacity may increase by up to three times the resting rate, because blood is returning to the lungs severely desaturated and thus there is a greater partial pressure gradient from the alveoli to the blood. In fact, rates of more than 80 ml/min have been observed among highly trained athletes.

The increase in oxygen diffusion capacity from rest to exercise is caused by a relatively inefficient, sluggish circulation through the lungs at rest, which results primarily from limited perfusion of the upper regions of the lungs attributable to gravity. If the lung is divided into three zones as depicted in figure 7.8, at rest only the bottom third (zone 3) of the lung is perfused with blood. During exercise, however, blood flow through the lungs is greater, primarily as a result of elevated blood pressure, which increases lung perfusion.

Carbon Dioxide Exchange

Carbon dioxide, like oxygen, moves along a partial pressure gradient. As shown in figure 7.6, the blood passing from the right side of the heart through the alveoli has a PCO_2 of about 46 mmHg. Air in the alveoli has a PCO_2 of about 40 mmHg. Although this results in a relatively small pressure gradient of only about 6 mmHg, it is more than adequate to allow for exchange of CO_2. Carbon dioxide's diffusion coefficient is 20 times greater than that of oxygen, so CO_2 can diffuse across the respiratory membrane much more rapidly.

Summary of Pulmonary Gas Diffusion

The partial pressures of gases involved in pulmonary diffusion are summarized in table 7.1. Note that the total pressure in the venous blood is only 706 mmHg, 54 mmHg lower than the total pressure in dry air and alveolar air. This is the result of a much greater decrease in PO_2 compared with the increase in PCO_2 as the blood goes through the body's tissues.

TABLE 7.1 Partial Pressures of Respiratory Gases at Sea Level

		Partial pressure (mmHg)				
Gas	% in dry air	Dry air	Alveolar air	Arterial blood	Venous blood	Diffusion gradient
H_2O	0	0	47	47	47	0
O_2	20.93	159.1	105	100	40	60
CO_2	0.03	0.2	40	40	46	6
N_2	79.04	600.7	568	573	573	0
Total	100.00	760	760	760	706[a]	0

[a]See text for an explanation of the decrease in total pressure.

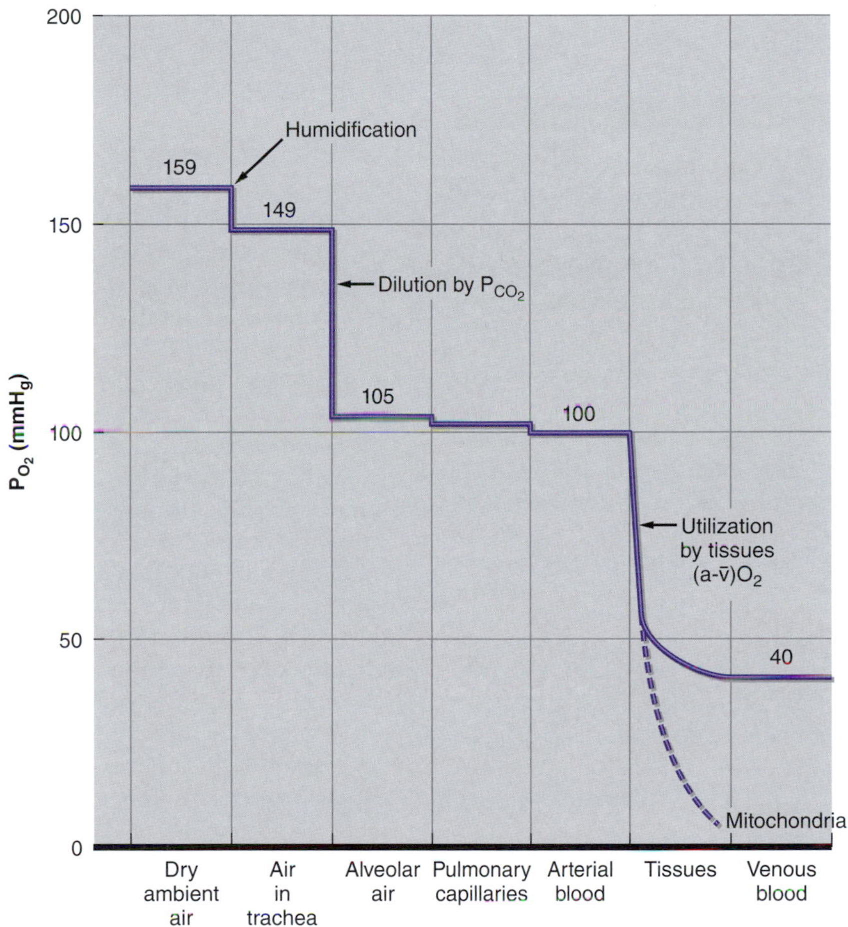

FIGURE 7.9 The oxygen cascade depicts the dropping partial pressures of oxygen (in this depiction, at sea level) from dry ambient air to the tissues and into the venous circulation draining those tissues.

Figure 7.9 shows the dropping partial pressures of oxygen at sea level from dry ambient air to the tissues and into the venous circulation draining those tissues. This is referred to as the **oxygen cascade**. At a sea level barometric pressure (P_B) of 760 mmHg, PO_2 in the ambient air (if it were completely devoid of moisture, which does not occur in nature) would be

$$0.2093 \times 760 \text{ mmHg} = 159 \text{ mmHg}.$$

As dry air moves through the nose and mouth and becomes humidified water vapor (which has a partial pressure, PH_2O, of 47 mmHg at body temperature), air in the trachea has a partial pressure of

$$0.2093 \times (760 - 47) = 149 \text{ mmHg}.$$

In the alveoli, air now becomes a mixture combining PCO_2 in blood returning from the systemic circulation and PO_2 from the tracheal air and equilibrates at approximately 105 mmHg. As oxygen diffuses from the alveoli into the pulmonary capillaries and into arterial blood, PO_2 continues to drop slightly down diffusion gradients, since pulmonary capillary blood is a mixture of arterial and venous blood, a so-called admixture.

At the tissue (e.g., muscle) level, cells extract O_2 from the arterial supply for aerobic metabolism, and the drop in PO_2 from arterial blood to venous blood flowing away from the tissues represents the arterial–venous oxygen difference, or (a-v)O_2 difference. Note that the PO_2 at the mitochondrial level is extremely low, approximately 1 to 2 mmHg. This ensures optimal O_2 delivery to these organelles, the ultimate destination of oxygen where it is used in oxidative phosphorylation.

In Review

- Pulmonary diffusion is the process by which gases are exchanged across the respiratory membrane in the alveoli.
- Dalton's law states that the total pressure of a mixture of gases equals the sum of the partial pressures of the individual gases in that mixture.
- The amount and rate of gas exchange that occur across the membrane depend primarily on the partial pressure of each gas, although other factors are also important, as shown by Fick's law. Gases diffuse along a pressure gradient, moving from an area of higher pressure to one of lower pressure. Thus, oxygen enters the blood and carbon dioxide leaves it.
- Oxygen diffusion capacity increases as one moves from rest to exercise. When exercising muscles require more oxygen to be used in the metabolic processes, venous oxygen is depleted and oxygen exchange at the alveoli is facilitated.
- The pressure gradient for carbon dioxide exchange is less than for oxygen exchange, but carbon dioxide's diffusion coefficient is 20 times greater than that of oxygen, so carbon dioxide crosses the membrane readily without a large pressure gradient.

Transport of Oxygen and Carbon Dioxide in the Blood

We have considered how air moves into and out of the lungs via pulmonary ventilation and how gas exchange occurs via pulmonary diffusion. Next we consider how gases are transported in the blood to deliver oxygen to the tissues and remove the carbon dioxide that the tissues produce.

Oxygen Transport

Oxygen is transported by the blood either (1) combined with hemoglobin in the red blood cells (greater than 98%) or (2) dissolved in the blood plasma (less than 2%). Only about 3 ml of oxygen is dissolved in each liter of plasma. Assuming a total plasma volume of 3 to 5 L, only about 9 to 15 ml of oxygen can be carried in the dissolved state. This limited amount of oxygen cannot adequately meet the needs of even resting body tissues, which generally require more than 250 ml of oxygen per minute (depending on body size). However, hemoglobin, a protein contained within each of the body's 4 to 6 billion red blood cells, allows the blood to transport nearly 70 times more oxygen than can be dissolved in plasma.

Hemoglobin Saturation

As just noted, over 98% of oxygen is transported in the blood bound to hemoglobin. Each molecule of hemoglobin can carry four molecules of oxygen. When oxygen binds to hemoglobin, it forms oxyhemoglobin; hemoglobin that is not bound to oxygen is referred to as deoxyhemoglobin. The binding of oxygen to hemoglobin depends on the PO_2 in the blood and the bonding strength, or affinity, between hemoglobin and oxygen. The curve in figure 7.10 is an oxygen–hemoglobin dissociation curve, which shows the amount of hemoglobin saturated with oxygen at different PO_2 values. The shape of the curve is extremely important for its function in the body. The relatively flat upper portion means that at high PO_2 concentrations, such as in the lungs, large drops in PO_2 result in only small changes in hemoglobin saturation. This is called the "loading" portion of the curve. A high blood PO_2 results in almost complete hemoglobin saturation, which means that the maximal amount of oxygen is bound. But as the PO_2 decreases, so does hemoglobin saturation.

The steep portion of the curve coincides with PO_2 values typically found in the tissues of the body. Here, relatively small changes in PO_2 result in large changes in saturation. This is advantageous because this is the "unloading" portion of the curve where hemoglobin loses its oxygen to the tissues.

Many factors determine the hemoglobin saturation. If, for example, the blood becomes more acidic, the dissociation curve shifts to the right. This indicates that more oxygen is being unloaded from the hemoglobin at the tissue level. This rightward shift of the curve (see figure 7.11*a*), attributable to a decline in pH, is referred to as the Bohr effect. The pH in the lungs is generally high, so hemoglobin passing through the lungs has a strong affinity for oxygen, encouraging high saturation. At the tissue level, especially during exercise, the pH is

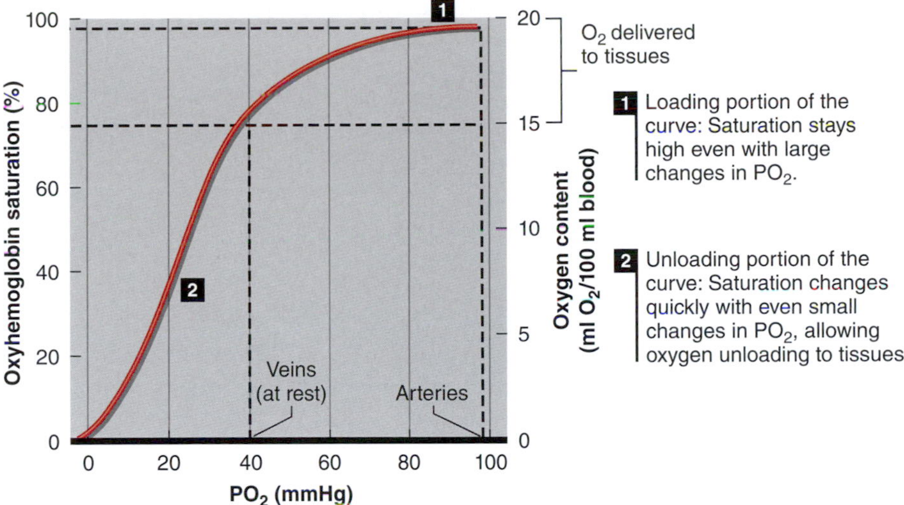

FIGURE 7.10 Oxyhemoglobin dissociation curve.

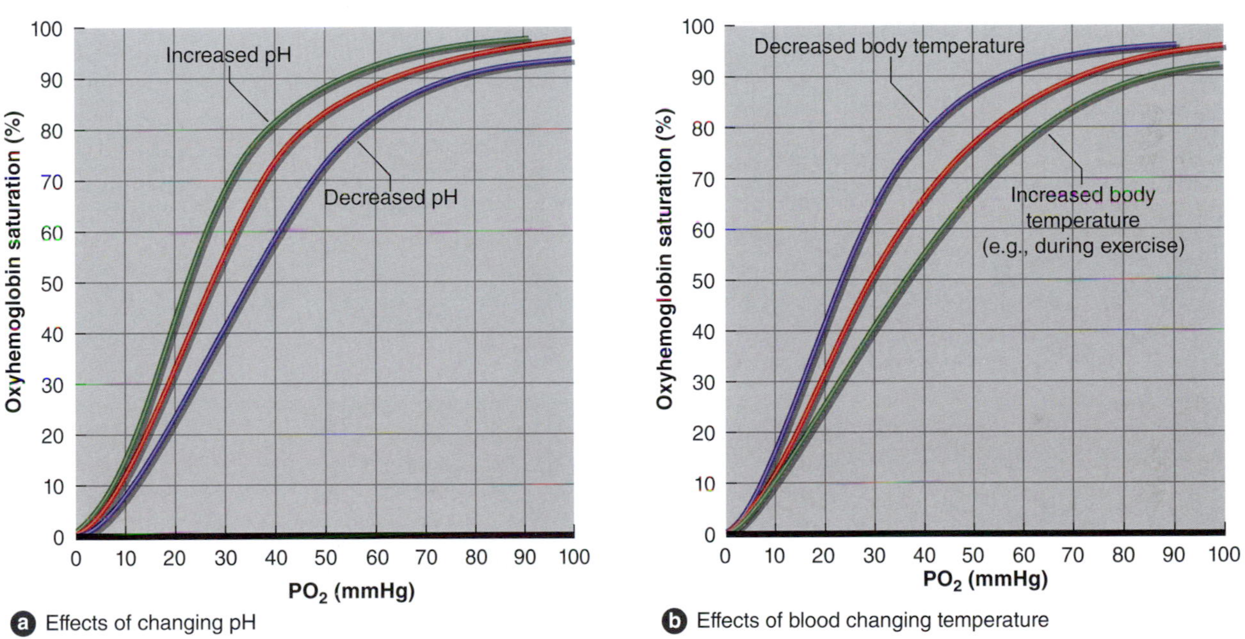

FIGURE 7.11 The effects of *(a)* changing blood pH and *(b)* blood temperature on the oxyhemoglobin dissociation curve.

lower, causing oxygen to dissociate from hemoglobin, thereby supplying oxygen to the tissues. With exercise, the ability to unload oxygen to the muscles increases as the muscle pH decreases.

Blood temperature also affects oxygen dissociation. As shown in figure 7.11*b*, increased blood temperature shifts the dissociation curve to the right, indicating that oxygen is unloaded from hemoglobin more readily at higher temperatures. Because of this, the hemoglobin unloads more oxygen when blood circulates through the metabolically heated active muscles.

Blood Oxygen-Carrying Capacity

The oxygen-carrying capacity of blood is the maximal amount of oxygen the blood can transport. It depends primarily on the blood hemoglobin content. Each 100 ml of blood contains an average of 14 to 18 g of hemoglobin in men and 12 to 16 g in women. Each gram of hemoglobin can combine with about 1.34 ml of oxygen, so the oxygen-carrying capacity of blood is approximately 16 to 24 ml per 100 ml of blood when blood is fully saturated with oxygen. At rest, as the blood passes through the lungs, it is in contact with the alveolar air for approximately 0.75 s. This is sufficient time for hemoglobin to become 98% to 99% saturated. At high intensities of exercise, the contact time is greatly reduced, which can reduce the binding of hemoglobin to oxygen and slightly decrease the saturation, although the unique "S" shape of the curve guards against large drops.

People with low hemoglobin concentrations, such as those with anemia, have reduced oxygen-carrying capacities. Depending on the severity of the condition, these people might feel few effects of anemia while they are at rest because their cardiovascular system can compensate for reduced blood oxygen content by increasing cardiac output. However, during activities in which oxygen delivery can become a limitation, such as highly intense aerobic effort, reduced blood oxygen content limits performance.

Carbon Dioxide Transport

Carbon dioxide also relies on the blood for transportation. Once carbon dioxide is released from the cells, it is carried in the blood primarily in three forms:

- As bicarbonate ions resulting from the dissociation of carbonic acid
- Dissolved in plasma
- Bound to hemoglobin (called carbaminohemoglobin)

Bicarbonate Ion

The majority of carbon dioxide is carried in the form of bicarbonate ion. Bicarbonate accounts for the transport of 60% to 70% of the carbon dioxide in the blood. Carbon dioxide and water molecules combine to form carbonic acid (H_2CO_3). This reaction is catalyzed by the enzyme carbonic anhydrase, which is found in red blood cells. Carbonic acid is unstable and quickly dissociates, freeing a hydrogen ion (H^+) and forming a bicarbonate ion (HCO_3^-):

$$CO_2 + H_2O \rightarrow H_2CO_3 \rightarrow H^+ + HCO_3^-$$

The H^+ subsequently binds to hemoglobin, and this binding triggers the Bohr effect, mentioned previously, which shifts the oxygen–hemoglobin dissociation curve to the right. The bicarbonate ion diffuses out of the red blood cell and into the plasma. In order to prevent electrical imbalance from the shift of the negatively charged bicarbonate ion into the plasma, a chloride ion diffuses from the plasma into the red blood cell. This is called the chloride shift.

Additionally, the formation of hydrogen ions through this reaction enhances oxygen unloading at the level of the tissue. Through this mechanism, hemoglobin acts as a buffer, binding and neutralizing the H^+ and thus preventing any significant acidification of the blood. Acid–base balance is discussed in more detail in chapter 8.

When the blood enters the lungs, where the PCO_2 is lower, the H^+ and bicarbonate ions rejoin to form carbonic acid, which then dissociates into carbon dioxide and water:

$$H^+ + HCO_3^- \rightarrow H_2CO_3 \rightarrow CO_2 + H_2O$$

The carbon dioxide that is thus re-formed can enter the alveoli and be exhaled.

Dissolved Carbon Dioxide

Part of the carbon dioxide released from the tissues is dissolved in plasma, but only a small amount, typically just 7% to 10%, is transported this way. This dissolved carbon dioxide comes out of solution where the PCO_2 is low, as in the lungs. There it diffuses from the pulmonary capillaries into the alveoli to be exhaled.

Carbaminohemoglobin

Carbon dioxide transport also can occur when the gas binds with hemoglobin, forming carbaminohemoglobin. The compound is so named because carbon dioxide binds with amino acids in the globin part of the hemoglobin molecule, rather than with the heme group as oxygen does. Because carbon dioxide binding occurs on a different part of the hemoglobin molecule than does oxygen binding, the two processes do not compete. However, carbon dioxide binding varies with the oxygenation of the hemoglobin (deoxyhemoglobin binds carbon dioxide more easily than oxyhemoglobin) and the partial pressure of CO_2. Carbon dioxide is released from hemoglobin when PCO_2 is low, as it is in the lungs. Thus, carbon dioxide is readily released from the hemoglobin in the lungs, allowing it to enter the alveoli to be exhaled.

In Review

- Oxygen is transported in the blood primarily bound to hemoglobin (as oxyhemoglobin), although a small part of it is dissolved in plasma.
- To better respond to increased oxygen demand, hemoglobin unloading of oxygen (desaturation) is enhanced (i.e., the curve shifts to the right) when
  - PO_2 decreases,
  - pH decreases, or
  - temperature increases.
- Because of the sigmoid shape of the curve, loading of hemoglobin with oxygen in the lungs is only minimally affected by the shift.
- In the arteries, hemoglobin is usually about 98% saturated with oxygen. This is a higher oxygen content than our bodies require, so the blood's oxygen-carrying capacity seldom limits performance in healthy individuals.
- Carbon dioxide is transported in the blood primarily as bicarbonate ion. This prevents the formation of carbonic acid, which can cause H^+ to accumulate and lower the pH. Smaller amounts of carbon dioxide are either dissolved in the plasma or bound to hemoglobin.

Gas Exchange at the Muscles

We have considered how the respiratory and cardiovascular systems bring air into our lungs, exchange oxygen and carbon dioxide in the alveoli, and transport oxygen to the muscles and carbon dioxide to the lungs. We now consider the delivery of oxygen from the capillary blood to the muscle tissue.

Arterial–Venous Oxygen Difference

At rest, the oxygen content of arterial blood is about 20 ml of oxygen per 100 ml of blood. As shown in figure 7.12*a*, this value decreases to 15 to 16 ml of oxygen per 100 ml after the blood has passed through the capillaries into the venous system. This difference in oxygen content between arterial and venous blood is referred to as the **arterial–mixed venous oxygen difference, or $(a\text{-}\bar{v})O_2$ difference**. The term *mixed venous* ($\bar{v}$) refers to the oxygen content of blood in the right atrium, which comes from all parts of the body, both active and inactive.

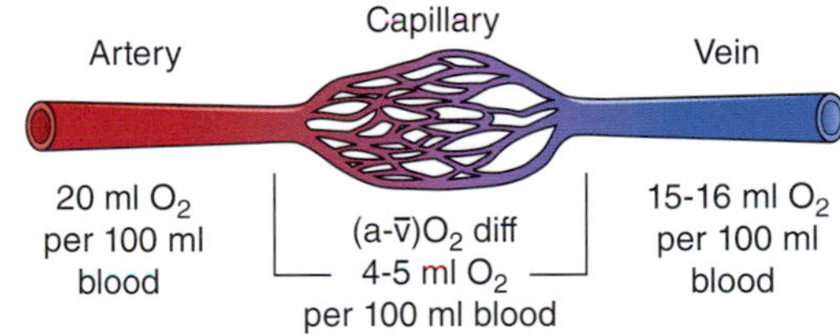

a Muscle at rest

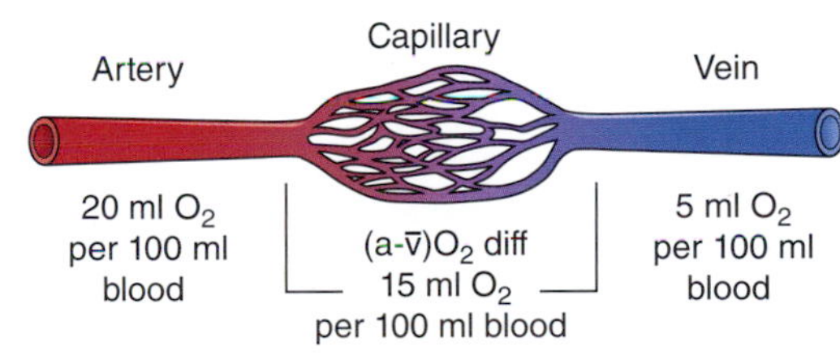

b Muscle during intense aerobic exercise

FIGURE 7.12 The arterial–mixed venous oxygen difference, or $(a\text{-}\bar{v})O_2$ difference, across the muscle *(a)* at rest and *(b)* during intense aerobic exercise.

The difference between arterial and mixed venous oxygen content reflects the 4 to 5 ml of oxygen per 100 ml of blood taken up by the tissues. The amount of oxygen taken up is proportional to its use for oxidative energy production. Thus, as the rate of oxygen use increases, the $(a\text{-}\bar{v})O_2$ difference also increases. It can increase to 15 to 16 ml per 100 ml of blood during maximal levels of endurance exercise (figure 7.12*b*). However, at the level of the contracting muscle, the **arterial–venous oxygen difference, or $(a\text{-}v)O_2$ difference**, during intense exercise can increase to 17 to 18 ml per 100 ml of blood. Note that there is not a bar over the *v* in this instance because we are now looking at local muscle venous blood, not mixed venous blood in the right atrium. During intense exercise, more oxygen is unloaded to the active muscles because the PO_2 in the muscles is substantially lower than in arterial blood.

Oxygen Transport in the Muscle

Before oxygen can be used in oxidative metabolism, it must be transported in the muscle to the mitochondria by a molecule called **myoglobin**. Myoglobin is similar in structure to hemoglobin, but myoglobin has a much greater affinity for oxygen than hemoglobin. This concept is illustrated in figure 7.13. At PO_2 values less than 20 mmHg, the myoglobin dissociation curve is much steeper than the dissociation curve for hemoglobin. Myoglobin releases its oxygen content only under conditions in which the PO_2 is very low. Note from figure 7.13 that at a PO_2 at which venous blood is unloading oxygen, myoglobin is loading oxygen. It is estimated that the PO_2 in the mitochondria of an exercising muscle may be as low as 1 mmHg; thus, myoglobin readily delivers oxygen to the mitochondria.

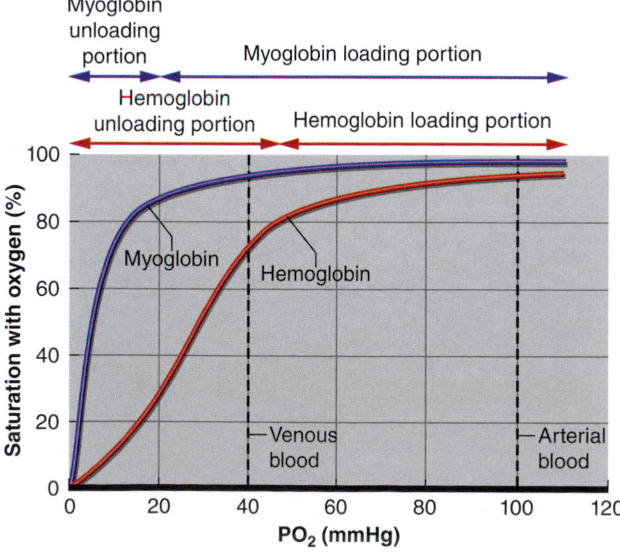

FIGURE 7.13 A comparison of the dissociation curves for myoglobin and hemoglobin.

Factors Influencing Oxygen Delivery and Uptake

The rates of oxygen delivery and uptake depend on three major variables:

- Oxygen content of blood
- Blood flow
- Local conditions (e.g., pH, temperature)

With exercise, each of these variables is adjusted to ensure increased oxygen delivery to active muscle. Under normal circumstances, hemoglobin is about 98% saturated with oxygen. Any reduction in the blood's normal oxygen-carrying capacity would hinder oxygen delivery and reduce cellular uptake of oxygen. Likewise, a reduction in the PO_2 of the arterial blood would lower the partial pressure gradient, limiting the unloading of oxygen at the tissue level. Exercise increases blood flow through the muscles. As more blood carries oxygen through the muscles, less oxygen must be removed from each 100 ml of blood (assuming the demand is unchanged). Thus, increased blood flow improves oxygen delivery.

Many local changes in the muscle during exercise affect oxygen delivery and uptake. For example, muscle activity increases muscle acidity because of lactate production. Also, muscle temperature and carbon dioxide concentration both increase because of increased metabolism. All these changes increase oxygen unloading from the hemoglobin molecule, facilitating oxygen delivery and uptake by the muscles.

Carbon Dioxide Removal

Carbon dioxide exits the cells by simple diffusion in response to the partial pressure gradient between the tissue and the capillary blood. For example, muscles generate carbon dioxide through oxidative metabolism, so the PCO_2 in muscles is relatively high compared with that in the capillary blood. Consequently, CO_2 diffuses out of the muscles and into the blood to be transported to the lungs.

In Review

- The $(a\text{-}\bar{v})O_2$ difference is the difference in the oxygen content of arterial and mixed venous blood throughout the body. This measure reflects the amount of oxygen taken up by the tissues, active and inactive.

- The $(a\text{-}\bar{v})O_2$ difference increases from a resting value of about 4 to 5 ml per 100 ml of blood up to values of 18 ml per 100 ml of blood during intense exercise. This increase reflects an increased extraction of oxygen from arterial blood by active muscle, thus decreasing the oxygen content of the venous blood.
- Oxygen delivery to the tissues depends on the oxygen content of the blood, blood flow to the tissues, and local conditions (e.g., tissue temperature and PO_2).
- Within muscle, oxygen is transported to the mitochondria by a molecule called myoglobin. Compared to the oxyhemoglobin dissociation curve, the myoglobin-O_2 dissociation curve is much steeper at low PO_2 values.
- Myoglobin releases its oxygen only at a very low PO_2. This is compatible with the PO_2 found in exercising muscle, which may be as low as 1 mmHg.
- Carbon dioxide exchange at the tissues is similar to oxygen exchange, except that carbon dioxide leaves the muscles, where it is formed, and enters the blood to be transported to the lungs for clearance.

Regulation of Pulmonary Ventilation

Maintaining homeostatic balance in blood PO_2, PCO_2, and pH requires a high degree of coordination between the respiratory, muscular, and circulatory systems. Much of this coordination is accomplished by involuntary regulation of pulmonary ventilation. This control is not yet fully understood, although many of the intricate neural controls have been identified.

The respiratory muscles are under the direct control of motor neurons, which are in turn regulated by **respiratory centers** (inspiratory and expiratory) located within the brain stem (in the medulla oblongata and pons). These centers establish the rate and depth of breathing by sending out periodic impulses to the respiratory muscles. The cortex can override these centers if voluntary control of respiration is desired. Additionally, input from other parts of the brain occurs under certain conditions.

The inspiratory area of the brain (dorsal respiratory group) contains cells that intrinsically fire and control the basic rhythm of ventilation. The expiratory area is quiet during normal breathing (recall that expiration is a passive process at rest). However, during forceful breathing such as during exercise, the expiratory area

RESEARCH PERSPECTIVE 7.3

Ventilation During Exercise in Asthma

Asthma, a condition in which the airways are inflamed and narrowed, changes airway function and makes breathing difficult. Because these changes in airway function are variable, asthmatics experience daily fluctuations in airway inflammation, airway hyper-responsiveness, pulmonary function, and clinical symptoms. Regular aerobic exercise is recommended for asthmatics, and asthmatics who are physically active show improvements in exercise capacity. However, despite a large body of literature characterizing exercise-induced bronchoconstriction in asthmatics, a significant gap in knowledge exists with regard to the influences of variable airway function at rest on the responses to aerobic exercise.

A recent study sought to determine the effects of both improved and worsened preexercise airway mechanical function on the ventilatory responses to aerobic exercise in asthmatic and nonasthmatic adults.[5] All subjects completed four separate exercise bouts of 3 min of cycling at 70% of their peak workload, followed by continuous exercise at 85% of peak workload until volitional exhaustion. Each exercise bout was preceded by one of four different interventions: (1) inhalation of a fast-acting β_2-agonist to improve airway function, (2) a eucapnic voluntary hyperpnea challenge to worsen airway function, (3) a sham version of the hyperpnea, and (4) a control trial. Pulmonary function was assessed using an automated spirometer.

Surprisingly, despite markedly different preexercise pulmonary function (experimentally manipulated by each intervention) in asthmatic adults, exercise ventilation was nearly identical among the four conditions. Moreover, there were no differences in exercise ventilation between asthmatic and nonasthmatic adults during any of the four different interventions. These data demonstrate that the pulmonary system of asthmatic adults is capable of adequately responding to the acute demand for increased airflow necessitated by high-intensity aerobic exercise. Clinically, the findings of this study support the notion that habitual aerobic exercise is beneficial for adults with asthma.

actively sends signals to the muscles of expiration. Two other brain centers aid in the control of respiration. The apneustic area has an excitatory effect on the inspiratory center, resulting in prolonged firing of the inspiratory neurons. Finally, the pneumotaxic center inhibits or switches off inspiration, helping to regulate inspiratory volume.

The respiratory centers do not act alone in controlling breathing. Breathing also is regulated and modified by the changing chemical environment in the body. For example, sensitive areas in the brain respond to changes in carbon dioxide and H^+ levels. The central chemoreceptors in the brain are stimulated by an increase in H^+ ions in the cerebrospinal fluid. The blood–brain barrier is relatively impermeable to H^+ ions or bicarbonate. However, CO_2 readily diffuses across the blood–brain barrier and then reacts to increase H^+ ions. This, in turn, stimulates the inspiratory center, which then activates the neural circuitry to increase the rate and depth of respiration. This increase in respiration, in turn, increases the removal of carbon dioxide and H^+.

Chemoreceptors in the aortic arch (the aortic bodies) and in the bifurcation of the common carotid artery (the carotid bodies) not only are sensitive primarily to blood changes in PO_2 but also respond to changes in H^+ concentration and PCO_2. The carotid chemoreceptors are more sensitive to changes in H^+ concentrations and PCO_2. Overall, PCO_2 appears to be the strongest stimulus for the regulation of breathing. When carbon dioxide levels become too high, carbonic acid forms, then quickly dissociates, giving off H^+. If H^+ accumulates, the blood becomes too acidic (pH decreases). Thus, an increased PCO_2 stimulates the inspiratory center to increase respiration—not to bring in more oxygen but to rid the body of excess carbon dioxide and limit further pH changes.

In addition to the chemoreceptors, other neural mechanisms influence breathing. The pleurae, bronchioles, and alveoli in the lungs contain stretch receptors. When these areas are excessively stretched, that information is relayed to the expiratory center. The expiratory center responds by shortening the duration of an inspiration, which decreases the risk of overinflating the respiratory structures. This response is known as the Hering-Breuer reflex.

Many control mechanisms are involved in the regulation of breathing, as shown in figure 7.14. Such simple stimuli as emotional distress or an abrupt change in the temperature of the surroundings can affect breathing. But all these control mechanisms are essential. The goal of respiration is to maintain appropriate levels of the blood and tissue gases as well as proper pH for normal cellular function. Small changes in any of these, if not carefully controlled, could impair physical activity and jeopardize health.

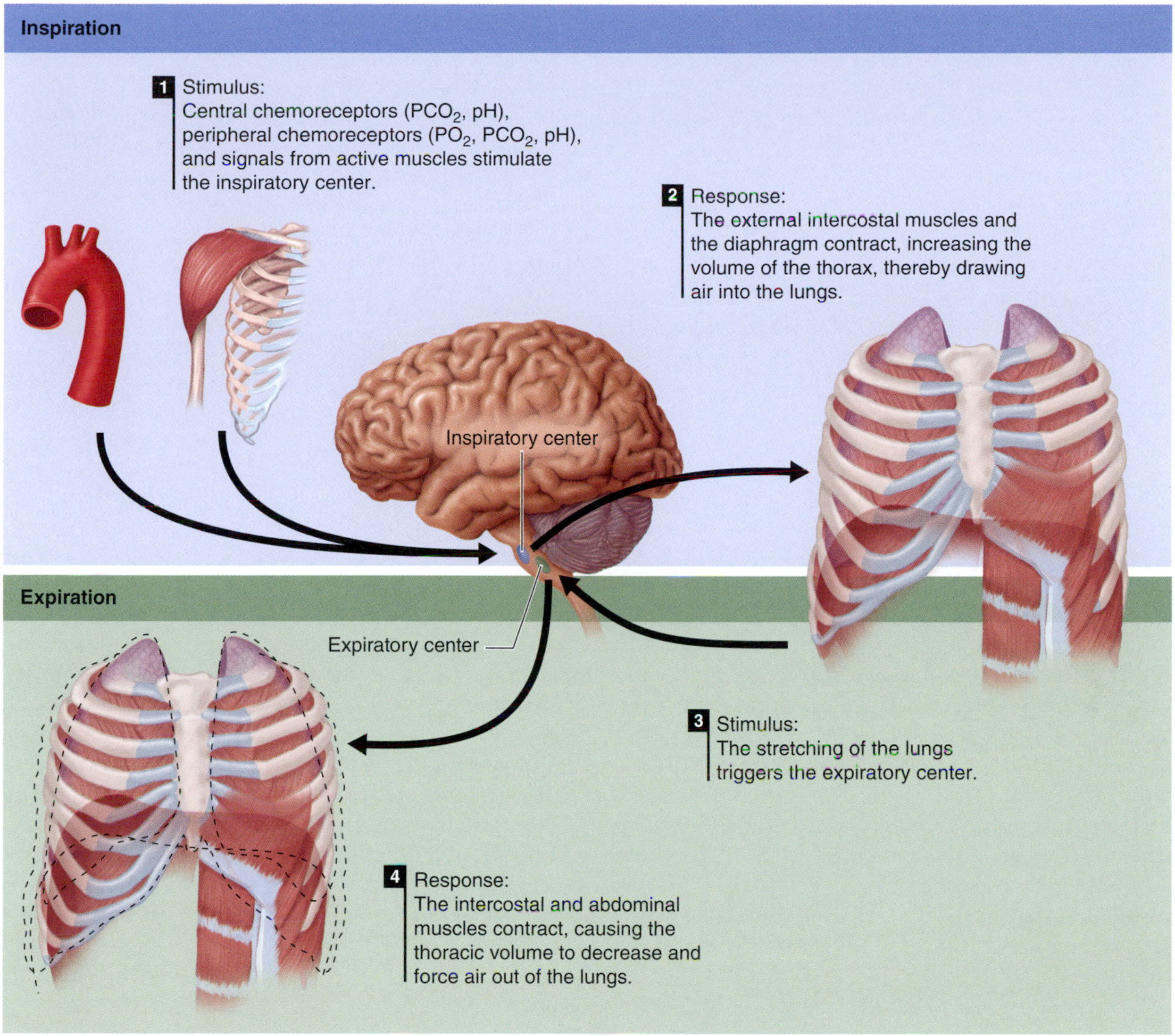

FIGURE 7.14 An overview of the processes involved in respiratory regulation.

Afferent Feedback From Exercising Limbs

The respiratory system responds almost immediately to increased ventilation at the initiation of exercise, even before there is a significant increase in the metabolic demand from exercising muscle. The fast initiation of the drive to breathe results from a combination of central command (the brain's feedforward mechanism) and afferent neural feedback from the working limbs.

In addition to those physiological mechanisms, it has been shown that the fast drive to breathe at the beginning of exercise is proportional to the frequency of limb movement. In attempting to separate the contributions to the control of ventilation from central command and afferent feedback from locomotor muscles, ventilation was measured in a group of subjects as they ran at two different speeds on a treadmill.[1] When the subjects started running at a given constant speed, their ventilation immediately increased in proportion to the treadmill speed. However, when subjects began running at a lower speed, but with the grade elevated to match the workload of the faster flat (0 grade) condition, their ventilation first increased to match the slower speed and then gradually drifted up to meet their actual oxygen demand. The immediate increase in ventilation was partially controlled by afferent feedback from the limbs, but the subsequent gradual increase in ventilation suggested that increased ventilation is a response to metabolic changes and increased metabolic demand from the exercising muscle.

RESEARCH PERSPECTIVE 7.4

Regular Exercise Reduces Respiratory Disease Mortality

Pneumonia, an infection that causes inflammation of the air sacs in the lungs, is the leading cause of infection-related death in the United States. The risk of pneumonia increases with age and comorbid conditions such as heart disease, chronic lung disease, and use of immunosuppressive drugs. Multiple health benefits have been attributed to regular physical activity; however, it remains unclear whether these benefits extend to decreased risk of respiratory disease. Certainly, reductions in the risk of pneumonia would be consistent with concept that regular exercise prevents age-related declines in function.

A 2014 report examined the association of running and walking with mortality due to respiratory disease in the National Walkers' and Runners' Health Studies, a prospective epidemiological cohort of over 150,000 adults.[6] This large cohort was used to test the hypothesis that greater exercise energy expenditure would be associated with a lower risk for respiratory diseases in general and pneumonia in particular. The results provided strong support for a reduction in the risk for respiratory diseases and pneumonia as underlying and contributing causes of mortality with greater exercise energy expenditure. Not surprisingly, this relation was dose dependent, with more substantial reductions in risk occurring in those with greater levels of habitual activity. Interestingly, this risk reduction was not different between walkers and runners. In addition, these effects appear to be independent of the effects of exercise on cardiovascular disease risk. These findings add to the compelling evidence for the health benefits of regular aerobic exercise.

More recently, scientists have been interested in whether afferent neural feedback from the limbs continues throughout exercise. Investigators at the University of Toronto had subjects independently alter either their pedal cadence or resistance while cycling during two different trials.[2] During one, they varied their pedal speed in a sinusoidal manner while keeping their total workload constant, and during the other one, they kept their speed constant while varying their pedal workload sinusoidally (see figure 7.15). During the trial in which pedal speed varied (figure 7.15*a*), there was a much faster increase in ventilation that preceded any changes in heart rate. In contrast, when subjects altered their workload (figure 7.15*b*) but kept their pedal speed constant, there was a greater lag time before the increase in ventilation, such that the metabolic changes preceded changes in ventilation. The results from these unique experiments suggest that limb movement frequency influences ventilation at the start of, and throughout, exercise. Continued afferent neural feedback from the limbs influences the drive to breathe during exercise.

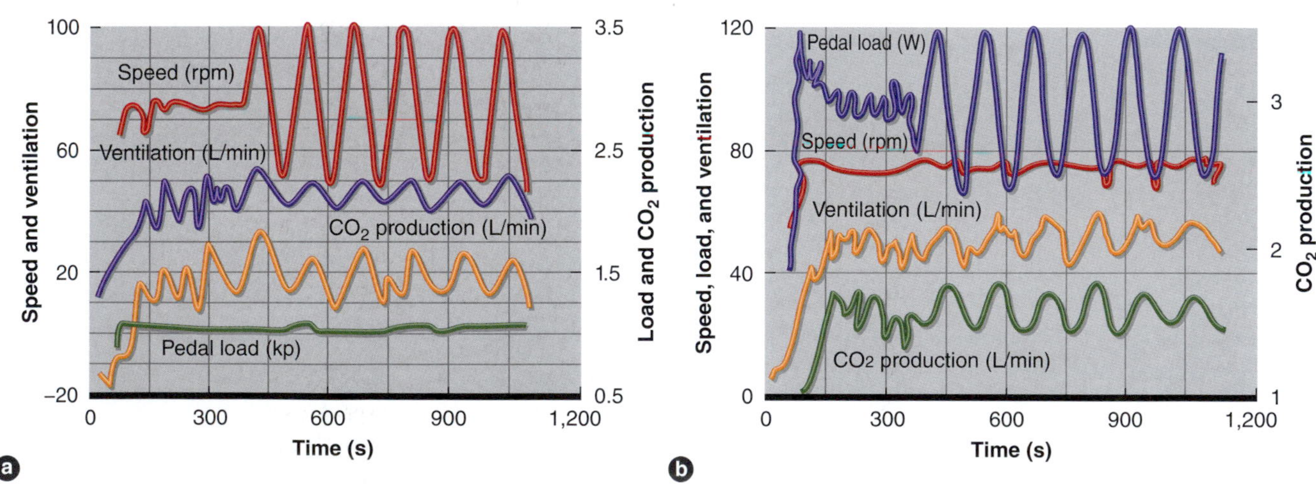

FIGURE 7.15 Sine wave exercise experiments. *(a)* Breath-by-breath variables measured during an exercise test with the subject varying pedaling speed (cadence) while pedal loading remains constant. The solid lines are fitted sine waves. *(b)* Breath-by-breath variables measured during an exercise test with varying pedal loading while pedaling speed (cadence) remains constant.

IN CLOSING

In chapter 6, we discussed the role of the cardiovascular system during exercise. In this chapter, we looked at the role played by the respiratory system. The entire process of respiration involves pulmonary ventilation (inspiration and expiration), diffusion of gases at the alveoli, transport of gases through the blood, and gas exchange at the tissues. In the next chapter, we examine how the cardiovascular and respiratory systems respond to an acute bout of exercise.

KEY TERMS

alveoli
arterial–mixed venous oxygen difference, or $(a\text{-}\bar{v})O_2$ difference
arterial–venous oxygen difference, or $(a\text{-}v)O_2$ difference
Boyle's gas law
Dalton's law
dead space
expiration
external respiration
Fick's law
Henry's law
inspiration
internal respiration
myoglobin
oxygen cascade
oxygen diffusion capacity
partial pressure
pulmonary diffusion
pulmonary ventilation
residual volume (RV)
respiratory centers
respiratory membrane
respiratory pump
spirometry
tidal volume
total lung capacity (TLC)
vital capacity (VC)

STUDY QUESTIONS

1. Describe and differentiate between external and internal respiration.
2. Describe the mechanisms involved in inspiration and expiration.
3. What is a spirometer? Describe and define the lung volumes measured using spirometry.
4. Explain the concept of partial pressures of respiratory gases—oxygen, carbon dioxide, and nitrogen. What is the role of gas partial pressures in pulmonary diffusion?
5. Where in the lung does the exchange of gases with the blood occur? Describe the role of the respiratory membrane.
6. How are oxygen and carbon dioxide transported in the blood?
7. Describe the oxygen cascade from dry ambient air to the tissues and into the venous circulation. Provide appropriate values for the various partial pressures of oxygen at each level.
8. How is oxygen unloaded from the arterial blood to the muscle and carbon dioxide removed from the muscle into the venous blood?
9. What is meant by the arterial–mixed venous oxygen difference? How and why does this change from resting conditions to exercise conditions?
10. Describe how pulmonary ventilation is regulated. What are the chemical stimuli that control the depth and rate of breathing? How do they control respiration during exercise?

STUDY GUIDE ACTIVITIES

In addition to the activities listed in the chapter opening outline, two other activities are available in the web study guide, located at

www.HumanKinetics.com/PhysiologyOfSportAndExercise

The **KEY TERMS** activity reviews important terms, and the end-of-chapter **QUIZ** tests your understanding of the material covered in the chapter.

Cardiorespiratory Responses to Acute Exercise

In this chapter and in the web study guide

Cardiovascular Responses to Acute Exercise 200

AUDIO FOR FIGURE 8.2 describes the use of a submaximal exercise test to estimate maximal exercise capacity.

VIDEO 8.1 presents Ben Levine on physiological differences in trained versus untrained people and the relationship between cardiac output and oxygen use.

AUDIO FOR FIGURE 8.9 describes an example of cardiovascular drift.

ANIMATION FOR FIGURE 8.12 details the integrated cardiovascular response to exercise.

ACTIVITY 8.1 *Cardiovascular Response to Exercise* reviews cardiovascular changes occurring during exercise.

ACTIVITY 8.2 *Cardiovascular Response Scenarios* explores how cardiovascular responses contribute to real-life situations.

Respiratory Responses to Acute Exercise 216

ACTIVITY 8.3 *Pulmonary Ventilation During Exercise* investigates the response of pulmonary ventilation to exercise and the factors that affect the phases of pulmonary ventilation.

ACTIVITY 8.4 *Pulmonary Ventilation and Energy Metabolism* reviews the key terms related to pulmonary ventilation and energy metabolism.

In Closing 224

Completing a full 26.2 mi (42 km) marathon is a major accomplishment, even for those who are young and extremely fit. On May 5, 2002, Greg Osterman completed the Cincinnati Flying Pig Marathon, his sixth full marathon, finishing in a time of 5 h and 16 min. This is certainly not a world record time, or even an exceptional time for fit runners. However, in 1990 at the age of 35, Greg had contracted a viral infection that went right to his heart and progressed to heart failure. In 1992, he received a heart transplant. In 1993, his body started rejecting his new heart and he also contracted leukemia, not an uncommon response to the antirejection drugs given to transplant patients. He miraculously recovered and started his quest to get physically fit. He ran his first race (15K) in 1994, followed by five marathons in Bermuda, San Diego, New York, and Cincinnati in 1999 and 2001. Greg is an excellent example of both human resolve and physiological adaptability.

After reviewing the basic anatomy and physiology of the cardiovascular and respiratory systems, this chapter looks specifically at how these systems respond to the increased demands placed on the body during acute exercise. With exercise, oxygen demand by the active muscles increases significantly. Metabolic processes speed up and more waste products are created. During prolonged exercise or exercise in a hot environment, body temperature increases. In intense exercise, H^+ concentration increases in the muscles and blood, lowering their pH.

Cardiovascular Responses to Acute Exercise

Numerous interrelated cardiovascular changes occur during dynamic exercise. The primary goal of these adjustments is to increase blood flow to working muscle; however, cardiovascular control of virtually every tissue and organ in the body is also altered. To better understand the changes that occur, we must examine the function of both the heart and the peripheral circulation. In this section, we examine changes in all components of the cardiovascular system from rest to acute exercise, looking specifically at the following:

- Heart rate
- Stroke volume
- Cardiac output
- Blood pressure
- Blood flow
- The blood

We then see how these changes are integrated to maintain adequate blood pressure and provide for the exercising body's needs.

Heart Rate

Heart rate (HR) is one of the simplest physiological responses to measure and yet one of the most informative in terms of cardiovascular stress and strain. Measuring HR involves simply taking the subject's pulse, usually at the radial or carotid artery. Heart rate is a good indicator of relative exercise intensity.

Resting Heart Rate

Resting heart rate (RHR) averages 60 to 80 beats/min in most individuals. In highly conditioned, endurance-trained athletes, resting rates as low as 28 beats/min have been reported. This is mainly due to an increase in parasympathetic (vagal) tone that accompanies endurance exercise training. Resting heart rate can also be affected by environmental factors; for example, it increases with extremes in temperature and altitude.

Just before the start of exercise, preexercise HR usually increases above normal resting values. This is called the anticipatory response. This response is mediated through release of the neurotransmitter norepinephrine from the sympathetic nervous system and the hormone epinephrine from the adrenal medulla. Vagal tone also decreases. Because preexercise HR is elevated, reliable estimates of the true RHR should be made only under conditions of total relaxation, such as early in the morning before the subject rises from a restful night's sleep.

Heart Rate During Exercise

When exercise begins, HR increases directly in proportion to the increase in exercise intensity (figure 8.1), until near-maximal exercise is achieved. As maximal exercise intensity is approached, HR begins to plateau even as the exercise workload continues to increase. This indicates that HR is approaching a maximal value. The **maximum heart rate (HR_{max})** is the highest HR value achieved in an all-out effort to the point of volitional fatigue. Once accurately determined, HR_{max} is a highly reliable value that remains constant from day to day. However, this value changes slightly from year to year due to a normal age-related decline.

HR_{max} is often estimated based on age because HR_{max} shows a slight but predictable decrease of about one beat per year beginning at 10 to 15 years of age. Subtracting one's age from 220 beats/min provides a reasonable approximation of one's predicted HR_{max}. However, this is only an estimate—individual values vary considerably from this average value. To illustrate, for a 40-year-old woman, HR_{max} would be estimated to be 180 beats/min (HR_{max} = 220 – 40 beats/min). However, 68% of all 40-year-olds have actual HR_{max} values between 168 and 192 beats/min (mean ± 1 standard

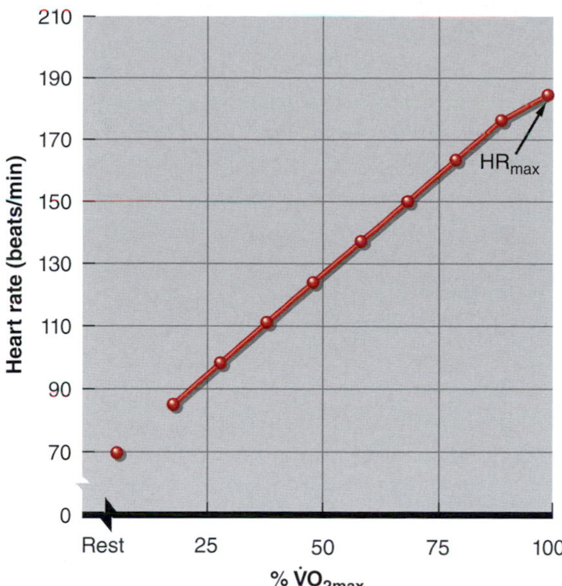

FIGURE 8.1 Changes in heart rate (HR) as a subject progressively walks, jogs, and then runs on a treadmill as intensity is increased. Heart rate is plotted against exercise intensity shown as a percentage of the subject's $\dot{V}O_{2max}$, at which point the rise in HR begins to plateau. The HR at this plateau is the subject's maximal HR or HR_{max}.

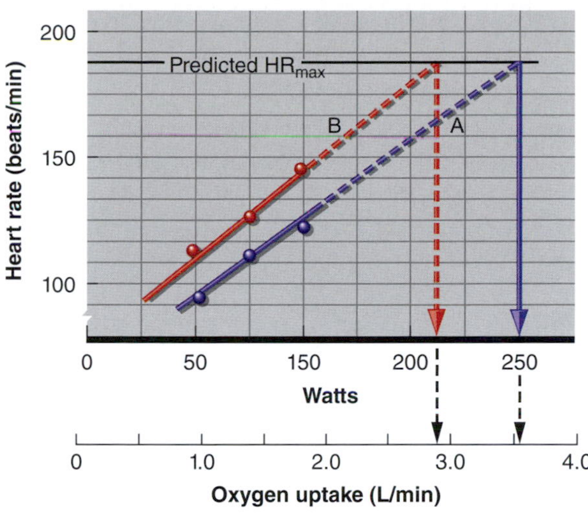

FIGURE 8.2 The increase in heart rate with increasing power output on a cycle ergometer and oxygen uptake is linear within a wide range. The predicted maximal oxygen uptake can be extrapolated using the subject's estimated maximum heart rate as demonstrated here for two subjects with similar estimated maximum heart rates but quite different maximal workloads and $\dot{V}O_{2max}$ values.

Reprinted by permission from P.O. Åstrand et al., *Textbook of Work Physiology: Physiological Bases of Exercise*, 4th ed. (Champaign, IL: Human Kinetics, 2003), 285.

deviation), and 95% fall between 156 and 204 beats/min (mean ± 2 standard deviations). This demonstrates the potential for error in estimating a person's HR_{max}. A similar but more accurate equation has been developed to estimate HR_{max} from age. In this equation, $HR_{max} = 208 - (0.7 \times \text{age})$.[16]

When the exercise intensity is held constant at any submaximal workload, HR increases fairly rapidly until it reaches a plateau. This plateau is the **steady-state heart rate**, and it is the optimal HR for meeting the circulatory demands at that specific rate of work. For each subsequent increase in intensity, HR will reach a new steady-state value within 3 min. However, the more intense the exercise, the longer it takes to achieve this steady-state value.

The concept of steady-state heart rate forms the basis for simple exercise tests that have been developed to estimate cardiorespiratory (aerobic) fitness. In one such test, individuals are placed on an exercise device, such as a cycle ergometer, and then perform exercise at two or three standardized exercise intensities. Those with better cardiorespiratory endurance capacity will have a lower steady-state HR at each exercise intensity than those who are less fit. Thus, a lower steady-state HR at a fixed exercise intensity is a valid predictor of better cardiorespiratory fitness.

Figure 8.2 illustrates results from a submaximal graded exercise test performed on a cycle ergometer by two different individuals of the same age. Steady-state HR is measured at three or four distinct workloads, and a line of best fit is drawn through the data points. Because there is a consistent relation between exercise intensity and energy demand, steady-state HR can be plotted against the corresponding energy ($\dot{V}O_2$) required to do work on the cycle ergometer. The resultant line can be extrapolated to the age-predicted HR_{max} to estimate an individual's maximal exercise capacity. In this figure, subject A has a higher fitness level than subject B because (1) at any given submaximal intensity, this subject's HR is lower and (2) extrapolation to age-predicted HR_{max} yields a higher estimated maximal exercise capacity ($\dot{V}O_{2max}$).

Heart Rate Variability

Heart rate variability is a measure of the rhythmic fluctuation in HR that occurs because of continuous changes in the sympathetic–parasympathetic balance that controls sinus rhythm. Analysis of HR variability has been used as a method of noninvasively evaluating the relative contributions of the sympathetic and parasympathetic nervous systems at rest and during exercise. During acute aerobic exercise, many different factors contribute to increasing HR variability, including increases in body core temperature, sympathetic nerve activity, and respiratory rate. After a bout of acute

RESEARCH PERSPECTIVE 8.1

HUNTing for a Better Prediction of Maximal Heart Rate

Maximal heart rate (HR_{max}) is commonly used in clinical exercise testing and to prescribe exercise intensity in physical training and rehabilitation settings. HR_{max} can be determined with an individual exercise test to exhaustion and is verified by a plateau of heart rate despite an increase in exercise intensity. However, an exercise test to maximal exertion may not always be feasible, especially in clinical settings where maximal exercise may not be safe or in field tests where advanced equipment (such as a treadmill or stationary bicycle with adjustable grade or resistance) may not be available. Because of these limitations, there is a need for accurate equations to predict HR_{max}.

HR_{max} declines linearly with age and is estimated using the common formulas in the text of this chapter. However, scientists have suggested that adding other factors, including sex, body mass index (BMI), smoking, and physical activity, to prediction equations may increase their accuracy. In 2013, a group of researchers in Norway set out to develop a new, more accurate prediction formula for HR_{max}.[8] To do this, the research team studied a subpopulation of participants who were enrolled in the HUNT Fitness Study, a large cohort designed to measure $\dot{V}O_{2max}$ in healthy Norwegian adults. To create a new formula for HR_{max}, the researchers analyzed HR_{max} measured during a peak $\dot{V}O_2$ test, then investigated the relations between HR_{max} and age, sex, physical activity status, BMI, and objectively measured aerobic fitness.

HR_{max} was linearly related to age and was best predicted by the formula

$$HR_{max} = 211 - 0.64 \times \text{age}$$

whereas the traditionally used prediction equation of

$$HR_{max} = 220 - \text{age}$$

(1) overestimated HR_{max} in young individuals, (2) best predicted actual HR_{max} around age 40, and (3) increasingly underestimated HR_{max} as people aged. Unexpectedly, the study team found that HR_{max} was adequately predicted by age alone—accounting for body mass index, sex, smoking status, physical activity, or $\dot{V}O_{2max}$ did not improve the equation's accuracy. This study concluded that the new prediction equation $HR_{max} = 211 - 0.64 \times age$ most accurately described HR_{max} as a function of age. However, like all prediction formulas, the standard error of ±11 beats/min must still be taken into consideration. Furthermore, although sex, body mass index, smoking, physical activity, and fitness did not influence the age-related decline in HR_{max} across the sample of subjects they surveyed, these factors may still influence HR_{max} on an individual basis.

This new equation may be better than the quick-and-easy standard *220 – age*, but direct measurement of HR_{max} using a maximal exercise test is always preferable when possible.

exercise, HR variability gradually increases compared to preexercising values due to greater vagal tone. Moreover, changes in HR variability can be used to assess the impact of exercise training (discussed in chapter 11), the occurrence of overtraining[15] (discussed in chapter 14), and even as a diagnostic tool in certain clinical populations[12] (discussed in chapter 20).

Heart rate, like other signals that repeat periodically over time, can be represented by a power spectrum, which describes how much of the signal occurs at each different frequency. HR signals are analyzed with respect to frequency, rather than time, using a mathematical technique called *spectral analysis.* In spectral analysis, the variability around the mean HR is separated into the contributing frequency domains. There are many physiological influences on HR variability frequency domains.[5] Mathematically separating these different elements of HR variability allows researchers to examine the impact of exercise training or disease on each one of the individual contributors. For example, with aerobic exercise training, there is an increase in the parasympathetic control of HR, characterized by greater vagal tone and reduced resting sympathetic nerve activity, that affects the high-frequency domain of HR variability.

Stroke Volume

Stroke volume (SV) also changes during acute exercise to allow the heart to meet the demands of exercise. At near-maximal and maximal exercise intensities, as HR approaches its maximum, SV is a major determinant of cardiorespiratory endurance capacity.

Stroke volume is determined by four factors:

1. The volume of venous blood returned to the heart (the heart can only pump what returns)
2. Ventricular distensibility (the capacity to enlarge the ventricle, to allow maximal filling)

3. Ventricular contractility (the inherent capacity of the ventricle to contract forcefully)
4. Aortic or pulmonary artery pressure (the pressure against which the ventricles must contract)

The first two factors influence the filling capacity of the ventricle, determining how much blood fills the ventricle and the ease with which the ventricle is filled at the available pressure. Together, these factors determine the end-diastolic volume (EDV), sometimes referred to as the **preload**. The last two characteristics influence the ventricle's ability to empty during systole, determining the force with which blood is ejected and the pressure against which it must be expelled into the arteries. The latter factor, the aortic mean pressure, which represents resistance to blood being ejected from the left ventricle (and to a less important extent, the pulmonary artery pressure resistance to flow from the right ventricle), is referred to as the **afterload**. These four factors combine to determine the SV during acute exercise.

Stroke Volume During Exercise

Stroke volume increases above resting values during exercise. Most researchers agree that SV increases with increasing exercise intensity up to intensities somewhere between 40% and 60% of $\dot{V}O_{2max}$. At that point, SV typically plateaus, remaining essentially unchanged up to and including the point of exhaustion, as shown in figure 8.3. However, other researchers have reported that SV continues to increase beyond 40% to 60% $\dot{V}O_{2max}$, even up through maximal exercise intensities, as discussed shortly.

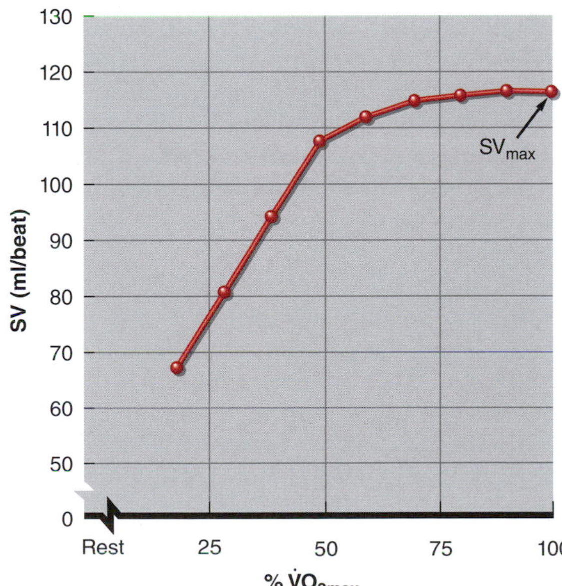

FIGURE 8.3 Changes in stroke volume (SV) as a subject exercises on a treadmill at increasing intensities. Stroke volume is plotted as a function of percent $\dot{V}O_{2max}$. The SV increases with increasing intensity up to approximately 40% to 60% of $\dot{V}O_{2max}$, before reaching a maximum (SV_{max}).

When the body is in an upright position, SV can approximately double from resting to maximal values. For example, in active but untrained individuals, SV increases from about 60 to 70 ml/beat at rest to 110 to 130 ml/beat during maximal exercise. In highly trained endurance athletes, SV can increase from 80 to 110 ml/beat at rest to 160 to 200 ml/beat during maximal exercise. During supine exercise, such as recumbent cycling, SV also increases but usually by only about 20% to 40%—not nearly as much as in an upright position. Why does body position make such a difference?

When the body is in the supine position, blood does not pool in the lower extremities. Blood returns more easily to the heart in a supine posture, which means that resting SV values are higher in the supine position than in the upright position. Thus, the increase in SV with maximal exercise is not as great in the supine position as in the upright position because SV starts out higher. Interestingly, the highest SV attainable in upright exercise is only slightly greater than the resting value in the reclining position. The majority of the SV increase during low to moderate intensities of exercise in the upright position appears to be compensating for the force of gravity that causes blood to pool in the extremities.

Although researchers agree that SV increases as exercise intensity increases up to approximately 40% to 60% $\dot{V}O_{2max}$, reports about what happens after that point differ. A few studies have shown that SV continues to increase beyond that intensity. Part of this apparent disagreement might result from differences among studies in the mode of exercise testing. Studies that show plateaus in the 40% to 60% $\dot{V}O_{2max}$ range typically have used cycle ergometers as the mode of exercise. This makes intuitive sense since blood is pooled in the legs during cycle ergometer exercise, resulting in decreased venous return of blood from the legs. Thus, the plateau in SV might be unique to cycling exercise.

Alternatively, in those studies in which SV continued to increase up to maximal exercise intensities, subjects were generally highly trained athletes. Many highly trained athletes, including highly trained cyclists tested on a cycle ergometer, can continue to increase their SV beyond 40% to 60% $\dot{V}O_{2max}$, perhaps because of adaptations caused by aerobic training. One such adaptation is an increased venous return, which leads to better ventricular filling, and an increased force of contraction (Frank-Starling mechanism). The increases in cardiac output and SV with increasing work as represented by increasing HR, in elite athletes, trained university distance runners, and untrained university students, are illustrated in figure 8.4.

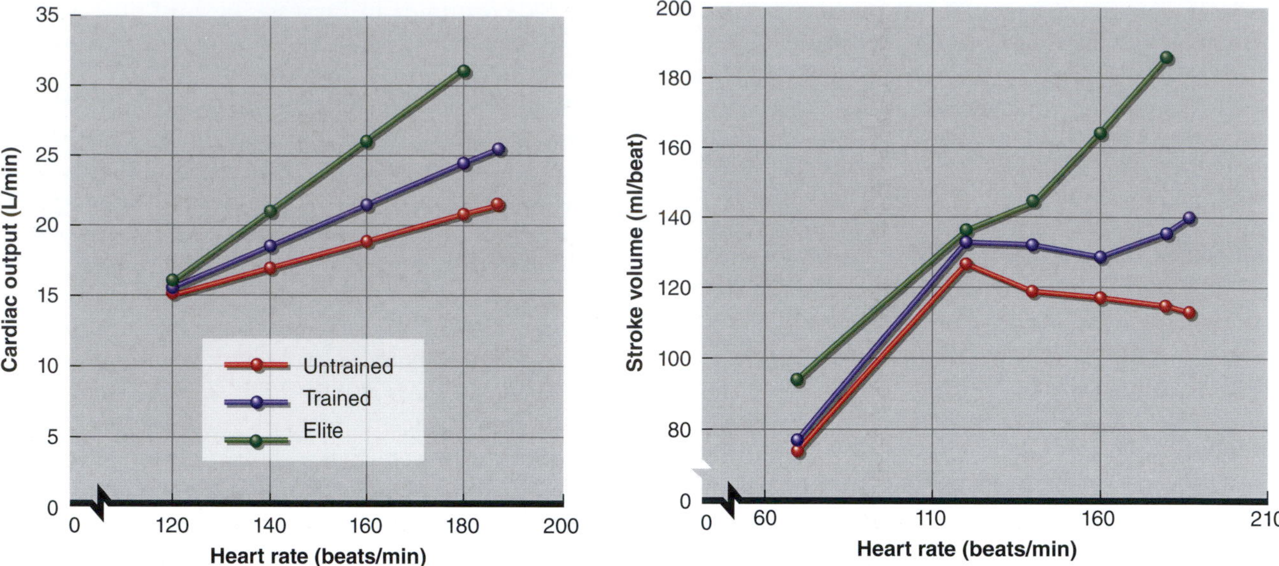

FIGURE 8.4 Cardiac output and stroke volume responses to increasing exercise intensities measured in untrained subjects, trained distance runners, and elite runners.

Adapted by permission from B. Zhou et al., "Stroke Volume Does Not Plateau During Graded Exercise in Elite Male Distance Runners," *Medicine and Science in Sports and Exercise* 33 (2001): 1849-1854.

Importance of Stroke Volume to $\dot{V}O_{2max}$

$\dot{V}O_{2max}$ is widely regarded as the single best measure of cardiorespiratory endurance, as discussed in chapter 5. At a maximal exercise intensity, $\dot{V}O_{2max}$ defines the upper limit of cardiovascular function, that is,

$$\dot{V}O_{2max} = HR_{max} \times SV_{max} \times (a\text{-}v)O_{2max}.$$

Table 8.1 shows the stark difference in $\dot{V}O_{2max}$ between an elite athlete, a normal age-matched subject, and a cardiac patient with mitral stenosis (a narrowing of the mitral valve). Because differences in HR_{max} and $(a\text{-}v)O_{2max}$ among these three groups are small, it is the ability to increase SV during maximal exercise that primarily determines $\dot{V}O_{2max}$.

How Does Stroke Volume Increase During Exercise?

Stroke volume increases during exercise despite the fact that there is less time for ventricular filling, especially at high heart rates. For example, at a resting HR of 70 beats/min, filling time between beats is 0.55 sec. At a HR of 195 beats/min, this interval decreases to 0.12 sec.[13] How does SV increase in light of less time to fill?

One explanation for the increase in SV with exercise is that the primary factor determining SV is increased preload, or the extent to which the ventricle stretches as it fills with blood, that is, the EDV. When the ventricle stretches more during filling, it subsequently contracts more forcefully. For example, when a larger volume of blood enters and fills the ventricle during diastole, the ventricular walls stretch to a greater extent. To eject that greater volume of blood, the ventricle responds by contracting more forcefully. This is referred to as the **Frank-Starling mechanism**. At the level of the muscle fiber, the greater the stretch of the myocardial cells, the more actin–myosin cross-bridges are formed, and greater force is developed.

Additionally, SV will increase during exercise if the ventricle's contractility (an inherent property of the ventricle) is enhanced. Contractility can increase by increasing sympathetic nerve stimulation or circulating catecholamines (epinephrine, norepinephrine), or both. An improved force of contraction can increase SV with or without an increased EDV by increasing the

TABLE 8.1 **The Importance of Stroke Volume in Determining $\dot{V}O_{2max}$**

Group	$\dot{V}O_{2max}$ (ml/min)	HR_{max} (beats/min)	SV_{max} (ml/beat)	$(a\text{-}v)O_{2max}$ (ml/100 ml)
Athletes	6,250	190	205	16
Normal subjects	3,500	195	112	16
Cardiac patients	1,400	190	43	17

ejection fraction. Finally, when mean arterial blood pressure is low, SV is greater since there is less resistance to outflow into the aorta. These mechanisms all combine to determine the SV at any given intensity of dynamic exercise.

Stroke volume is much more difficult to measure than HR. Some clinically used cardiovascular diagnostic techniques have made it possible to determine exactly how SV changes with exercise. Echocardiography (using sound waves) and radionuclide techniques (tagging red blood cells with radioactive tracers) have elucidated how the heart chambers respond to increasing oxygen demands during exercise. With either technique, continuous images of the heart can be taken at rest and up to near-maximal intensities of exercise.

Figure 8.5 illustrates the results of one study of normally active but untrained subjects.[9] In this study, participants were tested during both supine and upright cycle ergometry at rest and at three exercise intensities, which are depicted on the *x*-axis of figure 8.5.

When one goes from resting conditions to exercise of increasing intensity, there is an increase in left ventricular EDV (a greater filling or preload), which serves to increase SV through the Frank-Starling mechanism. There is also a decrease in the left ventricular ESV (greater emptying), indicating an increased force of contraction.

Figure 8.5 shows that both the Frank-Starling mechanism and increased contractility are important in increasing SV during exercise. The Frank-Starling mechanism appears to have its greatest influence at lower exercise intensities, and improved contractile force becomes more important at higher exercise intensities.

Recall that HR also increases with exercise intensity. The plateau or small decrease in left ventricular EDV at high exercise intensities could be caused by a reduced ventricular filling time due to the high HR. One study showed that ventricular filling time decreased from about 500 to 700 ms at rest to about 150 ms at HRs between 150 and 200 beats/min.[17] Therefore, with increasing intensities approaching $\dot{V}O_{2max}$ (and HR_{max}), the diastolic filling time could be shortened enough to limit filling. As a result, EDV might plateau or even start to decrease.

For the Frank-Starling mechanism to increase SV, left ventricular EDV must increase, necessitating an increased venous return to the heart. As discussed in chapter 6, the muscle pump and respiratory pump both aid in increasing venous return. In addition, redistribution of blood flow and volume from inactive tissues such as the splanchnic and renal circulations increases the available central blood volume.

To review, two factors that can contribute to an increase in SV with increasing intensity of exercise are increased venous return (preload) and increased ventricular contractility. The third factor that contributes to the increase in SV during exercise—a decrease in afterload—results from a decrease in total peripheral resistance. **Total peripheral resistance (TPR)** decreases because of vasodilation of the blood vessels in exercising skeletal muscle. This decrease in afterload allows

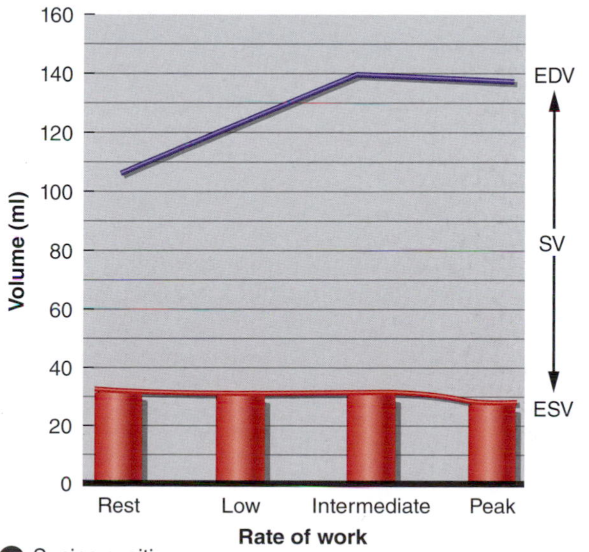

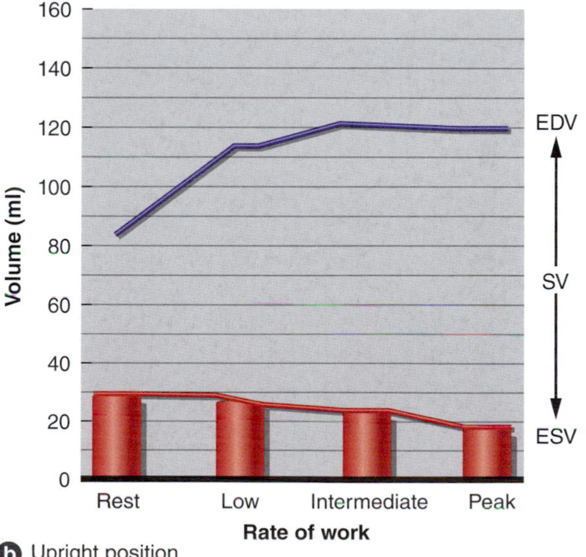

FIGURE 8.5 Changes in left ventricular end-diastolic volume (EDV), end-systolic volume (ESV), and stroke volume (SV) at rest and during low-, intermediate-, and peak-intensity exercise when the subject is in the *(a)* supine and *(b)* upright positions. Note that SV = EDV − ESV.

Adapted from Poliner et al. (1980).

the left ventricle to expel blood against less resistance, facilitating greater emptying of the ventricle.

Cardiac Output

Since cardiac output is the product of heart rate and stroke volume ($\dot{Q}$ = HR × SV), cardiac output predictably increases with increasing exercise intensity (figure 8.6). Resting cardiac output is approximately 5.0 L/min but varies in proportion to the size of the person. Maximal cardiac output varies between less than 20 L/min in sedentary individuals to 40 or more L/min in elite endurance athletes. Maximal $\dot{Q}$ is a function of both body size and endurance training. The linear relationship between cardiac output and exercise intensity is expected because the major purpose of the increase in cardiac output is to meet the muscles' increased demand for oxygen. Like $\dot{V}O_{2max}$, when cardiac output approaches maximal exercise intensity, it may reach a plateau (figure 8.6). In fact, it is likely that $\dot{V}O_{2max}$ is ultimately limited by the inability of cardiac output to increase further.

VIDEO 8.1 Presents Ben Levine on physiological differences in trained versus untrained people and the relationship between cardiac output and oxygen use.

The Fick Equation

In the 1870s, a cardiovascular physiologist by the name of Adolph Fick developed a principle critical to our understanding of the basic relationship between metabolism and cardiovascular function. In its simplest form, the Fick principle states that the oxygen consumption of a tissue is dependent on blood flow to that tissue and the amount of oxygen extracted from the blood by the tissue. This principle can be applied to the whole body or to regional circulations. Oxygen consumption is the product of blood flow and the difference in concentration of oxygen in the blood between the arterial blood supplying the tissue and the venous blood draining out of the tissue—the $(a\text{-}\bar{v})O_2$ difference. Whole-body oxygen consumption ($\dot{V}O_2$) is calculated as the product of the cardiac output ($\dot{Q}$) and $(a\text{-}\bar{v})O_2$ difference.

Fick equation:

$$\dot{V}O_2 = \dot{Q} \times (a\text{-}\bar{v})O_2 \text{ difference},$$

which can be rewritten as

$$\dot{V}O_2 = HR \times SV \times (a\text{-}\bar{v})O_2 \text{ difference}.$$

This basic relationship is an important concept in exercise physiology and comes up frequently throughout the remainder of this book.

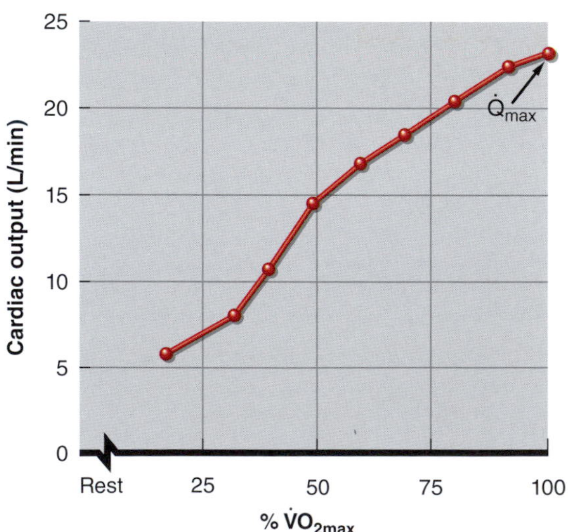

FIGURE 8.6 The cardiac output ($\dot{Q}$) response to walking-running on a treadmill at increasing intensities plotted as a function of percent $\dot{V}O_{2max}$. Cardiac output increases in direct proportion to increasing intensity, eventually reaching a maximum ($\dot{Q}_{max}$).

The Cardiac Response to Exercise

To see how HR, SV, and $\dot{Q}$ vary under various conditions of rest and exercise, consider the following example. An individual first moves from a reclining position to a seated posture and then to standing. Next the person begins walking, then jogging, and finally breaks into a fast-paced run. How does the heart respond?

In a reclining position, HR is ~50 beats/min; it increases to about 55 beats/min during sitting and to about 60 beats/min during standing. When the body shifts from a reclining to a sitting position and then to a standing position, gravity causes blood to pool in the legs, which reduces the volume of blood returning to the heart and thus decreases SV. To compensate for the reduction in SV, HR increases in order to maintain cardiac output; that is, $\dot{Q}$ = HR × SV.

During the transition from rest to walking, HR increases from about 60 to about 90 beats/min. Heart rate increases to 140 beats/min with moderate-paced jogging and can reach 180 beats/min or more with a fast-paced run. The initial increase in HR—up to about 100 beats/min—is mediated by a withdrawal of parasympathetic (vagal) tone. Further increases in HR are mediated by increased activation of the sympathetic nervous system. Stroke volume also increases with exercise, further increasing cardiac output. These relationships are illustrated in figure 8.7.

During the initial stages of exercise in untrained individuals, increased cardiac output is caused by an increase in both HR and SV. When the level of exer-

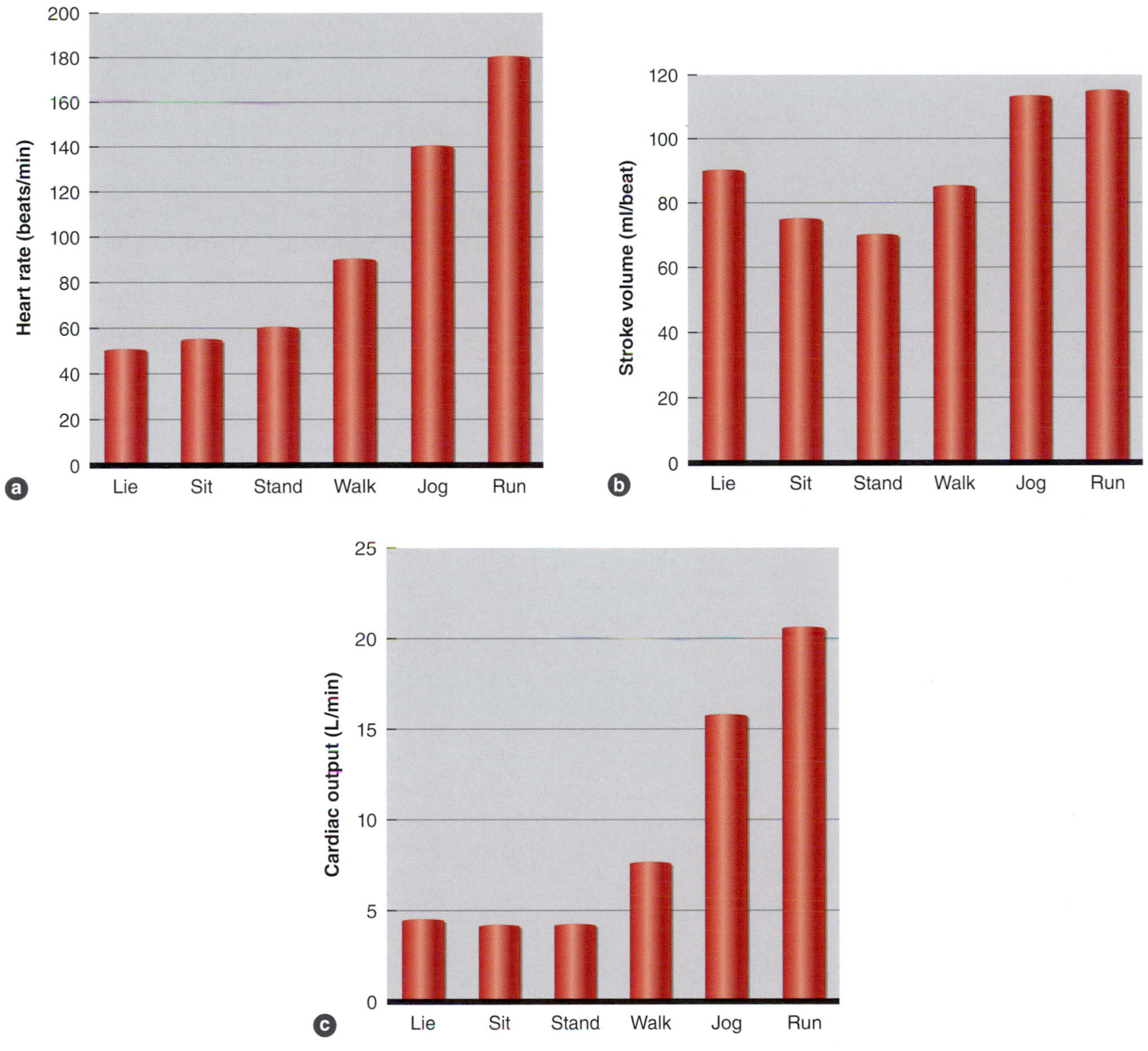

FIGURE 8.7 Changes in *(a)* heart rate, *(b)* stroke volume, and *(c)* cardiac output with changes in posture (lying supine, sitting, and standing upright) and with exercise (walking at 5 km/h [3.1 mph], jogging at 11 km/h [6.8 mph], and running at 16 km/h [9.9 mph]).

cise exceeds 40% to 60% of the individual's maximal exercise capacity, SV either plateaus or continues to increase at a much slower rate. Thus, further increases in cardiac output are largely the result of increases in HR. Further SV increases contribute more to the rise in cardiac output at high intensities of exercise in highly trained athletes.

Blood Pressure

During endurance exercise, systolic blood pressure increases in direct proportion to the increase in exercise intensity. However, diastolic pressure does not change significantly and may even decrease. As a result of the increased systolic pressure, mean arterial blood pressure increases. A systolic pressure that starts out at 120 mmHg in a normal healthy person at rest can exceed 200 mmHg at maximal exercise. Systolic pressures of 240 to 250 mmHg have been reported in normal, healthy, highly trained athletes at maximal intensities of aerobic exercise.

Increased systolic blood pressure results from the increased cardiac output that accompanies increasing rates of work. This increase in pressure helps facilitate the increase in blood flow through the vasculature. Also, blood pressure (that is, hydrostatic pressure) in large part determines how much plasma leaves the capillaries, entering the tissues and carrying needed supplies. Thus increased systolic pressure aids substrate delivery to working muscles.

After increasing initially, mean arterial pressure reaches a steady state during submaximal steady-state endurance exercise. As work intensity increases, so does systolic blood pressure. If steady-state exercise is prolonged, the systolic pressure might start to decrease gradually, but diastolic pressure remains constant. The slight decrease in systolic blood pressure, if it occurs, is a normal response and simply reflects increased vasodilation in the active muscles, which decreases the total peripheral resistance (since mean arterial pressure = cardiac output × total peripheral resistance).

Diastolic blood pressure changes little during submaximal dynamic exercise; however, at maximal exercise intensities, diastolic blood pressure may increase slightly. Remember that diastolic pressure reflects the pressure in the arteries when the heart is at rest (diastole). With dynamic exercise there is an overall increase in sympathetic tone to the vasculature, causing overall vasoconstriction. However, this vasoconstriction is blunted in the exercising muscles by the release of local vasodilators, a phenomenon called functional sympatholysis (discussed in chapter 6). Thus, because there is a balance between vasoconstriction to inactive regional circulations and vasodilation in active skeletal muscle, diastolic pressure does not change substantially. However, in some cases of cardiovascular disease, increases in diastolic pressure of 15 mmHg or more occur in response to exercise and are one of several indications for immediately stopping a diagnostic exercise test.

Upper body exercise causes a greater blood pressure response than leg exercise at the same absolute rate of energy expenditure. This is most likely attributable to the smaller exercising muscle mass of the upper body compared with the lower body, plus an increased energy demand to stabilize the upper body during arm exercise. This difference in the systolic blood pressure response to upper and lower body exercise has important implications for the heart. Myocardial oxygen uptake and myocardial blood flow are directly related to the product of HR and systolic blood pressure (SBP). This value is referred to as the **rate–pressure product (RPP)**, or double product (RPP = HR × SBP). With static or dynamic resistance exercise or upper body dynamic exercise, the RPP is elevated, indicating increased myocardial oxygen demand. The use of RPP as an indirect index of myocardial oxygen demand is important in clinical exercise testing.

Periodic blood pressure increases during resistance exercise, such as weightlifting, can be extreme. With high-intensity resistance training, blood pressure can briefly reach 480/350 mmHg. Very high pressures like these are more commonly seen when the exerciser performs a **Valsalva maneuver** to aid heavy lifts. This maneuver occurs when a person tries to exhale while the mouth, nose, and glottis are closed. This action causes an enormous increase in intrathoracic pressure. Much of the subsequent blood pressure increase results from the body's effort to overcome the high internal pressures created during the Valsalva maneuver.

In Review

- Preexercise HR is not a reliable estimate of RHR because of the anticipatory HR response.
- As exercise intensity increases, HR increases proportionately, approaching HR_{max} near the maximal exercise intensity.
- To estimate HR_{max}:

$$HR_{max} = 220 - \text{age in years, or}$$

$$HR_{max} = 208 - (0.7 \times \text{age in years})$$

- Stroke volume (the amount of blood ejected with each contraction) also increases proportionately with increasing exercise intensity but usually achieves its maximal value at about 40% to 60% of $\dot{V}O_{2max}$ in untrained individuals. Highly trained individuals can continue to increase SV, sometimes up to maximal exercise intensity.
- Increases in HR and SV combine to increase cardiac output. Thus, more blood is pumped during exercise, ensuring that an adequate supply of oxygen and metabolic substrates reaches the exercising muscles and that the waste products of muscle metabolism are cleared away.
- During exercise, cardiac output increases in proportion to exercise intensity to match the need for increased blood flow to exercising muscles.
- According to the Fick equation, whole-body oxygen consumption ($\dot{V}O_2$) is calculated as the product of the cardiac output ($\dot{Q}$) and $(a\text{-}\bar{v})O_2$ difference.
- The ability to increase cardiac output, predominantly driven by increases in stroke volume, is the primary determinant of $\dot{V}O_{2max}$.

Blood Flow

Acute increases in cardiac output and blood pressure during exercise allow for increased total blood flow to the body. These responses facilitate increased blood to areas where it is needed, primarily the exercising muscles. Additionally, sympathetic control of the cardiovascular system redistributes blood so that areas with the greatest metabolic need receive more blood than areas with low demands.

Redistribution of Blood During Exercise

Blood flow patterns change markedly in the transition from rest to exercise. Through the vasoconstrictor action of the sympathetic nervous system on local arterioles, blood flow is redirected away from areas where elevated flow is not essential to those areas that are active during exercise (see figure 6.11). Only 15% to 20% of the resting cardiac output goes to muscle, but during high-intensity exercise, the muscles may receive 80% to 85% of the cardiac output. This shift in blood flow to the muscles is accomplished primarily by reducing blood flow to the kidneys and the so-called splanchnic circulation (which includes the liver, stomach, pancreas, and intestines). Figure 8.8 illustrates a typical distribution of cardiac output throughout the body at rest and during heavy exercise. Because cardiac output increases greatly with increasing intensity of exercise, the values are shown both as the relative percentage of cardiac output and as the absolute cardiac output going to each regional circulation at rest and at three intensities of exercise.

Although several physiological mechanisms are responsible for the redistribution of blood flow during exercise, they work together in an integrated fashion. To illustrate this, consider what happens to blood flow during exercise, focusing on the primary driver of the response, namely the increased blood flow requirement of the exercising skeletal muscles.

As exercise begins, active skeletal muscles rapidly require increased oxygen delivery. This need is partially met through sympathetic stimulation of vessels in those areas to which blood flow is to be reduced (e.g., the splanchnic and renal circulations). The resulting vasoconstriction in those areas allows for more of the (increased) cardiac output to be distributed to the exercising skeletal muscles. In the skeletal muscles, sympathetic stimulation to the constrictor fibers in the arteriolar walls also increases; however, local dilator substances are released from the exercising muscle and overcome sympathetic vasoconstriction, producing an overall vasodilation in the muscle (functional sympatholysis).

Many local dilator substances are released in exercising skeletal muscle. As the metabolic rate of the muscle tissue increases during exercise, metabolic waste products begin to accumulate. Increased metabolism causes an increase in acidity (increased hydrogen ions and lower pH), carbon dioxide, and temperature in the muscle tissue. These are some of the local changes that trigger vasodilation of, and increasing blood flow through, the arterioles feeding local capillaries. Local vasodilation is also triggered by the low partial pressure of oxygen in the tissue or a reduction in oxygen bound to hemoglobin (increased oxygen demand), the act of muscle contraction, and possibly other vasoactive

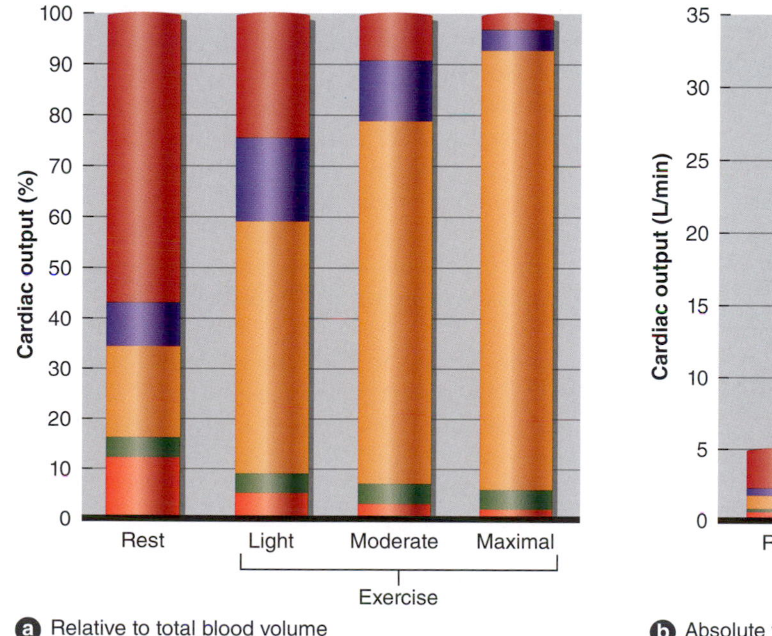

ⓐ Relative to total blood volume

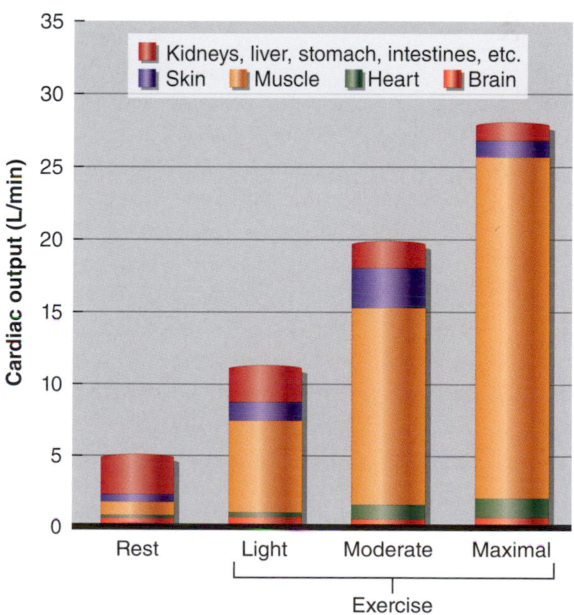

ⓑ Absolute values

FIGURE 8.8 The distribution of cardiac output at rest and during exercise *(a)* as a percentage of the total cardiac output and *(b)* as absolute volumes.

Data from Vander, Sherman, and Luciano (1985).

substances (including adenosine) released as a result of skeletal muscle contraction.

When exercise is performed in a hot environment, there is also an increase in blood flow to the skin to help dissipate the body heat. The sympathetic control of skin blood flow is unique in that there are sympathetic vasoconstrictor fibers (similar to skeletal muscle) and sympathetic active vasodilator fibers interacting over most of the skin surface area. During dynamic exercise, as body core temperature rises, there is initially a reduction in sympathetic vasoconstriction, causing a passive vasodilation. Once a specific body core temperature threshold is reached, skin blood flow begins to dramatically increase by activation of the sympathetic active vasodilator system. The increase in skin blood flow during exercise promotes heat loss, because metabolic heat from deep in the body can be released only when blood moves close to the skin. This limits the rate of rise in body temperature, as discussed in more detail in chapter 12.

Cardiovascular Drift

With prolonged aerobic exercise or aerobic exercise in a hot environment at a steady-state intensity, SV gradually decreases and HR increases. Cardiac output is well maintained, but arterial blood pressure also declines. These alterations, illustrated in figure 8.9, have been referred to collectively as **cardiovascular drift**. Cardiovascular drift has traditionally been associated with a

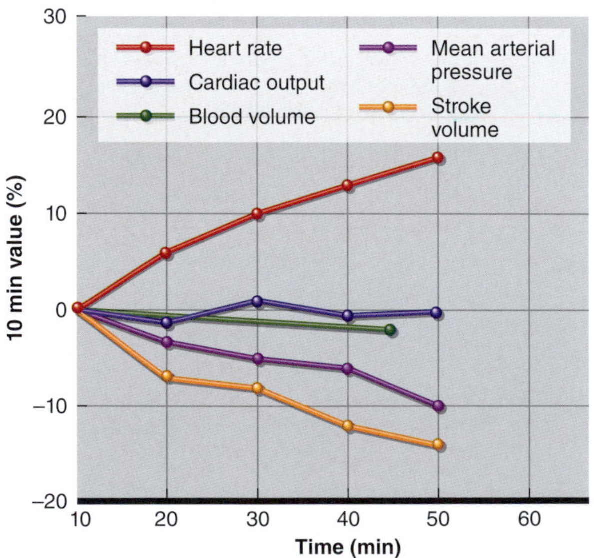

FIGURE 8.9 Circulatory responses to prolonged, moderately intense exercise in the upright posture in a thermoneutral 20 °C environment, illustrating cardiovascular drift. Values are expressed as the percentage of change from the values measured at the 10 min point of the exercise.

Adapted by permission from L.B. Rowell, *Human Circulation: Regulation During Physical Stress* (New York: Oxford University Press, 1986), 230.

progressive increase in the fraction of cardiac output directed to the vasodilated skin to facilitate heat loss and attenuate the increase in body core temperature. With more blood in the skin for the purpose of cooling the body, less blood is available to return to the heart, thus decreasing preload. There is also a small decrease in blood volume resulting from sweating and from a generalized shift of plasma across the capillary membrane into the surrounding tissues. These factors combine to decrease ventricular filling pressure, which decreases venous return to the heart and reduces the EDV. With the reduction in EDV, SV is reduced (SV = EDV – ESV). In order to maintain cardiac output ($\dot{Q}$ = HR × SV), HR increases to compensate for the decrease in SV.

A more recent hypothesis has been put forth to explain cardiovascular drift. As HR increases, there is less filling time for the ventricles. This exercise tachycardia may lower SV under the conditions of prolonged exercise even without peripheral displacement of blood volume. From the available research, it is not possible to pinpoint a single hypothesis that fully explains cardiovascular drift, and it is likely that the two mechanisms may interact.

Competition for Blood Supply

When the demands of exercise are added to blood flow demands for all other systems of the body, competition for a limited available cardiac output can occur. This competition for available blood flow can develop among several vascular beds, depending on the specific conditions. For example, there may be competition for available blood flow between active skeletal muscle and the gastrointestinal system following a meal. McKirnan and coworkers[7] studied the effects of feeding versus fasting on the distribution of blood flow during exercise in miniature pigs. The pigs were divided into two groups. One group fasted for 14 to 17 h before exercise. The other group ate their morning ration in two feedings: Half the ration was fed 90 to 120 min before exercise and the other half 30 to 45 min before exercise. Both groups of pigs then ran at approximately 65% of their $\dot{V}O_{2max}$.

Blood flow to the hindlimb muscles during exercise was 18% lower and gastrointestinal blood flow was 23% higher in the fed group than in the fasted group. Similar results in humans suggest that the redistribution of gastrointestinal blood flow to the working muscles is attenuated after a meal. As a practical application, these findings suggest that athletes should be cautious in timing their meals before competition to maximize blood flow to the active muscles during exercise.

Another example of the competition for blood flow is seen in exercise in a hot environment. In this scenario, competition for available cardiac output can occur between the skin circulation for thermoregulation and the exercising muscles. This is discussed in more detail in chapter 12.

Blood

We have now examined how the heart and blood vessels respond to exercise. The remaining component of the cardiovascular system is the blood: the fluid that carries oxygen and nutrients to the tissues and clears away waste products of metabolism. As metabolism increases during exercise, several aspects of the blood itself become increasingly critical for optimal performance.

Oxygen Content

At rest, the blood's oxygen content varies from 20 ml of oxygen per 100 ml of arterial blood to 14 ml of oxygen per 100 ml of venous blood returning to the right atrium. The difference between these two values (20 ml − 14 ml = 6 ml) is referred to as the arterial–mixed venous oxygen difference, or $(a\text{-}\bar{v})O_2$ difference. This value represents the extent to which oxygen is extracted, or removed, from the blood as it passes through the body.

With increasing exercise intensity, the $(a\text{-}\bar{v})O_2$ difference increases progressively and can almost triple from rest to maximal exercise intensities (see figure 8.10). This increased difference really reflects a decreasing venous oxygen content, because arterial oxygen content changes little from rest up to maximal exertion. With exercise, more oxygen is required by the active muscles; therefore, more oxygen is extracted from the blood. The venous oxygen content decreases, approaching zero in the active muscles. However, mixed venous blood in the right atrium of the heart rarely decreases below 4 ml of oxygen per 100 ml of blood because the blood returning from the active tissues is mixed with blood from inactive tissues as it returns to the heart. Oxygen extraction by the inactive tissues is far lower than in the active muscles.

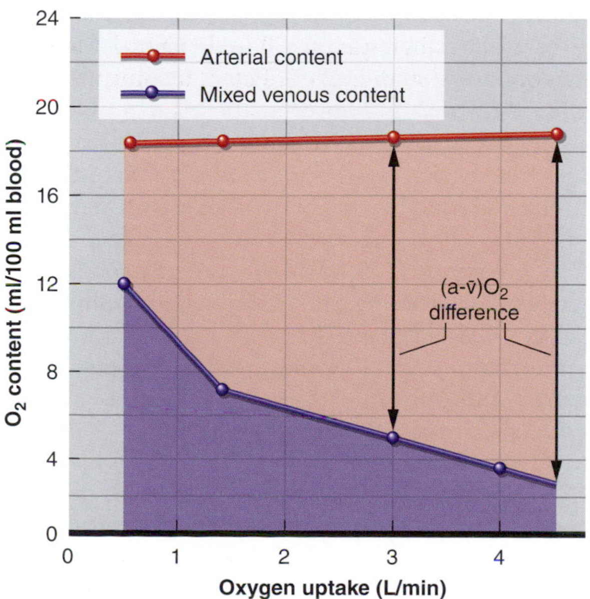

FIGURE 8.10 Changes in the oxygen content of arterial and mixed venous blood and the $(a\text{-}\bar{v})O_2$ difference (arterial–mixed venous oxygen difference) as a function of exercise intensity.

Plasma Volume

Upon standing, or with the onset of exercise, there is an almost immediate loss of plasma from the blood to the interstitial fluid space. The movement of fluid out of the capillaries is dictated by the pressures inside the capillaries, which include the **hydrostatic pressure** exerted by increased blood pressure and the **oncotic pressure**, the pressure exerted by the proteins in the blood, mostly albumin. The pressures that influence fluid movement outside the capillaries are the pressure provided by the surrounding tissue as well as the oncotic pressures from proteins in the interstitial fluid (figure 8.11). Osmotic pressures, those exerted by electrolytes in solution on both sides of the capillary wall, also play a role. As blood pressure increases with exercise, the hydrostatic pressure within the capillaries increases. This increase in blood pressure forces water from the intravascular

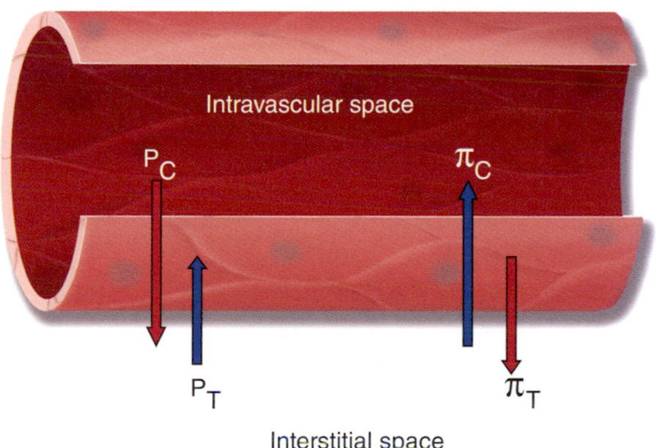

FIGURE 8.11 Filtration of plasma from the microvasculature. Both the blood pressure (P_C) inside the blood vessel and the oncotic pressure (π_T) in the tissue cause plasma to flow from the intravascular space to the interstitial space. The pressure that the tissue (P_T) exerts on the blood vessel and the oncotic pressure of the blood (π_C) inside the blood vessel cause plasma to be reabsorbed. Net filtration of plasma can be determined by summing the outward forces ($P_C + \pi_T$) and subtracting the inward forces ($P_T - \pi_C$); net capillary filtration = $(P_C + \pi_T) - (P_T - \pi_C)$.

compartment to the interstitial compartment. Also, as metabolic waste products build up in the active muscle, intramuscular osmotic pressure increases, which draws fluid out of the capillaries to the muscle.

Approximately a 10% to 15% reduction in plasma volume can occur with prolonged exercise, with the largest falls occurring during the first few minutes. During resistance training, the plasma volume loss is proportional to the intensity of the effort, with similar transient losses of fluid from the vascular space of 10% to 15%.

If exercise intensity or environmental conditions cause sweating, additional plasma volume losses may occur. Although the major source of fluid for sweat formation is the interstitial fluid, this fluid space will be diminished as sweating continues. This increases the oncotic (since proteins do not move with the fluid) and osmotic (since sweat has fewer electrolytes than interstitial fluid) pressures in the interstitial space, causing even more plasma to move out of the vascular compartment into the interstitial space. Intracellular fluid volume is impossible to measure directly and accurately, but research suggests that fluid is also lost from the intracellular compartment during prolonged exercise and even from the red blood cells, which may shrink in size.

A reduction in plasma volume can impair performance. For long-duration activities in which dehydration occurs and heat loss is a problem, blood flow to active tissues may be reduced to allow increasingly more blood to be diverted to the skin in an attempt to lose body heat. Note that a decrease in muscle blood flow occurs only in conditions of dehydration and only at high intensities. Severely reduced plasma volume also increases blood viscosity, which can impede blood flow and thus limit oxygen transport, especially if the hematocrit exceeds 60%.

In activities that last a few minutes or less, body fluid shifts are of little practical importance. As exercise duration increases, however, body fluid changes and temperature regulation become important for performance. For the football player, the Tour de France cyclist, or the marathon runner, these processes are crucial, not only for competition but also for survival. Deaths have occurred from dehydration and hyperthermia during, or as a result of, various sport activities. These issues are discussed in detail in chapter 12.

Hemoconcentration

When plasma volume is reduced, hemoconcentration occurs. When the fluid portion of the blood is reduced, the cellular and protein portions represent a larger fraction of the total blood volume; that is, they become more concentrated in the blood. This hemoconcentration increases red blood cell concentration substantially—by up to 25%. Hematocrit can increase from 40% to 50%. However, the total number and volume of red blood cells do not change substantially.

The net effect, even without an increase in the total number of red blood cells, is to increase the number of red blood cells per unit of blood; that is, the cells are more concentrated. As the red blood cell concentration increases, so does the blood's per-unit hemoglobin content. This substantially increases the blood's oxygen-carrying capacity, which is advantageous during exercise and provides a distinct advantage at altitude, as discussed in chapter 13.

The Integrated Cardiovascular Response to Exercise

As is evident from all of the changes in cardiovascular function that take place during exercise, the cardiovascular system is extremely complex but responds exquisitely to deliver oxygen to meet the demands of exercising muscle. Figure 8.12 is a simplified flow

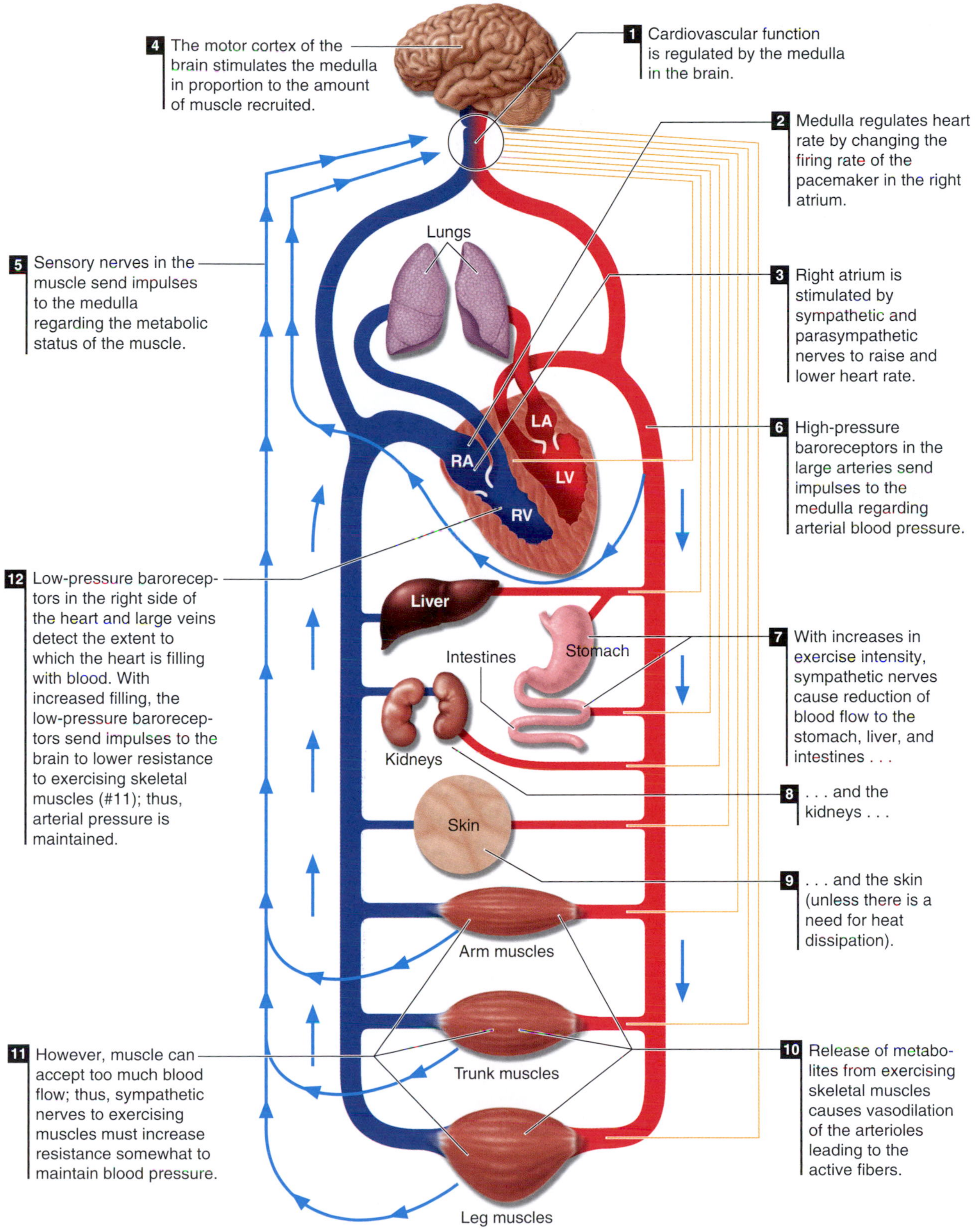

FIGURE 8.12 The integrated cardiovascular response to exercise.

Adapted by permission from E.F. Coyle, "Cardiovascular Function During Exercise: Neural Control Factors," *Sports Science Exchange* 4, no. 34 (1991): 1-6. Adapted with permission of Stokely-Van Camp, Inc.

RESEARCH PERSPECTIVE 8.2

Is Recovery a Distinct Cardiovascular State?

Exercise recovery refers to the time period immediately following a bout of exercise. This period continues until the system has completely recovered, or returned to a resting state, and can last anywhere from seconds to hours depending on the mode and intensity of the exercise. Exercise recovery can also refer to the specific physiological state that exists after exercise, which is distinctly different from the physiology of exercise or the physiology at rest. Some of these physiological changes during recovery may be necessary for long-term adaptation to exercise training, yet some can lead to cardiovascular instability during recovery. Over the last 20 years, the scientific understanding of exercise recovery as a distinct physiological state has grown immensely, mainly through human studies of cardiovascular variables such as blood pressure, heart rate, and cardiac output immediately following aerobic exercise or resistance exercise.[11]

In general, there is a dose-dependent effect of exercise intensity and duration on the cardiovascular changes that follow aerobic exercise. In general, the increase in vascular conductance (or decrease in resistance due to vasodilation of the blood vessels in the muscle) is greater than the increase in cardiac output following a bout of aerobic exercise. This means that peripheral vasodilation is the driving force that lowers blood pressure after exercise. This reduced blood pressure after exercise is called *postexercise hypotension*, and can last for several hours following a bout of aerobic exercise. The sustained postexercise vasodilation occurs largely within the previously active skeletal muscle, with a smaller but still relevant vasodilation in the nonactive skeletal muscle beds. Blood flow to the other tissues (e.g., brain, gut) reverts more quickly to resting values. Vasodilation of the nonactive skeletal muscle probably occurs due to a resetting of the blood pressure set point at the brain, while vasodilation of the previously active skeletal muscle is due to the release of local vasodilatory molecules. Recently, it has been demonstrated that the one important molecule released by the previously active muscle is histamine. Histamine is elevated in the muscle following exercise, and postexercise vasodilation is reduced by 80% when the actions of histamine are inhibited. While the lasting effects of histamine improve our understanding of what causes postexercise hypotension during recovery, the exercise-related trigger for histamine release from the muscle remains unknown.

The cardiovascular changes that occur during recovery following a bout of resistance exercise are distinctly different from those following acute aerobic exercise. Like aerobic exercise, blood pressure is reduced following resistance exercise. However, in contrast to aerobic exercise, postexercise hypotension following resistance exercise is due to decreases in cardiac output, not to vasodilation in the vascular beds of the previously active muscle. It is unclear why the mechanisms controlling blood pressure during recovery are different between aerobic and resistance exercise, but these differences are likely due to both central regulation (how the sympathetic nervous system controls blood pressure) and local cellular changes in the muscle. Interestingly, knee-extension exercise that replicates resistance training does not generate a local increase in histamines, while knee-extension exercise that replicates aerobic exercise does. Because combined aerobic and resistance exercise programs do not further reduce postexercise blood pressure compared to aerobic exercise alone, there is probably some overlap in the central mechanisms. Overall, there are fewer studies of the control of postexercise hypotension following resistance exercise. Although the recent research points to a larger role for changes in the central control of blood pressure during recovery, this is an area that requires further study.

Exercise recovery can be viewed as both a window of opportunity for the positive adaptations to training to be manipulated and a vulnerable period in which individuals are at heightened risk for adverse events such as fainting. Fully understanding this period may provide insight into when the cardiovascular system has recovered from prior training and is physiologically ready for additional training stress. The future may include training methods that take advantage of the exercise recovery state to avoid negative consequences of overtraining and to optimize training and health outcomes.

diagram that illustrates how the body integrates all these cardiovascular responses to provide for its needs during exercise. Key areas and responses are labeled and summarized to help illustrate how these complex control mechanisms are coordinated. It is important to note that although the body attempts to meet the blood flow needs of the muscle, it can do so only if blood pressure is not compromised. Maintenance of arterial blood pressure appears to be the highest priority of the cardiovascular system, regardless of exercise, the environment, or other competing needs.

The cardiovascular and respiratory adjustments to dynamic exercise are profound and rapid. Within 1 s of the initiation of muscle contraction, HR dramatically increases by vagal withdrawal and respiration increases. Increases in cardiac output and blood pressure increase blood flow to the active skeletal muscle to meet its metabolic demands. What causes these extremely rapid early changes in the cardiovascular system, since they take place well before metabolic needs of working muscle occur?

Over the years there has been considerable debate over what causes the cardiovascular system to be turned on at the onset of exercise. One explanation is the theory of **central command**, which involves parallel coactivation of both the motor and the cardiovascular control centers of the brain. Activation of central command rapidly increases HR and blood pressure. In addition to central command, the cardiovascular responses to exercise are modified by mechanoreceptors, chemoreceptors, and baroreceptors. As discussed in chapter 6, baroreceptors are sensitive to stretch and send information back to the cardiovascular control centers about blood pressure. Signals from the periphery are sent back to the cardiovascular control centers through the stimulation of mechanoreceptors that are sensitive to the stretch of the skeletal muscle and through the chemoreceptors that are sensitive to an increase in metabolites in the muscle. Feedback about blood pressure and the local muscle environment helps to fine-tune and adjust the cardiovascular response. These relationships are illustrated in figure 8.13.

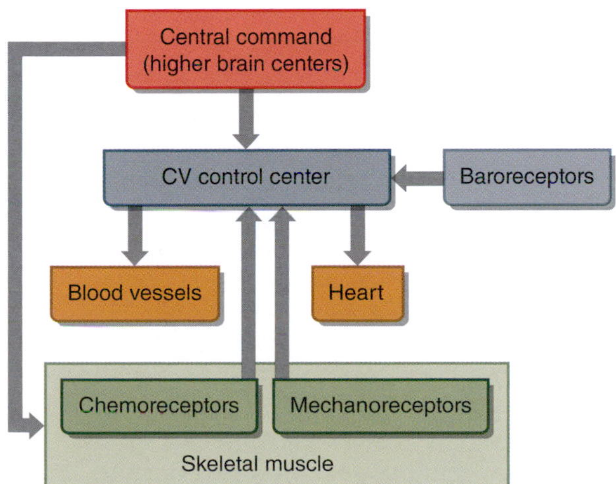

FIGURE 8.13 A summary of cardiovascular (CV) control during exercise.

Adapted by permission from S.K. Powers & E.T. Howley, *Exercise Physiology: Theory and Application to Fitness and Performance*, 5th ed. (New York, McGraw-Hill, 2004), 188. © The McGraw-Hill Education.

In Review

- Mean arterial blood pressure increases immediately in response to exercise, and the magnitude of the increase is proportional to the intensity of exercise. During whole-body endurance exercise, this is accomplished primarily by an increase in systolic blood pressure, with minimal changes in diastolic pressure.
- Systolic blood pressure can exceed 200 to 250 mmHg at maximal exercise intensity, the result of increases in cardiac output. Upper body exercise causes a greater blood pressure response than leg exercise at the same absolute rate of energy expenditure, likely due to the smaller muscle mass involved and the need to stabilize the trunk during dynamic arm exercise.
- Blood flow is redistributed during exercise from inactive or low-activity tissues of the body like the liver and kidneys to meet the increased metabolic needs of exercising muscles.
- With prolonged aerobic exercise, or aerobic exercise in the heat, SV gradually decreases and HR increases proportionately to maintain cardiac output. This is referred to as cardiovascular drift and is associated with a progressive increase in blood flow to the vasodilated skin and losses of fluid from the vascular space.
- The changes that occur in the blood during exercise include the following:
  1. The $(a\text{-}\bar{v})O_2$ difference increases as venous oxygen concentration decreases, reflecting increased extraction of oxygen from the blood for use by the active tissues.
  2. Plasma volume decreases. Plasma is pushed out of the capillaries by increased hydrostatic pressure as blood pressure increases, and fluid is drawn into the muscles by the increased oncotic and osmotic pressures in the muscle tissues, a by-product of metabolism. With prolonged exercise or exercise in hot environments, increasingly more plasma volume is lost through sweating.
  3. Hemoconcentration occurs as plasma volume (water) decreases. Although the actual number of red blood cells stays relatively constant, the relative number of red blood cells per unit of blood increases, which increases oxygen-carrying capacity.

Respiratory Responses to Acute Exercise

Now that we have discussed the role of the cardiovascular system in delivering oxygen to the exercising muscle, we examine how the respiratory system responds to acute dynamic exercise.

Pulmonary Ventilation During Dynamic Exercise

The onset of exercise is accompanied by an immediate increase in ventilation. In fact, like the HR response, the marked increase in breathing may occur even before the onset of muscular contractions—that is, it may be an anticipatory response. This is shown in figure 8.14 for light, moderate, and heavy exercise. Because of its rapid onset, this initial respiratory adjustment to the demands of exercise is undoubtedly neural in nature, mediated by respiratory control centers in the brain (central command), although neural signals also come from receptors in the exercising muscle.

The more gradual second phase of the respiratory increase shown during heavy exercise in figure 8.14 is controlled primarily by changes in the chemical status of the arterial blood. As exercise progresses, increased metabolism in the muscles generates more CO_2 and H^+. Recall that these changes shift the oxyhemoglobin saturation curve rightward, enhancing oxygen unloading in the muscles, which increases the $(a\text{-}\bar{v})O_2$ difference. Increased CO_2 and H^+ are sensed by chemoreceptors primarily located in the brain, carotid bodies, and lungs, which in turn stimulate the inspiratory center, increasing rate and depth of respiration. Chemoreceptors in the muscles themselves might also be involved. In addition, receptors in the right ventricle of the heart send information to the

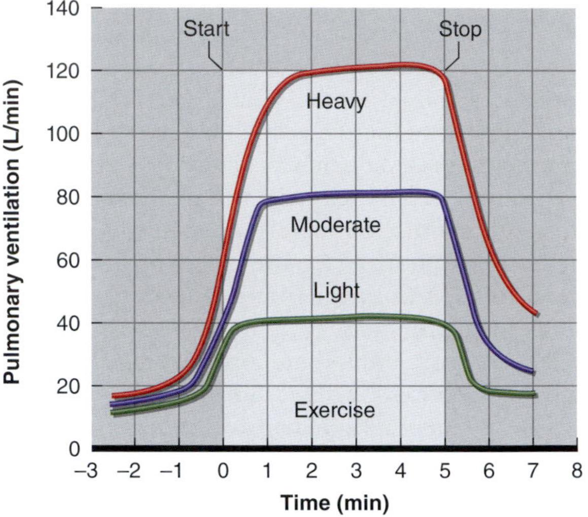

FIGURE 8.14 The ventilatory response to light, moderate, and heavy exercise. The subject exercised at each of the three intensities for 5 min. After an initial steep increase, the ventilation rate tended to plateau at a steady-state value at the light and moderate intensities but continued to increase somewhat at the heavy intensity.

RESEARCH PERSPECTIVE 8.3

Posture Affects Ventilation During Recovery After Exercise

Body posture affects cardiopulmonary function due to the effects of gravity. For example, in the upright posture, the mechanical actions of the inspiratory muscles expand the chest wall and elevate the rib cage against gravity, while changing to the supine posture increases abdominal pressure on the pleural cavity and inspiration is achieved predominantly through abdominal expansion. Despite reports that cardiopulmonary function is affected by body position, few exercise physiologists have investigated the effect of posture during recovery from aerobic exercise.

A 2017 study conducted in Korea examined cardiopulmonary function in relation to body position during recovery from a maximal exercise test.[4] Subjects were randomly assigned to one of three recovery postures: supine, sitting, or sitting with the trunk leaning forward. Each subject performed a maximal exercise test to exhaustion, then immediately assumed their assigned recovery position. Oxygen uptake, minute ventilatory volume, respiration rate, and heart rate were measured during the assigned posture at rest before the test and at 1, 3, and 5 min of recovery. No differences in these variables were seen preexercise. While there were no differences in heart or respiratory rate between recovery postures, the $\dot{V}O_2$ and minute ventilatory volume were significantly lower during recovery in the group assigned to the trunk-leaning-forward posture. This forward-leaning posture improves ventilatory capacity during recovery from maximal exercise, which, in turn, enables rapid recovery of the respiratory system after exercise. The study team concluded that the forward-leaning position has a positive effect on pulmonary ventilation after exercise and may be the most effective posture to promote recovery of breathing after maximal exertion.

inspiratory center so that increases in cardiac output can stimulate breathing during the early minutes of exercise. The influences of CO_2 and H^+ concentrations in the blood on breathing rate and pattern serve to fine-tune the neutrally mediated respiratory response to exercise in order to precisely match oxygen delivery with aerobic demands without overtaxing respiratory muscles.

Pulmonary ventilation increases during exercise in direct proportion to the metabolic needs of exercising muscle. At low exercise intensities, this is accomplished by increases in tidal volume (the amount of air moved in and out of the lungs during regular breathing). At higher intensities, the rate of respiration also increases. Maximal rates of pulmonary ventilation depend on body size. Maximal ventilation rates of approximately 100 L/min are common for smaller individuals but may exceed 200 L/min in larger individuals.

At the end of exercise, the muscles' energy demands decrease almost immediately to resting levels. But pulmonary ventilation returns to normal at a slower rate. If the rate of breathing perfectly matched the metabolic demands of the tissues, respiration would decrease to the resting level within seconds after exercise. But respiratory recovery takes several minutes, which suggests that postexercise breathing is regulated primarily by acid–base balance, the partial pressure of dissolved carbon dioxide (PCO_2), and blood temperature.

Breathing Irregularities During Exercise

Ideally, breathing during exercise is regulated in a way that maximizes aerobic performance. However, respiratory dysfunction during exercise can hinder performance.

Dyspnea

The sensation of **dyspnea** (shortness of breath) during exercise is common among individuals with poor aerobic fitness levels who attempt to exercise at intensities that significantly elevate arterial CO_2 and H^+ concentrations. As discussed in chapter 7, both stimuli send strong signals to the inspiratory center to increase the rate and depth of ventilation. Although exercise-induced dyspnea is sensed as an inability to breathe, the underlying cause is an inability to adjust breathing to blood PCO_2 and H^+.

Failure to reduce these stimuli during exercise appears to be related to poor conditioning of respiratory muscles. Despite a strong neural drive to ventilate the lungs, the respiratory muscles fatigue easily and are unable to reestablish normal homeostasis.

Exercise-Induced Asthma

In healthy humans, the respiratory system in general and the ability to conduct efficient gas exchange at the lungs in particular do not normally limit exercise performance. However, it is estimated that up to 55% of elite athletes participating in endurance winter sports and swimming experience symptoms of exercise-induced asthma (EIA), exercise-induced bronchospasm (EIB), or both.[1,6] Exercise-induced asthma is defined as a lower airway obstruction with symptoms that include coughing, wheezing, or dyspnea that is induced by exercise in individuals with underlying asthma. In addition to EIA, EIB is a reduction in lung function measured by the forced expiratory volume in one second (FEV_1) performed after a standardized exercise test. Many athletes experience these respiratory symptoms, and the onset can occur during childhood or later in life during their sport careers.

Physiologically, there are several different mechanisms by which EIA and EIB may occur in athletes. The classic reasoning has been that hyperventilation during

intense exercise leads to increased evaporation of water from the airway surface. This is a result of having to humidify and warm the air coming into the lung coupled with an increased rate of ventilation during intense exercise. The evaporation of water leads to an increase in osmolality, providing a stimulus for water to move from inside cells to the extracellular fluid. This shrinkage of cells then induces inflammation, in turn causing the airways to constrict.

Other proposed contributors to EIA and EIB in athletes include a disruption to the airway epithelium and microvasculature injury induced by strenuous exercise and airway cooling. Airway cooling causes a reflex increase in parasympathetic nerve activity, causing bronchoconstriction and vasoconstriction of the blood vessels in the bronchioles in order to conserve heat.

Some aspects of EIA and EIB in elite athletes relate to the specific environmental and sport-specific training conditions in which symptoms occur. For example, the cold and dry air that accompanies winter sports,[6] the ultrafine airborne particles emitted from ice resurfacing machines in indoor ice rinks,[14] the pollen and pollutant exposure in athletes practicing outdoors,[2] and chemical exposure in chlorine-rich atmospheres for swimmers have all been implicated as causal factors in breathing problems of athletes.

Hyperventilation

The anticipation of or anxiety about exercise, as well as some respiratory disorders, can cause an increase in ventilation in excess of that needed to support exercise. Such overbreathing is termed **hyperventilation**. At rest, hyperventilation can decrease the normal PCO_2 of 40 mmHg in the alveoli and arterial blood to about 15 mmHg. As arterial CO_2 concentrations decrease, blood pH increases. These effects combine to reduce the ventilatory drive. Because the blood leaving the lungs is almost always about 98% saturated with oxygen, an increase in the alveolar PO_2 does not increase the oxygen content of the blood. Consequently, the reduced drive to breathe—along with the improved ability to hold one's breath after hyperventilating—results from carbon dioxide unloading rather than increased blood oxygen. This is sometimes referred to as "blowing off CO_2." Even when performed for only a few seconds, such deep, rapid breathing can lead to light-headedness and even loss of consciousness. This phenomenon reveals the sensitivity of the respiratory system's regulation by carbon dioxide and pH.

Valsalva Maneuver

The Valsalva maneuver is a potentially dangerous respiratory procedure that frequently accompanies certain types of exercise, in particular the lifting of heavy objects. This occurs when the individual

- closes the glottis (the opening between the vocal cords),
- increases the intra-abdominal pressure by forcibly contracting the diaphragm and the abdominal muscles, and
- increases the intrathoracic pressure by forcibly contracting the respiratory muscles.

As a result of these actions, air is trapped and pressurized in the lungs. The high intra-abdominal and intrathoracic pressures restrict venous return by collapsing the great veins. This maneuver, if held for an extended period of time, can greatly reduce the volume of blood returning to the heart, decreasing cardiac output and lowering arterial blood pressure. Although the Valsalva maneuver can be helpful in certain circumstances, the maneuver can be dangerous and should be avoided.

Ventilation and Energy Metabolism

During long periods of mild steady-state activity, ventilation matches the rate of energy metabolism, varying in proportion to the volume of oxygen consumed and the volume of carbon dioxide produced ($\dot{V}O_2$ and $\dot{V}CO_2$, respectively) by the body.

Ventilatory Equivalent for Oxygen

The ratio between the volume of air expired or ventilated ($\dot{V}_E$) and the amount of oxygen consumed by the tissues ($\dot{V}O_2$) in a given amount of time is referred to as the **ventilatory equivalent for oxygen ($\dot{V}_E/\dot{V}O_2$)**. It is typically measured in liters of air breathed per liter of oxygen consumed per minute.

At rest, the $\dot{V}_E/\dot{V}O_2$ can range from 23 to 28 L of air per liter of oxygen. This value changes very little during mild exercise, such as walking. But when exercise intensity increases to near-maximal levels, the $\dot{V}_E/\dot{V}O_2$ can be greater than 30 L of air per liter of oxygen consumed. In general, however, the $\dot{V}_E/\dot{V}O_2$ remains relatively constant over a wide range of exercise intensities, indicating that the control of breathing is properly matched to the body's demand for oxygen.

Ventilatory Threshold

As exercise intensity increases, at some point ventilation increases disproportionately to oxygen consumption. The point at which this occurs, typically between ~55% and 70% of $\dot{V}O_{2max}$, is called the **ventilatory threshold**, illustrated in figure 8.15. At approximately the same intensity as the ventilatory threshold, more lactate starts to appear in the blood. This may result from greater production of lactate or less clearance of lactate or both. This lactic acid combines with sodium bicarbonate (which buffers acid) and forms sodium lactate, water, and carbon dioxide. As we know, the increase in carbon dioxide stimulates chemoreceptors that signal the inspiratory center to increase ventilation. Thus, the ventilatory threshold reflects the respiratory response to increased carbon dioxide levels. Ventilation increases dramatically beyond the ventilatory threshold, as seen in figure 8.15.

The disproportionate increase in ventilation without an equivalent increase in oxygen consumption led to early speculation that the ventilatory threshold might be related to the lactate threshold (that point at which blood lactate production exceeds lactate reuptake and clearance as described in chapter 5). The ventilatory threshold reflects a disproportionate increase in the volume of carbon dioxide produced per minute ($\dot{V}CO_2$) relative to the oxygen consumed. Recall from chapter 5 that the respiratory exchange ratio (RER) is the ratio of carbon dioxide production to oxygen consumption. Thus, the disproportionate increase in carbon dioxide production also causes RER to increase.

The increased $\dot{V}CO_2$ was thought to result from excess carbon dioxide being released from bicarbonate buffering of lactic acid. Wasserman and McIlroy[18] coined the term **anaerobic threshold** to refer to this phenomenon because they assumed that the sudden increase in CO_2 reflected a shift toward more anaerobic metabolism. They believed that this was a good noninvasive alternative to blood sampling for detecting the onset of anaerobic metabolism. It should be noted that a number of scientists objected to their use of the term *anaerobic threshold* to refer to this respiratory phenomenon.

Over the years, the anaerobic threshold concept has been refined considerably to provide a relatively accurate estimate of lactate threshold. One of the more accurate techniques for identifying this threshold involves monitoring both the ventilatory equivalent for oxygen ($\dot{V}_E/\dot{V}O_2$) and the **ventilatory equivalent for carbon dioxide ($\dot{V}_E/\dot{V}CO_2$)**, which is the ratio of the volume of air expired ($\dot{V}_E$) to the volume of carbon dioxide produced ($\dot{V}CO_2$). Using this technique, the threshold is defined as that point where there is a systematic increase in $\dot{V}_E/\dot{V}O_2$ without a concomitant increase in $\dot{V}_E/\dot{V}CO_2$. This is illustrated in figure 8.16. Both the $\dot{V}_E/\dot{V}CO_2$ and $\dot{V}_E/\dot{V}O_2$ decline with increasing exercise intensity at the lower intensities. However, the $\dot{V}_E/\dot{V}O_2$ starts to increase at about 75 W while the $\dot{V}_E/\dot{V}CO_2$ continues to decline. This indicates that the increase in ventilation to remove CO_2 is disproportionate to the body's need to provide O_2. In general, this respiratory threshold technique provides a reasonably close estimate of the lactate threshold, eliminating the need for repeated blood sampling.

Respiratory Limitations to Performance

Like all tissue activity, respiration requires energy. Most of this energy is used by the respiratory muscles during pulmonary ventilation. At rest, the respiratory muscles account for only about 2% of the total oxygen uptake. As the rate and depth of ventilation increase, so does the energy cost of respiration. The diaphragm, the intercostal muscles, and the abdominal muscles can account for up to 11% of the total oxygen consumed

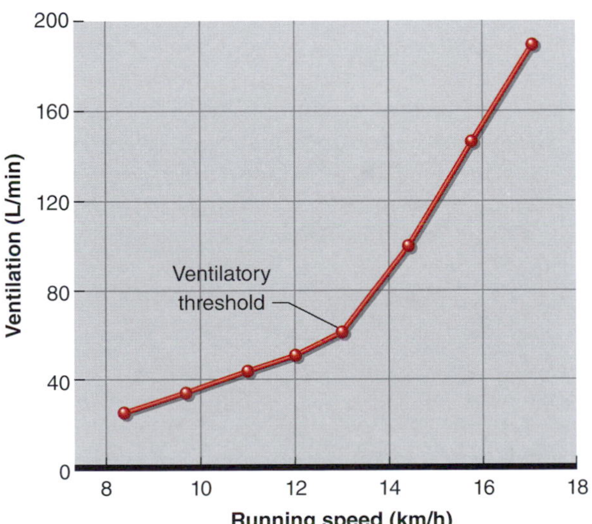

FIGURE 8.15 Changes in pulmonary ventilation ($\dot{V}_E$) during running at increasing velocities, illustrating the concept of ventilatory threshold.

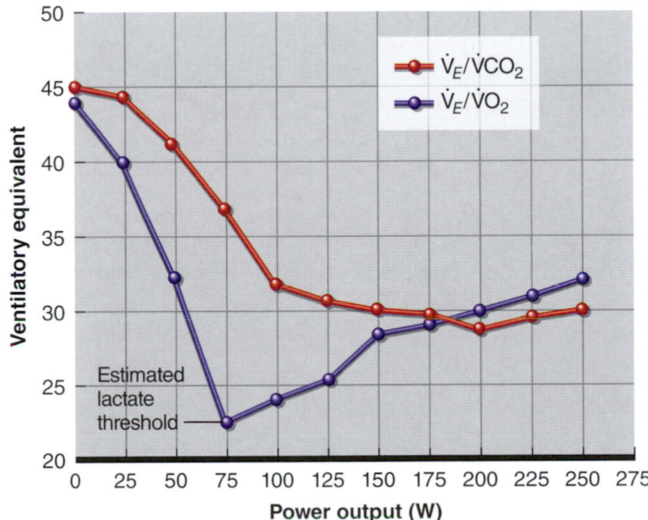

FIGURE 8.16 Changes in the ventilatory equivalent for carbon dioxide ($\dot{V}_E/\dot{V}CO_2$) and the ventilatory equivalent for oxygen ($\dot{V}_E/\dot{V}O_2$) during increasing intensities of exercise on a cycle ergometer. Note that the breakpoint of the estimated lactate threshold at a power output of 75 W is evident only in the $\dot{V}_E/\dot{V}O_2$ ratio.

during heavy exercise and can receive up to 15% of the cardiac output. During recovery from dynamic exercise, sustained elevations in ventilation continue to demand increased energy, accounting for 9% to 12% of the total oxygen consumed postexercise.

In Review

- During exercise, ventilation shows an almost immediate increase due to increased inspiratory center stimulation. This is caused by both central command and neural feedback from muscle activity itself. This phase is followed by a plateau (during light exercise) or a much more gradual increase in respiration (during heavy exercise) that results from chemical changes in the arterial blood resulting from exercise metabolism.
- Altered breathing patterns and sensations associated with exercise include dyspnea, exercise-induced asthma or bronchospasm, hyperventilation, and performance of the Valsalva maneuver.
- During mild, steady-state exercise, ventilation increases to match the rate of energy metabolism; that is, ventilation parallels oxygen uptake. The ratio of air ventilated to oxygen consumed is the ventilatory equivalent for oxygen ($\dot{V}_E/\dot{V}O_2$).
- At low exercise intensities, increased ventilation is accomplished by increases in tidal volume (the amount of air moved in and out of the lungs during regular breathing). At higher intensities, the rate of respiration also increases.
- Maximal rates of pulmonary ventilation depend on body size. Maximal ventilation rates of approximately 100 L/min are common for smaller individuals but may exceed 200 L/min in larger individuals.
- The ventilatory threshold is the point at which ventilation begins to increase disproportionately to the increase in oxygen consumption. This increase in $\dot{V}_E$ reflects the need to remove excess carbon dioxide.
- We can estimate lactate threshold with reasonable accuracy by identifying that point at which $\dot{V}_E/\dot{V}O_2$ starts to increase while $\dot{V}_E/\dot{V}CO_2$ continues to decline.

Although the muscles of respiration are heavily taxed during exercise, ventilation is sufficient to prevent an increase in alveolar PCO_2 or a decline in alveolar PO_2 during activities lasting only a few minutes. Even during maximal effort, ventilation usually is not pushed to its maximal capacity to voluntarily move air in and out of the lungs. This capacity is called the **maximal voluntary ventilation** and is significantly greater than ventilation at maximal exercise. However, considerable evidence suggests that pulmonary ventilation might be a limiting factor during exercise of very high intensity (95%-100% $\dot{V}O_{2max}$) in highly trained subjects.

Can heavy breathing for several hours (such as during marathon running) cause glycogen depletion and fatigue of the respiratory muscles? Animal studies have shown a substantial sparing of their respiratory muscle glycogen compared with muscle glycogen in exercising muscles. Although similar data are not available for humans, our respiratory muscles are better designed for long-term activity than are the muscles in our extremities. The diaphragm, for example, has two to three times more oxidative capacity (oxidative enzymes and mitochondria) and capillary density than other skeletal muscle. Consequently, the diaphragm can obtain more energy from oxidative sources than can skeletal muscles.

Similarly, airway resistance and gas diffusion in the lungs do not limit exercise in a normal, healthy individual. The volume of air inspired can increase 20- to 40-fold with exercise—from ~5 L/min at rest up to 100 to 200 L/min with maximal exertion. Airway resistance, however, is maintained at near-resting levels by airway dilation (through an increase in the laryngeal aperture and bronchodilation). During submaximal and maximal efforts in untrained and moderately trained individuals, blood leaving the lungs remains nearly saturated with oxygen (~98%). However, with maximal exercise in some highly trained elite endurance athletes, there is too large a demand on lung gas exchange, resulting in a decline in arterial PO_2 and arterial oxygen saturation (i.e., **exercise-induced arterial hypoxemia [EIAH]**). Approximately 40% to 50% of elite endurance athletes experience a significant reduction in arterial oxygenation during exercise approaching exhaustion.[10] Arterial hypoxemia at maximal exercise is likely the result of a mismatch between ventilation and perfusion of the lung. Since cardiac output is extremely high in elite athletes, blood is flowing through the lungs at a high rate and thus there may not be sufficient time for that blood to become saturated with oxygen. Thus, in healthy individuals, the respiratory system is well designed to accommodate the demands of heavy breathing during short- and long-term physical effort. However, some highly trained individuals who consume unusually large amounts of oxygen during exhaustive exercise can face respiratory limitations.

The respiratory system also can limit performance in patient populations with restricted or obstructed airways. For example, asthma causes constriction of the bronchial tubes and swelling of the mucous membranes. These effects cause considerable resistance to ventilation, resulting in a shortness of breath. Exercise is known to bring about symptoms of asthma or to worsen those symptoms in select individuals. The mechanism or mechanisms through which exercise induces airway obstruction in individuals with so-called exercise-induced asthma remain unknown, despite extensive study.

In Review

- Respiratory muscles can account for up to 10% of the body's total oxygen consumption and 15% of the cardiac output during heavy exercise.
- Pulmonary ventilation is usually not a limiting factor for performance even during maximal effort, although it can limit performance in some elite endurance athletes.
- The respiratory muscles are well designed to avoid fatigue during long-term activity.
- Airway resistance and gas diffusion usually do not limit performance in normal, healthy individuals exercising at sea level.
- The respiratory system can, and often does, limit performance in people with various types of restrictive or obstructive respiratory disorders.

Respiratory Regulation of Acid–Base Balance

As noted earlier, high-intensity exercise results in the production and accumulation of lactate and H^+. Although regulation of acid–base balance involves more than control of respiration, it is discussed here because the respiratory system plays such a crucial role in rapid adjustment of the body's acid–base status during and immediately after exercise.

Acids, such as lactic acid and carbonic acid, release hydrogen ions (H^+). As noted in the preceding chapters, the metabolism of carbohydrate, fat, or protein produces inorganic acids that dissociate, increasing the H^+ concentration in body fluids, thus lowering the pH. To minimize the effects of free H^+, the blood and muscles contain base substances that combine with, and thus buffer or neutralize, the H^+:

$$H^+ + \text{buffer} \rightarrow \text{H-buffer}$$

Under resting conditions, body fluids have more bases (such as bicarbonate, phosphate, and proteins) than acids, resulting in a slightly alkaline tissue pH that ranges from 7.1 in muscle to 7.4 in arterial blood. The tolerable limits for arterial blood pH extend from 6.9 to 7.5, although the extremes of this range can be tolerated only for a few minutes (see figure 8.17). An

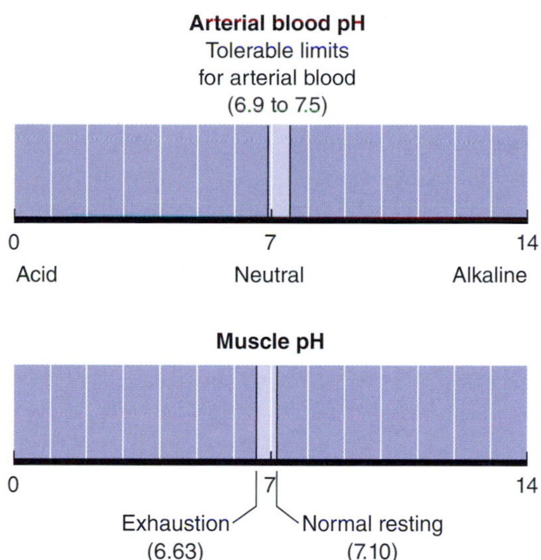

FIGURE 8.17 Tolerable limits for arterial blood pH and muscle pH at rest and at exhaustion. Note the small range of physiological tolerance for both muscle and blood pH.

H^+ concentration above normal (low pH) is referred to as acidosis, whereas a decrease in H^+ below the normal concentration (high pH) is termed alkalosis.

The pH of intra- and extracellular body fluids is kept within a relatively narrow range by

- chemical buffers in the blood,
- pulmonary ventilation, and
- kidney function.

The three major chemical buffers in the body are bicarbonate (HCO_3^-), inorganic phosphates (P_i), and proteins. In addition to these, hemoglobin in the red blood cells is also a major buffer. Table 8.2 illustrates the relative contributions of these buffers in handling acids in the blood. Recall that bicarbonate combines with H^+ to form carbonic acid, thereby eliminating the acidifying influence of free H^+. The carbonic acid in turn forms carbon dioxide and water in the lungs. The CO_2 is then exhaled and only water remains.

TABLE 8.2 **Buffering Capacity of Blood Components**

Buffer	Slykes[a]	%
Bicarbonate	18.0	64
Hemoglobin	8.0	29
Proteins	1.7	6
Phosphates	0.3	1
Total	**28.0**	**100**

[a]Milliequivalents of hydrogen ions taken up by each liter of blood from pH 7.4 to 7.0.

The amount of bicarbonate that combines with H^+ equals the amount of acid buffered. When lactic acid decreases the blood's pH from 7.4 to 7.0, more than 60% of the bicarbonate initially present in the blood has been used. Even under resting conditions, the acid produced by the end products of metabolism would use up a major portion of the bicarbonate from the blood if there were no other way of removing H^+ from the body. Blood and chemical buffers are required only to transport metabolic acids from their sites of production (the muscles) to the lungs or kidneys, where they can be removed. Once H^+ is transported and removed, the buffer molecules can be reused.

In the muscle fibers and the kidney tubules, H^+ is primarily buffered by phosphates, such as phosphoric acid and sodium phosphate. Less is known about the capacity of the buffers intracellularly, although cells contain more protein and phosphates and less bicarbonate than do the extracellular fluids.

As noted earlier, any increase in free H^+ in the blood stimulates the respiratory center to increase ventilation. This facilitates the binding of H^+ to bicarbonate and the removal of carbon dioxide. The end result is a decrease in free H^+ and an increase in blood pH. Thus, both the chemical buffers and the respiratory system provide short-term means of neutralizing the acute effects of exercise acidosis. To maintain a constant buffer reserve, the accumulated H^+ is removed from the body via excretion by the kidneys and eliminated in urine. The kidneys filter H^+ from the blood along with other waste products. This provides a way to eliminate H^+ from the body while maintaining the concentration of extracellular bicarbonate.

During sprint exercise, muscle glycolysis generates a large amount of lactate and H^+, which lowers the muscle pH from a resting level of 7.1 to less than 6.7. As shown in table 8.3, an all-out 400 m sprint decreases leg muscle pH to 6.63 and increases muscle lactate from a resting value of 1.2 mmol/kg to almost 20 mmol/kg of muscle. Such disturbances in acid–base balance can impair muscle contractility and its capacity to generate adenosine triphosphate (ATP). Lactate and H^+ accumulate in the muscle, in part because they do not freely diffuse across the skeletal muscle fiber membranes. Despite the great production of lactate and H^+ during the ~60 s required to run 400 m, these by-products diffuse throughout the body fluids and reach equilibrium after only about 5 to 10 min of recovery. Five minutes after the exercise, the runners described in table 8.3 had blood pH values of 7.10 and blood lactate concen-

TABLE 8.3 **Blood and Muscle pH and Lactate Concentration 5 min After a 400 m Run**

		Muscle		Blood	
Runner	Time (s)	pH	Lactate (mmol/kg)	pH	Lactate (mmol/L)
1	61.0	6.68	19.7	7.12	12.6
2	57.1	6.59	20.5	7.14	13.4
3	65.0	6.59	20.2	7.02	13.1
4	58.5	6.68	18.2	7.10	10.1
Average	60.4	6.64	19.7	7.10	12.3

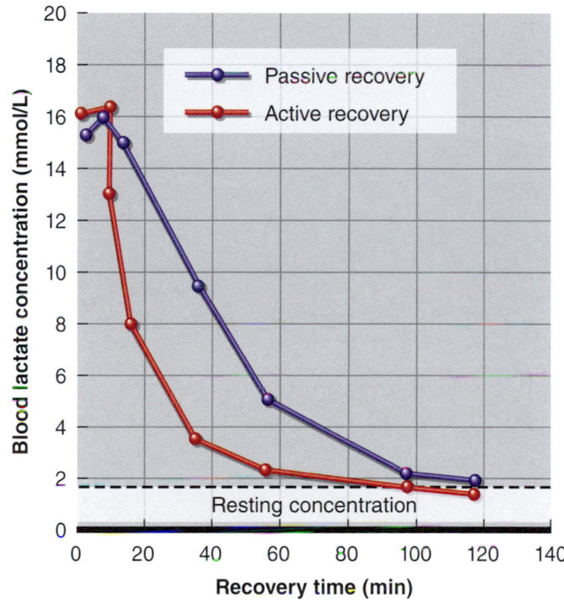

FIGURE 8.18 Effects of active and passive recovery on blood lactate concentrations after a series of exhaustive sprint bouts. Note that the blood lactate removal rate is faster when the subjects perform exercise during recovery than when they rest during recovery.

trations of 12.3 mmol/L, compared with a resting pH of 7.40 and a resting lactate level of 1.5 mmol/L.

Reestablishing normal resting concentrations of blood and muscle lactate after such an exhaustive exercise bout is a relatively slow process, often requiring 1 to 2 h. As shown in figure 8.18, recovery of blood lactate to the resting level is facilitated by continued lower-intensity exercise, called active recovery.[3] After a series of exhaustive sprint bouts, the participants in this study either sat quietly (passive recovery) or exercised at an intensity of 50% $\dot{V}O_{2max}$. Blood lactate is removed more quickly during active recovery because the activity maintains elevated blood flow through the active muscles, which in turn enhances both lactate diffusion out of the muscles and lactate oxidation.

Although blood lactate remains elevated for 1 to 2 h after highly anaerobic exercise, blood and muscle H^+ concentrations return to normal within 40 min of recovery. Chemical buffering, principally by bicarbonate, and respiratory removal of excess carbon dioxide are responsible for this relatively rapid return to normal acid–base homeostasis.

In Review

- Excess H^+ (decreased pH) impairs muscle contractility and ATP generation.
- The respiratory and renal systems play integral roles in maintaining acid–base balance. The renal system is involved in more long-term maintenance of acid–base balance through the secretion of H^+.
- Whenever H^+ concentration starts to increase, the inspiratory center responds by increasing the rate and depth of respiration. Removing carbon dioxide is an essential means of reducing H^+ concentrations.
- Carbon dioxide is transported in the blood primarily bound to bicarbonate. Once it reaches the lungs, carbon dioxide is formed again and exhaled.
- Whenever H^+ concentration begins to increase, whether from carbon dioxide or lactate accumulation, bicarbonate ion can buffer the H^+ to prevent acidosis.

IN CLOSING

In this chapter, we discussed the responses of the cardiovascular and respiratory systems to exercise. We also considered the limitations that these systems can impose on abilities to perform sustained aerobic exercise. The next chapter presents basic principles of exercise training, allowing us to better understand in the subsequent chapters how the body adapts to resistance training as well as aerobic and anaerobic training.

KEY TERMS

afterload
anaerobic threshold
cardiovascular drift
central command
dyspnea
exercise-induced arterial hypoxemia (EIAH)
Frank-Starling mechanism
hydrostatic pressure
hyperventilation
maximal voluntary ventilation
maximum heart rate (HR_{max})
oncotic pressure
preload
rate–pressure product (RPP)
resting heart rate (RHR)
steady-state heart rate
total peripheral resistance (TPR)
Valsalva maneuver
ventilatory equivalent for carbon dioxide ($\dot{V}_E/\dot{V}CO_2$)
ventilatory equivalent for oxygen ($\dot{V}_E/\dot{V}O_2$)
ventilatory threshold

STUDY QUESTIONS

1. Describe how heart rate, stroke volume, and cardiac output respond to increasing rates of work. Illustrate how these three variables are interrelated.
2. How do we determine HR_{max}? What are alternative methods using indirect estimates? What are the major limitations of these indirect estimates?
3. What information can be learned from measuring heart rate variability?
4. Describe two important mechanisms for returning blood back to the heart during exercise in an upright position.
5. Explain why the ability to increase stroke volume is important in determining maximal oxygen consumption.
6. What is the Fick principle, and how does this apply to our understanding of the relationship between metabolism and cardiovascular function?
7. Define the Frank-Starling mechanism. How does this work during exercise?
8. How does blood pressure respond to exercise?
9. What are the major cardiovascular adjustments that the body makes when someone is overheated during exercise?
10. What is cardiovascular drift? What two theories have been proposed to explain this phenomenon?
11. What changes occur in the plasma volume and red blood cells with increasing levels of exercise? With prolonged exercise in the heat?
12. How does pulmonary ventilation respond to increasing intensities of exercise?
13. Define the terms *dyspnea*, *hyperventilation*, *Valsalva maneuver*, and *ventilatory threshold*.
14. What causes exercise-induced asthma in some athletes? What athletes are most prone to being affected?
15. What role does the respiratory system play in acid–base balance?
16. What is the normal resting pH for arterial blood? For muscle? How are these values changed as a result of exhaustive sprint exercise?
17. What are the primary buffers in the blood? In muscles?

STUDY GUIDE ACTIVITIES

In addition to the activities listed in the chapter opening outline, two other activities are available in the web study guide, located at

www.HumanKinetics.com/PhysiologyOfSportAndExercise

The KEY TERMS activity reviews important terms, and the end-of-chapter QUIZ tests your understanding of the material covered in the chapter.

PART III

Exercise Training

The study of exercise physiology relies heavily on the understanding of (1) how the body responds during acute bouts of exercise and (2) how it adapts to repeated exercise sessions (i.e., training responses). In the two previous sections of the book, we examined the control and function of skeletal muscle during acute exercise (part I) and the roles of the cardiovascular and respiratory systems in supporting those functions (part II). In part III, we examine how these systems adapt when exposed to repeated bouts of exercise (i.e., adaptations to training). Chapter 9, Principles of Exercise Training, lays the groundwork for subsequent chapters by discussing the terminology and training principles used by exercise physiologists. The principles presented in this chapter can be used to optimize the physiological adaptations to a training program. In chapter 10, Adaptations to Resistance Training, we consider the mechanisms through which muscular strength and muscular endurance improve in response to resistance training. Finally, in chapter 11, Adaptations to Aerobic and Anaerobic Training, we discuss the changes in various systems of the body that result from performing regular physical activity involving a wide variety of combinations of exercise intensity and duration. Training adaptations that ultimately lead to improvements in exercise capacity and athletic performance are specific to all aspects of the training to which those physiological systems are exposed.

USA

Principles of Exercise Training

In this chapter and in the web study guide

American Ashton Eaton won the gold medal in the decathlon at the 2012 Olympic Games in London, accumulating 8,869 points over the grueling 2-day competition. At the U.S. Olympic Trials in June of that year, Eaton had broken the 9,000-point barrier as well as the world record, previously held by Roman Šebrle of the Czech Republic, whose mark had stood for 11 years. Decathletes are considered by many to be the ultimate athletes, since they have to compete in events that test their speed, strength, power, agility, and endurance. The decathlon is a 2-day event made up of the 100 m sprint, long jump, shot put, high jump, and 400 m run on the first day, and the 110 m hurdles, discus, pole vault, javelin, and 1,500 m run on the second day. Because training is very specific to the sport or event, intense muscular power training to increase the distance one can heave a 16 lb (~7 kg) shot put does little to improve one's 1,500 m run time. Decathletes spend countless hours training specifically for each of their 10 events, fine-tuning their training techniques to maximize performance in each event.

Previous chapters examining the acute response to exercise covered the body's immediate response to a single exercise bout. We now investigate how the body responds to repeated bouts of exercise performed over a period of time—exercise training. When one performs regular exercise over a period of days, weeks, and months, a variety of physiological adaptations occur. The positive adaptations that accompany proper training principles lead to improvement in both exercise capacity and sport performance. With resistance training, muscles become stronger. With aerobic training, the heart and lungs become more efficient at oxygen delivery, and exercise endurance increases. With high-intensity anaerobic training, the neuromuscular, metabolic, and cardiovascular systems adapt to generate more adenosine triphosphate (ATP) per unit of time, thus increasing muscular endurance and speed of movement over short periods of time. These adaptations are highly specific to the type of training performed. Before examining specific adaptations to training, this chapter first looks at the basic terminology and general principles used in exercise training and then gives an overview of the elements of proper training programs.

Terminology

Before discussing the principles of exercise training, we first define key terms that will be used throughout the rest of this book.

Muscular Strength

Strength is defined as the maximal force that a muscle or muscle group can generate. Someone with a maximal capacity to bench press 100 kg (220 lb) has twice the strength of someone who can bench press 50 kg (110 lb). In this example, strength is defined as the maximal weight the individual can lift with one single effort. This is referred to as **1-repetition maximum (1RM)**. To determine 1RM in the weight room or fitness center, people select a weight that they know they can lift at least one time. After a proper warm-up, they try to execute several repetitions. If they can perform more than one repetition, they add weight and try again to execute several repetitions. This continues until the person is unable to lift the weight more than a single repetition. This last weight that can be lifted only once is the 1RM for that particular exercise. The 1RM is commonly used in the laboratory or weight room as a measure of strength.

Muscular strength can also be accurately measured in the research laboratory through use of specialized equipment that allows quantification of static strength and dynamic strength at various speeds and at various angles in the joint's range of motion (see figure 9.1). Gains in muscular strength involve changes in both the structure of the muscle and its neural control. These are discussed in chapter 10.

FIGURE 9.1 An isokinetic testing and training device.

Muscular Power

Power is defined as the rate at which work is performed, thus the product of force and velocity. Unlike strength, it has a speed component. Maximal muscular power, generally referred to simply as power, is the explosive aspect of strength, the product of strength and the velocity of movement.

$$\text{Power} = \text{force} \times \text{distance} / \text{time},$$
$$\text{where force} = \text{strength}$$
$$\text{and distance} / \text{time} = \text{velocity}$$

Consider an example. Two individuals can each bench press 200 kg (441 lb), moving the weight the same distance, from where the bar touches the chest to full extension of the arms. But the person who can do it in 1 s has twice the power of the individual who takes 2 s to perform the lift. This is illustrated in table 9.1.

Although absolute strength is an important component of performance, muscular power is the functional application of both strength and speed of movement. It is a key component in almost every sport and competitive activity. In football, for example, an offensive lineman with a bench press 1RM of 200 kg (441 lb) may be unable to control a defensive lineman with a bench press 1RM of only 150 kg (330 lb) if the defensive lineman can move his 1RM at a much faster speed. The offensive lineman is 50 kg (110 lb) stronger, but the defensive lineman's faster speed coupled with adequate strength could give him the performance edge. Although simple field tests are available to estimate power, these tests are generally not very specific to power because their results are affected by other factors. Power can be measured, however, through use of more sophisticated electronic devices, such as the one depicted in figure 9.1.

Throughout this book, the primary concern is with issues of muscular strength, with only brief mention of muscular power. Recall that power has two components: strength and speed. Speed is a more innate quality that changes little with training. Thus, improvements in power follow improvements in strength gained through traditional resistance training programs. However, high power output exercises, such as vertical jump training and some types of resistance training, have been shown to increase power for those specific movements.[1]

Muscular Endurance

Many sporting activities depend on the muscles' ability to *repeatedly* develop or sustain submaximal forces or to do both. The capacity to perform repeated muscle contractions, or to sustain a contraction over time, is termed **muscular endurance**. Examples of muscular endurance include performing sit-ups or push-ups or sustaining force in an attempt to pin an opponent in wrestling. Although several valid laboratory techniques are available to directly measure muscular endurance, a simple way to estimate it is to assess the maximum number of repetitions one can perform at a given percentage of 1RM. For example, a man who has a 1RM for the bench press of 100 kg (220 lb) could evaluate his muscular endurance independent of his muscular strength by measuring how many repetitions he could perform at, for example, 75% of that 1RM (75 kg, or 165 lb). Muscular endurance is increased through gains in muscular strength and through changes in local blood flow and metabolic function. Metabolic and circulatory adaptations that occur with training are discussed in chapter 11.

Table 9.1 illustrates the functional differences between strength, power, and muscular endurance in three athletes. The actual values have been exaggerated for the purpose of illustration. From this table we can see that although athlete A has half the strength of athletes B and C, he has twice the power of athlete B and is equal in power to athlete C. Therefore, because of his fast speed of movement, his lack of strength

TABLE 9.1 Strength, Power, and Muscular Endurance of Three Athletes Performing the Bench Press

Component	Athlete A	Athlete B	Athlete C
Strength[a]	100 kg	200 kg	200 kg
Power[b]	100 kg lifted 0.6 m in 0.5 s = 120 kg · m/s = 1,177 J/s or 1,177 W	200 kg lifted 0.6 m in 2.0 s = 60 kg · m/s = 588 J/s or 588 W	200 kg lifted 0.6 m in 1.0 s = 120 kg · m/s = 1,177 J/s or 1,177 W
Muscular endurance[c]	10 repetitions with 75 kg	10 repetitions with 150 kg	5 repetitions with 150 kg

[a]Strength was determined by the maximum amount of weight the athlete could bench press just once (i.e., the 1RM).

[b]Power was determined as the athlete performed the 1-repetition maximum (1RM) test as explosively as possible. Power was calculated as the product of force (weight lifted) times the distance lifted from the chest to full arm extension (0.6 m, or about 2 ft), divided by the time it took to complete the lift.

[c]Muscular endurance was determined by the greatest number of repetitions that could be completed using 75% of the 1RM.

does not seriously limit his power output. Also, for purposes of designing training programs, the analysis of these three athletes indicates that athlete A should focus training on developing strength, without losing speed; athlete B should focus on training explosively to improve speed of movement (although this is unlikely to change much); and athlete C should focus training on developing muscular endurance. These recommendations are made assuming that each athlete needs to optimize performance in each of these three areas.

Aerobic Power

Aerobic power is defined as the rate of energy release by cellular metabolic processes that depend on the continued availability of oxygen. It is synonymous with the terms *aerobic capacity* and *maximal oxygen uptake* ($\dot{V}O_{2max}$). Maximal aerobic power is the highest oxygen uptake that an individual can obtain during dynamic exercise using large muscle groups for a few minutes. It depends on the maximal capacity for aerobic resynthesis of ATP. In most healthy individuals, maximal aerobic power is limited primarily by the central cardiovascular system and to a lesser extent by respiration and metabolism. The best laboratory test of aerobic power is a graded exercise test to exhaustion during which $\dot{V}O_2$ is measured and $\dot{V}O_{2max}$ is determined, as discussed in detail in chapter 5. A number of field tests, most often measuring the time needed to walk or run a set distance, or the distance covered in a given time, have been developed to estimate $\dot{V}O_{2max}$ without the need to actually measure it in the laboratory.

Anaerobic Power

Anaerobic power is defined as the rate of energy release by cellular metabolic processes that function without the involvement of oxygen. Maximal anaerobic power, or *anaerobic capacity*, is defined as the maximal capacity of the anaerobic systems (ATP-phosphocreatine [PCr] system and anaerobic glycolytic system) to produce ATP. Unlike the situation with aerobic power, there is no universally accepted laboratory test to determine anaerobic power. Several tests provide estimates of maximal anaerobic power, including the maximal accumulated oxygen deficit test, the critical power test, and the Wingate anaerobic test.

The commonly used Wingate test involves 30 s of all-out pedaling against a constant resistance on a cycle ergometer. The resistance, or braking force, is determined by the person's weight, sex, age, and level of training. Given a 5 s countdown, subjects begin to pedal as fast as they can, and the resistance is increased instantaneously and held constant for the duration of the test. Peak anaerobic power is determined from the number of revolutions performed in the first 5 s, while anaerobic capacity is measured as the total work performed during the 30 s.

In Review

- Muscular strength refers to the ability of a muscle or muscle group to exert force.
- Muscular power is the rate of performing work, or the product of force and velocity.
- Muscular endurance is the capacity to sustain a static contraction or to perform repeated muscle contractions.
- Maximal aerobic power, or aerobic capacity, is the highest oxygen uptake that an individual can obtain during sustained dynamic exercise using large muscle groups.
- Maximal anaerobic power, or anaerobic capacity, is defined as the maximal capacity of the anaerobic energy systems to produce ATP.

General Principles of Training

Chapters 10 and 11 present in detail the specific physiological adaptations that result from resistance training, aerobic training, and anaerobic training. Several principles, however, apply to all forms of exercise training.

Principle of Individuality

Individuals do not all possess the same inherent ability to respond to an acute exercise bout or the same capacity to adapt to exercise training. Heredity plays a major role in determining the body's response to a single bout of exercise, as well as the chronic changes that result from a training program. This is the **principle of individuality**. Except for identical twins, no two people have exactly the same genetic characteristics, so individuals are unlikely to exhibit the same responses. Variations in cellular growth rates, metabolism, cardiovascular and respiratory regulation, and neural and endocrine regulation lead to tremendous individual variation. Such individual variation likely explains why some people show great improvement after participating in a given program ("high responders") whereas others experience little or no change after following the same program ("low responders"). We discuss this phenomenon of high and low responders in more detail in chapter 11. For these reasons, any training program must take into account the specific needs and abilities of the individuals for whom it is designed. Do not expect all individuals to have exactly the same degree of improvement, even when they train exactly the same.

Principle of Specificity

Training adaptations are highly specific to the type of activity being performed and to the volume and intensity of the exercise. To improve muscular power, for example, a shot-putter would not emphasize distance running or slow, low-intensity resistance training. The shot-putter needs to develop explosive power. Similarly, the marathon runner would not focus on sprint training. This is likely the reason that athletes who train for strength and power, such as weightlifters, often have great strength but don't have highly developed aerobic endurance when compared to untrained people. According to the **principle of specificity**, exercise adaptations are specific to the mode, intensity, and duration of training, and the training program must stress the physiological systems that are critical for optimal performance in a given sport in order to achieve specific training adaptations and goals.

Principle of Reversibility

Resistance training improves muscle strength and the capacity to resist fatigue. Likewise, endurance training improves the ability to perform aerobic exercise at higher intensities and for longer periods. But if training is decreased or stopped (detraining), the physiological adaptations that caused those improvements in performance will be reversed. Any gains achieved with training will eventually be lost. The **principle of reversibility** lends scientific support to the saying "Use it or lose it." All effective training programs must include a maintenance plan that sustains the physiological adaptations gained by training. In chapter 14, we examine specific physiological changes that occur when the training stimulus stops.

RESEARCH PERSPECTIVE 9.1

Can Aerobic Exercise Increase Muscle Size?

The principle of specificity states that training adaptations are highly specific to the type of training performed. Aerobic exercise training is associated with improvements in aerobic capacity and cardiorespiratory function. However, there is debate among exercise physiologists about the impact of aerobic exercise training on skeletal muscle mass. Historically, it has been assumed that aerobic exercise has little effect on skeletal muscle hypertrophy. With the development of high-resolution imaging techniques, accumulating evidence now suggests that aerobic exercise training can improve muscle mass in sedentary individuals across the life span. The first of these studies established that 6 months of walking or running training could increase the cross-sectional area of the thigh by 9% in older men.[13] In that study, while the older men experienced a robust increase in muscle size, a group of young men did not show any changes in muscle size with the training. (This result may have been because the younger men attended fewer exercise sessions than the older men did throughout the study. The effectiveness of aerobic exercising training to induce skeletal muscle hypertrophy likely depends on obtaining sufficient exercise intensities, duration, and frequency to accumulate a large number of muscle contractions at this lower load.) Studies that have compared aerobic training with resistance training have found that, on average, both modalities increase muscle size by approximately the same percentage (~7%-9%) from baseline.

Aside from just increasing whole muscle size, aerobic training increased slow- and fast-twitch myofiber cross-sectional area of the exercised muscle in the majority of studies. Similarly, studies of the metabolic turnover of muscle proteins showed that aerobic exercise acutely and chronically stimulated skeletal muscle protein synthesis, resulting in a positive muscle protein balance and increased myofiber size, even in older men and women who might otherwise have age-related anabolic impairments. Despite a lack of standard methodology to measure muscle protein breakdown, most studies examining muscle protein breakdown and aerobic training agree that aerobic training results in reduced catabolic factors, leading to skeletal muscle hypertrophy.

Overall, the existing research indicates that aerobic exercise training can produce skeletal muscle hypertrophy.[11] Aerobic exercise-induced changes in the molecular regulation and protein metabolism of skeletal muscle increase both individual myofiber and whole muscle size in sedentary individuals. These data show that aerobic exercise should be acknowledged for its ability to increase skeletal muscle mass and considered an effective countermeasure for age-related muscle atrophy.

Principle of Progressive Overload

Two important concepts, overload and progressive training, form the foundation of all training programs. According to the **principle of progressive overload**, systematically increasing the demands on the body is necessary for continued improvement. For example, when undergoing a strength training program, in order to gain strength the muscles must be overloaded, which means they must be loaded beyond the point to which they are normally loaded. Progressive resistance training implies that as the muscles become stronger, either increased resistance or increased repetitions or both are required to stimulate further strength increases.

As an example, consider a young woman who can perform only 10 repetitions of a bench press before reaching fatigue, using 30 kg (66 lb) of weight. With a week or two of resistance training, she should be able to increase to 14 or 15 repetitions with the same weight. She then adds 2.3 kg (5 lb) to the bar, and her repetitions decrease to 8 or 10. As she continues to train, the repetitions continue to increase, and within another week or two, she is ready to add another 2.3 kg of weight. Thus, improvement depends on a progressive increase in the amount of weight lifted. In a similar way, training volume (intensity and duration) must be increased progressively with anaerobic and aerobic training for further improvements to occur.

Principle of Variation

The **principle of variation**, also called the **principle of periodization**, first proposed in the 1960s, has become very popular in the area of resistance training. Periodization is the systematic process of changing one or more variables in the training program—mode, volume, or intensity—over time to allow for the training stimulus to remain challenging and effective.[1] Training intensity and volume of training are the most commonly manipulated aspects of training to achieve peak levels of fitness for competition. Classical periodization involves high initial training volume with low intensity; then, as training progresses, volume decreases and intensity gradually increases. Undulating periodization uses more frequent variation within a training cycle.

For sport-specific training, the volume and intensity of training are varied over a macrocycle, which is generally up to a year of training. A macrocycle is composed of two or more mesocycles that are dictated by the dates of major competitions. Each mesocycle is subdivided into periods of preparation, competition, and transition. This principle is discussed in greater detail in chapter 14.

In Review

- According to the principle of individuality, each person responds uniquely to training, and training programs must be designed to allow for individual variation.
- According to the principle of specificity, to maximize benefits, training must be specifically matched to the type of activity or sport the person engages in. An athlete involved in a sport that requires tremendous strength, such as weightlifting, would not expect great strength gains from endurance running.
- According to the principle of reversibility, training benefits are lost if training is either discontinued or reduced abruptly. To avoid this, all training programs must include a maintenance program.
- According to the principle of progressive overload, as the body adapts to training at a given volume and intensity, the stress placed on the body must be increased progressively for the training stimulus to remain effective in producing further improvements.
- According to the principle of variation (or periodization), one or more aspects of the training program should be altered over time to maximize effectiveness of training. The systematic variation of volume and intensity is most effective for long-term progression.

Resistance Training Programs

Over the past 75 years, research has provided a substantial knowledge base concerning resistance training and its application to health and sport. The health aspects of resistance training are discussed in chapter 20. This section concerns primarily the use of resistance training for sport.

Recommendations for Resistance Training Programs

Resistance training programs can be designed and prescribed in terms of

- the exercises that will be performed;
- the order in which they will be performed;
- the number of sets for each exercise;
- the rest periods between sets and between exercises; and
- the amount of resistance, the number of repetitions, and the velocity of movement to be used.

In 2009, the American College of Sports Medicine (ACSM) revised its position stand on progressive resistance training for healthy adults (table 9.2).[1] Previous statements specified, for all adults, a minimum of one set of 8 to 12 reps for each of 8 to 10 different exercises that together involve all of the major muscle groups. The new position stand recommends resistance training models specific to desired outcomes, that is, improvements in strength, muscle **hypertrophy**, power, local muscular endurance, or gross motor performance.

Resistance programs aimed at improving strength should involve repetitions with both concentric (muscle shortening) and eccentric (muscle lengthening) actions. Isometric contractions play a beneficial, but secondary, role and may be included as well. Concentric strength improvement is greatest when eccentric exercises are included, and eccentric training has been shown to produce specific benefits for those action-specific movements. Large muscle groups should be stressed before smaller groups, multiple-joint exercises before single-joint exercises, and higher-intensity efforts before those of lower intensity. Table 9.2 provides a summary of the ACSM recommendations on loading, volume (sets and reps), velocity of movements, and frequency of training.

It is recommended that rest periods of 2 to 3 min or more be used between heavy loads for novice and intermediate lifters; for advanced lifters, 1 or 2 min may suffice. Once an individual can perform the current workload at or above the desired number of reps for two consecutive training sessions, a 2% to 10% increase in load should be applied. While both machine-based

TABLE 9.2 **American College of Sports Medicine's Recommendations for Resistance Training Programs[a]**

Primary goal of resistance training program	Training level	Loading	Volume	Velocity	Frequency (times per week)
Strength development	Novice	60%-70% 1RM	1-3 sets, 8-12 reps	Slow, moderate	2-3
	Intermediate	70%-80% 1RM	Multiple sets, 6-12 reps	Moderate	3-4
	Advanced	80%-100% 1RM	Multiple sets, 1-12 reps	Unintentionally slow to fast	4-6
Muscle hypertrophy	Novice	70%-85% 1RM	1-3 sets, 8-12 reps	Slow, moderate	2-3
	Intermediate	70%-85% 1RM	1-3 sets, 6-12 reps	Slow, moderate	4
	Advanced	70%-100% 1RM; emphasis on 70-85%	3-6 sets, 1-12 reps	Slow, moderate, fast	4-6
Development of muscle power	Novice	0%-60% 1RM—lower body; 30%-60% 1RM—upper body	1-3 sets, 3-6 reps	Moderate	2-3
	Intermediate	0%-60% 1RM—lower body; 30%-60% 1RM—upper body	1-3 sets, 3-6 reps	Fast	3-4
	Advanced	85%-100% 1RM	3-6 sets, 1-6 reps, various strategies[b]	Fast	4-5
Increased local muscular endurance	Novice	Light	1-3 sets, 10-15 reps	Slow—moderate reps Moderate—high reps	2-3
	Intermediate	Light	1-3 sets, 10-15 reps	Slow—moderate reps Moderate—high reps	3-4
	Advanced	30%-80% 1RM	Various strategies,[b] 10-25 reps or more	Slow—moderate reps Moderate—high reps	4-6

[a]These recommendations also include the type of muscle action (eccentric and concentric), single-joint versus multiple-joint exercises, order or sequence of exercises, and the rest intervals. See text for further information.

[b]Periodized—see text for explanation of periodization.

Adapted from information in ACSM (2009).

exercises and free weights can be used for novice and intermediate lifters, for advanced lifters, the emphasis should be placed on free weights.

When muscle hypertrophy (in bodybuilders, for example) or development of muscular power is the goal, recommendations for sequencing, rest periods, and so on are the same as those for strength development. However, as shown in table 9.2, other aspects of the program differ.

Types of Resistance Training

Resistance training can use static contractions, dynamic contractions, or both. Dynamic contractions can include concentric or eccentric contractions, or both. Typical resistance training can be performed using free weights, variable-resistance devices, isokinetic devices, and plyometrics.

Static-Contraction Resistance Training

Static-contraction resistance training, also called **isometric training**, gained great popularity in the mid-1950s as a result of research by several German scientists. These studies indicated that static resistance training caused tremendous strength gains and that those gains exceeded the gains resulting from dynamic-contraction procedures. Subsequent studies were unable to reproduce the original studies' results, and training programs based heavily on isometric contractions have generally fallen out of favor. However, static contractions remain an important form of training, particularly for core stabilization (discussed later in the chapter) and for enhancing grip strength.[1] Additionally, in postsurgical rehabilitation when a limb is immobilized and thus incapable of dynamic contractions, static contractions facilitate recovery and reduce muscle atrophy and strength loss.

Free Weights Versus Machines

With **free weights**, such as barbells and dumbbells, the resistance or weight lifted remains constant throughout the dynamic range of movement. If a 50 kg (110 lb) weight is lifted, it will always weigh 50 kg. In contrast, a variable-resistance contraction involves varying the resistance to try to match it to the strength curve. Figure 9.2 illustrates how strength varies throughout the range of motion in a two-arm curl. Maximal strength production by the elbow flexors occurs at approximately 100° in the range of movement. These muscles are weakest at 60° (elbows fully flexed) and at 180° (elbows fully extended). In these positions, one is able to generate only 67% and 71%, respectively, of the maximal force-producing capabilities at the optimal angle of 100°.

When one is using free weights, the range of motion is less restricted than with machines, and the resistance or weight used to train the muscle is limited by the weakest point in that range of motion. If the person in figure 9.2 had the capacity to lift only 45 kg (100 lb) at the optimal angle of 100°, then he would be able to lift only 32 kg (71 lb) at the fully extended position of 180°. Therefore, if he is starting with a barbell loaded with 32 kg, he can just barely move it from the fully extended position to start his lift. However, by the time he gets to an angle of 100° in his full range of motion, he is lifting only 70% of what he could maximally lift at that angle. Thus, free weights maximally tax the weakest points in the range of motion and provide moderate resistance at the midrange (90°-140°). Individuals performing the two-arm curl tend to greatly reduce their range of motion as they start to fatigue (referred to as "cheating"). They are simply trying to stay out of the weakest portion of their range of motion. The bottom

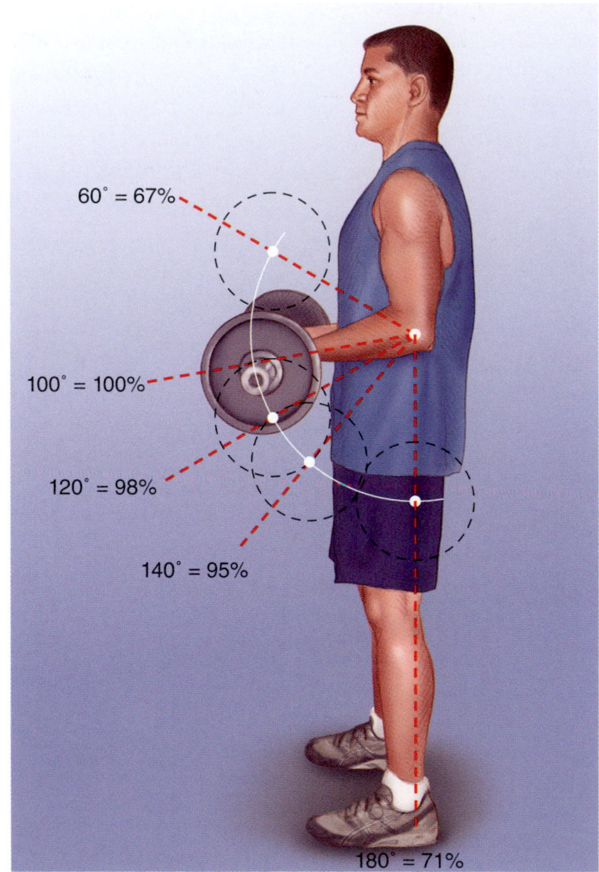

FIGURE 9.2 The variation in strength relative to the angle of the elbow during the two-arm curl. Strength is optimized at an angle of 100°. The maximal force-development capacity of the muscle group at a given angle is given as a percentage of the capacity at the optimal angle of 100°.

line is that with free weights, the maximum weight one can lift is limited by the weakest portion in the range of motion, which means that the strongest position in the range of motion is never maximally taxed! However, free weights do offer some distinct advantages, especially for the expert lifter.

Starting in the 1970s, a number of resistance training machines or devices were introduced that used stacked weights, variable-resistance, and isokinetic techniques. Variable-resistance machines use cams, pulleys, and levers to vary the weight throughout the range of movement. Such machines have been regarded as safer; they are easy to use and allow performance of some exercises that are difficult to do with free weights. Machines help stabilize the body, especially for novice lifters, and limit the muscle action to that desired without extraneous muscle groups firing.

On the other hand, free weights offer some advantages that resistance machines do not provide. The lifter must control the weight being lifted. A lifter must recruit more motor units—not only in the muscles being trained but also in supporting muscles—to gain control of the bar, stabilize the weight lifted, and maintain body balance. The lifter must both balance and stabilize the weight. In that regard, when an athlete is training for a sport such as football, the experience with free weights more closely resembles actions associated with actual sport competition. Also, because free weights do not limit the range of motion of a particular exercise, optimal training specificity can be achieved. Whereas a bicep curl on a machine may be done only in the vertical plane, an athlete using free weights can perform the curl in any plane, choosing one, for example, that reflects a sport-specific motion. And finally, data show that if significant strength gains are to be achieved over a shortened training period, free weights may provide greater strength gains than many types of weight machines.

Both machine-based resistance programs and free-weight training programs result in measurable gains in strength, hypertrophy, and power. Free-weight programs result in greater improvements in free-weight tests, and machine training results in greater gains in machine-based tests. The choice to use weight machines versus free weights depends on the experience of the lifter and the desired outcomes.

Eccentric Training

Another form of dynamic-contraction resistance training, called **eccentric training**, emphasizes the eccentric phase. With eccentric contractions, the muscle's ability to resist force is considerably greater than with concentric contractions (see chapter 1). Subjecting the muscle to this greater training stimulus theoretically produces greater strength gains. A number of studies have shown the importance of including the eccentric phase of muscle contraction along with the concentric phase to maximize gains in muscle strength and size. Further, eccentric contraction is important to stimulate muscle hypertrophy, as discussed in the next chapter.

Variable-Resistance Training

With a variable-resistance device, the resistance is decreased at the weakest points in the range of movement and increased at the strongest points. **Variable-resistance training** is the basis for several popular resistance training machines. The underlying theory is that the muscle can be more fully trained if it is forced to act at higher constant percentages of its capacity throughout each point in its range of movement. Figure 9.3 illustrates a variable-resistance device in which a cam alters the resistance through the range of motion. As noted earlier, there are advantages and disadvantages to training using such machines.

Isokinetic Training

Isokinetic training is conducted with equipment that keeps movement speed constant. Whether one applies

FIGURE 9.3 A variable-resistance training device that uses a cam to alter the resistance through the range of motion.

very light force or an all-out maximal muscle contraction, the speed of movement does not vary. Using electronics, air, or hydraulics, the device can be preset to control the speed of movement (angular velocity) from 0°/s (static contraction) to 300°/s or higher. An isokinetic device is illustrated in figure 9.1. Theoretically, if properly motivated, the individual can contract the muscles at maximal force at all points in the range of motion.

Plyometrics

Plyometrics, or stretch–shortening cycle exercise, became popular during the late 1970s and early 1980s, primarily for improving jumping ability. As an example, to develop knee extensor muscle strength and power, a person goes from standing upright to a deep squat position (eccentric contraction) and then jumps up onto a box (concentric contraction), landing in a squat position on top of the box. The person then jumps off the box onto the ground, landing in a squat position, and repeats the sequence with the next box (see figure 9.4).

Proposed to bridge the gap between speed and strength training, plyometrics uses the stretch reflex to facilitate recruitment of motor units. It also stores energy in the elastic and contractile components of muscle during the eccentric contraction (stretch) that can be recovered during the concentric contraction.

Electrical Stimulation

One can stimulate a muscle by passing an electric current directly across it or its motor nerve. This technique, called **electrical stimulation**, has proven effective in clinical settings to reduce the loss of strength and muscle size during periods of immobilization and to restore strength and size during rehabilitation. Electrical stimulation training also has been used experimentally in healthy subjects (including athletes). Athletes have used this technique to supplement their regular training programs, but no evidence shows any additional gains in strength, power, or performance from this supplementation.

Core Training

In recent years, a significant emphasis has been placed on core stability and strengthening exercises. While there are varying opinions on what anatomical features constitute the core, the general consensus is that the core is the group of trunk muscles that surround the spine and abdominal viscera and include the abdominal, gluteal, hip girdle, paraspinal, and other accessory muscles.

Initially, this type of core-specific exercise training was explored in rehabilitation settings, specifically for the treatment of lower back pain, but its benefits have also been recognized in sport performance. Theoretically, greater core stability could benefit sport performance by providing a foundation for greater force production and force transfer to the extremities. For example, having the core stabilized and engaged in the simple action of throwing a ball allows for greater biomechanical efficiency in the limb transmitting the force to throw the ball and for the activation of stabilizing muscles in the contralateral arm. The principle of core stabilization promotes proximal stability for distal mobility.

There has been little definitive research on the benefits of core stability and core strengthening for athletic performance. One reason is that there are no standardized tests for evaluating core strength and stability. Further, the studies that have been done have

FIGURE 9.4 Plyometric box jumping (see the text for a detailed explanation).

been mainly with injured populations and not specific to athletic performance. However, the limited research does show that this type of training decreases the likelihood of injury, especially in the lower back and the lower extremities, during sport performance. The physiological explanation for this finding is that core stability training increases the sensitivity of the muscle spindles, thereby permitting a greater state of readiness for loading joints during movement[15] and protecting the body from injury.

The many different types of core stability and strengthening training include balance and instability resistance (e.g., physioball). It is thought that because the core is composed mainly of type I muscle fibers, the core musculature may respond well to multiple sets of exercises with high repetitions.[4] Yoga, Pilates, tai chi, and the physioball are commonly incorporated into athletes' training programs to promote core stability and strength. Further research is needed to determine the benefits of core training and the underlying mechanisms.

In Review

- Low-repetition, high-resistance training enhances strength development, whereas high-repetition, low-intensity training optimizes the development of muscular endurance.
- Variation (or periodization), through which various aspects of the training program are altered, is important to optimize results and prevent overtraining or burnout.
- Resistance programs aimed at improving strength should involve repetitions with both concentric (muscle shortening) and eccentric (muscle lengthening) actions. Isometric contractions play a beneficial, but secondary, role and may be included as well.
- Large muscle groups should be stressed before smaller groups, multiple-joint exercises before single-joint exercises, and higher-intensity efforts before those of lower intensity.
- Rest periods of 2 to 3 min or more should be incorporated between heavy loads for novice and intermediate lifters; for advanced lifters, 1 to 2 min may suffice.
- The ability of a muscle or muscle group to generate force varies throughout the full range of movement.
- While both machine-based exercises and free weights can be used for novice and intermediate lifters, for advanced lifters, the emphasis should be placed on free weights.
- When neutral testing devices are used, strength gains from free-weight programs and machine-based programs are similar.
- Electrical stimulation can be successfully used in rehabilitating athletes but has no additional benefits when used to supplement resistance training in healthy athletes.
- Exercises aimed at improving core stability may benefit sport performance by providing a foundation for greater force production and force transfer to the extremities while stabilizing other parts of the body. However, direct evidence of such a benefit is lacking.

Anaerobic and Aerobic Power Training Programs

Anaerobic and aerobic power training programs, while quite different at the extremes (e.g., training for the 100 m dash versus the 42.2 km [26.2 mi] full marathon), are designed along a continuum. Table 9.3 illustrates how training requirements vary in competitive running events as one goes from short sprints to long distances. With this table serving as an example that can be applied to all sports, the primary emphasis for the short sprints is on training the ATP-PCr system. For longer sprints and middle distances, the primary emphasis is on the glycolytic system, and for the longer distances, the primary emphasis is on the oxidative system. Anaerobic power is represented by the ATP-PCr and anaerobic glycolytic systems, while aerobic power is represented by the oxidative system. Note, however, that even at the extremes, more than one energy system must be trained.

Different types of training programs can be used to meet the specific training requirements of each event, such as in running and swimming, and each sport. This section describes some of the more popular types of training programs and how they are used to improve the specific energy systems.

Group Exercise Training

The first description of group fitness can be found in the 1968 book *Aerobics* by Dr. Kenneth Cooper. His mission was to encourage people to exercise with the goal of disease prevention rather than disease treatment. One suggested method was a new form of exercise that utilized dance movements, primarily from hip-hop and jazz, choreographed with music and led by an instructor. Currently, group fitness options focus on varying types of cardiovascular, strength, and flexibility training. For instance, cardio classes include mixed martial arts, plyometric training, indoor cycling, and aquatic activities.

TABLE 9.3 **Percentage of Emphasis on the Three Metabolic Energy Systems in Training for Various Running Events**

Running event	Anaerobic speed (ATP-PCr system)	Anaerobic endurance (anaerobic glycolytic system)	Aerobic endurance (oxidative system)
100 m (109 yd)	95	3	2
200 m (218 yd)	95	2	3
400 m (436 yd)	80	15	5
800 m (872 yd)	30	65	5
1,500 m (0.93 mi)	20	55	25
3,000 m (1.86 mi)	20	40	40
5,000 m (3.10 mi)	10	20	70
10,000 m (6.20 mi)	5	15	80
Marathon (42.20 km; 26.20 mi)	5	5	90

Adapted from F. Wilt, Training for Competitive Running, in *Exercise Physiology*, edited by H.B. Falls (Amsterdam, Netherlands: Elsevier, 1968).

Multiple strength training formats exist that range from high-repetition barbell classes to boot camps with more traditional powerlifting techniques to core-based functional actions. Finally, flexibility is the emphasis in a range of yoga disciplines as well as in fall or injury prevention sessions.

Group fitness can provide the equivalent health benefits of independent exercise, increasing oxygen consumption, high-density lipoprotein (HDL), and lean muscle mass while decreasing fasting blood glucose, low-density lipoprotein (LDL), triglycerides, and fat mass. These positive physiological results occur in parallel to the improvement of many psychological variables such as satisfaction, enjoyment, challenge, and motivation. Because of these health benefits and the prevalence of many class styles, durations, and intensities, group fitness can be an ideal recommendation for all ages and abilities.

Interval Training

Interval training consists of repeated bouts of high- to moderate-intensity exercise interspersed with periods of rest or reduced-intensity exercise. Research has shown that athletes can perform a considerably greater total volume of exercise by breaking the overall exercise period into shorter, more intense bouts, with rest or active recovery intervals inserted between the intense bouts.

The vocabulary used to describe an interval training program is similar to that used in resistance training and includes the terms *sets, repetitions, training time, training distance* and *frequency, exercise interval,* and *rest* or *active recovery interval.* Interval training is frequently prescribed in these terms, as illustrated in the following example for a middle-distance runner:

- Set 1: 6 × 400 m at 75 s (90 s slow jog)
- Set 2: 6 × 800 m at 180 s (200 s jog-walk)

For the first set, the athlete would run six repetitions of 400 m each, completing the exercise interval in 75 s and recovering for 90 s between exercise intervals with slow jogging. The second set consists of running six repetitions of 800 m each, completing the exercise interval in 180 s, and recovering for 200 s with walking-jogging.

While interval training is traditionally associated with track, cross country running, and swimming, it is appropriate for all sports and activities. One can adapt interval training procedures for each sport or event by first selecting the form or mode of training and then manipulating the following primary variables to fit the sport and athlete:

- Rate of the exercise interval
- Distance of the exercise interval
- Number of repetitions and sets during each training session
- Duration of the rest or active recovery interval
- Type of activity during the active recovery interval
- Frequency of training per week

Exercise Interval Intensity

One can determine the intensity of the exercise interval either by establishing a specific duration for a set distance, as illustrated in our previous example for set 1 (i.e., 75 s for 400 m), or by using a fixed percentage of the athlete's maximal heart rate (HR_{max}). Setting a specific duration is more practical, particularly for short sprints. One typically determines this by using the athlete's best time for the set distance and then adjusting the duration according to the relative intensity that the athlete wants to achieve, with 100% equal to the athlete's best time. As an example, to develop the ATP-PCr system, the intensity should be near maximal (e.g., 90%-98%); to develop the anaerobic glycolytic system, it should be high (e.g., 80%-95%); and to develop the aerobic system, it should be moderate to high (e.g., 75%-85%). These estimated percentages are only approximations and are dependent on the athlete's genetic potential and fitness level, duration of the interval (e.g., 10 s versus 10 min), number of repetitions and sets, and duration of the active recovery interval.

Using a fixed percentage of the athlete's HR_{max} might provide a better index of the physiological stress experienced by the athlete. Heart rate monitors are now readily available and relatively inexpensive (see figure 9.5). HR_{max} can be determined during a maximal exercise test in the laboratory as described in chapter 8 or during an all-out run on the track using the heart rate monitor. Training the ATP-PCr system requires training at very high percentages of the athlete's HR_{max} (e.g., 90%-100%), as does training to develop the anaerobic glycolytic system (e.g., 85%-100% HR_{max}). To develop the aerobic system, the intensity should be moderate to high (e.g., 70%-90% HR_{max}).

FIGURE 9.5 A runner outfitted with a heart rate monitor. The receiving unit, attached to the chest strap, picks up and transmits electrical impulses from the heart to the digital monitor and memory device worn on the wrist. After the workout, the contents of the memory device can be downloaded to a computer.

Figure 9.6 illustrates changes in blood lactate concentration in a runner using interval training at three different intensities corresponding to those intensities needed to train the ATP-PCr system, the glycolytic system, and the oxidative system. The runner performed a single set consisting of five repetitions at each intensity on different days, and the lactate concentrations were obtained from a blood sample taken after the last repetition of each intensity. Monitoring blood lactate concentrations can verify the energy system that is primarily being trained.

Distance of the Exercise Interval

The distance of the exercise interval is determined by the requirements of the event, sport, or activity. Athletes who run or sprint short distances, such as track sprinters, basketball players, and soccer players, use short intervals of 30 to 200 m (33-219 yd), although a 200 m sprinter frequently runs over distances of 300 to 400 m (328-437 yd). A 1,500 m runner may run intervals as short as 200 m to increase speed, but most of this athlete's training would be at distances of 400 to 1,500 m (437-1,640 yd), or even longer distances, to increase endurance and decrease fatigue or exhaustion in the race.

Number of Repetitions and Sets During Each Training Session

The number of repetitions and sets is also largely determined by the needs of the sport, event, or activity.

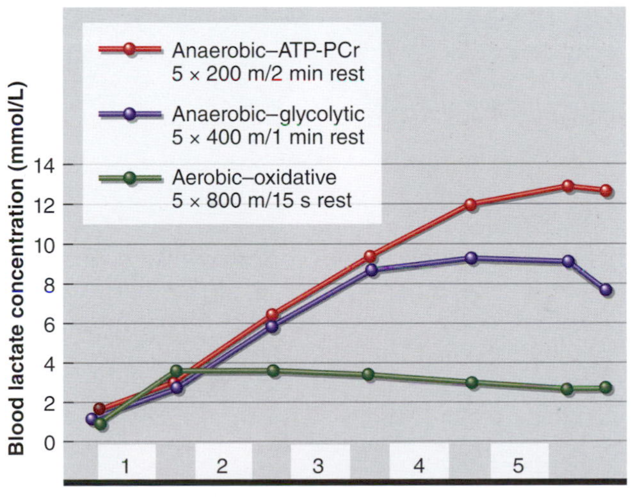

FIGURE 9.6 Blood lactate concentrations in a single runner following a set of five repetitions of interval training at three different paces, each on different days, corresponding to the appropriate pace for training each energy system.

Generally, the shorter and more intense the interval, the greater the number of repetitions and sets should be. As the training interval is lengthened in both distance and duration, the number of repetitions and sets is correspondingly reduced.

Duration of the Rest or Active Recovery Interval

The duration of the rest or active recovery interval depends on how rapidly the athlete recovers from the exercise interval. The extent of recovery is best determined by the reduction of the athlete's heart rate to a predetermined level during the rest or active recovery period. For younger athletes (30 years of age or younger), heart rate is generally allowed to drop to between 130 and 150 beats/min before the next exercise interval begins. For those over 30 years, since HR_{max} decreases ~1 beat/min per year, we subtract the difference between the athlete's age and 30 years from both 130 and 150. So, for a 45-year-old, we would subtract 15 beats/min to obtain the athlete's recovery range of 115 to 135 beats/min. The recovery interval between sets can be established in a similar manner, but generally the heart rate should be below 120 beats/min.

Type of Activity During the Active Recovery Interval

The type of activity performed during the active recovery interval for land-based training can vary from slow walking to rapid walking or jogging. In the pool, slow swimming using alternative strokes or the primary stroke is appropriate. In some cases, usually in the pool, total rest can be used. Generally, the more intense the exercise interval, the lighter or less intense the activity performed in the recovery interval. As athletes become better conditioned, they are able to increase the intensity of the exercise interval, decrease the duration of the rest interval, or both.

Frequency of Training per Week

The frequency of training depends largely on the purpose of the interval training. A world-class sprinter or middle-distance runner typically works out 5 to 7 days a week, although not every workout will include interval training. Swimmers use interval training almost exclusively. Team sport athletes can benefit from 2 to 4 days of interval training per week when interval training is used only as a supplement to a general conditioning program.

Continuous Training

Continuous training involves continuous activity without rest intervals. This can vary from **long, slow distance (LSD) training** to high-intensity endurance training. Continuous training is structured primarily to affect the oxidative and glycolytic energy systems. High-intensity continuous activity is usually performed at intensities representing 85% to 95% of the athlete's HR_{max}. For swimmers and track and cross country athletes, this could be above, at, or near race pace. This pace would likely match or exceed the pace associated with the athlete's lactate threshold. Scientific evidence has clearly demonstrated that marathon runners typically race at, or very close to, their lactate threshold.

Long, slow distance (LSD) training became extremely popular in the 1960s. With this form of training, introduced in the 1920s by Dr. Ernst van Aaken, a German physician and coach, the athlete typically trains at relatively low intensities, between 60% and 80% of HR_{max}, which is approximately the equivalent of 50% to 75% of $\dot{V}O_{2max}$. Distance, rather than speed, is the main objective. Distance runners may train 15 to 30 mi (24-48 km) per day using LSD techniques, with weekly distances of 100 to 200 mi (161-322 km). The pace of the run is substantially slower than the runner's maximal pace. While less stressful to the cardiovascular and respiratory systems, extreme distances can result in overuse injuries and general breakdown of muscles and joints. Further, the serious runner needs to train at or near race pace on a regular basis to develop leg speed and strength. Thus, most runners vary their workout from one day to the next, from week to week, and from month to month.

Long, slow distance training is probably the most popular and safest form of aerobic endurance conditioning for the nonathlete who just wants to get into shape and stay in shape for health-related purposes. More vigorous or burst types of activity generally are not encouraged in older, sedentary people. Long, slow distance is also a good training program for athletes in team sports for maintaining aerobic endurance during the season as well as the off-season.

Fartlek training, or speed play, is another form of continuous exercise that has some components of interval training. This form of training was developed in Sweden in the 1930s and is used primarily by distance runners. The athlete varies the pace from high speed to jogging speed at his or her discretion. This is a free form of training in which fun is the primary goal, and distance and time are not even considered. Fartlek training is normally performed in the countryside where there are hills of various inclines. Many coaches have used Fartlek training to supplement either high-intensity continuous training or interval training, since it provides variety to the normal training routine.

Interval-Circuit Training

Introduced in the Scandinavian countries in the 1960s and 1970s, **interval-circuit training** combines interval and circuit training into one workout. The circuit may be 3,000 to 10,000 m in length, with stations every 400 to 1,600 m (437-1,750 yd). The athlete jogs, runs, or sprints the distance between stations; stops at each station to perform a strength, flexibility, or muscular endurance exercise in a manner similar to that in actual circuit training; and continues on jogging, running, or sprinting to the next station. These courses are typically located in parks or in the country where there are many trees and hills. Such a training regimen can benefit almost any type of athlete and provide diversity to what might be an otherwise monotonous training regimen.

RESEARCH PERSPECTIVE 9.2

Tabata Training: The Original HIIT

Tabata, named for Dr. Izumi Tabata from Ritsumeikan University in Japan, is a high-intensity interval training (HIIT) protocol that incorporates brief, supramaximal (170% of $\dot{V}O_{2max}$) 20 s intervals followed by 10 s of rest into a 4 min exercise bout. In 1996, Dr. Tabata published the results of a study that found that—although just 4 min in duration—subjects who followed this HIIT protocol 4 days a week showed notable improvements in aerobic and anaerobic fitness, higher than those observed in subjects who performed classical endurance training (70% of $\dot{V}O_{2max}$) for 60 min a day.[14] Subsequent research has shown that Tabata and other HIIT programs can be effectively used to increase aerobic and anaerobic fitness, promote fat loss, and improve health outcomes in a relatively short period of time.[12]

The HIIT model of brief bouts of aerobic conditioning at close to maximal intensity has become increasingly popular with competitive athletes to enhance both aerobic and anaerobic endurance while mimicking the swings in intensity that typically occur in competition. However, HIIT protocols can induce similarly large increases in fitness in recreationally active adults as well. Although the intensity of the intervals in the Tabata study was supramaximal, more recent studies have demonstrated that modified versions of the 20 s on/10 s off protocol can yield similar results with submaximal intensities that are still very high intensity, ranging from 80% to 95% of $\dot{V}O_{2max}$. These aerobic intervals are more appropriate for the general population and are frequently utilized in group fitness and personal training settings.

Although somewhat counterintuitive because of the high heart rates achieved, a growing number of fitness experts are suggesting that HIIT should be considered as a strategy to improve cardiovascular and metabolic health. Studies of nonathletes training with high-intensity intervals have documented improved metabolic rate and fat oxidation, decreased abdominal fat, greater insulin sensitivity, improved blood glucose, and reduced blood pressure after training. A report from a research team in the United Kingdom showed that doing 20 s Tabata-style intervals for just 3 min a week improved insulin sensitivity in young men.[3] The authors reasoned that when insulin works more effectively, the muscles utilize more fatty acid oxidation for fuel, which in turn may result in greater utilization of fat even at rest. Indeed, the postexercise oxygen consumption following a 4 min bout of HIIT was double what it was just before exercise. The extra calories and potential for increased fat utilization resulting from HIIT may be an untapped resource to safely improve health outcomes, even in people who are not regularly physically active.

High-Intensity Interval Training (HIIT)

Traditionally, exercise physiologists have recommended one of three regimens to improve aerobic power: continuous exercise at a moderate to high intensity; long, slow (low-intensity) exercise; or interval training. However, a growing body of research suggests that **high-intensity interval training (HIIT)** is a time-efficient way to induce many adaptations normally associated with traditional endurance training. Scientists at McMaster University in Canada have studied the effects of training using short bursts of very intense cycling, interspersed with up to a few minutes of rest or low-intensity cycling for recovery.[6] A common training mode employed is based on the Wingate test, a test that consists of 30 s of all-out cycling and generally produces mean power outputs that are two to three times higher than what subjects typically generate during a maximal oxygen uptake test.

A typical HIIT workout consists of four to six bouts of 30 s all-out cycling separated by a few minutes of recovery. Therefore, the total exercise time is as little as 2 min spread over a 20 min total time period. Several studies have now confirmed that performing six or so sessions of this type of interval training over a 2-week span can dramatically improve aerobic capacity in previously untrained individuals. The best feature of this type of training for busy exercisers is that such a regimen involves only 15 min total of all-out cycling within a total time commitment of 2.5 h![5]

In addition to improving $\dot{V}O_{2max}$, HIIT has been proven to have additional health benefits. Similar to continuous aerobic training programs, HIIT improves glucose control and insulin sensitivity, especially in individuals with (or at risk for) type 2 diabetes. HIIT has also been shown to improve vascular endothelial function, a measure of blood vessel health. In fact, studies have demonstrated that HIIT may be more effective than continuous, long-duration training in promoting metabolic and cardiovascular adaptations.[10]

HIIT for Athletes

Can highly fit individuals and endurance athletes also benefit from HIIT? In sedentary individuals, exercise training affects both the cardiovascular system and the muscles' oxidative enzyme capacity, resulting in an increase in $\dot{V}O_{2max}$. In contrast, in already trained individuals, an increase in exercise intensity close to or even slightly above $\dot{V}O_{2max}$ is often necessary to elicit improvements in $\dot{V}O_{2max}$ and performance.

There is growing evidence that inserting HIIT into an already high-volume traditional aerobic training program can further enhance performance.[5] For example,

RESEARCH PERSPECTIVE 9.3

Exploring the Mechanisms That Increase $\dot{V}O_{2max}$ with HIIT

High-intensity interval training (HIIT) significantly improves maximal oxygen uptake ($\dot{V}O_{2max}$). In untrained individuals, HIIT can increase $\dot{V}O_{2max}$ to the same extent as moderate-intensity continuous training, despite the much shorter duration of exercise bouts. The mechanisms by which moderate-intensity aerobic training increases $\dot{V}O_{2max}$ (e.g., greater blood volume, higher cardiac output, increases in stroke volume) are well characterized and discussed in detail in this chapter. However, despite the widely seen increases in $\dot{V}O_{2max}$ in response to HIIT, the specific adaptations that underlie this outcome have not been clarified. According to the Fick equation, increases in $\dot{V}O_{2max}$ are mediated by increases in cardiac output or arterial–venous oxygen difference, or (a-v)O_2 difference. However, the scientific evidence is not clear whether HIIT improves one, the other, or both.

Recently, a study conducted by a team of scientists from Cal State San Marcos, SUNY Stony Brook, and the National College of Natural Medicine examined the cardiovascular adaptations to 6 weeks of HIIT in 71 healthy, active, young subjects.[2] The aims of this study were (1) how HIIT improves $\dot{V}O_{2max}$ and (2) whether there is an optimal HIIT regimen that would produce the most benefit. In order to test this, the subjects were divided into four groups: a sprint interval-training group (SIT), a high-volume interval-training group ($HIIT_{HI}$), a periodized interval-training group (PER), and a control group that did not exercise (CON). For the first 10 exercise sessions, all subjects in the exercising groups performed the same HIIT protocol of 8 to 10 bouts of 60 s of cycling at 90% to 110% of peak power with 75 s of recovery between bouts. After this initial training, the subjects were randomized into their specific regimens for the remainder of the study. $\dot{V}O_{2max}$, maximal cardiac output, stroke volume, heart rate, and the (a-v)O_2 difference were measured during progressive cycling exercise at the beginning of the study, at the midway point, and at the end of the study. Compared to the control group, all HIIT groups had a significant increase in $\dot{V}O_{2max}$, and the magnitude of the increase in $\dot{V}O_{2max}$ was not different between regimens. In all HIIT groups, maximal cardiac output and stroke volume increased, while maximal heart rate and the (a-v)O_2 difference did not change. Because HIIT increased maximal cardiac output but did not affect extraction, the study team concluded that HIIT increases $\dot{V}O_{2max}$ by improving oxygen delivery through increased blood flow rather than by increasing the muscles' ability to extract oxygen.

when a group of well-trained cyclists replaced 15% of their normal training time with HIIT, they improved their peak power and speed during a 4 km time trial. Such improvement was seen after only six HIIT sessions inserted over a 4-week period.

Another study used a type of HIIT training called the *10-20-30 concept* (5 min intervals alternating between low speed for 30 s, moderate speed for 20 s, and close to maximal speed for 10 s) to assess whether 7 weeks of HIIT training could improve endurance performance, cardiovascular fitness, and overall physical health in a group of already well-trained individuals.[7] The athletes who underwent the HIIT training increased their $\dot{V}O_{2max}$ by 4% and increased their performance in both 1,500 m and 5 km runs, despite a ~50% reduction in their total training time. They also had a significant reduction in their total blood cholesterol, cholesterol fractions, and resting blood pressure. These results suggest that high-intensity interval training is capable of improving $\dot{V}O_{2max}$, exercise performance, and overall markers of cardiovascular health in previously trained individuals. This same group of researchers had previously demonstrated an increase in peripheral muscle membrane proteins and transporters and changes in muscle oxidative enzyme capacity in previously trained athletes who followed a traditional HIIT regimen.[9]

For busy athletes, HIIT training is easily achievable and effective at improving cardiovascular health as well as athletic performance. It may be also useful for athletes who wish to reduce their training time or volume before competition without sacrificing continued improvements in $\dot{V}O_{2max}$ and performance.

Gibala and Jones recommend that for endurance athletes, 75% of total training volume be performed at continuous low intensities with 10% to 15% done using high-intensity intervals.[5] While each bout of activity is anaerobic, the overall effect of HIIT is to stimulate adaptations similar to those with endurance training but in a shorter period of time and with less total work performed. Studies that have compared HIIT to a much higher total volume of traditional continuous endurance training have shown similar improvements in $\dot{V}O_{2max}$ and cellular markers of improved aerobic capacity in untrained individuals. Adaptations were different in highly trained athletes. These adaptations are discussed in more detail in chapter 11.

High-Intensity Interval Training in Team Sports

The science of training athletes depends heavily on the concept of specificity of training. However, designing a training regimen that provides sport-specific performance benefits while maintaining overall speed, fitness, and athletic skills—without overtraining—is often a difficult task. Adding HIIT to traditional endurance workouts has gained popularity for its benefits in improving sport-specific athletic performance,[7,9] but most of the studies examining the effects of HIIT have been performed on athletes competing in individual sports, such as runners and cyclists, not in athletes who participate in team sports.

In order to determine whether HIIT training would be beneficial for performance, a group of elite soccer players were tested before, and again after, a 5-week HIIT intervention. The Danish Second Division players averaged 2.7 training sessions per week, with each session lasting 3.6 h, and played one match per week. The HIIT, accomplished by carrying out drills without the ball, consisted of six to nine 30 s intervals per week at an intensity of 90% to 95% of $\dot{V}O_{2max}$. The number of HIIT intervals increased each week. The performance evaluation included a sprint test, an agility test, and repeated 20 m shuttle runs at progressively increasing speeds. After the HIIT intervention, the elite soccer players performed better on the shuttle runs by 11%, but performance on the sprint and agility tests was unchanged. Interestingly, after the HIIT interventions, the $\dot{V}O_{2max}$ was unchanged, but there was a reduction in the athletes' $\dot{V}O_2$ during running at a fixed speed of 10 km/h, indicating that running economy was improved by HIIT in these elite soccer players.

In Review

- Anaerobic and aerobic power training programs are designed to train the three metabolic energy systems: the ATP-PCr system, the anaerobic glycolytic system, and the oxidative system.
- Interval training consists of repeated bouts of high- to moderate-intensity exercise interspersed with periods of rest or reduced-intensity exercise. For short intervals, the rate or pace of activity and the number of repetitions are usually high, and the recovery period is usually short. Just the opposite is the case for long intervals.
- Both the exercise rate and the recovery rate can be closely monitored with use of a heart rate monitor.
- Interval training is appropriate for all sports. The length and intensity of intervals can be adjusted based on the sport requirements.
- Continuous training has no rest intervals and can vary from LSD training to high-intensity training. Long, slow distance training is very popular for general fitness training.
- Fartlek training, or speed play, is an excellent activity for recovering from several days or more of intense training.
- Interval-circuit training combines interval training and circuit training into one workout.
- High-intensity interval training is a time-efficient way to induce many adaptations normally associated with traditional endurance training. In addition to consuming less time, it can be used to provide variety to the training.
- High-intensity interval training has been shown to improve performance in already-trained individuals, including those participating in team sports such as soccer.

IN CLOSING

In this chapter, we reviewed general principles of training and the terminology used to describe these principles. We then learned the essential ingredients of successful resistance training and anaerobic and aerobic power training programs. With this background, we can now focus on how the body adapts to these different types of training programs. In the next chapter, we will see how the body responds to resistance training.

KEY TERMS

1-repetition maximum (1RM)
aerobic power
anaerobic power
continuous training
eccentric training
electrical stimulation
Fartlek training
free weights
high-intensity interval training (HIIT)
hypertrophy
interval-circuit training
interval training
isokinetic training
isometric training
long, slow distance (LSD) training
muscular endurance
plyometrics
power
principle of individuality
principle of periodization
principle of progressive overload
principle of reversibility
principle of specificity
principle of variation
static-contraction resistance training
strength
variable-resistance training

STUDY QUESTIONS

1. Define and differentiate the terms *strength*, *power*, and *muscular endurance*. How does each component relate to athletic performance?
2. Define aerobic and anaerobic power. How does each relate to athletic performance?
3. Describe and provide examples for the principles of individuality, specificity, reversibility, progressive overload, and variation.
4. What factors need to be considered when one is designing a resistance training program?
5. What would be the appropriate range for resistance and repetitions when one is designing a resistance training program targeted to develop strength? Muscular endurance? Muscular power? Hypertrophy?
6. Describe the various types of resistance training, and explain the advantages and disadvantages of each.
7. What type of training program would likely provide the greatest improvement for sprinters? Marathon runners? Football players?
8. What are some advantages of exercising in a group setting rather than alone?
9. Describe the various forms of interval and continuous training programs, and discuss the advantages and disadvantages of each. Indicate the sport or event most likely to benefit from each one.
10. High-intensity interval training has been shown to cause beneficial adaptations leading to improved performance. Describe those physiological adaptations.

STUDY GUIDE ACTIVITIES

In addition to the activities listed in the chapter opening outline, two other activities are available in the web study guide, located at

www.HumanKinetics.com/PhysiologyOfSportAndExercise

The **KEY TERMS** activity reviews important terms, and the end-of-chapter **QUIZ** tests your understanding of the material covered in the chapter.

Adaptations to Resistance Training

In this chapter and in the web study guide

When he died on September 13, 2013, at the age of 84, few sport fans had heard of Jim Bradford. Bradford, an African American, spent much of his life working quietly behind the scenes at the Library of Congress as a researcher and bookbinder. At the 1952 Olympic Games in Helsinki and again at the 1960 Games in Rome, Bradford won silver medals in the weightlifting heavyweight division. Yet he was hardly known in his hometown of Washington in those decades, let alone nationally. Although it's hard to imagine in today's world of professional athletes, Bradford had to take unpaid leave from the Library of Congress to compete in the Olympics. "I come back to my job and that is it. That was par for the course then."[2]

Mr. Bradford was a self-proclaimed "butterball" during high school who started lifting weights after reading inspirational stories in a weightlifting magazine. He started with a set of dumbbells in his second-floor bedroom before moving his training to a nearby YMCA at his parents' request. There, he developed a unique lifting style—keeping his legs together and bending his back only as he lifted the bar overhead—for the simple reason that he feared dropping the weights, scuffing the floor, and getting kicked out of the gym![2]

With any type of effective chronic exercise, multiple adaptations occur in the neuromuscular system. The type and extent of the adaptations depend on the type of training: Aerobic training, such as running, cycling, or swimming, results in little gain in muscle size and strength, but major neuromuscular adaptations occur with **resistance training**.

Resistance training was once considered inappropriate for athletes except those in competitive weightlifting, throwing events in track and field, and wrestling and boxing. Women typically avoided the weight room for fear of becoming masculine looking! But in the late 1960s and early 1970s, coaches and researchers discovered that strength and power training are beneficial for almost all sports and activities, and for women as well as men. It was not until the late 1980s and early 1990s that health professionals began to recognize the importance of resistance training to overall health, fitness, and rehabilitation.

Most athletes now include resistance training as an important component of their overall training program. Much of this attitude change is attributable to research that has proven the performance benefits of resistance training and to innovations in training techniques and equipment. Resistance training is now an important part of the exercise prescription for all those who seek the health-related benefits of exercise.

Resistance Training and Gains in Muscular Fitness

Throughout this book, we see how important muscular fitness is to health, quality of life, and athletic performance. How do we get stronger and how do we increase muscle power and muscle endurance? Maintaining an active lifestyle is important in maintaining muscular fitness, but resistance training is necessary to increase muscular strength and power. In this section, we briefly review the changes that result from resistance training. We focus on strength, with only a brief mention of power and muscular endurance—topics that are discussed in more detail later in this book.

The neuromuscular system is one of the most responsive systems in the body to the repeated stimulation of training. Resistance training programs can produce substantial strength gains. In 3 to 6 months, one can see from 25% to 100% improvement, sometimes even more. These estimates of percentage gains in strength are, however, somewhat misleading. Most subjects in strength training research studies have never lifted weights or participated in any other form of resistance training. Most of their early gains in strength are the result of learning how to more effectively produce force and produce a true maximal movement, such as moving a barbell from the chest to a fully extended position in the bench press. This learning effect can account for as much as 50% of the early strength gains.

Muscle is very plastic, increasing in size and strength with training and decreasing in size and strength when immobilized. The remainder of this chapter details the physiological adaptations that allow people to become stronger.

Mechanisms of Gains in Muscle Strength

For many years, strength gains were assumed to result directly from increases in muscle size (hypertrophy). This assumption was logical because many people who regularly strength trained developed visually larger muscles. Also, muscles associated with a limb immobilized in a cast for weeks or months start to decrease in size (**atrophy**) and lose strength almost immediately. Gains in muscle size are generally paralleled by gains in strength, and losses in muscle size correlate highly with losses in strength. Thus, it is tempting to conclude that a direct cause-and-effect relationship exists between muscle size and muscle strength. While there is an association between size and strength, muscle strength involves far more than mere muscle size.

This does not mean that muscle size is unimportant in the ultimate strength potential of the muscle. The ability to generate force depends on the number of cross-bridges within sarcomeres, which in turn depends

RESEARCH PERSPECTIVE 10.1

Aerobic Benefits From Resistance Exercise Training

Resistance exercise training is a well-established method for increasing muscle size and strength. The classical skeletal muscle responses to resistance exercise training (e.g., hypertrophy, changes in muscle fiber type, increases in neural activation) are described in detail in this chapter. However, the long-held dogma that resistance exercise training results in distinctly different adaptations from aerobic exercise training (which are discussed in detail in chapter 11) has limited many studies to a strict focus on the mechanisms of muscle hypertrophy and gains in strength.

Conversely, increases in skeletal muscle mitochondrial number and function and increases in the number and density of capillaries in the skeletal muscle are both well-known adaptations to aerobic exercise training. These adaptations increase ATP production and oxygen and nutrient delivery to the exercising skeletal muscle; as such, they contribute to the improvements in muscle health and maximal oxygen uptake that occur with aerobic exercise training. In addition to improving fitness, increases in mitochondrial function and skeletal muscle capillarization also enhance overall health by improving muscle bioenergetics, insulin sensitivity, and glucose tolerance at rest. Therefore, strategies to induce these adaptations may have benefits for many people, not just athletes.

Recently, exercise physiologists have begun to explore whether resistance exercise training may also exert some of these aerobic benefits. Research studies conducted at the University of Texas Medical Branch[17] and Maastricht University in the Netherlands[26] are two of the first to explore this possibility. Both of these studies utilized a 12-week program of resistance exercise training to improve muscular fitness and collected muscle biopsy samples from the vastus lateralis before and after the training program (see figure). In Texas, the research team compared mitochondrial respiratory capacity, measured as citrate synthase activity, in the muscle biopsy samples from young subjects. They found that resistance exercise increased mitochondrial protein expression and respiratory capacity, suggesting that this resistance training protocol improved the oxidative capacity of the trained muscle. The Netherlands group examined the number of capillary contacts and the ratio of capillaries to muscle fibers in muscle biopsy samples from young and older men at baseline, and in the older men again after the resistance training protocol. (They chose to study resistance training in the older men only, opining that these subjects might have the most clinical benefit from resistance exercise to fight sarcopenia.) They found that older men had fewer capillary contacts and a lower ratio of capillaries to muscle fibers compared to young men, but, importantly, both of these aerobic-type variables increased in the older men after 12 weeks of resistance training. This led the researchers to conclude that resistance exercise can increase skeletal muscle capillarization.

Taken together, these studies show that the adaptations to resistance exercise may have more in common with aerobic exercise adaptations than we previously thought. Resistance exercise training improves mitochondrial function and capillarization in the skeletal muscle. These findings increase our understanding of how resistance exercise improves multiple fitness domains and gives us new knowledge about the health benefits of resistance training.

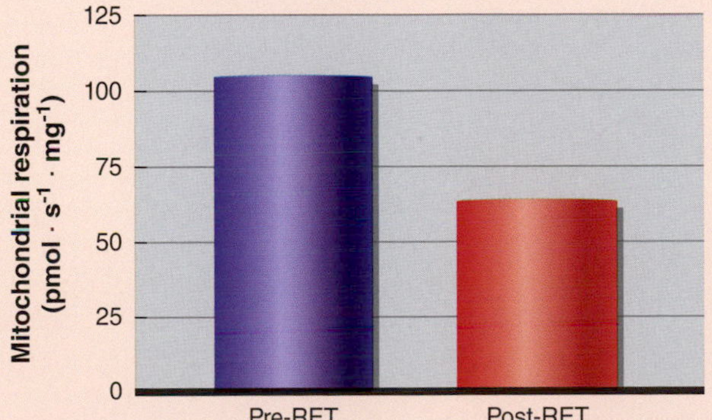

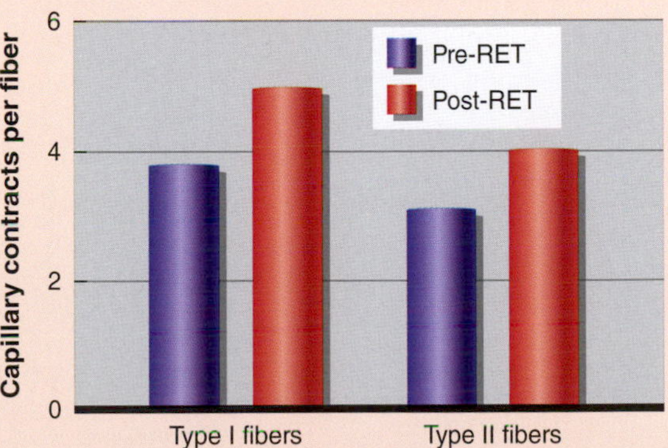

Twelve weeks of resistance training increases mitochondrial respiration[17] and the number of capillary contacts per muscle fiber[26] in trained skeletal muscle. These adaptations were previously thought to occur with aerobic exercise training only. However, resistance exercise training can induce these aerobic adaptations in the trained muscle as well.

on the amount of actin and myosin. Size is extremely important, as revealed by the existing men's and women's world records for competitive weightlifting, shown in figure 10.1. As weight classification increases (implying increased muscle mass), so does the record for the total weight lifted. However, the mechanisms associated with strength gains are more complex. What, in addition to increased size of the muscle, explains strength gains with training? Let's first consider the strong evidence that neural control of the muscle is altered with resistance training, allowing for a greater force production.

Neural Control of Strength Gains

An important component of the strength gains that result from resistance training, especially in the early stages, are neural adaptations. Enoka has made a convincing argument that strength gains can be achieved without structural changes in muscle but not without neural adaptations.[7] Thus, strength is not solely a property of muscle but rather a property of the neuromotor system. Motor unit recruitment, frequency of motor nerve firing rates, better synchronization of motor units during a particular movement, and other neural factors are important to strength gains. Removal of neural inhibition may also play a role. These neural factors may well explain most, if not all, strength gains that occur in the absence of hypertrophy, as well as episodic superhuman feats of strength.

Synchronization and Recruitment of Additional Motor Units

Motor units are generally recruited asynchronously; they are not all engaged at the same instant. They are

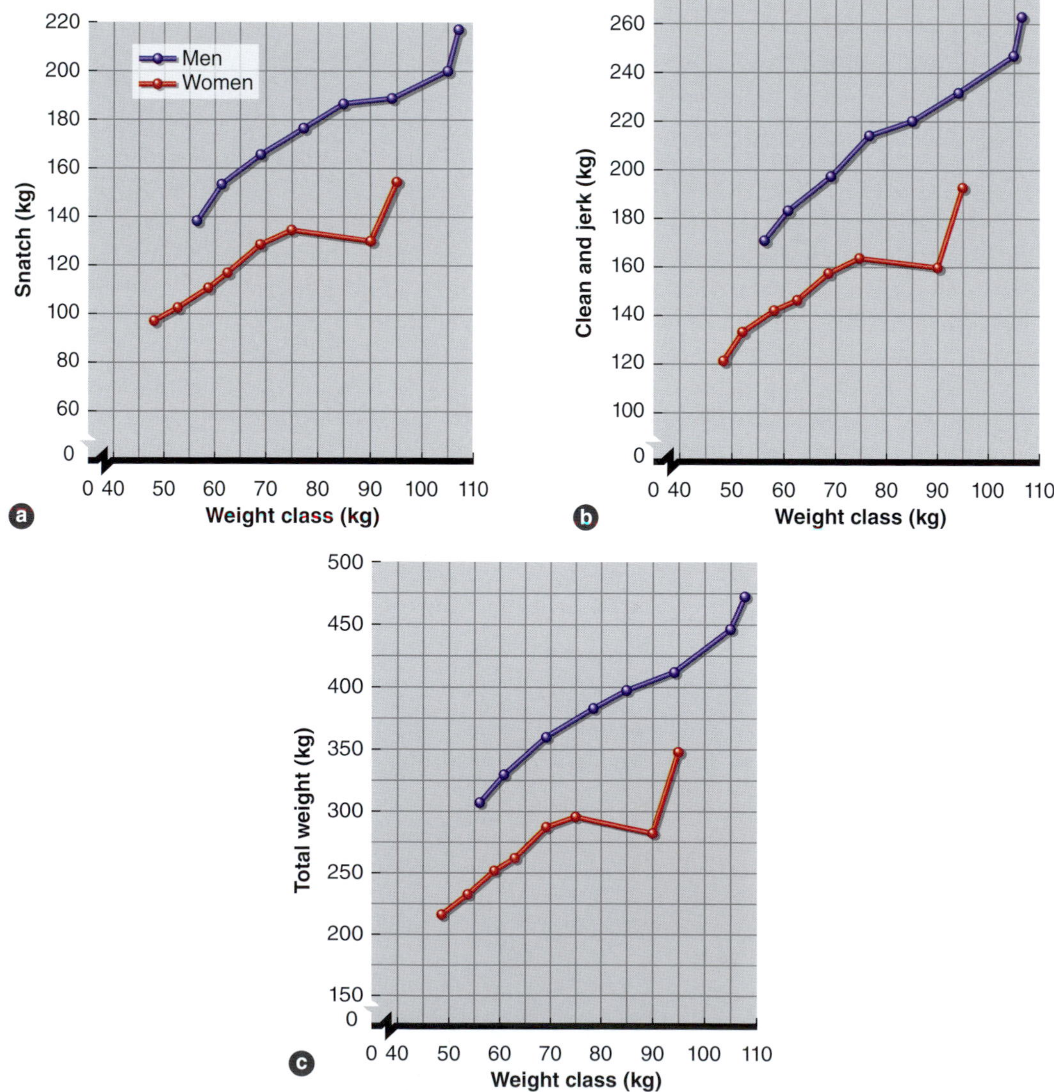

FIGURE 10.1 World records for *(a)* the snatch, *(b)* the clean and jerk, and *(c)* total weight for men and women through 2016.

controlled by a number of different neurons that can transmit either excitatory or inhibitory impulses (see chapter 3). Whether the muscle fibers contract or stay relaxed depends on the summation of the many impulses received by the given motor unit at any one time. The motor unit is activated and its muscle fibers contract only when the incoming excitatory impulses exceed the inhibitory impulses and the threshold is met or exceeded.

Strength gains may result from changes in the connections between motor neurons located in the spinal cord, allowing motor units to act more synchronously. This increased synchronicity means that a greater number of motor units will be firing at any one time, facilitating contraction and increasing the muscle's ability to generate force. There is good evidence to support increased motor unit synchronization with resistance training, but controversy still exists as to whether synchronization of motor unit activation produces a more forceful contraction. It is clear, however, that synchronization does improve the rate of force development and the capability to exert steady forces.[6]

Increased Rate Coding of Motor Units

The increase in neural drive of α-motor neurons could also increase the frequency of discharge, or rate coding, of their motor units. Recall from chapter 1 that as the frequency of stimulation of a given motor unit increases, the muscle eventually reaches a state of tetanus, producing the absolute peak force or tension of the muscle fiber or motor unit (see figure 1.14). There is limited evidence that rate coding is increased with resistance training. Rapid movement or ballistic-type training appears to be particularly effective in stimulating increases in rate coding.

Increased Neural Drive

Neural drive refers to the combination of motor unit recruitment and rate coding of the units. Neural drive starts in the central nervous system and is spread to muscle fibers through peripheral nerves. Electromyography (EMG) using surface electrodes over the muscle measures the total activity within the nerve and muscle and therefore is a good measure of neural drive.

An alternate explanation for neutrally mediated strength gains is simply that more motor units are recruited to perform the given task, independent of whether these motor units act in unison. Such improvement in recruitment patterns could result from an increase in neural drive to the α-motor neurons during maximal contraction. Trained muscles generate a given amount of submaximal force with less EMG activity, suggesting a more efficient motor unit recruitment pattern. This increase in neural drive could increase the frequency of discharge (rate coding) of the motor units or reduce inhibitory impulses, allowing more motor units to be activated or to be activated at a higher frequency. Additionally, maximal neural drive appears to increase with resistance training.

Autogenic Inhibition

Inhibitory mechanisms in the neuromuscular system, such as the Golgi tendon organs, might be necessary to prevent the muscles from exerting more force than the bones and connective tissues can tolerate. This control is referred to as **autogenic inhibition**. However, under extreme situations when larger forces are sometimes produced, significant damage can occur to these structures, suggesting that the protective inhibitory mechanisms can be overridden.

The function of Golgi tendon organs is discussed in chapter 3. When the tension on a muscle's tendons and internal connective tissue structures exceeds the threshold of the embedded Golgi tendon organs, motor neurons to that muscle are inhibited; that is, autogenic inhibition occurs. Both the reticular formation in the brain stem and the cerebral cortex function to initiate and propagate inhibitory impulses.

Resistance training can gradually reduce or counteract these inhibitory impulses, allowing the muscle to achieve a greater force production independent of increases in muscle mass. Thus, strength gains may be achieved by reduced neurological inhibition. This theory is attractive because it can at least partially explain superhuman feats of strength and strength gains in the absence of hypertrophy.

Other Neural Factors

In addition to increasing motor unit recruitment or decreasing neurological inhibition, other neural factors can contribute to strength gains with resistance training. One of these is referred to as coactivation of agonist and antagonist muscles (the agonist muscles are the primary movers, and the antagonist muscles act to impede the agonists). If we use forearm flexor concentric contraction as an example, the biceps is the primary agonist and the triceps is the antagonist. If the two were contracting with equal force development, no movement would occur. Thus, to maximize the force generated by an agonist, it is necessary to minimize the amount of coactivation. Reduction in coactivation could explain a portion of strength gains attributed to neural factors, but its contribution likely would be small.

Changes also have been noted in the morphology of the neuromuscular junction, with both increased and decreased activity levels that might be directly related to the muscle's force-producing capacity.

Muscle Hypertrophy

How does a muscle's size increase? Two types of hypertrophy can occur: transient and chronic. **Transient hypertrophy** is the increased muscle size that develops during and immediately following a single exercise bout. This results mainly from fluid accumulation (edema) in the interstitial and intracellular spaces of the muscle that comes from the blood plasma. Transient hypertrophy, as its name implies, lasts only for a short time. The fluid returns to the blood within hours after exercise.

Chronic hypertrophy refers to the increase in muscle size that occurs with long-term resistance training. This reflects actual structural changes in the muscle that can result from an increase in the size of existing individual muscle fibers (**fiber hypertrophy**), in the number of muscle fibers (**fiber hyperplasia**), or in both. Controversy surrounds the theories that attempt to explain the underlying cause of this phenomenon. Of importance, however, is the finding that the eccentric component of training is important in maximizing increases in muscle fiber cross-sectional area. A number of studies have shown greater hypertrophy and strength resulting solely from eccentric contraction training as compared to concentric contraction or combined eccentric and concentric contraction training. Further, higher-velocity eccentric training appears to result in greater hypertrophy and strength gains than slower-velocity training.[20] These greater increases appear to be related to disruptions in the sarcomere Z-lines. This disruption was originally labeled muscle damage but is now thought to represent fiber protein remodeling.[20] Thus, training with only concentric actions could limit muscle hypertrophy and increases in muscle strength.

Intensity and Hypertrophy

With traditional training methods, the prevailing opinion has been that an intensity of 60% to 85% of 1RM or higher is needed to achieve substantial increases in muscle size.

More recently, however, research has suggested that low-intensity exercise at <50% of 1RM can lead to gains in muscle size equal to those seen at high intensities, provided that the training is performed to volitional muscle fatigue.[18] This theory holds that fatiguing contractions at light loads lead to metabolic stimuli that result in maximal muscle fiber recruitment.

Is there a minimal intensity for resistance training that will lead to muscle hypertrophy, provided that resistance exercises are performed to volitional fatigue? It has been reported that from intensities as low as 30% and as high as 90% of 1RM, load played a minimal role in stimulating muscle protein synthesis, muscle hypertrophy, and strength gains in novice exercisers.[3] High-repetition (HR) and low-repetition (LR)—low and high load, respectively—training caused similar increases in skeletal muscle mass when resistance exercise was performed until volitional failure. Increases in lean body mass, as an indirect measure of muscle mass, and muscle fiber CSA, a direct measure of muscle area, occurred in both LR and HR groups with no differences between groups. There was a significant increase in 1RM strength for the leg press, knee extension, and shoulder press exercises, again with no differences between groups. These effects do not seem to depend on training status because similar results occurred in men with previous strength-training experience.[16]

The following section examines the two postulated mechanisms for increasing muscle size with resistance training: fiber hypertrophy and fiber hyperplasia.

Fiber Hypertrophy

Early research suggested that the number of muscle fibers in each of a person's muscles was established by birth or shortly thereafter and that this number remained fixed throughout life. If this were true, then whole-muscle hypertrophy could result only from

individual muscle fiber hypertrophy. This could be explained by

- more myofibrils,
- more actin and myosin filaments,
- more sarcoplasm,
- more connective tissue, or
- any combination of these.

As seen in the micrographs in figure 10.2, effective resistance training can significantly increase the cross-sectional area of muscle fibers. Such dramatic enlargement of muscle fibers does not occur, however, in all cases of muscle hypertrophy.

Muscle fiber hypertrophy is probably caused by increased numbers of myofibrils and actin and myosin filaments, which would provide more cross-bridges for force production during maximal contraction. The size of existing myofibrils does not appear to change. The increase in muscle cross-sectional area results from adding new sarcomeres in parallel to each other.

Individual muscle fiber hypertrophy from resistance training appears to result from a net increase in muscle protein synthesis. The muscle's protein content is in a continual state of flux. Protein is always being synthesized and degraded. But the rates of these processes vary with the demands placed on the body. During exercise, protein synthesis decreases, while protein degradation increases. After exercise, although protein degradation continues, protein synthesis increases three- to fivefold more, leading to a net synthesis of myofibrillar (myosin and actin) protein. A single bout of resistance exercise can elevate net protein synthesis for up to 24 h.

Hormones and Hypertrophy

The prevailing perspective in muscle physiology is that the hormone changes induced by resistance exercise facilitate increases in muscle mass that in turn increase muscular strength. The hormones that are typically associated with this response include the anabolic hormones testosterone, growth hormone (GH), and insulin-like growth factor 1 (IGF-1). The hormone testosterone has traditionally been thought to be at least partly responsible for these changes because one of its primary functions is promoting muscle growth. Testosterone is a steroidal hormone with major anabolic functions, and men experience a significantly greater increase in muscle growth starting at puberty, which is largely due to a 10-fold increase in testosterone production. Furthermore, it has been well established that massive doses of anabolic steroids coupled with resistance training markedly increase muscle mass and strength (see chapter 16).

While it is true that acute resistance training transiently increases the concentrations of these hormones, it has been shown experimentally that acute increases in these hormones are not required for increases in muscle mass or strength.[19] Researchers at McMaster University designed a series of studies to examine whether exercise-induced elevations in testosterone, GH, and IGF-1 were necessary for, or could enhance, muscle anabolism. They examined the elbow flexor muscles when exposed (1) to low hormone concentrations during a small muscle mass exercise consisting of isolated arm curls and (2) to high circulating hormone concentrations induced by an intense lower body exercise routine in addition to arm curls.[29,30,31] In the low hormone trials, myofibrillar protein synthesis was elevated after acute exercise bouts, and there were gains in strength and hypertrophy after training—even though testosterone, GH, and IGF-1 all remained near baseline concentrations. This implies that postexercise increases in these hormones are not necessary to stimulate muscle anabolism. Furthermore, when these hormones

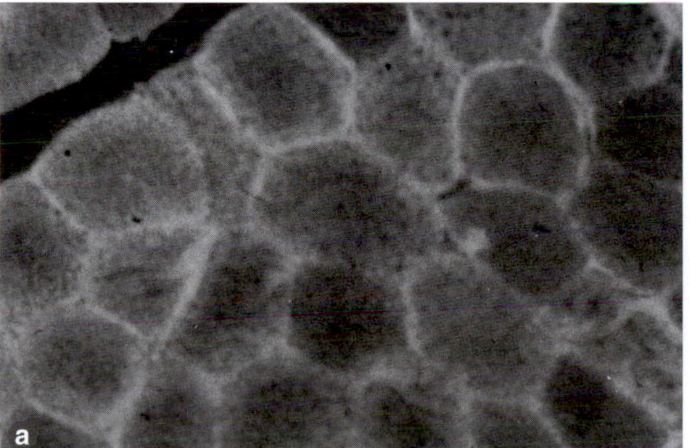

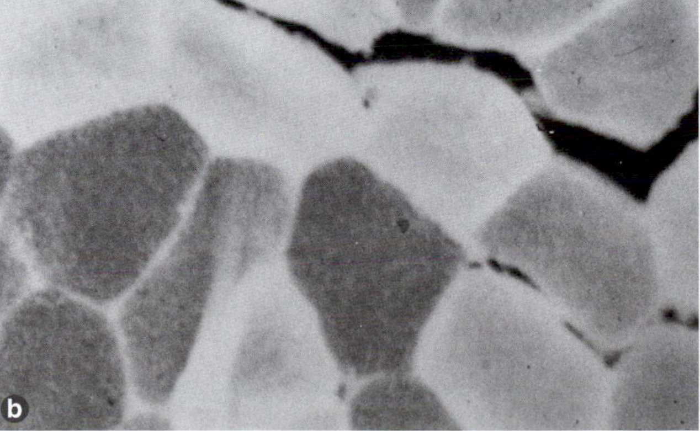

FIGURE 10.2 Microscopic views of muscle cross sections taken from the leg muscle of a man who had not trained during the previous 2 years, *(a)* before he resumed training and *(b)* after he completed 6 months of dynamic strength training. Note the significantly larger fibers (hypertrophy) after training.

were elevated postexercise, there was no further enhancement in myofibrillar protein synthesis or gain in strength and hypertrophy with training.

In addition to these studies, researchers have compared men's and women's responses to resistance training. Women have a 45-fold lower postexercise testosterone response compared to men even after their 20-fold lower baseline testosterone concentration is accounted for.[28] Despite having drastically lower postexercise increases in testosterone, the women were able to robustly increase rates of myofibrillar protein synthesis. Moreover, as in most exercise training studies, the men and women trained in this study had variable individual responses in terms of muscle hypertrophy. Despite these variable responses, there was no relation between each subject's exercise-induced increases in testosterone, GH, and IGF-1 and his or her muscle growth or strength gains.[32]

These new data provide strong new evidence that postexercise elevations in testosterone, GH, and IGF-1 are not required to increase muscle anabolism and strength. An alternate hypothesis is that the muscle hypertrophy and strength gains occurring with resistance training are mediated by changes in intrinsic intramuscular properties.

Fiber Hyperplasia

Research on animals suggests that hyperplasia, an increase in the total number of fibers within a muscle, may also be a factor in the hypertrophy of whole muscles. Studies on cats provide fairly clear evidence that fiber splitting occurs with extremely heavy weight training.[8] Cats were trained to move a heavy weight with a forepaw to get their food (figure 10.3). With the use of food as a powerful incentive, they learned to generate considerable force. With this intense strength training, selected muscle fibers appeared to actually split in half, and each half then increased to the size of the parent fiber.

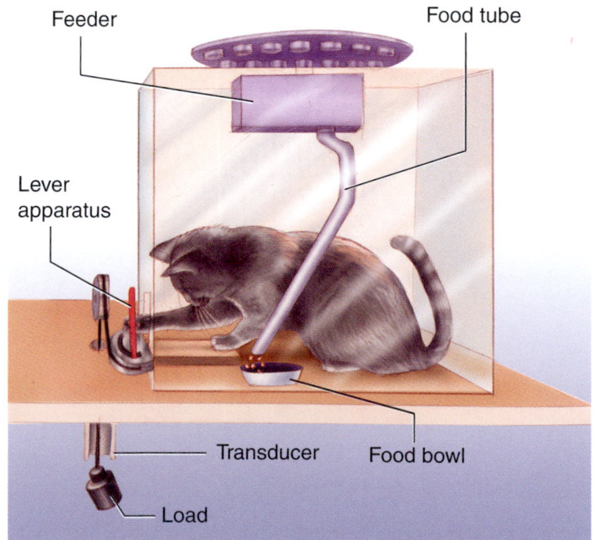

FIGURE 10.3 Heavy resistance training in cats.

Subsequent studies, however, demonstrated that hypertrophy of selected muscles in chickens, rats, and mice resulting from chronic exercise overload was attributable solely to hypertrophy of existing fibers, not hyperplasia. In these studies, each fiber in the whole muscle was actually counted. These direct fiber counts revealed no change in fiber number.

This finding led the scientists who performed the initial cat experiments to conduct an additional resistance training study with cats. This time they used actual fiber counts to determine if total muscle hypertrophy resulted from hyperplasia or fiber hypertrophy.[9] Following a resistance training program of 101 weeks, the cats were able to perform one-leg lifts of an average of 57% of their body weight, resulting in an 11% increase in muscle weight. Most important, the researchers found a 9% increase in the total number of muscle fibers, confirming that muscle fiber hyperplasia did occur.

The difference in results between the cat studies and those with rats and mice most likely is attributable to differences in the manner in which the animals were trained. The cats were trained with a pure form of resistance training: high resistance and low repetitions. The other animals were trained with more endurance-type activity: low resistance and high repetitions.

One additional animal model has been used to stimulate muscle hypertrophy associated with hyperplasia. Scientists have placed the anterior latissimus dorsi muscle of chickens in a state of chronic stretch by attaching weights to it, with the other wing serving as the normal control condition. In many of the studies that have used this model, the chronic stretch has resulted in substantial hypertrophy and hyperplasia.

Researchers are still uncertain about the roles played by hyperplasia and individual fiber hypertrophy in increasing *human* muscle size with resistance training. Most evidence indicates that individual fiber hypertrophy accounts for most whole-muscle hypertrophy. However, results from selected studies indicate that hyperplasia is possible in humans. It is possible that only very high intensity in resistance training can result in fiber hyperplasia, and even then, the percentage of the total muscle size increase due to this phenomenon is small, perhaps 5% to 10%. Whether strength training results in muscle fiber hyperplasia in humans remains unresolved.

In a cadaver study of seven previously healthy young men who had suffered sudden accidental death, the investigators compared cross sections of autopsied right and left tibialis anterior muscles (lower leg). Right-hand dominance is known to lead to greater hypertrophy of the left leg. In fact, the average cross-sectional area of the left muscle was 7.5% larger than that of the right. This was associated with a 10% greater number of fibers in the left muscle. There was no difference in mean fiber size.[21]

The differences between these studies might be explained by the nature of the training load or stimulus. Training at high intensities or high resistances is thought to cause greater fiber hypertrophy, particularly of the type II (fast-twitch) fibers, than training at lower intensities or resistances.

Only one longitudinal study demonstrated the possibility of hyperplasia in men who had previous recreational resistance training experience.[14] Following 12 weeks of intensified resistance training, the muscle fiber number in the biceps brachii of several of the 12 subjects appeared to increase significantly. It appears from this study that hyperplasia can occur in humans but possibly only in certain subjects or under certain training conditions.

From the preceding information, it appears that fiber hyperplasia can occur in animals and possibly in humans. How are these new cells formed? It is postulated that individual muscle fibers have the capacity to divide and split into two daughter cells, each of which can then develop into a functional muscle fiber. Importantly, satellite cells, which are the myogenic stem cells involved in skeletal muscle regeneration, are likely involved in the generation of new muscle fibers. These cells are typically activated by muscle stretching and injury; as we see later in this chapter, muscle injury results from intense training, particularly eccentric-action training. Muscle injury can lead to a cascade of responses in which satellite cells become activated and proliferate, migrate to the damaged region, and fuse to existing myofibers or combine and fuse to produce new myofibers.[13] This is illustrated in figure 10.4.

Satellite cells provide additional nuclei within muscle fibers. The added genetic machinery (DNA) is necessary to provide the increased muscle protein content and related materials to facilitate hypertrophy (and theoretically, hyperplasia).

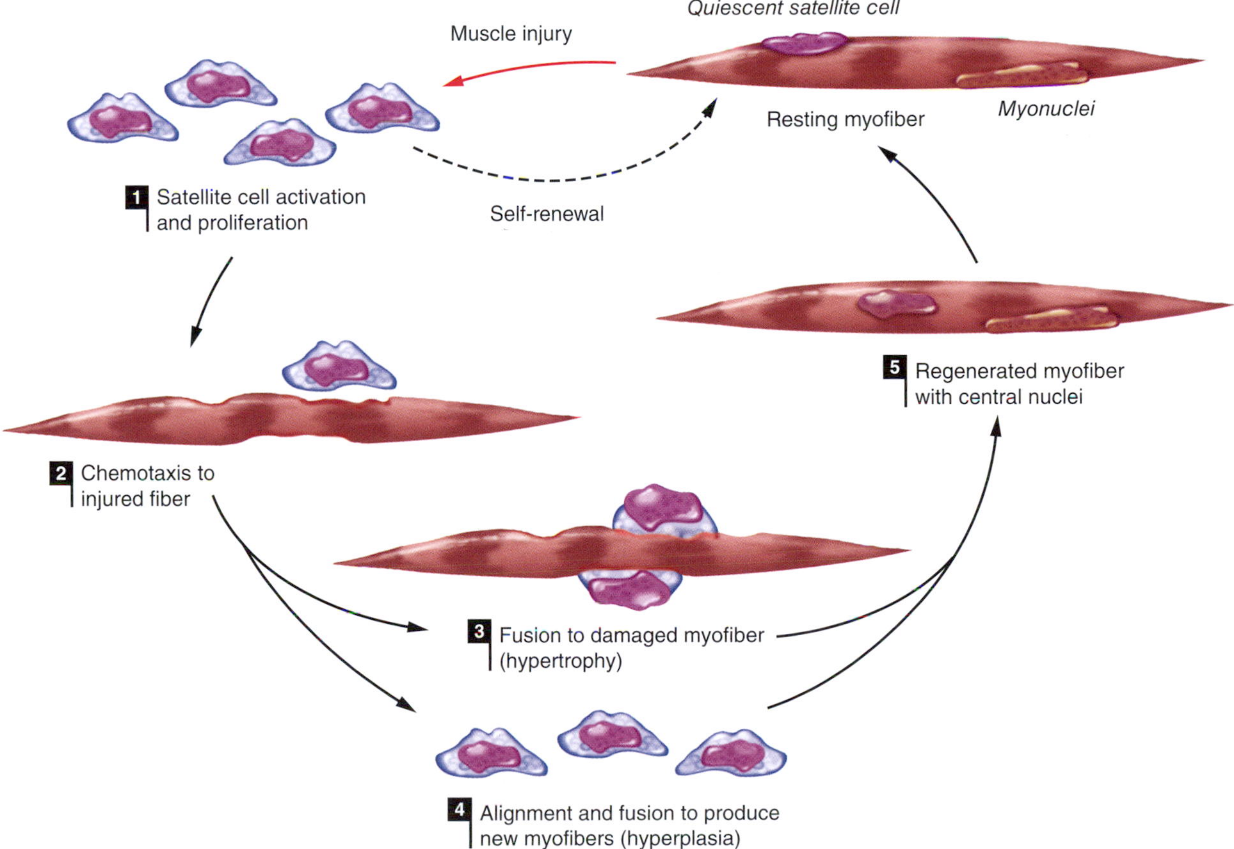

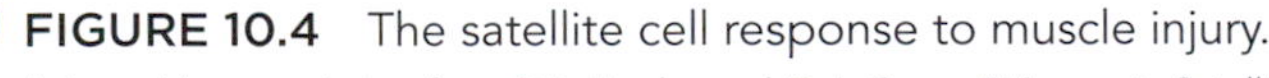
FIGURE 10.4 The satellite cell response to muscle injury.

Adapted by permission from T.J. Hawke and D.J. Garry, "Myogenic Satellite Cells: Physiology to Molecular Biology," *Journal of Applied Physiology* 91 (2001): 534-551.

Integration of Neural Activation and Fiber Hypertrophy

Research on resistance training adaptations indicates that early increases in voluntary strength, or maximal force production, are associated primarily with neural adaptations resulting in increased voluntary activation of muscle. This was clearly demonstrated in a study of both men and women who participated in an 8-week, high-intensity resistance training program, training twice per week.[22] Muscle biopsies were obtained at the beginning of the study and every 2 weeks during the training period. Strength, measured according to the 1RM, increased substantially over the 8 weeks of training, with the greatest gains coming after the second week. Muscle biopsies, however, revealed only a small, insignificant increase in muscle fiber cross-sectional area by the end of the 8 weeks of training. Thus, the strength gains were largely the result of increased neural activation.

Long-term increases in strength generally result from hypertrophy of the trained muscle. However, because it takes time to build protein through a decrease in protein degradation, an increase in protein synthesis, or both, early strength gains are typically due to changes in the pattern by which nerves activate the muscle fibers. Most research shows that neural factors contribute prominently to strength gains during the first 8 to 10 weeks of training. Hypertrophy contributes little during these initial weeks of training but progressively increases its contribution, becoming the major contributor after 10 weeks of training. However, not all studies concur with this pattern of strength development. One 6-month study of strength-trained athletes showed that neural activation explained most of the strength gains during the most intensive training months and that hypertrophy was not a major factor.[12]

Muscle Atrophy and Decreased Strength with Inactivity

When a normally active or highly trained person reduces his or her level of activity or ceases training altogether, changes occur in both muscle structure and function. This is illustrated by the results of two types of studies: studies in which entire limbs have been immobilized and studies in which highly trained people stop training—so-called detraining.

Immobilization

When a trained muscle suddenly becomes inactive through immobilization, major changes are initiated within that muscle in a matter of hours. During the first 6 h of immobilization, the rate of protein synthesis starts to decrease. This decrease likely initiates muscular atrophy, which is the wasting away or decrease in the size of muscle tissue. Atrophy results from lack of muscle use and the consequent loss of muscle protein that accompanies the inactivity. Strength decreases are most dramatic during the first week of immobilization, averaging 3% to 4% per day. This is associated not only with the atrophy but also with decreased neuromuscular activity of the immobilized muscle.

Immobilization appears to affect both type I and type II fibers. From various studies, researchers have observed disintegrated myofibrils, streaming Z-disks (discontinuity of Z-disks and fusion of the myofibrils), and mitochondrial damage. When muscle atrophies, the cross-sectional fiber area decreases. Several studies have shown the effect to be greater in type I fibers, including a decrease in the percentage of type I fibers, thereby increasing the relative percentage of type II fibers.

Muscles can and often do recover from immobilization when activity is resumed. The recovery period is substantially longer than the period of immobilization.

Cessation of Training

Similarly, significant muscle alterations can occur when people stop training. In one study, women resistance trained for 20 weeks and then stopped training for 30 to 32 weeks. The training program focused on the lower extremity, using a full squat, leg press, and leg extension. Finally, the participants retrained for 6 weeks.[23] Strength increases were dramatic, as seen in figure 10.5. Compare the women's strength after their initial training period (post-20) with their strength after detraining (pre-6). This represents the strength loss they experienced with cessation of training. During the two training periods, increases in strength were accompanied by increases in the cross-sectional[23] area of all fiber types and a decrease in the percentage of type IIx fibers. Detraining had relatively little effect on fiber cross-sectional area, although the type II fiber areas tended to decrease (figure 10.6).

To prevent losses in the strength gained through resistance training, basic maintenance programs must be established once the desired goals for strength development have been achieved. Maintenance programs are designed to provide sufficient stress to the muscles to maintain existing levels of strength while allowing a reduction in intensity, duration, or frequency of training.

In one study, men and women resistance trained with knee extensions for either 10 or 18 weeks and then spent an additional 12 weeks with either no training or

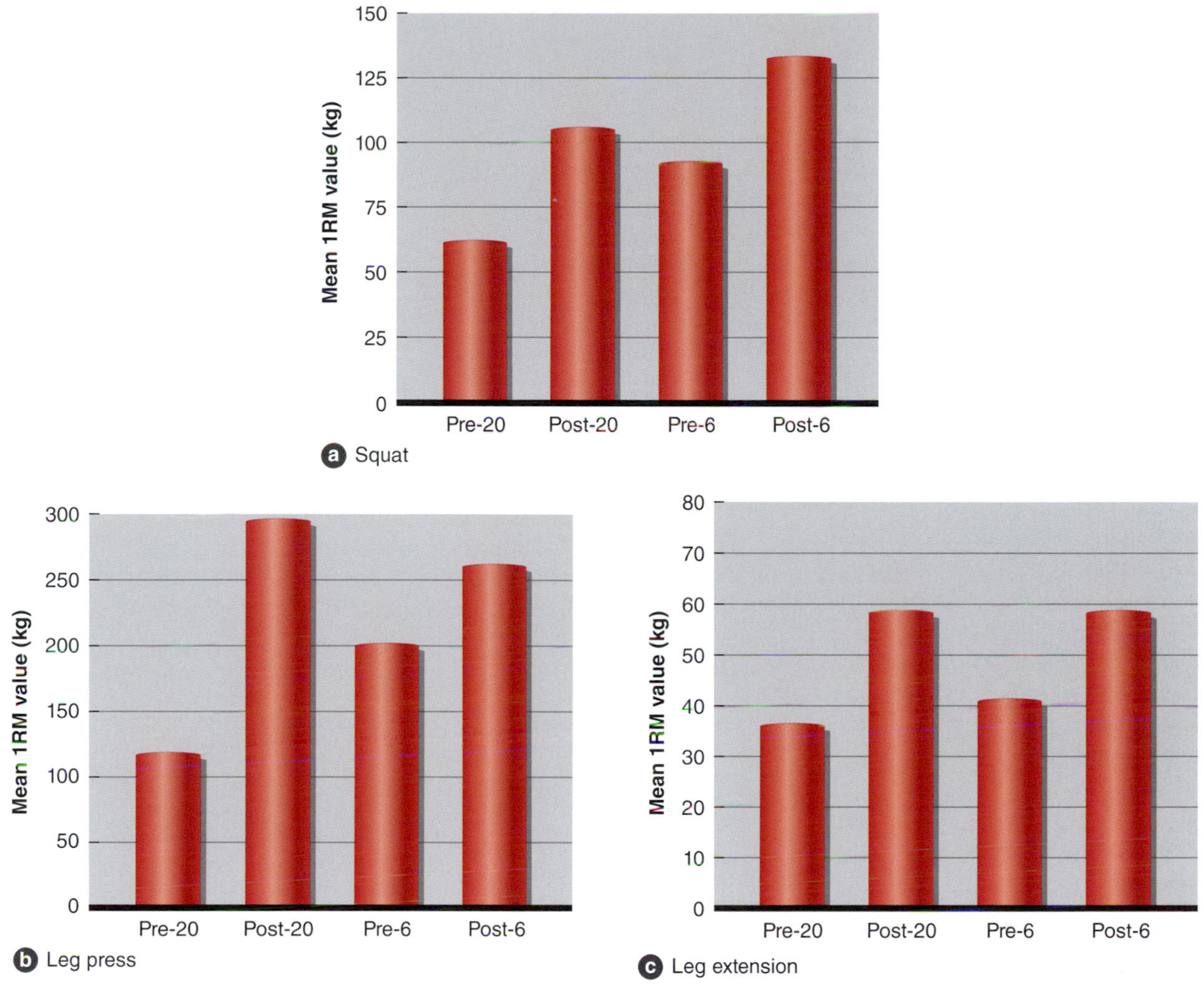

a Squat

b Leg press

c Leg extension

FIGURE 10.5 Changes in muscle strength for 3 different resistance exercises, *(a)* squat, *(b)* leg press, and *(c)* leg extension, with resistance training in women. Pre-20 values indicate strength before starting training, post-20 values indicate the changes following 20 weeks of training, pre-6 values indicate the changes following 30 to 32 weeks of detraining, and post-6 values indicate the changes following 6 weeks of retraining.

Adapted by permission from R.S. Staron et al., "Strength and Skeletal Muscle Adaptations in Heavy-Resistance-Trained Women After Detraining and Retraining," *Journal of Applied Physiology* 70 (1991): 631-640.

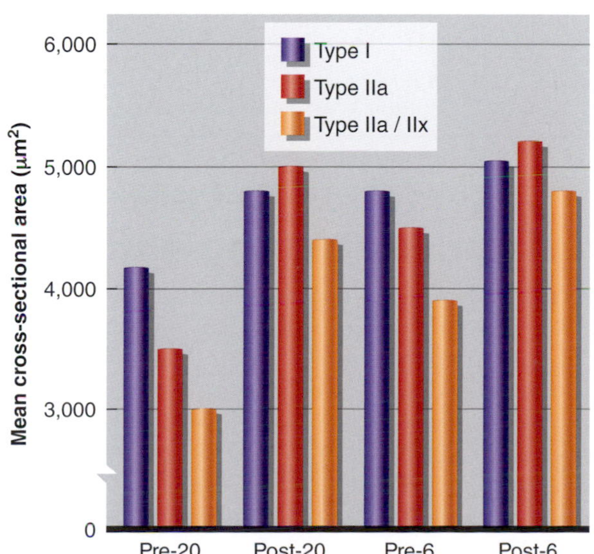

FIGURE 10.6 Changes in mean cross-sectional areas for the major fiber types with resistance training in women over periods of training (post-20), detraining (pre-6), and retraining (post-6). Type IIa/IIx is an intermediate fiber type. See figure 10.5 caption for more details.

reduced training.[10] Knee extension strength increased 21.4% during the training period. Subjects who then stopped training lost 68% of their strength gains during the weeks they didn't train. But subjects who reduced their training (from 3 days per week to 2, or from 2 to 1) did not lose strength. Thus, it appears that strength can be maintained for at least up to 12 weeks with reduced training frequency.

Fiber Type Alterations

Can muscle fibers change from one type to another through resistance training? The earliest research concluded that neither speed (anaerobic) nor endurance (aerobic) training could alter the basic fiber type, specifically from type I to type II or from type II to type I. These early studies did show, however, that fibers began to take on certain characteristics of the opposite fiber type if the training was of the opposite kind (e.g., type II fibers might become more oxidative with aerobic training).

Research using animal models has shown that fiber type conversion is indeed possible under conditions of cross-innervation, in which a type II motor unit is experimentally innervated by a type I motor neuron or a type I motor unit is experimentally innervated by a type II motor neuron. Also, chronic, low-frequency nerve stimulation transforms type II motor units into type I motor units within a matter of weeks. Muscle fiber types in rats have changed in response to 15 weeks of high-intensity treadmill training, resulting in an increase in type I and type IIa fibers and a decrease in type IIx fibers.[11] The transition of fibers from type IIx to type IIa and from type IIa to type I was confirmed by several different histochemical techniques.

Staron and coworkers found evidence of fiber type transformation in women as a result of heavy resistance training.[24] Substantial increases in static strength and in the cross-sectional area of all fiber types were noted following a 20-week heavy resistance training program for the lower extremity. The mean percentage of type IIx fibers decreased significantly, but the mean percentage of type IIa fibers increased. The transition of type IIx fibers to type IIa fibers with resistance training has been consistently reported in a number of subsequent studies. Further, other studies demonstrate that a combination of high-intensity resistance training and short-interval speed work can lead to a conversion of type I to type IIa fibers.

In Review

- Neural adaptations always accompany the strength gains that result from resistance training, but hypertrophy may or may not take place.
- Neural mechanisms leading to strength gains can include an increase in frequency of stimulation, or rate coding; recruitment of more motor units; more synchronous recruitment of motor units; and decreases in autogenic inhibition from the Golgi tendon organs.
- Early gains in strength appear to result more from changes in neural factors, but later long-term gains are largely the result of muscle hypertrophy.
- Transient muscle hypertrophy is the temporary enlargement of muscle resulting from edema immediately after an exercise bout.
- Chronic muscle hypertrophy occurs from repeated resistance training and reflects actual structural changes in the muscle.
- Most muscle hypertrophy results from an increase in the size of individual muscle fibers (fiber hypertrophy).
- Fiber hypertrophy increases the numbers of myofibrils and actin and myosin filaments, which provides more cross-bridges for force production.
- Muscle fiber hyperplasia has been clearly shown to occur in animal models with the use of resistance training to induce muscle hypertrophy. Only a few studies suggest evidence of hyperplasia in humans.
- Muscles atrophy (decrease in size and strength) when they become inactive, as with injury, immobilization, or cessation of training.
- Atrophy begins very quickly if training is stopped, but training can be reduced, as in a maintenance program, without resulting in atrophy or loss of strength.
- With resistance training there is a transition of type IIx to type IIa fibers.
- Evidence indicates that one fiber type can actually be converted to the other type (e.g., type I to type II, or vice versa) as a result of cross-innervation or chronic stimulation, and possibly with training.

Interaction Between Resistance Training and Diet

Muscle hypertrophy in response to resistance training can be either limited or enhanced by nutrition. As mentioned previously, a net positive protein balance (more synthesis than breakdown) is the necessary condition under which muscle hypertrophy occurs. Without adequate protein in the diet, protein synthesis is compromised and muscles cannot increase their protein content and hypertrophy. Ingesting protein within a few hours after a bout of resistance exercise increases the rate of protein synthesis and thus adds to the net positive protein balance. Increased protein intake over the subsequent 24 h period will continue to support muscle anabolism. Therefore, nutrition and exercise are powerful stimulators of skeletal muscle protein synthesis.[5]

VIDEO 10.1 Presents Luc von Loon on the role of protein in adaptations to resistance training.

Recommendations for Protein Intake

An international group of researchers recently performed a systematic analysis of 49 published studies (1,863 subjects) to determine if dietary protein supplementation enhances the gains in muscle mass and strength with resistance training.[15] They surveyed randomized controlled trials in which subjects performed resistance training for at least 6 weeks and took various amounts of dietary protein supplementation. Their analysis of this large sample showed that dietary protein supplementation significantly enhanced changes in strength as measured by 1-repetition maximum tests and muscle size (fiber cross-sectional area and whole-muscle cross-sectional area). However, protein intakes greater than ~1.6 g/kg of body weight per day did not further contribute to these gains. So, although the current U.S. Dietary Reference Intake (DRI) for protein for people over 18 years of age, regardless of physical activity status, is 0.8 g/kg per day, athletes engaged in resistance training may require protein intakes in the diet as high as 1.7 g/kg per day. Although ingestion of relatively small amounts of protein (5-10 g) can stimulate muscle protein synthesis in young men and women, to make muscles larger, one should consume larger amounts of protein, on the order of 20 to 25 g, immediately after resistance exercise.[1]

What type of protein should be ingested and how much? The best forms of protein for muscle hypertrophy are easily and rapidly digested and rich in essential amino acids, especially leucine. Whey protein found in milk is one source that meets both of these goals. In practice, after resistance training, athletes should consume a small amount of high-quality protein along with adequate carbohydrate in order to stimulate muscle proteins and replenish muscle glycogen stores after exercise. This can be accomplished with either a recovery beverage or foods such as milk or yogurt, a small sandwich, or a protein-rich energy bar. Adding carbohydrate to postexercise protein ingestion does not markedly affect muscle protein balance but does have other benefits, including aiding in the resynthesis of muscle glycogen.

Is there an optimal timing of protein ingestion when an individual is trying to optimize the hypertrophic response to successive exercise sessions? A single bout of exercise stimulates muscle protein synthesis rates for several hours, and intake of protein further enhances postexercise muscle protein synthesis. The protein synthesis–stimulating effect of a single dose of amino acids is transient and lasts only 1 to 2 h. Ingesting repeated small doses of protein during recovery from resistance training may be more effective in increasing muscle hypertrophy than eating just one large meal. However, elevated muscle protein synthesis rates are not totally limited to the few hours of acute postexercise recovery. The so-called window of opportunity lasts from just before the start of resistance exercise to several hours postexercise. Providing protein before or during exercise can enhance muscle protein synthesis during exercise and is a good strategy for prolonged or repeated workouts.

Mechanism of Protein Synthesis with Resistance Training and Protein Intake

The rate of protein synthesis within the myofibrils is controlled primarily by an enzyme, or kinase, known as **mTOR** (mechanistic target of rapamycin). mTOR integrates the input from upstream pathways, including insulin and growth factors and amino acids (figure 10.7), and controls transcription of messenger RNA (mRNA). If mTOR is blocked experimentally, resistance exercise does not result in muscle hypertrophy. The primary stimulus for protein synthesis is the mechanical stretch applied to the muscle, which activates mTOR. mTOR senses cellular nutrient and oxygen levels, so it is also activated by the proper timing of protein intake, specifically proteins rich in leucine. So, delivering leucine to muscles during the window of opportunity will increase mTOR more than acute exercise alone and lead to enhanced protein synthesis and muscle hypertrophy.

The increased protein synthesis with enhanced dietary amino acid availability occurs not only because of the greater net supply of amino acids but also because of changes in hormonal concentrations that create a more favorable anabolic environment. Insulin serves as a strong anabolic stimulus for skeletal muscle hypertrophy, as shown in figure 10.7. In the presence of adequate substrate, insulin (which rises after a meal) is capable of stimulating skeletal muscle protein synthesis and hypertrophy in young muscles.

Insulin stimulates protein synthesis from available amino acids by promoting a more efficient conversion of genetic codes carried by mRNA into proteins, a process known as translation. This process is accomplished by cellular organelles known as ribosomes, so it stands to reason that increasing ribosome content in the muscle fibers (that is, increasing the translational capacity) will also result in more protein being synthesized. Ribosome biogenesis, the creation of new ribosomes, appears to be an important mechanism regulating muscle size in response to resistance exercise. In fact, when the synthesis of new ribosomes is blocked biochemically, muscles fail to undergo hypertrophy. Notably, recent studies show that mTOR is involved in the synthesis of ribosomes in the cell nucleus in addition to its role in regulating translation in the cytoplasm (figure 10.7),

RESEARCH PERSPECTIVE 10.2

Lifting Before Bedtime for Enhanced Muscle Protein Synthesis

Ingesting protein after a bout of resistance exercise stimulates muscle protein synthesis and inhibits muscle breakdown, resulting in an overall increase in muscle protein content during the acute phase of recovery. Because of this phenomenon, postexercise protein ingestion is a widely used strategy for increasing muscle hypertrophy and speeding recovery following resistance exercise training.

Various factors can affect postexercise protein synthesis, including the amount, type, and timing of postexercise protein ingestion. Studies have shown that protein ingestion before bed increases overnight amino acid availability and stimulates muscle protein synthesis during overnight sleep. Presleep protein supplementation increases strength and hypertrophy gains over a prolonged resistance exercise training program, and protein ingestion before sleep may be a practical way to support muscle mass and maximize hypertrophy during training.

A 2016 study conducted in the Netherlands[25] examined whether an acute bout of resistance exercise performed in the evening could further increase the muscle protein synthesis response to presleep protein ingestion. Researchers hypothesized that by combining the powerful stimulus for protein synthesis immediately following exercise with presleep protein ingestion, they would see an even larger increase in new protein synthesis overnight. To study this, the researchers recruited 24 healthy young men and divided them into two groups: presleep protein ingestion plus evening exercise (PRO+EX) or presleep protein alone (PRO). After a standardized meal, the PRO+EX subjects performed 60 min of lower-body resistance exercise while the PRO subjects rested. After the exercise or rest session, all of the subjects consumed the same drink containing 20 g of protein. Muscle biopsy samples were taken, and a labeled amino acid isotope was continually infused for the measurement of protein turnover. Just before bed, each subject ingested another 30 g of labeled protein. Overnight, blood samples were collected at 30, 60, 90, 150, 210, 330, and 450 min while the subject slept, and a second muscle biopsy was obtained the following morning.

The results showed that overnight protein synthesis of the myofibrils was ~35% greater in the PRO+EX subjects compared to the PRO subjects. In addition, much more of the labeled dietary protein-derived amino acids were incorporated into the new myofibrils of the PRO+EX overnight. These findings led the study team to conclude that resistance exercise performed in the evening increases the overnight muscle protein synthesis response to presleep protein ingestion. Therefore, combining protein ingestion before sleep with resistance exercise may be a useful strategy when trying to maximize skeletal muscle reconditioning overnight.

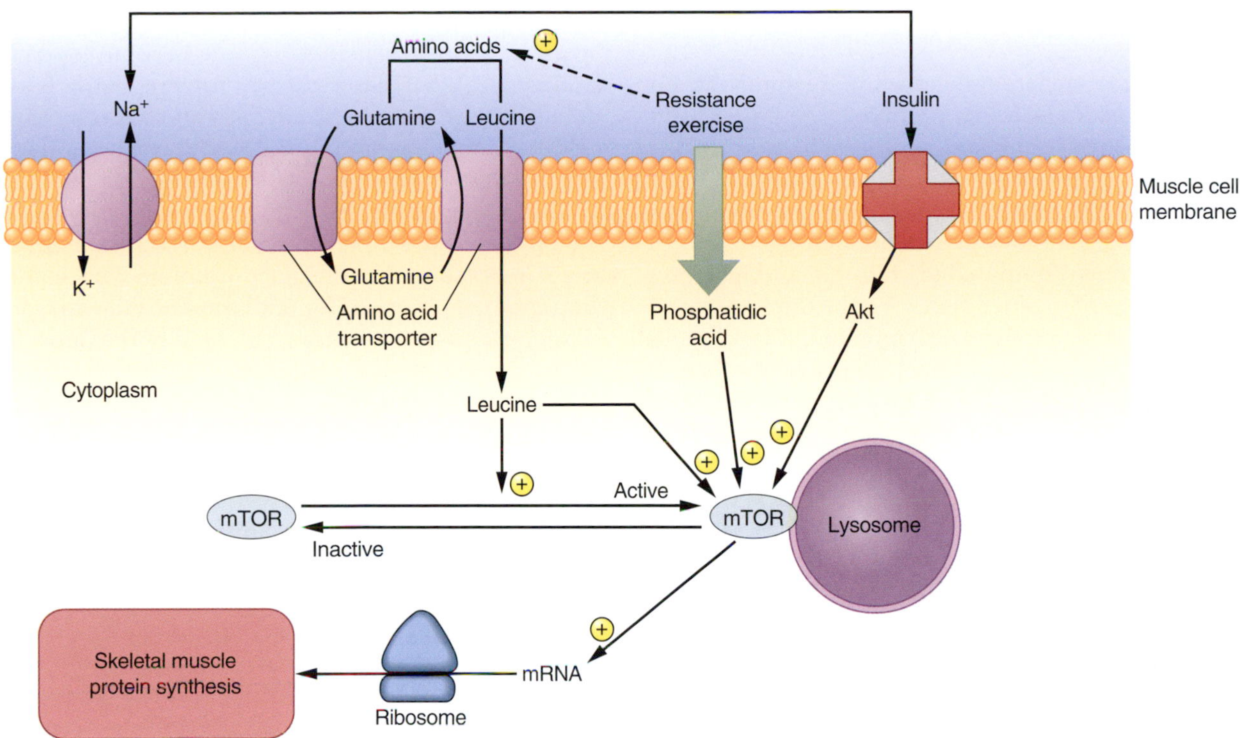

FIGURE 10.7 Schematic representation of the separate and combined roles of resistance training, insulin, and amino acid intake on skeletal muscle protein synthesis.

Redrawn from Dickinson et al. (2013).

which puts this kinase at center stage in the muscle growth process.

In Review

- Resistance exercise and protein intake are powerful stimulators of skeletal muscle protein synthesis.
- Resistance-trained athletes should consume an adequate amount of high-quality protein (as high as 1.7 g per kg of body weight per day) along with carbohydrate in order to stimulate muscle protein synthesis and also replenish muscle glycogen stores after exercise.
- The rate of protein synthesis within the myofibrils is controlled primarily by an enzyme known as mTOR. The primary stimulus for protein synthesis is the mechanical stretch applied to the muscle, which activates mTOR through a signaling pathway involving IGF-1.
- Ribosome biogenesis, the creation of new ribosomes, appears to be another important mechanism regulating muscle hypertrophy in response to resistance exercise.

Resistance Training for Special Populations

Until the 1970s, resistance training was widely regarded as appropriate only for young, healthy male athletes. This narrow concept led many people to overlook the benefits of resistance training when planning their own activities. In recent years, considerable interest has focused on training for women, children, and people who are elderly. As mentioned earlier in this chapter, the widespread use of resistance training by women, either for sport or for health-related benefits, is rather recent. Substantial knowledge has developed since the early 1970s revealing that women and men have the same ability to develop strength but that, on average, women may not be able to achieve peak values as high as those attained by men. This difference in strength is attributable primarily to muscle size differences related to sex differences in anabolic hormones. Resistance training techniques developed for and applied to men's training seem equally appropriate for women's training. Issues of strength and resistance training for women are covered in more detail in chapter 19. In this section, we first consider age, and then we summarize the importance of this form of training for all athletes, regardless of their sex, age, or sport.

Relative gains in strength appear to be similar when we compare women to men, children to adults, and elderly people to young and middle-aged adults when these gains are expressed as a percentage of their initial strength. However, the increase in the absolute weight lifted is generally greater in men compared to women, in adults compared to children, and in young adults compared to older adults. For example, after 20 weeks of resistance training, assume that a 12-year-old boy and a 25-year-old man each improves his bench press strength by 50%. If the man's initial bench press strength (1-repetition maximum, 1RM) were 50 kg (110 lb), he would have improved by 25 kg (55 lb) to a new 1RM of 75 kg (165 lb). If the boy's initial 1RM were 25 kg, he would have improved by 12.5 kg (28 lb) to a new 1RM of 37.5 kg (83 lb).

Resistance Exercise for Older Adults

Interest in resistance training for elderly people has increased since the 1980s. A substantial loss of fat-free body mass accompanies aging, a condition known as **sarcopenia**. This loss reflects mainly the loss of muscle mass, largely because most people become less active as they age. When a muscle isn't used regularly, it loses function, with predictable atrophy and loss of strength.

Can resistance training in elderly people prevent or reverse this process, and does nutrition play the same role in older people as it does in young individuals? Elderly exercisers can indeed gain strength and muscle mass in response to resistance training. This fact has important implications for both their health and quality of life (discussed in chapter 18). One of the important benefits is that, with maintained or improved strength, they are less likely to fall. This is significant because falls are a major source of injury and debilitation for elderly people and often lead to death.

Under basal conditions, fractional protein synthesis and breakdown are not much different between young and aged people. Rather, sarcopenia results from aged muscle's inability to respond appropriately to anabolic stimuli. An acute bout of resistance exercise does not appear to elicit the same hypertrophic response in skeletal muscles of older individuals. This anabolic resistance has been attributed to the inability of resistance exercise to appropriately increase mTOR signaling in the elderly.[5] Resistance training is certainly capable of increasing strength and muscle mass in elderly persons; the response is simply blunted. With resistance training, large increases in strength are often accompanied by only small increases in myofibrillar protein and muscle size. In this age group, strength increases depend significantly on neural adaptations.

The impact of protein intake on muscle hypertrophy in the elderly is likewise blunted. Whereas as little as 5 g of protein in combination with resistance training stimulates skeletal muscle protein synthesis in young individuals, larger amounts must be ingested to cause the same effect in the elderly. This may be attributable to changes in the sensitivity of aged muscle to branched-chain amino acids. Studies indicate that ingestion of 25 to 30 g of high-quality protein or greater than 2 g of leucine is necessary to stimulate aged muscle protein synthesis to a similar degree as in young muscle. Aging is also associated with a resistance of skeletal muscle to the influence of insulin on protein synthesis, which could be a key factor in the etiology of sarcopenia.

In human aging, there is significant variation in regional body composition, and age-associated dysfunction and disability have been associated with sex and race. For example, women have greater muscle fat infiltration and higher subcutaneous fat and lower limb mass compared to men, and African Americans exhibit greater limb muscle mass that is accompanied by greater subcutaneous fat and inter- and intramuscular fat than Caucasians of the same sex.

Are there sex- and race-based differences in the physiological adaptations to resistance training in middle-aged and older adults? A study conducted at the University of Maryland used a unique one-leg strength training protocol to examine the influence of sex and race on thigh muscle volume, subcutaneous fat, and intermuscular fat changes in response to resistance training.[27] Subjects, which included Caucasian and African American men and women aged 50 to 85 years, completed 10 weeks of unilateral knee extension training.

All groups had an increase in thigh muscle volume of the exercised leg. While the men experienced a greater absolute increase in muscle size, when the data were represented as a percentage increase, changes in muscle volume were similar between men and women. Nor were there any sex differences with respect to changes in subcutaneous fat or intermuscular fat. There were no differences in muscle or fat adaptations to training between Caucasian and African American exercisers. The results of this study indicate that strength training does not alter subcutaneous or intermuscular fat, regardless of sex or race. Given that there do appear to be racial differences in the incidence of metabolic disorders and functional measures of muscle quality among the elderly, other unexplored factors likely explain the racial disparity.

Resistance Training for Children

The wisdom of resistance training for children and adolescents has long been debated. The potential for injury, particularly growth plate injuries from the use of free weights, has caused much concern. Many people once believed that children would not benefit from resistance training, based on the assumption that the hormonal changes associated with puberty are necessary for gaining muscle strength and mass. We now know that children and adolescents can train safely with minimal risk of injury if appropriate safeguards are implemented. Furthermore, they can indeed gain both muscular strength and muscle mass.

Resistance training programs for children should be prescribed in much the same way as for adults, but with a special emphasis on teaching proper lifting technique. Specific guidelines have been established by a number of professional organizations, including the American Orthopaedic Society for Sports Medicine, the American Academy of Pediatrics, the American College of Sports Medicine, the National Athletic Trainers' Association, the National Strength and Conditioning Association, the President's Council on Physical Fitness and Sports, and the U.S. Olympic Committee. Basic guidelines for the progression of resistance exercise in children are presented in table 10.1.

Resistance Training for Athletes

Gaining strength, power, or muscular endurance simply for the sake of being stronger, being more powerful, or having greater muscular endurance is of relatively little importance to athletes unless it also improves their athletic performance. Resistance training by field-event athletes and competitive weightlifters makes intuitive sense. The need for resistance training by the gymnast,

TABLE 10.1 **Basic Guidelines for Resistance Exercise Progression in Children**

Age	Considerations
7 years or younger	Introduce child to basic exercises using little or no weight; develop the concept of a training session; teach exercise technique; progress from body weight calisthenics, partner exercises, and lightly resisted exercises; keep volume low.
8-10 years	Gradually increase the number of exercises; practice exercise technique in all lifts; start gradual progressive loading of exercises; keep exercises simple; gradually increase training volume; carefully monitor tolerance of the exercise stress.
11-13 years	Teach all basic exercise techniques; continue progressive loading of each exercise; emphasize exercise techniques; introduce more advanced exercises with little or no resistance. Progress to more advanced youth programs in resistance exercise; add sport-specific components; emphasize exercise techniques; increase volume.
14-15 years	Progress to more advanced youth programs in resistance exercise; add sport-specific components; emphasize exercise techniques; increase volume.
16 years or older	Move child to entry-level adult programs after all background knowledge has been mastered and a basic level of training experience has been gained.

Note. If a child of any age begins a program with no previous experience, start the child at the level for the previous age category and move him or her to more advanced levels as exercise toleration, skill, amount of training time, and understanding permit.

Reprinted by permission from W.J. Kraemer and S.J. Fleck, *Strength Training for Young Athletes*, 2nd ed. (Champaign, IL: Human Kinetics, 2005), 5.

RESEARCH PERSPECTIVE 10.3

Resistance Training Can Improve Health Without Changing BMI

Childhood obesity has increased dramatically over the last decade, and obese adolescents are significantly more likely to have metabolic and cardiovascular disease. Regular physical activity improves metabolic and cardiovascular health, and increasing aerobic activity is often recommended to reduce the risk of disease in obese individuals. It is well documented that aerobic training improves blood flow responses, reduces resting blood pressure, reduces inflammation, increases insulin sensitivity, and improves body composition in overweight and obese individuals. However, adherence to aerobic training programs is very low in this unfit population. Resistance training could be an alternative strategy to improve health while increasing adherence in obese individuals. Most studies to date that have investigated resistance training to improve cardiovascular and metabolic outcomes in obesity have combined resistance training with aerobic training; because of this, little is known about the isolated effects of resistance exercise training on cardiovascular and metabolic health in obesity.

In 2015, a group of exercise physiologists in Brazil conducted a study to investigate the effects of a supervised resistance exercise training program on measures of metabolic and cardiovascular health in obese adolescents.[4] Twenty-four obese adolescents (mean age of 14 years) performed 12 weeks of supervised resistance exercise training that included all major muscle groups. Body mass index, body composition, blood pressure, endothelial function (a measure of blood vessel health), inflammation, and insulin resistance were measured before and after training. Despite that fact that the body mass index (BMI) did not change, study participants had significantly lower body fat and waist circumferences after the 12 weeks of training. Blood pressure, endothelial function, inflammation, insulin resistance, and performance on a submaximal exercise test all improved as well.

Overall, the findings from this study led to the conclusion that resistance training improves cardiovascular and metabolic health in obese adolescents, even if BMI does not change. Although there were no changes in body mass, endothelial function, blood pressure, and metabolic profiles all improved. Resistance training programs may be an effective alternative to aerobic training to reduce the risk of cardiovascular and metabolic disease and may increase adherence to exercise programs in obese adolescents.

distance runner, baseball player, high jumper, or ballet dancer is less obvious.

We do not have extensive research to document the specific benefits of resistance training for every sport or for every event within a sport. But clearly, each has basic strength, power, and muscular endurance requirements that must be met to achieve optimal performance. Training beyond these requirements, however, may be unnecessary.

Training is costly in terms of time, and athletes can't afford to waste time on activities that won't result in better athletic performances. Thus, some performance measurement is imperative to evaluate any resistance training program's efficacy. To resistance train solely to become stronger, with no associated improvement in performance, is of questionable value. However, it should also be recognized that resistance training to improve muscular endurance can reduce the risk of injury for most sports because fatigued individuals are at an increased risk of injury.

In Review

- Resistance training can benefit almost everyone, regardless of the person's sex, age, or athletic involvement.
- In elderly people, resistance training can slow or reverse the age-associated loss of muscle mass known as sarcopenia.
- Aged skeletal muscle retains the ability to respond to exercise, insulin, and enhanced protein intake to substantially increase net protein synthesis. However, older muscles have a blunted response compared to young muscles.
- Most athletes in most sports can benefit from resistance training if an appropriate program is designed for them. But to ensure that the program is working, performance should be assessed periodically and the training regimen adjusted as needed.

IN CLOSING

In this chapter, we have considered the role of resistance training in increasing muscular strength and improving performance. We have examined how muscle strength is gained through both muscular and neural adaptations, the importance of dietary protein intake in muscle hypertrophy, how resistance training can slow the impact of sarcopenia in the elderly, and how resistance training is of importance for both health and sport, regardless of age or sex. In the next chapter, we turn our attention away from resistance training and begin exploring how the body adapts to aerobic and anaerobic training.

KEY TERMS

atrophy
autogenic inhibition
chronic hypertrophy
fiber hyperplasia
fiber hypertrophy
mTOR
resistance training
sarcopenia
transient hypertrophy

STUDY QUESTIONS

1. What is a reasonable expectation for percentage strength gains following a 6-month resistance training program? How do these percentage gains differ by age, sex, and previous resistance training experience?
2. What is the suggested minimal intensity for resistance training that will lead to muscle hypertrophy when the exercises are performed to volitional fatigue?
3. Discuss the different theories that have attempted to explain how muscles gain strength with training.
4. What is autogenic inhibition? How might it be important to resistance training?
5. Differentiate between transient and chronic muscle hypertrophy.
6. What is fiber hyperplasia? How might it occur? How might it be related to gains in size and muscle strength with resistance training?
7. What is the physiological basis for hypertrophy?
8. Describe the respective effects of intensity and circulating hormones on muscle adaptation to fatiguing resistance training.
9. What is the physiological response to muscle immobilization?
10. To support protein synthesis during resistance training, what type of protein should be ingested and how much?
11. Is there an optimal timing of protein ingestion when an individual is trying to optimize the hypertrophic response to successive exercise sessions?
12. What is the role of mTOR in protein synthesis? How are ribosomes involved in the process?
13. How do the basic guidelines for prescribing resistance exercise for children differ from those for adults?

STUDY GUIDE ACTIVITIES

In addition to the activities listed in the chapter opening outline, two other activities are available in the web study guide, located at

www.HumanKinetics.com/PhysiologyOfSportAndExercise

The **KEY TERMS** activity reviews important terms, and the end-of-chapter **QUIZ** tests your understanding of the material covered in the chapter.

1
KB
SCOTT

Adaptations to Aerobic and Anaerobic Training

11

In this chapter and in the web study guide

VIDEO 11.1 presents Ben Levine on the significance of $\dot{V}O_2max$ for sport performance.

AUDIO FOR FIGURE 11.7 describes how the variables of maximal cardiac output and red blood cell volume impact $\dot{V}O_{2max}$ values in individuals.

AUDIO FOR FIGURE 11.8 describes the increases in total blood volume and plasma volume with endurance training.

ACTIVITY 11.1 *Adaptations* reviews the cardiovascular, respiratory, and metabolic responses to training.

AUDIO FOR FIGURE 11.15 describes a twin study on the effect of heredity on $\dot{V}O_{2max}$.

ACTIVITY 11.2 *Individual Response* considers the factors affecting individual response to training.

ACTIVITY 11.3 *Aerobic Training* explores adaptations in response to aerobic training by applying them to real-life situations.

ACTIVITY 11.4 *Anaerobic Training* explores adaptations in response to anaerobic training by applying them to real-life situations.

ACTIVITY 11.5 *Putting It All Together* reviews all concepts related to adaptations to aerobic and anaerobic training.

On October 8, 2016, the Ironman World Championships were held in Kona, on the Big Island of Hawaii, for the 40th time. Organized by the World Triathlon Corporation, professional triathletes qualified for the race based on a point system and a total of $650,000 in prize money was awarded. German athlete Jan Frodeno completed this grueling event in 8:06:30 to win his second World Championship in as many years, completing the 3.9 km (2.4 mi) swim through tough ocean waves in just over 48 min, biking 180 km (112 mi) through hot lava fields in under 4.5 h, then running 42 km (26.2 mi) in 2:45:34. In the women's division, Daniela Ryf of Switzerland earned her second (and back-to-back) Ironman title, finishing almost 24 min ahead of the next closest woman competitor in 8:46:46—the sole woman's performance under 9 h. How are these athletes able to compete in this race? While there is little doubt that they are genetically gifted—including a high $\dot{V}O_{2max}$—rigorous sport-specific training is also required to develop their cardiorespiratory endurance capacities.

During a single bout of aerobic exercise, the human body precisely adjusts its cardiovascular and respiratory function to meet the energy and oxygen demands of actively contracting muscle. When these systems are challenged repeatedly, as happens with regular exercise training, they adapt in ways that allow the body to improve $\dot{V}O_{2max}$ and overall endurance performance. **Aerobic training**, or cardiorespiratory endurance training, improves cardiac function and peripheral blood flow and enhances the capacity of the muscle fibers to generate greater amounts of adenosine triphosphate (ATP). In this chapter, we examine adaptations in cardiorespiratory function in response to endurance training and how such adaptations improve an athlete's endurance capacity and performance. Additionally, we examine adaptations to anaerobic training. **Anaerobic training** improves anaerobic metabolism; short-term, high-intensity exercise capacity; tolerance for acid–base imbalances; and in some cases, muscle strength. Both aerobic and anaerobic training induce a variety of adaptations that benefit exercise and sport performance.

The effects of training on cardiovascular and respiratory, or aerobic, endurance are well known to endurance athletes like distance runners, cyclists, cross-country skiers, and swimmers but are often ignored by other types of athletes. Training programs for many nonendurance athletes often ignore the aerobic endurance component. This is understandable, because for maximum improvement in performance, training should be highly specific to the particular sport or activity in which the athlete participates, and endurance is frequently not recognized as important to nonendurance activities. The reasoning is, why waste valuable training time if the result is not improved performance?

The problem with this reasoning is that most nonendurance sports do indeed have an endurance, or aerobic, component. For example, in football, players and coaches might fail to recognize the importance of cardiorespiratory endurance as part of the total training program. From all outward appearances, American football is an anaerobic, or burst-type, activity consisting of repeated bouts of high-intensity work of short duration. Seldom does a run exceed 40 to 60 yd (37-55 m), and even this is usually followed by a substantial rest interval. The need for endurance may not be readily apparent. What athletes and coaches might fail to consider is that this burst-type activity must be repeated many times during the game. With a higher aerobic endurance capacity, an athlete could maintain the quality of each burst activity throughout the game and would still be relatively fresh (less drop-off in performance, fewer feelings of fatigue) during the all-important closing minutes of the game.

Chapters 9 and 14 cover the principles of training for sport performance—the "how," "when," and "how much" questions about how training improves athletic performance. The focus here is on those physiological changes that occur within the body systems when aerobic or anaerobic exercise is repeated regularly to induce a training response.

Adaptations to Aerobic Training

Improvements in endurance that accompany regular (e.g., daily, every other day) aerobic training, such as running, cycling, or swimming, result from multiple adaptations to the training stimuli. Some adaptations occur in the cardiovascular system, improving circulation to and within the muscles. Still other important changes occur within the muscles themselves, promoting more efficient utilization of oxygen and fuel substrates. Pulmonary adaptations, as will be noted later, occur to a lesser extent.

Muscular Versus Cardiorespiratory Endurance

Endurance is a term that refers to two separate but related concepts: muscular endurance and cardiorespiratory endurance. Each makes a unique contribution to athletic performance, and each differs in its importance to different athletes.

For sprinters, endurance is the quality that allows them to sustain a high speed over the full distance of, for example, a 100 m or 200 m dash. This component of fitness is termed muscular endurance, the ability of a single muscle or muscle group to maintain high-intensity, repetitive, or static contractions. This type of endurance is also exemplified by a weightlifter doing multiple repetitions, a boxer, or a wrestler. The exercise or activity can be rhythmic and repetitive in nature, such as multiple repetitions of the bench press for the weightlifter and jabbing for the boxer. Or the activity can be more static, such as a sustained muscle action when a wrestler attempts to pin an opponent. In either case, the resulting fatigue is confined to a specific muscle group, and the activity's duration is usually no more than 2 min. Muscular endurance is highly related to muscular strength and to anaerobic power development.

While muscular endurance is specific to individual muscles or muscle groups, cardiorespiratory endurance relates to the ability to sustain prolonged, dynamic whole-body exercise using large muscle groups. Cardiorespiratory endurance is related to the development of the cardiorespiratory systems' ability to maintain oxygen delivery to working muscles during prolonged exercise, as well as the muscles' ability to use energy aerobically (discussed in chapters 2 and 5). This is why the terms *cardiorespiratory endurance* and *aerobic endurance* are sometimes used synonymously.

Evaluating Cardiorespiratory Endurance Capacity

Studying the effects of training on endurance requires an objective, repeatable means of measuring an individual's cardiorespiratory endurance capacity. In that way, the exercise scientist, coach, or athlete can monitor improvements as physiological adaptations occur during the training program.

Maximal Endurance Capacity: $\dot{V}O_{2max}$

Most exercise scientists regard $\dot{V}O_{2max}$, sometimes called maximal aerobic power or maximal aerobic capacity, as the best objective laboratory measure of cardiorespiratory endurance. Recall from chapter 5 that $\dot{V}O_{2max}$ is defined as the highest rate of oxygen consumption attainable during maximal or exhaustive exercise. $\dot{V}O_{2max}$ as defined by the Fick equation is determined by maximal cardiac output (delivery of oxygen and blood flow to working muscles) and the maximal (a-$\bar{v}$) O_2 difference (the ability of the active muscles to extract and use the oxygen).

As exercise intensity increases, oxygen consumption eventually either plateaus or decreases slightly, even with further increases in workload, indicating that a true maximal $\dot{V}O_{2max}$ has been achieved.

With endurance training, more oxygen can be delivered to, and used by, active muscles than in an untrained state. Previously untrained subjects demonstrate average increases in $\dot{V}O_{2max}$ of 15% to 20% after a 20-week training program. These improvements allow individuals to perform endurance activities at a higher intensity, improving their performance potential. Figure 11.1 illustrates the increase in $\dot{V}O_{2max}$ after 12 months of aerobic training in a previously untrained individual. In this example, $\dot{V}O_{2max}$ increased by about 30%. Note that the $\dot{V}O_2$ cost of running at a certain submaximal intensity (referred to as running economy) did not change, but higher running speeds could be attained after training.

VIDEO 11.1 Presents Ben Levine on the significance of $\dot{V}O_{2max}$ for sport performance.

Submaximal Endurance

In addition to increasing maximal endurance capacity, endurance training also increases **submaximal endurance**, which is more difficult to evaluate. A lower steady-state heart rate at the same submaximal exercise intensity is one physiological variable that can be used to objectively quantify the effect of training. Additionally, one could measure the average peak absolute power output a person can maintain over a fixed period of time on a cycle ergometer. For running, the average peak speed or velocity a person can maintain for a set period of time would be a similar test of submaximal endurance. Generally, these tests last 30 min to an hour

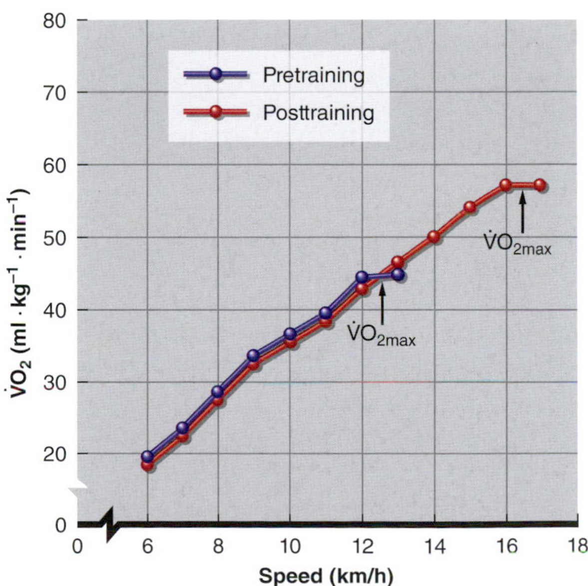

FIGURE 11.1 Changes in $\dot{V}O_{2max}$ with 12 months of endurance training. $\dot{V}O_{2max}$ increased from 44 to 57 ml · kg^{-1} · min^{-1}, a 30% increase. Peak speed during the treadmill test increased from 13 km/h (8 mph) to 16 km/h (~10 mph).

RESEARCH PERSPECTIVE 11.1

How Much Can $\dot{V}O_{2max}$ Improve?

In 1968, the Dallas bed-rest study demonstrated that $\dot{V}O_{2max}$ could be doubled (from roughly 25 ml · kg^{-1} · min^{-1} to 50 ml · kg^{-1} · min^{-1}) within a few weeks of training after a period of detraining.[30] Despite this huge increase in $\dot{V}O_{2max}$ following bed-rest induced detraining, 50 ml · kg^{-1} · min^{-1} is a rather typical $\dot{V}O_{2max}$ for a recreational endurance athlete, and it is unlikely that an average active adult can increase $\dot{V}O_{2max}$ from this average to values even remotely close to double.[21] Meanwhile, elite endurance athletes typically have $\dot{V}O_{2max}$ values approaching 80 ml · kg^{-1} · min^{-1}, with the highest value ever published at an incredible 90.6 ml · kg^{-1} · min^{-1} in an Olympic gold medalist cross-country skier. It is unlikely that any ordinary human can achieve such astounding values even with rigorous training programs, so how, and by how much, can $\dot{V}O_{2max}$ actually be enhanced?

Large $\dot{V}O_{2max}$ changes may take years to achieve. Prospective training studies are challenging to undertake in the laboratory, but the longest published study showed only a 21% increase in $\dot{V}O_{2max}$ over 12 months of training at moderate to high intensities every other day. Other training studies lasting 4 to 6 months show even more modest increases of 9% to 17%, and overall it appears that average endurance training improves $\dot{V}O_{2max}$ ~0.5 L/min. High-intensity interval training has shown the largest increase in $\dot{V}O_{2max}$ (44%), but it should be noted that the training intensities and volumes in all of these studies are far lower than the training load of world-class athletes. In contrast to the average-fit subjects in these longitudinal studies, young (15-25 years of age) world-class athletes can substantially increase their $\dot{V}O_{2max}$ from an already high 55 to 60 ml · kg^{-1} · min^{-1} to 75 to 80 ml · kg^{-1} · min^{-1} with years of intense training. This phenomenon suggests that large increases in $\dot{V}O_{2max}$ can be seen in athletes undergoing intense training and that training in early life is likely a determinant of very high $\dot{V}O_{2max}$ values recorded in endurance champions.

A high $\dot{V}O_{2max}$ is the product of a high maximal cardiac output ($\dot{Q}_{max}$) and a high oxygen-carrying capacity of the blood. This maximal blood flow and oxygen carrying leads to increased oxygen delivery to the exercising muscles. The high $\dot{Q}_{max}$ in elite athletes is the result of increased stroke volume, since maximal heart rate does not change with training. Endurance trained athletes achieve this higher stroke volume through changes in the left ventricle of the heart such that it has a larger mass and is more distensible and therefore easier to fill with each heartbeat. But what about nonelite athletes? It remains unknown if average individuals can ever reach $\dot{Q}_{max}$ values observed in elite athletes. The initial increase in stroke volume observed with training healthy, normal volunteers is due to an increased blood volume rather

and reflect the concept of critical power discussed in chapter 5. Submaximal endurance, like critical power, is more closely related to actual competitive endurance performance than $\dot{V}O_{2max}$. With endurance training, submaximal endurance increases.

Cardiovascular Adaptations to Training

Multiple cardiovascular adaptations occur in response to exercise training, including changes in the following:

- Heart size
- Stroke volume
- Heart rate
- Cardiac output
- Blood flow
- Blood and red cell volumes

Not surprisingly, these variables are interrelated. For example, training-induced increases in stroke volume depend on increases in both heart size and blood volume. To fully understand adaptations in these variables, it is important to review how these components relate to oxygen transport.

Oxygen Transport System

Cardiorespiratory endurance is related to the cardiovascular and respiratory systems' ability to deliver sufficient oxygen to meet the needs of metabolically active tissues.

Recall from chapter 8 that the ability of the cardiovascular and respiratory systems to deliver oxygen to active tissues is defined by the **Fick equation**, which states that whole-body oxygen consumption is determined by both the delivery of oxygen via blood flow (cardiac output) and the amount of oxygen extracted by the tissues, the $(a\text{-}\bar{v})O_2$ difference. The product of cardiac output and the $(a\text{-}\bar{v})O_2$ difference determines the rate at which oxygen is being consumed:

$$\dot{V}O_2 = \text{stroke volume} \times \text{heart rate} \times (a\text{-}\bar{v})O_2 \text{ difference}$$

and

$$\dot{V}O_{2max} = \text{maximal stroke volume} \times \text{maximal heart rate} \times \text{maximal } (a\text{-}\bar{v})O_2 \text{ difference}$$

than changes in the myocardium. After 1 year of exercise training in previously untrained individuals, left ventricle mass has been shown to increase. However, these changes do not result in very large increases in $\dot{Q}_{max}$. These findings suggest that it is unlikely that individuals with average cardiac function can ever reach values observed in elite athletes, but it may be that exercise training during childhood and early adulthood may favor the development of these advantageous cardiac characteristics.

Given the limited potential for training to increase $\dot{Q}_{max}$, the adaptation for improved oxygen delivery is an important factor for the improvement of $\dot{V}O_{2max}$ with endurance training. Oxygen-carrying capacity is directly related to the number of hemoglobin available to bind to oxygen, and hemoglobin mass correlates tightly with exercise performance. There is little doubt that exercise training increases hemoglobin mass or total red blood cell volume ~20%. It is unknown if long-term endurance training can increase hemoglobin mass from normal values (~700 g) to that observed in elite athletes (~1,200 g), but it appears unlikely. There may be genetic determinants of total hemoglobin mass, but research has thus far failed to find a genetic polymorphism to explain extremely high hemoglobin mass in elite athletes.

Finally, improvements in oxygen extraction, i.e., increases in (a-v)O_2 difference, may also contribute to the increase in $\dot{V}O_{2max}$ with training. Athletes have a more homogeneous blood flow distribution during submaximal exercise, which results in a higher oxygen extraction compared to untrained individuals. Systemic maximal oxygen extraction can be improved with training in healthy volunteers, from 72% up to 84%. Although this is a meaningful improvement, it is still nowhere near oxygen extraction reported in elite endurance athletes (93%). It is unlikely that average individuals can achieve the high oxygen extractions observed in elite athletes; however, the possibility that already trained elite endurance athletes can further improve oxygen extraction has yet to be studied.

Overall, although endurance training leads to improvements in the mechanisms contributing to $\dot{V}O_{2max}$, the overall increases observed in healthy, normal individuals rarely exceed 0.5 L/min and never reach the extraordinarily high values observed in elite endurance athletes. Even highly trained athletes seem to plateau after age 25, and increases in performance after that are due to increases in other mechanisms such as mechanical efficiency or critical power. $\dot{V}O_{2max}$ is a powerful determinant of endurance performance, but the magnitude of improvements that can be achieved through training are relatively small, even in elite endurance athletes.[20]

Because maximal heart rate (HR_{max}) either stays the same or decreases slightly with training, increases in $\dot{V}O_{2max}$ depend on adaptations in maximal stroke volume and maximal (a-$\bar{v}$)O_2 difference.

The oxygen demand of exercising muscles increases with increasing exercise intensity. Aerobic endurance depends on the cardiorespiratory system's ability to deliver sufficient oxygen to these active tissues to meet their heightened demands for oxygen for oxidative metabolism. As maximal levels of exercise are achieved, heart size, blood flow, blood pressure, and blood volume can all potentially limit the maximal ability to transport oxygen. Endurance training elicits numerous changes in these components of the **oxygen transport system** that enable it to function more effectively.

Heart Size

The measurement of heart size has been of interest to cardiologists for years because a hypertrophied, or enlarged, heart is typically a pathological condition indicating the presence of cardiovascular disease. Clinicians and scientists commonly use echocardiography to accurately measure the size of the heart and its chambers. Echocardiography involves the technique of ultrasound, which uses high-frequency sound waves directed through the chest wall to the heart. These sound waves are emitted from a transducer placed on the chest; once they contact the various structures of the heart, they rebound back to a sensor, which is able to capture the deflected sound waves and provide a moving picture of the heart. A trained physician or technician can visualize the size of the heart's chambers, thicknesses of its walls, and heart valve action. There are several forms of echocardiography: M-mode echocardiography, which provides a one-dimensional view of the heart; two-dimensional echocardiography; and Doppler echocardiography, which is used more often to measure blood flow through large arteries.

As an adaptation to the increased work demand, cardiac muscle mass and ventricular volume increase with training. Cardiac muscle, like skeletal muscle, undergoes morphological adaptations as a result of chronic endurance training. At one time, **cardiac hypertrophy** induced by exercise—**athlete's heart**, as it was called—

was viewed with concern because experts incorrectly believed that enlargement of the heart always reflected a pathological state, as sometimes occurs with severe hypertension. Training-induced cardiac hypertrophy, on the other hand, is now recognized as a normal adaptation to chronic endurance training.

The left ventricle, as discussed in chapter 6, does the most work and thus undergoes the greatest adaptation in response to endurance training. The type of ventricular adaptation depends on the type of exercise training performed. For example, during resistance training, the left ventricle must contract against increased afterload from the systemic circulation. From chapter 8 we learned that blood pressure during resistance exercise can exceed 480/350 mmHg. This presents a considerable resistance that must be overcome by the left ventricle. To overcome this high afterload, the heart muscle compensates by increasing left ventricular wall thickness, thereby increasing its contractility. Thus, the increase in its muscle mass is in direct response to repeated exposure to the increased afterload with resistance training. However, there is little change in ventricular volume.

With endurance training, left ventricular chamber size increases. This allows for increased left ventricular filling and consequently an increase in stroke volume. The increase in left ventricular dimensions is largely attributable to a training-induced increase in plasma volume (discussed later in this chapter) that increases left ventricular end-diastolic volume (increased preload). In concert with this, a decrease in heart rate at rest caused by increased parasympathetic tone, and during exercise at the same rate of work, allows a longer diastolic filling period. The increases in plasma volume and diastolic filling time increase left ventricular chamber size at the end of diastole. This effect of endurance training on the left ventricle is often called a volume loading effect.

It was originally hypothesized that this increase in left ventricular dimensions was the only change in the left ventricle caused by endurance training. Additional research has revealed that, similar to what happens in resistance training, myocardial wall thickness increases with endurance training. Highly trained endurance athletes (competitive cross-country skiers, endurance cyclists, and long-distance runners) have greater left ventricular masses than non–endurance-trained men and women. Furthermore, left ventricular mass is highly correlated with $\dot{V}O_{2max}$.

Fagard[12] conducted the most extensive review of the existing research literature in 1996, focusing on long-distance runners (135 athletes and 173 controls), cyclists (69 athletes and 65 controls), and strength athletes (178 athletes, including weight- and powerlifters, bodybuilders, wrestlers, throwers, and bobsledders, and 105 controls). For each group, the athletes were matched by age and body size with a group of sedentary control subjects. For each group of runners, cyclists, and strength athletes, the internal diameter of the left ventricle (LVID, an index of chamber size) and the total left ventricular mass (LVM) were greater in the athletes compared with their age- and sized-matched controls (figure 11.2). Thus, data from this review support the hypothesis that both left ventricular chamber size and wall thickness increase with endurance training.

Most studies of heart size changes with training have been cross-sectional, comparing trained individuals with sedentary, untrained individuals. Certainly a portion of the differences that we see in figure 11.2 can be attributed to genetics, not training. However, a number of longitudinal studies have followed individuals from an untrained state to a trained state, and others have followed individuals from a trained state to an untrained state. These studies have reported increases in heart size with training and decreases with detraining. Therefore, training does bring about changes, but they are likely not as large as the differences shown in figure 11.2.

In Review

- *Cardiorespiratory endurance* (also called *maximal aerobic power*) refers to the ability to perform prolonged, dynamic exercise using a large muscle mass.
- $\dot{V}O_{2max}$—the highest rate of oxygen consumption obtainable during maximal or exhaustive exercise—is the best single measure of cardiorespiratory endurance.
- Cardiac output, the product of heart rate and stroke volume, represents how much blood leaves the heart each minute, whereas $(a-\bar{v})O_2$ difference is a measure of how much oxygen is extracted from the blood by the tissues. According to the Fick equation, the product of these values is the rate of oxygen consumption: $\dot{V}O_2$ = stroke volume × heart rate × $(a-\bar{v})O_2$ difference.
- Of the chambers of the heart, the left ventricle adapts the most in response to endurance training.
- With endurance training, the internal dimensions of the left ventricle increase, mostly in response to an increase in ventricular filling secondary to an increase in plasma volume.
- Left ventricular wall thickness and mass also increase with endurance training, allowing for a greater force of contraction.

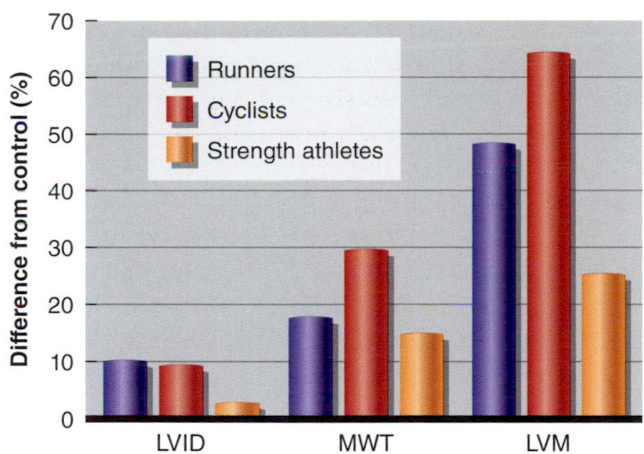

FIGURE 11.2 Percentage differences in heart size of three groups of athletes (runners, cyclists, and strength athletes) compared with their age- and size-matched sedentary controls (0%). Percentage differences are presented for left ventricular internal diameter (LVID), mean wall thickness (MWT), and left ventricular mass (LVM).

Data are from Fagard (1996).

Stroke Volume

Stroke volume at rest is substantially higher after an endurance training program than it is before training. This endurance training–induced increase is also seen at a given submaximal exercise intensity and at maximal exercise. This increase is illustrated in figure 11.3, which shows the changes in stroke volume of a subject who exercised at increasing intensities up to a maximal intensity before and after a 6-month endurance training program. Typical values for stroke volume at rest and during maximal exercise in untrained, trained, and highly trained athletes are listed in table 11.1. The wide range of stroke volume values for any given cell within this table is largely attributable to differences in body size. Larger people typically have larger hearts and a greater blood volume, and thus higher stroke volumes—an important point when one is comparing stroke volumes of different people.

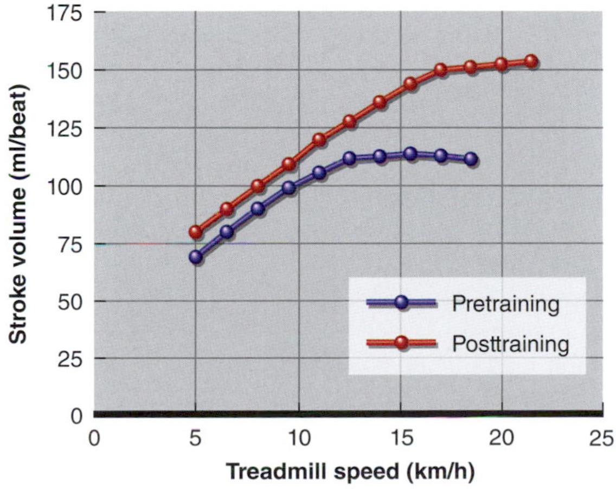

FIGURE 11.3 Changes in stroke volume with endurance training during walking, jogging, and running on a treadmill at increasing velocities.

TABLE 11.1 **Stroke Volumes at Rest (SV_{rest}) and During Maximal Exercise (SV_{max}) for Different States of Training**

Subjects	SV_{rest} (ml/beat)	SV_{max} (ml/beat)
Untrained	50-70	80-110
Trained	70-90	110-150
Highly trained	90-110	150-220+

After aerobic training, the left ventricle fills more completely during diastole. Plasma volume expands with training, which allows for more blood to enter the ventricle during diastole, increasing end-diastolic volume (EDV). The heart rate of a trained heart is also lower at rest and at the same absolute exercise intensity than that of an untrained heart, allowing more time for the increased diastolic filling. More blood entering the ventricle increases the stretch on the ventricular walls; by the Frank-Starling mechanism (see chapter 8), this results in an increased force of contraction.

The thickness of the posterior and septal walls of the left ventricle also increases slightly with endurance training. Increased ventricular muscle mass results in increased contractile force, in turn causing a lower end-systolic volume (ESV).

The decrease in ESV is facilitated by the decrease in peripheral resistance that occurs with training. Increased contractility resulting from an increase in left ventricular thickness and greater diastolic filling (Frank-Starling mechanism), coupled with the reduction in systemic peripheral resistance, increases the ejection fraction [equal to (EDV – ESV) / EDV] in the trained heart. More blood enters the left ventricle, and a greater percentage of what enters is forced out with each contraction, resulting in an increase in stroke volume.

Adaptations in stroke volume during endurance training are illustrated by a study in which older men trained aerobically for 1 year.[10] Their cardiovascular function was evaluated before and after training. The subjects performed running, treadmill, and cycle ergometer exercise for 1 h each day, 4 days per week. They exercised at intensities of 60% to 80% of $\dot{V}O_{2max}$, with brief bouts of exercise exceeding 90% of $\dot{V}O_{2max}$. End-diastolic volume increased at rest and throughout submaximal exercise. The ejection fraction increased, which was associated with a decreased ESV, suggesting increased contractility of the left ventricle. $\dot{V}O_{2max}$ increased by 23%, a substantial improvement in endurance.

To summarize, increased left ventricular dimensions, reduced systemic peripheral resistance, and a greater blood volume account for the increases in resting, submaximal, and maximal stroke volume after an endurance training program.

In Review

- Following endurance training, stroke volume (SV) is increased at rest and during submaximal and maximal exercise.
- A major factor leading to the SV increase is an increased end-diastolic volume (EDV) caused by an increase in plasma volume and a greater diastolic filling time secondary to a lower heart rate.
- Another contributing factor to increased SV is an increased left ventricular force of contraction. This is caused by hypertrophy of the cardiac muscle and increased ventricular stretch resulting from an increase in diastolic filling (increased preload), leading to greater elastic recoil (Frank-Starling mechanism).
- Reduced systemic vascular resistance (decreased afterload) also contributes to the increased volume of blood pumped from the left ventricle with each beat.

Heart Rate

Aerobic training has a major impact on heart rate at rest, during submaximal exercise, and during the postexercise recovery period. The effect of aerobic training on maximal heart rate is rather negligible.

Resting Heart Rate Resting heart rate decreases markedly as a result of endurance training. Some studies have shown that a sedentary individual with an initial resting heart rate of 80 beats/min can decrease resting heart rate by approximately 1 beat/min with each week of aerobic training, at least for the first few weeks. After 10 weeks of moderate endurance training, resting heart rate can decrease from 80 to 70 beats/min or lower. On the other hand, well-controlled studies with large numbers of subjects have shown much smaller decreases in resting heart rate, that is, fewer than 5 beats/min following up to 20 weeks of aerobic training.

Recall from chapter 6 that bradycardia is a term indicating a heart rate of fewer than 60 beats/min. In untrained individuals, bradycardia can be the result of abnormal cardiac function or heart disease. However, highly conditioned endurance athletes often have resting heart rates lower than 40 beats/min, and some have values lower than 30 beats/min. Therefore, it is necessary to differentiate between training-induced bradycardia, which is a normal response to endurance training, and pathological bradycardia, which can be cause for concern.

The low resting heart rate (HR) of well-trained endurance athletes is most often attributed to an elevated parasympathetic (vagal) tone. However, a 2013 review of the available evidence casts doubt on this mechanism.[6] The two alternative explanations for the resting bradycardia of athletes are a diminished sympathetic tone and a lower intrinsic heart rate. Recall from chapter 6 that the intrinsic heart rate is the rate of sinoatrial (SA) node firing in the absence of any neural or hormonal input. In studies that have blocked parasympathetic activity to the heart using the drug atropine, there is still a significant resting bradycardia in athletes. In fact, the difference in HR after parasympathetic blockade is greater than the difference in the normal HR, suggesting that the bradycardia is not the result of elevated vagal tone.

Other studies have blocked both branches of the autonomic nervous system, that is, used a complete autonomic blockade. The HR after complete autonomic blockade is a measure of the intrinsic HR. In studies showing a lowered resting HR after endurance training, the bradycardia persists after complete autonomic blockade. Thus, the resting bradycardia seen in athletes is at least partially, and perhaps completely, the result of a decreased intrinsic HR.

A decreased intrinsic HR can result from a remodeling of the SA node. The SA node serves as the pacemaker of the heart due to properties of ion channels and Ca^{2+}-handling proteins in the SA node cells. Changes in these properties cause the well-known bradycardias associated with SA node disease, heart failure, atrial fibrillation, and even aging itself. In fact, the age-associated decrease in resting HR has been attributed to a downregulation of ryanodine receptors (see chapter 6) that are involved in Ca^{2+} flux. If these mechanisms are involved in bradycardias associated with these other processes and diseases, it is likely that they are involved in training-induced bradycardia as well.

Submaximal Heart Rate During submaximal exercise, aerobic training results in a lower heart rate at any given absolute exercise intensity. This is illustrated in figure 11.4, which shows the heart rate of an individual exercising on a treadmill before and after training. At each walking or running speed, the posttraining heart rate is lower than the heart rate before training. The training-induced decrease in heart rate is typically greater at higher intensities.

While maintaining a cardiac output appropriate to meet the needs of working muscle, a trained heart performs less work (lower heart rate, higher stroke

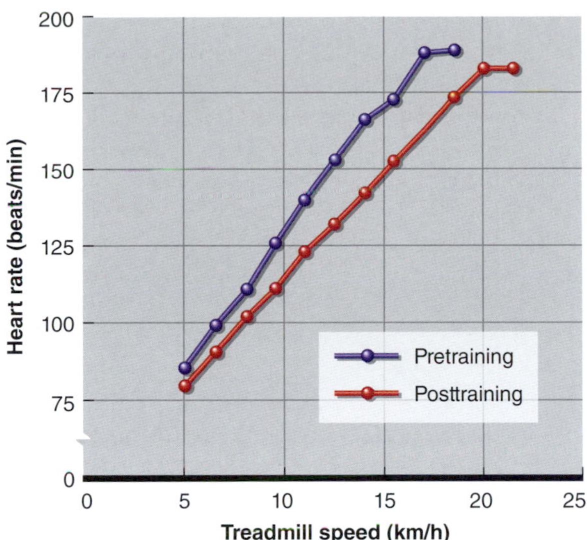

FIGURE 11.4 Endurance training-induced changes in heart rate during progressive walking, jogging, and running on a treadmill at increasing speeds.

volume) than an untrained heart at the same absolute workload.

Maximum Heart Rate A person's maximal heart rate (HR_{max}) tends to be stable and typically remains relatively unchanged after endurance training. However, several studies have suggested that for people whose untrained HR_{max} values exceed 180 beats/min, HR_{max} might be slightly lower after training. Also, highly conditioned endurance athletes often have lower HR_{max} values than untrained individuals of the same age, although this is not always the case. Athletes over 60 years old sometimes have higher HR_{max} values than untrained people of the same age.

Interactions Between Heart Rate and Stroke Volume During exercise, the product of heart rate and stroke volume provides a cardiac output appropriate to the intensity of the activity being performed. At maximal or near-maximal intensities, heart rate may change to provide the optimal combination of heart rate and stroke volume to maximize cardiac output. If heart rate is too fast, diastolic filling time is reduced, and stroke volume might be compromised. For example, if HR_{max} is 180 beats/min, the heart beats three times per second. Each cardiac cycle thus lasts for only 0.33 s. Diastole is as short as 0.15 s or less. This fast heart rate allows very little time for the ventricles to fill. As a consequence, stroke volume may decrease at high heart rates when filling time is compromised.

However, if the heart rate slows, the ventricles have longer to fill. This has been proposed as one reason highly trained endurance athletes tend to have lower HR_{max} values: Their hearts have adapted to training by drastically increasing their stroke volumes, so lower HR_{max} values can provide optimal cardiac output.

Which comes first? Does increased stroke volume result in a decreased heart rate, or does a lower heart rate result in an increased stroke volume? This question remains unanswered. In either case, the combination of increased stroke volume and decreased heart rate is a more efficient way for the heart to meet the metabolic demands of the exercising body. The heart expends less energy by contracting less often but more forcefully than it would if contraction frequency were increased. Reciprocal changes in heart rate and stroke volume in response to training share a common goal: to allow the heart to pump the maximal amount of oxygenated blood at the lowest energy cost.

Heart Rate Recovery During exercise, as discussed in chapter 6, heart rate must increase to increase cardiac output to meet the blood flow demands of active muscles. When the exercise bout is finished, heart rate does not instantly return to its resting level. Instead, it remains elevated for a while, slowly returning to its resting rate. The time it takes for heart rate to return to its resting rate is called the heart rate recovery period.

After endurance training, as shown in figure 11.5, heart rate returns to its resting level much more quickly after an exercise bout than it does before training. This is true after both submaximal and maximal exercise.

Because the heart rate recovery period is shorter after endurance training, this measurement has been proposed as an indirect index of cardiorespiratory fitness. In general, a more fit person recovers faster after a standardized rate of work than a less fit person, so this measure may have some utility in field settings when more direct measures of endurance capacity are

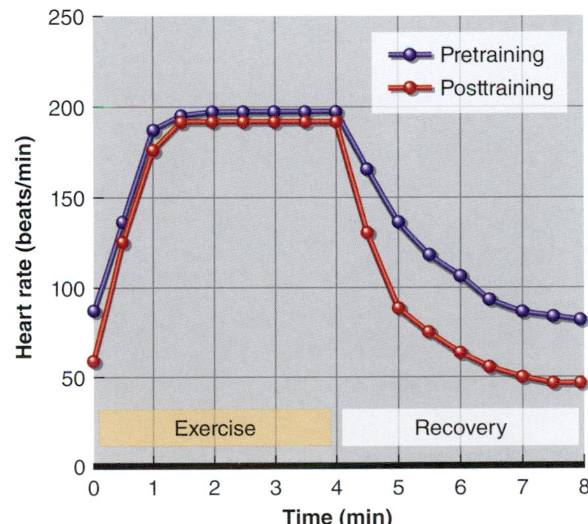

FIGURE 11.5 Changes in heart rate during recovery after a 4 min, all-out bout of exercise before and after endurance training.

not possible or feasible. However, factors other than training can also affect heart rate recovery time. For example, an elevated core temperature or an enhanced sympathetic nervous system response can prolong heart rate elevation.

The heart rate recovery curve is a useful tool for tracking a person's progress during a training program. But because of the potential influence of other factors, it should not be used to compare individuals.

Cardiac Output

We have looked at the effects of training on the two components of cardiac output: stroke volume and heart rate. While stroke volume increases with training, heart rate generally decreases at rest and during exercise at a given absolute intensity.

Because the magnitude of these reciprocal changes is similar, cardiac output at rest and during submaximal exercise at a given exercise intensity does not change much following endurance training. In fact, cardiac output can decrease slightly. This is likely the result of an increase in the $(a\text{-}\bar{v})O_2$ difference (reflecting greater oxygen extraction by the tissues) or a decrease in the rate of oxygen consumption (reflecting an increased mechanical efficiency). Generally, cardiac output matches the oxygen consumption required for any given intensity of effort.

Maximal cardiac output, however, increases considerably in response to aerobic training, as seen in figure 11.6, and is largely responsible for the increase in $\dot{V}O_{2max}$. This increase in cardiac output must result from an increase in maximal stroke volume, because HR_{max} changes little, if any. Maximal cardiac output ranges from 14 to 20 L/min in untrained individuals and from 25 to 35 L/min in trained individuals, and can be 40 L/min or more in highly conditioned endurance athletes. These absolute values, however, are greatly influenced by body size.

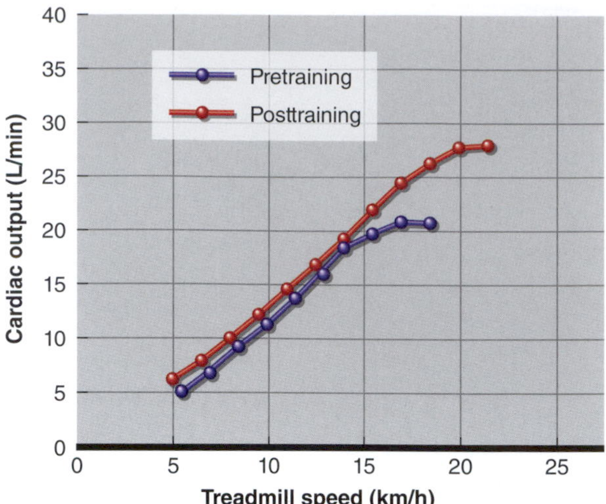

FIGURE 11.6 Changes in cardiac output with endurance training during walking, then jogging, and finally running on a treadmill as velocity increases.

Lundby and colleagues[19] have argued that variability in $\dot{V}O_{2max}$ among individuals is primarily determined by differences in two variables: maximal cardiac output and red blood cell volume (figure 11.7). (Red cell volume changes are discussed later in this chapter.) Therefore, the response of $\dot{V}O_{2max}$ to endurance training reflects relative changes in these two important determinants.

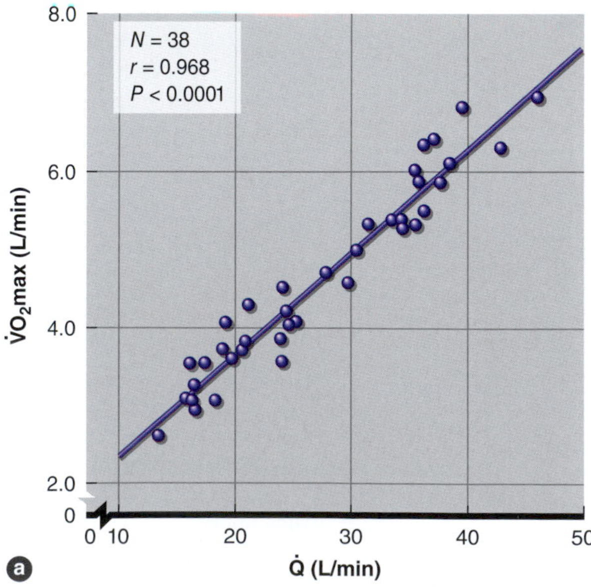

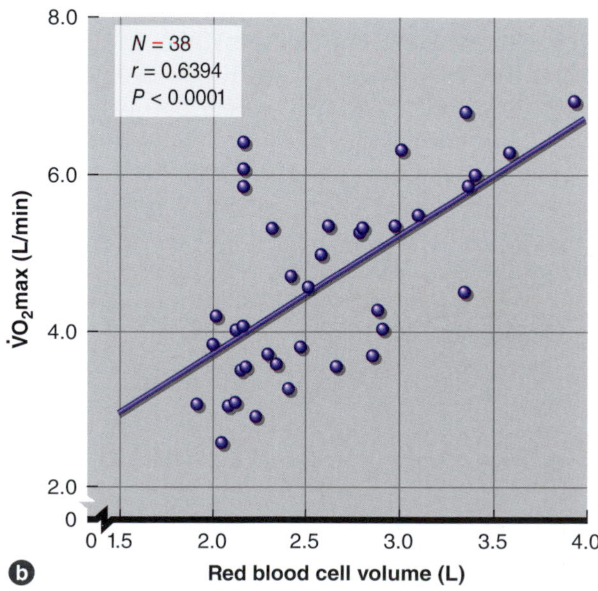

FIGURE 11.7 Correlations between $\dot{V}O_{2max}$ and *(a)* maximal cardiac output and *(b)* red blood cell volume.

Reprinted by permission from C. Lundby, D. Montero, and M. Joyner, "Biology of VO2 Max: Looking Under the Physiology Map," *Acta Physiologica* 220, no. 2 (2017): 218-228.

In Review

- Resting heart rate decreases as a result of endurance training. In a sedentary person, the decrease is typically about 1 beat/min per week during the initial weeks of training, but smaller decreases have been reported. Highly trained endurance athletes may have resting heart rates of 40 beats/min or lower.
- The mechanisms responsible for the sinus bradycardia associated with endurance training remain controversial, but likely involve both extrinsic (autonomic neural balance) and intrinsic (SA node function) components.
- Heart rate during submaximal exercise is also lower, with larger decreases seen at higher exercise intensities.
- Maximal heart rate either remains unchanged or decreases slightly with training.
- Heart rate during the recovery period decreases more rapidly after training, making it an indirect but convenient way of tracking the adaptations within an individual that occur with training. However, this value is not useful for comparing fitness levels of different people.
- Cardiac output at rest and at submaximal levels of exercise remains unchanged (or may decrease slightly) after endurance training.
- Cardiac output during maximal exercise increases considerably and is largely responsible for the increase in $\dot{V}O_{2max}$. The increased maximal cardiac output is the result of the substantial increase in maximal stroke volume, made possible by training-induced changes in blood volume and cardiac structure and function.

Blood Flow

Active muscles need substantially more oxygen and fuel substrates than inactive ones. To meet these increased needs, more blood must be delivered to these muscles during exercise. With endurance training, the cardiovascular system adapts to increase blood flow to exercising muscles to meet their higher demand for oxygen and metabolic substrates. In addition to changes in the heart that allow for better pumping and increased stroke volume, four factors account for this enhanced blood flow to muscle following training:

- Increased capillarization
- Greater recruitment of existing capillaries
- More effective blood flow redistribution away from inactive regions
- Increased total blood volume

To permit increased blood flow, new capillaries develop in trained muscles. This allows the blood flowing into skeletal muscle from arterioles to more fully perfuse the active fibers. This increase in capillaries usually is expressed as an increase in the number of capillaries per muscle fiber, or the **capillary-to-fiber ratio**. Table 11.2 illustrates the differences in capillary-to-fiber ratios between well-trained and untrained men, both before and after exercise.[15]

In all tissues, including muscle, not all capillaries are open at any given time. In addition to new capillary formation, existing capillaries in trained muscles can be recruited and open to flow, which also increases blood flow to muscle fibers. The increase in new capillaries with endurance training and increased capillary recruitment combine to increase the overall area for diffusion of oxygen between the vascular system and the metabolically active muscle fibers.

A more effective redistribution of cardiac output also can increase blood flow to the active muscles. Blood flow is directed to the active musculature and shunted away from areas that do not need high flow. Blood flow can increase to the more active fibers even within a specific muscle group. Armstrong and Laughlin[2] first demonstrated that endurance-trained rats could redistribute blood flow to their most active tissues during

TABLE 11.2 **Muscle Fiber Capillarization in Well-Trained and Untrained Men**

Stage	Capillaries per mm^2	Muscle fibers per mm^2	Capillary-to-fiber ratio	Diffusion distance[a]
		Well trained		
Preexercise	640	440	1.5	20.1
Postexercise	611	414	1.6	20.3
		Untrained		
Preexercise	600	557	1.1	20.3
Postexercise	599	576	1.1	20.5

Note. This table illustrates the larger size of the muscle fibers in the well-trained men in that they had fewer fibers for a given area (fibers per mm^2). They also had an approximately 50% higher capillary-to-fiber ratio than the untrained men.

[a]Diffusion distance is expressed as the average half-distance between capillaries on the cross-sectional view expressed in micrometers.

Adapted by permission from L. Hermansen and M. Wachtlova, "Capillary Density of Skeletal Muscle in Well Trained and Untrained Men," *Journal of Applied Physiology* 30 (1971): 860-863.

exercise better than untrained rats could. The total blood flow to the exercising hindlimbs did not differ between the trained and untrained rats. However, the trained rats distributed more of their blood to the most oxidative muscle fibers, effectively redistributing the blood flow away from the glycolytic muscle fibers. These findings are difficult to replicate in humans because of measurement challenges, as well as the fact that human skeletal muscle is a mosaic with mixed fiber types among individual muscles.

Finally, the body's total blood volume increases with endurance training, providing more blood to meet the body's many blood flow needs during endurance activity without compromising venous return, as discussed next in this chapter.

Blood Volume

Endurance training increases total blood volume, and this effect is larger with higher training intensities. Furthermore, the effect occurs rapidly. This increased blood volume results primarily from an increase in plasma volume, but there is also an increase in the volume of red blood cells. The time course and mechanism for the increase of each of these components of blood are quite different.

Plasma Volume The increase in plasma volume with training is thought to result from two mechanisms. The first mechanism, which has two phases, results in increases in plasma proteins, particularly albumin. Recall from chapter 8 that plasma proteins are the major driver of oncotic pressure in the vasculature. As plasma protein concentration increases, so does oncotic pressure, and fluid is reabsorbed from the interstitial fluid into the blood vessels. During an intense bout of exercise, proteins leave the vascular space and move into the interstitial space. They are then returned in greater amounts through the lymph system. It is likely that the first phase of rapid plasma volume increase is the result of the increased plasma albumin, which is noted within the first hour of recovery from the first training bout. In the second phase, protein synthesis is turned on (upregulated) by repeated exercise, and new proteins are formed. With the second mechanism, exercise increases the release of antidiuretic hormone and aldosterone, hormones that cause reabsorption of water and sodium in the kidneys, which increases blood plasma. That increased fluid is kept in the vascular space by the oncotic pressure exerted by the proteins. Nearly all of the increase in blood volume during the first 2 weeks of training can be explained by the increase in plasma volume. This early blood volume expansion allows stroke volume to increase despite the fact that changes in the structure and function of the heart itself take longer to develop.

Red Blood Cells An increase in red blood cell volume with endurance training also contributes to the overall increase in blood volume (figure 11.7*b*) and red cell volume, like cardiac output, is correlated to $\dot{V}O_{2max}$. Although the actual number of red blood cells may increase, the hematocrit—the ratio of the red blood cell volume to the total blood volume—may actually decrease. Figure 11.8 illustrates this apparent paradox. Notice that the hematocrit is reduced even though there has been a slight increase in red blood cells. A trained athlete's hematocrit can decrease to such an extent that the athlete appears to be anemic on the basis of a relatively low concentration of red cells and hemoglobin ("pseudoanemia").

The increased ratio of plasma to cells resulting from a greater increase in the fluid portion reduces the blood's viscosity, or thickness. Reduced viscosity may aid the smooth flow of blood through the blood vessels, particularly through the smaller vessels such as the capillaries. One of the physiological benefits of decreasing blood viscosity is that it enhances oxygen delivery to the active muscle mass.

Both the total amount (absolute values) of hemoglobin and the total number of red blood cells are typically elevated in highly trained athletes. This ensures that the blood has more than adequate oxygen-carrying capacity; that is, the blood's ability to deliver oxygen to exercising muscle is not a limiting factor in exercise. The turnover rate of red blood cells also may be higher with intense training.

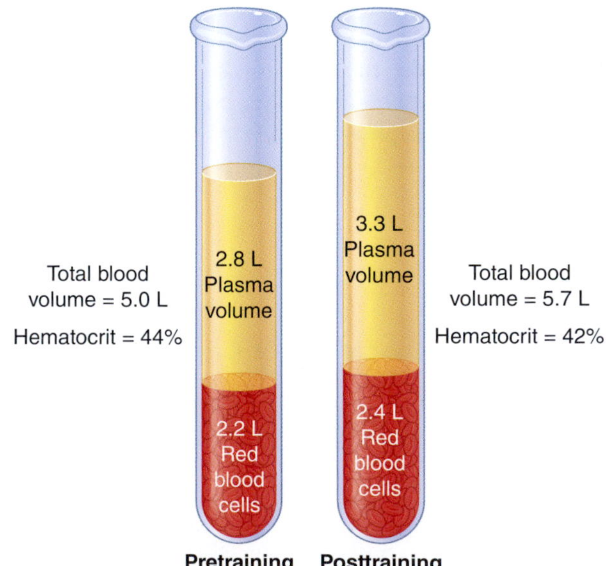

FIGURE 11.8 Increases in total blood volume and plasma volume occur with endurance training. Note that although the hematocrit (percentage of red blood cells) decreased from 44% to 42%, the total volume of red blood cells increased by 10%.

In Review

- Blood flow to active muscle is increased by endurance training.
- Increased muscle blood flow results from four factors:
  1. Increased capillarization
  2. Greater opening of existing capillaries (capillary recruitment)
  3. More effective blood flow distribution
  4. Increased blood volume
- Blood volume increases as a result of endurance training.
- Plasma volume is expanded through increased protein content (returned from lymph and upregulated protein synthesis). This effect is maintained and supported by fluid-conserving hormones.
- Red blood cell volume also increases, but the increase in plasma volume is typically higher. This decreases blood viscosity, which can improve tissue perfusion and oxygen availability.

Respiratory Adaptations to Training

No matter how proficient the cardiovascular system is at supplying blood to exercising muscle, endurance would be hindered if the respiratory system were not able to deliver enough oxygen to fully oxygenate red blood cells. Respiratory system function does not usually limit performance because ventilation can be increased to a much greater extent than cardiovascular function. But, as with the cardiovascular system, the respiratory system undergoes specific adaptations to endurance training to maximize its efficiency.

Pulmonary Ventilation

After training, pulmonary ventilation is essentially unchanged at rest. Although endurance training does not change the structure or basic physiology of the lung, it does decrease ventilation during submaximal exercise by as much as 30% at a given submaximal intensity. Maximal pulmonary ventilation is substantially increased from a rate of about 100 to 120 L/min in untrained sedentary individuals to about 130 to 150 L/min or more following endurance training. Breathing rates typically increase to about 180 L/min in highly trained athletes and can exceed 200 L/min in very large, highly trained endurance athletes. Two factors can account for the increase in maximal pulmonary ventilation following training: increased tidal volume and increased respiratory frequency at maximal exercise.

Ventilation is not usually a limiting factor for endurance exercise performance. However, in some very highly trained athletes, the pulmonary system's capacity for oxygen transport may not be able to meet the demands of exercising muscle and the cardiovascular system. This results in what has been termed *exercise-induced arterial hypoxemia*, in which arterial oxygen saturation decreases below 96%. This desaturation in highly trained elite athletes likely results from the large right heart cardiac output directed to the lung during exercise and consequently a decrease in the time the blood spends in the lung.

Pulmonary Diffusion

Pulmonary diffusion, or gas exchange occurring in the alveoli, is unaltered at rest and during submaximal exercise following training. However, it increases at maximal exercise intensity. Pulmonary blood flow (blood coming from the right side of the heart to the lungs) increases following training, particularly flow to the upper regions of the lungs when a person is sitting or standing. This increases lung perfusion. More blood is brought into the lungs for gas exchange, and at the same time ventilation increases so that more air is brought into the lungs. This means that more alveoli will be involved in pulmonary diffusion. The net result is that pulmonary diffusion increases.

Arterial–Venous Oxygen Difference

It is clear that stroke volume adapts with endurance training, but peripheral adaptations also contribute to the increase in $\dot{V}O_{2max}$. The oxygen content of arterial blood changes very little with endurance training. Even though total hemoglobin is increased, the amount of hemoglobin per unit of blood is the same or even slightly reduced. The $(a\text{-}\bar{v})O_2$ difference, however, does increase with training, particularly at submaximal exercise intensities. This increase results from a lower mixed venous oxygen content, reflecting both greater oxygen extraction by active tissues and a more effective distribution of blood flow to active tissues. The increased extraction results in part from an increase in oxidative capacity of active muscle fibers, as described later in this chapter.

This was demonstrated in a unique longitudinal study involving both exercise training and a bed-rest deconditioning model.[24] Five 20-year-old men were tested (baseline values), placed on bed rest for 20 days (deconditioning), and then trained for 60 days, starting immediately at the conclusion of bed rest. These same five men were restudied 30 years later at the age of 50; they were tested at baseline in a relatively sedentary state and after 6 months of endurance training. The average percentage increases in $\dot{V}O_{2max}$ were similar for the subjects at age 20 (18%) and at age 50 (14%). However, the increase in $\dot{V}O_{2max}$ at age 20 was explained by increases in both maximal cardiac output and maximal $(a-\bar{v})O_2$ difference; at age 50, the increase was explained primarily by an increase in $(a-\bar{v})O_2$ difference, while maximal cardiac output was unchanged. Maximal stroke volume was increased after training at both age 20 and age 50 but to a lesser degree at age 50 (+16 ml/beat at age 20 versus +8 ml/beat at age 50).

While most studies have shown an increase in maximal $(a-\bar{v})O_2$ difference after aerobic training, a 2015 analysis of the literature challenged this long-held notion.[25] That study reported that, based on a survey of 13 studies that measured both cardiac output and $(a-\bar{v})O_2$ difference before and after training, improvements in $\dot{V}O_{2max}$ following 5 to 13 weeks of training were associated with increases in cardiac output, but not in $(a-\bar{v})O_2$ difference. That an increase in maximal cardiac output is the predominant factor associated with increases in $\dot{V}O_{2max}$ is not surprising, given the close relation between these variables shown in figure 11.7a. However, the training period in the studies analyzed was relatively short, so training adaptations may not have been complete. In longer term endurance training studies, maximal $(a-\bar{v})O_2$ differences were enhanced by 1% to 29%[25].

In summary, the respiratory system is quite adept at bringing adequate oxygen into the body. For this reason, the respiratory system seldom limits endurance performance. Not surprisingly, the major training adaptations noted in the respiratory system are apparent mainly during maximal exercise, when all systems are being maximally stressed.

In Review

- Unlike what happens with the cardiovascular system, endurance training has little effect on lung structure and function.
- To support increases in $\dot{V}O_{2max}$, there is an increase in pulmonary ventilation during maximal effort following training as both tidal volume and respiratory rate increase.
- Pulmonary diffusion at maximal intensity increases, especially to upper regions of the lung that are not normally perfused.
- Although the largest part of the increase in $\dot{V}O_{2max}$ results from the increases in cardiac output and muscle blood flow, an increase in $(a-\bar{v})O_2$ difference also plays a key role.
- This increase in $(a-\bar{v})O_2$ difference is attributable to a more effective distribution of arterial blood away from inactive tissue to the active tissue and an increased ability of active muscle to extract oxygen.

Adaptations in Muscle

Repeated excitation and contraction of muscle fibers during endurance training stimulate changes in their structure and function. Our main interest here is in aerobic training and the changes it produces in muscle fiber type, mitochondrial function, and oxidative enzymes.

Muscle Fiber Type

As noted in chapter 1, low- to moderate-intensity aerobic activities rely extensively on type I (slow-twitch) fibers. In response to aerobic training, type I fibers become larger. More specifically, they develop a larger cross-sectional area, although the magnitude of change depends on the intensity and duration of each training bout and the length of the training program. Increases in cross-sectional area of up to 25% have been reported. Fast-twitch (type II) fibers, because they are not being recruited to the same extent during endurance exercise, generally do not increase cross-sectional area.

Most early studies showed no change in the percentage of type I versus type II fibers following aerobic training, but subtle changes were noted among the different type II fiber subtypes. Type IIx fibers have a low oxidative capacity and are recruited less often than type IIa fibers during aerobic exercise. However, during long-duration exercise, these fibers may eventually be recruited to perform in a manner resembling type IIa fibers. This can cause some type IIx fibers to take on the characteristics of the more oxidative type IIa fibers. Recent evidence suggests that not only is there a transition of type IIx to IIa fibers but also there can be a transition of type II to type I fibers. The magnitude of change is generally small, not more than a few percent. As an example, in the HERITAGE Family Study,[28] a 20-week program of aerobic training increased type I fibers from 43% pretraining to almost 47% posttraining and decreased type IIx fibers from 20% to 15%, with type IIa remaining essentially unchanged. These more recent studies have included larger numbers of subjects and have taken advantage of improved measurement technology; both might explain why fiber type composition changes within a muscle are now recognized.

Capillary Supply

One of the most important adaptations to aerobic training is an increase in the number of capillaries surrounding each muscle fiber. Table 11.2 illustrates that endurance-trained men have considerably more capillaries in their leg muscles than sedentary individuals.[15] With long periods of aerobic training, the number of capillaries may increase by more than 15%.[28] Having more capillaries allows for greater exchange of gases, heat, nutrients, and metabolic by-products between the blood and contracting muscle fibers. In fact, the increase in capillary density (i.e., increase in capillaries per muscle fiber) is potentially one of the most important alterations in response to training that causes the increase in $\dot{V}O_{2max}$. It is now clear that the diffusion of oxygen from the capillary to the mitochondria is a major factor limiting the maximal rate of oxygen consumption by the muscle. Increasing capillary density facilitates this diffusion, thus maintaining an environment well suited to energy production and repeated muscle contractions.

Myoglobin Content

When oxygen enters the muscle fiber, it binds to myoglobin, a molecule similar to hemoglobin. This iron-containing molecule shuttles the oxygen molecules from the cell membrane to the mitochondria. Type I fibers contain large quantities of myoglobin, which gives these fibers their red appearance (myoglobin is a pigment that turns red when bound to oxygen). Type II fibers, on the other hand, are highly glycolytic, so they contain (and require) little myoglobin—hence their whiter appearance. More important, their limited myoglobin supply limits their oxidative capacity, resulting in poor endurance for these fibers.

Myoglobin transports oxygen and releases it to the mitochondria when oxygen becomes limited during muscle action. This oxygen reserve is used during the transition from rest to exercise, providing oxygen to the mitochondria during the lag between the beginning of exercise and the increased cardiovascular delivery of oxygen. Endurance training has been shown to increase muscle myoglobin content by 75% to 80%. This adaptation clearly supports a muscle's increased capacity for oxidative metabolism after training.

Mitochondrial Function

As noted in chapter 2, oxidative energy production takes place in the mitochondria. Not surprisingly, aerobic training also induces changes in mitochondrial function that improve the muscle fibers' capacity to produce ATP. The ability to use oxygen and produce ATP via oxidation depends on the number and size of the muscle mitochondria. Both increase with aerobic training.

During one study that involved endurance training in rats, the number of mitochondria increased approximately 15% during 27 weeks of exercise.[16] Average mitochondrial size also increased by about 35% over that training period. As with other training-induced adaptations, the magnitude of change depends on training volume.

Not all mitochondria within a muscle fiber are equally efficient, as new mitochondria are constantly being formed (biogenesis) and old, weakened mitochondria are being cleared (mitophagy) (see figure 11.9). Regulation of this mitochondrial turnover cycle determines not only the number of mitochondria in a fiber but also the overall quantity and function of those mitochondria,[36] which in turn determine overall metabolic function and performance of skeletal muscles. There has been an explosion of new research aimed at understanding the underlying molecular mechanisms that regulate mitochondrial biogenesis, the process by which new mitochondria are formed. These efforts resulted in the discovery of *peroxisome proliferator-activated receptor-γ coactivator-1α (PGC-1α)*, a key regulator protein

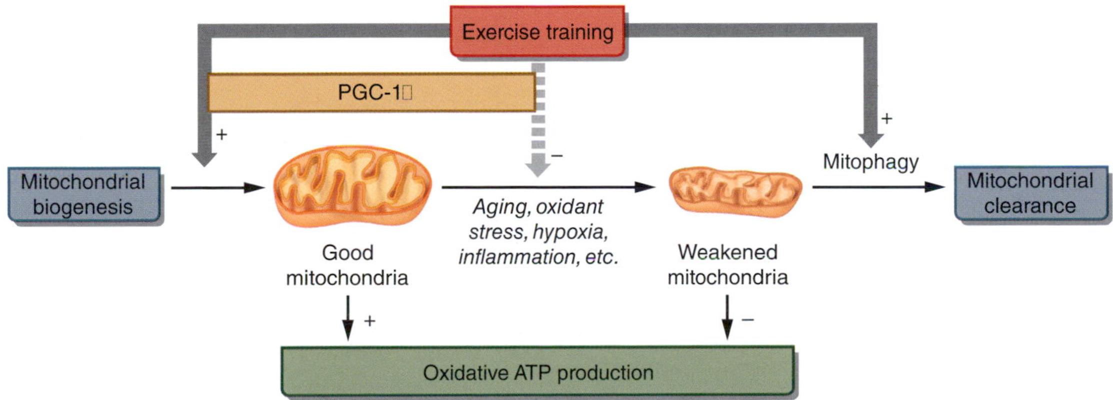

FIGURE 11.9 Endurance exercise training affects the quality of muscle mitochondria by increasing the production of new, healthy mitochondria (biogenesis), decreasing the degradation of mitochondria, and clearing away damaged mitochondria (mitophagy). The first two processes are controlled by the regulator protein PGC-1α. Solid arrows indicate a positive effect while dotted arrows indicate a negative effect.

that is integrally involved in mitochondrial biogenesis in skeletal muscle. Because of its multiple important roles in enhancing metabolic function, PGC-1α is often called the master regulator or master switch. It is also now well established that both acute exercise and exercise training—both endurance and resistance exercise—enhance PGC-1α expression.

As shown in figure 11.9, exercise training promotes biogenesis of new mitochondria, slows the decline in mitochondrial function by remodeling mitochondria through processes of fusion and fission, and helps maintain mitophagy in skeletal muscle. Thus, mitochondrial quality control is an important exercise-induced adaptation.[36]

Increased expression of PGC-1α protein can be measured in skeletal muscle even after a single bout of exercise; after two or three repeated bouts, markers for mitochondrial biogenesis can be observed. Increased PGC-1α not only increases mitochondrial biogenesis but also controls the replacement of old weakened mitochondria with new healthy mitochondria. Mitochondrial damage induced by such insults as hypoxia, inflammation, or increased oxidant stress can lead to the accumulation of metabolic by-products that impair mitochondrial function. Although addition of new mitochondria is of extreme importance, the maintenance of a healthy population of mitochondria is equally critical for optimal metabolic capacity. Continuous removal of damaged mitochondria is likewise important for optimal function of skeletal muscle.

Oxidative Enzymes

Regular endurance exercise has been shown to induce major adaptations in skeletal muscle, including an increase in the number and size of the muscle fiber mitochondria as just discussed. These changes are further enhanced by an increase in mitochondrial capacity. The oxidative breakdown of fuels and the ultimate production of ATP depend on the action of **mitochondrial oxidative enzymes**, the specialized proteins that catalyze (i.e., speed up) the breakdown of nutrients to form ATP. Aerobic training increases the activity of these important enzymes.

Figure 11.10 illustrates the changes in the activity of succinate dehydrogenase (SDH), a key muscle oxidative enzyme, over 7 months of progressive swim training. While the rate of increases in $\dot{V}O_{2max}$ slowed after the first 2 months of training, activity of this key oxidative enzyme continued to increase throughout the entire training period. This suggests that training-induced increases in $\dot{V}O_{2max}$ might be limited more by the circulatory system's ability to transport oxygen than by the muscles' oxidative potential.

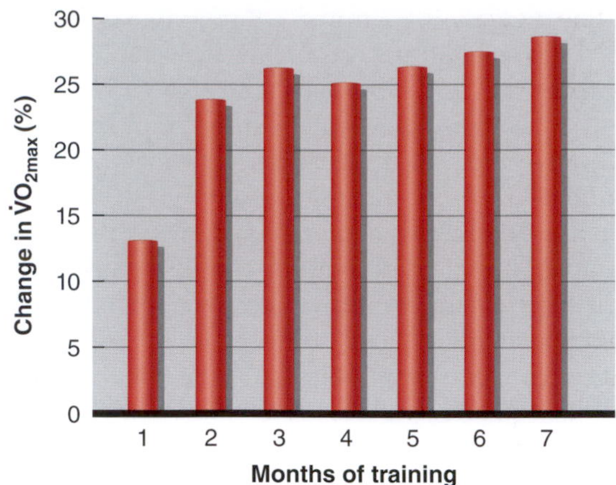

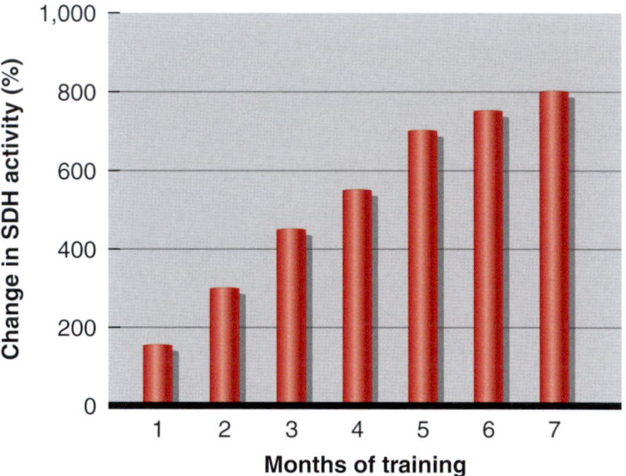

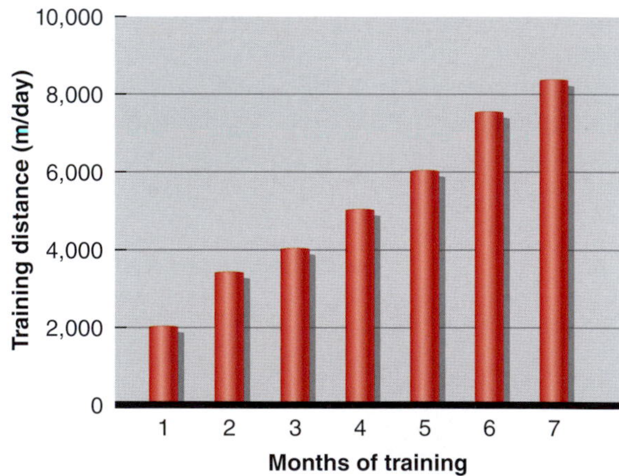

FIGURE 11.10 The percentage change in maximal oxygen uptake ($\dot{V}O_{2max}$) and the activity of succinate dehydrogenase (SDH), one of the muscles' key oxidative enzymes, during 7 months of swim training. Interestingly, although this enzyme activity continues to increase with increasing levels of training, the swimmers' maximal oxygen uptake appears to level off after the first 8 to 10 weeks of training. This implies that mitochondrial enzyme activity is not a direct indication of whole-body endurance capacity.

The activities of muscle enzymes such as SDH and citrate synthase are dramatically influenced by aerobic training. This is seen in figure 11.11, which compares the activities of these enzymes in untrained people, moderately trained joggers, and highly trained runners.[9] Even moderate daily exercise increases the activity of these enzymes and thus the oxidative capacity of the muscle. For example, jogging or cycling for as little as 20 min per day has been shown to increase SDH activity in leg muscles by more than 25%. Training more vigorously—for example, for 60 to 90 min per day—produces a two- to threefold increase in this enzyme's activity.

One metabolic consequence of mitochondrial changes induced by aerobic training is **glycogen sparing**, a slower rate of utilization of muscle glycogen and enhanced reliance on fat as a fuel source at a given exercise intensity. Enhanced glycogen sparing with endurance training most likely improves the ability to sustain a higher exercise intensity, such as maintaining a faster race pace in a 10 km run.

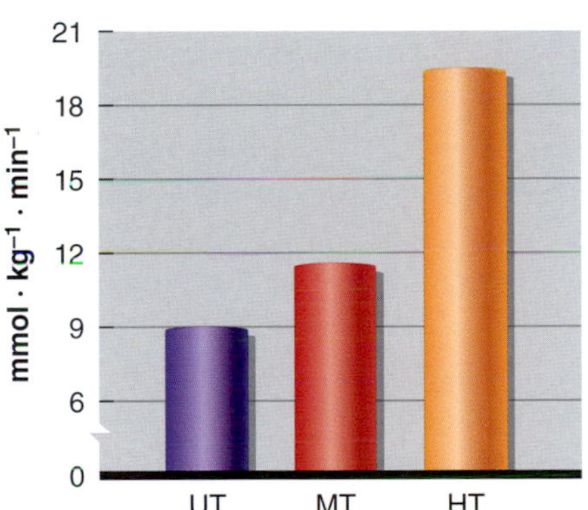

a Succinate dehydrogenase

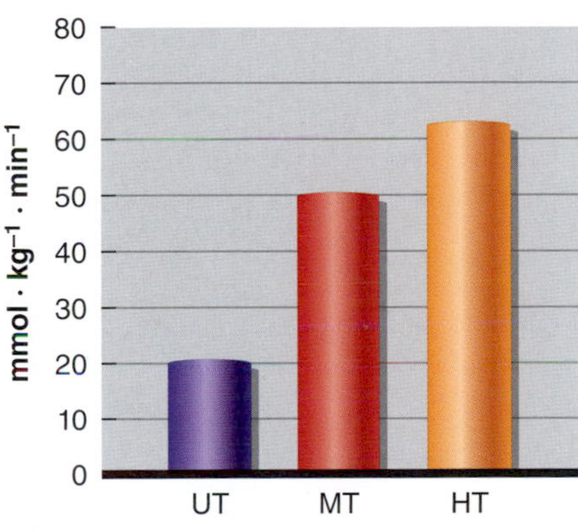

b Citrate synthase

FIGURE 11.11 Leg muscle (gastrocnemius) enzyme activities of untrained (UT) subjects, moderately trained (MT) joggers, and highly trained (HT) marathon runners. Enzyme levels are shown for two of many key enzymes that participate in the oxidative production of adenosine triphosphate.
Adapted from Costill, Fink, Lesmes, et al. (1979); Costill, Coyle, Fink, et al. (1979).

In summary, endurance exercise training causes a wide variety of phenotypic adaptations in skeletal muscle, including angiogenesis (creation of new capillaries), transformation of fiber types from glycolytic to oxidative, increased ability to mobilize and use fats as a substrate, and increased glucose uptake by muscle fibers, which increases the number of mitochondria and improves the overall quality of the existing mitochondrial pool.

In Review

- Aerobic training selectively recruits type I muscle fibers and fewer type II fibers. Consequently, the type I fibers increase their cross-sectional area with aerobic training.
- After training, there appears to be a small increase in the percentage of type I fibers, as well as a transition of some type IIx to type IIa fibers.
- Aerobic training increases both the number of capillaries per muscle fiber and the number of capillaries for a given cross-sectional area of muscle. These changes improve blood perfusion through the muscles, enhancing the diffusion of oxygen, carbon dioxide, nutrients, and by-products of metabolism between the blood and muscle fibers.
- Aerobic training increases muscle myoglobin content by as much as 80%. Myoglobin transports oxygen from cell membranes to the mitochondria.
- Aerobic training increases both the number and the size of muscle fiber mitochondria, providing the muscle with an increased capacity for oxidative metabolism.
- Endurance exercise training also improves the overall quality of the existing mitochondrial pool.
- Activities of many oxidative enzymes are increased with aerobic training.
- These changes occurring in the muscles, combined with adaptations in the oxygen transport system, enhance the capacity of oxidative metabolism and improve endurance performance.

Metabolic Adaptations to Training

Now that we have discussed training changes in both the cardiovascular and respiratory systems, as well as skeletal muscle adaptations, we are ready to examine how these integrated adaptations are reflected by changes in three important physiological variables related to metabolism:

- Lactate threshold
- Respiratory exchange ratio
- Oxygen consumption

Lactate Threshold

Lactate threshold, discussed in chapter 5, is a physiological marker that is closely associated with endurance performance—the higher the lactate threshold, the better the performance capacity. Figure 11.12*a* illustrates the difference in lactate threshold between an endurance-trained individual and an untrained individual. This figure also accurately represents the changes in lactate threshold that would occur following a 6- to 12-month program of endurance training. In either case, in the trained state, one can exercise at a higher percentage of one's $\dot{V}O_{2max}$ before lactate begins to accumulate in the blood. In this example, the trained runner could sustain a race pace of 70% to 75% of $\dot{V}O_{2max}$, an intensity that would result in continued lactate accumulation in the blood of the untrained runner. This translates into a much faster race pace (see figure 11.12*b*). Above the lactate threshold, the lower lactate at a given rate of work is likely attributable to a combination of reduced lactate production and increased lactate clearance. As athletes become better trained, their postexercise blood lactate concentrations are lower for a given rate of work.

Respiratory Exchange Ratio

Recall from chapter 5 that the respiratory exchange ratio (RER) is the ratio of carbon dioxide released to oxygen consumed during metabolism. The RER reflects the composition of the mixture of substrates being used as an energy source, with a lower RER reflecting an increased reliance on fats for energy production and a higher RER reflecting a higher contribution of carbohydrates.

After training, the RER decreases at both absolute and relative submaximal exercise intensities. These changes are attributable to a greater utilization of free fatty acids instead of carbohydrate at these work rates following training.

Resting and Submaximal Oxygen Consumption

Oxygen consumption ($\dot{V}O_2$) at rest is unchanged following endurance training. While a few cross-sectional comparisons have suggested that training elevates resting $\dot{V}O_2$, the HERITAGE Family Study—with a large number of subjects and with duplicate measures of resting metabolic rate both before and after 20 weeks of training—showed no evidence of an increased resting metabolic rate after training.[35]

During submaximal exercise at a given intensity, $\dot{V}O_2$ is either unchanged or slightly reduced following training. In the HERITAGE Family Study, training reduced submaximal $\dot{V}O_2$ by 3.5% at a work rate of 50 W. There was a corresponding reduction in cardiac output at 50 W, reinforcing the strong interrelationship between $\dot{V}O_2$ and cardiac output.[34] This small decrease in $\dot{V}O_2$ during submaximal exercise, not seen in many studies, could have resulted from an increase in exercise economy (performing the same exercise intensity with less extraneous movement).

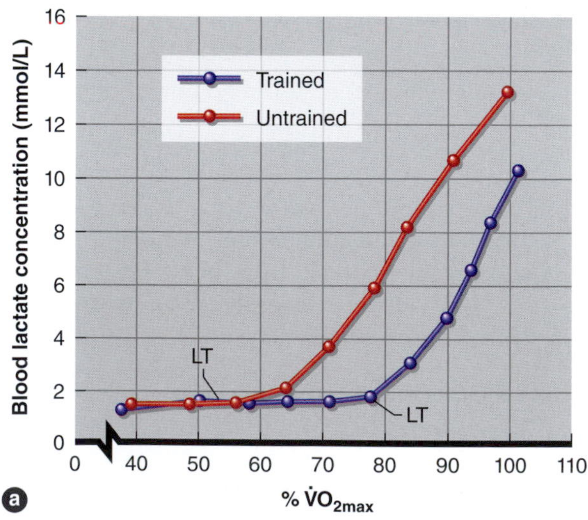

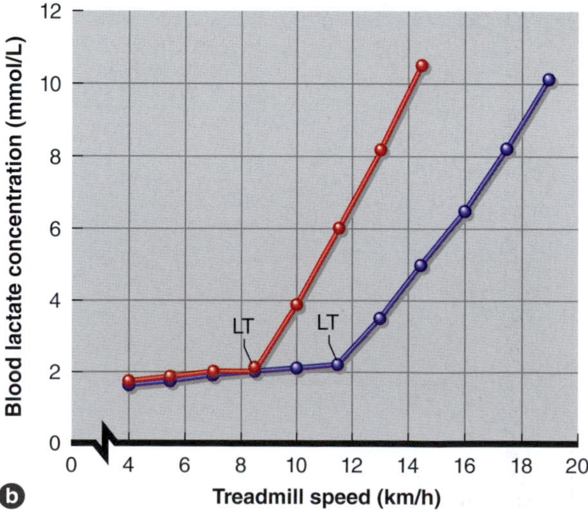

FIGURE 11.12 Changes in lactate threshold (LT) with training expressed as *(a)* a percentage of maximal oxygen uptake (% $\dot{V}O_{2max}$) and *(b)* an increase in speed on the treadmill. Lactate threshold occurs at a speed of 8.4 km/h (5.2 mph) in the untrained state and at 11.6 km/h (7.2 mph) in the trained state.

Maximal Oxygen Consumption

$\dot{V}O_{2max}$ is the best indicator of cardiorespiratory endurance capacity and increases substantially in response to endurance training. While small and very large increases have been reported, an increase of 15% to 20% is typical for a previously sedentary person who trains at 50% to 85% of his or her $\dot{V}O_{2max}$ three to five times per week, 20 to 60 min per day, for 6 months. For example, the $\dot{V}O_{2max}$ of a sedentary individual could reasonably increase from 35 ml · kg^{-1} · min^{-1} to 42 ml · kg^{-1} · min^{-1} as a result of such a program. This is far below the values we see in world-class endurance athletes, whose values generally range from 70 to 94 ml · kg^{-1} · min^{-1}. The more sedentary an individual is when starting an exercise program, the larger the increase in $\dot{V}O_{2max}$.

Integrated Adaptations to Chronic Endurance Exercise

It should now be clear that the adaptations that accompany endurance training are many and that they affect multiple physiological systems. Physiologists commonly establish models to help explain how various physiological factors or variables work together to affect a specific outcome or component of performance. Dr. Donna H. Korzick, an exercise physiologist at Pennsylvania State University, has created a unifying figure to model the factors that contribute to the cardiovascular adaptation to chronic endurance training (see figure 11.13).

What Limits Aerobic Power and Endurance Performance?

A number of years ago, exercise scientists were divided on what major physiological factor or factors actually limit $\dot{V}O_{2max}$. Two contrasting theories had been proposed.

One theory held that endurance performance was limited by the lack of sufficient concentrations of oxidative enzymes in the mitochondria. Endurance training programs substantially increase these oxidative enzymes, allowing active tissue to use more of the available

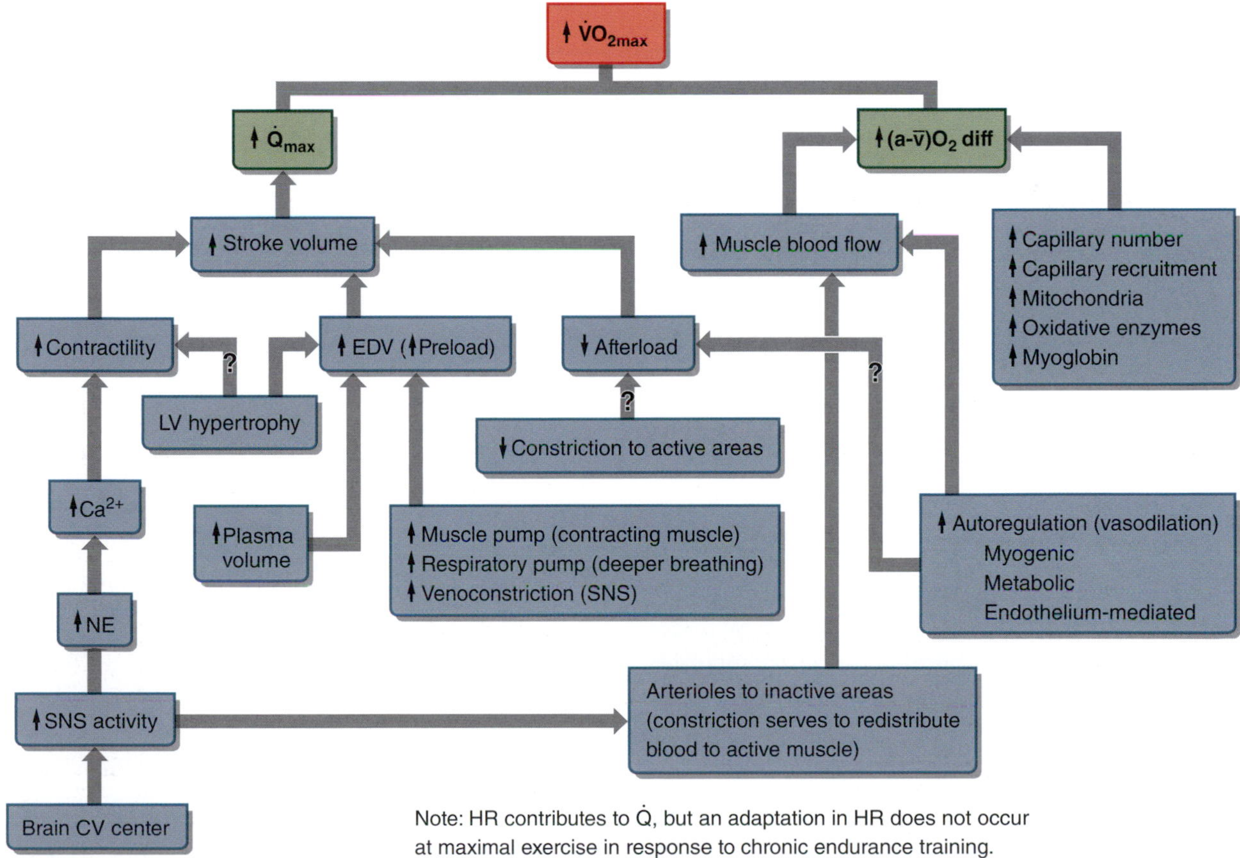

FIGURE 11.13 Cardiovascular adaptations to chronic endurance exercise.

Adapted by permission from Donna H. Korzick, Pennsylvania State University, 2006.

oxygen, resulting in a higher $\dot{V}O_{2max}$. In addition, endurance training increases both the size and number of muscle mitochondria. Thus, this theory argued, the main limitation of maximal oxygen consumption is the inability of the existing mitochondria to use the available oxygen beyond a certain rate. This theory was referred to as the utilization theory.

The second theory proposed that central and peripheral cardiovascular factors limit endurance capacity. These circulatory influences would preclude delivery of sufficient amounts of oxygen to the active tissues. Taking into account the observation that improvement in $\dot{V}O_{2max}$ following endurance training results from increased blood volume, increased cardiac output (via stroke volume), and a better perfusion of active muscle with blood, this theory proposed that these cardiovascular factors are the limiting factor for $\dot{V}O_{2max}$.

Evidence strongly supports the latter theory. In one study, subjects breathed a mixture of carbon monoxide (which irreversibly binds to hemoglobin, limiting hemoglobin's oxygen-carrying capacity) and air during exercise to exhaustion.[26] $\dot{V}O_{2max}$ decreased in direct proportion to the percentage of carbon monoxide breathed. The carbon monoxide molecules bonded to approximately 15% of the total hemoglobin; this percentage agreed with the percentage reduction in $\dot{V}O_{2max}$. In another study, approximately 15% to 20% of each subject's total blood volume was removed.[11] $\dot{V}O_{2max}$ decreased by approximately the same relative amount. Reinfusion of the subjects' packed red blood cells approximately 4 weeks later increased $\dot{V}O_{2max}$ well above baseline or control conditions. In both studies, the reduction in the oxygen-carrying capacity of the blood—via either blocking hemoglobin or removing whole blood—resulted in the delivery of less oxygen to the active tissues and a corresponding reduction in $\dot{V}O_{2max}$. Similarly, studies have shown that breathing oxygen-enriched mixtures, in which the partial pressure of oxygen in the inspired air is substantially increased, increases endurance capacity.

These and subsequent studies indicated that the available oxygen supply is the major limiter of endurance performance. Oxygen transport to the working muscles, not the available mitochondria and oxidative enzymes, limits $\dot{V}O_{2max}$. The argument was that increases in $\dot{V}O_{2max}$ with training are largely attributable to increased maximal blood flow and increased muscle capillary density in the active tissues. Skeletal muscle adaptations (including increased mitochondrial content and respiratory capacity of the muscle fibers) contribute importantly to the ability to perform prolonged, high-intensity, submaximal exercise.

Table 11.3 summarizes the typical physiological changes that occur with endurance training. The values (pre- and posttraining) for a previously inactive man are compared with values for a world-class male endurance runner.

In Review

- Lactate threshold increases with endurance training, allowing performance of higher exercise intensities without significantly increasing blood lactate concentration.
- With endurance training, the RER decreases at submaximal work rates, indicating greater utilization of free fatty acids as an energy substrate (carbohydrate sparing).
- Oxygen consumption generally remains unchanged at rest and remains unaltered or decreases slightly during submaximal exercise following endurance training.
- $\dot{V}O_{2max}$ increases substantially following endurance training, but the extent of increase possible is genetically limited in each individual. The major limiting factor appears to be oxygen delivery to the active muscles.

Long-Term Improvement in Aerobic Power and Cardiorespiratory Endurance

Although an individual's highest attainable $\dot{V}O_{2max}$ is usually achieved within 12 to 18 months of intense endurance training, endurance *performance* can continue to improve. Improvement in endurance performance without improvement in $\dot{V}O_{2max}$ is likely attributable to improvements in the ability to perform at increasingly higher percentages of $\dot{V}O_{2max}$ for extended periods. Consider, for example, a young male runner who starts training with an initial $\dot{V}O_{2max}$ of 52.0 ml · kg^{-1} · min^{-1}. He reaches his genetically determined peak $\dot{V}O_{2max}$ of 71.0 ml · kg^{-1} · min^{-1} after 2 years of intense training, after which no further increases occur, even with more frequent or more intense workouts. At this point, as shown in figure 11.14, the young runner is able to run at 75% of his $\dot{V}O_{2max}$ (0.75 × 71.0 = 53.3 ml · kg^{-1} · min^{-1}) in a 10 km (6.2 mi) race. After an additional 2 years of intensive training, his $\dot{V}O_{2max}$ is unchanged, but he is now able to compete at 88% of his $\dot{V}O_{2max}$ (0.88 × 71.0 = 62.5 ml · kg^{-1} · min^{-1}). Obviously, by being able to sustain an oxygen uptake of 62.5 ml · kg^{-1} · min^{-1}, he is able to run at a much faster race pace.

This ability to sustain exercise at a higher percentage of $\dot{V}O_{2max}$ is partly the result of an increase in the ability to buffer lactate, because race pace is directly related to the $\dot{V}O_2$ value at which lactate begins to accumulate.

TABLE 11.3 **Typical Effects of Endurance Training in a Previously Inactive Man, Contrasted with Values for a Male World-Class Endurance Athlete**

Variables	Pretraining, sedentary male	Posttraining, sedentary male	World-class endurance athlete
	Cardiovascular		
HR_{rest} (beats/min)	75	65	45
HR_{max} (beats/min)	185	183	174
SV_{rest} (ml/beat)	60	70	100
SV_{max} (ml/beat)	120	140	200
$\dot{Q}$ at rest (L/min)	4.5	4.5	4.5
$\dot{Q}_{max}$ (L/min)	22.2	25.6	34.8
Heart volume (ml)	750	820	1,200
Blood volume (L)	4.7	5.1	6.0
Systolic BP at rest (mmHg)	135	130	120
Systolic BP_{max} (mmHg)	200	210	220
Diastolic BP at rest (mmHg)	78	76	65
Diastolic BP_{max} (mmHg)	82	80	65
	Respiratory		
$\dot{V}_E$ at rest (L/min)	7	6	6
$\dot{V}_{E\,max}$ (L/min)	110	135	195
TV at rest (L)	0.5	0.5	0.5
TV_{max} (L)	2.75	3.00	3.90
VC (L)	5.8	6.0	6.2
RV (L)	1.4	1.2	1.2
	Metabolic		
$(a\text{-}\bar{v})O_2$ diff at rest (ml/100 ml)	6.0	6.0	6.0
$(a\text{-}\bar{v})O_2$ diff max (ml/100 ml)	14.5	15.0	16.0
$\dot{V}O_2$ at rest $(ml \cdot kg^{-1} \cdot min^{-1})$	3.5	3.5	3.5
$\dot{V}O_{2max}$ $(ml \cdot kg^{-1} \cdot min^{-1})$	40.7	49.9	81.9
Blood lactate at rest (mmol/L)	1	1	1
Blood lactate max (mmol/L)	7.5	8.5	9.0
	Body composition		
Weight (kg)	79	77	68
Fat weight (kg)	12.6	9.6	5.1
Fat-free weight (kg)	66.4	67.4	62.9
Fat (%)	16.0	12.5	7.5

Note. HR = heart rate; SV = stroke volume; $\dot{Q}$= cardiac output; BP = blood pressure; $\dot{V}_E$ = ventilation; TV = tidal volume; VC = vital capacity; RV = residual volume; $(a\text{-}\bar{v})O_2$ diff = arterial–mixed venous oxygen difference; $\dot{V}O_2$ = oxygen consumption.

Factors Affecting an Individual's Response to Aerobic Training

We have discussed general trends in adaptations that occur in response to endurance training. However, we must always remember that we are talking about adaptations in individuals and that everyone does not respond in the same manner. Several factors that can affect individual response to aerobic training must be considered.

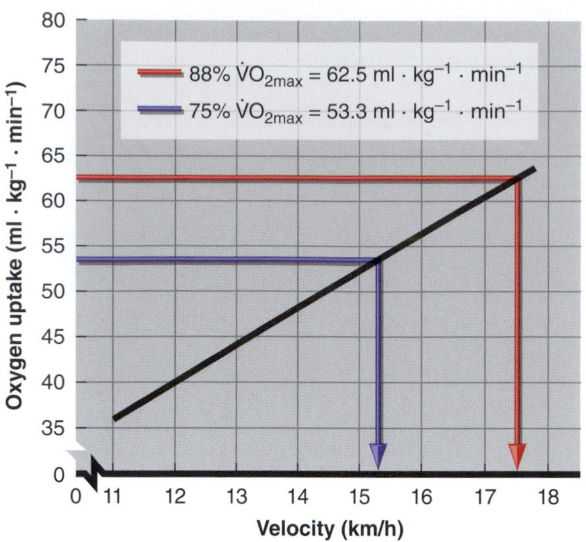

FIGURE 11.14 Change in race pace with continued training after maximal oxygen uptake stops increasing beyond 71 ml · kg^{-1} · min^{-1}.

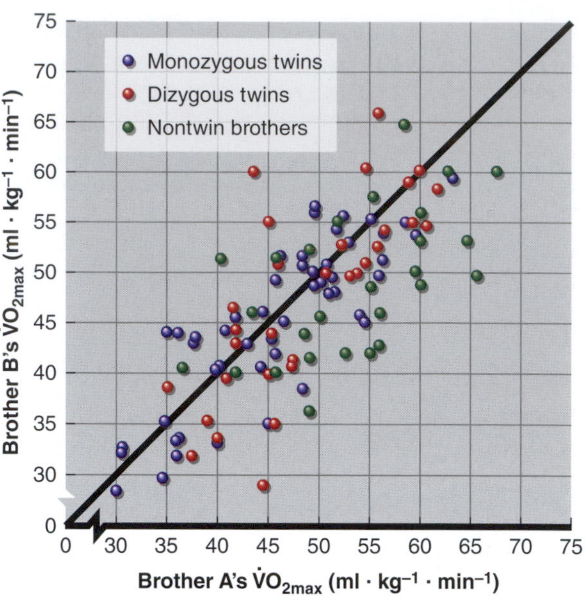

FIGURE 11.15 Comparisons of $\dot{V}O_{2max}$ in twin (monozygous and dizygous) and nontwin brothers.

Adapted by permission from C. Bouchard et al., "Aerobic Performance in Brothers, Dizygotic and Monozygotic Twins," *Medicine and Science in Sports and Exercise* 18 (1986): 639-646.

Training Status and $\dot{V}O_{2max}$

The higher the initial state of conditioning, the smaller the relative improvement for the same volume of training. For example, if two people, one sedentary and the other partially trained, undergo the same endurance training program, the sedentary person will show the greatest relative (%) improvement.

In fully mature athletes, the highest attainable $\dot{V}O_{2max}$ is reached within 8 to 18 months of intense endurance training, indicating that each athlete has a finite maximal attainable level of oxygen consumption. This finite range is genetically determined but may potentially be influenced by training in early childhood during the development of the cardiovascular system.

Heredity

The ability to increase maximal oxygen consumption levels is genetically limited. This does not mean that each individual has a preprogrammed $\dot{V}O_{2max}$ that cannot be exceeded. Rather, a range of $\dot{V}O_{2max}$ values seems to be predetermined by an individual's genetic makeup, with that individual's highest attainable $\dot{V}O_{2max}$ somewhere in that range. Each individual is born into a predetermined genetic window, and the person can shift up or down within that window with exercise training or detraining, respectively.

Research on the genetic basis of $\dot{V}O_{2max}$ began in the late 1960s and early 1970s. Recent research has shown that identical (monozygous) twins have similar $\dot{V}O_{2max}$ values, whereas the variability for dizygous (fraternal) twins is much greater (see figure 11.15).[5] Each symbol represents a pair of brothers. Brother A's $\dot{V}O_{2max}$ value is indicated by the symbol's position on the *x*-axis, and brother B's $\dot{V}O_{2max}$ value is on the *y*-axis. Similarity in the siblings' $\dot{V}O_{2max}$ values is noted by comparing the *x* and *y* coordinates of the symbol (i.e., how close it falls to the diagonal line $x = y$ on the graph). Similar results were found for endurance capacity, determined by the maximal amount of work performed in an all-out, 90 min ride on a cycle ergometer.

Bouchard and colleagues[4] concluded that heredity accounts for between 25% and 50% of the variance in $\dot{V}O_{2maxx}$ values. This means that of all factors influencing $\dot{V}O_{2max}$, heredity alone is responsible for one-quarter to one-half of the total influence. World-class athletes who have stopped endurance training continue for many years to have high $\dot{V}O_{2max}$ values in their sedentary, deconditioned state. Their $\dot{V}O_{2max}$ values may decrease from 85 to 65 ml · kg^{-1} · min^{-1}, but this deconditioned value is still very high compared with that of the general population.

Heredity also potentially explains the fact that some people have relatively high $\dot{V}O_{2max}$ values yet have no history of endurance training. In a study that compared untrained men who had $\dot{V}O_{2max}$ values below 49 ml · kg^{-1} · min^{-1} with untrained men who had $\dot{V}O_{2max}$ values above 62.5 ml · kg^{-1} · min^{-1}, those with high values were distinguished by having higher blood volumes, which contributed to higher stroke volumes and cardiac outputs at maximal intensities. The higher blood volumes in the high $\dot{V}O_{2max}$ group were most likely genetically determined.[11]

Thus, both genetic and environmental factors influence $\dot{V}O_{2max}$ values. The genetic factors probably establish the boundaries for the athlete, but endurance training can push $\dot{V}O_{2max}$ to the upper limit of these boundaries. Dr. Per-Olof Åstrand, one of the most highly recognized exercise physiologists during the second half of the 20th century, stated on numerous occasions that the best way to become a champion Olympic athlete is to be selective when choosing one's parents!

Sex

Healthy untrained girls and women have significantly lower $\dot{V}O_{2max}$ values (20%-25% lower) than healthy untrained boys and men. Highly conditioned female endurance athletes have values much closer to those of highly trained male endurance athletes (i.e., only about 10% lower). This is discussed in greater detail in chapter 19. Representative ranges of $\dot{V}O_{2max}$ values for athletes and nonathletes are presented in table 11.4 by age, sex, and sport.

High Responders and Low Responders

For years, researchers have found wide variations in the amount of improvement in $\dot{V}O_{2max}$ with endurance training. Studies have demonstrated individual improvements in $\dot{V}O_{2max}$ ranging from 0% to 50% or more, even in similarly fit subjects completing exactly the same training program.

TABLE 11.4 **Maximal Oxygen Uptake Values ($ml \cdot kg^{-1} \cdot min^{-1}$) for Nonathletes and Athletes**

Group or sport	Age group (years)	Males	Females
Nonathletes	10-19	47-56	38-46
	20-29	43-52	33-42
	30-39	39-48	30-38
	40-49	36-44	26-35
	50-59	34-41	24-33
	60-69	31-38	22-30
	70-79	28-35	20-27
Baseball and softball	18-32	48-56	52-57
Basketball	18-30	40-60	43-60
Bicycling	18-26	62-74	47-57
Canoeing	22-28	55-67	48-52
Football	20-36	42-60	
Gymnastics	18-22	52-58	36-50
Ice hockey	10-30	50-63	
Jockey	20-40	50-60	
Orienteering	20-60	47-53	46-60
Racquetball	20-35	55-62	50-60
Rowing	20-35	60-72	58-65
Skiing, alpine	18-30	57-68	50-55
Skiing, Nordic	20-28	65-94	60-75
Ski jumping	18-24	58-63	
Soccer	22-28	54-64	50-60
Speed skating	18-24	56-73	44-55
Swimming	10-25	50-70	40-60
Track and field, discus	22-30	42-55	*
Track and field, running	18-39	60-85	50-75
	40-75	40-60	35-60
Track and field, shot put	22-30	40-46	*
Volleyball	18-22		40-56
Weightlifting	20-30	38-52	*
Wrestling	20-30	52-65	

*Data not available.

In the past, exercise physiologists have assumed that these variations result from differing degrees of compliance with the training program. People who comply with the program should, and do, have the highest percentage of improvement, and poor compliers should show little or no improvement. However, given the same training stimulus and full compliance with the program, substantial variations still occur in the percent improvement in $\dot{V}O_{2max}$ for different people.

It is now evident that some of the response to a training program is also genetically determined. This is illustrated in figure 11.16. Ten pairs of identical twins completed a 20-week endurance training program; the improvements in $\dot{V}O_{2max}$, expressed as percentages, are plotted for each twin pair—twin A on the *x*-axis and twin B on the *y*-axis.[27] Notice the similarity in response of each twin pair. Yet across twin pairs, improvement in $\dot{V}O_{2max}$ varied from 0% to 40%. These results, and those from other studies, indicate that there will be **high responders** (showing large improvement) and **low responders** (showing little or no improvement) among groups of people who participate in identical training programs. However, while genetic variants may be involved, such variants appear to be associated with the physiological mechanisms (increased cardiac output, expanded blood volume, improved muscle oxygen extraction) that underpin such differences.[20]

Results from the HERITAGE Family Study also support a strong genetic component in the magnitude of increase in $\dot{V}O_{2max}$ with endurance training. Families, including the biological mother and father and three or more of their children, trained 3 days a week for 20 weeks, initially exercising at a heart rate equal to 55% of their $\dot{V}O_{2max}$ for 35 min per day and progressing to a heart rate equal to 75% of their $\dot{V}O_{2max}$ for 50 min per day by the end of the 14th week, which they maintained for the last 6 weeks.[3] The average increase in $\dot{V}O_{2max}$ was about 17% but varied from 0% to more than 50%. Figure 11.17 illustrates the improvement in $\dot{V}O_{2max}$ for each subject in each family. Maximal heritability was estimated at 47%. Note that subjects who are high responders tend to be clustered in the same families, as are those who are low responders.

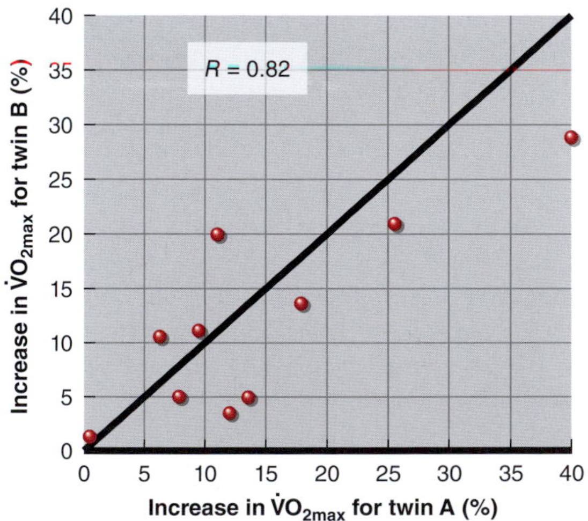

FIGURE 11.16 Variations in the percentage increase in $\dot{V}O_{2max}$ for identical twins undergoing the same 20-week training program.

Reprinted by permission from D. Prud'homme et al., "Sensitivity of Maximal Aerobic Power to Training is Genotype-Dependent," *Medicine and Science in Sports and Exercise* 16, no. 5 (1984): 489-493.

It is clear that this is a genetic phenomenon, not a result of compliance or noncompliance. One must consider this important point when conducting training studies and designing training programs. Individual differences must always be accounted for.

Cardiorespiratory Endurance in Nonendurance Sports

Many people regard cardiorespiratory endurance as the most important component of physical fitness. Low endurance capacity leads to fatigue, even in activities that are not aerobic. For any athlete, regardless of the sport or activity, fatigue represents a major deterrent to optimal performance. Even minor fatigue can hinder the athlete's total performance:

- Muscular strength is decreased.
- Reaction and movement times are prolonged.
- Agility and neuromuscular coordination are reduced.
- Whole-body movement speed is slowed.
- Concentration and alertness are reduced.

The decline in concentration and alertness associated with fatigue is particularly important. The athlete can become careless and more prone to serious injury, especially in contact sports. Even though these decrements in performance might be small, they can be just enough to cause an athlete to miss the critical free throw in basketball, the strike zone in baseball, or the 20 ft (6 m) putt in golf.

All athletes can benefit from improving their cardiorespiratory endurance. Even golfers, whose sport demands little in the way of aerobic endurance, can benefit. Improved endurance can allow golfers to complete a round of golf with less fatigue and to better withstand long periods of walking and standing.

For the sedentary, middle-aged adult, numerous health factors indicate that cardiovascular endurance should be the primary emphasis of training. Training

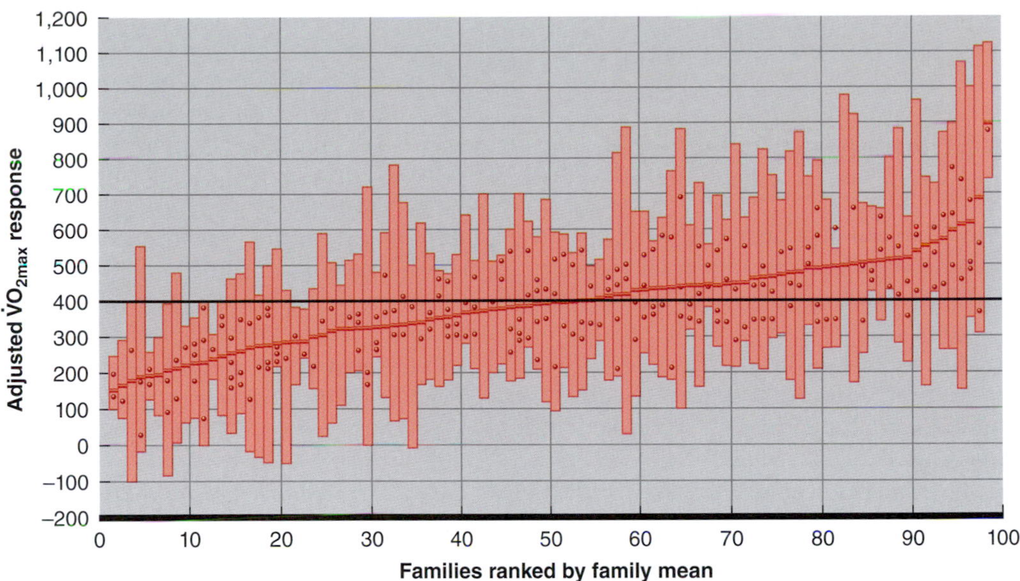

FIGURE 11.17 Variations in the improvement in $\dot{V}O_{2max}$ following 20 weeks of endurance training by families. Values represent the changes in $\dot{V}O_{2max}$ in ml/min, with an average increase of 393 ml/min. Data for each family are enclosed within a bar, and each family member's value is represented as a dot within the bar.

Adapted by permission from C. Bouchard, P. An, T. Rice, J.S. Skinner, J.H. Wilmore et al., "Familial Aggregation of $\dot{V}O_{2max}$ Response to Exercise Training. Results from HERITAGE Family Study," *Journal of Applied Physiology* 87 (1999): 1003-1008.

for health and fitness is discussed at length in part VII of this book.

The extent of endurance training needed varies considerably from one sport to the next and from one athlete to the next. It depends on the athlete's current endurance capacity and the endurance demands of the chosen activity. However, adequate cardiovascular conditioning must be the foundation of any athlete's general conditioning program.

In Review

- Although improvements in $\dot{V}O_{2max}$ eventually plateau, endurance performance can continue to improve for years with continued training.
- An individual's genetic makeup predetermines a range for that person's $\dot{V}O_{2max}$ and accounts for 25% to 50% of the variance in $\dot{V}O_{2max}$ values.
- Heredity also largely explains individual variations in response to identical training programs.
- Highly conditioned female endurance athletes have $\dot{V}O_{2max}$ values only about 10% lower than those of highly conditioned male endurance athletes.
- All athletes, regardless of their sport or event, can benefit from maximizing their cardiorespiratory endurance.

Aerobic Deconditioning

Issues related to deconditioning are particularly relevant to bed rest associated with diseases and disability as well as to the space program, since weightlessness and bed rest cause similar declines in $\dot{V}O_{2max}$. According to a recent analysis, 80 studies with a total of 949 participants have been published since 1949 that reported the effects of total bed rest on $\dot{V}O_{2max}$.[29] The studies were conducted mainly in young (age range 22-34 years), male (>90%) subjects with bed rest lasting from 1 to 90 days. Declines in $\dot{V}O_{2max}$ were fairly linear throughout periods of prolonged bed rest.

Surprisingly, while body weight and lean body mass both decline in response to bed rest, those changes were unrelated to the decline in $\dot{V}O_{2max}$. The most important predictor of how much $\dot{V}O_{2max}$ dropped was the subjects' fitness level at the beginning of the bed rest period. Higher initial $\dot{V}O_{2max}$ levels were associated with larger declines in $\dot{V}O_{2max}$.

Adaptations to Anaerobic Training

In muscular activities that require near-maximal force production for relatively short periods of time, such as sprinting, much of the energy needs are met by the ATP-phosphocreatine (PCr) system and the anaerobic breakdown of muscle glycogen (glycolysis). The following sections focus on the trainability of these two systems.

Changes in Anaerobic Power and Anaerobic Capacity

Exercise scientists have had difficulty agreeing on an appropriate laboratory or field test to measure anaerobic power. Unlike the situation with aerobic power, for which $\dot{V}O_{2max}$ is generally agreed to be the gold standard measurement, no single test adequately measures anaerobic power. Most research has been conducted through use of three different tests of either anaerobic power, anaerobic capacity, or both: the Wingate anaerobic test, the critical power test, and the maximal accumulated oxygen deficit test. Of these three, the Wingate test has been the most widely used. Despite the limitations inherent in each of these methods, they remain our only indirect indicators of the metabolic potential of anaerobic capacity.

As described in chapter 9, the Wingate anaerobic test is commonly used to measure anaerobic power. Peak power output, the highest mechanical power achieved during the first 5 to 10 s, is considered an index of anaerobic power. The mean power output is computed as the average power output over the total 30 s period, and one obtains total work simply by multiplying the mean power output by 30 s. Mean power output and total work have both been used as indexes of anaerobic capacity.

With anaerobic training, such as sprint training on the track or on a cycle ergometer, there are increases in both peak anaerobic power and anaerobic capacity. However, results have varied widely across studies, from those that showed only minimal increases to those showing increases of up to 25%.

Adaptations in Muscle with Anaerobic Training

With anaerobic training, which includes sprint training and resistance training, there are changes in skeletal muscle that specifically reflect muscle fiber recruitment for these types of activities. As discussed in chapter 1, at higher intensities, type II muscle fibers are recruited to a greater extent, but not exclusively, because type I fibers continue to be recruited. Overall, sprint and resistance activities use the type II muscle fibers significantly more than do aerobic activities. Consequently, both type IIa and type IIx muscle fibers undergo an increase in their cross-sectional areas. The cross-sectional area of type I fibers also is increased but usually to a lesser extent. Furthermore, with sprint training there appears to be a reduction in the percentage of type I fibers and an increase in the percentage of type II fibers, with the greatest change in type IIa fibers. In two of these studies, in which subjects performed 15 to 30 s all-out sprints, the type I percentage decreased from 57% to 48% and type IIa increased from 32% to 38%.[17,18] This shift of type I to type II fibers is not typically seen with resistance training.

Adaptations in the Energy Systems

Just as aerobic training produces changes in the aerobic energy system, anaerobic training alters the ATP-PCr and anaerobic glycolytic energy systems. These changes are not as obvious or predictable as those that result from endurance training, but they do improve performance in anaerobic activities.

Adaptations in the ATP-PCr System

Activities that emphasize maximal muscle force production, such as sprinting and weightlifting events, rely most heavily on the ATP-PCr system for energy. Maximal effort lasting less than about 6 s places the greatest demands on the breakdown and resynthesis of ATP and PCr. Costill and coworkers reported their findings from a study of resistance training and its effects on the ATP-PCr system.[8] Their participants trained by performing maximal knee extensions. One leg was trained using 6 s maximal work bouts that were repeated 10 times. This type of training preferentially stressed the ATP-PCr energy system. The other leg was trained with repeated 30 s maximal bouts, which instead preferentially stressed the glycolytic system.

The two forms of training produced the same muscular strength gains (about 14%) and the same resistance to fatigue. As seen in figure 11.18, the activities of the anaerobic muscle enzymes creatine kinase and myokinase increased as a result of the 30 s training bouts but were almost unchanged in the leg trained with repeated 6 s maximal efforts. This finding leads us to conclude that maximal sprint bouts (6 s) might improve muscular strength but contribute little to the mechanisms responsible for ATP and PCr breakdown. Data have been published, however, that show improvements in ATP-PCr enzyme activities with training bouts lasting only 5 s.

Regardless of the conflicting results, these studies suggest that the major value of training bouts that last only a few seconds (sprints) is the development of muscular strength. Such strength gains enable the individual to perform a given task with less effort, which reduces the risk of fatigue. Whether these changes allow the muscle to perform more anaerobic work remains unanswered, although a 60 s sprint-fatigue test suggests that short sprint-type anaerobic training does not enhance anaerobic endurance.[8]

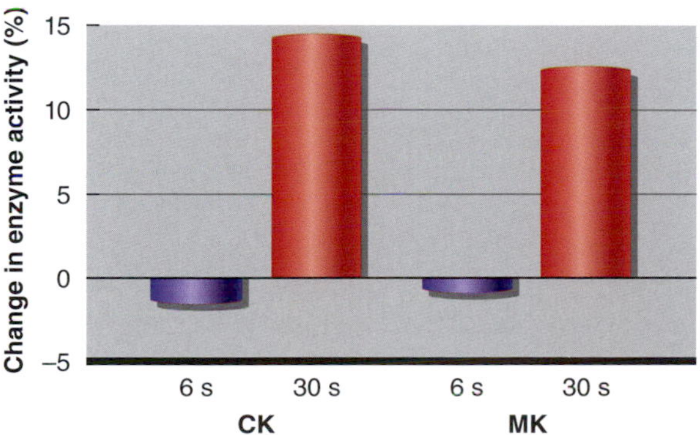

FIGURE 11.18 Changes in creatine kinase (CK) and muscle myokinase (MK) activities as a result of 6 s and 30 s bouts of maximal anaerobic training.

Adaptations in the Glycolytic System

Anaerobic training (30 s bouts) increases the activities of several key glycolytic enzymes. The most frequently studied glycolytic enzymes are phosphorylase, phosphofructokinase (PFK), and lactate dehydrogenase (LDH). The activities of these three enzymes increased 10% to 25% with repeated 30 s training bouts but changed little with short (6 s) bouts that stress primarily the ATP-PCr system.[8] In another study, 30 s maximal all-out sprints significantly increased hexokinase (56%) and PFK (49%) but not total phosphorylase activity or LDH.[21]

Because both PFK and phosphorylase are essential to the anaerobic yield of ATP, such training might enhance glycolytic capacity and allow the muscle to develop greater tension for a longer period of time. However, as seen in figure 11.19, this conclusion is not supported by results of a 60 s sprint performance test, in which the subjects performed maximal knee extension and flexion. Power output and the rate of fatigue (shown by a decrease in power production) were affected to the same degree after sprint training with either 6 s or 30 s training bouts. Thus, we must conclude that performance gains with these forms of training result from improvements in strength rather than improvements in the anaerobic yield of ATP.

Adaptations to High-Intensity Interval Training

In chapter 9 we introduced a special form of training using short bursts of very intense cycling, interspersed

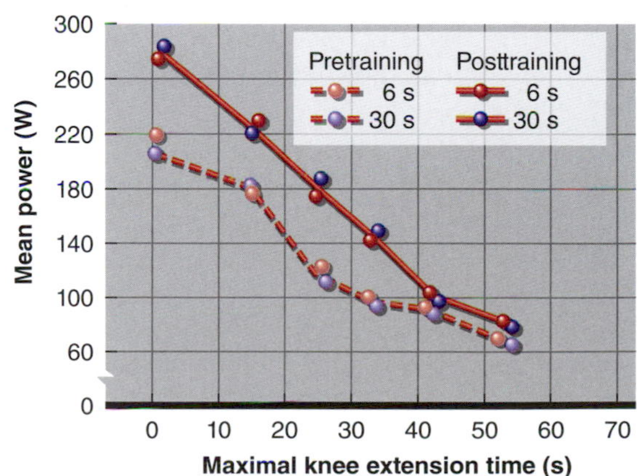

FIGURE 11.19 Performance in a 60 s sprint bout after training with 6 s and 30 s anaerobic bouts. Subjects are the same as in figure 11.18.

with up to a few minutes of rest or low-intensity cycling for recovery.[14] High-intensity interval training (HIIT) is a time-efficient way to induce many aerobic training benefits normally associated with continuous running, cycling, or swimming.

Adaptations to HIIT mirror those associated with more traditional aerobic training. In one study, untrained young subjects performed four to six 30 s sprints separated by 4 min of recovery, three times a week. These men showed the same beneficial changes in their heart, blood vessels, and muscles as another group who underwent a traditional training program involving up to an hour of continuous cycling, 5 days per week. Improvements in exercise performance—whether measured as cycling time to exhaustion at a fixed work intensity or in time trials that more closely resemble normal athletic competition—were comparable between groups, despite considerable differences in training time commitment.[14] High-intensity interval training appears to stimulate some of the same molecular signaling pathways that regulate skeletal muscle remodeling in response to endurance training, including mitochondrial biogenesis and changes in the capacity for carbohydrate and fat transport and oxidation.

In Review

- Anaerobic training bouts improve both anaerobic power and anaerobic capacity.
- The performance improvement noted with sprint-type anaerobic training appears to result more from strength gains than from improvements in the functioning of the anaerobic energy systems.
- Anaerobic training increases the ATP-PCr and glycolytic enzymes but has no effect on the oxidative enzymes.
- Conversely, aerobic training increases the oxidative enzymes but has little effect on the ATP-PCr or glycolytic enzymes.
- Adaptations to HIIT mirror those associated with more traditional aerobic training. High-intensity interval training appears to stimulate some of the same molecular signaling pathways that regulate skeletal muscle remodeling in response to endurance training, including mitochondrial biogenesis and changes in the capacity for carbohydrate and fat transport and oxidation.

RESEARCH PERSPECTIVE 11.2

Brief, Intense Stair Climbing

Low cardiorespiratory fitness is a strong predictor of cardiovascular disease and death. Public health guidelines generally recommend 150 min/week of moderate-intensity physical activity to achieve health benefits, but less than 15% of North Americans meet that recommendation. Lack of time and lack of necessary equipment are the two most commonly cited reasons for not achieving the recommended daily physical activity. Because of this, public health researchers and exercise physiologists are interested in finding easily accessible and briefer exercise protocols that achieve the same health benefits as the current 150 min/week of moderate-intensity exercise recommendation.

High-intensity interval training, or HIIT, which involves brief intermittent bursts of high-intensity exercise separated by recovery periods, improves cardiorespiratory fitness. Sprint interval training has been shown to improve fitness and insulin sensitivity to the same extent as a moderate-intensity continuous exercise protocol that required five times as much time to complete. Knowing this, a research team at McMaster University in Canada recently conducted a series of studies to see if brief, intense stair climbing, a readily available high-intensity activity, could improve cardiorespiratory fitness.[1] In these studies, young, sedentary women performed an acute exercise comparison to ensure that the stair-climbing protocol elicited the same acute physiological responses as a classical sprint interval training bout on a stationary bike, and a 6-week training intervention. In the intervention period, subjects were instructed to perform three 20 s bursts of all-out intensity going up the stairs as fast as possible with 2 min of recovery between bouts, 3 days/week.

Work output, heart rate, blood lactate, and RPE responses were the same between the brief intense stair-climbing protocol and the classical sprint interval training protocol. Following 6 weeks of stair-climbing training, cardiorespiratory fitness ($\dot{V}O_{2peak}$) had improved by 12%, which is similar to other studies using a cycle ergometer to administer sprint interval training. Importantly, the study team also reported that the participants completed 99% of all the training sessions and that the average time required to complete the training was ≤9 min/week. Overall, brief, intense stair climbing is a time-efficient and easily accessible mode of exercise that can increase cardiorespiratory fitness in sedentary adults.

RESEARCH PERSPECTIVE 11.3

Do Ice Baths Increase Recovery and Endurance Performance?

Postexercise cold-water immersion has become increasingly popular in athletic training programs because of the belief that it speeds up recovery. However, few studies have scientifically investigated the effect of postexercise cold-water immersion therapy on the adaptive responses to endurance training, and no studies have examined these effects with sprint interval training. Of the research studies that have been conducted, the results have been conflicting. Some suggest that cold-water immersion may stimulate muscle mitochondria biogenesis and allow for better recovery and harder training in subsequent bouts of exercise, while other say that postexercise cold-water immersion may counteract the molecular processes related to vascular remodeling and have detrimental long-term effects on skeletal muscle adaptations to endurance training. Overall, the utility of postexercise cold-water immersion to enhance recovery and subsequent performance is still unclear.

A recent study conducted at Victoria University in Australia investigated the effects of cold-water immersion on mitochondrial content and function within the muscle (1) following a single bout of sprint interval (HIIT) exercise and (2) after a 6-week HIIT intervention.[7] The researchers recruited healthy, recreationally active men and split them into two groups. One group received postexercise cold-water immersion after each training bout, while the other group performed the same training but had a passive recovery without cold water. After a baseline familiarization trial, subjects performed a single bout of HIIT training followed by a skeletal muscle biopsy. After the 6-week training intervention, another muscle biopsy was done along with posttraining time-trial and $\dot{V}O_{2max}$ testing. The investigators analyzed the skeletal muscle biopsies for p-AMPK, p-p38 MAPK, p-p53, and PGC-1α, which are markers of mitochondrial content and function. In short, the investigators did not find any effect of cold-water immersion on any of their measurements. There were no differences between the group of participants who were treated with cold-water immersion after each exercise bout and the control group, who performed the exercise training without cold-water immersion during recovery. This program of HIIT increased $\dot{V}O_{2max}$ and time trial performance without any effect on markers of mitochondrial content or function.

While these findings suggest that cold-water immersion is not detrimental to endurance adaptations after sprint interval training, they also suggest that cold-water immersion does not provide any benefit to endurance training adaptations or improvements in fitness and performance.

As discussed in chapter 9, athletes who already train vigorously can likewise improve performance by integrating HIIT into their training regimens. However, the mechanisms for these improvements appear to differ.[13] The rapid increases in skeletal muscle oxidative enzymes seen in previously untrained exercisers are not apparent in already trained individuals who add HIIT to their workouts. The underlying adaptations for improved performance in these athletes are not well understood.

Specificity of Training and Cross-Training

Physiological adaptations in response to physical training are highly specific to the nature of the training activity. Furthermore, the more specific the training program is to a given sport or activity, the greater the improvement in performance in that sport or activity. As discussed in chapter 9, the concept of **specificity of training** is very important for all physiological adaptations.

This concept is also important in testing of athletes. As an example, to accurately measure endurance improvements, athletes should be tested while engaged in an activity similar to the sport or activity in which they usually participate. Consider one study of highly trained rowers, cyclists, and cross-country skiers. Their $\dot{V}O_{2max}$ was measured while they performed two types of work: uphill running on a treadmill and maximal performance of their sport-specific activity.[33] A key finding, shown in figure 11.20, was that the $\dot{V}O_{2max}$ attained by all the athletes during their sport-specific activity was as high as or higher than the values obtained on the treadmill. For many of these athletes, $\dot{V}O_{2max}$ was substantially higher during their sport-specific activity.

A highly creative design for studying the concept of specificity of training involves one-legged exercise training, with the untrained opposite leg used as the control. In one study, subjects were placed into three groups: a group that sprint trained one leg and endurance trained the other, a group that sprint trained one leg and did not train the other, and a group that endurance trained one leg and did not train the other.[30] Improvement in $\dot{V}O_{2max}$ and lowered heart rate and

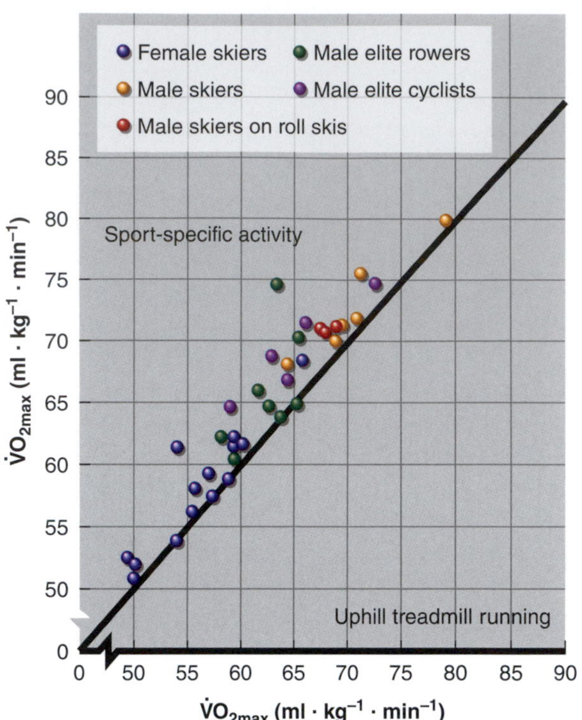

FIGURE 11.20 $\dot{V}O_{2max}$ values measured during uphill treadmill running versus sport-specific activities in selected groups of athletes.

Adapted by permission from S.B. Strømme, F. Ingjer, and H.D. Meen, "Assessment of Maximal Aerobic Power in Specifically Trained Athletes," *Journal of Applied Physiology* 42 (1977): 833-837.

blood lactate response at submaximal work rates were found only when exercise was performed with the endurance-trained leg.

Much of the training response occurs in the specific muscles that have been trained, possibly even in individual motor units in a specific muscle. This observation applies to metabolic as well as cardiorespiratory responses to training. Table 11.5 shows the activities of selected muscle enzymes from the three energy systems for untrained, anaerobically trained, and aerobically trained men. The table shows that aerobically trained muscles have significantly lower glycolytic enzyme activities. Thus, they might have less capacity for anaerobic metabolism or might rely less on energy from glycolysis. More research is needed to explain the implications of the muscular changes accompanying both anaerobic and aerobic training, but this table clearly illustrates the high degree of specificity to a given training stimulus.

Cross-training refers to training for more than one sport at the same time or training several different fitness components (such as endurance, strength, and flexibility) at one time. The athlete who trains by swimming, running, and cycling in preparation for competing in a triathlon is an example of the former, and the athlete involved in heavy resistance training and high-intensity cardiorespiratory training at the same time is an example of the latter.

TABLE 11.5 **Selected Muscle Enzyme Activities (mmol · g^{-1} · min^{-1}) for Untrained, Anaerobically Trained, and Aerobically Trained Men**

	Untrained	Anaerobically trained	Aerobically trained
	Aerobic enzymes		
Oxidative system			
Succinate dehydrogenase	8.1	8.0	20.8a
Malate dehydrogenase	45.5	46.0	65.5a
Carnitine palmityl transferase	1.5	1.5	2.3a
	Anaerobic enzymes		
ATP-PCr system			
Creatine kinase	609.0	702.0[a]	589.0
Myokinase	309.0	350.0[a]	297.0
Glycolytic system			
Phosphorylase	5.3	5.8	3.7[a]
Phosphofructokinase	19.9	29.2[a]	18.9
Lactate dehydrogenase	766.0	811.0	621.0

[a]Significant difference from the untrained value.

RESEARCH PERSPECTIVE 11.4

Age and Responses to HIIT

Maximal oxygen consumption ($\dot{V}O_{2max}$) is one of the strongest predictors of cardiovascular health span and mortality. Even with regular aerobic activity, $\dot{V}O_{2max}$ declines ~1% per year with age, and this decline accelerates in older age. Consequently, older adults, who already have a higher risk for cardiovascular disease and mortality, could benefit the most from interventions that increase $\dot{V}O_{2max}$. High-intensity interval training (HIIT) yields effective improvements in aerobic fitness and cardiovascular heath in healthy young and middle-aged adults. Because of the relatively short time commitment and significant improvements in fitness achieved, HIIT may be an especially valuable strategy for improving $\dot{V}O_{2max}$ in older adults. However, few research studies have examined how age affects the aerobic training response to HIIT in older adults.

A recent study of 94 healthy men and women ranging from 20 to 83 years of age sought to determine how age affected improvements in $\dot{V}O_{2max}$ following HIIT training.[32] In this study, participants with similar pretest $\dot{V}O_{2max}$ values relative to age were tested immediately before and immediately following an 8-week HIIT intervention. During the intervention, the study participants completed supervised HIIT training with a targeted intensity of 90% to 95% of maximal heart rate, three times a week. After the HIIT intervention, all of the subjects improved their $\dot{V}O_{2max}$. In order to examine age differences, the subjects were separated into six age groups (20-29 years, 30-39 years, 40-49 years, 50-59 years, 60-69 years, and 70+ years). All of the groups improved their $\dot{V}O_{2max}$ with no differences between age groups. In contrast to age, the percentage improvements in $\dot{V}O_{2max}$ were predicted by their baseline $\dot{V}O_{2max}$, with the least fit people showing the greatest improvements regardless of their age group. Healthy individuals of average fitness can improve $\dot{V}O_{2max}$ through HIIT, regardless of their age, and HIIT may be a useful strategy to improve fitness and slow the decline in $\dot{V}O_{2max}$ in healthy aging.

For the athlete training for cardiorespiratory endurance and strength at the same time, the studies conducted to date indicate that gains in strength, power, and endurance can result. However, the gains in muscular strength and power are less when strength training is combined with endurance training than when strength training alone is done. The opposite does not appear to be true: Improvement of aerobic power with endurance training does not appear to be attenuated by inclusion of a resistance training program. In fact, short-term endurance can be increased with resistance training. Although earlier studies supported the conclusion that concurrent strength and endurance training limits gains in strength and power, McCarthy and colleagues[23] reported similar gains in strength, muscle hypertrophy, and neural activation in a group of previously untrained subjects who underwent concurrent high-intensity strength training and cycle endurance training compared with a group who performed only high-intensity strength training.

In Review

- For athletes to maximize cardiorespiratory gains from training, the training should be specific to the type of activity that an athlete usually performs.
- The program must be carefully matched with the athlete's individual needs to maximize the physiological adaptations to training, thereby optimizing performance.
- Resistance training in combination with endurance training does not appear to restrict improvement in aerobic power and may increase short-term endurance, but it can limit improvement in strength and power when compared with gains from resistance training alone.

IN CLOSING

In this chapter, we examined how the cardiovascular, respiratory, and metabolic systems adapt to aerobic and anaerobic training, as well as HIIT training. The focus was on how these adaptations can improve both aerobic and anaerobic performance. This chapter concludes our review of how body systems respond to both acute and chronic exercise. Now that we have completed our examination of how the body responds to internal challenges induced by various types, durations, and intensities of exercise, we turn our attention to the external environment. In the next part of the book, we focus on the body's adaptations to varying environmental conditions, beginning in the next chapter by considering how external temperature can affect performance.

KEY TERMS

aerobic training
anaerobic training
athlete's heart
capillary-to-fiber ratio
cardiac hypertrophy
cross-training
Fick equation
glycogen sparing
high responders
low responders
mitochondrial oxidative enzymes
oxygen transport system
specificity of training
submaximal endurance

STUDY QUESTIONS

1. Differentiate between muscular endurance and cardiovascular endurance.
2. What is maximal oxygen uptake ($\dot{V}O_{2max}$)? How is it defined physiologically, and what determines its limits?
3. Of what importance is $\dot{V}O_{2max}$ to endurance performance? Why does the competitor with the highest $\dot{V}O_{2max}$ not always win?
4. Describe the changes in the oxygen transport system that occur with endurance training.
5. What is possibly the most important adaptation that the body makes in response to endurance training, which allows for an increase in both $\dot{V}O_{2max}$ and performance? Through what mechanisms do these changes occur?
6. What are the theoretical reasons given for the resting bradycardia that accompanies endurance exercise training?
7. What metabolic adaptations occur in response to endurance training?
8. Explain the two theories that have been proposed to account for limitations to aerobic performance that may be altered by endurance training. Which of these has the greatest validity today?
9. What is the most important predictor of how much $\dot{V}O_{2max}$ will decline with inactivity or bed rest?
10. How important is genetic potential in a developing young athlete?
11. What adaptations have been shown to occur in muscle fibers with anaerobic training?
12. Discuss specificity of anaerobic training with respect to enzyme changes in muscle.
13. Can athletes who already train vigorously still improve performance by integrating HIIT into their training regimens? In what way are adaptive mechanisms different from those seen in untrained individuals who undergo HIIT?
14. Why is cross-training beneficial to endurance athletes? How does it benefit sprint and power athletes?

STUDY GUIDE ACTIVITIES

In addition to the activities listed in the chapter opening outline, two other activities are available in the web study guide, located at

www.HumanKinetics.com/PhysiologyOfSportAndExercise

The **KEY TERMS** activity reviews important terms, and the end-of-chapter **QUIZ** tests your understanding of the material covered in the chapter.

PART IV

Environmental Influences on Performance

In previous sections of this book, we discussed the physiological adjustments and coordination of systems (muscular, neural, cardiovascular, respiratory) that allow us to perform physical activity. We also saw how these systems adapt when exposed to the repeated stress of training. In part IV, we turn our attention to how the body responds and adapts when challenged to exercise under extreme environmental conditions. In chapter 12, Exercise in Hot and Cold Environments, we examine the mechanisms by which the body regulates its internal temperature at rest and during exercise. Then we consider how the body responds and adapts to exercise in the heat and cold, along with the health risks associated with physical activity in hot and cold environments. In chapter 13, Exercise at Altitude, we discuss the unique challenges that the body faces when performing physical activity under conditions of reduced atmospheric pressure (altitude) and how the body adapts to time spent at altitude. We then discuss the best way to prepare for competing at altitude and whether altitude training might help people perform better at sea level. Finally, health risks associated with ascent to high altitude are discussed.

Exercise in Hot and Cold Environments

In this chapter and in the web study guide

Organizers of the 2014 Australian (Tennis) Open were criticized for forcing players to compete in intense heat as temperatures hit 42 °C (108 °F) in Melbourne on January 13. The day's peak temperature of 42 °C was just short of Melbourne's January record of 45.6 °C (114 °F), which occurred in 1936. The Olympic Movement Medical Code states, "In each sports discipline, minimal safety requirements should be defined and applied with a view to protecting the health of the participants and the public during training and competition. Depending on the sport and the level of competition, specific rules should be adopted regarding sports venues [and] safe environmental conditions."[5] All major sporting bodies abide by this code and have comprehensive management strategies in place. However, the Australian Open Extreme Heat Policy (EHP) is applied only at the referee's discretion, and only minimal changes were enacted to protect the players in Melbourne.

Several prominent competitors, including Scotland's Andy Murray, unsuccessfully called on Australian Open organizers to reconsider their decision to make players compete in such oppressive temperatures. Canadian player Frank Dancevic passed out during the second set of his first-round match with France's Benoit Paire on an unshaded court, as did a ball boy. It was so hot that Danish player Caroline Wozniacki's plastic water bottle melted on court and Serbia's Jelena Jankovic burned her backside and hamstrings sitting on an uncovered seat, then fell during her first-round victory when her rubber-soled shoe stuck to the court. Yet play continued.

The stresses of physical exertion are often complicated by environmental conditions. Performing exercise in extreme heat or cold places an additional burden on the mechanisms that regulate body temperature while supporting continued exercise. Although these mechanisms are amazingly effective in regulating body temperature under normal conditions, mechanisms of **thermoregulation** can be inadequate when we are subjected to extreme heat or cold. Fortunately, the body is able to adapt to such environmental stresses with continued exposure over time, a process known as *acclimation* (which refers to a short-term adaptation, e.g., days to weeks) or *acclimatization* (the proper term when we are referring to natural adaptations gained over long periods of time, e.g., months to years).

In the following discussion, we focus on the physiological responses to acute and chronic exercise in both hot and cold environments. Specific health risks are associated with exercise in both temperature extremes, so we also discuss the prevention of temperature-related illness and injuries during exercise.

Body Temperature Regulation

Humans are homeotherms, which means that our internal body temperature is physiologically regulated to keep it nearly constant even when environmental temperature changes. In physiology, temperatures are expressed as degrees Centigrade. To convert from °F to °C and vice versa, use the following transformations:

To go from °F to °C: Subtract 32°, then divide by 1.8.

To go from °C to °F: Multiply by 1.8, then add 32°.

Although a person's temperature varies from day to day, and even from hour to hour, these fluctuations are usually no more than about 1.0 °C (1.8 °F). Only during prolonged heavy exercise, fever due to illness, or extreme conditions of heat or cold do body temperatures deviate from the normal baseline range of 36.1 to 37.8 °C (97.0-100.0 °F). Body temperature reflects a careful balance between heat production and heat loss. Whenever this balance is disturbed, body temperature changes.

Metabolic Heat Production

Only a small part (usually less than 25%) of the energy (adenosine triphosphate, ATP) the body produces is used for physiological functions such as muscle contraction; the rest is converted to heat. All active tissues produce metabolic heat (M) that must be precisely offset by heat loss to the environment to maintain the internal temperature of the body. If the body's heat production exceeds its heat loss, as it often does during moderate- to heavy-intensity aerobic activity, the body stores the excess heat, and internal temperature increases. People's ability to maintain a constant internal temperature depends on their ability to balance the metabolic heat they produce and the heat they gain from the environment with the heat the body loses. This balance is depicted in figure 12.1.

Transfer of Body Heat to and From the Environment

Let's examine the mechanisms by which heat is transferred between a person and the surroundings. For the body to transfer heat to the environment, the heat produced in the body must first move from deep in the body (the core) to the skin (the shell), where it has access to the outside environment. Heat is primarily moved from the core to the skin by the blood. Only when heat reaches the skin can it be transferred to the environment by any of four mechanisms: conduction, convection, radiation, and evaporation. These are illustrated in figure 12.2.

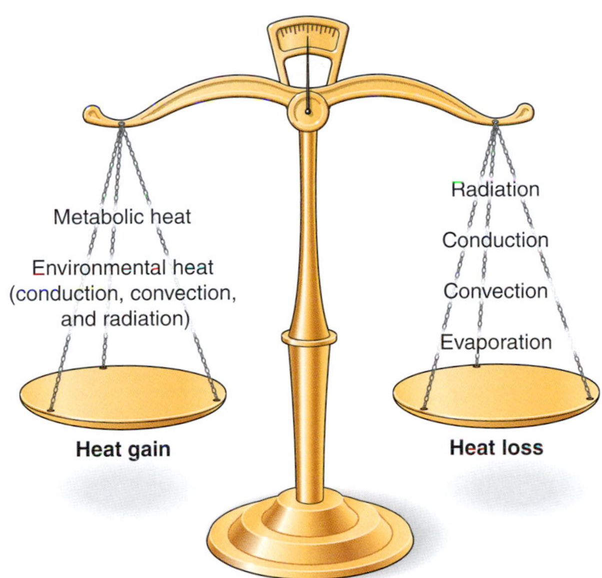

FIGURE 12.1 To maintain a steady-state core temperature, the body must balance the heat gained from metabolism and external environmental factors with the heat lost through the avenues of radiation, conduction, convection, and evaporation.

Conduction and Convection

Heat **conduction (K)** involves the transfer of heat from one solid material to another through direct molecular contact. As an example, heat can be lost from the body when the skin is in contact with a cold object, as when one sits on cold metal bleachers watching a soccer match on a chilly day. Conversely, if a hot object is pressed against the skin, heat from the object is conducted to the skin and heat is gained by the body. If the contact is prolonged, heat from the skin surface can be transferred to the blood as it flows through the skin and transferred to the core, raising internal (core) temperature. During exercise, conduction is usually negligible as a source of heat exchange because the body surface area in contact with solid objects (for example, soles of the feet on hot playing fields) is small. Therefore, many environmental physiologists treat conductive heat exchange as negligible in their calculations of heat balance and exchange.

Convection (C), on the other hand, involves transferring heat by the motion of a gas or a liquid across the heated surface. When the body is still and there is little air movement, a thin unstirred boundary layer of air surrounds the body. However, the air around us is usually in constant motion, especially so during exercise as we move either the whole body or body segments (e.g., the arms pumping as we run) through the air. As air moves around us, passing over the skin, heat is exchanged with the air molecules. The greater the movement of the air (or liquid, such as water), the greater the rate of heat exchange by convection.

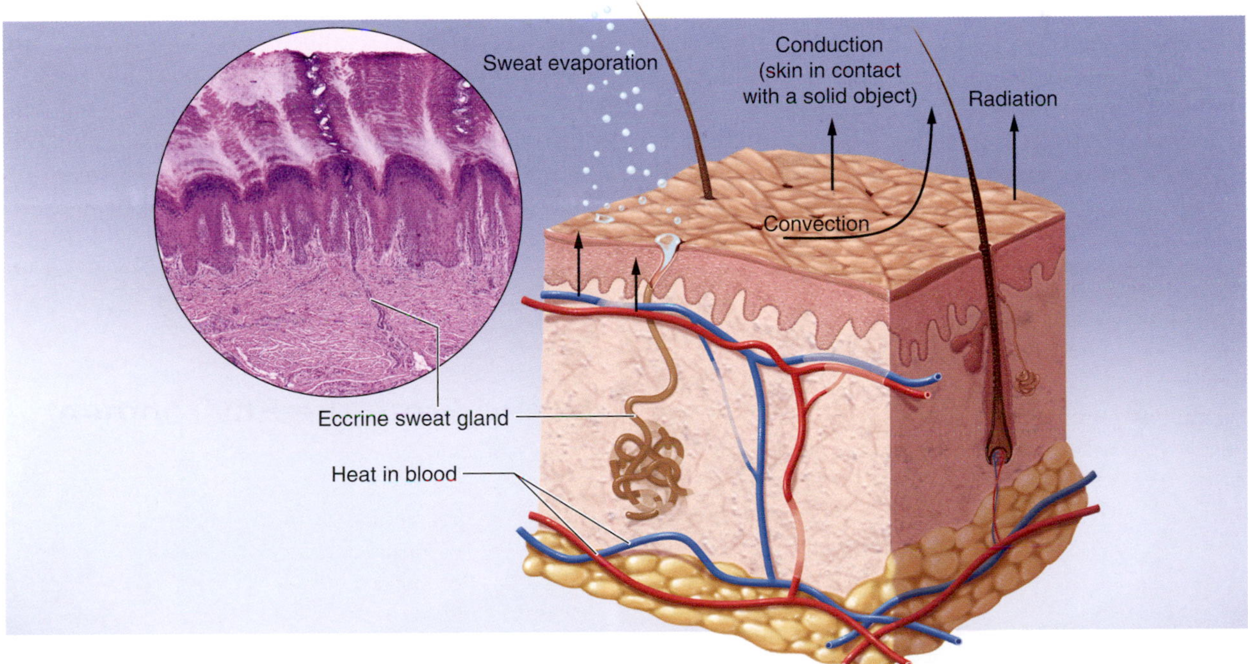

FIGURE 12.2 The removal of heat from the skin. Heat is delivered to the body surface via the arterial blood and to a lesser extent by conduction through the subcutaneous tissue. When the temperature of the skin is greater than that of the environment, heat is removed by conduction (if the skin is in contact with an object), convection, radiation, and sweat evaporation; when the environmental temperature exceeds skin temperature, heat can be removed only by evaporation.

Thus, in an environment in which air temperature is cooler than the skin temperature, convection permits the transfer of heat from the skin to the air (heat loss); however, if air temperature is higher than skin temperature, heat is gained by the body through convection. We often consider these processes to be mechanisms of heat loss, forgetting that when the environmental temperature exceeds skin temperature, the gradient works in the opposite direction.

Convection is important on a daily basis, since it constantly removes the metabolic heat we generate at rest and during activities of daily living, as long as the air temperature is lower than the skin temperature. However, if a person is submerged in cold water, the amount of heat dissipated from the body to the water by convection can be nearly 26 times greater than when the person is exposed to a similar cold air temperature.

Radiation

At rest, **radiation (R)** and convection are the primary methods for eliminating the body's excess heat. At normal room temperature (typically 21-25 °C, or ~70-77 °F), the nude body loses about 60% of its excess heat by radiation. The heat is given off in the form of infrared rays, which are a type of electromagnetic wave. Figure 12.3 shows two infrared thermograms of an individual.

The skin constantly radiates heat in all directions to objects around it, such as clothing, furniture, and walls, but it also can receive radiant heat from surrounding objects that are warmer. If the temperature of the surrounding objects is greater than that of the skin, the body experiences a net heat gain via radiation. A tremendous amount of radiant heat is received from exposure to the sun.

Taken together, conduction, convection, and radiation are considered avenues of **dry heat exchange**. Resistance to dry heat exchange is called **insulation**, a concept known to everyone as it relates to clothing and home heating and cooling. The ideal insulator is a layer of still air (remember that moving air causes convective heat loss), which we achieve by trapping layers of air in fibers (down, fiberglass, and so on). Adding insulation in this way minimizes unwanted heat loss in cold environments. However, during exercise we want to dissipate heat to the environment, which is best accomplished by wearing thin, light-colored clothing (to limit radiant heat absorption) that allows for a maximally exposed skin surface area. The effect of clothing on sweat evaporation is discussed later.

Evaporation

By far, **evaporation (E)** is the primary avenue for heat dissipation during exercise. In fact, when air temperature is close to skin temperature, the only available means of cooling is evaporation. As a fluid evaporates and turns into its gaseous form, heat is lost. An important concept to remember is that sweat must evaporate for any heat loss to occur. Some sweat drips off the body or stays on the skin or in the clothing, especially if the air is humid. This unevaporated sweat contributes nothing to body cooling and simply represents a wasteful loss of body water. Evaporation accounts for about 80% of the total heat loss when one is physically active and is therefore an extremely important avenue for heat loss. Even at rest, evaporation accounts for 10% to 20% of body heat loss, since some evaporation occurs without our awareness (termed *insensible water loss*).

As body core temperature increases, a threshold core temperature is reached, and production increases dramatically. As sweat reaches the skin, it is converted from a liquid to a vapor, and heat is lost from the skin

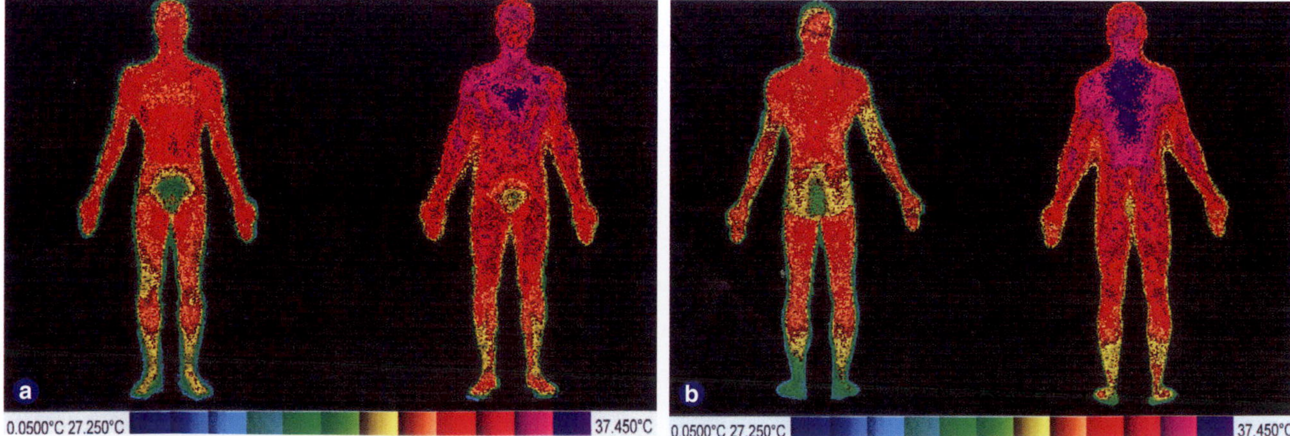

FIGURE 12.3 Thermograms of the body showing the variations in radiant (infrared) heat leaving the *(a)* front and *(b)* rear surfaces of the body both before (left) and after (right) running outside at a temperature of 30 °C and 75% humidity. The color scale at the bottom of each photo provides the temperature variation for the changes in color.

in the process, the latent heat of vaporization. Thus, sweat evaporation becomes increasingly important as body temperature increases.

Evaporation of 1 L of sweat in an hour results in the loss of 680 W (2,428 kJ) of heat. To put the tremendous cooling power of sweat evaporation into perspective, a 70 kg (154 lb) marathon runner who runs a 2:30 marathon will produce about 1,000 W of metabolic heat. If every drop of sweat he produced evaporated, he would need to sweat at a rate of about 1.5 L per hour. Because inevitably some sweat drips off, a better estimated sweating rate necessary to maintain core temperature would be 2 L per hour. Without replacing fluids, however, our hypothetical runner would lose about 5 L (5.3 qt) of water, or over 7% of his body weight!

Analogous to insulation, which limits dry heat exchange, clothing also adds resistance to sweat evaporation. While some cooling of the skin does occur as sweat evaporates from wet clothing surfaces, the cooling power is less than for evaporation directly from skin to air. Clothing that fits loosely and comprises fabrics that promote wicking or free movement of water vapor molecules through the fabric enhances evaporative cooling.

Figure 12.4 shows the complex interaction between the mechanisms of body heat balance (production and loss) and environmental conditions.[6] Using the symbols defined in the previous paragraphs, we can represent the state of heat balance by a simple equation:

$$M - W \pm R \pm C \pm K - E = 0,$$

where W represents any useful work being performed as a result of muscle contraction. Notice that while R, C, and K can be either positive (heat gain) or negative (heat loss), E can only be negative. When $M - W \pm R \pm C \pm K - E > 0$, heat is stored in the body and core temperature rises.

Humidity and Heat Loss

The water vapor pressure of the air (the pressure exerted by water vapor molecules suspended in the air) plays a major role in evaporative heat loss. *Relative humidity* is a more commonly used term that relates the water vapor pressure of the air to that of fully saturated air (100% humidity). When humidity is high, the air already contains many water molecules. This decreases its capacity to accept more water because the vapor pressure gradient between the skin and the air is decreased. Thus, high humidity limits sweat evaporation and heat loss, while low humidity offers an ideal opportunity for sweat evaporation and heat loss. But this efficient cooling mechanism can also pose a problem. If sweating is prolonged without adequate fluid replacement, dehydration can occur.

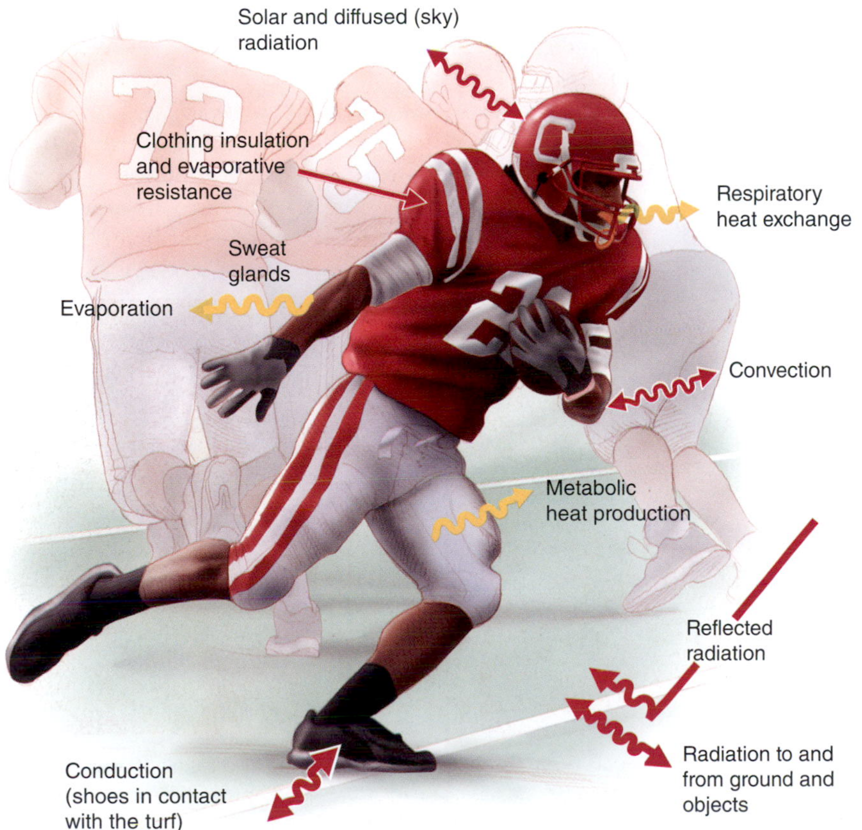

FIGURE 12.4 The complex interaction between the body's mechanisms for heat balance and environmental conditions.

Thermoregulatory Control

We live our entire lives within a very small, fiercely protected range of internal body temperatures. If sweating and evaporation were unlimited, we could withstand extreme ambient heat (e.g., even oven temperatures >200 °C for short periods!) if we were protected from contact with hot surfaces. On the other hand, the temperature limits for living cells range from about 0 °C (where ice crystals form) to about 45 °C (where intracellular proteins start to unravel), and humans can tolerate *internal* temperatures below 35 °C or above 41 °C for only very brief periods of time. To maintain internal temperature within these limits, we have developed very effective and, in some instances specialized, physiological responses to heat and cold. These responses involve the finely controlled coordination of several body systems.

Internal body temperature at rest is regulated at approximately 37 °C (98.6 °F). During exercise, the body is often unable to dissipate heat as rapidly as it is produced. In rare circumstances, a person can reach internal temperatures exceeding 40 °C (104 °F), with a temperature above 42 °C (107.6 °F) in active muscles. The muscles' energy systems become more chemically efficient with a small increase in muscle temperature, but internal body temperatures above 40 °C can adversely affect the nervous system and reduce further efforts to unload excess heat. How does the body regulate its internal temperature? The hypothalamus plays a central role (see figure 12.5).

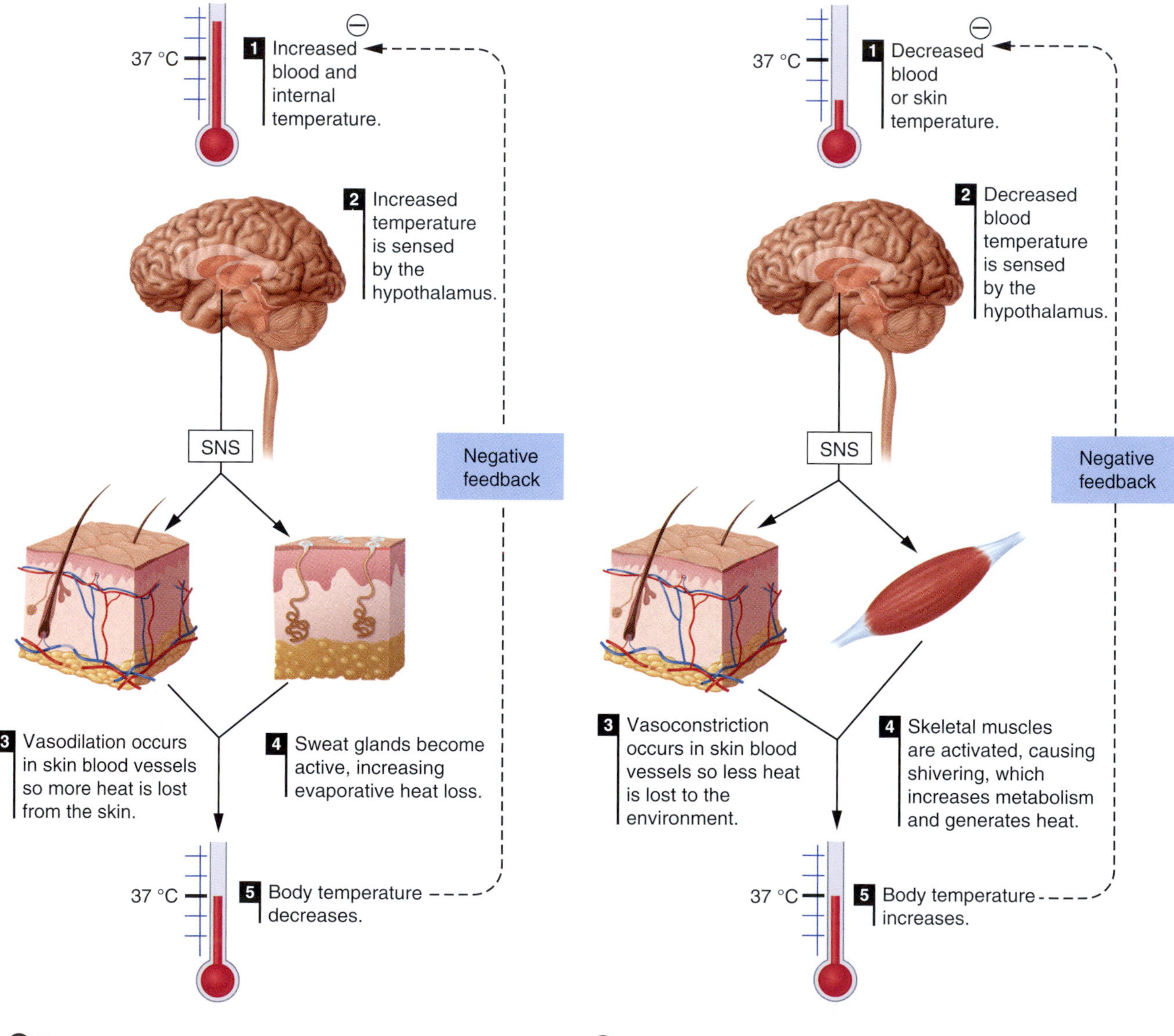

FIGURE 12.5 A simplified overview of the role of the hypothalamus in controlling body temperature during *(a)* heat stress and *(b)* cold stress.

The Preoptic-Anterior Hypothalamus: The Body's Thermostat

A simple way to envision the mechanisms that control internal body temperature is to compare them to the thermostat that controls the air temperature in a home, although the body's mechanisms function in a more complex manner and generally with greater precision than a home heating and cooling system. Sensory receptors called **thermoreceptors** detect changes in temperature and relay this information to the body's thermostat, located in a region of the brain called the **preoptic-anterior hypothalamus (POAH)**. In response, the hypothalamus activates mechanisms that regulate the heating or cooling of the body. Like a home thermostat, the hypothalamus has a predetermined temperature, or set point, that it tries to maintain. This is the normal body temperature. The smallest deviation from this set point signals this thermoregulatory center to readjust the body temperature.

Thermoreceptors are located throughout the body but especially in the skin and central nervous system. The peripheral receptors located in the skin monitor the skin temperature, which varies with changes in the temperature around a person. They provide information not only to the POAH but also to the cerebral cortex, which allows one to consciously perceive temperature and voluntarily control one's exposure to heat or cold. Because the skin temperature changes long before core temperature, these receptors serve as an early warning system for impending thermal challenges.

Central receptors are located in the hypothalamus, other brain regions, and the spinal cord and monitor the temperature of the blood as it circulates through these sensitive areas. These central receptors are responsive to blood temperature changes as small as 0.01 °C (0.018 °F) and to the rate of change as well. Because of this exquisite sensitivity, very small changes in the temperature of the blood passing through the hypothalamus quickly trigger reflexes that help one conserve or eliminate body heat as needed.

Environmental challenges to thermal homeostasis present challenges in parallel to many other bodily control systems, such as those that control blood pressure, body fluid and electrolyte balance, and circadian rhythms. In testimony to the intricate design of the human brain, these other hypothalamus-based controllers are located in close proximity to the POAH, with neural connections that finely coordinate control across all of these systems.

Thermoregulatory Effectors

When the POAH senses temperatures above or below normal, signals are sent through the sympathetic nervous system to four sets of effectors:

- **Skin arterioles.** When skin or core temperature changes, the POAH sends signals via the sympathetic nervous system (SNS) to the smooth muscle in the walls of the arterioles that supply the skin, causing them to dilate or constrict. This either increases or decreases skin blood flow. Skin vasoconstriction results primarily from SNS release of the neurotransmitter norepinephrine, although other neurotransmitters are involved as well, and facilitates heat conservation by minimizing dry heat exchange. Skin vasodilation in response to heat stress is a more complex and less understood process. Increased skin blood flow aids in heat dissipation to the environment through conduction, convection, and radiation (and, indirectly, evaporation as skin temperature goes up). Fine-tuning of skin blood flow is the mechanism by which minute-to-minute adjustments are made in heat balance and exchange. These adjustments are rapid and occur with no real energy cost to the body.

• **Eccrine sweat glands.** When either the skin or core temperature is elevated sufficiently, the POAH also sends impulses through the SNS to the **eccrine sweat glands**, resulting in active secretion of sweat onto the skin surface. The primary neurotransmitter involved is acetylcholine; thus we refer to sweat gland activation as sympathetic cholinergic stimulation. Like skin arterioles, sweat glands are about 10 times more responsive to increases in core temperature than to similar increases in skin temperature. The evaporation of this moisture, as discussed earlier, removes heat from the skin surface.

• **Skeletal muscle.** Skeletal muscle is called into action when a person needs to generate more body heat. In a cold environment, thermoreceptors in the skin or core sense cold and send signals to the hypothalamus. In response to this integrated neural input, the hypothalamus activates the brain centers that control muscle tone. These centers stimulate shivering, which is a rapid, involuntary cycle of contraction and relaxation of skeletal muscles. This increased muscle activity is ideal for generating heat to either maintain or increase the body temperature, because no useful work results from the shivering, only heat production.

• **Endocrine glands.** The effects of several hormones cause the cells to increase their metabolic rates. Increased metabolism affects heat balance because it increases heat production. Cooling the body stimulates the release of thyroxine from the thyroid gland. Thyroxine can elevate the metabolic rate throughout the body by more than 100%. Also, recall that epinephrine and norepinephrine (the catecholamines) mimic and enhance the activity of the SNS. Thus, they directly affect the metabolic rate of virtually all body cells.

In Review

- Humans are homeothermic, meaning that internal body temperature is regulated through physiological mechanisms, usually keeping it in the resting range of 36.1 to 37.8 °C (97.0-100.0 °F) despite changes in environmental temperatures.
- Body heat is transferred by conduction, convection, radiation, and evaporation. At rest, most heat is lost via radiation and convection, but during exercise, evaporation becomes the most important avenue of heat loss.
- When air temperature is close to skin temperature, the only available means of cooling is evaporation of sweat. Evaporation of 1 L of sweat in an hour results in the loss of 680 W (2,428 kJ) of heat.
- At any given air temperature, higher humidity (i.e., a higher water vapor pressure of the surrounding air) decreases evaporative heat loss.
- The preoptic-anterior area of the hypothalamus (POAH) is the primary thermoregulatory center. It acts as a thermostat, monitoring temperature and accelerating heat loss or heat production as needed.
- Two sets of thermoreceptors provide temperature information to the POAH. Peripheral receptors in the skin relay information about the temperature of the skin and the environment around it. Central receptors in the hypothalamus, other brain areas, and the spinal cord transmit information about the internal body temperature. Central thermoreceptors are far more sensitive to temperature change than peripheral receptors. The main role of the peripheral receptors is anticipatory, allowing for early adjustments to be made.
- Effectors stimulated by the hypothalamus through the SNS can alter the body temperature. Increased skeletal muscle activity (voluntary or involuntary as in the case of shivering) increases the temperature by increasing metabolic heat production. Increased sweat gland activity decreases the temperature by increasing evaporative heat loss. Smooth muscle in the skin arterioles can cause these vessels to dilate to direct blood to the skin for heat transfer or to constrict to retain heat deep in the body. Metabolic heat production can also be stimulated by the actions of hormones such as thyroxine and the catecholamines.

Physiological Responses to Exercise in the Heat

Heat production is beneficial during exercise in a cold environment because it helps maintain normal body temperature. However, even when exercise is performed in a cool environment, the metabolic heat load places a considerable burden on the mechanisms that control body temperature. In this section, we examine some physiological changes that occur in response to exercise while the body is exposed to heat stress and the impact that these changes can have on performance. For this discussion, heat stress means any environmental condition that increases body temperature and jeopardizes homeostasis.

Cardiovascular Function

As we learned in chapter 8, exercise increases the demands on the cardiovascular system. When the need to regulate body temperature is added during exercise in the heat, the burden placed on the cardiovascular system is enhanced. During exercise in hot conditions, the circulatory system has to continue to transport blood not only to working muscle but also to the skin, where the tremendous heat generated by the muscles can be transferred to the environment. To meet this dual demand during exercise in the heat, two changes occur. First, cardiac output increases further (above the level associated with a similar exercise intensity in cool conditions) by increasing both heart rate and contractility. Second, blood flow is shunted away from nonessential areas (like the gut, liver, and kidneys) and to the skin.

Consider what happens during prolonged running on a hot day. The aerobic exercise increases both metabolic heat production and the demand for blood flow and oxygen delivery to the working muscles. This excess heat can be dissipated only if blood flow increases to the skin.

In response to the elevated core temperature (and to a lesser extent, the higher skin temperature), the SNS signals sent from the POAH to the skin arterioles cause these blood vessels to dilate, delivering more metabolic heat to the body surface. Sympathetic nervous system signals also go to the heart to increase heart rate and cause the left ventricle to pump more forcefully. However, the ability to increase stroke volume is limited as blood pools in the periphery and less returns to the left atrium. To maintain cardiac output under such circumstances, the heart rate gradually creeps upward to help compensate for the decrease in stroke volume.

The causes of this phenomenon, known as cardiovascular drift, were discussed in chapter 8. Because blood volume stays constant or even decreases (as fluid is lost in sweat), another phase of cardiovascular adjustment occurs simultaneously. Sympathetic nerve signals to the kidneys, liver, and intestines cause vasoconstriction of those regional circulations, which allows more of the available cardiac output to reach the skin without compromising muscle blood flow.

What Limits Exercise in the Heat?

Seldom are records set in endurance events, such as distance running, when the environmental heat stress is great. The factors that cause early fatigue when heat stress is superimposed on prolonged exercise have been the topic of some debate, and several theories have

RESEARCH PERSPECTIVE 12.1

Dehydration Challenges the Cardiovascular System During Exercise in the Heat

Dehydration commonly occurs when people exercise for prolonged periods of time without adequate fluid replacement. This fall in total body fluid causes a reduction in sweat rate and skin blood flow, resulting in significant increases in body heat storage, hyperthermia, and physiological strain. A compromised blood flow to the active skeletal muscle has been proposed as an important factor in the chain of events that lead to fatigue during exercise in the heat, and the overall fall in total body fluid poses a significant challenge to thermoregulation and cardiovascular control during exercise.

Dehydration and the resulting increases in core temperature that occur during prolonged submaximal exercise in the heat (e.g., distance running, cycling) lead to increases in heart rate and total peripheral resistance, small reductions in blood volume and mean arterial pressure, and large reductions in stroke volume and cardiac output. These changes lead to marked reductions in blood flow to active and nonactive muscles, skin, and cerebral circulations. Importantly, these exercise-induced alterations in cardiovascular function and blood flow during dehydration are prevented with oral rehydration, or when dehydrated individuals exercise in the cold, preventing hyperthermia. Thus, the impact of dehydration on cardiovascular function and performance occurs only when dehydration is coupled with hyperthermia during prolonged exercise.

A progressive fall in cardiac output and stroke volume and a compensatory, although possibly detrimental, rise in heart rate are common features of the dehydration-induced cardiovascular strain observed during prolonged aerobic exercise in the heat. Progressive dehydration reduces total blood volume and venous return; these factors, combined with a reduced diastolic filling time (due to elevated heart rate), compound to reduce stroke volume. Importantly, dehydration and hyperthermia do not influence the contractile properties of the myocardium, and it is the reduction in filling during diastole that is primarily responsible for the lower cardiac output observed with dehydration. This reduction in cardiac output is, in turn, responsible for reductions in perfusion pressure and limb blood flow. It is unlikely that changes in mean arterial pressure are responsible for the reductions in cerebral blood flow. Recent research suggests that these reductions may be explained by hyperventilatory reductions in the partial pressure of CO_2; however, future studies will need to provide more evidence before these mechanisms are clear.

been proposed. None of these theories captures every situation but taken together they demonstrate the multiple control systems at work during thermoregulation.

At some point, the cardiovascular system can no longer compensate for the increasing demands of continuing endurance exercise and efficiently regulating the body's heat. Consequently, any factor that tends to overload the cardiovascular system or interfere with heat dissipation can drastically impair performance, increase the risk of overheating, or both. Exercise in the heat becomes limited when heart rate approaches maximum, especially in untrained or nonheat-acclimated exercisers, as illustrated in figure 12.6. Interestingly, working muscle blood flow is well maintained even at very high core temperatures unless significant dehydration occurs.

Another theory that helps explain limitations to exercise in the heat—especially in well-trained, acclimated athletes—is the **critical temperature theory**. This theory proposes that, regardless of the rate at which core temperature (and thus brain temperature) increases, the brain will send signals to stop exercise when some critical brain temperature is reached, usually between 40 and 41 °C (104 and 105.8 °F).

Precooling and Sport Performance

Many athletes now artificially lower their body core temperature before exercise, a practice known as *precooling*. Various methods are used, including cold-water immersion, sitting in a cold room, cold showers, cooling packs or vests, and ingesting a cold drink or ice slurry. If an athlete starts at a lower core temperature and his or her core temperature subsequently rises at the same rate, his or her core temperature will be lower at any given point in the exercise bout or competition. In laboratory studies, this often allows subjects to exercise longer at a given intensity under hot conditions. But does this practice improve sport performance?

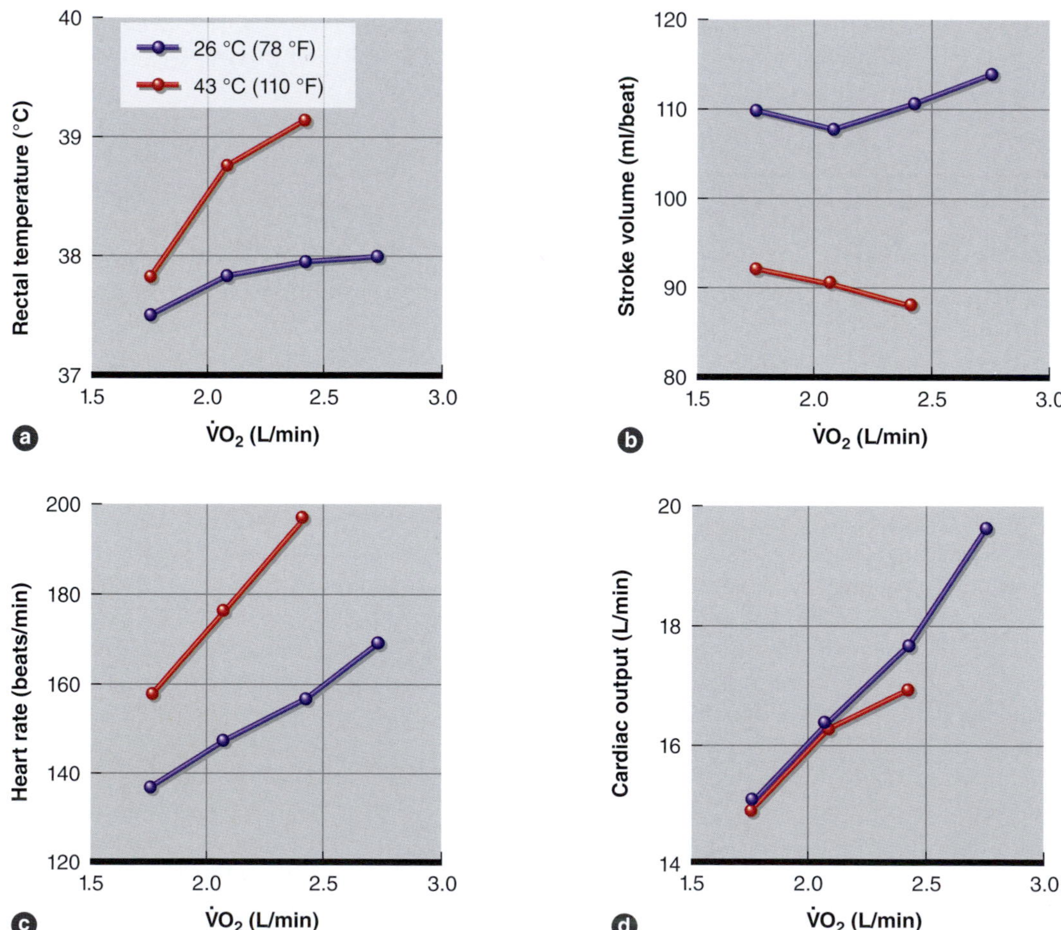

FIGURE 12.6 *(a)* Rectal temperature and *(b-d)* cardiovascular responses to graded exercise in thermoneutral (26 °C, or 78 °F; blue circles) and hot (43 °C, or 110 °F; red circles) environments. Note that, in addition to directional changes in these variables caused by heat stress, the maximal intensity is decreased when exercise is performed in a hot environment.

Based on Rowell (1974).

Two recent meta-analyses (statistically based review of existing studies) addressed that important question,[14,15] concurring that precooling *does* effectively enhance sport performance, particularly in hot environments. As expected, precooling had a larger effect during endurance events like cycling time trials, but improvement was also seen for intermittent running or cycling sprints. The effectiveness of precooling on short sprints was minimal. Of the methods compared, ingesting cold drinks was the most promising procedure. Further, the best results occurred in athletes with the highest $\dot{V}O_{2max}$.[12]

The second analysis[11] concluded that precooling actually impaired sprint performance, which might be expected if active muscle temperature fell. Intermittent activity and endurance performance were both improved with precooling, however. Similarly, cooling maneuvers used during exercise were also successful at improving performance. Whether used before exercise or during exercise, the effectiveness of body cooling procedures depends on both (1) the amount of cooling (intensity and duration) and (2) how much thermal strain is involved in the exercise.

Body Fluid Balance: Sweating

On hot summer days, it is not uncommon for the temperature of the environment to exceed both the skin and deep body temperatures. As mentioned earlier, this makes evaporation far more important for heat loss, because radiation, convection, and conduction now become avenues of heat gain from the environment. Increased dependence on evaporation means an increased demand for sweating.

The eccrine sweat glands are controlled by stimulation of the POAH. Elevated blood temperature causes this region of the hypothalamus to transmit impulses through the sympathetic nerve fibers to the millions of eccrine sweat glands distributed over the body's surface. The sweat glands are fairly simple tubular structures extending through the dermis and epidermis, opening onto the skin, as illustrated in figure 12.7.

A second type of sweat gland, the apocrine gland, is clustered in particular regions of the body, including the face, axilla, and genital regions. These are the glands associated with nervous perspiration, and they do not contribute significantly to heat loss by evaporation. The eccrine sweat glands, on the other hand, play a purely thermoregulatory role. They are located over most of the skin surface, with ~2 to 5 million covering the whole body. They are most densely distributed on the palms of the hands, the soles of the feet, and the forehead. The lowest densities are found on the forearms, lower legs, and thighs. When sweating begins, there are large regional variations in sweating rate. During exercise, the highest local sweating rates are typically measured on the middle and lower back and on the forehead, while the lowest sweating rates are observed on the hands and feet.

Sweat is formed in the coiled secretory portion of the sweat gland and at this stage has an electrolyte composition similar to that of the blood, since plasma is the source of sweat formation. As this filtrate of plasma passes through the uncoiled duct of the gland, sodium and chloride are reabsorbed back into the surrounding tissues and then into the blood. As a result, the final sweat that is extruded onto the skin surface through the sweat gland pores is hypotonic to (has less electrolytes than) plasma. During light sweating, the filtrate sweat travels slowly enough through the duct that there is time for reabsorption of sodium and chloride. Thus, the sweat that forms during light sweating contains very little of these electrolytes by the time it reaches the skin. However, when the sweating rate increases during exercise, the filtrate moves more quickly through the tubules, allowing less time for reabsorption, and the sodium and chloride content of sweat can be considerably higher.

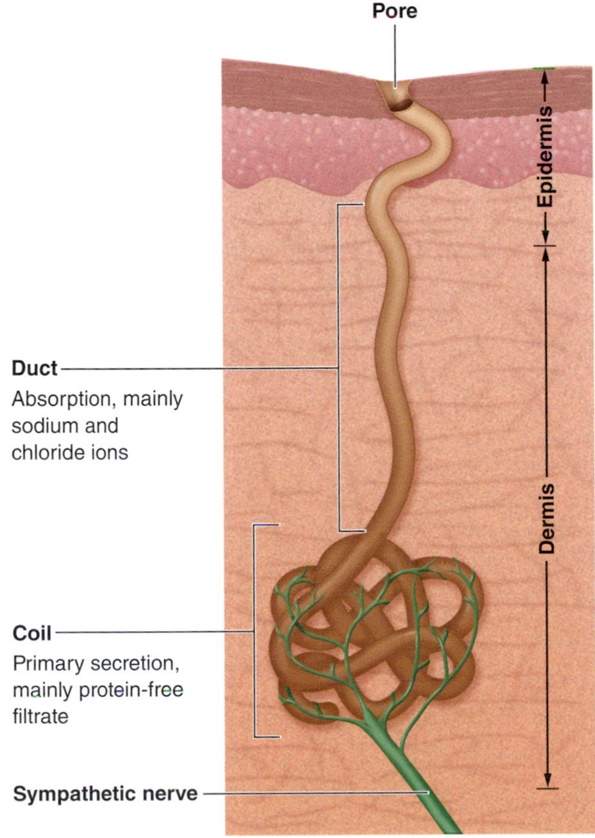

FIGURE 12.7 Anatomy of an eccrine sweat gland that is innervated by a sympathetic cholinergic nerve.

As seen in table 12.1, the electrolyte concentration of trained and untrained subjects' sweat is significantly different. With training and repeated heat exposure (acclimation), more sodium is reabsorbed and the sweat is more dilute, in part because the sweat glands become more sensitive to the hormone aldosterone. Unfortunately, the sweat glands apparently do not have a similar mechanism for conserving other electrolytes. Potassium, calcium, and magnesium are not reabsorbed by the sweat glands and are therefore found in similar concentrations in sweat and plasma. In addition to heat acclimation and aerobic training, genetics is a major determinant of both sweating rate and sweat sodium losses.

While performing heavy exercise in hot conditions, the body can lose more than 1 L of sweat per hour per square meter of body surface. This means that during intense effort on a hot and humid day (high level of heat stress), an average-sized female athlete (50-75 kg, or 110-165 lb) might lose 1.6 to 2.0 L of sweat, or about 2.5% to 3.2% of body weight, each hour. A person can lose a critical amount of body water in only a few hours of exercise in these conditions.

VIDEO 12.1 Presents Caroline Smith on research methods for measuring sweat rates over different parts of the body and the findings of this research.

A high rate of sweating maintained for a prolonged time ultimately reduces blood volume. This limits the volume of blood returning to the heart, increasing heart rate and eventually decreasing cardiac output, which in turn reduces performance potential, particularly for endurance activities. In long-distance runners, sweat losses can approach 6% to 10% of body weight. Such severe dehydration can limit subsequent sweating and make the individual susceptible to heat-related illnesses. Chapter 15 provides a detailed discussion of dehydration and the value of fluid replacement.

Loss of both electrolytes and water in the sweat triggers the release of both aldosterone and antidiuretic hormone (ADH), also known as **vasopressin** or **arginine vasopressin**. Recall that aldosterone is responsible for maintaining appropriate sodium concentrations in the blood and that ADH plays a key role in maintaining fluid balance (chapter 4). Aldosterone is released from the adrenal cortex in response to stimuli such as decreased blood sodium content, reduced blood volume, or reduced blood pressure. During acute exercise in the heat and during repeated days of exercise in the heat, this hormone limits sodium excretion from the kidneys. More sodium is retained by the body, which in turn promotes water retention. This allows the body to retain water and sodium in preparation for additional exposures to the heat and subsequent sweat losses.

Similarly, exercise and body water loss stimulate the posterior pituitary gland to release ADH. This hormone stimulates water reabsorption from the kidneys, which further promotes fluid retention in the body. Thus, the body attempts to compensate for loss of electrolytes and water during periods of heat stress and heavy sweating by reducing their loss in urine. Also, recall that blood flow to the kidneys is substantially reduced during exercise in the heat, which likewise aids in fluid retention.

In Review

- During exercise in the heat, the skin competes with the active muscles for a limited cardiac output. Muscle blood flow is well maintained (sometimes to the detriment of skin blood flow) unless severe dehydration occurs. A series of cardiovascular adjustments shunt blood away from nonessential regions such as the liver, gut, and kidneys to the skin to aid in heat dissipation.
- At a given exercise intensity in the heat, cardiac output may remain reasonably constant or decrease slightly at higher intensities, as a gradual upward drift in heart rate helps offset a lower stroke volume.
- Prolonged heavy sweating can lead to dehydration and excessive electrolyte loss. To compensate, increased release of aldosterone and ADH enhances sodium and water retention.

TABLE 12.1 **Example of Sodium, Chloride, and Potassium Concentrations in the Sweat of Trained and Untrained Subjects During Exercise**

Subjects	Sweat Na (mmol/L)	Sweat Cl^- (mmol/L)	Sweat K^+ (mmol/L)
Untrained men	90	60	4
Trained men	35	30	4
Untrained women	105	98	4
Trained women	62	47	4

Individual sweat electrolyte concentrations are highly variable, but training and heat acclimation decrease sodium losses in the sweat.

Data from the Human Performance Laboratory, Ball State University.

- Sweating rates as high as 4 L per hour have been observed in highly trained, well-acclimated athletes, but these rates cannot be sustained for more than several hours. Maximal daily sweating rates can be in the range of 10 to 15 L but only with adequate fluid replacement.

Health Risks During Exercise in the Heat

Despite the body's defenses against overheating, excessive heat production by active muscles, heat gained from the environment, and conditions that prevent the dissipation of excess body heat may elevate the internal body temperature to levels that impair normal cellular functions. Under such conditions, excessive heat gains pose a risk to one's health, as we highlighted in the opening anecdote of this chapter. Air temperature alone is not an accurate index of the total physiological stress imposed on the body in a hot environment. Six variables must be taken into account:

- Metabolic heat production
- Air temperature
- Ambient water vapor pressure (humidity)
- Air velocity
- Radiant heat sources
- Clothing

All of these factors influence the degree of heat stress that a person experiences. The contribution of each factor to the total heat stress under various environmental conditions can be predicted mathematically using advanced heat balance equations.

An individual exercising on a bright, sunny day with an air temperature of 23 °C (73.4 °F) and no measurable wind experiences considerably more heat stress than someone exercising in the same air temperature but under cloud cover and with a slight breeze. At

RESEARCH PERSPECTIVE 12.2

Tattoos and Sweating

The practice of tattooing dates back to ancient civilization but has become increasingly popular in recent times. Approximately 45 million Americans have at least one tattoo, and 40% of young adults (20-40 years) have one or more tattoos compared to only 10% of older adults.[10] Tattoos are particularly popular among college and professional athletes as well as military personnel, who are regularly physically active and rely on increasing skin blood flow and eccrine sweating to dissipate metabolic heat produced during physical activity.

The tattooing process involves puncturing the skin with a cluster of needles to deposit dye 3 to 5 mm below the skin's surface into the dermal layer. This process initiates an inflammatory response within the dermal layer of the skin, which contributes to the permanent color incorporated in the skin. The dermal layer of the skin is composed of collagen fibers, nerves, blood vessels, and eccrine sweat glands. Eccrine sweat glands draw fluid from the extracellular fluid into the secretory coils. As isosmotic primary sweat is extruded through the skin surface, sodium chloride is reabsorbed, resulting in a very low sweat sodium concentration. Because the eccrine sweat glands are located in the dermal layer, where the primary effects of tattooing occur, it is possible that tattooing will affect the sweat gland function (sweat production or sodium reabsorption) in the tattooed skin.

A recent study conducted at Alma College in Michigan examined the sweating response in tattooed compared to nontattooed skin.[7] In this study, the researchers stimulated sweating by delivering pilocarpine nitrate (a pharmacological agonist that stimulates muscarinic receptors on the sweat gland) to a 5.2 cm area of tattooed skin and a 5.2 cm area of nontattooed skin on the contralateral anatomical location on the same subjects. The researchers chose to measure the contralateral anatomical location to account for regional differences in sweat rate across the body. They measured the sweat rate by weighing an absorbent disk placed on the skin before and after the sweating stimulus, and collected the sweat from each site to measure sodium concentration by flame photometry later. The study team found that sweat rate in the tattooed skin was significantly lower compared to the nontattooed skin and that the sodium concentration was much higher in the tattooed skin. These data suggest that tattooing alters the function of the sweat gland in the area of skin that is tattooed. This study is the first of its kind and will likely open the door for future research exploring the mechanisms by which these alterations occur and whether there is any way to restore sweat gland function in tattooed skin.

temperatures above skin temperature, which is normally 32 to 33 °C (92 °F), radiation, conduction, and convection substantially add to the body's heat load rather than acting as avenues for heat loss. How, then, can we estimate the amount of heat stress to which an individual may be exposed?

Measuring Heat Stress

It has become common to hear about the heat index on local weather channels. The heat index, a complex equation involving air temperature and relative humidity, is a measure of how hot it *feels*, that is, how we perceive the heat. However, the heat index does not do a good job of reflecting the physiological stress on humans, so its use is limited in exercise physiology. Through the years, efforts have been made to quantify atmospheric variables into a single index that would reflect the physiological heat stress on an individual. In the 1970s, **wet-bulb globe temperature (WBGT)** was devised to simultaneously account for conduction, convection, evaporation, and radiation (see figure 12.8). It is based on three different thermometer readings and provides a single temperature reading to estimate the cooling capacity of the surrounding environment.

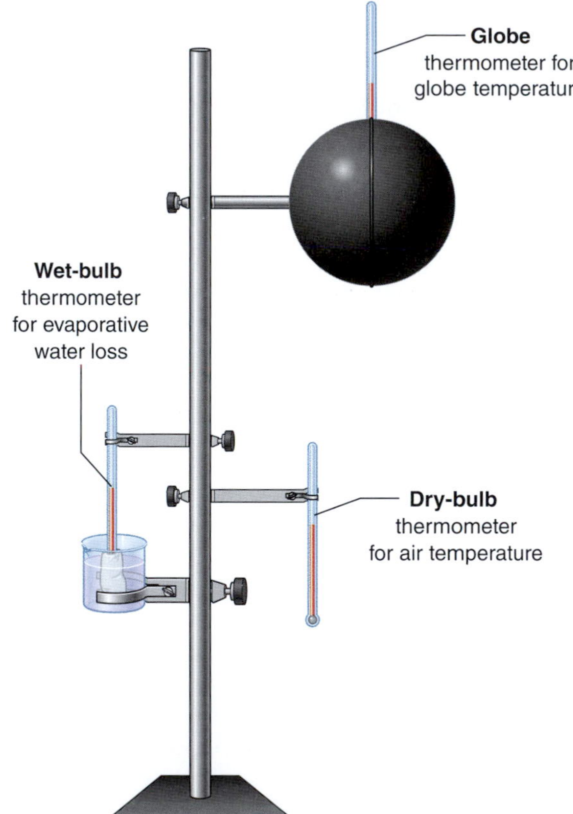

FIGURE 12.8 A wet-bulb globe temperature stand showing the three separate thermometers for air (dry-bulb) temperature, wet-bulb temperature reflecting the cooling effect of evaporation, and globe temperature, which measures additional effects of radiant heat.

The dry-bulb temperature (T_{db}) is the actual air temperature one would measure with a typical thermometer. A second thermometer has a bulb that is kept moist by a wetted cotton "sock" dipped in distilled water. As water evaporates from this wet bulb, its temperature (T_{wb}) will be lower than the T_{db}, reflecting the effect of sweat evaporating from the skin. The difference between the wet- and dry-bulb temperatures indicates the environment's capacity for cooling by evaporation. In still air with 100% relative humidity, these two bulb temperatures are the same because evaporation is impossible. Lower ambient water vapor pressure and increased air movement promote evaporation, increasing the difference between these two bulb temperatures. The third thermometer, placed inside a black globe, typically shows a temperature higher than T_{db} as the globe, painted a flat black, maximally absorbs radiant heat. Thus, its temperature (T_g) is a good indicator of the environment's radiant heat load.

The temperatures from these three thermometers can be combined into the following equation to estimate the overall atmospheric challenge to body temperature in outdoor environments:

$$WBGT = 0.1\ T_{db} + 0.7\ T_{wb} + 0.2\ T_g$$

The fact that the coefficient for T_{wb} is the largest reflects the importance of sweat evaporation in the physiology of heat exchange. Also note that WBGT reflects only the environment's impact on heat stress and is most effectively used along with a measure or estimate of metabolic heat production. Clothing further influences heat stress.

Wet-bulb globe temperature, as an index of **thermal stress**, is now routinely used by coaches, team physicians, and athletic trainers to anticipate the health risks associated with athletic practices and competitions in thermally stressful environments.

Heat-Related Disorders

Exposure to the combination of external heat stress and metabolically generated heat can lead to three heat-related disorders (see figure 12.9): heat cramps, heat exhaustion, and heatstroke.

Heat Cramps

Heat cramps, the least serious of the three heat disorders, are characterized by severe and painful cramping of large skeletal muscles. They involve primarily the muscles that are most heavily used during exercise, and such instances of athletes "locking up" are different from cramps everyone has experienced in small muscles. Heat cramps are brought on by sodium losses and

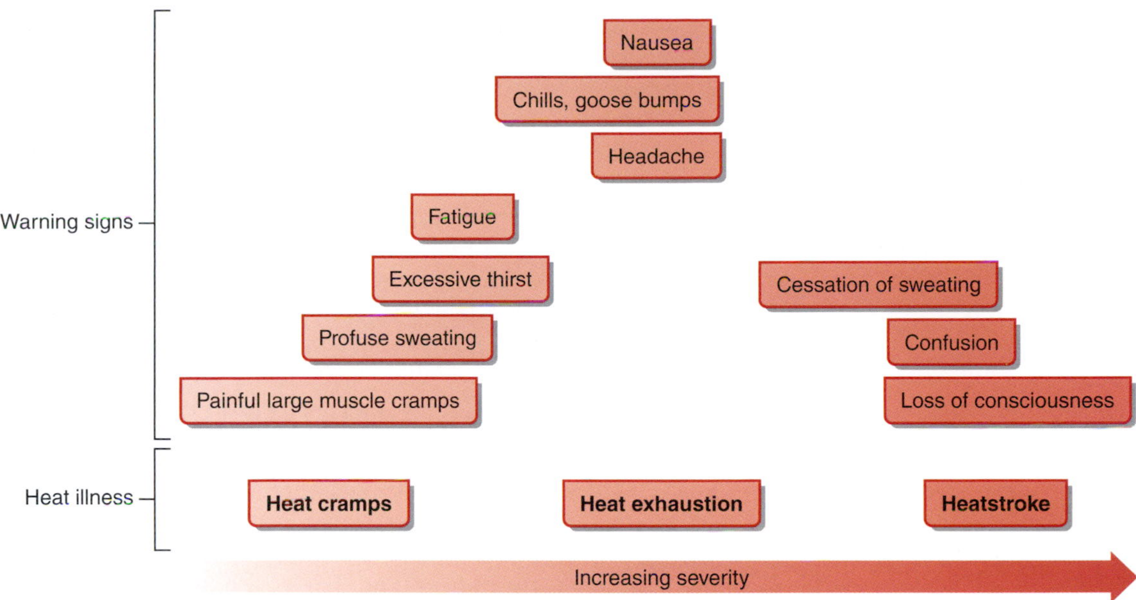

FIGURE 12.9 Schematic diagram of heat illnesses of increasing severity, showing some of the typical warning signs that accompany these illnesses. Not all of these warning signs are present in every case of heat illness, nor do they always occur in this, or any predictable, order. Any of the three heat illnesses shown here can occur suddenly without prior symptoms; that is, there is not always a progression from heat exhaustion to heatstroke.

Adapted by permission of All Sport, Inc.

dehydration that accompany high rates of sweating, and thus are most common in heavy sweaters who lose a lot of sodium in their sweat. (A common misconception is that potassium is involved in cramping and that eating potassium-rich foods like bananas will prevent heat cramps.) Heat cramps can be prevented or minimized in susceptible athletes by proper hydration practices involving liberal salt intake with foods and in beverages consumed during exercise. Treatment for these cramps involves moving the stricken individual to a cooler location and administering a saline solution, either orally or intravenously if necessary.

Heat Exhaustion

Heat exhaustion typically is accompanied by such symptoms as extreme fatigue, dizziness, nausea, vomiting, fainting, and a weak, rapid pulse. It is caused by the cardiovascular system's inability to adequately meet the needs of the body as it becomes severely dehydrated. Recall that during exercise in heat, active muscles and skin compete for a share of a limited, and decreasing, blood volume. Heat exhaustion may result when these simultaneous demands cannot be met, and it typically occurs when blood volume decreases as a result of excessive fluid loss from profuse sweating. A second form of heat exhaustion, from sodium depletion, is rare in athletes. Therefore, heat exhaustion can be thought of as a syndrome of dehydration and should be treated as such.

With heat exhaustion, the thermoregulatory mechanisms are functioning but cannot dissipate heat quickly enough because insufficient blood volume is available to allow adequate blood flow to the skin. Although the condition often occurs during moderate to heavy exercise in the heat, it is not necessarily accompanied by extremely high core temperatures. Some people who collapse from heat exhaustion have core temperatures well below 39 °C (102.2 °F). People who are unfit or not acclimated to the heat are more susceptible than others to heat exhaustion.

Treatment for victims of heat exhaustion involves rest in a cooler environment with their feet elevated to facilitate return of blood to the heart. If the person is conscious, administration of salt water is usually recommended. If the person is unconscious, medically supervised intravenous administration of saline solution is recommended.

Heatstroke

As we saw in our opening story, **heatstroke** is a life-threatening heat disorder that requires immediate medical attention. Heatstroke is caused by failure of the body's thermoregulatory mechanisms. It is characterized by

- an increase in internal body temperature to a value exceeding 40 °C (104 °F) and
- confusion, disorientation, or unconsciousness.

The latter element—altered mental status—is the key to recognizing impending heatstroke because neural tissues in the brain are particularly sensitive to extreme heat. In heatstroke, cessation of active sweating may also occur, but sweat may remain on the skin. The notion that heatstroke is always accompanied by dry, red skin is outdated, and this sign should never be used to distinguish heatstroke from heat exhaustion.

If heatstroke is left untreated, core temperature will continue to rise, progressing to coma and ultimately death. Appropriate treatment always involves cooling the body as rapidly as possible. In the field, this can best be accomplished by immersing the victim—as much of the body as possible excluding the head—in a bath of cold water or ice water. While cold-water immersion provides the fastest cooling rates, in cases in which cold or ice water is not available, temperate-water immersion is the next best option. Where immersion is not feasible, the combination of wrapping the entire body in cold, wet sheets and fanning vigorously may be used. Cooling methods that place ice bags on small areas, like the armpits, neck, and groin, are not effective in rapidly lowering core temperature because of the small surface area covered.

For the athlete, heatstroke is not just a problem associated with extreme conditions. Studies have reported rectal temperatures above 40.5 °C (104.9 °F) in marathon runners who successfully completed races conducted in temperate and even in cool conditions.

Sickle Cell Trait Complications

Sickle cell anemia is a genetic disease of the red blood cells that causes them to become oblong in shape and inefficient at carrying oxygen. In order to functionally display the disease of sickle cell anemia, a person must have two copies of the recessive sickle cell gene. People who inherit only one copy of the gene have **sickle cell trait**. Individuals with sickle cell trait do not display the sickling of red blood cells but are at an increased risk for several pathologies that are exacerbated by exercise, dehydration, or both.

As demonstrated in a recent review of the literature, college football players who have sickle cell trait are at a 15-fold increased risk for exertional death when compared to athletes without sickle cell trait.[4] The precise physiological mechanisms responsible for the increased risk associated with sickle cell trait are unclear; however, both defects in ryanodine receptors (see figure 6.2) and other genetic variations that affect the kidney's ability to concentrate urine and limit body water loss may contribute to this increased risk. The ryanodine receptor mediates the release of calcium from the sarcoplasmic reticulum, and mutations in this receptor are also responsible for *malignant hyperthermia*, a life-threatening condition that dramatically increases metabolic heat production during surgical anesthesia. In sickle cell trait, most sudden death events occur during high-intensity exercise plus high environmental heat and humidity, conditions that contribute to the sometimes fatal combination of severe dehydration and increased cardiovascular strain.[8] Sickle cell trait is now routinely screened for in National Collegiate Athletic Association student athletes,[12] an effort endorsed by many other sports medicine governing bodies.

Preventing Hyperthermia

We can do little about the prevailing weather conditions. In threatening conditions, athletes must either move the exercise session to a less stressful environment (for example, moving practice indoors) or decrease their effort (and thus their metabolic heat production) in order to reduce their risk of overheating. Athletes, coaches, and sport organizers should all be able to recognize the symptoms of heat illness.

To prevent heat disorders, several simple precautions should be taken. Competition and practice should not be held outdoors when the WBGT is more than 28 °C (82.4 °F) unless special precautions are taken. Scheduling practices and events either in the early morning or in the late evening avoids the severe heat stress of midday. Fluids should be readily available, and drink breaks should be scheduled every 15 to 30 min, with a goal of matching fluid intake to sweat loss. Because individual sweating rates and sweat sodium losses vary tremendously and cannot easily be predicted across individual athletes, athletes should customize their fluid intake based on their individual sweating rate. This is best accomplished by having athletes weigh themselves before and after exercise sessions and learn to estimate their approximate fluid needs.

Several organizations have set forth guidelines for practicing and competing under conditions of heat stress. As a summary,

1. Athletic events (distance races, tennis matches, sport team practices, and so on) should be scheduled to avoid the hottest times of the day. As a general rule, if the WBGT is above 28 °C (82-83 °F), consider cancelling, moving indoors, decreasing intensity of practice, or otherwise altering the event.
2. An adequate supply of palatable fluid must be available. Athletes should be educated and encouraged to prevent excessive (>2%) weight loss—that is, to replace their sweat losses to prevent dehydration but not overdrink to the point where they gain weight during the event.

3. Because individual sweating rates and sweat sodium losses vary tremendously, athletes should customize their fluid intake based on their individual sweating rate. Sweating rate can be estimated by measuring body weight before and after exercise. Fluids containing electrolytes and carbohydrate can provide benefits over water alone.
4. Athletes should be aware of signs and symptoms of heat illness. Cold-water immersion is the most efficient method for cooling hyperthermic athletes in the field.
5. Organizers of events and medical personnel should have the right to cancel or terminate events and to stop individual athletes who exhibit clear signs of heat exhaustion or heatstroke.

Clothing is another important consideration. Obviously, the more clothing worn, the less body area exposed to the environment to allow for direct heat loss. The foolish practice of exercising in a rubberized suit to promote weight loss is an excellent illustration of how a dangerous microenvironment (the isolated environment inside the suit) can be created in which temperature and humidity are sufficiently high to block virtually all heat loss from the body. This can rapidly lead to heat exhaustion or heatstroke. Football uniforms are another example of clothing that impedes heat loss. Coaches and athletic trainers should avoid practice sessions in full uniforms whenever possible, especially early in the season when temperatures tend to be hottest and players tend to be less fit and not well acclimated.

Distance athletes should wear as little clothing as possible when heat stress is a potential limitation to thermoregulation. They should tend toward underdressing rather than overdressing, because the metabolic heat generated will soon make extra clothing an unnecessary burden. Clothing should be loosely woven to allow sweat to be absorbed and wicked away from the skin and should be light colored to reflect heat back to the environment. Hats should be worn during exercise in bright sunlight or when cloud cover is limited.

It is also important to maintain adequate hydration, since the body loses considerable water through sweating. This is discussed in detail in chapter 15. Briefly, drinking fluid both before and during exercise can greatly reduce the negative effects of exercising in the heat. Adequate fluid intake will attenuate the increase in core body temperature and heart rate normally seen when a person exercises in the heat and will allow exercise to be continued longer. This is illustrated in figure 12.10.

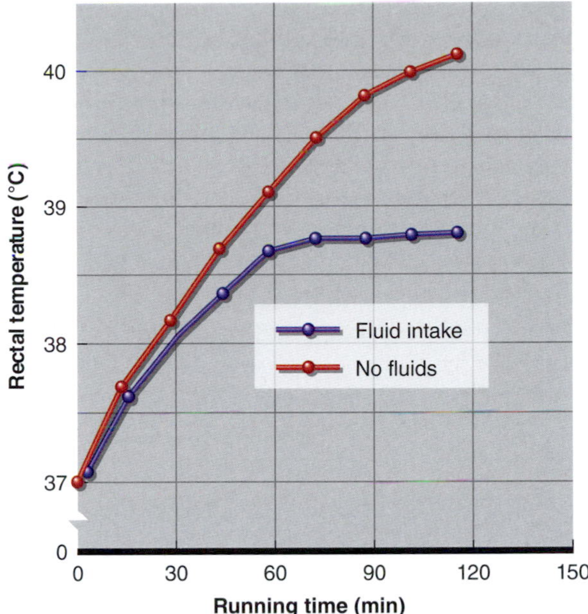

FIGURE 12.10 Effects of fluid intake on core (rectal) body temperature during a 2 h run. Subjects were given fluid during one trial, and on a separate day they completed a second trial without fluid. Note that fluid intake did not have much influence until about 45 min, after which time body heat storage was reduced compared with that in the no-fluid trial.

Based on Costill, Kammer, and Fisher (1970).

In Review

- Heat stress on the body is not accurately reflected by air temperature alone. Exercise intensity (metabolic heat), humidity, air velocity (or wind), radiation, and clothing also contribute to the total heat stress experienced during exercise in the heat.
- Perhaps the most widely used measurement of the combined physiological effects of heat stress is the WBGT, which measures air temperature and accounts for the heat exchange potential through convection, evaporation, and radiation in a specific environment. Exercise intensity and clothing must be considered separately along with the WBGT.
- To calculate WBGT:
  - Outdoor WBGT = $0.1\ T_{db} + 0.7\ T_{wb} + 0.2\ T_g$
  - Indoor WBGT = $0.7\ T_{wb} + 0.3\ T_g$
- Heat cramps are caused by the loss of fluids and salt (sodium) that results from excessive sweating in susceptible athletes. High dietary sodium intakes and proper hydration strategies can prevent heat cramps.

- Heat exhaustion results from the inability of the cardiovascular system to adequately meet the blood flow needs of the active muscles and the skin. It is often brought on by dehydration caused by excessive loss of fluids and electrolytes, which results in reduced blood volume. Although it is not in itself life threatening, heat exhaustion can deteriorate to heatstroke if untreated.
- Heatstroke is caused by failure of the body's thermoregulatory mechanisms. If it is not treated, core temperature continues to rise quickly and can be fatal. In addition to a severely elevated core temperature, altered mental status or cognitive function is a hallmark sign of heatstroke.

Acclimation to Exercise in the Heat

How can athletes prepare for prolonged activity in the heat? Does repeated exercise in the heat make us better able to tolerate thermal stress? Many studies have addressed these questions and have concluded that repeated exercise in the heat causes a series of relatively rapid adaptations that enable us to perform better, and more safely, in hot conditions. When these physiological changes occur over short periods of time, like days to weeks, or if they are artificially induced as in a climatic chamber, those adaptations are termed **acclimation (heat acclimation)**. A similar but much more gradual set of adaptations occurs in people who adapt to hot conditions by living in hot environments for months to years. This is known as **acclimatization** (note that the word *climate* is part of this latter term).

Effects of Heat Acclimation

Repeated bouts of prolonged, low-intensity exercise in the heat cause a relatively rapid improvement in the ability to maintain cardiovascular function and eliminate excess body heat, which reduces physiological strain. This process, termed heat acclimation, involves changes in plasma volume, cardiovascular function, sweating, and skin blood flow that allow for subsequent exercise bouts in the heat to be performed with a lower core temperature and heart rate response (figure 12.11). Because the body's heat loss capacity at a given rate of work is enhanced by acclimation, core temperature during exercise increases less than before acclimation (figure 12.11*a*), and heart rate increases less in response to standardized submaximal exercise after heat acclimation (figure 12.11*b*). In addition, after heat acclimation, more work can be done before adverse symptoms occur or a maximal tolerable core temperature or heart rate is reached.

This series of positive adaptations typically takes a period of 9 to 14 days of exercise in the heat to fully occur, as shown in figure 12.12. Well-trained individuals need fewer exposures than untrained individuals to fully acclimate. A critical physiological adjustment that

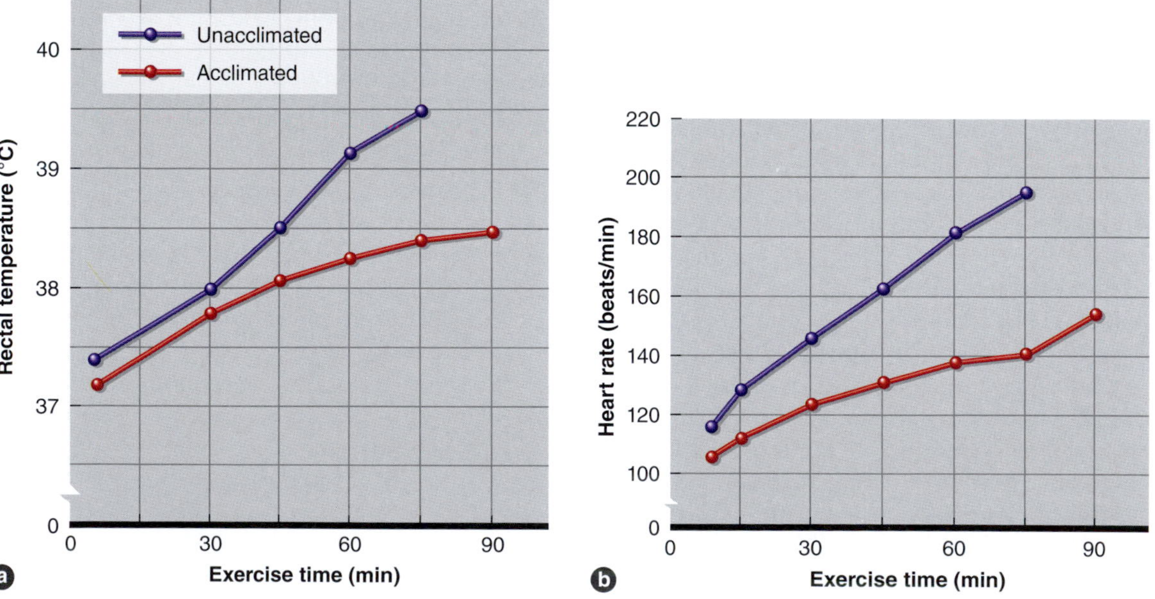

FIGURE 12.11 Typical *(a)* rectal temperature and *(b)* heart rate responses during an acute bout of exercise at the same intensity before and after heat acclimation. Note that, in addition to lower physiological strain, exercise time is usually longer after acclimation.

Data from King et al. (1985).

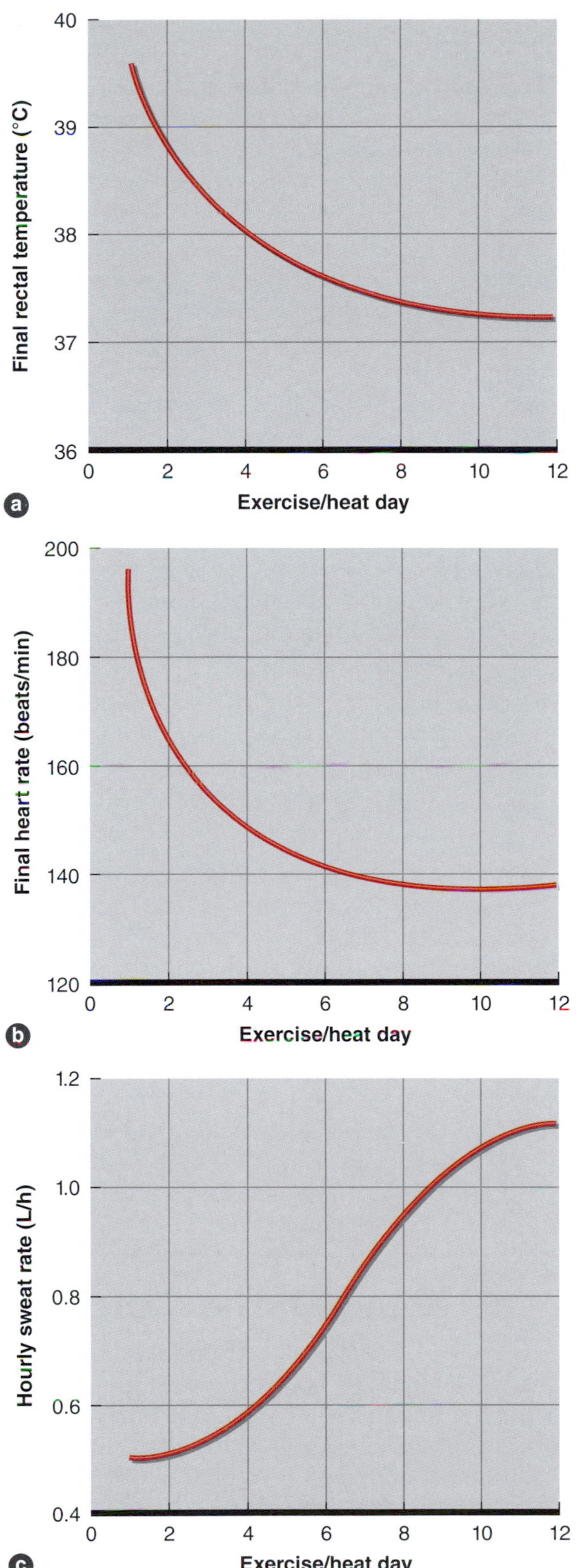

FIGURE 12.12 Changes in *(a)* rectal temperature, *(b)* heart rate, and *(c)* hourly sweating rate in a group of men exercising for 100 min per day for 12 consecutive days in a hot environment.

occurs over the first 1 to 3 days of acclimation is the expansion of plasma volume. The exact mechanism by which plasma volume expands after these initial exercise–heat exposures is not universally agreed upon. The process likely involves (1) proteins being forced out of the circulation as muscles contract, (2) these same proteins then being returned to the blood through the lymph, and (3) fluid moving into the blood because of the oncotic pressure exerted by the increased protein content. However, this change is temporary, and blood volume usually returns to original levels within 10 days. This early expansion of blood volume is important because it supports stroke volume, allowing the body to maintain cardiac output while additional physiological adjustments are made.

As shown in figure 12.12, end-exercise heart rate and core temperature decrease early in the acclimation process, while the increase in sweating rate during exercise in the heat occurs somewhat later. An additional adaptation is a more even distribution of sweat over the body, with increased sweating on the most exposed body areas such as the arms and legs, areas that are most effective at dissipating body heat. At the beginning of exercise, sweating also starts earlier in an acclimated person, which improves heat tolerance, and the sweat that is produced becomes more dilute, conserving sodium. This latter effect occurs in part because the eccrine sweat glands become more sensitive to the effects of circulating aldosterone.

Achieving Heat Acclimation

Heat acclimation requires more than merely resting in a hot environment. The benefits of acclimation, as well as the rate at which we acclimate, depend on

- the environmental conditions during each exercise session,
- the duration of exercise–heat exposure, and
- the rate of internal heat production (exercise intensity).

An athlete must exercise in a hot environment to attain full acclimation that sustains exercise in the heat. Simply sitting in a hot environment, such as a sauna or steam room, for long periods each day will not fully or adequately prepare the individual for physical exertion in the heat, at least not to the same extent as will exercising in the heat.

Because body temperature is elevated and sweating occurs, athletes gain partial heat tolerance simply by training, even in a cooler environment. Therefore, athletes are "preacclimated" to heat and need fewer exercise heat exposures to fully acclimate. To gain maximal benefits, athletes who train in environments cooler than those in which they will compete must achieve heat acclimation before the contest or event. Heat acclimation will improve their performance and reduce the associated physiological stress and risk of heat injury.

With a full 14-day heat acclimation schedule (90 min cycling, 40 °C, 20% relative humidity), improvements in whole-body heat loss can be demonstrated using direct calorimetry.[9] Direct calorimetry is described in chapter 5. After 14 days, whole-body heat loss was improved by 11% since sweat rate, and subsequently sweat evaporation, occurred to a greater extent over a shorter period of time. Approximately 70% of the improvements occurred within the first 7 days. While heat loss responses improved, there was no effect on core temperature in this study. Once the acclimation protocol was stopped, benefits persisted for up to 2 weeks.

Sex Differences

Exercise in the heat, in the cold, or at altitude confers additional stress or challenge to the body's adaptive abilities Many early studies indicated that women are less tolerant to heat than men are, particularly when exercising. Much of this difference, however, was the result of lower fitness levels of the women included in these studies, because the men and women were tested at the same absolute rate of work. When the rate of work is adjusted relative to individual $\dot{V}O_{2max}$ values, women's responses are almost identical to men's. Women have a delayed onset of sweating and dilation of the skin arterioles during the luteal phase of the menstrual cycle (i.e., onset occurs at a higher core temperature). However, this should not affect performance until core temperature approaches 40 °C. Women generally have lower sweat rates for the same exercise and heat stress. Although they possess a larger number of active sweat glands than men do, women produce less sweat per gland. This is a slight disadvantage in hot, dry environments but is a slight advantage in humid conditions in which sweat evaporation is minimal and reduced sweating delays dehydration.

After acclimatization, the internal temperature at which sweating and vasodilation begin is lowered to a similar extent in women and men. Also, the sensitivity of the sweating response per unit increase in internal temperature increases by a similar amount in the two sexes following both physical training and heat acclimatization. Therefore, most differences noted between women and men in the early studies can be attributed to initial differences in their physical conditioning and acclimation status and not to their sex.

In Review

- Repeated exercise exposure to heat stress gradually improves one's ability to support cardiovascular function and lose excess heat during subsequent bouts of exercise heat stress. This process is called heat acclimation.
- With heat acclimation, people start to sweat earlier and sweating rate increases, particularly on areas that are well exposed and are the most efficient at promoting heat loss. This reduces skin temperature, which increases the thermal gradient from the skin to the environment and promotes heat loss.
- Both core temperature and heart rate during exercise in the heat are reduced at any given intensity after heat acclimation. Plasma volume increases early in the process, contributing to an increase in stroke volume that supports the delivery of blood to both active muscles and skin.
- Full heat acclimation requires exercise in a hot environment, not merely exposure to heat.
- The rate of heat acclimation depends on training status, the conditions to which one is exposed during each session, the duration of the exposure, and the rate of internal heat production.
- With moderate-intensity exercise in the heat, people generally acclimate in 9 to 14 exposure days. Cardiovascular adjustments, which depend on expansion of plasma volume, occur first, followed by changes in sweating.
- When exercise intensity is adjusted relative to an individual's $\dot{V}O_{2max}$, women and men respond similarly to heat stress. Most differences noted are likely attributable to different initial levels of conditioning. Women tend to have lower sweating rates, which is advantageous in hot, humid conditions and disadvantageous in hot, dry conditions.

Exercise in the Cold

Humans can be thought of as tropical animals. Most of our adjustments to heat stress are physiological, whereas many adjustments to cold environments involve behavior, like putting on more clothing or seeking shelter. Increased year-round participation in sport has sparked new interest in, and concerns about, exercise in the cold. In addition, certain occupations and military endeavors require people to work in cold conditions—conditions that often limit performance. For these reasons, the physiological responses and health risks associated with cold stress are important issues in exercise science. We define cold stress here as any environmental condition causing a loss of body heat that threatens homeostasis. In the following discussion we focus on the two major cold environments: air and water.

The hypothalamus has a temperature set point of about 37 °C (98.6 °F), but daily fluctuations in the body temperature can be as much as 1 °C. A decrease in either skin or blood temperature provides feedback to the thermoregulatory center (POAH) to activate mechanisms that conserve body heat and increase heat production. The primary means by which our bodies avoid excessive heat loss (in the order in which they are invoked) are peripheral vasoconstriction, nonshivering thermogenesis, and shivering. Because these mechanisms or effectors of heat production and conservation are often inadequate, we also must rely on behavioral responses such as huddling behavior (decreasing exposed body surface area) and putting on more clothing to help insulate our deep body tissues from the environment.

Peripheral vasoconstriction occurs as a result of sympathetic stimulation of the smooth muscle layers of the arterioles in the skin. This stimulation causes the vascular smooth muscle to contract, which constricts the arterioles, reduces the blood flow to the shell of the body, and minimizes heat loss. Even at thermoneutral temperatures, there is tonic (constant baseline) skin vasoconstriction, and continuous adjustment of skin vascular tone occurs at all times to offset small heat imbalances in the body. When changing skin blood flow alone is not adequate to prevent heat loss, **nonshivering thermogenesis**—stimulation of metabolism by the SNS—is increased. Increasing the metabolic rate increases heat production. The next line of defense of body temperature during cold stress is **shivering**, a rapid, involuntary cycle of contraction and relaxation of skeletal muscles, which can cause a four- to fivefold increase in the body's rate of heat production. The overall adjustments in blood flow and metabolism that serve to maintain body core temperature are shown in figure 12.13.

Habituation and Acclimation to Cold

Whether humans truly acclimate to cold environments in the physiological sense—and if so, how this occurs—is far less clear than the process of heat acclimation. When one looks at studies of people who undergo repeated daily cold exposures, the results seem controversial. However, Dr. Andrew Young of the U.S. Army Research Institute for Environmental Medicine and others have proposed a scheme to explain the development of different patterns of cold adaptation that are observed in humans.[16] People who are regularly exposed to cold environments under which significant body heat loss does not occur typically undergo **cold habituation**, in which skin vasoconstrictor and shivering responses are blunted, and core temperature is allowed to fall to a greater extent than before the chronic cold exposures. This pattern of adaptation often occurs when small areas of skin—often the hands and face—are exposed repeatedly to cold air.

However, when heat loss is more severe or occurs at a faster rate, total body heat loss may occur. In cases in which increased metabolic heat production alone can sufficiently minimize heat loss, enhanced nonshivering and shivering thermogenesis develop (**metabolic acclimation**). Yet a third distinct pattern of cold adaptation, called **insulative acclimation**, tends to occur in situations in which increased metabolism is unable to maintain core temperature. In insulative acclimation, enhanced skin vasoconstriction occurs, which increases peripheral insulation and minimizes heat loss.

Other Factors Affecting Body Heat Loss

The mechanisms of conduction, convection, radiation, and evaporation, which usually perform effectively in dissipating metabolically produced heat during exercise

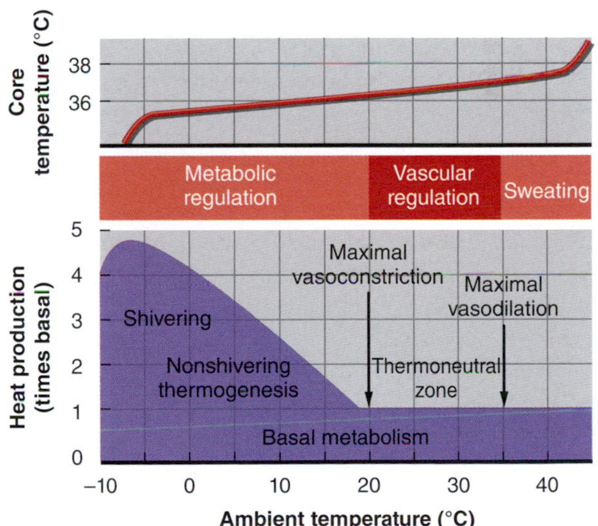

FIGURE 12.13 Thermoregulatory mechanisms by which humans strive to maintain a relatively constant body core temperature. In the thermoneutral zone, minor adjustments to skin blood flow minimize heat loss or gain. When maximal vasoconstriction is not sufficient to maintain core temperature, metabolic regulation, first in the form of nonshivering thermogenesis and then through shivering, serves to increase metabolic heat production.

RESEARCH PERSPECTIVE 12.3

Fuel for Shivering

During environmental cold exposure when behavioral changes (e.g., putting on additional layers, moving to a warmer area) are impossible or ill advised, humans rely on involuntary processes such as shivering to increase metabolic heat production. The muscle contractions associated with shivering are fueled by the oxidation of carbohydrates, lipids, and proteins, with the precise mix of fuels determined by the availability of each substrate and the muscle recruitment patterns and intensity of the shivering. Adult humans are able to sustain shivering and the metabolic heat production associated with it using a wide variety of fuels over a span of a few hours, and research has shown that when one fuel store is reduced, utilization of other fuels is increased to maintain shivering thermogenesis. Until recently, detailed studies of fuel sources for shivering had never been conducted in trials lasting longer than 3 h. Because of this, the long-term metabolic demands of cold exposure were largely unknown, and it was unclear if substrate depletion could limit shivering thermogenesis during extended cold exposures, such as during survival situations.

Recently, a team of research scientists from Canada conducted a 24 h cold-survival simulation to study the oxidative fuel selection of shivering during prolonged cold exposure in underfed and noncold-acclimatized men.[3] In this study, participants came to the laboratory in the morning following an overnight fast and were dressed in light cotton clothing, fleece mittens, and neoprene boots with their extremities protected to minimize the risk of nonfreezing cold injury. The subjects swallowed a telemetry pill to measure core temperature throughout the experiment and were fitted with thermal and electromyography sensors to measure skin temperature and muscle contractions. After baseline measurements, participants completed a 5 km treadmill walk to simulate physical activities associated with survival in an emergency. The subjects were then moved to an environmental chamber where they were exposed to 7.5 °C (45.5 °F) and 50% relative humidity for 24 h. Over the 24 h cold exposure, the subjects were fed 1,641 kcals divided into rations provided at 0, 3, 6, 9, 12, 15, 18, and 21 h. Metabolic and thermal responses as well as electromyography activity were measured in a seated position for 1 h at 6, 12, and 24 h of cold exposure. Changes in thermogenesis, fuel selection, and fuel availability were measured periodically using indirect calorimetry and metabolic isotope tracers.

Cold exposure decreased core temperature ~ 0.8 °C (1.4 °F) and mean skin temperature ~6.1 °C (11 °F). Total heat production initially increased 1.3 to 1.5 times compared to that at baseline and remained constant throughout the trial. This rise in metabolic heat production and shivering intensity was fueled by large modifications in fuel selection throughout the cold exposure. Total carbohydrate oxidation decreased after 6 h and remained low while lipid oxidation progressively doubled by 24 h (see figure). Protein utilization was low and did not change throughout the protocol. The research team concluded that these changes in fuel selection over an extended period of cold exposure dramatically reduced the utilization of limited muscle glycogen stores, extending the predicted time to muscle glycogen depletion as much as 10 times. Future research will be necessary to determine how nutritional substrate delivery or more extreme cold exposure may affect these responses.

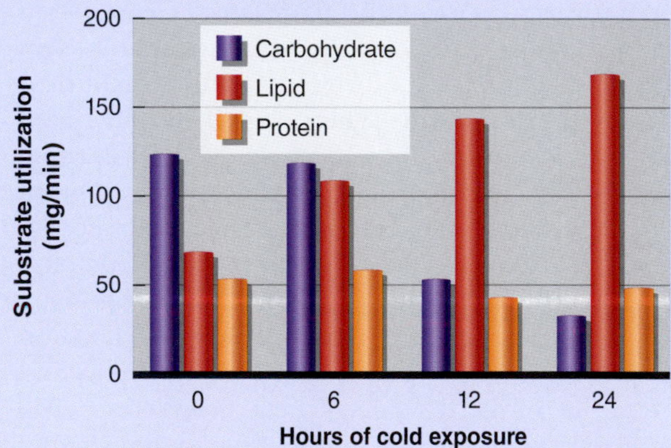

Representation of fuel sources for shivering thermogenesis after 0, 6, 12, and 24 h of cold exposure (7.5 °C, 50% relative humidity) in an environmental chamber. Carbohydrate oxidation decreased after 6 h while lipid oxidation progressively increased over the 24 h exposure. Protein utilization remained mostly unchanged throughout.

in warm conditions, can dissipate heat to the environment faster than the body produces it in extremely cold environments.

Pinpointing the exact conditions that permit excessive body heat loss and eventual **hypothermia** (low body core temperature) is difficult. Thermal balance depends on a wide variety of factors that affect the balance between body heat production and heat loss. Generally speaking, the larger the thermal gradient between the skin and the cold environment, the greater the heat loss. When exercising in the cold, people should not overdress. Overdressing can cause the body to become hot and initiate sweating. As the sweat soaks through the clothing, evaporation removes the heat, and heat is lost at an even faster rate.

A number of additional factors can also influence the rate of heat loss.

Body Size and Composition

Insulating the body against the cold is the most obvious form of protection against hypothermia. Recall that insulation is defined as resistance to dry heat exchange through radiation, convection, and conduction. Both inactive peripheral muscles and subcutaneous fat are excellent insulators. Skinfold measurements of subcutaneous fat thickness are a good indicator of an individual's tolerance for cold exposure. The thermal conductivity of fat (its capacity for transferring heat) is relatively low, so fat impedes heat transfer from the deep tissues to the body surface. People who have more fat mass conserve heat more efficiently in the cold than smaller, leaner individuals.

The rate of heat loss also is affected by the ratio of body surface area to body mass. Larger individuals have a small surface area-to-mass ratio, which makes them less susceptible to hypothermia.

Sex and Age Differences

Women have a slight advantage over men during cold exposure because they have more subcutaneous body fat. But their smaller muscle mass is a disadvantage in extreme cold because shivering is the major adaptation for generating body heat. The greater the active muscle mass, the greater the heat generation. Muscle also provides an additional insulating layer.

True sex differences in cold tolerance are minimal. Some studies have shown that the added subcutaneous fat in women may give them an advantage during cold-water immersion, but when men and women of similar body fat mass and size are compared, no real difference is noted in body temperature regulation during exposure to the cold.

As shown in table 12.2, small children have a large surface area-to-mass ratio compared with adults, leading to proportionately greater heat loss. This makes it more difficult for them to maintain normal body temperature in the cold. On the other end of the age spectrum, elderly individuals often tend to lose overall muscle mass, decreasing tissue insulation and making them more susceptible to hypothermia.

Windchill

As with heat, the air temperature alone does not provide a valid index of the heat loss experienced by the individual. Air movement, or wind, increases convective heat loss and therefore increases the rate of cooling. **Windchill** is an index based on the cooling effect of wind and is an often misunderstood and misused concept. Windchill is typically presented in charts of windchill equivalent temperatures showing various combinations of air temperature and wind speed that result in the same cooling power as that seen with no wind (figure 12.14). It is important to remember that windchill is not the temperature of the wind or the air (windchill does *not* change air temperature). True windchill refers to the cooling power of the environment. As windchill increases, so does the risk of freezing of tissues (figure 12.14).

Heat Loss in Cold Water

More research has been conducted on cold-water exposure than on exposure to cold air. Whereas radiation and sweat evaporation are the primary mechanisms for heat loss in air, convection allows the greatest heat transfer during immersion in water (recall that convection involves heat loss to moving fluids, which include both liquids and gases). As mentioned earlier, water has a thermal conductivity about 26 times greater than air. This means that heat loss by convection is 26 times faster in cold water than in cold air. When all heat-transfer mechanisms are considered (radiation, conduction, convection, and evaporation), the body generally loses heat four times faster in water than it does in air of the same temperature.

At rest, humans generally maintain a constant internal temperature in water at temperatures down to about 32 °C (89.6 °F). But when the water temperature decreases further, they may become hypothermic. Because of the large loss of heat from a body immersed in cold water, prolonged exposure or unusually cold water can lead to extreme hypothermia and death. Individuals immersed in water at 15 °C (59 °F) experience

TABLE 12.2 **Body Weight, Height, Surface Area, and Surface Area/Mass Ratio for an Average-Sized Adult and Child**

Person	Weight (kg)	Height (m)	Surface area (m^2)	Area/mass ratio (m^2/kg)
Adult	85	1.83	2.07	0.024
Child	25	1.00	0.79	0.032

Wind speed (km/h)	Air temperature (°C) −10	−15	−20	−25	−30	−35	−40	−45	−50
5	−13	−19	−24	−30	−36	−41	−47	−53	−58
10	−15	−21	−27	−33	−39	−45	−51	−57	−63
15	−17	−23	−29	−35	−41	−48	−54	−60	−66
20	−18	−24	−30	−37	−43	−49	−56	−62	−68
25	−19	−25	−32	−38	−44	−51	−57	−64	−70
30	−20	−26	−33	−39	−46	−52	−59	−65	−72
35	−20	−27	−33	−40	−47	−53	−60	−66	−73
40	−21	−27	−34	−41	−48	−54	−61	−68	−74
45	−21	−28	−35	−42	−48	−55	−62	−69	−75
50	−22	−29	−35	−42	−49	−56	−63	−69	−76
55	−22	−29	−36	−43	−50	−57	−63	−70	−77
60	−23	−30	−36	−43	−50	−57	−64	−71	−78
65	−23	−30	−37	−44	−51	−58	−65	−72	−79
70	−23	−30	−37	−44	−51	−58	−65	−72	−80
75	−24	−31	−38	−45	−52	−59	−66	−73	−80
80	−24	−31	−38	−45	−52	−60	−67	−74	−81

Risk	Description
Very low	Freezing is possible but unlikely
Likely	Freezing is likely > 30 min
High	Freezing risk < 30 min
Severe	Freezing risk < 10 min
Extreme	Freezing risk < 3 min

FIGURE 12.14 Windchill equivalent temperature chart showing various combinations of temperature and wind speed that result in the same cooling power as that seen with no wind. For example, a wind speed of 20 km/h at −10 °C would result in the same heat loss as −30 °C with no wind. Also shown in the figure is the risk of tissues freezing as windchill—the cooling power of the environment—increases.

a decrease in rectal temperature of about 2.1 °C (3.8 °F) per hour. In 1995, four U.S. Army rangers died of hypothermia after exposure to 11 °C (52 °F) swamp water in Florida, tragically publicizing the fact that hypothermia can occur when water and air temperatures are well above freezing.

If the water temperature were lowered to 4 °C (39.2 °F), rectal temperature would decrease at a rate of 3.2 °C (5.8 °F) per hour. The rate of heat loss is further accelerated if the cold water is moving around the individual because heat loss by convection increases. As a result, survival time in cold water under these conditions is quite brief. Victims can become weak and lose consciousness within minutes.

If the metabolic rate is low, as when the person is at rest, then even moderately cool water can cause hypothermia. But exercise in water increases the metabolic rate and offsets some of the heat loss. For example, although heat loss increases when one is swimming at high speeds (because of convection), the swimmer's accelerated rate of metabolic heat production more than compensates for the greater heat transfer. For competition and training, water temperatures between 23.9 and 27.8 °C (75-82 °F) are appropriate.

Physiological Responses to Exercise in the Cold

We have seen how the body adapts to maintain its internal temperature when exposed to a cold environment. Now consider what happens when the demands of physical performance are added to those of thermoregulation in the cold. How does the body respond to exercise in cold environmental conditions?

Muscle Function

Cooling a muscle causes it to contract with less force. The nervous system responds to muscle cooling by altering the normal muscle fiber recruitment patterns for force development, which may decrease the efficiency of the muscle's actions. Both muscle shortening velocity and power decrease significantly when muscle temperature is lowered. Luckily, large deep muscles seldom experience such low temperatures because they are protected from heat loss by a continuous supply of warm blood flow.

If clothing insulation and exercise metabolism are sufficient to maintain an athlete's body temperature in the cold, aerobic exercise performance may be unimpaired. However, as fatigue sets in and intensity decreases, so does metabolic heat production. Long-distance running, swimming, and skiing in the cold can expose the participant to such conditions. At the beginning of these activities, the athlete can exercise at an intensity that generates sufficient metabolic heat to maintain core temperature. However, late in the activity, when the energy reserves have diminished, exercise intensity declines, and this reduces metabolic heat production. The resulting decrease in core temperature causes the individual to become even more fatigued and less capable of generating heat. Under these conditions, the athlete may be confronted with a potentially dangerous situation.

Cold conditions affect muscle function in another way. As small muscles in the periphery like the fingers become cold, muscle function can be severely affected. This results in a loss of manual dexterity and limits the ability to perform fine motor skills like writing and manual labor tasks.

Metabolic Responses

Prolonged exercise increases the mobilization and oxidation of free fatty acids (FFAs) as a fuel source. The primary stimulus for this increased lipid metabolism is the release of catecholamines (epinephrine and norepinephrine). Exposure to cold markedly increases epinephrine and norepinephrine secretion, but FFA levels increase substantially less than during prolonged exercise in warmer conditions. Cold exposure triggers vasoconstriction in the vessels supplying not only the skin but also fatty subcutaneous tissues as well. The subcutaneous fat is a major storage site for lipids (as adipose tissue), so this vasoconstriction reduces the blood flow to an important area from which the FFAs would be mobilized. Thus, FFA levels do not increase as much as the elevated levels of epinephrine and norepinephrine would predict.

Blood glucose plays an important role in both cold tolerance and exercise endurance. For example, hypoglycemia (low blood sugar) suppresses shivering. The reasons for these changes are unknown. Fortunately, blood glucose concentrations are maintained reasonably well during cold exposure. Muscle glycogen, on the other hand, is used at a somewhat higher rate in the cold than in warmer conditions. However, studies on exercise metabolism in the cold are limited, and our knowledge regarding hormonal regulation of metabolism in the cold is too limited to support any definitive conclusions.

In Review

- Peripheral vasoconstriction decreases the transfer of heat from the skin to the air, thus decreasing heat loss to the environment. This is the body's first line of defense in the cold.
- The body's insulating shell consists of two regions: the superficial skin and subcutaneous fat and the underlying muscle. Increased skin vasoconstriction, increased subcutaneous fat thickness, and increased inactive muscle mass, especially in the limbs, can all increase total body insulation.
- Nonshivering thermogenesis increases metabolic heat production through activation of the SNS and by the action of hormones. Shivering thermogenesis increases metabolic heat production further to help maintain or increase body temperature.
- There are three distinct patterns of adaptation to repeated cold exposure: cold habituation, metabolic acclimation, and insulative acclimation.
- Body size is an important consideration for heat loss. Both a higher ratio of surface area to mass and a lower peripheral muscle mass or subcutaneous fat increase the loss of body heat to the environment.

- Because they have more insulating subcutaneous fat, women have a slight advantage over men during cold exposure, but their smaller muscle mass limits their ability to generate body heat.
- Wind increases heat loss by convection. The cooling power of the environment, known as windchill, is typically expressed as equivalent temperatures.
- Immersion in cold water tremendously increases heat loss through convection. In some cases, exercise may generate enough metabolic heat to offset some of this loss.
- When muscle is cooled, it is less able to produce force, and fatigue can occur more rapidly.
- During prolonged exercise in the cold, as fatigue causes exercise intensity to decline, metabolic heat production decreases and exercisers may become susceptible to hypothermia.
- Exercise triggers release of catecholamines, which increase the mobilization and use of FFAs for fuel. But in the cold, vasoconstriction impairs circulation to peripheral fat stores, so this process is attenuated.

Health Risks During Exercise in the Cold

If humans had retained the ability of lower animals, such as reptiles, to tolerate low body temperatures, we could survive extreme hypothermia. Unfortunately, the evolution of thermoregulation in humans has been accompanied by a diminished ability of tissues to function effectively outside a narrow range of temperatures. This section deals briefly with the health risks associated with cold stress. The American College of Sports Medicine published a comprehensive position stand, "Prevention of Cold Injuries During Exercise," in 2006 that addresses these topics in much greater detail.[1]

Hypothermia

Individuals immersed in near-freezing water will die within a few minutes when their rectal temperature decreases from a normal level of 37 °C (98.6 °F) to 24 or 25 °C (75.2 or 77 °F). Cases of accidental hypothermia, as well as data obtained from surgical patients who are intentionally made hypothermic, reveal that the lethal lower limit of body temperature is usually

RESEARCH PERSPECTIVE 12.4

The Yukon Arctic Ultramarathon

The Yukon Arctic Ultramarathon is considered the longest and coldest ultra-endurance event in the world. This 430 mi (692 km) event, which follows the Yukon Quest Trail beginning in Whitehorse (the capitol of the Canadian province of Yukon), is made up a series of nonstop multiday races of mountain biking, cross-country skiing, or walking. Moreover, it is run in February of each year. All participants must be self-reliant and pull a sled or attach a pack to their bike that contains all provisions and gear (including food, water, and tent) needed for the event. Because of the grueling conditions caused by the time of year and the extreme length, the Yukon Arctic Ultra is often ranked as the most difficult physical competition in the world. Cold exposure and exercise have been reported to influence circulating concentrations of myokines, adipokines, and hepatokines that may alter the regulation of metabolism. Specifically, changes in these metabolic signaling molecules may initiate changes in white or brown adipose tissue to promote thermogenesis.

In 2016, a team of research scientists from the University of Alaska Fairbanks and the Center for Space Medicine and Extreme Environments in Berlin, Germany, followed eight participants in the Yukon Arctic Ultramarathon.[2] Blood samples were collected at preevent, midevent, and postevent checkpoints. The temperature during the event ranged from −45 °C (−49 °F) to −8 °C (17.6 °F) that year. Because of these exceptionally harsh conditions, four of the eight study participants withdrew from competition by the 300 mi (483 km) mark. The four who did pass the 300 mi mark all lost significant body weight and fat mass. Analysis of the blood samples collected indicated that serum irisin, a myokine that has been purported to promote the breakdown of lipids and support brown adipose tissue formation, increased over the event. However, there were no changes in other myokines, adipokines, or hepatokines, and serum metabolites remained stable throughout the event. The data from the four participants who finished the Yukon Arctic Ultramarathon provide a glimpse into the physiology of these athletes and may suggest that the combination of cold exposure and extreme physical exertion promotes alterations in substrate metabolism that preserve skeletal muscle and increase physiological resilience in the face of these excessive demands. In the future, these researchers plan to obtain measurements of energy expenditure, dietary intake, and sleep quality to determine their contribution to the relative preservation of lean body mass in these athletes.

between 23 and 25 °C (73.4-77 °F), although patients have recovered after having rectal temperatures below 18 °C (64.4 °F).

Once core temperature falls below about 34.5 °C (94.1 °F), the hypothalamus begins to lose its ability to regulate body temperature. This ability is completely lost when the internal temperature decreases to about 29.5 °C (85.1 °F). This loss of function is associated with a slowing of metabolic reactions. For each 10 °C (18 °F) drop in cellular temperature, the metabolism of the cell decreases by half. As a result, low core temperatures can cause drowsiness, lethargy, and even coma.

Mild hypothermia can be treated by removing the affected person from the cold, providing dry clothing and blankets for insulation, and providing warm beverages. Moderate to severe cases of hypothermia require gentle handling to avoid initiating a cardiac arrhythmia. This requires slowly rewarming the victim. Severe cases of hypothermia require hospital facilities and medical care. Recommendations to prevent cold exposure–related injuries were outlined by the American College of Sports Medicine in its 2006 position stand on cold illness during exercise.[1]

Cardiorespiratory Effects

The hazards of excessive cold exposure include potential injury to both peripheral tissues and the cardiovascular and respiratory systems. Death from hypothermia has resulted from cardiac arrest while respiration was still functional. Cooling primarily influences the sinoatrial node, the heart's pacemaker, leading to a substantial decrease in heart rate and, ultimately, cardiac arrest.

People have questioned whether rapid, deep breathing of cold air can damage or freeze the respiratory tract. In fact, the cold air that passes into the mouth and trachea is rapidly warmed, even when the temperature of inhaled air is less than −25 °C (−13 °F).[6] Even at this temperature, when a person is at rest and breathing primarily through the nose, the air is warmed to about 15 °C (59 °F) by the time it has traveled about 5 cm (2 in.) into the nasal passage. As shown in figure 12.15, extremely cold air entering the nose is quite warm by the time it reaches the back of the nasal passage, thereby posing no threat of damage to the throat, trachea, or lungs. Mouth breathing, which often occurs during exercise, may result in cold irritation to the

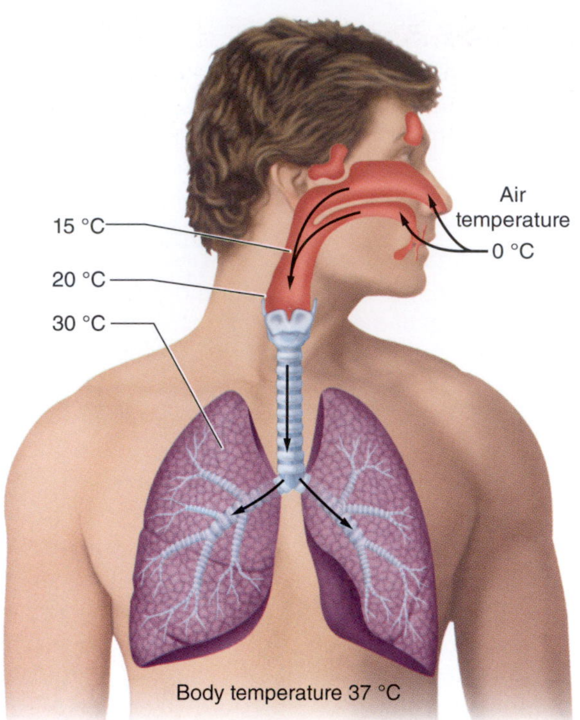

FIGURE 12.15 The warming of inspired air as it moves through the respiratory tract.

mouth, pharynx, trachea, and even bronchi when the air temperature is below −12 °C (10 °F). Excessive cold exposure also affects respiratory function by decreasing respiratory rate and volume.

Frostbite

Exposed skin can freeze when its temperature is lowered just a few degrees below the freezing point (0 °C, or 32 °F). Because of the warming influence of circulation and metabolic heat production, the environmental air temperature (including windchill; see figure 12.14) required to freeze exposed fingers, nose, and ears is about −29 °C (−20 °F). Recall from our earlier discussion that peripheral vasoconstriction helps the body retain heat. Unfortunately, during exposure to extreme cold, the circulation in the skin can decrease to the point that the tissue dies from lack of oxygen and nutrients. This is commonly called **frostbite**. If not treated early, frostbite injuries can be serious, leading to gangrene and loss of tissue. Frostbitten parts should be left untreated until they can be thawed, preferably in a hospital, without risk of refreezing.

Exercise-Induced Asthma

While not strictly a cold-related illness, exercise-induced asthma is a common problem that affects as many as 50% of winter sport athletes (see chapter 8). The main cause of this syndrome is the drying of the airways due to the combination of the high respiration rate associated with exercise and the extremely dry air as temperature drops, although multiple environmental and genetic factors may determine whether a given athlete will experience asthma symptoms as a result. The resultant airway narrowing often leaves athletes gasping for breath. Luckily, there are preventive medications available such as β-agonists plus inhalers that can quickly deliver corticosteroids and bronchodilating medications to relieve symptoms.

In Review

- The hypothalamus begins to lose its ability to regulate body temperature when core temperature drops below about 34.5 °C (94.1 °F).
- Hypothermia critically affects the heart's sinoatrial node, decreasing the heart rate, which in turn reduces cardiac output.
- Breathing cold air does not freeze the respiratory passages or the lungs because the inspired air is progressively warmed as it moves through the respiratory tract.
- Exposure to extreme cold decreases respiratory rate and volume.
- Frostbite occurs as a consequence of the body's attempt to prevent heat loss by skin vasoconstriction. If vasoconstriction is prolonged, the skin cools rapidly and the reduced blood flow, combined with the lack of oxygen and nutrients, ultimately can cause cutaneous tissue to die.
- Because cold air is inherently dry, many athletes experience the symptoms of exercise-induced asthma during high-intensity exercise in cold environments.

IN CLOSING

In this chapter, we began our examination of how the external environment affects the body's ability to perform physical work. We looked at the effects of extreme heat and cold stress and the body's responses to these conditions. We considered the health risks associated with these temperature extremes and how the body adapts to these conditions through acclimation. In the next chapter, we examine additional environmental extremes associated with exercise at altitude.

KEY TERMS

acclimation (heat acclimation)
acclimatization
arginine vasopressin
cold habituation
conduction (K)
convection (C)
critical temperature theory
dry heat exchange
eccrine sweat glands
evaporation (E)
frostbite
heat cramps
heat exhaustion
heatstroke
hypothermia
insulation
insulative acclimation
irisin
metabolic acclimation
nonshivering thermogenesis
peripheral vasoconstriction
preoptic-anterior hypothalamus (POAH)
radiation (R)
shivering
sickle cell trait
thermal stress
thermoreceptors
thermoregulation
vasopressin
wet-bulb globe temperature (WBGT)
windchill

STUDY QUESTIONS

1. What are the four major avenues for loss of body heat?
2. Which of these four pathways is most important for controlling body temperature at rest? During exercise?
3. What happens to the body temperature during exercise and why?
4. Why is water vapor pressure in the air an important factor when one is performing in the heat? Why are wind and cloud cover important?
5. What factors may limit the ability to continue to exercise in hot environments?
6. What is the purpose of the wet-bulb globe temperature (WBGT)? What does it measure?
7. Differentiate between heat cramps, heat exhaustion, and heatstroke.
8. What physiological adaptations occur that allow a person to acclimate to exercise in the heat?
9. How does the body minimize excessive heat loss during cold exposure?
10. What are the three patterns of cold adaptation and when might each be expected to occur?
11. What factors should be considered to provide maximal protection when people are exercising in the cold?

STUDY GUIDE ACTIVITIES

In addition to the activities listed in the chapter opening outline, two other activities are available in the web study guide, located at

www.HumanKinetics.com/PhysiologyOfSportAndExercise

The **KEY TERMS** activity reviews important terms, and the end-of-chapter **QUIZ** tests your understanding of the material covered in the chapter.

SFR
TF1

Exercise at Altitude

In this chapter and in the web study guide

On October 14, 2012, Austrian skydiver Felix Baumgartner set several world records for skydiving. On that date, Baumgartner piloted a helium balloon into the stratosphere at an altitude of 39 km (128,000 ft) (a balloon ascent record) and, leaving the balloon, jumped to Earth from that height (a record for parachute jumping from the highest altitude). In doing so, he became the first human to break the sound barrier outside a vehicle, reaching a velocity of 1,343 km/h (833 mph) during a 4 min and 19 s free fall.

At such altitudes, without a pressurized suit and breathing equipment, Baumgartner's body would have been unable to supply his tissues with oxygen, and he would have quickly lost consciousness. This specialized suit was also designed to protect him from the low ambient pressure, extreme cold (−52 °C, or −61 °F), friction, and other dangers. Had the suit failed, above 19.2 km he would have experienced the fatal condition known as ebullism, the formation of gas bubbles in his bodily fluids.

Instead, he deployed his parachute and landed safely in the New Mexico desert, illustrating not only an exceptional feat of human performance but also the power of technology in overcoming—for a brief period of time—the dangers of high altitude.

Our previous discussions of the physiological responses to exercise have all been based on the conditions that exist at or near sea level, where the **barometric pressure (P_b)** averages about 760 mmHg. Recall from chapter 7 that barometric pressure is a measure of the total pressure that all of the gases composing the atmosphere exert on the body (and everything else). Regardless of the P_b, oxygen molecules always make up 20.93% of the air. The **partial pressure of oxygen (PO_2)** is that portion of P_b exerted only by the oxygen molecules in the air. At sea level, PO_2 is therefore 0.2093 times 760 mmHg, or 159 mmHg. Partial pressure is an important concept in understanding altitude physiology, because it is primarily the low PO_2 at altitude that limits exercise performance. Although the human body tolerates small fluctuations in PO_2, large variations pose problems. This is evident when mountain climbers ascend to high altitudes where significantly reduced PO_2 can substantially impair physical performance and can even jeopardize life.

The reduced barometric pressure at altitude is referred to as a **hypobaric** environment or simply hypobaria (low atmospheric pressure). The lower atmospheric pressure also means a lower PO_2 in the inspired air, which limits pulmonary diffusion of oxygen from the lungs and oxygen transport to the tissues. The low PO_2 in the air is termed **hypoxia** (low oxygen), while the resulting low PO_2 in the blood is called **hypoxemia**.

In this chapter, we examine the unique characteristics of hypobaric, hypoxic environments and how these conditions alter physiological responses at rest and during physical activity, training, and sport performance. We cover these changes upon acute ascent to altitude, the ways in which these responses change as humans acclimate to high altitude, and specialized training strategies used by athletes to improve performance at altitude. We also examine several specific health risks associated with hypoxic environments.

Environmental Conditions at Altitude

Clinical problems associated with altitude were reported as early as 400 BC. However, most of the early concerns about ascent to high altitudes focused on the cold conditions at altitude rather than the limitations imposed by low air pressure. The initial landmark discoveries that led to our current understanding of the reduced P_b and PO_2 at altitude can be credited primarily to four scientists spanning the 17th through the 19th centuries. Torricelli (ca. 1644) developed the mercury barometer, an instrument that permitted the accurate measurement of atmospheric pressure. Only a few years later (1648), Pascal demonstrated a reduction in barometric pressure at high altitudes. Nearly 130 years after that (1777), Lavoisier described oxygen and the other gases that contribute to the total barometric pressure. Finally, in 1801, John Dalton set forth the principle (called Dalton's law of partial pressures) stating that the total pressure exerted by a mixture of gases is equal to the sum of the partial pressures of those individual gases.

The deleterious effects of high altitude on humans that are caused by low PO_2 (hypoxia) were recognized in the late 1800s. More recently, a team of scientists led by the late John Sutton performed an intricate series of laboratory studies in the hypobaric chamber at the U.S. Army Institute of Environmental Medicine. These experiments, known collectively as *Operation Everest II*, have significantly added to our understanding of exercise at altitude.[23]

Based on the effects of altitude on performance, the following definitions are useful:[1]

- Near sea level [below 500 m (1,640 ft)]: There are no effects of altitude on well-being or exercise performance.
- Low altitude [500-2,000 m (1,640-6,560 ft)]: There are no effects on well-being but performance may be diminished, especially in athletes

performing above 1,500 m (4,921 ft). These performance decrements may be overcome with acclimation.

- Moderate altitude [2,000-3,000 m (6,560-9,840 ft)]: Effects on well-being in unacclimated individuals and decreased maximal aerobic capacity and performance are likely. Optimal performance may or may not be restored with acclimation.
- High altitude [3,000-5,500 m (9,840-18,000 ft)]: There are adverse health effects (including acute mountain sickness, discussed later in this chapter) in a large percentage of individuals and significant performance decrements, even after full acclimation.
- Extreme altitude [above 5,500 m (~18,000 ft)]: Severe hypoxic effects are experienced. The highest permanent human settlements are at 5,200 to 5,800 m (17,000-19,000 ft).

For our discussion, the term *altitude* refers to elevations above 1,500 m (4,921 ft), because few negative physiological effects on exercise or sport performance are seen below that altitude.

While the major impact of altitude on exercise physiology is attributable to the low PO_2 that ultimately limits oxygen availability to the tissues, the atmosphere at altitude also differs in other ways from sea-level conditions.

Atmospheric Pressure at Altitude

At sea level, the air extending to the outermost reaches of the earth's atmosphere (approximately 38.6 km, or 24 mi) exerts a pressure equal to 760 mmHg. At the summit of Mount Everest, the highest point on Earth (8,848 m, or 29,029 ft), the pressure exerted by the air above is only about 250 mmHg. These and related altitude differences are depicted in figure 13.1.

The barometric pressure on Earth does not remain constant. Rather, it varies somewhat with changes in climatic conditions, time of year, and the specific site at which the measurement is taken. On Mount Everest, for example, the mean barometric pressure varies from 243 mmHg in January to nearly 255 mmHg in June and July. These minor variations are of little interest at sea level, except to meteorologists because of their effect on weather patterns, but are of considerable physiological importance for a climber attempting to ascend Mount Everest without supplemental oxygen.

Although barometric pressure varies, the percentages of gases in the air that we breathe remain

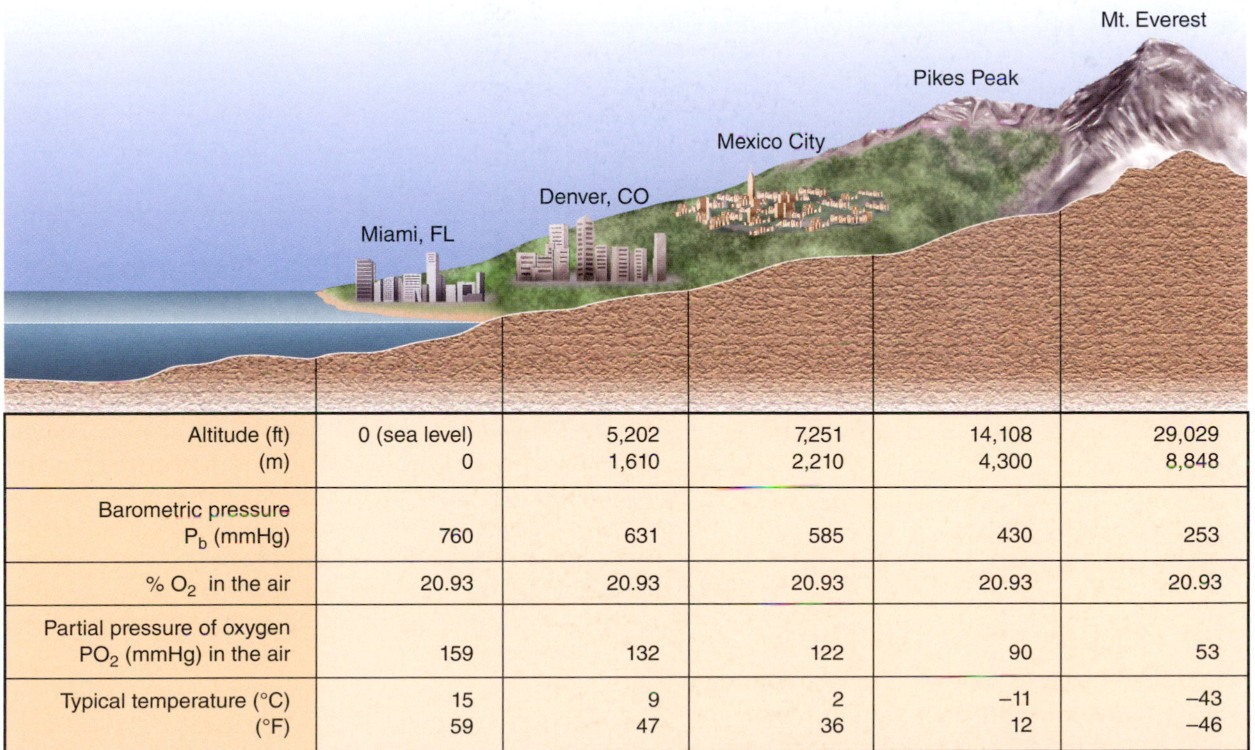

Altitude (ft) (m)	0 (sea level) 0	5,202 1,610	7,251 2,210	14,108 4,300	29,029 8,848
Barometric pressure P_b (mmHg)	760	631	585	430	253
% O_2 in the air	20.93	20.93	20.93	20.93	20.93
Partial pressure of oxygen PO_2 (mmHg) in the air	159	132	122	90	53
Typical temperature (°C) (°F)	15 59	9 47	2 36	−11 12	−43 −46

FIGURE 13.1 Differences in atmospheric conditions at sea level as altitude increases and barometric pressure falls accordingly. Note that the partial pressure of oxygen in the air decreases from 159 mmHg at sea level to only 53 mmHg at the summit of Mount Everest.

unchanged from sea level to high altitude. At any elevation, the air always contains 20.93% oxygen, 0.03% carbon dioxide, and 79.04% nitrogen. Only the partial pressures of these gases change. As shown in figure 13.1, the pressure that oxygen molecules in the air exert at various altitudes drops proportionally with decreases in the barometric pressure. The consequent changes in PO_2 have significant effects on the partial pressure of oxygen that reaches the lungs, as well as the gradients between the alveoli of the lungs and the blood (where oxygen is loaded) and between the blood and the tissues (where oxygen is unloaded). These effects are discussed in detail later in this chapter.

Air Temperature and Humidity at Altitude

Clearly, the low PO_2 at altitude has the greatest impact on exercise physiology. However, other environmental factors contribute to the ability to perform exercise as well. For example, air temperature decreases at a rate of about 1 °C (1.8 °F) for every 150 m (about 490 ft) of ascent. The average temperature near the summit of Mount Everest is estimated to be about −40 °C (−40 °F), whereas at sea level the temperature would be about 15 °C (59 °F). The combination of low temperatures, low ambient water vapor pressure, and high winds at altitude poses a serious risk of cold-related disorders, such as hypothermia, frostbite, and nonfreezing cold injuries.

Water vapor has its own partial pressure, also known as the water vapor pressure (P_{H_2O}). Because of the cold temperatures at altitude, the water vapor pressure in the air is extremely low. Cold air holds very little water. Thus, even if air is fully saturated with water (100% relative humidity), the actual vapor pressure of water contained in the air is low. The extremely low P_{H_2O} at high altitude promotes evaporation of moisture from the skin (or clothing) surface, because of the high gradient between skin and air, and can lead quickly to dehydration. In addition, a large volume of water is lost through respiratory evaporation due to a combination of (a) a large vapor pressure gradient between warmed air leaving the mouth and nose and the dry air in the environment and (b) an increased respiration rate (discussed later) experienced at altitude.

Solar Radiation at Altitude

The intensity of solar radiation increases at high altitude for two reasons. First, at high altitudes, light travels through less of the atmosphere before reaching the earth. For this reason, less of the sun's radiation, especially ultraviolet rays, is absorbed by the atmosphere at higher altitudes. Second, because atmospheric water normally absorbs a substantial amount of the sun's radiation, the low water vapor in the air at altitude also increases radiant exposure. Solar radiation may be further amplified by reflective light from snow, which is usually found at higher elevations.

In Review

- Altitude presents a hypobaric environment (one in which the atmospheric barometric pressure is reduced). Altitudes of 1,500 m (4,921 ft) or higher have a notable physiological impact on exercise performance.
- Although the percentages of the gases in the air that we breathe remain constant regardless of altitude, the partial pressure of each of these gases decreases with the decreased barometric pressure at higher altitudes.
- The low partial pressure of oxygen (PO_2) in the air at altitude is the environmental condition with the most profound physiological impact. Because the PO_2 in the lungs is low, PO_2 gradients between the alveoli of the lungs and the blood (where oxygen is loaded) and between the blood and the tissues (where oxygen is unloaded) are decreased.
- Air temperature typically decreases as altitude increases. Cold air can hold little water, so the air at altitude is dry. These two factors combine to increase susceptibility to cold-related disorders and dehydration.
- Because the atmosphere is thinner and drier at altitude, solar radiation is more intense at higher elevations. This effect is magnified when the ground is snow covered.

Physiological Responses to Acute Altitude Exposure

This section deals with how the human body responds to acute altitude exposure, emphasizing responses that can affect exercise and sport performance. The main concerns are respiratory, cardiovascular, and metabolic responses. Most physiological studies have been performed on healthy, fit young men; unfortunately, few studies on the effects of altitude have included women, children, or the elderly—populations whose responses to the conditions of altitude might differ from those described here.

Respiratory Responses to Altitude

Adequate oxygen supply to exercising muscles is essential to physical performance and, as seen in chapter 8, depends on an adequate supply of oxygen being brought into the body, moved from the lungs to the blood, transported to the muscles, and adequately taken up into the muscles. A limitation in any of these steps can impair performance.

Pulmonary Ventilation

The sequence of steps leading to the transport of oxygen to working muscle begins with pulmonary ventilation, the active movement of gas molecules into the alveoli of the lungs (breathing). Ventilation increases within seconds of exposure to high altitude, both at rest and during exercise, because chemoreceptors in the aortic arch and carotid arteries are stimulated by the low PO_2 and signals are sent to the brain to increase breathing. The increased ventilation is primarily associated with an increased tidal volume and an even greater increase in respiratory rate. Over the next several hours and days, ventilation remains elevated to a level proportional to the altitude.

Increased ventilation acts much the same as hyperventilation at sea level. The amount of carbon dioxide in the alveoli is reduced. Carbon dioxide follows the pressure gradient, so more diffuses out of the blood, where its pressure is relatively high, and into the lungs to be exhaled. This "blowing off" of CO_2 causes blood PCO_2 to fall and blood pH to increase, a condition known as **respiratory alkalosis**. This alkalosis has two effects. First, it causes the oxyhemoglobin saturation curve to shift to the left (as discussed in the next section). Second, it helps keep the rise in ventilation caused by the hypoxic (low PO_2) drive from increasing even further. At a given submaximal exercise intensity, ventilation is higher at altitude than at sea level, but maximal exercise ventilation is similar.

In an effort to offset respiratory alkalosis, the kidneys excrete more bicarbonate ion, the ions that buffer the carbonic acid formed from carbon dioxide. Thus, a reduction in bicarbonate ion concentration reduces the blood's buffering capacity. More acid remains in the blood, and the alkalosis is minimized.

Pulmonary Diffusion

Under resting conditions, pulmonary diffusion (diffusion of O_2 from the alveoli to the arterial blood) does not limit the exchange of gases between the alveoli and the blood. If gas exchange were limited or impaired at altitude, less oxygen would enter the blood, so the arterial PO_2 would be much lower than the alveolar PO_2. Instead, these two values are almost equal (figure 13.2). Therefore, the low arterial blood PO_2, or hypoxemia, is a direct reflection of the low alveolar PO_2 and not a limitation of oxygen diffusion from the alveoli to the arterial blood.

As introduced in chapter 7, the oxygen cascade is a way of depicting the changes in PO_2 from the air through the tissues and into the venous circulation.

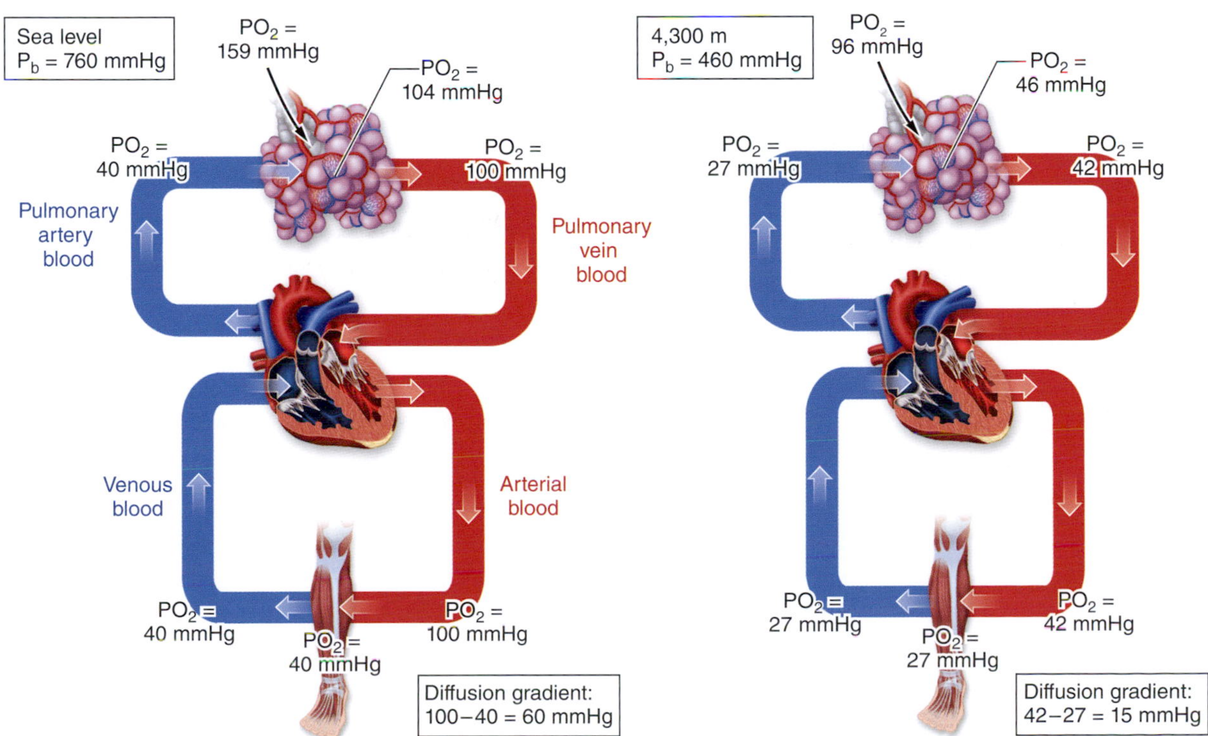

FIGURE 13.2 Comparison of the partial pressure of oxygen (PO_2) in the inspired air and in body tissues at sea level and at 4,300 m (14,108 ft) of altitude, the altitude of Pikes Peak, Colorado. As inspired PO_2 decreases, so does alveolar PO_2. Arterial PO_2 is similar to that in the lungs, but the gradient for diffusion of O_2 into tissues, including muscle, is greatly reduced.

Figure 13.3 shows the differences at various points in the oxygen cascade between sea level and an altitude of 5,800 m (19,029 ft).

Oxygen Transport

As shown in figures 13.2 and 13.3, the inspired PO_2 at sea level is 159 mmHg; however, it decreases to about 105 in the alveoli, primarily because of the addition of water vapor molecules (P_{H_2O} = 47 mmHg at 37 °C). When the alveolar PO_2 drops at altitude, fewer binding sites on the hemoglobin in the blood perfusing the lungs become saturated with O_2. As depicted in figure 13.4, the oxygen-binding (or oxyhemoglobin dissociation) curve for hemoglobin has a distinct S shape. At sea level, when alveolar PO_2 is about 104 mmHg, 96% to 97% of hemoglobin has O_2 bound to it. When PO_2 in the lungs is decreased to 46 mmHg at 4,300 m (14,108 ft), only about 80% of hemoglobin sites are saturated with O_2. If the oxygen-loading portion of the curve were not relatively flat, far less O_2 would be taken up by the blood as it passes through the lungs, and binding would be extremely limited at altitude. Therefore, while arterial blood is still desaturated at altitude, the inherent shape of the oxyhemoglobin dissociation curve serves to minimize this problem.

A second adaptation occurs very early in altitude exposure that also aids in preventing the fall in arterial oxygen content. As mentioned earlier, a respiratory alkalosis accompanies the increased ventilation caused by acute altitude exposure. This increase in blood pH actually shifts the oxyhemoglobin dissociation curve to the left, as shown in figure 13.3. The result is that, rather than 80% binding of oxygen to hemoglobin, 89% of hemoglobin is saturated with O_2. Because of this shift, more oxygen binds to hemoglobin in the lungs and more oxygen is unloaded to the tissues at higher altitudes, where PO_2 is lower in both tissues.

Gas Exchange at the Muscles

Figures 13.2 and 13.3 illustrate that arterial PO_2 at sea level is about 100 mmHg, and the PO_2 in body tissues is

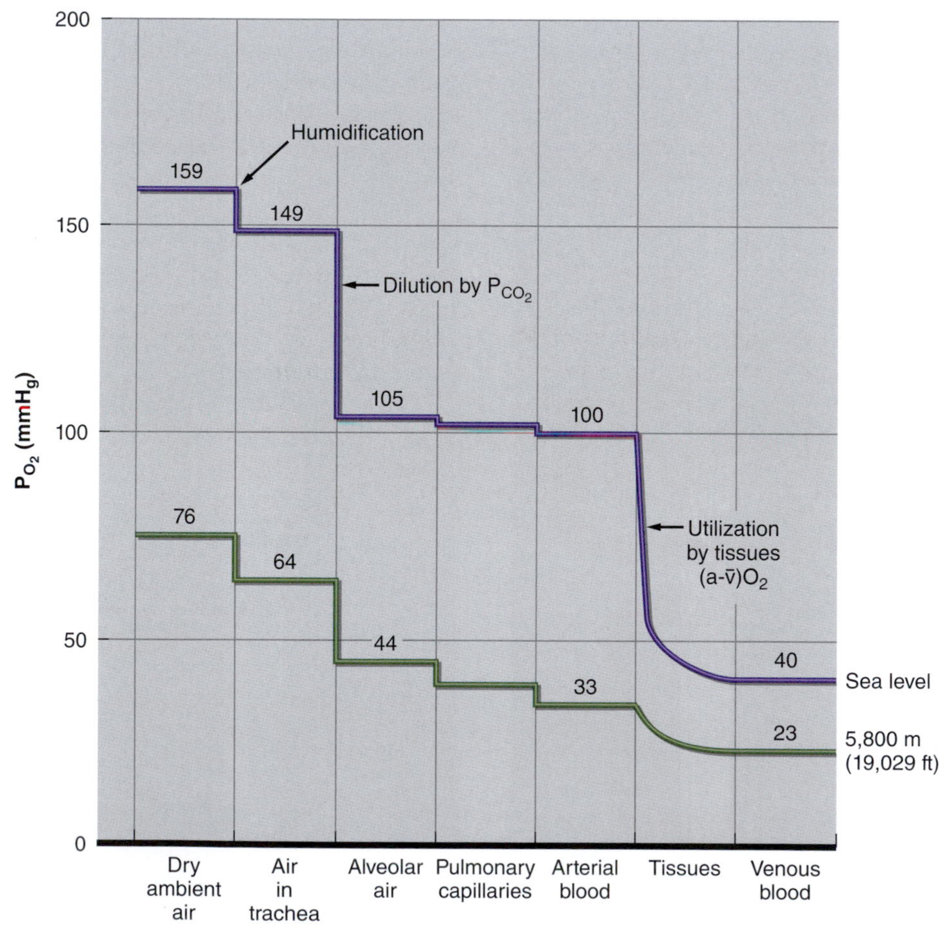

FIGURE 13.3 Comparison of the oxygen cascade at sea level and at an altitude of 5,800 m. Each step, from inspired air to tissue and venous blood shows a remarkable decrease in PO_2, reducing the gradient for diffusion of O_2 into tissues, including muscle.

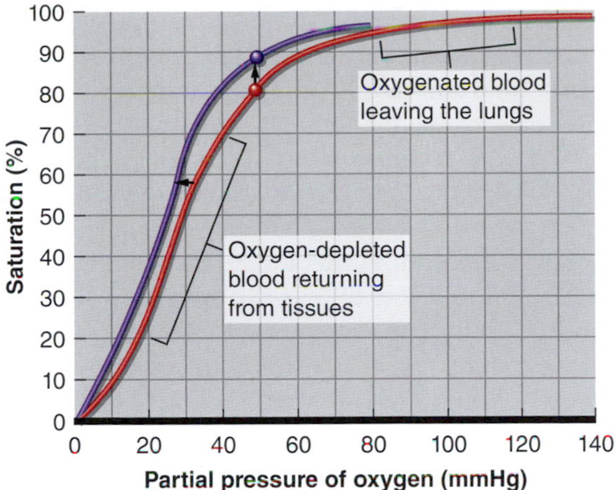

FIGURE 13.4 The S-shaped oxyhemoglobin dissociation curve at sea level (red line). When alveolar PO_2 is about 104 mmHg, 96% to 97% of hemoglobin is saturated with O_2. The respiratory alkalosis with acute altitude exposure shifts the oxyhemoglobin dissociation curve leftward (blue line), partially compensating for the desaturation resulting from a drop in PO_2.

consistently about 40 mmHg at rest; so the difference, or the pressure gradient, between the arterial PO_2 and the tissue PO_2 at sea level is about 60 mmHg. However, when one moves to an elevation of 4,300 m (14,108 ft), arterial PO_2 decreases to about 42 mmHg and the tissue PO_2 drops to 27 mmHg. Thus, the pressure gradient decreases from 60 mmHg at sea level to only 15 mmHg at the higher altitude. This represents a 75% reduction in the diffusion gradient. At 5,800 m (19,029 ft), the gradient shrinks to only 10 mmHg. Because the diffusion gradient is responsible for driving the oxygen from the hemoglobin in the blood into the tissues, this change in arterial PO_2 at altitude is a much greater consideration for exercise performance than the small reduction in hemoglobin saturation that occurs in the lungs.

Cardiovascular Responses to Altitude

As the respiratory system becomes increasingly limiting at altitude, the cardiovascular system likewise undergoes substantial changes to attempt to compensate for the decrease in arterial PO_2 that accompanies hypoxia.

Blood Volume

Within the first few hours of arrival at altitude, plasma volume begins to progressively decrease, and the decline reaches a plateau by the end of the first few weeks. This decrease in plasma volume is the result of both respiratory water loss and increased urine production. This combination of respiratory water loss and the excretion of fluid can reduce total plasma volume by up to 25%. Initially, the result of the plasma loss is an increase in the hematocrit, the percentage of the blood volume composed of red blood cells (containing hemoglobin). This adaptation—more red blood cells for a given blood flow—allows more oxygen to be delivered to the muscles for a given cardiac output. Over a period of weeks at altitude, this diminished plasma volume eventually returns to normal if adequate fluids are ingested.

Continued exposure to high altitude triggers the release of erythropoietin (EPO) from the kidneys; this is the hormone responsible for stimulating erythrocyte (red blood cell) production. This increases the total number of red blood cells and creates a greater total blood volume, which allows the person to partially compensate for the lower PO_2 experienced at altitude. However, this compensation is slow, taking weeks to months to fully restore red cell mass.

Cardiac Output

The preceding discussion clearly illustrates that the amount of oxygen delivered to the muscles by a given volume of blood is limited at altitude because of the reduced arterial PO_2. A logical means to compensate for this is to increase the volume of blood delivered to the active muscles. At rest and during submaximal exercise, this is accomplished by increasing cardiac output. Since cardiac output is the product of stroke volume and heart rate, increasing either or both of these variables will increase cardiac output. Upon ascent to altitude, the sympathetic nervous system is stimulated, releasing norepinephrine and epinephrine, the main hormones that alter cardiac function. The increase in norepinephrine in particular persists for several days of acute altitude exposure.

When submaximal exercise is performed during the first few hours at altitude, stroke volume is decreased compared to sea-level values (attributable to the reduced plasma volume). Fortunately, heart rate is increased disproportionately not only to compensate for the decrease in stroke volume but also to slightly increase cardiac output. However, this extra cardiac workload is not an efficient way to ensure sufficient oxygen delivery to the body's active tissues for prolonged periods. Consequently, after a few days at altitude, the muscles begin extracting more oxygen from the blood (increasing the arterial–venous oxygen difference), which reduces the demand for increased cardiac output, since $\dot{V}O_2 = \dot{Q} \times (a\text{-}\bar{v})O_2$ difference. The increase in heart rate and cardiac output peaks after about 6 to 10 days at high altitude, after which cardiac output and heart rate during a given exercise bout start to decrease.

At maximal or exhaustive work levels at higher altitudes, both maximal stroke volume and maximal heart rate are decreased, as is cardiac output. The decrease in stroke volume is directly related to the decrease in plasma volume. Maximal heart rate may be somewhat lower at high altitude as a consequence of a decrease in the response to sympathetic nervous system activity, possibly attributable to a reduction in β-receptors (receptors in the heart that respond to sympathetic nerve activation, thus increasing the heart rate). With a decreased diffusion gradient to move oxygen from the blood into the muscles, coupled with this reduction in maximal cardiac output, it is apparent why both $\dot{V}O_{2max}$ and submaximal aerobic performance are hindered at altitude. In summary, hypobaric conditions significantly limit oxygen delivery to the muscles, reducing the capacity to perform high-intensity or prolonged aerobic activities.

Table 13.1 summarizes the acute responses to altitude at rest and during submaximal exercise. Given the hypoxic conditions at altitude, and because any fixed amount of work at altitude represents a higher percentage of $\dot{V}O_{2max}$, we would expect anaerobic metabolism to be increased. If this occurs, we would expect lactic acid production to increase at any given work rate above the lactate threshold. This is in fact the case upon arrival at altitude. However, with longer exposure to altitude, the lactate concentration in the muscles and venous blood at a given intensity of exercise (including maximal exertion) is lower, despite the fact that muscle $\dot{V}O_{2max}$ does not change with adaptation to altitude. To date, there is no universally accepted explanation for this so-called lactate paradox.[4]

Metabolic Responses to Altitude

Ascent to altitude increases the basal metabolic rate, possibly due to increases in both thyroxin and catecholamine concentrations. This increased metabolism must be balanced by an increased food intake to prevent body weight from decreasing, a common occurrence during the first few days at altitude because appetite declines as well. In individuals who maintain their body weight at altitude, there is an increased reliance on carbohydrate for fuel, both at rest and during submaximal exercise. Because glucose yields more energy than fat or protein per liter of oxygen, this adaptation is beneficial.

Nutrition Needs at Altitude

In addition to the alterations in physiological systems and processes already described, several other considerations are important to note with ascent to altitude. At altitude, the body has a natural tendency to lose fluids through the skin (insensible water loss), the respiratory system, and the kidneys. This water loss is exaggerated with exercise as sweat evaporation increases from the wetted skin to the relatively dry air. These avenues of fluid loss dramatically increase the risk of dehydration, and one should pay careful attention to staying hydrated. A rule of thumb at altitude is to consume at

TABLE 13.1 **Effects of Acute Hypoxia (Initial 48 h) on Physiological Responses at Rest and During Submaximal Exercise**

System	Acute hypoxic effect at rest	Acute hypoxic effect at a given submaximal exercise intensity
Respiratory and oxygen transport	Immediate increase in ventilation (increased frequency > increased tidal volume) Decreased 2,3-diphosphoglycerate concentration Leftward shift in the oxyhemoglobin dissociation curve Stimulation of peripheral chemoreceptors Respiratory alkalosis	Increased ventilation
Cardiovascular	Decreased plasma volume Increased heart rate Decreased stroke volume Increased cardiac output Increased blood pressure	Increased heart rate Decreased stroke volume (due to decreased plasma volume) Increased cardiac output Increased $\dot{V}O_2$
Metabolic	Increased basal metabolic rate Decreased $(a-\bar{v})O_2$ difference	Greater utilization of carbohydrates for energy Increased lactate production initially, then lower Decreased blood pH
Renal	Diuresis Excretion of bicarbonate ions Increased release of erythropoietin	

least 3 L of fluid per day; however, this must be tailored to individual needs. It may seem counterproductive to increase fluid intake when the decrease in plasma volume is taking place to help pack red blood cells. However, dehydration can negatively alter the body water balance among fluid compartments, so staying well hydrated and allowing the natural decrease in plasma volume to occur is sound advice.

Appetite decreases at altitude, and decreased food intake often accompanies that decline. This decreased energy consumption coupled with increased metabolic rates can lead to daily energy deficits of up to 500 kcal/day, resulting in weight loss over time. Consuming adequate calories to support exercise and recreational activities is important, and climbers should be instructed to eat more calories than their appetite suggests.

Finally, successful acclimation and acclimatization to high altitude depend on adequate iron stores in the body. Iron deficiency may prevent the increase in red blood cell production that occurs progressively over the first 4 weeks or so at altitude. Therefore, consumption of iron-rich foods and perhaps even iron supplements is recommended before and during altitude exposure.

In Review

- Altitude causes hypobaric hypoxia, resulting in a decreased partial pressure of oxygen in the inspired air, in the alveoli, in the blood, and at the tissue level.
- With acute exposure to altitude, a series of adaptations occur in an attempt to maximize oxygen delivery to the tissues. Pulmonary ventilation increases and pulmonary diffusion is reasonably well maintained, but oxygen transport is slightly impaired because hemoglobin saturation at altitude is reduced.
- Ventilation increases almost immediately upon exposure to hypoxia because the decreased PO_2 stimulates peripheral chemoreceptors. The increased rate and depth of breathing help offset even larger decreases in PO_2 in the body.
- The diffusion gradient that allows oxygen exchange between the blood and active tissue is substantially reduced at moderate and high altitudes; thus, oxygen uptake by muscle is impaired.
- A decrease in plasma volume initially increases red blood cell concentration, allowing more oxygen to be transported per unit of blood, partially compensating for the impaired oxygen binding to hemoglobin.
- Upon initial ascent to altitude, cardiac output increases during submaximal work to compensate for the decreased oxygen content per liter of blood. It does so by increasing heart rate, because stroke volume falls with the fall in plasma volume.
- During maximal work at altitude, stroke volume and heart rate are both lower, which reduces cardiac output. This reduced cardiac output, combined with the decreased pressure gradient, severely impairs oxygen delivery to tissues.
- Ascent to altitude increases metabolic rate by increasing sympathetic nervous system activity. There is an increased reliance on carbohydrate for fuel, both at rest and during submaximal exercise.
- The exaggerated fluid loss and general loss of appetite at altitude increase the risk of dehydration.
- Decreased energy intake coupled with the increased energy expenditure of activity at altitude can lead to daily energy deficits and weight loss.

Exercise and Sport Performance at Altitude

The difficulty of demanding physical exertion at high altitude has been described by many climbers. In 1925, E.G. Norton[17] gave the following account of climbing without supplemental oxygen at 8,600 m (28,208 ft): "Our pace was wretched. My ambition was to do 20 consecutive paces uphill without a pause to rest and pant, elbow on bent knee, yet I never remember achieving it—13 was nearer the mark." In this section we briefly consider how exercise and sport performance is affected by altitude.

Maximal Oxygen Uptake and Endurance Activity

Maximal oxygen uptake decreases as altitude increases (see figure 13.5). $\dot{V}O_{2max}$ decreases little until the atmospheric PO_2 drops below 131 mmHg. This decline generally begins at an altitude of about 1,500 m (approximately 5,000 ft)—about the elevation of Denver, Colorado, and Albuquerque, New Mexico. At altitudes between 1,500 m and 5,000 m (5,000 and 16,400 ft), the decreased $\dot{V}O_{2max}$ is due primarily to the reduced arterial PO_2; at higher elevations, a decreased maximal cardiac output further limits $\dot{V}O_{2max}$. $\dot{V}O_{2max}$ decreases approximately 8% to 11% for every 1,000 m increase (or 3% for every 1,000 ft increase) in altitude above 1,500 m. The rate of decline may become even steeper at very high altitudes (figure 13.6). Endurance athletes with a high $\dot{V}O_{2max}$ at sea level have a competitive advantage at altitude if everything else is equal. As $\dot{V}O_{2max}$ declines on arrival at altitude, competition at any

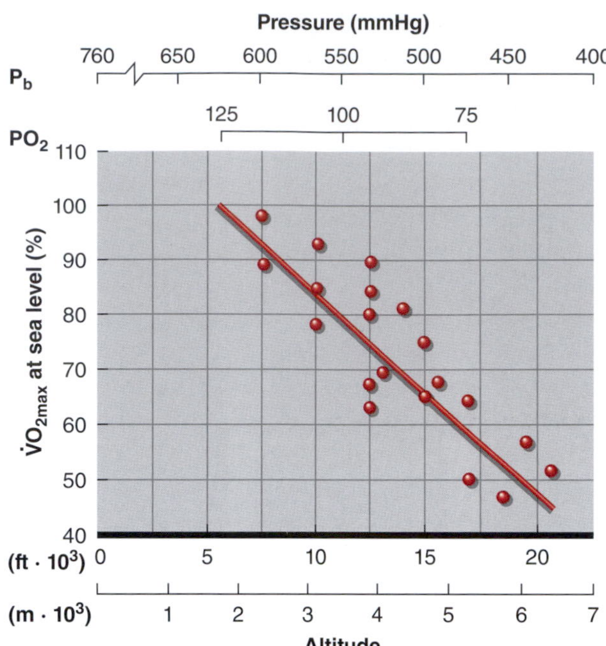

FIGURE 13.5 Changes in maximal oxygen uptake ($\dot{V}O_{2max}$) with decreases in barometric pressure (P_b) and partial pressure of oxygen (PO_2). Values for $\dot{V}O_{2max}$ are recorded as percentages of $\dot{V}O_{2max}$ attained at sea level (P_b = 760 mmHg). Note that the decline in $\dot{V}O_{2max}$ begins at about 1,500 m and is fairly linear. At the altitudes of Mexico City (2,240 m), Leadville, Colorado (3,180 m), and Nuñoa, Peru (4,000 m), one's $\dot{V}O_{2max}$ would be significantly below one's capacity at sea level or in Denver, Colorado (1,600 m).[6]

Data from Buskirk et al. (1967).

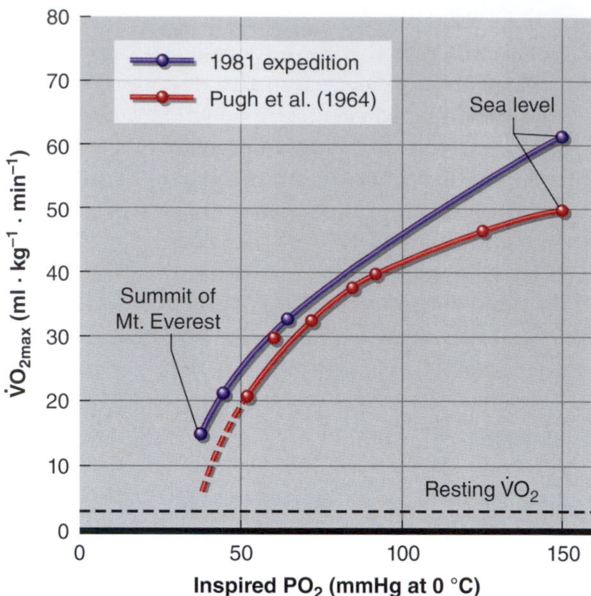

FIGURE 13.6 $\dot{V}O_{2max}$ relative to the partial pressure of oxygen (PO_2) of the inspired air for two expeditions to Mount Everest.

Adapted by permission from J.B. West et al., "Maximal Exercise at Extreme Altitudes on Mount Everest," *Journal of Applied Physiology* 55 (1983): 688-698.

given pace will be performed at a lower percentage of $\dot{V}O_{2max}$.

When men and women are matched for their initial aerobic fitness level, there are no sex differences in the rate of decline in $\dot{V}O_{2max}$.

As shown in figure 13.6, men climbing Mount Everest in a 1981 expedition experienced a change in $\dot{V}O_{2max}$ from about 62 ml · kg^{-1} · min^{-1} at sea level to only about 15 ml · kg^{-1} · min^{-1} near the mountain's peak. Because resting oxygen requirements are about 3.5 ml · kg^{-1} · min^{-1}, without supplemental oxygen these men had little capacity for physical effort at this elevation. A study by Pugh and coworkers,[18] also shown in figure 13.5, showed that men with $\dot{V}O_{2max}$ values of 50 ml · kg^{-1} · min^{-1} at sea level would be unable to exercise, or even to move, near the peak of Mount Everest because their $\dot{V}O_{2max}$ values at that altitude would decrease to 5 ml · kg^{-1} · min^{-1}. Thus, most normal people with sea-level $\dot{V}O_{2max}$ values below 50 ml · kg^{-1} · min^{-1} would not be able to survive without supplemental oxygen at the summit of Mount Everest because their $\dot{V}O_{2max}$ values at such an altitude would be too low to sustain their body tissues. Enough oxygen would be consumed to barely meet their resting requirements.

Obviously, activities of long duration that place considerable demands on oxygen transport and uptake by the tissues are those that are most severely affected by the hypoxic conditions at altitude. At the summit of Mount Everest, $\dot{V}O_{2max}$ is reduced to 10% to 25% of its sea-level value. This severely limits the body's exercise capacity. Because $\dot{V}O_{2max}$ is reduced by a certain percentage, individuals with larger aerobic capacities can perform a standard work task with less perceived effort and with less cardiovascular and respiratory stress at altitude than those with a lower $\dot{V}O_{2max}$. This may explain how noted mountaineers Reinhold Messner of Italy and Peter Habeler of Austria were able to reach the summit of Everest without supplemental oxygen in 1978—they obviously possessed high sea-level $\dot{V}O_{2max}$ values.

Anaerobic Sprinting, Jumping, and Throwing Activities

Whereas endurance events are impaired at altitude, anaerobic sprint activities that last less than a minute (such as 100 to 400 m track sprints) are generally not impaired by moderate altitude and can sometimes be improved. Such activities place minimal demands on the oxygen transport system and aerobic metabolism. Instead, most of the energy is provided through the adenosine triphosphate (ATP), phosphocreatine, and glycolytic systems.

In addition, the thinner air at altitude provides less aerodynamic resistance to athletes' movements. At the

1968 Olympic Games, for example, the thinner air of Mexico City clearly aided the performances of certain athletes. At Mexico City, world or Olympic records were set or tied in the men's 100 m, 200 m, 400 m, 800 m, long jump, and triple jump events and in the women's 100 m, 200 m, 400 m, 800 m, 4 × 100 relay, and long jump events. Because similar results occurred in swimming events up to 800 m, some exercise scientists have questioned the role of lower air density in improved sprint performance. Interestingly, while shot put performance was not affected at the altitude of Mexico City, discus performance declined because there is less lift at low barometric pressures.

In Review

- Prolonged endurance performance suffers the most at high altitude because oxidative energy production is limited.
- Maximal oxygen consumption decreases in proportion to the decrease in atmospheric pressure, beginning to decline at about 1,500 m (4,921 ft).
- Anaerobic sprint activities that last 2 min or less are generally not impaired at moderate altitude. In some instances, sprint performance may be improved because the thinner air at altitude provides less resistance to movement.

Acclimation: Chronic Exposure to Altitude

When people are exposed to altitude over days, weeks, and months, their bodies gradually adjust to the lower oxygen partial pressure in the air. But, however well they acclimate to the conditions at high altitude, they never fully compensate for the hypoxia. Even endurance-trained athletes who live at altitude for years never attain the level of performance or the $\dot{V}O_{2max}$ values that they might achieve at sea level. In this regard, acclimation to altitude is similar to heat acclimation discussed in chapter 12. Heat acclimation improves performance and attenuates physiological strain during exercise in the heat compared to that experienced during the first few days; however, performance is still poorer than in cooler environments.

The following sections cover some of the physiological adaptations that occur with prolonged altitude exposure. These include changes at the pulmonary, cardiovascular, and muscle tissue (cellular) level. In general, these adaptations take longer to fully develop (several weeks to several months) than those associated with heat acclimation (typically 1-2 weeks). Generally, about 3 weeks are needed for full acclimation to even moderate altitude. For each additional 600 m (1,970 ft) altitude increase, another week is needed on average. All of these beneficial effects are lost within a month of return to sea level. Many of these adjustments in resting and maximal exercise variables are shown in figure 13.7.

Pulmonary Adaptations

One of the most important adaptations to altitude is an increase in pulmonary ventilation, both at rest and during exercise. Within 3 or 4 days at 4,000 m (13,123 ft), the increased resting ventilation rate levels off at a value about 40% higher than at sea level. Submaximal exercise ventilatory rate also plateaus at about 50% higher but over a longer time frame. Increases in ventilation during exercise remain elevated at altitude and are more pronounced at higher exercise intensities.

Blood Adaptations

During the first 2 weeks at altitude, the number of circulating erythrocytes increases. The lack of oxygen at altitude stimulates the renal release of erythropoietin or EPO. Within the first 3 h after arrival at a high elevation, the blood's EPO concentration increases; it then continues to increase for 2 or 3 days. Although blood EPO concentrations return to baseline levels in about a month, the **polycythemia** (increased red blood cells) may be evident for 3 months or more. After a person lives at 4,000 m (13,123 ft) for about 6 months, total blood volume (composed mainly of the red cell volume and the plasma volume) increases by about 10%, not only as a result of the altitude-induced stimulation of erythrocyte production but also because of plasma volume expansion.[18]

The percentage of total blood volume composed of erythrocytes is referred to as the hematocrit. Residents in the central Andes of Peru (4,540 m, or 14,895 ft) have an average hematocrit of 60% to 65% (properly termed an acclimatization rather than an acclimation response; see chapter 12). This is considerably higher than the average hematocrit of sea-level residents, which is only 45% to 48%. However, during 6 weeks of exposure to the Peruvian altitude, sea-level residents have shown remarkable increases in their hematocrit levels, up to an average of 59%.

As the volume of erythrocytes increases, so does the blood's hemoglobin (Hb) content (and concentration, after an initial decline; see figure 13.7). As noted in figure 13.8, blood hemoglobin concentration tends to increase proportionately with increases in the elevation at which people reside. The data presented are for men. For women, however, the limited available data show a similar trend but with a lower concentration than for men at a given altitude. These adaptations improve oxygen-carrying capacity of a fixed volume of blood.

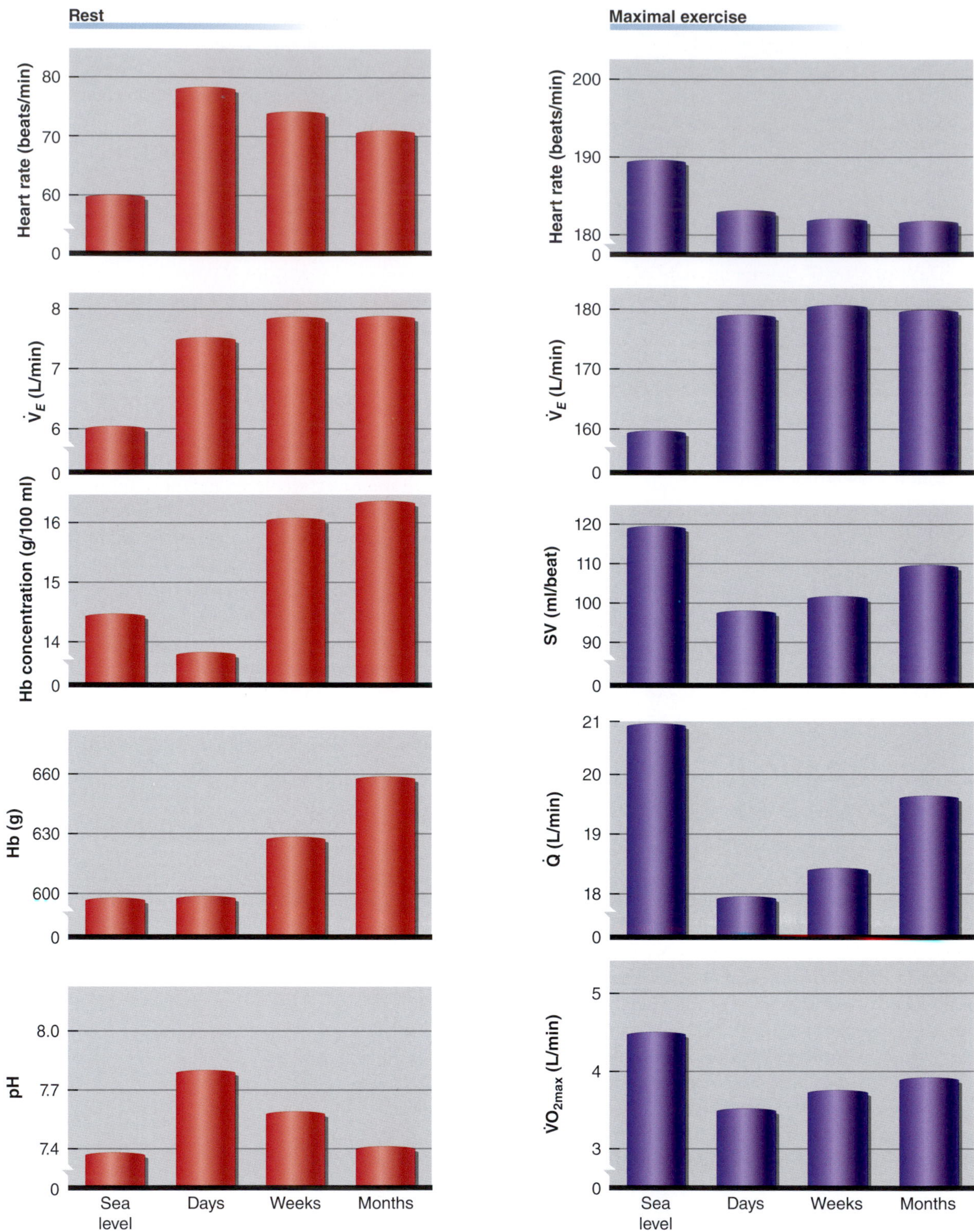

FIGURE 13.7 Physiological variables measured at sea level, after 2 or 3 days at altitude, and after weeks and months at altitude (3,000-3,500 m, or 9,843-11,483 ft). Both resting (left) and maximal exercise (right) variables are shown.

Drawn from data presented in Bartsch and Saltin (2008).

The reduction in plasma volume during acute altitude exposure reduces total blood volume, thus reducing submaximal and maximal cardiac output. But with acclimatization, as plasma volume increases over several weeks at altitude and as red blood cells continue to increase, maximal cardiac output increases. However, it does not return to sea-level values, as shown in figure 13.7. Thus, overall oxygen delivery capacity is increased

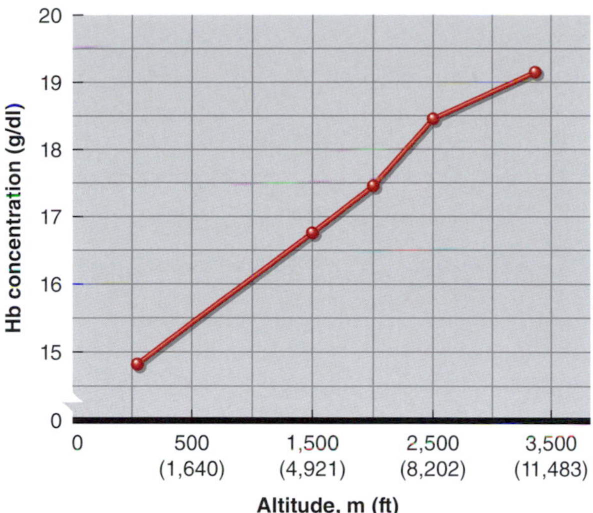

FIGURE 13.8 Hemoglobin (Hb) concentrations of men living at, and acclimatized to, various altitudes.

with acclimatization but not to the extent needed to achieve sea-level $\dot{V}O_{2max}$ values.

There is some debate about whether acclimation alters oxygen transport in the blood by changing the shape and position of the oxyhemoglobin dissociation curve (figure 13.4). The concentration of 2,3-diphosphoglycerate (2,3-DPG) increases in red blood cells, which shifts the curve to the right. This would favor the unloading of oxygen at the tissues (because more oxygen would be unloading from hemoglobin at any given low arterial PO_2), but this effect opposes the loading benefit of the respiratory alkalosis, a leftward shift. The net effect of both mechanisms is variable.

Muscle Adaptations

Although few attempts have been made to study muscle changes that occur during exposure to altitude, sufficient muscle biopsy data exist to indicate that muscles undergo significant structural and metabolic changes during ascent to altitude. In a study of climbers experiencing 4 to 6 weeks of chronic hypoxia on expeditions, muscle fiber cross-sectional area decreased, thus decreasing total muscle area. Capillary density in the muscles increased, which allowed more blood and oxygen to be delivered to the muscle fibers. Muscles' inability to meet exercise demands at high altitude might be related to a decrease in their mass and their ability to generate ATP.

The cause of the decreased muscle cross-sectional area within the first days and weeks at altitude is not fully understood. As previously discussed, prolonged exposure to high altitude frequently causes a loss of appetite and a noticeable weight loss. During a 1992 expedition to climb Mount McKinley, six men experienced an average weight loss of 6 kg, or 13 lb (D.L. Costill et al., unpublished data). Although part of this loss represents a general decrease in body weight and extracellular water, all of the men experienced a noticeable decrease in muscle mass. It seems logical to assume that much of this decrease in muscle mass is associated with loss of appetite and a wasting of muscle protein. Perhaps future studies on nutrition and body composition in mountain climbers will provide a more thorough explanation of the incapacitating influences of high altitude on muscle structure and function.

Several weeks at altitudes above 2,500 m (8,202 ft) reduce the metabolic potential of muscle, although this may not occur at lower elevations. Both mitochondrial function and glycolytic enzyme activities of the leg muscles (vastus lateralis and gastrocnemius) are significantly reduced after 4 weeks at altitude. This suggests that, in addition to receiving less oxygen, muscles lose some of their capacity to perform oxidative phosphorylation and generate ATP. Unfortunately, no muscle biopsy data have been obtained from long-term residents at high altitudes to determine whether those individuals experience any muscular adaptations as a consequence of lifelong residence at these elevations.

Cardiovascular Adaptations

Studies conducted on endurance-trained runners in the late 1960s indicated that the decrement in $\dot{V}O_{2max}$ when they first reached high altitude improved little for the duration of their exposure to hypoxia. Aerobic capacity remained unchanged for up to 2 months at altitude.[6] Although the runners who previously had been exposed to altitude were more tolerant of hypoxia, their $\dot{V}O_{2max}$ values and running performance were not significantly improved with acclimation. Because of the many adaptations that occur during acclimation to altitude, this lack of improvement in aerobic capacity and performance was unexpected. Perhaps these trained subjects had already attained maximal training adaptations and were unable to further adapt in response to altitude exposure. Or perhaps the reduced PO_2 of altitude made it more difficult for them to train at the same intensity and volume as at sea level. Both possibilities have some merit based on the available literature. While a ceiling effect on performance is more difficult to prove, athletes who train at altitude routinely have a difficult time maintaining their sea-level training intensities or volumes.

In Review

- Hypoxic conditions stimulate the renal release of EPO, which increases erythrocyte (red blood cell) production in bone marrow. More red blood cells means more hemoglobin. Although plasma volume decreases initially, which helps concentrate the hemoglobin, it eventually returns to normal. Normal plasma volume plus additional red blood cells increases total blood volume. All these changes increase the blood's oxygen-carrying capacity.

- Total muscle mass decreases after a few weeks at altitude, as does total body weight. Part of this decrease is from dehydration and appetite suppression. However, there is also protein breakdown in the muscles.
- Other muscle adaptations include decreased fiber area, increased capillary supply, and decreased oxidative enzyme activities.
- While work capacity improves with altitude acclimation, the decrease in $\dot{V}O_{2max}$ with initial exposure to altitude does not improve much over several weeks of exposure and typically never returns to sea-level values.

Altitude: Optimizing Training and Performance

We have considered the major changes that occur as the human body becomes acclimated to altitude and how these adaptations affect performance at altitude. But is there any advantage to training at altitude to improve performance at sea level? Are there advan-

RESEARCH PERSPECTIVE 13.1

Human Adaptation to High Altitude: Tibetan and Sherpa Physiology

Three main regions of the world host human populations at an altitude over 4,000 m. These are the Tibetan plateau and Himalayan valleys, the South American Andes, and the Ethiopian Highlands. Of these regions, the Tibetan plateau is both the largest and the highest, and the human population there has successfully lived and reproduced at this high altitude for hundreds of generations. With hypoxia as a constant evolutionary pressure, these populations have adapted to life at high altitude, which is evident in the physiology of the Tibetan people who live at these altitudes compared to lowlanders.[13]

Since the early expeditions to the summit of Mount Everest, tales of Sherpas' extraordinary tolerance and exercise capacity at high altitude have provided anecdotal evidence of their physiological adaptations to the hypoxic environment. More recently, research studies aimed at exploring these unique populations have characterized a number of physiological adaptations that increase oxygen absorption and delivery in the face of hypoxia. Matched for age, height, weight, and smoking history, Tibetans demonstrate larger chest circumferences, total lung capacity, vital capacity, residual volumes, and tidal volumes compared to lowlanders while Sherpas also demonstrate a greater expiratory flow rates and forced vital capacity. These morphological and mechanical respiratory adaptations allow for greater respiratory volumes and improved lung diffusion capacity. Aiding in this improved diffusion capacity at the lung is a blunted hypoxic pulmonary vasoconstriction in Tibetan people. Hypoxic pulmonary vasoconstriction aids in ventilation and perfusion matching in healthy lowlanders but leads to pulmonary hypertension in lowlanders exposed to hypoxic conditions. This response is blunted in Tibetans, resulting in greater pulmonary perfusion for oxygen absorption at the alveoli. Interestingly, Tibetans do not have higher hemoglobin concentrations compared to lowlanders; in fact, they have a different set point for hypoxia-induced erythropoiesis and a blunted rise in hemoglobin with altitude.

At altitude, exercising Tibetans and Sherpas are able to generate greater maximal heart rates and increased cardiac outputs and to maintain stroke volume when compared to lowlanders. This enhanced cardiac capacity likely facilitates superior work at altitude and does not induce right ventricular hypertrophy, as would be expected in lowlanders exposed to altitude for a prolonged period. This preservation of cardiac function despite chronic exposure to hypoxia may be due to metabolic adaptations that drive a preference for glucose as a substrate for the myocardium. Under resting conditions, the preferred substrate for cardiac muscle is normally fatty acids. However, Sherpas have an elevated myocardial glucose uptake compared to lowlanders under the same conditions. This switch toward glucose as a preferred fuel makes sense under hypoxic conditions, given that the ATP yield per oxygen molecule is much higher with glucose than with fatty acids.

Sherpas possess a greater number of capillaries and a lower muscle fiber cross-sectional area, which increases their ratio of capillaries to muscle fibers in skeletal muscle. This serves to increase convective and diffusive blood flow to deliver oxygen to the working muscle. Despite a 25% lower muscle mitochondrial volume density, Tibetans demonstrate a larger ratio of maximal oxygen consumption to mitochondrial volume. Similar to the myocardium, this seems to be achieved through an adaptive preference for glucose

tages to training at altitude versus training at sea level when one must compete at altitude? And what about the relatively new concept of "live high, train low" to optimize performance?

Does Altitude Training Improve Sea-Level Performance?

Athletes have hypothesized for decades that training under hypoxic conditions, for example in an altitude chamber or by simply breathing low-oxygen gas mixtures, can improve sea-level endurance performance. Since many of the beneficial changes associated with altitude acclimation are similar to those conferred by aerobic training, can combining the two be even more beneficial? Can altitude training improve sea-level performance?

A strong theoretical argument can be made for altitude training. First, altitude training evokes substantial tissue hypoxia (reduced oxygen supply). This is thought to be essential for initiating the conditioning response. Second, the altitude-induced increase in red blood cell mass and hemoglobin content improves oxygen delivery on return to sea level. Although evidence suggests that these latter changes are transient, lasting only several days, in theory this still should provide an advantage for the athlete.

metabolism in the skeletal myocytes. Tibetans also have elevated myoglobin and antioxidant proteins, which both contribute to increased oxygen inflow and consumption under hypoxic conditions. Finally, Sherpas and Tibetans have a slight dominance of type I muscle fibers. This increased prevalence may explain the lactate paradox, in which people acclimatized to higher altitudes show lower than expected rise in blood lactate for a given work rate. Tibetan's predominance of slow type I muscle fibers seems to favor improved coupling between ATP demand and ADP supply, limiting lactate accumulation.

Overall, the unique physiology of Tibetan and Sherpa populations provides insight into the successful adaptation to hypobaric and hypoxic conditions. More recently, advancements in technological and analytical research methods have opened the door for the identification of the genetic and molecular mechanisms that contribute to these phenotypes[14] (see figure). While these studies are in their infancy, it has been proposed that epigenetic changes (changes in heritable traits that are not explained by changes in the DNA sequence but are instead modulated by external environmental factors) may explain the ability of populations living at high altitude to rapidly acquire these adaptive features. Further studies in this area will continue to provide insight into the cellular and molecular mechanisms underlying human adaptive potential and will not only clarify the physiological adaptations of the Tibetans, but may also hold the key to further understanding in biomedical research and the study of human evolutionary theory.

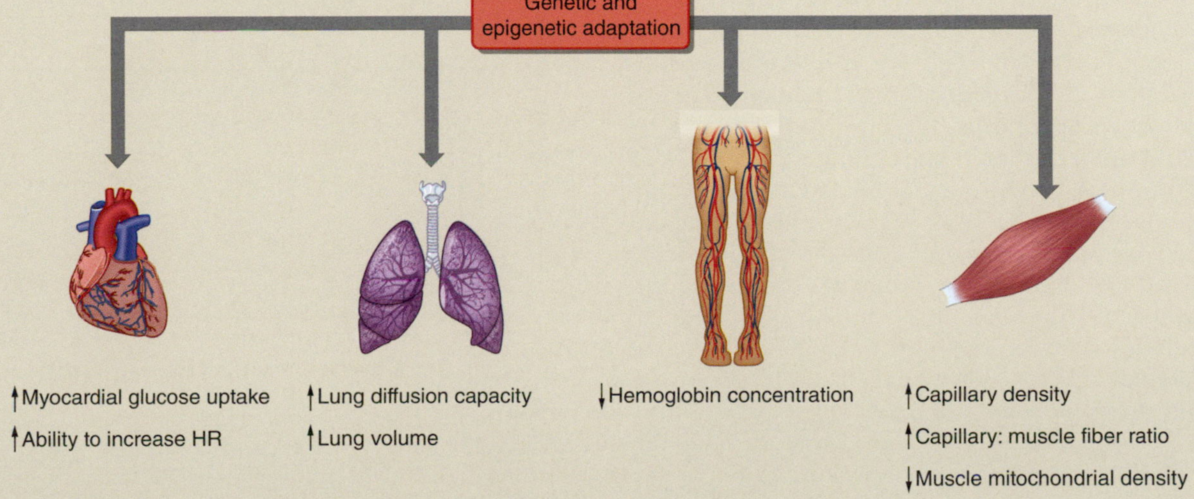

Genetic and epigenetic adaptations lead to key physiological differences between Sherpas and Tibetans and lowlanders.

Studying athletes at altitude poses additional problems because they are often unable to train at the same volume and intensity of effort as when at sea level. This was demonstrated in a group of elite female cyclists who performed self-selected maximum power outputs during high-intensity interval training. They completed trials under the following conditions: breathing atmospheric air (normoxia) and breathing a hypoxic gas mixture simulating 2,100 m (6,888 ft). The athletes' sustained (10 min) and short-term (15 s) power outputs at maximal intensity were reduced under hypoxic conditions.[5] Training at even higher elevations, where acclimatization effects would be even more beneficial, causes even greater disruptions in training.

In addition, living and training at moderate to high altitude often causes athletes to dehydrate and to lose blood volume and muscle mass. These and other side effects tend to diminish the athletes' fitness and their motivation and tolerance for intense training. As a result, studies are difficult to interpret, but the value of altitude training for optimal sea-level performance has not been validated.

Live High, Train Low

When living and training at altitude, athletes are faced with the problem that the intensity of training is reduced because aerobic capacity and cardiorespiratory function are reduced at altitude. Thus, although athletes gain certain physiological benefits from being at altitude, they lose training adaptations associated with higher intensities of training. One way to get around this problem is to have athletes *live* at moderate altitude but *train* at low altitude, where training intensity is not compromised.

Researchers at the Institute for Exercise and Environmental Medicine in Dallas, Texas, conducted a series of studies in the mid-1990s to investigate altitude training for enhancing endurance performance. In one study,[15] researchers divided 39 competitive runners into three equal groups: One (the high–low group) lived at moderate altitude (2,500 m, or 8,202 ft) and trained at low altitude (1,250 m, or 4,101 ft); one group (high–high) lived and trained at moderate altitude (2,500 m); and one group (low–low) lived and trained at low altitude (150 m, or 490 ft). Using a 5,000 m time trial as the primary performance outcome measure, the researchers found that the high–low group was the only group to significantly improve their running performance, even though both the high–low and high–high groups increased their $\dot{V}O_{2max}$ values by 5% in direct proportion to their increase in red cell mass. Thus, there appear to be performance benefits from living at moderate altitude but going to lower elevations to maximize training intensity.

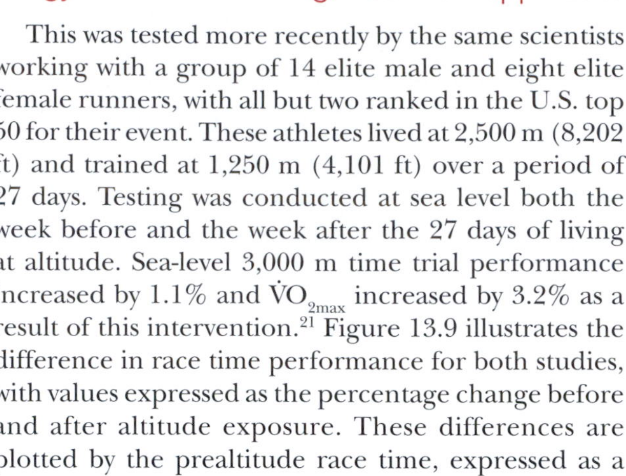

VIDEO 13.1 Presents Ben Levine on the physiology behind the live high–train low approach. 

This was tested more recently by the same scientists working with a group of 14 elite male and eight elite female runners, with all but two ranked in the U.S. top 50 for their event. These athletes lived at 2,500 m (8,202 ft) and trained at 1,250 m (4,101 ft) over a period of 27 days. Testing was conducted at sea level both the week before and the week after the 27 days of living at altitude. Sea-level 3,000 m time trial performance increased by 1.1% and $\dot{V}O_{2max}$ increased by 3.2% as a result of this intervention.[21] Figure 13.9 illustrates the difference in race time performance for both studies, with values expressed as the percentage change before and after altitude exposure. These differences are plotted by the prealtitude race time, expressed as a percentage of the existing U.S. record in the event at the time of the time trial.

Multiple studies have now shown a benefit of living and sleeping at high altitudes but training at low altitude for increasing sea-level $\dot{V}O_{2max}$ or improving sea-level aerobic performance in elite endurance athletes. These improvements have been linked to an increase in the oxygen-carrying capacity of the blood. Chronic living above 2,500 m (8,202 ft) induces such hematological acclimatization features in most athletes, although there is some variability. A recent study examined whether there is a minimum threshold "living" altitude for improvements in sea-level performance.[8] That is, do athletes who live at relatively higher altitudes

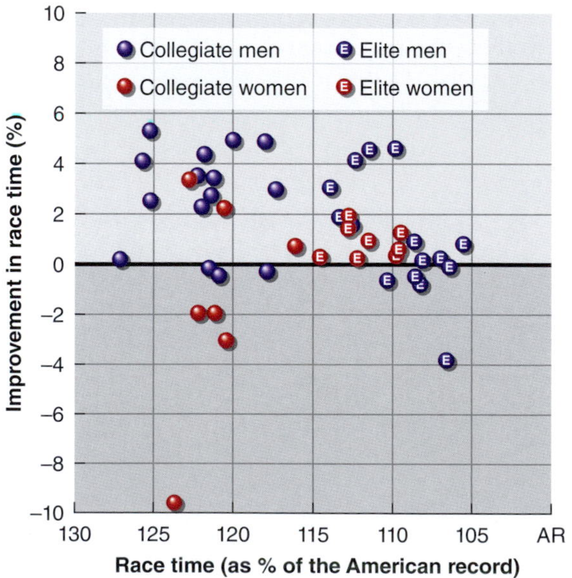

FIGURE 13.9 Improvement in race time (%) in elite male and female runners[21] and collegiate male and female runners[15] following 4 weeks of living at altitude but training at 1,250 m (4,100 ft). See the text for details.

demonstrate greater performance improvements in a live high–train low setting?

In the study, 48 trained distance runners were randomly assigned to live at one of four altitudes: 1,780 m (5,840 ft), 2,085 m (6,841 ft), 2,454 m (8,051 ft), or 2,800 m (9,186 ft). The athletes all trained together at altitudes between 1,250 m (4,101 ft) and 3,000 m (9,842 ft). Upon return from the live high–train low altitude camp, red blood cell mass and EPO concentrations increased similarly in all four groups, but EPO returned to baseline sooner in the 1,780-m group. Sea-level $\dot{V}O_{2max}$ was improved for all groups as expected (figure 13.10). However, performance on a 3 km (1.9 mi) run at sea level was significantly improved only in the groups that had lived at the middle two altitudes. It appears that increased red blood cell mass is necessary, but not sufficient, for improved performance. Also, there appears to be an optimal living altitude to optimize performance gains in such a setting.

It appears that team sports performance can likewise be enhanced using the live high–train low strategy. In a study of 32 elite male field hockey players,[3] 14 days of living high (2,800-3,000 m) and training low as a team increased hemoglobin mass by 3% and shuttle-run performance by 21%. Moreover, these effects lasted for at least 3 weeks posttraining.

Optimizing Performance at Altitude

What can athletes who normally train at sea level but must compete at altitude do to prepare most effectively for competition? Although not all combinations have been attempted and research thus far is not conclusive, it appears that athletes have two viable options. One option is to compete as soon as possible after arriving at altitude, that is, within 24 h of arrival. This strategy does not provide the beneficial effects of acclimation, but the altitude exposure is brief enough that the classic symptoms of altitude sickness are not yet totally manifest. After the first 24 h, the athlete's physical condition often worsens because of the untoward effects of acute altitude exposure, such as dehydration, headache, and sleep disturbances. However, one laboratory study that asked whether there was a performance advantage to competing only 2 h after arrival at altitude showed minimal differences in cycling performance and no physiological changes that would provide an advantage[11] compared to a 14 h exposure to the simulated altitude (2,500 m).

Another option is to train at higher altitudes for a minimum of 2 weeks before competing. But not even 2 weeks is sufficient for *total* acclimation, which would require a minimum of 3 weeks, and usually even longer. As previously mentioned, several weeks of intense aerobic training at sea level to increase the athletes' $\dot{V}O_{2max}$ will allow them to compete at altitude at a lower relative intensity ($\%\dot{V}O_{2max}$) than if they had not trained aerobically.

Extended training for optimal performance at altitude requires an elevation between 1,500 m (4,921 ft), which is considered the lowest level at which an effect will be noticed, and 3,000 m (9,842 ft), which is the highest level for efficient conditioning. Work capacity is reduced during the initial days at altitude. For this reason, when first reaching higher altitudes, athletes should reduce workout intensity to between 60% and 70% of sea-level intensity, gradually working up to full intensity within 14 days.

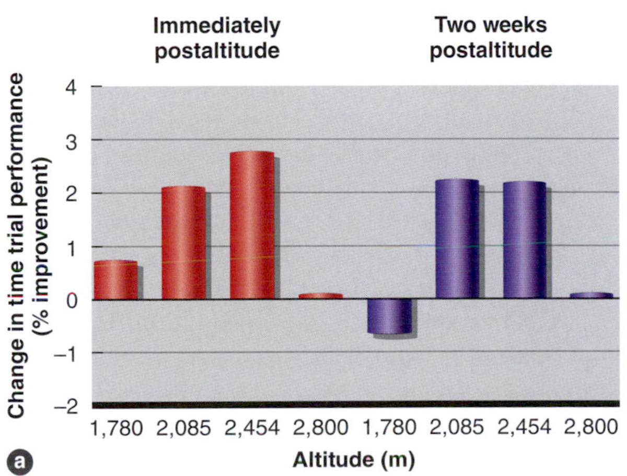

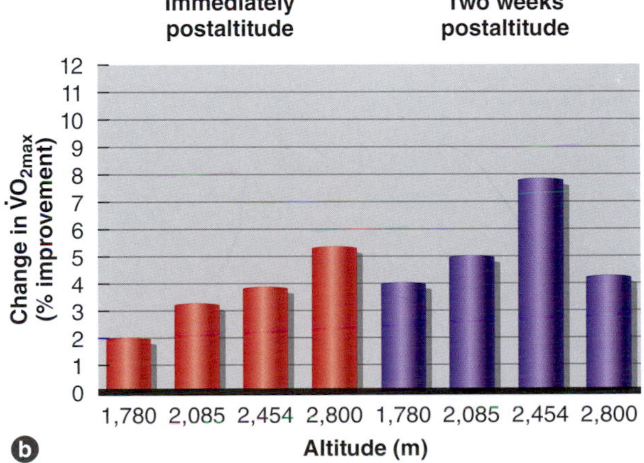

FIGURE 13.10 *(a)* Improvement in 3K time trial (%) and *(b)* $\dot{V}O_{2max}$ immediately upon return to sea level and 2 weeks later in elite collegiate distance runners[8] after 4 weeks of living at one of the four altitudes shown but training together as a group. See the text for details.

Adapted by permission from R.F. Chapman et al., "Defining the "Dose" of Altitude Training: How High to Live for Optimal Sea Level Performance Enhancement," *Journal of Applied Physiology* 116, no. 6 (2014): 595-603.

RESEARCH PERSPECTIVE 13.2

Altitude Training for Swimmers

The classical "live high, train high" approach to altitude training has been used by athletes since the late 1960s to improve performance at sea level. In a series of studies published in the 1990s, the "live high, train low" strategy was shown to improve running performance in collegiate athletes by increasing red cell mass (the altitude adaptation) and maintaining high-intensity training velocities (possible only at low altitude). Since that time, altitude training has been adopted by endurance athletes of all disciplines, including elite swimmers. However, no studies have examined the effectiveness of altitude training for improved swim performance. Because of this disconnect between a lack of research evidence and common practical use of altitude training in swimmers, an international group of investigators recently conducted a multidisciplinary research project to study the effects of altitude training on swimmers' performance and health.

In this study (The Altitude Project), scientists examined the effects of four different altitude training paradigms on performance, $\dot{V}O_{2max}$, oxygen kinetics, and total hemoglobin mass.[20] To do this, 61 swimmers were divided into four groups: (1) living and training at moderate altitude (2,320 m) for 4 weeks (Hi-Hi4), (2) living and training at moderate altitude for 3 weeks (Hi-Hi3), (3) living at altitude (2,320 m) and training at both moderate and low (690 m) altitude for 4 weeks (Hi-HiLo), and (4) living and training near sea level (190 or 655 m) for 4 weeks (Lo-Lo). Performance and physiological variables were tested before, once a week during, immediately after, and throughout a 4-week sea-level recovery from these training regimens. The researchers found that all four training regimens improved time trial performance by ~3.5% following 1 week of recovery, while Hi-HiLo achieved the greatest improvements in time trial performance compared to the other training regimens. Total hemoglobin mass increased in both Hi-Hi groups but did not increase in the Hi-HiLo or Lo-Lo groups. $\dot{V}O_{2max}$ and oxygen kinetics were unchanged by any of the training regimens, a result that is likely explained by the elite status of these athletes and the limitations to improving $\dot{V}O_{2max}$ beyond an already very high value.

Overall, the study findings led to the conclusion that the performance enhancement observed in the Hi-HiLo group was not linked to changes in $\dot{V}O_{2max}$, hemoglobin mass, or oxygen kinetics, but that "living high, training high and low" for 4 weeks has the potential to improve swimming performance above altitude or sea-level training alone. The physiological alterations responsible for altitude acclimatization and subsequent training effects are complex, and a deeper understanding will require future investigations targeting the specific mechanisms and mediators.

Artificial Altitude Training

The largest and most important adaptations to altitude are physiological changes caused by the hypoxia experienced there, so we might anticipate that people could achieve similar adaptations simply by breathing gases with a low PO_2. But no evidence supports the idea that brief periods (1-2 h per day) of breathing hypoxic gases or hypobaric mixtures induce even a partial adaptation similar to that observed at altitude. Nor does such artificial hypoxic training confer any advantage for endurance performance at sea level.[19] On the other hand, alternating periods (lasting between 5 and 14 days) of training at 2,300 m (7,546 ft) and at sea level adequately stimulated altitude acclimation in a group of elite middle-distance runners.[9] Staying at sea level for up to 11 days did not interfere with the usual adjustments to altitude as long as training was maintained.

The favorable results of the studies on living high and training low have stimulated considerable interest in how this concept might be applied without having to send athletes to altitude to live. One approach has been to develop a hypoxic apartment where athletes sleep and live. The gas mixture inside the apartment is adjusted so that nitrogen represents a higher percentage of the inspired air, reducing the percentage of oxygen in the inspired air as well as its partial pressure. Pioneered by Finnish sport scientists, these apartments can simulate altitudes between 2,000 m (6,562 ft) and 3,000 m (9,842 ft) as the nitrogen and oxygen percentages in the inspired air are adjusted to reduce the partial pressure of oxygen to those levels associated with 2,000 to 3,000 m altitude. Hypoxic sleeping devices or tents have also been proposed.

Unfortunately, at this time, few carefully controlled scientific studies exist to confirm whether these apartments or sleeping devices actually improve performance and physiological function. One recent meta-analysis (a statistical approach that combines data from multiple studies to draw conclusions) reported that natural live high–train low approaches provide the best results for enhancing performance in elite athletes, while some nonelite exercisers seem to benefit from the artificial approaches.[2] However, the authors cautioned that improvements seen in these subelite athletes could have been due to a placebo effect. Ethical issues have also been raised about the use of such devices.

In Review

- Living at high altitudes and training at low altitudes currently appears to be the best practice for improving subsequent sea-level performance.
- Athletes who must perform at altitude should do so as early after arrival as possible, certainly within 24 h of arrival, before the detrimental side effects that occur at altitude become too great.
- Alternatively, athletes who must perform at altitude could train at an altitude between 1,500 m (4,921 ft) and 3,000 m (9,842 ft) for a minimum of 2 weeks (more is better) before performing.
- There is no evidence that brief periods (1-2 h per day) of breathing hypoxic gases or hypobaric gas mixtures induce even a partial adaptation similar to that observed at altitude.

Health Risks of Acute Exposure to Altitude

A large proportion of people who ascend to moderate and high altitudes experience symptoms of **acute altitude (mountain) sickness**. This disorder is characterized by symptoms such as headache, nausea, vomiting, dyspnea (difficult breathing), and insomnia. These symptoms can begin anywhere from 6 to 48 h after arrival at high altitude and are most severe on days 2 and 3. Although not life threatening, acute altitude sickness can be incapacitating for several days or longer. In some cases, the condition can worsen. The victim can develop the more lethal altitude-related illnesses of high-altitude pulmonary edema or high-altitude cerebral edema.

Acute Altitude (Mountain) Sickness

The incidence of acute altitude sickness varies with the altitude, the rate of ascent, and the individual's experience and susceptibility. Several studies have been conducted to determine the incidence of acute altitude sickness in groups of novice hikers (tourists) and more experienced climbers. Results vary widely, ranging from a frequency of less than 1% to almost 60% at altitudes of 3,000 to 5,500 m (9,842-18,045 ft) (figure 13.11). Forster,[10] however, reported that 80% of those who ascended to the top of Mauna Kea (4,205 m, or 13,796 ft) on the island of Hawaii experienced some symptoms of acute altitude sickness. Another study showed that at elevations of 2,500 to 3,500 m (8,202-11,483 ft), altitudes commonly experienced by recreational skiers and hikers, the incidence of acute altitude sickness was about 7% for men and 22% for women, but the reason for this sex difference is unclear.[22]

Although the precise underlying cause of acute altitude sickness is not fully understood, it appears that people who experience the greatest distress also have a low ventilatory response to hypoxia. This inadequate ventilation allows PO_2 to decrease further and carbon dioxide to accumulate in the tissues, and these two factors may induce most of the symptoms associated with altitude sickness.

Headache is the most common symptom associated with ascent to high altitude. Headache is rarely experienced below 2,500 m (~8,000 ft), but ascent to 3,600 m (~12,000 ft) results in headache in the majority of people. The headache at altitude, which many sufferers describe as continuous and throbbing, is typically worse in the morning and after exercise. Alcohol consumption

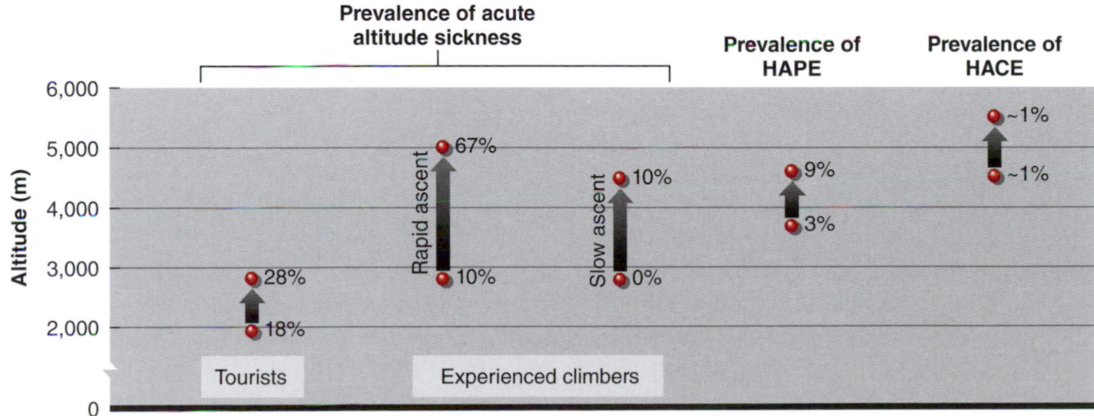

FIGURE 13.11 Change in the reported prevalence of acute altitude sickness, high-altitude pulmonary edema (HAPE), and high-altitude cerebral edema (HACE) as a function of altitude, experience, and, in the case of acute altitude sickness, rate of ascent.

Adapted from compiled data presented in Bartsch and Saltin (2008).

RESEARCH PERSPECTIVE 13.3

Should Athletes Live *Extra* High and Train Low?

Endurance exercise performance is impaired with acute altitude exposure, and athletes who normally live and train at sea level commonly plan to live and train for some period of time at altitude before competing in events scheduled to take place at altitude. This period of acclimatization reduces altitude-mediated declines in performance, but from a practical standpoint, the athletes must consider both how long to arrive at altitude before the event and at what altitude they plan to reside relative to the altitude of the event. Research data suggest that athletes competing at altitude should arrive 14 days prior to competition. Performance improves significantly at altitude over the first 14 days with minimal improvements after that. However, the best altitude to reside at during that time remains unclear. Studies exploring the effects of different altitudes on performance at altitude have utilized the same acclimatization and performance altitudes or lower acclimatization altitudes for preacclimatization before high-altitude exposure. To date, only one study has explored whether living at a higher altitude than the competition altitude is more advantageous than the previously studied paradigms.

A recent study examined the decline in competitive distance running performance between sea level and low altitude (1,780 m) as a function of living altitudes and days of residence at altitude in 48 collegiate track and cross country athletes.[7] The research team hypothesized that athletes living at higher altitudes than the altitude of the competition would have smaller declines in performance. To test this hypothesis, all subjects completed 4 weeks of sea-level training and were then randomized into one of four groups: live at 1,780 m, live at 2,085 m, live at 2,454 m, or live at 2,800 m. Each group was transported by car to their designated training altitude camp and lived and trained there for 28 days. Once a week, every training group performed a high-intensity training bout at low altitude (1,250 m). Subjects completed 3,000 m performance trials on the track at sea level, at baseline, 6 days before departure for altitude camp, and at 1,780 m on days 5, 12, 19, and 26 of the altitude camp. $\dot{V}O_{2max}$ plasma volume, total blood volume, and red blood cell volume were measured once at sea level and once after the 28 days of altitude training.

All of the groups living at altitude had increases in red blood cell volume, but there were no differences between groups. $\dot{V}O_{2max}$ increased in the two highest altitude groups (2,085 and 2,454 m) but did not improve in the 1,780 or 2,800 m groups. In terms of performance, athletes living at 2,454 and 2,800 m had a larger decline in performance compared to the 1,780 m group after 5 days at altitude. The 1,780 m group did not show changes in performance across the 26 days at altitude, while all of the groups living higher than the competition altitude showed improvements from day 5 to day 19 but no further improvement at day 26. The study team concluded that an endurance athlete competing acutely at low to moderate altitude should live at the altitude of the competition and not higher. For athletes living ~300 to 1,000 m higher than the competition altitude, acute altitude performance may be much worse and may require much longer (up to 19 days) acclimatization to minimize performance decrements.

worsens the symptoms. The precise mechanism is unsubstantiated, but hypoxia causes dilation of the cerebral blood vessels, so stretching of pain receptors in these structures is a likely cause.

Another consequence of acute altitude sickness is an inability to sleep even if the individual is markedly fatigued. Studies have shown that the inability to achieve satisfying sleep at altitude is associated with an interruption in the sleep stages. Additionally, some people suffer from a pattern of interrupted breathing, called **Cheyne-Stokes breathing**, which prevents them from falling asleep and remaining asleep. Cheyne-Stokes breathing is characterized by alternating periods of rapid breathing and slow, shallow breathing, usually including intermittent periods in which breathing completely stops. The incidence of this irregular breathing pattern increases with altitude, occurring in 24% of people at 2,440 m (8,005 ft), 40% of people at 4,270 m (14,009 ft), and almost everyone at altitudes above 6,300 m (20,669 ft).[24]

The first few days at high altitude are associated with reduced exercise performance and increased risk of acute mountain sickness (AMS). How can exercisers and athletes prevent acute altitude sickness? Even athletes who are highly endurance trained before altitude exposure seem to have little protection against the effects of hypoxia, and it is difficult to determine which athletes may be susceptible to these symptoms unless this is suggested by a prior history of acute altitude sickness.

People can usually prevent acute altitude sickness by ascending to altitude gradually, spending periods of a few days at lower elevations. A gradual ascent of no more than 300 m (984 ft) per day at elevations above 3,000 m (9,842 ft) has been suggested to minimize the risks of altitude sickness. Scientists at the U.S. Army Research Institute for Environmental Medicine have conducted studies examining the efficacy of various staging and hypoxia treatments as preacclimatization strategies.[12,16] To minimize symptoms associated with AMS and preserve performance capabilities at altitudes above 4,000 m (13,123 ft), evidence suggests the following:

- Ascending to moderate altitudes of >1,500 m for 1 to 2 days induces ventilatory acclimatization.
- Six days at 2,200 m substantially reduces AMS and improves exercise performance after rapid ascent to >4,000 m.
- Five or more days at 3,000 m within the last 2 months significantly reduces AMS.
- The longer the residence at moderate altitudes, the better the ventilatory adaptations and prevention of AMS at high altitude.
- Exercise training during these preexposures may improve physical performance even more.
- Strategies involving hypobaric hypoxia (actual altitude exposures, chambers, and tents) are much more effective that normobaric hypoxia (breathing low-oxygen mixtures at sea level).

Of the drugs that have been used to reduce symptoms in those who develop acute altitude sickness, acetazolamide started the day before ascent is the only established preventive measure. Acetazolamide is sometimes combined with a steroid such as dexamethasone. Both drugs must be used with medical supervision. Of course, the definitive treatment for severe acute mountain sickness is a retreat to lower altitude, but high-flow oxygen and the use of hyperbaric rescue bags are also effective in severe cases.

High-Altitude Pulmonary Edema

Unlike AMS, **high-altitude pulmonary edema (HAPE)**, which is the accumulation of fluids in the lungs, is life threatening. The cause of HAPE is likely related to the pulmonary vasoconstriction resulting from hypoxia, causing blood clots to form in the lungs. Remaining tissue becomes overperfused, and fluid and protein leak out of the capillaries. This seems to occur most frequently in unacclimatized people who rapidly ascend to altitudes above 2,500 m (8,202 ft). The disorder occurs in otherwise healthy people and has been reported more often in children and young adults. The fluid accumulation interferes with air movement into and out of the lungs, leading to shortness of breath, a persistent cough, chest tightness, and excessive fatigue. Disruption of normal breathing impairs blood oxygenation and, if it is severe enough, cyanosis (a bluish tint) of the lips and fingernails, mental confusion, and loss of consciousness may occur. High-altitude pulmonary edema is treated via administration of supplemental oxygen and movement of the victim to a lower altitude.

High-Altitude Cerebral Edema

Rare cases of **high-altitude cerebral edema (HACE)**, which is fluid accumulation in the cranial cavity, have been reported. The condition is often a subsequent complication of HAPE. This neurological condition is characterized by mental confusion, lethargy, and ataxia (difficulty walking), progressing to unconsciousness and death. Most cases have been reported at altitudes greater than 4,300 m (14,108 ft). Similar to that for HAPE, the cause of HACE involves hypoxia-induced leakage of fluids from cerebral capillaries, causing edema and a resultant pressure buildup in the confined intracranial space. Treatment involves administration of supplemental oxygen, a hyperbaric bag, and prompt descent to a lower altitude. If descent is delayed, permanent impairment may ensue.

In Review

- Acute altitude sickness, often called acute mountain sickness, typically causes symptoms such as headaches, nausea, dyspnea, and insomnia. Symptoms usually appear 6 to 48 h after arrival at altitude.
- The exact cause of acute altitude sickness is not known, but many researchers suspect that the symptoms may result from the combination of hypoxia and carbon dioxide accumulation in the tissues.
- Acute altitude sickness usually can be avoided by a slow, gradual ascent to altitude, climbing no more than 300 m (984 ft) per day at elevations above 3,000 m (9,842 ft).
- High-altitude pulmonary and cerebral edema (HAPE and HACE)—which involve accumulation of fluid in the lungs and cranial cavity, respectively—are life-threatening conditions. Both are treated by oxygen administration, hyperbaric bags, and descent.

IN CLOSING

Activities are seldom conducted under ideal environmental conditions. Heat, cold, humidity, and altitude—alone or in combination—present unique problems that are superimposed on the physiological demands of exercise. This chapter and the preceding chapter summarized the characteristics of these common environmental stresses and how we can cope with them.

Much of our discussion thus far has dealt with how physiological variables and environmental stress can hinder our performance. In the next part of the book, we examine various ways to optimize performance. We begin by looking at the importance of the amount of training, considering what happens when we train either too much or too little.

KEY TERMS

acute altitude (mountain) sickness

barometric pressure (P_b)

Cheyne-Stokes breathing

high-altitude cerebral edema (HACE)

high-altitude pulmonary edema (HAPE)

hypobaric

hypoxemia

hypoxia

partial pressure of oxygen (PO_2)

polycythemia

respiratory alkalosis

STUDY QUESTIONS

1. Describe the conditions at altitude that could limit the ability to perform physical activity.
2. What types of exercise are detrimentally influenced by exposure to high altitude and why?
3. When someone ascends to an altitude of over 1,500 m, describe the physiological adjustments that occur within the first 24 h.
4. Differentiate the physiological adjustments that accompany acclimation to altitude over a period of days, weeks, and months.
5. Would an endurance athlete who trained at altitude be able to perform better during subsequent sea-level performance? Why or why not?
6. Describe the theoretical advantage of living high and training low.
7. What are the best strategies for preparing athletes for high-altitude competition?
8. What are the health risks associated with acute exposure to high altitude and how can they be minimized?
9. How can the likelihood of experiencing acute altitude sickness be minimized? Once it occurs, how is it treated?

STUDY GUIDE ACTIVITIES

In addition to the activities listed in the chapter opening outline, two other activities are available in the web study guide, located at

www.HumanKinetics.com/PhysiologyOfSportAndExercise

The **KEY TERMS** activity reviews important terms, and the end-of-chapter **QUIZ** tests your understanding of the material covered in the chapter.

PART V

Optimizing Performance in Sport

Previous chapters explained how the body responds to an acute bout of exercise, how it adapts to chronic training, and how it adjusts to environmental extremes. We can now apply that knowledge to optimizing performance specific to sport. In part V, we focus on how athletes can best prepare for competition from a physiological standpoint. In chapter 14, Training for Sport, we discuss how to optimize athletes' training regimen and explore how too little or too much training can impair performance. In chapter 15, Body Composition and Nutrition for Sport, we address the issues of assessing body composition, how body composition relates to sport performance, and sports that have weight standards. We then evaluate athletes' dietary needs and consider how changing the diet or using nutritional supplementation may improve performance. In chapter 16, Ergogenic Aids in Sport, we discuss many of the pharmacological, hormonal, and physiological agents that have been proposed to improve performance. We examine the potential benefits, proven effects, and health risks associated with their use.

Training for Sport

In this chapter and in the web study guide

ACTIVITY 14.1 *Overtraining Syndrome* considers the effects of overtraining on two swimmers and explores symptoms of sympathetic and parasympathetic overtraining.

AUDIO FOR FIGURE 14.6 describes a model for the development of overtraining and the roles of the brain and immune system.

VIDEO 14.1 presents Scott Trappe on tapering for peak sport performance.

ACTIVITY 14.2 *Tapering for Peak Performance* investigates the effects of tapering on the performance of two distance runners.

ACTIVITY 14.3 *Detraining* explores how physiological responses to detraining change.

Throughout his college career, Eric had trained at swimming 4 h each day, covering as much as 13.7 km (8.5 mi) per day. Despite this effort, his performance time for the 200 yd (183 m) butterfly event had not improved since his freshman year. With a best performance of 2 min 15 s for the event, he was seldom given a chance to compete because several teammates could perform the event in less than 2 min 5 s. During Eric's senior year, his coach made a major change in the team's training plan. The swimmers trained only 2 h per day and swam an average of 4.5 to 4.8 km (2.8-3.0 mi) per day. Further, they swam each interval at a faster pace and had a longer rest period between intervals. Suddenly Eric's performance began to improve. After 3 months, his time dropped to 2 min 10 s, still not good enough to make him a major contender. But as a reward for Eric's improvement, the coach chose him to swim the 200 yd butterfly event at the conference championship meet, which was preceded by 3 weeks of reduced training (tapering) of only 1.6 km (1 mi) per day. Subsequently, with less volume of training than in previous years and well rested after the taper, Eric was able to make the finals of the event at the championship meet. His preliminary time was 2 min 1 s. In the finals, he improved even further, posting a third-place finish with a time of 1 min 57.7 s, an impressive performance for a swimmer who performed better with lower-volume but higher-quality training.

Repeated days and weeks of training can be considered a positive stress because the adaptations caused by training improve the capacity for energy production, oxygen delivery, muscle contraction, and other mechanisms that enhance exercise performance. The major changes associated with training occur within the first 6 to 10 weeks. The magnitude of these adaptations depends on the volume and intensity of exercise performed during training, which has led many coaches and athletes to believe erroneously that the athlete who trains the longest and hardest will be the best performer. However, the quantity and the quality of training are two separate things. Too often, training sessions are judged by the total volume (e.g., the distance run, cycled, or swum) performed in each training session, leading coaches to design training programs that are not optimal for improving performance and often impose unrealistic demands on the athlete.

The rate at which an individual adapts to training is genetically limited. Too much training can reduce the athlete's optimal potential for improvement and in some cases can cause a breakdown in the adaptation process, eventually reducing performance. When training is taken to extremes, serious illness or injury can occur.

Although the volume of work performed during training is an important stimulus for physiological improvements in performance, a proper balance should be established between volume and intensity. Training can be overdone, leading to fatigue, illness, overuse injury, overtraining syndrome, and performance decrements. In contrast, proper rest and achieving the proper balance between training volume and intensity can, and will, enhance performance. Much effort has been directed toward determining the appropriate volume and intensity required to achieve optimal adaptation. Exercise physiologists have tested numerous training regimens to determine both the minimal and maximal stimuli needed for cardiovascular and muscular improvements. The next section examines those factors that can affect the response to a given training program, developing a model for optimizing the training stimulus.

Optimizing Training

All well-designed training programs incorporate the principle of progressive overload. As discussed in chapter 9, this principle states that to continue to provide the benefits of training, the training stimulus must be progressively increased as the body adapts to the current stimulus. The only way to continue to improve with training is to progressively increase the training stimulus. However, when this concept is carried too far, training may become excessive, pushing the body beyond its ability to adapt, producing no additional improvement in conditioning or performance and leading to performance decrements. Conversely, if the volume or the intensity of training is too low, the resulting physiological changes will be hindered and optimal performance will not be achieved. Thus, the coach and athlete face the challenge of determining the optimal training stimulus for each particular athlete, recognizing that what works for one athlete might not work for another.

Figure 14.1 provides a model demonstrating the continuum of training stages. In this model, **undertraining** represents the type of training an athlete would undertake between competitive seasons or during active rest. Generally, physiological adaptations will be minor, and there will be no improvement in performance during this stage. **Acute overload** represents what might be considered an average training load, whereby the athlete is stressing the body to the extent necessary to improve both physiological function and performance. **Overreaching** is a relatively new term that refers to a brief period of heavy overload without adequate recovery, thus exceeding the athlete's adaptive capacity. There will be a brief performance decrement, from several days to several weeks, but eventually performance will improve. Finally, **overtraining** refers to that point at which an athlete experiences physiological maladap-

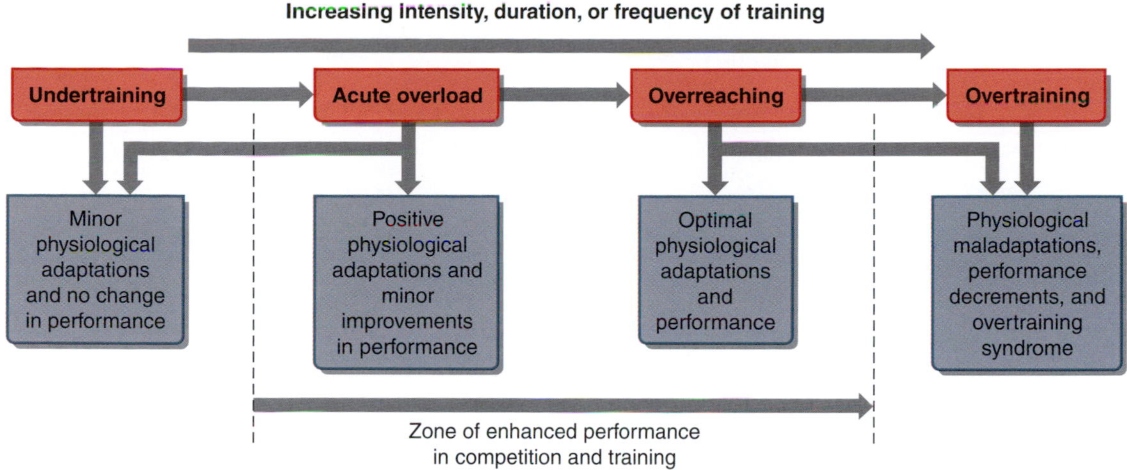

FIGURE 14.1 Model of the continuum of training stages.

Adapted by permission from L.E. Armstrong and J.L. VanHeest, "The Unknown Mechanism of the Overtraining Syndrome," *Sports Medicine* 32, no. 1 (2002): 185-209.

tations and chronic performance decrements. This generally leads to the **overtraining syndrome**.[1]

Overreaching

Overreaching is a systematic attempt to intentionally overstress the body for a short period of training. Done correctly, this allows the body to adapt to the increased training stimulus, that is, beyond the level of adaptation attained during normal overload. As with overtraining, there is a brief decrement in performance lasting several days to several weeks, followed by increased physiological function and increased performance. Obviously, this is the critical phase of training—on the edge, leading either to improved physiological function and performance or to overtraining if one goes too far. With overreaching, the period of full recovery from training takes several days to several weeks; however, with overtraining, recovery can take many months or, in some cases, years. The key to overreaching is to push the athlete hard enough to accomplish the desired positive physiological and performance improvements but to avoid going into the stage of overtraining. This is not an easy task to accomplish!

Excessive Training

Although not shown in the model in figure 14.1, **excessive training** refers to training that is well above what is needed for peak performance but does not strictly meet the criteria for either overreaching or overtraining. With excessive training, either the volume or the intensity of training, or both, are increased beyond the optimal level. A "more is better" philosophy often drives the training schedule. For many years, athletes were undertrained. As coaches and athletes became bolder and started to push the envelope by increasing both training volume and intensity, they found that athletes responded well, and world records began to tumble. However, one can take this philosophy only so far. At a certain point, performance begins to either plateau or decline.

Most research on excessive training has been conducted on swimmers, but the principles also apply to most other forms of training. Swim training 3 to 4 h per day, 5 or 6 days each week, provides no greater benefits than training only 1 to 1.5 h per day.[6] In fact, such excessive training has been shown to significantly decrease muscular strength and sprint swimming performance.

Studies conducted to date reveal no scientific evidence that multiple daily training sessions enhance fitness and performance more than a single daily session. This is illustrated by the data in figure 14.2, which show the responses of two groups of swimmers who trained once per day (group 1) or twice per day (group 2) for weeks 5 through 10 of a 25-week training program.[6] All swimmers began the program following the same training regimen: one session per day. But from the beginning of the 5th week through the end of the 10th week, group 2 trained twice per day, doubling their volume of training. After 6 weeks on the different regimens, both groups returned to the once-daily program. All the swimmers' heart rates and blood lactate values decreased dramatically when training began, and no significant differences were seen in the two groups' results in response to the change in training volume. The swimmers who trained twice per day showed no additional improvements over those who trained only once per day. In fact, their blood lactate concentrations (figure 14.2*a*) and heart rates (figure 14.2*b*) appeared to be slightly higher for the same fixed-pace swim.

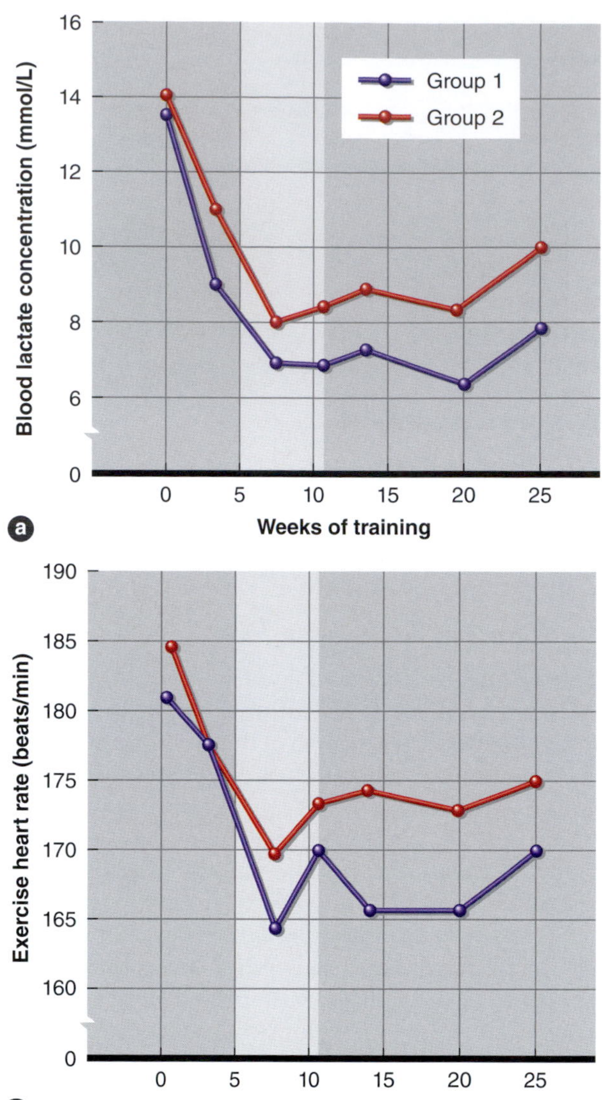

FIGURE 14.2 Changes in swimmers' *(a)* blood lactate concentrations and *(b)* heart rates during a standardized 366 m (400 yd) swim over 25 weeks of training. From the beginning of the fifth through the end of the 10th week, one group trained once per day (group 1) whereas the other group trained twice each day (group 2).

To determine the influence of long-term excessive training, performance improvements of swimmers who trained twice daily for a total distance of more than 10,000 m (10,936 yd) per day (the LS, or long-swim, group) were compared with improvements of those who swam approximately half that distance in a single session each day (the SS, or short-swim, group).[6] Changes in performance time for the 100 yd (91 m) front crawl were examined over a 4-year period for both groups. The LS swimmers and SS swimmers experienced an identical improvement of 0.8% per year. The concept of training specificity (see chapter 9) implies that several hours of daily training will not provide the adaptations needed for athletes who participate in events of short duration. Most competitive swimming events last less than 2 min. How can training for 3 to 4 h per day at speeds that are markedly slower than competitive pace prepare the swimmer for the maximal efforts of competition? Such a large training volume prepares the athlete to tolerate a high volume of training but likely does little to benefit actual performance.

The need for long daily workouts (high *volume*) is now being questioned by researchers. For many sports, it appears that training volume could be reduced significantly, possibly by as much as one-half in some sports, without reducing the benefits and with less risk of overtraining athletes to the point of decreased performance. The principle of training specificity suggests that low-intensity, high-volume training does not improve sprint-type performance.

Training *intensity* is also an important factor and refers to both the relative force of muscle action (i.e., resistance training) and the relative stress placed on the metabolic and cardiovascular systems (i.e., anaerobic and aerobic training). There is a strong interaction between training intensity and training volume: As intensity is reduced, training volume must be increased to achieve adaptation. Training at very high intensities requires substantially less training volume, but the adaptations that occur will be significantly different from those achieved with low-intensity, high-volume training. This concept applies to all three types of training, that is, resistance, anaerobic, and aerobic.

High-intensity, low-volume training can be tolerated only for brief periods. While this type of training does increase muscular strength and anaerobic capacity in high-intensity interval training, it provides little or no improvement in aerobic capacity. Conversely, low-intensity, high-volume training stresses the oxygen transport and oxidative metabolic systems, causing greater gains in aerobic capacity, but has little or no effect on muscular strength, anaerobic capacity, or sprint speed.

Attempts to perform large volumes of high-intensity training can have negative effects on adaptation. The energy needs of high-intensity exercise place greater demands on the glycolytic system, rapidly depleting muscle glycogen. If such training is attempted too often—for example, daily—the muscles can become chronically depleted of their energy reserves, and the person might demonstrate signs of overtraining, a topic discussed later in this chapter.

Periodization of Training

Based on the model of training shown in figure 14.1, effective training is accomplished by applying the principle of **periodization**, dividing the entire sport training season into smaller periods of time and training units. Periodization allows for a varied training load over time that enables acute overload and overreaching without overtraining the athlete. Traditional periodization began in the 1960s and gained popularity among coaches and athletes over the next five decades. More recently, alternative approaches have appeared, but little research has been performed to validate either the physiological adaptations or sport performance improvements that accompany them.

Traditional Periodization

Periodization, in its traditional form, involved the planned sequencing of training units (longer and shorter sessions and cycles, higher and lower intensity periods) so that athletes could achieve their desired performance. For decades, periodization was based on a hierarchy that included

- multiyear preparation (for example, for the 4-year Olympic cycle),
- macrocycles (monthlong training cycles),
- mesocycles (weekly training cycles),
- microcycles (cycle lasting a few days), and finally,
- individual workouts.

One such periodized training program is illustrated in figure 14.3.

Traditional periodization programs have two major disadvantages.[14] First, models such as that shown in figure 14.3 are best used to prepare athletes to participate successfully in a single planned competition. The recent trend for high-caliber athletes to compete in multiple competitions throughout the year adds too much complexity to the traditional design of macro-, meso-, and microcycles. Secondly, the traditional periodization scheme assumes that all physiological systems and skills involved in the sport are being developed simultaneously. This approach works best for sports like running, swimming, and cycling but falls short for ball sports, combat sports, and aesthetic sports that require more than the development of general attributes like aerobic capacity, muscular strength, or sprint speed.

An effective variation of this model involves the implementation of general and specific exercises with the goal of stimulating specific motor skills that will stress both metabolic and sport-specific movement patterns. General exercises aim at developing basic motor abilities (e.g., general strength, speed, or endurance) regardless of the mechanical or metabolic demands of the sports, for example, squats for soccer. Specific exercises are designed to mimic mechanical gestures (movement patterns) specific to a sport, for example,

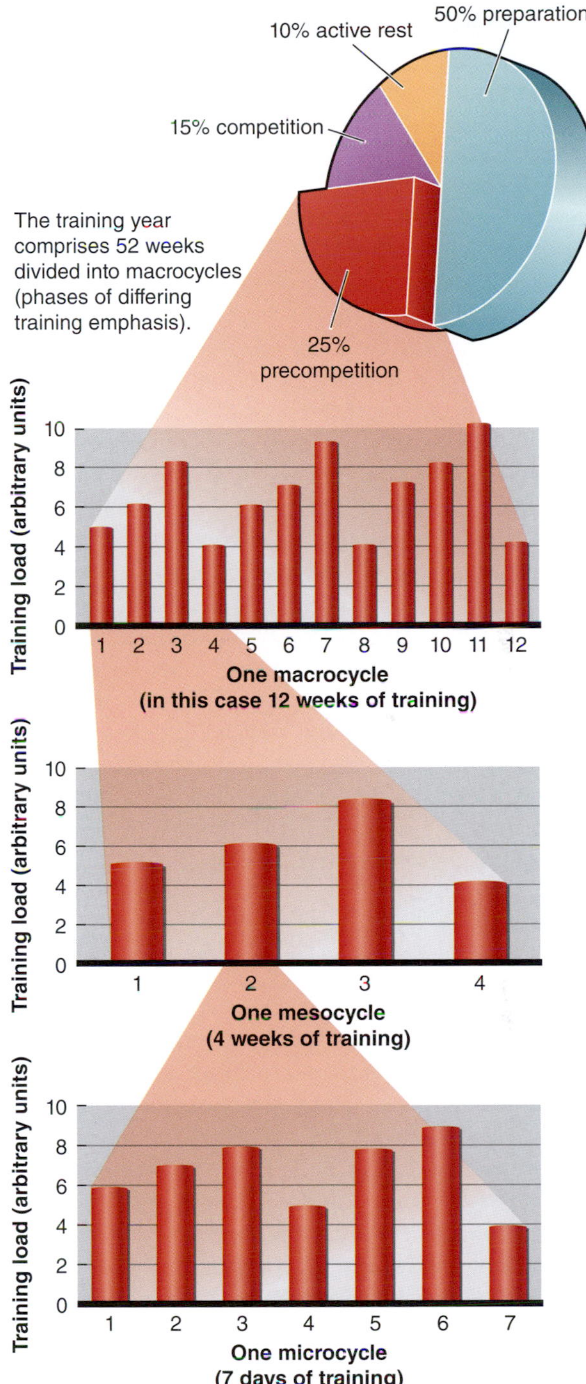

FIGURE 14.3 The structure of a traditional periodized training program. This model varies the training load over time to achieve acute overload and some overreaching while avoiding overtraining.

Adapted by permission from R.W. Fry, A.R. Morton, and D. Keast, "Overtraining in Athletes: An Update," *Sports Medicine* 12 (1991): 32-65.

lunges for fencing. Specificity can be also achieved by stressing the *most relevant* physiological systems for a given sport, with special emphasis on intensity and duration. By combining the neuromuscular and mechanical

RESEARCH PERSPECTIVE 14.1

Periodization of High-Intensity Training Models and Endurance Adaptation

The training term *periodization* is widely used to describe and quantify the planning process of training. Periodization plans add structure, with well-defined training periods that stimulate specific physiological adaptations or performance qualities for optimal performance development. Such endurance training models manipulate different training sessions periodized over time, scales ranging from microcycles (2-7 days), mesocycles (3-6 weeks), and macrocycles (6-12 months). Training intensity distribution is also considered, with most studies suggesting a high volume of low-intensity training, substantial moderate-intensity training, and less high-intensity training. The optimization of the duration and intensity of individual training sessions, the frequency of training sessions, and the organizational pattern of these stimulus variables will maximize physiological adaptations and performance capability in elite athletes.

Research has recently focused on the periodization of high-intensity training. Although these limited studies have advanced our understanding of the intensity and duration of high-intensity training sessions and endurance performance, the cumulative effects of the intensity and duration of such training sessions is not completely understood. Thus, investigators undertook a study to compare the effects of three different high-intensity training models, balanced for total load but periodized in a specific mesocycle order or in a mixed distribution, on endurance adaptations during a 12-week training period in endurance athletes.[21] Three separate groups were assigned to organized interval sessions in a specific periodized mesocycle order (increasing or decreasing intensity) or in a mixed distribution during mesocycles. This training intervention is illustrated in the accompanying figure. Pre- and postintervention testing assessed both physiological (e.g., $\dot{V}O_2$) and performance (e.g., power) responses.

The results showed that all three groups demonstrated moderate-to-large improvements in physiological and performance variables, indicating that the basic load features of all three training models were effective. Perhaps surprisingly, the specific high-intensity training periodized mesocycle order or mixed distribution did not have any effect on the adaptive responses to endurance training. Although the authors found a tendency toward less adaptation in the mixed distribution group, this was not statistically different. The authors argue that many different soundly designed periodization structures can achieve the same training goals.

with the metabolic components, sport-specific exercise can be optimized for almost every sport.

A key to determining the effectiveness of periodization programs is to test performance improvements following each cycle using sport-specific drills. Tests need to be specific to the sport for which training is implemented so that a true performance improvement can be assessed. Implementing this approach will serve as the basis for the coming training cycles. With appropriate testing following each training cycle, it is possible to determine not only performance improvements but also any transfer effects from the general to the specific training modules or cycles.

Block Periodization

Beginning in the 1980s, the term *training blocks* became popular among highly trained athletes and their coaches.[14] A block is a training cycle of highly specialized and concentrated work. While each sport differs, block periodized training involves several general principles.[14]

- Each block focuses on a minimal number of desired outcomes and targeted abilities.
- The number of blocks is small (three or four), unlike traditional mesocycles that may contain 9 to 11 different types of smaller cycles.
- A single block typically lasts 2 to 4 weeks. This allows for adaptation without excessive fatigue.
- Putting training stages into the proper sequence leads to optimal competitive sport performance.

The major purported benefit of block periodization is the consecutive development of many abilities related to the given sport, although the published data on this topic are scarce because it is relatively new and results are difficult to compare across sports and studies.

In Review

- Every athlete's rate of adaptation to training is genetically determined. Each individual responds differently to the same training stress, so what might be excessive training for one person could be well below the capacity of another for optimal adaptations. Therefore, it is important to recognize individual differences when designing training programs.

- Optimal training involves following a model that incorporates the principles of periodization, because the body needs to systematically go through stages of undertraining, acute overload, and overreaching to maximize performance.
- Excessive training refers to training that is done with an unnecessarily high volume, intensity, or both. It provides few or no additional improvements in conditioning or performance and can lead to decreased performance and health problems.
- Training volume can be increased through increase in the duration or frequency of training bouts or both. Many studies have shown no significant differences in improvement between athletes who train with typical training volumes and those who train with twice the volume (training conducted for twice the duration or twice a day instead of once a day).
- Training intensity determines the specific adaptations that occur in response to the training stimulus. As training intensity increases, training volume must be reduced, and vice versa.
- Periodization allows for a varied training load over time that allows for acute overload and overreaching without overtraining the athlete.

Overtraining

With overly intense training, athletes may experience an unexplained decline in performance and physiological function that extends over weeks, months, or years. This condition is termed overtraining, and the precise cause or causes of the resulting performance decrement are not fully understood. Research has pointed to both psychological and physiological causes. Furthermore, overtraining can occur with each of three major forms of training—resistance, anaerobic, and aerobic—so it is likely that the cause or causes and symptoms vary by the type of training.

All athletes experience some degree of fatigue during repeated days and weeks of training, so not all fatigue-producing situations can be classified as overtraining (as noted previously with overreaching). Fatigue that follows one or more exhaustive training sessions is typically relieved by a few days of reduced training or rest and a proper calorie- and carbohydrate-rich diet. Overtraining, on the other hand, is characterized by a sudden decline in performance and physiological function that cannot be remedied by a few days of reduced training, rest, or dietary intervention.

The Overtraining Syndrome

Most of the symptoms that result from overtraining, collectively referred to as the overtraining syndrome, are subjective and identifiable only after the individual's performance and physiological function have suffered. Unfortunately, these symptoms can be highly individualized, making it very difficult for athletes, trainers, and coaches to recognize that performance decrements were indeed caused by overtraining. A decline in physical performance with continued training is usually the first indication of the overtraining syndrome (see figure 14.4). The athlete senses a loss of muscular strength,

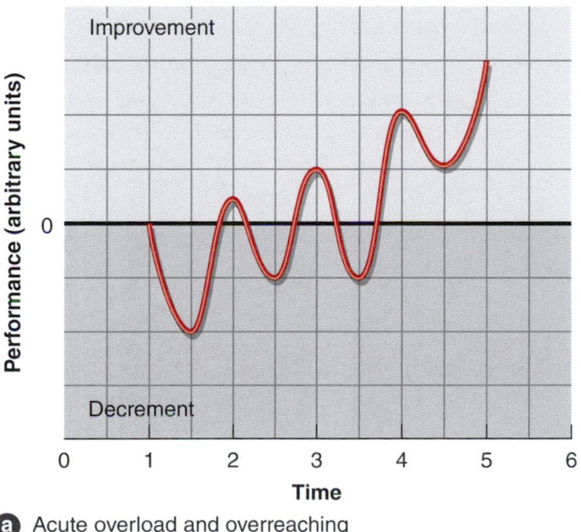

ⓐ Acute overload and overreaching

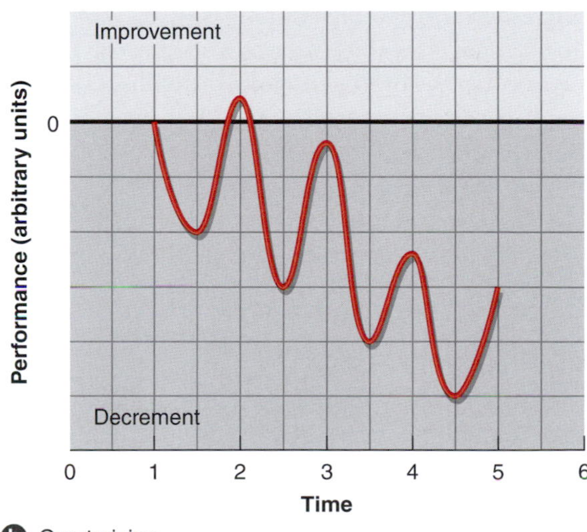

ⓑ Overtraining

FIGURE 14.4 Typical pattern of the expected improvement in performance with *(a)* acute overload and overreaching in contrast to the pattern seen with *(b)* overtraining.

Adapted by permission from M.L. O'Toole, Overreaching and Overtraining in Endurance Athletes, in *Overtraining in Sport*, edited by R.B. Kreider, A.C. Fry, and M.L. O'Toole (Champaign, IL: Human Kinetics, 1998), 10, 13.

coordination, and exercise capacity and generally feels fatigued. Other primary signs and symptoms of the overtraining syndrome include[1]

- change in appetite;
- body weight loss;
- sleep disturbances;
- irritability, restlessness, excitability, and anxiousness;
- loss of motivation and vigor;
- lack of mental concentration;
- feelings of depression; and
- lack of appreciation for things—including exercise—that normally are enjoyable.

The underlying causes of overtraining syndrome are often a complex combination of emotional and physiological factors. The emotional demands of competition, the desire to win, the fear of failure, unrealistically high goals, and others' expectations can be sources of intolerable emotional stress. Because of this, overtraining is typically accompanied by a loss of competitive desire and a loss of enthusiasm for training. Furthermore, Armstrong and VanHeest[1] make the important observation that the overtraining syndrome and clinical depression involve remarkably similar signs and symptoms, brain structures, neurotransmitters, endocrine pathways, and immune responses, suggesting that they may have similar etiologies.

The physiological factors responsible for the detrimental effects of overtraining are likewise not fully understood. However, research suggests that overtraining is associated with alterations in the nervous, endocrine, and immune systems. Although a cause-and-effect relationship between these changes and the symptoms of overtraining has not been clearly established, these symptoms can help determine whether an individual is overtrained. In the following discussion, we focus on some of the observed changes associated with overtraining and on potential causes of the overtraining syndrome.

Autonomic Nervous System Responses to Overtraining

Some studies suggest that overtraining is associated with abnormal responses of the autonomic nervous system. Physiological symptoms accompanying the decline in performance often reflect changes in those organs or systems that are controlled by either the sympathetic or the parasympathetic branch of the autonomic nervous system (see chapter 3). Sympathetic nervous system abnormalities due to overtraining can lead to

- increased resting heart rate,
- increased blood pressure,
- loss of appetite,
- decreased body mass,
- sleep disturbances,
- emotional instability, and
- elevated basal metabolic rate.

This form of overtraining occurs predominantly among athletes who emphasize resistance training methods of very high intensity.

Other studies suggest that the parasympathetic nervous system might be dominant in some cases of overtraining, usually in endurance athletes. In these cases, the performance decrements markedly differ from those associated with sympathetic overtraining. Signs of parasympathetic overtraining, assumed to be the result of high training volume overload, include

- early onset of fatigue,
- decreased resting heart rate,
- rapid heart rate recovery after exercise, and
- decreased resting blood pressure.

Thus, it appears that athletes in different sports or events will likely exhibit unique signs and symptoms of the overtraining syndrome that are related to their training regimens. In fact, some have differentiated intensity-related and volume-related forms of overtraining, recognizing that specific training stressors result in unique signs and symptoms when applied excessively.

The symptoms of overtraining syndrome are highly individualized and subjective, so they cannot be universally applied. The presence of one or more of these symptoms is sufficient to alert the coach or trainer that an athlete might be overtraining.

Some of the symptoms associated with autonomic nervous system overtraining are also seen in people who are not overtrained. For this reason, the presence of these symptoms cannot be used to confirm overtraining. Although there is not strong scientific evidence to support the autonomic nervous system overtraining theory, the autonomic nervous system is definitely affected by overtraining.

Hormonal Responses to Overtraining

Measurements of various blood hormone concentrations during periods of overreaching suggest that marked disturbances in endocrine function accompany excessive stress. As shown in figure 14.5, when swimmers increase their training 1.5- to 2-fold, blood concentrations of thyroxine and testosterone usually decrease and blood concentrations of cortisol increase. The ratio of testosterone to cortisol is thought to regulate anabolic processes in recovery, so a change in this ratio is considered an important indicator, and perhaps a cause, of the overtraining syndrome. Decreased tes-

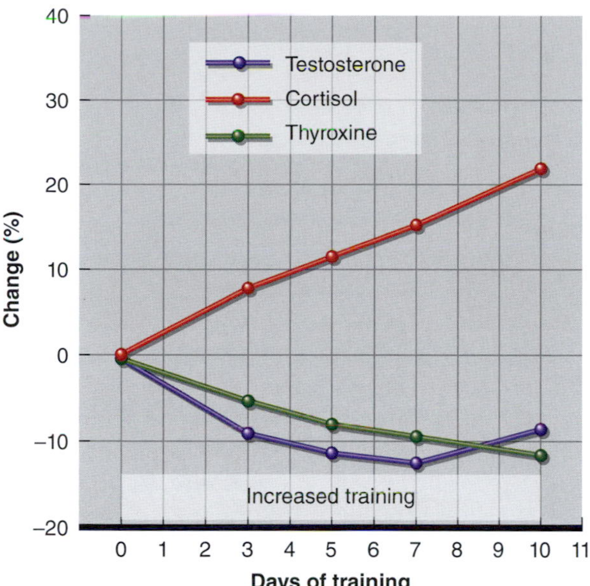

FIGURE 14.5 Changes in resting blood concentrations of testosterone, cortisol, and thyroxine over a period of intensified training. During the 10-day period shown here, swimmers increased their training from about 4,000 m (4,374 yd) to 8,000 m (8,749 yd) per day. These data show that resting cortisol concentrations increased in response to the added stress, whereas testosterone and thyroxine showed an unwanted decline during this period.

tosterone coupled with increased cortisol might lead to more protein catabolism than anabolism in the cells. Other research, however, suggests that although cortisol concentrations increase with overreaching and the early stages of overtraining, both resting and exercise cortisol concentrations eventually decrease in the overtraining syndrome. Further, most overtraining studies have been conducted on aerobically trained endurance athletes. Fewer studies exist on anaerobically trained and resistance-trained athletes. Using the terminology introduced in the preceding section, intensity-related overtraining (anaerobic and resistance training) does not appear to alter resting hormonal concentrations.

Overtrained athletes often have higher blood concentrations of urea because urea is produced by the breakdown of protein, that is, protein catabolism. This protein loss is the mechanism that is thought to be responsible for the weight loss seen in many overtrained athletes.

Elevated resting blood concentrations of epinephrine and norepinephrine have also been reported during periods of intensified aerobic training. These two hormones elevate heart rate and blood pressure. This has led some exercise physiologists to suggest that the blood concentrations of these catecholamines could be measured to confirm overtraining. However, other studies have found no change in these catecholamines during intensified training, and some have even found decreased resting concentrations.

Acute overload training and overreaching can produce most of the same hormonal changes reported in overtrained athletes. For this reason, measuring these and other hormones does not provide valid confirmation of overtraining. Athletes whose hormone concentrations appear to be abnormal may simply be experiencing the normal effects of training. Further, the time interval between the last training bout and the resting blood sample is very important. Some potential markers remain elevated for more than 24 h and might not reflect a true resting state. These hormonal changes might simply reflect the stress of training rather than a breakdown in the adaptive process. Consequently, experts generally agree that no blood marker conclusively defines the overtraining syndrome.

Armstrong and VanHeest[1] proposed that the various stressors associated with the overtraining syndrome act primarily through the hypothalamus. They postulated that these stressors activate the following two predominant hormonal axes involved in the body's response to stressors:

- The sympathetic–adrenal medullary axis (SAM), involving the sympathetic branch of the autonomic nervous system
- The hypothalamic-pituitary-adrenocortical axis (HPA)

This is illustrated in figure 14.6*a*. Figure 14.6*b* illustrates the brain and immune system interactions with these two axes. These two figures are quite complex and go well beyond the scope of an introductory-level exercise physiology text. However, a cursory study of the interactions depicted in these figures will give one an appreciation of the complexity of this syndrome. Importantly, note that the stressors have their initial effect on the brain (hypothalamus). Thus, it is highly likely that brain neurotransmitters play an important role in the overtraining syndrome. Serotonin is a major neurotransmitter that is suspected to play a significant role in the overtraining syndrome. Unfortunately, plasma concentrations of this important neurotransmitter do not accurately reflect concentrations in the brain. Advances in technology should provide the necessary tools to help us better understand what is going on inside the brain.

Overtraining syndrome appears to be associated with systemic inflammation and increased synthesis of cytokines,[20] providing support for the model depicted in figure 14.6*b*. Elevated circulating cytokines result from infection as well as from skeletal muscle, bone, and joint trauma associated with overtraining. They appear to be a normal part of the body's inflammatory

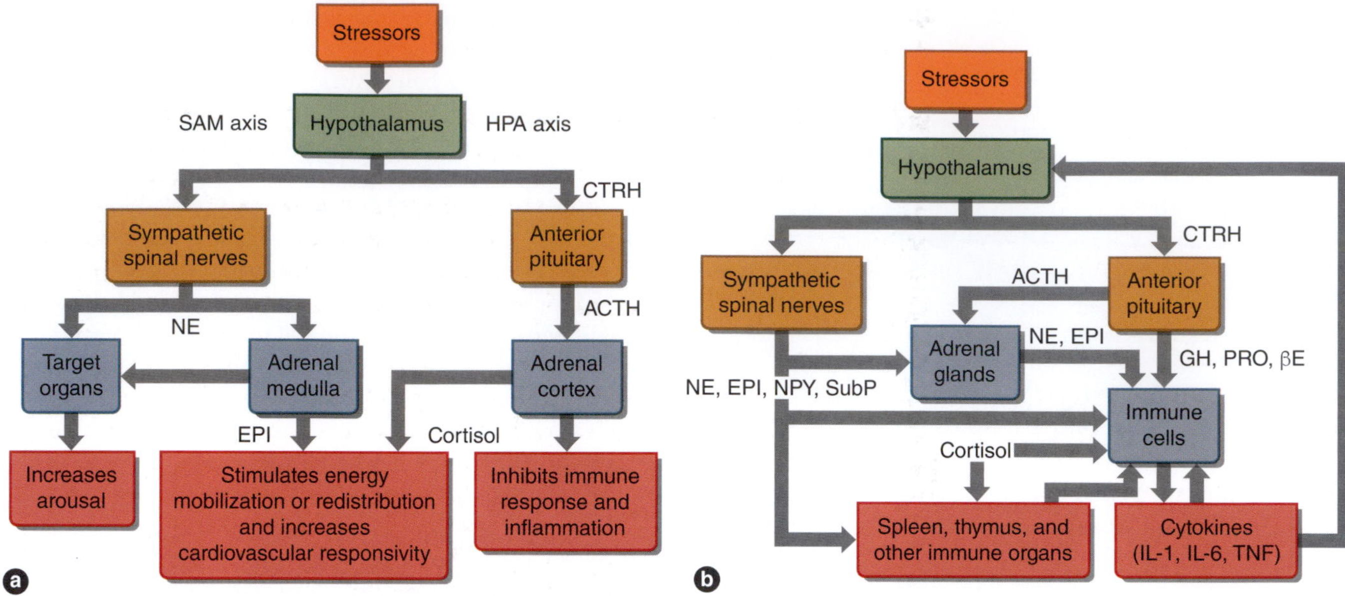

FIGURE 14.6 *(a)* The role of the hypothalamus and the sympathetic–adrenal medullary (SAM) and hypothalamic-pituitary-adrenocortical (HPA) axes as possible mediators of the overtraining syndrome. *(b)* The brain–immune system interactions in this model, with the cytokines playing a potentially major role in mediating overtraining. ACTH = adrenocorticotropin; EPI = adrenaline (epinephrine); CTRH = corticotrophin-releasing hormone; GH = growth hormone; IL-1 = interleukin-1; IL-6 = interleukin-6; NE = noradrenaline; SubP = substance P; PRO = prolactin; TNF = tumor necrosis factor; NPY = neuropeptide Y; βE = β-endorphin.

Adapted by permission from L.E. Armstrong and J.L. VanHeest, "The Unknown Mechanism of the Overtraining Syndrome," *Sports Medicine* 32 (2002): 185-209.

response to infection and injury. It is theorized that excessive musculoskeletal stress, coupled with insufficient rest and recovery, sets up a cascade of events whereby a local acute inflammatory response evolves into chronic inflammation and eventually into systemic inflammation. Systemic inflammation activates circulating monocytes, which can then synthesize large quantities of cytokines. Cytokines then act on most of the brain and body functions in a manner consistent with symptoms expressed in the overtraining syndrome.[20]

Immunity and Overtraining

The immune system provides a line of defense against invading bacteria, parasites, viruses, and tumor cells. This system depends on the actions of specialized cells (such as lymphocytes, granulocytes, and macrophages) and antibodies. These primarily eliminate or neutralize foreign invaders that might cause illness (pathogens). Unfortunately, one of the most serious consequences of overtraining is the negative effect it has on the body's immune system. In fact, according to the model proposed in figure 14.6, compromised **immune function** is potentially a major factor in the initiation of the overtraining syndrome.

Many studies have shown that excessive training suppresses normal immune function, increasing the overtrained athlete's susceptibility to infections. This is illustrated in figure 14.7. Studies also show that short bouts of intense exercise can temporarily impair the immune response, and successive days of heavy training

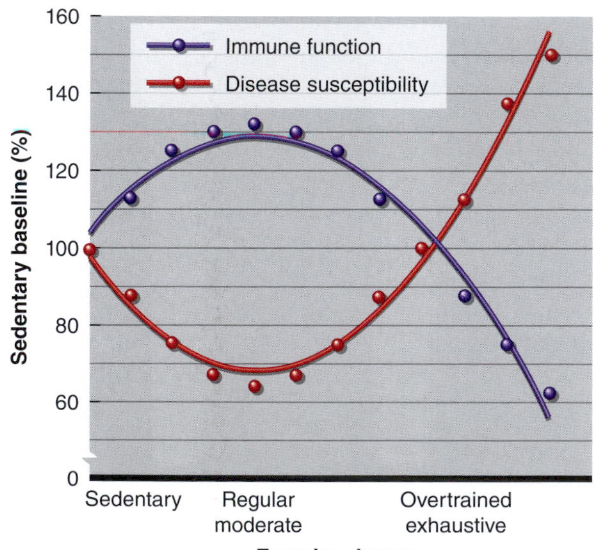

FIGURE 14.7 The inverted J-shaped model of the relationship between amount of exercise and immune function. This model suggests that moderate exercise may lower the risk of infection or disease, whereas overtraining may increase the risk.

Data from Nieman (1997).

can amplify this suppression. Several investigators have reported an increased incidence of illness following a single, exhaustive exercise bout, such as running a full 42 km (26.2 mi) marathon. This immune suppression is characterized by abnormally low concentrations of both lymphocytes and antibodies. Invading organisms or substances are more likely to cause illness when these concentrations are low. Also, intense exercise during illness might decrease one's ability to fight off the infection and increase the risk of even greater complications.[18]

Joint Consensus Statement on Overtraining

The importance of preventing, diagnosing, and treating the overtraining syndrome was addressed by two international organizations (the American College of Sports Medicine and the European College of Sport Science) in the form of a consensus statement.[17] The balance between providing the overload needed for training adaptations and avoiding the combination of excessive overload and inadequate recovery is a difficult one to achieve. The consensus statement delineates between nonfunctional overreaching (NFOR) and overtraining syndrome (OTS). Nonfunctional overreaching leads to a state of stagnated or decreased performance, but eventually the athlete can recover with sufficient rest. Overtraining syndrome, on the other hand, results in a "long-term decrement in performance capacity with or without related physiological and psychological signs and symptoms of maladaptation in which restoration of performance capacity may take *several weeks or months*" [emphasis added] (p. 187). Survey data suggest that the prevalence of NFOR or OTS is approximately 10% in endurance athletes, but some estimates have been even higher.

The key distinction is that OTS results in prolonged maladaptation, but in reality it is extremely difficult to diagnose OTS. Symptoms of OTS are usually (but not always) more severe than those of NFOR and include fatigue, performance decline, and mood disturbances. Several markers of OTS have been proposed, including hormones, psychological tests, biochemical markers, and indices of immune function, but none of them meet the criteria to make their use generally accepted. Measuring heart rate variability as an index of autonomic balance has shown some promise, but varied measurement techniques have limited its utility and acceptance. One necessary but tedious task in diagnosing OTS is to rule out other causes of performance decline such as diseases and infections, negative energy balance, insufficient carbohydrate or protein intakes, allergies, and so on.

Finally, it is important to consider that demanding, intensified training is not the only element in the development of OTS. Overtraining syndrome is complex and involves psychological factors such as excessive expectations, competitive stress, social and family involvement, personal and emotional problems, training monotony, and extra-sport demand from work or school. The two organizations propose the following checklist to aid in the diagnosis of OTS.

Performance-Fatigue

Is the athlete experiencing the following:

- ☐ Unexplainable underperformance
- ☐ Persistent fatigue
- ☐ Increased sense of effort in training
- ☐ Sleep disorders

Exclusion Criteria

Are there confounding diseases?

- ☐ Anemia
- ☐ Epstein-Barr virus
- ☐ Other infectious diseases
- ☐ Muscle damage (high CK)
- ☐ Lyme disease
- ☐ Endocrine diseases (diabetes, thyroid, adrenal gland, . . .)
- ☐ Major disorders of eating behavior
- ☐ Biological abnormalities (increased erythrocyte sedimentation rate, C-reactive protein, creatinine, or liver enzymes, decreased ferritin, . . .)
- ☐ Injury (musculoskeletal system)
- ☐ Cardiovascular symptoms
- ☐ Adult-onset asthma
- ☐ Allergies

Are there training errors?

- ☐ Training volume increased (>5%) ($h \cdot wk^{-1}$, $km \cdot wk^{-1}$)
- ☐ Training intensity increased significantly
- ☐ Training monotony present
- ☐ High number of competitions
- ☐ In endurance athletes: decreased performance at "anaerobic" threshold
- ☐ Exposure to environmental stressors (altitude, heat, cold, . . .)

Other confounding factors:

- ☐ Psychological signs and symptoms (disturbed POMS, RESTQ-Sport, RPE, . . .)
- ☐ Social factors (family relationships, financial, work, coach, team, . . .)
- ☐ Recent or multiple time zone travel

Exercise test:

- ☐ Are there baseline values to compare with (performance, heart rate, hormonal, lactate, . . .)?
- ☐ Maximal exercise test performance
- ☐ Submaximal or sport-specific test performance
- ☐ Multiple performance tests

Reprinted by permission from R. Meeusen et al., "Prevention, Diagnosis, and Treatment of the Overtraining Syndrome: Joint Consensus Statement of the European College of Sport Science and the American College of Sports Medicine," *Medicine and Science in Sports and Exercise* 45 (2013): 186-205.

Predicting the Overtraining Syndrome

We must remember that the underlying cause or causes of the overtraining syndrome are not fully known, although it is likely that physical or psychological (or emotional) overload, or a combination of the two, might trigger this condition. Trying not to exceed an athlete's stress tolerance by regulating the amount of physiological and psychological stress experienced during training is difficult. Most coaches and athletes use intuition to determine training volume and intensity, but few can accurately assess the true impact of a workout on the athlete. No preliminary symptoms warn athletes that they are on the verge of becoming overtrained. By the time coaches realize that they have pushed an athlete too hard, it is often too late. The damage done by repeated days of excessive training or stress can be repaired only by days, and in some cases weeks or months, of reduced training or complete rest.

Numerous investigators have tried to identify markers of the overtraining syndrome in its early stages by using assorted physiological and psychological measurements. A list of potential markers is provided in table 14.1. Unfortunately, none has proven totally effective. It is often difficult to determine whether the measure-

TABLE 14.1 **Potential Markers of Overreaching (OR), Overtraining (OT), and Overtraining Syndrome (OTS)**

Physiological and psychological marker	Response	OR	OT	OTS
HR_{rest} and HR_{max}	Decreased		✓	✓
HR_{submax} and $\dot{V}O_{2submax}$	Increased	✓		✓
$\dot{V}O_{2max}$	Decreased			✓
Anaerobic metabolism	Impaired		✓	
Basal metabolic rate	Increased			✓
RER_{submax} and RER_{max}	Decreased		✓	✓
Nitrogen balance	Negative			✓
Nerve excitability	Increased			✓
Sympathetic nervous response	Increased			✓
Psychological mood states	Altered	✓		
Risk of infection	Increased	✓		
Hematocrit and hemoglobin	Decreased		✓	
Leukocytes and immunophenotypes	Decreased		✓	
Serum iron and ferritin	Decreased		✓	
Serum electrolyte levels	Decreased			✓
Serum glucose and free fatty acids	Decreased		✓	
Plasma lactate concentration, submax, max	Decreased		✓	✓
Ammonia	Increased		✓	✓
Serum testosterone and cortisol	Decreased	✓		
ACTH, growth hormone, prolactin	Decreased			✓
Catecholamines, rest, night	Decreased			✓
Creatine kinase	Increased			✓

Note. HR = heart rate; RER = respiratory exchange ratio; ACTH = adrenocorticotropic hormone.

From Armstrong and VanHeest (2002).

ments obtained are related to overtraining or whether they simply reflect normal responses to overload or overreaching training.

One good method to identify the overtraining syndrome is to monitor the athlete's heart rate during a standardized workout, such as a fixed-paced run or swim, using a digital heart rate monitor (figure 9.5). The data presented in figure 14.8 illustrate a runner's heart rate response during a 1 mi (1.6 km) run performed at a fixed pace of 6 min/mi (3.7 min/km), or 10 mph (16 km/h). This response was monitored when the runner was untrained (UT), after the runner had trained (TR), and during a period when the runner demonstrated symptoms of overtraining syndrome (OT). This figure shows that heart rate was higher when the runner was in the overtrained state than when the runner was responding well to training. Such a test provides a simple and objective way to monitor training and can possibly provide an early warning sign of overtraining.

Exercise Addiction

The condition of exercise addiction may not sound like a bad thing to many people who understand all of the health benefits associated with regular exercise. However, exercise addiction is maladaptive, so instead of improving the person's life, it causes more problems. This form of addictive behavior can threaten one's overall health, cause overuse and other injuries, ruin personal lives and careers, and, in some cases, contribute to nutrition problems and eating disorders. Not surprisingly, exercise addiction has received recent attention among exercise scientists, including exercise and sport psychologists. Surprisingly, despite the known harmful consequences, exercise addiction is not listed as a disorder in the latest version of the *Diagnostic and Statistical Manual of Mental Disorders (DSM-V)*.

The distinction between healthy levels of exercise and exercise addiction is difficult to make. People who are addicted to exercise will continue to exercise regardless of physical injury, personal inconvenience, disruption to other areas of their lives (e.g., marriage or career), and lack of time for other important activities.[15] Committed exercisers find a variety of rewards in their physical activity and view their exercise routines as an important but not all-defining part of their lives. Addicted exercisers identify exercise as a central part of their lives and exercise to avoid unpleasant emotions that would arise if they missed their exercise (e.g., guilt, anxiety). Addicted exercisers may even consider exercise *the* most important part of their lives and the only way to elevate their mood or feel better. A 2013 review on the topic of exercise addiction[15] notes that "exercisers must acquire a sense of life-balance while embracing an attitude conducive to sustainable long-term physical, psychological and social health outcomes".

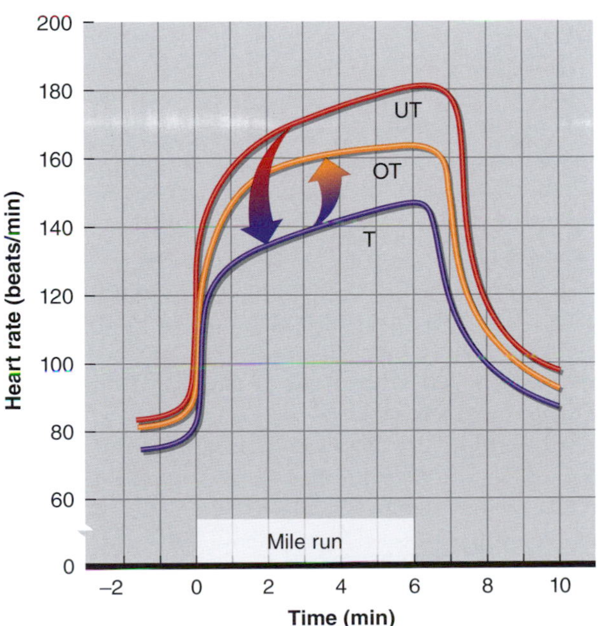

FIGURE 14.8 A runner's heart rate responses during a fixed-pace treadmill run at 10 mph (16 km/h) performed before training (UT), after training (T), and when the runner showed symptoms of overtraining (OT).

Prevention and Recovery

Recovery from the overtraining syndrome is possible with a marked reduction in training intensity or complete rest. Although most coaches recommend a few days of easy training, overtrained athletes require considerably more time for full recovery. This might necessitate the total cessation of training for a period of weeks or months. In some cases, counseling might be needed to help athletes cope with other stress in their lives that might contribute to this condition.

The best way to minimize the risk of overtraining is to follow periodization training procedures, alternating easy, moderate, and hard periods of training. Although individual tolerance varies tremendously, even the strongest athletes have periods when they are susceptible to the overtraining syndrome. As a rule, 1 or 2 days of intense training should be followed by an equal number of easy training days. Likewise, a week or two of hard training should be followed by a week of reduced effort with little or no emphasis on anaerobic exercise.

Endurance athletes (such as swimmers, cyclists, and runners) should pay particular attention to their caloric and carbohydrate intakes. Repeated days of hard training gradually reduce muscle glycogen. Unless these athletes consume extra carbohydrate during these periods, muscle and liver glycogen reserves can be depleted. As a consequence, the most heavily recruited muscle fibers are not able to generate the energy needed for exercise.

In Review

- Addicted exercisers identify exercise as a central part of their lives and exercise to avoid unpleasant emotions that would arise if they missed their exercise (e.g., guilt, anxiety).
- Athletes often overtrain in the erroneous belief that more training always produces further improvement. As their performance declines with overtraining, they train even harder in an effort to compensate. The importance of designing training programs to include rest and variations in training intensity and volume to avoid overtraining syndrome cannot be overstated.
- Overtraining stresses the body beyond its capacity to adapt, decreasing physiological capacity and performance.
- The symptoms of overtraining syndrome are subjective and vary from individual to individual; many also accompany regular training, making prevention or diagnosis of the overtraining syndrome difficult.
- Possible explanations for the overtraining syndrome include changes in function of the autonomic nervous system, altered endocrine responses, suppressed immune function, and altered brain neurotransmitters.
- Many potential signs and symptoms of overtraining have been proposed to diagnose overtraining in its earliest stages. However, at this time, the heart rate response to a fixed-pace exercise bout appears to be the easiest and most accurate technique. Performance decrements are also good indicators but often appear later in the syndrome.
- Overtraining syndrome is treated by a marked reduction in training intensity or complete rest, for weeks or months. Prevention can best be accomplished through use of periodization training procedures that vary training intensity and volume.
- For endurance athletes, it is important to ensure adequate calorie and carbohydrate intake to meet energy needs.

Tapering for Peak Performance

Peak performance requires maximal physical and psychological tolerance for the stress of the activity. But periods of intense training can reduce muscular strength, decreasing athletes' performance capacity. For this reason, to compete at their peak, many athletes reduce their training intensity and volume before a major competition to give their bodies a break from the rigors of intense training, a practice commonly referred to as **tapering**. The **taper period**, during which intensity and volume are reduced, provides adequate time for healing of tissue damage caused by intense training and for the body's energy reserves to be fully replenished. Tapering periods range from 4 to 28 days or longer, depending on the sport, the event, and the athlete's needs. Tapering is not appropriate for all sports, particularly those in which competition occurs once a week or more frequently. Still, rested athletes generally perform better.

The most notable change during the taper period is a marked increase in muscular strength, which explains at least part of the performance improvement that occurs. It is difficult to determine whether strength improvements result from changes in the muscles' contractile mechanisms or improved muscle fiber recruitment. However, examination of individual muscle fibers taken from swimmers' arms before and after 10 days of intensified training showed that the type II (fast-twitch) fibers exhibited a significant reduction in their maximal shortening velocity.[8] This change has been attributed to changes in the fibers' myosin molecules. In these cases, the myosin in the type II fibers became more like that in the type I fibers. We assume from these data that such changes in the muscle fibers cause the power loss that swimmers and runners experience during prolonged periods of intense training. We can also assume that the recovery of strength and power that occurs with tapering is linked to modifications of the muscles' contractile mechanisms. Tapering also allows time for the muscle to repair any damage incurred during intense training and for the energy reserves (i.e., muscle and liver glycogen) to be restored.

Although tapering is widely practiced in a variety of sports, many coaches fear that reduced training for such a long period before a major competition will decrease conditioning and impair performance. But numerous studies clearly show that this fear is unwarranted. Developing optimal $\dot{V}O_{2max}$ initially requires a considerable amount of training, but once it has been developed, much less training is needed to maintain it at its highest level. In fact, the training level of $\dot{V}O_{2max}$ can be maintained even when training frequency is reduced by two-thirds.[11]

Runners and swimmers who reduce their training by about 60% for 15 to 21 days show no losses in $\dot{V}O_{2max}$ or endurance performance.[5,12] One study showed that swimmers' blood lactate concentrations after a standard swim were lower after a taper period than before. More important, the swimmers experienced a 3% improvement in performance as a result of the reduced training and demonstrated an 18% to 25% increase in arm strength and power.[5] In a study of distance runners, those runners who went through a 7-day taper decreased their running time in a 5 km time trial by 3% compared to no improvement in those who did not taper. Submaximal oxygen uptake during running at 80% $\dot{V}O_{2max}$ was decreased by 6% in those who tapered, indicating a greater economy of effort. Blood lactate concentrations at 80% of $\dot{V}O_{2max}$ were unchanged, as were $\dot{V}O_{2max}$ and leg extension peak force.[13]

The effectiveness of various tapering strategies on improving sport performance is difficult to assess from any single study. One way to analyze data across many studies is to do a meta-analysis, which focuses on combining the results from a number of research studies in the hope of identifying patterns and contrasts among the various results; the goal is to identify interesting relationships and conclusions from multiple studies. The effects of different types of tapering on competitive performance were assessed by meta-analysis of data from 27 studies.[3] The authors analyzed tapering by determining the decrease in training intensity, volume, and frequency, as well as the pattern of tapering and duration of the tapering period. Most of the studies involved competitive swimming, running, and cycling. The optimal tapering strategy that emerged from this

RESEARCH PERSPECTIVE 14.2

Peak Performance During the Taper Phase

Coaches of elite athletes have a common goal of designing a well-controlled training program to ensure that the maximal performance is achieved at major competitions. The best competition performances often occur following a taper phase, which is a progressive, nonlinear reduction of the training load in the period before competition. The main purpose of the taper is to reduce physiological and psychological stressors of previous training and remove residual fatigue so that sport performance can be optimized. However, there is relatively little scientific evidence available to guide coaches in prescribing appropriate tapering strategies for individual athletes. Most studies have used mathematical modeling simulations to predict the most ideal tapering strategies. Despite this limited support, it is common for athletes to undertake a period of overload training before tapering in an effort to maximize performance gains.

A 2014 study[2] described the relations between performance and training before and during a simulated taper. The investigators examined whether well-trained triathletes would demonstrate greater performance improvements than a control group during a simulated 4-week taper after completing 3 weeks of overload training. The investigators tested the hypothesis that completing overload training before the taper would allow larger performance gains, particularly in overreached athletes (those demonstrating decreased performance), but this would require a longer taper for performance compensation. The training of each triathlete was monitored for 11 weeks, which was divided into four distinct phases. The first two phases were identical for both the overload training and control groups: phase I was 3 weeks during which the subjects completed their own usual training and phase II was 1 week of moderate training load during which the subjects reduced habitual training volume by 30% while maintaining training intensity. During phase III, the control group repeated the training program as in phase I; however, the overload training group completed 3 weeks of training designed to deliberately overreach the subjects (duration of each training session was increased by 30%). During phase IV, all participants completed a 4-week taper, where their normal training load was decreased by 40% each week. At the end of phases II and III, and each week during phase IV, subjects reported to the laboratory for a maximal oxygen consumption test.

In the overload training group of athletes, 10 triathletes developed clear symptoms of being functionally overreached (transient reduced performance associated with high perceived fatigue followed by the occurrence of a performance supercompensation), whereas 12 triathletes did not. The investigators found that greater gains in performance and maximal oxygen consumption were achieved when a higher training load was prescribed before the taper but not if functional overreaching occurs. Functional overreaching was associated with poor performance supercompensation during tapering and presented a greater risk for training maladaptation, including increased risk of injury or infection.

Practically, these results highlight the importance of carefully monitoring athletes during training, as well as during the taper. The investigators suggest that future studies should examine the effectiveness of different training models using early markers of functional overreaching to inform adjustments in training loads in an attempt to maximize training, tapering, and performance responses.

analysis was a tapering duration of 2 weeks over which the training volume is decreased by 41% to 60% without any change in intensity or frequency.

Unfortunately, little information is available to demonstrate the influence of tapering on performance in team sports and in long-duration endurance events such as cycling and marathon running. Before guidelines can be offered for athletes in these sports, research is needed to demonstrate that similar benefits can be generated by such periods of reduced training.

VIDEO 14.1 Presents Scott Trappe on tapering for peak sport performance.

In Review

- Many athletes decrease their training intensity and volume before a competition to increase strength, power, and performance capacity. This practice is called tapering.
- In some sports, tapering for competition is crucial to optimal performance. Reduced training volume and intensity, coupled with quality rest, are needed to allow muscle to repair itself and to restore its energy reserves for competition.
- The optimal duration of the taper is between 4 and 28 days, or longer, and is dependent on the sport or event and the athlete's needs.
- Muscular strength increases significantly during the tapering period.
- Tapering allows time for the muscle to repair any damage incurred during intense training and for the energy reserves (i.e., muscle and liver glycogen) to be restored.
- Less training is needed to maintain previous gains than was originally needed to attain them, so tapering does not decrease conditioning.
- Aerobic performance may improve by an average of about 3% with proper tapering.

Detraining

Tapering, by reducing the training stimulus, can facilitate performance. What happens to highly conditioned athletes who have fine-tuned their performance abilities to a peak level but then come to the end of their competitive season? Many athletes welcome the opportunity to completely relax, purposely avoiding any strenuous physical activity. But how does physical inactivity affect highly trained athletes from a physiological standpoint?

Detraining is defined as the partial or complete loss of training-induced adaptations in response to either the cessation of training or a substantial decrement in the training load—in contrast to tapering, which is a gradual reduction of the peak training load over only a few days to a few weeks. Some of our knowledge about detraining comes from clinical research with patients who have been forced into inactivity because of injury or surgery. Most athletes fear that all they have gained through hard training will be lost during even a brief period of inactivity. But recent studies reveal that a few days of rest or reduced training will not impair and might even enhance performance, similar to what we see with tapering. Yet at some point, training reduction or complete inactivity will decrease physiological function and performance.

Muscular Strength and Power

When a broken limb is immobilized in a rigid cast, changes begin almost immediately in both the bone and the muscles. Within only a few days, the cast that was applied tightly around the injured limb is loose. After several weeks, a large space separates the cast and the limb. Skeletal muscles undergo a substantial decrease in size, a process known as atrophy, when they remain inactive. Not surprisingly, muscle atrophy is accompanied by considerable loss of muscular strength and power. Total inactivity leads to rapid losses, but even prolonged periods of reduced activity lead to gradual losses that eventually can become quite significant.

Similarly, research confirms that muscular strength and power both are reduced once athletes stop training. The rate and magnitude of loss vary with the level of training. Highly skilled, accomplished weightlifters appear to have a rather rapid decline in strength within a few weeks of discontinuing intense training.[9] With previously untrained people, strength gains can be maintained from several weeks up to over 7 months. In a study of young (20-30 years) and older (65-75 years) men and women who trained for 9 weeks, the increase in strength (1-repetition maximum) averaged 34% for the younger subjects and 28% for the older subjects, with no differences between men and women. After 12 weeks of detraining, none of the four groups had significant losses in strength from the end of the 9-week training program values. After 31 weeks of detraining, there was only an 8% loss in the younger subjects and a 13% decrease in the older subjects.[16]

A study with collegiate swimmers revealed that even with up to 4 weeks of inactivity, terminating training did not affect arm or shoulder strength.[4] No strength changes were seen in these swimmers, whether they spent 4 weeks at complete rest or whether they reduced their training frequency to one or three sessions per week. But swimming power was reduced by 8% to about 14% during the 4 weeks of reduced activity, whether the swimmers underwent complete rest or merely reduced training frequency. Although muscular strength might not have diminished during the 4 weeks of rest or reduced training, the swimmers might have lost their ability to apply force during swimming, possibly attributable to a loss of skill.

The physiological mechanisms responsible for the loss of muscular strength as a consequence of either immobilization or inactivity are not clearly understood. Muscle atrophy causes a noticeable decrease in muscle mass and water content, which could partly account for a loss in the development of maximal muscle fiber tension. Changes occur in the rates of protein synthesis and degradation as well as in specific fiber type characteristics. When muscles aren't used, the frequency of their neurological stimulation is reduced, and normal fiber recruitment is disrupted. Thus, part of the strength loss associated with detraining could result from an inability to activate some muscle fibers.

The retention of muscular strength, power, and size is extremely important for the injured athlete. The athlete can save much time and effort during rehabilitation by performing even a low level of exercise with the injured limb, starting in the first few days of recovery. Simple isometric contractions are very effective for rehabilitation because their intensity can be graded and they do not require joint movement. Any program of rehabilitation, however, must be designed in cooperation with the supervising physician and physical therapist.

Muscular Endurance

Muscular endurance decreases after only 2 weeks of inactivity. At this time, not enough evidence is available to establish whether this performance decrement results from changes in the muscle or from changes in cardiovascular capacity. In this section, we examine muscle changes that are known to accompany detraining and that could decrease muscular endurance.

The localized muscle adaptations that occur during periods of inactivity are well documented, but the exact role that these changes play in the loss of muscular endurance is unclear. We know from postsurgery cases that after a week or two of cast immobilization, the activities of oxidative enzymes such as succinate dehydrogenase (SDH) and cytochrome oxidase decrease by 40% to 60%. Data collected from swimmers, shown in figure 14.9, indicate that the muscles' oxidative potential decreases much more rapidly than the subjects' maximal oxygen uptake with detraining. Reduced oxidative enzyme activity would be expected to impair muscular endurance, and this most likely relates to submaximal endurance capacity rather than to maximal aerobic capacity, or $\dot{V}O_{2max}$.

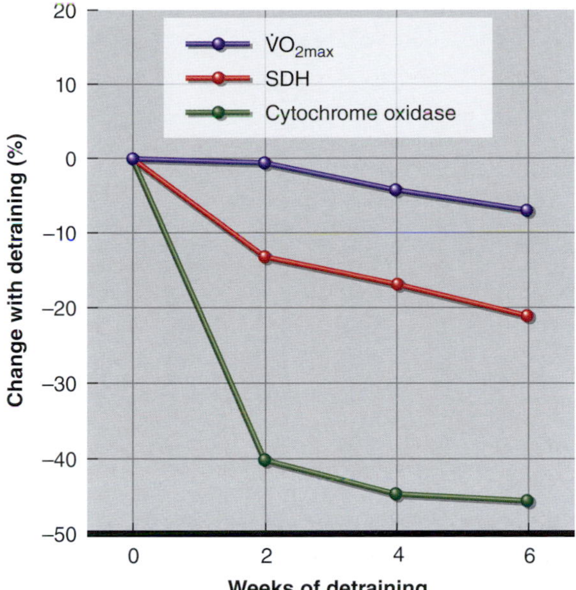

FIGURE 14.9 Percentage decreases in $\dot{V}O_{2max}$, muscle succinate dehydrogenase (SDH) activity, and cytochrome oxidase activity during 6 weeks of detraining. These findings suggest that muscles experience a decline in metabolic potential, although tests of $\dot{V}O_{2max}$ show little change over this period of detraining.

In contrast, when athletes stop training, the muscles' glycolytic enzymes, such as phosphorylase and phosphofructokinase, change little, if at all, for at least 4 weeks. In fact, Coyle and colleagues[7] observed no change in glycolytic enzyme activities with up to 84 days of detraining compared with a nearly 60% decrease in the activities of various oxidative enzymes. This might at least partly explain why performance times in sprint events are unaffected by a month or more of inactivity, but the ability to perform longer endurance events may decrease significantly with as little as 2 weeks of detraining.

One notable change in muscle during detraining is a change in its glycogen content. Endurance-trained muscle tends to increase glycogen storage. But 4 weeks of detraining has been shown to decrease muscle glycogen by as much as 40%. Figure 14.10 illustrates the decrease in muscle glycogen accompanying 4 weeks of detraining in competitive collegiate swimmers and in untrained subjects (serving as a time control). The untrained people showed no change in muscle glycogen content after 4 weeks of inactivity, but the swimmers' values, which were high initially, decreased until they were about equal to those of the untrained people. This indicates that the trained swimmers' improved capacity for muscle glycogen storage was reversed during detraining.

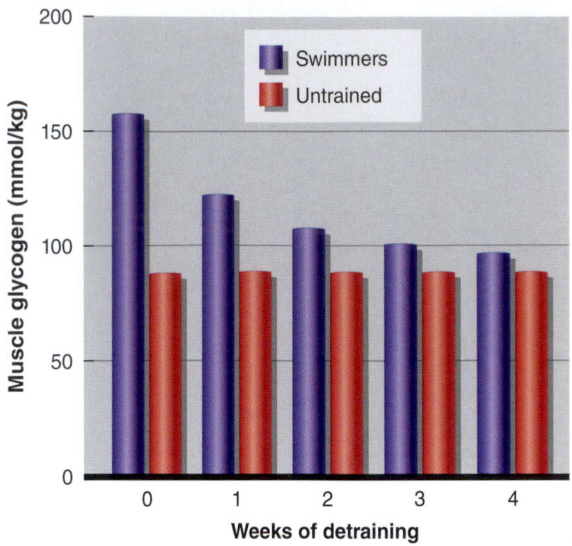

FIGURE 14.10 Changes in glycogen content of the deltoid muscle in competitive swimmers during 4 weeks of detraining. Muscle glycogen returned almost to the untrained concentration at the end of this period.

Measurements of blood lactate and pH after a standard work bout have been used to assess the physiological changes that accompany training and detraining. For example, a group of collegiate swimmers were required to perform a standard-paced, 200 yd (183 m) swim at 90% of their seasonal best following 5 months of training and then to repeat this test at the same absolute pace once a week for the following 4 weeks of detraining. The results are shown in table 14.2. Blood lactate concentrations, measured immediately after this standard swim, increased from week to week during a month of inactivity. At the end of the fourth week of detraining, the swimmers' acid–base balance was significantly disturbed. This was reflected by a significant increase in blood lactate concentrations and a significant decrease in the concentrations of bicarbonate (a buffer).

Speed, Agility, and Flexibility

Training produces less improvement in speed and agility than it does in strength, power, muscular endurance, flexibility, and cardiorespiratory endurance. Consequently, losses of speed and agility that occur with inactivity are relatively small, and peak speed and agility can be maintained with a limited amount of training. But this does not imply that the track sprinter can get by with training only a few days a week. Success in competition relies on factors other than basic speed and agility, such as correct form, skill, and the ability to generate a strong finishing sprint. Many hours of practice are required to tune performance to its optimal level, but most of this time is spent developing performance qualities other than speed and agility.

Flexibility, on the other hand, is lost rather quickly during inactivity. Stretching exercises should be incorporated into both in-season and off-season training programs. It has been proposed that reduced flexibility increases athletes' susceptibility to serious injury.

Cardiorespiratory Endurance

The heart, like other muscles in the body, is strengthened by endurance training. Inactivity, on the other hand, can substantially decondition the heart and the cardiovascular system. The most dramatic example of this is seen in a classic study conducted on subjects undergoing long periods of total bed rest; they weren't allowed to leave their beds, and physical activity was kept to an absolute minimum.[19] Cardiovascular and metabolic function were assessed at a constant submaximal rate of work and at maximal rates of work both before and after the 20-day period of bed rest. The cardiovascular effects that accompanied bed rest included

- a considerable increase in submaximal heart rate,
- a 25% decrease in submaximal stroke volume,

TABLE 14.2 **Blood Lactate, pH, and Bicarbonate (HCO_3-) Values in Collegiate Swimmers Undergoing Detraining**

	Weeks of detraining			
Measurement	0	1	2	4
Lactate (mmol/L)	4.2	6.3	6.8	9.7*
pH	7.26	7.24	7.24	7.18*
HCO_3^- (mmol/L)	21.1	19.5*	16.1*	16.3*
Swim time (s)	130.6	130.1	130.5	130.0

Note. Measurements were taken immediately after a fixed-pace swim. The values at week 0 represent the measurements taken at the end of 5 months of training. The values for weeks 1, 2, and 4 are the results obtained after 1, 2, and 4 weeks of detraining, respectively.

*Significant difference from the value at the end of training.

- a 25% reduction in maximal cardiac output, and
- a 27% decrease in maximal oxygen consumption.

The reductions in cardiac output and $\dot{V}O_{2max}$ appear to result from reduced stroke volume; this seems to be largely attributable to a decreased plasma volume, with reductions in heart volume and ventricular contractility playing a smaller role.

It is interesting that the two most highly conditioned subjects in this study (the two who had the highest $\dot{V}O_{2max}$ values) experienced greater decrements in $\dot{V}O_{2max}$ than the three less fit people, as shown in figure 14.11. Furthermore, the untrained subjects regained their initial conditioning levels (before bed rest) in the first 10 days of reconditioning, but the well-trained subjects needed about 40 days for full recovery. This suggests that highly trained individuals cannot afford long periods with little or no endurance training. The athlete who totally abstains from endurance training at the completion of the season will experience great difficulty getting back into top aerobic fitness when the new season begins.

Inactivity can significantly reduce $\dot{V}O_{2max}$. How much activity is needed to prevent such considerable losses of physical conditioning? Although a decrease in training frequency and duration reduces aerobic capacity, the losses are significant only when frequency and duration are reduced by two-thirds of the regular training load.

However, training intensity apparently plays a more crucial role in maintaining aerobic power during periods of reduced training. Training at 70% $\dot{V}O_{2max}$ appears to be necessary to maintain maximal aerobic capacity.[11]

Detraining During Spaceflight

As astronauts orbit the earth, they are in an environment where the gravitational forces are considerably less than those on Earth, that is, microgravity. While astronauts experience a sense of weightlessness in orbit, the gravitational forces do not reach 0 *g*. During an extended stay in microgravity, astronauts undergo physiological changes that are nearly identical to those of detraining. But what could be perceived as maladaptations on Earth might in fact be necessary adaptations to accommodate the microgravity of space. Let's briefly review the changes that take place when astronauts leave the 1 *g* environment on Earth to spend weeks or months in space.

Muscle mass and strength decline in microgravity, particularly in the postural muscles, that is, those muscles that maintain the body upright countering the force of gravity. The cross-sectional area of both type I and type II muscle fibers is decreased as well. The extent of decline depends on the muscle group, the duration of the flight, and the type and extent of the in-flight exercise program. Microgravity also affects bone, with average bone mineral losses of about 4% in the weight-bearing bones; the magnitude of this loss depends on the length of exposure to microgravity.

The cardiovascular system also undergoes major adaptations to microgravity. When the body is in microgravity, there is a reduction in hydrostatic pressure so that blood no longer pools in the lower extremities as it does at 1 *g*. As a consequence, more blood returns to the heart, which leads to transient increases in stroke volume. Reductions in plasma volume occur over time but are likely due to reduced fluid intake rather than increased production of urine (diuresis) by the kidneys. Transcapillary fluid shifts between the microcirculation and surrounding tissues can also account for some of the reduction in plasma volume, most likely upper body capillary filtration; for example, blood relocates into the facial tissues, creating a puffy facial appearance. Red cell mass also decreases, so total blood volume is reduced as well. The reduced blood volume serves astronauts

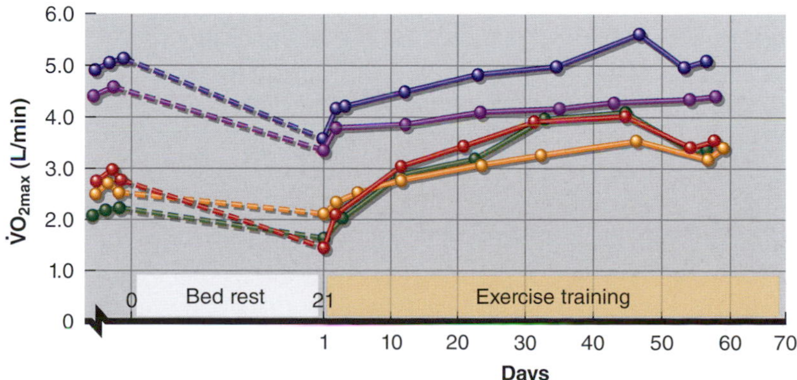

FIGURE 14.11 Changes in $\dot{V}O_{2max}$ with 20 days of bed rest for five individual subjects. Note that the subjects who were least fit (lowest $\dot{V}O_{2max}$ values) at the start of bed rest showed smaller decrements with inactivity and greater gains when they trained after bed rest. Highly fit individuals, on the other hand, were far more affected by the period of inactivity.

Adapted by permission from B. Saltin et al., "A Longitudinal Study of Adaptive Changes in Oxygen Transport and Body Composition," *Circulation* 38, no. VII (1968): 1-78.

RESEARCH PERSPECTIVE 14.3

Disturbed Sleep and Increased Illness in Overreached Athletes

In an attempt to improve the physiological adaptations to training and maximize sport performance, most athletes increase exercise intensity and volume at some point during training. However, this practice may also lead to functional overreaching and compromises short-term performance outcomes, especially when the recovery time is insufficient. Although the causes of functional overreaching remain incompletely understood, it is well known that adequate passive rest and sufficient sleep promote enhanced recovery and reduced fatigue. The first study to look at the link between disruptions in sleep and functional overreaching was conducted in 2014 to determine whether changes in objective sleep parameters were evident between a healthy group of athletes and a group of athletes developing functional overreaching following overload training.[10]

During the study, performance was assessed by measurement of $\dot{V}O_{2max}$ and sleep was monitored using actigraphy. As expected, there was a progressive decline in indices of sleep quality and small reductions in sleep quantity during the overload training period in athletes who developed functional overreaching. In these athletes, alterations in sleep metrics were reversed during the subsequent taper phase of training. In the control group, no athletes developed any symptoms of training intolerance. However, it remains unclear whether poor sleep was a consequence or a cause of functional overreaching; for these determination, additional studies must be conducted.

well while they remain in microgravity. But it presents a serious problem on their return to a 1 *g* environment, where the body is again subjected to the hydrostatic pressure effect. Astronauts have experienced postural (orthostatic) hypotension and fainting during their first few hours back in a normal 1 *g* environment because their blood volume was insufficient to meet all their circulatory needs.

Maximal aerobic power ($\dot{V}O_{2max}$) is generally reduced immediately postflight, likely due to the reduction in plasma volume and leg strength during flight. However, data are limited on directly measured $\dot{V}O_{2max}$ in astronauts preflight, during flight, and postflight. Head-down tilt (−6°) bed rest, used as an Earth-based model of spaceflight, shows consistent reductions in $\dot{V}O_{2max}$ that are associated with reductions in total blood volume, plasma volume, and, consequently, maximal stroke volume. The head-down tilt model has been shown to provide $\dot{V}O_{2max}$ (or $\dot{V}O_{2peak}$) data comparable to actual pre- to postflight changes.[22]

Importantly, having an understanding of the general decline in physiological function that occurs during spaceflight has led the scientific and medical community to the realization that in-flight exercise programs are essential to preserve the long-term health of astronauts. Research is now under way to design the most appropriate programs and exercise equipment to meet this objective.

In Review

- The body rapidly loses many of the benefits of training if training is discontinued. Some minimal level of training is necessary to prevent these losses. Research indicates that at least three training sessions per week at an intensity of at least 70% $\dot{V}O_{2max}$ are needed to maintain aerobic conditioning.
- Detraining is defined as the partial or complete loss of training-induced adaptations in response to either the cessation of training or a substantial decrement in the training load.
- Detraining causes muscle atrophy, which is accompanied by losses of muscular strength and endurance. However, muscles require only minimal stimulation to retain these qualities during periods of reduced activity.
- Muscular endurance decreases after only 2 weeks of inactivity. Possible explanations for this are
  1. decreased oxidative enzyme activity,
  2. decreased muscle glycogen storage, or
  3. disturbance of acid–base balance.
- Detraining-induced losses in speed and agility are small, but flexibility seems to be lost quickly.
- With detraining, losses of cardiorespiratory endurance are much greater than losses of muscular strength, power, and endurance over the same time period.
- To maintain cardiorespiratory endurance, training must be conducted at least three times per week, and training intensity should be at least 70% of $\dot{V}O_{2max}$.
- Prolonged exposure to microgravity in space causes physiological changes that are nearly identical to those of detraining. While these are generally maladaptations on Earth, in space they serve as positive physiological adaptations that lead to a better ability to tolerate microgravity.

IN CLOSING

In this chapter, we have examined how the quantity of training can affect performance. We saw that too much training, in the form of either excessive training or overtraining, can actually impair performance. Then we looked at the effects of too little training—detraining—as a result of either inactivity or immobilization after an injury. Finally, we saw that with detraining, many of the gains achieved during regular training are quickly lost, especially cardiovascular endurance.

Now that we have dispelled the myth that more training always means better performance, it makes sense to ask about other ways in which athletes can try to optimize their performance. In the next chapter, we turn our attention to optimal body composition and nutrition for the serious athlete.

KEY TERMS

acute overload
detraining
excessive training
immune function
overreaching
overtraining
overtraining syndrome
periodization
tapering
taper period
undertraining

STUDY QUESTIONS

1. Describe the model used to optimize training. Define the terms *undertraining*, *acute overload*, *overreaching*, and *overtraining*.
2. What is excessive training? How does it relate to the model for optimizing training?
3. Distinguish between traditional and block forms of training periodization. What are the advantages and disadvantages of each?
4. Define and describe the overtraining syndrome. What are the general symptoms of the overtraining syndrome? How do these differ between sympathetic and parasympathetic overtraining?
5. What distinguishes a committed exerciser from one who may be addicted to exercise?
6. How might the hypothalamus be involved in the overtraining syndrome? What role might cytokines play?
7. Describe the relationship between physical activity and immune function and disease susceptibility.
8. What appears to be the best predictor of the overtraining syndrome?
9. How do we treat the overtraining syndrome?
10. What physiological changes occur during the taper period that can be credited with improvements in performance?
11. What alterations occur in strength, power, and muscular endurance with physical detraining?
12. What alterations occur in speed, agility, and flexibility with physical detraining?
13. What changes occur in cardiovascular function as one becomes deconditioned?
14. What similarities do we see between spaceflight and detraining?

STUDY GUIDE ACTIVITIES

In addition to the activities listed in the chapter opening outline, two other activities are available in the web study guide, located at

www.HumanKinetics.com/PhysiologyOfSportAndExercise

The **KEY TERMS** activity reviews important terms, and the end-of-chapter **QUIZ** tests your understanding of the material covered in the chapter.

Body Composition and Nutrition for Sport

15

In this chapter and in the web study guide

Defensive linemen in the National Football League are large men, especially the defensive tackles who play the so-called 0-technique position in the middle of the line. By nature of their role, these behemoths must be hard to push off the line of scrimmage, yet agile enough to fill running lanes and rush the opposing quarterback. Dontari Poe is one of the best at his position and was rewarded by signing a 1-year $8 million contract with the Atlanta Falcons in 2017. However, there was a catch. Five hundred thousand dollars of that amount was tied to a weight clause that could be earned only if he reportedly weighed in at 149.7 kg (330 lb), a weight that may optimize his athletic performance. During the 2016 season, Poe was listed by the Chiefs as weighing 157.3 kg (346 lb), so he had some work to do to reach his goal. Similarly, a weight clause was included in his $27 million deal in 2018 with the Carolina Panthers, which included quarterly bonuses for meeting successful weight targets. Attesting to his overall athleticism, although he is a defensive player, Poe holds the records for being the heaviest NFL player to run for a touchdown and the heaviest player to throw a pass for a touchdown! Many athletes compete in sports for which management of weight—and body composition—is critical, including wrestling, boxing, gymnastics, and distance running. Yet extreme changes may be detrimental to an athlete's performance and overall health. Achieving optimal body weight and composition is a critical component of successful sport performance.

Coaches and athletes today are acutely aware of the importance of achieving and maintaining optimal body weight for peak performance in sport. Appropriate size, physique, and body composition are critical to success in almost all athletic endeavors. Compare the specific performance requirements of the 152 cm, 45 kg (5 ft, 100 lb) Olympic gymnast and those of the 206 cm, 147 kg (6 ft 9 in., 325 lb) defensive lineman in professional football. Although size and body build can be altered only slightly, body composition can change substantially with dieting and exercise. Resistance training can substantially increase muscle mass, and sound dieting combined with vigorous exercise can significantly decrease body fat. Such changes can be of major importance in achieving optimal athletic performance.

Peak performance also requires a careful dietary balance of the essential nutrients. The U.S. government has established standards for optimal nutrient intake that are termed Dietary Reference Intakes (DRIs). The DRIs provide estimates of the range of intakes of various food substances needed to maintain good health. However, these guidelines were not written for extremely active people or athletes, and the nutrition needs of athletes can exceed the DRIs considerably. Individual caloric needs are extremely variable, depending on the athlete's size, sex, training volume, and sport. Cyclists competing in the Tour de France and Norwegian cross-country skiers during training have been reported to expend up to 9,000 kcal per day. One ultra-endurance runner expended an average of 10,750 kcal per day over a 5.2-day 600 mi (966 km) race![29] Also, some competitive sports require adherence to rigid weight standards. Athletes who participate in these sports must closely monitor their weight and thus their caloric intake. Too often, this leads to poor nutrition, reliance on supplements and drugs, dehydration, and serious health risks. In addition, the dietary tactics used by some athletes to achieve excessive weight loss are of increasing concern because of the potential association with eating disorders, such as anorexia nervosa and bulimia nervosa.

Assessing Body Composition

Body composition refers to the body's tissue composition. Figure 15.1 illustrates three models of body composition. The first two divide the body into its various chemical or anatomical components; the last one simplifies body composition into two components, the fat mass and the fat-free mass, which is the model used in this book. **Fat mass** is often discussed in terms of **relative body fat**, which is the percentage of the total body mass that is composed of fat. **Fat-free mass** simply refers to all body tissue that is not fat, including bone, muscle, organs, and connective tissue.

Assessment of body composition provides additional information to both the coach and the athlete, beyond the basic measures of height and weight. As an example, if the center fielder of a Major League Baseball team is 188 cm (6 ft 2 in.) tall and weighs 91 kg (200 lb), is he at his ideal playing weight? Knowing that 4.5 kg (10 lb) of a total weight of 91 kg (200 lb) is fat weight and that the remaining 86.5 kg (190 lb) is fat-free weight provides considerably more insight than knowing height and weight alone. In this example, only 5% of his body weight is fat, which is about as low as any athlete should go. Armed with this knowledge, both athlete and coach realize that this athlete's body composition is ideal. They should not be concerned with weight loss, even though standard height–weight charts indicate that the athlete is overweight. However, another baseball player of the same height and weight who has 23 kg (50 lb) of fat would be 25% fat. This would constitute a problem for an elite athlete: He would be overfat.

In most sports, the higher the percentage of body fat, the poorer the performance. An accurate assessment of the athlete's body composition provides valuable insight

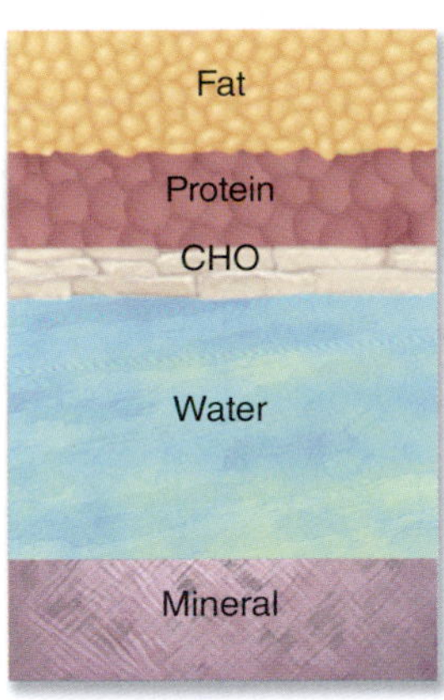

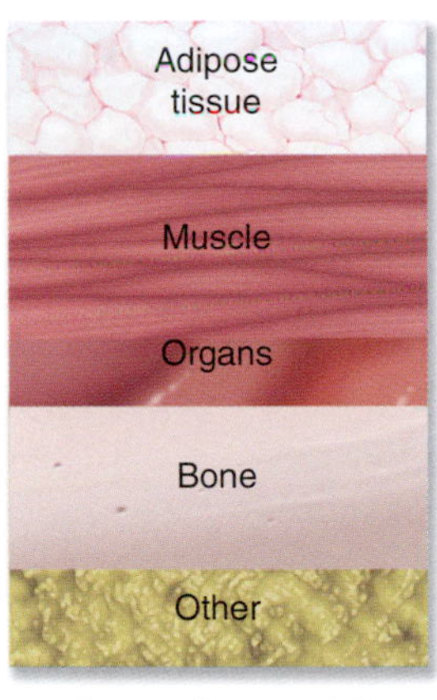

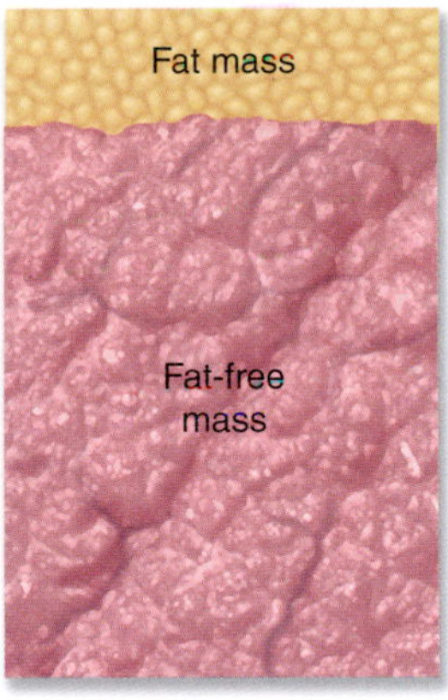

FIGURE 15.1 Three models of body composition (see text for description).

Adapted by permission from J.H. Wilmore, Body Weight and Body Composition, in *Eating, Body Weight, and Performance in Athletes: Disorders of Modern Society*, edited by K.D. Brownell, J. Rodin, and J.H. Wilmore (Philadelphia, PA: Lippincott, Williams, and Wilkins, 1992), 77-93.

into the weight that allows optimal performance. But how do we determine an athlete's body composition?

Densitometry

Densitometry involves measuring the density of the athlete's body in order to assess body composition. Density (D) is defined as mass (M) divided by volume (V):

$$D_{body} = M_{body} / V_{body}$$

The mass of the body is the athlete's scale weight. Body volume can be obtained by several different techniques, but the most common is **hydrostatic weighing** (also called *underwater weighing*), in which the athlete is weighed while totally immersed in water. The difference between the athlete's scale weight and underwater weight, when corrected for the density of water, equals the body's volume. This volume must be further corrected to account for air trapped in the body. The amount of air trapped in the gastrointestinal tract is small, difficult to measure, and usually ignored. The gas trapped in the lungs, however, must be measured because its volume is generally large.

Figure 15.2 shows the hydrostatic weighing technique. The density of the fat-free mass is higher than the density of water, while that of fat is lower than the density of water.

Densitometry has long been the technique of choice for assessing body composition. New techniques typically are often compared against densitometry to determine their accuracy. If body weight, underwater weight, and lung volume during underwater weighing are measured correctly, the resulting **body density** value is accurate. However, densitometry has its limitations. Densitometry's major weakness is in the conversion of body density to an estimate of relative body fat. Accurate estimates of the individual densities of fat mass and fat-free mass are required when the two-component model of body composition is used. The equation most often used to convert body density to an estimate of relative or percentage body fat is the standard equation of Siri:

$$\% \text{ Body fat} = (495 / D_{body}) - 450.$$

FIGURE 15.2 A simple hydrostatic weighing system (underwater weighing technique) to determine the density of the body.

This equation assumes that the densities of the fat mass and the fat-free mass are relatively constant in all people. Indeed, the density of fat at different sites is very consistent in the same individual and relatively consistent between people. The value generally used is 0.9007 g/cm. But determining the density of the fat-free mass (D_{FFM}), which the equation of Siri assumes is 1.100, is more problematic. To determine this density, we make assumptions that (1) the density of each tissue constituting the fat-free mass is known and remains constant and (2) each tissue type represents a constant proportion of the fat-free mass (e.g., we assume that bone always represents 17% of the fat-free mass). Exceptions to either of these assumptions cause error when we convert body density to relative body fat.

Other Laboratory Techniques

Many other laboratory techniques are available for assessing body composition. These include radiography, computed tomography (CT), magnetic resonance imaging (MRI), hydrometry (for measuring total body water), total body electrical conductivity, and neutron activation. Most of these techniques are complex and require expensive equipment. None of them is likely to be used for the general assessment of athletic populations, so we will not discuss them further in this chapter; however, we do discuss CT in chapter 22. Two other techniques hold considerable promise—dual-energy X-ray absorptiometry and air displacement.

Dual-energy X-ray absorptiometry (DEXA) evolved from the earlier single- and dual-photon absorptiometry techniques used between 1963 and 1984. The earlier techniques were used to estimate regional bone mineral content and bone mineral density, primarily in the spine, pelvis, and femur. The newer DEXA technique (see figure 15.3) allows the quantification of not only bone but also soft tissue composition. Furthermore, it is not limited to regional estimates but can provide total body estimates. Research to date suggests that DEXA provides precise and reliable estimates of body composition. DEXA is becoming more commonly accepted as the gold standard technique. The advantages of DEXA over the underwater weighing technique include the ability to estimate bone density and bone mineral content in addition to fat and fat-free mass. Furthermore, it is a passive technique whereby the athlete simply lies on a table during the scan, as opposed to having to be submerged underwater multiple times. The disadvantage is the cost of the equipment and technical support.

Air plethysmography, similar to hydrostatic weighing, is a densitometric technique. Volume is determined by air displacement rather than by water immersion. Commercial models of air plethysmographs have made this technique popular and widely available (see figure 15.4). The principle of operation is rather simple. It involves a closed chamber of room air at atmospheric pressure, which has a known volume. The individual to be tested opens the chamber door, is seated in the chamber, then closes the chamber door, forming an airtight seal. The new volume of the air in the chamber is determined, which is then subtracted from the total volume of the chamber to provide an estimate of the person's volume.

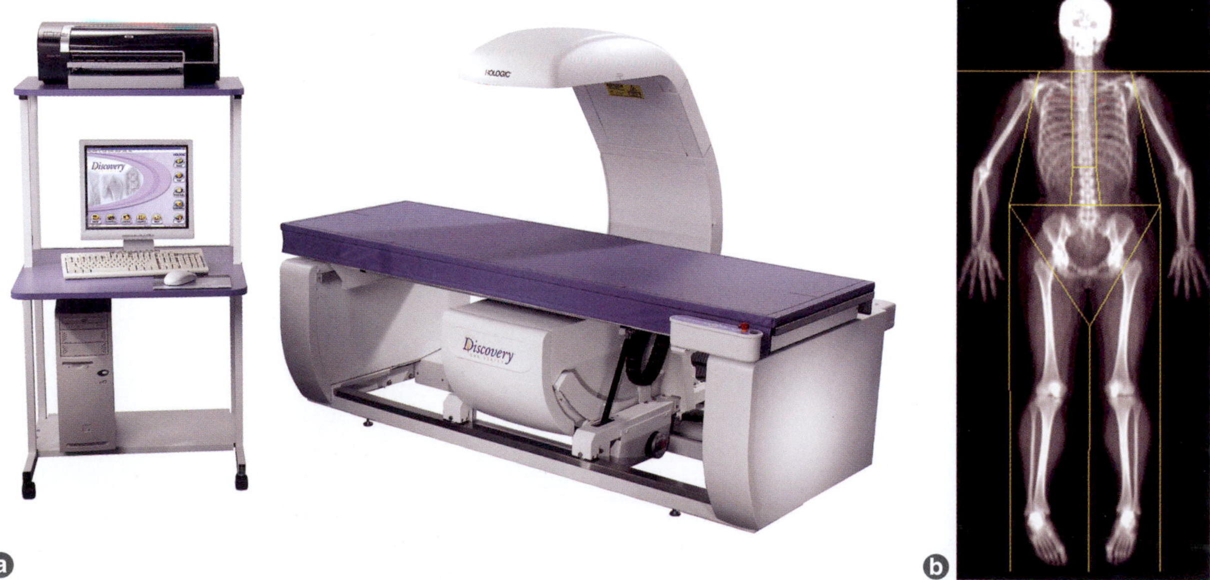

FIGURE 15.3 *(a)* The dual-energy X-ray absorptiometry (DEXA) machine used to estimate bone density and bone mineral content as well as total body composition (fat mass and fat-free mass). *(b)* A regional scan of the body.

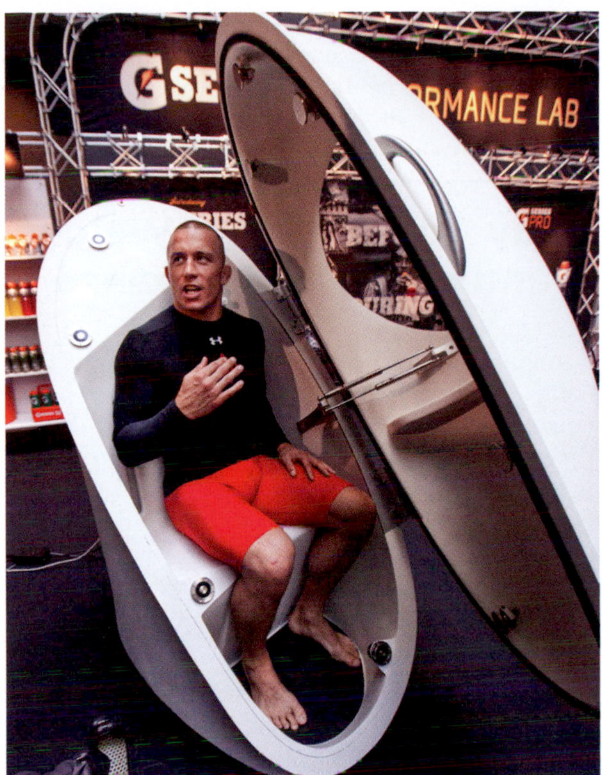

FIGURE 15.4 The Bod Pod air plethysmography device uses the air displacement technique to estimate total body volume.

Although this is a relatively simple technique for the subject, it requires considerable accuracy in controlling for changes in temperature, gas composition, and the subject's breathing while in the chamber. Studies have confirmed the accuracy of this technique under most conditions. It appears to provide a relatively precise measurement of body volume. Just as with the underwater weighing technique, one must still use the subject's body density in an equation to estimate relative body fat, recognizing the uncertainties of the D_{FFM} for that subject.

Field Techniques

Several field techniques are also available for assessing body composition. These techniques are more accessible than laboratory techniques because the equipment is less costly and cumbersome; they are techniques that therefore can be used more easily outside the laboratory by the coach, the trainer, or even the athlete.

Skinfold Fat Thickness

The most widely applied field technique involves measuring the **skinfold fat thickness** (see figure 15.5) at one or more sites and using the values obtained to estimate body composition. It generally is recommended that

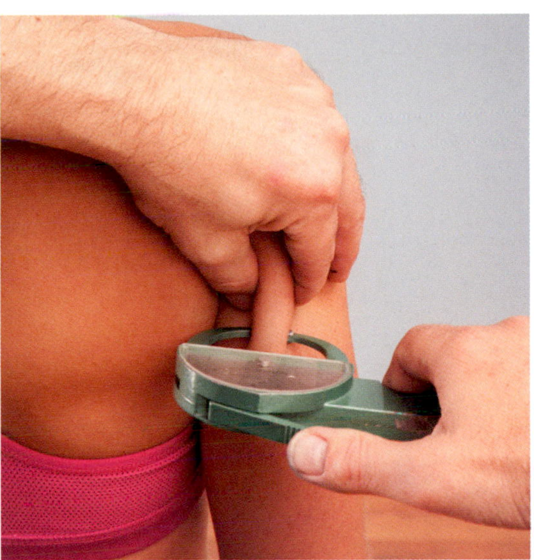

FIGURE 15.5 Measuring skinfold fat thickness at the triceps skinfold site.

the sum of the measurements from three or more skinfold sites be used in a quadratic, curvilinear equation to estimate body density. A curvilinear equation more accurately describes the relationship between the sum of skinfold measurements and body density. Linear equations underestimate the density of lean people and overestimate the density of obese people. Skinfold fat thickness measurements that use quadratic equations provide reasonably accurate estimates of total body fat or relative fat.

Bioelectric Impedance

Bioelectric impedance is a simple procedure introduced in the 1980s that takes just a few minutes to perform. Four electrodes are attached to the body—at the ankle, the foot, the wrist, and the back of the hand. Newer technology utilizes foot plate and handheld electrodes—as shown in figure 15.6. An undetectable current is passed through the electrodes on one side of the body (hand and foot), while the other side receives the current flow. Electrical conduction through the tissues between the electrodes depends on the water and electrolyte distribution in that tissue. Fat-free mass contains almost all the body water and the conducting electrolytes, so conductivity is much greater in the fat-free mass than in the fat mass. The fat mass has a much greater impedance, meaning that it is much more difficult for the current to flow through the fat mass. Thus, the amount of current flow through the tissues reflects the relative amount of fat contained in those tissues.

With the bioelectric impedance technique, measurements of the impedance, the conductivity, or both are transformed into estimates of relative body

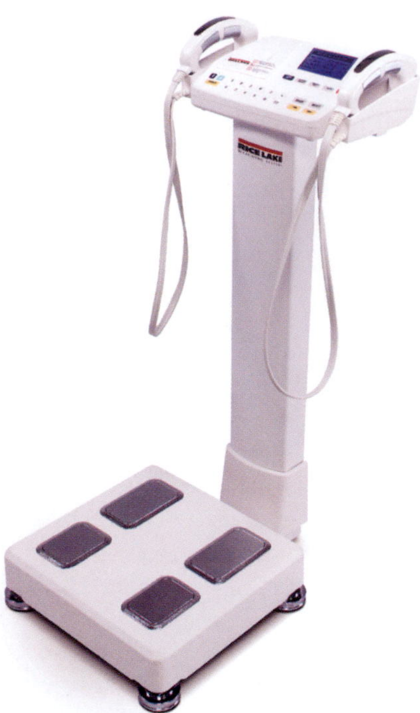

FIGURE 15.6 A commercially available bioelectric impedance device marketed for assessing relative body fat.

fat. Estimates of relative body fat based on bioelectric impedance highly correlate with body fat measurements obtained through hydrostatic weighing. However, the relative body fat in lean athletic populations tends to be overestimated with bioelectric impedance because of the nature of the equations used. Furthermore, hydration alters the bioimpedance and therefore must be tightly controlled.

In Review

- Knowing a person's body composition is more valuable for predicting performance potential than merely knowing height and weight.
- Densitometry is one of the best methods for assessing body composition and has long been considered the most accurate, although it does carry certain risks of error. It involves calculating the density of the athlete's body by dividing body mass by body volume, which is typically determined via hydrostatic weighing or air displacement. Body composition can be calculated, although there is some margin of error.
- Dual-energy X-ray absorptiometry, originally developed for estimating bone density and bone mineral content, is now capable of providing accurate estimates of not only total body composition—fat mass and fat-free mass—but also segmental body composition and bone mass.
- Field techniques for assessing body composition include measuring skinfold fat thickness and bioelectric impedance. These techniques are less costly and more accessible to the athlete and the coach than are laboratory techniques.

Body Composition, Weight, and Sport Performance

In certain sports (e.g., football and basketball), a large size traditionally has been associated with performance quality: The bigger the athlete, the better the performance. But big does not always mean better. In certain other sports, smaller and lighter are considered better for performance (e.g., gymnastics, figure skating, and diving). Yet this can be taken to extremes, compromising the athlete's health and performance. In the following sections, we consider how performance can be affected by weight and body composition.

Fat-Free Mass and Relative Body Fat

Rather than be concerned with total body size or weight, most athletes should be concerned specifically with fat-free mass. Maximizing fat-free mass is desirable for athletes involved in activities that require strength, power, and muscular endurance. But increased fat-free mass may be undesirable for the endurance athlete, such as a distance runner, for whom a higher fat-free mass is an additional load that must be carried and might impair the athlete's performance. This might also be true for the high jumper, long jumper, triple jumper, and pole-vaulter, who must maximize their vertical or horizontal distances or both. Additional weight, even though it is active fat-free mass, could decrease rather than facilitate performance in these events.

Relative body fat is a major concern of athletes. Adding more fat to the body just to increase the athlete's weight and overall size is generally detrimental to performance. This is true of all activities in which the body weight must be moved through space, such as running and jumping. It is less important for more stationary activities, such as archery and shooting. Endurance athletes try to minimize their fat stores because excess weight has been proven to impair their performance.

Heavyweight weightlifters might be exceptions to the general rule that less fat is better. These athletes add large amounts of fat weight just before competition under the premise that the additional weight will lower their center of gravity and give them a greater mechanical advantage in lifting. The sumo wrestler is another notable exception to the theory that overall size is not the major determinant of athletic success. In this sport, the larger individual has a decided advantage; but even

RESEARCH PERSPECTIVE 15.1

Exercise Type and Body Composition

Despite the well-established benefits of exercise on health and chronic disease, the precise role of exercise in weight loss is still controversial, likely resulting from the differences in outcome measures in the variety of studies that test whether regular exercise is associated with decreased body weight. Also, changes in outcome measures such as body weight or body mass index (BMI) may not reflect changes in body composition, which may be more likely to occur in response to adherence to prolonged exercise training.

Exercise interventions for weight loss or weight maintenance have often focused on aerobic exercise because of the higher energy expenditure as compared to resistance exercise. However, resistance exercise is associated with increased generalized functional capacity, which may affect total daily energy expenditure by increasing total physical activity. Furthermore, many exercise intervention trials have found that participants in an exercise program reduce their daily nonexercise activity or increase energy intake, both of which may minimize the potential weight reduction. Examining the effects of habitual exercise, rather than a prescribed intervention, may eliminate these compensatory metabolic and behavioral changes and enrich our scientific understanding of the long-term effects of exercise on body weight and composition.

In 2015, a team of researchers examined the effects of various self-selected exercise programs on measures of body composition.[15] Researchers followed 430 young adults who were participants in the *Energy Balance Study*, a long-term observational study that examines determinants of weight change over the span of 1 year. All of the participants were healthy and had BMIs between 20 and 35 kg/m^2. At the start of the study, height, weight, fat mass, fat-free mass, and lean tissue mass were measured on all of the participants. Percent body fat was calculated and participants were classified into three groups: normal fat (%BF = 22±7), overfat (%BF = 30±8), and obese (%BF = 35±7). Every 3 months, the participants reported their habitual self-selected engagement (frequency and time) in different exercise types. Based on these reports, time spent engaging in endurance exercise, resistance exercise, and other exercise (e.g., sports, group fitness, or aerobics) was calculated for each participant. At the end of the study, 348 of the 430 original participants had provided sufficient data to be included in the 1-year analysis.

The authors found that neither total exercise nor specific exercise type had an impact on BMI. Resistance exercise increased lean body mass and decreased fat mass, while aerobic exercise decreased fat mass but had no effect on lean mass. All exercise types positively affected lean body mass in the normal-fat participants. Interestingly, in overfat and obese participants, fat mass was reduced by increasing resistance exercise but was not changed with increasing aerobic exercise. The research team concluded that despite the limited effects on BMI, exercise was associated with beneficial changes in body composition such as increased lean mass in normal-fat participants and reduced fat mass in overfat and obese participants. Additionally, adults with excess body fat may benefit more from resistance exercise training, contrary to the commonly held belief that they should focus on aerobic exercise training.

so, the wrestler with the higher fat-free mass should have the best overall success. Performance in swimming also seems to be an exception to this general rule. Body fat might provide some advantage to the swimmer because it improves buoyancy, which can reduce body drag in the water and reduce the metabolic cost of staying on the surface.

Weight Standards

Weight standards are now used in several sports to ensure that athletes have the optimal body size and composition for maximal performance. Unfortunately, this is not always the result.

Theoretically, the elite athlete's genetic foundation and years of intense training have combined to provide the ultimate athletic profile for the given sport. These elite athletes set the standards toward which others aspire. However, this can be misleading, as we see in figure 15.7.[36] The figure displays the percentage body fat values of elite female track-and-field athletes. If we look just at the distance runners, many of the best were below 12% body fat and the top two had only about 6% fat. However, one of the best distance runners in the United States at that time had 17% body fat. Furthermore, one of the women in this study had a relative body fat of 37%, and she set the best time in the world for the 50 mi (80 km) run within 6 months of her evaluation! Neither of these women would have been successful if she had been forced to decrease her weight to achieve 12% body fat or lower.

Furthermore, weight standards have been abused over the years by coaches, athletes, and parents who have adopted the philosophy that if small weight losses improve performance a little, then major weight losses should improve it even more.

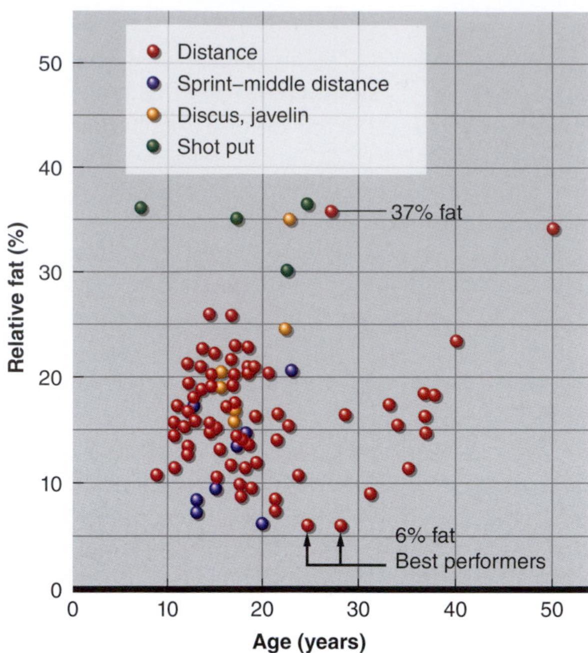

FIGURE 15.7 Relative body fat in elite female track-and-field athletes (see text for explanation).

Data from Wilmore, Brown, and Davis (1977).

Risks Associated with Severe Weight Loss

Many schools, districts, or state- and national-level organizations organize sports (e.g., wrestling) on the basis of size, with weight as the predominant factor. Athletes in these sports often attempt to achieve the lowest possible weight to gain an advantage over opponents. In so doing, many have jeopardized their health. In the following sections, we examine a few of the consequences of severe weight loss in athletes, both male and female.

Dehydration Fasting or very low-calorie diets lead to large amounts of weight loss, primarily through dehydration. As we discuss later in this chapter, for every gram of carbohydrate stored, there is an obligatory gain of 2.6 g of water. When carbohydrate is used for energy, that water is lost. With fasting and very low-calorie diets, carbohydrate stores are substantially depleted during the first few days. This results in a significant loss of weight attributable to the loss of body water.

Additionally, athletes trying to make weight might exercise in rubberized sweat suits, sit in steam and sauna baths, chew on towels to lose saliva, and minimize their fluid intake. Such severe water losses compromise kidney and cardiovascular function and are potentially dangerous. Losses of as little as 2% of the athlete's weight through dehydration can impair performance, even skilled performance in intermittent, high-intensity sports like tennis, soccer, and basketball.[17]

Fatigue Pushing body weight too low can have major repercussions. When weight drops below a certain level, the athlete is likely to experience performance decrements and increased incidence of illness and injury. The performance decrements can be attributable to many factors, including the prolonged feelings of fatigue that often accompany excessive weight loss. The causes of this fatigue have not been established, but there are several likely possibilities. This type of fatigue could be attributed to substrate depletion. Both neural and hormonal components seem to be involved. In some cases, the balance between the sympathetic nervous system and the parasympathetic system may be altered in favor of the latter. In addition, effects of severe weight loss on the hypothalamus can impair immune function.

The Female Athlete Triad The constant attention given to achieving and maintaining a prescribed weight goal, particularly if the weight goal is inappropriate, can lead to energy imbalances, especially in female athletes. The precipitating cause of these disturbances is the failure of the athlete to consume an adequate number of calories to meet the energy expenditure needs associated with exercise training.[1] As a result, the athlete is in a state of energy deficiency or low energy availability. A female athlete who experiences energy imbalance for prolonged periods (low energy availability compared to high energy expenditure associated with training, for example) is prone to disordered eating, menstrual dysfunction, and bone mineral loss.[1] Together, these disorders are referred to as the female athlete triad. These important problems are discussed in more detail in chapter 19.

Establishing Appropriate Weight Standards

The potential for abuse of weight standards is clearly established. Thus, it is critically important to properly set weight standards. Body weight standards for sports should be based on an athlete's body composition and not purely the body mass. Thus, establishing weight standards should translate into establishing standards of relative body fat for each sport and, where appropriate, for each event within a sport. With this in mind, what is the recommended relative body fat for an elite athlete in any given sport? For each sport, an optimal range of values for relative body fat should be established, outside of which the athlete's performance is likely impaired. And because fat distribution shows definite sex differences, the weight standards should be sex specific. Representative ranges for men and women in various sports are presented in table 15.1. In most cases, these values represent the elite athletes in those sports.

TABLE 15.1 **Ranges of Relative Body Fat Values for Male and Female Athletes in Various Sports**

	% fat	
Group or sport	Men	Women
Baseball or softball	8-14	12-18
Basketball	6-12	10-16
Bodybuilding	5-8	6-12
Canoeing or kayaking	6-12	10-16
Cycling	5-11	8-15
Fencing	8-12	10-16
Football	5-25	
Golf	10-16	12-20
Gymnastics	5-12	8-16
Horse racing (jockey)	6-12	10-16
Ice or field hockey	8-16	12-18
Orienteering	5-12	8-16
Pentathlon	*	8-15
Racquetball	6-14	10-18
Rowing	6-14	8-16
Rugby	6-16	*
Skating	5-12	8-16
Skiing (alpine and Nordic)	7-15	10-18
Ski jumping	7-15	10-18
Soccer	6-14	10-18
Swimming	6-12	10-18
Synchronized swimming		10-18
Tennis	6-14	10-20
Track and field, field events	8-18	12-20
Track and field, running events	5-12	8-15
Triathlon	5-12	8-15
Volleyball	7-15	10-18
Weightlifting	5-12	10-18
Wrestling	5-16	

*Data not available.

The recommended values may not be appropriate for all athletes who engage in a specific activity. The existing techniques for measuring body composition include inherent errors, as we discussed earlier in this chapter. Even with the best laboratory techniques, measurement of body composition is associated with a 1% to 3% error or higher. In addition, we must understand the concept of individual variability. Not every female distance runner will have her best performance at 12% body fat or lower. While some will improve performance with these low values, others won't be able to get down to such low relative fat values, or they will find that their performance starts to decline before they reach the suggested values. For these reasons, a range of values should be set for men and women in specific activities, recognizing individual variability, methodological error, and sex differences.

Achieving Optimal Weight

Many athletes find themselves considerably above their assigned playing weight with only a few weeks remaining before they report to training camp. Consider the NFL lineman introduced at the beginning of this chapter. He had to weigh 149.7 kg (330 lb) to achieve peak performance and receive a financial incentive but weighed 157.3 kg (346 lb). If that scenario took place 1 month before the competitive season started, he would have some work to do to reach his goal. Exercise alone is of little value because he would need 6 to 8 months or more to lose this much weight through that means. Will this athlete accomplish his goal?

VIDEO 15.1 Presents Louise Burke on strategies for in-season diet changes for athletes who need to lose or gain weight.

Avoiding Fasting and Crash Dieting

Consider a wrestler who decides that he wants to wrestle at a much lower weight class in order to be more competitive. He wants to lose 3 kg (6.6 lb) per week for the next 4 weeks, so he decides to embark on a crash diet, selecting whatever diet is in vogue, knowing that a person can lose weight quickly with such diets. But much of this weight loss would be from body water and very little from stored fat. Several studies have shown that substantial weight losses occur with very low-calorie diets (of 500 kcal per day or less) over the first several weeks, but that of the weight lost, more than 60% comes from the body's fat-free tissue and less than 40% from fat depots.

While much of the player's weight is lost from water stores, a substantial amount of protein is lost as well. Also, most crash diets are based on a major reduction in carbohydrate intake. This reduced intake is insufficient to supply the body's needs for carbohydrate, and as a result the body's carbohydrate stores become depleted. Because water storage accompanies carbohydrate storage, the water stores are also reduced as the carbohydrate stores diminish, as discussed earlier in this chapter.

In addition, the body relies more heavily on free fatty acids for energy because its carbohydrate stores are depleted. As a result, ketone bodies, a by-product of fatty acid metabolism, accumulate in the blood, causing a condition known as ketosis. This condition further increases water loss, especially during the first week of the diet. Athletes who have attempted this ill-advised shortcut to rapid weight loss have lost substantial weight, but, because much of the weight loss was from the fat-free mass, their performance was severely compromised.

Optimal Weight Loss: Decreasing Fat Mass and Increasing Fat-Free Mass

The sensible approach to reducing body fat stores is to combine moderate dietary restriction with increased physical activity. Athletes should work toward achieving their goal weight slowly, losing no more than 1 kg (less than 2.2 lb) per week. Losing more weight than that per week leads to losses in fat-free mass. The rate of this loss should be reduced still more, or the weight loss program terminated, if performance is affected or if medical symptoms are noted.

Decreasing caloric intake by 200 to 500 kcal per day will allow weight losses of about 0.5 kg (1.1 lb) per week, particularly if combined with a sound exercise program. This is a realistic goal, and such losses add up to a substantial weight loss over time. When trying to reduce weight, athletes should consume their total daily calories over at least three meals per day. Many athletes make the mistake of eating only one or two meals per day, skipping breakfast, lunch, or both, and then consuming a large dinner. Research in animals has shown that, given the same number of total calories, the animals that eat their daily food ration in one or two meals gain more weight than those that nibble their ration throughout the day. Human research is less clear.

The purpose of weight loss programs is to lose body fat, not fat-free mass. Therefore, the combination of diet and exercise is the preferred approach. Combining increased activity with caloric reduction prevents any significant loss of fat-free mass. Resistance training promotes gains in fat-free mass, and both resistance training and endurance training promote loss of fat mass. To lose weight, athletes should combine a moderate resistance and endurance training program with modest caloric restriction.

In Review

- Maximizing fat-free mass is desirable for athletes in sports that require strength, power, and muscular endurance but could be a hindrance to endurance athletes, who must be able to move their total body mass for extended periods, and jumpers, who must move their body mass vertically or horizontally for distance.
- The degree of fatness has more influence on performance than does total body weight. In general, the greater the relative body fat, the poorer the performance. Possible exceptions include heavyweight weightlifters, sumo wrestlers, and swimmers.
- Severe weight loss in athletes can cause potential health problems, such as dehydration, chronic fatigue, disordered eating, menstrual dysfunction, and bone mineral disorders.
- Body weight standards should be based on body composition and should emphasize relative body fat rather than total body mass.
- For each sport, a range of values should be established, recognizing the importance of individual variation, methodological error, and sex differences.
- When severe (very low calorie) diets are followed, much of the weight loss that occurs is from water, not fat.
- Most severe diets limit carbohydrate intake, depleting carbohydrate stores. Water is lost along with the carbohydrate, exacerbating the problem of dehydration. Also, the increased reliance on free fatty acids can lead to ketosis, which further increases water loss.
- The combination of diet and exercise is the preferred approach to optimal weight loss.
- Athletes should lose no more than about 1 kg (about 2 lb) per week until reaching the upper end of the desired weight range. After that, weight loss should be less than 0.5 kg (1 lb) per week until goal weight is reached. More rapid weight losses result in loss of fat-free mass. People can accomplish weight loss at the recommended rate by reducing dietary intake 200 to 500 kcal per day, especially in combination with a sound exercise program.
- For fat loss, moderate resistance and endurance training is most effective. Resistance training promotes gains in fat-free mass.

Classification of Nutrients

We now turn our attention to nutrition aspects of preparing the athlete for optimal performance. As we will see, it is important to maintain a diet that will provide general health benefits, maintain an appropriate weight and body composition, and maximize athletic performance.

A person's diet should contain a relative balance of carbohydrate, fat, and protein. Of the total calories consumed, the recommended balance for most people is

- carbohydrate—55% to 60%;
- fat—no more than 35% (less than 10% saturated); and
- protein—10% to 15%.

Interestingly, this recommended percentage distribution of total calories consumed appears to be optimal

RESEARCH PERSPECTIVE 15.2

Meal Timing and the Aerobic Exercise Window

In 1982, a study published in the *American Journal of Physiology* demonstrated that moderate-intensity physical activity (treadmill walking) that began 30 min after a meal rapidly returned blood glucose levels to fasting levels.[24] Over the decades that have passed since that finding, research has shown that blood glucose concentrations are sensitive to multiple exercise conditions, including timing, intensity, duration, frequency, and sequencing of exercise modes. A large number of studies have examined exercise and meal timings; however, a wide variety in exercise intensities, mode, and meal timings have led to conflicting reports. For example, light to moderate exercise before a meal raises postprandial glucose, but similar activity after a meal lowers blood glucose. However, changing to a high-intensity interval exercise bout gives the opposite result, with premeal exercise improving the glycemic response. Despite these conflicting results, the consensus is that postmeal exercise of moderate intensity is superior to premeal exercise if your goal is to limit hyperglycemia. Moderate exercise during the mid-postprandial period consistently lowers the postmeal spike in blood glucose, and exercise starting 30 min after the first bite into the meal shows the most effective blunting of the glucose surge. These data are especially important in meal timing and exercise plans for individuals with diabetes, who are challenged to manage glycemia closely.[10]

Building on these data surrounding meal consumption and exercise timing, a 2015 study examined the role of postexercise protein consumption on skeletal muscle protein synthesis rate and recovery after aerobic exercise in 12 aerobically trained men.[28] On three separate trials, each subject completed high-intensity cycling then ingested either (1) a protein–carbohydrate blend with 5 g of leucine, (2) a protein–carbohydrate blend with 15 g of leucine, or (3) a non-nitrogenous, isocaloric control beverage. The leucine doses were chosen based on previous studies showing that 15 g of leucine improved recovery and increased muscle protein synthesis after resistance exercise, as well as to see if aerobic athletes could confer similar benefits with lower doses. The research team collected muscle biopsies to determine muscle protein fractional synthetic rate during recovery. They also collected blood samples to measure plasma amino acids throughout recovery.

The researchers found that the beverages containing protein increased muscle protein fractional synthetic rate compared to the control. The addition of 5 mg leucine increased muscle protein fractional synthetic rate by 33% and represented near-maximum stimulation for protein synthesis. Further increasing leucine did not increase muscle protein synthesis above what was observed with the lower dose beverage. Plasma amino acids and leucine were decreased during recovery following the control beverage, but increased in a dose-dependent manner with ingestion of the protein–leucine beverages. The study concluded that ingestion of a protein-containing meal improves muscle synthesis and recovery following a bout of aerobic exercise; however, the addition of 5 mg of leucine is sufficient to reach maximal fraction protein synthesis. Although additional supplementation after that does increase plasma amino acids, it does not result in additional muscle synthesis. To address this, future studies should attempt to determine if these customized protein–leucine recovery beverages also translate to improvements in performance.

for both athletic performance and health. However, when planning diets for athletes, using grams per kilogram of body weight for carbohydrate and protein results in a more precise prescription for training and competition needs.[27] A similar distribution of caloric intake is recommended for the prevention of cardiovascular disease, diabetes, obesity, and cancer, as discussed in greater detail later in this chapter. Although all foods ultimately can be broken down to carbohydrate, fat, or protein, these nutrients are not all that the body needs. Food can be categorized into six classes of nutrients, each with specific functions in the body:

- Carbohydrate
- Fat
- Protein
- Vitamins
- Minerals
- Water

The following discussion examines the physiological importance to the athlete of each class of nutrients.

Carbohydrate

A carbohydrate (CHO) is classified as either a monosaccharide, a disaccharide, or a polysaccharide. Monosaccharides are the simple one-unit sugars, such as glucose, fructose, and galactose, that cannot be reduced to a simpler compound. Disaccharides (such as sucrose, maltose, and lactose) are made up of two monosaccharides. For example, sucrose (table sugar) consists of glucose and fructose. Oligosaccharides are short chains of 3 to 10 monosaccharides linked together. Polysaccharides are composed of large chains of linked monosaccharides. Glycogen is the energy-providing polysaccharide found in animals, including man, and is stored in muscle and liver. Starch and fiber are the two plant polysaccharides and are commonly referred to as complex carbohydrates. Simple carbohydrates are carbohydrates derived from processed foods or foods high in sugar. All carbohydrates must be broken down to monosaccharides before the body can use them.

Carbohydrate serves many functions in the body:

- It is a major energy source, particularly during high-intensity exercise.
- Its presence regulates fat and protein metabolism.
- The nervous system relies exclusively on carbohydrate for energy.
- Muscle and liver glycogen are synthesized from ingested carbohydrates.

Major sources of carbohydrate in the diet include grains, fruits, vegetables, milk, and concentrated sweets. Refined sugar, syrup, and cornstarch are nearly pure carbohydrates, as are candy, honey, jellies, and molasses. Some soft drinks contain few if any nutrients other than carbohydrates.

Carbohydrate Consumption and Glycogen Storage

The body stores excess carbohydrate, primarily in the muscles and liver, as glycogen. Because of this, carbohydrate consumption directly influences muscle glycogen storage and the ability to train and compete in endurance events. As shown in figure 15.8, athletes who trained intensely over 3 consecutive days and ate a low-carbohydrate diet (40% of total calories) experienced a progressive decrease in muscle glycogen.[12] When these same athletes consumed a high-carbohydrate diet (70% of total calories), their muscle glycogen levels recovered almost completely within the 22 h between training bouts.

Early studies demonstrated that when men ate a diet containing a normal amount of carbohydrate (about 55% of total calories ingested), their muscles stored about 100 mmol of glycogen per kg of muscle. One study showed that diets containing less than 15% carbohydrate led to storage of only 53 mmol/kg, but carbohydrate-rich diets (60%-70% CHO) led to storage of 205 mmol/kg. When subjects exercised to exhaustion at 75% of their maximal oxygen uptake, their exercise times were proportional to the amount of muscle glycogen stored before the test, as shown in figure 15.9.

Most studies have shown that glycogen storage replacement is not determined simply by carbohydrate intake. Exercise with an eccentric (muscle lengthening) component, such as running and weightlifting, can induce some muscle damage and impair glycogen resynthesis. In these situations, muscle glycogen levels can appear quite normal during the first 6 to 12 h after exercise, but glycogen resynthesis slows or stops completely as muscle repair begins.

The precise cause for this response is unknown, but conditions within the muscle could inhibit muscle glucose uptake and glycogen storage. For example, within 12 to 24 h after intense eccentric exercise, damaged muscle fibers are infiltrated with inflammatory cells (leukocytes, macrophages) that remove cellular debris

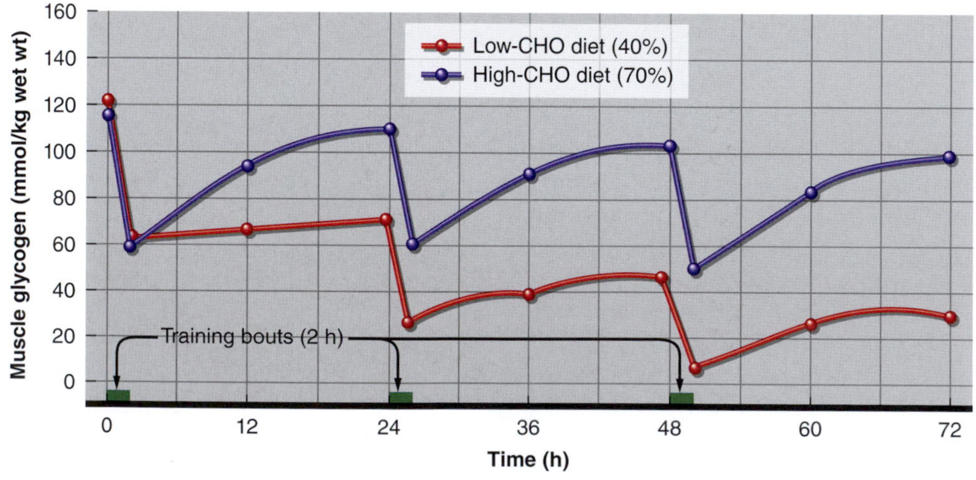

FIGURE 15.8 The influence of dietary carbohydrate (CHO) on muscle glycogen stores during repeated days of training. Note that when a low-CHO diet was consumed, muscle glycogen gradually declined over the 3 days of study, whereas the CHO-rich diet was able to return the glycogen to near normal each day.

Data from Costill and Miller (1980); Costill et al. (1971).

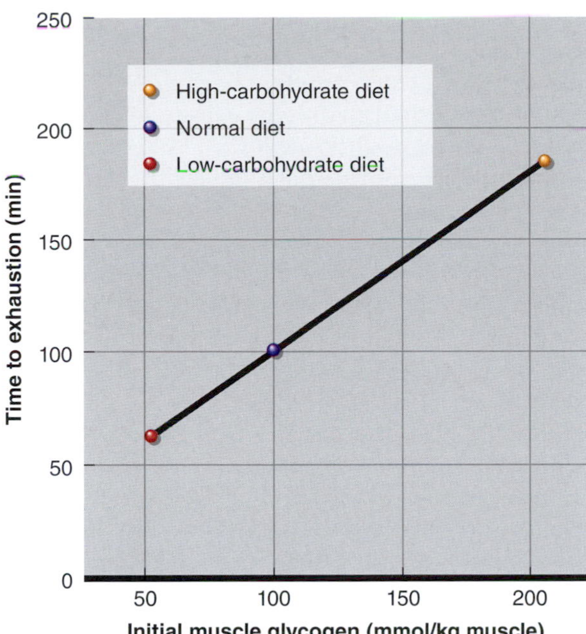

FIGURE 15.9 The relation between preexercise muscle glycogen content and exercise time to exhaustion. The exercise time to exhaustion and muscle glycogen were nearly four times greater when the subjects ate a carbohydrate-rich diet than when the diet was composed mostly of fat and protein.

resulting from damage to the cells' membranes (see chapters 5 and 14). This repair process can require a significant amount of the blood glucose, reducing the amount of glucose available for resynthesizing muscle glycogen. In addition, some evidence suggests that eccentrically exercised muscle is less sensitive to insulin, which would limit muscle fiber uptake of glucose. Perhaps future studies will more fully explain why eccentric-type activities delay glycogen storage. But for now, we can only observe that glycogen recovery from various forms of exercise can differ and that this should be considered for optimal diet, training, and competition.

Carbohydrate is the primary fuel source for most athletes and should constitute at least 50% of their total caloric intake. For endurance athletes, carbohydrate intake as a percentage of total caloric intake might need to be higher: 55% to 65%. However, most important is the total number of grams of carbohydrate ingested. It appears that athletes need from 3 to 12 g/kg of body weight per day in order to maintain glycogen stores. This wide range is necessary to account for the training intensity and total daily energy expenditure, the person's sex, and environmental conditions. For example, during periods of moderate-intensity training, 5 to 7 g/kg per day should be adequate. However, with long-duration high- and extremely high-intensity training, intake should be increased to 6 to 10 g/kg per day and 8 to 12 g/kg per day, respectively.[9,23]

The Glycemic Index

It has long been thought that the rapid increase in blood sugar concentration (hyperglycemia) with the intake of carbohydrate usually is associated with simple carbohydrates, such as glucose, sucrose, fructose, and high-fructose corn syrup. However, this is not always the case. Scientists have discovered that the glycemic response (i.e., increase in blood sugar) to carbohydrate intake varies considerably for both simple and complex carbohydrates. This has led to the use of what has been termed the *glycemic index* of foods (GI). The ingestion of glucose or white bread leads to a rapid increase in blood sugar. This response is used as a standard and has been arbitrarily assigned a GI of 100. The glycemic response for all other foods is referenced against the response for glucose or white bread, using 50 g of both the test food and glucose or white bread as the standard. The GI is calculated as follows: GI = 100 ü (blood glucose response over 2 h to 50 g of test food/blood glucose response over 2 h to 50 g of glucose or white bread). Three categories of GI have been established:[23]

- High glycemic index foods (GI >70) such as sport drinks, jelly beans, baked potato, French fries, popcorn, cornflakes, Corn Chex, and pretzels
- Moderate glycemic index foods (GI 56-70) such as pastry, pita bread, boiled white rice, bananas, most popular soft drinks, and regular ice cream
- Low glycemic index foods (GI ≤55) such as white boiled spaghetti, kidney and baked beans, milk, grapefruit, apples, pears, peanuts, M&M's, and yogurt

Food items were classified according to the 2002 International Table of Glycemic Index and Glycemic Load Values.

While the GI is a useful tool for rating foods, it is not without controversy. First, the GI for a given food can vary considerably between individuals as well as between similar foods. Second, some complex carbohydrates have high GIs. Third, adding small amounts of fat to a high-GI carbohydrate can greatly reduce the GI of that food. Finally, GI values differ substantially depending on whether glucose or white bread is used as the reference food, with white bread producing substantially higher values.[23] An additional index has been proposed that might be of importance during exercise. The glycemic load (GL) considers both the GI and the amount of carbohydrate (CHO) in a single serving and is calculated as follows: GL = (GI ü CHO, g)/100.

Before exercise, low-GI foods would be preferred to reduce the likelihood of hyperinsulinemia. However, high-GI foods should be an advantage during exercise by helping maintain blood glucose levels. This should also be the case during recovery from intense and prolonged exercise, since the higher blood sugar level should increase muscle and liver glycogen storage.

Carbohydrate Intake and Performance

As noted earlier, muscle glycogen provides a major source of energy during exercise. Muscle glycogen depletion has been shown to be a major cause of fatigue and ultimate exhaustion in high-intensity exercise of short duration or in moderate-intensity exercise lasting more than an hour. This is clearly illustrated in figure 15.10, which shows the marked depletion of muscle glycogen at very high intensities (150% and 120% of $\dot{V}O_{2max}$) for durations of less than 30 min, as well as at lower intensities (83%, 64%, and 31% of $\dot{V}O_{2max}$) for durations of 1 to 3 h.

On the basis of these results, scientists speculated that loading the muscle with extra glycogen before starting exercise should enhance performance. Studies in the 1960s demonstrated that men who ate a carbohydrate-rich diet for 3 days stored nearly twice their normal amounts of muscle glycogen.[4] When they were asked to exercise to exhaustion at 75% of $\dot{V}O_{2max}$, their exercise times significantly increased (see figure 15.9). This practice, called **glycogen loading** or **carbohydrate loading**, is widely used by distance runners, cyclists, and other athletes who must perform exercise for several hours. We discuss this practice in greater detail later in this chapter.

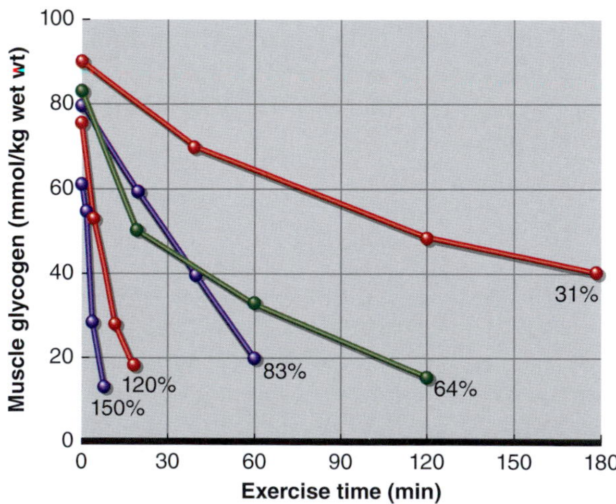

FIGURE 15.10 The influence of exercise intensity (31%, 64%, 83%, 120%, and 150% of $\dot{V}O_{2max}$) on the reduction in muscle glycogen stores. At relatively high intensities, the rate of muscle glycogen use is extremely high compared to that at the moderate and lower intensities.

Adapted by permission from A. Jeukendrup and M. Gleeson, *Sport Nutrition: An Introduction to Energy Production and Performance* (Champaign, IL: Human Kinetics, 2004). Original data from Gollnick, Piehl, and Saltin (1974).

Blood glucose concentrations become low (hypoglycemia) during exhaustive high-intensity, long-duration exercise, and this might contribute to fatigue. Numerous studies have shown that subjects' performances improve when they are given carbohydrate feedings during exercise lasting 1 to 4 h. Comparisons of subjects receiving carbohydrate feedings and those receiving placebos generally reveal no performance differences during the early stages of exercise, but during the final stages, performance is greatly improved with carbohydrate feedings.

Although all of the precise mechanisms by which carbohydrate feedings improve performance are not fully understood, maintaining blood glucose near normal concentrations allows the muscles to obtain more energy from this source. Carbohydrate feedings during exercise generally do not spare muscle glycogen use. Instead, they may preserve liver glycogen or even promote glycogen synthesis during exercise, enabling the exercising muscles to rely more on blood glucose for energy late in the exercise. Carbohydrate feedings might also enhance central nervous system function, reducing the perception of effort. Endurance performance (more than 1 h) can be enhanced when carbohydrate is consumed within 5 min before the exercise begins, more than 2 h before exercise (such as during the precompetition meal), and at frequent intervals during the activity.

VIDEO 15.2 Presents Asker Jeukendrup on carbohydrate intake during exercise. 

An athlete should use caution when ingesting carbohydrate-rich foods during the period from 15 to 45 min before exercise, because this could cause hypoglycemia shortly after exercise begins, which could lead to early exhaustion by depriving the muscle of its primary energy sources. Carbohydrate ingested during that period stimulates insulin secretion, elevating insulin when the activity begins. In response to the elevated insulin level, glucose uptake by the muscles reaches an abnormally high rate, leading to hypoglycemia and early fatigue (figure 15.11). Not everyone experiences this reaction, but sufficient evidence indicates that high-GI carbohydrates (those that cause a large increase in blood insulin) should be avoided or moderated in the period from 15 to 45 min before exercise.

Why don't carbohydrate feedings during exercise produce the same hypoglycemic effects observed with preexercise feedings? Sugar intake during exercise results in smaller increases in both blood glucose and insulin, lessening the threat of an overreaction that leads to a sudden decrease in blood glucose. This finer control of blood glucose during exercise might

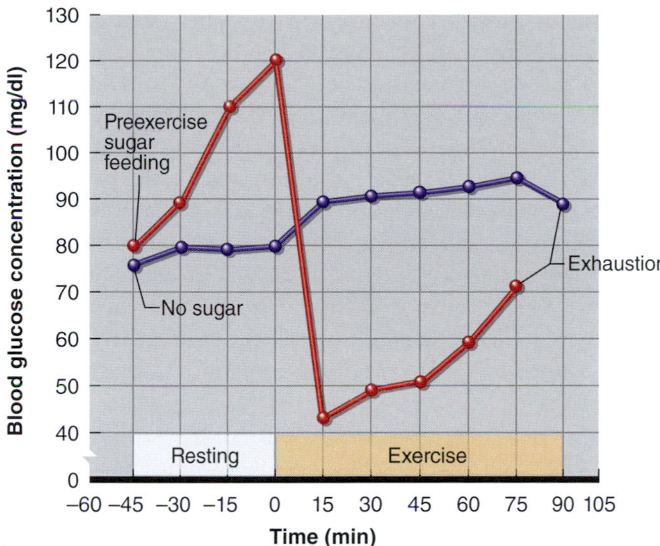

FIGURE 15.11 The effects of preexercise carbohydrate (sugar) feeding on blood glucose levels during exercise. Note the decrease in blood glucose to hypoglycemic levels with sugar feeding 45 min before exercise. Also, subjects during the sugar-feeding trial were unable to complete the full 90 min at 70% of $\dot{V}O_{2max}$, achieving only 75 min.

Based on Costill et al. (1977).

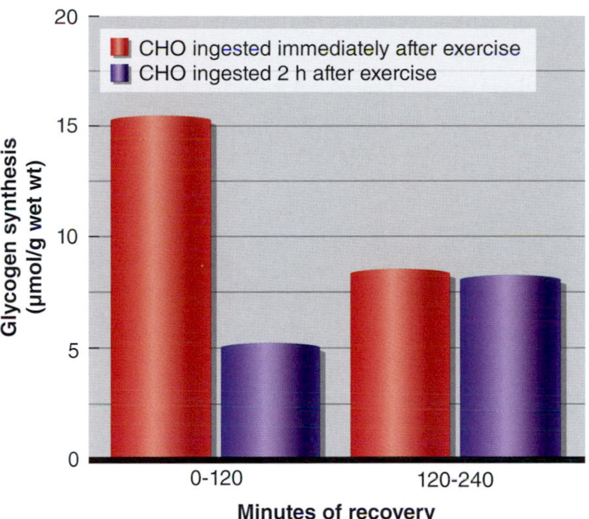

FIGURE 15.12 Replenishment of muscle glycogen stores after 70 min of muscle glycogen-depleting exercise using two different regimens of carbohydrate replacement. In the trial in which the carbohydrate solution was provided immediately following exercise (left), muscle glycogen storage was three times higher during the first 2 h of recovery compared to the other trial, in which the carbohydrate solution was not given until after 2 h of recovery (right). There was no difference in muscle glycogen storage during the next 2 h.

Adapted by permission from J.L. Ivy et al., "Muscle Glycogen Synthesis After Exercise: Effect of Time of Carbohydrate Ingestion," *Journal of Applied Physiology* 64 (1988): 1480-1485.

be caused by increased muscle fiber permeability that decreases the need for insulin, or insulin-binding sites may be altered during muscular activity. Regardless of the cause, carbohydrate intake during exercise appears to supplement the carbohydrate supply needed for muscular activity.

Finally, it is important to consume carbohydrate immediately after high-intensity and long-duration exercise during which carbohydrate stores have been reduced or depleted. Rates of glycogen resynthesis are very high during the first 2 h of recovery and progressively decrease thereafter. In a study by Ivy and colleagues,[21] cyclists exercised continuously for 70 min on a cycle ergometer on two separate occasions, a week apart, at moderate to high work rates to deplete the active muscles' glycogen stores. During one trial, a 25% carbohydrate solution was ingested immediately after exercise, while in the other trial the solution was ingested after 2 h of recovery. In the trial in which the solution was provided immediately after exercise, glycogen storage rates during the first 2 h were three times higher compared to the trial in which the solution was not provided until after 2 h of recovery. The storage rates were the same for the two trials during the second 2 h (see figure 15.12). More recently, it has been shown that adding protein to the carbohydrate supplement enhances the replenishment of muscle glycogen stores during the recovery period. Adding protein to the carbohydrate supplement maximizes glycogen synthesis with less frequent supplementation and less carbohydrate.[23] Further, it also appears to stimulate muscle tissue repair.

The importance of maximizing carbohydrate storage in the liver and muscles before exercise, and of providing carbohydrate during and immediately following exercise, has led food and nutrition companies to develop products to meet these needs, including sports drinks, as discussed at the end of this chapter.

Fat

Fat, also termed lipid, is a class of organic compounds with limited water solubility. It exists in the body in many forms, such as triglycerides, free fatty acids (FFAs), phospholipids, and sterols. The body stores most of its fat as triglycerides, composed of three molecules of fatty acids and one molecule of glycerol. Triglycerides are our most concentrated source of energy.

Excessive dietary fat, especially cholesterol and triglycerides, plays a major role in cardiovascular disease (discussed in chapter 21), and excessive fat intake also has been linked to other diseases such as cancer, diabetes, and obesity. But despite the negative effects of too much fat intake, fat is necessary in the diet and serves many vital functions:

- It is an essential component of cell membranes and nerve fibers.
- It is a primary energy source, providing up to 70% of our total energy in the resting state.
- It supports and cushions vital organs.
- All steroid hormones in the body are produced from cholesterol.
- Fat-soluble vitamins gain entry into, are stored in, and are transported through the body via fat.
- Body heat loss is minimized by the insulating effect of subcutaneous fat.

The most basic unit of fat is the fatty acid, the component used for energy production. Fatty acids occur in two forms: saturated and unsaturated. Unsaturated fats contain one (monounsaturated) or more (polyunsaturated) double bonds between carbon atoms, and each double bond takes the place of two hydrogen atoms. A saturated fatty acid possesses no double bonds, so it has the maximum amount of hydrogen bound to the carbons. Excessive saturated fat consumption is a risk factor for numerous diseases.

Fats derived from animal sources generally contain more saturated fatty acids than fats derived from plants. Fats that are more highly saturated tend to be solids at room temperature (e.g., bacon fat, coconut oil), whereas less saturated fats tend to be liquid (e.g., olive and canola oils). And although many vegetable oils are low in saturated fats, they are often used in foods as hydrogenated shortening. The process of hydrogenation adds hydrogen to the fat, increasing its saturation.

Fat Consumption

Fat can enhance food's palatability by absorbing and retaining flavors and by affecting the food's texture. For this reason, it is quite common in our diets, and extremely low-fat diets are unpalatable. The dietary fat

RESEARCH PERSPECTIVE 15.3

Low-Carb and Low-Fat Diets

Trends in dieting strategies for weight loss often become popular before the scientific community has an opportunity to appropriately test their effects not only on weight loss and maintenance but also on numerous other health outcomes as well. To date, only the *Dietary Approaches to Stop Hypertension (DASH)* diet, which is low in fat and sodium and high in whole grains, fruits, and vegetables, has been scientifically evaluated and shown to improve cardiovascular health. Low-carbohydrate diets, in which the person adhering to the diet consumes <40 g of carbohydrates per day, have become increasingly popular in recent years. However, the cardiovascular effects of this approach are relatively unknown. Prospective studies (which apply an intervention or stimulus and then follow the participants into the future to measure outcomes) of low-carbohydrate diets have produced conflicting results, and few randomized controlled studies have examined the effects of carbohydrate restriction on cardiovascular disease risk factors in a diverse population.

In 2014, a research team from Tulane University conducted a randomized, parallel group trial to examine the effects of a 12-month low-carbohydrate diet compared to a low-fat diet.[6] In this study, 148 men and women ages 22 to 75 years with a body mass index (BMI) of 30 to 45 kg/m^2 from diverse socioeconomic and ethnic backgrounds were assigned to one of two diet groups: low fat or low carbohydrate. The 73 participants assigned to the low-fat diet were instructed to maintain less than 30% of their daily energy intake from fat and 55% from carbohydrates. The 75 participants assigned to the low-carbohydrate diet were instructed to eat less than 40 g of carbohydrate daily. Subjects met with a dietician weekly for the first 4 weeks, followed by group counseling sessions every other week for the next 5 weeks, and then monthly for the last 6 months of the study. The study team collected data on weight, cardiovascular disease risk factors, and dietary composition at baseline, and 3, 6, and 12 months into the intervention. By the end of the yearlong intervention, 60 participants in the low-fat group and 59 in the low-carbohydrate group had completed the entire year.

After 12 months, participants in the low-carbohydrate group had greater decreases in weight, fat mass, and low-density lipoprotein cholesterol and greater increases in high-density lipoprotein cholesterol compared to the low-fat diet group. These data indicate that low-carbohydrate diets may be more effective for weight loss and reducing cardiovascular disease risk factors compared to a low-fat diet, even in a very large and diverse group of participants. The study concluded that restricting carbohydrates may be a good option for people seeking to lose weight and reduce cardiovascular disease risk.

intake for both men and women was as high as 45% of total calories consumed in 1965 but has decreased to about 33% to 34% as reported in the most recent National Health and Nutrition Examination Survey (NHANES).[33] Most likely this decrease was attributable to recent media attention on the health risks of dietary fat. Most nutritionists recommend that fat consumption not exceed 35% of total calories consumed. *Dietary Guidelines for Americans, 2015-2020,* produced by the U.S. Departments of Health and Human Services and Agriculture, advises us to limit saturated fats to less than 10% of total caloric intake and replace with polyunsaturated and monounsaturated fats, to limit cholesterol to less than 300 mg per day, and to eliminate trans fatty acids.[16] Recent research has shown that not all fats categorized by degree of saturation have the same health effect. For example, not all saturated fats are "bad" fats; the saturated fat stearic acid does not negatively affect serum cholesterol levels.[34,35]

Fat Intake and Performance

For the athlete, fat is especially important as an energy source. Muscle and liver glycogen stores in the body are limited, so the use of fat (or FFA) for energy production can delay exhaustion. Clearly, any change that allows the body to use more fat would be an advantage, particularly for endurance performance. In fact, one adaptation that occurs in response to endurance training is an increased ability to use fat as an energy source. Unfortunately, merely eating fat does not stimulate the muscles to burn fat. Instead, eating fatty foods tends only to elevate plasma triglycerides, which then must be broken down before the FFA can be used for energy production. To increase the use of fat, the FFA levels in the blood, not the triglyceride levels, must be increased.

Highly trained athletes can adapt to a high-fat diet. But is this beneficial to overall performance? We have just seen how glycogen loading can improve endurance performance. Does fat loading have similar benefits? While several studies have demonstrated limited benefit of a high-fat versus a high-carbohydrate diet, most studies have shown either no benefit or decreased performance. The body adapts to a high-fat diet by increasing the supply of fat and the capacity for fat oxidation in the muscle, thus increasing fat oxidation during exercise. But this usually occurs at the expense of a reduction in muscle glycogen stores, thus negating any beneficial effects. Drawing conclusions from the studies conducted so far is complicated by the fact that the type of fat (medium-chain versus long-chain triglycerides) and the duration of the high-fat diet intervention (less than a week versus several weeks or more) have varied widely in these studies.

Protein

Protein is a class of nitrogen-containing compounds formed by amino acids. Protein serves numerous functions in our bodies:

- It is the major structural component of the cell.
- It is used for growth, repair, and maintenance of body tissues.
- Hemoglobin, enzymes, and many hormones are produced from protein.
- It is one of the three primary buffers in the control of acid–base balance.
- Proteins in the plasma help maintain normal blood osmotic pressure.
- Antibodies for disease protection are formed from protein.
- Energy can be produced from protein.

Twenty amino acids have been identified as necessary for human growth and metabolism (table 15.2). Of these, 11 (for children) or 12 (for adults) are termed **nonessential amino acids**, meaning that our bodies synthesize them, so we don't rely on dietary intake for their supply. The remaining eight or nine are termed **essential amino acids** because we cannot synthesize them; thus, they are an essential part of our daily diets. Absence of one of these essential amino acids from the diet precludes formation of any proteins that contain that amino acid, and thus any tissue requiring those proteins cannot be maintained.

A dietary protein source that contains all the essential amino acids is called a complete protein. Meat, fish,

TABLE 15.2 **Essential and Nonessential Amino Acids**

Essential	Nonessential
Isoleucine	Alanine
Leucine	Arginine
Lysine	Asparagine
Methionine	Aspartic acid
Phenylalanine	Cysteine
Threonine	Glutamic acid
Tryptophan	Glutamine
Valine	Glycine
Histidine (children)[a]	Proline
	Serine
	Tyrosine
	Histidine (adult)[a]

[a]Histidine is not synthesized in infants and young children, so it is an essential amino acid for children but not for adults.

poultry, eggs, and milk are examples. The proteins in vegetables and grains are called incomplete proteins because they do not supply all the essential amino acids. This concept is important for people on vegetarian diets (discussed later in this chapter). However, combining several incomplete protein sources at one meal should resolve this problem.

Protein Consumption

Protein accounts for approximately 15% of the total calories consumed per day in the United States. The Recommended Dietary Allowances (RDAs) for protein are 0.95 g/kg of body weight per day for 4- to 13-year-olds, 0.85 g/kg per day for 14- to 18-year-olds, and 0.80 g/kg per day for adults. Men typically require more protein than women because men generally weigh more and have greater muscle mass. However, men generally eat more food per day to sustain their weight and muscle mass. Thus, an allowance of 0.8 g/kg of body weight is considered appropriate for both men and women.

Protein Intake and Performance

Should athletes who are training for muscular strength and endurance and aerobic endurance increase their protein intake? Amino acids are the body's building blocks, so protein is essential for the growth and development of body tissues. Over the years, nutritionists and physiologists argued against the need for supplementing proteins for optimal sport performance. It was generally believed that the RDA of 0.8 g of protein per kilogram of body weight each day would adequately meet the demands of hard training. However, in sports that rely on the maintenance or building of muscle mass, the RDA is probably inadequate.

Studies using metabolic-tracer and nitrogen-balance technologies have shown that the overall protein and specific amino acid requirements are higher for individuals in training than for normally active people. The role of protein differs for endurance- and strength-trained athletes. It appears that strength-training individuals need up to 2.1 times the RDA, or about 1.6 to 1.7 g of protein per kilogram of body weight per day, whereas athletes engaging in endurance training need 1.2 to 1.4 g of protein per kilogram of body weight per day.[2] Most athletes cross-train (endurance athletes do some strength training and power athletes perform some aerobic exercise); therefore, a general recommendation of 1.2 to 1.7 g of protein per kilogram of body weight per day has been recommended,[27] which is close to the 1.4 to 1.6 g/kg per day supported by empirical studies. While endurance exercise places greater demand on protein to increase mitochondrial content and as an auxiliary fuel, strength training requires additional amino acids as the building blocks

RESEARCH PERSPECTIVE 15.4

The Myth of High-Protein Diets

Trends in dieting for weight loss come and go. For many years, dietary fat received a lot of the blame for increasing rates of overweight and obesity, and fat-free or low-fat diets were popular. However, despite this emphasis on reducing fat intake, rates of overweight and obesity continued to climb. Recently, some people have been making the counterargument that Americans have grown fatter because they consume too many starches and sugars and not enough meat and fat. In 2015, the *Dietary Guidelines Advisory Committee* lifted their recommendation that dietary cholesterol should be restricted, citing evidence that dietary cholesterol does not influence blood cholesterol concentrations all that much. So, is a high-fat, high-protein regimen the new healthy diet du jour?

In an opinion piece published in *The New York Times*, Dean Ornish, a medical doctor and clinical professor of medicine at the University of California San Francisco, makes the argument that animal fats and proteins are not health foods.[25] Despite being told for decades to reduce fats and animal proteins in their diets, Americans consumed 67% more added fat and 41% more meat in 2000 than they did in 1950. Additionally, consumption of refined sugars has gone up 39%. Increasing consumption of animal protein increases risk for premature death from all causes, and heavy consumption of saturated fats and trans fats doubles the risk for Alzheimer's disease. Dr. Ornish and his colleagues at the Preventative Research Institute at UCSF have conducted clinical research trials showing the benefits of a plant-based diet for reducing circulating free fatty acids and insulin and reducing inflammation associated with animal protein and fat consumption. Dr. Ornish and others have found that diets rich in fruits, vegetables, whole grains, and polyunsaturated fats and low in saturated fats, refined carbohydrates, and animal proteins actually reverses the progression of cardiovascular diseases and can reduce the need for chronic medications. The more people adhered to the dietary recommendations, the more improvement Dr. Ornish measured, even in the oldest patients. Overall, despite the popularity of high-protein diets, only a whole-foods, plant-based diet has been tested clinically and shown to reverse disease trends.

for muscle protein synthesis. Of course, there could be exceptions to this for athletes who are just starting a new, rigorous training program or who are doing very intense, long-duration workouts. Protein timing is important for athletes, and current advice is to consume about 20 g of high-quality protein (one that contains all of the essential amino acids) in the hour or two after resistance exercise.[8,32]

Protein Intake and Resistance Exercise

In order to increase muscle mass, the rate of protein synthesis (anabolism) must exceed the rate of protein degradation (catabolism). While both processes are important in building muscle mass, for healthy exercisers, increasing the synthesis of new protein plays a much larger role than decreasing degradation. As summarized by Stuart Phillips,[26] both resistance training and protein (or amino acid) consumption can independently promote protein synthesis. However, combining the two—that is, consuming protein after a bout of resistance exercise—synergistically stimulates protein synthesis. Based on years of elegant research, we now have better answers to the specific questions of what to eat, how much is necessary, and when to consume proteins in relation to exercise if the goal is an increased muscle mass.

Rapidly digested proteins that contain high proportions of essential amino acids are more effective in stimulating muscle protein synthesis than are other proteins. The key amino acid appears to be leucine, since peak muscle protein synthesis correlates with peak leucine concentration. This finding has led to the concept of leucine as a trigger for promoting protein synthesis.

There is no evidence that commercially available protein supplements are necessary to achieve desired goals. Food protein, particularly high-quality dairy protein—containing all essential amino acids and high in leucine—may be more effective than other protein sources. In particular, high-quality milk-based proteins like casein and whey are effective in promoting gains in muscle mass during periods of intense resistance training.

Consuming protein in close temporal proximity to exercise leads to greater rates of protein synthesis and greater muscle hypertrophy. Current evidence favors consumption of protein after resistance exercise, but the optimal timing after exercise is not known. The anabolic effect of resistance exercise is long lasting, but likely diminishes with increasing time postexercise, so current evidence supports a 1 to 2 h postexercise window for optimal synthesis rates. In addition, athletes should consume protein as part of each meal to promote optimal protein synthesis throughout the day.

Intakes in the range of 1.4 to 1.6 g of protein per kilogram per day should be consumed for the maintenance and building of muscle mass. This represents a higher overall daily protein intake than the RDA, which is currently 0.8 g of protein/kg body mass per day. Consuming amounts of protein in excess of 2.0 g/kg per day provides no additional benefit. Individual meals should include at least 0.25 g of protein/kg body mass.

In Review

- Carbohydrates are sugars and starches. They exist in the body as monosaccharides, disaccharides, oligosaccharides, and polysaccharides. All carbohydrates must be broken down into monosaccharides before the body can use them as a fuel.
- Insufficient intake of carbohydrate during periods of intense training can lead to depletion of glycogen stores. Conversely, glycogen loading by consumption of a diet rich in carbohydrate offers major benefits to performance.
- Endurance performance can be enhanced when carbohydrate is consumed an hour or more before beginning exercise, within 5 min of starting exercise, and during exercise. Exercisers can replenish carbohydrate stores rapidly by ingesting carbohydrate during the first 2 h of recovery. This can be facilitated by the addition of protein to the carbohydrate supplement.
- Fats, or lipids, exist in the body as triglycerides, FFAs, phospholipids, and sterols. They are stored primarily as triglycerides, which are the body's most concentrated energy source. A triglyceride molecule can be broken down into one glycerol and three fatty acid molecules. Only the FFAs are used by the body for energy production.
- Although fat is a major energy source, the use of high-fat diets to enhance endurance performance by sparing glycogen has generally been unsuccessful.
- The smallest unit of protein is an amino acid. All proteins must be broken down to amino acids before the body can use them. Only nonessential amino acids can be synthesized in the body; the essential amino acids must be attained through diet.
- Protein is not a primary energy source in our bodies, but it can be used for energy production during endurance exercise.
- The current RDA for protein (0.8 g/kg per day) may be too low for athletes involved in intense resistance training or for endurance athletes. In most cases, athletes should consume 1.2 to 1.7 g/kg per day. During the initial days of training,

during periods of very intense training, or when insufficient energy is consumed, the requirement might be higher. However, extremely high-protein diets offer no additional benefits and could pose a health risk to normal kidney function and contribute to risk of dehydration.

- Protein supplementation during recovery from resistance training can stimulate muscle protein synthesis.

Vitamins

Vitamins are a group of organic compounds that perform specific functions in the body. We need them in relatively small quantities to fully use the other nutrients we ingest. Vitamins act primarily as catalysts or cofactors in chemical reactions. They are essential for energy release, tissue building, and metabolic regulation. Vitamins can be classified into one of two major categories: fat soluble and water soluble. The fat-soluble vitamins, A, D, E, and K, are absorbed from the digestive tract bound to lipids (fats). These vitamins are stored in the body, so excessive intake can cause toxic accumulations. Vitamin C and the B-complex vitamins biotin, pantothenic acid, and folate are water soluble. They are absorbed from the digestive tract along with water. Any excess of these vitamins is excreted, mostly in the urine, but vitamin toxicity has been reported with high doses of some of these. Table 15.3 lists the various vitamins and their RDA values, or Adequate Intake (AI) values when the RDA values are not available.

Most vitamins have some functions important to the athlete:

- Vitamin A is crucial for normal growth and development because it plays an integral role in bone development.
- Vitamin D is essential for intestinal absorption of calcium and phosphorus and thus for bone development and strength. By regulating calcium absorption, this vitamin also has a key role in neuromuscular function.
- Vitamin K is an intermediate in the electron transport chain, making it important for oxidative phosphorylation.

Of all the vitamins, the B-complex vitamins and vitamins C and E have been most extensively studied for their potential to facilitate athletic performance. In the following sections, we briefly consider these vitamins.

B-Complex Vitamins

The B-complex vitamins were once thought to be a single vitamin. Now more than a dozen B-complex vitamins have been identified. These vitamins play essential roles in cellular metabolism, primarily by serving as cofactors in various enzyme systems involved in the oxidation of food and the production of energy. Vitamin B_1 (thiamin) is needed for the conversion of pyruvic acid to acetyl coenzyme A. Vitamin B_2 (riboflavin) becomes flavin adenine dinucleotide (FAD), which acts as a hydrogen acceptor during oxidation. Vitamin B_3 (niacin) is a component of nicotinamide adenine dinucleotide phosphate (NADP), a coenzyme in glycolysis. Vitamin B_{12} has a role in amino acid metabolism and is also needed for the production of red blood cells, which transport oxygen to the cells for oxidation. The B-complex vitamins have such a close interrelationship that a deficiency in one can impair utilization of the others. Symptoms of deficiencies vary with the vitamins involved.

A few studies have shown that supplementation of one or more of the B-complex vitamins facilitates performance. However, most researchers agree that this is true only if the individual being studied has a preexisting B-complex deficiency. Creating a deficiency in one or more of the B-complex vitamins usually impairs performance, but this is reversed when the deficiency is corrected with supplementation. No compelling evidence supports supplementation when there is no deficiency.

Vitamin C

Vitamin C (ascorbic acid) is common in our foods, but deficiencies can occur in people who smoke, use oral contraceptives, have surgery, or run a fever. This vitamin is important for the formation and maintenance of collagen, a crucial protein found in connective tissue, so it is essential for healthy bones, ligaments, and blood vessels. Vitamin C also functions in

- the metabolism of amino acids;
- synthesis of some hormones, such as the catecholamines (epinephrine and norepinephrine) and the anti-inflammatory corticoids; and
- promotion of iron absorption from the intestines.

Some people also believe that vitamin C assists healing, combats fever and infection, and prevents or cures the common cold, but evidence of these attributes is inconclusive.

Vitamin C supplementation to enhance performance has produced equivocal findings in the research conducted to date. As with other vitamins, vitamin C supplementation does not improve performance when no deficiency exists.

Vitamin E

Vitamin E is stored in muscle and fat. This vitamin's functions are not well established, although it is known to enhance the activity of vitamins A and C by preventing their oxidation. Indeed, the most important role

TABLE 15.3 **RDAs or AIs for Vitamins and Minerals**

		9-13 years		14-18 years		19-50 years		51-70 years	
Vitamin or Mineral	**Dose**	**Male**	**Female**	**Male**	**Female**	**Male**	**Female**	**Male**	**Female**
Vitamins									
A (retinol)	μg/day	600	600	900	700	900	700	900	700
B_1 (thiamine)	mg/day	0.09	0.09	1.2	1.0	1.2	1.1	1.2	1.2
B_2 (riboflavin)	mg/day	0.9	0.9	1.3	1.0	1.3	1.1	1.3	1.1
B_3 (niacin)	mg/day	12	12	16	14	16	14	16	14
B_6	mg/day	1.0	1.0	1.3	1.2	1.3	1.3	1.7	1.5
B_{12}	μg/day	1.8	1.8	2.4	2.4	2.4	2.4	2.4	2.4
C	mg/day	45	45	75	65	90	75	90	75
D	μg/day	15[a]	15[a]	15[a]	15[a]	15[a]	15[a]	15[a]	15[a]
E	mg/day	11	11	15	15	15	15	15	15
Biotin (vitamin H)	μg/day	20[a]	20[a]	25[a]	25[a]	30[a]	30[a]	30[a]	30[a]
K	μg/day	60[a]	60[a]	75[a]	75[a]	120[a]	90[a]	120[a]	90[a]
Folate	μg/day	300	300	400	400	400	400	400	400
Pantothenic acid	mg/day	4[a]	4[a]	5[a]	5[a]	5[a]	5[a]	5[a]	5[a]
Minerals									
Calcium	mg/day	1,300[a]	1,300[a]	1,300[a]	1,300[a]	1,000[a]	1,000[a]	1,000[a]	1,200[a]
Chloride	g/day	2.3[a]	2.3[a]	2.3[a]	2.3[a]	2.3[a]	2.3[a]	2.0[a]	2.0[a]
Chromium	μg/day	25[a]	21[a]	35[a]	24[a]	35[a]	25[a]	30[a]	20[a]
Copper	μg/day	700	700	890	890	900	900	900	900
Fluoride	mg/day	2[a]	2[a]	3[a]	3[a]	4[a]	3[a]	4[a]	3[a]
Iodine	μg/day	120	120	150	150	150	150	150	150
Iron	mg/day	8	8	11	15	8	18	8	8
Magnesium	mg/day	240	240	410	360	410[b]	315[b]	420	320
Manganese	mg/day	1.9[a]	1.6[a]	2.2[a]	1.6[a]	2.3[a]	1.8[a]	2.3[a]	1.8[a]
Molybdenum	μg/day	34	34	43	43	45	45	45	45
Phosphorus	mg/day	1,250	1,250	1,250	1,250	700	700	700	700
Potassium	g/day	4.5[a]	4.5[a]	4.7[a]	4.7[a]	4.7[a]	4.7[a]	4.7[a]	4.7[a]
Selenium	μg/day	40	40	55	55	55	55	55	55
Sodium	g/day	1.5[a]	1.5[a]	1.5[a]	1.5[a]	1.5[a]	1.5[a]	1.3[a]	1.3[a]
Zinc	mg/day	8	8	11	9	11	8	11	8

[a]Adequate Intake (AI). RDA is not available.

[b]Men: age 19 to 30 years = 400; age 31 to 50 years = 420. Women: age 19 to 30 years = 310; age 31 to 50 years = 320.

Note. Values are also available for infants and small children and for women during pregnancy and lactation.

From USDA National Agricultural Library. Available: http://fnic.nal.usda.gov/dietary-guidance/dietary-reference-intakes/dri-tables

of vitamin E is its action as an antioxidant. It disarms **free radicals** (highly reactive molecules) that could otherwise severely damage cells and disrupt metabolic processes. Exercise has been shown to cause DNA damage to the cell, whereas supplementing the intake of vitamin E reduces DNA damage induced by exercise. However, investigators found no benefit of 30 days of vitamin E supplementation with respect to the muscle damage that resulted from 240 maximal isokinetic eccentric knee flexion–extension actions (24 sets of 10 repetitions each) when compared with a placebo control condition.[7]

Many athletes have consumed supplementary doses of vitamin E since it has been postulated to benefit performance through its relationship with oxygen use and energy supply. However, research reviews generally conclude that vitamin E supplementation does not improve athletic performance.

Minerals

A number of inorganic substances known as minerals are essential for normal cellular functions. Minerals account for approximately 4% of body weight. Some are present in high concentrations in the skeleton and teeth, but minerals are also found throughout the body, in and around every cell, dissolved in the body's fluids. They can be present either as ions or combined with various organic compounds. Mineral compounds that can dissociate into ions in the body are called **electrolytes**.

By definition, **macrominerals** are those minerals needed by the body in amounts greater than 100 mg per day. **Microminerals**, or **trace elements**, are those needed in smaller amounts. Table 15.3 lists the essential minerals and their RDAs or AIs.

Mineral intake is less likely to be supplemented by athletes than vitamin intake, possibly because far fewer claims have been made about the performance-enhancing qualities of specific minerals. Of the minerals, calcium and iron have been most frequently investigated.

Calcium

Calcium is the most abundant mineral in the body, constituting approximately 40% of the total mineral content. Calcium is well known for its importance in building and maintaining healthy bones, and that is where most of it is stored. But it is also essential for nerve impulse transmission. Calcium plays major roles in enzyme activation and regulation of cell membrane permeability, both important for metabolism. This mineral is also essential for normal muscle function: Recall from chapter 1 that calcium is stored in the sarcoplasmic reticulum of muscles and released when the muscle fibers are stimulated. It is required for formation of the actin–myosin cross-bridges that cause the fibers to contract.

Sufficient calcium intake is critical to our health. If we do not consume enough calcium, it is removed from its storage sites in the body, especially the bones. This condition is called osteopenia. It weakens the bones and can lead to osteoporosis, a common problem in postmenopausal women and aging men and women. Unfortunately, few studies have been conducted on calcium supplementation, but their results suggest that supplementation is of no value in the presence of an adequate (RDA) dietary intake of calcium.

Phosphorus

Phosphorus is closely linked to calcium. It constitutes approximately 22% of the body's total mineral content. About 80% of this phosphorus is combined with calcium (calcium phosphate), providing strength and rigidity to the bones. Phosphorus is an essential part of metabolism, cell membrane structure, and the buffering system to maintain constant blood pH. Phosphorus plays a major role in bioenergetics: It is an essential component of adenosine triphosphate. There is no evidence to suggest the need for supplementation in athletes.

Iron

Iron—a micromineral—is present in the body in relatively small amounts (35-50 mg/kg of body weight). It plays a crucial role in oxygen transportation: Iron is required for the formation of both hemoglobin and myoglobin. Hemoglobin, located in the red blood cells, binds with oxygen in the lungs and then transports it to the body tissues via the blood. Myoglobin, found in muscle, combines with oxygen and stores it until needed.

Iron deficiency is prevalent throughout the world. By some estimates, as much as 25% of the world's population is iron deficient. In the United States, approximately 20% of women and 3% of men are iron deficient, as are 50% of pregnant women. The major problem associated with this condition is iron-deficiency anemia, in which hemoglobin levels are reduced, decreasing the blood's oxygen-carrying capacity. This causes fatigue, headaches, and other symptoms. Iron deficiency is a more common problem in women than in men because both menstruation and pregnancy cause iron losses that must be replenished. This problem is compounded by the fact that women generally consume less food, and thus less iron, than men.

Iron has received much attention in the research literature. Women are considered anemic when their hemoglobin concentration is below 10 g per 100 ml of blood, and men are considered to be anemic below values of 12 g per 100 ml of blood. Studies suggest that 22% to 25% of female athletes and 10% of male athletes are iron deficient, but these numbers may be conservative. Hemoglobin concentration is not the only marker of anemia or necessarily the best. Plasma ferritin concentration provides a good marker of the body's iron stores. Values below 20 to 30 μg/L indicate low stores of body iron.

When iron supplements are given to those who are iron deficient (i.e., with low plasma ferritin levels), endurance performance typically improves. However, supplementation of iron in those who are not deficient has little or no benefit. In fact, iron supplements can be a health risk, because excess iron is toxic for the liver, and ferritin levels higher than 200 μg/L are associated with an increased risk for coronary artery disease.

Sodium, Potassium, and Chloride

Sodium, potassium, and chloride are the important electrolytes that are distributed throughout all body fluids and tissues. Sodium and chloride are found primarily in the fluid outside of the cells and in the blood plasma, but potassium is located mainly inside the cells. This selective distribution of these three minerals establishes the separation of electrical charge across neuron and muscle cell membranes. Thus, these minerals enable neural impulses to control muscle activity (see chapter 3). In addition, they are responsible for maintaining the body's water balance and distribution, normal osmotic equilibrium, acid–base balance (pH), and normal cardiac rhythm (chapter 6).

Western diets are high in sodium, so sodium deficiency is an unlikely occurrence. However, minerals are lost with sweating, so excessive sweating during extreme exertion or exercise in a hot environment can deplete these minerals. When discussing mineral imbalances, we often focus on deficiencies. However, many of these minerals also have negative effects when taken in excess. In fact, excess potassium can cause heart failure! Individual needs vary, but megadoses are never advisable.

To conclude this section on vitamins and minerals, we can say that while physical activity increases vitamin and mineral requirements, the need is generally met by adequate food intake. For athletes who eat balanced meals that meet their caloric needs, it is highly likely that all vitamin and mineral needs will be met and supplementation will have no performance benefits. However, for those who are intentionally consuming a low-energy or unbalanced diet, supplementation may be necessary to maintain health and performance. If there is any question about the adequacy of an athlete's diet, a low-dose multivitamin and mineral supplement may be appropriate.

The previous discussion on nutrient intakes provides several RDAs. In the early 1940s, the Food and Nutrition Board of the National Academy of Sciences established the United States Recommended Daily Allowance (RDA) guidelines for all food nutrients. The latest edition of the RDAs in their original form was released in 1989. The RDAs provide estimates of safe and adequate daily dietary intakes and estimated minimum requirements for selected vitamins and minerals. A major revision of the RDAs was initiated in the early 1990s. The RDAs have been replaced by new recommendations called Dietary Reference Intakes (DRIs). The DRIs reflect a joint effort between the United States and Canada to provide dietary intake recommendations grouped by nutrient function and classification.

The new DRIs were released in a series of reports starting in 1997 and continuing through 2005. They include four different reference values:

- Estimated Average Requirement (EAR)—the intake value estimated to meet the requirement for 50% of healthy individuals in an age- and sex-specific group
- Recommended Dietary Allowance (RDA)—the intake value sufficient to meet the nutrient requirements of nearly all (97%-98%) individuals in a specific group
- Tolerable Upper Intake Level (UL)—the highest level of daily nutrient intake that is unlikely to pose no risk of adverse health effects to almost all individuals in a specific group
- Adequate Intake (AI)—a recommended intake value based on observed or experimentally determined approximations or estimates of nutrient intake by healthy individuals in a specific group that are assumed to be adequate; this is used when an RDA cannot be determined

For further information on DRIs and specific recommendations for each of the nutrient classifications by sex and age, refer to the U.S. Department of Agriculture's Food and Nutrition Information Center at www.nal.usda.gov/fnic.

In Review

- Vitamins perform numerous functions in the body and are essential for normal growth and development. Many are involved in metabolic processes, such as those leading to energy production.
- Vitamins A, D, E, and K are fat soluble. These can accumulate to toxic levels in the body. Vitamin C and the B-complex vitamins biotin, pantothenic acid, and folate are water soluble. Excesses of these are excreted, so toxicity is rarely a problem. Several of the B-complex vitamins are involved in the processes of energy production.
- Macrominerals are minerals required in amounts of more than 100 mg per day. Microminerals (trace elements) are those we require in smaller amounts.
- Minerals are required for numerous physiological processes, such as muscle contraction, oxygen transport, fluid balance, and bioenergetics. Minerals can dissociate into ions, which can participate in numerous chemical reactions. Minerals that can dissociate into ions are called electrolytes.
- Vitamins and minerals do not appear to have any special performance-enhancing value. Taking them in amounts greater than the RDA will not improve performance and may have unwanted effects.

Water

Water is often not thought of as a nutrient because it has no caloric value. Yet its importance in maintaining life is perhaps second only to that of oxygen. Water constitutes about 60% of a typical young man's and 50% of a typical young woman's total body weight, but this varies with body composition, since the fat-free mass has a much higher water content (~73% water) than the fat mass (~10% water). It has been estimated that we can survive losses of up to 40% of our body weight in fat, carbohydrate, and protein. But a water loss of only 9% to 12% of body weight can be fatal.

Approximately two-thirds of the water in our bodies is contained in our cells and is referred to as **intracellular fluid**. The remainder is outside the cells, referred to as the **extracellular fluid**. Extracellular fluid includes the interstitial fluid surrounding the cells, the blood plasma, lymph, and other body fluids.

Water plays several critical roles in exercise. Among its most important functions, water provides transportation between and delivery to the body's various tissues, regulates body temperature, and maintains blood pressure for proper cardiovascular function. In the next sections, we more closely examine the role of water in exercise and performance.

Water and Electrolyte Balance

For optimal performance, the body's water and electrolyte contents should remain relatively constant. Unfortunately, this doesn't always happen during exercise. In the next sections, we examine water content and electrolyte balance at rest, how exercise affects these, and the impact on performance when water or electrolyte balance is disturbed.

Water Balance at Rest

Under normal resting conditions, the body's water content is relatively constant: Water intake equals water output. About 60% of our daily water intake is obtained from the fluids we drink and about 30% is from the foods we consume. The remaining 10% is produced in our cells during metabolism (recall from chapter 2 that water is a by-product of oxidative phosphorylation). Metabolic water production varies from 150 to 250 ml per day, depending on the rate of energy expenditure: Higher metabolic rates produce more water. The total daily water intake from all sources averages about 33 ml per kilogram of body weight per day. For a 70 kg (154 lb) person, average intake is 2.3 L per day. Water output, or water loss, occurs from four sources:

- Evaporation from the skin
- Evaporation from the respiratory tract
- Excretion from the kidneys
- Excretion from the large intestine

Human skin is permeable to water. Water diffuses to the skin's surface, where it evaporates into the environment. In addition, the gases we breathe are constantly being humidified by water as they pass through the respiratory tract. These two types of water loss (from the skin and respiration) occur without our sensing them. Thus, they are termed insensible water losses. Under cool, resting conditions, these losses account for about 30% of daily water loss.

The majority of our daily water loss—60% at rest—occurs from our kidneys, which excrete water and waste products as urine. Under resting conditions, the kidneys excrete about 50 to 60 ml of water per hour. Another 5% of the water is lost by sweating (although this is often considered along with insensible water loss), and the remaining 5% is excreted from the large intestine in the feces. The sources of water gain and water loss at rest are depicted in figure 15.13.

Water Balance During Exercise

Water loss accelerates during exercise, as seen in table 15.4. The ability to lose the heat generated during exercise depends primarily on the formation and evaporation of sweat. As body temperature increases, sweating increases in an effort to prevent overheating (see chapter 12). But at the same time, more water is produced during exercise because of increased oxida-

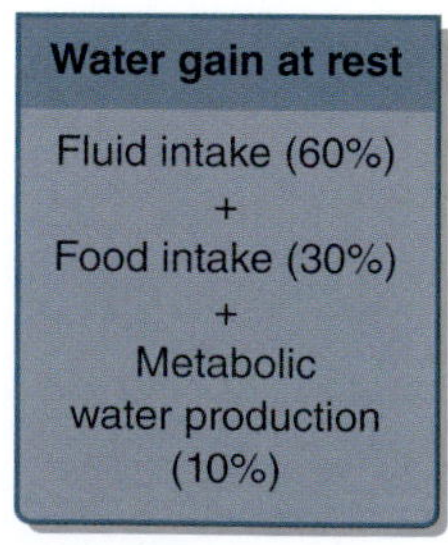

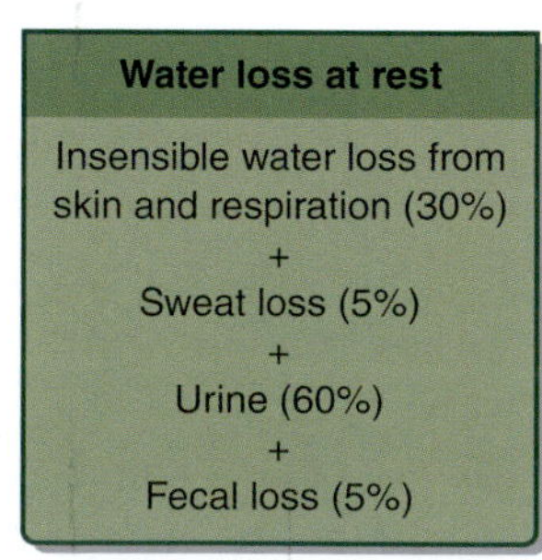

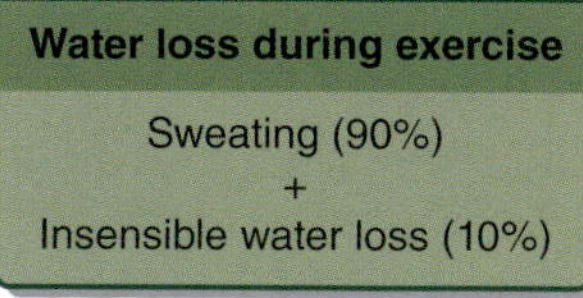

FIGURE 15.13 Sources and approximate percentages of body water gains and losses at rest and during exercise (see text for more detail).

TABLE 15.4 **Typical Values of Water Loss From the Body at Rest in a Cool Environment and During Prolonged Exhaustive Exercise**

	Resting		Prolonged exercise	
Source of loss	ml/h	% total	ml/h	% total
Skin (insensible loss)	15	15	15	1
Respiration (insensible loss)	15	15	100	7
Sweating	4	5	1,200	91
Urine	58	60	10	1
Feces	4	5		0
Total	96	100	1,325	100

tive metabolism. Unfortunately, the amount produced even during the most intense effort has only a small impact on the **dehydration**, or water loss, that results from heavy sweating.

In general, the volume of sweat produced during exercise is determined by metabolic rate, environmental temperature, radiant heat load, humidity, air velocity, clothing, and body size.

These factors influence the body's heat storage and temperature regulation. Heat is transferred from warmer areas to cooler ones, so heat loss from the body is impaired by high environmental temperatures, radiation, high humidity, and still air. Body size, specifically the ratio between surface area and mass, is important because large individuals generally expend more energy to do a given task, so they typically have higher metabolic rates and produce more heat. But they also have more surface area (skin), which allows more sweat formation and evaporation. As exercise intensity increases, so does the metabolic rate. This increases body heat production, which in turn increases sweating. To conserve water during exercise, blood flow to the kidneys decreases in an attempt to prevent dehydration; but like the increase in metabolic water production, this may be insufficient. Athletes commonly lose between 1% and 6% of their body water during intense, prolonged exercise. During high-intensity exercise under environmental heat stress, sweating can cause losses of as much as 3 L of water per hour. (Chapter 12 contains additional information about body water losses during exercise in warm environments.) In cold, dry environments or at altitude, water loss from respiration contributes to the overall loss of body water as well.

Dehydration and Exercise Performance

Even minimal changes in the body's water content can impair endurance performance. Without adequate fluid replacement, an athlete's exercise tolerance shows a pronounced decrease during long-term activity because of water loss through sweating. The impact of dehydration on the cardiovascular and thermoregulatory systems is quite predictable. Fluid loss decreases plasma volume. This decreases blood pressure, which in turn reduces blood flow to the muscles and skin. In an effort to overcome this, heart rate increases. Because less blood reaches the skin, heat dissipation is hindered, and the body retains more heat. Thus, when a person is dehydrated by 2% of body weight or more, both heart rate and body temperature are elevated during exercise above values observed when normally hydrated.

As one might expect, these physiological changes decrease exercise performance. Figure 15.14 illustrates the effects of an approximate 2% decrease in body

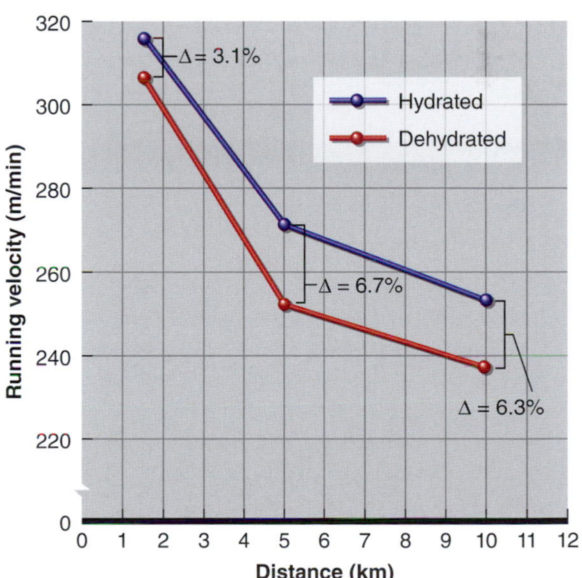

FIGURE 15.14 The decline in running velocity (meters per minute) with dehydration of about 2% of body weight for 1,500 m, 5,000 m, and 10,000 m time trials compared with velocity in the normally hydrated condition.

Reprinted by permission from L.E. Armstrong, D.L. Costill, and W.J. Fink, "Influence of Diuretic-Induced Dehydration on Competitive Running Performance," *Medicine and Science in Sports and Exercise* 17 (1985): 456-461.

weight attributable to dehydration from the use of a diuretic on distance runners' performance in 1,500 m, 5,000 m, and 10,000 m time trials on an outdoor track.[3] The dehydration condition resulted in plasma volume decreases between 10% and 12%. Although the average $\dot{V}O_{2max}$ did not differ between the normally hydrated and dehydrated trials, mean running velocity decreased by 3% in the 1,500 m run and by more than 6% in the 5,000 and 10,000 m runs. The greater the duration of the performance, the greater is the expected decline in performance for the same degree of dehydration. These trials were conducted in relatively cool weather. The higher the temperature, humidity, and radiation, the greater the expected decrement in performance for the same degree of dehydration. The decrement in performance would be progressively greater with greater levels of dehydration.

The effect of dehydration on performance in muscular strength, muscular endurance, and anaerobic types of activities is not as clear. Decrements have been seen in some studies, whereas other studies have shown no change in performance. In one of the best-controlled studies, researchers at Penn State University reported that 2% dehydration resulted in significant deterioration of basketball skills in 12- to 15-year-old boys who were skilled basketball players.[17]

Wrestlers and other weight-category athletes commonly dehydrate to get a weight advantage during the weigh-in for a competition. Most rehydrate after the weigh-in before the competition and experience only small decrements in performance. A summary of the effects of dehydration on exercise performance is shown in table 15.5.

In Review

- Water balance depends on electrolyte balance, and vice versa.
- At rest, water intake equals water output. Water intake includes water ingested from foods and fluids and produced as a metabolic by-product. The majority of water output at rest occurs from the kidneys, but water also is lost from the skin, from the respiratory tract, and in feces.
- During exercise, metabolic water production increases as metabolic rate increases.
- Water loss during exercise increases because as heat in the body increases, more water is lost with increased sweating. Sweat becomes the primary avenue for water loss during exercise. In fact, the kidneys decrease urine production in an effort to prevent dehydration.
- When dehydration reaches 2% of body weight, aerobic endurance performance—and even skilled performance in sports such as shooting free throws in basketball—is notably impaired. Heart rate and body temperature increase in response to dehydration.

Electrolyte Balance During Exercise

Normal body function depends on a balance between water and electrolytes. We have discussed the effects of water loss on performance. Now we turn our attention to the effects of the other component of this delicate balance: electrolytes. When large amounts of water are lost from the body, as during exercise, the balance between water and electrolytes can be disrupted quickly. In the next sections, we examine the effects of exercise on electrolyte balance. Our focus is on the two major routes for electrolyte loss: sweat and urine production.

Electrolyte Loss in Sweat Human sweat is a filtrate of blood plasma, so it contains many substances found there, including sodium (Na^+), chloride (Cl^-), potassium (K^+), magnesium (Mg^{2+}), and calcium (Ca^{2+}). Although sweat tastes salty, it contains far fewer minerals than the plasma and other body fluids. In fact, sweat is 99% water.

Sodium and chloride are the predominant ions in sweat and blood. As indicated in table 15.6, the concentrations of sodium and chloride in sweat are about one-third those found in plasma and five times those found in muscle. Each of these three fluids' **osmolarity**, which is the ratio of solutes (such as electrolytes) to fluid, is also shown. Sweat's electrolyte concentration can vary considerably between individuals. It is strongly influenced by genetics, the rate of sweating, the state of training, and the state of heat acclimatization.

At the elevated rates of sweating reported during endurance events, sweat contains large amounts of sodium and chloride but little potassium, calcium, and magnesium. Based on estimates of the athlete's total body electrolyte content, such losses would lower the body's sodium and chloride content by only about 5% to 7%. Total body levels of potassium and magnesium, two ions principally confined to the insides of cells, would decrease by about 1%. These losses probably have no measurable effect on an athlete's performance.

As electrolytes are lost in sweat, the remaining ions are redistributed among the body tissues. Consider potassium. It diffuses from active muscle fibers as they contract, entering the extracellular fluid. The increase this causes in extracellular potassium levels does not equal the amount of potassium that is released from active muscles, because potassium is taken up by inactive muscles and other tissues while the active muscles are losing it. During recovery, intracellular potassium levels normalize quickly. Some researchers suggest that

TABLE 15.5 **Alterations in Physiological Function and Performance During Dehydration of 2% or Greater**

Variable	Dehydration
Cardiovascular	
Blood volume/Plasma volume	Decreased
Cardiac output	Decreased
Stroke volume	Decreased
Heart rate	Increased
Metabolic	
Aerobic capacity ($\dot{V}O_{2max}$)	No change or decreased
Anaerobic power (Wingate test)	No change or decreased
Anaerobic capacity (Wingate test)	No change or decreased
Blood lactate, peak value	Decreased
Buffer capacity of the blood	Decreased
Lactate threshold, velocity	Decreased
Muscle and liver glycogen	Decreased
Blood glucose during exercise	Possibly decreased
Protein degradation with exercise	Possibly decreased
Thermoregulation and fluid balance	
Electrolytes, muscle and blood	Decreased
Exercise core temperature	Increased
Sweat rate	Decreased, delayed onset
Skin blood flow	Decreased
Performance	
Muscular strength	No change or decreased
Muscle endurance	No change or decreased
Muscular power	Unknown
Speed of movement	No change or decreased
Run time to exhaustion	Decreased
Total work performed	Decreased
Attention and focus	Decreased
Some skill aspects of performance	Decreased

Note. Data for this table were derived from the following reviews: M. Fogelholm, 1994, "Effects of bodyweight reduction on sports performance," *Sports Medicine* 18: 249-267; C.A. Horswill, 1994, "Physiology and nutrition for wrestling," in D.R. Lamb, H.G. Knutten, & R. Murray (Eds.), *Physiology and nutrition for competitive sport* (Vol. 7, pp. 131-174); H.L. Keller, S.E. Tolly, & P.S. Freedson, 1994, "Weight loss in adolescent wrestlers," *Pediatric Exercise Science* 6: 211-224; and R. Opplinger, H. Case, C. Horswill, G. Landry, & A. Shelter, 1996, "Weight loss in wrestlers: An American College of Sports Medicine position stand," *Medicine and Science in Sports and Exercise* 28: ix-xii.

TABLE 15.6 **Electrolyte Concentrations and Osmolarity in Sweat, Plasma, and Muscle of Men Following 2 h of Exercise in the Heatw**

	Electrolytes (mEq/L)				
Site	Na+	Cl^-	K^+	$Mg2^+$	Osmolarity (mOsm/L)
Sweat	40-60	30-50	4-6	1.5-5	80-185
Plasma	140	101	4	1.5	295
Muscle	9	6	162	31	295

Note. mEq/L = milliequivalents per liter (thousandths of 1 g of solute per 1 L of solvent).

these muscle potassium disturbances during exercise might contribute to fatigue by altering the membrane potentials of neurons and muscle fibers, making it more difficult to transmit impulses.

Electrolyte Loss in Urine In addition to clearing wastes from the blood and regulating water levels, the kidneys also regulate the body's electrolyte content. Urine production is the other major source of electrolyte loss. At rest, electrolytes are excreted in the urine as necessary to maintain homeostatic levels, and this is the primary route for electrolyte loss. But as the body's water loss increases during exercise, urine production rate decreases considerably in an effort to conserve water. Consequently, with very little urine being produced, electrolyte loss by this avenue is minimized.

The kidneys play another role in electrolyte management. If, for example, a person ingests 250 mEq of salt (NaCl), the kidneys will normally excrete 250 mEq of these electrolytes to keep the body's NaCl content constant. Heavy sweating and dehydration, however, trigger the release of the hormone aldosterone from the adrenal gland. This hormone stimulates renal reabsorption of sodium. Consequently, the body retains more sodium than usual during the hours and days after a prolonged exercise bout. This elevates the body's sodium content and increases the osmolality of the extracellular fluids.

This increased sodium content triggers thirst, compelling the person to consume more water, which is then retained in the extracellular compartment. The increased water consumption reestablishes normal osmolality in the extracellular fluids but leaves these fluids expanded, which dilutes the other substances present there. This expansion of the extracellular fluids has no negative effects and is temporary. In fact, this is one of the major mechanisms for the increase in plasma volume that occurs with training and with acclimatization to exercise in the heat. Fluid levels return to normal within 48 to 72 h after exercise, provided there are no subsequent exercise bouts.

In Review

- The loss of large volumes of sweat can disrupt electrolyte balance, although the concentrations of electrolytes in sweat are much lower than in plasma.
- Electrolyte loss during exercise occurs primarily with water loss from sweating. Sodium and chloride are the most abundant electrolytes in sweat.
- There is substantial variability among individuals in both sweating rate and the electrolyte composition of the sweat. This makes it virtually impossible to create a "one size fits all" plan for replacing fluids and electrolytes.
- At rest, excessive electrolytes are excreted in the urine by the kidneys. But urine production declines substantially during exercise, so little electrolyte loss occurs by this route.
- Dehydration induces the hormones ADH (antidiuretic hormone) and aldosterone to promote renal retention of water and sodium. The increased sodium and decreased fluid volume trigger thirst (see chapter 4).

Replacement of Body Fluid Losses

The body loses more water than electrolytes during heavy sweating. This raises the osmotic pressure in the body fluids because the electrolytes become more concentrated. The need to replace body water is greater than the need for electrolytes because only by replenishing water content can the electrolytes return to normal concentrations. But how does the body know when this is necessary?

Thirst When people feel thirsty, they drink. The thirst sensation is regulated largely by the osmoreceptors in the hypothalamus. Sensory signals of thirst are invoked when the plasma's osmolality is increased above a threshold value. A second set of signals arises from the low-pressure baroreceptors when low blood volume is sensed. However, relative to osmolar control of thirst, a large blood volume loss is necessary for this backup control system to be activated. Unfortunately, the body's **thirst mechanism** doesn't precisely gauge its state of dehydration. It does not sense thirst until well after dehydration begins. Even when dehydrated, people might desire fluids only at intermittent intervals.

The control of thirst is not fully understood. When permitted to drink water as their thirst dictates, people can require 24 to 48 h to completely replace water lost through heavy sweating. Because of our sluggish drive to replace body water and to prevent chronic dehydration, we are advised to drink more fluid than our thirst dictates. Because of the increased water loss during exercise, it is imperative that athletes' water intake be sufficient to meet their bodies' needs, and it is essential that they rehydrate during and after an exercise bout.

Benefits of Fluids During Exercise Drinking fluids during prolonged exercise, especially in hot weather, has obvious benefits. Water intake minimizes dehydration, increases in body temperature, cardiovascular stress, and declines in performance. As seen in figure 15.15, when subjects became dehydrated during several hours of treadmill running in the heat (40 °C, or 104 °F) without fluid replacement, their heart rates

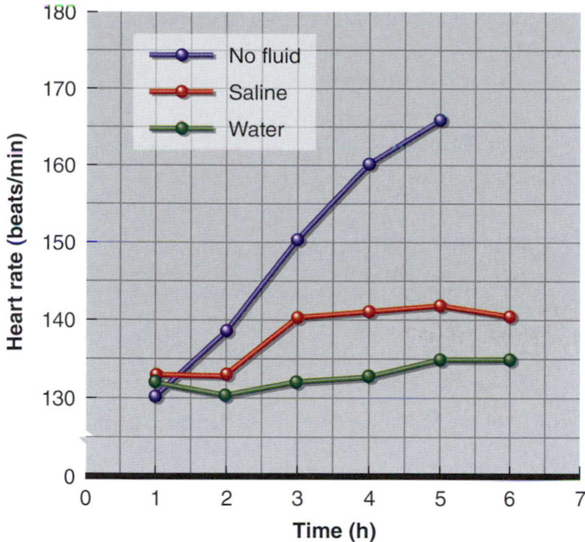

FIGURE 15.15 Heart rate during 6 h of treadmill running in the heat. Subjects received no fluid, a saline solution, or water. Subjects deprived of fluids became exhausted and could not complete the 6 h exercise.

Data from Barr, Costill, and Fink (1991).

increased steadily throughout the exercise.[5] When they were deprived of fluids, the subjects became exhausted and couldn't complete the 6 h exercise. Ingesting either water or a saline solution in amounts equal to weight loss prevented dehydration and kept subjects' heart rates lower. Even warm fluids (near body temperature) provide some protection against overheating, but cold fluids enhance body cooling because some of the deep body heat is used to warm cold drinks to body temperature.

Hyponatremia

Fluid replacement is beneficial, but too much of a good thing is problematic. In the 1980s, the first cases of hyponatremia were reported in endurance athletes. **Hyponatremia** is clinically defined as a serum sodium concentration below the normal range of 135 to 145 mmol/L. Symptoms of hyponatremia generally appear when serum sodium levels drop below 130 mmol/L. Early signs and symptoms include bloating, puffiness, nausea, vomiting, and headache. As the severity increases, due to increasing cerebral edema (swelling of the brain), the symptoms include confusion, disorientation, agitation, seizures, pulmonary edema, coma, and death. How likely is hyponatremia to occur?

The processes that regulate fluid volumes and electrolyte concentrations are highly effective, so consuming enough water to dilute plasma electrolytes is difficult under normal circumstances. Marathoners who lose 3 to 5 L of sweat and drink 2 to 3 L of water maintain normal plasma concentrations of sodium, chloride, and potassium. And distance runners who run 25 to 40 km (15.5-24.9 mi) per day in warm weather and do not salt their food don't develop electrolyte deficiencies.

Some research has suggested that during ultramarathon running (more than 42 km, or 26.2 mi), athletes can experience hyponatremia. A case study of two runners who collapsed after an ultramarathon race (160 km, or 100 mi) in 1983 revealed that their blood sodium concentrations had decreased from a normal value of 140 mmol/L to values of 123 and 118 mmol/L.[19] One of the runners experienced a grand mal seizure; the other became disoriented and confused. Examining the runners' fluid intakes and estimating their sodium intakes during the run suggested that they had diluted their sodium contents by consuming too much fluid that contained too little sodium.

The ideal resolution to prevent hyponatremia would be to replace water at the exact rate at which it is being lost or to add sodium to the ingested fluid. The problem with the latter approach is that most sport drinks contain no more than 25 mmol/L of sodium and are apparently too weak to prevent sodium dilution alone, but very strong concentrations cannot be tolerated. Exercise hyponatremia appears to be the result of a fluid overload due to overconsumption, underreplacement of sodium losses, or both. Only a small number of cases have been reported. Thus, it is probably inappropriate to form conclusions from this information to design a fluid replacement regimen for people who must exercise for long periods in the heat.

Recommendations for Fluid Replacement Before, During, and After Exercise

Dehydration is a potential problem for athletes who train and compete for long periods of time, as well as in hot and humid environments. The American College of Sports Medicine (ACSM), the American Dietetic Association, and the Dieticians of Canada have all published guidelines for adequate fluid intake before, during, and following exercise. Key recommendations include the following:

- Two hours before exercise, the athlete should consume 400 to 600 ml (14 to 22 oz.) of fluid to provide hydration and allow time for excretion of excess ingested water.
- During exercise, the athlete should drink enough fluid to keep fluid losses to less than 2% of body weight. Weight gain from overdrinking should be avoided.
- After exercise, the athlete should consume adequate fluids to fully replace sweat losses that occurred during exercise.

- Sport drinks containing carbohydrate concentrations of 4% to 8%, and sodium in concentrations between 0.5 and 0.7 g/L, are recommended during intense exercise events lasting longer than 1 h.
- Including sodium in drinks or eating high-sodium foods during the recovery period can help the rehydration process.[2]

In Review

- The need to replace lost body water is greater than the need to replace lost electrolytes, because sweat is very dilute.
- The thirst mechanism does not exactly match the body's hydration state, so often more fluid must be consumed than thirst dictates to maintain body weight.
- Fluid intake during prolonged exercise minimizes dehydration and optimizes cardiovascular and thermoregulatory function.
- In some rare cases, drinking too much fluid with too little sodium has led to hyponatremia (a low plasma concentration of sodium), which can cause confusion, disorientation, and even seizures, coma, and death.

Nutrition and Athletic Performance

Athletes place considerable demands on their bodies every day they train and compete. Optimal performance must include optimal nutrition. Too often, athletes spend considerable time and effort perfecting skills and attaining top physical condition while ignoring proper nutrition and sleep. Performance deterioration often can be traced to poor nutrition.

VIDEO 15.3 Presents Nancy Williams on reasons why some athletes might not eat enough.

Evidence-Based Nutrition Guidelines for Athletes

In 2016, experts from the American College of Sports Medicine, the American Dietetic Association, and the Dieticians of Canada collaborated to prepare a joint position statement on nutrition strategies for athletes.[2] This publication provides guidelines for the appropriate type, amount, and timing of foods, fluids, and dietary supplements. Based on an evidence-based analysis, their conclusions are summarized as follows:

1. Athletes need to consume energy that is adequate in amount and timing of intake during periods of high-intensity or long-duration training to maintain health and maximize training outcomes.
2. The primary goal of a training diet is to allow an athlete to stay healthy and injury-free while maximizing the functional and metabolic adaptations to training.
3. The optimal physique, including body size, shape, and composition (e.g., muscle mass and body fat levels), depends upon the sex, age, and heredity of the athlete, and may be sport and event specific. Where significant manipulation of body composition is required, it should ideally take place well before the competitive season to minimize the influence on event performance or reliance on rapid weight loss techniques.
4. Body carbohydrate stores provide an important fuel source for the brain and muscle during exercise and are manipulated by exercise and dietary intake. Recommendations for carbohydrate intake typically range from 3 to 10 g/kg body weight per day (and up to 12 g/kg body weight per day for extreme and prolonged activities), but should be individualized to the athlete and situation.
5. Recommendations for protein intake typically range from 1.2 to 2.0 g/kg body weight per day, but have more recently been expressed in terms of the regular spacing of intakes of modest amounts of high-quality protein (0.3 g/kg body weight) after exercise and throughout the day. Such intakes can generally be met from food sources.
6. For most athletes, fat intakes associated with eating styles that accommodate dietary goals typically range from 20% to 35% of total energy intake. Consuming greater than or equal to 20% of energy intake from fat does not benefit performance, and extreme restriction of fat intake may limit the food range needed to meet overall health and performance goals.
7. Athletes should consume diets that provide at least the RDA or AI for all micronutrients.
8. A primary goal of competition nutrition is to address nutrition-related factors that may limit performance by causing fatigue and a deterioration in skill or concentration over the course of the event.

9. Foods and fluids consumed in the 1 to 4 h before an event should contribute to body carbohydrate stores, ensure appropriate hydration status, and maintain gastrointestinal comfort throughout the event.
10. Dehydration or hypohydration can increase the perception of effort and impair exercise performance; thus, appropriate fluid intake before, during, and after exercise is important for health and optimal performance. Individualized fluid plans should be developed.
11. An additional nutrition-related strategy for events of less than 60 min duration is to consume carbohydrate according to its potential to enhance performance.
12. Rapid restoration of performance between physiologically demanding training sessions or competitive events requires appropriate intake of fluids, electrolytes, energy, and carbohydrates to promote rehydration and restore muscle glycogen.
13. In general, vitamin and mineral supplements are unnecessary for athletes who consume a diet providing high energy availability from a variety of nutrient-dense foods. A multivitamin and mineral supplement may be appropriate in some cases when these conditions do not exist.
14. Athletes should be counseled regarding the appropriate use of ergogenic aids. These products (typically supplements) should be used only after careful evaluation for safety, efficacy, potency, and compliance with relevant antidoping codes and legal requirements.
15. Vegetarian athletes may be at risk for low intakes of energy, protein, fat, creatine, carnosine, n-3 fatty acids, and key micronutrients such as iron, calcium, riboflavin, zinc, and vitamin B_{12}.

Much of the scientific rationale for these recommendations is provided throughout this chapter and in the full position stand.[2] There are, however, special situations in which additional information is needed. We will now look at the precompetition meal and muscle glycogen replacement and loading.

Precompetition Meal

Although the meal ingested a few hours before competition might contribute little to muscle glycogen stores, which reflect long-term carbohydrate intake, it can ensure a normal blood glucose level and prevent hunger. This meal should contain only about 200 to 500 kcal and consist mostly of carbohydrate foods that are easily digested. Foods such as cereal, milk, juice, and toast are digested rather quickly and won't leave the athlete feeling full during competition. In general, this meal should be consumed at least 2 h before competition. The rates at which food is digested and nutrients are absorbed into the body are quite individual, so timing the precompetition meal might depend on prior experience. In one study of endurance cyclists, a prolonged cycling exercise trial to exhaustion at 70% of the subject's $\dot{V}O_{2max}$ was performed under two different conditions, with 14 days between trials: a breakfast containing 100 g of carbohydrate fed 3 h before exercise (fed) and no feeding before exercise (fasted). Subjects tested under the fed condition exercised 136 min before reaching exhaustion compared with 109 min in the fasted trial, indicating the importance of the precompetition meal.[29]

A liquid precompetition meal might be less likely to result in nervous indigestion, nausea, vomiting, and abdominal cramps. Such feedings are commercially available and generally have been found useful both before and between events. Finding time for athletes to eat is often difficult when they must perform in multiple preliminary and final events. Under these circumstances, a liquid feeding that is low in fat and high in carbohydrate might be the only solution.

Muscle Glycogen Replacement and Loading

Earlier in this chapter, we established that different diets can markedly influence muscle glycogen stores and that endurance performance depends largely on these stores. The theory is that the greater the amount of glycogen stored, the better the potential endurance performance, because fatigue will be delayed. Thus, an athlete's goal is to begin an exercise bout or competition with as much stored glycogen as possible.

On the basis of muscle biopsy studies conducted in the mid-1960s, Åstrand[4] proposed a plan to help runners store the maximum amount of glycogen. This process is known as glycogen or carbohydrate loading. According to Åstrand's regimen, athletes should begin to prepare for an aerobic endurance competition by completing an exhaustive training bout 7 days before the event. For the next 3 days, they should eat fat and protein almost exclusively to deprive the muscles of carbohydrate, which increases the activity of glycogen synthase, an enzyme responsible for glycogen synthesis and storage. Athletes should then eat a carbohydrate-rich diet for the remaining 3 days before the event. Because glycogen synthase activity is increased, increased carbohydrate intake results in greater muscle glycogen storage. Training intensity and volume during this 6-day period should be markedly reduced to prevent additional muscle glycogen depletion, thus maximizing liver and muscle glycogen reserves. Originally, an additional intense training bout was performed 4 days before competition.

This regimen has been shown to elevate muscle glycogen stores to twice the normal level, but it is somewhat impractical for most highly trained competitors. During the 3 days of low carbohydrate intake, athletes generally find training difficult. They are also often irritable and unable to perform mental tasks, and they typically show signs of low blood sugar, such as muscle weakness and disorientation. In addition, the exhaustive depletion bouts of exercise performed 7 days before the competition have little training value and can impair glycogen storage rather than enhance it. This depletion exercise also exposes athletes to possible injury or overtraining.

Considering these limitations, many now propose that the depletion exercise and the low-carbohydrate aspects of Åstrand's regimen be eliminated. Instead, the athlete should simply reduce training intensity a week before competition and eat a normal, mixed diet containing 55% of the calories from carbohydrate until 3 days before the competition. For these days, training should be reduced to a daily warm-up of 10 to 15 min of activity and accompanied by a carbohydrate-rich diet. With this plan, as seen in figure 15.16, glycogen will be elevated to nearly 200 mmol/kg of muscle, the same level attained with Åstrand's depletion regimen, and the athlete will be better rested for competition.

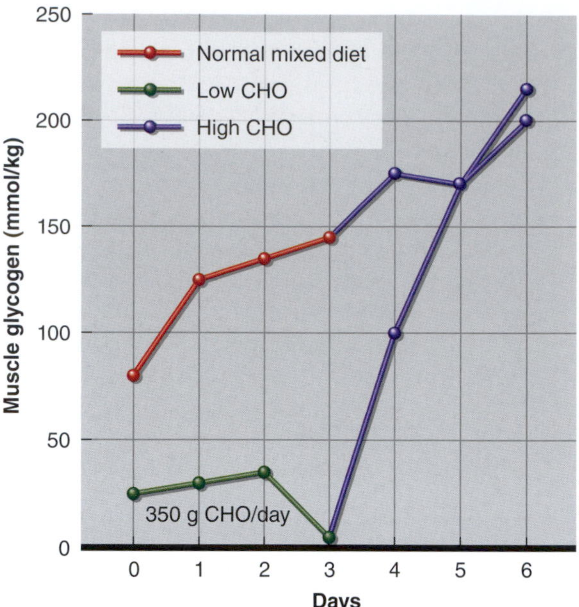

FIGURE 15.16 Two regimens for muscle glycogen loading. In one regimen, the subjects were depleted of muscle glycogen (day 0) and then ate a low-carbohydrate (CHO) diet for 3 days. They then switched to a CHO-rich diet, which caused muscle glycogen to increase to about 200 mmol/kg. In the other dietary regimen, the subjects ate a normal, mixed diet and reduced their training volume for the first 3 days; they then changed to a high-CHO diet and a further reduction in training volume for 3 days, which also resulted in muscle glycogen of about 200 mmol/kg.

Data from Åstrand (1979).

It is possible to increase carbohydrate stores rapidly after even a very short near-maximal-intensity bout of exercise. In a study of seven endurance athletes, scientists found that cycling for 150 s at 130% of $\dot{V}O_{2peak}$ followed by 30 s of all-out cycling and 24 h of high carbohydrate intake was sufficient to nearly double muscle glycogen stores in just one day.[18]

Diet is also important in preparing the liver for the demands of endurance exercise. Liver glycogen stores decrease rapidly when a person is deprived of carbohydrate for only 24 h, even when at rest. With only 1 h of strenuous exercise, liver glycogen decreases by 55%. Thus, hard training combined with a low-carbohydrate diet can empty the liver glycogen stores. A single carbohydrate meal, however, quickly restores liver glycogen to normal. Clearly, a carbohydrate-rich diet in the days preceding competition will maximize the liver glycogen reserve and minimize the risk of hypoglycemia during the event.

Water is stored in the body at a rate of about 2.6 g of water with each gram of glycogen. Consequently, an increase or decrease in muscle and liver glycogen generally produces a change in body weight of 0.5 to 1.4 kg (1-3 lb). Some scientists have proposed monitoring changes in muscle and liver glycogen stores via recording the athlete's early morning weight immediately after rising—after emptying the bladder but before eating breakfast. A sudden decrease in weight might reflect a failure to replace glycogen, a deficit in body water, or both.

Athletes who must train or compete in exhaustive events on successive days should replace muscle and liver glycogen stores as rapidly as possible. Although liver glycogen can be depleted totally after 2 h of exercise at 70% $\dot{V}O_{2max}$, it is replenished within a few hours when a carbohydrate-rich meal is consumed. Muscle glycogen resynthesis, on the other hand, is a slower process, taking several days to return to normal after an exhaustive exercise bout such as the marathon (see figure 15.17). Studies in the late 1980s revealed that muscle glycogen resynthesis was most rapid when individuals were fed at least 50 g (about 0.7 g/kg body weight) of glucose every 2 h after the exercise.[22] Feeding subjects more than this amount did not appear to accelerate the replacement of muscle glycogen. During the first 2 h after exercise, the rate of muscle glycogen resynthesis is much higher than later in recovery,

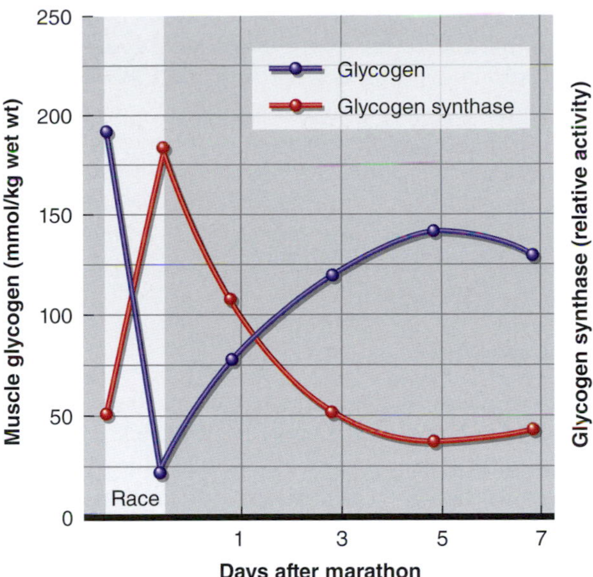

FIGURE 15.17 Muscle glycogen resynthesis is a slow process, requiring several days to restore normal muscle glycogen storage following exhaustive exercise. Note that when muscle glycogen decreases with hard exercise (race), muscle glycogen synthase is markedly elevated. This triggers the muscle to store glycogen when carbohydrates are eaten, returning glycogen synthase to the baseline level.

as discussed earlier in this chapter. Thus, an athlete recovering from an exhaustive endurance event should ingest sufficient carbohydrate as soon after exercise as is practical. Adding protein and amino acids to the carbohydrate ingested during the recovery period enhances muscle glycogen synthesis above that achieved with carbohydrate alone.

Sport Drinks

We mentioned earlier that ingesting carbohydrate before, during, and following exercise benefits performance by ensuring adequate fuel for energy production during exercise and replenishing glycogen stores following exercise. While selecting a wise diet can provide for most of the athlete's nutrition needs, nutritional supplements can also be of great value. In addition, adequate fluid intake is necessary for preexercise hydration, hydration during exercise, and rehydration following exercise. Sport drinks are uniquely designed to meet both the energy and fluid needs of the athlete. Performance benefits from these drinks have been clearly documented, not only in endurance activities but also in burst activities as well (e.g., soccer and basketball).[10,17]

Sport drinks differ from one another in a number of ways besides in taste. Of major concern, however, is the rate at which energy and water are delivered. Energy delivery is primarily determined by the concentration of the carbohydrates in the drink, and fluid replacement is influenced by both taste and the sodium concentration of the drink.

In Review

- Some athletes have adopted vegetarian diets and appear to perform well. However, careful consideration must be given to protein sources and to consuming adequate levels of iron, zinc, calcium, and several vitamins.
- The precompetition meal should be eaten no less than 2 h before competition, and it should be low in fat, high in carbohydrate, and easily digestible. A liquid precompetition meal low in fat and high in carbohydrate has advantages.
- Carbohydrate loading increases muscle glycogen content, which in turn increases endurance performance.
- After endurance competition or training, it is important to consume substantial carbohydrate to replace the glycogen used during activity. Replacing glycogen during the first few hours after training or competing is optimal because glycogen synthase levels are at their peak.

Energy Delivery: The Carbohydrate Concentration

A major concern is how rapidly the drink leaves the stomach, or the rate of **gastric emptying**. In general, carbohydrate solutions empty more slowly from the stomach than either water or a weak sodium chloride (salt) solution. Research suggests that a solution's caloric content, a reflection of its concentration, might be a major determinant of how quickly it empties from the stomach and is absorbed from the intestine. Since carbohydrate solutions remain in the stomach longer than either water or weak solutions, increasing the glucose concentration of a sport drink significantly reduces the gastric emptying rate. For example, 400 ml (14 oz) of a weak glucose solution (139 mmol/L) is almost completely emptied from the stomach in 20 min, but emptying a similar volume of a strong glucose solution (834 mmol/L) can require nearly 2 h.[14] However, when even a small amount of a strong glucose drink leaves the stomach, it can contain more sugar than a larger amount of a weaker solution simply because of its higher concentration. But, if an athlete is trying to prevent dehydration, this would deliver less water and thus be counterproductive.

What is the optimal CHO ingestion rate, that is, what dose of CHO is most beneficial to athletic performance? The ACSM and the National Athletic Trainers'

Association both recommend CHO ingestion at a rate of 30 to 60 g/h for endurance athletes, but both lower and higher CHO doses have also been shown to improve performance. The optimal concentration would depend on several factors, including emptying from the stomach, absorption into the blood from the intestines, and the rate at which muscle can use the CHO (about 60-90 g/h depending on the type of CHO ingested). Concentrations of CHO that are too high can cause stomach discomfort and slow absorption of both CHO and fluid from the intestines.

A 2013 study[30] investigated the relation between CHO concentration in sport drinks and cycling time-trial performance. Fifty-one young cyclists and triathletes each completed four trials. After a 2 h ride at about 70% $\dot{V}O_{2peak}$, subjects performed a computer-simulated 20 km time trial, which they were asked to complete as quickly as possible. Twelve different beverages were consumed (1 L/h) during the 2 h ride in a double-blind manner (neither the person giving the athlete the beverage nor the athlete knew which beverage was which). The beverages provided subjects with 10, 20, 30, 40, 50, 60, 70, 80, 90, 100, 110, or 120 g of mixed CHO (1:1:1 glucose-fructose-maltodextrin) per hour or a noncaloric placebo. Subsequent time-trial performance improved incrementally with CHO concentrations up to 78 g/h, with diminishing performance enhancement at CHO levels greater than 78 g/h. That is, the performance-enhancing qualities of the sport drinks started to plateau at CHO intakes above 78 g/h, suggesting that in competitive race settings, athletes should strive to consume a mixed-CHO sport drink at a CHO ingestion rate of 68 to 88 g/h.

Most sport drinks on the market contain about 6 to 8 g of carbohydrate per 100 ml (3.5 oz) of fluid (6% to 8%). The carbohydrate source is generally glucose, glucose polymers, or a combination of glucose and glucose polymers, although fructose or sucrose has also been used.[22] Research studies have confirmed enhanced endurance performance with use of solutions in this range of concentration and with these sources of carbohydrate when compared to water.[2] Carbohydrate solutions above ~6% slow gastric emptying and limit the immediate availability of fluid. However, they can provide a greater amount of carbohydrate in a given period of time to meet the increased energy needs.[2,23]

Rehydration with Sport Drinks: The Sodium Concentration

Just adding fluid to the body during exercise lessens the risk of serious dehydration. But research indicates that adding glucose and sodium to sport drinks, aside from supplying an energy source, stimulates both water and sodium absorption. Sodium increases both thirst and the palatability of the drink. Recall that when sodium is retained, this causes more water to be retained. For rehydration purposes, both during and following exercise, the sodium concentration should range between 20 and 60 mmol/L.[23] There is an important loss of sodium from the body with sweating. With high rates of sweating and large volumes of water intake, this can lead to critical reductions in the sodium concentration of the blood and possibly to hyponatremia, as discussed earlier in this chapter.

Palatability

Athletes will not drink solutions that taste bad. Unfortunately, we all have different taste preferences. To further confound the issue, what tastes good before and after a long, hot bout of exercise will not necessarily taste good during the event. Studies of taste preferences of runners and cyclists during 60 min of exercise showed that most chose a drink with a light flavor and no strong aftertaste. But, will athletes drink more if given a sport drink as compared to water? In one study, runners ran on a treadmill for 90 min and then recovered while seated for an additional 90 min. Both exercise and recovery conditions were controlled in an environmental chamber at a temperature of 32 °C (86 °F), 50% humidity. Three trials were conducted, two with two different sport drinks (6% and 8% carbohydrate) and one with water. Subjects were encouraged to drink throughout each trial. The volume consumed during exercise was similar for all three drinks, but during recovery, the runners drank about 55% more of each of the two sport drinks than water.[37]

In Review

- Sport drinks have been shown to reduce the risk of dehydration and provide an important source of energy. They also can improve the performance of the athlete in both endurance and sudden burst activities such as soccer and basketball.
- To maximize both CHO and fluid intake, the carbohydrate concentration of a sport drink generally should not exceed 6% to 8%.[2]
- In competitive race settings, athletes should strive to consume a mixed-CHO sport drink at a CHO ingestion rate of 68 to 88 g/h.
- The inclusion of sodium in a sport drink facilitates the intake and storage of water.
- Taste is an important factor when one is considering a sport drink. Most athletes prefer a light flavor without a strong aftertaste. Each athlete should select the drink that tastes best, provided that the nutritional ingredients are the same.

IN CLOSING

In this chapter, we examined the body composition and nutrition needs of the athlete, considering the importance of optimizing body composition for peak performance and eating wisely to enhance athletic performance. We discovered the importance of each of the six nutrient categories and how they can be adjusted to meet the athlete's training and performance needs. We looked at the precompetition meal, ways to effectively replenish and load muscle glycogen stores, and the effectiveness of commercial sport drinks. Now that we have a better understanding of the importance of an appropriate weight and a balanced diet, we turn our attention to another aspect of the athlete's quest for success. In the next chapter, we evaluate substances that have been proposed to enhance athletic performance—ergogenic aids.

KEY TERMS

air plethysmography
bioelectric impedance
body composition
body density
carbohydrate loading
dehydration
densitometry
dual-energy X-ray absorptiometry (DEXA)
electrolyte
essential amino acids
extracellular fluid
fat
fat-free mass
fat mass
free radicals
gastric emptying
glycogen loading
hydrostatic weighing
hyponatremia
intracellular fluid
macrominerals
microminerals
nonessential amino acids
osmolarity
protein
relative body fat
skinfold fat thickness
thirst mechanism
trace elements
vitamin

STUDY QUESTIONS

1. Differentiate between body size and body composition. What tissues of the body constitute the fat-free mass?
2. What is densitometry? How is it used to assess the body composition of the athlete? What is the major weakness of densitometry with respect to its accuracy?
3. What are several field techniques for estimating body composition? What are their strengths and weaknesses?
4. What is the relationship of relative leanness and fatness to performance in sport?
5. What guidelines should be used to determine the athlete's goal weight?
6. What are the six categories of nutrients?
7. What role does dietary carbohydrate play in endurance performance? How about fat? Protein?
8. When consuming carbohydrates during competition, what is the optimal dose that athletes should strive to consume?
9. What is an appropriate protein allowance for a normally active adult man? For a woman? Discuss the value of using protein supplements to enhance performance in strength and endurance events.
10. Should the athlete supplement vitamins and minerals?
11. How does dehydration affect exercise performance? What effect does dehydration have on exercise heart rate and body temperature?
12. Describe the recommended precompetition meal.
13. Describe the method used to maximize muscle glycogen storage (glycogen loading).
14. Discuss the value of consuming carbohydrate during and after endurance exercise. What are the potential benefits of sport drinks?

STUDY GUIDE ACTIVITIES

In addition to the activities listed in the chapter opening outline, two other activities are available in the web study guide, located at

www.HumanKinetics.com/PhysiologyOfSportAndExercise

The **KEY TERMS** activity reviews important terms, and the end-of-chapter **QUIZ** tests your understanding of the material covered in the chapter.

I.M. Only
1 ML
100 mg.

Ergogenic Aids in Sport

In this chapter and in the web study guide

In the summer of 2013, two world-class track sprinters failed their in-competition urine tests and were provisionally suspended until each of their urine B samples was analyzed in the presence of the athletes or their representatives. The B-sample testing confirmed the results of the A-sample testing: The athletes had ingested oxilofrine, a banned stimulant. Each sprinter faced a suspension of up to 4 years, a ban that would effectively end their competitive careers. Both sprinters professed their innocence, claiming that they must have consumed a dietary supplement that was contaminated with the drug. Subsequent analysis of one of the many supplements the athletes admitted consuming showed that the supplement did indeed contain oxilofrine. Athletes are held fully responsible—strictly liable—for everything they ingest, so even when athletes inadvertently ingest a banned substance, ignorance of supplement contamination is not a valid defense. Unfortunately for the athletes, it is entirely possible that they had no prior knowledge that the supplement they willingly ingested on the advice of a personal trainer contained a banned substance.

Some dietary supplements purported to improve muscle strength, sexual performance, or weight loss have been found to contain banned substances—often prescription medications or altered versions thereof. Surprising to many people is that some supplements even have banned substances listed as ingredients on the product labels. Setting aside the accidental contamination of supplements, the never-ending quest to improve performance, lose weight, or alter body image has led some athletes to purposefully use banned substances or techniques. Decades of purposeful attempts at cheating have led to the implementation of sophisticated drug testing administered by the World Anti-Doping Agency (WADA) and other sports organizations to catch athletes who try to gain an unfair advantage over their competitors. Unfortunately, some athletes fail sport drug tests because they ingested supplements that, unbeknownst to them, contained prohibited ingredients. Even though they had no intent to cheat, those athletes also risk long suspensions from their sports, large legal fees, and loss of their good reputations.

Some athletes are willing to try anything to improve their performance and believe that special dietary supplements can be the deciding factor. Others choose to use prohibited techniques such as blood doping. Still others try prescription medications or various formulations of hormones in an attempt to create a competitive advantage.

Ergogenic aids are substances or practices (such as hypnosis or a coach's pep talk) that improve an athlete's performance. Some ergogenic aids, such as proper training and basic sport nutrition, are a legitimate part of sport, and their benefits have been extensively researched. However, the proposed benefits of many **ergogenic** (work-producing) substances or techniques lack the scientific support required to verify their effectiveness. Athletes often receive tips about ergogenic aids from friends or coaches or have seen advice on the Internet and assume that such guidance is safe and accurate. Some athletes experiment with dietary supplements, special clothing and equipment, or unique training techniques hoping for even a slight performance improvement, regardless of possible negative consequences.

The list of ergogenic aids is long, but the list of those with substantiated ergogenic properties is much shorter. In fact, some allegedly ergogenic substances or practices can actually impair performance. These are usually drugs—beta-blockers are one example—and one author has termed them **ergolytic** (work-decreasing) agents.[22] Ironically and sometimes tragically, several ergolytic agents have been promoted as ergogenic aids.

Many athletes indiscriminately take prescription drugs and dietary supplements in the belief that these will improve their performance. According to one study of 53 Division I university coaches and trainers, 94% provided their athletes with dietary supplements, despite the fact that the National Collegiate Athletic Association (NCAA) and the National Athletic Trainers' Association (NATA) encourage nutrition education and a food-first approach rather than reliance on dietary supplements.[14,47] Use of dietary supplements might seem like a legitimate attempt to improve performance, but some dietary supplements are purposefully or accidentally contaminated with prohibited substances.

The focus of this chapter is on selected nutritional supplements with possible ergogenic effects, as well as banned substances and techniques with confirmed ergogenic effects. More general sport nutrition practices are addressed in chapter 15. Psychological practices and mechanical factors are beyond the scope of this book but are reviewed in depth elsewhere.[56]

Before we cover specific ergogenic aids, it is important to take a quick look at some of the challenges associated with studying the proposed effects of performance-enhancing agents. For example, the inherent limitations of designing studies on ergogenic aids, including the power of the placebo effect, directly affect our ability to draw practical and scientific conclusions. For those reasons, many well-designed studies are usually required before it is possible to confidently draw conclusions about the ergogenic benefit of any intervention.

Researching Ergogenic Aids

Anyone can claim that a certain substance or technique is ergogenic, but before an intervention can be legitimately classified as truly performance enhancing, it must be shown by competent research to consistently enhance performance in a variety of relevant circumstances. High-quality scientific studies are essential to differentiate between a true ergogenic response and a placebo response, in which performance improves simply because research subjects (or athletes) expect improvement.

The Placebo Effect

Assume that an athlete consumes a dietary supplement several hours before game time and then has a successful performance. The athlete might attribute the success to the supplement, even though there is no evidence that ingesting the supplement could have had any effect whatsoever. This is a good example of why anecdotal accounts are not a substitute for valid and reliable scientific data.

As discussed in the introductory chapter, the phenomenon by which one's expectations affect the body's response in some measurable way is known as the **placebo effect**. From a practical standpoint, a placebo effect can improve performance simply because the athlete expects an improvement. For that reason, placebo effects can be beneficial. For example, if a coach tells a young athlete that his or her new shoes will definitely help with running faster, there is a good chance the athlete will run faster. The placebo effect can be helpful from a practical perspective, but from a scientific perspective, the placebo effect can complicate studies of ergogenic aids because researchers must be able to separate the placebo effect from the true physiological responses to the intervention being tested.

The power of the placebo effect was clearly demonstrated in one of the earliest studies of **anabolic steroids** in sport.[4] Fifteen male athletes who had been involved in heavy weightlifting for the previous 2 years volunteered for a strength training experiment using anabolic steroids. They were told that those who made the greatest strength gains over a preliminary 4-month weight training period would be selected for the second phase of the study, in which they would receive anabolic steroids.

Following the initial period, eight of these 15 subjects were randomly selected to enter the treatment phase. Only six of these subjects passed all medical screening tests and were allowed to continue the study. The next phase of the study consisted of a 4-week period in which the subjects were told that they would receive 10 mg per day of Dianabol (a potent anabolic steroid), when in fact they received a placebo—an inactive substance typically provided in a form that looks identical to the genuine drug.

Strength data were collected over the last 7 weeks of the 4-month pretreatment training period and over all 4 weeks of the treatment (placebo) period (see figure 16.1). Even though the subjects were experienced weightlifters, they continued to gain impressive amounts of strength during the pretreatment training period. However, strength gains while subjects were taking the placebo were even greater than during the pretreatment period! The group improved by an average of 11 kg (24 lb) in combined 1-repetition maximum (1RM) lift totals (squat, bench press, military press, sitting press) during the 7-week pretreatment period but improved 45 kg (~100 lb) during the 4-week treatment (placebo) period. This represents an average gain in strength of 1.6 kg (3.5 lb) per week during the pretreatment training period and 11.3 kg (25 lb) per week during the placebo period—an increase in the rate of strength gain more than seven times greater during the placebo (supposed steroid) period over the pretreatment training period.

From a practical standpoint, a placebo effect can certainly be a good thing for sport performance, provided that the related risks are acceptable. For example, football players may breathe from an oxygen tank on the sidelines, but there is no scientific evidence that breathing oxygen-enriched air will speed recovery

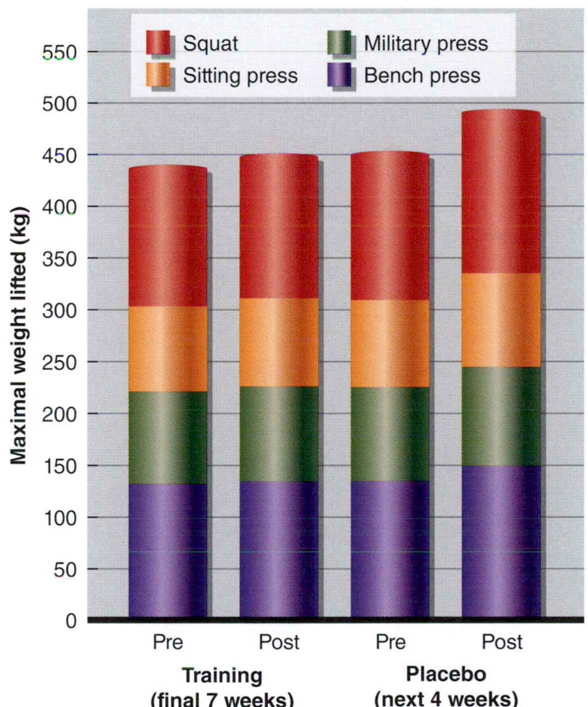

FIGURE 16.1 The placebo effect on muscular strength gains. The increase in total strength and strength in each of four maximum lifts over the last 7 weeks of an intense 4-month pretreatment training period is compared with strength increases during a subsequent 4-week treatment period, in which the subjects took placebos that they thought were anabolic steroids and continued intense resistance training.

Data from Ariel and Saville (1972).

RESEARCH PERSPECTIVE 16.1

The "Nocebo" Effect on Sport Performance

A placebo effect is defined as a positive psychobiological response to a purported beneficial treatment. A "nocebo" effect, conversely, is a negative psychobiological response to a purported harmful treatment, for example, when a patient is told to anticipate a side effect of a medication or that a certain exercise might cause pain. There is empirical support for both placebo and nocebo effects in sport performance. However, those studies have generally been too small to reliably identify the variables associated with the placebo and nocebo responses.

A recent study sought to estimate the relative magnitude of placebo and nocebo effects on sport performance.[34] Approximately 600 adults participated and were randomized to three different treatments: positive-belief treatment (participants were deceptively administered an inert capsule described as a potent supplement to improve spring performance), negative-belief treatment (participants were deceptively administered an inert capsule described as a potent supplement to negatively affect spring performance), or no-treatment control (participants received no instruction or capsule). Participants completed an exercise performance test consisting of five 20 m sprints before and after treatment.

In contrast to previous studies, there was no improvement in sprint performance following the ingestion of a purportedly beneficial supplement. However, the authors did observe a significant worsening in sprint performance in those individuals receiving a purportedly harmful supplement (i.e., a nocebo effect). Second, within the positive-belief treatment, the performance of participants reporting that they were likely to use a supplement was improved to a greater degree than in those participants not intending to use a supplement. In other words, the intention to use sports supplements might relate to placebo responding. This has practical implications for sports practitioners in that the effectiveness of a given treatment may relate to the athlete's intentions to use it.

during football games. But if athletes believe that they will recover faster, that very belief is likely to make them feel better. Such is the power of the placebo effect, an often effective approach to reducing feelings of pain and fatigue.

Rawdon and colleagues[42] conducted a meta-analysis of 37 studies and confirmed the presence of a strong placebo effect associated with sport supplements that claimed to improve performance. This finding may not be surprising, but it does confirm that the mere anticipation of a benefit from a dietary supplement is often associated with a benefit. When subjects received either the test supplement or a placebo, measures of muscular performance improved by a similar amount when compared to values in the control groups. The finding that the supplement treatments did not differ from the placebo treatments lends further support to the authors' conclusion that "any improvements seen with supplementation should not be attributed solely to the supplement, as it is acting in synergy with a training stimulus".

Controlling for the Placebo Effect

All studies of potential ergogenic substances must include a placebo group so that researchers can compare responses resulting from the test substance with those resulting from a placebo. In many studies, a *double blind* experimental design is used, in which neither the subject nor the experimenter knows who is getting the proposed ergogenic aid and who is getting the placebo. This is done to eliminate experimenter bias, whereby the experimenter's beliefs might affect the outcome of the study. In this design, the substances are assigned a code, and only an independent person not associated with the project has access to the codes until all data have been collected and analyzed.

Placebos have such a potentially powerful effect that sport scientists often recommend that studies of ergogenic aids include not only a placebo treatment (or group) but also a control treatment to help identify (control for) the placebo effect. For example, in a study designed to determine if consuming oak leaf extract improves muscle strength after 8 weeks of training, at least three experimental treatments would be needed: (1) oak leaf extract pill, (2) a placebo pill, and (3) no pill. If strength gains after training were similar for all three treatments, the researchers would conclude that oak leaf extract has no ergogenic effect. If both the oak leaf extract treatment and the placebo treatment were associated with greater strength gains than the control treatment, the conclusion would be that the gain in strength was caused by a placebo effect. If strength gains on the oak leaf extract treatment were significantly greater than on both the placebo and the control treatments, only then could the researchers conclude that there was something in oak leaf extract that improves adaptations to strength training. In studying ergogenic aids, the addition of a control group improves the ability to understand the possible impact of the placebo effect.

Limitations of Ergogenic Aid Research

To establish the effectiveness (efficacy) of a possible ergogenic aid, scientists rely on accepted research methods to control for the placebo effect. However, even the most rigorously conducted studies can miss small effects that might actually be meaningful to athletes, whose success is often defined in fractions of a second. Such small differences in performance can sometimes be obscured by the statistical analyses that research requires.

Research can also be limited by the accuracy of lab equipment and the variability of research techniques. All research methods have some margin of error as well as inherent variability, and if the results fall within that margin of error, the researcher will be unable to isolate the effect of the intervention being tested. Because of measurement error, equipment variability, differences in individual responses, and the day-to-day variability of subjects' responses, a potential ergogenic aid must exert a major effect before scientific studies can confirm that it is ergogenic.

It is also important to keep in mind that performance in a laboratory is considerably different from performance in the athletic environment, so laboratory results do not always accurately reflect natural athletic results. Yet an advantage of laboratory testing is that the test environment can be carefully controlled. This is usually not possible in field studies, where several uncontrollable variables—such as temperature, humidity, wind, and distractions—can affect the results.

The ergogenic potential of any intervention cannot be confirmed by one study or, for that matter, even by a dozen studies. Dozens and sometimes hundreds of studies are required to draw a confident conclusion about whether or not an intervention is ergogenic. The ergogenic effect of sport drinks is a good example of the need for multiple studies to confirm efficacy. Before the late 1980s, the prevailing opinion among sport scientists was that sport drinks did not provide hydration or performance benefits differing from those seen with plain water. This opinion was based on the scientific literature at the time, which contained little evidence that sport drinks were ergogenic. It was only after the publication of dozens of competent research studies conducted over many years that scientific opinion changed.

Another challenge in studying the potential benefits of dietary supplements is that researchers often have to assume that the contents of the supplement are accurately portrayed on the product label. This may not be the case. The values for active ingredients in the supplement may actually be considerably more or less than the values listed on the product label, or the supplement may be contaminated with a banned substance that could be responsible for whatever effect might be found. For this reason, some sport scientists argue

RESEARCH PERSPECTIVE 16.2

Analgesic Use in Sport

For many years, clinicians have debated the use, misuse, and abuse of analgesics—painkillers like ibuprofen, naproxen, and aspirin—in sport. In 1967, the original International Olympic Committee prohibited narcotic analgesics from sport. The 2017 World Anti-Doping Agency prohibited list bans narcotics and cannabinoids. More commonly used analgesics, including nonsteroidal anti-inflammatory drugs, local anesthetics, and some weak opioids, however, are *not* prohibited. This leaves a gray area because no clear definition separates narcotics and cannabinoids from more commonly used analgesics in terms of their health risks or ergogenic potential.

This leads to one obvious question: What are the key determinants for including a substance on the prohibited list? The current criteria for the prohibition of a substance require consideration of the potential for, or enhancement of, sport performance, as well as the potential for, or actual, health risk. Whether use constitutes a violation of the spirit of sport (that is, the concept of fair, ethical, and respectful competition) is also considered. However, the specific health risks of narcotics and cannabinoids relative to other analgesics have not been rigorously assessed. In fact, there is even an argument that *all* opioids and cannabinoids should be at the physician's discretion in treating an athlete, but, currently, the window for using these substances in elite athletes is extremely narrow.[53]

There are often passionate views of the role that the anti-doping code should play in regulating opioids and cannabinoids. Some suggest that the social use of drugs might be better addressed by a code of conduct approach in which substance abuse counseling and treatment strategies would be employed. Others believe that the use of opioids and cannabinoids is a completely unacceptable behavior in the sport setting and potentially so dangerous that every effort should be taken to prevent their use; thus, drug testing and anti-doping sanctioning should continue. Moving forward, any approach to quell the debate surrounding the aforementioned gray area must be simultaneously grounded in the principles and best practices guiding pain management and treatment strategies in the sport setting.

that studies on dietary supplements should include third-party verification of the purity and potency of the supplement being tested.

Realizing that science progresses slowly, does so in small steps, and is limited in its ability to unequivocally determine the efficacy of performance-related interventions, we now turn our attention to examples of ergogenic aids.

Ergogenic Nutrition Aids

We continue this chapter with an overview of a few selected nutrition interventions that have been studied for their ergogenic effects, and we end the chapter by reviewing prohibited substances and techniques. Although basic concepts of nutrition and the specific performance-related properties of carbohydrate, fat, protein, vitamins, and minerals are discussed in chapter 15, many nutrition interventions have been proposed to have specific ergogenic properties. We intentionally provide only a brief overview of a handful of nutrition interventions and encourage interested readers to seek additional information from the references provided. This section on ergogenic nutrition aids also highlights the necessity for dozens of well-designed research studies to be conducted on a particular topic before it is possible to confidently draw scientific conclusions and make practical recommendations.

It is impossible to cover all ergogenic nutrition aids being used by athletes, so the information in this section of the chapter is limited to the following:

- Bicarbonate
- β-alanine
- Leucine
- Caffeine
- Cherry juice polyphenols
- Creatine
- Nitrate

Bicarbonate

Recall from chapter 7 that bicarbonate is an important part of the buffering system necessary to maintain the acid–base balance of body fluids. In highly anaerobic events, in which large amounts of lactic acid are formed, enhancing the body's buffering capacity through elevation of the blood's bicarbonate concentrations could improve performance capacity.

Proposed Ergogenic Benefits of Bicarbonate

Through ingestion of sodium bicarbonate (baking soda) or sodium citrate, the buffering capacity for metabolic acids such as lactic acid can be increased, allowing extracellular pH to be maintained even as active muscles release greater amounts of lactic acid. Theoretically, this could allow high-intensity exercise to continue for a longer time before fatigue sets in. Not surprisingly, oral intake of sufficient amounts of sodium bicarbonate does elevate plasma bicarbonate concentration. However, this has little effect on intracellular concentrations of bicarbonate. Therefore, the potential benefits of bicarbonate ingestion would theoretically be limited to anaerobic bouts of exercise lasting longer than 2 min, because bouts less than 2 min would be too brief to allow enough hydrogen ions (H^+, from the lactic acid) to diffuse out of the muscle fibers into the extracellular fluid, where they could be buffered by the extra bicarbonate ions.

Demonstrated Effects of Bicarbonate

In 1990, Roth and Brooks[44] described a cell membrane lactate transporter that operates in response to the pH gradient between extracellular fluid and muscle cells. Increasing the extracellular buffering capacity by ingesting bicarbonate or citrate increases the extracellular pH, which in turn increases transport of lactate from the muscle fiber via the membrane transporter.

The theory underlying bicarbonate or citrate ingestion as a way to increase buffering capacity and improve high-intensity performance is sound, although the existing research literature is conflicting, in part because of the relatively limited number of published studies. Linderman and Fahey,[38] in their review of the literature, found several important patterns in the research that might explain these conflicts. They concluded that bicarbonate ingestion had little or no effect on performances less than 1 min or more than 7 min in duration but that for performances between 1 and 7 min, the ergogenic effects were evident. Furthermore, they found that the dose was important. Most studies that used a dose of 300 mg/kg of body mass showed a benefit, whereas most studies using a lower dosage showed little or no benefit. Thus, it appears that bicarbonate ingestion of 300 mg/kg of body mass can enhance the performance of all-out, maximal anaerobic activities of 1 to 7 min duration. (Preliminary research indicates a possible synergistic effect of combining β-alanine and sodium bicarbonate,[46] although this possibility must be confirmed by additional research.)

An example of a study supporting this conclusion is illustrated in figure 16.2. In this study, blood bicarbonate concentrations were artificially elevated by bicarbonate ingestion before and during five sprint-cycling bouts, each lasting 1 min (see figure 16.2*a*).[20] Performance on the final trial improved by 42%. This elevation in blood bicarbonate reduced the concentration of free H^+ both during and after exercise (see figure 16.2*b*), thereby elevating blood pH (keep in mind that

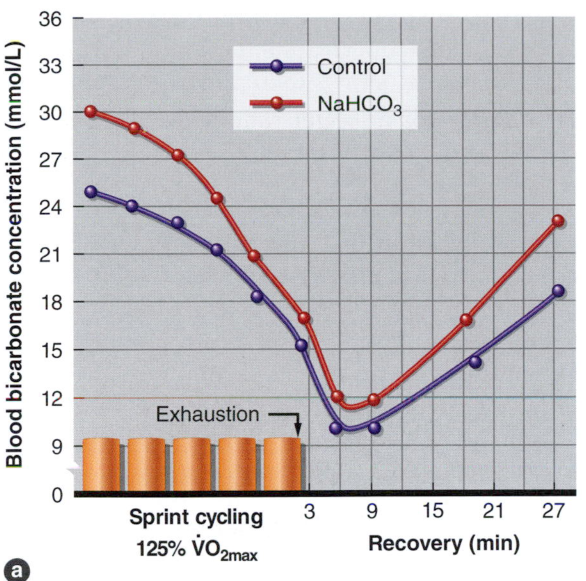

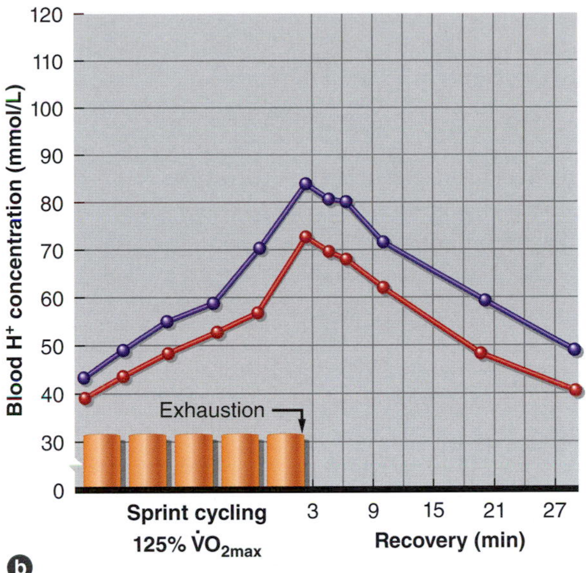

FIGURE 16.2 Concentrations of *(a)* blood bicarbonate (HCO_3^-) and *(b)* blood hydrogen ion (H^+) before, during, and after five sprint-cycling bouts with and without ingestion of sodium bicarbonate ($NaHCO_3$). The fifth sprint was performed to exhaustion. The elevated blood HCO_3^- concentrations caused attenuation of the elevation in blood H^+, a smaller decline in blood pH, a 42% increase in performance to exhaustion during the fifth sprint, and faster recovery after the sprints.

Adapted by permission from D.L. Costill et al., "Acid–base Balance During Repeated Bouts of Exercise: Influence of HCO3," *International Journal of Sports Medicine* 5 (1984): 228-231. © Georg Thieme Verlag KG.

pH is the negative logarithm of the H^+ concentration, so when H^+ falls, pH increases). The authors concluded that in addition to improving buffering capacity, the extra bicarbonate appeared to speed the removal of H^+ ions from the muscle fibers, thereby blunting the decrease in intracellular pH. This conclusion essentially predicted the presence of a lactate transporter in the muscle cell membrane that Roth and Brooks[44] reported 6 years later.

Risks Associated with Bicarbonate

Although smaller amounts of sodium bicarbonate have long been used as a remedy for indigestion, many authors studying bicarbonate loading have reported severe gastrointestinal discomfort in some of their subjects, including diarrhea, cramps, and bloating, from the high doses of sodium bicarbonate or citrate required for a buffering effect. These symptoms can be reduced by ingesting ample water and by dividing the total bicarbonate or citrate into five equal parts over a 1 to 2 h period (e.g., five doses of 400 mg for a 70 kg athlete).[38]

β-Alanine

Consuming fairly large amounts of sodium bicarbonate or sodium citrate increases the buffering capacity of the *extracellular* fluid compartment and is associated with improved high-intensity performance. But is it possible to increase the buffering capacity of the *intracellular* compartment, particularly in muscle cells? Research on β-alanine loading suggests that improved intracellular buffering may be possible.

Proposed Ergogenic Benefits of β-Alanine

β-alanine is an amino acid and as such is used by cells to create a variety of different proteins. One of those proteins is carnosine, a small, simple molecule found in high concentrations in muscle and nerve cells. Carnosine is composed of β-alanine and histidine, another amino acid. Carnosine functions as an intracellular buffer of the hydrogen ions (H^+) produced during intense exercise, especially as a result of lactic acid production.

Demonstrated Effects of β-Alanine

Research shows that β-alanine loading (e.g., 3 g/day for 4 weeks as a loading dose, followed by 1.5 g/day as a maintenance dose) results in increased muscle cell carnosine levels and improved high-intensity performance.[33] In addition, combining β-alanine loading with bicarbonate loading may have an additive effect on performance, perhaps in events lasting an hour or more.[6] Although some results look promising, other studies report no performance benefit of β-alanine loading, so more research is needed before we have a better understanding of the mechanisms by which, the extent to which, and the consistency with which β-alanine improves performance.

Risks Associated with β-Alanine

More research is required to fully understand the side effects and possible risks associated with β-alanine loading. For example, one study reported that subjects were able to identify the β-alanine treatment because it caused paresthesia (tingling of the skin).[17] Additional research is needed to determine if other side effects are associated with β-alanine loading.

Leucine

Leucine is a *branched-chain amino acid*, as are isoleucine and valine. Leucine is also an essential amino acid that must be supplied by the diet. Soybeans, beef, peanuts, fish, chicken, and almonds are examples of food sources with relatively high leucine content. During recovery from exercise training, adequate food and fluid intake helps muscle cells restore intracellular glycogen levels, reestablish intracellular hydration, repair damaged proteins, and build new proteins.

Proposed Ergogenic Benefits of Leucine

Athletes, coaches, and sport scientists are all interested in ways to optimize these processes, and leucine intake appears to play an important role as a trigger to stimulate muscle protein synthesis as well as acting as a substrate for newly synthesized protein. Leucine is also catabolized by muscle cells, producing metabolites that may also have anabolic properties. For these reasons, leucine is an interesting candidate as a dietary supplement whose ergogenic properties may mimic the muscle-building effects of illegal anabolic agents.

Demonstrated Effects of Leucine

Wilkinson and colleagues[55] studied the changes in muscle protein synthesis (MPS) and muscle protein breakdown (MPB) in 15 young men at rest in response to the acute ingestion of 3.42 g of either leucine or a metabolite of leucine that has also been proposed as a dietary supplement: β-hydroxy-β-methylbutyrate (HMB). For 2.5 h after ingestion, MPS and MPB were calculated from blood samples and muscle biopsies of the vastus lateralis in combination with leucine and phenylalanine tracers.

The researchers concluded that consuming small doses (<4 g) of leucine and HMB resulted in similar increases in MPS, responses that are comparable to those that have been reported following the consumption of a mixed meal. In addition, HMB ingestion was associated with a decrease in MPB. (MPB was not measured in the leucine treatment group.)

These results lend support to prior reports that leucine and HMB supplementation are associated with increases in MPS. While HMB and leucine both increased the anabolic signaling molecule *mechanistic target of rapamycin* (mTOR), the effect was more pronounced with leucine. The authors concluded that exogenous HMB induces acute muscle anabolism (increased MPS and reduced MPB) but through different mechanisms than the branched-chain amino acid leucine.

Risks Associated with Leucine

As with all dietary supplement research, more studies are needed before a clear picture emerges as to the anabolic potential of leucine and HMB supplementation. However, research does suggest a benefit, and there appears to be little to no risk associated with supplementation with leucine or HMB or both.

VIDEO 16.1 Presents Nicholas Burd on the role of leucine in muscle repair and growth. 

In Review

- An ergogenic aid is any substance or phenomenon that enhances performance. An ergolytic agent is one that has a detrimental effect on performance. Some substances generally thought to be ergogenic are actually ergolytic.
- Although the placebo effect has a psychological origin, the body's physiological and physical responses to a placebo can be quite real. The placebo effect clearly illustrates how effective our mental state can be in altering physiological status.
- Bicarbonate is an important component of the blood's buffering system, needed to maintain normal pH by neutralizing excess acid.
- Bicarbonate (and citrate) loading is proposed to increase the blood's alkalinity, thus increasing the buffering capacity so that more H^+ can be neutralized. This delays the onset of fatigue.
- Bicarbonate ingestion of 300 mg/kg of body weight can delay fatigue and increase performance in high-intensity bouts of exercise lasting approximately 1 to 7 min.
- Bicarbonate (and citrate) loading can cause gastrointestinal distress, including cramping, bloating, and diarrhea.
- β-alanine is an amino acid that is part of the carnosine molecule, a dipeptide that acts as an intracellular buffer (among other functions).

- β-alanine loading is associated with an increase in the carnosine content of muscle cells.
- The impact of β-alanine loading on high-intensity performance is equivocal; some studies report positive effects and others do not.
- Intake of the essential branched-chain amino acid leucine intake can play an important role as a trigger to stimulate muscle protein synthesis.

Caffeine

Caffeine, as with sodium bicarbonate, is not a nutrient, although it is commonly ingested in beverages such as coffees, teas, soft drinks, energy drinks, and energy shots. Caffeine is also a common ingredient in several over-the-counter medications, often in aspirin and other analgesic products. Caffeine is a central nervous system (CNS) stimulant that acts on adenosine receptors in the brain; its sympathomimetic effects are similar to those of **amphetamines**, although considerably weaker. As a stimulant capable of improving exercise performance, caffeine deserves special attention, in part because it is one of the most widely consumed drugs in the world.

Proposed Benefits of Caffeine

As with sympathomimetic amines, caffeine ingestion is linked to improved alertness, concentration, reaction time, and perceived energy level. Caffeine is known to have metabolic effects on adipose tissue and skeletal muscle as well as on the CNS.

Demonstrated Effects of Caffeine

Because of its effects on the CNS, the general effects of caffeine include

- increased mental alertness,
- increased concentration,
- elevated mood,
- decreased fatigue and delayed onset,
- decreased reaction time (i.e., faster response),
- enhanced catecholamine release,
- increased free fatty acid mobilization, and
- increased use of muscle triglycerides.

The ergogenic benefits of caffeine were first reported by scientists in the late 1970s.[19,35] Caffeine ingested before exercise increased endurance times in fixed-pace work bouts and decreased times in fixed-distance tasks.

Although several subsequent studies were unable to replicate these results, recent studies have consistently demonstrated the ergogenic effects of caffeine ingestion.[30,39] It was initially thought that this improvement was the result of an increased mobilization of free fatty acids, sparing muscle glycogen for later use. But the mechanisms by which caffeine improves endurance performance appear to be more complex, since glycogen sparing does not always occur, and also because caffeine improves performance in tasks in which muscle glycogen is not limiting. A growing number of studies now demonstrate that caffeine has its effect directly on the CNS.[49] It is now well documented that caffeine

RESEARCH PERSPECTIVE 16.3

Caffeine Use in Cycling

Caffeine can enhance endurance exercise performance and short-term high-intensity anaerobic exercise performance, as discussed in this chapter. However, the mechanism underlying caffeine's ergogenic effects are not completely understood. In recent years, focus has shifted from caffeine's promotion of fat oxidation (glycogen sparing) to its actions within the central and peripheral nervous systems, which could alter sensations of effort and muscle pain. Yet few studies have systematically assessed neuromuscular function (maximal voluntary contraction, motor-unit recruitment, and muscle contractile properties), muscle pain, perceived effort, and endurance performance after caffeine consumption.

Investigators recently tested two hypotheses: (1) caffeine would improve cycling performance because of increased strength due to enhanced motor-unit recruitment and (2) caffeine would have no effect on the perception of pain and effort during leg cycling performance.[11] Using a novel study design that incorporated both leg cycling and arm crank cycling allowed investigators to manipulate caffeine's effects on force production and motor-unit recruitment. Caffeine ingestion enhanced strength and motor-unit recruitment in the knee extensors but not the elbow flexors. Caffeine ingestion also reduced muscle pain and ratings of perceived exertion during moderate-intensity exercise, but this effect disappeared at high exercise intensities, like those exerted during all-out time trials. Lastly, caffeine ingestion improved endurance performance during leg cycling but not arm crank cycling. These findings cast doubt on the ability of caffeine to be hypoalgesic (decrease the sensitivity to pain) during high-intensity exercise. Instead, the data support the concept that enhanced strength is a likely mechanism, at least in part, by which caffeine is ergogenic.

lowers the perception of effort, allowing the athlete to perform at a higher intensity with the same perceived effort. The dose of caffeine that is commonly associated with improved performance is 3 g/kg body weight consumed before exercise. That equates to about 200 mg of caffeine for a 70 kg athlete. As a single dose, that is greater than the amount in a 12 oz energy drink (150 mg) or a can of cola (40 mg), but similar to the amount in a large serving of strong coffee.

Caffeine has also been shown to improve performance in sprint and strength types of activities and in high-intensity team sports. Unfortunately, fewer studies have investigated the mechanisms by which caffeine might improve performance in high-intensity activity. However, in addition to CNS stimulation, caffeine might facilitate calcium exchange at the sarcoplasmic reticulum and increase the activity of the sodium–potassium pump, better maintaining the muscle membrane potential.[30]

Risks Associated with Caffeine Use

In people who are not accustomed to using caffeine, who are sensitive to it, or who consume high doses, caffeine can produce nervousness, restlessness, insomnia, headache, gastrointestinal problems, and tremors. Caffeine also acts as a mild diuretic at rest, but this effect plays no significant role during exercise because renal blood flow and urine production decreases with exercise. Regular caffeine use does not jeopardize daily hydration, especially among habitual caffeine users, because the diuretic effect of caffeine is muted and people generally consume enough water and other beverages to maintain 24 h hydration status. Caffeine can disrupt normal sleep patterns in some people, contributing to fatigue. Caffeine is also physically addictive; abrupt discontinuation of intake can result in headache, fatigue, irritability, and gastrointestinal distress. At one time, caffeine was on the WADA list of prohibited substances, but in 2004 it was removed from the list, although its presence in urine samples is still monitored.

Cherry Juice Polyphenols

Fruit and vegetable juices contain hundreds, if not thousands, of compounds with biological activities and health-related effects that are poorly understood simply because of a lack of research. Of interest to athletes and coaches is emerging research on how tart cherry juice might be a potential nutrition intervention with analgesic (pain-reducing) effects that could speed recovery from training bouts causing muscle damage and related pain.

Proposed Benefits of Cherry Juice Polyphenols

Cherries, much like blueberries, cranberries, raspberries, purple grapes, pomegranate, acai berries, and other dark fruits, contain polyphenolic compounds such as flavonoids, secondary plant metabolites that provide the pigmentation characteristics of the fruit and also help protect the fruit from microbes and insects. Daily polyphenol intake is one of the reasons dietitians recommend several servings of fruits and vegetables each day. (There are estimated to be more than 4,000 different flavonoids in foods, and we typically eat about 1,500 mg of flavonoids every day.) Eating dark fruits such as cherries exposes our cells to flavonoids such as gallic acid, kaempferol, quercetin, resveratrol, and other compounds with antioxidant and anti-inflammatory properties. Unlike taking an antioxidant supplement, eating a dark fruit provides a large array of micronutrients that play important roles in the life cycle of plants. Whether or not that array of micronutrients (or single micronutrients) provides meaningful benefits beyond good nutrition is a question that scientists are beginning to address. The premise for the use of cherry juice as an ergogenic aid is that the anti-inflammatory and antioxidant characteristics of cherries may have a beneficial effect for pain reduction following muscle-damaging exercise.

Demonstrated Effects of Cherry Juice Polyphenols

Studies in which subjects consumed cherry juice or placebo for a week or more before a bout of eccentric exercise have reported a reduction in oxidative stress, strength loss, and subjective perceptions of muscle pain associated with the cherry juice treatment.[12,18,37] These studies suggest that cherry consumption might be associated with real benefits, associations that will eventually be supported or rejected through further research. At this time, it is impossible to state with any confidence whether cherry juice consumption is associated with consistent analgesic effects.

Possible Risks Associated with Cherry Juice Polyphenols

In terms of basic nutrition, cherries are wholesome, healthy, and packed with a wide variety of good-for-us nutrients. For that reason, there is no obvious downside to including cherries (or cherry juice) in an athlete's diet; and, optimistically speaking, the anti-inflammatory upside of cherries just might do muscles some good. However, it will take years of research and dozens of studies to determine whether or not that optimism can be scientifically justified.

Creatine

Creatine has become a popular ergogenic aid among athletes, especially those participating in team sports and other activities that involve repeated, high-intensity movements. Creatine is a molecule found inside every muscle cell. The body is capable of synthesizing creatine in the liver, with additional creatine supplied by the diet (mostly from meat and fish). Consuming creatine supplements to boost daily creatine intake has been shown to increase the creatine content of muscles.

Proposed Benefits of Creatine

The primary use of creatine is based on its role in skeletal muscle, where approximately 60% of the total creatine content is in the form of phosphocreatine (PCr). Increasing the creatine content of skeletal muscle cells through creatine supplementation is theorized to increase muscle PCr levels, thus enhancing the adenosine triphosphate (ATP)-PCr energy system by better maintaining muscle ATP levels. This would theoretically enhance peak power production during intense exercise and possibly facilitate recovery from high-intensity exercise. Creatine also serves as an intracellular buffer, helping to regulate acid–base balance, and plays a role in oxidative metabolic pathways. These are two other possible ways in which creatine loading might improve performance.

Demonstrated Effects of Creatine

Because of the popularity of creatine supplementation in the 1990s and the wide-ranging claims of its ergogenic properties, the American College of Sports Medicine (ACSM) published a consensus statement, "The Physiological and Health Effects of Oral Creatine Supplementation," in 2000.[2] A group of scientists reviewed the research literature on creatine and performance to develop a list of conclusions supported by scientific research. The following conclusions were drawn:

- Creatine supplementation can increase muscle PCr content, although not in all individuals.
- Exercise performance in short periods of intense, high-power exercise can be enhanced, particularly with repeated bouts, consistent with the role of PCr in this type of activity.
- Maximal isometric strength, the rate of maximal force production, and maximal aerobic capacity are not enhanced by creatine supplementation.
- Creatine supplementation often leads to weight gain within the first few days, likely attributable to water accumulation with creatine uptake in the muscle.
- In combination with resistance training, creatine supplementation is associated with greater gains in strength, perhaps in relation to the increased ability to train at higher intensities.

RESEARCH PERSPECTIVE 16.4

Creatine Supplementation Plus Resistance Exercise to Prevent Sarcopenia

Sarcopenia is the loss of muscle mass (myopenia) and strength (dynapenia) that occurs with normal aging. It is characterized by type II muscle fiber atrophy, loss of myofibers, and increased intramuscular lipid and connective tissue. Resistance training increases both muscle mass and muscle strength and is therefore an important countermeasure to sarcopenia. In addition to resistance training, protein consumption also stimulates muscle protein synthesis and may be an additional strategy for mitigating age-related losses in muscle mass and strength. Does the use of protein supplementation in the form of creatine and resistance training together have a greater effect on muscle strength and functional performance than either resistance training or creatine supplementation alone? In 2014, investigators performed a meta-analysis (systemically combining findings from several different studies) to determine whether the addition of creatine supplementation to resistance training improves body composition and increases strength and functional performance in older adults.[21]

Ten studies with a total of 357 participants met the inclusion criteria and were included in the meta-analysis. As hypothesized, in older adults, creatine supplementation during resistance training (≥6 weeks) improved body composition, strength, and functional performance to a greater extent than resistance training alone. The mechanisms by which creatine improves muscle strength are likely multifactorial and may potentially include increased phosphocreatine energy stores, increased phosphocreatine synthesis, and reduced muscle damage—all of which enhance the ability to perform resistance training, thereby having a greater effect on body composition, strength, and functional performance.

When considered collectively, the results of this meta-analysis support a role for concurrent resistance training and creatine supplementation in older adults to slow sarcopenia.

- The high expectations for performance enhancement exceed the true ergogenic benefits of creatine loading.

Subsequent scientific reviews have been published since the release of the ACSM consensus statement and are generally in agreement. Creatine is one of the few supplements that, in combination with resistance exercise, is associated with increases in both fat-free mass and strength.[40] With respect to improving athletic performance, the studies are mixed. This is likely due to two factors: the physiological demands of the sport or event and the individual variability in response to the supplement. Performance enhancement is more likely to occur in sports involving brief periods of repeated high-intensity exercise.[7] Concerning individual variability, we discussed in chapter 9 the principle of individuality—the fact that there are high responders and low responders to any given intervention. In studies involving only a few subjects (e.g., <10), it is possible that more high than low responders are represented in the study sample or vice versa.

Risks Associated with Creatine

There do not appear to be any short-term health risks associated with appropriate levels of creatine supplementation, a conclusion based on a number of research articles and the widespread use of creatine by athletes. One remaining note of caution relates to long-term creatine supplementation, especially in young, growing athletes, because there have been no long-term studies on the topic. It may well be that creatine supplementation presents no long-term health risks, but verification of that conclusion awaits further research. There are several reported cases of kidney damage in young athletes who used massive doses of creatine.

Nitrate

Inorganic nitrate (NO_3^-) is a molecule found in all plants (as part of the nitrogen cycle that you might remember from high school biology class); vegetables such as spinach, celery, and beets contain higher levels of nitrate than other vegetables. Bacteria on the tongue (and other enzymes in the body) convert nitrate to nitrite (NO_2^-). Cells can then convert nitrite to nitric oxide (NO). Nitric oxide can also be produced from the amino acid arginine via nitric oxide synthase enzyme (NOS). Citrulline, another amino acid and a precursor to arginine, is included in some dietary supplements as an endogenous source of NO. Our diets provide an estimated 50% of the NO produced in the body, and dietary intake of nitrate represents one way in which NO production might be increased. Nitric oxide molecules play a critical role in a wide variety of functions, including control of blood flow, white blood cell chemotaxis, mitochondrial function, and cellular apoptosis.

Proposed Benefits of Nitrate Ingestion

Because NO affects muscle blood flow and mitochondrial function, scientists have been interested in determining if ingestion of nitrate, arginine, and citrulline (and a few other compounds) is associated with ergogenic benefits. For example, if nitrate ingestion results in increased delivery of oxygen and nutrients to active skeletal muscles, that response could be the basis for improved performance in a variety of exercise settings ranging in intensity and duration.

Nitrate (and nitrite) ingestion is also associated with a reduction in resting systolic blood pressure and other positive cardiovascular changes that might alter the course of cardiovascular aging and therefore have important implications for overall cardiovascular health. As with nitrate ingestion and performance, this is a promising area of research that requires more studies using subjects of both sexes and a range of ages as well as a variety of intake doses consumed over varying durations and intensities of exercise.

Demonstrated Effects of Nitrate Ingestion

Early research on nitrate ingestion (often in the form of beet juice) showed performance benefits, a finding that many other studies—although not all—have corroborated. As is always the case with research on ergogenic aids, many studies are needed before a clear picture emerges, and nitrate ingestion is no exception. However, a growing body of well-done research supports an ergogenic effect with nitrate ingestion, at least in untrained or modestly fit young male subjects. It appears that the effect of nitrate ingestion might be considerably less than the benefits associated with being highly fit, because studies using very fit subjects often report no benefit of nitrate ingestion.[9]

Research has consistently shown that when subjects consume beet juice for a week before an exercise challenge, or even a few hours before exercise, plasma nitrite concentration rises, resting blood pressure decreases, exercise oxygen consumption falls slightly, and exercise performance is significantly improved. It is thought that these responses are related to enhanced muscle cell contractility, improved efficiency of ATP production in mitochondria, and increased blood flow to muscle cells. Most of these studies have examined a single dose of beet juice; little work has been done to determine the dose of beet juice (i.e., the dose of nitrate) that is associated with the greatest responses.

In one study, scientists reported that eight male subjects who ingested 500 ml (16 oz) of beet juice per day for 6 days (an average of 486 mg of sodium nitrate per day) improved their time to exhaustion on a cycle ergometer test even though oxygen consumption was reduced, a finding consistent with other studies.[5]

The hypothesis is that nitrate ingestion increases NO production and improves mitochondrial efficiency for ATP production, reducing oxygen consumption even as performance improves. It appears that acute consumption of beet juice before exercise might also be ergogenic.[52]

In another study,[58] 10 young, recreationally active men ingested 70, 140, or 280 ml (roughly 2.5, 5.0, and 10.0 oz) of beet juice at rest and immediately before exercise. Following beet juice ingestion, plasma nitrate concentration rose in a dose-dependent manner, with peak levels at 2 to 3 h postingestion. The lowest volume of beet juice did not affect exercise responses, but the 140 and 280 ml volumes were associated with lower average steady-state oxygen consumption (by 1.7% and 3.0%, respectively) and improved exercise performance (by 14% and 12%, respectively) compared to the control treatment (water). Performance was assessed as time to failure on a cycle ergometer as the subjects repeated bouts of moderate- and high-intensity exercise until they could not maintain a preset pedaling cadence.

Although the exact mechanisms by which nitrate ingestion alters exercise performance are not yet entirely clear, many research studies have reported similar results, at least with recreationally active subjects. More work is needed to corroborate these early findings, further understand dose–response relationships, identify mechanisms of action, and provide additional recommendations to coaches and athletes interested in using beet juice or other nitrate sources as part of their training and competitive strategies.

Possible Risks Associated with Nitrate Ingestion

Adverse health consequences do not appear to be associated with nitrate intake, with the possible exception of people with heart disease who are taking prescription medications (e.g., nitroglycerine) that also affect NO metabolism. However, little is known about the possible side effects of consuming large amounts of nitrate supplements.

In Review

- Caffeine can enhance performance in endurance sports and may even be of benefit in activities of much shorter duration (e.g., 1 to 6 min). However, some athletes may experience a negative reaction to caffeine, in which case caffeine could be ergolytic.
- There is some evidence that consuming tart cherry juice is associated with reduced pain and faster recovery of strength from eccentric exercise compared to a placebo treatment.
- Creatine supplementation appears to have ergogenic benefits, particularly for improving performance in repeated, high-intensity exercise bouts between 30 and 150 s in duration.
- Creatine supplementation has been shown to increase muscle creatine levels and increase performance in sports involving brief, repeated periods of high-intensity exercise.
- Ingesting nitrate (usually in the form of beet juice) is associated with improved exercise performance in sedentary or recreationally active subjects but is less commonly evident in well-trained subjects. Nitrate ingestion has been shown to lower systolic blood pressure and may have other benefits to cardiovascular health.

Anti-Doping Codes and Drug Testing

Performance-enhancing drugs (PEDs), used by athletes who choose to cheat in order to gain an unfair advantage over their competitors, have a long history in sport. Several of these prohibited substances and techniques are discussed in more detail later in the chapter. Increased PED use among athletes has led to the formation of regulatory organizations, increased testing of supplement products and ingredients for contamination, and the proliferation of designer drugs. In this section, we introduce prohibited substances from this regulatory perspective.

The World Anti-Doping Code

In 1968, the International Olympic Committee (IOC) began drug testing athletes during the Summer and Winter Games. As concerns over the use of PEDs grew through the 1970s and 1980s, an IOC-led initiative established the World Anti-Doping Agency (WADA; www.wada-ama.org) in 1999. WADA is composed of and funded by the Olympic Movement and various governments.

The actions of WADA are guided by the World Anti-Doping Code, adopted in 2003. The Code is the core document that provides the framework for harmonized anti-doping policies, rules, and regulations within sport organizations and among public authorities. More than 600 athletic governing organizations have now adopted the Code.

An important component of the Code is the Prohibited Substances List, which is updated annually. Substances or practices are considered for inclusion if they meet two of the following three criteria:

1. There is scientific or medical evidence that the substance or practice has the potential to

enhance, or has been shown to enhance, sport performance.

2. There is scientific or medical evidence that use of the substance or practice presents a health risk to athletes.
3. The World Anti-Doping Agency determines that the substance or practice violates the spirit of sport.

Possession, trafficking, or administration of PEDs and banned practices are also considered violations of the Code. Banned practices include tampering with or substituting urine or blood samples. For instance, athletes have tried to substitute someone else's urine as their own or have attempted to add proteases or other substances to a urine sample to mask banned substances. Intravenous infusions of saline, glucose, and other solutions are also prohibited in competitive settings. Even gene doping is now a banned practice in anticipation of what future technologies might allow.

Compliance with the Code is monitored through a testing program administered in conjunction with various sport governing bodies. Athletes must report their whereabouts and be available for randomized testing. Those who test positive for a banned substance or practice may be subject to a variety of sanctions ranging from increased monitoring up to a lifetime ban from their sport.

The Code is based on the principle of strict liability; that is, athletes are responsible for any substance in their body, even if they were unaware of it. In other words, even if an athlete accidentally ingests a banned substance, the athlete is held fully responsible. Athletes who have medical conditions that require the medicinal use of a banned substance (for example, an athlete who is prescribed an asthma medication) may apply for a **therapeutic use exemption**. Athletes who believe that a positive test or sanction is incorrect or unfair may appeal the result through their sport's governing body and the Court of Arbitration for Sport.

Other sport organizations such as the NCAA, the National Football League, and NASCAR have their own banned substance lists but all are similar in many respects to the WADA list. The IOC, the United States Olympic Committee (USOC), the International Association of Athletics Federations (IAAF), and the NCAA all publish extensive lists of banned substances, most of which are pharmacological agents. The IOC and the USOC use the standards established by WADA. In the United States, these standards are administered through the United States Anti-Doping Agency (USADA; www.usada.org). The list of banned substances is updated annually. Table 16.1 lists prohibited substances as of 2017 and the mechanisms of action by which these ergogenic aids have been proposed to work.

Each athlete, coach, athletic trainer, and team physician must know which drugs are prescribed for and taken by the athlete, and they must check these drugs periodically against the relevant listing of banned substances because lists change frequently. Athletes who require prescription medications can apply for and be granted therapeutic use exemptions so they can continue to take their medications without risk of running afoul of anti-doping regulations.

Contamination of Dietary Supplements

Many athletes take one or more types of nutritional supplements, ranging from vitamin and mineral tablets to elaborate mixtures of plant parts and herbs. Most athletes assume that they are ingesting a substance that exactly matches the ingredients listed on the product container. Unfortunately, that may not be the case, since accidental or purposeful contamination of dietary supplements with banned substances has occurred repeatedly. Although most supplement manufacturers are diligent in ensuring the purity and potency of their products, some are not. This inconsistency requires all supplement consumers to follow a *caveat emptor*—buyer beware—approach to purchasing and consuming supplements.

Starting in 1999, researchers have investigated the potency and purity of selected sport supplements. The findings were sobering. In some cases, product potency varied from negligible values to up to 150% of the dose listed on the product label. In addition, many supplements were confirmed to be contaminated with prohibited substances[27] that could lead to positive doping results and subsequent severe penalties for athletes. Common contaminants have included anabolic steroids, stimulants, and diuretics. Numerous studies have substantiated the extent and the critical nature of this problem. The bottom line: Athletes are responsible for what they ingest; those who use supplements are taking what could be an extremely high risk.

Supplement consumers—and especially athletes—should be concerned about three critical aspects with any supplement they are considering.

1. Purity—does the supplement contain any harmful or prohibited substances?
2. Potency—does the supplement actually contain the amounts of ingredients listed on the label?
3. Efficacy—does the supplement actually provide any benefits?

Unfortunately, there are no easy answers to any of these questions. Some supplement manufacturers enlist third parties to analyze their products for potency and purity and display a certification mark on the product label.

TABLE 16.1 **WADA 2017 Prohibited Substances List, with Possible Mechanisms of Action (Where Appropriate)**

Agent	Influence heart, blood, circulation, and aerobic endurance	Increase oxygen delivery	Supply fuel for muscle and general muscle function	Act on muscle mass and strength	Result in weight loss or weight gain	Counteract or delay onset or sensation of fatigue	Counteract central nervous system inhibition	Aid in relaxation and stress reduction	Speed recovery
Anabolic agents				✓	✓				✓
Peptide hormones	✓	✓		✓	✓				✓
β-2 agonists	✓					✓	✓		
Hormone and metabolic modulators	✓		✓		✓	✓	✓		
Diuretics and other masking agents	✓			✓	✓				
Blood manipulation	✓	✓	✓			✓			
Chemical and physical manipulation (tampering with samples)									
Chemical and physical manipulation: large intravenous infusions	✓	✓	✓			✓			✓
Gene doping	✓	✓	✓	✓	✓	✓	✓	✓	✓
Stimulants	✓		✓			✓	✓		✓
Narcotics						✓		✓	
Cannabinoids								✓	
Glucocorticosteroids	✓		✓	✓		✓			
Alcohol								✓	
β-blockers	✓					✓		✓	

Doing so reflects a good-faith effort on the part of the manufacturer to assure consumers that their manufacturing processes and product formulations adhere to the most current standards. But as rigorous as testing for banned substances can be, it is impossible to ensure that any product is completely free of banned substances because it is impossible to test for all banned substances. However, reputable manufacturers that make the extra effort and incur the expense of third-party testing are doing the right thing on behalf of their consumers.

Designer Steroids and Drug Testing

In an effort to circumvent sport drug testing, some chemists have become adept at synthesizing variations of steroids—or their metabolic precursors—that cannot be detected by current testing procedures. That does not mean that these illicit compounds are undetectable—all compounds can be detected by the right instruments and analytical techniques. But sport drug testing typically analyzes for an established array of steroids, and if a new steroid is not in that array, it will not be detected until someone begins to look for it. This is exactly what happened with tetrahydrogestrinone (THG)—the powerful designer steroid illegally used by athletes in the 1990s. Tetrahydrogestrinone remained undetected by sport drug testing until a sample of the substance was analyzed by forensic toxicologists and subsequently added to the WADA banned substance list.

The World Anti-Doping Agency recognizes two classes of steroids: (1) endogenous steroids such as **testosterone**, estrogen, and their metabolic precursors including epitestosterone, androstenedione (Andro), and dehydroepiandrosterone (DHEA); and (2) exogenous steroids—in other words, synthetic steroids such as nandrolone, trenbolone, methyltestosterone, THG, and other designer steroids.

Most sport drug testing is accomplished with urine samples; in the case of steroids, the test methodology requires steps to extract, separate, and detect steroid metabolites in the urine. After the urine sample has been treated to make the metabolites measurable, the sample is typically run through gas or liquid chromatographs to separate the compounds and through mass spectrometers to identify the compounds.

In addition to random drug testing, the Athlete Biological Passport (ABP) will see more use in future years. The ABP is a way to determine a young athlete's normal profile for measures such as testosterone, epitestosterone, the ratio between those two steroids (T/E ratio), hemoglobin, hematocrit, and other parameters that can be used to establish a normal baseline for individual athletes. Major changes in those parameters reflect possible use of banned substances. For example, the T/E ratio has long been used as a marker of steroid use. The average T/E ratio in men is 1:1. Use of illegal steroids increases that ratio. Because there is a large variation in T/E ratio among people, sport governing bodies have adopted cut-off values such as 4:1 or 6:1 to allow for that normal variation. Of course, this does not help athletes whose normal T/E ratio is 10:1, nor does it prevent athletes from using a combination of testosterone and epitestosterone to benefit from an anabolic effect without altering the T/E ratio. However, it is now possible to distinguish exogenous testosterone from endogenous testosterone, another analytical weapon that can be used to catch cheaters.

In Review

- The World Anti-Doping Code, first adopted in 2003, is the core document that provides the framework for harmonized anti-doping policies, rules, and regulations within sport organizations and among public authorities. More than 600 athletic governing organizations have now adopted the Code.
- Substances are considered for inclusion on the WADA banned substance list if they meet any two of the following three criteria based on scientific evidence: (1) They have the potential to enhance, or have been shown to enhance, sport performance, (2) they present a health risk to athletes, or (3) they violate the spirit of sport.
- A substantial risk is associated with using nutritional supplements due to potential contamination of the products and the challenges associated with ensuring that supplements are free of banned substances.

Prohibited Substances and Techniques

Table 16.1 introduced a list of the substances, agents, and methods included in the WADA 2017 Prohibited Substances List. In this section, we consider the following prohibited substances and practices in detail:

- Stimulants
- Anabolic steroids
- Human growth hormone
- Diuretics and other masking agents
- β-blockers
- Blood doping

Stimulants

Stimulants include amphetamines and related compounds. Stimulants are often referred to as *sympathomimetic* amines, which means that their activity mimics that of the sympathetic nervous system. For many years, stimulants have been used as appetite suppressants in medically supervised weight loss programs. During World War II, army troops used amphetamines to combat fatigue and to improve endurance. Stimulants are now used to treat attention deficit hyperactivity disorder (ADHD) to activate parts of the brain that help reduce the symptoms associated with ADHD and similar disorders. Long ago, amphetamines found their way into sport. Other sympathomimetic amines such as **ephedrine** and **pseudoephedrine** have been used as ergogenic aids. Ephedrine is derived from herbs (often known as ma huang) and has been used as a decongestant and as a bronchodilator in the treatment of asthma. Pseudoephedrine is used in over-the-counter medications primarily as a decongestant and in the illicit manufacturing of methamphetamine.

Proposed Benefits of Stimulants

Stimulants have physiological and psychological effects that reduce appetite, increase concentration and alertness, boost metabolic rate, and reduce the sensation of fatigue. Athletes who have used amphetamines report a sense of indestructibility, which they feel spurs them on to higher performance levels. Others rely on the sympathetic stimulation to raise metabolic rate and promote fat loss.

In terms of actual performance, amphetamines are thought to help athletes run faster, throw farther, jump higher, and delay the onset of fatigue. Similar claims and expectations have been associated with the use of ephedrine and pseudoephedrine.

Demonstrated Effects of Stimulants

Not surprisingly, some studies report that amphetamines have no effect on performance, others demonstrate a strong ergogenic effect, and still others conclude that amphetamines have an ergolytic effect. As potent CNS stimulants, amphetamines do increase the state of arousal, which leads to a sense of increased energy, self-confidence, and faster decision-making. Athletes who take amphetamines experience a decreased sense of fatigue, along with increased heart rate, systolic and diastolic blood pressure, and blood flow to skeletal muscles, as well as an elevation of blood glucose and free fatty acids.

The key question is, do these effects aid physical performance? Although studies are not in total agreement, research that has used better experimental designs and controls shows that amphetamine ingestion is associated with the types of responses that would be expected from a CNS stimulant, including

- weight loss;
- improved reaction time, acceleration, and speed;
- greater strength, power, and muscular endurance;
- possibly improved aerobic endurance, although not $\dot{V}O_{2max}$;
- higher maximum heart rate and peak lactate concentrations at exhaustion;
- improved focus; and
- enhanced fine motor coordination.

The results are not as clear for less-potent stimulants such as synephrine, ephedrine, and pseudoephedrine. While studies have shown small improvements in markers of athletic performance with use of these substances, the current thinking is that performance benefits are inconsistent for speed, strength, power, and endurance.[1,16]

Risks Associated with Stimulants

Deaths have been attributed to excessive amphetamine and ephedrine use, likely because heart rate and blood pressure are increased, placing greater stress on the cardiovascular system. Stimulants can trigger cardiac arrhythmias in susceptible individuals. Also, rather than delaying the onset of fatigue, amphetamines likely delay the sensation of fatigue, enabling some athletes to push dangerously beyond the normal limits of exhaustion.

Amphetamines can be psychologically addictive because of the euphoria and energized feelings they cause. But the drugs also can be physically addictive if taken regularly, and a person's tolerance to them builds with continued use, requiring increasingly larger doses over time to obtain the same effects. Amphetamines also can be toxic. Extreme nervousness, acute anxiety, aggressive behavior, and insomnia are side effects of regular use. Ephedrine has side effects similar to those of amphetamines and is associated with risks of cardiovascular events and heat illness.

Anabolic Steroids

Anabolic steroids are prohibited substances because of their well-established ergogenic benefit of increasing muscle mass beyond what can be normally achieved by the combination of strength training and diet. Steroids and similar substances have been used by athletes in sports ranging from endurance cycling to baseball to bodybuilding. Many male sex hormones have both muscle-building and masculinizing effects and are therefore referred to as anabolic-androgenic steroids. The use of steroids and other hormonal agents as ergogenic aids began in the late 1940s or early 1950s, if not sooner. Anabolic agents are banned for all sports, and the medical risks associated with their use are high.

Anecdotal stories suggest that anywhere from 20% to 90% of athletes in certain sports are using, or have used, anabolic steroids. Scientific surveys, however, suggest a much lower estimate of 6% across sports.[8] Anabolic steroid use has even been reported in the general high school population in the United States, varying from 4% to 11% in boys and up to 3% in girls.[16,43]

Steroids are a large class of substances that include male and female sex hormones such as testosterone and estradiol. Interestingly, cholesterol is also classified as a steroid. Although cholesterol has no anabolic or androgenic effects, its structure is similar to that of all steroid hormones. Plants also produce steroids (phytosterols) that, like cholesterol, do not interact with androgen receptors in cells and therefore have no anabolic effects. Steroids that do interact with androgen receptors, however, accelerate growth by increasing the rate of bone maturation and the development of muscle mass. For years, anabolic steroids have been given to children with delayed growth patterns to normalize their growth curves. Dozens of steroids have been synthesized with slightly altered chemical structures that reduce the androgenic and anabolic properties of the hormones. As a result, synthetic steroid hormones differ in their anabolic and androgenic (masculinizing) effects.

Proposed Benefits of Anabolic Steroids

Steroid administration is known to increase muscle mass, and therefore strength, and to reduce fat mass.

For those reasons, steroids are understandably classified as a PED. In addition, anabolic steroids have been postulated to facilitate recovery from exhaustive training bouts, allowing athletes to train hard on subsequent days.

Athletes who take steroids have become very good at beating the system. Diuretics and other masking agents and techniques have been used to reduce the risk of detection. And chemists continue to develop designer steroids to help cheating athletes avoid detection. Randomized drug testing (both in competition and out of season), coupled with improved analytical techniques, has increased the likelihood that PED users will be caught. Those actions, along with improved education efforts, will hopefully be successful in deterring most athletes from using steroids.

Demonstrated Effects of Anabolic Steroids

The ergogenic benefits of anabolic steroids are clear. A strong dose–response relationship between steroids and lean body mass, muscle mass, and strength has now been demonstrated, making steroids one of the most powerful prohibited ergogenic aids. However, that was not always the case. Results of early studies on anabolic steroids were almost evenly divided among those studies that showed no significant change in body size or physical performance and those that reported significant improvements in muscle mass and strength.

One basic problem with almost all research conducted in this area is the researcher's inability to use the large drug dosages commonly consumed by athletes. It is estimated that some athletes use 5 to 20 times (or more) the recommended maximum daily medicinal dosage.[31] However, some researchers have been able to observe athletes when they are taking high doses of steroids and when they are off them, aiding our understanding of how anabolic steroids affect muscle size and strength.

In one of the first studies involving athletes who were illegally taking steroids, the effects of relatively high doses were observed in seven male weightlifters.[32] Two treatment periods, each lasting 6 weeks, were separated by a 6-week interval without treatment. Half the subjects received a placebo during the first treatment period and then the steroid during the second treatment period. The other half received the medications in reverse order: steroid first, then placebo. When the data from all subjects were analyzed, results showed that while on the steroid, the weightlifters had significant increases in

- body mass and fat-free mass,
- total body potassium and total body nitrogen (markers of fat-free mass),
- muscle size, and
- leg strength.

These increases did not occur during the placebo period. Results of this study are summarized in figure 16.3.

In a second study, Forbes[26] observed body composition changes in a professional bodybuilder and a competitive weightlifter. Both were on self-prescribed high doses of steroids. The bodybuilder continued on

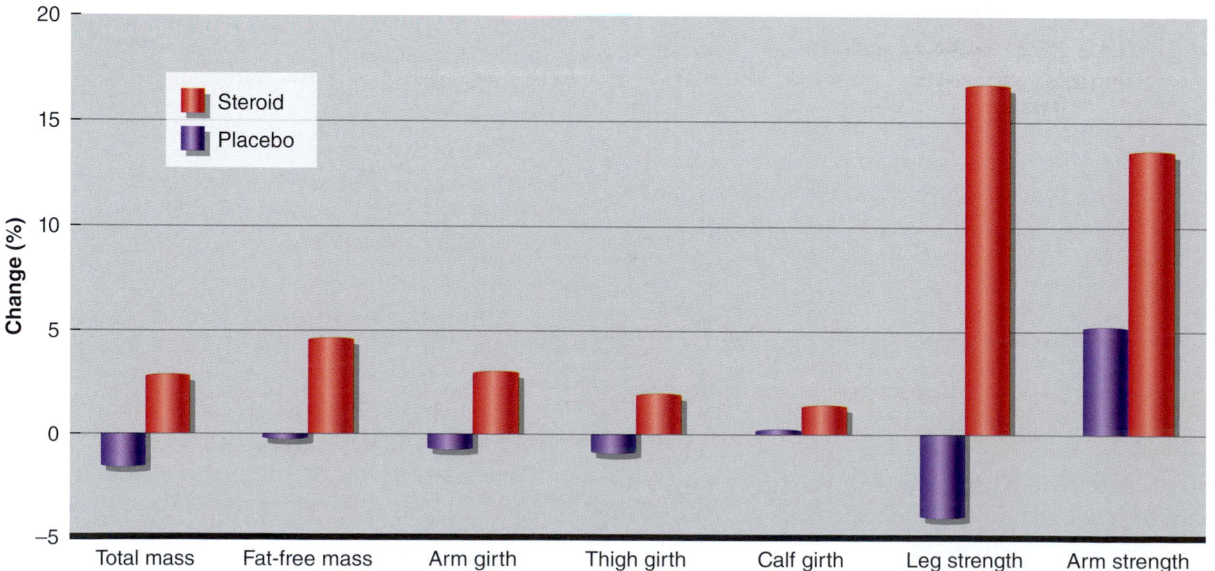

FIGURE 16.3 Percentage changes in body size, body composition, and strength when athletes used anabolic steroids or a placebo.

Adapted from Hervey et al. (1981).

the high dose for 140 days and the weightlifter for 125 days. Fat-free body mass increased an average of 19.2 kg (42.3 lb), and fat mass decreased almost 10 kg (22 lb). Forbes[26] plotted the results of a number of studies that used different dosages (figure 16.4). He observed that minimal increases of 1 to 2 kg (2.2-4.4 lb) in fat-free body mass occur with low doses of anabolic steroids. But with high doses, fat-free body mass increases markedly. These results suggest a threshold level for steroid doses, with only regular high doses resulting in substantial increases in fat-free body mass. Similarly, brief increases in testosterone, such as those that might occur after strength training or protein intake, do not appear to have a major effect on muscle mass or strength.

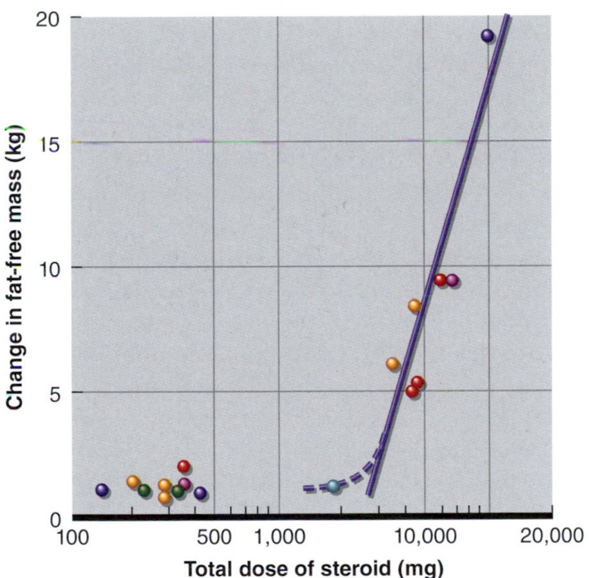

FIGURE 16.4 The relationship between the total dose of steroid (mg/day) and the change in fat-free mass in kilograms. The colors represent different anabolic steroid drugs. Steroid dose plotted logarithmically.

Adapted by permission from G.B. Forbes, *"The Effect of Anabolic Steroids on Lean Body Mass: The Dose Response Curve," Metabolism*, 34 (1985): 571-573, Copyright 1985, with permission from Elsevier. http://www.sciencedirect.com/science/journal

A third study looked at supraphysiological doses of testosterone in relation to muscle size and strength in men who were nonathletes but were experienced with weightlifting.[10] Forty-three men completed the study and were assigned to one of the following groups: placebo with no exercise, placebo with exercise, testosterone with no exercise, and testosterone with exercise. The men received either 600 mg of testosterone enanthate or placebo intramuscularly each week for 10 weeks. The exercise groups lifted weights 3 days each week for 10 weeks. Body composition was measured by underwater weighing, triceps and quadriceps muscle size by magnetic resonance imaging, and arm and leg strength by 1RM bicep curl and squat testing. In the no-exercise groups, testosterone injections resulted in increased arm and leg size and strength. Subjects who strength trained and received testosterone injections had the greatest increases in arm and leg size, mass, and strength (see table 16.2).

The increase in muscle mass is generally associated with increases in the cross-sectional areas of both type I and type II muscle fibers and an increase in the number of muscle nuclei. These increases are dose dependent and likely due to increased muscle protein synthesis resulting from continued stimulation of muscle androgen receptors by the exogenous steroids.[25]

The effects of steroids on cardiovascular function are not well studied. Early research reported improvements in $\dot{V}O_{2max}$ with the use of anabolic steroids, possibly as a result of increases in red blood cell production and total blood volume. However, later studies failed to confirm these results, and little research has been done on the effects of steroid administration on the training responses to endurance exercise.

Some athletes use steroids to speed recovery from hard training in the belief that anabolic steroids accelerate recovery and repair of muscle and other tissues. Anecdotal information from professional cyclists and other athletes seems to support this contention. Tamaki and colleagues[50] reported less muscle fiber damage after a single exhaustive bout of weightlifting

TABLE 16.2 **Average Changes in Muscle Size and Strength in Men Taking Testosterone or a Placebo, With and Without Exercise**

Variable	Placebo, no exercise	Testosterone, no exercise	Placebo with exercise	Testosterone with exercise
Fat-free mass	+0.8 kg	+3.2 kg	+2 kg	+6.1 kg
Triceps area	−82 mm^2	+424 mm^2	+57 mm^2	+501 mm^2
Quadriceps area	−131 mm^2	+607 mm^2	+534 mm^2	+1,174 mm^2
Bench press strength	No change	+9 kg	+10 kg	+22 kg
Squat strength	+3 kg	+13 kg	+25 kg	+38 kg

Data from Bhasin et al. (1996).

in a group of rats that received a single injection of the long-acting androgenic-anabolic steroid nandrolone decanoate compared to a control group that received a placebo. They also found that the rats in the steroid group had an increased rate of protein synthesis during recovery when compared to the control group. A fuller understanding of the effects of anabolic steroid use on recovery processes will likely develop as more research is conducted. However, regardless of the results, use of anabolic steroids in sport is prohibited.

Risks Associated with Anabolic Steroids

Steroid use in sport is cheating. It is not ethical for athletes to use drugs to improve their chances because doing so violates the very spirit of competition. Most athletes feel that it is wrong for their competitors to artificially improve their performance, yet many of these same athletes recognize that they are unable to compete at the same level if they do not use steroids. Fair competition is not possible if some athletes use illegal substances. This is one of the guiding principles of the World Anti-Doping Code.

The medical risks associated with steroid use are large, especially with the massive doses often used by athletes. And the risks associated with steroid use are not negated by stacking, cycling, and other techniques of steroid use. Children who use steroids before puberty risk early closure of the epiphyses of the long bones, so adult stature is reduced. Use of anabolic steroids suppresses the secretion of gonadotropic hormones, which control the development and function of the gonads (testes and ovaries). In boys and men, decreased gonadotropin secretion can cause testicular atrophy, decreased secretion of testosterone, a reduced sperm count, and impotence long after steroid administration ends. Excess testosterone can lead to higher estrogen production as well, causing enlargement of the male breasts (gynecomastia). In girls and women, gonadotropins are required for ovulation and secretion of estrogens, so a decrease in these hormones disrupts those processes and leads to irregular menstrual cycles. In addition, these hormonal disturbances in women can lead to masculinization: breast regression, enlargement of the clitoris, deepening of the voice, and growth of facial hair. Prostate gland enlargement in men is another possible side effect of steroid use, elevating the risk for prostate cancer.

Many steroids are metabolized by the liver. The tendency for steroids and their metabolites to accumulate in the liver can produce a form of chemical hepatitis and can lead to liver tumors. Abnormal cardiac hypertrophy (enlarged heart), cardiomyopathy (diseased heart muscle), myocardial infarction (heart attack), thrombosis, arrhythmia, and hypertension have been reported in chronic steroid users. Scientists have found markedly depressed high-density lipoprotein cholesterol (HDL-C) levels—reductions of 30% or more—in athletes on even moderate steroid doses. High-density lipoprotein cholesterol has antiatherogenic properties, meaning that it prevents the development of atherosclerosis. Low levels of HDL-C are associated with a high risk for both coronary artery disease and heart attack (see chapter 21). Furthermore, low-density lipoprotein cholesterol (LDL-C), which has atherogenic properties, appears to be increased with steroid use.

Substantial personality changes have occurred with steroid use, changes that can be enhanced by abuse of alcohol and other drugs. The most notable change is a marked increase in aggressive behavior, or "'roid rage." Some teens have become extremely violent and have attributed these drastic mood changes to steroid use. Evidence also suggests that drug dependency can result from steroid use.

It is important to note that not everyone using anabolic steroids is an athlete. In fact, it appears that the majority of those taking steroids are nonathletes who use steroids for cosmetic purposes. Further, many self-inject steroids, and when needles are shared, there is a dramatic increase in the risk of contracting infections such as hepatitis and human immunodeficiency virus and acquired immunodeficiency syndrome (HIV/AIDS).

Neither scientists nor physicians know the potential long-term effects of chronic steroid use. One study of male mice that received four different anabolic steroids of the kinds and doses taken by athletes showed that their life span was markedly decreased.[13] A number of birth defects have been reported by former East German athletes who were unknowingly given steroids during their athletic careers, although the true cause and incidence of these abnormalities remain unknown. Most diseases begin many years before symptoms develop, and it is possible that the more serious health risks of steroid use will not be apparent for 20 or 30 years.

Recent scientific review articles provide more detail on the potential ergogenic effects and health risks associated with anabolic steroid use.[16,25,31,36,41,60] Most sport governing bodies have developed educational materials for their athletes in hopes of preventing the use of steroids and other PEDs. Also, national governing bodies for most sports have instituted aggressive year-round testing programs in which athletes are tested randomly for steroid use.

Human Growth Hormone

For years, the medical treatment for hypopituitary dwarfism has been administration of **human growth**

hormone (hGH), a hormone secreted by the anterior pituitary gland. Before 1985, this hormone was obtained from cadaver pituitary extracts, and the supply was limited. Since the introduction of genetically engineered hGH in the mid-1980s, availability is no longer an issue.

During the 1980s, realizing this hormone's numerous functions, athletes started using hGH as a possible substitute for, and as a complement to, their use of anabolic steroids. As drug testing for anabolic steroids became more sophisticated, some athletes were looking for an alternative for which there was no drug test at the time.

Proposed Benefits of Growth Hormone

The six normal functions of growth hormone attracted the interest of some coaches and athletes:

- Stimulation of protein and nucleic acid synthesis in skeletal muscle
- Stimulation of bone growth (elongation) if bones are not yet fused
- Stimulation of insulin-like growth factor (IGF-1) synthesis
- Increase in lipolysis, leading to an increase in free fatty acids and an overall decrease in body fat
- Increase in blood glucose levels
- Enhancement of healing after musculoskeletal injuries

Demonstrated Effects of Growth Hormone

Administration of hGH to older men (>60 years) has been shown to increase fat-free mass, decrease fat mass, and increase bone density.[45] However, in studies of young men and experienced weightlifters, there appear to be few significant benefits.[59] More consistently, an increase in collagen synthesis and a reduction in fat mass are associated with hGH, suggesting that hGH is more useful as a cutting (fat-reducing) agent than an anabolic PED.

There is little doubt that exogenous testosterone is a powerful anabolic agent, but hGH does not have the same effects, except in individuals with hGH deficiencies. Exercise and certain amino acid supplements stimulate hGH release from the pituitary, but there is little evidence to suggest that this normal increase in hGH affects muscle protein synthesis or muscle mass and strength. In fact, exercise-related increases in hGH and testosterone are not needed to stimulate muscle protein synthesis and do not influence muscle mass or strength development.[54]

Risks Associated with Growth Hormone Use

As with steroids, potential medical risks are associated with hGH use. Acromegaly can result from taking hGH after the bones have fused. This disorder results in bone thickening, which causes broadening of the hands, feet, and face; skin thickening; and unwanted soft tissue growth. Internal organs typically enlarge. Ultimately, the victim suffers muscle and joint weakness and often heart disease. Cardiomyopathy is the most common cause of death with hGH use. Glucose intolerance, diabetes, and hypertension also can result from hGH use. Athletes who use PEDs often ingest or inject multiple substances, theoretically increasing the risks of adverse health effects.

In Review

- Stimulants such as amphetamines can improve performance in certain sports or activities, but, in addition to being illegal, these drugs carry risks that far outweigh their benefits. Amphetamines can be addictive, and they may mask important afferent and efferent signals that are designed to prevent injury and overexertion.
- Anabolic steroids are more appropriately termed androgenic-anabolic steroids or androgens, because in their natural state they include both androgenic (masculinizing) and anabolic (building) properties. Synthetic steroids have been designed to maximize the anabolic effects while minimizing androgenic effects.
- Anabolic steroid use increases muscle mass and strength and reduces body fat, which can improve sport performance. Aerobic endurance appears to be unaffected by steroid use, but additional research is needed to confirm or contradict that conclusion. Steroid use is illegal in sport and is banned by all governing bodies. Furthermore, the health risks can be considerable.
- Potential risks associated with use of anabolic steroids include personality changes, "'roid rage," testicular atrophy in men, reduced sperm count in men, breast enlargement in men, breast regression in women, prostate gland enlargement in men, masculinization in women, menstrual cycle disruption in women, liver damage, and cardiovascular diseases.
- Growth hormone does not appear to have strong anabolic or ergogenic effects. Research supports the ability of hGH to increase fat-free mass and decrease fat mass in older men, but hGH appears to have little or no effect on increasing muscle

mass and strength in younger people. Most of the increase in mass and fat-free mass is associated with increased water retention.

- It appears that hGH has no anabolic properties in young, healthy athletes. However, major health risks are associated with the use of growth hormone. Risks associated with hGH use include acromegaly, hypertrophy of internal organs, muscle and joint weakness, diabetes, hypertension, and heart disease.

Diuretics and Other Masking Agents

Diuretics such as desmopressin and acetazolamide affect the kidneys, increasing the loss of electrolytes and therefore water in the urine. Used under the supervision of a physician, diuretics reduce blood volume and total body water. They are generally prescribed to control hypertension and reduce edema (water retention) associated with congestive heart failure or other conditions.

Proposed Benefits of Diuretic Use

Diuretic use by athletes is often for weight control in weight-class sports such as wrestling and among athletes like gymnasts who believe that a lighter body will enhance their performance. Athletes who knowingly consume banned substances often use diuretics to increase their urine volume in the hope that the practice will flush the banned substance out of their bodies and dilute its concentration in the urine, making it more difficult to detect. Other masking agents such as glycerol, dextran, and albumin have been used in an attempt to expand blood volume and dilute the concentration of banned substances. All are on WADA's list of prohibited substances and techniques.

Demonstrated Effects of Diuretic Use

Diuretics lead to significant temporary weight loss, but no evidence suggests any other potential ergogenic effects. In fact, several side effects make diuretics ergolytic. The fluid loss results primarily from losses in extracellular fluid, including plasma, lowering blood volume. For athletes, particularly those who depend on moderate to high levels of aerobic endurance, this reduction in plasma volume reduces maximal cardiac output, which in turn reduces oxygen delivery and aerobic capacity, impairing performance. In other words, diuretics ensure dehydration.

Risks Associated with Diuretic Use

In addition to reducing plasma volume, the dehydrating effects of diuretics may hinder thermoregulation. As heat production increases during exercise, blood is diverted to the skin so that heat can be lost to the environment. However, when blood plasma volume is diminished, as with diuretic use, skin blood flow is reduced so that blood remains in the central regions of the body to maintain cardiac filling pressure and adequate blood supply and blood pressure to vital organs. Thus, less blood is available to be shunted to the skin, and heat loss may be impaired.

Electrolyte imbalance also can occur with prolonged use of diuretics. Many diuretics cause fluid loss by ensuring electrolyte loss. For example, a diuretic called furosemide inhibits sodium reabsorption in the kidneys, thus allowing more sodium to be excreted in the urine. Because fluid follows the sodium, more fluid is excreted. Electrolyte imbalances can occur with losses of either sodium or potassium. These imbalances can cause fatigue and muscle cramping. More serious imbalances can lead to exhaustion, cardiac arrhythmias, and even cardiac arrest. Some athletes' deaths have been attributed to electrolyte imbalances caused by diuretic use.

β-Blockers

The sympathetic nervous system influences bodily functions through adrenergic nerves: nerves that use norepinephrine as their neurotransmitter. Neural impulses traveling through these nerves trigger the release of norepinephrine, which crosses the synapses and binds to adrenergic receptors on the target cells. These adrenergic receptors are classified into two groups: α-adrenergic receptors and β-adrenergic receptors.

β-Adrenergic blockers, or **β-blockers**, are a class of drugs that block the β-adrenergic receptors, preventing norepinephrine from binding to its receptors on the target cells. Both nonspecific and specific (i.e., cardioselective) forms of β-blockers exist in many different prescription formulations. β-Blockers generally are prescribed for the treatment of hypertension, angina pectoris, and certain cardiac arrhythmias. They also are prescribed as a preventive treatment for migraine headaches, to reduce the symptoms of anxiety and stage fright, and for initial recovery from heart attacks.

Proposed Benefits of β-Blockers

Beta-blocker use by athletes has generally been limited to sports in which anxiety and tremor could impair performance, such as shooting sports, and by golfers seeking to steady their stroke during putting. When a person stands on a force platform (a device that measures mechanical forces), measurable body movement is detected each time the heart beats. These small movements are enough to affect a shooter's aim. Accuracy in shooting sports improves if the rifle or pistol can be shot—or the arrow released—between heartbeats. β-Blockers can slow a shooter's heart rate, allowing more time to stabilize the aim before the next heartbeat. Because of this effect, WADA has banned the use of β-blockers for sports such as archery, billiards, golf, and shooting.

Demonstrated Effects of β-Blockers

Because β-blockers decrease the effects of sympathetic nervous system activity, there is a marked reduction in maximum heart rate. For example, a 20-year-old male athlete with a normal maximum heart rate of 190 beats/min might have a maximum heart rate of only 130 beats/min when taking β-blocking drugs. These drugs also reduce resting heart rate and heart rate during submaximal exercise.

Risks Associated with β-Blockers

Most risks from β-blockers are associated with prolonged use, not isolated use as might happen with athletes. By blocking the sympathetic nervous system's relaxation effect on smooth muscle, β-blockers can induce bronchospasm in people with asthma and can cause cardiac failure in people who have underlying problems with cardiac function. In people with bradycardia, β-blocker medications can lead to heart block. In addition, the decreased blood pressure caused by β-blockers can result in light-headedness. Some people with type 2 diabetes can become hypoglycemic on β-blockers because insulin secretion is no longer limited. For most athletes, β-blockers are ergolytic because of the reduction in maximal heart rate and feelings of pronounced fatigue. For athletes who are prescribed β-blockers for a medical condition such as hypertension or a heart arrhythmia, β-1-selective blockers are usually preferred because they have fewer negative effects on performance.

Blood Doping

Blood doping refers to practices that alter the oxygen-carrying capacity of blood. These include infusions of red blood cells, artificial hemoglobin, or erythropoietin to stimulate the body to produce more red blood cells. Erythropoietin is the naturally occurring hormone that stimulates red blood cell production. It increases the number of red blood cells and therefore the blood's oxygen-carrying capacity.

Blood doping is prohibited because transfusion of red blood cells, previously donated by either the recipient (autologous transfusions) or someone else with the same blood type (homologous transfusions), increases red blood cell mass and the oxygen-carrying capacity of blood. Use of erythropoietin (EPO) or EPO-stimulating substances that increase red blood cell mass is prohibited for the same reason and also because numerous deaths have been associated with EPO use.

Proposed Benefits of Blood Doping

The premise underlying blood doping is simple. Because most oxygen is carried in blood bound to hemoglobin, increasing the number of red blood cells (and hemoglobin) available to ferry the oxygen to muscles could benefit endurance performance.

Demonstrated Effects of Blood Doping

In a landmark study in the early 1970s, Ekblom and associates[24] withdrew between 800 and 1,200 ml of blood from their subjects, refrigerated the blood, then reinfused the red blood cells into the subjects about 4 weeks later. Results showed a considerable improvement in $\dot{V}O_{2max}$ (9%) and treadmill performance time (23%) after reinfusion. Over the next few years, several studies confirmed these original findings, although several others failed to demonstrate an ergogenic effect.

Thus, the research literature was divided on the effectiveness of blood doping until a major breakthrough occurred in 1980 as a result of a study by Buick and colleagues.[15] Eleven highly trained distance runners were tested at different times during the study: (1) before blood withdrawal, (2) following blood withdrawal after allowing adequate time to reestablish normal red blood cell levels but before reinfusion of the removed blood, (3) following a sham reinfusion of 50 ml of saline (a placebo), (4) following reinfusion of 900 ml of the subject's own blood that originally had been withdrawn and preserved by freezing, and (5) after the elevated red blood cell levels had returned to normal.

As shown in figure 16.5, the researchers found a substantial increase in $\dot{V}O_{2max}$ and treadmill running time to exhaustion after the reinfusion of the red blood cells and no change after the sham reinfusion. This increase in $\dot{V}O_{2max}$ persisted for up to 16 weeks, although the improvement in treadmill time was lost within the first 7 days.

Why was the Buick study a major breakthrough? Gledhill[28] helped explain the controversy arising from the early studies. Many early studies that showed no improvement with blood doping had reinfused only small volumes of red blood cells, and the reinfusion was conducted within 3 to 4 weeks of the blood withdrawal. First, it appears 900 ml or more of whole blood must be reinfused to have an effect on performance. Increases in $\dot{V}O_{2max}$ and performance are not as great when smaller volumes are used. In fact, some studies using smaller volumes failed to show any differences in $\dot{V}O_{2max}$ or performance.

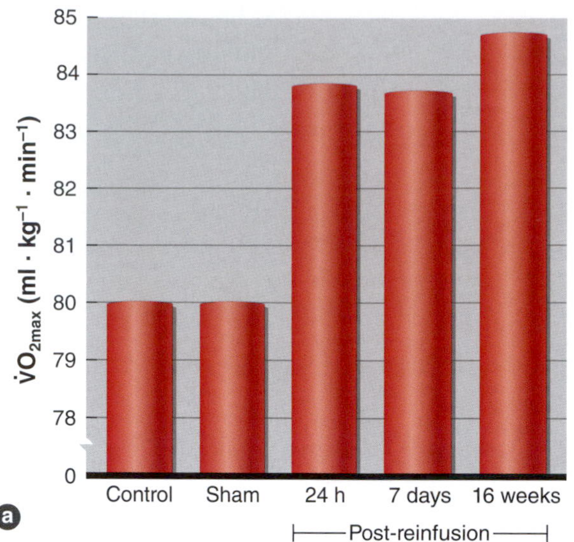

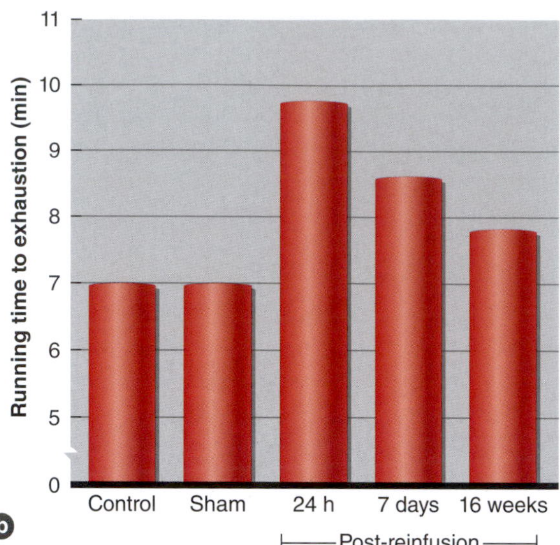

FIGURE 16.5 Changes in *(a)* $\dot{V}O_{2max}$ and *(b)* running time to exhaustion following reinfusion of red blood cells.

(a) Based on Buick et al. (1980). *(b)* Adapted by permission from F.J. Buick et al., "Effect of Induced Erythocythemia on Aerobic Work Capacity," *Journal of Applied Physiology* 48 (1980): 636-642.

Second, it appears that it is necessary to wait for at least 5 weeks and possibly as long as 10 weeks before reinfusion. This is based on the time it takes to reestablish the blood's original, prewithdrawal red blood cell mass and hematocrit.

Finally, researchers who conducted early studies refrigerated the withdrawn blood. When blood is refrigerated, approximately 40% of the red blood cells are destroyed; maximal storage time for blood under refrigeration is approximately 5 weeks. Later studies used frozen storage. Freezing allows an almost unlimited storage time, and only about 15% of the red blood cells are lost. Gledhill[28] concluded that blood doping significantly improves $\dot{V}O_{2max}$ and endurance performance when done under optimal conditions:

- A minimum of 900 ml of blood reinfusion
- A 5- to 6-week minimum interval between withdrawal and reinfusion
- Blood storage by freezing

The researchers also showed that the improvements in $\dot{V}O_{2max}$ and performance are the direct result of the blood's increased hemoglobin content, not an increased cardiac output caused by an expanded plasma volume.

Does an increase in $\dot{V}O_{2max}$ and treadmill time as a result of blood doping translate into improved endurance performance? Several studies have addressed this issue. Researchers in one study observed 5 mi (8 km) treadmill run times in a group of 12 experienced distance runners.[59] Their times were checked before and after saline (placebo) infusion and before and after blood infusion; the run times on the treadmill were significantly faster following blood infusion, but this difference became clear only over the second half of the trial. The blood infusion trials were 33 s (3.7%) faster over the last 2.5 mi (4 km) and 51 s (2.7%) faster over the full distance, compared to the placebo trials. A second study looked at 3 mi (4.8 km) run times in a group of six trained distance runners and reported an average decrease of 23.7 s following blood doping when compared to the runners' blind control trials.[29] Subsequent studies confirmed improvements in distance running and cross-country skiing performance with blood doping.[23,48] Figure 16.6 illustrates the improvement in run time with blood doping for distances of up to 11 km (6.8 mi).

Risks Associated with Blood Doping

Although blood doping is relatively safe in the hands of competent physicians, it has inherent dangers.[3] Adding more red blood cells to the cardiovascular system can cause the blood to become too viscous, which can lead to clotting and possibly heart failure. In an effort to control blood doping, some sport governing bodies, such as professional cycling, do not allow athletes to compete if their hemoglobin concentrations or hematocrit levels are too high (e.g., >50%). With autologous blood transfusions, in which the recipients receive their own blood, mislabeling of the blood can occur. With homologous blood transfusions, in which blood is received from a matched donor, several other complications can occur. The reinfused blood could be mismatched and an allergic reaction could be triggered. The athlete

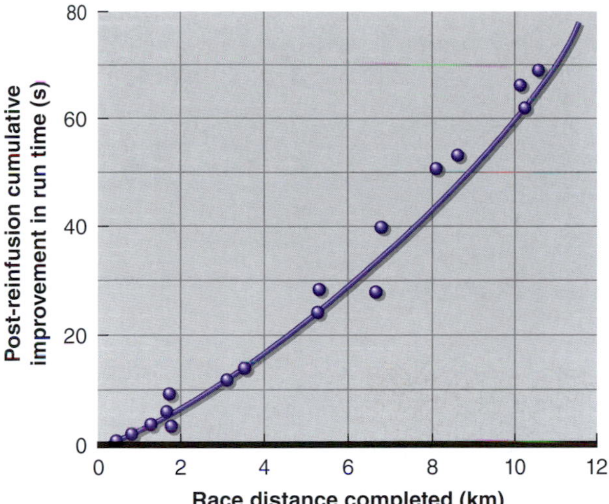

FIGURE 16.6 Improvements in running times for distances of up to 11 km (6.8 mi) following reinfusion of red blood cells from two units of blood preserved by freezing. Values on the *y*-axis reflect the reduction in time to run a specific distance on the *x*-axis. As an example, for a 10 km (6.2 mi) race, one could expect to run 60 s faster after reinfusion.

Adapted by permission from L.L Spriet, Blood Doping and Oxygen Transport, in *Ergogenics: Enhancement of Performance in Exercise and Sport*, edited by D.R. Lamb and M.H. Williams (Dubuque, IA: Brown & Benchmark, 1991), 213-242. Copyright 1991 Cooper Publishing Group, Carmel, IN.

also risks infection from hepatitis or HIV pathogens.[51]

The outcome of EPO use is less predictable than that of red blood cell reinfusion (blood doping). Once EPO has been administered, it is difficult to predict how much red blood cell production will occur. This places the athlete at great risk of substantial increases in blood viscosity and associated problems such as thrombosis (blood clot), myocardial infarction (heart attack), congestive heart failure, hypertension, stroke, and pulmonary embolism.

The potential risks of blood doping, even without consideration of the legal and ethical issues involved, outweigh any potential benefits.

In Review

- Diuretics affect the kidneys, increasing urine production. They often are used by athletes for temporary weight reduction and also by those trying to mask the use of other drugs in anticipation of drug testing.
- Weight loss (dehydration) is the only substantiated potential ergogenic effect of diuretics, but this weight loss is primarily from the extracellular fluid compartment, including blood plasma. This can lead to dehydration, volume depletion, increased cardiac strain, and electrolyte imbalances.
- β-blocker medications are banned in sports in which a slowed heart rate can provide a competitive advantage. The World Anti-Doping Agency maintains a list of specific sports in which β-blockers are prohibited substances.
- β-blockers block β-adrenergic receptors, limiting the binding of catecholamines. They slow the resting heart rate, which is a distinct advantage for shooters who try to release the arrow or squeeze the trigger between heartbeats to minimize the slight tremor associated with each beat, and could be an advantage for golfers when chipping and putting.
- β-blockers can cause heart block, hypotension, bronchospasm, pronounced fatigue, and decreased motivation. Selective β-blockers have fewer side effects than nonselective blockers.
- Blood doping and administration of EPO can enhance aerobic capacity and the performance of aerobic sports or activities. This occurs through an increase in the oxygen-carrying capacity of blood, primarily attributable to increased red blood cells. Both procedures involve risk.
- Studies have shown major increases in maximal oxygen uptake, time to exhaustion, and actual performance in cross-country skiing, cycling, and distance running as a result of blood doping.
- Erythropoietin is the naturally occurring hormone that stimulates red blood cell production in bone marrow. Studies have clearly demonstrated increased maximal oxygen consumption and increased exercise time to exhaustion after administration of EPO.
- Serious risks associated with blood doping include blood clotting, heart failure, and, if blood from another donor is used by accident or intentionally, transfusion reactions and transmission of hepatitis and HIV.
- Because the magnitude of the response to EPO administration cannot be accurately predicted, EPO use can be dangerous. The hormone can lead to death if red blood cells are overproduced, increasing the blood's viscosity. Known risks include thrombosis, myocardial infarction, congestive heart failure, hypertension, stroke, and pulmonary embolism.

IN CLOSING

All forms of ergogenic aids are associated with potential benefits and possible risks. That risk-to-benefit ratio is greater for some interventions than for others. For example, there are potential benefits to creatine loading for some athletes, but some risk is involved if the supplement contains banned substances. In addition, athletes must recognize the legal, ethical, and medical consequences of using any ergogenic agent, especially those that are expressly prohibited (e.g., steroid use, blood doping) or that may contain banned substances (e.g., contaminated substances.) Athletes who use banned substances or procedures risk disqualification from a particular competition and can be banned from competition in their sport for a year or more. We have discussed pharmacological, hormonal, physiological, and some specific ergogenic nutrition aids. In the next part of the book, we shift our focus to younger, older, and female athletes within the broader categories of growth and development, aging, and sex differences in exercise performance. We begin in chapter 17 by examining special considerations in the child and adolescent.

KEY TERMS

amphetamines
anabolic steroids
β-blockers
blood doping
caffeine
creatine
diuretics
ephedrine
ergogenic
ergolytic
human growth hormone (hGH)
placebo effect
pseudoephedrine
testosterone
therapeutic use exemption

STUDY QUESTIONS

1. What is the meaning of the term *ergogenic aid*? What is an ergolytic effect?
2. Why is it important to include control groups and placebos in studies of the ergogenic properties of any substance or phenomenon?
3. How do β-alanine and sodium bicarbonate function as possible ergogenic aids?
4. How might caffeine improve athletic performance?
5. How might the polyphenols in cherry juice benefit recovery from intense exercise?
6. What are the potential ergogenic properties of creatine supplementation?
7. What roles do the amino acid leucine play in building muscle mass?
8. What effect does nitrate ingestion have on the response to exercise?
9. By what criteria are substances considered for the banned list, according to the World Anti-Doping Code?
10. What is presently known about the use of amphetamines in athletic competition? What are the potential risks of using amphetamines?
11. What are the effects of anabolic steroid use on athletic performance? What are some of the medical risks of steroid use?
12. What is known about hGH as a potential ergogenic aid? What are the risks associated with its use?
13. Can diuretics be ergogenic? What are some risks associated with their use?
14. Under what circumstances might β-blockers be ergogenic aids?
15. What is blood doping? Does blood doping improve athletic performance?
16. By what mechanism is erythropoietin theorized to benefit performance?

STUDY GUIDE ACTIVITIES

In addition to the activities listed in the chapter opening outline, two other activities are available in the web study guide, located at

www.HumanKinetics.com/PhysiologyOfSportAndExercise

The **KEY TERMS** activity reviews important terms, and the end-of-chapter **QUIZ** tests your understanding of the material covered in the chapter.

PART VI

Age and Sex Considerations in Sport and Exercise

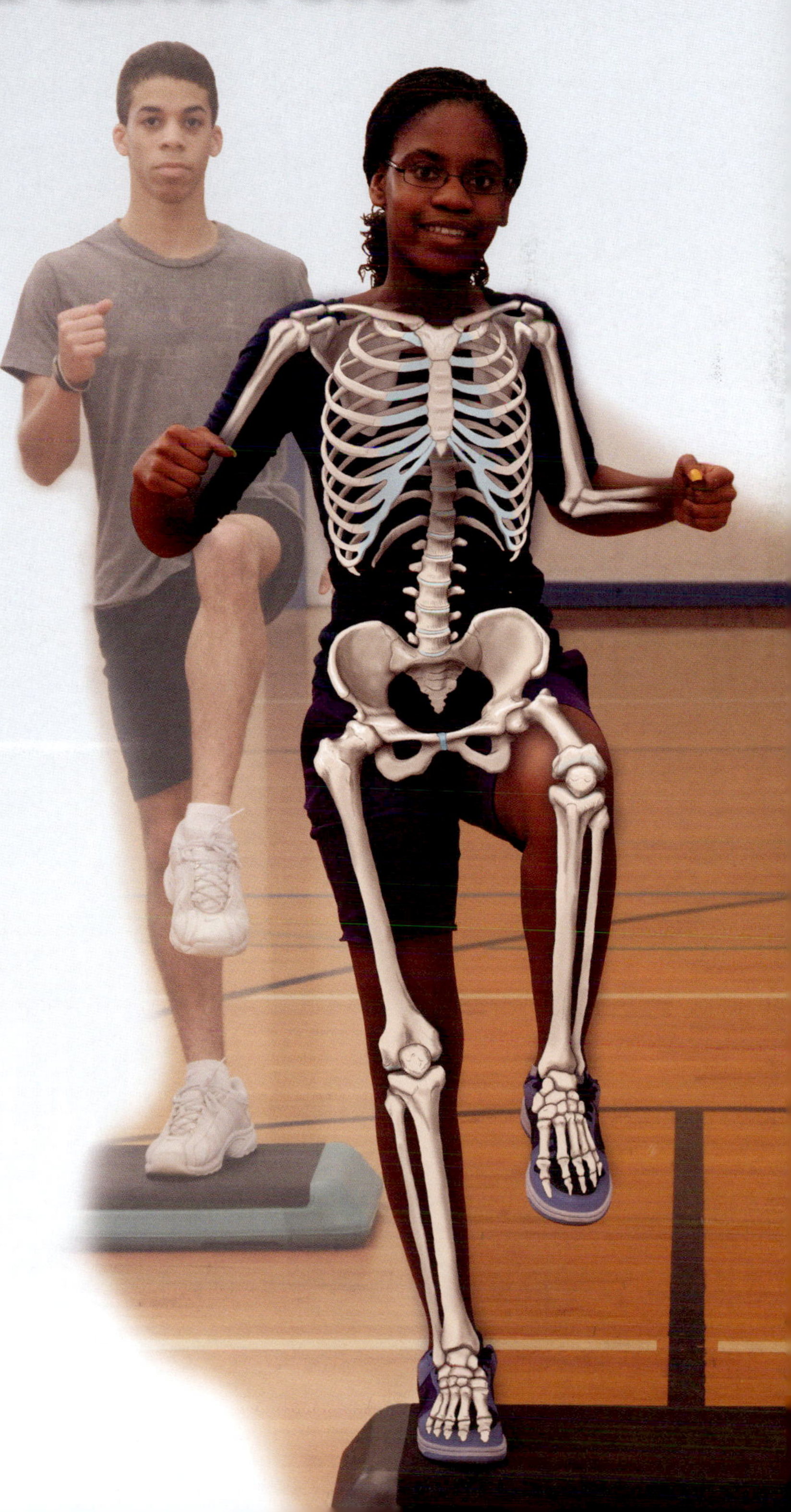

From the previous parts of this book, we gained a solid understanding of the general principles of exercise and sport physiology. Historically, much of the basic and applied exercise physiology literature has been based on responses of young men. Now we turn our attention to how these principles are applied to children and adolescents, older individuals, and women. In chapter 17, Children and Adolescents in Sport and Exercise, we examine the processes of human growth and development and how different developmental stages affect a child's physiological capacity and exercise performance. We also consider how these stages of growth and development might alter our strategies for training young athletes for competition. In chapter 18, Aging in Sport and Exercise, we discuss how physiological changes affect exercise and sport performance as people age beyond the middle years, focusing on which of these changes are attributable to the aging process, per se, and what changes might be attributable to an increasingly sedentary lifestyle. We discuss the important role that continued training can play in minimizing the loss of performance capacity and physical conditioning that accompanies aging. In chapter 19, Sex Differences in Sport and Exercise, we examine differences between women's and men's responses to acute exercise and exercise training and the extent to which these differences are biologically determined. We also focus on physiological and clinical issues specific to female athletes, including menstrual function, pregnancy, osteoporosis, and the higher prevalence of eating disorders in female athletes.

17 Children and Adolescents in Sport and Exercise

In this chapter and in the web study guide

The fastest man and the fastest woman in the world come from the same small country of only 2.8 million people. Jamaican runners Usain Bolt and Shelly-Ann Fraser-Pryce both won track-and-field gold medals in the 100 m race in Beijing in 2008 and again in London in 2012. In 2016, Elaine Thompson won the gold in both the 100 m and 200 m races in Rio, with Bolt again winning the 100 m and 200 m races for men. How can a tiny island known for sun, beaches, and reggae music produce so many track champions? While theories are plentiful, one thing that separates Jamaican athletes is their early childhood interest in track and field, nurtured by a culture that promotes, supports, and rewards childhood exercise and sport.

Although sustaining an in-school program of regular physical activity in the United States and other countries is currently challenging, such is not the case in Jamaica. The Jamaican education system, through a rigorous physical education syllabus and dedicated physical education teachers, has paved the way for the island's Olympic track tradition. Exercise—and, in particular, running—is engrained in the culture and widely promoted among the children of the island.

Competitions encourage children to stay active, exercise regularly, and test their athletic skills against their schoolmates. As early as age 3, while children are in preschool, they begin training and preparing for one of every school's most eagerly anticipated occasions, Sports Day. Sports Days are held at virtually every school and continued throughout high school and college. Boys and girls begin to enter nationally sponsored races when they are as young as 5, and by the time they are teenagers, top sprinters are competing before huge crowds at the National Stadium at the Inter-Secondary Schools Sports Association (ISSA) Boys' and Girls' Athletic Championship, or simply "Champs," held during the first week of April every year.

Not every child will become the next Elaine Thompson or Usain Bolt, but we can learn from the Jamaican experience: Regular physical activity is good for all children, whether they grow to become world-class athletes or simply healthy and fit adults.

Throughout the previous chapters of this book, we have examined the physiological responses of adults to acute bouts of exercise and their adaptations to training and the environment. For many years, it was assumed that children and adolescents responded and adapted identically to adults, but few scientists actually had studied children and adolescents. It is especially important to understand how children and adolescents respond to exercise because physical activity is vital in battling childhood obesity and in teaching children to develop lifelong healthy habits. We now have a better understanding and appreciation of both the differences and similarities between adults and children and adolescents, which we discuss in this chapter.

Growth, Development, and Maturation

Growth, development, and maturation are terms used to describe changes that occur in the body starting at conception and continuing through adulthood. **Growth** refers to an increase in the size of the body or any of its parts. **Development** refers to differentiation of cells along specialized lines of function (e.g., organ systems), so it reflects the functional changes that occur with growth. Finally, **maturation** refers to the process of taking on adult form and becoming fully functional, and it is defined by the system or function being considered. For example, skeletal maturity refers to having a fully developed skeletal system in which all bones have completed normal growth and ossification, whereas sexual maturity refers to having a fully functional reproductive system. The state of a child's or adolescent's maturity can be defined by

- chronological age,
- skeletal age, and
- stage of sexual maturation.

Throughout this chapter we refer to the child and the adolescent. The period of life from birth to the start of adulthood is generally divided into three phases: infancy, childhood, and adolescence. **Infancy** is defined as the first year of life. **Childhood** spans the period of time between the end of infancy (the first birthday) and the beginning of adolescence and is usually divided into early childhood (preschool) and middle childhood (elementary school). The period of **adolescence** is more difficult to define in chronological years, because it varies in both its onset and its termination. Its onset generally is defined as the onset of **puberty**, when secondary sex characteristics develop and sexual reproduction becomes possible, and its termination as the completion of growth and development processes, such as attaining adult height. For most girls, adolescence ranges from 8 to 19 years, and for most boys from 10 to 22 years.

Given the increasing popularity of youth sport and an emphasis on increasing children's physical fitness to combat childhood obesity, we must understand the impact of exercise on the physiological aspects of growth and development. Children and adolescents must not be regarded as mere miniature versions of adults. The growth and development of their bones, muscles, nerves, and organs largely dictate their physiological and performance capacities. As a child's size increases, so do almost all of his or her functional capacities. This is true of motor ability, strength, cardiovascular and respiratory function, and aerobic and anaerobic capacity. In the following sections, we examine age-related changes in a growth and development.

Height and Weight

Specialists in the field of growth and development have spent considerable time analyzing the changes in height and weight that accompany growth and—importantly—the rates at which growth occurs over time. Figure 17.1 shows that height increases rapidly during the first 2 years of life, with children reaching about 50% of adult height by age 2. After this, height increases at a progressively slower rate throughout childhood as shown by the decline in the rate of its change. Just before puberty, the rate of change in height increases markedly, followed by an exponential decrease in growth rate until full height is attained at a mean or average age of about 16 years in girls and 18 years in boys (although some boys do not reach their full height until their early 20s). The peak rate of change in height occurs at approximately 12 years in girls and 14 years in boys. The peak rate of change in body weight occurs at approximately 12.5 years in girls and 14.5 years in boys—slightly later than for height.

Bone

Bones, joints, cartilage, and ligaments form the body's structural support. Bones provide points of attachment for the muscles, protect delicate tissues, and act as reservoirs for calcium and phosphorus, and some are involved in blood cell formation. During fetal development, as well as during the initial 14 to 22 years of life, membranes and cartilage are transformed into bone through the process of **ossification**, or bone formation. The line of cartilage in our bones is also known as the growth plate. The average age at which the growth plate closes and ossification is complete is highly variable, but bones typically begin to fuse in the preteens, and all are fused by the early 20s. On average, girls achieve full bone maturity several years before boys. This is due to the role of different hormones, including estrogen, in signaling the growth plate to close.

In general, bone structural integrity and bone health are evaluated by examining **bone mineral density (BMD)** as well as blood markers of bone formation and resorption. During childhood and throughout adolescence, BMD increases significantly, generally peaking sometime in the second decade and falling thereafter throughout the life span. This concept is illustrated for women across the life span in figure 17.2. Therefore, adolescence is a prime window for increasing BMD with proper nutrition and physically stressing bone through weight-bearing exercise.[22]

A recent longitudinal study incorporating jumping exercises (box jumping) for 8- to 9-year-old prepubescent boy and girls showed that simple short-term high-impact exercise confers long-term benefits. The boys and girls who participated in box jumping exercises during their school physical education class saw an increase in their BMD after 7 months, and this benefit was sustained for 4 years after the intervention. The increase was over and above what is observed with normal growth and development. Furthermore, if the benefits of this type of exercise are sustained until BMD plateaus in early adulthood, this type of exercise could have substantial effects on reducing fracture risks later in life when BMD decreases.[10]

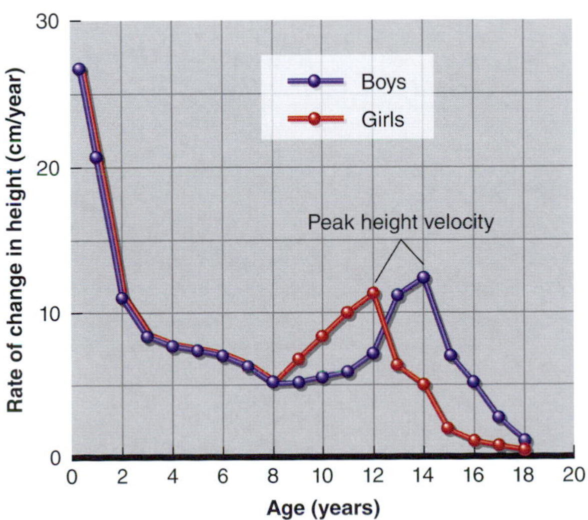

FIGURE 17.1 Changes with age in the rate of increase in height (cm/year).

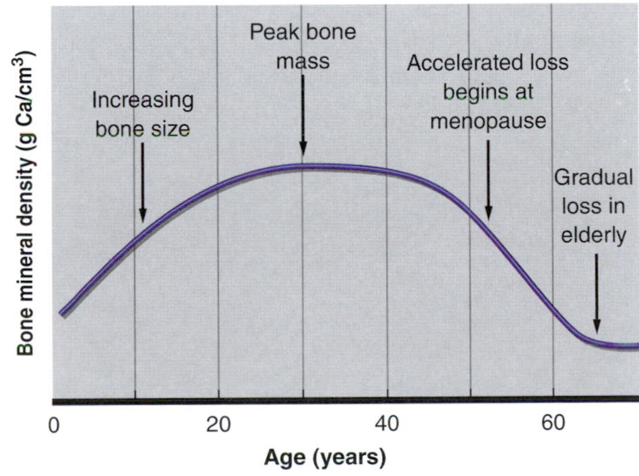

FIGURE 17.2 Changes in bone mineral density throughout the female life span. The decline after age 50 is less precipitous for men.

In Review

- Growth in height is very rapid during the first 2 years of life, with a child reaching 50% of adult stature by age 2. After that, the rate is slower throughout childhood until a marked increase occurs near puberty.
- The peak rate of height growth occurs at age 12 in girls and age 14 in boys. Full height is typically attained at age 16 in girls and age 18 in boys.
- Growth in weight follows the same trend as height. The peak rate of weight increase occurs at age 12.5 in girls and age 14.5 in boys.
- Bone mineral density increases significantly throughout childhood and adolescence, peaking in early adulthood. High-impact load-bearing exercise can substantially increase BMD.

Muscle

From birth through adolescence, the body's muscle mass steadily increases as weight increases. In males, the skeletal muscle mass increases from 25% of total body weight at birth to about 40% to 45% or more in young men (20-30 years). Much of this gain occurs when the rate of muscle development peaks during puberty. This peak corresponds to a sudden, almost 10-fold increase in testosterone. Girls don't experience such rapid acceleration of muscle growth at puberty, but their muscle mass does continue to increase (although more slowly than boys') to about 30% to 35% of their total body weight as young adults. This rate difference is largely attributed to hormonal differences at puberty (see chapter 19).

Increases in muscle mass with growth appear to result primarily from hypertrophy (increase in size) of existing fibers, with little or no hyperplasia (increase in fiber number). Fiber hypertrophy results from increases in the myofilaments and myofibrils. Increases in muscle length as young bones elongate result from increases in the number of sarcomeres (which are added at the junction of the muscle and the tendon) and from increases in the length of existing sarcomeres. Muscle mass peaks in girls at age 16 to 20 years and in boys at 18 to 25 years, unless it is increased further through exercise, diet, or both.

Fat

Fat cells form and fat deposition starts in these cells early in fetal development, and this process continues indefinitely thereafter. Each fat cell can increase in size at any age from birth to death. The amount of fat that accumulates with growth and aging depends on

- diet,
- exercise habits, and
- heredity.

Heredity is unchangeable, but both diet and exercise can be altered to either increase or decrease fat stores.

At birth, 10% to 12% of total body weight is fat. At **physical maturity**, the fat content reaches approximately 15% of total body weight for men and approximately 25% for women. This sex difference, like that seen in muscle growth, is primarily attributable to hormonal differences. When girls reach puberty, their estrogen concentrations and tissue exposure increase, promoting the deposition of body fat. Figure 17.3 illustrates the changes in percent body fat, fat mass, and fat-free mass for both boys and girls from ages 8 to 20 years.[16] It is important to realize that both fat mass and fat-free mass increase during this time period, so an increase in absolute fat does not necessarily mean an increase in relative fat percentage.

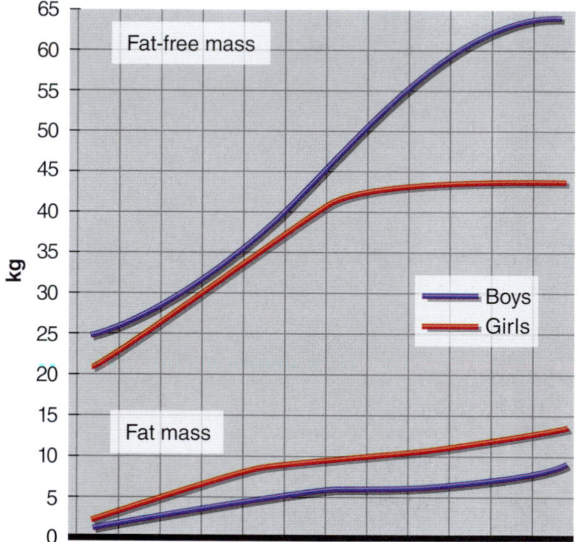

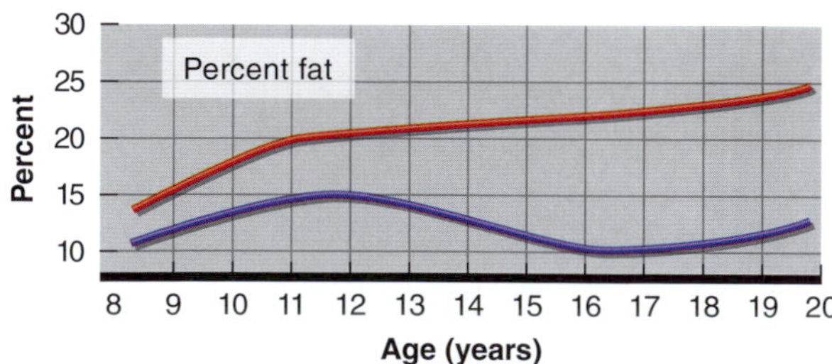

FIGURE 17.3 Changes in percent fat, fat mass, and fat-free mass for girls and boys from age 8 to 20.

Reprinted by permission from R.M. Malina, C. Bouchard, and O. Bar-Or, *Growth, Maturation, and Physical Activity*, 2nd ed. (Champaign, IL: Human Kinetics, 2004), 114.

Nervous System

As children grow, they develop better balance, agility, and coordination as their nervous systems develop. Myelination of the nerve fibers must be completed before fast reactions and skilled movement can occur because conduction of an impulse along a nerve fiber is considerably slower if myelination is absent or incomplete (discussed in chapter 3). **Myelination** of the cerebral cortex occurs most rapidly during childhood but continues well beyond puberty. Although practicing an activity or skill can improve performance to a certain extent, the full development of that activity or skill depends on full maturation (and myelination) of the nervous system. The development of strength is also likely influenced by myelination.

In Review

- Muscle mass increases steadily along with weight from birth through adolescence.
- In boys, the rate of muscle mass increase peaks at puberty, when testosterone production increases dramatically. Girls do not experience this sharp increase in muscle mass.
- Muscle mass increases in boys and girls result primarily from fiber hypertrophy with little or no hyperplasia.
- Muscle mass peaks in girls between ages 16 and 20 and in boys between ages 18 and 25, although it can be further increased through diet and exercise.
- Fat cells can increase in size and number throughout life.
- The amount of fat accumulation depends on diet, exercise habits, and heredity.
- At physical maturity, the body's fat content averages 15% in young men and 25% in young women. The differences are caused primarily by higher testosterone levels in men and higher estrogen levels in women.
- Balance, agility, and coordination improve as children's nervous systems develop.
- Myelination of nerve fibers is accompanied by faster reactions and more skilled movements because myelination speeds the transmission of electrical impulses.

Physiological Responses to Acute Exercise

The function of almost all physiological systems continues to improve until full maturity is reached, or shortly before. After that, physiological function plateaus for a period of time before starting to decline beginning in middle age. In this section, we focus on some of the changes in children and adolescents during an acute bout of exercise during growth and development.

Strength

Strength improves as muscle mass increases with growth and development. Peak strength usually is attained by age 20 in women and between age 20 and 30 in men. The hormonal changes that accompany puberty lead to marked increases in strength in pubescent boys because of increased muscle mass. The extent of development and the performance capacity of muscle also depend on the concurrent maturation of the nervous system. High levels of strength, power, and skill are impossible if the child has not reached neural maturity. Myelination of many motor nerves is incomplete until sexual maturity, so the neural control of muscle function is limited before that time.

Figure 17.4 illustrates changes in leg strength in a group of boys from the Medford Boys' Growth Study.[4] The boys were followed longitudinally from age 7 to 18. The rate of strength gain (slope of line) increased noticeably around age 12, the typical age for onset of puberty. Similar longitudinal data for girls over this same age span are not available. Cross-sectional data, however, indicate that girls experience a more gradual and linear increase in absolute strength and do not exhibit any significant change in strength relative to body weight after puberty,[9] as shown in figure 17.5.

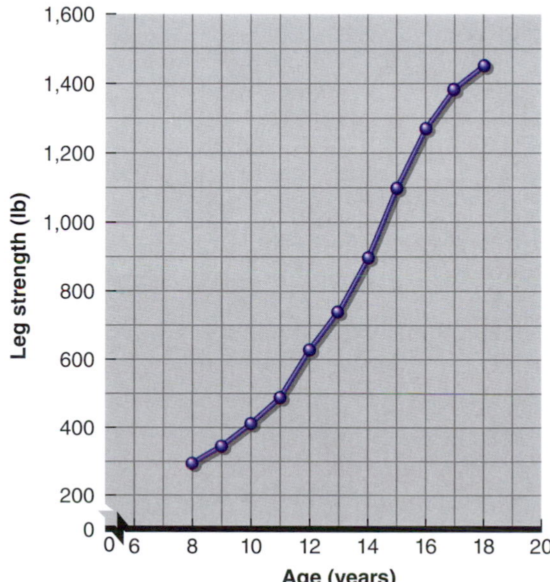

FIGURE 17.4 Gains with age in leg strength of boys followed longitudinally over 12 years. Note the increase in the slope of the curve from 12 to 16 years of age.

Data from Clarke (1971).

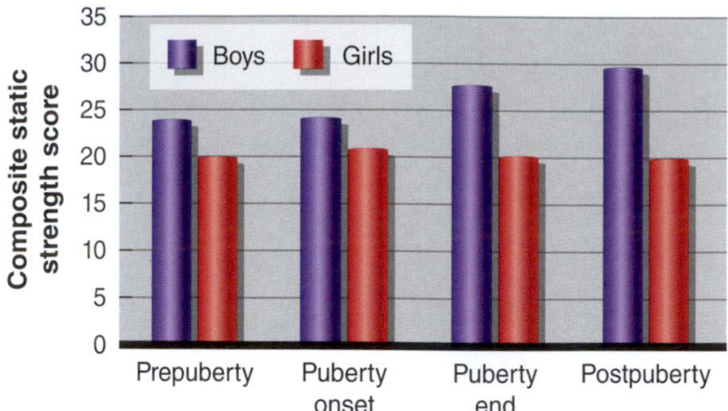

FIGURE 17.5 Changes in strength with developmental status in boys and girls. Strength is expressed as a composite static strength score from several strength testing sites, and the data are expressed per kilogram of body mass to account for differences in size between boys and girls. PHV = peak height velocity.

Cardiovascular and Respiratory Function

Cardiovascular function undergoes considerable change as children grow and age. Because of significant increases in aerobic power over the course of growth and development, we need to consider these changes at rest and during submaximal exercise, as well as during maximal exercise.

Rest and Submaximal Exercise

Blood pressure at rest and during submaximal exercise is lower in children than in adults but progressively increases to adult values during the late teen years. Blood pressure is directly related to body size. Larger people generally have bigger hearts and higher blood pressures, so size is at least partially responsible for children's lower blood pressure values. In addition, blood flow to active muscles during exercise in children can be greater for a given volume of muscle than in adults because children have less peripheral resistance. Thus, in children, for a given submaximal workload, blood pressure is lower and the muscle is relatively overperfused.

Recall that cardiac output is the product of heart rate and stroke volume. A child's smaller heart size and total blood volume result in a lower stroke volume, both at rest and during exercise, than in an adult. In an attempt to compensate for the lower stroke volume to maintain cardiac output, the child's heart rate at a given absolute submaximal intensity is higher than an adult's. As the child ages and heart size and blood volume increase along with body size, stroke volume increases as body size increases and heart rate decreases for the same absolute work intensity.

However, a child's higher submaximal heart rate cannot completely compensate for the lower stroke volume, so the child's cardiac output is also somewhat lower than the adult's for a given oxygen consumption. To maintain adequate oxygen uptake during these submaximal levels of work, the child's arterial–mixed venous oxygen difference, or (a-$\bar{v}$)O_2 difference, increases to further compensate for the lower cardiac output. The increase in (a-$\bar{v}$)O_2 difference is most likely attributable to a greater percentage of the cardiac output going to the active muscles.[28] These cardiovascular responses during submaximal exercise are illustrated in figure 17.6.

Maximal Exercise

Maximum heart rate (HR_{max}) is higher in children than in adults but decreases linearly with age. Children under age 10 frequently have maximum heart rates exceeding 210 beats/min, whereas the average 20-year-old has a maximum heart rate of approximately 195 beats/min.

As with submaximal exercise, the child's smaller heart and blood volume limit the maximal stroke volume that he or she can achieve. Again, the high HR_{max} cannot fully compensate for this, leaving the child with a lower maximal cardiac output than the adult. This limits the child's performance at high absolute workloads during some activities (e.g., cycling) because the child's capacity for oxygen delivery is less than an adult's. However, for high relative workloads in which the child is responsible for moving only his or her body mass (e.g., running), this lower maximal cardiac output is not as serious a limitation. In running, for example, a 25 kg (55 lb) child requires (in proportion to body size) considerably less oxygen than a 90 kg (198 lb) man would require, yet the rate of oxygen consumption when scaled for body size is about the same for both.

In Review

- Strength improves in children as muscle mass increases with growth and development.
- Gains in strength with growth also depend on neural maturation because neuromuscular control is limited until myelination is complete, usually around sexual maturity.
- Blood pressure is directly related to body size: It is lower in children than in adults but increases to adult levels in the late teen years, both at rest and during exercise.
- During both submaximal and maximal exercise, a child's smaller heart and blood volume result in a lower stroke volume than in adults. In partial compensation, a child's heart rate is higher than an adult's for the same exercise intensity. Maximal

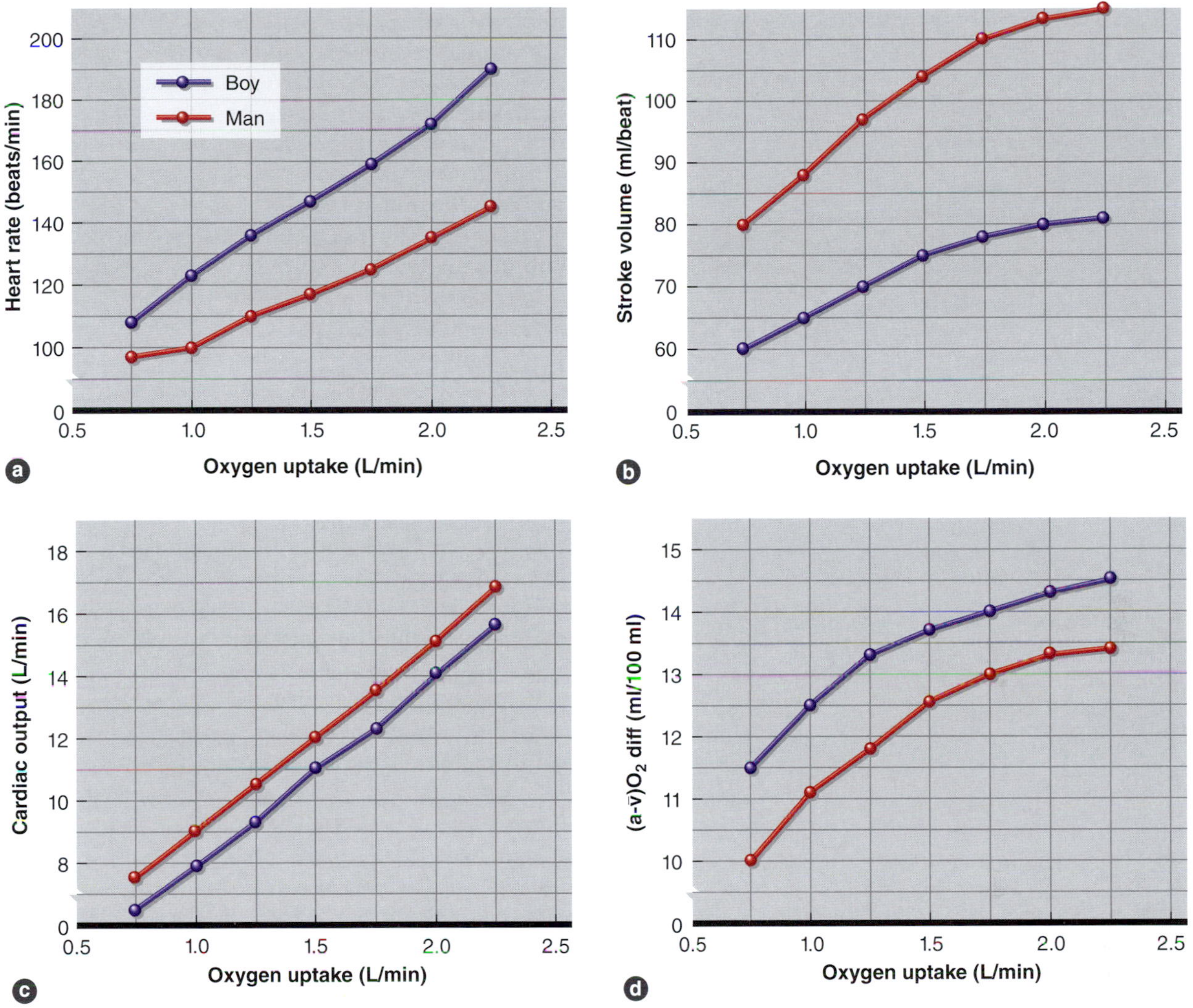

FIGURE 17.6 Submaximal *(a)* heart rate, *(b)* stroke volume, *(c)* cardiac output, and *(d)* arterial–mixed venous oxygen difference, or (a-v̄)O_2 difference, in a 12-year-old boy and a fully mature man at the same rates of oxygen uptake.

cardiac output is lower than that of an equally trained adult.

- Even with increased heart rate, a child's cardiac output remains less than an adult's. In submaximal exercise, an increase in the (a-v̄)O_2 difference ensures adequate oxygen delivery to the active muscles. But at maximal work rates, oxygen delivery limits performance in activities other than those in which the child merely needs to move body mass, such as running.

Metabolism

Metabolism and substrate utilization at rest and during exercise also change as the child and adolescent grow larger, as we would expect from the changes that we have just reviewed in muscle mass and strength and cardiorespiratory function.

Maximal Oxygen Consumption

The purpose of the cardiovascular and respiratory adjustments that occur in response to exercise is to accommodate the exercising muscles' need for oxygen. Thus, increases in cardiovascular and respiratory function that accompany growth support corresponding increases in aerobic capacity ($\dot{V}O_{2max}$). In 1938, Robinson[24] demonstrated this phenomenon in a cross-sectional sample of boys and men ranging in age from 6 to 91 years. He found that $\dot{V}O_{2max}$ peaks between ages 17 and 21 and then decreases linearly with age. Other studies have subsequently confirmed these observations. Studies of girls and women have shown essentially the same trend, although the decrease begins at a much younger age, generally age 12 to 15 years, in part attributable to an earlier assumption of a more sedentary lifestyle. The changes in $\dot{V}O_{2max}$ with age, expressed in liters per minute, are illustrated in figure 17.7*a*.

Scaling Physiological Data to Account for Size Differences

Expressing $\dot{V}O_{2max}$ relative to body weight (ml · kg^{-1} · min^{-1}) provides a considerably different picture, as shown in figure 17.7*b*. Values change little in boys from age 6 to young adulthood. For girls, little change occurs from age 6 to 13, but after age 13, $\dot{V}O_{2max}$ shows a gradual decrease. These observations may not accurately reflect the development of the cardiorespiratory system as children grow or changes in their physical activity levels. Several concerns have been raised about the validity of using $\dot{V}O_{2max}$ relative to body weight to account for changes in the cardiorespiratory and metabolic systems during periods of growth and development. Rather, the sex differences that begin to emerge around puberty may reflect differences in increasing body mass and changing body composition. Because girls tend to increase fat mass with estrogen exposure at puberty, decreases in $\dot{V}O_2$ per kilogram may not be significant when normalized for fat-free mass.

There are several arguments against using body weight to scale $\dot{V}O_{2max}$ for differences in size. First, although $\dot{V}O_{2max}$ values expressed relative to body weight remain relatively stable or decline with age, endurance performance steadily improves. The average 14-year-old boy can run the mile (1.6 km) almost twice as fast as the average 5-year-old boy, yet the two boys' $\dot{V}O_{2max}$ values expressed relative to body weight are similar.[25] Second, although the increases in $\dot{V}O_{2max}$ that accompany endurance training in children are relatively small compared with those in adults, the performance increases in these children can be larger. Therefore, body weight is likely not the most appropriate way to scale $\dot{V}O_{2max}$ values for differences in body size in children and adolescents. The relations between $\dot{V}O_{2max}$, body dimensions, and system functions during growth are extraordinarily complex and are discussed in greater detail later in the chapter.

Strong cases have been made for scaling $\dot{V}O_2$, cardiac output, stroke volume, and other size-related physiological variables relative to body surface area, measured in square meters, or relative to weight, expressed to the 0.67 power or the 0.75 power (wt$^{0.67}$ or wt$^{0.75}$). For years, cardiologists have expressed heart volumes relative to body surface area. Research suggests that using body surface area (ml · m^{-2} · min^{-1}) or wt$^{0.75}$ (ml · kg$^{-0.75}$ · min^{-1}) provides the best means by which to express the data to reduce the effect of body size. One study followed young boys longitudinally from 12 to 20 years of age; one group remained untrained but active, and the other group trained.[27] There was little or no increase with run training in $\dot{V}O_{2max}$ expressed in ml · kg^{-1} · min^{-1}, whereas submaximal $\dot{V}O_2$ expressed in the same manner decreased with age, suggesting no change in aerobic capacity but an improvement in running economy. When these same data were expressed in ml · kg$^{-0.75}$ · min^{-1}, the boys who were training showed increased aerobic capacity with increased training and age but no change in running economy. This latter finding intuitively makes more sense, suggesting that the use of wt$^{0.75}$ is the best way of expressing the data.

Running Economy

How do growth-related changes in aerobic capacity affect a child's performance? For any activity that

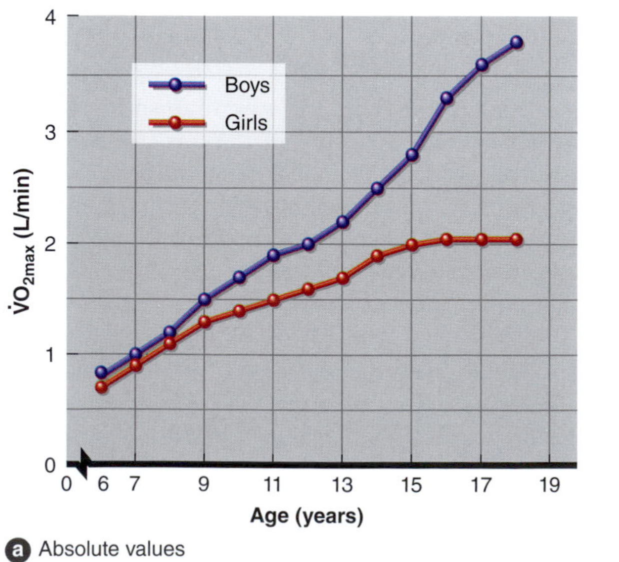

(a) Absolute values

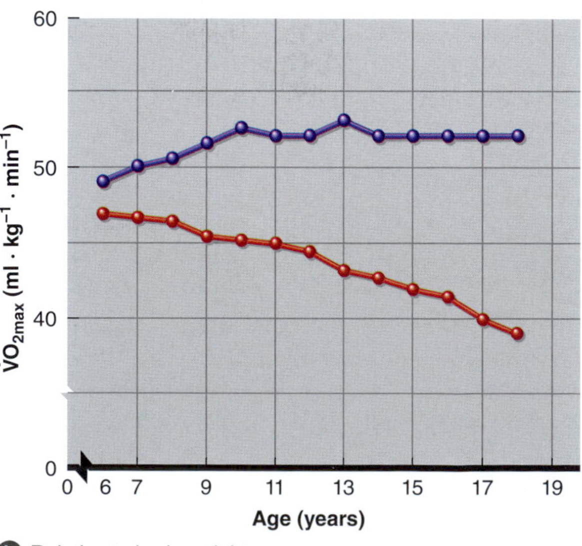

(b) Relative to body weight

FIGURE 17.7 Changes in *(a)* absolute and *(b)* relative maximal oxygen uptake with age in children and adolescents.

requires a fixed rate of work, such as cycling on an ergometer, the child's lower $\dot{V}O_{2max}$ limits endurance performance. But as noted earlier, for activities in which body weight is the major resistance to movement, such as distance running, children should not be at a disadvantage, because their $\dot{V}O_{2max}$ values expressed relative to body weight are already at or near adult values.

Yet children cannot maintain a running pace as fast as adults can because of basic differences in economy of effort. At a given speed on a treadmill, a child has a substantially higher submaximal oxygen consumption when expressed relative to body weight than an adult. As children age, their legs lengthen, their muscles become stronger, and their running skills improve. Running economy increases, and this improves their distance-running pace, even if they are not training and their $\dot{V}O_{2max}$ values don't increase.[6,12] While an increased stride frequency as children and adolescents grow is likely the most important factor in explaining these changes in running economy, inappropriate methods of scaling oxygen consumption to body weight during growth and development, as discussed in the previous section, add to the controversy.

In Review

- As pulmonary and cardiovascular function improve with continued development, so does aerobic capacity.
- $\dot{V}O_{2max}$, expressed in liters per minute, peaks between ages 17 and 21 years in boys and between 12 and 15 years in girls, after which it plateaus for several years and then steadily decreases.
- When $\dot{V}O_{2max}$ is expressed relative to body weight, it plateaus in boys from ages 6 to 25 years before it begins to decline. In girls, the decline in $\dot{V}O_{2max}$ is small from ages 6 to 12 years but becomes more substantial starting at about age 13. However, expressing $\dot{V}O_{2max}$ relative to body weight might not provide an accurate estimate of aerobic capacity. Such $\dot{V}O_{2max}$ values do not reflect the significant gains in endurance performance capacity that are noted with both maturation and training.
- The child's lower $\dot{V}O_{2max}$ value (L/min) limits endurance performance unless body weight is the major resistance to movement, as in distance running.
- $\dot{V}O_{2max}$, when expressed in liters per minute, is lower in children than in adults at similar levels of training. This is attributable primarily to the child's lower maximal cardiac output. When $\dot{V}O_{2max}$ values are normalized for the differences in body size between children and adults, there is little or no difference in aerobic capacity, yet in activities such as distance running, a child's performance is far inferior to adult performance.
- Running economy is lower in children compared with adults when $\dot{V}O_{2max}$ is expressed relative to body weight. One factor has been identified to explain this difference: the difference between children and adults in stride frequency for the same fixed-pace run.

Anaerobic Capacity

Children have a limited ability to perform anaerobic activities due to a lower glycolytic capacity. Muscle glycogen content in children is about 50% to 60% that of adults, and children do not achieve the same concentrations of lactate in either muscle or blood at maximal exercise intensities as their adult counterparts. The lower lactate concentrations might reflect a lower concentration of phosphofructokinase, the key rate-limiting enzyme of anaerobic glycolysis, and significantly lower (~3.5-fold) lactate dehydrogenase activity.[11] Lower blood lactate concentrations in children after exhaustive exercise may also reflect their lower relative muscle mass, higher lactate clearance, a greater reliance on aerobic metabolism, or some combination of these. In terms of the other anaerobic metabolic pathways, children's resting stores of adenosine triphosphate (ATP) and phosphocreatine (PCr) are similar to those of adults, so activities less than 10 to 15 s in duration should not be compromised. Only activities that tax the anaerobic glycolytic system—those from 15 s up to 2 min in duration—seem to be limited.

Anaerobic mean and peak power output, as determined by the Wingate anaerobic power test (a 30 s, all-out maximal effort on a cycle ergometer), is also lower in children than in adults. Figure 17.8 illustrates the results of a similar cycle ergometer anaerobic power test.[26] In this figure, peak power is statistically adjusted for body mass to account for differences in body size when we compare values for preteenagers, teenagers, and adults. This figure demonstrates the very low peak power outputs for preteenagers (9-10 years of age) compared with both teenagers (14-15 years of age) and adults (mean age of 21 years). Teenagers were much closer to the values for adults than the preteenagers were.

Bar-Or[1] summarized the development of both the aerobic and anaerobic characteristics of boys and girls from ages 9 through 16, using 18 years of age as the criterion for 100% of the adult value. The changes with age are shown in figure 17.9. Aerobic power is represented by the child's $\dot{V}O_{2max}$, whereas anaerobic power is represented by the child's performance on

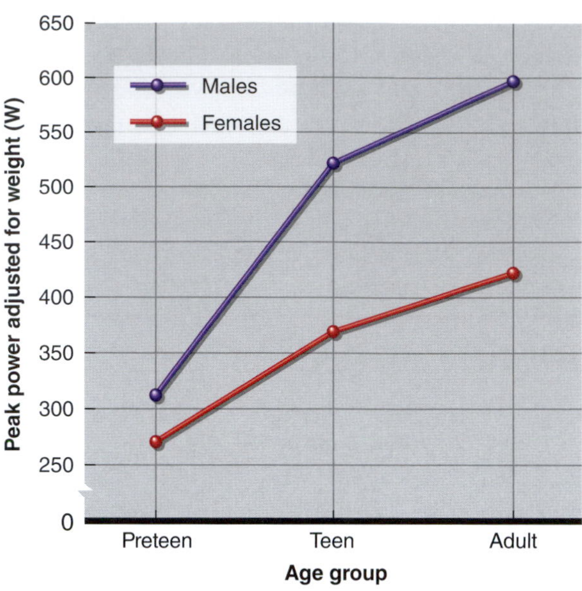

FIGURE 17.8 Optimal peak power output (anaerobic power) statistically adjusted for body mass in preteenagers (9-10 years old), teenagers (14-15 years old), and adults (mean age of 21 years). These values represent anaerobic power independent of body size.

Data from Santos et al. (2002).

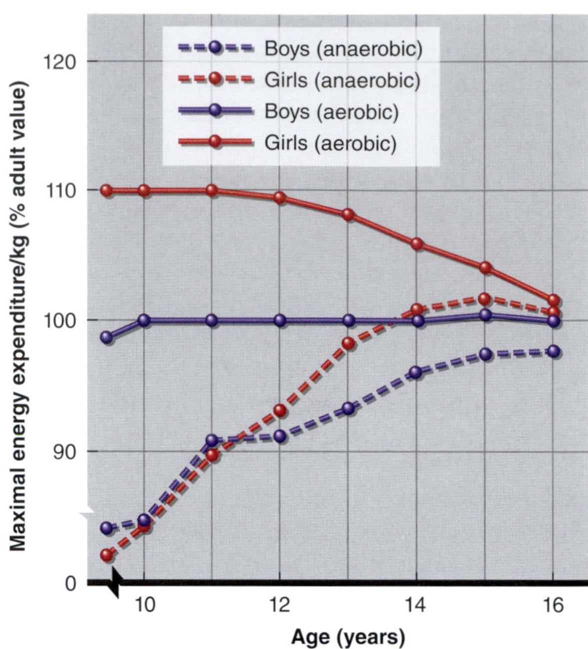

FIGURE 17.9 Development of aerobic and anaerobic characteristics in boys and girls ages 9 to 16 years. Values are expressed as a percentage of adult values (values at age 18).

Adapted by permission from O. Bar-Or, *Pediatric Sports Medicine for the Practitioner: From Physiologic Principles to Clinical Applications* (New York: Springer-Verlag, 1983). By permission of Marilyn Bar-Or.

RESEARCH PERSPECTIVE 17.1

Cognitive Benefits of Exercise for Children

Over the past decade, there has been a steady increase in classroom teaching time in an effort to improve standardized testing and academic achievement scores, involving predominately sedentary learning activities. In order to provide this increased teaching time, extensive cuts to nonacademic areas, especially those involving physical activity such as physical education and recess, have become the norm. However, a growing body of literature suggests that the exact opposite approach should likely be considered.

Physical activity/fitness and nutrition/overweight are two focus areas for *Healthy People 2020*, a comprehensive health promotion and disease prevention initiative by the Department of Health and Human Services. These are important objectives because childhood and youth physical inactivity is a growing concern and the percentage of youth considered overweight has more than tripled in the past 20 years. These statistics are alarming because of the propensity for physical inactivity and obesity to lead to chronic diseases. Schools remain an important setting in which to promote national health objectives, including lifetime physical activity. This is especially important given that physical activity habits develop very early in childhood and track over time from youth to adulthood. Thus, school programs can create expectations for regular physical activity that can persist into adulthood.

The health benefits of regular physical activity are clear. Moreover, the *Centers for Disease Control and Prevention* has published substantial evidence that physical activity can help improve academic achievement, including grades and standardized test scores. Additionally, physical activity also affects cognitive skills and attitudes as well as academic behavior, all of which improve overall academic performance, likely via improvements in concentration, attention, and classroom behavior. This scientific evidence can, and perhaps should, be used to implement policy change, gain support and funding for physical activity programs at schools and in the community, change lifestyle habits, reduce medical and financial burdens, and improve overall quality of life. Thus, physical and health education provides the foundation for healthy and active lifestyles throughout life.[7]

the Margaria step-running test (a field test). Maximal energy expenditure per kilogram represents the maximal energy-generating capacities of the aerobic and anaerobic systems, scaled to body weight to account for body size differences with growth. Notice that aerobic fitness remains constant for the boys but declines for the girls from 12 to 16 years. Nine- to twelve-year-old girls have a higher aerobic capacity than the eighteen-year-old reference adult value; thus, their values are 110% of the adult value. For both boys and girls, anaerobic capacity increases from 9 through 15 years of age.

Endocrine Responses and Substrate Utilization During Exercise

As discussed in previous chapters, physical activity causes the release of several key metabolic regulatory hormones to mobilize carbohydrates and fats to be used as fuels. Many of the hormones that regulate metabolism during exercise can also influence growth and development. For example, exercise is a potent stimulus for the growth hormone (GH) and insulin-like growth factor axis. High-intensity exercise in children can cause dramatic peaks in GH and influence the normal daily cycling of this hormone.

In general, pediatric studies suggest that the insulin response to exercise differs with pubertal stage and sex and that children have a higher stress response to exercise. This results in differences in blood glucose control. At the start of exercise, children have a relative hypoglycemia. The reasons for this are unclear, but in addition to lower muscle glycogen content, it is thought that children have an immature capacity for hepatic glycogenolysis. It is therefore not surprising that children rely more heavily on fat oxidation for fuel during exercise. However, exogenous glucose oxidation appears to be relatively high, potentially due to reduced endogenous glucose production. This fuel utilization profile changes throughout puberty such that adolescents have a decreased relative rate of fat oxidation, more like that in adults. The change in substrate utilization during exercise may have an impact on body composition throughout development, and from a practical point of view it may also affect nutrition needs for peak performance in children.

In Review

- Children's ability to perform anaerobic activities is limited. A child has a lower glycolytic capacity, possibly because of a limited amount of the rate-limiting enzyme phosphofructokinase or lactate dehydrogenase.
- Children have lower lactate concentrations in both blood and muscle at maximal and supramaximal rates of work.
- Anaerobic mean and peak power outputs are lower in children than in adults, even when scaled for body mass.
- Children have a different insulin and stress hormone response than adults and rely on greater fat oxidation for fuel during exercise.

Physiological Adaptations to Exercise Training

Training can improve body composition, strength, aerobic capacity, and anaerobic capacity of children. Generally, children adapt well to the same types of training regimens used by adults. But training programs for children and adolescents should be designed specifically for each age group, keeping in mind the developmental factors associated with that age.

Body Composition

The child and adolescent respond to physical training similarly to adults with respect to changes in body weight and composition. With both resistance and aerobic training, both boys and girls decrease body weight and fat mass and increase fat-free mass, although the increase in fat-free mass is less in the child. As we have already observed, there is also evidence of significant bone growth as a result of high-impact weight-bearing exercise training,[10] above that seen with normal growth.

Strength

For many years, the use of resistance training to increase muscular strength and endurance in prepubescent and adolescent boys and girls was controversial. Boys and girls were discouraged from using free weights for fear that they might injure themselves or interfere with the growth process. Furthermore, many scientists speculated that resistance training would have little or no effect on the muscles of prepubescent boys because their circulating androgen concentrations were still low. It is now widely accepted that properly designed resistance training programs are safe and effective for youth and adolescents, and the risk of injury with resistance training in youth is very low.

A number of studies conducted on both children and adolescents have clearly demonstrated that resistance training is effective in increasing strength. The increase is largely dependent on the volume and intensity of training, but in general, the percentage increases for children and adolescents are similar to those for young adults.

RESEARCH PERSPECTIVE 17.2

Physical Activity and Obesity in Children Around the World

While generic physical activity guidelines for adults were modified in 2018, global physical activity guidelines for school-aged children suggest at least 60 min/day of moderate- to vigorous-intensity activity. These guidelines are based on evidence linking physical activity to health outcomes, such as physical fitness, bone health, and markers of cardiovascular and metabolic health in children. Recently, studies have also demonstrated a negative relation between sedentary behaviors and health outcomes. Further, there is a direct association between physical activity and adiposity. However, there is a relative lack of comparative objective data on the relation between physical activity, sedentary behavior, and obesity in international samples. Thus, a group of researchers undertook a study in order to examine the association between moderate- to vigorous-intensity physical activity, vigorous physical activity, and sedentary behavior with obesity in 9- to 11-year-old children from countries that differ in stages of epidemiologic transition and represent a range of economic development (The International Study of Childhood Obesity, Lifestyle and the Environment, ISCOLE).[12]

The ISCOLE study included over 7,000 children from 12 different countries around the globe. The amount of time spent in moderate- to vigorous-intensity physical activity, vigorous physical activity, and sedentary behavior was estimated from 24 h waist-worn accelerometers used for at least 4 days. Not surprisingly, the amount of time spent in moderate- to vigorous-intensity and vigorous physical activity were significantly associated with obesity (determined from body mass index measurements). The authors report an optimal threshold of 55 min/day of moderate- to vigorous-intensity physical activity, which is consistent with the global recommendations of 60 min/day. The relation between sedentary behavior and obesity was much weaker. Taken together, these data, which were remarkably consistent between countries at different stages of economic and epidemiologic transition, strongly suggest that physical activity is directly related to obesity; this relation is evident across different cultures, races, and geographical settings. However, given the study design, it remains unclear whether low levels of physical activity are the cause or consequence of obesity in children; this remains an active area of research. Nevertheless, these data very clearly highlight a robust association between physical activity and obesity in children.

The mechanisms allowing strength changes in children are similar to those for adults, with one minor exception: Prepubescent strength gains are accomplished largely without any changes in muscle size and likely involve improvements in neural mechanisms, including improved motor skill coordination, increased motor unit activation, and other undetermined neurological adaptations.[23] Strength gains in the adolescent result primarily from both neural adaptations and increases in muscle size and specific tension.

Maximal Oxygen Consumption

Do prepubescent boys and girls benefit from aerobic training to improve their cardiorespiratory function and $\dot{V}O_{2max}$? This also has been a controversial area because several early studies indicated that training prepubescent children did not change their $\dot{V}O_{2max}$ values. Interestingly, even without significant increases in $\dot{V}O_{2max}$, the running performance of the children studied did improve substantially.[25] Other studies have shown small increases in aerobic capacity with training in prepubescent children, but these increases are less than would be expected for adolescents or adults—about 5% to 15% in children compared with about 15% to 25% in adolescents and adults.

More substantial changes in $\dot{V}O_{2max}$ appear to occur once children have reached puberty, although the reason for this is unknown. Because stroke volume appears to be the major limitation to aerobic performance, it is quite possible that further increases in $\dot{V}O_{2max}$ depend on heart growth. Also, as discussed earlier in this chapter, scaling of these variables is an issue.

Anaerobic Capacity

Anaerobic training appears to improve children's anaerobic capacity. Following training, children have

- increased resting levels of PCr, ATP, and glycogen;
- increased phosphofructokinase activity;
- increased maximal blood lactate concentrations;
- and higher ventilatory thresholds.[14]

When designing aerobic and anaerobic training programs for children and adolescents, standard training principles for adults can be applied. Again, it is prudent to be conservative to reduce the risk of injury, overtraining, and loss of interest in sport.

In Review

- Body composition changes with training in children and adolescents are similar to those seen in adults—loss of total body weight and fat mass and increase in fat-free mass.
- The risk of injury from resistance training in young athletes is relatively low, and the programs they should follow are much like those for adults.
- Strength gains achieved from resistance training in preadolescents result primarily from improved motor skill coordination, increased motor unit activation, and other neurological adaptations. Unlike adults, preadolescents who resistance train experience little change in muscle size. Mechanisms of strength gains in adolescents are similar to those for adults.
- Aerobic training in preadolescents does not alter $\dot{V}O_{2max}$ as much as would be expected for the training stimulus, possibly because $\dot{V}O_{2max}$ depends on heart size. But endurance performance improves with aerobic training. Adolescents are similar to adults in their improvement.
- A child's anaerobic capacity increases with anaerobic training.

Physical Activity Patterns Among Youth

Because physical activity patterns that are established in childhood track through adolescence and into adulthood, it is imperative to increase physical activity among our youth. However, current intervention strategies aimed at getting children more active have been mostly ineffective (discussed in the following section). Figure 17.10 depicts the percentage (as of 2014) of U.S. boys and girls aged 12 to 15 who were physically active by number of days per week. Physical activity was defined as any type of moderate- to vigorous-intensity activity, both in and out of school, sufficient to increase heart rate and breathing for at least 60 min. In 2012, only about one in four U.S. children engaged in this level of activity. In the boys surveyed, as weight status increased, the percentage of boys who were active decreased significantly. A similar trend was noted among the girls, with fewer obese girls engaging in physical activity (although the difference was not statistically significant). Among active boys, basketball was the most commonly reported activity, followed by running, football, biking, and walking. Among the girls, running was the most common activity, followed by walking, basketball, dancing, and biking.[8]

Children

Overweight children tend to be less physically active than children of normal body weight, but is physical inactivity the cause or the result of obesity? Being overweight as a child is a social and psychological impediment to physical activities that are enjoyable for normal-weight children. Therefore, for a variety of reasons, overweight children often shy away from such activities. In turn, a lack of physical activity may lead to a decrease in fitness and impaired motor skills that discourage the child even further from engaging in physical activity. It is reasonable to assume, then, that obesity in childhood may be associated with a sedentary lifestyle and poor cardiorespiratory fitness later in life.

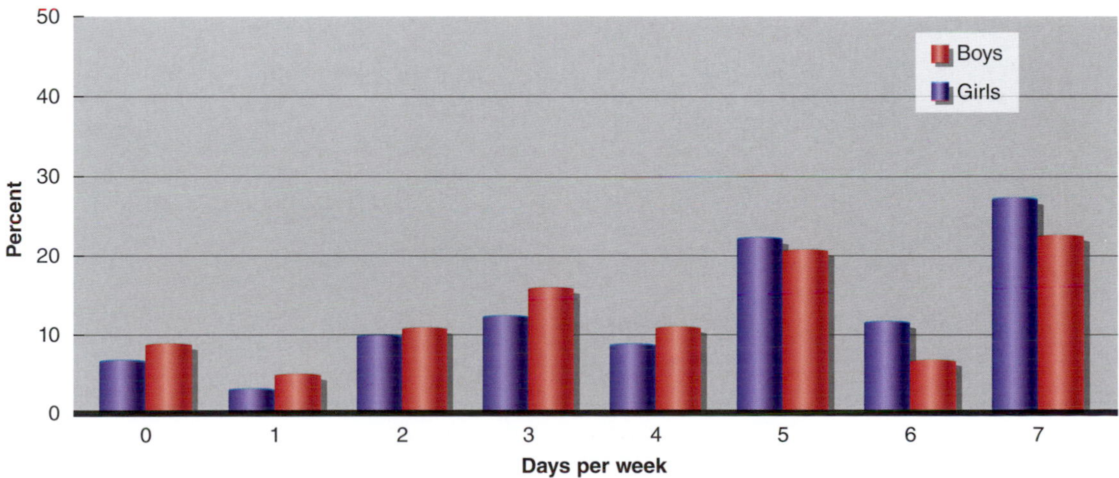

FIGURE 17.10 Percentage of U.S. boys and girls aged 12 to 15 who were physically active by number of days per week.

From U.S. Department of Health and Human Services, 2014, *HCHS Data Brief No. 141,* Available: http://www.cdc.gov/nchs/data/databriefs/db141.htm; Data from CDC/NCHS, National Health and Nutrition Examination Survey and National Youth Fitness Survey, 2012.

An ongoing study in Finland[21] is examining the association between early childhood weight status and cardiorespiratory fitness and leisure-time physical activity (LTPA) during adolescence and whether the patterns of LTPA and fitness seen during childhood persist through adolescence. This is one of the few longitudinal studies on children and adolescents in which children have been followed from birth through childhood to adolescence. A high preschool BMI was associated with low fitness in adolescence, independent of sex and regardless of adolescent LTPA. On a positive note, children who had a high preschool BMI but lost weight as they grew older had fitness levels in adolescence similar to those of the children with low BMIs. And regardless of fitness level in childhood, if LTPA increased from age 9 to 17, a similar fitness level was seen at age 17.

The data show the importance of maintaining a healthy body weight and a physically active lifestyle from childhood through adolescence if the goal is to improve fitness during adolescence. Clearly, obesity, physical inactivity, and poor cardiorespiratory fitness in childhood and adolescence are major public health concerns, since these patterns tend to persist through adulthood.

Are Current Physical Activity Interventions for Children Working?

Children, in large numbers, fail to meet current national physical activity guidelines, despite many initiatives and interventions aimed at making our children more active. Physical activity programs have had little impact on the growing childhood obesity epidemic. According to a 2013 review and meta-analysis, the fact that current interventions to promote physical activity achieve only "a small to negligible" increase in total activity provides one explanation.[17] Researchers from the United Kingdom analyzed 30 studies from the literature involving over 14,000 participants under the age of 16 years. Accelerometers, a popular method for measuring physical activity, were used to measure activity in 43% of the children studied.

The investigators looked at both total physical activity and time spent in moderate or vigorous physical activity. The disappointing results indicated that physical activity interventions have had only a small effect (about 4 min more walking or running per day) on children's overall activity levels. Further data analysis indicated that this small intervention effect, unlikely to make any clinical difference on health variables, did not differ significantly between any of the subgroups in the study. The effect of the interventions was unaffected by age, body mass index, intervention duration, or type of program (home- or family-based intervention versus school-based intervention).

This finding may explain, in part, why physical intervention programs have had little impact on reducing the body mass index or body fat of children. The reasons that interventions have failed are less than clear. Investigators often point to flaws in the intervention

designs, including poor delivery of the program, failure of the children to comply with the intervention, or lack of sufficient duration of the intervention or intensity of the exercise to elicit the designed exercise change. An alternate explanation offered by the researchers is that time spent in the intervention

> may simply be replacing periods of equally intense activity. For example, after-school activity clubs may simply replace a period of time that children usually spend playing outdoors or replace a time later in the day/week when the child would usually be active. (p. 226)[17]

Physical Inactivity in Children and Risk of Injury

As we strive to increase physical activity in children, we must remember that unstructured free play, exercise, and organized sport all carry with them an inherent risk of injury. This risk of injury may limit some children's interest in, or enthusiasm for, healthy physical activity. It is therefore important to understand what factors increase the risk of injuries in children engaging in physical activities.

To describe the risk factors associated with injuries occurring during physical education classes, leisure-time activities, and organized sports, Bloemers and colleagues[2] studied a group of approximately 1,000 Dutch schoolchildren aged 9 to 12 years (the iPlay Study). The children were followed for one school year, and physical education classes were held twice per week for 45 min per class. Activity-related injuries were continuously monitored by physical education teachers. A total of 119 injuries were reported by 104 different children over the school year. Girls were at higher risk of injury as were older children. At odds with other similar studies, there was no relation between BMI and injury risk. However, remarkably, the most active children had the lowest injury risk despite spending more time engaged in physical activity, lending support to the idea that increasing activity in children need not increase the risk of injury!

Adolescents

As described previously, childhood patterns of obesity, cardiorespiratory fitness, and physical activity portend similar patterns throughout adolescence and most likely into adulthood. Inadequate physical activity and low cardiorespiratory fitness ($\dot{V}O_{2peak}$) are clearly established as important risk factors for cardiovascular morbidity and mortality, some cancers, type 2 diabetes, and longevity, yet the relation between the two is complex. Therefore, despite the importance of understanding trends in $\dot{V}O_{2peak}$ as they relate to physical activity and other determinants, few studies have addressed whether $\dot{V}O_{2peak}$ is closely associated with physical activity during the formative adolescent years. A 2013 Norwegian study[20] (Young-HUNT Study) examined the distribution of cardiorespiratory fitness within a large (n = 570) group of healthy teenagers between the ages of 13 and 18 years. The goal was to determine the associations between $\dot{V}O_{2peak}$, self-reported physical activity, and several markers for future cardiovascular health.

A total of 289 girls and 281 boys had their resting blood pressure and heart rate, height, weight, and waist circumference measured, then underwent $\dot{V}O_{2peak}$ testing (maximal treadmill graded exercise testing). The mean $\dot{V}O_{2peak}$ was 60 ml · kg^{-1} · min^{-1} for the boys and 49 ml · kg^{-1} · min^{-1} for the girls. Absolute $\dot{V}O_{2peak}$ (in L/min) increased with age in both sexes. Self-reported physical activity was positively associated with $\dot{V}O_{2peak}$ for the entire group of teens, regardless of sex or age. A high $\dot{V}O_{2peak}$ was associated with a low resting heart rate in both sexes but with a low body mass index and waist circumference only in the boys.

Adolescence is an important period of life with regard to establishing and maintaining good health habits. Although $\dot{V}O_{2peak}$ was generally high in this group of adolescents, $\dot{V}O_{2peak}$ was higher in physically active adolescents of both sexes. The strong and consistent relation between self-reported physical activity and $\dot{V}O_{2peak}$ in this study suggests that teenagers who engage in physical activity at the recommended level maintain or even increase $\dot{V}O_{2peak}$ throughout adolescence.

Young Adulthood

As adolescents transition into early adulthood, physical activity declines modestly. An analysis of 49 longitudinal studies in the literature showed that self-reported daily physical activity declined by an average of about 5 min per day.[5] When accelerometers were used to directly measure daily activity in nine other published studies, a decline of about 7.5 min per day was seen. Both men and women had lower physical activity in adulthood compared to their late teens, but the decline was slightly larger in the men. This may be related to the fact that the men were more physically active as adolescents.

Sport Performance and Specialization

Sport performance in children and adolescents improves with growth and maturation, as can be seen for age-group records in sports such as swimming and track and field. Figure 17.11 illustrates the improvement in American records for various age groups.

The figure gives values for the 100 and 400 m swim and the 100 and 1,500 m run. These events were

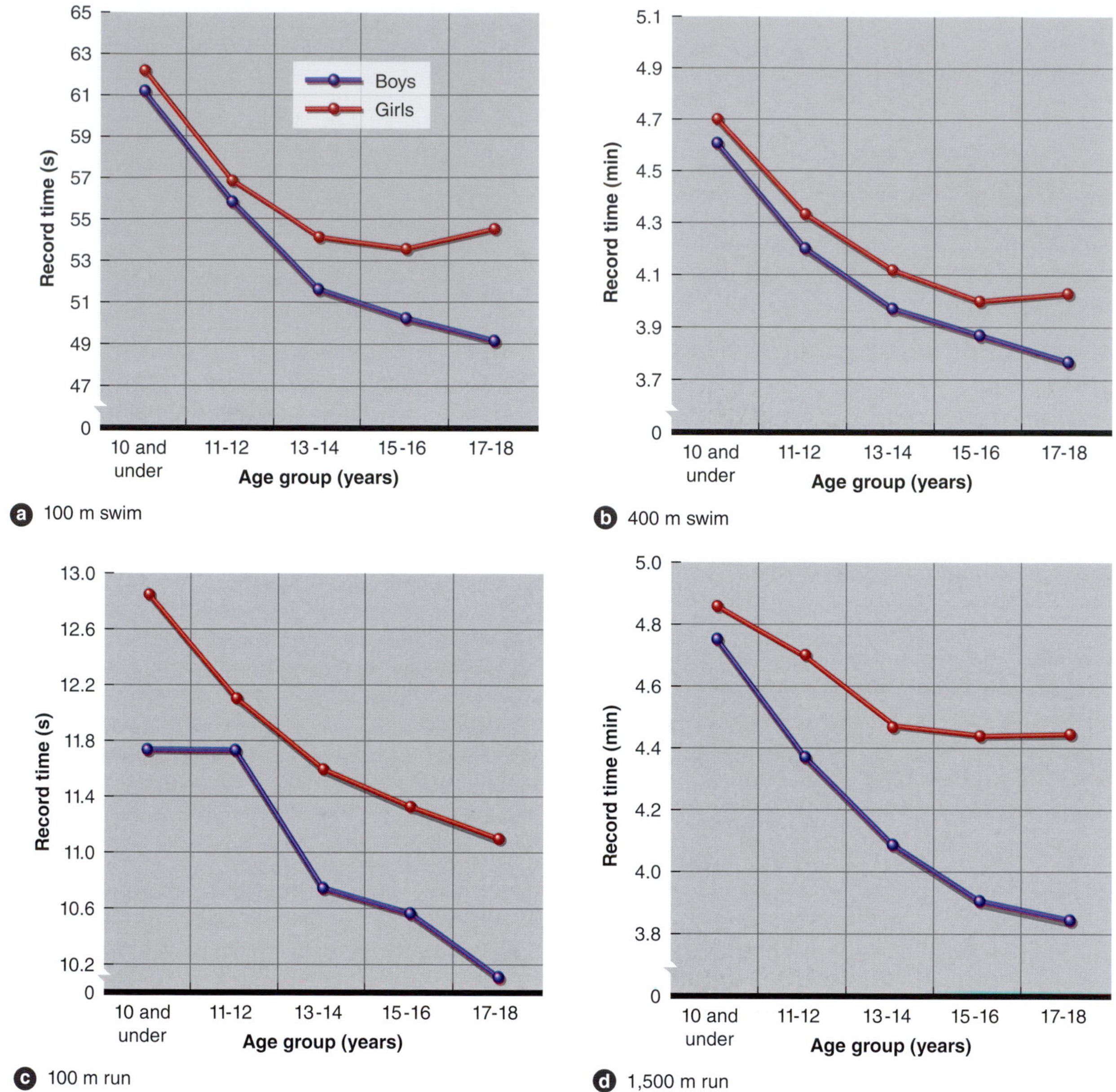

FIGURE 17.11 United States national record performances for boys and girls from 10 years of age and under through 17 to 18 years of age in *(a)* 100 m swim, *(b)* 400 m swim, *(c)* 100 m run, and *(d)* 1,500 m run.

(b) Data from USA Track & Field (2011).

selected because they represent a predominantly anaerobic event in swimming and running (100 m swim and run) and a predominantly aerobic activity (400 m swim and 1,500 m run). Both anaerobic and aerobic performances improve progressively with increasing age groups, with the exception of the 1,500 m run for 17- and 18-year-old girls. Similar age-group records for weightlifting do not appear to be available, because weightlifting competition is organized by weight in broad classifications such as 16 and under, 17 to 20 years of age, and then adult classifications. On the basis of normal strength gains with growth and development, it is assumed that weightlifting records would increase markedly from late childhood through adolescence, particularly in boys.

As participation in youth sport grows throughout the world, so does the prevalence of year-round sport-specific training, and many children and adolescents play for multiple teams in the same sport. What began in Eastern European countries in an attempt to develop Olympic athletes has evolved into single sport-specific programs worldwide. Many parents and coaches believe that the best way to develop their children and athletes to be elite in a sport is to have them play only that sport from an early age.[14]

A recent editorial[19] opined that sport specialization may contribute to reduced lifelong physical activity. Reduced participation in a variety of diverse sports and fun physical activities may lead to a limited development of lifetime sport and exercise skills and interests. Furthermore, restricting exercise time to a single sport may lead to diminished overall motor skill development and increased injury risk.

The National Association for Sport and Physical Education (NASPE) has also strongly encouraged delayed sport specialization for children and adolescents. In addition to the physical benefits of engaging in multiple sports, social and psychological benefits can be seen as well. Multiple-sport athletes develop more diverse relationships and experiences than their single-sport peers, skills that are helpful in adult life. It is perhaps not surprising, then, that the most elite athletes are most commonly those who waited until later in adolescence or young adulthood to specialize in their chosen sport.

Special Issues

During the period of growth and development from childhood through adolescence, potential concerns include children's susceptibility to thermal stress and the effect of exercise training on growth and maturation. These special issues are addressed in this section.

Thermal Stress

Laboratory experiments suggest that children are more susceptible to heat- and cold-induced illness or injury than adults. Yet there has been a minimal number of reported cases of thermal illness or injury among children.

RESEARCH PERSPECTIVE 17.3

Declines in Physical Activity During Adolescence

Physical activity declines from childhood through adulthood; however, this decline is most rapid during adolescence. Interestingly, physical activity appears to decline more rapidly in girls than in boys during this period. Given that physical inactivity is one of the most pressing public health issues, it is important to understand the underlying reasons for age- and sex-related declines in physical activity. Because adolescence is a period of rapid and dramatic change, not only in physical activity but also in physiological, psychological, and social factors, a question remains as to whether declines in physical activity are driven by biological changes or by changes in behavior and the environment.

Several studies have recently examined the adolescent fall in physical activity. One such study included data derived from the Physical Health Activity Study Team project, which ran for 5 years in the Canadian public school system and included children aged ~11 to 14 years. Participation in physical activity was quantified using a standard questionnaire. Findings from this study add to the growing body of literature documenting somewhat lower levels of overall activity and a more rapid decline in physical activity during adolescence in girls compared to boys. Interestingly, biological age (based on sexual maturity), and not chronological age (based on birth date), fully accounts for sex differences in the rate of decline but not for differences in overall levels of activity. Collectively, these findings underscore the importance of considering biological age, perhaps instead of chronological age or sex, as a main contributing factor to declines in physical activity during adolescence.

A nonintervention prospective cohort study, termed EarlyBird study, collected annual measures of accelerometry-based physical activity from ~300 children aged 5 to 15 years. The goal of the EarlyBird study was to reveal the start, extent, and nature of the decline in physical activity, determine the types of activity that contributed to the decline, and ascertain whether certain biological or environmental factors influenced the decline. The authors report that the decline in physical activity begins around 9 years of age and can be largely attributed to a reduction in light-intensity activity rather than a reduction in more intense activities.[17,18] This may have important clinical implications for the development of intervention strategies, since light activities are more likely to be habitual whereas moderate- to vigorous-intensity physical activity is more likely to be structured. Second, the age-related decline in physical activity in both sexes was unrelated to puberty, although puberty accounted for the steeper age-related decline in physical activity among girls. Finally, physical activity appears to track throughout childhood, such that those who are inactive in early childhood are more likely to remain inactive during adolescence. In broad agreement with the Physical Health Activity Study Team project, the inactivity of adolescence is driven, at least in part, by biological variables;[3] thus, these variables should be considered in the development of intervention strategies targeted to increase physical activity throughout adolescence.

A major concern is the child's apparently lower capacity during exercise in the heat to dissipate heat through evaporation. Compared with adults, children have a greater ratio of body surface area to mass, meaning that they have more skin surface area from which to gain or lose heat for each kilogram of body weight. Unless the environment is hot, this is an advantage, because children are better able to lose heat through radiation, convection, and conduction. However, once the environmental temperature exceeds the skin's temperature, children more readily gain heat from the environment, which is a distinct disadvantage. A child's lower capacity for evaporative heat loss is largely the result of a lower sweating rate. Individual sweat glands in children form sweat more slowly and are less sensitive to increases in the body's core temperature than those in adults. Although young boys can acclimatize to exercise in the heat, their rate of acclimatization is slower than that of adults. Acclimatization data are not available for girls.

Only a few studies have focused on children exercising in the cold. From the limited information available, children appear to have greater conductive heat loss than adults because of a larger ratio of surface area to mass. This should be expected to place them at higher risk for hypothermia and to necessitate more clothing layers during exercise in cold temperatures.

Few studies have been conducted on children in relation to either heat or cold stress, and conclusions from existing studies sometimes have been contradictory. More research is needed in this area to determine the risks faced by children who exercise in the heat and cold. In the meantime, a conservative approach is advisable.

Growth and Maturation with Training

Many people have wondered what effect physical training has on growth and maturation. Does hard physical training slow down or accelerate normal growth and development? In a comprehensive review of this area, Malina determined that regular training has no apparent effect on growth in terms of height.[15] It does, however, affect weight and body composition, as discussed earlier in this chapter.

As for maturation, the age at which peak height velocity occurs generally is not affected by regular training nor is the rate of skeletal maturation. But the data concerning the influence of regular training on indices of sexual maturation are not as clear. Although some data suggest that menarche (the initial onset of menstruation) is delayed in highly trained girls, these data are confounded by a number of factors that generally have not been properly controlled in each study's analysis. Menarche is discussed in more detail in chapter 19.

In Review

- Because physical activity patterns that are established in childhood track through adolescence and into adulthood, it is imperative to increase physical activity among our youth.
- As more children specialize early in only one sport, the reduced participation in a variety of diverse sports and fun physical activities may lead to a limited development of lifetime sport and exercise skills and interests.
- Laboratory studies suggest that children may be more susceptible to injury or illness from thermal stress because they have a greater ratio of body surface area to mass when compared to adults. However, the number of reported cases does not support this.
- Children are limited in their evaporative heat loss compared to adults because children sweat less (less sweat is produced by each active sweat gland).
- Young boys acclimatize to heat more slowly than adults do. Data are not available for girls.
- Physical training appears to have little or no negative effect on normal growth and development. Its effects on markers of sexual maturation are less clear.

IN CLOSING

In this chapter, we have discussed children and young athletes. We have seen how children gain more control of movements as their body systems grow and develop, how their developing systems can sometimes limit performance capacities, and how training can improve children's performances.

We have seen that, in general, the ability to perform increases as children approach physical maturity. We have also discussed the importance of being physically active during childhood and the adolescent period, since patterns developed during these stages follow through into adulthood. Having considered the developmental process, we are now ready to consider the aging process. How is performance affected as we move beyond our physiological prime? This is our focus in the next chapter as we turn our attention to aging and the older athlete.

KEY TERMS

adolescence
bone mineral density (BMD)
childhood
development
growth
infancy
maturation
myelination
ossification
physical maturity
puberty

STUDY QUESTIONS

1. Explain the concepts of growth, development, and maturation. How do they differ?
2. At what ages do height and weight reach their peak rate of growth in boys and in girls?
3. What typical changes occur in fat cells with growth and development?
4. How does pulmonary function change with growth?
5. What changes occur in heart rate and stroke volume for a fixed rate of work as a child grows? What factors explain these changes? What changes in these two variables occur with aerobic training?
6. What changes occur in cardiac output for a fixed rate of work as a child grows? What factors explain these changes? What changes occur with aerobic training?
7. What changes occur in maximal heart rate as a child grows?
8. What physiological variables account for the $\dot{V}O_{2max}$ increase from age 6 to 20?
9. What advice would you give to children if they wanted to improve their strength? Can they improve strength? If so, how does this occur?
10. What happens to aerobic capacity ($\dot{V}O_{2max}$) as a prepubescent child trains aerobically?
11. What is the evidence linking childhood patterns of obesity, cardiorespiratory fitness, and physical activity with similar patterns throughout adolescence and into adulthood?
12. What happens to anaerobic capacity as a prepubescent child trains anaerobically?
13. What are the major concerns associated with early youth sport specialization?
14. How do children differ from adults with respect to thermoregulation?
15. How do physical activity and regular training affect the growth and maturation processes?

STUDY GUIDE ACTIVITIES

In addition to the activities listed in the chapter opening outline, two other activities are available in the web study guide, located at

www.HumanKinetics.com/PhysiologyOfSportAndExercise

The **KEY TERMS** activity reviews important terms, and the end-of-chapter QUIZ tests your understanding of the material covered in the chapter.

Aging in Sport and Exercise

In this chapter and in the web study guide

Dara Torres is the first and only swimmer from the United States to compete in five different Olympic Games (1984, 1988, 1992, 2000, and 2008). More impressively, Torres is the first swimmer from any country to compete in the Olympic Games past the age of 40! On August 17, 2008, at the age of 41 years and 125 days, she won the silver medal in the women's 50 m freestyle, finishing in a new American record time of 24.07 s (a heartbreaking 0.01 s behind the winner). Only half an hour later, she won another silver medal as part of the U.S. 4 × 100 m medley relay team, swimming the fastest 100 m freestyle split (52.27 s) in relay history. She followed that up with yet a third silver medal, this time in the 4 × 100 m freestyle relay, swimming the anchor leg for the United States.

Years of training at such a high level, however, took its toll on her aching shoulders and arthritic knees. Following reconstructive knee surgery and rehabilitation, Torres started training once again in 2010 with the goal of competing in the 2012 London Olympics. At the 2012 U.S. Olympic Trials, at age 45, she placed fourth in her specialty event—the 50 m freestyle. Even so, she was only 0.09 s from qualifying for her sixth U.S. Olympic team. Finally, odds-defying Dara Torres had concluded her illustrious Olympic career.

The number of men and women over age 50 years who exercise regularly or participate in competitive sport has increased dramatically over the past 30 years. According to current population forecasts, the number of elderly people will increase worldwide from 6.9% of the population in 2000 to a projected 19.3% by 2050. In parallel with this overall increase in older adults, the number of middle-aged and older athletes is expected to increase as well. Many of these older competitors, often termed masters or senior athletes, engage in competition for fun, general recreation, and fitness, while others train with the same enthusiasm and intensity as Olympians. Opportunities are now available for older athletes to compete in activities ranging from marathon running to powerlifting. The success achieved and the performance records set by many older athletes are phenomenal. However, although these older athletes exhibit strength and endurance capacities that are far greater than those of untrained people their age, even the most highly trained older athlete experiences a decline in performance after the fourth or fifth decade of life.

In modern societies, voluntary physical activity begins to decline soon after people reach physical maturity. Technology has made many aspects of life less physically demanding. Voluntary participation in strenuous physical activity on a regular basis is an unusual pattern of behavior that is not observed in most aging laboratory animals. Studies have shown that humans and other animals tend to decrease their physical activity as they grow older. As shown in figure 18.1, rats that were allowed to eat ad libitum ran an average of more than 4,000 m (4,374 yd) per week in the early months of life but covered less than 1,000 m (1,094 yd) per week during their final months.

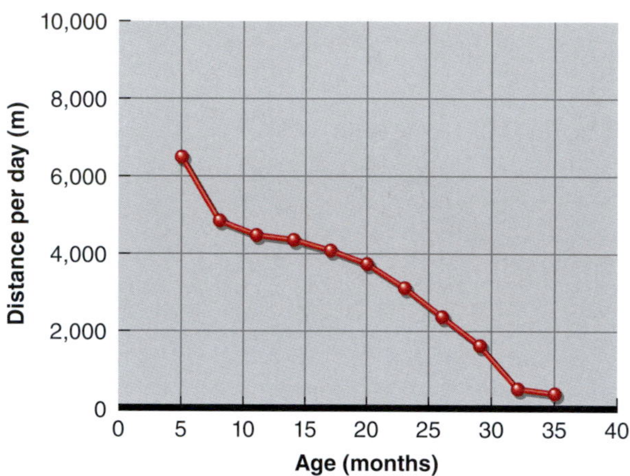

FIGURE 18.1 Voluntary running activity in rats throughout life.

Adapted by permission from J.O. Holloszy, "Mortality Rate and Longevity of Food-Restricted Exercising Male Rats: A Reevaluation," *Journal of Applied Physiology* 82 (1997): 399-403.

Considering the importance of exercise for maintaining muscular and cardiorespiratory fitness, it is not surprising that inactivity can lead to deterioration of one's capacity for strenuous effort. Because of this, it is difficult to distinguish between the effects of chronological aging alone (known as **primary aging**) and those of the reduced activity and comorbidities that often accompany aging. Researchers who study aging most commonly use cross-sectional designs, but these have important limitations compared to longitudinal study designs. For example, changes in medical care, diet and exercise, and other lifestyle variables may affect age cohorts differently. Selective mortality— that is, the fact that the subject population consists of the survivors of a cohort that has already experienced some degree of mortality—is an issue as well. Finally, when all but apparently healthy older individuals are excluded from exercise studies, it is difficult to apply research findings to the larger aged population, many of whom have underlying disease, take medications, or both. It is important to understand the impact of primary aging on physiological function, but the interpretation and applicability of the results are influenced by the study design, as well as the specific population being tested.

Height, Weight, and Body Composition

As we age, we tend to lose height and gain weight, as illustrated in figure 18.2.[35] The reduction in height gen-

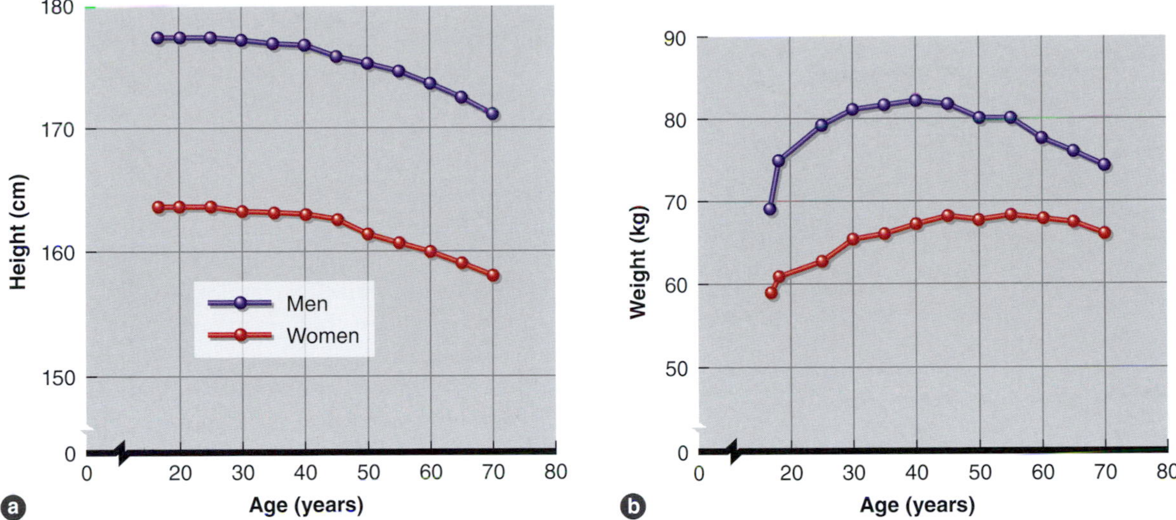

FIGURE 18.2 Changes in *(a)* height and *(b)* weight in men and women up to 70 years of age.
Reprinted by permission from W.W. Spirduso, *Physical Dimensions of Aging* (Champaign, IL: Human Kinetics, 1985), 59.

erally starts at about 35 to 40 years of age and is primarily attributable to compression of the intervertebral disks and poor posture early in aging. At about age 40 to 50 years in women, and 50 to 60 years in men, osteopenia and osteoporosis become a factor. **Osteopenia** is a reduction in bone mineral density below normal that occurs before **osteoporosis**, which is a severe loss of bone mass with deterioration of the microarchitecture of bone, leading to increased risk of bone fracture (see chapter 19). Genetic factors and poor diet and exercise habits throughout the life span contribute to the development of osteoporosis in both men and women, while decreased estrogen concentrations after menopause appear to be responsible for the greater rate of bone loss in women. During the adult life span, a gain in weight typically occurs between age 25 and 45 and is attributable to both a decrease in physical activity levels and excess caloric intake. Beyond the age of 45, weight stabilizes for about 10 to 15 years and then decreases as the body loses bone calcium and muscle mass. Many people over 65 years of age tend to lose their appetite and thus don't consume sufficient calories to maintain body weight. An active lifestyle, however, tends to help stimulate appetite so that caloric intake more closely approximates caloric expenditure, thereby maintaining weight and preventing frailty in old age.

Beginning at about 20 years of age, humans also tend to gain fat. This is largely attributable to changes in diet, physical inactivity, and a reduced ability to mobilize fat stores. However, the body fat content of physically active older people, including older athletes, is significantly lower than that of age-matched sedentary people. In addition, with primary aging there tends to be a shift in the location where body fat is stored, from the periphery toward the center of the body around the organs. This *central adiposity* is associated with cardiovascular and metabolic diseases. Although physical activity cannot fully counteract the age-related gain in fat mass, in active men and women there is less of a shift of the fat stores with aging, which is more advantageous for reducing the risk of cardiovascular and metabolic disease.

Fat-free mass decreases progressively in both men and women beginning at about the age of 40. This results primarily from decreased muscle and bone mass, with muscle having the greatest effect because it constitutes about 50% of the fat-free mass. *Sarcopenia* is the term used to describe the loss of muscle mass associated with the aging process. Figure 18.3 illustrates the changes in muscle mass with aging in a cross-sectional study of 468 men and women, aged 18 to 88 years.[21] There is almost no decline in muscle mass until about age 40, at which time the rate of decline increases, with a greater decline in men than women. Obviously, a reduction in physical activity is a major cause of this decline in muscle mass with aging, but there are other factors. For example, the rate of muscle protein synthesis is reduced while the rate of muscle protein breakdown is unchanged or accelerated with aging, leading to negative nitrogen balance and net loss of muscle. The rate of muscle protein synthesis in 60- to 80-year-olds is about 30% lower than in a 20-year-old. This reduction in muscle protein synthesis rate in older people is likely associated with declines in growth hormone, insulin-like growth factor 1,[14] and cell signaling. Longitudinal data suggest that loss of fat-free mass and gain of fat mass offset each other. As a result, percent fat is increasing while total body mass remains relatively stable.

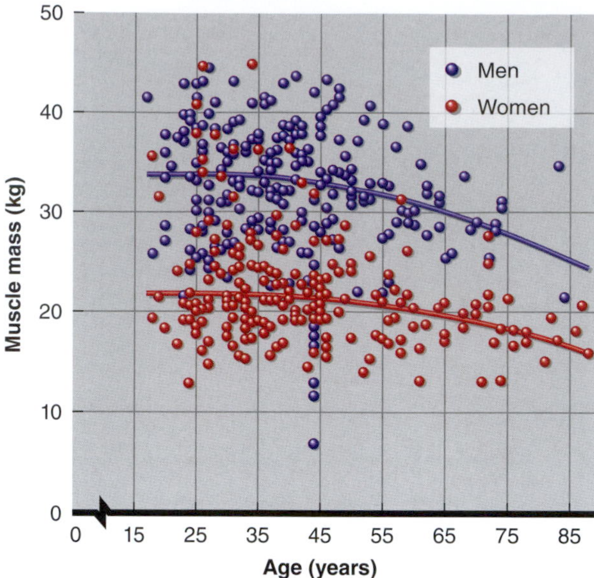

FIGURE 18.3 Changes in muscle mass with aging in 468 men and women 18 to 88 years of age. The rate of decline is greater in men than in women and is steeper after about 45 years of age.

Adapted by permission from I. Janssen et al., "Skeletal Muscle Mass and Distribution in 468 Men and Women Aged 18-88 yr.," *Journal of Applied Physiology* 89 (2000): 81-88.

VIDEO 18.1 Presents Scott Trappe on age-related sarcopenia.

There is also a significant decrease in bone mineral content, starting at about age 30 to 35 in women and at age 45 to 50 in men. Throughout the life cycle, bone is constantly being formed by osteoblasts and resorbed by osteoclasts. Early in life, resorption occurs at a slower rate than synthesis and bone mass increases. With aging, resorption exceeds synthesis, resulting in a net loss of bone. The loss of both muscle and bone mass is at least partially attributable to decreased physical activity, especially a lack of weight-bearing exercise. Since bone mineral accounts for less than 4% of total body mass in young adults, the contribution of osteopenia to the loss of total fat-free mass is small compared to that of sarcopenia.

These differences in weight, relative (%) body fat, fat mass, and fat-free mass with aging are illustrated in figure 18.4.[25] These data are from a study of young (18-31 years) and older (58-72 years) men and women who either were sedentary or were endurance-trained athletes. Body weight, relative body fat, and fat mass were higher in the older sedentary groups, whereas fat-free mass was lower. Similar trends, except for body weight, were noted for the endurance-trained athletes. However, the young and older endurance-trained athletes had much lower total body weight, relative body fat, and fat mass values and similar fat-free mass values compared to their sedentary age-matched counterparts.

With training, older men and women can reduce weight, percent body fat, and fat mass. Furthermore, they can increase their fat-free mass, but as in younger individuals, this is more likely with resistance training than with aerobic training. Men appear to experience greater changes in body composition than women, but the reasons for this have not been clearly established.

For older adults desiring to lose weight and body fat, the most significant changes in body composition result from a combination of diet and exercise, with a modest reduction in caloric intake (500-1,000 kcal/day) being the preferred approach. A more substantial reduction in caloric intake (>1,000 kcal/day) is likely to result in a loss of fat-free mass as well as fat mass. This is not desirable, since a loss in fat-free mass is associated with a reduction in resting metabolic rate, thus decreasing the rate of weight and fat loss. Exercise that increases fat-free mass will likely increase resting metabolic rate, which would increase the rate of weight loss. It appears that older adults experience changes in body composition due to exercise training that are similar to those for younger adults.

In Review

- Body weight tends to increase with aging, whereas height decreases.
- Body fat increases with age, primarily because of increased caloric intake, decreased physical activity, and a reduced ability to mobilize fat.
- Beyond age 45, fat-free mass decreases, primarily because of decreased muscle and bone mass, both resulting at least partly from decreased activity.
- The sarcopenia observed with aging is thought to be the result of decreased muscle protein synthesis in older people, accompanied by an unchanged or accelerated rate of muscle protein breakdown.
- Exercise training can help attenuate these changes in body composition, even in individuals as old as 80 to 90 years of age.

Physiological Responses to Acute Exercise

As we age, muscular and cardiovascular endurance and muscular strength tend to decrease, with the extent of decrease dependent on physical activity and genetics. As activity decreases, which appears to be a

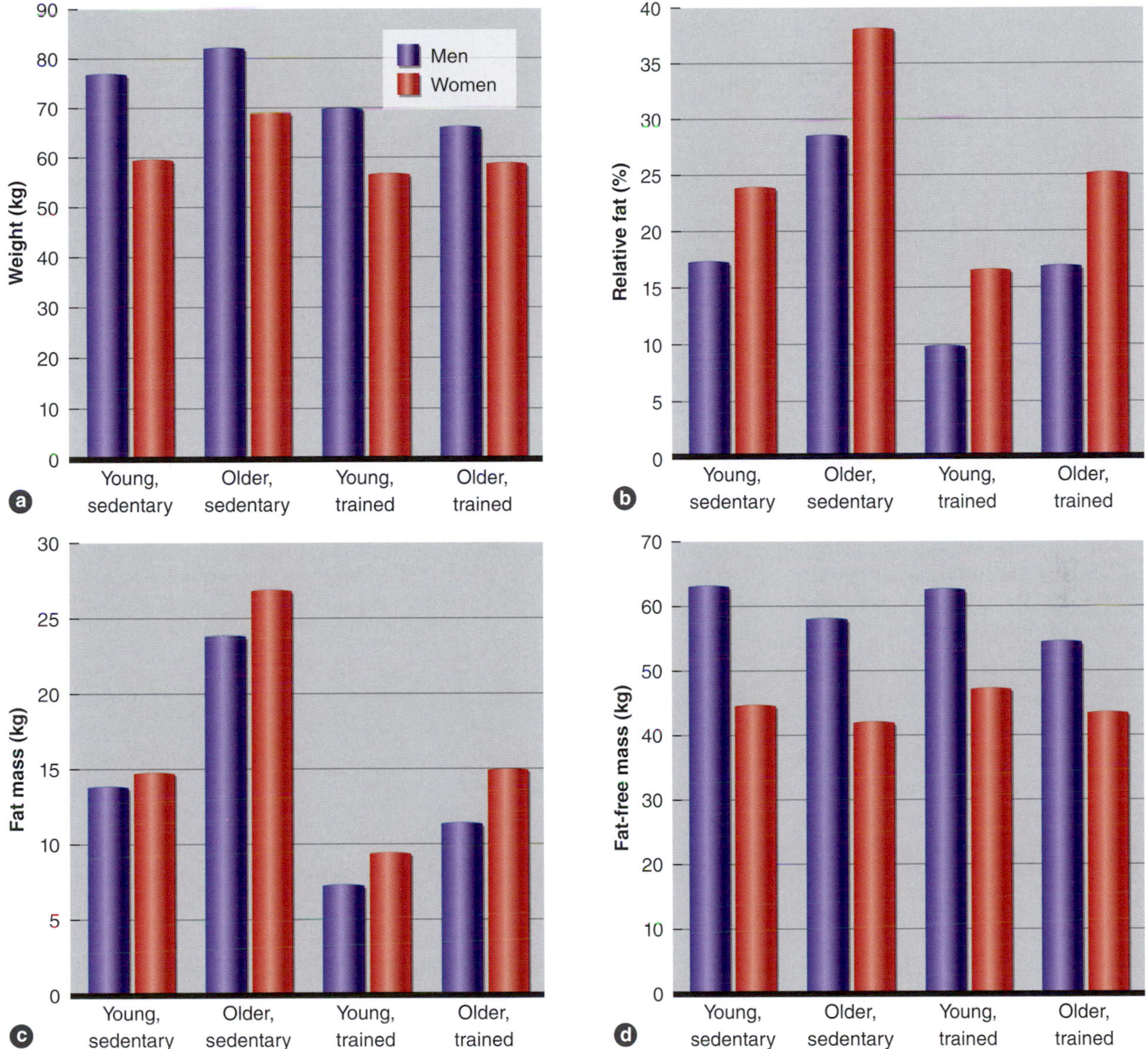

FIGURE 18.4 Differences in *(a)* total body weight, *(b)* relative body fat, *(c)* fat mass, and *(d)* fat-free mass in young and older men and women, sedentary and endurance trained.

Adapted by permission from W.M. Kohrt et al., "Body Composition of Healthy Sedentary and Trained, Young and Older Men and Women," *Medicine and Science in Sports and Exercise* 24 (1992): 832-837.

natural phenomenon in both animals and humans, these reductions in physiological function are much more substantial.

Strength and Neuromuscular Function

The strength needed to meet the activities of daily living remains unchanged throughout life. However, a person's maximal strength, generally well above the daily demands early in adulthood, decreases steadily with aging. Eventually, strength may decline to the point where previously simple activities become challenging. For example, the ability to stand up from a sitting position in a chair starts to be compromised at age 50, and before age 80 this task becomes impossible for some people (see figure 18.5*a*). As a further example, opening the lid on a jar is a task that can easily be accomplished by almost all men and women below the age of 60. After age 60, the failure rate for this task increases significantly.

Figure 18.5*b* shows leg strength changes with aging in men. Knee extension strength in normally active men and women starts to decrease by age 40. But training

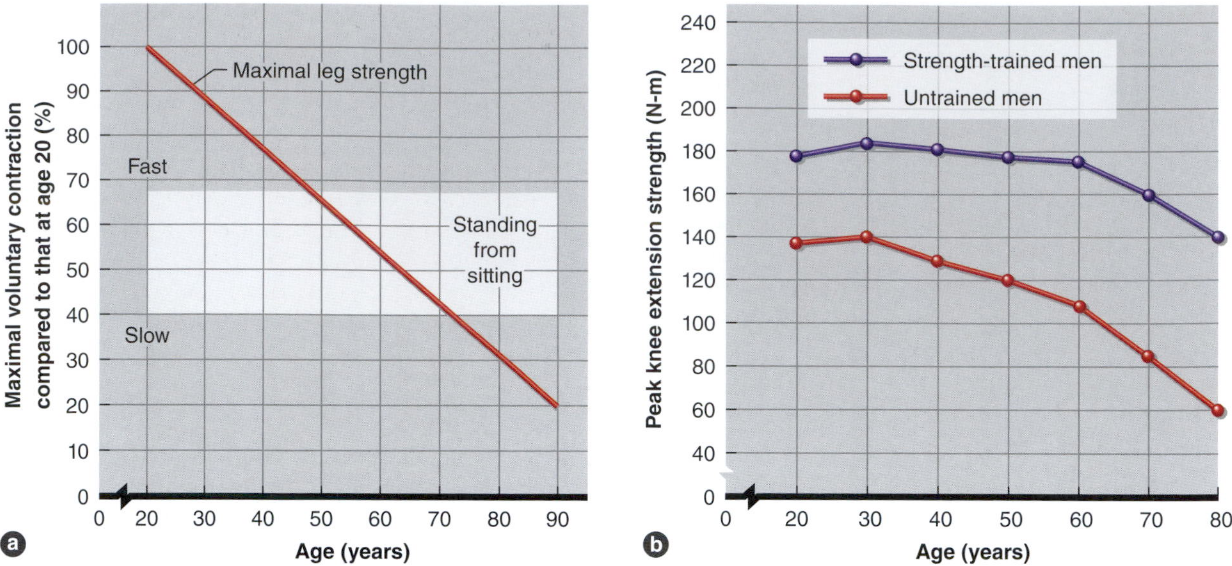

FIGURE 18.5 *(a)* The ability to stand from a sitting position is compromised at age 50, and by age 80 this task becomes impossible for some people. *(b)* Changes in peak knee extension strength in untrained and trained men at various ages. Note that older men (e.g., 60-80 years) who strength train can have knee extension strength equal to or greater than that of individuals who are only a third their age.

the knee extensor muscles with resistance exercises enables older men to perform better at age 60 than most normally active men half that age. The reduction in strength with aging was highly correlated with the reduction in the cross-sectional area of the involved muscles. The reductions in strength with aging appear to be modality specific, in that losses in isokinetic strength are greatest at high angular velocities and losses in concentric strength are greater than losses in eccentric strength.

Age-related loss of muscle strength results primarily from the substantial loss of muscle mass that accompanies aging or decreased physical activity (or both), as discussed earlier in this chapter. Figure 18.6 shows a computed tomographic (CT) scan of the upper arms of three 57-year-old men of similar body weight (about 78-80 kg, or 172-176 lb). Note that the untrained subject has substantially less muscle and more fat than the others. The swim-trained subject has less fat and a markedly larger triceps muscle than the untrained subject, but his biceps muscle, a muscle seldom used during swimming, is not much different. However, both of these muscles are larger in the resistance-trained subject. The differences between these three men are likely attributable to a combination of genetics and their volume and type of training.

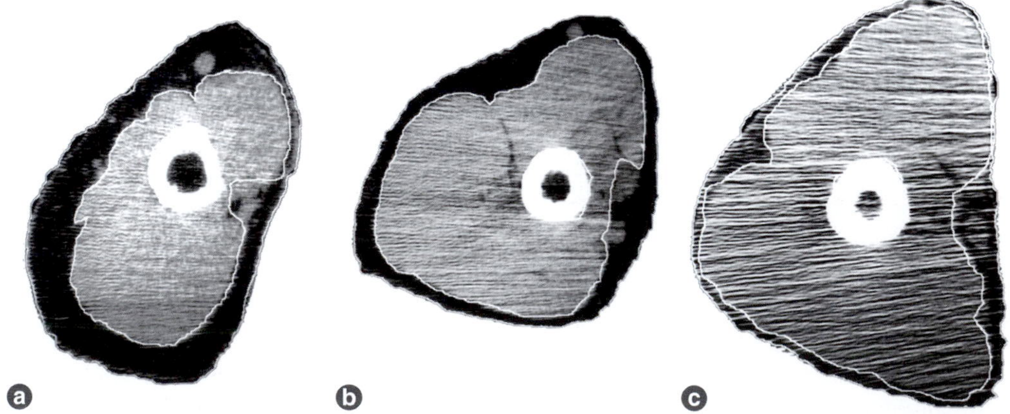

FIGURE 18.6 Computed tomography scans of the upper arms of three 57-year-old men of similar body weights. The scans show bone (dark center surrounded by white ring), muscle (striated gray area), and subcutaneous fat (dark perimeter). Note the difference in the muscle areas of *(a)* the untrained man, *(b)* the swim-trained man, and *(c)* the strength-trained man.

VIDEO 18.2 Presents Luc Von Loon on functional strength and the importance of strength training for both athlete and elderly populations.

Aging has a marked effect on muscle mass and strength, but what about muscle fiber type? There are conflicting results about the effects of aging on type I and type II fibers. Postmortem cross-sectional comparisons of the vastus lateralis (quadriceps) muscle in 15- to 83-year-old men and women have suggested that fiber type remains unchanged throughout life.[22] However, results from longitudinal studies conducted over a 20-year period indicate that the frequency or intensity of activity, or perhaps both, might play an important role in fiber type distribution with aging.[36,37] Muscle biopsy samples from the gastrocnemius (calf) muscles of a group of previously elite distance runners, obtained in 1970 through 1974 and again in 1992, demonstrated that the runners who had decreased their activity (fitness trained) or become sedentary (untrained) had a significantly greater proportion of type I fibers than when they were 18 to 22 years younger (figure 18.7). Those who continued to train had no change. Although some of the elite runners who still competed in distance running (highly trained) showed a small increase in the percentage of type I fibers, on average these highly trained runners showed no change in their calf muscle fiber composition over the 18- to 22-year span of this study.

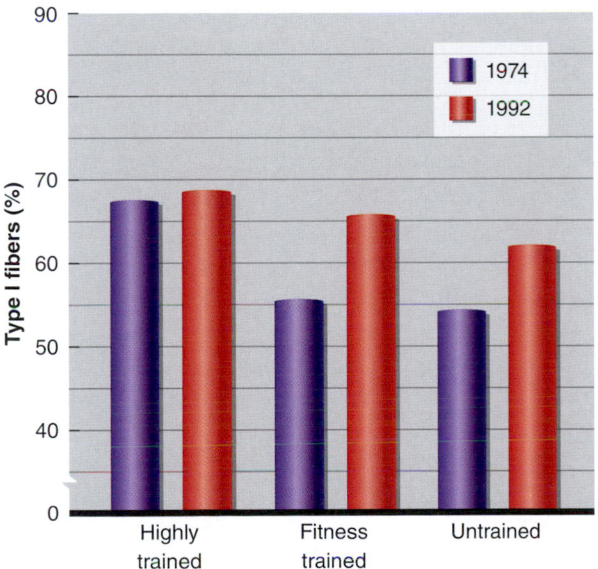

FIGURE 18.7 Changes in muscle fiber type composition of the gastrocnemius in elite distance runners who remained highly trained, stayed fitness trained, or became untrained during the 18 to 22 years between tests. Note that the runners who continued to compete showed almost no change in percentage of type I fibers, whereas the less fit and untrained individuals experienced an increase in the percentage of type I fibers.

It has been suggested that the apparent increase in type I fibers may be attributable to an actual decrease in the number of type II fibers, resulting in a greater *relative* proportion of type I fibers. Although the precise cause of type II fiber loss is unclear, it has been suggested that the number of type II motor neurons decreases during aging, which eliminates innervation of these muscle fibers.

Numerous investigations have shown a decrease in both the number and the size of muscle fibers with aging. One study reported a loss of approximately 10% of the total number of muscle fibers per decade after age 50.[27] This explains in part the muscle atrophy that occurs as we get older. It appears that the size of both type I and type II fibers decreases with aging. Endurance training (such as distance running) has little impact on the decline in muscle mass with aging. Strength training, on the other hand, reduces muscle atrophy in aging adults and can, in fact, cause older individuals to increase their muscle cross-sectional area.[27]

Aging is accompanied by substantial changes in the nervous system's capacity to process information and to activate muscles. Specifically, aging affects the ability to detect a stimulus and process the information to produce a response. Simple and complex movements are slowed with aging, although people who remain physically active are only slightly slower than younger, active individuals. Motor unit activation is lower in aged adults. For example, one study showed that older men (~80 years) had lower firing rates and longer twitch contraction durations, whereas younger men (~20 years) had relatively higher firing rates and shorter contraction times.[4] However, others have shown that older individuals retain their ability to maximally recruit skeletal muscle, suggesting that reduced strength is due to local muscle rather than neural factors.

Neuromuscular changes during aging are at least partially responsible for decreased strength and endurance, but active participation in exercise and sport tends to lessen the impact of aging on performance. This doesn't mean that regular physical activity can arrest biological aging, but an active lifestyle can markedly reduce many of the decrements in physical work capacity.

Despite the loss of muscle mass in active aging men, the structural and biochemical properties of the remaining muscle mass are well maintained. The number of capillaries per unit area is similar in young and old endurance runners. Oxidative enzyme activities in the muscles of endurance-trained older athletes are only 10% to 15% lower than in endurance-trained young athletes. Thus, oxidative capacity of skeletal muscle of endurance-trained older runners is only slightly less than that of young elite runners, which suggests that aging has little effect on skeletal muscle's adaptability to endurance training.

In Review

- Maximal strength decreases steadily with aging.
- Age-related losses of strength result primarily from a substantial loss of muscle mass.
- In general, normally active people experience a shift toward a higher percentage of type I muscle fibers as they age, possibly attributable to a reduction in type II fibers.
- The total number of muscle fibers and the fiber cross-sectional area decrease with age, but resistance training appears to lessen the decline in fiber area.
- Aging slows the nervous system's ability to respond to a stimulus, process the information, and produce a muscular contraction.
- Whereas endurance training does little to prevent the age-associated loss in muscle mass, strength training can maintain or increase the muscle fiber cross-sectional area in older men and women.

Cardiovascular and Respiratory Function

Changes in endurance capacity that accompany aging can be attributed to decrements in both central and peripheral cardiovascular function. Changes in respiratory function that accompany primary aging likely play a lesser role, although the incidence of respiratory dysfunction increases with age. In this section, we examine aging effects on both the cardiovascular and respiratory systems.

Cardiovascular Function

Similar to muscle function, cardiovascular function declines as we age. One of the most notable changes that accompanies aging is a decrease in maximal heart rate (HR_{max}). HR_{max} is estimated to decrease slightly less than 1 beat/min per year as we age. Traditionally, the average HR_{max} for any age has been estimated from the equation $HR_{max} = 220 - \text{age}$. This equation tends to overestimate the HR_{max} of children and young adults and underestimate the HR_{max} of older adults. Also, when the old equation ($HR_{max} = 220 - \text{age}$) is used, individual values can deviate by ±20 beats/min from the predicted value. For example, the old equation predicts that an *average* 60-year-old would have an HR_{max} of 160 beats/min; however, an individual's actual HR_{max} might be as low as 140 beats/min or as high as 180 beats/min.

Tanaka and colleagues proposed a more accurate equation that seems to be appropriate for all people and is not influenced by sex or activity level.[35]

$$HR_{max} = [208 - (0.7 \times \text{age})]$$

While this latter equation improves the prediction of the average individual's heart rate, there is still substantial interindividual variability. We will see in chapter 20 that overestimates and underestimates make a big difference when HR_{max} is used for exercise prescription.

The reduction in HR_{max} with aging is not affected by training; that is, it appears to be similar for sedentary and well-trained adults. This reduction in HR_{max} might be attributable to morphological and electrophysiological alterations in the cardiac conduction system, specifically in the sinoatrial (SA) node and in the atrioventricular (AV) bundle, which could slow cardiac conduction. Downregulation of the β_1-adrenergic receptors in the heart also decreases the heart's sensitivity to catecholamine stimulation.

The decrease in maximal cardiac output with aging in highly trained men and women is primarily attributable to decreased heart rate and, to a lesser extent, a decrease in stroke volume. Maximal stroke volume (SV_{max}) is modestly reduced (~10%-20% reduction) in highly trained older adults. The responses to catecholamine stimulation and myocardial contractility are reduced, and recent evidence provided with more sophisticated Doppler imaging techniques indicates that the Frank-Starling mechanism is less effective due to left ventricular and arterial stiffening. Studies of endurance runners have shown that the lower $\dot{V}O_{2max}$ values observed in older athletes result from a reduction in maximal cardiac output, despite the fact that heart volumes of older athletes are similar to those of young athletes, confirming that a decreased maximal heart rate is the primary cause of reduced $\dot{V}O_{2max}$. In untrained men and women, a number of studies have demonstrated a clear and larger decrease in maximal stroke volume—and maximal cardiac output—with aging.

VIDEO 18.3 Presents Ben Levine on the factors that affect $\dot{V}O_{2max}$ and cause its decline with aging.

Peripheral blood flow decreases with aging, even though capillary density in the muscles is unchanged. In middle-aged athletes, there is a 10% to 15% reduction in blood flow to working muscles in the leg at any given work rate when compared with well-trained young athletes (figure 18.8). This attenuation in blood flow is due to a number of peripheral factors including blunted functional sympatholysis (i.e., a greater sympathetic vasoconstriction in the exercising muscle) and a reduction in local vasodilators and their effect. But the reduced blood flow to the legs of these middle-aged and older endurance runners during submaximal exercise is apparently compensated for by a greater arterial–mixed venous oxygen difference, or $(a\text{-}\bar{v})O_2$ difference (more oxygen extracted by the muscles). As a result, although the blood flow is lower, oxygen uptake by the exercising muscles is similar at a given submaximal work intensity in the older age group. This was confirmed in a study of endurance-trained men, which compared subjects aged 22 to 30 years with subjects aged 55 to 68 years. Leg blood flow, vascular conductance, and femoral venous oxygen saturation were each 20% to 30% lower in the older men at each submaximal work rate, while leg $(a\text{-}\bar{v})O_2$ difference was higher in older subjects.[31]

It is hard to determine the extent to which the age-related changes in stroke volume, cardiac output, and peripheral blood flow result from primary aging and the extent to which they are attributable to the **cardiovascular deconditioning** that accompanies reduced activity. These declines in cardiovascular function with aging are largely responsible for the declines observed in $\dot{V}O_{2max}$, as discussed later in this chapter.

Additionally, sedentary aged humans have a decrease in cardiac and arterial compliance caused by stiffening of the heart and the large elastic arteries. The aged blood vessels also exhibit a change in local control of blood flow including a dysfunction in the ability of the endothelium to release and respond to vasodilators such as nitric oxide and prostaglandins, termed **endothelial dysfunction**. This change contributes to the inability of blood vessels to vasodilate and the reduction in peripheral muscle blood flow during exercise.

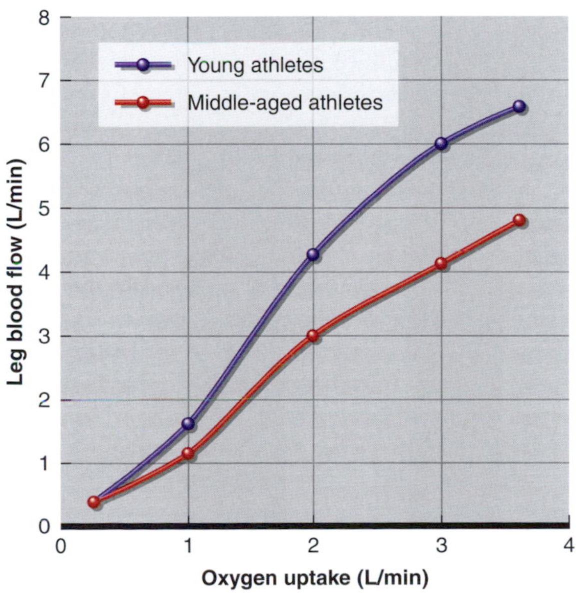

FIGURE 18.8 Leg blood flow during cycling exercise in young and middle-aged orienteers.

Adapted by permission from B. Saltin, The Aging Endurance Athlete, in *Sports Medicine for the Mature Athlete*, edited by J.R. Sutton and R.M. Brock (Indianapolis: Benchmark Press, 1986). Copyright 1986 Cooper Publishing Group, Carmel, IN.

Habitual exercise training in older men and women offsets some of the arterial stiffening and endothelial dysfunction. Part of the reason for this preservation in vascular function with habitual exercise training is a preservation or restoration in vasodilator signaling, including an increase in nitric oxide bioavailability. Research is ongoing to determine the appropriate dose of exercise (time, duration, and intensity) needed to see these positive cardiovascular benefits in healthy and clinical older populations.

Respiratory Function

Pulmonary function changes with aging, but these changes do not appear to limit exercise capacity. Both vital capacity (VC) and **forced expiratory volume in 1 s ($FEV_{1.0}$)** decrease linearly with age, starting at age 20 to 30. Whereas these decrease, residual volume (RV) increases, and the total lung capacity (TLC) remains essentially unchanged. As a result, the ratio of the residual volume to total lung capacity (RV/TLC) increases, meaning that less air can be exchanged. In our early 20s, RV accounts for 18% to 22% of the TLC, but this increases to 30% or more as we reach age 50. Smoking appears to accelerate this increase.

These changes are matched by changes in maximal ventilatory capacity during exhaustive exercise. **Maximal expiratory ventilation ($\dot{V}_{Emax}$)** increases during growth until people achieve physical maturity, and then it decreases with age. $\dot{V}_{Emax}$ values average about 40 L/min for 4- to 6-year-old boys, increase to 110 to 140 L/min for fully mature men, and then decrease to 70 to 90 L/min for 60- to 70-year-old men. Girls and women follow the same general pattern, although their absolute values are considerably lower at each age, primarily because of smaller body size. Recall from chapter 7 that TLC is directly proportional to height, which is why men often have higher values than women.

The changes in pulmonary function as adults get older result from several factors. The most important of these is a loss of elasticity of the lung tissue and chest wall, which increases the work involved in breathing. The resulting stiffening of the chest wall appears to be responsible for most of the reduction in lung function. But despite all these changes, the lungs still have a remarkable reserve and maintain an adequate diffusion capacity to permit maximal exertion, and do not appear to limit exercise capacity.

Endurance-trained older athletes have only slightly decreased pulmonary ventilation capacities, so the decreased aerobic capacity in older athletes cannot be attributed to changes in pulmonary ventilation. During strenuous exercise, both normally active older people and athletes can maintain near-maximal arterial oxygen saturation. Changes in the lungs and the blood's oxygen-carrying capacity are not responsible for the observed decrease in $\dot{V}O_{2max}$ seen in aging athletes. Rather, the primary limitation is apparently linked to the cardiovascular changes described earlier. Aging decreases maximum heart rate, which lowers maximal cardiac output and blood flow to the exercising muscles. Submaximal $(a-\bar{v})O_2$ difference is maintained in older exercisers, suggesting that O_2 extraction is well preserved with aging.

In Review

- Much of the decline in endurance performance associated with aging can be attributed to decreased cardiovascular function.
- Maximum heart rate decreases about 1 beat/min per year as we age. The average HR_{max} for a given age can be estimated by the following equation: $HR_{max} = [208 - (0.7 \times \text{age})]$.
- Maximal stroke volume is only slightly reduced in older athletes, but their cardiac output decreases with age primarily because of the reduction in HR_{max}. In untrained people, maximal stroke volume decreases due to left ventricular and arterial stiffening with aging.
- Peripheral blood flow also decreases with age; in trained older athletes, however, this is offset by an increased $(a-\bar{v})O_2$ difference.
- The decrease in $\dot{V}O_{2max}$ with aging and inactivity is largely explained by decreased $\dot{Q}_{max}$ due to decreased HR_{max}. Maximal stroke volume is modestly reduced (SV_{max}), and maximal $(a-\bar{v})O_2$ difference typically does not change much with aging.
- The decrease in HR_{max} is attributable largely to decreases in the heart's intrinsic rate but also could be caused by decreases in sympathetic nervous system activity and alterations in the cardiac conduction system.
- The decrease in $\dot{V}O_{2max}$ with aging is primarily a function of reduced blood flow to the active muscles, which is associated with the reduction in maximal cardiac output.
- Habitual exercise can partially reverse or prevent many of the detrimental vascular changes that occur with aging, including reducing arterial stiffness and improving endothelial function.
- It is unclear how much of the decrease in cardiovascular function with aging is attributable to aging alone and how much is attributable to deconditioning because of decreased activity. However, many studies indicate that these changes are attenuated in older athletes who continue to train, which indicates that inactivity plays a substantial role.
- Both vital capacity and forced expiratory volume decrease linearly with age. Residual volume increases, so total lung capacity remains unchanged. This increases the RV/TLC ratio, meaning that less air can be exchanged in the lung with each breath.
- Maximal expiratory ventilation also decreases with age.
- Pulmonary changes that accompany aging are primarily caused by a loss of elasticity in the lung tissue and the chest wall. However, trained older athletes have only slightly decreased pulmonary ventilation capacity.

Aerobic and Anaerobic Function

In investigating the effect of aging on aerobic and anaerobic function during exercise, the focus is on two key variables—$\dot{V}O_{2max}$ and the lactate threshold.

$\dot{V}O_{2max}$

To determine how $\dot{V}O_{2max}$ changes with aging, there are several important issues to consider. First, one must decide how to express the $\dot{V}O_{2max}$ values—in liters per minute (L/min) or in liters per minute per kilogram of body weight to adjust for changes in weight ($ml \cdot kg^{-1} \cdot min^{-1}$). While $\dot{V}O_{2max}$ expressed in liters per minute may decrease little over a 10- to 20-year period, when the same subjects' values are expressed relative to body weight, there may be a large decrease if the individuals gain weight over the interval between the initial test and the final test. For nonweight-bearing exercise, such as cycling, using liters per minute is usually most appropriate. For weight-bearing activities such as running, it is usually more appropriate to express the values per unit of body weight ($ml \cdot kg^{-1} \cdot min^{-1}$).

A second issue relates to whether change values in variables with aging should be expressed as an absolute change (L/min or $ml \cdot kg^{-1} \cdot min^{-1}$) or as a percentage change.

This might seem like a minor point, but it is not. As an example, a 30-year-old man had an initial $\dot{V}O_{2max}$ of 50 $ml \cdot kg^{-1} \cdot min^{-1}$, and at the age of 50 years his $\dot{V}O_{2max}$ decreased to 40 $ml \cdot kg^{-1} \cdot min^{-1}$. A 60-year-old man had an initial $\dot{V}O_{2max}$ of 35 $ml \cdot kg^{-1} \cdot min^{-1}$, and at the age of 80 years his vO_{2max} decreased to 25 ml · kg^{-1} · min^{-1}. In this example, both men have decreased their $\dot{V}O_{2max}$ by 10 ml · kg^{-1} · min^{-1} over a 20-year period, showing a decline of 0.5 ml · kg^{-1} · min^{-1} per year. However, the younger man has had a decrease of 20% (10/50 = 0.20) over 20 years, or 1% per year, whereas the older man has had a decrease of 29% (10/35 = 0.29), or 1.4% per year. Although the two men have identical decreases in $\dot{V}O_{2max}$ when expressed in ml · kg^{-1} · min^{-1}, the older man has a substantially greater decrease when expressed as a percentage decrease. Keeping this in mind, let's consider changes in $\dot{V}O_{2max}$ with aging, looking first at changes in normally active people and then at changes in highly trained endurance athletes.

Normally Active People The first studies of aging and physical fitness were performed by Sid Robinson[32] in the late 1930s. He demonstrated that $\dot{V}O_{2max}$ in normally active men declined steadily from ages 25 to 75 (table 18.1). His cross-sectional data show that aerobic capacity declines an average of 0.44 ml · kg^{-1} · min^{-1} per year up to age 75, which is about 1% per year or 10% per decade. A review of 11 cross-sectional studies on men, most under age 70, showed that the average rate of decrease in $\dot{V}O_{2max}$ was 0.41 ml · kg^{-1} · min^{-1} per year for men and 0.30 ml · kg^{-1} · min^{-1} per year for women, again close to 1% per year.[1]

In the mid-1990s, a large cross-sectional study of changes in $\dot{V}O_{2max}$ with aging was conducted at the NASA/Johnson Space Center in Houston, Texas. This study included 1,499 men and 409 women, all of whom were healthy and had performed a maximal treadmill test to exhaustion during which $\dot{V}O_{2max}$ was directly measured.[19,20] The authors reported a decline in $\dot{V}O_{2max}$ of 0.46 ml · kg^{-1} · min^{-1} per year in men (1.2% per year) and 0.54 ml · kg^{-1} · min^{-1} per year in women (1.7% per year).

Unfortunately, few longitudinal studies have been conducted in this area. Nevertheless, the rate of decline in $\dot{V}O_{2max}$ generally is agreed to be approximately 10% per decade or 1% per year (–0.4 ml · kg^{-1} · min^{-1} per year) in relatively sedentary men. The results are similar for women, although fewer subjects have been studied.

TABLE 18.1 Changes in $\dot{V}O_{2max}$ Among Normally Active Men

Age (years)	$\dot{V}O_{2max}$ (ml · kg^{-1} · min^{-1})	% change from 25 years
25	47.7	
35	43.1	–10
45	39.5	–17
52	38.4	–20
63	34.5	–28
75	25.5	–47

Data from Robinson (1938).

Older Athletes One of the most notable long-term studies of distance runners and aging was conducted by D.B. Dill and his colleagues from the Harvard Fatigue Laboratory.[6] Don Lash, world record holder for the 2 mi run (8 min 58 s) in 1936, was among the athletes studied by the Harvard group. Although few of the former runners continued to train after leaving college, Lash was still running about 45 min per day at age 49. Despite this activity, his $\dot{V}O_{2max}$ had declined from 81.4 ml · kg^{-1} · min^{-1} at age 24 to 54.4 ml · kg^{-1} · min^{-1} at age 49, a 33% decline. Runners who did not continue to train during middle age showed declines of about 43% from ages 23 to 50 (from 70 to 40 ml · kg^{-1} · min^{-1}). These data suggest that prior training offers little advantage to endurance capacity in later life unless a person continues to engage in some form of vigorous activity. However, due to their high initial values, these individuals have a large functional reserve, and this large decrease in aerobic capacity has little effect on their ability to carry out activities of daily living. In addition, there are large individual differences in the rate of decline in $\dot{V}O_{2max}$ with aging, and genetics is a major contributor.

More recent longitudinal studies of older male runners and rowers have shown a decline in aerobic capacity and cardiovascular function and changes in muscle fiber composition with aging. These athletes were studied for 20 to 28 years, during which time those athletes who continued high-volume and -intensity training experienced a 5% to 6% decline in $\dot{V}O_{2max}$ per decade. On the other hand, elite runners who stopped training experienced nearly a 15% decline in aerobic capacity per decade (1.5% per year).

Fewer studies have been published on women, but the results show the same trends. In one study of 86 male and 49 female masters endurance runners, the authors observed both cross-sectional and longitudinal (approximately 8.5 years) changes in $\dot{V}O_{2max}$ with age.[15] Their results are illustrated in figure 18.9. The average rate of decline, as indicated by the cross-sectional data regression line, was 0.47 ml · kg^{-1} · min^{-1} per year in men (0.8% per year) and 0.44 ml · kg^{-1} · min^{-1} in women (0.9% per year). However, this figure shows that the longitudinal changes are greater than the cross-sectional changes, particularly for the older ages. In a cross-sectional study of sedentary women (n = 2,256), active women (n = 1,717), and endurance-trained women (n = 911) aged 18 to 89 years, $\dot{V}O_{2max}$ declined by 0.35 ml · kg^{-1} · min^{-1} per year in the sedentary women (1.2% per year), 0.44 ml · kg^{-1} · min^{-1} per year in the active women (1.1% per year), and 0.62 ml · kg^{-1} · min^{-1} per year in the endurance-trained women (1.2%).[8]

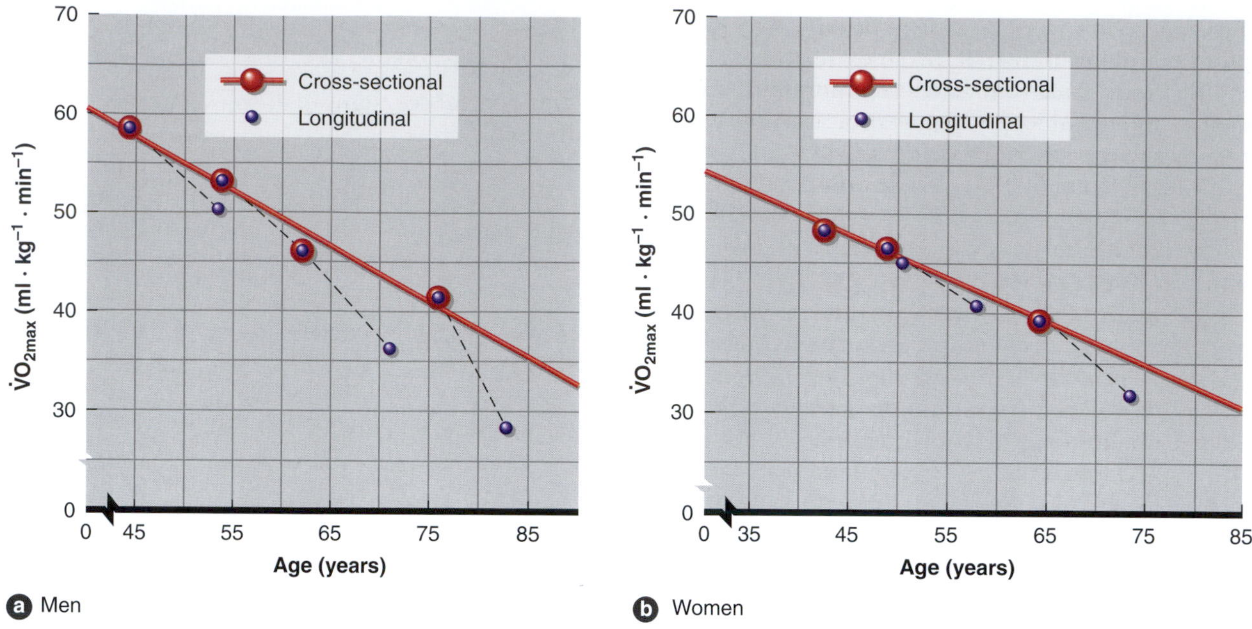

FIGURE 18.9 Cross-sectional and longitudinal declines in $\dot{V}O_{2max}$ with age in a group of *(a)* 86 male and *(b)* 49 female masters endurance athletes.

Adapted from Hawkins et al. (2001).

In the early 2000s, a 25-year follow-up study reexamined highly competitive, older male distance runners.[37,38] These men were initially tested at 18 to 25 years of age. During the interval between testing sessions, the runners trained at about the same relative intensity as they had when they were younger. As a consequence, their $\dot{V}O_{2max}$ values (L/min) declined only 3.6% over the 25-year period,[38] as shown in table 18.2. Although their maximal oxygen uptake decreased from 69.0 to 64.3 ml · kg^{-1} · min^{-1}, this is a decrease of only 0.19 ml · kg^{-1} · min^{-1} per year or 0.3% per year, and most of that change was attributable to a 2.1 kg (4.6 lb) increase in body weight.

This rate of decrease in these older runners' $\dot{V}O_{2max}$ values is significantly less than that of either sedentary people or those who fitness train at levels and intensities below those of these older runners. Scott Trappe and colleagues[39] had the unique opportunity to examine aerobic capacity in two cohorts of healthy, independently living men aged 80 to 91 years who differed dramatically in their lifelong physical activity habits. The authors recruited a group of nine elite Swedish masters cross-country skiers who were still actively engaged in the sport. These octogenarian athletes were lifelong exercisers who had never stopped training for more than 6 months at any point in the past 50 years! A second subject sample involved six healthy age-matched controls who had no history of regular exercise. Each subject performed a $\dot{V}O_{2max}$ test on a cycle ergometer and had a resting vastus lateralis muscle biopsy performed to measure such markers of aerobic capacity as the oxidative enzyme citrate synthase and the mitochondrial biogenesis initiator PGC-1α.

Not surprisingly, the lifelong endurance athletes exhibited a higher $\dot{V}O_{2max}$, whether expressed in absolute (2.6 versus 1.6 L/min) or relative (38 versus 21 ml · kg^{-1} · min^{-1}) terms. Skeletal muscle citrate synthase was 54% higher, and basal PGC-1α was 135% higher in the athletes. Both mitochondrial enzyme activity and PGC-1α were correlated with $\dot{V}O_{2max}$. The combined cardiorespiratory and muscle adaptations resulted in almost a 40% greater maximal power output on the cycle ergometer in the athletes. When one considers that some absolute $\dot{V}O_{2max}$ is associated with simply

TABLE 18.2 **Changes in Mean Aerobic Capacity and Maximal Heart Rate with Aging in a Group of 10 Highly Trained Masters Distance Runners**

		$\dot{V}O_{2max}$		
Age (years)	Weight (kg)	(L/min)	(ml · kg^{-1} · min^{-1})	HR_{max} (beats/min)
21.3	63.9	4.41	69.0	189
46.3	66.0	4.25	64.3	180

being able to complete activities of daily living (see figure 18.10), the authors of this study clearly show the expanded functional reserve between this minimal fitness level and the uniquely high aerobic capacity of the octogenarian athletes that reflects a lower risk for disability and mortality. As the authors point out, not only is the 11-MET (metabolic equivalent) aerobic capacity of these aged athletes the highest ever recorded in this age group, "it places them in the lowest all-cause mortality risk category for men *of any age*".

Although the number of studies on women is much smaller, a similar trend would be expected. Note that although intense training reduces the normal aging-related decline in $\dot{V}O_{2max}$, aerobic capacity still declines. Thus, it appears that highly intense training has a slowing effect on the rate of loss in aerobic capacity during the early and middle years of adult life (e.g., 30-50 years) but less effect after 50 years of age.

In summary, $\dot{V}O_{2max}$ declines with age, and the rate of decline is approximately 1% per year. Many factors influence this rate of decline, including the following:

- Genetics
- General activity level
- Intensity of training
- Volume of training
- Increased body weight and body fat mass, decreased fat-free mass
- Age range, with older individuals experiencing greater declines

There is not universal agreement on which of these factors are most important. An integrated view of the concepts of the physiological mechanisms contributing to the reductions in endurance performance with aging is presented in figure 18.11.

Lactate Threshold

In young endurance-trained adults, the lactate threshold predicts exercise performance in distance events ranging from 3.2 km (2 mi) to the marathon. Lactate threshold expressed as a percentage of $\dot{V}O_{2max}$ (LT-%

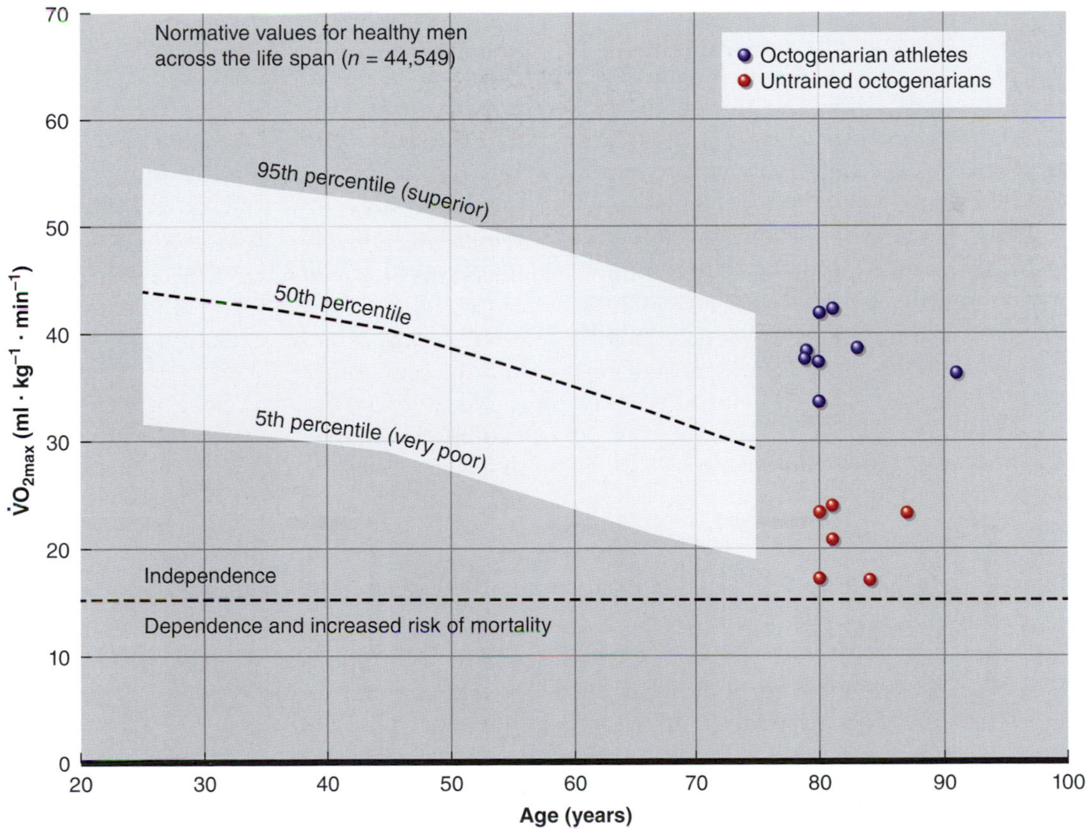

FIGURE 18.10 The figure shows individual $\dot{V}O_{2max}$ data from the lifelong endurance athletes and healthy untrained octogenarians studied by Trappe and colleagues. The dotted line at the bottom represents a 5-MET exercise capacity thought to be necessary for independently performing activities of daily living. The normative values for healthy men across the life span, shown in the shaded area, are derived from data from the Cooper Institute in Dallas, Texas.

Reprinted by permission from S. Trappe et al., "New Records in Aerobic Power Among Octogenarian Lifelong Endurance Athletes," *Journal of Applied Physiology* 114 (2013): 3-10.

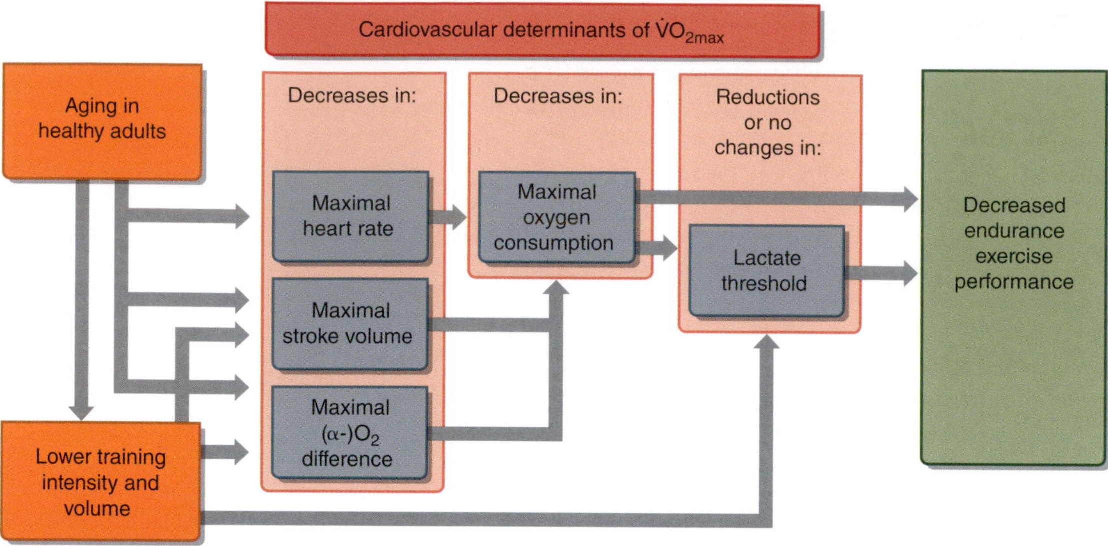

FIGURE 18.11 Factors and physiological mechanisms contributing to the reductions in endurance performance with advancing age in healthy humans. Primary aging directly contributes to the reduction in the cardiovascular determinants of $\dot{V}O_{2max}$; however, there tends to also be a decrease in training volume and intensity with aging (dotted lines), so the relative contributions of each to reduced endurance performance are difficult to determine.

$\dot{V}O_{2max}$) provides the best marker relative to endurance running performance in individuals with similar $\dot{V}O_{2max}$ values. Few studies have addressed the changes in either the lactate threshold or the anaerobic threshold, derived from ventilatory variables, with aging. Lactate threshold was determined in a cross-sectional study of a group of masters endurance runners, 40 to 70+ years of age (111 men and 57 women).[40] Interestingly, LT-% $\dot{V}O_{2max}$ did not differ between men and women, but it did increase with age. More recent longitudinal studies with masters athletes reported that the change in lactate threshold over a 6-year follow-up was not predictive of running performance when it was expressed as a percentage of $\dot{V}O_{2max}$.[28] Another study reported similar results in 152 untrained men and 146 untrained women.[29] However, in both studies, $\dot{V}O_{2max}$ was lower in the older groups, which helps explain the increase in LT-% $\dot{V}O_{2max}$. When the LT at the absolute $\dot{V}O_2$ is compared between age groups, LT declined with age.

RESEARCH PERSPECTIVE 18.1

Centenarian Athletes

Centenarians (people who live to or beyond the age of 100 years) are often considered models of healthy aging. Many reach this age while maintaining generally good health, and some of them preserve a remarkably high level of physical activity. Indeed, elite athletes live longer than the general population.[3] Although these individuals may be associated with having outstanding performance capacity, the age-related decline in performance in centenarians was only recently investigated.

A group of researchers identified and compared all of the best performances achieved by centenarians in athletics, swimming, and yoga.[26] Within each discipline, the decline in performance was expressed as a percentage of the world record for that discipline. The researchers noted that the average decline in performance of centenarian athletes was 78% in comparison with the initial world record, consistent with the concept that performance declines by ~10% during each decade of life after the age of 35 years. The best performance achieved by a centenarian was in track cycling, potentially owing to improvements in skeletal muscle cycling efficiency that offset limitations in oxygen transport. It is thought that as the number of centenarians continues to grow and as attitudes about exercise change among older adults, these types of exceptional performances by athletes will become more common.

Neuromuscular junctions in slow-twitch muscle fibers in the young rats demonstrated significant adaptations to exercise training, but those in the old rats showed no change. On the other hand, NMJs in many fast twitch-fibers were affected by age, but training did not induce any beneficial adaptations in the NMJs. Changes in muscle fiber size were unrelated to changes in NMJs. Taken together, these data suggest that aging interferes with the ability of the NMJ to adapt to exercise training and show how complex the interactions of different muscles and fiber types are in their response to both aging and training.

Age-related changes in the motor unit have a profound effect on motor function, especially among the expanding number of older adults. A recent review[18] presented evidence that age-related changes in motor unit morphology and properties lead to impaired motor performance. The resulting changes included reduced maximal strength and power, slower contractile velocity, and increased fatigability. Additionally, older subjects showed increased variability while performing motor tasks due to reduced and more variable synaptic inputs that drive motor neuron activation, fewer and larger motor units, less stable neuromuscular junctions, lower and more variable motor unit action potential discharge rates, and smaller and slower skeletal muscle fibers. The impact of regular physical activity on such variability in motor performance is unknown.

Aerobic and Anaerobic Capacity

Studies have shown that improvements in $\dot{V}O_{2max}$ with training are similar for younger (ages 21-25) and older (ages 60-71) men and women.[24,29] Although the pretraining $\dot{V}O_{2max}$ values were, on the average, lower for the older subjects, the absolute increases of 5.5 to 6.0 $ml \cdot kg^{-1} \cdot min^{-1}$ were similar in the two groups. Additionally, older men and women experienced similar increases in $\dot{V}O_{2max}$, averaging 21% for men and 19% for women, when they trained for 9 to 12 months by walking, running, or both walking and running about 4 mi (6 km) per day. Cardiovascular adaptation to training in previously sedentary older subjects appear to peak after 3 to 6 months of moderate training.[33] Together, these studies indicate that endurance training produces similar gains in aerobic capacity in healthy people throughout the age range of 20 to 70 years, and this adaptation is independent of age, sex, and initial fitness level. However, this does not mean that endurance training can enable older athletes to achieve the performance standards established by younger athletes.

The precise mechanisms that trigger the body's adaptations to training at any age are not fully understood, so we do not know if improvements from training are achieved in the same way throughout life. For example, much of the improvement in $\dot{V}O_{2max}$ seen in younger subjects is associated with an increase in maximal cardiac output. But older subjects show significantly greater gains in muscle oxidative enzyme activities, which suggests that peripheral factors in older subjects' muscles might play a greater role in aerobic adaptations to training than in younger subjects.

Very little is known about the trainability of anaerobic capacity in older people. We saw earlier in this chapter that lactate threshold, expressed as a percentage of a person's $\dot{V}O_{2max}$, increases with aging and is not associated with endurance running performance. In young and middle-aged adults, LT-% $\dot{V}O_{2max}$ is the best predictor of endurance performance—running, cycling, swimming, and cross-country skiing. As we have already seen, different rates of aging of the oxygen transport and lactic acid buffering systems are a likely explanation for this difference between older adults and young and middle-aged adults. A related point is that when one is comparing the lactate threshold of individuals with different $\dot{V}O_{2max}$ values, it is likely more appropriate to consider the absolute $\dot{V}O_2$ value to which that threshold corresponds when attempting to explain endurance performance.

Mobility

One of the major problems associated with aging is a loss of mobility, the ability to move freely and with minimal restrictions or pain. Loss of mobility is associated with increased disability, hospitalization, and mortality. The Lifestyle Interventions and Independence for Elders (LIFE) study sought to determine whether a structured physical activity program could reduce the risk of major mobility problems in a cohort of sedentary men and women aged 70 to 89 years.[30]

A sample of 1,635 sedentary men and women who had physical limitations but were able to walk 400 m were recruited, then randomized to either a structured, moderate-intensity physical activity program that included aerobic, resistance, and flexibility training, or a control group. After 2.6 years, they tested both groups' ability to walk 400 m, which they defined as a major mobility disability. Mobility disability occurred in 36% of the control group but only 30% of the physical activity group. The authors concluded that a structured, moderate-intensity physical activity program can reduce major mobility disability over 2.6 years among older adults at risk for disability.

Another recent study also showed improvement in mobility after 20 weeks of resistance training in individuals over the age of 65.[10] Interestingly, training older men and women with fast, unloaded movements improved mobility as well as conventional resistance training does. The results were attributed to neural

RESEARCH PERSPECTIVE 18.3

Age-Related Changes in Human Skeletal Muscle

The term sarcopenia, based on the Greek words *sarx* (flesh) and *penia* (poverty), defines the age-related loss of muscle mass. In addition to loss of muscle mass, there are also notable changes in muscle architecture and molecular structure and function in older adults, all of which ultimately manifest as limitations in muscle function. Reductions in muscle force production (strength) and endurance are features of sarcopenia and can be attributed to decreases in muscle fiber number as well as fiber cross-sectional area. Recently, mitochondria have been implicated as potential contributors to sarcopenia.[2]

Mitochondria are the main suppliers of cellular energy but are also primary contributors to the generation of reactive oxygen species (ROS). It appears that mitochondrial function, in general, is impaired in aging, including reductions in mitochondrial protein synthesis, respiration, and maximal rate of ATP production, partly as a result of increased uncoupling of oxygen consumption to ATP synthesis. Moreover, recent studies have suggested that mitochondrial dysfunction may contribute to muscle atrophy. This linkage to decrements in muscle performance indicates that altered mitochondrial function and morphology may promote intracellular oxidative stress, which can interfere with myofilament function via oxidative modification of proteins or disruptions in myofilament protein metabolism. Therefore, the relation between alterations in mitochondrial biology and deficits in muscle and myofilament structure and function likely contribute to the age-related impairments in skeletal muscle function. Interestingly, it appears that this relation may be modulated by aerobic exercise and habitual physical activity.

It is well known that exercise is a powerful stimulus for mitochondrial biogenesis in young adults. In sedentary older adults, a decline in mitochondrial content and function is evident at the mRNA, protein, morphological, and function (e.g., respiration) levels. In rodent models of aging, as well in humans, part of this loss of function can be alleviated by exercise training, suggesting that low physical activity likely contributes to mitochondrial dysfunction. In older adults, exercise increases mitochondrial content and provides a greater capacity to defend against oxidative stress. Habitual exercise has also been demonstrated to improve mitochondrial respiration in older adults. Collectively, it appears that exercise can be used as an effective intervention strategy to improve mitochondrial dysfunction in aged muscle, thereby restoring coupling efficiency and improving overall muscle function and performance.

adaptations that activate more type II muscle fibers. Older people do not have to lift heavy weights or use specialized equipment to achieve neuromuscular benefits that improve mobility—similar adaptations occurred when older subjects were trained to move their limbs as quickly as possible.

In Review

- Older adults appear to get the same benefits—maintenance of body weight, decreased percent body fat and fat mass, and increased fat-free mass—from exercise training as younger and middle-aged adults.
- It appears that aging does not impair a person's ability to increase muscle strength or muscle hypertrophy. Individual muscle fibers of older individuals also have the ability to increase in size.
- Endurance exercise training produces similar absolute gains in healthy people, regardless of their age, sex, or initial level of fitness. However, percent improvement is larger in those with low initial baseline levels.
- With endurance training, an increased $\dot{V}O_{2max}$ in older exercisers results mostly from improvement in the muscles' oxidative enzyme activities (peripheral adaptation), whereas improvement in younger people is largely attributable to increased maximal cardiac output (central adaptation).
- Age-related changes occur in motor unit structure and function that lead to impaired motor performance.
- Loss of mobility with aging can be offset by a variety of exercise programs and do not necessarily require lifting heavy weights.

Sport Performance

World and national records in running, swimming, cycling, and weightlifting suggest that people are in their physical prime during their 20s or early 30s. If we use a cross-sectional approach, comparing these records with national and world records for older athletes in these events allows us to examine the effects aging has on the best performers. Unfortunately, we have little longitudinal information about the effects of aging on performance because few studies have enabled us to follow physical performance in selected individuals over the span of their athletic careers. However, we can look historically at performance times in certain athletic events to gain insight into the influence of physiological function on performance with aging. In the following sections, we consider how aging affects certain types of sport performance.

Running Performance

In 1954, Roger Bannister, a 21-year-old medical student, stunned the sporting world when he became the first person to run the mile (1.61 km) in less than 4 min (3 min 59.4 s). The current record for the mile is 3:43:13, set by Hicham El Guerrouj of Morocco in 1999, which is more than 16 s faster than Bannister's record—a gap that would have placed Bannister more than 100 m (109 yd) behind today's record holder. In 1954, it seemed inconceivable that a sub-4 min mile could have been accomplished by someone over the age of 30 years. The oldest individual to record a sub-4 min mile was Eamonn Coghlan, who was 41 years old when he ran it indoors in 3:58:13. The oldest individual to record a sub-5 min mile was 65 years old.

Although older runners have achieved some exceptional records, running performance in general declines with age, and the rate of this decline appears to be relatively independent of distance. Longitudinal studies of elite distance runners indicate that despite a high level of training, performance in events from the mile (1.61 km) to the marathon (42 km) declines at a rate of about 1% per year from the age of 27 to 47 years.[37,38] It is interesting to note that age-group masters records for both 100 m and 10 km runs decrease fairly linearly from age 25 to age 70 or so, as shown in figure 18.12. Beyond age 75 to 80, however, the records for

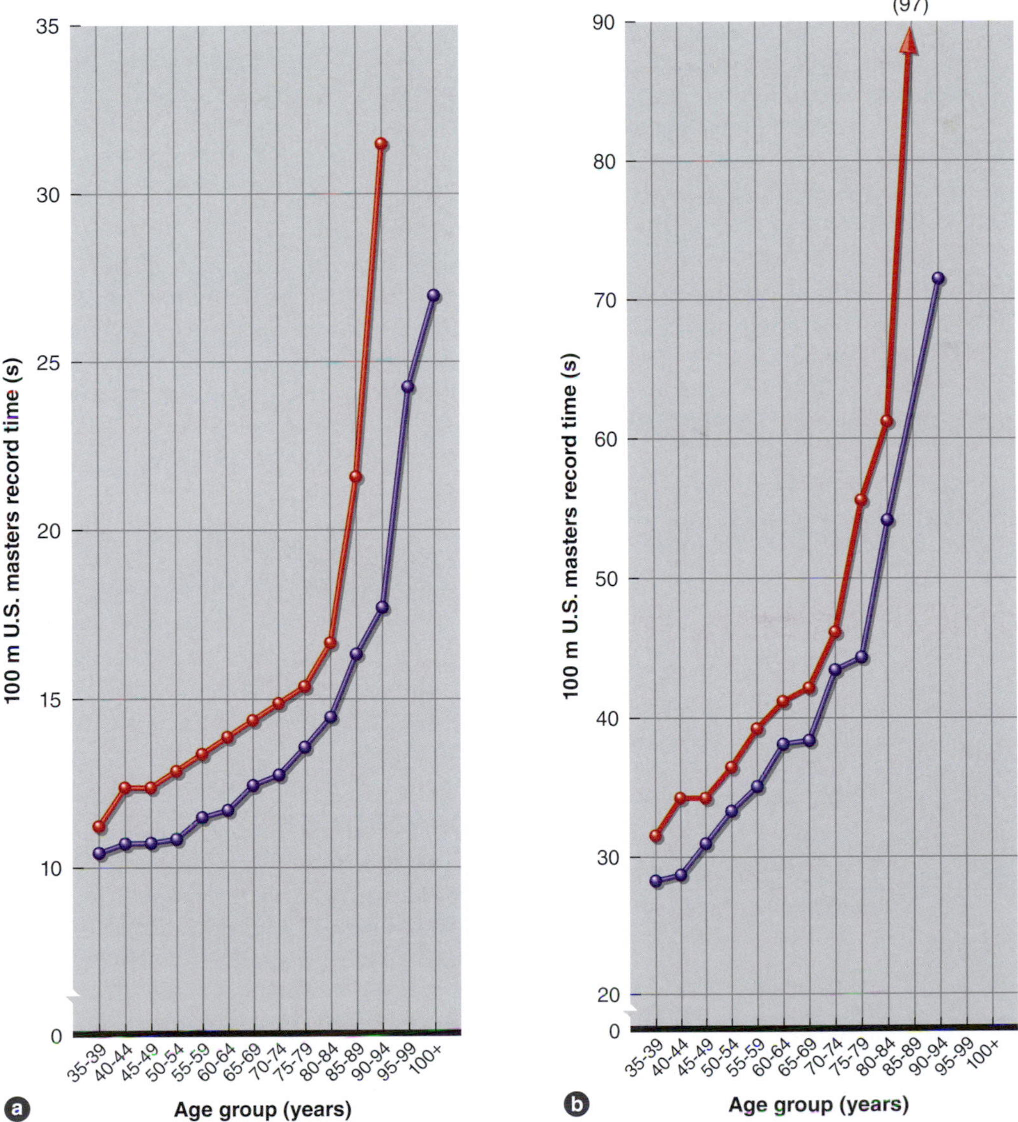

FIGURE 18.12 Men's and women's U.S. masters records across age groups for *(a)* 100 m and *(b)* 10 km runs.

men and women drop more precipitously. A sprint-running test of 560 women between ages 30 and 70 revealed a steady decrease in maximal running velocity of 8.5% per decade (0.85% per year).[29] The patterns of change are about the same in sprint- and endurance-running performances.

Swimming Performance

A retrospective study of freestyle performances at the U.S. Masters swimming championships between 1991 and 1995 revealed that both men's and women's performances in the 1,500 m declined steadily from age 35 to about 70 years, after which swimming times slowed at a faster rate.[34] However, the rate and magnitude of the declines in both 50 m and 1,500 m performances with age were found to be greater for women than for men.

Cycling Performance

As with other strength and endurance sports, record-setting cycling performances are generally achieved in the age range of 25 to 35. Male and female cyclists' records for 40 km (24.9 mi) races decrease at about the same rate with age, an average of 20 s (approximately 0.6%) per year. The U.S. national cycling records for 20 km (12.4 mi) show a similar pattern for both men and women. For this distance, speed decreases by about 12 s (approximately 0.7%) per year from age 20 to nearly age 65.

In Review

- As we age, peak performances in both endurance and strength events decrease by about 1% to 2% per year, starting between ages 25 and 35.
- Records in running, swimming, cycling, and weightlifting indicate that we are in our physical and physiological prime during our 20s and early 30s.
- In all of these sports, performance generally declines with aging beyond age 30 to 35.
- Most athletic performances decline steadily during middle and older age, primarily because of decrements in endurance and strength.

Weightlifting

In general, maximal muscle strength is achieved between the ages of 25 and 35. Beyond that age range, as shown in figure 18.13, international records for the sum of two power lifts decline at a steady rate. Of course, as with other measurements of human performance, individual strength performances vary considerably.

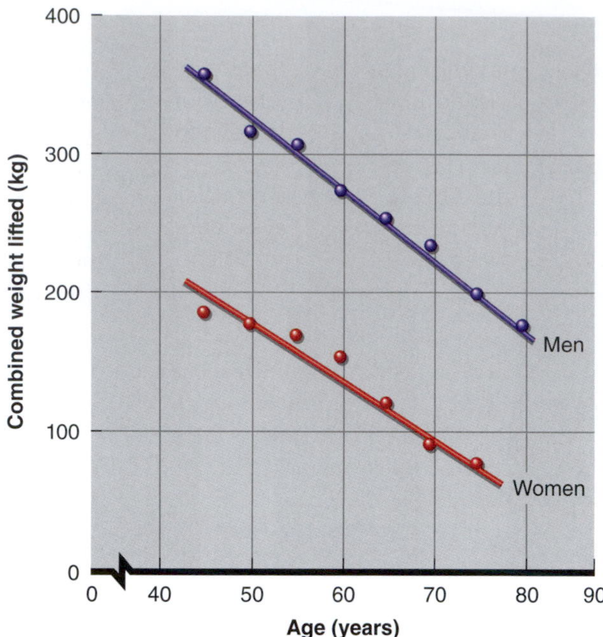

FIGURE 18.13 Changes in International Weightlifting Federation records with age among male and female weightlifters. The values reported are combined weight totals for the snatch and the clean and jerk.

Some individuals, for example, exhibit greater strength at age 60 than people half their age.

Most athletic performances decline at a steady rate during middle and advanced ages. These decreases result from decrements in both muscular and cardiovascular endurance and strength, as discussed earlier in this chapter.

Special Issues

As we age, we need to consider several special issues that can directly affect us when exercising or performing various sport activities. Here we look briefly at environmental stresses and then consider the issues of longevity, injury, and risk of death resulting from exercise and sport.

Environmental Stress

Because a variety of physiological control processes become less effective with aging, we can logically assume that older people are less tolerant of environmental stress than their younger counterparts. As we have seen elsewhere in this chapter, it is difficult to determine the separate effects of aging and physical fitness. In the following discussion, we compare the responses of younger and older adults to exercise in the heat and comment on cold stress and altitude exposure in older athletes.

Exposure to Heat

Exposure to heat stress presents a problem for older people. A preponderance of deaths during environmental heat waves occurs in people over the age of 70. The rate of metabolic heat production is related to the absolute exercise intensity, while heat loss mechanisms are related to the relative exercise intensity, so matching young and older subjects for $\dot{V}O_{2max}$ is important. When subjects are matched for body composition and $\dot{V}O_{2max}$, there is no difference in core temperature during exercise in the heat. However, when older subjects with a normal $\dot{V}O_{2max}$ for their age are compared to young subjects, they have a higher core temperature (see figure 18.14).[23]

VIDEO 18.4 Presents Lacy Alexander on the effects of age on thirst and temperature regulation.

These results indicate that physical training affects certain thermoregulatory responses. Sweat gland density does not appear to decline with aging but sweat gland output does, and research indicates that sweating rate is much more closely related to $\dot{V}O_{2max}$ than it is to age. As we saw in chapter 12, an increase in skin blood flow is necessary for the transfer of heat from the core to the shell for dissipation via evaporation of sweat. Even when young and older subjects are matched for $\dot{V}O_{2max}$, skin blood flow is lower in the older subjects but higher in older highly fit compared to older normally fit subjects, indicating that regular aerobic training can improve heat dissipation. Also, exercise in the heat requires significant blood flow to both the skin and exercising muscle, which is accomplished through increased cardiac output and reduced blood flow to renal and splanchnic regions. This redistribution of blood flow is less effective in older individuals, but similar to the situation with skin blood flow, improving aerobic fitness can improve this response. One study demonstrated that 4 weeks of endurance training in older men, which improved $\dot{V}O_{2max}$ by ~25%, improved regional redistribution so that renal and splanchnic blood flow decreased ~200 ml more than before training, which reduced cardiovascular strain.

In short, although aging reduces our ability to adapt to exercise in the heat, this deficit is largely due to reduced aerobic capacity. Increasing aerobic capacity can increase skin blood flow, increase sweating rate by increasing sweat gland output, and improve redistribution of blood to the skin and exercising muscles.

Exposure to Cold and to Altitude

In contrast to heat exposure, exercise in the cold typically poses less of a health risk. Due to declining aerobic fitness and loss of muscle mass, older individuals do have a reduced ability to generate metabolic heat. Also, the ability of the cutaneous vasculature to constrict is impaired with aging, which may increase heat loss. As a result of these changes, older individuals fail to appropriately maintain core temperature during cold stress.

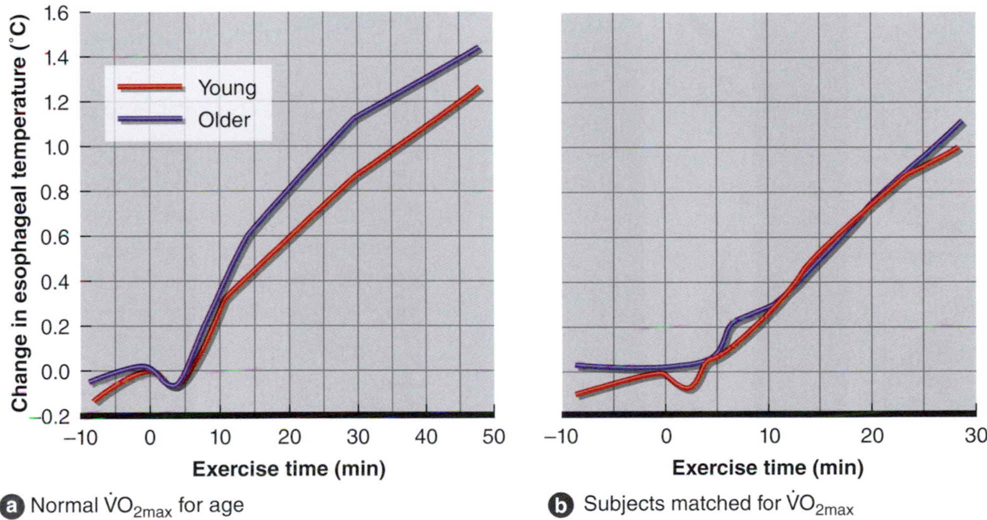

FIGURE 18.14 Changes in body core temperature in response to young (red line) and older (blue line) subjects exercising in a warm environment. When subjects in each age group have normal fitness levels for their age *(a)*, core temperature rises more sharply in the older individuals. However, when the two age groups are selected to have the same $\dot{V}O_{2max}$, the difference in core temperature disappears *(b)*, suggesting that $\dot{V}O_{2max}$ is more important than chronological age in determining this response.

Adapted by permission from W.L. Kenney, "Thermoregulation at Rest and During Exercise in Healthy Older Adults," *Exercise and Sport Sciences Reviews* 25 (1997): 41-77.

This is the case even when elderly subjects are exposed to what we would consider a mild cold stress.[7] However, people can easily offset these decrements by wearing clothing that is appropriate for the environmental conditions and activity level. By performing so-called behavioral thermoregulation, older athletes can offset the decrements in physiological thermoregulation and continue to exercise safely in cold environments.

During exposure to high altitude, there is little reason to expect older athletes to respond differently than their younger counterparts. Unfortunately, data are lacking regarding aging and the rate and magnitude of acclimatization to altitude. Likewise, it is unclear whether aging per se increases the incidence of any of the altitude illnesses. We can expect the performance of an older athlete at altitude to be similar to that of a younger athlete of comparable physical fitness.

Longevity

Regular physical activity is an important contributor to good health. Does training throughout adulthood affect longevity? Because the aging rate in rats is more rapid than in humans, they have been used as subjects in studies conducted to determine the influence of chronic exercise (training) on **longevity** (the length of one's life). A study by Goodrick[11] demonstrated that rats that exercised freely lived about 15% longer than sedentary rats. But an investigation at Washington University in St. Louis showed no significant increase in the life span of rats that voluntarily ran on an exercise wheel.[17] More of the active rats lived to old age, but on average, they still died at the same age as their sedentary counterparts. It is interesting that rats that had restricted food intake and maintained a lower body weight lived 10% longer than the freely eating, sedentary rats. Although exercise training is a key component to energy balance, the only known way to increase longevity is through caloric restriction.

Of course, we cannot directly apply these findings to humans, but these results raise some interesting questions that might be relevant to our health and longevity. Although it is true that an endurance exercise program can reduce a number of the risk factors associated with cardiovascular disease, only limited information supports the contention that people will live longer if they exercise regularly. Data collected from the alumni at Harvard University and the University of Pennsylvania and from participants at the Aerobic Center in Dallas suggest that there is a decrease in mortality rate and a small increase in longevity (about 2 years) among people who remain physically active throughout life. At a minimum, regular physical activity can increase the **health span**—the number of years of generally healthy living, free from serious disease or chronic disability.

In Review

- The impaired ability of older individuals to tolerate exercise in the heat is due to reduced $\dot{V}O_{2max}$ and impaired cardiovascular adaptations rather than a direct effect of aging on thermoregulatory control or sweating.
- Regular exercise training can increase skin blood flow and sweating rate and improve the redistribution of cardiac output in older individuals as well as young men and women.
- Older people generally have an impaired ability to tolerate cold, but they can compensate by wearing appropriate clothing.
- Adaptation to altitude appears to be independent of age.
- An active lifestyle appears to be associated with a small increase in longevity. Just as important, an active lifestyle leads to a higher quality of life!
- There is an increased risk of injury from exercise as people age, and injuries tend to be slower to heal.
- The risk of death during exercise is not increased in those who are regularly active but is increased in those who seldom exercise.

IN CLOSING

In this chapter, we examined the effects of aging on physical performance. We evaluated changes in cardiorespiratory endurance and strength with age. We considered the effect of aging on body composition, which we know can affect performance. And yet, in the course of our discussion, it became clear that much of the change that occurs with aging is attributable to the inactivity that often accompanies aging. When older people participate in training, most of the changes associated with aging are lessened, and the resulting degree of change is similar to that seen in young and middle-aged adults. Thus, we have dispelled many of the myths about the capacity for physical activity of older people.

In the next chapter, we turn our attention to women, who as a group are often considered less capable of physical activity than men. We consider the physiology of girls and women, the impact of this physiology on athletic ability, how performances of female athletes compare with those of male athletes, and special issues associated with being female.

KEY TERMS

cardiovascular deconditioning
endothelial dysfunction
forced expiratory volume in 1 s ($FEV_{1.0}$)
health span
longevity
maximal expiratory ventilation ($\dot{V}O_{2max}$)
osteopenia
osteoporosis
peripheral blood flow
primary aging

STUDY QUESTIONS

1. What changes occur in height, weight, and body composition with aging? What accounts for these changes? How do these changes affect maximal oxygen uptake?
2. What changes occur in muscle with aging? How do they affect strength and athletic performance?
3. Describe the changes in HR_{max} with age. How does training alter this relationship?
4. How does aging affect maximal stroke volume and maximal cardiac output? What mechanisms can potentially explain these changes?
5. How does the respiratory system change with aging? What happens to vital capacity, $FEV_{1.0}$, residual volume, total lung capacity, and the ratio RV/TLC?
6. $\dot{V}O_{2max}$ declines with age across the entire population. Describe the physiological mechanisms that account for this decline. How do trained older individuals maintain a relatively high $\dot{V}O_{2max}$?
7. How do aging and habitual exercise affect blood vessel function?
8. How does age affect anaerobic function?
9. Differentiate between biological aging and physical inactivity.
10. What influence do aging and training have on body composition?
11. Describe the trainability of the older individual for both strength and aerobic endurance.
12. What can be done to minimize age-related losses in motor function and mobility?
13. Describe the changes in strength and endurance performance records with aging.
14. What concerns should we have about older people exercising in hot and cold environments or at altitude?
15. What is the difference between life span and health span?

STUDY GUIDE ACTIVITIES

In addition to the activities listed in the chapter opening outline, two other activities are available in the web study guide, located at

www.HumanKinetics.com/PhysiologyOfSportAndExercise

The **KEY TERMS** activity reviews important terms, and the end-of-chapter **QUIZ** tests your understanding of the material covered in the chapter.

Sex Differences in Sport and Exercise

In this chapter and in the web study guide

Amber Miller accomplished two amazing feats during the same weekend in October 2011. The 27-year-old marathoner ran the Chicago Marathon while nearly 39 weeks pregnant—and then gave birth to a baby girl hours later! Because of her training and running history, Miller received clearance from her doctor to participate, provided that she took a half-run, half-walk approach, drank plenty of fluids, and ate food along the course. She finished in just under 6.5 h. Miller had intended to stop halfway, but once she and her husband started running, they just kept going. After the race, she grabbed some food and headed to the hospital.

In 2014 and 2015, Olympian Alysia Montaño ran an 800 m race and runner Amy Keil competed in the Boston Marathon, both nearly 8 months into their pregnancies. Although these feats may seem extreme and should not be attempted by most pregnant women, physicians generally agree that if a woman is healthy and running regularly before she got pregnant, running during pregnancy is fine. Marathon world record holder Paula Radcliffe ran 14 mi (22.5 km) a day while pregnant and resumed training weeks after the birth of her first child. She won the New York City Marathon in 2007 just 10 months after giving birth.

It is unfathomable now that before 1972, girls and women were banned from official participation in most mainstream marathons, including the Boston Marathon!

In the past, it was not uncommon for young girls to be discouraged from participating in vigorous physical activity, while young boys climbed trees, raced against each other, and participated in various sports. The underlying notion was that boys were meant to be active and athletic but girls were less suited to physical activity and competition. Physical education classes furthered this idea by having girls perform nonstrenuous exercise activities. In sports, girls and women were not allowed to run in long-distance races, and basketball was limited to half-court, with each team having only offensive or defensive players.

With the advent of Title IX, athletic activities and programming must be equally accessible to girls and women, and the results have been amazing. Their athletic accomplishments parallel those of boys and men, with performance differences of 15%-17% or less for most sports and events. This is illustrated in table 19.1, where men's and women's world records are compared for representative events in both track and field and swimming. Do these performance differences represent true biological differences, or are there other factors involved? The focus of this chapter is on the extent to which biological differences between women and men affect performance capacity.

Sex Versus Gender in Exercise Physiology

The literature examining physiological differences between men and women refers to these comparisons as *sex differences* or *gender differences*, with the terms often used interchangeably. The dual terminology can cause confusion with regard to literature searches, data reporting, and interpretation of results, leading researchers to consider whether there is a distinction between the two terms and when each should be used.

With respect to research on male and female physiology, sex and gender are not synonyms. Rather, sex is biologically determined and gender is culturally determined.[25] In other words, the term *sex* refers to the physiological, genetic, and biological traits that define a human as being male or female. Gender, by contrast, is the social, cultural, and psychological influences that

TABLE 19.1 **Selected Men's and Women's World Records Through 2018***

Event	Men	Women	Difference
Track and field			
100 m	9.58 s	10.49 s	9%
1,500 m	3:26.00 min:s	3:50.07 min:s	11%
10,000 m	26:17.53 min:s	29:17.45 min:s	11%
High jump	2.45 m	2.09 m	16%
Long jump	8.95 m	7.52 m	17%
Swimming (freestyle)			
100 m	46.91 s	51.71 s	10%
400 m	3:40.07 min:s	3:56.46 min:s	3%
1,500 m	14:31.02 min:s	15:20.48 min:s	5.5%

*Records current through June 2018, per iaaf.org and fina.org.

construe an individual's self-representation as male or female. For example, it is possible for one to have the structural and functional characteristics of a man but to identify and live as a woman, and possibly to have undergone surgical transformation to associate with the female gender. Consequently, most basic physiological comparisons to date between men and women have established sex differences, unless the investigation was specifically targeted at physiological outcomes associated with one or more of the cultural, social, behavioral, or psychological influences of different gender states. Although this seems to be a theoretical concept, real-life applications are observed in research and competitive sport.

For example, in 2011, the Internal Association of Athletics Federations (IAAF) devised a new policy in response to the case of Caster Semenya, a South African female runner who won the 800 m race at the 2009 Berlin World Championships. Semenya was criticized for her masculine build and forced to undergo sex testing that revealed unusually high levels of testosterone (a condition known as hyperandrogenism). The IAAF policy on hyperandrogenism thus states that women who are suspected of being too masculine must undergo testing and treatment in order to remain eligible for competition. This policy has been criticized on many points, among them that the development of the policy was confused and confounded by lack of distinction between sex and gender, as well as the difficulty of distinguishing easily between the two for the sake of competitive eligibility.[16]

Why the reluctance to use the term *sex differences* in modern-day physiology? It may be that the term *gender* avoids the innuendo associated with the term *sex*, and therefore seems less suggestive, more polite, or politically correct. In addition, even the concept of sex itself is not as dichotomous as one might wish, since genetic determinants of sex and biological structure and function can be discrepant so as to create a spectrum of male or female traits known as intersex characteristics in the same individual. In any case, though, referring to comparisons of the basic biological and physiological male and female attributes as sex differences establishes in the literature a basis for better classifying and interpreting research studies.

Body Size and Composition

Body size and composition are similar in boys and girls during early childhood. During late childhood, as we saw in chapter 17, girls begin to accumulate more fat than boys. Starting in early adolescence, boys begin to increase their fat-free mass (FFM) at a much higher rate than girls (see figure 17.3).

Body composition differences between the sexes occur primarily because of endocrine changes that begin during puberty. Before puberty, the anterior pituitary gland secretes very small amounts of the gonadotropic hormones: follicle-stimulating hormone (FSH) and luteinizing hormone (LH). During puberty, however, the anterior pituitary begins to secrete significantly greater amounts of both of these hormones. In girls, when sufficient quantities of FSH and LH are secreted, the ovaries develop and estrogen secretion begins. In boys, these same hormones trigger development of the testes and, in turn, testosterone secretion.

Figure 19.1 illustrates these changes in estrogen (estradiol—the most potent form of estrogen) and testosterone from the beginning of puberty (S1) to the end of puberty (S5). Testosterone increases bone formation, which leads to larger bones, and increased protein synthesis, which in turn leads to increased muscle mass. As a result, adolescent boys are larger and more muscular than girls, and these characteristics continue into adulthood. At full maturity, not only do men have a greater muscle mass but also the distribution of the muscle mass differs from that of women. Men carry approximately 3% more muscle mass in the upper body than women. Testosterone also stimulates erythropoietin production by the kidneys, which leads

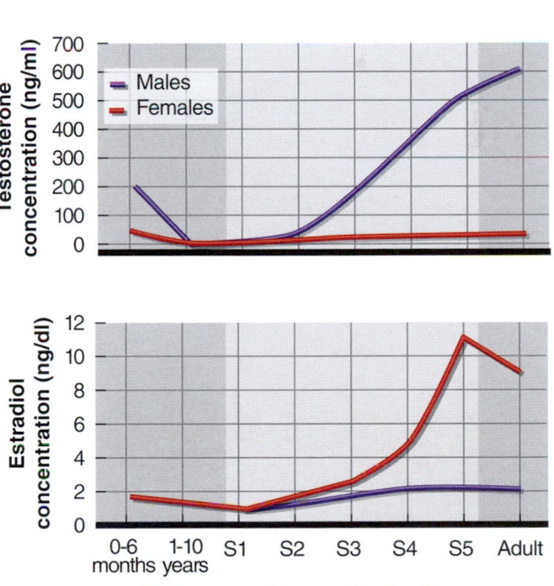

FIGURE 19.1 Changes in blood concentrations of testosterone and estrogen (estradiol) from birth to adulthood. S1 through S5 represent stages of puberty based on secondary sex characteristics, with S1 representing the initial stages of puberty and S5 the final stages.

Reprinted by permission from R.M. Malina, C. Bouchard, and O. Bar-Or, *Growth, Maturation, and Physical Activity,* 2nd ed. (Champaign, IL: Human Kinetics, 2004), 414; Data from Esoterix, Inc. *Endocrinology: Expected Values and S.I. Unit Conversion Tables,* 5th edition. (Calabasas Hills, CA: Esoteric, Inc., 2000).

to increased red blood cell production, as discussed later in this chapter.

Estrogen has a significant influence on women's body growth by broadening the pelvis, stimulating breast development, and increasing fat deposition, particularly in the thighs and hips. This increase in fat deposition in the thighs and hips is the result of increased **lipoprotein lipase** activity in these areas. This enzyme is considered the gatekeeper for storing fat in adipose tissue. Lipoprotein lipase is produced in the fat cells (adipocytes) but is bound to the walls of the capillaries, where it exerts its influence on the chylomicrons, which are the major transporters of triglycerides in the blood. When lipoprotein lipase activity in any area of the body is high, chylomicrons are trapped and their triglycerides are hydrolyzed and transported into the adipocytes in that area for storage.

Many women are constantly fighting fat deposition on the thighs and hips, but they are usually fighting a losing battle. Lipoprotein lipase activity is very high and lipolytic activity (fat breakdown) is low in the hips and thighs of women compared with women's other fat storage areas and compared with the hips and thighs of men. This results in a rapid storage of fat in women's thighs and hips, and the decreased lipolytic activity makes it difficult for women to lose fat from these areas. During the last trimester of pregnancy and throughout lactation, the activity of lipoprotein lipase decreases and lipolytic activity increases dramatically, which suggests that fat is stored in the hips and thighs for reproductive purposes.

Estrogen also accelerates the growth rate of bone, allowing the final bone length to be reached within 2 to 4 years following the onset of puberty. As a result, girls grow very rapidly for the first few years following puberty and then cease to grow. Boys have a much longer growth phase, allowing them to attain a greater height. Because of these differences, compared with fully mature men, fully mature women are on average about

- 13 cm (5 in.) shorter,
- 14 to 18 kg (30-40 lb) lighter in total weight,
- 3 to 6 kg (7-13 lb) heavier in fat mass, and
- 6% to 10% higher in relative body fat.

In Review

- The term *sex* refers to physiological, genetic, and biological traits. Gender, by contrast, is the social, cultural, and psychological influences that construe an individual's self-representation as male or female.
- Until puberty, girls and boys do not differ significantly in most measurements of body size and composition.
- At puberty, because of the influences of estrogen and testosterone, body composition begins to change markedly.
- Testosterone increases bone formation and protein synthesis, leading to a larger FFM. It also stimulates the production of erythropoietin, which increases red blood cell production.
- Estrogen causes increased fat deposition in women, particularly in the hips and thighs, and an increased rate of bone growth, such that bones in women reach their final length earlier than in men.

Physiological Responses to Acute Exercise

When women and men undergo an acute bout of exercise, whether an all-out run to exhaustion on the treadmill or a single attempt to lift the maximal weight possible, there are response differences between the sexes. Differences between children and adolescent boys and girls are discussed in chapter 17. Here we focus on these differences in adults in the areas of strength and of cardiovascular, respiratory, and metabolic responses to exercise.

Strength

In terms of absolute strength, women are approximately 40% to 60% weaker than men in upper body strength but only 25% to 30% weaker in lower body strength. However, because of the considerable size difference between the average man and the average woman, it is more appropriate to express strength either relative to body weight (absolute strength divided by body weight) or relative to FFM, as a reflection of muscle mass (absolute strength divided by FFM). When lower body strength is expressed relative to body weight, women are still 5% to 15% weaker than men, but when it is expressed relative to FFM, this difference disappears. This suggests that the innate qualities of muscle and mechanisms of neuromuscular control are similar for women and men, a fact that has been confirmed by computed tomography (CT) scans of the thighs and upper arms, that is, knee extensor muscles (figure 19.2*a*) and elbow flexor muscles (figure 19.2*b*).

Muscle biopsies have become more common among female athletes, allowing fiber type comparisons with male athletes in the same sport or event. From these data, men and women have similar fiber type distributions, although there is some evidence that men may reach greater extremes (greater than 80% type I or greater than 80% type II). Among elite distance runners,[7,12] women had a mean value of 69% type I fibers compared with 79% for the men. The women predictably had much smaller type I and type II fibers (mean values of less than 4,500 μm^2 in women and greater than 8,000 μm^2 in men). Despite smaller fiber size in women, capillarization appears to be similar between men and women.

Research indicates that women have a greater resistance to fatigue compared with men. Fatigue is often tested by having subjects maintain a constant force at a given percentage of their maximal voluntary static action for as long as possible. As an example, women would be able to maintain a constant force output at 50% of their maximal static action for a longer period of time than men at the same 50% of their maximal static action. Men, being stronger, have to apply a greater absolute amount of force to achieve the same 50% relative force. The reason for this greater resistance to fatigue is not yet known but could be related to the amount of muscle mass recruited and the compression of blood vessels, substrate utilization, muscle fiber type, and neuromuscular activation.

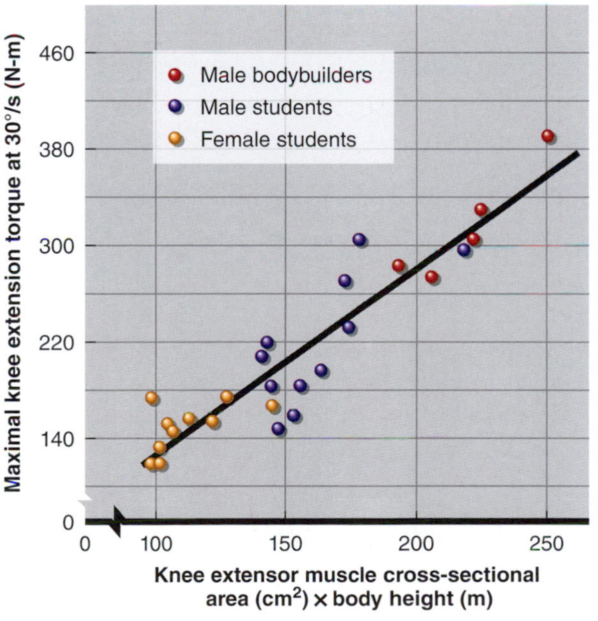

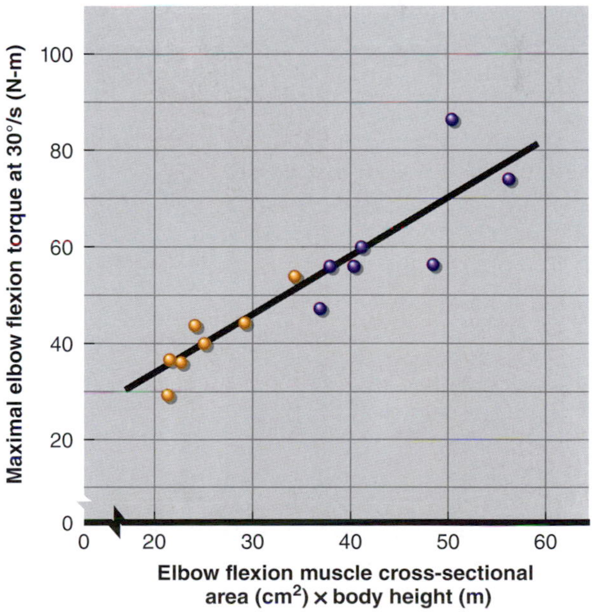

FIGURE 19.2 No sex differences between men and women in strength [(*a*) maximal knee extension torque or (*b*) maximal elbow flexion torque] are seen when strength is expressed per unit of muscle cross-sectional area.

Reprinted by permission from P. Schantz et al., "Muscle Fibre Type Distribution, Muscle Cross-Sectional Area and Maximal Voluntary Strength in Humans," *Acta Physiologica Scandinavica* 117 (1983): 219-226.

Cardiovascular and Respiratory Function

When exercising on a cycle ergometer at a fixed power output (independent of body weight), women generally have a higher heart rate (HR) response. Maximal heart rate (HR_{max}) is generally the same in both sexes. During submaximal exercise, stroke volume (SV) is lower in women, but HR is higher; therefore, cardiac output ($\dot{Q}$) is usually the same in men and women at any absolute submaximal power output. Remember that the *average* woman may also be less aerobically active and therefore less aerobically conditioned. However, even in fitness-matched men and women, the lower SV can be attributed to the following:

- Women have smaller hearts and therefore smaller left ventricles because of their smaller body size and possibly lower testosterone concentrations.
- Women have a smaller blood volume, which also is related to their size and lower FFM.

When power output is controlled to provide the same *relative* intensity of exercise, (% $\dot{V}O_{2max}$), women's heart rates are still slightly higher than men's, and their stroke volumes are markedly lower. At 60% $\dot{V}O_{2max}$, for example, a woman's cardiac output, stroke volume, and oxygen consumption are generally less than a man's, and her heart rate is slightly higher. These relationships between HR, SV, and $\dot{Q}$ for the same absolute power output (50 W) and the same relative power output (60% $\dot{V}O_{2max}$) are illustrated in figure 19.3. These data were derived from the HERITAGE Family Study.[27]

While several early studies reported $\dot{Q}$ to be higher in women than in men at identical submaximal power outputs, possibly compensating for their lower hemoglobin concentrations, more recent studies have consistently shown no differences between sexes.[27] Apparently, women can compensate for their lower hemoglobin concentrations with a steeper increase in their arterial–mixed venous oxygen difference, or (a-$\bar{v}$) O_2 difference, for a given power output. Women may also have less potential for increasing their peak (a-$\bar{v}$) O_2 difference with training due to their lower hemoglobin content, which results in lower arterial oxygen content and reduced muscle oxidative potential. Lower hemoglobin content is an important contributor to **sex-specific differences** in $\dot{V}O_{2max}$ because less oxygen is delivered to the active muscle for a given volume of blood.

Sex differences in respiratory responses to exercise are largely attributable to body size differences (figure 19.4). Breathing frequency during exercise at the same relative power output (e.g., 60% $\dot{V}O_{2max}$) differs little between the sexes. However, at the same absolute power output, women tend to breathe more rapidly than men, probably because when men and women are working at the same absolute power output, women are working at a higher percentage of their $\dot{V}O_{2max}$. Tidal volume and ventilatory volume are generally lower in women at the same relative and absolute power outputs, up to and including maximal levels. Again, these differences are related to their smaller body size.

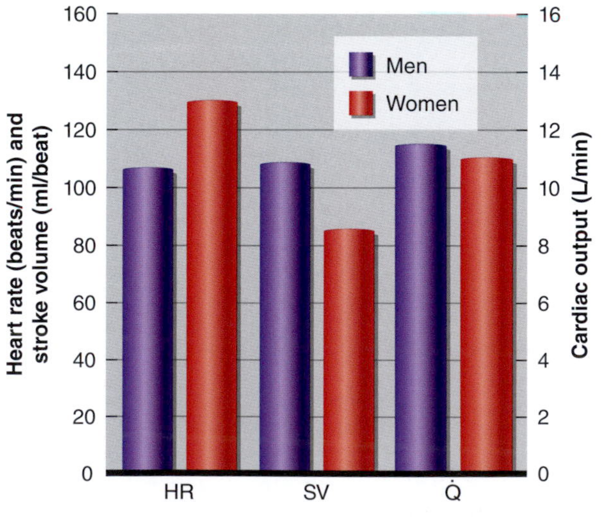

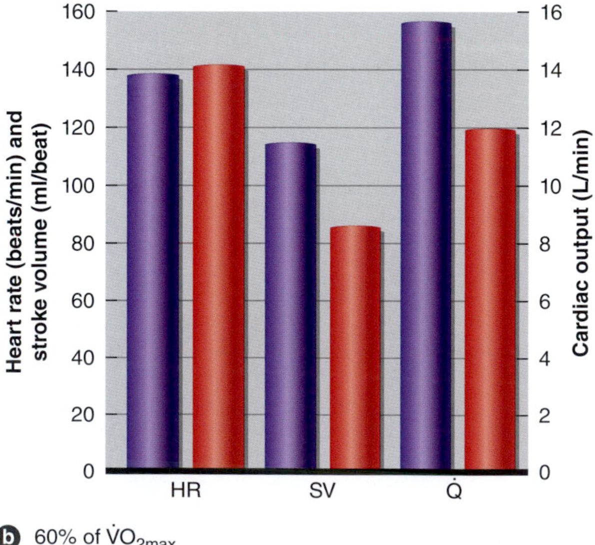

FIGURE 19.3 Comparison of submaximal heart rate (HR), stroke volume (SV), and cardiac output ($\dot{Q}$) between men and women at *(a)* the same absolute power output (50 W) and *(b)* the same relative power output (60% $\dot{V}O_{2max}$).
Based on Wilmore et al. (2001).

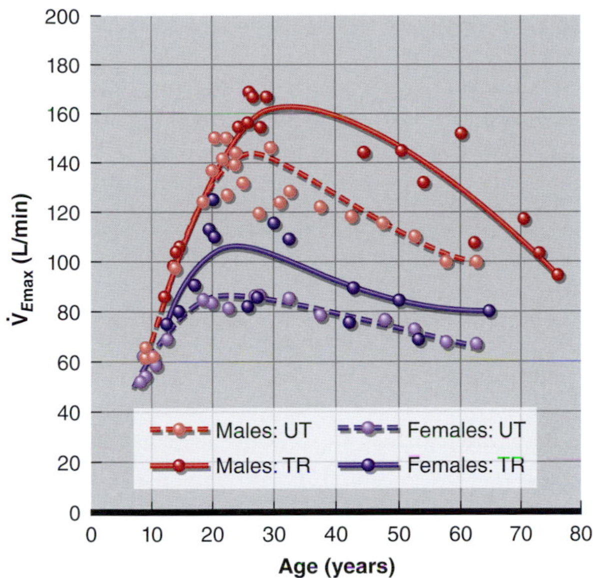

FIGURE 19.4 Differences in maximal ventilatory volumes with age in untrained (UT) and trained (TR) women and men.

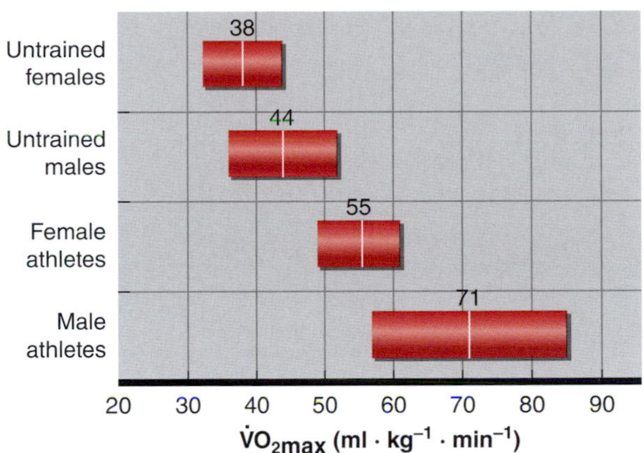

FIGURE 19.5 Range of $\dot{V}O_{2max}$ values (mean ± 2 standard deviations from the mean) for female nonathletes, male nonathletes, elite female athletes, and elite male athletes. The mean $\dot{V}O_{2max}$ value is presented above each bar. This figure illustrates that although differences in the average $\dot{V}O_{2max}$ between groups can be substantial, there can be considerable overlap of one group with another.

Data from Hermansen and Andersen (1965).

Maximal Oxygen Consumption

$\dot{V}O_{2max}$ is regarded by most exercise scientists as the single best index of a person's cardiorespiratory endurance capacity. The average woman's $\dot{V}O_{2max}$ is only 70% to 75% of the value of the average man's, but sex differences in $\dot{V}O_{2max}$ must be interpreted with caution. A classic study published in 1965, involving both elite athletes and nonathletes, showed that while mean $\dot{V}O_{2max}$ was lower in women, there is considerable variability and overlap in $\dot{V}O_{2max}$ within and between sexes.[13] For example, when the subjects' physiological responses to submaximal and maximal exercise were compared, 76% of the female nonathletes overlapped 47% of the male nonathletes, and 22% of the female athletes overlapped 7% of the male athletes, as illustrated in figure 19.5.

Comparisons of $\dot{V}O_{2max}$ values of normal nonathletic women and men beyond puberty are confounded by differences in the level of conditioning (i.e., nonathlete men are typically more well trained than women at the same age) as well as possible sex-specific differences. Saltin and Åstrand[21] sought to minimize the effect of differential physical activity between sexes by comparing $\dot{V}O_{2max}$ values of highly trained female and male athletes from Swedish national teams. In comparable events, the women had 15% to 30% lower $\dot{V}O_{2max}$ values. More recent data, however, suggest a smaller difference of 8% to 12% between sexes, as shown in figure 19.6, where $\dot{V}O_{2max}$ values for a group of elite female distance runners are compared with values for elite male distance runners and average, nonathletic women and men.

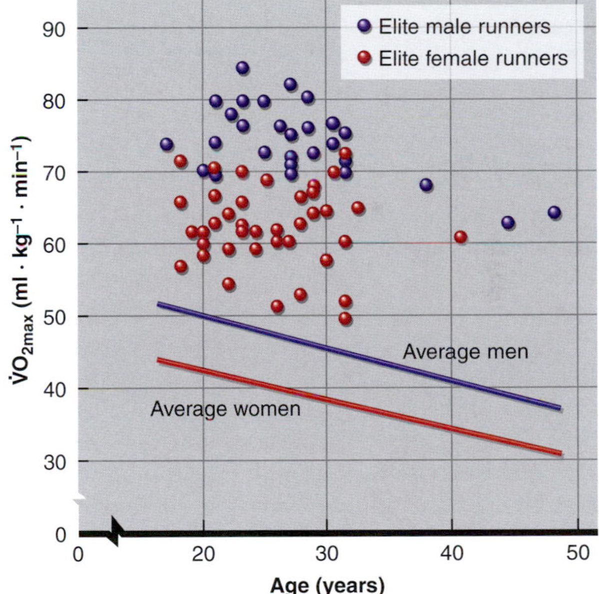

FIGURE 19.6 $\dot{V}O_{2max}$ values, compiled from the literature, for elite female and male distance runners compared with average values in untrained women and men.

Differences between the sexes may disappear when $\dot{V}O_{2max}$ is expressed relative to FFM or active muscle mass. For example, manipulating body weight in trained men (by adding external weight to the torso) to match the greater fat weight being carried by similarly trained women reduced the mean sex differences in submaximal $\dot{V}O_2$ (expressed in milliliters per kilogram of FFM) and $\dot{V}O_{2max}$ by 38% and 65%, respectively.[9] These results suggest that women's greater sex-specific

RESEARCH PERSPECTIVE 19.1

Do Men Lose More Weight Than Women with Regular Exercise?

Exercise is recommended by multiple public health organizations and professional societies as a primary method for weight control and weight loss, with the overall goal of improving health. Exercise improves health outcomes independent of weight loss, and the efficacy of an exercise intervention to produce weight loss without an accompanying diet plan is still in question (discussed in Research Perspective 15.1); however, there has been a prevailing assumption that women do not lose as much weight as men do in response to exercise. Studies that examine sex differences in weight loss often explain a lower weight loss in women by theorizing about a stronger defense of body fat resulting in greater energy intake. However, common methodological issues, including compliance to the exercise regimen, lack of actual energy expenditure data, differences in the duration and intensity of the exercise programs, and variability in subject characteristics make these studies difficult to interpret and compare.

In 2014, a research team of scientists from the University of Leeds in the United Kingdom and the Karolinska Research Institute in Sweden conducted a systematic review of studies that had examined weight loss differences between men and women.[6] This review included studies involving short-, medium-, and long-term interventions; supervised versus unsupervised programs; and normal weight versus overweight individuals. The review also examined exercise-induced energy expenditure and compensatory increases in energy intake. Their meta-analysis indicated that when energy expenditure was equal, weight loss was the same between men and women. Additionally, the variability in weight loss responses to exercise was driven by individual variability in energy intake, independent of sex. Furthermore, there was no evidence to suggest that women compensated for exercise-induced energy-expenditure by increasing caloric intake any more than men did. These researchers speculate that previous studies showing a sex difference in exercise-induced weight loss were confounded by inattention to such variables as body size and energy intake, skewing their results.

essential body fat stores are major determinants of sex-specific differences in $\dot{V}O_{2max}$ during running.

Women's lower hemoglobin concentrations have also been proposed as a factor contributing to their lower $\dot{V}O_{2max}$ values. However, removing a small amount of blood from men to equalize their hemoglobin concentrations to those of women reduces men's $\dot{V}O_{2max}$ values, but accounts for only a relatively small portion of the sex difference.[8]

It is also important to understand that a woman's lower maximal cardiac output is a limitation to achieving a high $\dot{V}O_{2max}$ value. A woman's smaller heart size and lower plasma volume greatly limit her maximal stroke volume. When we consider submaximal oxygen consumption ($\dot{V}O_2$), little if any difference is found between women and men for the same absolute power output. But since women usually are working at a higher percentage of their $\dot{V}O_{2max}$ at the same absolute submaximal work rate, blood lactate concentrations are higher, and the lactate threshold occurs at a lower absolute power output. Lactate threshold is similar between men and women when expressed as $\%\dot{V}O_{2max}$.

In Review

- The innate properties of muscle and the mechanisms of motor control are similar for women and men.
- Women and men do not differ in lower body strength expressed relative to body weight or to FFM. But women have less upper body strength expressed relative to body weight or FFM than men, largely because more of women's muscle mass is below the waist and women use their lower body muscles more than their upper body muscles.
- At submaximal exercise intensities, women have higher heart rates than men, but women's submaximal cardiac outputs are similar for the same intensity. This indicates that women have lower stroke volumes, primarily because they have smaller hearts and less blood volume.
- Women also have a lesser capacity to increase $(a\text{-}\bar{v})O_2$ difference, probably because of lower hemoglobin content, so less oxygen is delivered to their active muscles per unit of blood.
- Sex differences in respiratory responses are primarily attributable to body size differences.
- Beyond puberty, the average woman's $\dot{V}O_{2max}$ is only 70% to 75% that of the average man, although these differences may be attributable to discrepant physical activity levels between men and women. Research with highly trained athletes reveals an 8% to 15% difference, and much of this difference is attributable to women's greater

fat mass, lower hemoglobin levels, and lower maximal cardiac output.

- At a given relative exercise intensity, little or no difference in lactate threshold is found between the sexes.

Physiological Adaptations to Exercise Training

Basic physiological function, both at rest and during exercise, changes substantially with exercise training. In this section, we investigate how women adapt to regular exercise, emphasizing areas in which their responses might differ from that of men.

Body Composition

With either cardiorespiratory endurance training or resistance training, both women and men experience

- a decrease in total body mass,
- a decreased fat mass,
- a decreased relative (%) fat, and
- increased FFM.

The magnitude of the change in body composition appears to be related more to the total energy expenditure associated with the training activities than to the participant's sex. More FFM is gained in response to strength training than with endurance training, and the magnitude of these gains is similar between sexes.

Bone and connective tissue undergo alterations with training. In adults, weight-bearing exercise is critical for maintaining bone mass and bone mineral density. This adaptation appears to be independent of sex. Connective tissue appears to be strengthened with endurance training, and sex-specific differences in this response have not been identified.

Strength

Until the early 1970s, prescribing strength training programs for girls and women was considered inappropriate. Women were not believed capable of gaining strength because of their innately low levels of male anabolic hormones. During the 1960s and 1970s, however, it became evident that many of the United States' best female athletes were not doing well in international competition, largely because they were weaker than their competitors. Slowly, research began to demonstrate that women can benefit considerably from strength training programs even though the resulting strength gains are usually not accompanied by large increases in muscle bulk.

In part because of their lower testosterone, women have less total muscle mass than men. If muscle mass is the major determinant of strength, then women have a distinct disadvantage. But if neural factors are as important as or more important than size, women's potential for absolute strength gains is considerable. Also, some women can attain significant muscle hypertrophy. This has been demonstrated in female bodybuilders who have remained free of anabolic steroids. A number of studies have shown similar increases for men and women in FFM and muscle volume, as well as hypertrophy of type I, type IIa, and type IIx muscle fibers following periods of resistance training. Accordingly, women can experience major increases in strength (20%-40%) as a result of resistance training, and the magnitude of these changes is similar to that seen in men.

Cardiovascular and Respiratory Function

Major cardiovascular and respiratory adaptations result from cardiorespiratory endurance training, as discussed in chapter 11, and these adaptations do not appear to be sex specific. Increases in maximal cardiac output ($\dot{Q}_{max}$) accompany training despite the fact that maximal heart rate usually does not change (or decreases slightly), so the increase in $\dot{Q}_{max}$ is the result of a large increase in maximal stroke volume. This results from two factors. End-diastolic volume (the amount of blood in the ventricles before contraction) increases with training because blood volume increases. In addition, end-systolic volume (the amount of blood remaining in the ventricles after contraction) is reduced with training because the stronger myocardium and increased sympathetic nerve activity produces a more forceful contraction, ejecting more blood.

At submaximal work rates, cardiac output shows little or no change with training, although stroke volume is considerably higher for the same absolute rate of work. Consequently, heart rate for any given rate of work is reduced after training. These changes are independent of sex.

Maximal Oxygen Consumption

With cardiorespiratory endurance training, women experience the same percent increase (15% to 25% on average) in $\dot{V}O_{2max}$ that has been observed in men. The magnitude of change depends on the intensity and duration of the training sessions, the frequency of training, and the length of the study. There are also similar changes in lactate production and accumulation: Blood lactate concentrations are reduced for given absolute submaximal rates of work, peak lactate concentrations generally are increased, and the lactate threshold increases with training.

In Review

- With training, women and men experience similar changes in body composition as determined by total energy expenditure during training.
- Women, similar to men, gain considerable strength through resistance training, an effect associated with increases in FFM and muscle volume as well as hypertrophy of type I, type IIa, and type IIx muscle fibers.
- Cardiovascular and respiratory changes that accompany cardiorespiratory endurance training do not appear to be sex specific.
- Women experience the same relative increases in $\dot{V}O_{2max}$ as men with cardiorespiratory endurance training.

Sport Performance

Women are outperformed by men in all athletic activities in which performance can be precisely and objectively measured by distance or time. The difference is most pronounced in activities such as the shot put, where high levels of upper body strength are crucial to successful performance. It would appear that the gap between the sexes narrowed as more women began to participate in sport. In the 400 m freestyle swim, however, the winning time for women in the 1924 Olympic Games was 19% slower than that for men, but this difference decreased to only 7% in the 1984 Olympics. That gap was narrowed even further by Katie Ledecky, whose 2016 Olympic record time of 3:56.46 is less than 3% away from the men's record of 3:40.14 set in 2012 by Sun Yang. As we see from table 19.1, the difference between the men's and women's world records for the for the 1,500 m freestyle is only 5.5%. Differences are larger for track-and-field events. Unfortunately, making valid comparisons through the years has been difficult because the degree to which sport has been emphasized and women's access to coaching, facilities, and advanced training techniques have differed considerably over the years.

As noted at the beginning of this chapter, large numbers of girls and women did not start entering competitive sport until the 1970s. Once girls and women started training as hard as boys and men, their performance improved dramatically. This is illustrated in figure 19.7, which shows world track records from 1960 to 2016 for women and men in six running events in track and field. For distances of 100 m through the marathon, women's present world records are consistently 8% to 9% slower than men's. Furthermore, as can be seen in this figure, the improvement in women's records, which was initially quite dramatic, is beginning to level off and to parallel the curves for men's records.

It is firmly established that there are sex differences in anaerobic and aerobic performance attributable to sex differences in muscle mass and strength, maximal oxygen uptake, and metabolism. This results in men outperforming women at every distance and event. However, trends in the performance gap between men and women have changed nonuniformly over the past 50 years. For example, whereas sex differences in ultra-endurance performance have been decreasing, the gap between male and female performance has actually increased in the sprint events. These observations suggest that factors that mediate the sex difference in performance are not solely physiological.

In the endurance events, by contrast, it may be a participation issue. For example, the gap between the top female and male 24 h ultramarathoners has decreased over the last 35 years, indicating that as increasing numbers of women enter ultra-endurance sports, performance gaps may shrink.[19] Moreover, marathon data indicate that sex differences in performance widen with age and increasing place of finisher (1st through 10th), and these relationships are mediated predominantly by lower numbers of women in competition.[14]

In any case, evidence to date indicates that differences between men and women in sport performance have physiological, social, environmental, and psychological influences; thus, predicting the performance

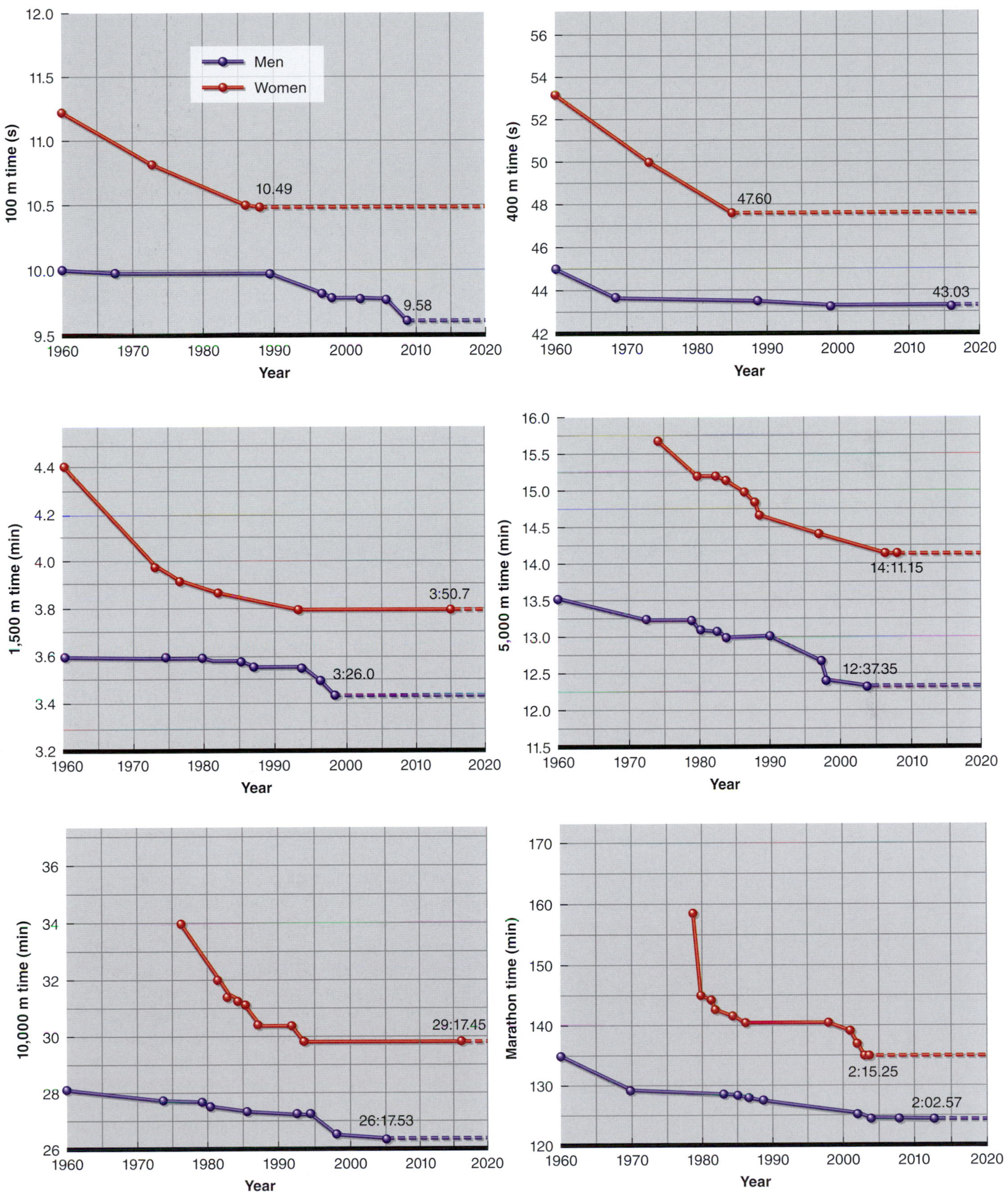

FIGURE 19.7 Women's and men's world records in six running events between 1960 and 2016.

RESEARCH PERSPECTIVE 19.2

Men Are More Likely Than Women to Slow Down During a Marathon

Optimal pacing strategies in endurance races has been a topic of interest for coaches, athletes, and exercise physiologists for some time. Traditionally, these studies have been limited to elite athletes. However, recently available data from mass-participation events have allowed investigators to study marathon running across a broader range of athletes, including nonelite distance runners. Based on these newly available data, new pacing patterns have emerged, including a finding that men are more likely than women to slow their pace as the marathon proceeds.

A 2015 study investigated these apparent sex differences in pacing in 91,921 runners across 14 marathons all around the United States.[10] Researchers obtained data on sex, age, race completion time, and time to the halfway point, along with measures of racing experience (total previous races, previous marathons, fastest previous marathon, fastest 5K, and earliest year with a recorded race) when available. Pace maintenance was calculated for each participant as the percentage change in the pace observed in the second half of the marathon relative to the first half. The research team made a 12% adjustment to women's performances to account for the fact that women, on average, are 10% to 12% slower compared to similarly conditioned men.

Across the entire study, men slowed their pace 15.6% while the women slowed their pace by an average of only 11.7%. This sex difference was significant for the entire subject group across marathons, for each individual marathon, and was not influenced by finishing times (indicating that overall speed did not influence this sex difference). Slower finishing times were associated with greater slowing, especially in men. Finally, although more experience was associated with less slowing for both sexes, controlling for experience did not eliminate the sex differences in pacing.

The reason for this apparent sex difference in pacing is unclear. It may reflect physiological differences (such as a greater susceptibility to glycogen depletion and muscle fatigue in men) or simply differences in decision-making. Men, for example, may be more likely to adopt a risky pace early, i.e., they begin the race with a fast pace that they are not able to maintain across the entire course.

gap over future generations may be more complex than simply generating a regression equation based on past performance.

Special Issues

Although the sexes respond to acute exercise and adapt to chronic exercise in much the same manner, several additional areas that are unique to women must be considered. Specifically, we look at

- menstruation and menstrual dysfunction,
- pregnancy,
- bone health across the life span,
- eating disorders,
- relative energy deficiency and the female athlete triad, and
- menopause.

Menstruation and Menstrual Dysfunction

How does the menstrual cycle influence exercise capacity and performance? How do physical activity and competition influence the menstrual cycle? These are two questions of interest to exercising women, particularly female athletes.

The three major phases of the **menstrual cycle** are illustrated in figure 19.8. The first is the menstrual (flow) phase, or **menses**, which lasts 3 to 5 days, during which time the uterine lining (endometrium) is shed and menstrual flow, or bleeding, occurs. The second is the proliferative phase, which prepares the uterus for fertilization and lasts about 10 days. During this phase, the endometrium begins to thicken, and some of the ovarian follicles that house the ova mature. These follicles secrete estrogen. The proliferative phase ends when a mature follicle ruptures, releasing its ovum (ovulation). The menstrual and proliferative phases correspond to the follicular phase of the ovarian cycle.

The third and final phase of the menstrual cycle is the secretory phase, which corresponds to the luteal phase of the ovarian cycle. This phase lasts 10 to 14 days, during which the endometrium continues to thicken, its blood and nutrient supply increases, and the uterus prepares itself for pregnancy. During this time, the remnants of the ruptured follicle (now termed a corpus luteum, hence the term *luteal phase*) secretes progesterone, and estrogen secretion also continues. The complete menstrual cycle averages 28 days; however, there is considerable variation in the cycle length among healthy women, from 23 to 36 days.

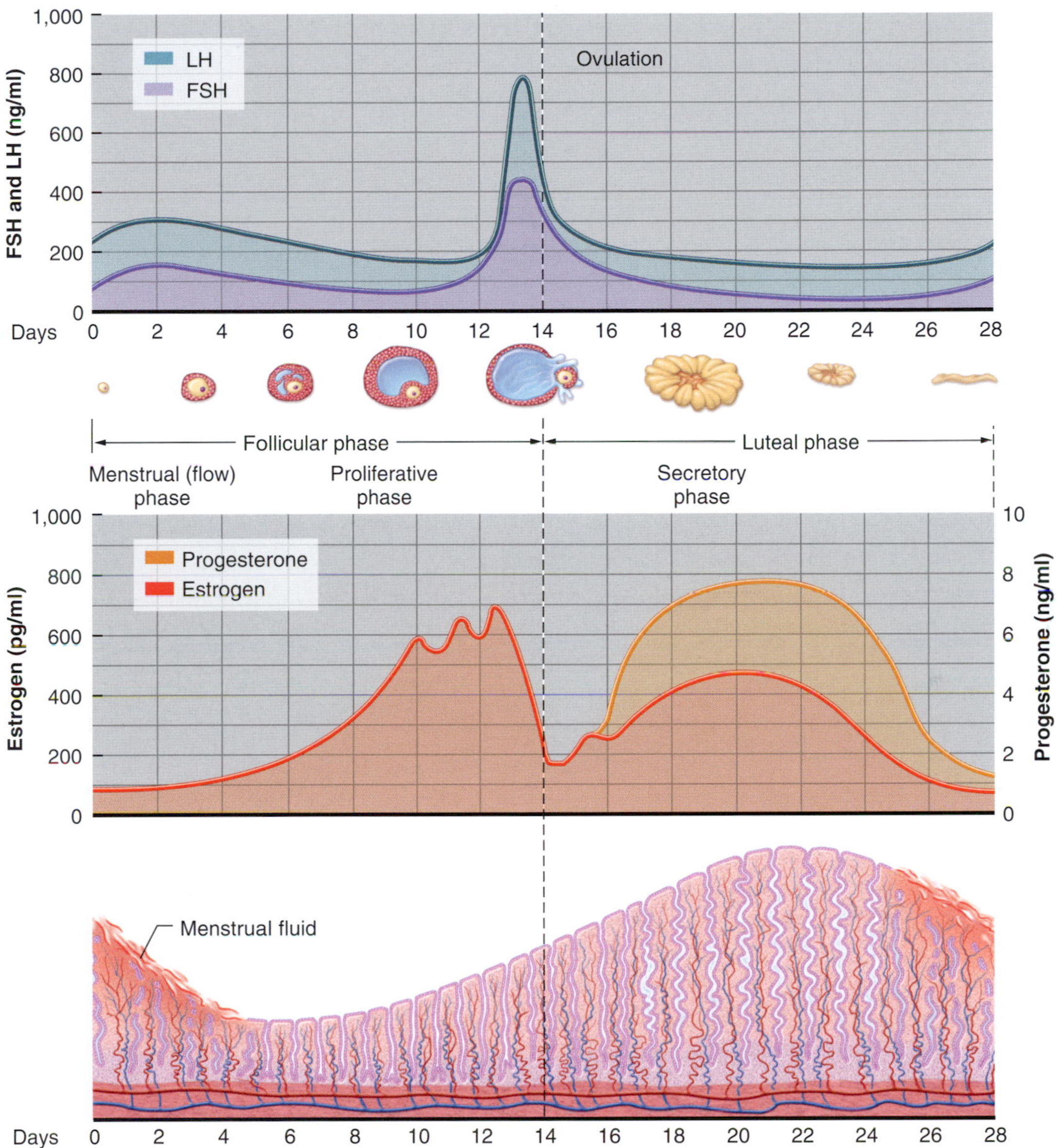

FIGURE 19.8 Phases of the menstrual cycle and the concomitant changes in progesterone and estrogen (middle) and follicle-stimulating hormone (FSH) and luteinizing hormone (LH) (top). For most purposes, the cycle is divided into the follicular phase, which starts with the initiation of bleeding during menstrual flow, and the luteal phase, which starts with ovulation.

Menstruation and Performance

Alterations in athletic performance experienced during different phases of the menstrual cycle are subject to considerable individual variability. There are no reliable data that demonstrate any significant physiological differences across phases of the cycle or changes in athletic performance at any time during the menstrual cycle. Elite-level performances have been achieved by female athletes during all phases of the menstrual cycle. Therefore, data gathered in the laboratory and during athletic competition concur that neither the physiological responses nor the performance of most women is substantially affected by the menstrual cycle, although women may be affected by factors such as premenstrual syndrome, bloating, and cramping.

Menarche

Menarche, which refers to the occurrence of the first menstrual period, has been reported to be delayed in some young athletes involved in certain sports and activities, such as gymnastics and ballet. A delay in menarche is defined as a menarche after the age of 14; the median age of menarche for American girls is between 12.4 to 13.0 years, depending on the population sampled. For gymnasts, the median age appears to be closer to 14.5

RESEARCH PERSPECTIVE 19.3

Should Female Athletes Be Tested for Iron Deficiency and Anemia?

Iron deficiency anemia impairs athletic performance, particularly in endurance sports, due to the reduced oxygen-carrying capacity of red blood cells. Iron deficiency anemia has been reported to be higher in female (<1%-18%) than in male (<1%-7%) athletes competing in a wide variety of sports. Traditionally, the higher prevalence of iron deficiency and anemia among athletes has been attributed to inadequate iron intake (e.g., restricting caloric intake in sports that emphasize leanness) and increased iron losses (e.g., via sweating, hemolysis, or menstruation).

A recent study conducted at the University of Wisconsin Madison examined the prevalence of anemia among incoming female college athletes at an NCAA Division 1 institution.[18] The aims of the study were to determine what proportion of these athletes had iron deficiency anemia and what the expenses of iron-related testing would be in this setting. Medical records for all athletes on varsity team rosters were examined over a 12-year period. Anemia was defined as hemoglobin concentration below 11.6 g/dl and iron deficiency as ferritin below 20 ng/ml. A total of 5,674 total laboratory reports were obtained from 2,749 individuals (56% female). The prevalence of low hemoglobin in female athletes at their initial physical when joining a team was approximately 6%. Across all female athletes, 2% had iron deficiency anemia and 31% had iron deficiency without anemia. Interestingly, the prevalence of anemia was not different from the average for any individual sport.

Currently, the International Olympic Committee recommends serum iron screenings as part of routine health evaluations for elite athletes. However, screening practices in the National Collegiate Athletic Association (NCAA) are inconsistent, with institutions using different biochemical indices and cutoffs to define iron deficiency anemia. In clinical testing, hemoglobin concentration and levels of serum ferritin and serum transferrin receptor are obtained in order to diagnose iron deficiency anemia. Although routine clinical screening would facilitate earlier detection and intervention in NCAA athletes, it comes with a financial cost. The median cost of iron screening exceeded $20,000 annually for the institution.

years. These data imply that exercise training causes delayed menarche. However, the alternative hypothesis is that late-maturing girls, such as those with a later age of menarche, are more likely to be successful in sports such as gymnastics because of their small, lean bodies. This implies that those who naturally experience a later menarcheal age have an advantage in, and thus are likely to be involved in, certain sports rather than that their sport involvement delays menarche. At this time, evidence is insufficient to strongly support either viewpoint.

Menstrual Dysfunction

Female athletes can experience disruptions of their normal menstrual cycle. These disruptions are collectively referred to as **menstrual dysfunction**, of which there are several types. **Eumenorrhea** is the term for normal menstrual function, referring to consistent menstrual cycle lengths of 26 to 35 days. **Oligomenorrhea** refers to inconsistent and irregular menses at intervals longer than 36 days and up to 90 days. **Amenorrhea** refers to the absence of menstruation; **primary amenorrhea** refers to the absence of menarche in girls and women 15 years of age and older—women who have not yet begun menstruating. When athletes with previously normal menstrual function report the absence of menstruation for 90 days or longer, the condition is called secondary amenorrhea. Thus, **secondary amenorrhea** is the absence of menses for 90 days or more in girls and women who were previously menstruating. Women involved in any sport and even recreationally active women can experience amenorrhea, as it occurs independent of intensity of exercise training.

The prevalence of secondary amenorrhea and oligomenorrhea among female athletes is well documented and is estimated to be anywhere between 5% and 66% or higher, depending on the sport or activity, the level of competition, and the definition used for amenorrhea. Nonetheless, the prevalence of amenorrhea in female athletes is considerably higher than the estimated 2% to 5% prevalence for amenorrhea and the 10% to 12% prevalence for oligomenorrhea in the general (nonathlete) population. The prevalence appears to be greater in athletes who participate in activities that emphasize a lean physique, such as ballet, gymnastics, and distance running.

Since the 1970s, scientists have been conducting experiments to determine the primary cause of secondary amenorrhea in exercising girls and women and female athletes. Some of the factors that have been proposed as potential causes are the following:

- A history of menstrual dysfunction
- The acute effects of stress
- A high volume or intensity of training
- A low body weight or percent body fat

- Hormonal alterations
- An energy deficit through inadequate nutrition, disordered eating, or both

Considerable research has been conducted on each of these proposed factors and five of the six have largely been eliminated for consideration as the primary cause. It is tempting to surmise that high-volume or high-intensity training (or a combination of the two) leads to menstrual dysfunction, but this factor is likely not involved.

Current evidence indicates that inadequate nutrition resulting in an energy deficit is the primary cause of secondary amenorrhea, that is, an inadequate intake of calories such that the body is not matching caloric intake to caloric expenditure over an extended period of time. For example, research by Dr. Anne Loucks[20] at Ohio University and Dr. Nancy Williams at Penn State University[26] has clearly demonstrated that inducing an energy deficit in eumenorrheic women results in significant alterations in hormones (estrogen, progesterone, leptin, and triiodothyronine [T_3]) associated with menstrual dysfunction, including amenorrhea. Food deprivation may trigger signals that inhibit LH secretion and menstrual function, and the greater the energy deficit, the more severe the impact on menstrual function.[26]

Thus, exercise training, per se, is likely not directly associated with menstrual dysfunction at all, other than through its contribution to an energy deficit. Rather, an energy deficit, in either the absence or presence of exercise training, is associated with these hormonal alterations. Therefore, high-intensity or high-volume training most likely is not associated with menstrual dysfunction as long as energy intake matches or exceeds energy expenditure over days, weeks, and months.

In Review

- The consequences of competing during different phases of the menstrual cycle on performance are subject to considerable individual variability. In general, there is no research evidence that demonstrates a consistent effect of phase of the menstrual cycle on sport performance.
- Menarche can occur late in some young athletes in certain sports. However, an alternative explanation for this is that late maturers, because of lean body build, are more likely to participate successfully in these activities, not that these activities cause delayed menarche.
- Female athletes can experience menstrual dysfunction, most often in the form of secondary amenorrhea or oligomenorrhea. Current evidence implicates inadequate nutrition in combination with exercise, rather than high exercise-related energy expenditure alone, as leading to a prolonged energy deficit that causes secondary amenorrhea.
- Hormonal changes associated with an energy deficit may feedback on the hypothalamus or pituitary and disrupt the normal cycle. This, too, is associated with a prolonged energy deficit.

Pregnancy

The major concern of pregnancy for active women is whether there are potential adverse effects of exercise during **pregnancy**. Over the years, several risks of exercise during pregnancy have been postulated. More recently, these risks have become balanced by the suggested and documented benefits of maternal exercise while pregnant.

TABLE 19.2 **Risks and Postulated Benefits of Exercise During Pregnancy**

	Maternal health	Fetal health
Risks*	Acute hypoglycemia Chronic fatigue Musculoskeletal injury	Acute hypoxia Acute hyperthermia Acute reduction in glucose availability
Benefits	Increased energy level (aerobic fitness) Reduced cardiovascular stress Prevention of excessive weight gain Facilitation of labor Faster recovery from labor Promotion of good posture Prevention of lower back pain Prevention of gestational diabetes Improved mood state and body image	Fewer complications of a difficult labor Improved cardiovascular health[5] Decreased risk of chronic disease[5]

*Risks are typically low unless the patient has certain conditions, such as preeclampsia. Visit www.acog.org for details on conditions that would contraindicate exercise.

Benefits of and Recommendations for Exercise During Pregnancy

According to a 2013 review on the health benefits of exercise in pregnancy,[17] physical activity during pregnancy has beneficial impacts on gestational diabetes, hypertensive disorders, excessive gestational weight gain, birth weight, timing of delivery, and child body composition. In fact, physical inactivity and excessive weight gain have been recognized as independent risk factors for maternal obesity and related pregnancy complications.[1]

Unfortunately, much of the research on exercise in pregnancy has been observational, with physical activity assessed through self-report. Therefore, specific mechanisms through which exercise affects maternal and fetal well-being have not been elucidated. In any case, exercise during pregnancy can have associated risks (see table 19.2), but the benefits far outweigh the potential risks if caution is taken in designing the exercise program. In the absence of medical complications or contraindications, physical activity in pregnancy is safe and desirable, and pregnant women should be encouraged to continue or to start a program of safe physical activity. It is, however, important that the pregnant woman coordinate her exercise program with her obstetrician so that sound medical judgment can be used to determine the most appropriate mode, frequency, duration, and intensity of activity.

Despite compelling data about the substantial health benefits to being physically active during pregnancy, only about 15% of pregnant women meet the American College of Obstetricians and Gynecologists[1] (ACOG) recommendation of 150 min of moderate-intensity aerobic activity per week during pregnancy.[11] Consequently, an emerging area of research has focused on the types of interventions and approaches that can increase participation in physical activity immediately before and during pregnancy.

VIDEO 19.1 Presents Jim Pivarnik on the benefits of exercise during pregnancy.

The American College of Obstetricians and Gynecologists[1] makes the following recommendations for exercising during pregnancy:

1. Physical activity in pregnancy has minimal risks and has been shown to benefit most women, although some modification to exercise routines may be necessary because of normal anatomic and physiologic changes and fetal requirements.
2. A thorough clinical evaluation should be conducted before recommending an exercise program to ensure that a patient does not have a medical reason to avoid exercise.
3. Women with uncomplicated pregnancies should be encouraged to engage in aerobic and strength-conditioning exercises before, during, and after pregnancy.
4. Obstetrician–gynecologists and other obstetric care providers should carefully evaluate women with medical or obstetric complications before making recommendations on physical activity participation during pregnancy. Although frequently prescribed, bed rest is only rarely indicated; in most cases, allowing ambulation should be considered.
5. Regular physical activity during pregnancy improves or maintains physical fitness, helps with weight management, reduces the risk of gestational diabetes in obese women, and enhances psychologic well-being.
6. Additional research is needed to study the effects of exercise on pregnancy-specific outcomes and to clarify the most effective behavioral counseling methods and the optimal intensity and frequency of exercise. Similar work is needed to create an improved evidence base concerning the effects of occupational physical activity on maternal–fetal health.

Most activities are safe or can be modified to be safe. However, some athletic activities should be avoided,[1] including the following:

- Contact sports
- Activities with a high risk of falling
- Scuba diving
- Sky diving
- Hot yoga or Pilates

Athletic Competition and Elite-Level Training During Pregnancy

A small but growing subset of female athletes are remaining highly physically active during pregnancy and choosing to compete during pregnancy, to return to competition very quickly after delivery, or both. For example, Serena Williams won the 2017 Australian Open while pregnant and advanced to the final match of Wimbledon in 2018 just 10 months after a complicated childbirth, and U.S. Olympic discus thrower Aretha Thurmond competed in the U.S. National Championships 2 weeks after her son was born. Relatively little research exists, though, to support whether such vigorous, intense, and sustained training regimens are as safe and beneficial during pregnancy as more established patterns of moderate physical activity.

One study in 2005 followed 41 fit women through pregnancy as they participated in either high-volume

RESEARCH PERSPECTIVE 19.4

Land Versus Water Exercise in Pregnancy

Healthy lifestyle habits maintained or adopted during pregnancy improve maternal and fetal outcomes, and obstetricians recommend that their patients follow healthy nutrition and physical activity guidelines during their pregnancy. An unhealthy lifestyle during pregnancy has long-term consequences on the lifetime health of the mother and the offspring, and these consequences result in a significant public health problem. The benefits of exercise on maternal and fetal health are well established, and studies of both land-based and aquatic exercise programs have reported positive results on maternal and fetal outcomes during and immediately following pregnancy. However, few studies have compared modes of exercise with respect to efficacy, safety, and compliance in pregnant women. Promoting exercise during pregnancy can be challenging, especially in women who are not accustomed to regular exercise programs and many have concerns about safety. Because of these concerns, studies that compare pregnancy outcomes following different exercise modalities may provide pregnant women with a choice of which modality they are most comfortable, along with clinically based evidence of safety.

Researchers from Spain, Argentina, and Canada conducted a cross-sectional analysis of three separate randomized clinical trials in healthy pregnant women. This large study aimed to compare the results from studies that used (1) land, (2) water, or (3) land and water exercise programs during pregnancy. Importantly, all three of these studies used the same study design and the same exercising workloads, increasing the validity of making comparisons across studies. A total of 578 women completed the studies. For the cross-sectional analysis, there were 311 women in the nonexercise control group (pooled from all three studies), 107 women who completed the land-based exercise, 49 who completed the water exercise, and 101 who completed the land and water exercise. Total maternal weight gain, gestational diabetes rates, pregnancy-induced hypertension, gestational age, and birth weights were compared. Overall, there were no adverse effects of any of the exercise modalities. Total maternal weight gain and the percentage of women with excessive weight gain were lower in the land exercise group compared to nonexercising women. However, the number of women who developed gestational diabetes was lower in the water and water and land groups compared to no exercise. The study authors concluded that exercise performed on land is more effective for preventing excessive weight gain, whereas programs that include aquatic exercise may be beneficial for preventing the development of gestational diabetes.

More randomized controlled trials are needed to confirm these findings and explain the physiological mechanisms behind them. However, and possibly most importantly, while all exercise modalities showed some benefit, no adverse events were reported with any of the exercise modalities. These findings suggest that pregnant women can safely choose the exercise modality or combination of modalities that they find most comfortable and that they are most likely to continue throughout their pregnancy.

(8.5 h/week of exercise) or moderate-volume (6 h/week of exercise) training.[15] Maternal and fetal safety and adverse outcomes were not different between the two groups, and women participating in high-volume exercise improved $\dot{V}O_{2max}$ by 9.1% from 17 weeks of gestation to 12 weeks postpartum in contrast to the moderate-intensity exercisers, who only maintained $\dot{V}O_{2max}$. Therefore, it appears that higher-volume training can actually evoke fitness gains through pregnancy in athletes accustomed to exercise before conception. There are also case reports of pregnant endurance runners who maintained training exceeding 60 mi/week (96.5 km/week) of running without any adverse outcomes and relatively few decrements in performance. By contrast, though, a more recent study of six pregnant elite Olympic-level athletes 23 to 29 weeks pregnant revealed that endurance exercise bouts at >90% maximal heart rate evoked reductions in uterine artery blood flow and fetal heart rate, indicating a potential risk to fetal well-being at from exercise of very high intensity.[23] Since data are limited to just a handful of studies on high-level athletes who maintain competitive and rigorous training during pregnancy, additional research is needed to address with certainty the questions about safety and efficacy of sustained, high-intensity training during the gestational and postpartum periods. Competitive athletes should pay particular attention to avoiding hyperthermia, maintaining proper hydration, and sustaining adequate caloric intake to prevent weight loss, which may adversely affect fetal growth.[1]

More than 40% of Olympic athletes in the most recent Olympiads were women, so there is little doubt that this topic will be more thoroughly investigated as increasing numbers of female athletes seek to maintain or even expand fitness and training during pregnancy.

In Review

- During exercise, major concerns for the pregnant athlete include the possible risk of fetal hypoxia, fetal hyperthermia, reduced carbohydrate supply to the fetus, miscarriage, premature labor, low birth weight, and abnormal fetal development.
- The benefits of a properly prescribed exercise program during pregnancy outweigh the potential risks. Such an exercise program must be coordinated with the woman's obstetrician.

Bone Health Across the Life Span

Bone health is a particular concern for women that is affected both by physical activity and menstrual status. As they age, women are at greater risk for reduced bone mineral content and osteoporosis than men are, particularly after menopause. In addition, because bone health is connected with menstrual status in women, it is an important consideration for athletes who experience menstrual dysfunction.

Regular physical activity can play an important role in affecting bone health across the life span. In children, for example, regular physical activity is associated with greater bone mass, size, and bone mineral density, and large-scale intervention studies have shown substantial improvements in bone mineral density without an increase in fracture risk. It is particularly important for girls to engage in weight-bearing activities, given their risk of osteoporosis after menopause. The optimal dose–response and characteristics of an exercise program for bone health have not been fully described through research. School-based physical activity protocols have been successful, although they can conflict with demands on curricular time; and recent interventions showing that jumping (5-10 min of exercise performed several times a week) can increase bone mineral density by 5% to 10% have demonstrated that there are efficient ways to incorporate bone-specific physical activity into the school day. In any case, it is clear that weight-bearing physical activity can augment bone mineral density, which is particularly important for children and adolescents since growing bone has more capacity to adapt to mechanical loading than does mature bone.

Maximizing bone density early in life can help reduce the risk of osteoporosis in older women. Osteoporosis is characterized by decreased bone mineral content, which causes increased bone porosity (see figure 19.9). Osteopenia, as we noted in chapter 18, refers to a loss of bone mass that occurs with aging. Osteoporosis is a more severe loss of bone mass with deterioration of the microarchitecture of bone, leading to skeletal fragility and increased risk of bone fracture. These changes typically begin in the early 30s. The occurrence rate for fractures associated with osteoporosis increases by two to five times after the onset of menopause. Men also experience osteoporosis but to a lesser degree early in life because of a slower rate of bone mineral loss. Much remains to be learned about the etiology of osteoporosis; however, three major contributing factors common to postmenopausal women are estrogen deficiency, inadequate calcium intake, and inadequate physical activity.

Although the first of these is a direct result of menopause, the last two reflect dietary and exercise patterns throughout life. Evidence suggests that in addition to the positive relationship between exercise and bone growth in children described previously, routine physical activity is associated with the maintenance of bone mass in middle-aged and older adults. For example, physically active postmenopausal women typically

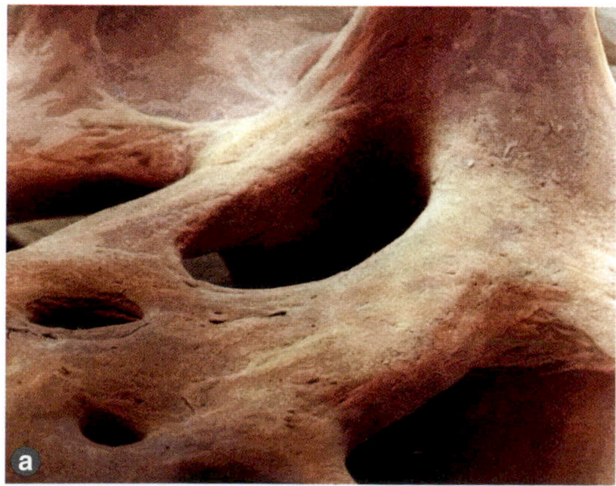

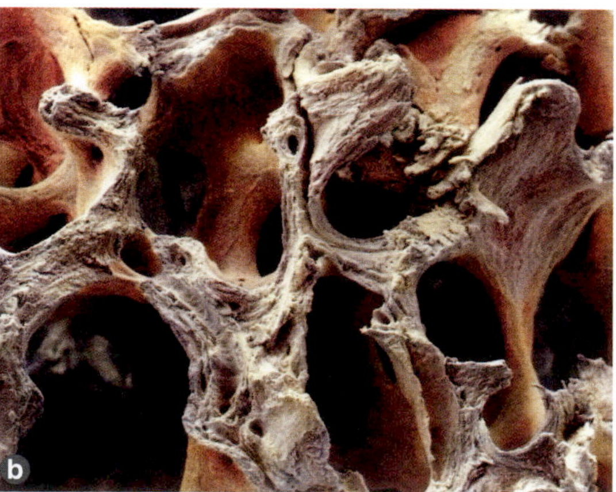

FIGURE 19.9 *(a)* Healthy bone and *(b)* bone showing increased porosity (decreased density, weakened and leached away) resulting from osteoporosis.

exhibit greater bone mineral density than sedentary age-matched women. Therefore, exercise in this age cohort protects against the normal age-related reductions in bone mineral density and thus is protective against osteoporosis and osteopenia across the life span.

In addition to postmenopausal women, women with amenorrhea and those with anorexia nervosa also suffer from low bone mass and osteoporosis attributable to insufficient calcium intake, low serum estrogen levels, or possibly both. In studies of women with anorexia, investigators found that their bone density was reduced significantly compared with that of controls. Amenorrheic athletes also demonstrate lower bone density than normally menstruating control athletes, suggesting that physical activity does not protect amenorrheic athletes from significant bone density losses. As shown in figure 19.10, the bone mineral content of normally menstruating runners tends to be higher than that of normally menstruating nonrunning controls, and female runners who are amenorrheic have higher bone mineral contents than untrained women who are amenorrheic. When we compare women of like menstrual status, those who are exercising have the higher bone mineral content.

Therefore, while evidence certainly suggests that increased physical activity and adequate calcium intake combined with adequate caloric intake are a sensible approach to preserving the integrity of bone at any age, an important point is that maintaining normal menstrual function is critical for those who have not reached menopause.

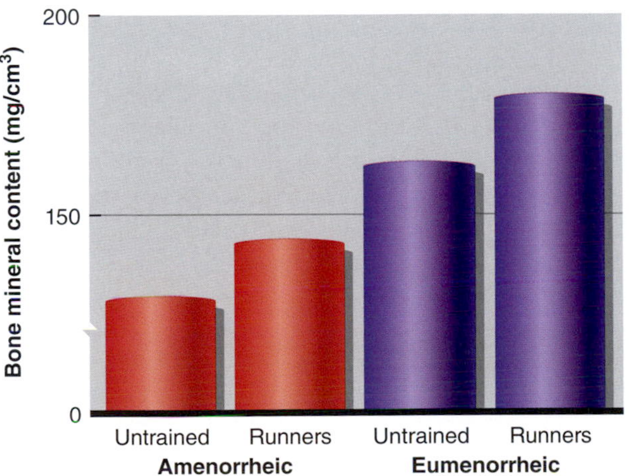

FIGURE 19.10 Bone mineral content of female runners and of untrained women who are amenorrheic and eumenorrheic. Note that when women with the same menstrual status are compared, the runners have higher bone mineral content than the untrained women.

Unpublished data from Dr. Barbara Drinkwater.

Eating Disorders

Eating disorders are a group of disorders that must meet specific criteria established by the American Psychiatric Association. The two most commonly diagnosed eating disorders are anorexia nervosa and bulimia nervosa. **Disordered eating**, on the other hand, refers to patterns of eating that are not considered normal but don't meet the specific diagnostic criteria for a given eating disorder.

Eating disorders in girls and women became the focus of considerable attention beginning in the 1980s. Men constitute about 10% or less of the reported cases. Anorexia nervosa has been considered a clinical syndrome since the late 19th century, but bulimia nervosa was first described in 1976.

Anorexia nervosa is a disorder characterized by

- refusal to maintain more than the minimal normal weight based on age and height,
- distorted body image,
- intense fear of fatness or gaining weight, and
- amenorrhea.

Girls and women from ages 12 to 21 are at greatest risk for this disorder. Its prevalence in this group is likely less than 1%.

Bulimia nervosa, originally termed bulimarexia, is characterized by

- recurrent episodes of binge eating;
- a feeling of lack of control during these binges; and
- purging behavior, which can include self-induced vomiting, laxative use, and diuretic use.

The prevalence of bulimia in the population at greatest risk, again adolescent girls and young women, is generally considered to be about 4% and possibly closer to 1%.

It is important to realize that a person might exhibit disordered eating and yet not meet the strict diagnostic criteria for either anorexia or bulimia. As an example, the diagnosis of bulimia requires that the individual average a minimum of two binge-eating and purging episodes a week for at least 3 months. What about the person who meets all the criteria except that bingeing and purging occur only once per week? Although this person cannot technically be diagnosed as having bulimia, her or his eating is certainly disordered and is a potential cause for concern. Thus, the term *disordered eating* has been used to refer to the behavior of those who do not meet the strict criteria for an eating disorder but who do have abnormal eating patterns.

The prevalence of eating disorders in athletes is controversial. Numerous studies have used either self-reports or at least one of two inventories developed to diagnose disordered eating: the Eating Disorders Inventory (EDI) and the Eating Attitudes Test (EAT). Results have varied because not all studies used the strict standard diagnostic criteria for either anorexia or bulimia. As in the general population, female athletes are typically at a much higher risk than male athletes, and certain sports carry higher risks than others. The high-risk sports can generally be grouped into three categories:

1. Appearance sports, such as diving, figure skating, gymnastics, bodybuilding, and ballet
2. Endurance sports, such as distance running and swimming
3. Weight-classification sports, such as horse racing (jockeys), boxing, and wrestling

Self-reports or inventories do not always provide accurate results. Secrecy is part of the behavior pattern associated with eating disorders, and people with these disorders often will not admit their problem even when anonymity is ensured. For the athlete, this need for secrecy might be heightened by fear that a coach or a parent will learn of the eating disorder and not allow the athlete to compete.

Even though research is limited, it seems appropriate to conclude that athletes are at a higher risk for eating disorders than the general population. Existing evidence likely does not reflect the seriousness of this problem in athletic populations. Although research data are not yet available, the prevalence might be as high as 60% or more in the specific high-risk athletic populations listed earlier.

The National Collegiate Athletic Association developed a list of warning signs for anorexia nervosa and bulimia nervosa (table 19.3). When an eating disorder is suspected, it is important to recognize the seriousness of the disorder and refer the athlete to a person specifically trained in dealing with eating disorders.

Female athletes are at higher risk than nonathletes for disordered eating and eating disorders for several reasons. Perhaps most important, there is tremendous pressure on athletes, particularly female athletes, to get their weight down to very low levels, often below what is appropriate. This weight limit can be imposed by the coach, trainer, or parent or can be self-imposed by the athlete. In addition, the personality of the typical elite female athlete closely matches the profile of someone at high risk for an eating disorder (competitive, perfectionistic, and under the tight control of a parent or other significant figure such as a coach). Furthermore, the nature of the sport or activity largely dictates those at high risk. As previously mentioned, athletes in three categories are at high risk: appearance sports, endurance sports, and weight-classification sports. Added to these risks are the normal pressures imposed by the media and culture on young women, whether they are athletes or not.

Female Athlete Triad

In the early 1990s, it became apparent that there is a reasonably strong connection of three interrelated clinical disorders observed in physically active women and female athletes; the interrelatedness has been referred to as the **female athlete triad**:

- Relative energy deficiency
- Secondary amenorrhea
- Low bone mass

It should be noted that disordered eating is not a necessary component of the female athlete triad. Rather, the common element is low energy availability that may or may not be a product of disordered eating. Either way, relative energy deficiency occurs as the athlete fails to consume an adequate volume of calories for her exercise energy expenditure. Over a period of time (the length of which has not been well established and might vary considerably from one athlete to another), an athlete who has low energy availability may start to

TABLE 19.3 **Warning Signs for Anorexia Nervosa and Bulimia Nervosa**

Anorexia nervosa	Bulimia nervosa
Dramatic loss in weight	A noticeable weight loss or gain
A preoccupation with food, calories, and weight	Excessive concern about weight
Wearing baggy or layered clothing	Bathroom visits after meals
Relentless, excessive exercise	Depressed mood
Mood swings	Strict dieting followed by eating binges
Avoiding food-related social activities	Increased criticism of one's body

Note. The presence of one or two of these signs does not necessarily indicate an eating disorder. Diagnosis should be made by appropriate health professionals.

Adapted from a poster distributed by the National Collegiate Athletic Association (1990).

experience abnormal menstrual function, which eventually may lead to secondary amenorrhea. Over time, this may lead to low bone mass. All of these conditions have been described previously within the chapter, but the female athlete triad refers specifically to the interrelatedness of the three disorders. The most recent revision of the American College of Sports Medicine position stand on the female athlete triad,[2] published in 2007, emphasizes that the three triad disorders can occur alone or in combination and should be addressed early before serious consequences develop.

VIDEO 19.2 Presents Nancy Williams on exercise and reproductive health and the causes of the female athlete triad.

Exercise and Menopause

Menopause is the permanent cessation of menses and is defined as occurring 12 months after the final menstrual period; it is typically reached between the ages of 45 and 55. The menopausal transition in women is driven by the declining and ultimately the loss of production of female sex hormones, leading to cessation of menses. Of much concern is the number of menopausal symptoms experienced by women over this transition, which can be severe and disruptive to the quality of life. It has been estimated that up to 75% of women exhibit symptoms such as hot flashes, vaginal bleeding, vaginal and urinary symptoms, and mood changes. While hormone therapy (i.e., estrogen alone or estrogen + progesterone) has typically been prescribed to women to reduce menopausal symptoms, data from major clinical trials including the Women's Health Initiative indicate that hormone therapy may be tied to an increased risk of heart disease, breast cancer, stroke, and blood clots. Therefore, women with risk factors for cancer and cardiovascular disease may not wish to exacerbate risk with hormone therapy.

Exercise has been investigated as a potential alternative treatment option for alleviating menopausal symptoms, including psychological, vasomotor, somatic, and sexual symptoms. Unfortunately, there are few rigorous, well-controlled randomized clinical trials of exercise for menopausal symptoms, and cross-sectional and self-report studies have multiple methodological flaws. However, to date, it appears that exercise is effective for improving mood, insomnia, and depression in menopausal women. Its effect on vasomotor symptoms, particularly hot flashes, has not been established in large-scale studies, although in general exercise is recommended during menopause for its impacts on quality of life, body composition, and disease risk.

In Review

- Postmenopausal women, amenorrheic women, and those who have anorexia nervosa are at greater risk of osteoporosis. Physical activity and adequate calcium and caloric intake are important to the preservation of bone at any age.
- Eating disorders, such as anorexia nervosa and bulimia nervosa, are much more common in women than in men and are especially common among athletes in appearance sports, endurance sports, and weight-classification sports. Athletes seem to be at a higher risk for eating disorders than the general population.
- The female athlete triad comprises interrelated disorders—nutritional inadequacy, secondary amenorrhea, and loss in bone mineral density—and can be associated with negative health outcomes leading to injury and inability to compete.
- The effect of regular exercise on menopausal symptoms such as hot flashes has not been established in large-scale studies; however, exercise is recommended during menopause because of its impact on quality of life, body composition, and disease risk.

IN CLOSING

In this chapter, we discussed sex-specific differences in performance. Most true differences between the sexes result from women's smaller body size, lower FFM, and greater relative and absolute body fat. Making valid comparisons in sport performances has been difficult because an event's popularity and other factors—such as opportunities to participate, coaching, facilities, and training techniques—have differed considerably between the sexes over the years. Despite these limitations, we have found that female and male athletes are not as different as many people believe.

With this chapter, we conclude our examination of age and sex considerations in sport and exercise. In the next part of the book, we turn our attention from athletics to a different application of exercise physiology: the use of physical activity for health and fitness. We begin with an examination of exercise prescription.

KEY TERMS

amenorrhea
anorexia nervosa
bulimia nervosa
disordered eating
eating disorders
estrogen
eumenorrhea
female athlete triad
lipoprotein lipase
menarche
menopause
menses
menstrual cycle
menstrual dysfunction
oligomenorrhea
pregnancy
primary amenorrhea
secondary amenorrhea
sex-specific differences

STUDY QUESTIONS

1. Most physiological distinctions between men and women should be referred to as "sex differences" rather than "gender differences." Explain.
2. How does the body composition of girls and women compare with that of boys and men? How do male and female athletes differ from male and female nonathletes?
3. What are the roles of testosterone and estrogen in the development of strength, fat-free mass, and fat mass?
4. How does women's upper body strength compare with men's? Lower body strength? Fat-free mass? Can women gain strength with resistance training?
5. What differences in $\dot{V}O_{2max}$ exist between average women and men? Between highly trained women and men? What can explain these differences?
6. What cardiovascular differences exist between women and men with respect to submaximal exercise? Maximal exercise?
7. How does the menstrual cycle influence athletic performance?
8. What is the primary reason that some female athletes in exercise training stop menstruating for intervals of several months to several years or more?
9. What risks are associated with training during pregnancy? How can these be avoided?
10. What are the effects of amenorrhea on bone mineral? How does exercise training affect bone mineral?
11. What are the two major eating disorders, and what is the level of risk for elite female athletes having these eating disorders? How does this vary by sport?
12. What is the female athlete triad? What factors are involved, and how does the triad develop?

STUDY GUIDE ACTIVITIES

In addition to the activities listed in the chapter opening outline, two other activities are available in the web study guide, located at

www.HumanKinetics.com/PhysiologyOfSportAndExercise

The **KEY TERMS** activity reviews important terms, and the end-of-chapter **QUIZ** tests your understanding of the material covered in the chapter.

PART VII

Physical Activity for Health and Fitness

In previous parts of this book, we focused on the physiological bases of physical activity and sport performance, describing the physiological responses to an acute bout of exercise, the adaptations to chronic training, and the means of improving performance in sport. In part VII, we shift our focus from athletic performance to a special area of exercise physiology: the role of physical activity in improving and maintaining overall physical fitness and health. In chapter 20, Prescription of Exercise for Health and Fitness, we discuss how to design an exercise program that can improve health-based physical fitness. We consider the essential components, ways to tailor the program to an individual's specific needs, and the unique role exercise plays in the rehabilitation of people with illness or disease. In chapter 21, Cardiovascular Disease and Physical Activity, we examine the major types of cardiovascular disease, their physiological bases, and how physical activity can help prevent or slow the progression of these diseases. Finally, in chapter 22, Obesity, Diabetes, and Physical Activity, we examine some of the causes of obesity and diabetes, the health risks associated with each, and the ways in which physical activity can be used to control both disorders.

Prescription of Exercise for Health and Fitness

20

In this chapter and in the web study guide

Statistics about the average adult's health and physical activity patterns do not paint a bright picture. The President's Council on Physical Fitness[15] lists the following:

- Only one in three children are physically active every day.
- Less than 5% of adults participate in 30 min of physical activity each day.
- Only 1 in 3 adults receive the recommended amount of physical activity each week.
- 80.2 million people above the age of 6 are physically inactive.
- Only 35% to 44% of adults 75 years or older and 28% to 34% of adults ages 65 to 74 are physically active.
- More than 80% of adolescents and adults do not meet the guidelines for both aerobic and muscle-strengthening activities.
- Children now spend more than 7.5 h/day in front of a TV, video game, or computer screen.
- Nearly one-third of high school students play video or computer games for ≥3 h on an average school day.

Although the human body is designed for movement, modern lifestyles have channeled the average American into an increasingly sedentary existence. Decades of research have clearly and unequivocally determined that an active lifestyle is important for optimal health. From a health perspective, we have not adapted well to this inactive lifestyle.

Health Benefits of Regular Physical Activity and Exercise

It seems rather ironic that it took research from clinicians and scientists to reach the conclusion that Galen identified in the first century that regular exercise was as important as breathing fresh air and eating the proper foods in promoting heath (see the Introduction chapter)!

The first acknowledgment from the modern medical profession came in 1992, when the American Heart Association proclaimed physical inactivity a major risk factor for coronary artery disease (CAD), placing it alongside smoking, abnormal blood lipids, and hypertension. Subsequently, the Centers for Disease Control and Prevention (CDC), in collaboration with the American College of Sports Medicine (ACSM), stated the importance of physical activity as a public health initiative and then in 1995 published a consensus statement by a panel of the National Institutes of Health (National Heart, Lung, and Blood Institute) advocating physical activity as important for cardiovascular health. Finally, in 1996, the Surgeon General of the United States released a written report on the health benefits of physical activity. This was a landmark report recognizing the importance of physical activity in reducing the risk for chronic degenerative diseases.

Much of the research supporting the benefits of physical activity in reducing the risk of developing chronic degenerative disease has come from the field of epidemiology, in which large populations are studied and associations between activity levels and disease risk determined. In the year 2000, exercise physiologists joined the war on what they termed the "sedentary death syndrome" by forming an action group advocating governmental support for research into the diseases and disorders associated with a sedentary lifestyle.[3]

Whether things have gotten better since that time is up for debate. Fewer adults now smoke cigarettes and fewer adolescents use alcohol or illicit drugs. And more adults are meeting physical activity guidelines, although we have a long way to go. Regular physical activity includes participation in moderate and vigorous physical activities and muscle-strengthening activities. Only one in three adults receives the recommended amount of physical activity each week, and more than 80 million Americans above the age of 6 are physically inactive.[15]

The health benefits associated with regular physical activity are undeniable, and more and more research is documenting these benefits.[1] There are well-documented effects of exercise on premature death, as well as cardiovascular disease, hypertension, and stroke. Other diseases that are minimized by regular physical activity include type 2 diabetes, osteoporosis, obesity and metabolic syndrome, many types of cancers, and depression. Further, physical and cognitive function are both improved by a program of regular exercise. These inverse associations between physical activity and disease and disability follow dose–response relations, i.e., more physical activity is associated with large impacts on the adverse health condition.

Physical Activity Recommendations

Despite these somewhat discouraging statistics, most Americans are aware that exercise is an integral part of preventive medicine. Yet people often equate exercise with jogging 8 km (5 mi) a day or lifting weights until their muscles can do no more. Many believe that high volume and intensity of exercise training are necessary

RESEARCH PERSPECTIVE 20.1

Sitting, Physical Activity, and Mortality

The association of sedentary behavior (i.e., sitting) with chronic disease and mortality outcomes is now apparent and has significant clinical implications. Using objectively collected accelerometry data, the U.S. National Health and Nutrition Examination Survey (NHANES) estimates that children and adults spend nearly 8 h per day being sedentary. A common question is whether the association between sedentary behavior and chronic disease is modulated by physical activity. A recent meta-analysis revealed that the deleterious impact of sitting on all-cause mortality was attenuated at increasingly higher levels of physical activity. In other words, participating in 60 to 75 min of moderate- to vigorous-intensity activity per day essentially eliminated the chronic disease risk attributed to sitting.

An important follow-up question then becomes whether the association between moderate- to vigorous-intensity physical activity and chronic disease and mortality is modulated by sedentary behavior. Researchers recently attempted to answer this question by re-analyzing the results of the meta-analysis mentioned previously.[8] Their results demonstrate a dose–response relation between moderate- to vigorous-intensity physical activity and mortality across all levels of sitting. In other words, even in individuals with little time spent sitting per day, more physical activity still reduces mortality risk. Thus, regardless of the amount of time spent sitting in one day, public health policy must focus on *both* reducing sedentary behavior and increasing moderate- to vigorous-intensity physical activity in order to maximize health benefits and reduce mortality risk.

This recommendation to simultaneously reduce sitting and increase physical activity is often referred to as the displacement hypothesis (i.e., displace activities with prolonged sitting with exercise or nonexercise physical activities). Implicit in this hypothesis is that replacing sitting time with physical activity will be associated with lower disease risk; however, this has not been experimentally evaluated. Another group of investigators[11] estimated the mortality benefits associated with the replacement of sitting time with different types of physical activity, including exercise and sports and nonexercise activity (e.g., household chores, daily walking). In this study, greater overall time spent sitting (≥12 versus <5 h/day) was associated with a graded increase in risk of all-cause mortality as well as greater risk for cardiovascular mortality. In less active adults, replacement of 1 h/day of sitting with an equal amount of either exercise or nonexercise activity was associated with lower mortality. Interestingly, for adults who engage in more overall activity, the majority of which is nonexercise, replacing sitting time with even more nonexercise activity does not confer additional benefits; instead, purposeful and planned exercise is required to further reduce risk for mortality in active adults. Taken together, the results of this study provide strong support for the public health recommendation to replace inactivity with more physical activity. This displacement is associated with greater longevity, particularly so for less-active adults.

to attain health-related benefits, yet this is not true. This myth was the primary focus of the CDC/ACSM report published in 1995, which concluded that significant health benefits can be obtained if one includes a moderate amount of physical activity, such as 30 min of brisk walking, 15 min of running, or 45 min of playing volleyball, on most, if not all, days of the week. The major emphasis of this report was that through a modest increase in daily activity, most people can improve their health and quality of life. In fact, a study of older adults (70-82 years) showed that just being more active greatly reduced the risk of mortality, independent of a formal exercise program.[10]

Additional health benefits can be gained through greater amounts of physical activity. Research suggests that people who can maintain a regular regimen of activity that is longer in duration or more vigorous in intensity are likely to derive greater benefit. It is now apparent that the appropriate exercise type and intensity vary depending on individual characteristics, current fitness level, and specific health concerns and individual goals.

ACSM's Guidelines for Exercise Testing and Prescription, Tenth Edition,[1] is a rich source of updated and compiled information concerning the health benefits of exercise. Based on combined efforts by the American College of Sports Medicine and the American Heart Association, the following recommendations have been in place since 2007:[7]

- All healthy adults aged 18 to 65 years should participate in moderate-intensity aerobic physical activity (PA) for a minimum of 30 min on 5 days per week or vigorous-intensity aerobic activity for a minimum of 20 min on 3 days per week.
- Combinations of moderate- and vigorous-intensity activity can be performed to meet this recommendation.

RESEARCH PERSPECTIVE 20.2

Revised Physical Activity Guidelines for Americans

In November 2018, the U.S. Department of Health and Human Services released a revision to the *Physical Activity Guidelines for Americans.* The original guidelines, published in 2008, highlighted the fact that being physically active is one of the most important actions that people of all ages can take to improve their health. Although the primary audiences for the guidelines are policy makers and health professionals, individuals can also benefit from the information. The revised edition, based on science from the intervening 10 years, reconfirmed the importance of daily physical activity to reduce the risk of many chronic diseases and reaffirmed the goal of 150 minutes of moderate-intensity aerobic activity per week. However, recognizing that many adults may not be able to achieve this goal, the guidelines now also state that any amount of daily activity—no matter how big or small—is beneficial to health. In that regard, the new guidelines eliminate the caveat that, in order to be beneficial, physical activity has to in bouts of at least 10 minutes. Other additions to the new guidelines include the following:

1. Information about the benefits of physical activity for older adults and people whose chronic conditions impede regular exercise
2. Guidance for children as young as 3 years of age
3. Contrasts between sedentary behavior and physical activity, and the risks associated with the former
4. New knowledge about the health benefits of physical activity for brain health, certain cancers, and fall-related injuries
5. Information about the benefits of exercise related to general well-being, daily functioning, and sleep
6. Science-based strategies to get the population as a whole more active

- Moderate-intensity aerobic activity can be accumulated to total the 30 min minimum by performing bouts each lasting ≥10 min.
- Every adult should perform activities that maintain or increase muscular strength and endurance for a minimum of 2 days per week.
- Because of the dose–response relationship between PA and health, individuals who wish to further improve their fitness, reduce risk for chronic diseases and disabilities, and prevent unhealthy weight gain may benefit by exceeding the minimum recommended amounts of exercise.

During the first decade of the 21st century, the ACSM in collaboration with the American Medical Association (AMA) launched a major initiative to encourage health care providers to counsel their patients on the importance of physical activity in promoting and maintaining health and preventing disease. With the recognition that many physicians and associated health care workers have not had extensive education and training in this area, a formal program was developed to educate and train them in the basics of exercise prescription. The website ExerciseIsMedicine.org offers the "Health Care Providers' Action Guide" to guide clinicians in including physical activity in their patients' overall health care plan.

VIDEO 20.1 Presents Bob Sallis with a physician's perspective on the prescription of exercise. 

Knowing the benefits of regular exercise, how should people begin exercise programs to improve their general health and fitness? The first step is deciding to take action. The next step is getting medical clearance.

Health Screening

Is a medical evaluation really necessary before one starts an exercise program? Although comprehensive medical screening is useful and desirable before exercise is prescribed, not all people need one. Many cannot afford the cost of such an evaluation, and the medical system is not prepared to provide this service for the total population, even if money were available. Also, medical evaluation before prescribing exercise for a population presumed healthy has not been proven to reduce the medical risks associated with exercise.

Most sports medicine organizations, including the ACSM, agree that a preparticipation health screening is encouraged for individuals who want to start or intensify an exercise program. However, medical evaluation is perceived as a significant barrier to starting an exercise program for many people. New guidelines clarify the level of screening suggested in the form of an algorithm, shown in table 20.1.

TABLE 20.1A **ACSM Preparticipation Screening Algorithm for Subjects Who Do Not Exercise Regularly**

	Medical clearance		Follow exercise prescription guidelines on progression*		
Health status	Recommended	Not necessary	Light-intensity exercise (30%-<40% HRR)	Moderate-intensity exercise (40%-<60% HRR)	Vigorous-intensity exercise (≥60% HRR)
No CV, metabolic, or renal disease and no signs or symptoms of these		x	Recommended	Recommended	May progress to this intensity*
Known CV, metabolic, or renal disease and asymptomatic	x		Recommended, after medical clearance	Recommended, after medical clearance	May progress to this intensity*
Any sign or symptom of CV, metabolic, or renal disease	x		Recommended, after medical clearance	Recommended, after medical clearance	May progress to this intensity*

CV = Cardiac, peripheral vascular, or cerebrovascular disease. Metabolic disease = type 1 and 2 diabetes mellitus. Regular exercise is defined as performing planned, structured physical activity for ≥30 min at a moderate or higher intensity for at least 3 days per wk over the past 3 mo or longer. *See chapter 11 for details on exercise prescription.

Reprinted by permission from M. Shipe, Health Risk Appraisal, in *Fitness Professional's Handbook*, 7th ed., edited by E.T. Howley and D.L. Thompson (Champaign, IL: Human Kinetics, 2017), 25; Adapted from Riebe et al., "Updating ACSM's Recommendations for Exercise Preparticipation Health Screening," *Medicine in Science in Sports and Exercise* 47 (2005): 2473-2479.

TABLE 20.1B **ACSM Preparticipation Screening Algorithm for Subjects Who Exercise Regularly**

	Medical clearance		Follow exercise prescription guidelines on progression*		
Health status	Recommended	Not necessary	Light-intensity exercise (30%-<40% HRR)	Moderate-intensity exercise (40%-<60% HRR)	Vigorous-intensity exercise (≥60% HRR)
No CV, metabolic, or renal disease and no signs or symptoms of these		x		May continue at this intensity	May continue or progress to this intensity*
Known CV, metabolic, or renal disease and asymptomatic		x		May continue at this intensity	
	x			May continue at this intensity	May continue this intensity after medical clearance
Any sign or symptom of CV, metabolic, or renal disease	x Stop exercising		May resume after medical clearance	May resume after medical clearance	May resume after medical clearance

CV = Cardiac, peripheral vascular, or cerebrovascular disease. Metabolic disease = type 1 and 2 diabetes mellitus. Regular exercise is defined as performing planned, structured physical activity for ≥30 min at a moderate or higher intensity for at least 3 days per wk over the past 3 mo or longer. *See chapter 11 for details on exercise prescription.

Reprinted by permission from M. Shipe, Health Risk Appraisal, in *Fitness Professional's Handbook*, 7th ed., edited by E.T. Howley and D.L. Thompson (Champaign, IL: Human Kinetics, 2017), 25; Adapted from Riebe et al., "Updating ACSM's Recommendations for Exercise Preparticipation Health Screening," *Medicine in Science in Sports and Exercise* 47 (2005): 2473-2479.

RESEARCH PERSPECTIVE 20.3

Exercise and the Brain

Over the past two centuries, life expectancy has nearly doubled, leading to an increasingly older population. Aging is the primary risk factor for chronic disease, including dementia and Alzheimer's disease, whose prevalence continues to steadily (and alarmingly) increase. Epidemiological studies suggest that the development of Alzheimer's disease may be related to modifiable risk factors, including physical activity and cardiovascular risk factors, in ~30% of patients. There is now convincing evidence linking habitual aerobic exercise to reductions in age-related cognitive decline. The precise mechanisms mediating exercise-induced improvements in brain structure and function are not clear. However, it appears likely that the cardiovascular adaptations that occur as a consequence of chronic physical activity may maintain brain health, reducing the risk of cognitive impairment and the development of dementia.

The cognitive benefits of exercise, including higher cognitive performance in memory and executive function, are greatest when regular exercise is started earlier in life and when aerobic exercise and strength training are combined.[17] Habitual aerobic exercise training also causes structural adaptations in the brain, reducing the deteriorations in gray and white matter functioning that typically occur with aging. In addition, exercise preserves brain volume, particularly in brain areas associated with cognitive impairment (e.g., parietal lobe and visual cortex).

In addition, the beneficial effects that habitual physical activity has on cardiovascular function also likely improve brain health. The brain lacks intracellular energy storage and thus relies entirely on the vascular supply of oxygen and nutrients; therefore, exercise-induced improvements in cardiovascular control of cerebral blood flow play a critical role in maintaining normal brain structure and function. The brain is particularly sensitive to changes in blood pressure because of the low resistance of the cerebral vasculature. In this regard, regular aerobic exercise attenuates the stiffening of the central elastic arteries (the aorta and carotid arteries). Reductions in central arterial stiffness are accompanied by reductions in systolic blood pressure, an established risk factor for stroke and cognitive impairment, and blunt the transmission of excessive pulsatile blood pressure waves into the brain. Exercise also improves endothelial function in large conduit arteries, cerebral autoregulation and vasomotor reactivity, and substrate exchange and waste clearance in the cerebral microcirculation, all of which ultimately improve brain health.

A 2014 epidemiological study also suggests that aerobic fitness reduces the risk of brain cancer mortality.[18] In this study, the risk of brain cancer mortality was prospectively examined in the National Walkers' and Runners' Health Study cohort (~153,000 individuals, ~50 years old, followed for ~12 years). In this sample, 100 deaths were attributed to brain cancer as the primary cause. Within these individuals, the risk for brain cancer mortality was significantly reduced with greater amounts of regular exercise, with the most benefit incurred with more activity (e.g., running 12 to 25 km/week).

Interestingly, the relation between exercise intensity and brain health may not be linear; instead, this link appears hormetic (i.e., a biphasic dose–response relation) (see figure). That is, emerging evidence suggests that strenuous endurance exercise may actually *cause* brain injury if not enough recovery is allowed. The mechanisms mediating the adverse effects of exercise on the brain remain unclear but conceivably could be related to the increased risk of systemic catabolic burden, inflammatory responses, and risks of cardiovascular injury or event. However, future studies are necessary in order to more precisely define the optimal dose of exercise training that may prevent or slow the age-related functional and structural deteriorations in the brain. Nevertheless, aerobic exercise is clearly an effective strategy to improve cardiovascular health, thereby slowing cognitive decline and the onset of dementia in older adults, improving the quality of life and extending the health span.

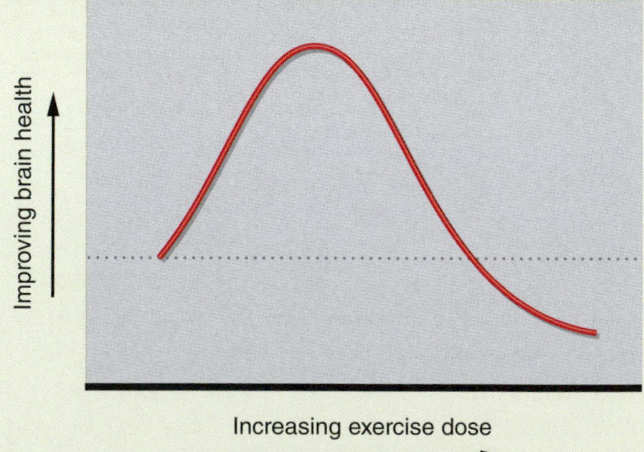

The potential hormetic relation between increases in exercise intensity and improvements in brain health. Aerobic exercise improves cognitive function and brain structure; however, strenuous exercise may have deleterious effects.

Health-Related Fitness Testing

Proper measurement of various aspects of physical fitness is common in preventive medicine, laboratory research, and rehabilitation and physical therapy programs.

While some of these tests are done for the purpose of diagnosing or documenting diseases such as coronary artery disease, the focus here is on the use of these tests to establish an individual's general health status and fitness level.

Body Composition

Increased adiposity is associated with a variety of diseases and disabilities. Central or abdominal adiposity has a particularly strong association. Height and weight measurements alone do not adequately capture this risk nor are they generally related to fitness level. Rather, determination of the components of the two-compartment model—fat mass and fat-free mass—can provide valuable information about health. Changes in these two components with exercise and diet are hallmarks of their effectiveness and health benefits. Techniques used for measurement of body composition are covered in detail in chapter 15.

Muscular Strength and Endurance

The definitions and principles of muscular strength and endurance are covered in chapter 9. These fitness components can be assessed in a number of valid ways in the laboratory or clinic. Muscular strength is defined as the maximal force that can be produced by a muscle or muscle group. Traditionally, the 1RM test has been the standard for measuring movement-specific strength, but 5RM and 10RM tests have also been proposed.[1] Grip strength is an important construct for activities of daily living and a good predictor of functional status and premature mortality in older individuals.

Muscular endurance, on the other hand, refers to the ability of a muscle or muscle group to either (1) sustain repeated contractions over a period of time or (2) maintain a given percentage of 1RM for a period of time. Perhaps the most common field test for muscular endurance is the push-up test, i.e., the maximal number of properly performed push-ups that can be done in a fixed period of time (usually 1 or 2 min). Sit-up or curl-up tests are not recommended because they only moderately relate to abdominal endurance and may cause lower back injury.[1]

Flexibility

The ability to move a joint through a complete range of motion is called flexibility, and tests of flexibility usually measure degrees of movement around that joint. While the flexibility of many joints can be measured, the most common flexibility test is the sit-and-reach test, or trunk flexion test. This test measures hamstring and lower back flexibility. Flexibility tests are unique in that women almost routinely score better than men of a similar age.

Cardiorespiratory Fitness

As mentioned throughout this book, the standard measure of cardiorespiratory fitness, or aerobic capacity, is maximal oxygen consumption, or $\dot{V}O_{2max}$. $\dot{V}O_{2max}$ can be estimated from field tests (such as the time it takes to complete a 12 min walk or run) or from a submaximal or maximal **graded exercise test (GXT)**. GXTs can be performed on a treadmill or a cycle ergometer or using a step test.

Submaximal tests typically involve measuring HR at several exercise intensities and extrapolating to HR_{max}. This process is illustrated in figure 8.2. Maximal GXTs push subjects to the point of volitional fatigue and often directly measure oxygen uptake at this maximal work load. GXTs are also used as clinical tools for the diagnosis or verification of coronary artery disease, as discussed next.

Clinical Graded Exercise Testing

Exercise testing alone can be used for determining readiness for exercise and activity or assessing the effectiveness of therapy as well as for other clinical decision-making situations, but it is rarely used in isolation for diagnostic purposes. When used for diagnostic purposes, a clinical GXT is usually used in conjunction with other procedures, such as echocardiography, pharmacological challenges using specific cardiovascular drugs or agents, or radionuclide scans. Diagnostic testing may also include echocardiography and radionuclide imaging to obtain anatomical and physiological images of the myocardium. These techniques improve the specificity and the sensitivity of the test. A 12-lead **exercise electrocardiogram (ECG)** (figures 20.1 and 20.2) usually accompanies the GXT. Exercise blood pressures and other measures of physiological and even psychological responses (rating of perceived exertion and pain scales, for example) may also be obtained during these assessment procedures.

The ECG and blood pressure are monitored as the person progresses from low-intensity exercise such as slow walking to more intense exercise and sometimes, but rarely, up to maximal-intensity exercise. Maximal-intensity exercise is, again, rarely used or seen with current methods of diagnostic testing. Exercise testing is, however, generally progressive; that is, the rate of work is progressively increased every 1 to 3 min until

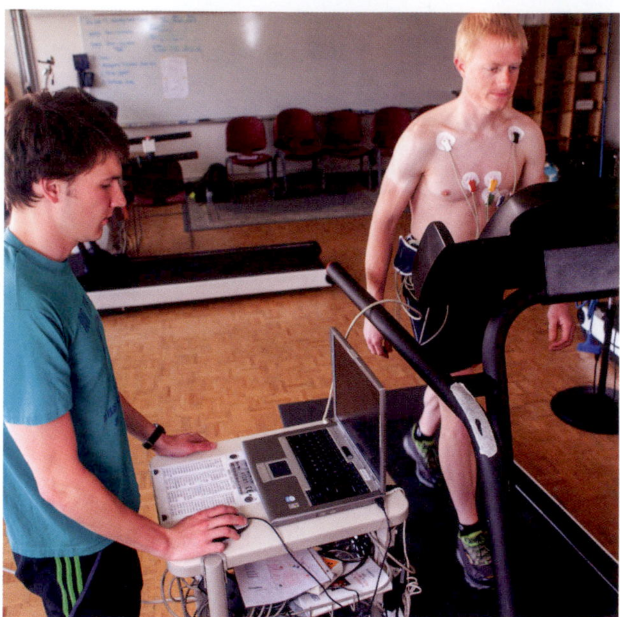

FIGURE 20.1 Obtaining an exercise electrocardiogram during a clinical graded exercise test.

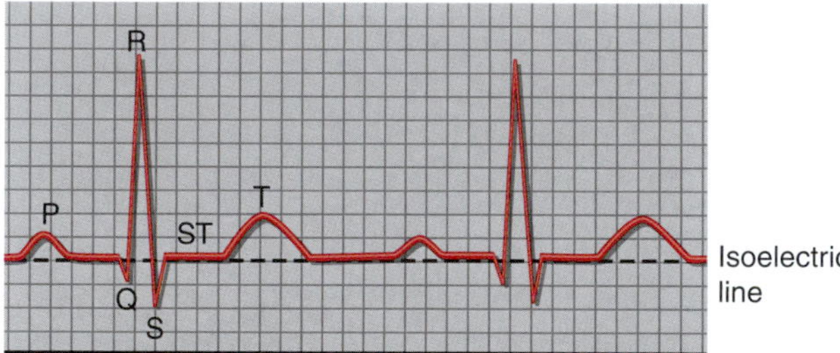

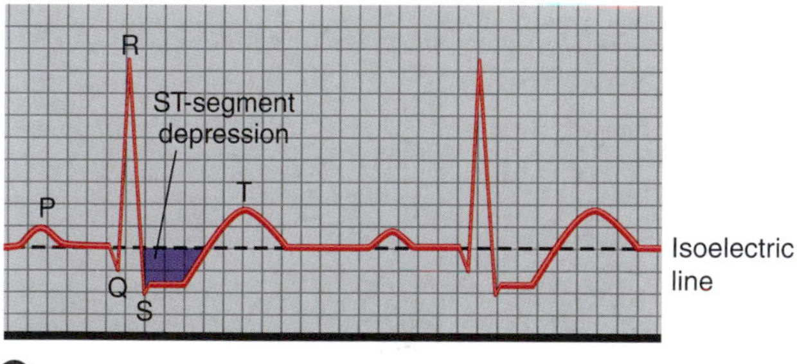

FIGURE 20.2 Illustration of *(a)* a normal electrocardiogram (ECG) and *(b)* an ECG with ST-segment depression, suggestive of the presence of coronary artery disease.

some usually predetermined heart rate (commonly 85% of estimated maximum) is achieved. The ECG is monitored to detect heart rhythm and electrical conductivity abnormalities. Blood pressure is monitored to determine whether there is a normal increase in systolic blood pressure and little or no change in diastolic blood pressure as the rate of work progresses from low intensity to maximal or near-maximal levels. It is also important for the clinician to interact with the subject, observing signs and symptoms during and immediately following the exercise test, such as chest pain or pressure (angina), unusual shortness of breath, light-headedness or dizziness, and inappropriate heart rate response.

Results from a diagnostic GXT, the ECG, the echocardiogram, the images, and the physiological measurements are all considered in assessing the test as either positive, negative, or equivocal. Positive or equivocal tests usually lead to subsequent diagnostic procedures such as coronary angiograms, other cardiovascular testing procedures, or medical management to determine the etiology and origin of the signs and symptoms that brought the person to the physician in the first place. Typically, a person with abnormal changes in a diagnostic stress test is referred for more direct clinical procedures.

To determine the accuracy of exercise ECG results, the sensitivity, specificity, and predictive value of an exercise test must be considered. **Sensitivity** refers to the exercise test's ability to correctly identify people who have the disease in question, such as CAD. **Specificity** refers to the test's ability to correctly identify people who do not have the disease. And the **predictive value of an abnormal exercise test** refers to the accuracy with which abnormal test results reflect presence of the disease.

Unfortunately, both the sensitivity and the predictive value of an abnormal exercise test for detecting CAD are relatively low in healthy people who have no symptoms of this disease. Past studies reveal that sensitivity averages about 66%, indicating that 66% of those with CAD are correctly identified by exercise ECGs as having the disease. Conversely, 34% of those with the disease are incorrectly diagnosed as disease-free based on exercise ECGs. Using exercise testing on a population that has a high probability of having CAD raises the sensitivity. Thus, it is now relatively rare to see a GXT performed for purely diagnostic purposes because GXTs are of limited value in screening young, apparently healthy individuals for coronary artery disease.

VIDEO 20.2 Presents Bob Sallis on whether a physician should be consulted before beginning moderate exercise. 

In Review

- Before beginning any exercise program, anyone who is considered to be at a high risk for CAD should be screened according to the ACSM screening algorithm.

- Health-related fitness tests include measures of body composition, muscular strength and endurance, flexibility, and cardiorespiratory fitness.
- The most recent ACSM guidelines should be followed for each phase of the evaluation, and the physician should be consulted about the proposed exercise activity in case there are any medical contraindications.
- Test sensitivity refers to the test's ability to correctly identify people with a given disease. Test specificity refers to its ability to correctly identify people who do not have the disease. The predictive value of an abnormal exercise test refers to the accuracy with which the test reflects presence of the disease in a given population.
- The most recent ACSM guidelines state that a medical examination and exercise test might not be necessary if moderate exercise is undertaken gradually, without competition, in people without symptoms of cardiovascular, pulmonary, and metabolic disease.

Exercise Prescription

The **exercise prescription** involves six basic factors:

1. Mode (type) of exercise
2. Frequency of participation
3. Intensity of the exercise bout
4. Duration (time) of each exercise bout
5. Volume of exercise
6. Progression

In the ACSM's newest guidelines,[1] the acronym FITT-VP is used to denote frequency, intensity, type, time, volume, and progression.

We outlined early in this chapter the recommended FITT-VP characteristics documented to promote health and minimize the risk of chronic disease. In this section, we assume that the goal of the exercise program is to improve cardiorespiratory endurance, including making small changes to sedentary behavior such as promoting breaks from sitting. Since the prescription of resistance training programs is discussed in detail in chapter 9, we will only briefly mention the inclusion of resistance training as a part of a total exercise program later in this chapter. The focus of this section is on aerobic training. Also, the information in this section is not appropriate for designing training programs for competitive endurance athletes, which is covered in chapter 9.

A minimum threshold for frequency, duration, and intensity of exercise must be reached before aerobic benefits are obtained. But, as we have discussed elsewhere, individual responses to any given training program are highly variable, so the threshold required differs from one person to the next. If we use exercise intensity as an example, the ACSM defines the lower end of moderate training intensity as 40% of one's heart rate reserve (HRR) or $\dot{V}O_{2max}$.[1] Although this recommendation is appropriate for most healthy adults, some low-fit individuals might improve their aerobic capacities at intensities below 40% of their $\dot{V}O_{2max}$. Each individual's threshold for frequency, duration, and intensity must be exceeded to achieve gains in cardiorespiratory endurance, and this threshold is likely to increase as cardiorespiratory endurance improves.

Exercise Mode

The prescribed exercise program should focus on one or more **modes**, or types, of cardiorespiratory endurance activities. Traditionally, the activities prescribed most frequently are

- walking,
- jogging,
- running,
- hiking,
- cycling,
- rowing, and
- swimming.

Because these activities do not appeal to everyone, alternative activities have been identified that promote similar cardiorespiratory endurance gains. Spinning, cardiovascular ("cardio") endurance group exercise, and combination resistance training–cardiorespiratory endurance training programs are excellent substitutes for the modes of cardiorespiratory endurance exercise listed. The benefits of HIIT training have been discussed throughout the book.

Individuals should select activities that they enjoy and are willing to continue throughout life. Exercise must be regarded as a lifetime pursuit because, as we saw in chapter 14, the benefits are soon lost if participation stops. Motivation is probably the most important factor in a successful exercise program. Selecting an activity that is fun, provides a challenge, and can produce needed benefits is one of the most crucial tasks in exercise prescription. Having several different activities to pursue is also wise in case of inclement weather, travel, or other barriers. Other considerations include geographic location, climate, and availability of equipment and facilities. Home exercise is useful when people are homebound either because of responsibilities, such as child rearing, or because of weather considerations such as heat, humidity, cold, rain, ice, and snow.

RESEARCH PERSPECTIVE 20.4

Golf Is (Probably) Good for Your Health

The potential of golf to provide an opportunity for physical activity, and the associated health benefits, is perhaps readily apparent. This may be particularly true for middle-aged and older adults, who are generally less active than younger adults. However, limited studies have comprehensively examined the relation between golf and health. Thus, a group of investigators recently undertook a scoping review to map the available evidence and identify the existing gaps in knowledge regarding the impact of golf on health.[12] Scoping reviews summarize a range of evidence in order to convey the breadth and depth of a field. In contrast to a systematic review, authors generally do not asses the quality of the included studies. Overall, this scoping review identified 301 eligible studies that assessed the effects of golf on physical and mental health. Although the scoping review cannot determine causative effects, the reports included provide consistent evidence for a positive association between golf and physical and mental health. Priority areas for future research include the effects of golf on mental well-being, the relation of golf to muscle strength and function, balance and fall prevention, and health behaviors among golfers.

Exercise Frequency

Research studies on exercise frequency show that 3 days per week of vigorous activity or 5 days per week of moderate activity are optimal. This does not mean that 6 or 7 days per week won't give additional benefits, but simply that for the health-related benefits, the optimal gain is achieved with a time investment of 3 to 5 days per week. Obviously, additional days above the 3- or 4-day frequency are beneficial for weight loss, but this frequency should not be encouraged until the exercise habit is firmly established and the injury risk is reduced.

Exercise Intensity

The intensity of the exercise bout appears to be the most important factor for developing cardiorespiratory endurance. How hard must people push themselves to gain benefits? Former athletes immediately recall the exhaustive workouts they endured to condition themselves for their sport. Unfortunately, this concept also gets carried over into the exercise programs they pursue for health benefits. Evidence now suggests that a modest training effect can be seen in some people through training at intensities of 40% of their $\dot{V}O_{2max}$. Generally, these intensities are most effective for people at lower levels of aerobic fitness. For most, however, the appropriate intensity appears to be between 40% and 90% of HRR. An upper level for intensity depends on the purposes and goals of the training program.

The ACSM defines a moderate training intensity of 40% to 59% of one's heart rate reserve (HRR) or $\dot{V}O_{2max}$ and vigorous exercise as 60% to 90% of HRR or $\dot{V}O_{2max}$.[1]

Multiple studies have clearly demonstrated that high-intensity interval training (HIIT) of low volume and very high intensity can markedly increase cardiorespiratory endurance. HIIT training has metabolic effects similar to longer, lower-intensity training. Substantial increases in muscle oxidative capacity, cardiometabolic variables associated with the protective effects of exercise, and cardiorespiratory endurance performance have been obtained in a training period as short as 10 min per day for 2 weeks.[6]

Exercise Duration

Duration and intensity of exercise are closely linked. Several studies have demonstrated improvement in cardiovascular conditioning with endurance exercise periods of 20 to 30 min per day at an appropriate intensity. Again, "optimal" is used here to reflect the greatest return for time invested, and the specified time refers to the time during which one is at one's appropriate exercise intensity. Similar improvements in cardiorespiratory endurance can be gained with either a short-duration, high-intensity program or a long-duration, low-intensity program if the minimal threshold for both duration and intensity is exceeded. Similar benefits are also gained whether the daily endurance training session is conducted in multiple shorter bouts (e.g., three 10 min bouts) or a single long one (e.g., one 30 min bout). Longer bouts are most often prescribed to facilitate weight loss or maintenance of weight loss.

Exercise Volume

It is appropriate to specify the weekly volume of exercise as MET-minutes (METs × total weekly minutes) or calories expended per week. MET stands for metabolic equivalent, a system for quantifying the intensity of activities that is discussed in more detail later in this chapter. Using MET-minutes per week as a single measure that combines intensity, frequency, and volume allows the exercise prescription to be progressed in a more deliberate, quantifiable manner (and more cautiously when appropriate). The ACSM recommends ≥500 to 1,000 MET-minutes/week or 150 min/week of moderate-intensity exercise. Caloric expenditures of

1,000 to 1,500 per week are close to those volume levels, depending on weight and the intensity of the exercise.

Progression

The rate at which an individual progresses through an exercise program is variable and must be individualized. Recall from chapter 9 that overload is a general principle of exercise training. Progression may involve changing one, or several, of the individual FITT-VP components.

Breaks in Prolonged Sitting

Prolonged sitting is a fact of life in our everyday world. Watching TV, sitting at work in front of computer screens all day, driving and commuting, eating, even having coffee with friends all involve sitting. Sedentary activities make up almost 60% of the day for most Americans and, if sleeping is included, may exceed 75% of the day.

Important research on sedentary behavior (defined as physical activity <1.5 METs) has described a sedentary physiology that promotes chronic disease. Pathophysiological conditions that are common to cardiovascular disease, obesity, diabetes, and other chronic diseases, in fact, are similar to those induced by being sedentary. As few as 30 min of continuous sitting increases insulin resistance, negatively affects lipid and carbohydrate metabolism, shifts ongoing metabolism more toward glucose, promotes storage of fat and changes in muscle fiber type, and, through those mechanisms, promotes a systemic, subacute inflammation.[2] All of these changes are consistent with and similar to those found in the chronic diseases mentioned.

Since it is clear that prolonged and continuous sitting causes these changes, the next question is how to moderate or prevent them. Does regular exercise do that, or do breaks in sitting do that? Here's what is known: Regular exercise, while clearly preventive with regard to chronic disease, does not change those effects of sedentary behavior. Breaks in sitting or sedentary behavior do, in fact, change that physiology. As little as 2 min of low-intensity (2.0-2.5 METs) activity can positively affect such physiological variables as insulin resistance and glucose metabolism.[14] The evidence is compelling enough that people with sedentary lifestyles should consider simply getting up and moving around periodically to break up the duration of sitting or other sedentary behavior.

In Review

- Six basic elements of an exercise program are exercise mode, frequency, intensity, duration, volume, and progression. A minimum threshold for the frequency, intensity, and duration must be met to yield significant cardiorespiratory endurance benefits, and this threshold is quite variable from one individual to another.
- The program should include one or more cardiorespiratory endurance activities. If the activity involves competition, preconditioning with a standard endurance activity is recommended before sport participation begins in order to bring the person up to an appropriate level of fitness.
- Activities must be matched with individual needs and likes so that motivation can be maintained.
- Optimally, exercise should be performed daily or on most days of the week. Exercise should begin with three or four sessions per week and then progress to more if desired.
- Exercise duration of 30 to 60 min working at moderate intensity or 20 to 60 min at vigorous intensity is recommended, but the key is reaching the threshold for combined duration and intensity, which can be expressed as MET-minutes.
- For most people, intensity should be at least 40% to 59% HRR (moderate) but can range from 40% to 90%. However, health benefits might occur at intensities lower than those needed for cardiorespiratory endurance conditioning in deconditioned people and can also occur at very high intensities.
- Research suggests that prolonged sitting has negative impacts on a person's health, even if the person does regular exercise. Frequent breaks from sitting are recommended as part of an exercise prescription for health.

Monitoring Exercise Intensity

Exercise intensity can be quantified on the basis of the training heart rate (THR), the metabolic equivalent (MET), or the rating of perceived exertion (RPE). Let's examine each of these and their strengths and weaknesses in quantifying exercise intensity.

Training Heart Rate

The concept of **training heart rate (THR)** is based on the linear relation between heart rate and $\dot{V}O_2$ with increasing rates of work, as shown in figure 20.3. When people are exercise tested, their heart rate and $\dot{V}O_2$ values are obtained each minute and plotted against each other. The THR is established through use of the heart rate that is equivalent to a set percentage of the $\dot{V}O_{2max}$. For example, if a training level of 75% $\dot{V}O_{2max}$ is desired, 75% of $\dot{V}O_{2max}$ is calculated ($\dot{V}O_{2max} \times 0.75$), and the heart rate corresponding to this $\dot{V}O_2$ then is selected as the THR. An important point is that the exercise intensity necessary to achieve a given percentage of $\dot{V}O_{2max}$ does not necessarily result in a matching percentage of maximal heart rate (HR_{max}). As an example, a THR that is set at 75% of the $\dot{V}O_{2max}$ represents an intensity of approximately 65% to 90% of the HR_{max} (see figure 20.3).

One can also establish the THR by using the heart rate reserve (HRR), also known as the Karvonen method. **Heart rate reserve** is defined as the difference between HR_{max} and the resting heart rate (HR_{rest}):

$$\text{Maximal HRR} = HR_{max} - HR_{rest}$$

With this method, we calculate the THR by taking a given percentage of the maximal HRR and adding it to the resting heart rate. Let's consider an example. For 75% of maximal HRR, the equation would be as follows:

$$THR_{75\%} = HR_{rest} + 0.75\ (HR_{max} - HR_{rest}).$$

Heart rate reserve adjusts the THR so that the THR as a specific percentage of the maximal HRR is nearly identical to the heart rate equivalent of that same percentage of $\dot{V}O_{2max}$ at moderate to high intensities. Thus, a THR computed as 75% of maximal HRR is approximately the same as the heart rate corresponding to 75% of the $\dot{V}O_{2max}$. However, there may be a substantial difference between the two at low intensities.[16]

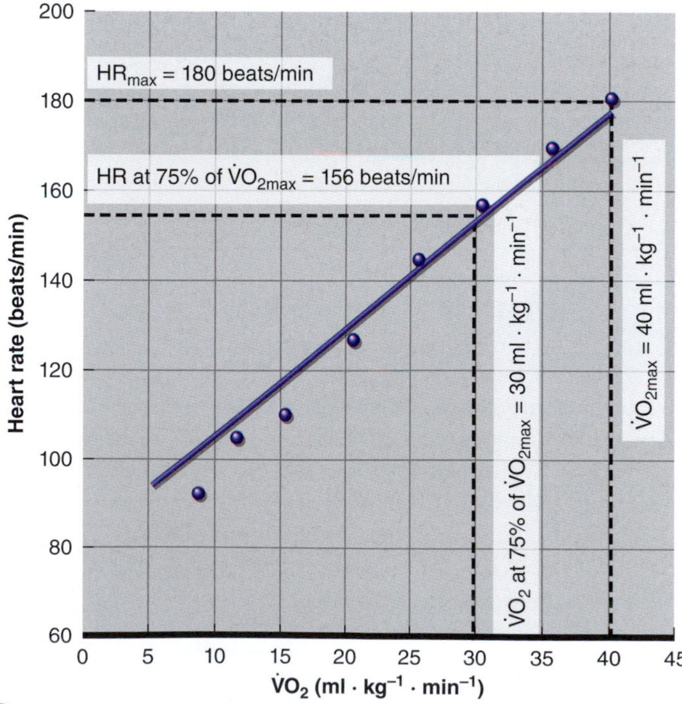

FIGURE 20.3 The linear relation between heart rate and oxygen consumption ($\dot{V}O_2$) with increasing rates of work and the heart rate equivalent to a set percentage (75%) of $\dot{V}O_{2max}$.

Training Heart Rate Range

An appropriate exercise intensity can be established by setting a heart rate range rather than a single THR value. This is a more sensible approach because exercising at a set percentage of $\dot{V}O_{2max}$ can place people above their lactate threshold, making it difficult for them to train for an extended period of time. With the THR range concept, low and high values are established that can ensure a training response. Individuals may choose to exercise at the low end of the HRR range or to progress through the range as comfort dictates. To illustrate this, using the HRR method for establishing the THR, consider the following example. A 40-year-old man has a resting heart rate of 75 beats/min and a maximum heart rate of 180 beats/min, and he is advised to exercise within a THR range of 50% to 75% HRR. His THR range would be as follows:

$$THR_{50\%} = 75 + 0.50\ (180 - 75) = 75 + 53 = 128 \text{ beats/min}$$

$$THR_{75\%} = 75 + 0.75\ (180 - 75) = 75 + 79 = 154 \text{ beats/min}$$

This same THR range method can be used if one estimates HR_{max} [208 − (0.7 × age)] without losing much accuracy if true HR_{max} has not been determined.

Heart rate is the preferred method for monitoring exercise intensity because it is highly correlated to the work of the heart (or stress on the heart) and allows for a progressive increase in the rate of training with improvements in fitness to maintain the same THR. The use of the HRR (Karvonen) method is the preferred method for establishing the THR range. When one is prescribing exercise intensity, it is appropriate to establish a THR range, with exercise starting at the low end of the range and progressing to the upper end of the range over time.

The $\dot{V}O_{2max}$ Reserve Method

The ACSM also advocates a slightly different approach to prescribing exercise intensity. Exercise intensity is prescribed based on what has been termed the $\dot{V}O_2$ reserve method ($\dot{V}O_2R$). Instead of prescribing exercise at a given percentage of $\dot{V}O_{2max}$, one bases the prescription on a given percentage of the $\dot{V}O_2R$, where $\dot{V}O_2R$ is defined as $\dot{V}O_{2max} - \dot{V}O_{2rest}$. This also can be thought of as the $\dot{V}O_{2max}$ reserve. As an example, with a $\dot{V}O_{2max}$ of 40 ml · kg^{-1} · min^{-1} and a $\dot{V}O_{2rest}$ of 3.5 ml · kg^{-1} · min^{-1},

$$\dot{V}O_2R = 40 - 3.5 \text{ ml} \cdot \text{kg}^{-1} \cdot \text{min}^{-1}$$
$$= 36.5 \text{ ml} \cdot \text{kg}^{-1} \cdot \text{min}^{-1}.$$

To prescribe an exercise intensity range between 60% and 75% of $\dot{V}O_2R$, we simply multiply $\dot{V}O_2R$ by 60% and 75%:

$$\dot{V}O_2R_{60\%} = 36.5 \text{ ml} \cdot \text{kg}^{-1} \cdot \text{min}^{-1} \times 0.60$$
$$= 21.9 \text{ ml} \cdot \text{kg}^{-1} \cdot \text{min}^{-1}$$
$$\dot{V}O_2R_{75\%} = 36.5 \text{ ml} \cdot \text{kg}^{-1} \cdot \text{min}^{-1} \times 0.75$$
$$= 27.4 \text{ ml} \cdot \text{kg}^{-1} \cdot \text{min}^{-1}$$

The major advantage of using the $\dot{V}O_2R$ technique is that one now has an equivalency between the percentage of the maximal HRR and the percentage of $\dot{V}O_{2max}$ reserve. There is a potential problem with this technique, however, in that using 3.5 ml · kg^{-1} · min^{-1} as a standard value for $\dot{V}O_{2rest}$ assumes that everyone has the same resting value. This, in fact, is not the case. Further, in one study, a large sample of women (n = 642) and men (n = 127) were found to have average $\dot{V}O_{2rest}$ values of 2.5 and 2.7 ml · kg^{-1} · min^{-1}, respectively. The range of values varied from 1.6 to 4.1 ml · kg^{-1} · min^{-1}.[5] Ultimately, however, the $\dot{V}O_2R$ technique mirrors the HRR method for prescribing exercise because the two values are so closely related.

Metabolic Equivalents

Exercise intensity also has been prescribed on the basis of the **metabolic equivalent (MET)** system. The amount of oxygen the body consumes is directly proportional to the energy expended during physical activity. In this system, it is assumed that the body uses approximately 3.5 ml of oxygen per kilogram (2.2 lb) of body weight per minute (3.5 ml · kg^{-1} · min^{-1}) at rest. As noted earlier, resting oxygen consumption varies among individuals, making this assumption not entirely accurate. The MET system, however, is based on this value, and the resting metabolic rate value of 3.5 ml · kg^{-1} · min^{-1} is referred to as 1 MET. All activities can be classified by intensity according to their oxygen requirements. An activity that is rated as a 2-MET activity would require two times the resting metabolic rate, or 7 ml · kg^{-1} · min^{-1}, and an activity that is rated at 4 METs would require approximately 14 ml · kg^{-1} · min^{-1}. Some activities and their MET values are presented in table 20.2.

These values are only approximations because of the potential error in using a standard value of 3.5 ml · kg^{-1} · min^{-1} as a constant resting value. Further, metabolic efficiency varies considerably from one person to the next, and even in the same individual. Although the

TABLE 20.2 **Selected Activities and Their Respective MET Values**

Activity	MET value	Activity	MET value
Rest and self-care activities			
Rest (supine)	1.0	Showering	2.0
Sitting	1.5	General grooming, standing	2.0
Eating	1.5	Dressing or undressing, standing	2.5
Bathing	1.5		
Home activities			
Knitting or hand sewing, light effort	1.3	Vacuuming (general, moderate effort)	3.3
Washing dishes	1.8	Making beds, changing linens	3.3
Ironing	1.8	Cleaning (scrubbing floor, washing car, washing windows)	3.5
Laundry, folding or hanging clothes	2.0-2.3	Sweeping, moderate effort	3.8
Cooking or food preparation	2.0-3.5	Moving furniture, carrying boxes	5.8
Machine sewing	2.8	Scrubbing floors on hands and knees, vigorous effort	6.5
Occupational			
Sitting tasks, office work, working at a computer	1.5	Construction (outside)	4.0
Driving a delivery truck, taxi, school bus, and so on	2.0	Hotel housekeeper	4.0
Cook, chef	2.5	Yard work	4.0
Standing tasks, light to moderate effort	3.0-4.5	Manual or unskilled labor	2.8-6.5

>continued

TABLE 20.2 *>continued*

Activity	MET value	Activity	MET value
Custodial work	2.5-4.0	Farming, light to vigorous effort	2.0-7.8
Carpentry (general, light to moderate effort)	2.5-4.3	Firefighter on the job	6.8-9.0
Physical conditioning			
Walking			
2.5 mph, level	3.0	4.5 mph, level	7.0
3.5 mph, level	4.3	5.0 mph, level	8.3
4.0 mph, level	5.0	5.0 mph, 3% grade	9.8
Jogging or running on level surface			
4.0 mph	6.0	10.0 mph	14.5
6.0 mph	9.8	12.0 mph	19.0
8.0 mph	11.8	14.0 mph	23.0
Swimming			
Freestyle, vigorous effort	9.8	Breaststroke, recreational or training and competition	5.3-10.3
Freestyle, slow to moderate	5.8	Sidestroke, general	7.0
Backstroke, recreational or training and competition	4.8/9.5		
Cycling			
Leisure, 5.5 mph	3.5	Leisure, 14.0-15.9 mph (vigorous effort)	10.0
Leisure, 10.0-11.9 mph (slow, light effort)	6.8	Racing, 16.0-19.0 mph (vigorous effort)	12.0
Leisure, 12.0-13.9 mph (moderate effort)	8.0	Racing, >20 mph (vigorous effort)	15.8
Recreational activities			
Aerobic dance	5.0-7.3	General resistance training	3.5-6.0
Video game activities	2.3-6.0	Rowing machines	4.8-12.0
Stationary cycle ergometer	3.5-14.0	Water aerobics	5.3
Circuit training	4.3-8.0	Video exercise workouts, light to vigorous	2.3-6.0
Sport activities			
Archery	4.3	Rock or mountain climbing	5.0-8.0
Badminton	5.5-7.0	Roller skating	7.0
Basketball	6.0-9.3	Rugby	6.3-8.3
Bowling, lawn bowling	3.0-3.8	Skateboarding	5.0-6.0
Football, flag or touch	4.0-8.0	Soccer	7.0-10.0
Golf	4.8	Softball	5.0-6.0
Handball	12.0	Squash	7.3-12.0
Hockey, field	7.8	Table tennis (Ping-Pong)	4.0
Hockey, ice	8.0-10.0	Tennis, singles	7.3-8.0
Horseback riding	5.8-7.3	Tennis, doubles	4.5-6.0
Lacrosse	8.0	Volleyball	3.0-4.0
Orienteering	9.0	Volleyball, competitive	8.0
Racquetball	7.0-10.0	Volleyball, competitive beach	6.0

Data from B.E. Ainsworth et al., "2011 Compendium of Physical Activities: A Second Update of Codes and MET Values," *Medicine and Science in Sports and Exercise* 43, no. 8 (2011):1575-1581.

MET system is useful as a guideline for training, it fails to account for changes in environmental conditions, and it does not allow for changes in physical conditioning as discussed in the previous section. The MET system for exercise prescription is, however, very useful for determining someone's potential ability to perform physical activity that may not be exercise. For example, a deconditioned person who has just started an exercise training program and is walking on a treadmill at 3.5 to 4.0 mph (4.3-5.0 METs) might ask whether it would be appropriate to incorporate tennis doubles into the program. Using the MET chart, you can see that doubles tennis is in the range of 4.5 to 6.0 METs. Thus you may advise the person that it is within his or her aerobic capacity, but that some of the MET levels for doubles tennis may be above that aerobic capacity. Additionally, activities that require some level of skill may also be more locally fatiguing. Though these facts may not preclude the activity, they may make it more fatiguing than it would be if the person were more fit.

Ratings of Perceived Exertion

Ratings of perceived exertion (RPE) also have been proposed for use in prescribing exercise intensity. With this method, individuals subjectively rate how hard they feel they are working. A given numerical rating corresponds to the perceived relative intensity of exercise. When the RPE scale is used correctly, this system for monitoring exercise intensity has proven very accurate. On the **Borg RPE scale**, which is a rating scale ranging from 6 to 20, moderate exercise corresponds to an RPE of 12 to 14 (somewhat hard) and vigorous exercise to an RPE of 14 to 16 (hard).[4] Initially, this sounds too simple. However, most people can use the RPE technique very accurately. Studies have shown that when people are asked to select a pace on a treadmill or a resistance on a cycle ergometer at a moderate or heavy intensity of exercise (see table 20.3), they are able to select a pace or resistance that gets their heart rates into the appropriate range. This is a more natural way to prescribe exercise and very efficient if the person is able to relate perceptions of intensity accurately.

One simple way of monitoring exercise intensity is referred to as the talk test, which has been used as an informal guideline for years. Scientists have now confirmed that the highest exercise intensity that just barely allows a person to talk comfortably while exercising is a very consistent method that correlates well with ventilatory threshold (see chapter 8) and is well within the THR range.[13]

Table 20.3 compares the various methods for rating exercise intensity. Let's use them to determine a moderate exercise intensity. As the table shows, one would want to work within a range of 40% to 60% if one is monitoring intensity using the HRR method. Using rating of perceived exertion, shown in the fourth column, this is equivalent to an RPE value of 12 to 13. All these values reflect moderate-intensity exercise. Perhaps the best use of RPE is to combine it with THR, allowing the exerciser to self-select exercise intensity by both heart rate and comfort level. Thus, self-selecting the comfort level is also part of the exerciser's repertoire. The relationship between HRR and RPE can be so close (assuming the exerciser uses the RPE scale honestly and correctly) that most people can control most of their exercise by RPE and not have to take heart rate.

In Review

- Exercise intensity can be monitored on the basis of THR, metabolic equivalent, or RPE.
- Training heart rate can be established through use of the heart rate equivalent to a certain percentage of $\dot{V}O_{2max}$. It can also be determined using the HRR method, which takes a given percentage of maximal HRR and adds it to resting heart rate. With this method, the percentage of maximal HRR used corresponds to approximately the same

TABLE 20.3 **Classification of Exercise Intensity Based on 20 to 60 min of Endurance Activity: Comparing Three Methods**

	Relative intensity		
Classification of intensity	HR_{max}	HRR, $\dot{V}O_2R$	Rating of perceived exertion
Very light	<57%	<30%	<9
Light	57%-64%	30%-40%	9-11
Moderate	64%-76%	40%-60%	12-13
Vigorous	76%-96%	60%-90%	14-17
Near maximal to maximal	≥96%	≥90%	≥18

Data from American College of Sports Medicine (2018).

percentage of $\dot{V}O_{2max}$ when a person is exercising at moderate to high intensities.

- A sensible approach is to establish a THR range to work within, rather than a single THR, attempting to establish the low end at an intensity below lactate threshold.
- The amount of oxygen consumed reflects the amount of energy expended during an activity. $\dot{V}O_2$ at rest has been assigned a value of 3.5 ml · kg^{-1} · min^{-1}, which equals 1 MET. Activity intensities can be classified by their oxygen requirements as multiples of the resting metabolic rate.
- The RPE method requires that a person subjectively rate how difficult the work is, using a numerical scale that is related to exercise intensity. The subject looks at the standard scale to determine the appropriate number.

Exercise Programming

Once the exercise prescription has been determined, it is integrated into a total exercise program, which is generally only part of an overall health improvement plan. Individual exercise capacity varies widely even among people of similar ages and physical builds. For this reason, each program must be individualized based on results of physiological and medical tests and, if possible, individual needs and interests.

The total exercise program consists of the following activities:

- Warm-up and stretching activities
- Endurance training
- Cool-down and stretching activities
- Flexibility training
- Resistance training
- Neuromotor exercise
- Recreational activities

Generally, the first three activities are performed three or four times each week. Flexibility training can be included as part of the warm-up, cool-down, or stretching exercises, or it can be done at a separate time during the week. Resistance training is usually done on alternate days when endurance training is not performed; however, the two can be combined into the same workout.

Warm-Up and Stretching Activities

The exercise session should begin with a low-intensity, dynamic warm-up that includes some of the cardiorespiratory endurance activities and perhaps functional (neuromotor) activities that are part of the exercise program itself but at a much reduced intensity. Stretching (flexibility) exercises may be included in the warm-up. Such a warm-up period gradually increases both heart rate and breathing, and prepares the exerciser for efficient and safe functioning of the heart, blood vessels, lungs, and muscles during the more vigorous exercise that follows. For example, someone who trains by running might start with 5 to 10 min of a light jog, some lunges or light-resistance band work, and then stretch before beginning the training run.

Endurance Training

Physical activities that develop cardiorespiratory endurance are the heart of the exercise program. They improve both the capacity and efficiency of the cardiovascular, respiratory, and metabolic systems. These activities may also help one control or reduce body weight. Activities such as walking, jogging, running, cycling, swimming, rowing, group cardiorespiratory endurance and other exercise activities, and hiking are good endurance activities. Sports such as handball, racquetball, tennis, badminton, and basketball also have aerobic potential if they are pursued vigorously. Activities such as golf, bowling, and softball are generally of little value for developing cardiorespiratory endurance, but they are fun, have definite recreational value, and may offer health-related benefits as physical activity. For these reasons, such activities certainly have a place in the overall exercise program.

Cool-Down and Stretching Activities

Every endurance exercise session should conclude with a cool-down period. The best way to accomplish cool-down is to slowly reduce the intensity of the endurance activity during the last several minutes of a workout. After running, for example, a slow, restful walk for several minutes helps prevent blood from pooling in the extremities. Stopping abruptly after an endurance exercise bout causes blood to pool in the legs and can result in dizziness or fainting. Also, catecholamine levels may be elevated during the immediate recovery period, which can lead to a heart dysrhythmia.

After the cool-down period, stretching exercises can be performed to facilitate increased flexibility.

Flexibility Training

Flexibility exercises usually supplement exercises performed during the warm-up or cool-down period and are useful for those who have poor flexibility or muscle and joint problems, such as lower back pain that may result from musculoskeletal imbalances. These exercises should be performed slowly. Quick stretching movements are potentially dangerous and can lead to muscle pulls or spasms. At one time it was recommended that these exercises be performed before the endurance conditioning period. However, recently it has been hypothesized that the muscles, tendons, ligaments, and joints are more adaptable and responsive to flexibility exercises when they are done after the endurance conditioning phase. Research has yet to confirm this hypothesis.

VIDEO 20.3 Presents Malachy McHugh on the timing and duration of stretching, and the role of stretching in injury prevention.

Resistance Training

The importance of resistance training as part of a general health and fitness exercise program has been clearly established. Many health-related benefits can be obtained from resistance training. The ACSM has included resistance training in its recommendations for a general health and fitness program.[1]

Recall from chapter 9 that the maximum amount of weight one can lift successfully only one time is the 1-repetition maximum, or 1RM. When people begin a resistance training program, they should start with a weight that is between 40% and 60% of their maximal strength, or 1RM, for each lift. The level will depend on age, conditioning level (especially previous resistance training), and goals. Generally, again depending on goals and experience, between 8 and 12 repetitions are recommended for improvement of strength and power in most adults. Up to 15 to 20 repetitions of a specific weight in two to four sets may be recommended if goals or conditioning levels are significantly different from those common to most adults.

When a given weight brings the exerciser to fatigue, that is called "repetition max" (or rep max). People should try to achieve as many repetitions as possible during the second and third sets, but the number of repetitions they can complete in these last sets may decrease as their muscles become fatigued. As muscular strength and endurance increase, the number of repetitions they can complete per set will increase. When people reach 15 repetitions on the first set, they are usually ready to progress to the next higher weight. This training technique is referred to as progressive resistance exercise, as discussed in chapter 9.

One can perform two or three sets of each lift per day, 2 or 3 days per week, for weight control purposes. Strength gains, however, appear to be fully achieved with just one set per day in previously untrained people.[14] People should select a variety of exercises that tax most or all of the major muscle groups of the upper body, trunk, and lower body. If time is a factor, it is better to reduce the number of sets to one or two and maintain a full-body workout.

Neuromotor Exercise

Neuromotor exercise includes such aspects of fitness as balance, neuromotor engagement of musculature, proprioceptive training, gait, and perhaps even core training.[1] This is also called *functional training* by many people. Many physical activities can be included; most are multijoint exercises using body weight or light hand weights and other devices (e.g., kettle bells, bands, weighted balls) that are incorporated into movements such as lunges, lifts, and bending-twisting. These exercises are aimed at improving our ability to perform everyday activities that we all do in order to live. They also train us to engage muscle groups—for example, gluteal or abdominal muscles—that we tend to become less able to use as we become sedentary and stop using a wide variety of muscles.

There is little research on the effects, the benefits, or the exact amount of functional exercise that is beneficial, though some initial results indicate that it may help in fall prevention in older adults and reduce injury in athletes. However, many professionals prescribe functional exercise for healthy adults, and these exercises should be considered for inclusion in programs for healthy adults and older adults.

Recreational Activities

Recreational activities are important to any comprehensive exercise program. Although people engage in these activities primarily for enjoyment and relaxation, many recreational activities can also improve health and fitness. Activities such as hiking, tennis, handball, squash, and certain team sports fall into this category. Guidelines for selecting these activities include the following:

- Can you learn or perform the activities with at least a moderate degree of success?
- Do the activities include opportunities for social development, if that is desired?
- Are the costs associated with participation reasonable and within your budget?
- Are the activities varied enough to maintain your continued long-term interest?
- Given your current age and health status, is the given activity safe for you?

Many excellent opportunities exist for people who have no recreational hobbies or activities but would like to become involved in some way. Local public recreation centers, park districts, YMCAs, YWCAs, churches, some public schools, community colleges, and universities offer classes in a wide variety of activities at little or no cost. Often the entire family can participate in these classes—an added bonus to a total health improvement program. Also, the number of commercial fitness centers is rapidly growing, and most now employ trained staff who can properly prescribe exercise programs and help individuals get started.

Exercise and Rehabilitation of People with Diseases

Exercise has become a major component in **rehabilitation programs** for a number of diseases. Cardiopulmonary rehabilitation programs, which began in the 1950s, have become the most visible (see chapter 21). Tremendous advances in cardiopulmonary rehabilitation have led to the formation of a professional association, the American Association of Cardiovascular and Pulmonary Rehabilitation, and a professional research journal, *Journal of Cardiopulmonary Rehabilitation and Prevention*.

Exercise is also an important part of the rehabilitation of people with

- cancer;
- obesity;
- diabetes;
- renal disease;
- osteoporosis;
- arthritis, chronic fatigue syndrome, and fibromyalgia; and
- cystic fibrosis.

Most recently, emphasis on the use of exercise in the rehabilitation of transplant patients, including those with heart, liver, or kidney transplants, has increased because exercise helps alleviate some drug side effects and improves general health.

Although the specific physiological mechanisms explaining the benefits of exercise training for each of these diseases have not been clearly defined, exercise training carries many general health benefits that appear to improve the patient's prognosis. The manner in which exercise is used in the rehabilitation of people with disease is highly specific to the nature and extent of the disease. It is therefore beyond the scope of this chapter to present specific details for any disease, but many resources are now available that provide extensive detail about establishing exercise programs for people with particular diseases and the clinical values of these programs.[1]

In Review

- An exercise session should begin with a low-intensity, dynamic warm-up of a type similar to the cardiorespiratory endurance exercise prescribed; functional and stretching exercises follow to prepare the cardiovascular, respiratory, and muscle systems to work more efficiently.
- Cardiorespiratory endurance activities should be performed three to five times each week.
- Each endurance session should be followed by cool-down and stretching to prevent blood pooling in the extremities and muscle soreness.
- Flexibility exercise should be performed slowly, and this phase of the program might be best included immediately after the endurance component.
- Resistance training should begin with a weight that is 40% to 60% of the person's 1RM. This is the proper weight if the person can lift it about 10 times; if he or she can lift it fewer than eight times, a lighter weight is required.
- Recreational activities should be included in an exercise program for enjoyment and relaxation.
- Exercise is a vital part of rehabilitation for most diseases. The type and details of the rehabilitation program depend on the patient, the specific disease involved, and its extent.

IN CLOSING

In this chapter, we have seen that the medical community now regards physical activity as vital to maintaining good health and reducing the risk of disease. We looked at the importance and practicality of both a medical examination and an exercise ECG in screening previously sedentary adults before prescribing exercise. We discussed the components of an exercise prescription and methods of monitoring exercise intensity. Finally, we reviewed the components of an exercise program and the role of exercise in rehabilitating patients with disease.

Now that we have seen the importance of exercise in disease prevention, we will look more closely at physical activity as it relates to specific disease states. In the next chapter, we turn our attention to cardiovascular diseases.

KEY TERMS

Borg RPE scale
exercise electrocardiogram (ECG)
exercise prescription
graded exercise test (GXT)
heart rate reserve
metabolic equivalent (MET)
mode
predictive value of an abnormal exercise test
rating of perceived exertion (RPE)
rehabilitation programs
sensitivity
specificity
training heart rate (THR)

STUDY QUESTIONS

1. How active are adult Americans today?
2. Discuss the concepts of sensitivity and specificity of exercise testing and the predictive value of an abnormal test. Of what value is this information in the establishment of policy mandating who should be exercise tested?
3. How can we get our population to be more active? What levels of exercise do we need to promote to help people gain the health-related benefits associated with exercise?
4. What six factors must be considered in the exercise prescription? Which of these is the most important?
5. Discuss the concept of a minimal threshold for initiating physiological changes with exercise training as it relates to the exercise prescription.
6. Discuss the various ways of monitoring exercise intensity, and give the advantages and disadvantages of each.
7. Describe the components of a good exercise program and their importance in the total program.

STUDY GUIDE ACTIVITIES

In addition to the activities listed in the chapter opening outline, two other activities are available in the web study guide, located at

www.HumanKinetics.com/PhysiologyOfSportAndExercise

The **KEY TERMS** activity reviews important terms, and the end-of-chapter **QUIZ** tests your understanding of the material covered in the chapter.

Cardiovascular Disease and Physical Activity

In this chapter and in the web study guide

On Saturday afternoon, June 22, 2002, St. Louis Cardinals pitcher Darryl Kile was found dead in his hotel room in Chicago. The Cardinals were in Chicago to play a three-game series against the Chicago Cubs. Kile was scheduled to start the last game of the series on Sunday night. He was considered one of the Cardinals' best pitchers and a clubhouse leader. Kile, who was only 33 years old, apparently died of a heart attack caused by coronary atherosclerosis—on autopsy, two of his three major coronary arteries were found to be narrowed by 80% to 90%. Although he had no medical history or symptoms of disease, his father had died of a stroke at the age of 44 years, and Kile had complained of shoulder pain and fatigue during dinner the previous night.

On November 3, 2007, Ryan Shay, a highly ranked distance runner (National Collegiate Athletic Association 10,000 m champion in 2001 and USA marathon champion in 2003), collapsed during the U.S. Olympic marathon trials in New York City after running just 5.5 mi (8.9 km). The report on autopsy results stated that his death resulted from "cardiac arrhythmia due to cardiac hypertrophy with patchy fibrosis of undetermined etiology." In October 2009, three men collapsed and died while running the 32nd Detroit Free Press/Flagstar Marathon, all within a span of 16 min. Several weeks earlier, two runners, one man and one woman in their mid-30s, died during the Rock 'n' Roll San Jose Half Marathon. Autopsy reports are not available for these five runners, but it is likely that their deaths were heart related since heat stress was likely not an issue. Covering the on-field collapse of a 14-year-old female soccer star who had to be revived with an AED and later underwent successful surgery for a cardiac abnormality, a 2015 CBS news report highlighted the fact that an estimated 20,000 young adults die each year from sudden cardiac arrest.

These tragedies illustrate the important fact that being a good or even an outstanding athlete during youth and young adulthood does not confer lifelong immunity to cardiovascular diseases.

Most of us consider ourselves to be healthy until we experience some overt sign of illness. With progressive diseases, such as heart disease, most people are unaware that the disease process is ongoing and could eventually cause major complications, including death. Fortunately, early detection and effective treatment of various chronic diseases can substantially reduce their severity and often avert disability and death. From a preventive standpoint, decreasing the risk factors for a disease often can either prevent the disease or delay its onset. In this chapter, we look at cardiovascular diseases, focusing primarily on coronary artery disease (CAD) and hypertension (HTN).

Prevalence of Cardiovascular Disease

Heart disease is the major cause of death in the United States (figure 21.1). The following statistics have been published by the American Heart Association:[4]

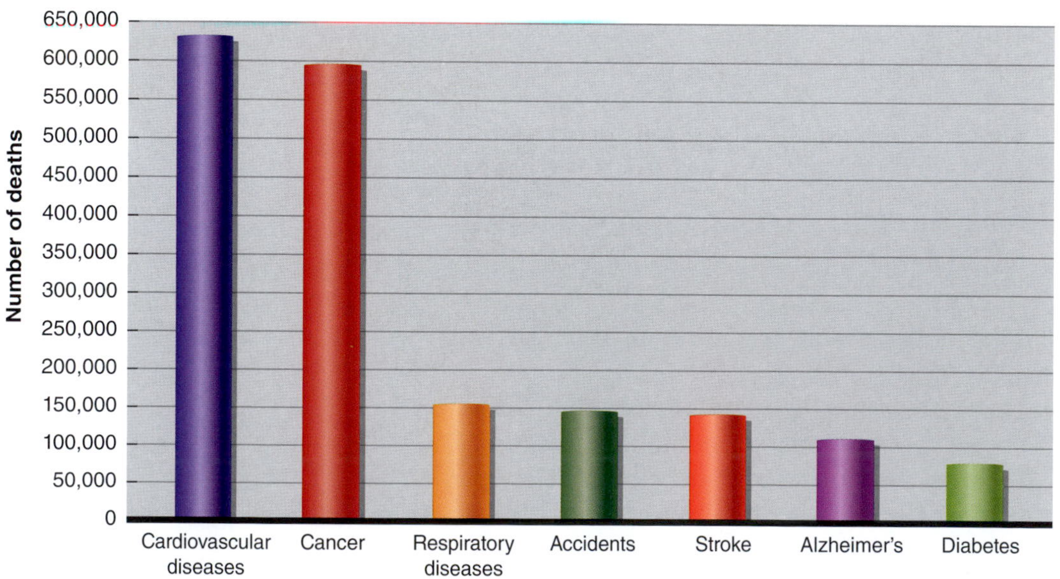

FIGURE 21.1 The leading causes of death in the United States in 2015.

Data from American Heart Association, "Heart Disease and Stroke Statistics--2014 Update," *Circulation* 129 (2014): e28-e292. http://circ.ahajournals.org/content/129/3/e28.full

- An estimated 92.1 million adults in the United States have at least one type of cardiovascular disease (CVD).
- By 2030, almost 44% of the U.S. population is projected to have some form of CVD.
- CVD was the most common cause of death worldwide, accounting for 17.3 million deaths, or 31.5% of total deaths around the world.
- Cardiovascular diseases claim more lives than all forms of cancer combined.
- The economic impact of CVD and stroke in the U.S. was estimated to be over $316 billion in 2013-14, accounting for 14% of all health care expenditures.

But there is some good news as well:

- From 2004 to 2014, CVD death rates dropped by 25.3% and the actual number of CVD deaths decreased by 6.7%.
- Continued improvements in overall cardiovascular health metrics are projected to reduce CHD by 30% between 2010 and 2020.

From the early 1900s to the mid-1960s, the relative number of heart disease deaths, expressed per 100,000 people, increased threefold. The population of the United States more than doubled during that time, so the absolute number of heart disease deaths increased even more dramatically than the relative rate indicates. In the 1970s, cardiovascular diseases accounted for well over 50% of all deaths in the United States. They remain the number one underlying cause of death in the United States. Cardiovascular diseases continue to be a major problem in the United States, accounting for about 775,000 deaths—one in every three deaths—as of 2016.[4]

As noted previously, CVD death rates dropped by about 25% over the decade from 2004 to 2014. The reasons for this decline have been heavily debated but likely include a greater focus on disease prevention. These are some examples:

- Improved public awareness of risk factors and symptoms
- Increased use of preventive measures, including lifestyle changes (e.g., nutrition, exercise, stress reduction, and smoking cessation) to reduce individual risk
- Earlier and more accurate diagnosis
- Greater awareness and use of cardiopulmonary resuscitation techniques

Another probable reason is better treatment of patients with disease, including these examples:

- Improved drugs for specific treatment
- Angioplasty, drug-coated stents, and bypass surgery
- Greater focus on secondary prevention

Cardiovascular disease incidence and mortality rates vary by sex, age, and race. Table 21.1 shows rates of all-cause deaths and death from cardiovascular disease in the United States by sex and race, adjusted for age.

Forms of Cardiovascular Disease

There are several different cardiovascular diseases. In this section, we focus primarily on those that are preventable and that affect the largest number of individuals each year. According to the World Health Organization, an estimated 17.7 million people died from CVD in 2015. Of these deaths, an estimated 7.4 million were due to coronary artery disease and 6.7 million were due to stroke.[43] Other types of cardiovascular diseases include congenital diseases of the heart, cardiomyopathy, valvular heart disease, venous thromboembolism, and peripheral artery disease.

Coronary Artery Disease

As most humans age, their coronary arteries (see figure 6.4), which supply blood to the myocardium (heart muscle), become progressively narrowed as a result of the formation of fatty **plaque** under the inner lining

TABLE 21.1 **U.S. Death Rates per 100,000 Adults by Sex and Race**

Category	All cause	Cardiovascular disease
All people	733	169
Men	863	212
White	862	211
Black or African American	1,040	258
Hispanic or Latino	629	146
Women	624	134
White	627	132
Black or African American	711	166
Hispanic or Latina	438	93

These data are age standardized (adjusted to account for younger or older average age of a population).

Data from World Health Organization, *Global Atlas on Cardiovascular Disease Prevention and Control*, edited by S. Mendis, P. Puska, and Bo Norrving (World Health Organization, Geneva, 2011). Available http://whqlibdoc.who.int/publications/2011/9789241564373_eng.pdf?ua=1

(the endothelium) of the artery, as seen in figure 21.2. This progressive narrowing of the arteries in general is referred to as atherosclerosis, and when the coronary arteries are involved, it is termed **coronary artery disease (CAD)**, or coronary heart disease. As the disease progresses and the coronary arteries become narrower, the capacity to supply blood to the myocardium is progressively reduced.

As the narrowing worsens, there is a mismatch between oxygen supply to the myocardium (heart muscle) and its demand, which increases with exercise. When this occurs, the area of the myocardium that is supplied by the narrowed artery or arteries becomes ischemic, meaning that it suffers a deficiency of blood (with oxygen and nutrients). **Ischemia** of the myocardium often causes mild to severe discomfort, referred to as *angina pectoris* or simply *angina*. Anginal symptoms are often first experienced during physical exertion or stress, when the oxygen demands of the heart increase, creating a mismatch between supply and demand.

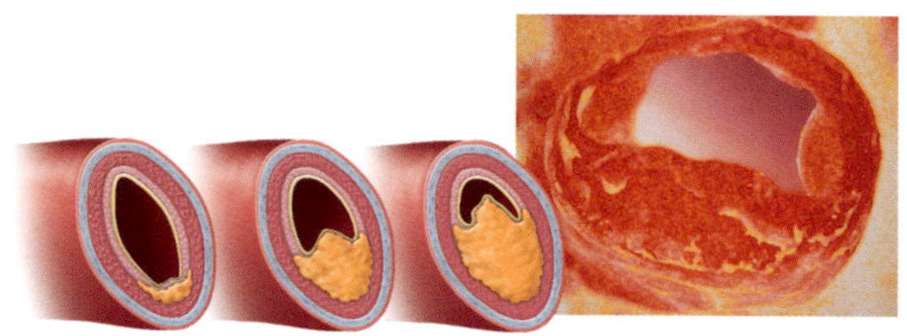

FIGURE 21.2 The progressive formation of plaque in a coronary artery.

When blood supply to a part of the myocardium is severely or totally restricted, ischemia can lead to a heart attack, or **myocardial infarction**, as cardiac muscle cells that are deprived of adequate blood flow (and thus deprived of oxygen) for several minutes undergo damage (injury) and sometimes necrosis (cellular death). The outcome of this event depends on the size of the affected myocardial area and the extent of the damage. Ischemic injury is reversible, while infarcted tissue is permanently damaged. Sometimes a heart attack is so mild that the victim is unaware that it has occurred. In such cases, the heart attack may only be discovered weeks, months, or even years later when an electrocardiogram is obtained during a routine medical examination.

Atherosclerosis is a progressive disease process, and the pathological changes that lead to atherosclerosis can begin in infancy and progress during childhood.[25]

RESEARCH PERSPECTIVE 21.1

Long-Term Marathon Running Reduces Coronary Artery Plaque Formation in Women

Regular endurance exercise leads to improvements in cardiovascular health and reduces cardiovascular disease risk factors, such as high blood pressure and high blood cholesterol. Typically, marathon training causes a large increase in cardiorespiratory fitness and a reduction in cardiovascular risk profiles. However, fairly little is known about the effects of long-term marathon participation on the coronary vasculature. Two studies of middle-aged men who were long-term marathon runners (classified in one study as having run more than five races in the past 3 years, and in the other study as having completed one or more marathons a year for 25 or more consecutive years) found that these men paradoxically had *greater* accumulations of plaque on their coronary arteries compared to sedentary age-matched men. Although these findings were surprising, until recently, no study had examined this phenomenon in women.

A recent study conducted at the Abbott Northwestern Hospital in Minneapolis, Minnesota, examined coronary artery plaque formation and the prevalence of cardiovascular disease risk factors in women who were long-term marathon runners.[36] In order to do this, researchers recruited 26 women who had run the Twin Cities Marathon for at least 10 consecutive years and 28 control women who were sedentary. Using coronary computed tomography angiography (CCTA), the researchers identified any plaque formation in the coronary arteries. At the time of the CCTA scan, participants also had a 12-lead ECG, resting heart rate and blood pressure measurements, and a blood draw for the measurement of serum lipids and creatinine. Five of the twenty-six marathon runners and fourteen of the twenty-eight controls had calcified plaques in their coronary arteries, a statistically lower incidence in the marathon running group. The five female runners with coronary plaque had run marathons for more years but were on average 12 years older than the runners without plaque. The marathon runners also had lower resting heart rate, body weight, blood pressure, and serum triglycerides and higher high-density lipoproteins compared to controls. Women who are long-term marathon runners are less likely to have coronary artery plaque formation and have a reduced cardiovascular disease risk profile compared to sedentary age-matched controls. Instead, the development of coronary plaque in women marathoners appears related to older age and increased cardiac risk.

Fatty streaks, or lipid deposits, which are thought to be the precursors of atherosclerosis, are commonly found in the aortas of children by age 3 to 5. These fatty streaks start to appear in the coronary arteries during the early teens, can develop into fibrous plaques during one's 20s, and can progress to unstable or complicated lesions during one's 40s and 50s.

The rate at which atherosclerosis progresses is determined largely by genetics and lifestyle factors, including smoking history, diet, and physical activity. For some people, the disease progresses rapidly, with a heart attack occurring at a relatively young age—in their 20s or 30s. For others, the disease progresses very slowly, with only minor evidence of atherosclerosis even late in life. Most people fall somewhere between these two extremes.

To illustrate this, a study of combat fatalities from the Korean War revealed that a large percentage of autopsied American soldiers, with an average age of 22 years, already had evidence of coronary atherosclerosis.[15] The extent of disease ranged from fibrous thickening to complete occlusion of one or more of the main branches of the coronary arteries. Evidence of coronary atherosclerosis also was found in almost half of the American fatalities from the Vietnam War, some with severe manifestations of the disease.[30]

Hypertension

Hypertension is the medical term for high blood pressure, a condition in which blood pressure is chronically elevated above levels considered desirable or healthy. Blood pressure depends in part on body size, so children and young adolescents have much lower blood pressures than adults. For this reason, defining what constitutes HTN in the growing child and adolescent is difficult. Clinically, HTN in these groups is defined as blood pressure values above the 90th or the 95th percentile for the youth's age. Hypertension is uncommon during childhood but can appear during mid-adolescence.

Historically, normal blood pressure and stages of HTN were classified by well-established blood pressure cutoffs. For example, prehypertension was defined as systolic blood pressure (SBP) between 120 and 139 mmHg or a diastolic blood pressure (DBP) between 80 and 89 mmHg and stage 1 HTN was defined as SBP of 140 to 159 or DBP of 90 to 99 mmHg. The Joint National Committee on Detection, Evaluation, and Treatment of High Blood Pressure has recently published new evidence-based guidelines (JNC8) for determining and treating HTN in adults.[24] The JNC8 guidelines eliminate categories of HTN (normal, pre-HTN, stage 1, and stage 2) based on blood pressure cutoffs and simply recommend treatment for specific population groups.

New treatment goals are summarized in table 21.2. These new recommendations also strongly endorse lifestyle modification guidelines, with diet and increased physical activity at the core of therapeutic management of HTN. Other major changes include altering the classes of medications recommended for the management of HTN and new recommendations for treating persons with chronic kidney disease. In hypertensive African Americans, for example, the guidelines recommend the use of thiazide-type diuretics or calcium channel blockers because these classes of drugs have higher efficacy rates in this population than other antihypertensive drugs, such as angiotensin receptor blockers or angiotensin-converting enzyme blockers.

The JNC8 panel's attempt to simplify the guidelines for treating HTN received some criticism but has generally been embraced by health care professionals. Perhaps the most significant and interesting point to be taken from JNC8 is that, unlike previous guidelines, these are evidence based, and the reason for the significant changes is the lack of evidence for some previously published treatment standards.

Hypertension causes the heart to work harder than normal because it has to pump blood from the left ventricle against a greater resistance (afterload). Furthermore, HTN places increased chronic strain on the systemic arteries and arterioles. Over time, the stress can cause the heart to enlarge and the arteries and arterioles to become less elastic. Eventually, this can lead to atherosclerosis, heart attacks, heart failure, stroke, and kidney failure.

In 2014, almost 86 million American adults, or about 34% of the adult population, were estimated to have high blood pressure (as defined by systolic ≥140 mmHg,

TABLE 21.2 **JNC8 Guidelines for the Management of Hypertension in Adults**

Population group	Treatment goal
Age ≥60 years with SBP ≥150 or DBP ≥90	SBP <150 and DBP <90
General population <60 years with SBP >140 or DBP >90	SBP <140 DBP <90
Diabetes ≥18 years	SBP <140 DBP <90

Note. All blood pressures are in mmHg. SBP = systolic blood pressure; DBP = diastolic blood pressure.

diastolic >90 mmHg, or both).[4] Among Americans age 60 and up, over 65% have high blood pressure. Like CAD, it is most prevalent (42%) among non-Hispanic black adults.[24] Compared with white Americans, black Americans develop high blood pressure at an earlier age, and it is more severe at any decade of life. Consequently, black Americans have a 1.3 times greater rate of nonfatal stroke, a 1.8 times greater rate of fatal stroke, a 1.5 times greater rate of heart disease deaths, and a 4.2 times greater rate of end-stage renal disease when compared with white Americans.[3] Sadly, of those diagnosed with high blood pressure, only about 50% have controlled HTN. This means that large numbers of Americans who are being seen by a physician for high blood pressure remain hypertensive.

Stroke

Stroke is a form of cardiovascular disease that affects the cerebral arteries, those that supply the brain. Globally, stroke was the cause of about 6.5 million deaths in 2013.[4] As with other forms of CVD, the death rate from stroke have decreased significantly in recent years.

Strokes generally fall into two categories: **ischemic stroke** and **hemorrhagic stroke**. Ischemic strokes are the most common (~87% of all cases) and result from an obstruction within a cerebral blood vessel that limits the flow of blood to that region of the brain. Obstructions have one of two causes:

- Cerebral thrombosis, the most common, in which a thrombus (blood clot) forms in a cerebral vessel, often at the site of atherosclerotic damage to the vessel. This thrombus is very similar in etiology to the blood clots formed in coronary arteries that cause myocardial ischemia, injury, and infarction.
- Cerebral embolism, in which an embolus (an undissolved mass of material, such as fat globules, or bits of tissue, a blood clot) breaks loose from another site in the body and lodges in a cerebral artery. An irregular heartbeat, atrial fibrillation, creates conditions in which clots can form in the heart, dislodge, and then travel to the brain.

In cases of ischemic stroke, blood flow distal to the blockage is restricted, and the part of the brain that relies on that supply becomes ischemic—oxygen deficient—and can die. Ischemic stroke is caused by the same vascular disease process present in atherosclerosis (see earlier) and other obstructive vascular diseases, including peripheral vascular disease.

Hemorrhagic strokes are of two major types:

- Intracerebral hemorrhage, in which one of the cerebral arteries ruptures in the brain
- Subarachnoid hemorrhage, in which one of the brain's surface vessels ruptures, dumping blood into the space between the brain and the skull

In both cases, blood flow beyond the rupture is diminished because as blood leaks from the vessel at the site of injury. Because the brain is enclosed in the rigid skull, as the blood accumulates outside of the vessel, it puts pressure on the fragile brain tissue, which can alter brain function. Brain hemorrhages often result from aneurysms, weak spots in the vessel wall that balloon outward, and aneurysms may arise because of HTN or atherosclerotic damage to the vessel wall. Arteriovenous malformations, clusters of abnormally formed blood vessels, are another cause of hemorrhagic strokes.

As with a heart attack, stroke can result in death of the affected brain tissue. The consequences depend largely on the location and extent of the stroke. Brain damage from a stroke can affect the senses, speech, body movement, thought patterns, and memory. Paralysis on one side of the body is common, as is the inability to verbalize thoughts. Most effects of a stroke are indicative of the side of the brain that was damaged. One side of the brain controls the functions of the opposite side of the body. For example, a stroke on the right side of the brain may have the following effects:

- Paralysis on the left side of the body
- Vision problems
- Quick, inquisitive behavioral style
- Memory loss

Alternatively, a stroke on the left side of the brain more often results in the following:

- Paralysis on the right side of the body
- Speech and language problems
- Slow, cautious behavioral style
- Memory loss

Heart Failure

Heart failure is a chronic and progressive clinical condition in which the heart muscle (myocardium) becomes too weak to maintain a cardiac output adequate to meet the body's blood and oxygen demands. When cardiac output is inadequate, blood may pool in the veins. This causes excess fluids to accumulate in the body, particularly in the legs and ankles. This fluid accumulation (edema) also can affect the lungs (pulmonary edema), disrupting breathing and causing shortness of breath. For that reason, this condition is commonly called *congestive heart failure* (*CHF*). It usually results from damage to the myocardium, which may have been caused by myocardial infarction, severe HTN, valvular heart disease, or cardiomyopathy.

Other Cardiovascular Diseases

Other cardiovascular diseases include peripheral vascular diseases, valvular heart diseases, and congenital heart diseases.

Peripheral vascular disease involves the systemic arteries and veins, as opposed to the coronary vessels. Peripheral artery disease (PAD) is atherosclerotic disease most often affecting arteries supplying blood to the legs, brain, or, more rarely, other organs. PAD in the legs can result in severe restrictions in activity due to pain. In arteries feeding the brain, PAD can result in *transient ischemic attacks* (*TIAs*; stroke-like symptoms caused by temporary restriction in blood flow to the brain).

Peripheral venous diseases include varicose veins and phlebitis. Varicose veins result from incompetency of the valves in the veins, allowing blood to back up in the veins and causing them to become enlarged, tortuous, and painful. Phlebitis is inflammation of a vein and is also very painful.

Valvular heart diseases involve one or more of the four valves that control the direction of blood flow into and out of the four heart chambers. **Rheumatic heart disease** is one form of valvular heart disease resulting from rheumatic fever, a streptococcal infection that typically occurs in children between age 5 and 15. Rheumatic fever is an inflammatory disease of the connective tissue and commonly affects the heart, specifically the heart valves. The damage to the valves can cause narrowing of the opening (*stenosis*); difficulty in valves opening, hindering blood flow out of the prevalve chamber; or incomplete valve closure (*prolapse* or *regurgitation*), allowing blood to leak back into the prevalve chamber. The disease is typically referred to by the valve affected and the type of problem, e.g., mitral stenosis.

Congenital heart disease includes any heart defects that are present at birth, which are also termed *congenital heart defects*. These defects occur when the heart itself, or the blood vessels near the heart, do not develop normally *in utero*. These include coarctation of the aorta, in which the aorta is abnormally constricted; valvular stenosis, in which one or more heart valves are narrowed; and septal defects, in which the septum or wall separating the right and left sides of the heart is defective, allowing blood from the systemic side to mix with that in the pulmonary side and vice versa.

The remainder of this chapter focuses on the two major diseases of the heart and blood vessels: CAD and HTN.

In Review

- Atherosclerosis is a process in which arteries become progressively narrowed. Coronary artery disease is atherosclerosis of the coronary arteries.
- When coronary artery blood flow is sufficiently blocked, the area of the myocardium supplied by the diseased artery suffers from lack of blood (ischemia). In severe cases, the resulting oxygen deprivation can cause a myocardial infarction, which results in tissue necrosis.
- Atherosclerotic changes in the arteries actually begin in young children, but the extent and progression of this disease process are quite variable.
- *Hypertension* is the clinical term for high blood pressure.
- Stroke affects the cerebral arteries so that the part of the brain they supply receives too little blood. Ischemic stroke is the most common form of stroke, usually resulting from cerebral thrombosis or embolism. The other cause of stroke is cerebral hemorrhage (cerebral and subarachnoid).
- Heart failure is a condition in which the cardiac muscle becomes too weak to maintain an adequate cardiac output, causing blood to back up in the veins.
- Peripheral vascular diseases involve systemic, rather than coronary, vessels and include peripheral artery disease, varicose veins, and phlebitis.
- Congenital heart disease includes all heart defects present at birth.

Understanding the Disease Process

Pathophysiology refers to the pathology and physiology of a specific disease process or disordered function. Understanding the pathophysiology of a disease gives us insight into how physical activity might affect or alter the disease process. Note that some common underlying pathophysiology such as systemic, subacute inflammation and insulin resistance is associated with many chronic diseases. In the following sections, we examine the pathophysiology of CAD and HTN.

Pathophysiology of Coronary Artery Disease

How does atherosclerosis develop in the coronary arteries? The walls of the coronary arteries are composed of three distinct layers, as shown in figure 21.3: the tunica intima (inner layer), the tunica media (middle layer), and the tunica adventitia (outer layer). These are referred to more simply as the intima, the media, and the adventitia. The innermost layer of the intima is called the **endothelium**. It is formed by a thin lining of endothelial cells that provides a smooth protective coating between the blood flowing through the artery and

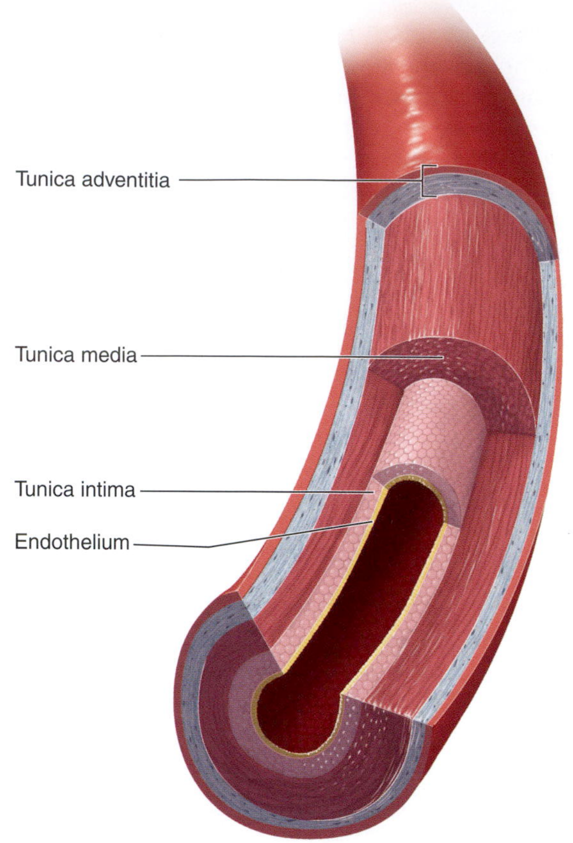

FIGURE 21.3 The wall of an artery has three layers: tunica intima, tunica media, and tunica adventitia.

the intimal layer of the vessel wall. The endothelium provides a protective barrier between toxic substances in the blood and the vascular smooth muscle cells. For vessels larger than 1 mm in diameter, the intima also includes a subendothelial layer, formed from connective tissue. The media consists mainly of the smooth muscle cells, which control the constriction and dilation of the vessel, and elastin. The adventitia is composed of collagen fibers that protect the vessel and anchor it to its surrounding structure.

The endothelium forms a protective barrier between the blood and outward layers of the artery wall. It functions to provide a smooth surface along which blood can travel, but it also has many significant physiological functions, including controlling blood flow within the vascular system by influencing vasodilation and vasoconstriction. It does this by producing physiologically active substances like **nitric oxide (NO)** and stimulating the production of similar substances in other tissues. These, in turn, cause the smooth muscle within the arterial wall to contract or dilate, thereby exerting control over the flow of blood though those channels.

Atherosclerosis is an inflammatory disease.[29] The initial injury—that is, the event that initiates the onset and progression of atherosclerosis (figure 21.4)—may result from any of several environmental stimuli. Such factors include elevated triglycerides or low-density lipoprotein (LDL) cholesterol after a high-fat meal

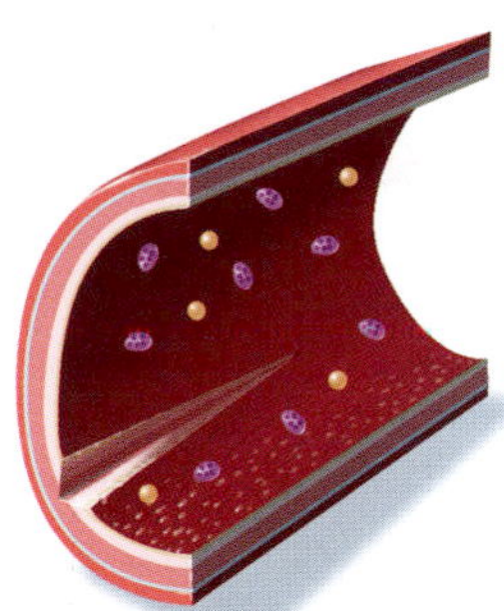
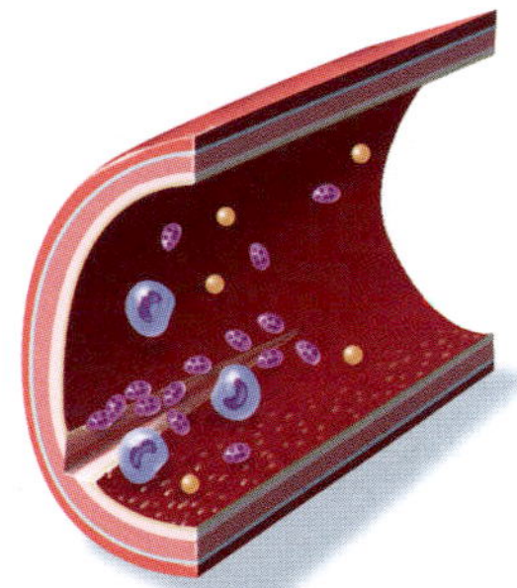
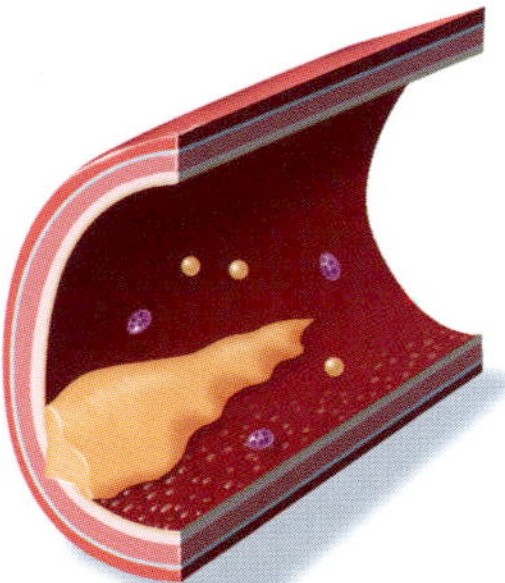
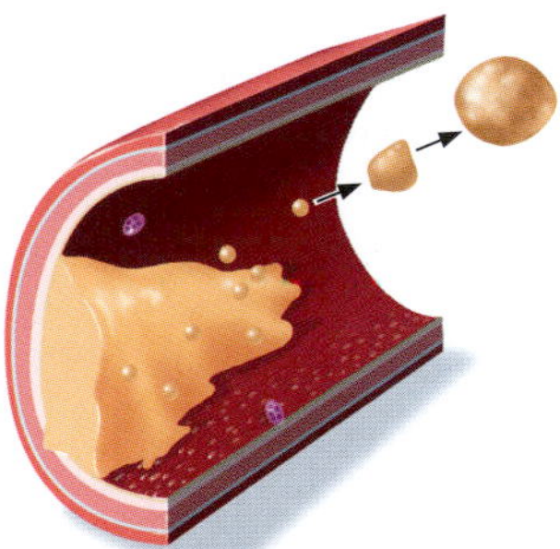

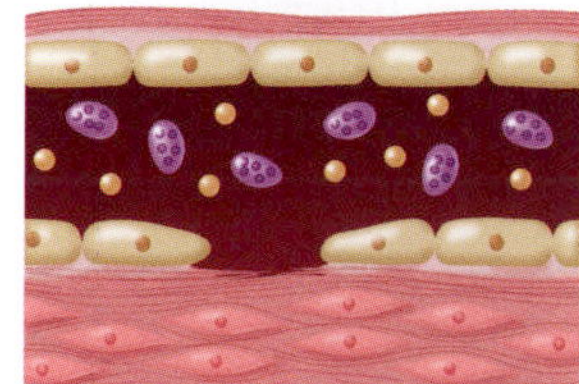
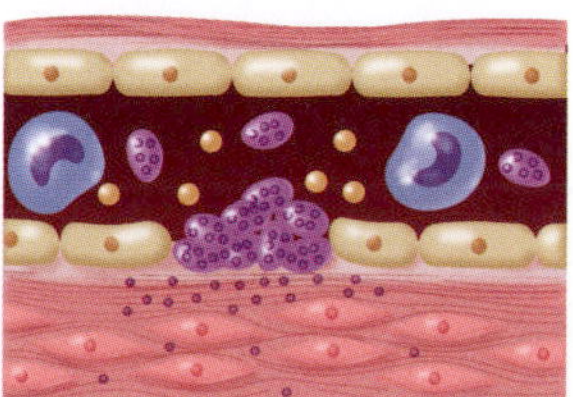
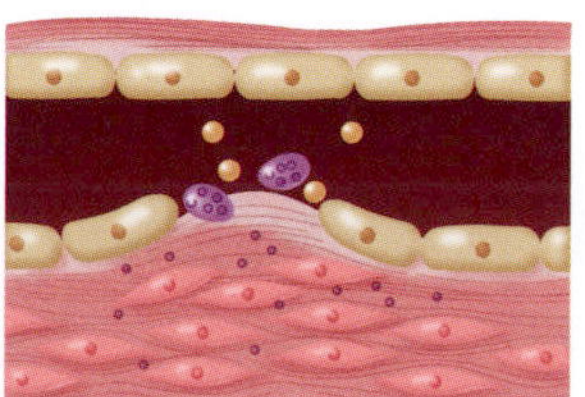
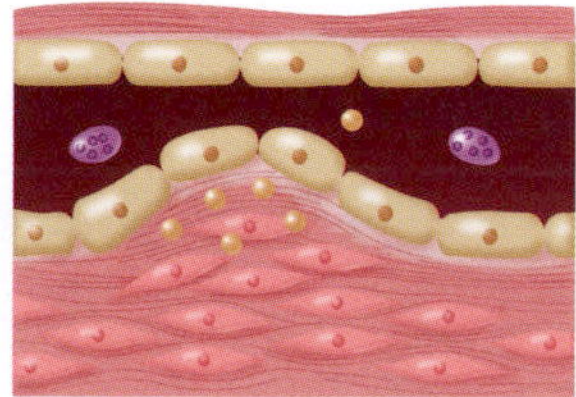

FIGURE 21.4 Changes in the arterial wall *(a-d)* with injury, illustrating the disruption of the endothelium and the subsequent alterations that lead to atherosclerosis.

(postprandial lipemia), compounds in cigarette smoke and inhaled environmental pollutants, **adipokines** and other inflammatory substances associated with excess adipose tissue, infectious microorganisms, and even emotional stress, all of which are associated with vascular inflammation. In turn, the inflammation causes endothelial dysfunction. Platelets in the blood, which may be attracted to the injury site at the endothelium, adhere to the exposed connective tissue (see figure 21.4*b*). These platelets release substances that promote migration of smooth muscle cells from the media into the intima. The intima normally contains few if any smooth muscle cells. A plaque, which is basically composed of smooth muscle cells, connective tissue, and debris, forms at the site of injury (see figure 21.4*c*). Eventually, lipids in the blood, specifically LDL cholesterol, are attracted to and deposited in the plaque (see figure 21.4*d*). Through this process, the initial injury and the subsequent inflammation result in a local deposition of lipids (fat), smooth muscle cells, and other substances within the wall of the artery.

The inflammatory process is driven by many physiologically active substances produced by the endothelium or actively stimulated by it (or both). These substances include **platelet-derived growth factor** (**PDGF**; figure 21.4*b*) and other substances that inhibit or drive endothelial NO production. The exact mechanisms and integration of these factors, as well as their individual contributions to the atherosclerotic process, are complex and still not fully understood.

Low-level, systemic inflammation is an underlying characteristic of many chronic diseases including atherosclerosis, obesity, and HTN (all discussed later in this and the next chapter). Both inflammation and insulin resistance are important forerunners of the pathophysiology of CAD.

Relatively recently, researchers have theorized that plaque forms when monocytes—white blood cells that are effector cells of the immune system—attach between endothelial cells. These monocytes differentiate into macrophages, which ingest oxidized LDL cholesterol. They slowly become large foam cells and form fatty streaks. Smooth muscle cells then accumulate under these foam cells. The endothelial cells then separate or are sloughed off, exposing the underlying connective tissue and allowing platelets to attach to it. The role of monocytes in atherosclerosis is the subject of ongoing research.

The plaque consists of a collection of smooth muscle cells and inflammatory cells (macrophages and T lymphocytes) with both intracellular and extracellular lipid. The plaque also contains a fibrous cap. The composition of the plaque and of its fibrous cap is critical to its stability. Unstable plaques are those that have thin fibrous caps and are heavily infiltrated by foam cells. These plaques are much more susceptible than others to rupture. When plaques rupture, proteolytic enzymes are released, causing a breakdown of the cellular matrix and leading to blood clotting (thrombus) as illustrated in figure 21.5. The thrombus, depending on its size, can occlude or block the artery, resulting in myocardial infarction. Even if a ruptured plaque does not directly cause an infarction, the thrombus may actually become part of the plaque and further enlarge it. In fact, this may be one reason for the unpredictably progressive nature of the disease. Plaque rupture and thrombosis account for up to 70% of myocardial infarctions and cardiac arrests. Interestingly, the plaques that do rupture are typically small, causing less than 50% stenosis or narrowing of a coronary artery.

There is good evidence that a plaque is a dynamic structure, undergoing cycles of erosion and repair that are responsible for its progressive growth. Ironically, smooth muscle cells are important to the stability of

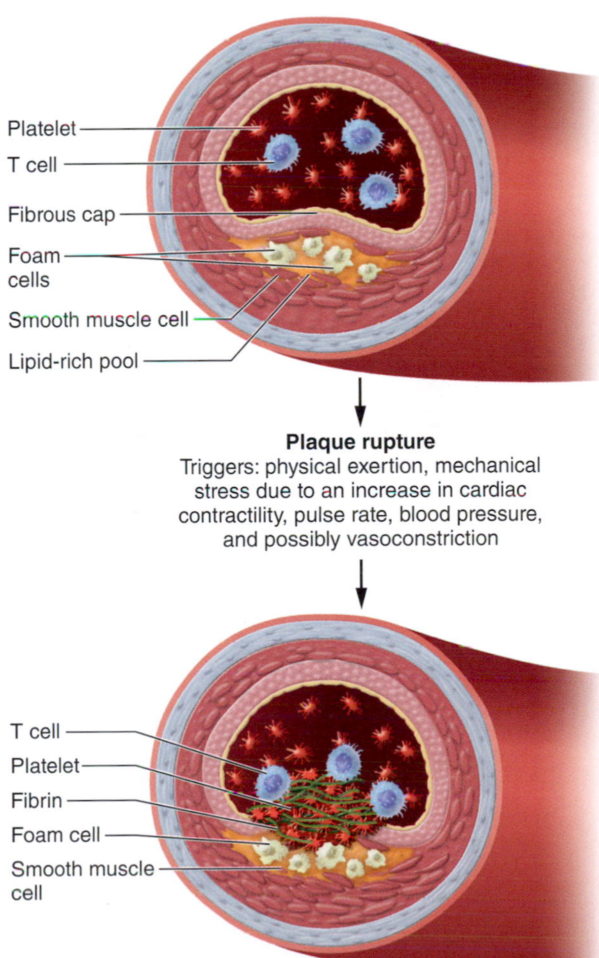

FIGURE 21.5 Illustration of fissure or rupture of an unstable plaque in a coronary artery, releasing its contents into the bloodstream and stimulating the formation of a thrombus (clot).

the plaque, and smooth muscle cell proliferation is potentially beneficial to maintaining the integrity of the plaque. Plaque rupture sites are characterized by a low density of smooth muscle cells.

Pathophysiology of Hypertension

The pathophysiology of HTN is not well understood. In fact, it is estimated that 90% to 95% of those identified with HTN are classified as having *essential* HTN, or hypertension of unknown origin. Essential HTN is also referred to as primary or idiopathic HTN. The remaining 5% to 10% are classified as having secondary HTN, meaning that the HTN is secondary to other health issues such as kidney disease, adrenal tumors (that may increase the release of epinephrine), or aortic defects. Although the pathophysiology of the most common type of HTN is largely unknown and undefined, it is clear that it is a multifactorial disease related to hormonal mediators, vascular reactivity (especially elasticity), blood viscosity, and neural stimulation of vascular resistance.

In Review

- *Pathophysiology* refers to the pathology and physiology of a specific disease process or disordered function.
- Early theories held that CHD can be initiated by injury to the smooth endothelial lining of the intimal layer of the arterial wall. This damage causes endothelial dysfunction and also attracts platelets to the area, which in turn release PDGF. The dysfunctional endothelium along with PDGF attracts smooth muscle cells, and a plaque (composed of smooth muscle cells, connective tissue, and debris) begins to form. Eventually, lipids, smooth muscle, collagenous tissue, and even blood clots may be deposited in the plaque.
- More recent research indicates that monocytes, involved with the immune system, can attach between endothelial cells in the intima and begin forming fatty streaks; this also leads to plaque formation. According to this theory, endothelial damage is not necessary for plaque formation, but endothelial dysfunction stimulated by environmental factors may also promote atherosclerosis. The details and role of this process still need to be clarified by further research.
- It is now clear that the composition of the plaque and its fibrous cap is critical with respect to myocardial infarctions and cardiac arrest. Softer plaques, where there is typically less than 50% occlusion of the artery, that have thin fibrous caps and are heavily infiltrated with foam cells are the most unstable and likely to rupture.
- The pathophysiology of HTN is poorly understood.
- More than 90% of people with HTN have essential hypertension, meaning that its cause is unknown.

Cardiovascular Disease Risk

Over the years, scientists have attempted to determine the basic etiology, or cause, of both CAD and HTN. Much of our understanding of these two diseases comes from the field of epidemiology, a science that studies the relationships of various factors to a specific disease or disease process. In a number of studies, selected members of various communities have been observed for extended periods of time. These observations include periodic medical examinations and clinical tests.

Eventually, some of the participants in such studies become diseased and, over time, many die as a result. All those who develop heart disease or HTN or who die from heart attacks or strokes are grouped accordingly. Then their previous medical and clinical tests are analyzed to determine shared attributes or factors. This approach provides researchers with valuable insights into the disease process and may, under some circumstances, provide indirect evidence of the causes of disease.

As identified in long-term longitudinal population studies, the factors that place individuals at risk for disease are referred to as **risk factors**. Often, having more risk factors imparts additional risk because the risk factors are additive and interact; thus, someone with three or four risk factors is at significantly greater risk than someone with one or two risk factors. Let's examine the risk factors for heart disease and HTN.

Risk Factors for Coronary Artery Disease

The factors associated with an increased risk for premature development of CAD can be classified into two groups: those over which a person has no control and those that can be altered through basic changes in lifestyle. Those that a person cannot control include family history of CAD (heredity), race, male sex, and advanced age. According to the American Heart Association (AHA), the **primary risk factors** that can be controlled or altered include

- exposure to or inhalation of tobacco smoke,
- hypertension,
- abnormal blood lipids and lipoproteins,

- physical inactivity,
- obesity and overweight, and
- diabetes.

Table 21.3 shows the levels of some of these risk factors that are associated with increased risk. However, these should be evaluated on an individual basis. One should note that recently published guidelines and practice standards for the management of lipids, HTN, and lifestyle to reduce cardiovascular risk, HTN, and overweight or obesity have significantly changed the practice of risk assessment and altered or even removed the thresholds and optimal values for many of these risk factors.

Other CAD risk factors have been proposed, but there is not yet sufficient evidence to support their inclusion as primary risk factors, that is, factors that add to the ability to predict CAD events. Those listed next have been added to the most recent AHA risk factor assessment guidelines. The AHA recommends that if, based on primary risk factors, a person's risk level is borderline or uncertain, other factors be considered to decide what treatment or changes are appropriate.[20]

- Family history, specifically first-degree relatives (a parent or sibling), of early CAD (men <55 years or women <65 years).
- C-reactive protein (hs-CRP) levels ≥2 mg/L. C-reactive protein is produced in the liver and smooth muscle cells within coronary arteries in response to injury or infection. C-reactive protein is a marker of inflammation.
- High coronary artery calcium (CAC) scoring. Coronary artery calcium is an indicator of coronary artery disease and can be assessed using conventional, noninvasive computed tomography scans. People with CAC ≥300 or ≥75th percentile for age, sex, and ethnicity are considered to be at risk.
- Low ankle–brachial index (ABI). This index compares blood pressure taken at the ankle and at the arm (brachial artery). It is a diagnostic test for peripheral artery disease. An ABI <0.9 is considered a risk factor.

Identification of Risk Factors

In July 1948, the National Institutes of Health (National Heart, Lung, and Blood Institute) started the Framingham Heart Study (FHS). The FHS was designed as a longitudinal investigation aimed at identifying those factors that influence the development of cardiovascular diseases. The original study population of 5,209 persons from Framingham, Massachusetts, was examined over a 4-year period starting in September 1948 and was reexamined every 2 years for over 48 years. The FHS pioneered the concept of risk factors associated with the development of CHD. Inclusion of the offspring, or second generation, of the original Framingham group began in 1971, and the third-generation study started in 2002. The FHS has been one of the most successful longitudinal studies in the history of medical research and resulted in over 2,473 research-based publications as of 2012. As just one example, the FHS was the first to indicate the importance of cholesterol as a risk factor for CHD, and subsequently to demonstrate that the true risk was associated with high levels of LDL cholesterol and low levels of high-density lipoprotein (HDL) cholesterol.

The Bogalusa Heart Study is another long-term epidemiological study (similar to the FHS) that began in 1972 and continues today. The Bogalusa study is a longitudinal study of cardiovascular disease risk factor development from birth through age 39. In 204 of the subjects who died prematurely (primarily from accidents, homicides, or suicides), the scientists found a strong relation between the risk factors and development of fatty streaks; the greater the number of risk factors, the greater the development of aortic

TABLE 21.3 **Level of Risk for Coronary Artery Disease Associated with Selected Risk Factors**

Risk factor		Level of risk: Optimal	Borderline	At risk
Blood pressure[a]				
Systolic	mmHg	<120		>140
Diastolic	mmHg	<80		>90
Overweight-obesity (body mass index)[a]	kg/m^2	18.5-24.9	25.0-29.9	> 30
Fasting plasma glucose[b]	mg/dl	<100	100-125	>126
Physical activity[a,c]	min/week	150-300		<150

[a]American Heart Association, 2010.

[b]Data from the American Diabetes Association: www.diabetes.org/diabetes-basics.

[c]Moderate to vigorous exercise on most days of the week.

and coronary artery fatty streaks.[5] In fact, the authors of the Bogalusa Heart Study were among the first to describe the presence of fatty streaks, the origins of atherosclerotic heart disease, in children.

Lipids and Lipoproteins

For many years, cholesterol and triglycerides were the only lipids observed in epidemiological studies. More recently, scientists have studied the manner in which lipids are transported in the blood. Lipids are insoluble in blood, so they are packaged with a protein to allow transport through the body. **Lipoproteins** are the proteins that carry the **blood lipids**. Two classes of lipoproteins of major concern for CHD are **low-density lipoprotein cholesterol (LDL-C)** and **high-density lipoprotein cholesterol (HDL-C)**. Low-density lipoprotein cholesterol has been implicated in plaque formation, whereas HDL-C is probably involved in plaque regression. High levels of LDL-C and low levels of HDL-C place a person at increased risk of having a heart attack. Conversely, a high level of HDL-C and a low level of LDL-C indicate a lower risk for heart attack. Thus, the ratio of the two lipoproteins is important. Simply assessing total cholesterol is not adequate to predict risk. A person with a moderately high level of total cholesterol (Total-C) coupled with a high concentration of HDL-C and low concentration of LDL-C is at a lower risk than a person with a moderately low level of Total-C but a higher level of LDL-C and a low HDL-C. Because of the important and distinct functions of these two lipoproteins, physicians and researchers now use the term *dyslipidemia* to refer to the lipoprotein-related risk factor for CHD, rather than "high cholesterol." The emphasis is on the relative levels of LDL-C and HDL-C.

Why are these two cholesterol carriers associated with different risk levels? Low-density lipoprotein cholesterol is thought to be responsible for depositing cholesterol in the arterial wall, whereas HDL-C removes cholesterol from the arterial wall and transports it to the liver to be metabolized. Because of these very different roles, it is essential to know the specific concentrations of both of these lipoproteins when one is determining individual risk. The new AHA/ACC (American College of Cardiology) guidelines for managing dyslipidemia no longer specify target treatment levels for lipids, but rather assess the need to treat with drugs and the effectiveness of any treatment based on the levels of LDL-C.[38] The preferred treatment strategy involves prescribing HMG-CoA reductase inhibitors, or statins. These are a class of drugs that block the formation of substances used to produce cholesterol in the liver. Thus, unlike the previous treatment guidelines (ATP3), the new guidelines give no specific targets for LDL, HDL, or even ratios of lipoprotein classes; rather, the therapy is guided by its effectiveness.

A new, long-awaited set of practice guidelines for the treatment of high cholesterol was published in 2013,[38] replacing previous guidelines from 2001. The ATP4

RESEARCH PERSPECTIVE 21.2

Maintaining Fitness Into Middle Age Reduces CVD Risk

Aerobic fitness has a dose–response relation with cardiorespiratory fitness, which confers many health benefits, including overall lower risk of cardiovascular disease, metabolic syndrome, diabetes, and total mortality. Studies that have examined the effects of changes in cardiorespiratory fitness suggest that those who maintain or improve fitness over time have lower risk for developing and dying of cardiovascular diseases, including hypertension, high cholesterol, and diabetes, compared to peers with longitudinal decreases in fitness. However, most of these longitudinal studies have focused on a white male population, reducing their generalizability, and have typically spanned only a 5- to 7-year period.

In 2015, an analysis of participants in the Coronary Artery Risk Development in Young Adults (CARDIA) study examined how changes in fitness over 20 years affected cardiometabolic risk factors in middle age.[11] The CARDIA study is an ongoing study of African American and Caucasian men and women that has collected fitness and cardiovascular health data over 25 years. For this investigation, the study team was interested in how fitness changed over the first 20 years of the study and how those changes in fitness were related to cardiometabolic outcomes. They used multiple linear regression and proportional hazards regression with adjustments for individual differences in baselines values to analyze data from 2,048 participants (43% men, ~50% African American). Across the 20-year span of the study, 20.6% of participants maintained their fitness level. The regression analysis revealed that maintaining fitness was associated with greater increases in high-density lipoproteins (HDLs) and smaller increases in low-density lipoproteins (LDLs) and triglycerides. Participants who maintained fitness had also gained less weight and waist circumference after 20 years. Overall, maintaining fitness over 20 years is associated with more favorable cardiometabolic risk factors in middle age. These findings extend previous findings to a larger and more generalizable cohort and underline the importance of maintaining fitness throughout the life span.

(Adult Treatment Panel) guidelines take an entirely different (and somewhat controversial) approach to treating high cholesterol. There are no overtly expressed treatment goals based on LDL or total cholesterol concentrations; rather, the recommendation is risk reduction through identification of statin benefit groups. These are groups of individuals who would benefit from pharmacological treatment with statins. People who fall within one of these specified groups should be treated with moderate- or high-intensity statin therapy. The benefit groups are identified according to risk as individuals

- with clinically diagnosed CHD;
- without clinically diagnosed CHD but with LDL ≥190 mg/dl;
- not in the preceding category, but between 40 and 75 years of age with type 1 or type 2 diabetes; and
- not in the preceding categories but aged 40 to 75 years with a 10-year risk ≥7.5% (by new risk assessment standards that use a complex formula based on age, sex, total and HDL cholesterol, smoking status, SBP, and use of antihypertensive medications).

The controversy about this particular set of guidelines arises from the potential increase in the use of statins among Americans, even those who do not have elevated cholesterol concentrations. Some estimate that statin prescriptions will double, increasing the use from 15 million to 30 million Americans. The cost of individual statin drugs ranges from about $0.50 to $4.00 per dose, depending on which statin is prescribed.

Important for exercise professionals, the expert panel prefaces the new guidelines with the following statement:

> It must be emphasized that lifestyle modification (i.e., adhering to a heart healthy diet, regular exercise habits, avoidance of tobacco habits, and maintenance of a healthy weight) remains a critical component of health promotions and ASCVD risk reduction, both prior to and in concert with the use of cholesterol-lowering drug therapies.

Risk Factors for Hypertension

The risk factors for HTN, like those for CHD, can be classified as those we can control and those we cannot. Those we cannot control are heredity (family history of HTN), sex, advanced age, and race (increased risk for people of African or Hispanic ancestry). Risk factors we can control are

- obesity and overweight,
- diet (sodium, alcohol),
- use of tobacco products,
- use of oral contraceptives,
- stress, and
- physical inactivity.

Although heredity is a risk factor for HTN, it probably plays a much smaller role than many of the other proposed factors. We must remember that lifestyle factors are often quite similar within a family.

Recently, scientists have shown great interest in a possible link between HTN, obesity, type 2 diabetes, and CHD through the common pathways of inflammation and insulin resistance. Obesity also has been established as an independent risk factor for HTN. Numerous studies have shown substantial reductions in blood pressure with weight loss in hypertensive patients. Also, although sodium intake traditionally has been linked to HTN, this relationship is controversial.

Physical inactivity is a risk factor for HTN. Its role has been conclusively established in epidemiological studies. Furthermore, substantial evidence indicates that increasing physical activity tends to reduce elevated blood pressure.

In Review

- Risk factors for CHD that we cannot control are heredity (and family history), male sex, and advanced age. Those that we can control are abnormal blood lipids and lipoproteins, HTN, smoking, physical inactivity, obesity, and diabetes.
- Inflammation, insulin resistance, and vascular dysfunction are common pathophysiological conditions found in atherosclerosis, type 2 diabetes mellitus, obesity, HTN, and other chronic diseases.
- Low-density lipoprotein cholesterol is thought to be responsible for depositing cholesterol in the arterial walls. However, HDL-C acts as a scavenger, removing cholesterol from the vessel walls. Thus, high HDL-C levels provide some degree of protection from CHD.
- Risk factors for HTN that cannot be controlled include heredity, advanced age, and race. Those we can control are obesity, diet (sodium and alcohol), use of tobacco products and oral contraceptives, stress, and physical inactivity.

Reducing Risk Through Physical Activity

The potential role of physical activity in preventing or delaying the onset of CHD and HTN has been of major interest to the medical community for many years. In

RESEARCH PERSPECTIVE 21.3

Physical Activity Reduces Cigarette Cravings

Cigarette smoking is a global health concern. In the United States, over 18% of adults smoke cigarettes and about one in five deaths is caused by smoking. Many smokers want to quit but only 3% to 5% of unaided attempts are successful after 1 year. Most people who quit smoking relapse within the first 8 days. Even with assistive behavioral or pharmacological therapy, fewer than 30% of smokers successfully quit smoking. Physical activity has been recommended as an aid for smoking cessation programs; however, there is little scientific evidence of the ability of physical activity to assist in smoking cessation.

Recently, an individual patient data meta-analysis investigated the acute effects of short bursts of physical activity on cigarette craving.[18] The meta-analysis reviewed 19 intervention studies that had examined the acute effects of physical activity on the desire to smoke. The interventions included moderate-intensity walking, running, light to vigorous cycling, and isometric exercise, with exercise bouts lasting anywhere from 5 to 40 min. Short bouts of physical activity resulted in large reductions in the desire to smoke. Furthermore, the effect size of the physical activity intervention was much larger when moderate-intensity exercise was considered, suggesting that the intensity of the exercise is likely related to the acute reduction in cravings. The researchers concluded that acute bouts of physical activity reduce cigarette cravings in individuals who are trying to quit smoking. Future studies should investigate if the mode, intensity, or duration of the activity can modulate the reduction in cravings and should determine if acute reductions in cravings can lead to long-term success in quitting smoking.

the following sections, we try to unravel this mystery by examining the following areas:

- Epidemiological evidence
- Physiological adaptations with training that might reduce risk
- Risk factor reduction with exercise training

Reducing the Risk of Coronary Artery Disease

Physical activity has been proven effective in reducing the risk of cardiovascular disease. In the following sections, we discover what is known about this topic and what physiological mechanisms are involved.

Lifestyle Management to Reduce the Risk of Cardiovascular Disease

New practice guidelines have been established by the AHA and the ACC to assist physicians and other health care professionals in evaluating the roles of physical activity and a healthy diet in the prevention and treatment of cardiovascular disease (CVD).[14] These are evidence-based guidelines focused on answering critical questions, an approach similar to that of practice guidelines and standards published jointly by the AHA and the ACC in 2013.[20,38] In the evidence-based approach, recommendations are based largely on analyses of randomized controlled trials (RCT) and on meta-analyses of those trials. The evidence present in the research is classified as strong, moderate, weak, or insufficient. (A final category, "recommend against," is also sometimes used.) The resulting guidelines replace similar previously published guidelines and practice standards.

The three critical questions addressed by the CVD guidelines are as follows:

- What are the effects of dietary pattern and macronutrient composition on risk factors and outcomes for CVD when compared to no treatment and to other types of treatment?
- What are the effects of dietary intake of sodium and potassium on CVD risk factors and outcomes when compared to no treatment and to other types of treatment?
- What is the effect of physical activity on blood pressure and serum lipids compared to no treatment and to other types of treatment?

Based on the existing research literature, the following summarized conclusions were reached:

Dietary pattern. This question centers on the efficacy of specific dietary changes for controlling lipids (specifically LDL cholesterol) and blood pressure. Strong evidence supported a dietary pattern emphasizing eating a largely plant-based diet (vegetables, fruits, and whole grains), low-fat dairy products, poultry, and fish and restricting sugar-sweetened beverages and red meats, as well as restricting saturated fat intake to 5% to 6% of total calories and reducing intake of trans fatty acids.

Dietary sodium and potassium. This question focuses on adults who may benefit from lowering resting blood pressure. Specifically, it highlights two micronutrients: sodium and potassium. The recommendations include a dietary pattern similar to that recommended for the first critical question. Strong evidence supported sodium restriction, either limiting sodium to the recommended daily intake (RDI) of 2,400 mg/day or reducing sodium intake by 1,000 mg/day if the RDI cannot be reached. A further reduction in sodium intake to 1,500 mg/day, if possible, is associated with further reductions in blood pressure and is recommended.

Physical activity. This question centers on the effects of exercise and physical activity on lowering lipids and blood pressure. Moderate evidence supported a role for aerobic exercise in lowering LDLs and total cholesterol but not in altering triglycerides or HDLs. The statement recommends aerobic activity for three or four sessions per week, 40 min per session, at a moderate to vigorous intensity. The evidence was strong in supporting aerobic exercise for the reduction of blood pressure for those with HTN, and similar exercise recommendations are suggested.

The guidelines recommend similar dietary changes and exercise for all adults with the goal of reducing risk factors, including 150 min/week of moderate-intensity exercise or 75 min/week of vigorous-intensity exercise. There is little doubt that a healthy dietary pattern and daily physical activity are healthy partners in a lifestyle to prevent chronic disease.

Inactivity as a Risk Factor: Epidemiological Evidence

Hundreds of research papers have dealt with the epidemiological relationship between physical inactivity and CAD. Generally, studies have shown the risk of heart attack in sedentary male populations to be about two to three times that of men who are physically active in either their jobs or their recreational pursuits. Early epidemiological studies from the 1950s were among the first to demonstrate this relation.[32] In these studies, sedentary bus drivers were compared with active bus conductors who worked on double-decker buses, and sedentary postal workers were compared with active postal carriers who walked their routes. The death rate from CAD was about twice as high in the sedentary groups as in the active groups. Many studies published over the subsequent 20 years showed essentially the same results: People who were occupationally sedentary were at approximately twice the risk for death from CAD as those who were active.

Most of these early epidemiological studies focused exclusively on occupational activity. Not until the 1970s did researchers start looking at leisure-time activity as well. Dr. Morris and his colleagues, who conducted the early studies of bus drivers and postal workers, were among the first to observe the relationship between leisure-time activity and the risk of CAD: The least active people were at two to three times greater risk. Subsequent studies have provided similar results.[8] Physical inactivity approximately doubles the risk of having a fatal heart attack. While most of these early studies were conducted on men, subsequent investigations have demonstrated similar results in women.[12]

Many extensive reviews of epidemiological studies published on physical inactivity and CAD have been published in the past 30 to 40 years. Many have applied stringent criteria for inclusion of studies and the quality of the research. Most indicate that inactive people have about twice the risk of dying from heart disease as more active people. Additionally, these studies have demonstrated that the relative risk from physical inactivity is similar to the risk associated with other major risk factors for CAD.[2] The results from these epidemiological studies have played a major role in leading the AHA and other organizations to declare physical inactivity a primary risk factor for CAD.

Exercise Type and Intensity

Another important concern was raised in the mid-1980s: What level of physical activity or fitness is necessary to reduce one's risk for CAD? It was not totally clear from the epidemiological studies what threshold level of fitness or activity was effective in reducing risk. In fact, during the mid-1980s, scientists had just started to differentiate between activity level (a pattern of behavior) and fitness (defined by a person's $\dot{V}O_{2max}$), a discussion that continues today.[7,33] In retrospect, distinguishing these two terms was crucial because a person can be active yet unfit (low $\dot{V}O_{2max}$) or fit (high $\dot{V}O_{2max}$) yet inactive. This area of research was subsequently redirected based on various epidemiological studies demonstrating that the levels of activity associated with a lower risk for CAD were generally low, well below the level that would be necessary to increase aerobic capacity. Subsequent studies have supported this concept.[8] Low levels of activity, such as walking and gardening, can provide considerable benefit in reducing the risk for CAD. However, it has also been demonstrated that more vigorous exercise may provide even greater benefit.

In 2002, a group of scientists from Harvard University reported the relationship of exercise type and intensity to CAD in more than 44,000 men enrolled in the Health Professionals Follow-Up Study.[39] These men were followed every 2 years from 1986 through 1998 to assess potential CAD risk factors, identify newly diagnosed cases of CAD, and assess levels of leisure-time physical activity. Men who ran 6 mph (9.7 km/h) or faster for 1 h or more per week had a 42% risk reduction compared with men who didn't run. Men who trained with weights for 30 min or more per week had a 23% risk reduction when compared with men who didn't weight train. Brisk walking for 30 min or more per day was associated with an 18% risk reduction, as was rowing for 1 h or more per week. Surprisingly, swimming and cycling were unrelated to risk. This study was the first to show the direct benefits of weight training on CAD risk and indicate that exercise intensity is also a critical consideration, with higher intensities providing greater risk reduction.

More recently, research has shown that the type and intensity of exercise is directly related to mortality and disease prevention.[42] The main outcome of this research was that people who didn't meet the physical activity guidelines recommended by the American College of Sports Medicine (ACSM) (150-250 MET-minutes per week) had a significantly higher mortality than those who met or exceeded those guidelines. In fact, after adjustment for medication use and body mass index, and even such variables as education and dietary pattern, those who exceeded the recommendations by two to three times had almost 25% lower mortality rates than those who just met the recommendations. There was also a modest dose–response relation between the amount of walking and the reduction in mortality[42] (figure 21.6).

Effects of Exercise on Risk Factors

The importance of regular physical activity and exercise in reducing the risk of CAD becomes apparent when we consider anatomical and physiological adaptations in response to exercise training. Many studies have investigated the role of exercise in altering risk factors associated with heart disease. Let's consider the major risk factors and how exercise might affect them.

Acute exercise and even lower levels of physical activity that may not induce significant training effects can directly and indirectly affect the risk for and the progression of CAD. Such basic and important factors as inflammation, endothelial function, insulin resistance, the production and secretion of adipokines, glucose metabolism, blood pressure, and even the influence of lipoproteins in the blood are positively influenced by exercise and physical activity.

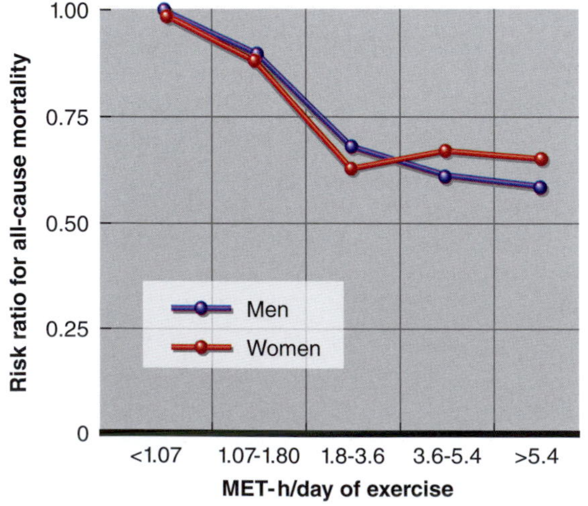

FIGURE 21.6 The relation between total daily exercise and all-cause mortality for men and women in the Williams study. This is a dose–response curve, indicating that the higher the dose, the better the response.

Data from Williams (2013).

Smoking There is little direct evidence to indicate that exercise leads to smoking cessation or reduces the number of cigarettes smoked.

Lipids According to the recently published AHA guidelines on lifestyle management, there is moderately strong evidence that aerobic (endurance) exercise decreases LDL-C, but the evidence for the effects of exercise in altering either triglycerides or HDL-C is low.[14] While many studies have demonstrated increases in HDL-C and decreases in triglycerides from training, others have reported little or no change.

Postprandial (after a meal) lipemia, the influx of fat into the blood after a fatty meal, increases the risk for CAD, stimulates plaque progression, and decreases plaque stability. Exercise before a high-fat meal moderates this effect by decreasing the amount of lipids entering the blood. High fat (or even moderate amounts of fat) in a meal can cause a significant postprandial outflow of lipoproteins that is atherogenic and, in fact, a risk factor.

We must consider two confounding factors when we evaluate lipid changes with exercise training because each can have a marked effect. Because plasma lipids are expressed as a concentration (milligrams of lipid per deciliter of blood), any change in plasma volume will affect plasma concentrations independently of the change in total lipid. Recall that training typically increases plasma volume (discussed in chapter 11). With this plasma expansion, the absolute amount of HDL-C could increase yet the HDL-C concentration might not change or even could decrease. In addition, plasma lipid levels are tightly coupled with changes in body weight. When we evaluate the effects of exercise training, the independent effects that a change in body weight could have on plasma lipids must be considered.

Hypertension Relatively strong data support the effectiveness of exercise in reducing blood pressure in those with mild to moderate HTN. Endurance training can reduce systolic (2-5 mmHg) and diastolic blood pressures (1-4 mmHg) in all individuals, including those with HTN.[14] The specific mechanisms responsible for the decreases in blood pressure with endurance training have yet to be fully determined.

Insulin Sensitivity and Glucose Exercise improves glucose uptake and utilization, as well as insulin sensitivity.[17] In fact, a single bout of exercise has been shown to improve insulin receptor sensitivity.

Endothelial Function Strong evidence demonstrates that exercise training improves endothelial function. Endothelial dysfunction is primarily the result of decreased NO bioavailability. Exercise training has been shown to increase NO bioavailability.[40]

Inflammation Moderate exercise is acutely anti-inflammatory.[19] Exercise has been shown to be associated with decreased levels of inflammatory markers and with increased immune system substances that slow the atherosclerotic process.

Other Factors With respect to the remaining risk factors, exercise plays an important role in weight reduction and control and in the management of diabetes. These topics are discussed in detail in chapter 22. Exercise also has been reported to be effective for reducing and controlling stress, reducing anxiety, and treating depression and anxiety.[9]

Endurance training produces other, long-term favorable anatomical and physiological changes that decrease the risk of heart attack, including larger coronary arteries, increased heart size, and increased pumping capacity. It also has a favorable effect on most of the risk factors for CAD. While not studied to the same extent, resistance training appears to provide many of these same benefits. This may be especially true when one combines aerobic training with resistance training, which may positively affect many other physiological variables pertinent to primary and secondary disease prevention.[41]

Sedentary Time and Disease Risk

In chapter 20, we mentioned sedentary death syndrome, a term coined by Dr. Frank Booth and colleagues. They have studied the physiology (and genetics) of a sedentary lifestyle, postulating that humans are genetically driven to be physically active and that being sedentary is a root cause of all chronic disease and pathophysiological dysfunction. More recently, an emerging research trend has been to look at sedentary time as reflected by the amount of time populations spend watching TV, working, playing or watching videos on a computer, and engaging in other seated activities. Most of us spend large amounts of time sitting—some more sedentary people may spend up to 12 to 15 h per day sitting.

The epidemiological research related to sedentary time and mortality from heart disease is sparse and not well developed. It is clear that sedentary time confers increased risk for fatal and nonfatal CVD.[16] This appears to be true regardless of whether someone is a regular exerciser or not.[13] The time spent sitting or engaging in a sedentary activity (irrespective of sleep time) appears to be a significant factor in a healthy lifestyle.

The National Health and Nutrition Examination Survey (NHANES) database (an ongoing analysis of a cross section of U.S. residents) provides further support for the direct relation between the amount of sedentary time and such risk indicators as HDL-C, triglycerides, insulin and insulin resistance, and inflammatory markers. Longer periods of sitting and higher daily sitting times are associated with harmful metabolic changes across the board, irrespective of whether moderate or vigorous exercise is interspersed.[26]

Reducing the Risk of Hypertension

Physical activity's role in reducing the risk of developing HTN has not been as well established as its role with respect to CAD and its role in lowering already elevated blood pressure. As we saw in the previous section, exercise training lowers blood pressure in those with moderate HTN, but the precise mechanisms allowing this reduction are not yet fully known. Let's consider what is known.

Epidemiological Evidence

Few epidemiological studies have dealt with the relation between physical inactivity and HTN. In the Tecumseh Community Health Study, 1,700 men (age 16 and older) completed questionnaires and interviews to provide estimates of their average daily energy expenditures, their peak daily energy expenditures, and the hours they spent in particular activities. The more active men had significantly lower systolic and diastolic blood pressures, irrespective of age.[31] In a follow-up of the participants from the Cooper Clinic study, the investigators reported a relative risk of 1.5 for the development of HTN in people with low levels of fitness compared with highly fit people.[6] The NHANES database revealed that newly identified HTN was associated with low fitness estimated from a submaximal treadmill test. Odds ratios of 2.12 for women and 1.83 for men (20-49 years) indicated that low fitness levels more than double the risk of HTN in women and almost double the risk in men.[10] (An odds ratio represents the odds that an outcome will occur given a particular risk factor. An odds ratio greater than 1.0 implies an association between the outcome and the risk factor.)

These limited studies indicate that active people and fit people are at reduced risk for developing HTN. Epidemiological studies also have shown that higher physical activity levels and aerobic fitness are related to a decreased risk for stroke in both men[28] and women.[23]

Training Adaptations That Might Reduce Risk

Specific mechanisms responsible for reductions in resting blood pressure with endurance training have not been clarified. The most likely mechanisms include neurohumoral adaptations, decreased sympathetic nervous system activity, changes in the renin–angiotensin system (a critical control system for blood pressure), and perhaps even structural changes in the vascular system itself.

The phenomenon known as postexercise hypotension (PEH) was noted almost 50 years ago. Subsequently, it has been confirmed in most exercising populations, and the evidence is overwhelming that PEH is a significant, acute effect of exercise that may in fact be linked to the chronic effect of exercise on blood pressure in people with HTN. Postexercise hypotension occurs in people with normal blood pressure as well as in people with HTN. It lasts from 1 to 2 h postexercise and sometimes up to 24 h postexercise. The greatest reductions in blood pressure seem to occur in those with the highest preexercise blood pressure. Reductions in SBP of 4 to 15 mmHg have been reported. The fact that this is an acute response that may last up to 24 h further supports the need for regular, daily physical activity, especially in those with HTN.

Risk Reduction with Exercise Training

In the previous section on CHD, we saw that exercise training lowers resting blood pressure in normotensives and in those with HTN. These reductions are unrelated to the duration of the training program but might be greater in response to low- to moderate-intensity activity compared with higher-intensity activity.

Not only does exercise reduce blood pressure in itself but also it affects other risk factors. Exercise is important in reducing body fat, can increase muscle mass, and promotes improved glycemic (blood sugar) control. Exercise training also has been associated with stress reduction.

In Review

- Epidemiological studies generally have found that the risk of CAD in sedentary male populations is about two to three times that of men who are physically active, and that physical inactivity approximately doubles a person's risk of a fatal heart attack.
- The levels of activity associated with a reduced risk for CAD can be lower than those needed to increase aerobic capacity.
- Physical training improves the heart's contractility, work capacity, and coronary circulation.
- Exercise has a significant impact on LDL-C. Endurance training decreases the ratios of LDL-C to HDL-C and of Total-C to HDL-C.
- Exercise is anti-inflammatory and improves endothelial function.
- Exercise can help control blood pressure, weight, and blood glucose levels and alleviate stress.
- People who are active and those who are fit have reduced risk for developing HTN.
- Exercise helps to reduce blood pressure in people with HTN.
- Resting blood pressure is decreased by training in individuals both with and without HTN; this is probably attributable to decreased peripheral resistance, coupled with neurohormonal mechanisms, but the actual mechanisms are not well defined.
- Exercise also reduces body fat, blood glucose levels, and insulin resistance, factors related to an increased risk for HTN.

Risk of Heart Attack and Death During Exercise

When someone dies while exercising, the incident usually makes headlines. Deaths during exercise don't happen often, but they are highly publicized. The stories at the beginning of this chapter offer a few examples. How safe, or how dangerous, is exercise? Death rates and mortality associated with exercise and physical activity among adults (not including youth and adult athletes) have been documented in the scientific literature. The incident rates depend on how the numbers are expressed, and they vary according to population (e.g., healthy adults, those with diagnosed CAD) and sex (women have higher rates than men for unknown reasons).[1] As summarized by the ACSM,[1] these rates vary from 1 cardiovascular event in almost 3,000,000 person-hours of activity, according to the YMCA data, to 1 fatality per 396,000 person-hours of jogging, to 1 fatality in 2.597 million workouts reported by a commercial health–fitness facility. Fatality rates in medically supervised exercise programs for patients with diagnosed CAD have been reported as 1 in 752,365 patient-hours. The extremely low rates in this population are undoubtedly the result of the medical supervision, available emergency equipment and assistance, and the pre-entry medical screening.

The overall risk of heart attack and death during exercise is very low. Further, although the risk of death increases during a period of vigorous exercise, habitual vigorous exercise is associated with an overall decreased risk of heart attack.[1,37] This is illustrated in figure 21.7. There is concern, however, regarding athletes who pursue ultra-endurance exercise, that is, training bouts or competition in excess of 4 h per session. Theoretically, they are at greater risk for cardiovascular disorders because of the high oxidative stress associated with this type of training or competition. Additionally, a recent

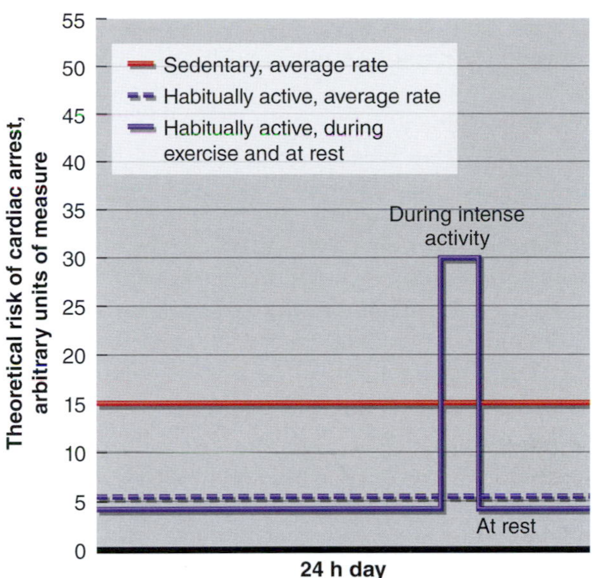

FIGURE 21.7 The risk of primary cardiac arrest during vigorous exercise and at other times throughout a 24 h period, comparing sedentary men with habitually active men.
Data from Siscovick et al. (1984).

study of ultra-endurance athletes (runners) found that "long-term excessive sustained exercise may be associated with coronary artery calcification, diastolic dysfunction, and large-artery wall stiffening" (p. 587).[34] This conclusion was somewhat speculative and requires further documentation. It is important to note, as the authors also state, that runners live longer than nonexercisers and have excellent physical fitness levels, which are also associated with lower mortality.

When death during exercise occurs in people age 35 or older, it usually results from a cardiac arrhythmia caused by atherosclerosis of the coronary arteries. On the other hand, those under age 35 are most likely to die from hypertrophic cardiomyopathy (enlarged diseased heart, usually genetically transmitted), congenital coronary artery anomalies, aortic aneurysm, or myocarditis (inflammation of the myocardium).

The risk of heart attack increases during the actual period of exercise. However, over the course of a 24 h period, those who exercise regularly have a much lower risk of a heart attack than those who do not exercise.

Exercise Training and Rehabilitation of Patients with Heart Disease

Can active participation in a cardiac rehabilitation program that has a strong aerobic and resistance exercise component help a heart attack survivor to either survive a subsequent attack or avoid one altogether? Endurance training leads to many physiological changes that reduce the work or oxygen demand of the heart and can be seen as preventive for future events. As we have seen, many of these are peripheral and do not involve the heart directly but are directly beneficial to the prevention of the pathophysiology of chronic diseases, specifically CAD. Exercise training increases the capillary/skeletal muscle fiber ratio and plasma volume and improves the efficiency of the working heart, thus resulting in a decreased workload on the left ventricle. Endurance training also improves vascular endothelial function, decreases insulin resistance through increased insulin receptor sensitivity, and increases glucose utilization; exercise training is anti-inflammatory and moderates postprandial lipemia (fat in the blood after a high-fat meal). All of these changes promote cardiovascular health and contribute to the preventive effects of regular exercise training.

VIDEO 21.1 Presents Ben Levine on the clinical implications of $\dot{V}O_{2max}$. 

Significant changes also may occur in the heart itself. Studies of heart disease patients at Washington University in St. Louis have provided dramatic evidence that high-intensity aerobic conditioning not only can substantially change peripheral factors but also can alter the heart itself, possibly increasing blood flow to the heart and increasing left ventricular function.

From our previous discussions in this chapter, it is clear that endurance exercise training can significantly reduce the risk of cardiovascular disease through its independent effect on the individual risk factors for CAD and HTN. Favorable changes in blood pressure, lipid levels, body composition, glucose control, and stress have been reported in patients undergoing exercise training for cardiac rehabilitation. We have every reason to believe that these changes are just as important to the health of a patient who has had a heart attack as they are for an apparently healthy person.

Resistance training also has substantial benefits when included as a part of a comprehensive cardiac rehabilitation program.[41] Table 21.4 provides a summary comparison of the benefits of resistance training and aerobic training on various physiological and clinical markers of health and fitness. It is obvious that combining the two types of training into a comprehensive rehabilitation program will maximize the overall training benefits.

A comprehensive cardiac rehabilitation program must consider all aspects of the patient's recovery, not just exercise and physical activity. Nutrition counseling is extremely important for everyone, dealing not only with the total calories consumed in a day but most importantly with food selection and dietary pattern to minimize risk and maximize health. Psychological and sexual counseling might also be necessary in some

TABLE 21.4 **Comparison of the Effects of Aerobic Endurance Training and Strength Training on Health and Fitness Variables**

Variable	Aerobic exercise	Resistance exercise
Bone mineral density	↑	↑↑
Body composition		
% fat	↓↓	↓
Lean body mass	↔	↑↑
Strength	↔	↑↑↑
Glucose metabolism		
Insulin response to glucose challenge	↓↓	↓↓
Insulin sensitivity	↑↑↑↑	↑↑↑↑
Serum lipids		
HDL-C	↑↔	↑↔
LDL-C	↓↔	↓↔
Resting heart rate	↓↓	↔
Stroke volume	↑↑	↔
Blood pressure at rest		
Systolic	↓↔	↔
Diastolic	↓↔	↓↔
$\dot{V}O_{2max}$	↑↑↑	↑↑
Endurance performance	↑↑↑	↑↑
Basal metabolism	↑	↑↑

Note. HDL-C = high-density lipoprotein cholesterol; LDL-C = low-density lipoprotein cholesterol; ↑ = increase; ↓ = decrease; ↔ = little or no change. The more arrows, the greater the change.

Reprinted from M.L. Pollock and K.R. Vincent, "Resistance Training for Health," *Research Digest: Presidents' Council on Physical Fitness and Sports* 2, no. 8 (1996): 1-6.

patients. It is not unusual for patients to become anxious about their heart, and their spouses sometimes fear that having sexual relations will further harm the heart of their recovering husband or wife. Many cardiac rehabilitation programs have patient support groups in which patients can openly discuss their concerns.

Patient education, however, is not (and cannot be) the sole answer to helping patients change behavior. Clearly, providing information does not cause lifestyle change nor is it even associated with long-term changes in an individual's health habits. Disease management programs that focus on health behavior change are becoming integral to rehabilitation programs for all kinds of chronic disease, including obesity, diabetes, heart disease, and chronic lung disease. Such programs require intensified patient contact, tools for behavior change and relapse prevention, and lots of support from within and particularly from outside the health care environment.[21]

A number of researchers have tried to determine whether participation in a cardiac rehabilitation program reduces the risk of a subsequent heart attack or of death from a subsequent heart attack. It is nearly impossible to design a study to answer these questions, primarily because it would be necessary to enroll several thousand people into one study to have a large enough sample to prove a statistically significant effect. Consequently, several published reports have combined the results of the most highly controlled of these studies and have used meta-analysis, a special type of statistical analysis, to examine these data. A recent report using 34 randomized, controlled trials of exercise-based cardiac rehabilitation programs with >6,000 patients concluded that exercise rehabilitation significantly reduces total mortality, all-cause mortality, and reinfarction rates among participants.[27]

The evidence that a comprehensive cardiac rehabilitation program is essential for almost all patients with diagnosed atherosclerotic vascular disease is clear. Evidence-based recommendations state that all eligible patients should be referred to a comprehensive cardiovascular rehabilitation program.[21] Comprehensive programs should provide a team-based approach to rehabilitation, risk reduction using appropriate behavioral therapies for changing health behaviors and lifestyle, ongoing educational assistance, and technology for tracking patient outcomes and providing feedback and patient support.

In Review

- Deaths during exercise are rare, although typically highly publicized.
- Deaths during exercise in people over age 35 usually are caused by a cardiac arrhythmia resulting from atherosclerosis.
- Deaths during exercise in people under age 35 usually are caused by hypertrophic cardiomyopathy, congenital coronary artery anomalies, aortic aneurysm, or myocarditis.
- Comprehensive cardiac rehabilitation programs facilitate rehabilitation for cardiovascular-related health problems, reduce the risk of subsequent heart attacks, and improve the lifestyle and health behaviors of individuals with cardiovascular disease.
- These programs should include both aerobic and resistance training components, as well as intensive health behavior change counseling and coaching.
- The benefits of comprehensive cardiac rehabilitation programs include favorable changes in body composition, glucose metabolism, plasma lipids and lipoproteins, heart function and cardiovascular dynamics, metabolism, health-related quality of life, and reduction in the risk of subsequent heart attacks and death from heart attacks.

IN CLOSING

In this chapter, we have seen how important physical activity is in reducing the risk for cardiovascular diseases, especially CAD and HTN. We discussed the prevalence of these disorders, the risk factors associated with each, and the ways in which physical activity can help reduce our personal risks. In the next chapter, we continue examining the effects of exercise on health as we turn our attention to obesity and diabetes.

KEY TERMS

adipokines
blood lipids
congenital heart disease
coronary artery disease (CAD)
endothelium
fatty streaks
heart failure
hemorrhagic stroke
high-density lipoprotein cholesterol (HDL-C)
hypertension
ischemia
ischemic stroke
lipoproteins
low-density lipoprotein cholesterol (LDL-C)
myocardial infarction
nitric oxide (NO)
pathophysiology
peripheral vascular disease
plaque
platelet-derived growth factor (PDGF)
primary risk factors
rheumatic heart disease
risk factor
stroke
valvular heart disease

STUDY QUESTIONS

1. What are currently the major causes of death in the United States?
2. What is atherosclerosis, how does it develop, and at what age does it begin?
3. What is hypertension, how does it develop, and at what age does it begin?
4. What is stroke? How does stroke occur? What are the results of stroke?
5. What are the basic risk factors for coronary artery disease? For hypertension?
6. What is the risk of death from coronary artery disease associated with a sedentary lifestyle as compared with an active lifestyle? How has this been established?
7. What are three basic physiological alterations resulting from exercise training that would reduce the risk of death from coronary artery disease?
8. In what ways does endurance exercise training alter risk factors for heart disease?
9. What are the areas covered by the 2013 AHA/ACC practice guidelines with respect to lifestyle management for people with cardiovascular disease?
10. What is a sedentary individual's risk of developing hypertension compared with an active individual's?
11. What are three basic physiological alterations resulting from exercise training that would reduce the risk for developing hypertension?
12. What changes in blood pressure would be expected to result from endurance exercise training in hypertensive individuals?
13. Of what value is cardiac rehabilitation in treating a patient who has had a heart attack?
14. What is the risk of death with endurance exercise training?

STUDY GUIDE ACTIVITIES

In addition to the activities listed in the chapter opening outline, two other activities are available in the web study guide, located at

www.HumanKinetics.com/PhysiologyOfSportAndExercise

The KEY TERMS activity reviews important terms, and the end-of-chapter QUIZ tests your understanding of the material covered in the chapter.

Obesity, Diabetes, and Physical Activity

22

In this chapter and in the web study guide

Dan Donlevie is a 10-time successful Ironman triathlete and a two-time ultramarathon finisher. He is also a type I diabetic. At age 40, Dan began experiencing a series of worsening symptoms indicative of diabetes—excessive thirst, frequent urination, and unexplained weight loss. At first, Dan chalked these symptoms up to the intense training regimen necessary for a competitive endurance athlete, but when his blood glucose concentration was found to be over 600 mg/dl (normal fasting values are 80 to 120 mg/dl), he was diagnosed with adult-onset type 1 diabetes. Overcoming the negative feelings associated with this surprise diagnosis, Dan did two amazing things. He trained, modified his diet, and managed his disease to the extent that he not only completed the U.S. Ironman Championship in New York City 5 months later, he set a personal record. Second, he became a certified diabetes lifestyle coach dedicated to helping other athletes take control of their diabetes while continuing their active, competitive lifestyles. Many notable athletes have type I diabetes, including Jay Cutler (American football), Chris Dudley (basketball), Gary Hall, Jr. (swimming), Wasim Akram (cricket), Missy Foy (marathon), and Ben Coker (soccer). Diabetes is indeed a disease; however, with the proper mindset, knowledge base, and approach to activity and diet, it can be successfully managed—even with the demanding life of an elite athlete.

While hundreds of millions of people around the world are malnourished, many Americans are dying as a direct or indirect result of *over*consumption of food. Billions of dollars are spent each year on diets, gadgets, and other weight loss methods and additional billions for the health care costs associated with obesity. Another common, and sometimes related, chronic disease is type 2 diabetes mellitus, which links obesity and insulin resistance. Further, there are common connections between insulin resistance and coronary artery disease, hypertension, and many other diseases or disorders.[9]

A sedentary lifestyle has been associated with an increased risk for both obesity and diabetes, and both are strongly linked to other diseases such as coronary artery disease and many types of cancer. Furthermore, approximately one-third of all Americans are obese, have type 2 diabetes, or both. In this chapter, we focus on obesity and diabetes, discussing their prevalence, their etiology, the health problems associated with each disease, and general treatment options. Finally, we consider the role that physical activity can play in their prevention and treatment.

Understanding Obesity

The terms *overweight* and *obesity* are often used interchangeably, but technically they have different meanings.

Terminology and Classification

Overweight and obesity are both conditions related to the accumulation of excess adipose (fat) tissue. In the past, these conditions were defined using standard height/weight tables to establish norms. These tables were established and published by insurance companies in the 1950s and based on actuarial rather than physiological data, but they are used less frequently today. Other, more objective measures of body composition—such as body mass index, percent fat, lean body mass, and hip-to-waist ratio—are now more commonly used. Perhaps the most commonly used measure of overweight and obesity is body mass index.

In the older standard tables, normative values for weight were based on population averages. For this reason, a person could be overweight according to these standards and yet have a lower-than-normal body fat content. For example, football players may be overweight according to both the old weight tables and body mass index standards, yet many are typically much leaner than other people of the same age, height, and weight. Still other individuals are within the normal range of body mass index yet are overweight or obese.

Obesity refers to the condition of having an excessive percentage of body fat. This implies that the actual amount of body fat or its percentage of the total weight must be assessed or estimated (see chapter 15 for assessment techniques). Exact standards for unhealthy fat percentages have not been established. However, men with more than 25% body fat and women with more than 35% are typically considered obese. Men with relative fat values of 20% to 25% and women with values of 30% to 35% are considered **overweight**. Allowances are higher for women because of sex-specific fat deposits such as breast tissue and hips, buttocks, and thighs, as will be discussed later in the chapter.

Despite its pitfalls, **body mass index (BMI)** is the most widely used clinical standard to estimate obesity because of its simplicity. To determine a person's BMI, body weight in kilograms is divided by the square of body height in meters. As an example, a man who weighs 104 kg (230 lb) and is 1.83 m (6 ft) tall would have a BMI of 31 kg/m^2 [$104\ kg/(1.83\ m)^2 = 104\ kg/3.35\ m^2 = 31\ kg/m^2$]. Generally, the BMI is highly correlated with body fat and usually provides a reasonable estimate of obesity. Table 22.1 provides a simple way of determining BMI from height and weight.

The classification system based on BMI proposed by the World Health Organization was adopted by

TABLE 22.1 **Body Mass Index**

	BMI																
BMI	19	20	21	22	23	24	25	26	27	28	29	30	31	32	33	34	35
Height (ft and in.)	Weight (lb)																
4'10" (58")	91	96	100	105	110	115	119	124	129	134	138	143	148	153	158	162	167
5' (60")	97	102	107	112	118	123	128	133	138	143	148	153	158	163	168	174	179
5'2" (62")	104	109	115	120	126	131	136	142	147	153	158	164	169	175	180	186	191
5'4" (64")	110	116	122	128	134	140	145	151	157	163	169	174	180	186	192	197	204
5'6" (66")	118	124	130	136	142	148	155	161	167	173	179	186	192	198	204	210	216
5'8" (68")	125	131	138	144	151	158	164	171	177	184	190	197	203	210	216	223	230
5'10" (70")	132	139	146	153	160	167	174	181	188	195	202	209	216	222	229	236	243
6' (72")	140	147	154	162	169	177	184	191	199	206	213	221	228	235	242	250	258
6'2" (74")	148	155	163	171	179	186	194	202	210	218	225	233	241	249	256	264	272
6'4" (76")	156	164	172	180	189	197	205	213	221	230	238	246	254	263	271	279	287

1 lb = 0.454 kg, 1 in. = 2.54 cm.

Reprinted from NIH/National Heart, Lung, and Blood Institute (NHLBI), Evidence Report of Clinical Guidelines on the Identification, Evaluation, and Treatment of Overweight and Obesity in Adults, 1998.

the National Institutes of Health in 1998 with several modifications and has been in use since 2000.[27] In table 22.2, BMI values are divided into five categories: underweight, normal weight, and overweight, along with three classes of obesity—classes I, II, and III. *Morbid obesity* usually refers to those persons with BMI >40.0.[18] The degree of disease risk in table 22.2 is determined by both BMI and waist circumference. Waist circumference is important because it reflects visceral fat, which imparts higher risk for mortality, as discussed later in the chapter. A larger waist circumference increases risk within any given BMI category. Racial and ethnic differences affect the relation between BMI and fatness, necessitating different BMI cut-points for overweight and obesity in specific groups. As an example, a number of studies have indicated that for a given BMI, Asians have a higher percentage body fat. Accordingly, for the same BMI, the health risk is higher in Asian men and women. However, a BMI of 30 or higher almost always indicates excessive adiposity or obesity across all populations, regardless of race or ethnicity.

Historically, the percentage of adults who were overweight or obese varied tremendously, depending on what cut-points or standards were used to classify people as overweight and obese. We now better understand the true prevalence of overweight and obesity and

TABLE 22.2 **Classification of Overweight and Obesity by BMI, Waist Circumference, and Associated Disease Risk[a]**

			Disease risk[b]	
Classification	BMI (kg/m²)	Obesity class	Men ≤37 in. (94 cm) Women ≤32 in. (80 cm)	Men >40 in. (102 cm) Women >35 in. (88 cm)
Underweight	<18.5			
Normal[c]	18.5-24.9			
Overweight	25.0-29.9		Increased	Substantially increased
Obesity	30.0-34.9	I	High	Very high
	35.0-39.9	II	Very high	Very high
Extreme obesity	≥40	III	Extremely high	Extremely high

[a]Disease risk for type 2 diabetes, hypertension, and cardiovascular disease.

[b]Relative to normal weight and waist circumference.

[c]Increased waist circumference also can be a marker for increased risk even in persons of normal weight.

Adapted from World Health Organization (2016).

how it has changed over time. Further, including the overweight category has proven useful, providing a gray area between normal weight and obesity. People in this category can either have an above-average fat-free body mass, such as the football players mentioned previously, or can be marginally overfat. However, as stated earlier, beyond a BMI of 30, individuals are likely to be obese.

Prevalence of Overweight and Obesity

The prevalence of overweight and obesity in the United States has increased dramatically since the 1960s. According to the Centers for Disease Control and Prevention,[28] the overall prevalence of obesity in the United States was just over 30% for the period from 2011 to 2014. As illustrated in figure 22.1, the prevalence of obesity is lower in 20- to 39-year-olds but increases beyond age 40 for both men and women.

If we include the category of overweight (those with BMI values of 25 and above consistent with the World Health Organization and National Institutes of Health classification systems), more than 70% of men and over 64% of women in the U.S. adult population are overweight or obese. The prevalence of overweight and obesity has increased by 68% and 142%, respectively, since 1976, and the prevalence of morbid obesity (class III) has increased by 360%.[23]

The most recent published data from the U.S. population show that there were no significant overall changes in either adults or youth with respect to overweight and obesity over the decade from 2003-2004 to 2011-2012. However, the prevalence of obesity across the U.S. population remains very high.

When we look at these data by race, it is apparent that the problem is much more significant in Hispanic men and women and in non-Hispanic black women (figure 22.2).

These trends are not unique to the United States. Canada, Australia, and most of Europe have seen similar increases in obesity prevalence.[39] The most recent data show that obesity is spreading to most regions of the world. Table 22.3 provides estimates of the rates of obesity (combined for men and women) for select countries around the world. These data may be somewhat misleading since there is a great variation in the dates of these surveys; however, indications are that rates of obesity have continued to skyrocket during the past 10 years, resulting in a worldwide obesity epidemic. Most recent estimates project that approximately 2 billion people worldwide are overweight or obese.

Unfortunately, this same trend of increasing prevalence of overweight and obesity has been reported in U.S. children and adolescents. Figure 22.3 illustrates the trends in the prevalence of overweight from 1971 through 2012 for preadolescent and adolescent boys and girls. Because BMI is much less precise with respect to estimating body fat in children and adolescents, scientists typically use the cut-point for BMI of greater than the 95th percentile, a value that likely indicates that the child is overfat. The prevalence of overweight remained relatively constant from 1971 through 1980, increased dramatically from 1980 through 2004, and now seems to be leveling off. The other good news,

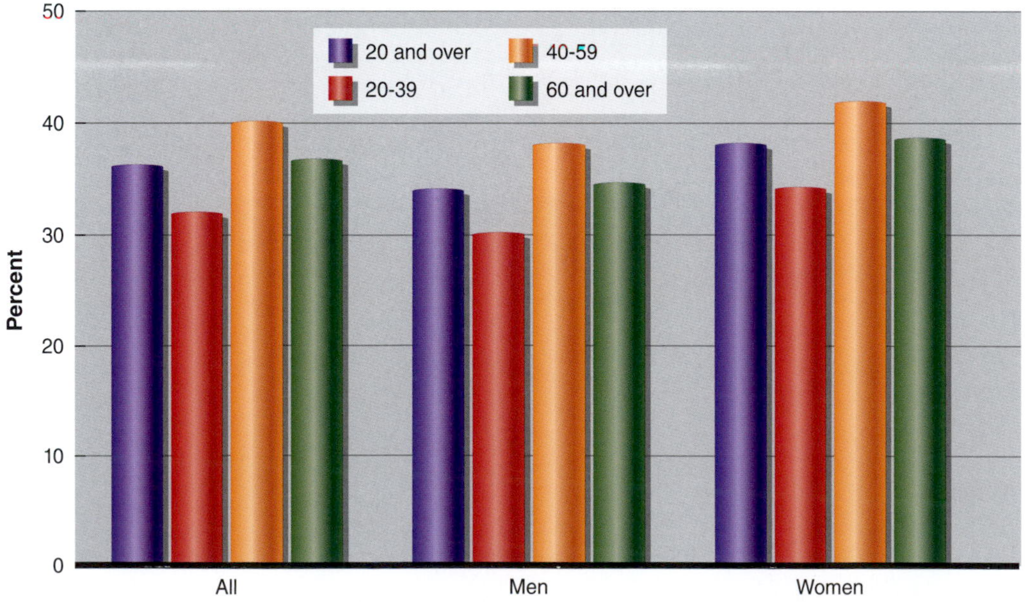

FIGURE 22.1 Prevalence of obesity in the United States among adults aged 20 and older, 2011 to 2014.
Data from Center for Disease Control and Prevention (2011).

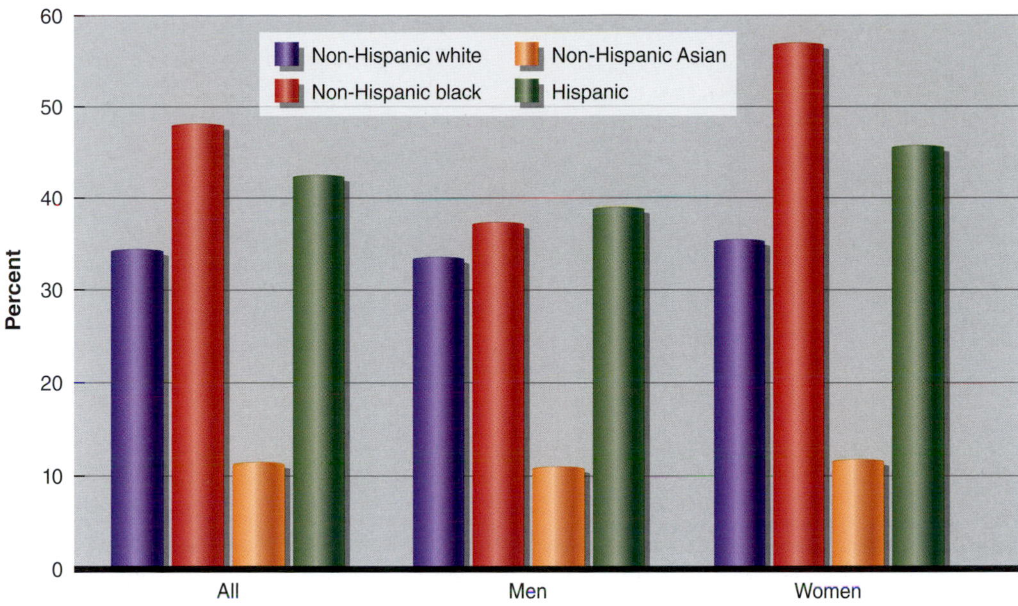

FIGURE 22.2 The prevalence of obesity in U.S. adults aged 20 and older by sex, race, and Hispanic origin, 2011 to 2014.
Data from Ogden et al. (2015).

TABLE 22.3 **Prevalence of Adult Obesity in Selected Countries**

Country	Obesity prevalence
North Korea	2.4%
Japan	3.3%
India	4.9%
South Korea	5.8%
Switzerland	19.4%
Netherlands	19.8%
Brazil	20.0%
Germany	20.1%
Italy	21.0%
Spain	23.7%
France	23.9%
Russian Federation	24.1%
Israel	25.3%
Canada	28.0%
Mexico	28.1%
United Kingdom	28.1%
Australia	28.6%
Turkey	29.5%
Saudi Arabia	34.7%
Kuwait	39.7%
Cook Islands	50.8%

Data from OECD (2014).

perhaps, is that the prevalence of obesity in 2- to 5-year-olds significantly decreased from 14% in 2003-2004 to 8.4% in 2011-2012.

The average person in the United States gains approximately 0.3 to 0.5 kg (0.7 to 1.1 lb) of weight each year after age 25. This seemingly small gain results in an average of 9 to 15 kg (15 to 33 lb) of excess weight by age 55. At the same time, bone and muscle mass decrease by approximately 0.1 kg (0.22 lb) per year due to the combined effects of reduced physical activity and normal aging processes. Taking this into account, an average person's fat mass actually increases by about 0.4 kg (0.9 lb) each year, which equates to a 12 kg (27 lb) fat gain over a 30-year period! (It should be noted that these values are approximations and that they vary by sex and race or ethnicity.)

However, population estimates of increases in calorie intake and the decreases in activity that have accompanied modern life do not match this average weight gain over a 30-year period.[16] The estimated increase in caloric intake coupled with decreases in physical activity should have accounted for annual weight gains of 39.1 kg (85.8 lb) and 37.2 kg (82 lb) for men and women, respectively. Thus, the adages often heard in weight management programs that "a calorie is a calorie" and 3,500 kcal = 1 lb does not accurately reflect real-life weight gains or the weight loss experienced by those in (or out of) weight management programs. Other factors must be at work that modify the energy balance equation.[15]

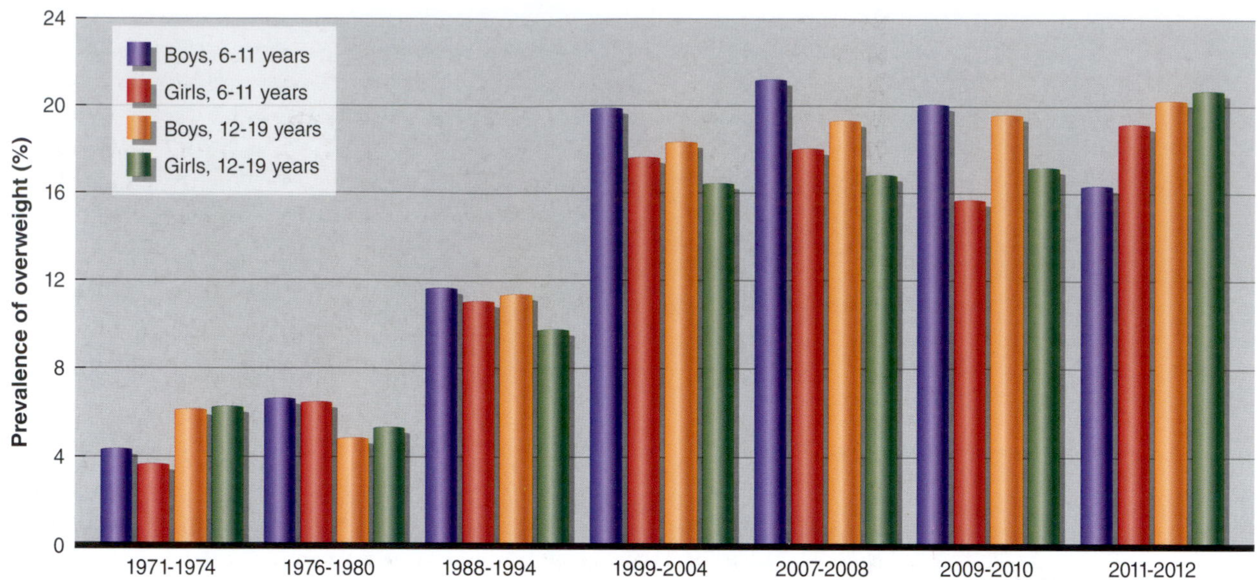

FIGURE 22.3 The increasing prevalence of overweight (>95th percentile body mass index [BMI] for age) in children and adolescents in the United States from 1971 through 2012.

Data from Flegal et al. (2002).

A major concern in light of the increasing rates of obesity, coupled with the earlier onset of obesity, is the impact that this will have on individual and national health care. With the onset of obesity occurring at younger ages will come an increased accumulative exposure to excess weight, and this will likely precipitate an earlier onset of obesity-related diseases such as diabetes.

Control of Body Weight

One must know how body weight is controlled or regulated to better understand how a person becomes obese. Body weight regulation has puzzled scientists for years. The average male consumes about 2,500 kcal per day, or nearly 1 million kcal per year. An average gain of 0.4 kg (0.9 lb) of fat each year represents an imbalance of only 3,111 kcal per year between energy intake and expenditure (3,500 kcal is the energy equivalent of 0.45 kg, or 1 lb, of adipose tissue). This translates to a surplus of less than 9 kcal per day. Even with a weight gain of 1.5 lb (0.7 kg) of fat per year, the body can balance caloric intake to within one potato chip per day of what is expended, a truly remarkable example of homeostasis.

In the past, the body's ability to balance its caloric intake and expenditure to within such a narrow range led scientists to believe that body weight is regulated around a given set point for body weight, similar to the way core temperature is regulated. More recently, research has centered on the regulation of caloric intake and expenditure, collectively called "energy balance." The physiological drive to control and regulate energy balance, as well as to conserve fuel in the form of stored fat, may be at the core of the obesity problem.[16,26] People become overweight when they have a positive energy balance. That is, when caloric intake exceeds caloric expenditure over an extended period of time, excess adipose tissue is accumulated.

Weight management programs center on losing weight (and fat) by establishing a negative energy balance in which caloric expenditure exceeds caloric intake over an extended period of time. It is critical to understand, however, that a physiological control system (and thus the control of energy balance) is affected by changes in, and regulation of, each of these components—energy intake and energy expenditure. So changes in dietary pattern and nutrient consumption, in physical activity, and ultimately in weight, will change the physiology of energy balance and probably the control system itself. Thus, sustained weight loss can be exceedingly difficult because of the physiological adaptations that occur with long-term negative caloric balance and physiological change that occur during any attempt to modify calorie expenditure and intake.

Energy balance is established and controlled within the three components of energy expenditure (see figure 22.4):

1. Resting metabolic rate (RMR)
2. The thermic effect of a meal (TEM)
3. The thermic effect of (physical) activity (TEA)

Recall that RMR, as discussed in chapter 5, is the body's metabolic rate early in the morning following an

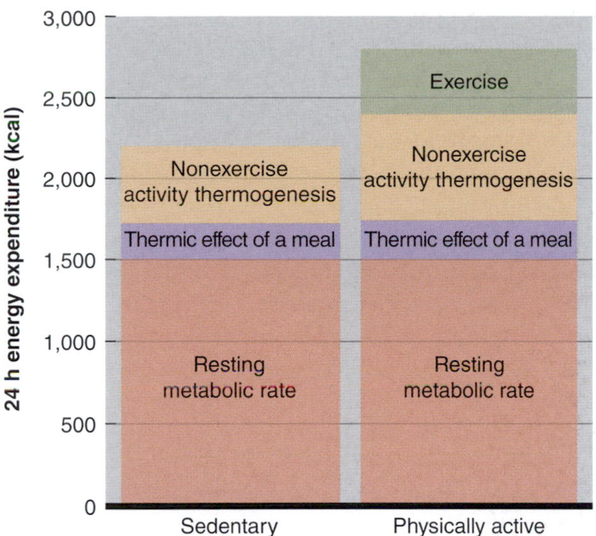

FIGURE 22.4 The components of energy expenditure, showing examples for a sedentary person and a physically active person. See the text for a detailed explanation.

overnight fast and 8 h of sleep. The term *basal metabolic rate (BMR)* also is used but generally implies that the person fasted for 12 to 18 h and that the BMR measurement was made immediately upon waking. Most research today reports the RMR represents the minimal amount of energy expenditure needed to support basic physiological processes. It accounts for about 60% to 75% of the total energy we expend each day.

The **thermic effect of a meal (TEM)** represents the increase in energy expenditure that is associated with the digestion, absorption, transport, metabolism, and storage of ingested food. While TEM peaks during and after food ingestion, on average the TEM accounts for approximately 8% to 10% of our total energy expenditure each day. This value also includes some energy waste, because the body can increase its metabolic rate above that necessary for food processing and storage. We seldom notice the TEM; however, after a very large holiday meal with family, people often start feeling warm and drowsy, perhaps indicating that the metabolic rate has increased perceptively. The TEM component of metabolism might be defective in some people with obesity, possibly attributable to a defect in the energy-wastage component, leading to a surplus of calories.

The **thermic effect of activity (TEA)** is simply the energy expended above the RMR to accomplish a given task or activity, whether it is sitting and typing on a computer or running a 10 km race. The TEA accounts for the remaining 15% to 30% of our energy expenditure. This is the most variable component of energy expenditure and the component that can most easily be purposefully modified.

The body adapts to major increases or decreases in energy intake by altering the energy expended by each of these three components—RMR, TEM, and TEA. With fasting or very low-calorie diets, all three components of total energy expenditure decrease in an attempt to conserve energy stores. This is dramatically illustrated by reported decreases in RMR of 20% to 30% or more within several weeks after initiating a fast or consuming a very low-calorie diet (<800 kcal/day). Conversely, all three components of energy expenditure increase somewhat with overeating. In this case, the body appears to be trying to prevent unnecessary storage of the surplus calories. These adaptations appear to be controlled by changes in the sympathetic nervous system, the endocrine system, and other physiological and biochemical systems that control both metabolism and appetite. The systems are extremely complex and difficult to study because of the large number of interacting variables and cross talk between systems.[15] This remains a critical area for ongoing and future research.

The composition of the diet (nutrients) and physical activity (including exercise and nonexercise physical activity) are primary control points for the energy balance equation. It is clear that changes in fat and carbohydrate (especially intake of simple or refined carbohydrates) content of the diet, coupled with a decrease in physical activity levels over an extended period of time, change our physiological processes markedly.

In Review

- Overweight is a body weight that exceeds the standard weight for a certain height and frame size. Obesity refers to excessive body fat, meaning more than 25% body fat for men and more than 35% body fat for women.
- To calculate a person's BMI, we divide body weight in kilograms by the square of height in meters. This value is highly correlated with relative body fat and provides a reasonable estimate of obesity. Body mass index values of 25.0 to 25.9 correspond to overweight, and values of 30.0 or higher correspond to obesity.
- Prevalence of obesity and overweight in the United States has increased dramatically since the 1970s.
- The average person gains 0.3 to 0.5 kg (0.7 to 1.1 lb) per year after age 25 but also loses 0.1 kg (0.22 lb) of fat-free mass per year, meaning a net gain of 0.4 kg (0.9 lb) of fat each year.
- Our bodies attempt to regulate body weight through a regulatory system based on energy balance and total energy expenditure.

- Daily energy expenditure is reflected by the sum of the RMR, the TEM, and the TEA. The body adapts to changes in energy intake by adjusting all of these components.

Etiology of Obesity

At various times throughout human history, obesity was thought to be caused by basic hormonal imbalances resulting from failure of one or more of the endocrine glands to properly regulate body weight. At other times, it has been believed that gluttony, rather than glandular malfunction, was the primary cause of obesity. In the first case, a person is perceived as having no control over the situation, yet in the second, he or she is held directly responsible. Findings from recent research show that obesity most likely results from a combination of many factors. Thus, the etiology, or cause, of obesity is extremely complex and multifactorial, rather than a simple matter of genes, hormones, or even personal choice.

Experimental studies on animals have linked obesity to heredity. Studies of humans have also shown a direct genetic influence on height, weight, and BMI. Studies by Dr. Claude Bouchard have provided possibly the strongest evidence yet of a significant genetic component for obesity.[4,5] He and his coinvestigators took 12 pairs of young adult male monozygotic (identical) twins and housed them in a closed section of a dormitory under 24 h observation for 120 consecutive days. The subjects' diets were monitored during the initial 14 days to determine their baseline caloric intake. Over the next 100 days, the subjects were fed 1,000 kcal above their baseline consumption for 6 of every 7 days. On the seventh day, the subjects were fed only their baseline diet. Thus, they were overfed by 1,000 kcal per day for 84 of the 100 days. Activity levels were also tightly controlled. At the end of the study period, the individual weight gained varied widely, from 4.3 to 13.3 kg (9.5-29.3 lb)—a threefold variation in weight gain for overconsumption of the same number of calories. However, the response of both twins in any twin pair was quite similar. The same association was found for gains in fat mass, percentage body fat, and subcutaneous fat. These and subsequent studies have demonstrated that genetics plays a major role in determining one's *susceptibility to becoming* obese. But the data demonstrate that other factors are also responsible since every subject in this study gained at least 4 kg.

Hormonal imbalances, emotional issues, and alterations in basic homeostatic mechanisms all have been shown to be either directly or indirectly related to the onset of obesity. Environmental factors, such as cultural habits, food availability, inadequate physical activity, and improper diets, are also major factors related to obesity.

While research confirms that there is a significant genetic component in the etiology of obesity, it is clearly possible to be obese based simply on lifestyle, with or without a family history (genetics) of obesity. It is also possible to be relatively lean, even with a genetic predisposition to obesity, through a healthy combination of diet and physical activity.[33]

Thus, the origin and etiology of obesity are highly complex, and the specific causes undoubtedly differ from one person to the next. Recognizing this is important for treating existing obesity and for preventing its onset. To attribute obesity solely to a single cause such as overeating or inactivity underestimates the cause and the magnitude of the problem.

Energy Balance and Obesity

In a 2012 review article, Dr. James Hill and colleagues[16] discuss the physiology (and neurophysiology) of weight gain, weight loss, and obesity. Energy balance, the sum of calorie expenditure and calorie intake, is the key to weight loss or the prevention of weight gain. Establishing a negative energy balance (caloric expenditure > caloric intake) is required for weight loss to occur. Our ancestors had a very high daily level of physical activity, coupled with a high drive to consume calories in order to establish that balance and to maintain a stable weight. Thus, humans seem to be wired to maintain weight (and energy balance) best with high levels of energy expenditure, not with large decreases in energy intake. Weight gain underlying the current obesity epidemic has been postulated as our body's attempt to reestablish energy balance in the face of excess food intake. With an increased body weight comes an increased metabolic rate and an increased caloric cost of most activities; however, with large weight gains and low activity levels, we cannot achieve energy balance, and the result is excess adipose tissue deposition. Thus, the adaptation that is most easily achieved is accumulating excess adipose tissue (thus added body weight) and becoming overweight and eventually obese. The fact is that becoming obese increases energy expenditure and eventually reestablishes the energy balance.[16]

Increases in caloric consumption with concurrent decreases in physical activity (both occupational and recreational) over the past 50 to 100 years have resulted in a high prevalence of overweight and obesity in the U.S. population. In 1960, the prevalence of overweight and obesity in the U.S. population was approximately 45%. By 2008, the prevalence was 68%. This large increase was due to a positive energy balance in a large portion of the population.[16]

Obesity is not simply a result of a problem with one of the components of the energy balance equation. Since 1960, the average individual food consumption has increased significantly, by 168 kcal/day for men and 335 kcal/day for women. There have been concomitant decreases in the average caloric expenditure of occupational activities, as well as decreased voluntary daily physical activity among men and women. When this decrease is combined with the average increase in calorie consumption, the positive energy balance would predict an average annual weight gain over the past 50 years that is much greater than that actually recorded. The actual average total weight gain for men and women has been about 30 and 25 lb (13.6 and 11.3 kg), respectively, not the amounts noted that would have been predicted by estimates of population-based changes in energy intake and expenditure. Thus, there must be additional physiological factors at work in the obesity epidemic. These are the adaptations that occur in the extremely complex physiology that controls energy balance and therefore body weight.

Jean Mayer was a preeminent Harvard physiologist who studied the physiology of nutrition and hunger. Mayer hypothesized that energy intake (caloric intake) is matched to calorie expenditure but only at high levels of energy expenditure. He further hypothesized that regulation of calorie intake is much less sensitive to energy expenditure at low levels of energy expenditure; thus, chronically maintaining a positive energy balance (calorie intake > calorie expenditure) is physiologically very difficult without weight gain. Therefore, the control of energy balance in humans works well at high levels of physical activity, but not at lower levels. So, in active and very active people, weight gain is often not a problem. But in sedentary individuals, the difficulty in sufficiently decreasing caloric intake to match very low levels of physical activity ultimately results in weight gain and obesity. It also appears that low metabolism (presumably meaning low RMR) does not significantly contribute to weight gain and obesity. Though this concept is not entirely supported by the literature, a number of well-accepted studies do support the plausibility of this view.

The most efficient way to prevent weight gain (and probably to lose weight) is to mimic those preindustrial high levels of energy expenditure in order to balance the higher average caloric intake. And most people simply cannot balance today's historically low levels of physical activity by consuming fewer calories. As a homeostatic mechanism, excess adipose tissue elevates RMR and increases the caloric cost of physical activity. Even with this physiological adaptation (increased stores of energy in the form of fat), energy balance remains positive and weight is gained as increased caloric expenditure simply cannot keep pace with the increased food intake.

Health Problems Associated with Overweight and Obesity

Overweight and obesity are associated with an increased overall mortality.[6] This relationship is curvilinear, as shown in figure 22.5. A major increase in the risk of death occurs when the BMI exceeds 30 kg/m^2, although BMI values between 25.0 and 29.9 are associated with an increased morbidity risk for many diseases. A number of more recent studies have reported that excess mortality is primarily associated with BMI values of 35.0 and above.

Excess morbidity and mortality associated with obesity and overweight are linked with a number or major diseases, including the following:[7]

- Hypertension
- Dyslipidemia
- Type 2 diabetes
- Coronary artery disease
- Stroke
- Gallbladder disease
- Osteoarthritis
- Sleep apnea and breathing problems
- Some cancers (endometrial, breast, colon, kidney, gallbladder, and liver)
- Mental illness such as clinical depression, anxiety, and other mental disorders

With the large increase in the prevalence of obesity in the United States since the 1970s, it is not surprising to also see a very high prevalence of metabolic syndrome in U.S. adults. **Metabolic syndrome** is a cluster of related conditions—increased blood pressure, elevated blood glucose, excess abdominal body fat, and elevated cholesterol or triglycerides—that occur together, increasing the risk of heart disease, stroke, and diabetes. Metabolic syndrome is defined as (1) waist circumference ≥102 cm (male adults) and ≥88 cm (female adults), (2) fasting plasma glucose ≥100 mg/dl, (3) blood pressure of ≥130/85 mmHg, (4) triglycerides ≥150 mg/dl, and (5) high-density lipoprotein cholesterol (HDL-C) <40 mg/dl (male adults) and <50 mg/dl (female adults). One-third of all adults in the United States meet these criteria. Data from the National Health and Nutrition Examination Survey (1988-2012) revealed that the prevalence of metabolic syndrome rose from 25% in 1988-1994 to 34% in 2007-2012.[25] Non-Hispanic black men were less likely than non-Hispanic white men to have metabolic syndrome, but the exact opposite relation is true for women.

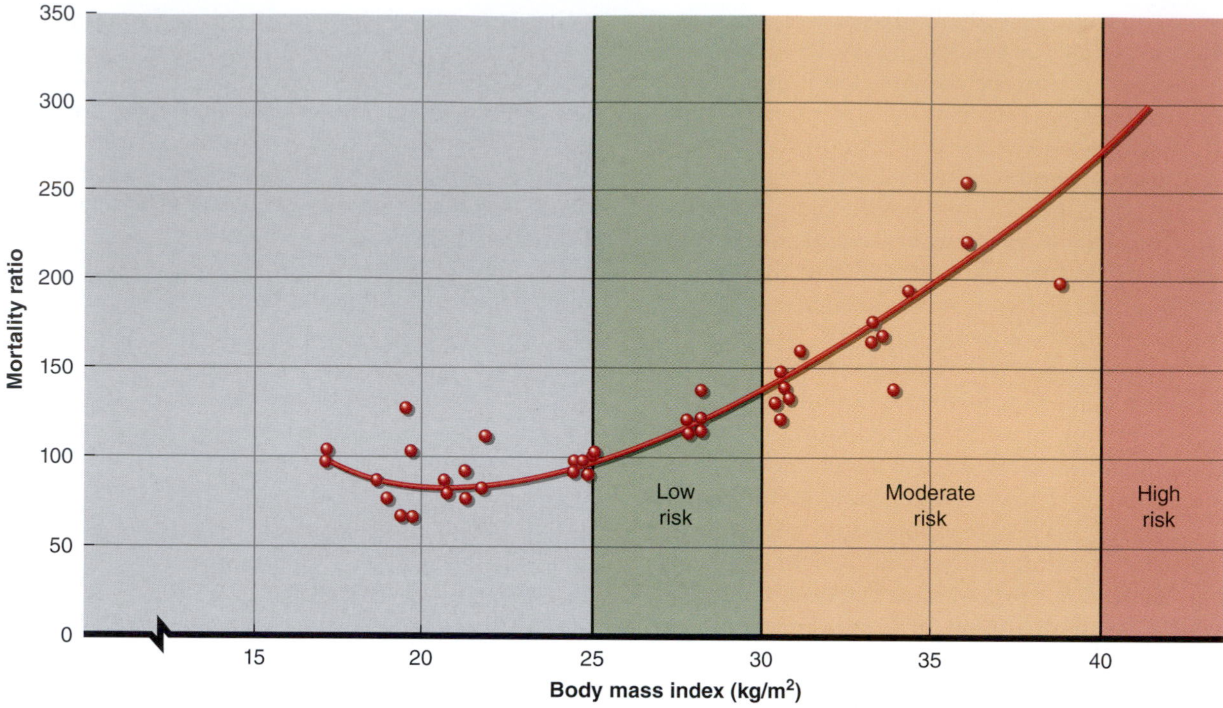

FIGURE 22.5 The relation between body mass index and excess mortality. A mortality ratio of 100 represents average mortality. The lowest portion of the curve (body mass indexes under 25) indicates very low risk, while mortality risk increases disproportionately with increasing BMI.

Bray GA. Obesity: Definition, diagnosis and disadvantages. *Med J Aust* 1985; 142: S2-S8. © Copyright 1985 The Medical Journal of Australia - reproduced with permission. The Medical Journal of Australia does not accept responsibility for any errors in translation.

Changes in Normal Body Function

The prevalence and extent of changes in physiological function associated with excess adiposity vary with the individual and with the degree of fatness. Conditions and diseases commonly associated with obesity (especially morbid obesity) include respiratory problems (sleep apnea, chronic pulmonary disease), abnormal blood clotting (thrombosis), enlargement of the heart, and congestive heart failure. Obese individuals (and often those who are overweight) typically have decreased exercise tolerance and a lower level of physical fitness due to decreased levels of physical activity. Additional weight gain (and excess adiposity) further reduces activity levels and exercise tolerance decreases. This is a vicious, positive feedback cycle of abnormal function associated with excess adiposity and decreased levels of physical activity and physical fitness, which leads to more inactivity and more pathophysiology. This cycle affects millions of people in the United States and around the world.

Increased Risk for Certain Diseases

An increased risk of developing certain chronic diseases also is associated with obesity. Both hypertension and atherosclerosis are directly linked to obesity (see chapter 21), as are impaired carbohydrate metabolism and type 2 diabetes. Obesity and type 2 diabetes are particularly common comorbid diseases.

Since the 1940s, major sex differences in the way in which fat is stored or patterned on the body have been recognized. As shown in figure 22.6, men tend to store fat in the upper body, particularly the abdominal area, whereas women tend to store fat in the lower body, particularly the hips, buttocks, and thighs. Abdominal obesity is closely associated with visceral obesity. Obesity that follows the male pattern is referred to as **upper body (android) obesity**, or apple-shaped obesity, and the female pattern is referred to as **lower body (gynoid) obesity**, or pear-shaped obesity. Apple-shaped obesity is more closely attributable to visceral fat and thus carries a higher risk.

Research beginning in the late 1970s and early 1980s established upper body obesity as a risk factor for the following conditions:

- Coronary artery disease
- Hypertension
- Stroke
- Elevated blood lipids
- Diabetes

FIGURE 22.6 Obesity patterns tend to differ by sex.

Furthermore, abdominal obesity appears to be more important than total body fatness as a risk factor for these diseases. Waist and hip circumference, or girth measurements, can be used to identify people with increased risk. A waist/hip girth ratio greater than 0.90 for men and greater than 0.85 for women indicates increased risk. With upper body obesity, the increased risk may result from the close proximity of visceral fat depots to the portal circulatory system (circulation to the liver). Figure 22.7 shows a young woman being placed into a computed tomography (CT) scanner in order to assess visceral abdominal fat (figure 22.7*a*) and MRI scans of the upper thigh of two men (figure 22.7*b*).[32] The subject in the lower two scans of figure 22.7*b* illustrates the effect of muscular atrophy and the comparison of visceral (deep) fat versus subcutaneous fat.

Effects of Obesity on Established Diseases

Though the effects of obesity on existing diseases are not totally clear, obesity clearly contributes to progression and exacerbation of some medical conditions, including type 2 diabetes, hypertension, and heart disease. Weight loss is typically prescribed as an integral part of the treatment plan. Conditions that generally benefit from weight reduction include

- angina pectoris,
- hypertension,
- congestive heart disease,
- myocardial infarction (reduced risk of recurrence),
- varicose veins,
- type 2 diabetes, and
- orthopedic problems.

Weight Loss

Healthy weight loss should generally not exceed 0.45 to 0.9 kg (1-2 lb) per week, and greater rates of weight loss should be managed only with medical supervision. Losing just 0.45 kg (1 lb) per week will result in the loss of 23.4 kg (52 lb) in only 1 year! Few people become

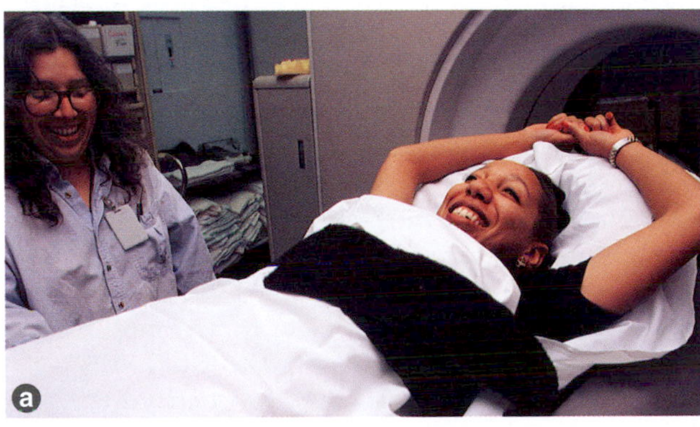

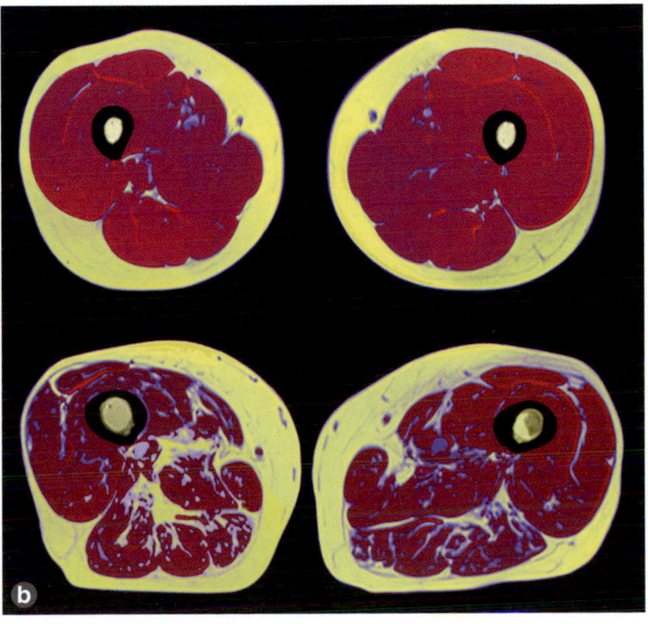

FIGURE 22.7 *(a)* A subject in the process of undergoing a computed tomography (CT) scan. *(b)* Colorized MRI scans of the upper thigh in two people. The subject in *the bottom scans* has considerably more fat infiltration (light areas).

RESEARCH PERSPECTIVE 22.1

The Fat-but-Fit Paradox

In the past 30 years, a multitude of longitudinal studies have convincingly demonstrated that low cardiorespiratory fitness (below first quintile = 20th percentile) increases all-cause mortality as well as cardiovascular disease mortality. Obesity is also a primary risk factor for all-cause and cardiovascular disease mortality. Interestingly, it has been suggested that being fit may alleviate some of the negative consequences of obesity, a concept known as the fat-but-fit paradox. That is, the mortality risk in obese individuals who are fit is not different from the risk in normal-weight and fit adults (theoretically, the healthiest group possible). In addition, several studies have now suggested that normal-weight but unfit individuals could be at a higher risk than obese but fit individuals.

Though this emerging evidence indicates that moderate-to-high cardiorespiratory fitness may mitigate some of the detrimental health consequences of obesity, many research questions remain.[30] For example, the available evidence supports a direct link between cardiorespiratory fitness and cardiometabolic risk factors in children and adolescents. Limited reports indicate that higher levels of fitness may attenuate the consequences of excess adiposity in children; however, these types of studies must be expanded to include more participants. This is especially important given the increasing public health concerns related to childhood obesity.

Perhaps the most important studies that remain to be conducted regarding the fat-but-fit paradox are randomized controlled trials. To date, all available evidence for the fat-but-fit paradox comes from observational studies. Although the information derived from these studies is valuable, it cannot determine cause and effect. Thus, randomized controlled trials are necessary to determine whether exercise interventions in obese individuals reduce cardiovascular disease risk even in the absence of weight loss. Such studies will further our understanding of the fat-but-fit paradox, because they will allow researchers to determine whether improvements in health without weight loss are driven by improvements in cardiorespiratory fitness. Public health strategies targeting obese adults should target both weight and fat reduction as well as increases in cardiorespiratory fitness, especially for unfit individuals.

obese that rapidly, so weight loss should be considered a long-term undertaking. Rapid weight loss is almost always short lived, and the weight is regained quickly. Rapid weight loss involves large losses of body water, and because the body has built-in mechanisms to prevent an imbalance in body fluid levels, the lost water eventually will be replaced. Thus, a person wishing to lose 9 kg (20 lb) is advised to attempt to attain this goal over a period of 3 to 5 months.

VIDEO 22.1 Presents Louise Burke on successful weight loss approaches and why some people have difficulty losing weight.

Many special diets have achieved popularity over the years, with each claiming to be the ultimate in effective and manageable weight loss. Some diets have been developed for use either in the hospital or at home under the supervision of a physician. These are often referred to as *very low calorie diets*, since they allow less than 800 kcal of food per day. Most of these have been formulated with proportions of protein and carbohydrate aimed at minimizing the loss of fat-free body mass. Many of these are effective, but no single diet has been shown to be more effective than any other. Again, the crucial factor for all diets should be the development of a caloric deficit while maintaining a complete, balanced diet that meets the body's nutrition needs. The best diet is the one that meets these simple criteria and is best suited to the individual.

Because improper eating habits are at least partially responsible for most cases of obesity, no diet should be viewed as a stand-alone solution or quick fix. A person should learn to make permanent changes in dietary habits, especially reducing the intake of fat and simple sugars. For most people, simply eating a low-fat diet will gradually reduce weight to a desirable level. One potential problem with low-fat diets, however, is that people mistakenly assume that low fat equates to low calorie, and this often is not the case. For most people, simply reducing total caloric intake by 250 to 500 kcal per day, combined with selecting low-fat foods and foods low in simple sugar, would be sufficient to accomplish their desired weight loss goals.

Pharmaceutical approaches to obesity have also been used to assist patients in weight loss by decreasing their appetite or increasing their RMR. Unfortunately, a number of side effects are associated with these drugs, some of which are very serious and can be life threatening. Surgical techniques also are used to treat extreme obesity but only as a last resort when other treatment procedures have failed and the obesity is life threatening. Gastric bypass surgery and gastric banding are the most common surgical procedures used to treat morbid obesity. Both restrict the amount of food that can enter the stomach. While highly effective, these procedures are expensive and have associated risks, although the average mortality rate is about 1% to 2%. These procedures have also been shown to moderate chronic diseases that are associated with obesity, including type 2 diabetes. Generally, surgery should be reserved for people who have morbid obesity or obesity with significant associated risk factors.

Behavior modification is one of the most effective techniques for helping people with weight problems. Major weight losses have been achieved through changes in basic behavior patterns associated with eating. Furthermore, these weight losses appear to be more permanent than those associated with other weight loss strategies. This approach appeals to most people because the techniques seem to make sense and are often easy to incorporate into a normal daily routine. For example, an individual might not have to consciously reduce the amount of food eaten but simply agree that all eating will be done in one location, which often cuts down on snacking. Or an individual might be allowed to take as much food as desired with the first helping but have no second helpings. Many such simple changes can help regulate eating behavior and result in substantial weight loss.

Management Guidelines for Overweight and Obesity

In 2014, new guidelines for the management of overweight and obesity were jointly published by the American Heart Association (AHA), the American College of Cardiology (ACC), and The Obesity Society.[18] These new guidelines, published in concert with three other sets of guidelines by the AHA and the ACC (assessment of cardiovascular risk, lifestyle modification to reduce cardiovascular risk, and the treatment of elevated cholesterol), were collectively designed to "promote optimal patient care and cardiovascular health." As with most recent guidelines published by highly respected professional organizations, these guidelines are evidence based, focusing on critical questions that can be answered with published, evidence-based studies in the literature.

In this new set of guidelines, the critical questions centered on (1) the criteria that guide appropriate weight loss programs; (2) the health benefits and risks of weight loss with respect to amount of loss and duration of a program; (3) which strategies for dietary interventions are effective; (4) the effectiveness of a comprehensive approach involving diet, exercise, and behavioral therapy; and (5) the safety and effectiveness of bariatric surgery as a primary therapeutic procedure. These documents are aimed at health care providers, primarily physicians and other gatekeepers, including advanced nurse practitioners and physician's assistants, but all allied health care professionals should be aware of such guidelines and employ them when appropriate.

Recommendations include the following:

- Assess height, weight, and waist circumference at least annually or more frequently if appropriate; determine BMI and advise therapeutic modality based on coronary heart disease risk.
- Counsel overweight and obese patients about cardiovascular risk factors and lifestyle changes that result in a sustained weight loss of 3% to 5% of body weight or more if possible. Greater amounts of weight loss result in greater benefits, but even modest amounts will reduce risk.
- Recommend a dietary pattern that restricts calories as part of a comprehensive lifestyle intervention. Prescribe 1,200 to 1,500 calories/day for women and 1,500 to 1,800 calories/day for men. There are effective, evidence-based diets that have been shown to promote healthy weight loss, but any intervention must reduce daily caloric intake. Among the effective diets are vegetarian and vegan, low fat (20% of calories from fat), low glycemic load, Mediterranean style, and low carbohydrate (<20 g/day without formal caloric restriction but with some total negative energy balance). Note that the common factor among these potentially effective diets is a negative energy balance.
- Because of individual variability, the strength of evidence (an important factor in determining both real-life efficacy and appropriateness of a diet for particular individuals) varies from high to low among these diets and other specific diet recommendations in these guidelines.
- A comprehensive program should include a prescribed, moderately reduced-calorie diet, increased physical activity, and behavioral strategies that facilitate compliance with the changes.
- Bariatric surgery referral is recommended for adults with a BMI >40 or in adults with a BMI >35 who have comorbid (accompanying) conditions—for example, diabetes, hypertension, or dyslipidemia.

The published guidelines are extremely comprehensive, with detailed discussions about each of these recommendations, as well as extensive references for each of the statements. Exercise professionals should be familiar with these recommendations and guidelines for populations that they work with on a regular basis.

In Review

- The etiology of obesity is not simple; the cause can be any one or a combination of many factors.
- Studies of animals and humans indicate that there is a genetic component to obesity. The disorder also has been linked to hormonal imbalances, emotional trauma, homeostatic imbalances, cultural influences, physical inactivity, and improper diets. It is unlikely that a single cause is responsible.
- Overweight and obesity are associated with increased risk of general excess mortality.
- Obesity increases the risk of certain chronic degenerative diseases. Abdominal (visceral) obesity increases the risk of developing coronary artery disease, hypertension, stroke, elevated blood lipids, diabetes, and the metabolic syndrome. Also, obesity can worsen preexisting health conditions and diseases.
- Emotional or psychological problems may contribute to obesity, and the disorder itself, with the stigma it carries, can be psychologically damaging.
- In the treatment of obesity, it is important to remember that people respond differently to a given intervention. Some have large losses of weight in a relatively short period of time, while others appear to be resistant to weight loss and see only small losses.
- Weight loss generally should not exceed 0.45 to 0.9 kg (1-2 lb) per week. Simple diet modification, reducing intake of fat and simple sugars, is sufficient to help most people lose weight. Behavior modification is also an effective method of weight loss.
- The use of drugs or surgery in the treatment of obesity is generally not recommended unless deemed necessary for the patient's health by a physician.

Role of Physical Activity in Weight Management and Risk Reduction

Inactivity is a major cause of obesity in the United States. In fact, a sedentary lifestyle may be just as important in the development of obesity as overeating. Thus, increasing physical activity levels must be recognized as an essential component in any program of weight reduction or control.

RESEARCH PERSPECTIVE 22.2

Sedentary Behavior, Physical Activity, and Adiposity

Obesity is a traditional risk factor for cardiovascular disease. The type of fat is meaningful: excess visceral adipose tissue (VAT), intermuscular adipose tissue (IMAT), and liver fat increase cardiovascular risk, whereas subcutaneous adipose tissue (SAT) may be relatively less important in the pathophysiology of disease development.

A lack of regular physical activity is a primary factor contributing to obesity, and a sedentary lifestyle also contributes to greater cardiovascular risk. In fact, some evidence suggests that meeting the guidelines for physical activity may not be sufficient for disease prevention if accompanied by high levels of sedentary time. Sedentary behavior likely influences fat deposition, which would, of course, potentially explain part of the link between sedentary behavior and cardiovascular risk. However, to date, no studies have specifically examined sedentary behavior and its relation to the type of fat deposition (e.g., VAT, IMAT). This critical gap in knowledge was recently addressed by a team of researchers who tested the hypothesis that sedentary time would be positively associated with abdominal fat depots, whereas physical activity level would be inversely associated.[36] The investigators also predicted that sedentary time and physical activity would be independent, and additive, in their prediction of abdominal fat deposition. The investigators were also interested in potential sex and racial differences between these relations.

Sedentary behavior and physical activity were assessed using self-report questionnaires, and abdominal fat depots were measured using CT scans. The findings provide evidence that sedentary behavior (particularly time spent watching television) and physical activity have distinct, independent associations with patterns of abdominal adipose tissue deposition. Specifically, there appears to be a positive relation between time spent watching television and VAT, independent of total adiposity. The strongest associations were observed in white men. These findings add to a growing body of literature examining the relations between sedentary behavior, physical activity, and obesity. Future longitudinal studies are necessary to track these associations over time.

Changes in Body Composition with Exercise Training

Regular exercise can alter body composition. Despite the belief that even vigorous exercise burns too few calories to lead to substantial body fat reductions, research has conclusively demonstrated the effectiveness of exercise training in promoting moderate alterations in body composition.

While it is relatively easy to calculate the number of calories a person will expend in an endurance training program and then estimate the total weight loss over a period of time, this type of calculation is rarely accurate because of the interaction of the physiology of energy balance discussed earlier. Changing one (or more) of the three components of energy balance affects the others. Increased energy expenditure in an exercise program that burns about 1,300 kcal per week calculates to a weight loss of approximately 0.15 kg (0.33 lb) per week. However, the simple act of increasing energy expenditure coupled with even small amounts of decreased food intake (or even weight loss) positively affects energy balance. So, simple calculations like the preceding one are rarely accurate, and over long periods of time, individual variation in weight loss from an exercise program can be considerable.

In addition, when significant body weight is lost, the caloric cost of activity (TEA) is lowered proportionately along with RMR. Physiology, appetite, neurohormonal stimuli for food consumption, and even the availability of food all affect the amount of weight loss for any given negative caloric balance. It is simply too difficult and inaccurate to try to calculate these kinds of outcomes over a short-term weight management program. However, most clients in weight management programs want and expect some kind of estimate. Providing that estimate with great caution is well advised.

The ACSM's 2011 position stand on weight management lists several important recommendations for weight loss and weight management.[11] One of these is that physical activity is an important adjunct to a weight management program. The ACSM suggests that 150 to 250 min per week of moderate-intensity physical activity is important to prevent weight gain. However, that volume of activity provides "only moderate weight loss," and greater volumes of activity are necessary for more significant weight loss. In fact, even more (>250 min per week) may be necessary to prevent weight regain after significant weight loss.

Additionally, examining the energy expended only during exercise does not give the full picture. Metabolic rate remains temporarily elevated after exercise ends. This phenomenon, as mentioned in chapter 5, is referred to as the excess postexercise oxygen consumption (EPOC). Returning the metabolic rate back to its preexercise level can require several minutes following light exercise, such as walking; several hours following very heavy exercise, such as playing a football game; and up to 12 to 24 h or even longer for prolonged, exhaustive exercise, such as running a marathon or ultra-endurance race.

The EPOC can require substantial energy expenditure when considered over the entire recovery period. If, for example, the oxygen consumption following exercise remains elevated by an average of only 0.05 L/min, this will amount to approximately 0.25 kcal/min, or 15 kcal/h. If the metabolism remained elevated for 5 h, this would provide an additional expenditure of 75 kcal that would not normally be included in the calculated total energy expenditure for that particular activity. This additional energy expenditure is ignored in most calculations of the energy costs of various activities. The person in this example, if exercising 5 days per week, would expend 375 kcal, or lose the equivalent of about 0.05 kg (0.1 lb) of fat in 1 week, or 0.45 kg (1.0 lb) in 10 weeks, from the additional caloric expenditure during the recovery period alone.

Studies have shown relatively small but significant changes in both weight and body composition with both aerobic and resistance training, which include

- total weight decrease,
- fat mass and relative body fat decrease, and
- either maintained or increased fat-free mass.

From a summary of hundreds of published studies that monitored body composition changes with aerobic training, the expected changes from a typical 1-year exercise training program (three times per week, 30 to 45 min per day, at 55% to 75% of $\dot{V}O_{2max}$) would be as follows: −3.2 kg (−7.1 lb) total body mass, −5.2 kg (−11.5 lb) fat mass, and +2.0 kg (+4.4 lb) fat-free mass.[37] Relative body fat would decrease by nearly 6% (e.g., from 30% body fat to 24% body fat).

Since the 1990s, abdominal visceral fat (figure 22.7) has been identified as a major independent risk factor for cardiovascular diseases and obesity. There is now substantial evidence that physical activity reduces the rate of accumulation of visceral fat and that exercise training actually reduces visceral fat stores.[34] This could be one of the most important health benefits of an active lifestyle.

Mechanisms for Change in Body Weight and Composition

When attempting to explain how exercise causes such changes in body weight and composition, it is necessary to consider both sides of the energy balance equation. Evaluating energy expenditure requires that we consider each of the three components of energy expenditure: RMR, TEM, and TEA. Evaluating energy intake requires that we also consider the energy that is lost in the feces (energy excreted), which is generally less than 5% of the total caloric intake. Keeping this balance in mind, in the next section we examine some of the possible mechanisms through which exercise might affect body weight and body composition.

Exercise and Appetite

The link between exercise and appetite remains controversial and unresolved. The interrelationships of energy intake, energy expenditure, appetite, and obesity are complex and multifactorial. Issues of hunger, the coupling of energy intake and energy expenditure, the hormones involved with hunger and satiety (the satisfied feeling after a meal), macronutrient intake (dietary pattern), and even differences in responses to exercise between men and women are difficult to sort out. At the present time, we can conclude that (1) exercise, by itself, probably does not significantly stimulate appetite, especially during a weight management program that includes exercise and diet and (2) even if exercise does cause minor increases in energy intake, this does not preclude its inclusion in weight management programs because the TEA is a major variable that can be altered in these programs.

A classic study by Jean Mayer led him to theorize that energy intake and energy expenditure are related, that is, "coupled within a certain zone" that he called "normal activity."[24] In a study of a large population of factory workers, he and his coauthors found that those who had the highest job-related energy expenditures increased energy intake appropriately to match the energy expenditure. However, the most sedentary workers, who were the heaviest group, had high energy intakes relative to their activity level. Thus, there was a zone of normal activity within which people increased their food intake to accommodate high levels of energy expenditure (work related), but outside that zone (where the relatively sedentary workers fell), energy intake was much higher than necessary by activity standards. Mayer further hypothesized that humans were evolutionarily active (not sedentary); thus, the role of physical activity, perhaps influenced by our genome, functions to increase energy intake only within a small range of caloric expenditure (Mayer's "normal activity range").

An additional influential factor in the appetite–weight loss issue is satiety (feeling satisfied after eating), which also affects hunger and energy intake. In a study on exercise, weight loss, and satiety, subjects who lost weight progressively increased their satiety during the training program. Subjects who lost less weight than expected progressively decreased their satiety and concurrently increased their caloric intake during the training program.[20] Combining exercise and diet in a weight management program may have some effect on both hunger and satiety, thus affecting how much weight is lost.

The relation between exercise and the hormones associated with appetite and energy intake is part of another complex and as yet unanswered question. Two hormones, leptin and ghrelin (discussed in chapter 4), have major influences on appetite and satiety, and are therefore significant players in the problem of obesity. These hormones provide acute and chronic feedback to the brain about nutrition status. Ghrelin is produced and secreted by the gastrointestinal system and provides acute input to the brain about food consumption. Leptin, on the other hand, provides signals related to the long-term energy status associated with energy stores, that is, adipose tissue. Greater fat deposits stimulate greater leptin production. Thus, there is immediate feedback from the gastrointestinal tract about what and how much we eat, plus long-term feedback from the adipocytes about the amount of stored energy. Some people with obesity have been shown to have leptin resistance. Leptin receptors in the central nervous system that were previously sensitive to leptin in the feedback loop become increasingly insensitive to this signaling hormone as adiposity increases.

Exercise and Resting Metabolic Rate

The effects of exercise on the components of energy expenditure became a major topic of interest among researchers in the late 1980s and early 1990s. Of obvious interest is how exercise training might affect the RMR, since RMR represents 60% to 75% of the total calories expended each day. For example, if a 25-year-old man's total daily caloric intake were 2,700 kcal and his RMR accounted for just 60% of that total (0.60 × 2,700 = 1,620 kcal RMR), a mere 1% increase in his RMR would require an extra 16 kcal expenditure each day, or 5,840 kcal per year. This small increase in RMR alone would account for the equivalent of a 0.8 kg (1.7 lb) fat loss per year.

The role of exercise training in increasing RMR has not been totally resolved. Several cross-sectional studies have shown that highly trained athletes have higher RMRs than untrained people of similar age and size. But other studies have not been able to confirm this.[31] Few longitudinal studies have been conducted to determine the change in RMR in untrained people who undergo training for a period of time, with some suggesting that RMR might increase following training. However, in a study of 40 men and women 17 to 62 years of age (HERITAGE Family Study), a 20-week aerobic training program (three times per week, 35-55 min per day, at 55%-75% of $\dot{V}O_{2max}$) failed to increase RMR even though $\dot{V}O_{2max}$ increased by nearly 18%.[38]

Because RMR is closely related to the fat-free mass of the body (fat-free tissue is more metabolically active), interest has increased in the use of resistance training to increase fat-free mass in an attempt to increase RMR.

Spot Reduction

Many people, including athletes, believe that exercising a specific area of the body will use the fat in that area, reducing the locally stored fat. Results of several early research studies tended to support this concept of **spot reduction**. But later research suggests that spot reduction is a myth and that exercise, even when localized, draws from almost all of the fat stores of the body, not just from local depots.

One such study used outstanding tennis players, theorizing that they would be ideal subjects for studying spot reduction because they could act as their own controls: Their dominant arms exercise vigorously for several hours every day, whereas their nondominant arms are relatively inactive.[14] The players' dominant arms had substantially greater girths attributable to exercise-induced muscle hypertrophy. But the subcutaneous skinfold fat thickness in the active and inactive arms showed absolutely no differences.

Another study examined the localized effects of a 27-day intense sit-up training program on subcutaneous fat in the abdomen. Researchers found no difference in the rate at which fat cell diameter changed in the abdomen, the subscapular region, and the gluteal region, which indicates a lack of adaptation at the training site (abdomen).[19] Decreases in girth can occur with exercise training, but these primarily result from increased muscle tone, not fat loss.

Low-Intensity Aerobic Exercise

As we discussed in earlier chapters, the higher the exercise intensity, the greater the body's reliance on carbohydrate as an energy source. With high-intensity aerobic exercise, carbohydrate can supply up to 90% or more of the body's energy needs. During the late 1980s, various groups promoted **low-intensity aerobic exercise** to increase the loss of body fat. These groups theorized that low-intensity aerobic training would allow the body to use more fat as the energy source, hastening the loss of body fat. Indeed, the body uses a higher percentage of fat for energy at lower exercise intensities. However, the total calories expended does not necessarily change as a result of the body's use of fat.

This is illustrated in table 22.4. In this hypothetical example, a 23-year-old woman with a $\dot{V}O_{2max}$ of 3.0 L/min exercises for 30 min at 50% of her $\dot{V}O_{2max}$ on one day and for 30 min at 75% of her $\dot{V}O_{2max}$ on another. The total calories from fat do not differ between the low- and high-intensity aerobic workouts: In both cases, she burns about 110 kcal of fat during 30 min. Most importantly, however, if she is fit enough to sustain the higher-intensity workout, she will expend about 50% more total calories for the same time period.

Scientists have determined that there is an optimal zone within which rates of fat oxidation are at their highest. The Fat_{max} zone, defined as that zone where fat oxidation rates are within 10% of the peak rate, was found to vary from between 55% and 72% of $\dot{V}O_{2max}$.[1] This is illustrated in figure 22.8. However, as already noted, this high rate of fat oxidation does not necessarily mean a higher total calorie expenditure or greater contribution to weight loss.

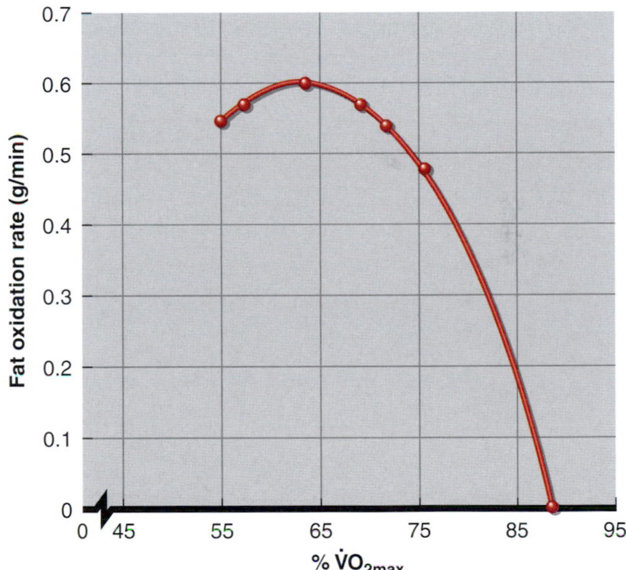

FIGURE 22.8 Rate of fat oxidation at various exercise intensities, expressed as a percentage of $\dot{V}O_{2max}$.

Reprinted by permission from J. Acten, M. Gleeson, and A.E. Jeukendrup, "Determination of the Exercise Intensity That Elicits Maximal Fat Oxidation," *Medicine and Science in Sports and Exercise* 34 (2002): 92-97.

TABLE 22.4 **Estimation of Kilocalories Used From Fat and Carbohydrate for a Low- and High-Intensity Aerobic Training Bout of 30 Minutes Duration**

Exercise intensity	Average $\dot{V}O_2$ (L/min)	Average RER	% kcal CHO	% kcal fat	kcal for 30 min CHO	kcal for 30 min fat	kcal for 30 min total
Low, 50%	1.50	0.85	50	50	110	110	220
High, 75%	2.25	0.90	67	33	222	110	332

Note. RER = respiratory exchange ratio; CHO = carbohydrate. Subject was a fit but not highly trained 23-year-old woman ($\dot{V}O_{2max}$ = 3.0 L/min).

In Review

- Inactivity is a major cause of obesity and type 2 diabetes in the United States; it is perhaps as significant as overeating.
- A minimum of 150 to 250 min per week of moderate-intensity physical activity is important to prevent weight gain. Greater volumes of activity are necessary for significant weight loss.
- The energy expended by activity includes the rate of energy expenditure during the activity and also the energy expended after the exercise, because the metabolic rate remains elevated for some time after the activity ends. This phenomenon is known as excess postexercise oxygen consumption, or EPOC.
- Dietary restriction alone may cause fat loss, but fat-free mass is also lost. With exercise, either alone or with diet, fat is lost, but fat-free mass is either maintained or increased. Possibly the greatest benefit of physical activity and formal exercise is their role in attenuating the accumulation of visceral fat or in reducing visceral fat stores.
- Simply being active, independent of a formal exercise program, is important in the prevention of obesity.
- Energy intake − energy expended = RMR + TEM + TEA when one is in energy balance.
- A certain amount of activity, Mayer's "normal zone," appears to be necessary for the body to balance energy intake and expenditure.
- Research indicates that exercise may suppress appetite, but this issue is controversial.
- A single bout of exercise increases the TEM.
- Exercise increases lipid mobilization from adipose tissue.
- Spot reduction is a myth.
- Low-intensity aerobic exercise burns no more fat than more vigorous exercise, and more total calories are burned in a more strenuous workout.

Understanding Diabetes

Diabetes mellitus, also called diabetes, is a disease characterized by high blood glucose concentrations (hyperglycemia) resulting from either an inadequate production of insulin by the pancreas, an inability of insulin to facilitate the transport of glucose into the cells, or both. Recall from chapter 4 that insulin is a hormone that reduces circulating glucose by facilitating its transport into the cells. We first address the terminology used in defining diabetes and disordered blood glucose control (glycemic control) and then look at the prevalence of diabetes in the United States.

Terminology and Classification

Diabetes mellitus is really two separate diseases—type 1 diabetes mellitus (previously known as juvenile-onset diabetes, or insulin-dependent diabetes mellitus) and type 2 diabetes mellitus (previously adult-onset diabetes).[22] Type 2 diabetes is closely associated with obesity, and as is the case with obesity, there has been a near epidemic of type 2 diabetes in the United States in children and adults.

Type 1 diabetes is caused by the failure of the **β-cells** in the pancreas to produce sufficient insulin. This condition is the result of the body's immune system destroying pancreatic β-cells (a process called apoptosis). As they decrease in number, less insulin is produced, and at some critical tipping point, there is insufficient insulin for proper control of blood glucose. Thus, this type has also been referred to as **insulin-dependent diabetes mellitus (IDDM)**. Type 1 diabetes accounts for only 5% to 10% of all cases of diabetes.

Type 2 diabetes is the result of the failure of insulin to facilitate the transport of glucose into the cells and is a result of insulin resistance. Type 2 diabetes accounts for 90% to 95% of all cases of diabetes. Insulin's primary function is to facilitate the transport of glucose from the blood across the cell membrane. *Insulin resistance* refers to the condition in which a normal insulin concentration in the blood produces a less than normal biological response. The body needs more insulin to transport a given amount of glucose into the cell. *Insulin sensitivity* is a related concept; it provides an index of the effectiveness of a given insulin concentration in the blood. Insulin resistance increases as type 2 diabetes develops and progresses.

A third type of diabetes, gestational diabetes, is a form of diabetes that develops in women and their fetuses during pregnancy and occurs in about 4% of all pregnancies. Fortunately, it usually disappears in both mother and baby after delivery. Unfortunately, gestational diabetes can lead to complications during pregnancy.

Prediabetes refers to the condition in those who have impaired fasting glucose, impaired glucose tolerance, or both. Both type 1 and type 2 diabetes are diagnosed on the basis of a plasma glucose level greater than 125 mg/dl following an 8 h fast. **Impaired fasting glucose** is defined as a plasma glucose level between 100 and 125 mg/dl, again following an 8 h fast. **Impaired glucose tolerance** is determined by a glucose tolerance test. This involves drinking a solution in which 75 g of anhydrous glucose is dissolved in water. Plasma glucose concentrations are measured 2 h later. A glucose concentration of

200 mg/dl or higher is diagnostic of diabetes. Values of 140 to 199 mg/dl represent impaired glucose tolerance, and values below 140 mg/dl are considered normal.

For research studies, an intravenous glucose tolerance test (IVGTT) can also be used. A catheter is placed in each arm, with the glucose solution injected into a vein in one arm, and blood samples are withdrawn from a vein in the other arm over the course of 3 h. Samples are taken more frequently during the first 15 to 45 min of the test and less frequently from 60 to 180 min. This allows one to develop a curve of both the glucose and insulin responses to the injected glucose load. The IVGTT is more precise than the oral glucose tolerance test.

There are common symptoms that can be used to identify those at risk for diabetes. They include the following:

Type 1 Diabetes

- Frequent urination
- Excessive or unusual thirst
- Unusual, unexplained weight loss
- Extreme hunger
- Extreme fatigue and irritability

Type 2 Diabetes

- Any of the type 1 symptoms
- Frequent infections
- Blurred vision or sudden vision changes
- Tingling or numbness in hands or feet
- Cuts, bruises, or sores that are slow to heal
- Recurring skin, gum, or bladder infections

RESEARCH PERSPECTIVE 22.3

Metformin or Exercise or Both to Treat Diabetes?

Type 2 diabetes and prediabetes are risk factors for the development of cardiovascular disease. Although the causes of type 2 diabetes are multifactorial, insulin resistance (and the resulting hyperinsulinemia) is likely a primary contributor. Thus, targeting insulin resistance is important for the management and prevention of type 2 diabetes. Diet and exercise are the first-line therapy for improving glycemic control; however, long-term adherence to exercise programs remains poor, and many individuals also require pharmacological therapy to maintain normal blood sugar concentrations.

The American Diabetes Association recently recommended that, in addition to lifestyle modifications, individuals with prediabetes should be considered for metformin therapy (the most widely used medication for diabetes treatment, which decreases liver glucose production). The underlying assumption behind this recommendation is that because metformin and exercise each lower blood glucose, the actions of metformin and the benefits of physical activity will be *additive*, such that combined treatment will be more beneficial than either treatment alone. However, because the generation of reactive oxygen species (ROS) is a primary factor promoting the metabolic adaptation to exercise and because metformin reduces ROS generation, metformin may actually *attenuate* metabolic adaptation to exercise via an oxidative stress mechanism. So, does combining metformin with exercise have additive, opposing, or no benefit for glucose regulation and cardiovascular disease risk reduction?

The answer to this seeming paradox was recently reviewed by Malin and Braun.[23] Work from a series of studies published by their laboratory suggests that combining the two metabolically beneficial treatments does not necessarily result in an additive outcome. They conducted a prospective, double-blind, randomized controlled trial to examine the long-term effects of combining exercise training and metformin on insulin sensitivity in men and women with prediabetes. Participants were assigned to four different treatment groups: placebo, metformin, exercise training plus placebo, or exercise training plus metformin. All treatments reduced fasting plasma insulin. Metformin alone or exercise training alone increased skeletal muscle insulin sensitivity. However, metformin blunted the effect of exercise training on skeletal muscle insulin sensitivity by ~30%, suggesting that metformin dampens some of the beneficial effects of exercise training in regards to glucose regulation.

This leaves clinicians with the question of whether they should coprescribe metformin with exercise in order to treat patients or whether exercise treatment alone is sufficient to lower diabetes risk. Not surprisingly, the answer is complicated. Research studies have suggested that metformin blunts the effects of exercise on improving insulin sensitivity, metabolic syndrome prevalence, and inflammation. Although these effects may seem concerning, it is important to realize that metformin has distinct effects on multiple tissues. Thus, the degree of additive benefit will depend, in large extent, on the tissue of interest. Nevertheless, the authors argue that exercise alone is the best first-line therapy for reducing the risk of type 2 diabetes.

Prevalence of Diabetes

According to the American Diabetes Association, in 2015, approximately 30 million Americans had diabetes and an estimated 7 million people were undiagnosed. Additionally, an estimated 84 million adults had prediabetes. The prevalence of diabetes increased from 4.9% of the U.S. population in 1990 to 9.4% in 2015. Approximately 25% of people over 65 have diabetes, and diabetes remains the seventh leading cause of death in the United States. Further, the prevalence of diagnosed diabetes in those aged 20 years or older was 7.4% for non-Hispanic whites, 8.0% for Asian Americans, 12.7% for non-Hispanic blacks, and 12.1% for Hispanics and Latinos.[2] These ethnic differences in diabetes rates parallel the differences in obesity rates discussed earlier in this chapter.

The true prevalence of type 2 diabetes in children is still low, but prevalence has been estimated to have increased by as much as 10-fold over the last 20 years. Furthermore, in 10- to 19-year-olds, studies have shown that type 2 diabetes accounted for between 33% and 46% of all diabetes.[41] This is very disturbing, considering that not too long ago we referred to type 2 diabetes as adult-onset diabetes.

Etiology of Diabetes

Heredity appears to play a major role in both type 1 and type 2 diabetes. With type 1 diabetes, the β-cells (insulin-secreting cells) of the pancreas are destroyed. This destruction may be caused by the body's immune system, increased β-cell susceptibility to viruses, or β-cell degeneration. Type 1 diabetes typically has a sudden onset during childhood or young adulthood. This leads to almost total insulin deficiency, and daily injections of insulin are usually required to control the disease.

In type 2 diabetes, the onset of the disease is more gradual, and the causes are more difficult to establish. Type 2 diabetes is usually characterized by three major metabolic abnormalities: insulin resistance (impaired insulin action), primarily in skeletal muscle tissue, increased insulin secretion (hyperinsulinemia in response to the increased insulin resistance), or excessive glucose output from the liver, or some combination of these.

Obesity plays a major role in the development of type 2 diabetes. With obesity, insulin resistance is commonly a primary initial problem. The β-cells of the pancreas respond to the increased blood glucose (secondary to the insulin resistance) by producing and secreting more insulin. Thus, it is not uncommon for people with recently diagnosed type 2 diabetes to have both hyperglycemia (high blood glucose) and hyperinsulinemia (high blood insulin). This is all a result of insulin receptors throughout the body (especially in skeletal muscle) that are less effective in facilitating the transport of glucose into the cell.

Health Problems Linked to Diabetes

Considerable health risks are associated with diabetes. People with this disease have a relatively high mortality rate. Diabetes places a person at increased risk for

- coronary artery or heart disease and peripheral vascular disease;
- hypertension, cerebrovascular disease, and stroke;
- kidney disease;
- nervous system disease;
- eye disorders, including blindness;
- dental disease;
- amputations; and
- complications during pregnancy.

Scientists have established the interrelatedness among coronary artery disease, hypertension, obesity, and type 2 diabetes. **Hyperinsulinemia** (high blood levels of insulin) and insulin resistance appear to be the important threads linking these disorders, possibly through insulin-mediated sympathetic nervous system stimulation (increased insulin levels cause increased sympathetic nervous system activity), inflammation, metabolic dysfunction, or some combination of these. Again, obesity seems to be the trigger setting off this reaction.

Treatment of Diabetes

The major modes of treatment for type 1 diabetes are insulin administration, diet, and exercise. The dosage of insulin is adjusted to allow normal carbohydrate, fat, and protein metabolism and normal levels of blood glucose. The type of insulin injected—short acting or intermediate acting—and the time of day at which the injections are administered are also individualized to maintain glycemic control throughout the day.

For type 2 diabetes, the focus has traditionally been on three factors: weight loss, a healthy dietary pattern, and exercise and physical activity. Treatment of diabetes (especially type 2 diabetes) with drugs has expanded significantly over the past 25 years. Drugs that stimulate production and release of insulin, inhibit production and release of glucose from the liver, or block enzymes in the stomach that break down carbohydrates and increase insulin sensitivity are in common use.

A balanced, healthy diet generally is prescribed for people with diabetes. In the past, a low-carbohydrate diet was recommended to better control blood sugar levels. However, a low-carbohydrate diet is usually associated with an increase in dietary fat, which can have a major negative effect on blood lipid levels. Because people with diabetes often have abnormal lipids and are already at greater risk for coronary artery disease, this is not desirable. Maintenance of proper blood sugar levels may be difficult in patients with obesity, so a reduced-calorie diet is used in order to facilitate significant loss of body fat. For many people with type 2 diabetes, weight loss alone can bring blood sugar levels back into the normal range. Additionally, since regular exercise and physical activity can increase insulin sensitivity, increased physical activity and a healthy dietary pattern are both very important aspects of the treatment plan for the person with type 2 diabetes.

During the mid-1990s, the National Institute of Diabetes and Digestive and Kidney Diseases (NIDDK) of the National Institutes of Health designed and initiated a study to determine whether either a lifestyle intervention (diet and exercise) or an oral diabetes drug (metformin) could prevent or delay the onset of type 2 diabetes in people with impaired glucose tolerance (prediabetes). A total of 3,234 persons 25 years of age and older were randomly assigned to placebo, drug, or lifestyle groups. The goals for the lifestyle modification program were to lose at least 7% of body weight and to participate in physical activity for at least 150 min per week. Both interventions proved very successful, delaying the development of diabetes by 11 years in the lifestyle group and by 3 years in the metformin group.[21] This landmark study clearly illustrates the power of lifestyle intervention in reducing the risk of a major debilitating disease and its serious consequences. The Finnish Diabetes Prevention Study confirmed those results and further reinforced the value of physical activity for prevention of diabetes.[35]

Role of Physical Activity in Diabetes

Indirect scientific evidence has clearly established that a physically active lifestyle reduces the risk of developing type 2 diabetes,[3] but there is little evidence to support this for type 1 diabetes. However, most physicians and scientists agree that physical activity is an important part of the treatment plan for both types of diabetes. Because there is such a disparity between the characteristics and responses of those with type 1 and type 2 diabetes, we discuss them separately.

Type 1 Diabetes

The role of regular exercise and physical training in improving glycemic control (regulation of blood sugar levels) in patients with type 1 diabetes has not been clearly defined and is controversial. The most distinguishing feature differentiating type 1 and type 2 diabetes is that those with type 1 have low blood insulin levels attributable to the inability or reduced ability of the pancreas to produce insulin. Those with type 1 diabetes are prone to hypoglycemia (low blood sugar

levels) during and immediately after exercise because the liver fails to release glucose at a rate that can keep up with glucose utilization. For these people, exercise can lead to excessive swings in plasma glucose levels that are unacceptable for the management of the disease. The degree of glycemic control during exercise varies tremendously among individuals with type 1 diabetes. As a result, exercise and exercise training can improve glycemic control in some patients, mainly those who are less prone to hypoglycemia, but not in others.

Although glycemic control is generally not improved by exercise in most people with type 1 diabetes, there are other potential benefits of exercise for these patients. These patients have a risk for coronary artery disease two to three times greater, and exercise may help reduce this risk. Exercise may also help reduce the risk for cerebrovascular and peripheral arterial diseases.

People with uncomplicated type 1 diabetes do not have to restrict physical activity provided that blood sugar levels are controlled appropriately. A number of athletes who have type 1 diabetes have trained and competed successfully. Monitoring blood sugar levels in an exercising person with type 1 diabetes is important so that diet and insulin dosages can be altered accordingly. This is particularly important for those who are competing at high intensities or for extended periods of time.

Special attention also should be given to the feet of people with diabetes, since it is common for them to experience peripheral neuropathy (diseased nerves) with some loss of sensation in the feet. Peripheral vascular disease is also more common in patients with diabetes, so the circulation to the extremities, especially the feet, often is impaired significantly. Ulcerations and other lesions on the feet account for more than half of all hospitalizations of diabetic patients. Because weight-bearing exercise places additional stress on the feet, the proper selection of footwear and appropriate preventive foot care are important.

Type 2 Diabetes

Exercise plays a major role in glycemic control for people with type 2 diabetes. Insulin production is generally not of concern in this group, particularly during the early stages of the disease, so the major problem with this form of diabetes is the lack of target cell response to insulin (insulin resistance). Because the cells become resistant to insulin, the hormone cannot perform its function of facilitating glucose transport across the cell membrane. The act of muscle contraction has an insulin-like effect. Membrane permeability to glucose increases with muscular contraction, likely attributable to an increase in the number of GLUT-4 glucose transporters associated with the plasma membrane.[17] Thus, acute bouts of exercise decrease insulin resistance and increase insulin sensitivity. This reduces the cells' requirements for insulin. Resistance and aerobic training appear to produce similar effects,[40] although some evidence suggests that combining resistance and aerobic exercise is the optimal exercise strategy for reducing insulin resistance.[10] This decrease in insulin resistance and increase in insulin sensitivity may primarily be a response to each individual bout of exercise rather than the result of a long-term change associated with training. Some studies have shown that the effect dissipates within 72 h.

The American Diabetes Association recently released an updated position statement on physical activity and exercise and diabetes.[8] This position statement provides evidence-based recommendations regarding physical activity and exercise for people with type 1 diabetes and type 2 diabetes. Recommendations from the ADA are summarized in table 22.5.

In Review

- Type 1 diabetes involves destruction of the β-cells in the pancreas, and it typically has an early onset, that is, in adolescence or at younger ages. Type 2 diabetes is characterized by impaired glucose tolerance because of insulin resistance.
- Diabetes is characterized by hyperglycemia. It develops from inadequate insulin utilization (insulin resistance), secretion, or both.
- Major modes of treatment for diabetes are drugs (if needed), diet, and, especially for type 2 diabetes, exercise and weight loss.
- In people with type 1 diabetes, glycemic control may or may not be improved with exercise. They have a greater risk for coronary artery disease, and exercise can decrease that risk.
- People with type 1 diabetes must carefully monitor blood glucose before, during, and after exercise so that diet and insulin dosage can be altered as needed.
- Peripheral neuropathy is a serious complication associated with type 1 and longstanding type 2 diabetes. Those with peripheral neuropathy should give their feet special attention because of the loss of sensation and impaired peripheral circulation that accompany it.
- Type 2 diabetes responds well to exercise. Glucose utilization and insulin sensitivity improve with exercise, most likely because of an increase in GLUT-4 receptors in skeletal muscle.

TABLE 22.5 **Summary of Recommendations by the American Diabetes Association for Reducing Sedentary Time and Increasing Physical Activity in People with Diabetes**

Recommendations for reduced sedentary time
All adults, and particularly those with type 2 diabetes, should decrease the amount of time spent in daily sedentary behavior.
Prolonged sitting should be interrupted with bouts of light activity every 30 min for blood glucose benefits, at least in adults with type 2 diabetes.
The above two recommendations are additional to, and not a replacement for, increased structured exercise and incidental movement.
Physical activity and type 2 diabetes
Daily exercise, or at least not allowing more than 2 days to elapse between exercise sessions, is recommended to enhance insulin action.
Adults with type 2 diabetes should ideally perform both aerobic and resistance exercise training for optimal glycemic and health outcomes.
Children and adolescents with type 2 diabetes should be encouraged to meet the same physical activity goals set for youth in general.
Structured lifestyle interventions that include at least 150 min/week of physical activity and dietary changes resulting in weight loss of 5% to 7% are recommended to prevent or delay the onset of type 2 diabetes in populations at high risk and with prediabetes.
Physical activity and type 1 diabetes
Youth and adults with type 1 diabetes can benefit from being physically active, and activity should be recommended to all.
Blood glucose responses to physical activity in all people with type 1 diabetes are highly variable based on activity type/timing and require different adjustments.
Additional carbohydrate intake or insulin reductions are typically required to maintain glycemic balance during and after physical activity. Frequent blood glucose checks are required to implement strategies for adjusting carbohydrate intake and insulin dose.
Insulin users can exercise using either basal-bolus injection regimens or insulin pumps, but there are advantages and disadvantages to both insulin delivery methods.
Continuous glucose monitoring during physical activity can be used to detect hypoglycemia when used as an adjunct rather than in place of capillary glucose tests.
Recommended physical activity participation for people with diabetes
Preexercise medical clearance is generally unnecessary for asymptomatic individuals prior to beginning low- or moderate-intensity physical activity not exceeding the demands of brisk walking or everyday living.
Most adults with diabetes should engage in 150 min or more of moderate- to vigorous-intensity activity weekly, spread over at least 3 days/week, with no more than 2 consecutive days without activity. Shorter durations (minimum 75 min/week) of vigorous-intensity activity or interval training may be sufficient for younger and more physically fit individuals.
Children and adolescents with type 1 or type 2 diabetes should engage in 60 min/day or more of moderate- or vigorous-intensity aerobic activity, with vigorous, muscle-strengthening, and bone-strengthening activities included at least 3 days/week.

IN CLOSING

In these last two chapters, we have concluded our look at the role of physical activity in preventing and treating coronary artery disease, hypertension, obesity, and diabetes. We have seen that exercise can decrease individual risk and also can be an integral part of treatment, improving overall health as well as alleviating some symptoms.

With this chapter, we also conclude our journey to an understanding of exercise and sport physiology. We began this book by reviewing how various body systems function during exercise and how they respond to chronic training. We saw how physical activity and performance are affected by the environment, such as extremes of heat, cold, and barometric pressure. We turned our attention to ways in which athletes can, or can attempt to, optimize their performance. Then we considered the unique differences for older and younger participants and male and female participants in sport and exercise. And finally, we examined the role of exercise in the maintenance of health and the development of fitness.

It has been a long journey from cover to cover, but we hope that you will close this book with a new appreciation of physical activity. Perhaps you will leave this book with a new awareness of how your body performs physical activity. Maybe, if you have not yet done so, you will feel compelled to commit to a personal exercise program. And we hope that you now feel some excitement about exercise and sport physiology, realizing that these areas of study affect so many aspects of our lives.

KEY TERMS

β-cells
body mass index (BMI)
diabetes mellitus
hyperinsulinemia
impaired fasting glucose
impaired glucose tolerance
insulin-dependent diabetes mellitus (IDDM)
lower body (gynoid) obesity
low-intensity aerobic exercise
metabolic syndrome
obesity
overweight
prediabetes
spot reduction
thermic effect of activity (TEA)
thermic effect of a meal (TEM)
type 1 diabetes
type 2 diabetes
upper body (android) obesity

STUDY QUESTIONS

1. What is the difference between overweight and obesity?
2. What are the methods for determining body composition, and what are the strengths and weakness of each?
3. What is body mass index? What is its significance?
4. What is the prevalence of obesity in the United States today? Is there a difference in the prevalence between men and women? Children and adults? Blacks and whites?
5. What are several health-related problems associated with obesity?
6. How are obesity, coronary artery disease, hypertension, insulin resistance, and diabetes interrelated?
7. Describe several methods for treating obesity. Which are the most effective?
8. What role does exercise play in the prevention and treatment of obesity?
9. By what mechanisms might exercise effect losses in total weight and fat weight?
10. How effective is spot reduction? Low-intensity aerobic exercise?
11. Describe the two major types of diabetes. How are they caused?
12. What health risks are associated with diabetes?
13. Discuss the major points published in the *Summary of Recommendations by the American Diabetes Association for Reducing Sedentary Time and Increasing Physical Activity in Diabetes* as they relate to exercise.
14. Describe the role of exercise in preventing diabetes and in treating patients with type 1 and type 2 diabetes.

STUDY GUIDE ACTIVITIES

In addition to the activities listed in the chapter opening outline, two other activities are available in the web study guide, located at

www.HumanKinetics.com/PhysiologyOfSportAndExercise

The **KEY TERMS** activity reviews important terms, and the end-of-chapter **QUIZ** tests your understanding of the material covered in the chapter.

Glossary

1-repetition maximum (1RM)—The maximal amount of weight that can be lifted just one time.

1RM—*See* 1-repetition maximum.

acclimation (heat acclimation)—Physiological adaptation to repeated environmental stresses, occurring over a relatively brief period of time (days to weeks). Acclimation often occurs in a laboratory environment.

acclimatization—Physiological adaptation to repeated environmental stress in a natural environment, occurring over months and years of living and exercising in that environment.

ACE—*See* angiotensin-converting enzyme.

acetylcholine—A primary neurotransmitter that transmits impulses across the synaptic cleft.

acetyl CoA—*See* acetyl coenzyme A.

acetyl coenzyme A (acetyl CoA)—The compound that forms the common entry point into the Krebs cycle for the oxidation of carbohydrate and fat.

actin—A thin protein filament that acts with myosin filaments to produce muscle action.

action potential—A rapid and substantial depolarization of the membrane of a neuron or muscle cell that is conducted through the cell.

activation energy—The initial energy required to start a chemical reaction or chain of reactions.

acute altitude (mountain) sickness—Illness characterized by headache, nausea, vomiting, dyspnea, and insomnia. It typically begins 6 to 96 h after one reaches high altitude and lasts several days.

acute exercise—A single bout of exercise.

acute muscle soreness—Soreness or pain felt during and immediately after an exercise bout.

acute overload—An average training load, whereby the athlete is stressing the body to the extent necessary to improve both physiological function and performance.

adenosine diphosphate (ADP)—A high-energy phosphate compound from which ATP is formed.

adenosine triphosphatase (ATPase)—An enzyme that splits the last phosphate group off ATP, releasing a large amount of energy and reducing the ATP to ADP and P_i.

adenosine triphosphate (ATP)—A high-energy phosphate compound from which the body derives its energy.

ADH—*See* antidiuretic hormone.

adipokines—Hormones that are released by, or act on, fat tissue.

adolescence—The period of life between the end of childhood and the beginning of adulthood. The onset of puberty marks the beginning of adolescence.

ADP—*See* adenosine diphosphate.

adrenaline—A chemical compound that serves as a neurotransmitter throughout the body. Also called epinephrine.

adrenergic—Refers to norepinephrine or epinephrine (also called noradrenaline and adrenaline, respectively).

aerobic metabolism—A process occurring in the mitochondria that uses oxygen to produce energy (ATP). Also known as cellular respiration.

aerobic power—Another term for maximal oxygen uptake, or $\dot{V}O_{2max}$.

aerobic training—Training that improves the efficiency of the aerobic energy-producing systems and can improve cardiorespiratory endurance.

afterload—The pressure against which the heart must pump blood, determined by the peripheral resistance in the large arteries.

air plethysmography—A procedure for assessing body composition by using air displacement to measure body volume, allowing the calculation of body density.

aldosterone—A mineralocorticoid hormone secreted by the adrenal cortex that prevents dehydration by promoting renal absorption of sodium.

α-motor neuron—A neuron innervating extrafusal skeletal muscle fibers.

alveoli—Grapelike clusters, or air sacs, at the ends of the terminal bronchioles.

amenorrhea—The absence (primary amenorrhea) or cessation (secondary amenorrhea) of normal menstrual function.

amphetamine—A central nervous system stimulant proposed to have ergogenic properties.

anabolic steroids—Prescription drugs with the anabolic (growth stimulating) characteristics of testosterone, taken by some athletes to increase body size, muscle mass, and strength.

anaerobic metabolism—The production of energy (ATP) in the absence of oxygen.

anaerobic power—Mean or peak power output in exercise lasting 30 s or less.

anaerobic threshold—The point at which the metabolic demands of exercise can no longer be met by available aerobic sources and at which an increase in anaerobic metabolism occurs, reflected by an increase in blood lactate concentration.

anaerobic training—Training that improves the efficiency of the anaerobic energy-producing systems and can increase muscular strength and tolerance for acid–base imbalances during high-intensity effort.

angiotensin-converting enzyme (ACE)—An enzyme that converts angiotensin I to angiotensin II.

anorexia nervosa—A clinical eating disorder characterized by distorted body image, intense fear of fatness or weight gain, amenorrhea, and refusal to maintain more than the minimal normal weight based on age and height.

antidiuretic hormone (ADH)—A hormone secreted by the pituitary gland that regulates fluid and electrolyte balance in the blood by reducing urine production.

arginine vasopressin—*See* antidiuretic hormone.

arterial–mixed venous oxygen difference, or (a-$\bar{v}$)O_2 difference—The difference in oxygen content between arterial and mixed venous blood, which reflects the amount of oxygen removed by the whole body.

arterial–venous oxygen difference, or (a-v)O_2 difference—The difference in oxygen content between arterial and venous blood at the tissue level.

arteries—Blood vessels that transport blood away from the heart.

arterioles—The smallest arteries that transport blood from larger arteries to the capillaries.

atherosclerosis—A condition that involves changes in the lining of the arteries and plaque accumulation, leading to progressive narrowing of the arteries.

athlete's heart—A nonpathological enlarged heart, often found in endurance athletes, that results primarily from left ventricular hypertrophy in response to training.

ATP—*See* adenosine triphosphate.

ATPase—*See* adenosine triphosphatase.

ATP-PCr system—The short-term anaerobic energy system that maintains ATP levels. Breakdown of phosphocreatine (PCr) frees P_i, which then combines with ADP to form ATP.

atrioventricular (AV) node—The specialized mass of conducting cells in the heart located at the atrioventricular junction.

atrophy—Loss of size, or mass, of body tissue, such as muscle atrophy with disuse.

autocrine—A locally acting messenger molecule, such as a prostaglandin.

autogenic inhibition—Reflex inhibition of a motor neuron in response to excessive tension in the muscle fibers it supplies, as monitored by the Golgi tendon organs.

AV node—*See* atrioventricular node.

(a-v)O_2 difference—*See* arterial–venous oxygen difference.

(a-$\bar{v}$)O_2 difference—*See* arterial–mixed venous oxygen difference.

axon hillock—A part of the neuron, between the cell body and the axon, that controls traffic down the axon through summation of excitatory and inhibitory postsynaptic potentials.

axon terminal—One of numerous branched endings of an axon. Also known as a terminal fibril.

barometric pressure (P_b)—The total pressure exerted by the atmosphere at a given altitude.

baroreceptor—Stretch receptor located within the cardiovascular system that senses changes in blood pressure.

basal metabolic rate (BMR)—The lowest rate of body metabolism (energy use) that can sustain life, measured after an overnight sleep in a laboratory under optimal conditions of quiet, rest, and relaxation and after a 12 h fast. *See also* resting metabolic rate.

β-blockers—A class of drugs that block transmission of neural impulses from the sympathetic nervous system, proposed to have ergogenic properties.

β-cells—Cells in the islets of Langerhans in the pancreas that secrete insulin.

β-oxidation—The first step in fatty acid oxidation, in which fatty acids are broken into separate two carbon units of acetic acid, each of which is then converted to acetyl CoA.

bioelectric impedance—A procedure for assessing body composition in which an electrical current is passed through the body. The resistance to current flow through the tissues reflects the relative amount of fat present.

bioenergetics—Term given to the study of metabolic processes that yield or consume energy.

bioinformatics—The science of analyzing complex biological data such as genetic codes.

blood doping—Any means by which a person's total volume of red blood cells is increased, typically via transfusion of red blood cells or use of erythropoietin.

blood lipids—Blood-borne fats, such as triglycerides and cholesterol.

BMI—*See* body mass index.

BMR—*See* basal metabolic rate.

body composition—The chemical composition of the body. The model used in this book considers two components: fat-free mass and fat mass.

body density—Body weight divided by body volume.

body mass index (BMI)—A measurement of body overweight or obesity determined by dividing weight (in kilograms) by height (in meters) squared. BMI is highly correlated with body composition.

bone mineral density—The mass of bone per unit volume. Decreased bone mineral density increases the risk of fractures.

Borg RPE scale—A numerical scale for rating perceived exertion.

Boyle's gas law—Law stating that at a constant temperature, the number of gas molecules in a given volume depends on the pressure.

bradycardia—A resting heart rate lower than 60 beats/min.

bulimia nervosa—A clinical eating disorder characterized by recurrent episodes of binge eating, a feeling of lack of control during these binges, and purging behavior, which may include self-induced vomiting and use of laxatives and diuretics. Sometimes the disorder also includes fasting or excessive exercise behaviors.

CAD—*See* coronary artery disease.

caffeine—A central nervous system stimulant believed by some athletes to have ergogenic properties.

calorie (cal)—A unit of measure of energy in biological systems, where 1 calorie is equal to the amount of heat energy needed to raise the temperature of 1 g of water 1 °C, from 15 to 16 °C.

calorimeter—A device for measuring the heat produced by the body (or by specific chemical reactions).

cAMP—*See* cyclic adenosine monophosphate.

capillaries—The smallest vessels transporting blood from the heart to the tissues and the actual sites of exchange between the blood and tissue.

capillary-to-fiber ratio—The number of capillaries per muscle fiber.

carbohydrate—An organic compound formed from carbon, hydrogen, and oxygen; includes starches, sugars, and cellulose.

carbohydrate loading—Increased dietary consumption of carbohydrate, a process used by athletes to increase carbohydrate stores in the body before prolonged endurance exercise.

cardiac cycle—The period that includes all events between two consecutive heartbeats.

cardiac hypertrophy—Enlargement of the heart by increases in muscle wall thickness or chamber size or both.

cardiac output ($\dot{Q}$)—The volume of blood pumped out by the heart per minute. $\dot{Q}$ = heart rate × stroke volume.

cardiorespiratory endurance—The ability of the body to sustain prolonged exercise.

cardiovascular deconditioning—A decrease in the cardiovascular system's ability to deliver sufficient oxygen and nutrients.

cardiovascular drift—An increase in heart rate during exercise to compensate for a decrease in stroke volume. This compensation helps maintain a constant cardiac output.

catabolism—The tearing down of body tissue; the destructive phase of metabolism.

catecholamines—Biologically active amines (organic compounds derived from ammonia), such as epinephrine and norepinephrine, that serve as neurotransmitters in the sympathetic nervous system and as hormones.

central command—Information originating in the brain that is transmitted to the cardiovascular, muscular, or pulmonary systems.

central governor theory—Theory proposing that processes occur in the brain that regulate power output by the muscles to prevent unsafe levels of exertion.

central nervous system (CNS)—System consisting of the brain and spinal cord.

chemoreceptor—A sensory organ capable of reacting to a chemical stimulus.

Cheyne-Stokes breathing—Alternating periods of rapid breathing and slow, shallow breathing, including periods in which breathing may actually cease temporarily. A symptom of acute mountain sickness.

childhood—The period of life between the first birthday and the onset of puberty.

cholecystokinin (CCK)—A hormone released by the gastrointestinal tract that signals the brain to suppress hunger.

cholinergic—Systems mediated by the neurotransmitter acetylcholine.

chronic adaptation—A physiological change that occurs when the body is exposed to repeated exercise bouts over weeks or months. These changes generally improve the body's efficiency at rest and during exercise.

chronic hypertrophy—An increase in muscle size that results from repeated long-term resistance training.

CNS—*See* central nervous system.

cold habituation—A response to repeated cold exposure, often to the hands and face, in which skin vasoconstrictor and shivering responses are blunted.

concentric contraction—Muscle shortening.

conduction (K)—(1) Transfer of heat through direct molecular contact with a solid object; (2) movement of an electrical impulse, such as through a neuron.

congenital heart disease—A heart defect present at birth that occurs from abnormal prenatal development of the heart or associated blood vessels. Also known as congenital heart defect.

continuous training—Training at a moderate to high intensity without stopping to rest.

control group—In an experimental design, the nontreated group to which the experimental group is compared.

convection (C)—The transfer of heat or cold via the movement of a gas or liquid across an object, such as the body.

coronary artery disease (CAD)—A disease characterized by pathological changes in the coronary arteries that supply blood flow to the myocardium.

cortisol—A corticosteroid hormone released from the adrenal cortex that stimulates gluconeogenesis, increases mobilization of free fatty acids, decreases use of glucose, and stimulates catabolism of protein. Also known as hydrocortisone.

creatine—A substance found in skeletal muscles most commonly in the form of PCr. Creatine supplements are often used as ergogenic aids because they are theorized to increase PCr levels, thus enhancing the ATP-PCr energy system by better maintaining muscle ATP levels.

creatine kinase—The enzyme that facilitates the breakdown of PCr to creatine and P_i.

critical power—The maximum power output or exercise intensity that can be sustained, at least theoretically, without exhaustion occurring.

critical temperature theory—Theory that prolonged exercise in hot environments is limited by attainment of a fixed elevated core temperature.

crossover concept—The exercise intensity at which fat and carbohydrate utilization intersect as the energy derived from fat decreases and the energy from carbohydrate increases.

crossover design—Experimental design in which the control group becomes the experimental group after the first experimental period, and vice versa.

cross-sectional research design—A research design in which a cross section of a population is tested at one specific time and then data from groups within that population are compared.

cross-training—Training for more than one sport at the same time, or training multiple fitness components (such as endurance, strength, and flexibility) within the same period.

cycle ergometer—An exercise device that uses cycling to measure physical work.

cyclic adenosine monophosphate (cAMP)—Intracellular second messenger that mediates hormone action.

cytochrome—A series of iron-containing proteins that facilitate the transport of electrons within the electron transport chain.

Dalton's law—A principle stating that the total pressure exerted by a mixture of gases is equal to the sum of the partial pressures of those individual gases.

DBP—*See* diastolic blood pressure.

dead space—The volume of air inhaled that does not take part in the gas exchange because it remains in the conducting airways.

dehydration—Loss of body fluids.

delayed-onset muscle soreness (DOMS)—Muscle soreness that develops a day or two after a heavy bout of exercise and is associated with actual injury within the muscle.

densitometry—The measurement of body density.

dependent variable—The physiological factor that is allowed to vary as another factor (the independent variable) is manipulated. Usually plotted on the *y*-axis.

depolarization—A decrease in the electrical potential across a membrane, as when the inside of a neuron becomes less negative relative to the outside.

detraining—Changes in physiological function in response to a reduction or cessation of regular physical training.

development—Changes that occur in the body starting at conception and continuing through adulthood; differentiation along specialized lines of function, reflecting changes that accompany growth.

diabetes mellitus—A disorder of carbohydrate metabolism characterized by hyperglycemia (high blood sugar levels) and glycosuria (presence of sugar in the urine). The disease develops when there is inadequate production of insulin by the pancreas or inadequate utilization of insulin by the cells.

diastolic blood pressure (DBP)—The lowest arterial pressure, resulting from ventricular diastole (the resting phase).

direct calorimetry—A method that gauges the body's rate and quantity of energy production by direct measurement of the body's heat production.

direct gene activation—The method of action of steroid hormones. They bind to receptors in the cell, and then the hormone–receptor complex enters the nucleus and activates certain genes.

disordered eating—Abnormal eating behavior that ranges from excessive restriction of food intake to pathological behaviors, such as self-induced vomiting and laxative abuse. Disordered eating can lead to clinical eating disorders, such as anorexia nervosa and bulimia nervosa.

diuretics—Substances that promote water excretion.

diurnal variation—Fluctuations in physiological responses that occur during a 24 h period.

DOMS—*See* delayed-onset muscle soreness.

dose–response relation—A relationship between two variables in which one changes predictably as the other increases or decreases.

downregulation—Decreased cellular sensitivity to a hormone, likely the result of a decreased number of cell receptors available to bind with the hormone.

dry heat exchange—Heat transfer by the combined avenues of convection, conduction, and radiation.

dual-energy X-ray absorptiometry (DEXA)—A technique used to assess both regional and total body composition through the use of X-ray absorptiometry.

dynamic contraction—Any muscle action that produces joint movement.

dyspnea—Labored or difficult breathing.

EAMCs—*See* exercise-associated muscle cramps.

eating disorders—A group of clinical disorders involving eating. *See also* anorexia nervosa, bulimia nervosa.

eccentric contraction—Any muscle action in which muscle lengthens.

eccentric training—Training that involves eccentric action.

eccrine sweat glands—Simple sweat glands dispersed over the body surface that respond to increases in core or skin temperature (or both) and facilitate thermoregulation.

ECG—*See* electrocardiogram and exercise electrocardiogram.

EDV—*See* end-diastolic volume.

EF—*See* ejection fraction.

effector (efferent) nerves—The motor division of the peripheral nervous system, carrying impulses from the CNS toward the periphery.

EIAH—*See* exercise-induced arterial hypoxemia.

ejection fraction (EF)—The fraction of blood pumped out of the left ventricle with each contraction, determined by dividing stroke volume by end-diastolic volume and expressed as a percentage.

electrical stimulation—Stimulation of a muscle by passing an electrical current through it.

electrocardiogram (ECG)—A recording of the heart's electrical activity.

electrocardiograph—A machine used to obtain an electrocardiogram.

electrolyte—A dissolved substance that can conduct an electrical current.

electron transport chain—A series of chemical reactions that convert the hydrogen ion generated by glycolysis and the Krebs cycle into water and produce energy for oxidative phosphorylation.

end branches—Branches coming off the ends of the axons leading to the axon terminals.

end-diastolic volume (EDV)—The volume of blood inside the left ventricle at the end of diastole, just before contraction.

endomysium—A sheath of connective tissue that covers each muscle fiber.

endothelial dysfunction—Negative changes in the cells that line the lumen of blood vessels resulting in a relative inability of those vessels to constrict or dilate.

endothelium—Layer of thin cells that line the lumen of the blood vessels.

end-systolic volume (ESV)—The volume of blood remaining in the left ventricle at the end of systole, just after contraction.

environmental physiology—Study of the effects of the environment (heat, cold, altitude, hyperbaria, and so on) on the function of the body.

enzyme—Protein molecules that speed up reactions by lowering their energy of activation.

ephedrine—A sympathomimetic amine derived from ephedra herbs (also known as ma huang) and used as a decongestant and as a bronchodilator in the treatment of asthma.

epigenetics—The study of changes in organisms caused by modifying gene expression rather than changing the genetic code itself.

epimysium—The outer connective tissue that surrounds an entire muscle, holding it together.

epinephrine—A catecholamine released from the adrenal medulla that, along with norepinephrine, prepares the body for a fight-or-flight response. It is also a neurotransmitter. *See also* catecholamines.

EPOC—*See* excess postexercise oxygen consumption.

EPSP—*See* excitatory postsynaptic potential.

ergogenic—Able to improve work or performance.

ergolytic—Able to impair work or performance.

ergometer—An exercise device that allows the amount and rate of a person's physical work to be controlled (standardized) and measured.

erythropoietin (EPO)—The hormone that stimulates erythrocyte (red blood cell) production.

essential amino acids—The eight or nine amino acids necessary for human growth that the body cannot synthesize and are thus essential parts of our diets.

estrogen—A female sex hormone.

ESV—*See* end-systolic volume.

eumenorrhea—Normal menstrual function.

evaporation (E)—Heat loss through the conversion of water (such as in sweat) to vapor.

excessive training—Training in which volume, intensity, or both are too great or are increased too quickly without proper progression.

excess postexercise oxygen consumption (EPOC)—Elevated oxygen consumption above resting levels after exercise; at one time referred to as oxygen debt.

excitation–contraction coupling—The sequence of events by which a nerve impulse reaches the muscle membrane and leads to cross-bridge activity and thus muscle contraction.

excitatory postsynaptic potential (EPSP)—A depolarization of the postsynaptic membrane caused by an excitatory impulse.

exercise-associated muscle cramps (EAMCs)—Painful prolonged contractions of muscles that accompany or result from muscle contractions.

exercise electrocardiogram (ECG)—A recording of the heart's electrical activity during exercise.

exercise-induced arterial hypoxemia (EIAH)—A decline in arterial PO_2 and arterial oxygen saturation during maximal or near-maximal exercise.

exercise physiology—The study of how body structure and function are altered by exposure to acute and chronic bouts of exercise.

exercise prescription—Individualization of the prescription of exercise duration, frequency, intensity, and mode.

expiration—The process by which air is forced out of the lungs through relaxation of the inspiratory muscles and elastic recoil of the lung tissue, which increase the pressure in the thorax.

external respiration—The process of bringing air into the lungs and the resulting exchange of gas between the alveoli and the capillary blood.

extracellular fluid—The 35% to 40% of the water in the body that is outside the cells, including interstitial fluid, blood plasma, lymph, cerebrospinal fluid, and other fluids.

extrinsic neural control—Redistribution of blood at the system or body level through neural mechanisms.

Fartlek training—Developed in the 1930s; the term comes from the Swedish for "speed play." This type of training combines continuous and interval training and stresses both the aerobic and anaerobic energy pathways.

fascicle—A small bundle of muscle fibers wrapped in a connective tissue sheath within a muscle.

fat—A class of organic compounds with limited water solubility that exists in the body in many forms, such as triglycerides, free fatty acids, phospholipids, and steroids.

fat-free mass—The mass (weight) of the body that is not fat, including muscle, bone, skin, and organs.

fatigue—General sensations of tiredness and accompanying decrements in muscular performance.

fat mass—The absolute amount or mass of body fat.

fatty streaks—Early lipid deposits within blood vessels.

female athlete triad—Three interrelated disorders—disordered eating, menstrual dysfunction, and bone mineral disorders—to which some female athletes are prone.

$FEV_{1.0}$—*See* forced expiratory volume in 1 s.

FFA—*See* free fatty acids.

fiber hyperplasia—An increase in the number of muscle fibers.

fiber hypertrophy—An increase in the size of existing individual muscle fibers.

Fick equation—$\dot{V}O_2 = \dot{Q} \times (a\text{-}\bar{v})O_2$ difference.

Fick's law—Law stating that the net diffusion rate of a gas across a fluid membrane is proportional to the difference in partial pressure, proportional to the area of the membrane, and inversely proportional to the thickness of the membrane.

force–velocity relation—The force generated by a muscle is a function of its velocity. Increasing the velocity of contraction while shortening reduces force, while increasing speed of contractions as the muscle is lengthening increases force.

forced expiratory volume in 1 s ($FEV_{1.0}$)—The volume of air exhaled in the first second after maximal inhalation.

Frank-Starling mechanism—The mechanism by which an increased amount of blood in the ventricle causes a stronger ventricular contraction to increase the amount of blood ejected.

free fatty acids (FFA)—The components of fat that are used by the body for metabolism.

free radicals—Univalent (unpaired) oxygen intermediates that leak out of the electron transport chain during metabolic processes and may damage tissues.

free weights—Traditional resistance training modality that uses only barbells, dumbbells, and so on to provide resistance.

frostbite—Tissue damage that occurs during cold exposure because circulation to the skin decreases, in an attempt to retain body heat, to the point that the tissue receives insufficient oxygen and nutrients.

functional sympatholysis—The process in which vasoactive molecules released from active skeletal muscle inhibit sympathetic vasoconstriction in order to increase blood flow to exercising muscle.

gastric emptying—The movement of food mixed with gastric secretions from the stomach into the duodenum.

genomics—The branch of molecular biology concerned with the structure, function, evolution, and mapping of genomes.

genotype—The genetic makeup of an individual.

ghrelin—A hormone secreted from the stomach and pancreas when the stomach is empty in order to stimulate hunger.

glucagon—A hormone released by the pancreas that promotes increased breakdown of liver glycogen to glucose (glycogenolysis) and increased gluconeogenesis.

glucocorticoid—A family of steroid hormones produced by the adrenal cortex that help maintain homeostasis through a variety of effects throughout the body.

gluconeogenesis—The conversion of protein or fat into glucose.

glucose—Six-carbon sugar that is the primary form of carbohydrate used for metabolism.

glycogen—The form of carbohydrate stored in the body, found predominantly in the muscles and liver.

glycogen loading—The manipulation of exercise and diet to optimize the body's glycogen storage.

glycogenolysis—The conversion of glycogen to glucose.

glycogen sparing—Increased reliance on fats for energy production during endurance activity, rather than stores of glycogen.

glycolysis—The breakdown of glucose to pyruvic acid.

Golgi tendon organ—A sensory receptor in a muscle tendon that monitors tension.

graded exercise test (GXT)—An exercise test in which the rate of work is increased gradually in 1 to 3 min increments, usually to the point of fatigue or exhaustion.

graded potential—A localized change (depolarization or hyperpolarization) in the membrane potential.

growth—An increase in the size of the body or any of its parts.

growth hormone—An anabolic agent that stimulates fat metabolism and promotes muscle growth and hypertrophy by facilitating amino acid transport into the cells.

GXT—*See* graded exercise test.

HACE—*See* high-altitude cerebral edema.

Haldane transformation—An equation allowing one to calculate the inspired air volume from expired air volume, or expired air volume from inspired air volume.

HAPE—*See* high-altitude pulmonary edema.

HDL-C—*See* high-density lipoprotein cholesterol.

health span—The number of years of generally healthy living, free from serious disease or chronic disability.

heart failure—A clinical condition in which the myocardium becomes too weak to maintain adequate cardiac output to meet the body's oxygen demands; heart failure usually results from the heart's being damaged or overworked.

heart murmur—A condition in which abnormal heart valve sounds are detected with the aid of a stethoscope.

heart rate reserve—The difference between maximal heart rate (HR_{max}) and resting heart rate (HR_{rest}).

heat acclimation—*See* acclimation.

heat cramp—Cramping of the skeletal muscles as a result of excessive dehydration and the associated salt loss.

heat exhaustion—A heat disorder resulting from an inability of the cardiovascular system to meet all the body tissues' needs while also shifting blood to the periphery for cooling, characterized by elevated body temperature, breathlessness, extreme tiredness, dizziness, and rapid pulse.

heatstroke—The most serious heat disorder, resulting from failure of the body's thermoregulatory mechanisms. Heatstroke is characterized by body temperature above 40.5 °C (105 °F), cessation of sweating, and total confusion or unconsciousness; it can lead to death.

hematocrit—The percentage of cells or formed elements in the total blood volume. More than 99% of the cells or formed elements are red blood cells.

hematopoiesis—Increased red blood cell concentration by increased production of cells.

hemoconcentration—A relative (not absolute) increase in the cellular content per unit of blood volume, resulting from a reduction in plasma volume.

hemodilution—An increase in blood plasma, resulting in a dilution of the blood's cellular contents.

hemoglobin—The iron-containing pigment in red blood cells that binds oxygen.

hemorrhagic stroke—Involves bleeding within the brain, which damages nearby brain tissue.

Henry's law—Law stating that gases dissolve in liquids in proportion to their partial pressures, depending also on their solubilities in the specific fluids and on the temperature.

hGH—*See* human growth hormone.

high-altitude cerebral edema (HACE)—A condition of unknown cause in which fluid accumulates in the cranial cavity at altitude; characterized by mental confusion that can progress to coma and death.

high-altitude pulmonary edema (HAPE)—A condition of unknown cause in which fluid accumulates in the lungs at altitude, interfering with ventilation, resulting in shortness of breath and fatigue, and characterized by impaired blood oxygenation, mental confusion, and loss of consciousness.

high-density lipoprotein cholesterol (HDL-C)—The cholesterol carried by HDL.

high-intensity interval training (HIIT)—Training that uses short bursts of very intense exercise interspersed with only a few minutes of rest or low-intensity exercise.

high responders—Those individuals within a population that show clear or exaggerated responses or adaptations to a stimulus.

HIIT—*See* high-intensity interval training.

homeostasis—Maintenance of a constant internal environment.

hormone—A chemical substance produced or released by an endocrine gland and transported by the blood to a specific target tissue.

HR_{max}—*See* maximum heart rate.

human growth hormone (hGH)—A hormone that promotes anabolism and is believed by some athletes to have ergogenic properties.

hydrostatic pressure—The pressure exerted by a stationary column of fluid in a tube.

hydrostatic weighing—A method of measuring body volume in which a person is weighed while submerged underwater. The difference between the scale weight on land and the underwater weight (corrected for water density) equals body volume. This value must be further corrected to account for any air trapped in the lungs and other parts of the body.

hyperglycemia—An elevated blood glucose level.

hyperinsulinemia—High levels of insulin in the blood.

hyperpolarization—An increase in the electrical potential across a membrane.

hypertension—Abnormally high blood pressure. In adults, hypertension is usually defined as a systolic pressure of 140 mmHg or higher or a diastolic pressure of 90 mmHg or higher.

hypertrophy—Increase in the size or mass of an organ or body tissue. *See also* fiber hypertrophy.

hyperventilation—A breathing rate or tidal volume greater than necessary for normal function.

hypobaric—Referring to an environment, such as that at high altitude, involving low atmospheric pressure.

hypoglycemia—A low blood glucose level.

hyponatremia—A blood sodium concentration below the normal range of 136 to 143 mmol/L.

hypothermia—Low body temperature; any temperature below the given person's normal temperature.

hypoxemia—A decreased oxygen content or concentration within the blood.

hypoxia—A decreased availability of oxygen to the tissues.

IDDM—*See* insulin-dependent diabetes mellitus.

immune function—The body's normal ability to fight infection and illness with antibodies and lymphocytes.

impaired fasting glucose—A plasma glucose level between 110 and 125 mg/dl following an 8 h fast.

impaired glucose tolerance—An abnormal glucose response to an oral glucose load (glucose tolerance test), sometimes seen as a precursor to diabetes.

independent variable—In an experiment, the variable that is manipulated by the experimenter to determine the response of the dependent variable. Usually plotted on the *x*-axis.

indirect calorimetry—A method of estimating energy expenditure by measuring respiratory gases.

infancy—The first year of life.

inhibiting factors—Hormones transmitted from the hypothalamus to the anterior pituitary that inhibit release of some other hormones.

inhibitory postsynaptic potential (IPSP)—A hyperpolarization of the postsynaptic membrane caused by an inhibitory impulse.

inspiration—The active process involving the diaphragm and the external intercostal muscles that expands the thoracic dimensions and thus the lungs. The expansion decreases pressure in the lungs, allowing outside air to rush in.

insulation—Resistance to dry heat loss.

insulative acclimation—A pattern of cold acclimation in which enhanced skin vasoconstriction increases peripheral insulation and minimizes heat loss.

insulin—A hormone produced by the β-cells in the pancreas that assists glucose entry into cells.

insulin-dependent diabetes mellitus (IDDM)—One of two major categories of diabetes mellitus that is caused by the inability of the pancreas to produce sufficient insulin as a result of failure of the β-cells in the pancreas. This is also known as type 1 diabetes.

insulin resistance—A physiological condition in which cells fail to respond to the normal actions of the hormone insulin.

insulin sensitivity—An index of the effectiveness of a given insulin concentration on the disposal of glucose.

integrative physiology—The study of organisms as functioning systems of molecules, cells, tissues, and organs with an emphasis on whole-body function.

intercalated disks—Specialized cell junctions in the myocardium where one muscle cell connects to the next.

internal respiration—The exchange of gases between the blood and tissues.

interval-circuit training—Training program that involves rapid movement from one exercise to another around a "circuit" or established set of exercises.

interval training—Repeated, brief, fast-paced exercise bouts with short rest intervals between bouts.

intracellular fluid—The approximately 60% to 65% of total body water that is contained in the cells.

IPSP—*See* inhibitory postsynaptic potential.

irisin—A muscle-derived hormone that signals conversion of white adipose tissue to brown adipose tissue.

ischemia—A temporary deficiency of blood to a specific area of the body.

ischemic stroke—Brain tissue damage resulting from insufficient oxygen supply to an area of the brain. May be caused by narrowing or blockage of blood vessels supplying the area.

isokinetic training—Resistance training in which the rate of movement is kept constant through the range of motion.

isometric training—Resistance training involving a static action.

kilocalorie (kcal)—The equivalent of 1,000 calories. *See also* calorie.

Krebs cycle—A series of chemical reactions that involve the complete oxidation of acetyl CoA and produce 2 mol of ATP (energy) along with hydrogen and carbon, which combine with oxygen to form H_2O and CO_2.

lactate threshold—The point during exercise of increasing intensity at which blood lactate begins to accumulate above resting levels, where lactate clearance is no longer able to keep up with lactate production.

LDL-C—*See* low-density lipoprotein cholesterol.

length–tension relation—The tension developed by a muscle is a function of its length. The highest force can be generated when there is optimal overlap between actin and myosin filaments.

leptin—A hormone primarily secreted by fat cells that acts on receptors in the hypothalamus to decrease hunger.

lipogenesis—The process of converting protein into fatty acids.

lipolysis—The process of breaking down triglyceride to its basic units to be used for energy.

lipoprotein lipase—The enzyme that breaks down triglycerides to free fatty acids and glycerol, allowing the free fatty acids to enter the cells for use as a fuel or for storage.

lipoproteins—The proteins that carry the blood lipids.

longevity—The length of a person's life.

longitudinal research design—A research design in which subjects are tested initially and then one or more times later to directly measure changes over time resulting from a given intervention.

long, slow distance (LSD) training—Endurance training involving long, slow distances.

low-density lipoprotein cholesterol (LDL-C)—The cholesterol carried by LDL.

lower body (gynoid) obesity—Obesity that follows the typically female pattern of fat storage, in which fat is stored primarily in the lower body, particularly in the hips, buttocks, and thighs.

low-intensity aerobic exercise—Aerobic exercise performed at low intensity, theoretically to cause the body to burn a higher percentage of fat.

low responders—Those individuals within a population who show little or no response or adaptation to a stimulus.

macrominerals—Those minerals of which the body needs more than 100 mg per day.

MAP—*See* mean arterial pressure.

maturation—The process by which the body takes on the adult form and becomes fully functional. It is often defined by the system or function being considered.

maximal expiratory ventilation ($\dot{V}_{Emax}$)—The highest ventilation that can be achieved during exhaustive exercise.

maximal oxygen uptake ($\dot{V}O_{2max}$)—The maximal capacity for oxygen consumption by the body during maximal exertion. It is also known as aerobic power, maximal oxygen intake, maximal oxygen consumption, and cardiorespiratory endurance capacity.

maximal voluntary ventilation—The maximal capacity to move air into and out of the lungs, usually measured for 12 s and extrapolated to a per minute value.

maximum heart rate (HR_{max})—The highest heart rate value attainable during an all-out effort to the point of exhaustion.

mean arterial pressure (MAP)—The average pressure exerted by the blood as it travels through the arteries. It is estimated as follows: $MAP = DBP + [0.333 \times (SBP - DBP)]$.

mechanoreceptors—An end organ that responds to changes in mechanical stress, such as stretch, compression, or distension.

menarche—The onset of menstruation; the first menses.

menopause—The permanent cessation of menses in women, typically occurring between ages 45 and 55; defined as occurring 12 months after the final menstrual period.

menses—The menstrual or flow phase of the menstrual cycle.

menstrual cycle—The cycle of uterine changes, averaging 28 days and consisting of the menstrual (flow) phase, the proliferative phase, and the secretory phase.

menstrual dysfunction—Disruption of the normal menstrual cycle; includes oligomenorrhea, primary amenorrhea, and secondary amenorrhea.

metabolic acclimation—A pattern of cold acclimation involving increased metabolic heat production through enhanced nonshivering and shivering thermogenesis.

metabolic equivalent (MET)—A unit used to estimate the metabolic cost (oxygen consumption) of physical activity.

One MET equals the resting metabolic rate of approximately 3.5 ml of $O_2 \cdot kg^{-1} \cdot min^{-1}$.

metabolic syndrome—A term that has been used to link coronary artery disease, hypertension, type 2 diabetes, and upper body obesity to insulin resistance and hyperinsulinemia. This syndrome has also been referred to as syndrome X and the civilization syndrome.

metabolism—All energy-producing and energy-using processes within the body.

microminerals (trace elements)—The minerals of which the body needs less than 100 mg per day.

mineralocorticoids—Steroid hormones released from the adrenal cortex that are responsible for electrolyte balance within the body—for example, aldosterone.

mitochondria—Cellular organelles that generate ATP through oxidative phosphorylation.

mitochondrial oxidative enzymes—Oxidative enzymes located in the mitochondria.

mode—Type of exercise.

motor nerves (motor neurons)—Efferent nerves that carry impulses to skeletal muscle.

motor neurons—*See* motor nerves.

motor reflex—An involuntary motor response to a given stimulus.

motor unit—The motor nerve and the group of muscle fibers it innervates.

mTOR—Enzyme that controls the rate of protein synthesis within the myofibrils after resistance training.

muscle fiber—An individual muscle cell.

muscle pump—The rhythmic mechanical compression of the veins that occurs during skeletal muscle contraction in many types of movement and exercise—for example, during walking and running—and assists the return of blood to the heart.

muscle spindle—A sensory receptor located in the muscle that senses how much the muscle is stretched.

muscular endurance—The ability of a muscle to resist fatigue.

musculoskeletal system—Body system composed of the skeleton and skeletal muscles that allows, supports, and helps control human movement.

myelination—The process of acquiring a myelin sheath.

myelin sheath—The outer covering of a myelinated nerve fiber, formed by a fatlike substance called myelin.

myocardial infarction—Death of heart tissue that results from insufficient blood supply to part of the myocardium.

myocardium—The muscle of the heart.

myofibril—The contractile element of skeletal muscle.

myoglobin—A compound similar to hemoglobin, but found in muscle tissue, that carries oxygen from the cell membrane to the mitochondria.

myosin—One of the proteins that form filaments that produce muscle action.

myosin cross-bridge—The protruding part of a myosin filament. It includes the myosin head, which binds to an active site on an actin filament to produce a power stroke that causes the filaments to slide across each other.

nebulin—A giant protein that coextends with actin and appears to play a regulatory role in mediating actin and myosin interactions.

negative feedback—A decrease in the amount or effect of a molecule or substance by its influence on the process giving rise to it, such as when a high level of a particular hormone in the blood may inhibit further secretion of that hormone.

nerve impulse—The electrical signal conducted along a neuron, which can be transmitted to another neuron or an end organ such as a group of muscle fibers.

neuromuscular junction—The site at which a motor neuron communicates with a muscle fiber.

neuron—A specialized cell in the nervous system responsible for generating and transmitting nerve impulses.

neurotransmitter—A chemical used for communication between a neuron and another cell.

nitric oxide (NO)—Important cellular signaling molecule involved in many physiological processes, including dilation of arterioles.

nonessential amino acids—The 11 or 12 amino acids that the body synthesizes.

nonshivering thermogenesis—The stimulation of metabolism by the sympathetic nervous system to generate more metabolic heat.

nonsteroid hormones—Hormones derived from protein, peptides, or amino acids that cannot easily cross cell membranes.

norepinephrine—A catecholamine released from the adrenal medulla that, along with epinephrine, prepares the body for a fight-or-flight response. It is also a neurotransmitter. *See also* catecholamines.

obesity—An excessive amount of body fat, generally defined as more than 25% in men and 35% in women; a body mass index of 30 or greater.

oligomenorrhea—Abnormally infrequent or scant menstruation.

oncotic pressure—The pressure exerted by the concentration of proteins in a solution, drawing water from regions with lower oncotic pressures.

osmolality—The number of solutes (such as electrolytes) dissolved in a fluid divided by the weight of that fluid; usually expressed in units of osmoles (or milliosmoles) per kilogram.

osmolarity—The number of solutes (such as electrolytes) dissolved in a fluid divided by the volume of that fluid; usually expressed in units of osmoles (or milliosmoles) per liter.

ossification—The process of bone formation.

osteopenia—The loss of bone mass with aging.

osteoporosis—Decreased bone mineral content that increases bone porosity.

overreaching—A systematic attempt to intentionally overstress the body, allowing the body to adapt to the training stimulus above and beyond the adaptation attained during a period of acute overload.

overtraining—The attempt to do more work than can be physically tolerated.

overtraining syndrome—A condition brought on by overtraining and characterized by performance decrements and a general breakdown in physiological function.

overweight—Body weight that exceeds the normal or standard weight for a particular individual based on sex, height, and frame size; a BMI of 25.0 to 29.9.

oxidative phosphorylation—Mitochondrial process that uses oxygen and high-energy electrons to produce ATP and water.

oxidative system—The body's most complex energy system, which generates energy by disassembling fuels with the aid of oxygen and has a very high energy yield.

oxygen cascade—The progressively dropping partial pressures of oxygen from dry ambient air as oxygen flows to the tissues and into the venous circulation draining those tissues.

oxygen deficit—The difference between the oxygen required for a given exercise intensity (steady state) and the actual oxygen consumption.

oxygen diffusion capacity—The rate at which oxygen diffuses from one place to another.

oxygen transport system—The components of the cardiovascular and respiratory systems involved in transporting oxygen.

partial pressure—The pressure exerted by an individual gas in a mixture of gases.

partial pressure of oxygen (PO_2)—The pressure exerted by oxygen in a mixture of gases.

pathophysiology—The physiology of a specific disease or disorder.

P_b—*See* barometric pressure.

PCr—*See* phosphocreatine.

PDGF—*See* platelet-derived growth factor.

peak oxygen uptake ($\dot{V}O_{2peak}$)—The highest oxygen uptake achieved during a graded exercise test when a subject reaches volitional fatigue before a plateau occurs in the $\dot{V}O_2$ response (the criterion for a true $\dot{V}O_{2max}$).

pericardium—A double-layered outer covering of the heart.

perimysium—The connective tissue sheath surrounding each muscle fasciculus.

periodization—The breaking down of an entire season's training program for sport into smaller periods or training units.

peripheral blood flow—Blood flow to the extremities and the skin.

peripheral nervous system (PNS)—That section of the nervous system through which motor nerve impulses are transmitted from the brain and spinal cord to the periphery and sensory nerve impulses are transmitted from the periphery to the brain and spinal cord.

peripheral vascular disease—Diseases of the systemic arteries and veins, especially those to the extremities, that impede adequate blood flow.

peripheral vasoconstriction—*See* vasoconstriction.

PFK—*See* phosphofructokinase.

phenotype—Observable characteristics of an individual resulting from the interaction of his or her genotype with the environment.

phosphocreatine (PCr)—An energy-rich compound that plays a critical role in providing energy for muscle action by maintaining ATP concentration.

phosphofructokinase (PFK)—A key rate-limiting enzyme of the anaerobic glycolytic energy system.

phosphorylation—The addition of a phosphate group (PO_4) to a molecule.

physical maturity—The point at which the body has attained the adult physical form.

physiology—The study of the function of organisms.

placebo effect—An effect produced by the subject's expectations after administration of an inactive substance (placebo).

placebo group—The group in an intervention study given a placebo rather than a test substance.

platelet-derived growth factor (PDGF)—A family of molecules released from platelets in the blood that help to heal wounds, repair damage to blood vessel walls, and help blood vessels grow.

plaque—A buildup of lipids, smooth muscle cells, connective tissue, and debris that forms at the site of injury to an artery.

plasmalemma—Plasma membrane, the selectively permeable lipid bilayer coated by proteins that composes the outer layer of a cell.

plyometrics—A type of dynamic-action resistance training based on the theory that use of the stretch reflex during jumping will recruit additional motor units.

PNS—*See* peripheral nervous system.

PO_2—*See* partial pressure of oxygen.

POAH—*See* preoptic-anterior hypothalamus.

polycythemia—Increased red blood cells.

power—The rate of performing work; the product of force and velocity. The rate of transformation of metabolic potential energy to work or heat.

power stroke—The tilting of the myosin head, caused by a strong intermolecular attraction between the myosin cross-bridge and the myosin head, that causes the actin and myosin filaments to slide across each other.

prediabetes—A term used to define those who have impaired fasting glucose, impaired glucose tolerance, or both, but are not truly diabetic.

predictive value of an abnormal exercise test—The accuracy with which abnormal test results reflect the presence of a disease.

pregnancy—The state of carrying an embryo or fetus in the body.

preload—The degree to which the myocardium is stretched before it contracts, determined by factors such as central blood volume.

premature ventricular contraction (PVC)—A common cardiac arrhythmia that results in the feeling of skipped or extra beats caused by impulses originating outside the SA node.

preoptic-anterior hypothalamus (POAH)—The area of the midbrain that is the primary controller of thermoregulatory function.

primary aging—The effect of chronological age alone, i.e., in the absence of other comorbidities that often accompany aging.

primary amenorrhea—The absence of menarche (the beginning of menstruation) beyond age 18.

primary risk factors—Risk factors that have been conclusively shown to have a strong association with a certain disease. Primary risk factors for coronary artery disease include smoking, hypertension, high blood lipid levels, obesity, and physical inactivity.

principle of individuality—The theory that any training program must consider the specific needs and abilities of the individual for whom it is designed.

principle of orderly recruitment—The theory that motor units generally are activated on the basis of a fixed order of recruitment, in which the motor units within a given muscle appear to be ranked according to the size of the motor neuron.

principle of periodization—The gradual cycling of specificity, intensity, and volume of training to achieve peak levels of fitness for competition. Also called the principle of variation.

principle of progressive overload—The theory that, to maximize the benefits of a training program, the training stimulus must be progressively increased as the body adapts to the current stimulus.

principle of reversibility—The theory that a training program must include a maintenance plan to ensure that the gains from training are not lost.

principle of specificity—The theory that a training program must stress the physiological systems critical for optimal performance in a given sport to achieve desired training adaptations in that sport.

principle of variation—The systematic process of changing one or more variables in an exercise training program—mode, volume, or intensity—over time to allow for the training stimulus to remain challenging and effective. Also called the principle of periodization.

prostaglandins—Substances derived from a fatty acid that act as hormones at the local level.

protein—A class of nitrogen-containing compounds formed by amino acids.

pseudoephedrine—A sympathomimetic amine that is used in over-the-counter medications primarily as a decongestant and in the illicit manufacturing of methamphetamine.

puberty—The point at which a person becomes physiologically capable of reproduction.

pulmonary diffusion—The exchange of gases between the lungs and the blood.

pulmonary ventilation—The movement of gases into and out of the lungs.

Purkinje fibers—The terminal branches of the AV bundle that transmit impulses through the ventricles six times faster than through the rest of the cardiac conduction system.

PVC—*See* premature ventricular contraction.

$\dot{Q}$—*See* cardiac output.

radiation (R)—The transfer of heat through electromagnetic waves.

rate coding—Refers to the frequency of impulses sent to a muscle. Increased force can be generated through increase in either the number of muscle fibers recruited or the rate at which the impulses are sent. Also called frequency coding.

rate-limiting enzyme—An enzyme found early in a metabolic pathway that determines the rate of the pathway.

rate–pressure product (RPP)—The mathematical product of heart rate × systolic blood pressure. Also called the double product.

rating of perceived exertion (RPE)—A person's subjective assessment of how hard he or she is exercising.

rehabilitation programs—Programs designed to reestablish health or fitness following a disability or illness.

relative body fat—The ratio of fat mass to total body mass, expressed as a percentage.

releasing factors—Hormones transmitted from the hypothalamus to the anterior pituitary that promote release of some other hormones.

renin—An enzyme formed by the kidneys to convert a plasma protein called angiotensinogen into angiotensin II. *See also* renin-angiotensin-aldosterone mechanism.

renin-angiotensin-aldosterone mechanism—The mechanism involved in renal control of blood pressure. The kidneys respond to decreased blood pressure or blood flow by forming renin, which converts angiotensinogen into angiotensin I, which is finally converted to angiotensin II. Angiotensin II constricts arterioles and triggers aldosterone release.

RER—*See* respiratory exchange ratio.

residual volume (RV)—The amount of air that cannot be exhaled from the lungs.

resistance training—Training designed to increase strength, power, and muscular endurance.

respiratory alkalosis—A condition in which increased carbon dioxide clearance allows blood pH to increase.

respiratory centers—Autonomic centers located in the medulla oblongata and the pons that establish breathing rate and depth.

respiratory exchange ratio (RER)—The ratio of carbon dioxide expired to oxygen consumed at the level of the lungs.

respiratory membrane—The membrane separating alveolar air and blood, composed of the alveolar wall, the capillary wall, and their basement membranes.

respiratory pump—Passive movement of blood through the central circulation as a function of pressure changes during breathing.

resting heart rate (RHR)—The heart rate at rest, averaging 60 to 80 beats/min.

resting membrane potential (RMP)—The potential difference between the electrical charges inside a cell and outside the cell, caused by a separation of charges across the membrane.

resting metabolic rate (RMR)—The body's metabolic rate early in the morning following an overnight fast and 8 h of sleep. Determining RMR does not require sleeping overnight in a laboratory or clinical facility. *See also* basal metabolic rate.

rheumatic heart disease—A form of valvular heart disease involving a streptococcal infection that has caused acute rheumatic fever, typically in children between ages 5 and 15.

RHR—*See* resting heart rate.

risk factor—A predisposing factor statistically linked to the development of a disease, such as coronary artery disease.

RMP—*See* resting membrane potential.

RMR—*See* resting metabolic rate.

RPE—*See* rating of perceived exertion.

RV—*See* residual volume.

saltatory conduction—The means of rapid nerve impulse conduction along myelinated neurons.

SA node—*See* sinoatrial node.

sarcolemma—A muscle fiber's cell membrane.

sarcomere—The basic functional unit of a myofibril.

sarcopenia—The loss of muscle mass associated with aging.

sarcoplasm—The gelatin-like cytoplasm in a muscle fiber.

sarcoplasmic reticulum (SR)—A longitudinal system of tubules that is associated with the myofibrils and stores calcium for muscle action.

satellite cells—Immature cells that can develop into mature cell types, such as myoblasts.

SBP—*See* systolic blood pressure.

secondary amenorrhea—The cessation of menstruation in a woman with previously normal menstrual function.

second messenger—A substance inside a cell that acts as a messenger after a nonsteroid hormone binds to receptors outside the cell.

sensitivity—A test's ability to correctly identify subjects who fit the criteria being tested, such as coronary artery disease.

sensory (afferent) nerves—Afferent nerves that carry impulses toward the central nervous system from the periphery.

sensory–motor integration—The process by which the sensory and motor systems communicate and coordinate with each other.

sex-specific differences—True physiological differences between females and males.

shivering—A rapid, involuntary cycle of contraction and relaxation of skeletal muscles that generates heat.

sickle cell trait—Inheritance of one gene for the disease sickle cell anemia leads to a condition in which those afflicted are at an increased risk for several pathologies that are exacerbated by exercise and dehydration.

single-fiber contractile velocity (V_o)—The rate at which an individual muscle cell can shorten and develop tension.

sinoatrial (SA) node—A group of specialized myocardial cells, located in the wall of the right atrium, that control the heart's rate of contraction; the pacemaker of the heart.

size principle—Principle asserting that the size of the motor neuron dictates the order of motor unit recruitment, with small-sized motor neurons being recruited first.

skinfold fat thickness—The most widely applied field technique used to estimate body density, relative body fat, and fat-free mass. It involves measurement with calipers of the skinfold fat at one or more sites.

sliding filament theory—A theory explaining muscle action: A myosin cross-bridge attaches to an actin filament, and then the power stroke drags the two filaments past one another.

sodium–potassium pump—An enzyme called Na^+-K^+-ATPase, which maintains the resting membrane potential in disequilibrium at −70 mV.

specificity—A test's ability to correctly identify subjects who do not fit the criteria being tested.

specificity of training—The principle that physiological adaptations in response to physical training are highly specific to the nature of the training activity. To maximize benefits, training should be carefully matched to an athlete's specific performance needs.

spirometry—The measurement of lung volumes and capacities.

sport physiology—The application of the concepts of exercise physiology to training athletes and enhancing sport performance.

spot reduction—The practice of exercising a specific area of the body, theoretically to reduce locally stored fat.

SR—*See* sarcoplasmic reticulum.

static-contraction resistance training—Resistance training that emphasizes static muscle action. Also known as isometric resistance training.

static (isometric) muscle contraction—Action in which the muscle contracts without moving, generating force while its length remains static (unchanged). Also known as isometric action.

steady-state heart rate—A heart rate that is maintained constant at submaximal levels of exercise when the rate of work is held constant.

steroid hormones—Hormones with chemical structures similar to cholesterol that are lipid soluble and that diffuse through cell membranes.

strength—The ability of a muscle to exert force—generally the maximal ability.

stroke—A cerebral vascular accident, a condition in which blood supply to some part of the brain is impaired, typically caused by infarction or hemorrhage, so that the tissue is damaged.

stroke volume (SV)—The amount of blood ejected from the left ventricle during contraction; the difference between the end-diastolic volume and the end-systolic volume.

submaximal endurance—The average absolute power output a person can maintain during a fixed period of time on a cycle ergometer, or the average speed or velocity a person can maintain during a fixed period of time. Generally, these tests last at least 30 min but usually not more than 90 min.

substrate—Basic fuel source, such as carbohydrate, protein, and fat.

summation—The summing of all individual changes in a neuron's membrane potential.

SV—*See* stroke volume.

synapse—The junction between two neurons.

systolic blood pressure (SBP)—The greatest arterial blood pressure, resulting from systole (the contracting phase of the heart).

T_3—*See* triiodothyronine.

T_4—*See* thyroxine.

tachycardia—A resting heart rate greater than 100 beats/min.

tapering—A reduction in training intensity before a major competition to give the body and mind a break from the rigors of intense training.

taper period—A period during which training intensity is reduced, allowing time for tissue damage from intense training to heal and for the body's energy reserves to be fully replenished.

target cells—Cells that possess specific hormone receptors.

TEA—*See* thermic effect of activity.

TEM—*See* thermic effect of a meal.

testosterone—The predominant male sex hormone.

tetanus—Highest tension developed by a muscle in response to stimulation of increasing frequency.

therapeutic use exemption—An exemption granted by the governing body of a sport that allows an athlete to use an otherwise banned substance if it is needed to treat a medical condition.

thermal stress—Stress imposed on the body by external temperature.

thermic effect of activity (TEA)—The energy expended in excess of the resting metabolic rate to accomplish a given task or activity.

thermic effect of a meal (TEM)—The energy expended in excess of resting metabolic rate associated with digestion, absorption, transport, metabolism, and storage of ingested food.

thermoreceptors—Sensory receptors that detect changes in body temperature and external temperature and relay this information to the hypothalamus. Also called thermoceptors.

thermoregulation—The process by which the thermoregulatory center, located in the hypothalamus, readjusts body temperature in response to small deviations from the set point.

thirst mechanism—A neural mechanism that triggers thirst in response to dehydration.

THR—*See* training heart rate.

threshold—A minimum amount of stimulus needed to elicit a response. Also, the minimum depolarization required to produce an action potential in neurons.

thyrotropin (TSH)—A hormone secreted by the anterior lobe of the pituitary gland that promotes the release of thyroid hormones.

thyroxine (T_4)—A hormone secreted by the thyroid gland that increases the rate of cellular metabolism and the rate and contractility of the heart.

tidal volume—The amount of air inspired or expired during a normal breathing cycle.

titin—A protein that positions the myosin filament to maintain equal spacing between actin filaments.

TLC—*See* total lung capacity.

total lung capacity (TLC)—The sum of vital capacity and residual volume.

total peripheral resistance (TPR)—The resistance to the flow of blood through the entire systemic circulation.

trace elements—*See* microminerals.

training effect—Physiological adaptation to repeated bouts of exercise.

training heart rate (THR)—A heart rate goal established by using the heart rate equivalent of a desired percentage of $\dot{V}O_{2max}$. For example, if a training level of 75% $\dot{V}O_{2max}$ is desired, 75% of $\dot{V}O_{2max}$ is calculated, and the heart rate corresponding to this $\dot{V}O_2$ is selected as the THR.

transient hypertrophy—The "pumping up" of muscle that happens during a single exercise bout, resulting mainly from fluid accumulation in the interstitial and intracellular spaces of the muscle.

translational physiology—The processes by which basic research findings are extended to the clinical research setting, then to the realm of clinical practice, and finally to health policy.

transverse tubules (T-tubules)—Extensions of the sarcolemma (plasma membrane) that pass laterally through the muscle fiber, allowing nutrients to be transported and nerve impulses to be transmitted rapidly to individual myofibrils.

treadmill—An ergometer in which a motor and pulley system drive a large belt that a person can either walk or run on.

triglycerides—The body's most concentrated energy source and the form in which most fats are stored in the body.

triiodothyronine (T_3)—A hormone released by the thyroid gland that increases the rate of cellular metabolism and the rate and contractility of the heart.

tropomyosin—A tube-shaped protein that twists around actin strands, fitting into the groove between them.

troponin—A complex protein attached at regular intervals to actin strands and tropomyosin.

TSH—*See* thyrotropin.

twitch—The smallest contractile response of a muscle fiber or a motor unit to a single electrical stimulus.

type 1 diabetes—A type of diabetes mellitus that generally has a sudden onset during childhood or young adulthood and leads to almost total insulin deficiency, usually requiring daily insulin injections. Also known as insulin-dependent diabetes mellitus (IDDM) or juvenile-onset diabetes.

type 2 diabetes—A type of diabetes mellitus in which disease onset is more gradual and the causes are more difficult to establish than in type 1 diabetes. Type 2 diabetes is characterized by impaired insulin secretion, impaired insulin action, or excessive glucose output from the liver. Also known as non-insulin-dependent diabetes mellitus (NIDDM).

type I fiber—A type of muscle fiber that has a high oxidative and a low glycolytic capacity, associated with endurance-type activities.

type II fiber—A type of muscle fiber with a low oxidative capacity and a high glycolytic capacity; associated with speed or power activities.

undertraining—The type of training an athlete would undertake between competitive seasons or during active rest. Generally, physiological adaptations will be minor, and there will be no improvement in performance.

upper body (android) obesity—Obesity that follows the typically male pattern of fat storage, in which fat is stored primarily in the upper body, particularly in the abdomen.

upregulation—An increased cellular sensitivity to a hormone, often caused by increased hormone receptors.

Valsalva maneuver—The process of holding the breath and attempting to compress the contents of the abdominal and thoracic cavities, causing increased intra-abdominal and intrathoracic pressure.

valvular heart disease—A disease involving one or more of the heart valves. Rheumatic heart disease is one example.

variable-resistance training—A technique that allows variation in the resistance applied throughout the range of motion in an attempt to match the ability of the muscle or muscle groups to apply force at any specific point in the range of motion.

vasoconstriction—The constriction or narrowing of blood vessels.

vasodilation—The dilation or widening of blood vessels.

vasopressin—*See* antidiuretic hormone.

VC—*See* vital capacity.

$\dot{V}_{Emax}$—*See* maximal expiratory ventilation.

$\dot{V}_E/\dot{V}CO_2$—*See* ventilatory equivalent for carbon dioxide.

$\dot{V}_E/\dot{V}O_2$—*See* ventilatory equivalent for oxygen.

veins—Blood vessels that transport blood back to the heart.

ventilatory equivalent for carbon dioxide ($\dot{V}_E/\dot{V}CO_2$)—The ratio of the volume of air ventilated ($\dot{V}_E$) to the amount of carbon dioxide produced ($\dot{V}CO_2$).

ventilatory equivalent for oxygen ($\dot{V}_E/\dot{V}O_2$)—The ratio between the volume of air ventilated ($\dot{V}_E$) and the amount of oxygen consumed ($\dot{V}O_2$); indicates breathing economy.

ventilatory threshold—Older name for the ventilatory breakpoint.

ventricular fibrillation—A serious cardiac arrhythmia in which the contraction of the ventricular tissue is uncoordinated, affecting the heart's ability to pump blood. *See also* ventricular tachycardia.

ventricular tachycardia—A serious cardiac arrhythmia consisting of three or more consecutive premature ventricular contractions. *See also* premature ventricular contraction and ventricular fibrillation.

venules—Small vessels that transport blood from the capillaries to the veins and then back to the heart.

vital capacity (VC)—The maximal volume of air expelled from the lungs after maximal inhalation.

vitamin—One of a group of unrelated organic compounds that perform specific functions to promote growth and to maintain health. Vitamins act primarily as catalysts in chemical reactions.

$\dot{V}O_2$—The volume of oxygen consumed per minute.

$\dot{V}O_2$ drift—A slow increase in $\dot{V}O_2$ during prolonged submaximal exercise at a constant power output.

$\dot{V}O_{2max}$—*See* maximal oxygen uptake.

wet-bulb globe temperature (WBGT)—A measurement of temperature that simultaneously accounts for conduction, convection, evaporation, and radiation, providing a single temperature reading to estimate the cooling capacity of the surrounding environment. The apparatus for measuring WBGT consists of a dry bulb, a wet bulb, and a black globe.

windchill—A chill factor created by the increase in the rate of heat loss via convection and conduction caused by wind.

References

Introduction

1. Åstrand, P.-O., & Rhyming, I. (1954). A nomogram for calculation of aerobic capacity (physical fitness) from pulse rate during submaximal work. *Journal of Applied Physiology,* **7,** 218-221.
2. Bainbridge, F.A. (1931). *The physiology of muscular exercise (3rd ed.).* London: Longmans, Green.
3. Bouchard, C. (2012). Overcoming barriers to progress in exercise genomics. *Exercise and Sports Science Reviews,* **39,** 212-217.
4. Bouchard, C. (2015). Exercise genomics—A paradigm shift is needed: A commentary. *British Journal of Sports Medicine,* **49,** 1492-1496.
5. Buford, T.W., & Pahor, M. (2012). Making preventive medicine more personalized: Implications for exercise-related research. *Preventive Medicine,* **55,** 34-36.
6. Buskirk, E.R., & Taylor, H.L. (1957). Maximal oxygen uptake and its relation to body composition, with special reference to chronic physical activity and obesity. *Journal of Applied Physiology,* **11,** 72-78.
7. Collins, F.S. (1999). The human genome project and the future of medicine. *Annals of the New York Academy of Sciences,* **882,** 42-55, discussion 56-65.
8. Collins, F.S. (2001). Contemplating the end of the beginning. *Genome Research,* **11,** 641-643.
9. Cooper, K.H. (1968). *Aerobics.* New York: Evans.
10. Dill, D.B. (1938). *Life, heat, and altitude.* Cambridge, MA: Harvard University Press.
11. Dill, D.B. (1985). *The hot life of man and beast.* Springfield, IL: Charles C Thomas.
12. Fletcher, W.M., & Hopkins, F.G. (1907). Lactic acid in amphibian muscle. *Journal of Physiology,* **35,** 247-254.
13. Flint, A., Jr. (1871). On the physiological effects of severe and protracted muscular exercise; with special reference to the influence of exercise upon the excretion of nitrogen. *New York Medical Journal,* **13,** 609-697.
14. Foster, M. (1970). *Lectures on the history of physiology.* New York: Dover.
15. Ginsburg, G.S., & Willard, H.F. (2009). Genomic and personalized medicine: Foundations and applications. *Translational Research,* **154,** 277-287.
16. Hamburg, M.A., & Collins, F.S. (2010). The path to personalized medicine. *New England Journal of Medicine,* **363,** 301-304.
17. Joyner, M.J., & Pedersen, B.K. (2011). Ten questions about systems biology. *Journal of Physiology,* **589,** 1017-1030.
18. Kerksick, C.M., Tsatsakis, A.M., Hayes, A.W., Kafantaris, I., & Kouretas, D. (2015). How can bioinformatics and toxicogenomics assist the next generation of research on physical exercise and athletic performance. *Journal of Strength and Conditioning Research,* **29,** 270-278.
19. LaGrange, F. (1889). *Physiology of bodily exercise.* London: Kegan Paul International.
20. Ling, C., & Ronn, T. (2014). Epigenetic adaptation to regular exercise in humans. *Drug Discovery Today,* **19,** 1015-1018.
21. Pérusse, L., Rankinen, T., Hagberg, J.M., Loos, R.J., Roth, S.M., Sarzynski, M.A., Wolfarth, B., & Bouchard, C. (2013). Advances in exercise, fitness, and performance genomics in 2012. *Medicine and Science in Sports and Exercise,* **45,** 824-831.
22. Petriz, B.A., Gomes, C.P., Rocha, L.A.O., Rezende, T.M.B., & Franco, O.L. (2012). Proteomics applied to exercise physiology: A cutting-edge technology. *Journal of Cellular Physiology,* **227,** 885-898.
23. Pitsiladis, Y.P., Durussel, J., & Rabin, O. (2014). An integrative "omics" solution to the detection of recombinant human erythropoietin and blood doping. *British Journal of Sports Medicine,* **48,** 856-861.
24. Robinson, S. (1938). Experimental studies of physical fitness in relation to age. *Arbeitsphysiologie,* **10,** 251-327.
25. Seals, D.R. (2013). Translational physiology: From molecules to public health. *Journal of Physiology,* **591,** 3457-3469.
26. Séguin, A., & Lavoisier, A. (1793). Premier mémoire sur la respiration des animaux. *Histoire et Mémoires de l'Academie Royale des Sciences,* **92,** 566-584.
27. Talmud, P.J., Hingorani, A.D., Cooper, J.A., Marmot, M.G., Brunner, E.J., Kumari, M., Kivimaki, M., & Humphries, S.E. (2010). Utility of genetic and non-genetic risk factors in prediction of type 2 diabetes: Whitehall II prospective cohort study. *BMJ,* **340,** b4838.
28. Taylor, H.L., Buskirk, E.R., & Henschel, A. (1955). Maximal oxygen intake as an objective measure of cardiorespiratory performance. *Journal of Applied Physiology,* **8,** 73-80.
29. Wang, L., McLeod, H.L., & Weinshilboum, R.M. (2011). Genomics and drug response. *New England Journal of Medicine,* **364,** 1144-1153.
30. Webborn, N., & Dijkstra, H.P. (2015). Twenty-first century genomics for sports medicine: What does it all mean? *British Journal of Sports Medicine,* **49,** 1481-1482.
31. Zuntz, N., & Schumberg, N.A.E.F. (1901). *Studien Zur Physiologie des Marches* (p. 211). Berlin: A. Hirschwald.

Chapter 1

1. Brooks, G.A., Fahey, T.D., & Baldwin, K.M. (2005). *Exercise physiology: Human bioenergetics and its applications* (4th ed.). New York: McGraw-Hill.
2. Bruusgaard, J.C., Egner, I.M., Larsen, T.K., Dupré-Aucouturier, S., Desplanches, D., & Gundersen, K. (2012). No change in myonuclear number during muscle unloading and reloading. *Journal of Applied Physiology,* **113**(2), 290-296.
3. Bruusgaard, J.C., Johansen, I.B., Egner, I.M., Rana, Z.A., & Gundersen, K. (2010). Myonuclei acquired by overload exercise precede hypertrophy and are not lost on detraining. *Proceedings of the National Academy of Sciences USA,* **107,** 15111-15116.
4. Costill, D.L., Daniels, J., Evans, W., Fink, W., Krahenbuhl, G., & Saltin, B. (1976). Skeletal muscle enzymes and fiber composition in male and female track athletes. *Journal of Applied Physiology,* **40,** 149-154.
5. Costill, D.L., Fink, W.J., Flynn, M., & Kirwan, J. (1987). Muscle fiber composition and enzyme activities in elite female distance runners. *International Journal of Sports Medicine,* **8,** 103-106.
6. Costill, D.L., Fink, W.J., & Pollock, M.L. (1976). Muscle fiber composition and enzyme activities of elite distance runners. *Medicine and Science in Sports,* **8,** 96-100.
7. Gallagher, I.J., Stephens, N.A., MacDonald, A.J., Skipworth, R.J.E., Husi, H., Greig, C.A., Ross, J.A., Timmons, J.A., & Fearon, K.C.H. (2012). Suppression of skeletal muscle turnover in cancer cachexia: Evidence from the transcriptome in sequential human muscle biopsies. *Clinical Cancer Research,* **18,** 2817-2827.
8. Haizlip, K.M., Harrison, B.C., & Leinwand, L.A. (2015). Sex-based differences in skeletal muscle kinetics and fiber-type composition. *Physiology (Bethesda),* **30,** 30-39.
9. Herzog, W., Duvall, M., & Leonard, T.R. (2012). Molecular mechanisms of muscle force regulation: A role for titin. *Exercise and Sports Sciences Reviews,* **40**(1), 50-57.
10. Herzog, W., Schappacher, G., DuVall, M., Leonard, T.R., & Herzog, J.A. (2016). Residual force enhancement following eccentric contractions: A new mechanism involving titin. *Physiology (Bethesda),* **31,** 300-312.
11. Lee, J.D., & Burd, N.A. (2012). No role of muscle satellite cells in hypertrophy: Further evidence of a mistaken identity? *Journal of Physiology,* **590,** 2837-2838.
12. MacIntosh, B.R., Gardiner, P.F., & McComas, A.J. (2006). *Skeletal muscle form and function* (2nd ed.). Champaign, IL: Human Kinetics.
13. Mahon, J.J., & Pearson, S. (2012). Changes in medial gastrocnemius fascicle-tendon behavior during single-leg hopping with increased joint stiffness. Poster communication. *Proceedings of the Physiological Society,* **26,** PC75.
14. Monroy, J.A., Powers, K.L., Gilmore, L.A., Uyeno, T.A., Lindstedt, S.L., & Nishikawa, K.C. (2012). What is the role of titin in active muscle? *Exercise and Sports Sciences Reviews,* **40**(2), 73-78.
15. Nishikawa, K.C., Monroy, J.A., Uyeno, T.A., Yeo, S.H., Pai, D.K., & Lindstedt, S.L. (2012). Is titin a "winding filament"? A new twist on muscle contraction. *Proceedings in Biological Sciences,* **279**(1730), 981-990.
16. Rana, M., Hamarneh, G., & Wakeling, J.M. (2014). 3D curvature of muscle fascicle in triceps surae. *Journal of Applied Physiology,* **117,** 1388-1397.
17. Timmins, R.G., Ruddy, J.D., Presland, J., Maniar, N., Shield, A.J., Williams, D., and Opar, D.A. (2016). Architectural changes of the biceps femoris long head after concentric or eccentric training. *Medicine and Science in Sports and Exercise,* **48,** 499-508.
18. Tskhovrebova, L., & Trinick, J. (2003). Titin: Properties and family relationships. *Nature Reviews Molecular Cell Biology,* **4,** 679-689.

Chapter 2

1. Brooks, G.A., & Mercier, J. (1994). Balance of carbohydrate and lipid utilization during exercise: The "crossover" concept. *Journal of Applied Physiology,* **76,** 2253-2261.
2. Chechi, K., van Marken Lichtenbelt, W.D., & Richard, D. (2017). Brown and beige adipose tissues: Phenotypes and metabolic potential in mice and men. *Journal of Applied Physiology,* Mar 16 [Epub ahead of print]. jap.00021.2017.
3. Cypess, A.M., Lehman, S., Williams, G., Tal, I., Rodman, D., Goldfine, A.B., Kuo, F.C., Palmer, E.L., Tseng, Y.H., Doria, A., Kolodny, G.M., & Kahn, C.R. (2009). Identification and importance of brown adipose tissue in adult humans. *New England Journal of Medicine,* **360,** 1509-1517.
4. Dubé, J.J., Broskey, N.T., Despines, A.A., Stefanovic-Racic, M., Toledo, F.G., Goodpaster, B.H., & Amati, F. (2016). Muscle characteristics and substrate energetics in lifelong endurance athletes. *Medicine and Science in Sports and Exercise,* **3,** 472-480.
5. Pathi, B., Kinsey, S.T., Howdeshell, M.E., Priester, C., McNeill, R.S., & Locke, B.R. (2012). The formation and functional consequences of heterogeneous mitochondrial distributions in skeletal muscle. *Journal of Experimental Biology,* **215,** 1871-1883.
6. Pathi, B., Kinsey, S.T., & Locke, B.R. (2011). Influence of reaction and diffusion on spatial organization of mitochondria and effectiveness factors in skeletal muscle cell design. *Biotechnology and Bioengineering,* **108,** 1912-1924.
7. Pathi, B., Kinsey, S.T., & Locke, B.R. (2013). Oxygen control of intracellular distribution of mitochondria in muscle fibers. *Biotechnology and Bioengineering,* **110,** 2513-2524.
8. van der Zwaard, S., de Ruiter, C.J., Noordhof, D.A., Sterrenburg, R., Bloemers, F.W., de Koning, J.J., Jaspers, R.T., & van der Laarse, W.J. (2016). Maximal oxygen uptake is proportional to muscle fiber oxidative capacity, from chronic heart failure patients to professional cyclists. *Journal of Applied Physiology,* **121,** 636-645.

Chapter 3

1. Distefano, L.J., Casa, D.J., Vansumeren, M.M., Karslo, R.M., Huggins, R.A., Demartini, J.K., Stearns, R.L., Armstrong, L.E., & Maresh, C.M. (2013). Hypohydration and hyperthermia impair neuromuscular control after exercise. *Medicine and Science in Sports and Exercise,* **45**(6), 1166-1173.
2. Gerstner, G.R., Thompson, B.J., Rosenberg, J.G., Sobolewski, E.J., Scharville, M.J., & Ryan, E.D. (2017). Neural and muscular contributions to the age-related reductions in rapid strength. *Medicine and Science in Sports and Exercise,* **49,** 1331-1339.
3. Girard, O., Millet, G.P., Micallef, J.P., & Racinais, S. (2012). Alteration in neuromuscular function after a 5 km running time trial. *European Journal of Applied Physiology,* **112,** 2323-2330.
4. Handschin, C. (2010). Regulation of skeletal muscle cell plasticity by the peroxisome proliferator-activated receptor gamma coactivator 1alpha. *Journal of Receptor and Signal Transduction Research,* **30,** 376-384.
5. Marieb, E.N. (1995). *Human anatomy and physiology* (3rd ed.). New York: Benjamin Cummings.
6. Martinez-Valdes, E., Falla, D., Negro, F., Mayer, F., & Farina, D. (2017). Differential motor unit changes after endurance or high-intensity interval training. *Medicine and Science in Sports and Exercise,* **49,** 1126-1136.
7. Pette, D., & Vrbova, G. (1985). Neural control of phenotypic expression in mammalian muscle fibers. *Muscle and Nerve,* **8,** 676-689.

Chapter 4

1. Bermon, S., & Garnier, P. (2017). Serum androgen levels and their relation to performance in track and field: Mass spectrometry results from 2127 observations in male and female elite athletes. *British Journal of Sports Medicine* July [Epub ahead of print].
2. Boden, B.P., Sheehan, F.T., Torg, J.S., & Hewett, T.E. (2010). Noncontact anterior cruciate ligament injuries: Mechanisms and risk factors. *Journal of the American Academy of Orthopaedic Surgeons,* **18,** 520-527.
3. Broom, D.R., Stensel, D.J., Bishop, N.C., Burns, S.F., & Miyashita, M. (2007). Exercise-induced suppression of acylated ghrelin in humans. *Journal of Applied Physiology,* **102,** 2165-2171.
4. Bruning, J.C., Gautam, D., Burks, D.J., Gillette, J., Schubert, M., Orban, P.C., Klein, R., Krone, W., Muller-Wieland, D., & Kahn, C.R. (2000). Role of brain insulin receptor in control of body weight and reproduction. *Science,* **289,** 2122-2125.
5. Jurimae, J., Maestu, J., Jurimae, T., Mangus, B., & von Duvillard, S.P. (2011). Peripheral signals of energy homeostasis as possible markers of training stress in athletes: A review. *Metabolism: Clinical and Experimental,* **60,** 335-350.
6. Kjellberg, S.R., Rudhe, U., & Sjöstrand, T. (1949). Increase of the amount of hemoglobin and blood volume in connection with physical training. *Acta Physiologica Scandinavica,* **19,** 146-151.
7. Lam, C.K., Chari, M., & Lam, T.K. (2009). CNS regulation of glucose homeostasis. *Physiology,* **24,** 159-170.
8. Lam, T.K., Gutierrez-Juarez, R., Pocai, A., & Rossetti, L. (2005). Regulation of blood glucose by hypothalamic pyruvate metabolism. *Science,* **309,** 943-947.
9. Laursen, T.L., Zak, R.B., Shute, R.J., Heesch, M.W.S., Dinan, N.E., Bubak, M.P., La Salle, D.T., & Slivka, D.R. (2017). Leptin, adiponectin, and ghrelin responses to endurance exercise in different ambient conditions. *Temperature* (Austin), **4,** 166-175.
10. Leidy, H.J., Gardner, J.K., Frye, B.R., Snook, M.L., Schuchert, M.K., Richard, E.L., & Williams, N.I. (2004). Circulating ghrelin is sensitive to changes in body weight during a diet and exercise program in normal-weight young women. *Journal of Clinical Endocrinology and Metabolism,* **89,** 2659-2664.
11. Magistretti, P.J., Pellerin, L., Rothman, D.L., & Shulman, R.G. (1999). Energy on demand. *Science,* **283,** 496-497.
12. Montero, D., Breenfeldt-Andersen, A., Oberholzer, L., Haider, T., Goetze, J.P., Meinild-Lundby, A.K., & Lundby, C. (2017). Erythropoiesis with endurance training: Dynamics and mechanisms. *American Journal of Physiology: Regulatory Integrative Comparative Physiology,* **312,** R894-R902.
13. Powell, J.W., & Barber-Foss, K.D. (2000). Sex-related injury patterns among selected high school sports. *American Journal of Sports Medicine,* **28,** 385-391.
14. Stensel, D. (2010). Exercise, appetite and appetite-regulating hormones: Implications for food intake and weight control. *Annals of Nutrition and Metabolism,* **57**(Suppl 2), 36-42.

Chapter 5

1. American College of Sports Medicine. (2018). *Guidelines for graded exercise testing and prescription* (10th ed.). Philadelphia: Lippincott, Williams and Wilkins.
2. Bergeron, M.F. (2008). Muscle cramps during exercise—is it fatigue or electrolyte deficit? *Current Sports Medicine Reports,* **7,** S50-S55.
3. Blanchfield, A.W., Hardy, J., De Morree, H.M., Staiano, W., & Marcora, S.M. (2014). Talking yourself out of exhaustion: The effects of self-talk on endurance performance. *Medicine and Science in Sports and Exercise,* **46,** 998-1007.
4. Bruckert, E., Hayem, G., Dejager, S., Yau, C., & Begaud, B. (2005). Mild to moderate muscular symptoms with high-dosage statin therapy in hyperlipidemic patients—the primo study. *Cardiovascular Drugs and Therapy,* **19,** 403-414.
5. Cannon, D.T., Bimson, W.E., Hampson, S.A., Bowen, T.S., Murgatroyd, S.R., Marwood, S., Kemp, G.J., & Rossiter, H.B. (2014). Skeletal muscle ATP turnover by 31P magnetic resonance spectroscopy during moderate and heavy bilateral knee extension. *Journal of Physiology,* **592,** 5287-5300.

6. Costill, D.L. (1986). *Inside running: Basics of sports physiology.* Indianapolis: Benchmark Press.
7. Craighead, D.H., Shank, S.W., Gottschall, J.S., Passe, D.H., Murray, B., Alexander, L.M., & Kenney, W.L. (2017). Ingestion of transient receptor potential channel agonists attenuates exercise-induced muscle cramps. *Muscle and Nerve* Feb 13 [Epub ahead of print]. doi: 10.1002/mus.25611.
8. DeBold, E.F. (2016). Decreased myofilament calcium sensitivity plays a significant role in muscle fatigue. *Medicine and Science in Sports and Exercise,* **43,** 144-149.
9. Eichner, E.R. (2007). The role of sodium in 'heat cramping'. *Sports Medicine,* **37,** 368-370.
10. Foure, A., Wegrzyk, J., Le Fur, Y., Mattei, J.-P., Boudinet, H., Vilmen, C., Bendahan, D., & Gondin, J. (2016). Impaired mitochondrial function and reduced energy cost as a result of muscle damage. *Medicine and Science in Sports and Exercise,* **47,** 1135-1144.
11. Gaesser, G.A., & Poole, D.C. (1996). The slow component of oxygen uptake kinetics in humans. *Exercise and Sport Sciences Reviews,* **24,** 35-70.
12. Galloway, S.D.R., & Maughan, R.J. (1997). Effects of ambient temperature on the capacity to perform prolonged cycle exercise in man. *Medicine and Science in Sports and Exercise,* **29,** 1240-1249.
13. Grassi, B., Rossiter, H.B., & Zoladz, J.A. (2015). Skeletal muscle fatigue and decreased efficiency: Two sides of the same coin? *Medicine and Science in Sports and Exercise,* **43,** 75-83.
14. Kent, J.A., Ortenblad, N., Hogan, M.C., Poole, D.C., & Muscj, T.I. (2016). No muscle is an island: Integrative perspectives on muscle fatigue. *Medicine and Science in Sports and Exercise,* **48,** 2281-2293.
15. Ludlow, L.W., & Weyand, P.G. (2016). Energy expenditure during level human walking: Seeking a simple and accurate predictive solution. *Journal of Applied Physiology,* **120,** 481-494.
16. Maughan, R.J., Otani, H., & Watson, P. (2012). Influence of relative humidity on prolonged exercise capacity in a warm environment. *European Journal of Applied Physiology,* **112,** 2313-2321.
17. Mikus, C.R., Boyle, L.J., Borengasser, S.J., Oberlin, D.J., Naples, S.P., Fletcher, J., Meers, G.M., Ruebel, M., Laughlin, M.H., Dellsperger, K.C., Fadel, P.J., & Thyfault, J.P. (2013). Simvastatin impairs exercise training adaptations. *Journal of the American College of Cardiology,* **62,** 709-714.
18. Neyroud, D., Maffiuletti, N.A., Kayser, B., & Place, N. (2012). Mechanisms of fatigue and task failure induced by sustained submaximal contractions. *Medicine and Science in Sports and Exercise,* **44,** 1243-1251.
19. Oosthuyse, T., & Bosch, A.N. (2017). The effect of gender and menstrual phase on serum creatine kinase activity and muscle soreness following downhill running. *Antioxidants,* **23,** E16.
20. Ortenblad, N., Westerblad, H., & Nielsen, J. (2013). Muscle glycogen stores and fatigue. *Journal of Physiology,* **591,** 4405-4413.
21. Pandolf, K.B., Givoni, B., & Goldman, R.F. (1977). Predicting energy expenditure with loads while standing or walking very slowly. *Journal of Applied Physiology,* **43,** 577-581.
22. Parker, B.A., & Thompson, P.D. (2012). Effect of statins on skeletal muscle: Exercise, myopathy, and muscle outcomes. *Exercise and Sport Sciences Reviews,* **40,** 188-194.
23. Phillips, P.S., Haas, R.H., Bannykh, S., Hathaway, S., Gray, N.L., Kimura, B.J., Vladutiu, G.D., England, J.D., & Scripps Mercy Clinical Research Center. (2002). Statin-associated myopathy with normal creatine kinase levels. *Annals of Internal Medicine,* **137,** 581-585.
24. Radak, Z., Naito, H., Taylor, A.W., & Goto, S. (2012). Nitric oxide: Is it the cause of muscle soreness? *Nitric Oxide: Biology and Chemistry,* **26,** 89-94.
25. Reid, M.B. (2016). Reactive oxygen species as agents of fatigue. *Medicine and Science in Sports and Exercise,* **48,** 2239-2246.
26. Schwane, J.A., Johnson, S.R., Vandenakker, C.B., & Armstrong, R.B. (1983). Delayed-onset muscular soreness and plasma CPK and LDH activities after downhill running. *Medicine and Science in Sports and Exercise,* **15,** 51-56.
27. Schwane, J.A., Watrous, B.G., Johnson, S.R., & Armstrong, R.B. (1983). Is lactic acid related to delayed-onset muscle soreness? *Physician and Sportsmedicine,* **11**(3), 124-131.
28. Vanhatalo, A., Jones, A.M., & Burnley, M. (2011). Application of critical power in sport. *International Journal of Sports Physiology and Performance,* **6,** 128-136.
29. Zuntz, N., & Hagemann, O. (1898). *Untersuchungen uber den Stroffwechsel des Pferdes bei Ruhe und Arbeit.* Berlin: Parey.

Chapter 6

1. Billman, G.E. (2017). Counterpoint: Exercise training induced bradycardia: The case for enhanced parasympathetic regulation. *Journal of Applied Physiology* Jul 6 [Epub ahead of print]. doi: 10.1152/japplphysiol.00605.2017.
2. Boyett, M.R., Wang, Y., Nakao, S., Ariyaratnam, J., Hart, G., Monfredi, O., & D'Souza, A. (2017). Point: Exercise training-induced bradycardia is caused by changes in intrinsic sinus node function. *Journal of Applied Physiology* Jul 6 [Epub ahead of print]. doi: 10.1152/japplphysiol.00604.2017.
3. Casey, D.P., Curry, T.B., Wilkins, B.W., & Joyner, M.J. (2011). Nitric oxide-mediated vasodilation becomes independent of beta-adrenergic receptor activation with increased intensity of hypoxic exercise. *Journal of Applied Physiology,* **110,** 687-694.
4. Casey, D.P., & Joyner, M.J. (2011). Local control of skeletal muscle blood flow during exercise: Influence of available oxygen. *Journal of Applied Physiology,* **111,** 1527-1538.

5. Casey, D.P., Mohamed, E.A., & Joyner, M.J. (2013). Role of nitric oxide and adenosine in the onset of vasodilation during dynamic forearm exercise. *European Journal of Applied Physiology,* **113,** 295-303.
6. Casey, D.P., Walker, B.G., Ranadive, S.M., Taylor, J.L., & Joyner, M.J. (2013). Contribution of nitric oxide in the contraction-induced rapid vasodilation in young and older adults. *Journal of Applied Physiology,* **115,** 446-455.
7. Dorfman, T.A., Rosen, B.D., Perhonen, M.A., Tillery, T., McColl, R., Peshock, R.M., & Levine, B.D. (2008). Diastolic suction is impaired by bed rest: MRI tagging studies of diastolic untwisting. *Journal of Applied Physiology,* **104,** 1037-1044.
8. Eijsvogels, T.M.H., Fernandez, A.B., & Thompson, P.D. (2016). Are there deleterious cardiac effects of acute and chronic endurance exercise? *Physiological Reviews,* **96,** 99-125.
9. Joyner, M.J., & Casey, D.P. (2014). Muscle blood flow, hypoxia and hypoperfusion. *Journal of Applied Physiology,* **116**(7), 852-857.
10. Moreau, K.L., & Ozemek, C. (2017). Vascular adaptations to habitual exercise in older adults: Time for the sex talk. *Exercise and Sport Sciences Review,* **45,** 116-123.
11. Notomi, Y., Martin-Miklovic, M.G., Oryszak, S.J., Shiota, T., Deserranno, D., Popovic, Z.B., Garcia, M.J., Greenberg, N.L., & Thomas, J.D. (2006). Enhanced ventricular untwisting during exercise: A mechanistic manifestation of elastic recoil described by Doppler tissue imaging. *Circulation,* **113,** 2524-2533.

Chapter 7

1. Casey, K., Duffin, J., Kelsey, C.J., & McAvoy, G.V. (1987). The effect of treadmill speed on ventilation at the start of exercise in man. *Journal of Physiology,* **391,** 13-24.
2. Duffin, J. (2014). The fast exercise drive to breathe. *Journal of Physiology,* **592,** 445-451.
3. Coffman, K.E., Carlson, A.R., Miller, A.D., Johnson, B.D., & Taylor, B.J. (2017). The effect of aging and cardiorespiratory fitness on the lung diffusing capacity response to exercise in healthy humans. *Journal of Applied Physiology,* **122,** 1425-1434.
4. Molino-Lova, R., Pasquini, G., Vannetti, F., Zipoli, R., Razzolini, L., Fabbri, V., Frandi, R., Cecchi, F., Gigliotti, F., & Macchi, C. (2013). Ventilatory strategies in the six-minute walk test in older patients receiving a three-week rehabilitation programme after cardiac surgery through median sternotomy. *Journal of Rehabilitation Medicine,* **45,** 504-509.
5. Rossman, M.J., Nader, S., Berry, D., Orsini, F., Klansky, A., & Haverkamp, H.C. (2014). Effects of altered airway function on exercise ventilation in asthmatic adults. *Medicine and Science in Sports and Exercise,* **46,** 1104-1113.
6. Williams, P.T. (2014). Dose-response relationship between exercise and respiratory disease mortality. *Medicine and Science in Sports and Exercise,* **46,** 711-717.
7. Wuthrich, T.U., Marty, J., Benaglia, P., Eichenberger, P.A., & Spengler, C.M. (2015). Acute effects of a respiratory sprint-interval session on muscle contractility. *Medicine and Science in Sports and Exercise,* **47,** 1979-1987.

Chapter 8

1. Helenius, I., & Haahtela, T. (2000). Allergy and asthma in elite summer sport athletes. *Journal of Allergy and Clinical Immunology,* **106,** 444-452.
2. Helenius, I.J., Tikkanen, H.O., & Haahtela, T. (1998). Occurrence of exercise induced bronchospasm in elite runners: Dependence on atopy and exposure to cold air and pollen. *British Journal of Sports Medicine,* **32,** 125-129.
3. Hermansen, L. (1981). Effect of metabolic changes on force generation in skeletal muscle during maximal exercise. In R. Porter & J. Whelan (Eds.), *Human muscle fatigue: Physiological mechanisms* (pp. 75-88). London: Pitman Medical.
4. Hwangbo, G., Lee, D.H., Park, S.H., & Han, J.W. (2017). Changes in cardiopulmonary function according to posture during recovery after maximal exercise. *Journal of Physical Therapy Science,* **29,** 1163-1166.
5. Karim, N., Hasan, J.A., & Ali, S.S. (2011). Heart rate variability—A review. *Journal of Basic and Applied Sciences,* **7,** 71-77.
6. Larsson, K., Ohlsen, P., Larsson, L., Malmberg, P., Rydstrom, P.O., & Ulriksen, H. (1993). High prevalence of asthma in cross country skiers. *BMJ,* **307,** 1326-1329.
7. McKirnan, M.D., Gray, C.G., & White, F.C. (1991). Effects of feeding on muscle blood flow during prolonged exercise in miniature swine. *Journal of Applied Physiology,* **70,** 1097-1104.
8. Nes, B.M., Janszky, I., Wisløff, U., Støylen, A., & Karlsen, T. (2013). Age-predicted maximal heart rate in healthy subjects: The HUNT fitness study. *Scandinavian Journal of Medicine and Science in Sports,* **23,** 697-704.
9. Poliner, L.R., Dehmer, G.J., Lewis, S.E., Parkey, R.W., Blomqvist, C.G., & Willerson, J.T. (1980). Left ventricular performance in normal subjects: A comparison of the responses to exercise in the upright and supine position. *Circulation,* **62,** 528-534.
10. Powers, S.K., Martin, D., & Dodd, S. (1993). Exercise-induced hypoxaemia in elite endurance athletes: Incidence, causes and impact on $\dot{V}O_{2max}$. *Sports Medicine,* **16,** 14-22.
11. Romero, S.A., Minson, C.T., & Halliwill, J.R. (2017). The cardiovascular system after exercise. *The Journal of Applied Physiology,* **122,** 925-932.
12. Routledge, F.S., Campbell, T.S., McFetridge-Durdle, J.A., & Bacon, S.L. (2010). Improvements in heart rate variability with exercise therapy. *Canadian Journal of Cardiology,* **26,** 303-312.
13. Rowell, L.B. (1993). *Human cardiovascular control.* New York: Oxford University Press.

14. Rundell, K.W. (2003). High levels of airborne ultrafine and fine particulate matter in indoor ice arenas. *Inhalation Toxicology,* **15,** 237-250.

15. Saboul, D., Pialoux, V., & Hautier, C. (2014). The breathing effect of the lf/hf ratio in the heart rate variability measurements of athletes. *European Journal of Sport Science,* **14**(Suppl 1), S282-S288.

16. Tanaka, H., Monahan, D.K., & Seals, D.R. (2001). Age-predicted maximal heart rate revisited. *Journal of the American College of Cardiology,* **37,** 153-156.

17. Turkevich, D., Micco, A., & Reeves, J.T. (1988). Noninvasive measurement of the decrease in left ventricular filling time during maximal exercise in normal subjects. *American Journal of Cardiology,* **62,** 650-652.

18. Wasserman, K., & McIlroy, M.B. (1964). Detecting the threshold of anaerobic metabolism in cardiac patients during exercise. *American Journal of Cardiology,* **14,** 844-852.

Chapter 9

1. American College of Sports Medicine. (2009). ACSM position stand: Progression models in resistance training for healthy adults. *Medicine and Science in Sports and Exercise,* **41,** 687-708.

2. Astorino, T.A., Edmunds, R.M., Clark, A., King, L., Gallant, R.A., Namm, S., Fischer, A., & Wood, K.M. (2017). High-intensity interval training increases cardiac output and $\dot{V}O_{2max}$. *Medicine and Science in Sports and Exercise,* **49,** 265-273.

3. Babraj, J.A., Vollaard, N.B., Keast, C., Guppy, F.M., Cottrell, G., & Timmons, J.A. (2009). Extremely short duration high intensity interval training substantially improves insulin action in young healthy males. *BMC Endocrine Disorders,* **9,** 3.

4. Behm, D.G., Drinkwater, E.J., Willardson, J.M., & Cowley, P.M. (2010). The use of instability to train the core musculature. *Applied Physiology, Nutrition, and Metabolism,* **35,** 91-108.

5. Gibala, M.J., & Jones, A.M. (2013). Physiological and performance adaptations to high-intensity interval training. *Nestlé Nutritional Institute Workshop Series,* **76,** 51-60.

6. Gibala, M.J., Little, J.P., van Essen, M., Wilkin, G.P., Burgomaster, K.A., Safdar, A., Raha, S., & Tarnopolsky, M.A. (2006). Short-term sprint interval versus traditional endurance training: Similar initial adaptations in human skeletal muscle and exercise performance. *Journal of Physiology,* **575,** 901-911.

7. Gunnarsson, T.P., & Bangsbo, J. (2012). The 10-20-30 training concept improves performance and health profile in moderately trained runners. *Journal of Applied Physiology,* **113,** 16-24.

8. Gunnarsson, T.P., Christensen, P.M., Holse, K., Christiansen, D., & Bangsbo, J. (2012). Effect of additional speed endurance training on performance and muscle adaptations. *Medicine and Science in Sports and Exercise,* **44,** 1942-1948.

9. Iaia, F.M., Thomassen, M., Kolding, H., Gunnarsson, T., Wendell, J., Rostgaard, T., Nordsborg, N., Krustrup, P., Nybo, L., Hellsten, Y., & Bangsbo, J. (2008). Reduced volume but increased training intensity elevates muscle Na+-K+ pump alpha1-subunit and NHE1 expression as well as short-term work capacity in humans. *American Journal of Physiology: Regulatory, Integrative and Comparative Physiology,* **294,** R966-R974.

10. Kilpatrick, M.W., Jung, M.E., & Little, J.P. (2017). High-intensity interval training: A review of physiological and psychological responses. *ACSM's Health and Fitness Journal,* **18,** 11-16.

11. Konopka, A.R., & Harber, M.P. (2014). Skeletal muscle hypertrophy after aerobic exercise training. *Exercise and Sport Science Reviews,* **42,** 53-61.

12. Olson, M. (2014). Tabata: It's a HIIT! *ACSM's Health and Fitness Journal,* **18,** 17-24.

13. Schwartz, R.S., Shuman, W.P., Larson, V., Cain, K.C., Fellingham, G.W., Beard, J.C., Kahn, S.E., Stratton, J.R., Cerqueira, M.D., & Abrass, I.B. (1991). The effect of intensive endurance exercise training on body fat distribution in young and older men. *Metabolism,* **45,** 545-551.

14. Tabata, I., Nishimura, K., Kouzaki, M., Hirai, Y., Ogita, F., Miyachi, M., & Yamamoto, K. (1996). Effects of moderate-intensity endurance and high-intensity intermittent training on anaerobic capacity and $\dot{V}O_{2max}$. *Medicine and Science in Sports and Exercise,* **28,** 1327-1330.

15. Willardson, J.M. (2007). Core stability training: Applications to sports conditioning programs. *Journal of Strength and Conditioning Research,* **21,** 979-985.

Chapter 10

1. Aguirre, N., van Loon, L.J., & Baar, K. (2013). The role of amino acids in skeletal muscle adaptation to exercise. *Nestlé Nutritional Institute Workshop Series,* **76,** 85-102.

2. Barnes, B. (2013). Jim Bradford, Olympic weightlifter, dies at 84. *Washington Post* October 13. Available: www.washingtonpost.com/local/obituaries/jim-bradford-dies-at-84-olympic-weightlifter/2013/10/13/abc758ba-302d-11e3-9ccc-2252bdb14df5_story.html [August 15, 2014].

3. Burd, N.A., West, D.W., Staples, A.W., Atherton, P.J., Baker, J.M., Moore, D.R., Holwerda, A.M., Parise, G., Rennie, M.J., Baker, S.K., & Phillips, S.M. (2010). Low-load high volume resistance exercise stimulates muscle protein synthesis more than high-load low volume resistance exercise in young men. *PLoS One,* **5,** e12033.

4. Dias, I., Farinatti, P., De Souza, M.G., Manhanini, D.P., Balthazar, E., Dantas, D.L., De Andrade Pinto, E.H., Bouskela, E., & Kraemer-Aguiar, L.G. (2015). Effects of resistance training on obese adolescents. *Medicine and Science in Sports and Exercise,* **47,** 2636-2644.

5. Dickinson, J.M., Volpi, E., & Rasmussen, B.B. (2013). Exercise and nutrition to target protein synthesis impairments in aging skeletal muscle. *Exercise and Sports Sciences Reviews,* **41,** 216-223.

6. Duchateau, J., & Enoka, R.M. (2002). Neural adaptations with chronic activity patterns in able-bodied humans. *American Journal of Physical Medicine and Rehabilitation,* **81**(Suppl 11), 517-527.

7. Enoka, R.M. (1988). Muscle strength and its development: New perspectives. *Sports Medicine,* **6,** 146-168.

8. Gonyea, W.J. (1980). Role of exercise in inducing increases in skeletal muscle fiber number. *Journal of Applied Physiology,* **48,** 421-426.

9. Gonyea, W.J., Sale, D.G., Gonyea, F.B., & Mikesky, A. (1986). Exercise induced increases in muscle fiber number. *European Journal of Applied Physiology,* **55,** 137-141.

10. Graves, J.E., Pollock, M.L., Leggett, S.H., Braith, R.W., Carpenter, D.M., & Bishop, L.E. (1988). Effect of reduced training frequency on muscular strength. *International Journal of Sports Medicine,* **9,** 316-319.

11. Green, H.J., Klug, G.A., Reichmann, H., Seedorf, U., Wiehrer, W., & Pette, D. (1984). Exercise-induced fibre type transitions with regard to myosin, parvalbumin, and sarcoplasmic reticulum in muscles of the rat. *Pflugers Archiv: European Journal of Physiology,* **400,** 432-438.

12. Hakkinen, K., Alen, M., & Komi, P.V. (1985). Changes in isometric force and relaxation-time, electromyographic and muscle fibre characteristics of human skeletal muscle during strength training and detraining. *Acta Physiologica Scandinavica,* **125,** 573-585.

13. Hawke, T.J., & Garry, D.J. (2001). Myogenic satellite cells: Physiology to molecular biology. *Journal of Applied Physiology,* **91,** 534-551.

14. McCall, G.E., Byrnes, W.C., Dickinson, A., Pattany, P.M., & Fleck, S.J. (1996). Muscle fiber hypertrophy, hyperplasia, and capillary density in college men after resistance training. *Journal of Applied Physiology,* **81,** 2004-2012.

15. Morton, R.W., Murphy, K.T., McKellar, S.R., Schoenfeld, B.J., Henselmans, M., Helms, E., Devries, M.C., Banfield, L., Krieger, J.W., & Phillips, S.M. (2017). A systematic review, meta-analysis and meta-regression of the effect of protein supplementation on resistance training-induced gains in muscle mass and strength in healthy adults. *British Journal of Sports Medicine* [Epub ahead of print]. http://dx.doi.org/10.1136/bjsports-2017-097608.

16. Morton, R.W., Oikawa, S.Y., Wavell, C.G., Mazara, N., McGlory, C., Quadrilatero, J., Baechler, B.L., Baker, S.K., & Phillips, S.M. (2016). Neither load nor systemic hormones determine resistance training-mediated hypertrophy or strength gains in resistance-trained young men. *Journal of Applied Physiology,* **121,** 129-138.

17. Porter, C., Reidy, P.T., Bhattarai, N., Sidossis, L.S., & Rasmussen, B.B. (2015). Resistance exercise training alters mitochondrial function in human skeletal muscle. *Medicine and Science in Sports and Exercise,* **47,** 1922-1931.

18. Schoenfield, B.J. (2013). Is there a minimum intensity threshold for resistance training-induced hypertrophic adaptations? *Sports Medicine,* **43**(12), 1279-1288.

19. Schroeder, E.T., Villanueva, M., West, D.D., & Phillips, S.M. (2013). Are acute post-resistance exercise increases in testosterone, growth hormone, and IGF-1 necessary to stimulate skeletal muscle anabolism and hypertrophy? *Medicine and Science in Sports and Exercise,* **45,** 2044-2051.

20. Shepstone, T.N., Tang, J.E., Dallaire, S., Schuenke, M.D., Staron, R.S., & Phillips, S.M. (2005). Short-term high- vs. low-velocity isokinetic lengthening training results in greater hypertrophy of the elbow flexors in young men. *Journal of Applied Physiology,* **98,** 1768-1776.

21. Sjöström, M., Lexell, J., Eriksson, A., & Taylor, C.C. (1991). Evidence of fibre hyperplasia in human skeletal muscles from healthy young men? A left-right comparison of the fibre number in whole anterior tibialis muscles. *European Journal of Applied Physiology,* **62,** 301-304.

22. Staron, R.S., Karapondo, D.L., Kraemer, W.J., Fry, A.C., Gordon, S.E., Falkel, J.E., Hagerman, F.C., & Hikida, R.S. (1994). Skeletal muscle adaptations during early phase of heavy resistance training in men and women. *Journal of Applied Physiology,* **76,** 1247-1255.

23. Staron, R.S., Leonardi, M.J., Karapondo, D.L., Malicky, E.S., Falkel, J.E., Hagerman, F.C., & Hikida, R.S. (1991). Strength and skeletal muscle adaptations in heavy-resistance-trained women after detraining and retraining. *Journal of Applied Physiology,* **70,** 631-640.

24. Staron, R.S., Malicky, E.S., Leonardi, M.J., Falkel, J.E., Hagerman, F.C., & Dudley, G.A. (1990). Muscle hypertrophy and fast fiber type conversions in heavy resistance-trained women. *European Journal of Applied Physiology,* **60,** 71-79.

25. Trommelen, J., Holwerda, A.M., Kouw, I.W., Langer, H., Halson, S.L., Rollo, I., Verdijk, L.B., & Van Loon, L.J. (2016). Resistance exercise augments postprandial overnight muscle protein synthesis rates. *Medicine and Science in Sports and Exercise,* **48,** 2517-2525.

26. Verdijk, L.B., Snijders, T., Holloway, T.M., Van Kranenburg, J., & Van Loon, L.J. (2016). Resistance training increases skeletal muscle capillarization in healthy older men. *Medicine and Science in Sports and Exercise,* **48,** 2157-2164.

27. Walts, C.T., Hanson, E.D., Delmonico, M.J., Yao, L., Wang, M.Q., & Hurley, B.F. (2008). Do sex or race differences influence strength training effects on muscle or fat? *Medicine and Science in Sports and Exercise,* **40,** 669-676.

28. West, D.W., Burd, N.A., Churchward-Venne, T.A., Camera, D.M., Mitchell, C.J., Baker, S.K., Hawley, J.A., Coffey, V.G., & Phillips, S.M. (2012). Sex-based comparisons of myofibrillar protein synthesis after resistance exercise in the fed state. *Journal of Applied Physiology,* **112,** 1805-1813.

29. West, D.W., Burd, N.A., Tang, J.E., Moore, D.R., Staples, A.W., Holwerda, A.M., Baker, S.K., & Phillips, S.M. (2010). Elevations in ostensibly anabolic hormones with resistance exercise enhance neither training-induced

muscle hypertrophy nor strength of the elbow flexors. *Journal of Applied Physiology,* **108,** 60-67.

30. West, D.W., Cotie, L.M., Mitchell, C.J., Churchward-Venne, T.A., MacDonald, M.J., & Phillips, S.M. (2013). Resistance exercise order does not determine postexercise delivery of testosterone, growth hormone, and IGF-1 to skeletal muscle. *Applied Physiology, Nutrition, and Metabolism,* **38,** 220-226.
31. West, D.W., Kujbida, G.W., Moore, D.R., Atherton, P., Burd, N.A., Padzik, J.P., De Lisio, M., Tang, J.E., Parise, G., Rennie, M.J., Baker, S.K., & Phillips, S.M. (2009). Resistance exercise-induced increases in putative anabolic hormones do not enhance muscle protein synthesis or intracellular signalling in young men. *Journal of Physiology,* **587,** 5239-5247.
32. West, D.W., & Phillips, S.M. (2012). Associations of exercise-induced hormone profiles and gains in strength and hypertrophy in a large cohort after weight training. *European Journal of Applied Physiology,* **112,** 2693-2702.

Chapter 11

1. Allison, M.K., Baglole, J.H., Martin, B.J., Macinnis, M.J., Gurd, B.J., & Gibala, M.J. (2016). Brief intense stair climbing improves cardiorespiratory fitness. *Medicine and Science in Sports and Exercise,* **49,** 298-307.
2. Armstrong, R.B., & Laughlin, M.H. (1984). Exercise blood flow patterns within and among rat muscles after training. *American Journal of Physiology,* **246,** H59-H68.
3. Bouchard, C., An, P., Rice, T., Skinner, J.S., Wilmore, J.H., Gagnon, J., Pérusse, L., Leon, A.S., & Rao, D.C. (1999). Familial aggregation of $\dot{V}O_{(2max)}$ response to exercise training: Results from the HERITAGE Family Study. *Journal of Applied Physiology,* **87,** 1003-1008.
4. Bouchard, C., Dionne, F.T., Simoneau, J.-A., & Boulay, M.R. (1992). Genetics of aerobic and anaerobic performances. *Exercise and Sport Sciences Reviews,* **20,** 27-58.
5. Bouchard, C., Lesage, R., Lortie, G., Simoneau, J.A., Hamel, P., Boulay, M.R., Pérusse, L., Theriault, G., & Leblanc, C. (1986). Aerobic performance in brothers, dizygotic and monozygotic twins. *Medicine and Science in Sports and Exercise,* **18,** 639-646.
6. Boyett, M.R., D'Souza, A.D., Zhang, H., Morris, G.M., Dobrzynski, H., & Monfredi, O. (2013). Viewpoint: Is the resting bradycardia in athletes the result of remodeling of the sinoatrial node rather than high vagal tone? *Journal of Applied Physiology,* **114,** 1351-1355.
7. Broatch, J.R., Petersen, A.C., & Bishop, D.J. (2017). Cold-water immersion following sprint interval training does not alter endurance signaling pathways or training adaptations in human skeletal muscle. *American Journal of Physiology. Regulatory, Integrative and Comparative Physiology,* **313**(4), R372-R384.
8. Costill, D.L., Coyle, E.F., Fink, W.F., Lesmes, G.R., & Witzmann, F.A. (1979). Adaptations in skeletal muscle following strength training. *Journal of Applied Physiology: Respiratory Environmental Exercise Physiology,* **46,** 96-99.
9. Costill, D.L., Fink, W.J., Ivy, J.L., Getchell, L.H., & Witzmann, F.A. (1979). Lipid metabolism in skeletal muscle of endurance-trained males and females. *Journal of Applied Physiology,* **28,** 251-255.
10. Ehsani, A.A., Ogawa, T., Miller, T.R., Spina, R.J., & Jilka, S.M. (1991). Exercise training improves left ventricular systolic function in older men. *Circulation,* **83,** 96-103.
11. Ekblom, B., Goldbarg, A.M., & Gullbring, B. (1972). Response to exercise after blood loss and reinfusion. *Journal of Applied Physiology,* **33,** 175-180.
12. Fagard, R.H. (1996). Athlete's heart: A meta-analysis of the echocardiographic experience. *International Journal of Sports Medicine,* **17,** S140-S144.
13. Gibala, M.J., & Jones, A.M. (2013). Physiological and performance adaptations to high-intensity interval training. *Nestlé Nutritional Institute Workshop Series,* **76,** 51-60.
14. Gibala, M.J., Little, J.P., van Essen, M., Wilkin, G.P., Burgomaster, K.A., Safdar, A., Raha, S., & Tarnopolsky, M.A. (2006). Short-term sprint interval versus traditional endurance training: Similar initial adaptations in human skeletal muscle and exercise performance. *Journal of Physiology,* **575,** 901-911.
15. Hermansen, L., & Wachtlova, M. (1971). Capillary density of skeletal muscle in well-trained and untrained men. *Journal of Applied Physiology,* **30,** 860-863.
16. Holloszy, J.O., Oscai, L.B., Mole, P.A., & Don, I.J. (1971). Biochemical adaptations to endurance exercise in skeletal muscle. In B. Pernow & B. Saltin (Eds.), *Muscle metabolism during exercise* (pp. 51-61). New York: Plenum Press.
17. Jacobs, I., Esbjörnsson, M., Sylvén, C., Holm, I., & Jansson, E. (1987). Sprint training effects on muscle myoglobin, enzymes, fiber types, and blood lactate. *Medicine and Science in Sports and Exercise,* **19,** 368-374.
18. Jansson, E., Esbjörnsson, M., Holm, I., & Jacobs, I. (1990). Increase in the proportion of fast-twitch muscle fibres by sprint training in males. *Acta Physiologica Scandinavica,* **140,** 359-363.
19. Lundby, C., Montero, D., & Joyner, M.J. (2017). Biology of $\dot{V}O_{2max}$: Looking under the physiology lamp. *Acta Physiologica,* **220,** 218-228.
20. Lundby, C., & Robach, P. (2015). Performance enhancement: What are the physiological limits? *Physiology,* **30,** 282-292.
21. MacDougall, J.D., Hicks, A.L., MacDonald, J.R., McKelvie, R.S., Green, H.J., & Smith, K.M. (1998). Muscle performance and enzymatic adaptations to sprint interval training. *Journal of Applied Physiology,* **84,** 2138-2142.
22. Martino, M., Gledhill, N., & Jamnik, V. (2002). High $\dot{V}O_2$max with no history of training is primarily due to high blood volume. *Medicine and Science in Sports and Exercise,* **34,** 966-971.
23. McCarthy, J.P., Pozniak, M.A., & Agre, J.C. (2002). Neuromuscular adaptations to concurrent strength and endurance training. *Medicine and Science in Sports and Exercise,* **34,** 511-519.

24. McGuire, D.K., Levine, B.D., Williamson, J.W., Snell, P.G., Blomqvist, C.G., Saltin, B., & Mitchell, J.H. (2001). A 30-year follow-up of the Dallas Bedrest and Training Study: II. Effect of age on cardiovascular adaptation to exercise training. *Circulation,* **104,** 1358-1366.

25. Montero, D., Diaz-Canestro, C., & Lundby, C. (2015). Endurance training and $\dot{V}O_{2max}$: Role of maximal cardiac output and oxygen extraction. *Medicine and Science in Sports and Exercise,* **47,** 2024-2033.

26. Pirnay, F., Dujardin, J., Deroanne, R., & Petit, J.M. (1971). Muscular exercise during intoxication by carbon monoxide. *Journal of Applied Physiology,* **31,** 573-575.

27. Prud'homme, D., Bouchard, C., LeBlanc, C., Landrey, F., & Fontaine, E. (1984). Sensitivity of maximal aerobic power to training is genotype-dependent. *Medicine and Science in Sports and Exercise,* **16,** 489-493.

28. Rico-Sanz, J., Rankinen, T., Joanisse, D.R., Leon, A.S., Skinner, J.S., Wilmore, J.H., Rao, D.C., & Bouchard, C. (2003). Familial resemblance for muscle phenotypes in the HERITAGE Family Study. *Medicine and Science in Sports and Exercise,* **35**(8), 1360-1366.

29. Ried-Larsen, M., Aarts, H., & Joyner, M.J. (2017). The effects of strict prolonged bed rest on cardiorespiratory fitness: Systematic review and meta-analysis. *Journal of Applied Physiology* Jul 13 [Epub ahead of print]. doi: 10.1152/japplphysiol.00415.2017.

30. Saltin, B., Blomqvist, G., Mitchell, J.H., Johnson, R.L., Jr., Wildenthal, K., & Chapman, C.B. (1968). Response to exercise after bed rest and after training. *Circulation,* **38,** VII1-78.

31. Saltin, B., Nazar, K., Costill, D.L., Stein, E., Jansson, E., Essen, B., & Gollnick, P.D. (1976). The nature of the training response: Peripheral and central adaptations to one-legged exercise. *Acta Physiologica Scandinavica,* **96,** 289-305.

32. Støren, Ø., Helgerud, J., Sæbø, M., Støa, E.M., Bratland-Sanda, S., Unhjem, R., Hoff, J., & Wang, E. (2017). The effect of age on the $\dot{V}O_{2max}$ response to high-intensity interval training. *Medicine and Science in Sports and Exercise,* **49,** 78-85.

33. 33.Strømme, S.B., Ingjer, F., & Meen, H.D. (1977). Assessment of maximal aerobic power in specifically trained athletes. *Journal of Applied Physiology,* **42,** 833-837.

34. Wilmore, J.H., Stanforth, P.R., Gagnon, J., Rice, T., Mandel, S., Leon, A.S., Rao, D.C., Skinner, J.S., & Bouchard, C. (2001). Cardiac output and stroke volume changes with endurance training: The HERITAGE Family Study. *Medicine and Science in Sports and Exercise,* **33,** 99-106.

35. Wilmore, J.H., Stanforth, P.R., Hudspeth, L.A., Gagnon, J., Daw, E.W., Leon, A.S., Rao, D.C., Skinner, J.S., & Bouchard, C. (1998). Alterations in resting metabolic rate as a consequence of 20 wk of endurance training: The HERITAGE Family Study. *American Journal of Clinical Nutrition,* **68,** 66-71.

36. Yan, Z., Lira, V.A., & Greene, N.P. (2012). Exercise training-induced regulation of mitochondrial quality. *Exercise and Sports Science Reviews,* **40,** 159-164.

Chapter 12

1. American College of Sports Medicine. (2006). Prevention of cold injuries during exercise. *Medicine and Science in Sports and Exercise,* **38**(11), 2012-2029.

2. Coker, R.H., Weaver, A.N., Coker, M.S., Murphy, C.J., Gunga, H.C., & Steinach, M. (2017). Metabolic responses to the Yukon Arctic Ultra: Longest and coldest in the world. *Medicine and Science in Sports and Exercise,* **49,** 357-362.

3. Haman, F., Mantha, O.L., Cheung, S.S., DuCharme, M.B., Taber, M., Blondin, D.P., McGarr, G.W., Hartley, G.L., Hynes, Z., & Basset, F.A. (2016). Oxidative fuel selection and shivering thermogenesis during a 12- and 24-h cold-survival simulation. *Journal of Applied Physiology,* **120,** 640-648.

4. Harmon, K.G., Drezner, J.A., Klossner, D., & Asif, I.M. (2012). Sickle cell trait associated with a RR of death of 37 times in National Collegiate Athletic Association Football athletes: A database with 2 million athlete-years as the denominator. *British Journal of Sports Medicine,* **46,** 325-330.

5. International Olympic Committee. (2009, September 28). *Olympic movement medical code.* Available: www.olympic.org/medical-commission?tab=medical-code [October 23, 2014].

6. King, D.S., Costill, D.L., Fink, W.J., Hargreaves, M., & Fielding, R.A. (1985). Muscle metabolism during exercise in the heat in unacclimatized and acclimatized humans. *Journal of Applied Physiology,* **59,** 1350-1354.

7. Luetkemeier, M.J., Hanisko, J.M., & Aho, K.M. (2017). Skin tattoos alter sweat rate and Na^+ concentration. *Medicine and Science in Sports and Exercise,* **49,** 1432-1436.

8. O'Connor, F.G., Deuster, P., & Thompson, A. (2013). Sickle cell trait: What's a sports medicine clinician to think? *British Journal of Sports Medicine,* **47,** 667-668.

9. Poirier, M.P., Gagnon, D., Friesen, B.J., Hardcastle, S.G., & Kenny, G.P. (2015). Whole-body heat exchange during acclimation and its decay. *Medicine and Science in Sports and Exercise,* **47,** 390-400.

10. Research Center for the People & The Press. (2007). *How young people view their lives, futures, and politics: A portrait of Generation Next.* Washington, D.C.: Author.

11. Rowell, L.B. (1974). Human cardiovascular adjustments to heat stress. *Physiological Reviews,* **54,** 75-159.

12. Tarini, B.A., Brooks, M.A., & Bundy, D.G. (2012). A policy impact analysis of the mandatory NCAA sickle cell trait screening program. *Health Services Research,* **47,** 446-461.

13. Trangmar, S.J., & González-Alonso, J. (2017). New insights into the impact of dehydration on blood flow and metabolism during exercise. *Exercise and Sport Sciences Reviews,* **45,** 146-153.

14. Tyler, C.J., Sunderland, C., & Cheung, S.S. (2014). The effect of cooling prior to and during exercise on exercise performance and capacity in the heat: A meta-analysis. *British Journal of Sports Medicine,* **49,** 7-13.
15. Wegmann, M., Oliver, F., Wigand, P., Hecksteden, A., Frohlich, M., & Meyer, T. (2012). Pre-cooling and sports performance: A meta-analytical review. *Sports Medicine,* **42,** 545-564.
16. Young, A.J. (1996). Homeostatic responses to prolonged cold exposure: Human cold acclimation. In M.J. Fregley & C.M. Blatteis (Eds.), *Handbook of physiology: Section 4. Environmental physiology* (pp. 419-438). New York: Oxford University Press.

Chapter 13

1. Bartsch, P., & Saltin, B. (2008). General introduction to altitude adaptation and mountain sickness. *Scandinavian Journal of Medicine and Science in Sports,* **18**(Suppl 1), 1-10.
2. Bonetti, D.L., & Hopkins, W.G. (2009). Sea-level exercise performance following adaptation to hypoxia: A meta-analysis. *Sports Medicine,* **39,** 107-127.
3. Brocherie, F., Millet, G.P., Hauser, A., Steiner, T., Rysman, J., Wehrlin, J.P., & Girard, O. (2015). "Live high-train low and high" hypoxic training improves team-sport performance. *Medicine and Science in Sports and Exercise,* **47,** 2140-2149.
4. Brooks, G.A., Wolfel, E.E., & Groves, B.M. (1992). Muscle accounts for glucose disposal but not blood lactate appearance during exercise after acclimatization to 4,300 m. *Journal of Applied Physiology,* **72,** 2435-2445.
5. Brosnan, M.J., Martin, D.T., Hahn, A.G., Gore, C.J., & Hawley, J.A. (2000). Impaired interval exercise responses in elite female cyclists at moderate simulated altitude. *Journal of Applied Physiology,* **89,** 1819-1824.
6. Buskirk, E.R., Kollias, J., Piconreatigue, E., Akers, R., Prokop, E., & Baker, P. (1967). Physiology and performance of track athletes at various altitudes in the United States and Peru. In R.F. Goddard (Ed.), *The effects of altitude on physical performance* (pp. 65-71). Chicago: Athletic Institute.
7. Chapman, R.F., Karlsen, T., Ge, R.L., Stray-Gundersen, J., & Levine, B.D. (2016). Living altitude influences endurance exercise performance change over time at altitude. *Journal of Applied Physiology,* **120,** 1151-1158.
8. Chapman, R.F., Karlsen, T., Resaland, G.K., Ge, R.-L., Harber, M.P., Witkowski, S., Stray-Gundersen, J., & Levine, B.D. (2014). Defining the "dose" of altitude training: How high to live for optimal sea level performance enhancement. *Journal of Applied Physiology,* **116**(6), 595-603.
9. Daniels, J., & Oldridge, N. (1970). Effects of alternate exposure to altitude and sea level on world-class middle-distance runners. *Medicine and Science in Sports,* **2,** 107-112.
10. Forster, P.J.G. (1985). Effect of different ascent profiles on performance at 4200 m elevation. *Aviation, Space, and Environmental Medicine,* **56,** 785-794.
11. Fulco, C.S., Beidleman, B.A., & Muza, S.R. (2013). Effectiveness of preacclimation strategies for high-altitude exposure. *Exercise and Sport Sciences Reviews,* **41,** 55-63.
12. Foss, J.L., Constantini, K., Mickleborough, T.D., & Chapman, R.F. (2017). Short-term arrival strategies for endurance exercise performance at moderate altitude. *Journal of Applied Physiology,* **123**(5), 1258-1265.
13. Gilbert-Kawai, E.T., Milledge, J.S., Grocott, M.P., & Martin, D.S. (2014). King of the mountains: Tibetan and Sherpa physiological adaptations for life at high altitude. *Physiology,* **29,** 388-402.
14. Julian, C.G. (2017). Epigenomics and human adaptation to high altitude. *Journal of Applied Physiology,* **123**(5), 1362-1370.
15. Levine, B.D., & Stray-Gundersen, J. (1997). "Living high–training low": Effect of moderate-altitude acclimatization with low-altitude training on performance. *Journal of Applied Physiology,* **83,** 102-112.
16. Muza, S.R., Beidleman, B.A., & Fulco, C.S. (2010). Altitude preexposure recommendations for inducing acclimatization. *High Altitude Medicine and Biology,* **11,** 87-92.
17. Norton, E.G. (1925). *The fight for Everest: 1924.* London: Arnold.
18. Pugh, L.C.G.E., Gill, M., Lahiri, J., Milledge, J., Ward, M., & West, J. (1964). Muscular exercise at great altitudes. *Journal of Applied Physiology,* **19,** 431-440.
19. Robach, P., Bonne, T., Fluck, D., Burgi, S., Toigo, M., Jacobs, R.A., & Lundby, C. (2014). Hypoxic training: Effect on mitochondrial function and aerobic performance in hypoxia. *Medicine and Science in Sports and Exercise,* **46,** 1936-1945.
20. Rodríguez, F.A., Iglesias, X., Feriche, B., Calderón-Soto, C., Chaverri, D., Wachsmuth, N.B., Schmidt, W., & Levine, B.D. (2015). Altitude training in elite swimmers for sea level performance (altitude project). *Medicine and Science in Sports and Exercise,* **47,** 1965-1978.
21. Stray-Gundersen, J., Chapman, R.F., & Levine, B.D. (2001). "Living high–training low" altitude training improves sea level performance in male and female elite runners. *Journal of Applied Physiology,* **91,** 1113-1120.
22. Sutton, J., & Lazarus, L. (1973). Mountain sickness in the Australian Alps. *Medical Journal of Australia,* **1,** 545-546.
23. Sutton, J.R., Reeves, J.T., Wagner, P.D., Groves, B.M., Cymerman, A., Malconian, M.K., Rock, P.B., Young, P.M., Walter, S.D., & Houston, C.S. (1988). Operation Everest II: Oxygen transport during exercise at extreme simulated altitude. *Journal of Applied Physiology,* **64,** 1309-1321.
24. West, J.B., Peters, R.M., Aksnes, G., Maret, K.H., Milledge, J.S., & Schoene, R.B. (1986). Nocturnal peri-

odic breathing at altitudes of 6300 and 8050 m. *Journal of Applied Physiology,* **61,** 280-287.

Chapter 14

1. Armstrong, L.E., & VanHeest, J.L. (2002). The unknown mechanism of the overtraining syndrome. *Sports Medicine,* **32,** 185-209.
2. Aubry, A., Hausswirth, C., Louis, J., Coutts, A.J., & Le Meur, Y. (2014). Functional overreaching: The key to peak performance during the taper? *Medicine and Science in Sport and Exercise,* **46,** 1769-1777.
3. Bosquet, L., Montpetit, J., Arvisais, D., & Mujika, I. (2007). Effects of tapering on performance: A meta-analysis. *Medicine and Science in Sports and Exercise,* **39,** 1358-1365.
4. Costill, D.L. (1998). Training adaptations for optimal performance. Paper presented at the VIII International Symposium on Biomechanics and Medicine of Swimming, June 28, University of Jyväskylä, Finland.
5. Costill, D.L., King, D.S., Thomas, R., & Hargreaves, M. (1985). Effects of reduced training on muscular power in swimmers. *Physician and Sportsmedicine,* **13**(2), 94-101.
6. Costill, D.L., Thomas, R., Robergs, R.A., Pascoe, D.D., Lambert, C.P., Barr, S.I., & Fink, W.J. (1991). Adaptations to swimming training: Influence of training volume. *Medicine and Science in Sports and Exercise,* **23,** 371-377.
7. Coyle, E.F., Martin, W.H., III, Sinacore, D.R., Joyner, M.J., Hagberg, J.M., & Holloszy, J.O. (1984). Time course of loss of adaptations after stopping prolonged intense endurance training. *Journal of Applied Physiology,* **57,** 1857-1864.
8. Fitts, R.H., Costill, D.L., & Gardetto, P.R. (1989). Effect of swim-exercise training on human muscle fiber function. *Journal of Applied Physiology,* **66,** 465-475.
9. Fleck, S.J., & Kraemer, W.J. (2004). *Designing resistance training programs* (3rd ed.). Champaign, IL: Human Kinetics.
10. Hausswirth, C., Louis, J., Aubry, A., Bonnet, G., Duffield, R., & Le Meur, Y. (2014). Evidence of disturbed sleep and increased illness in overreached athletes. *Medicine and Science in Sports and Exercise,* **46,** 1036-1045.
11. Hickson, R.C., Foster, C., Pollock, M.L., Galassi, T.M., & Rich, S. (1985). Reduced training intensities and loss of aerobic power, endurance, and cardiac growth. *Journal of Applied Physiology,* **58,** 492-499.
12. Houmard, J.A., Costill, D.L., Mitchell, J.B., Park, S.H., Hickner, R.C., & Roemmish, J.N. (1990). Reduced training maintains performance in distance runners. *International Journal of Sports Medicine,* **11,** 46-51.
13. Houmard, J.A., Scott, B.K., Justice, C.L., & Chenier, T.C. (1994). The effects of taper on performance in distance runners. *Medicine and Science in Sports and Exercise,* **26,** 624-631.
14. Issurin, V.B. (2010). New horizons for the methodology and physiology of training periodization. *Sports Medicine,* **40,** 189-206.
15. Landolfi, E. (2013). Exercise addiction. *Sports Medicine,* **43,** 111-119.
16. Lemmer, J.T., Hurlbut, D.E., Martel, G.F., Tracy, B.L., Ivey, F.M., Metter, E.J., Fozard, J.L., Fleg, J.L., & Hurley, B.F. (2000). Age and gender responses to strength training and detraining. *Medicine and Science in Sports and Exercise,* **32,** 1505-1512.
17. Meeusen, R., Duclos, M., Foster, C., Fry, A., Gleeson, M., Nieman, D., Raglin, J., Rietjens, G., Steinacker, J., & Urhausen, A. (2012). Prevention, diagnosis, and treatment of the overtraining syndrome: Joint consensus statement of the European College of Sport Science and the American College of Sports Medicine. *Medicine and Science in Sports and Exercise,* **45,** 186-205.
18. Nieman, D.C. (1994). Exercise, infection, and immunity. *International Journal of Sports Medicine,* **15,** S131-S141.
19. Saltin, B., Blomqvist, G., Mitchell, J.H., Johnson, R.L., Jr., Wildenthal, K., & Chapman, C.B. (1968). Response to submaximal and maximal exercise after bed rest and after training. *Circulation,* **38**(Suppl 5), VII-VII78.
20. Smith, L.L. (2000). Cytokine hypothesis of overtraining: A physiological adaptation to excessive stress? *Medicine and Science in Sports and Exercise,* **32,** 317-331.
21. Sylta, O., Tonnessen, E., Hammarstrom, D., Danielsen, J., Skovereng, K., Ravn, T., Ronnestad, B.R., Sandbakk, O., & Seiler, S. (2016). The effect of different high-intensity periodization models on endurance adaptations. *Medicine and Science in Sport and Exercise,* **48,** 2165-2174.
22. Trappe, T., Trappe, S., Lee, G., Widrick, J., Fitts, R., & Costill, D. (2006). Cardiorespiratory responses to physical work during and following 17 days of bed rest and spaceflight. *Journal of Applied Physiology,* **100,** 951-957.

Chapter 15

1. American College of Sports Medicine. Nattiv, A., Loucks, A.B., Manore, M.M., Sanborn, C.F., Sundgot-Borgen, J., & Warren, M.P. (2007). The female athlete triad. Position stand. *Medicine and Science in Sports and Exercise,* **39**(10), 1867-1882.
2. American College of Sports Medicine, American Dietetic Association, and Dietitians of Canada. (2016). Nutrition and athletic performance. Joint position statement. *Medicine and Science in Sports and Exercise,* **48,** 453-468.
3. Armstrong, L.E., Costill, D.L., & Fink, W.J. (1985). Influence of diuretic-induced dehydration on competitive running performance. *Medicine and Science in Sports and Exercise,* **17,** 456-461.
4. Åstrand, P.-O. (1967). Diet and athletic performance. *Federation Proceedings,* **26,** 1772-1777.
5. Barr, S.I., Costill, D.L., & Fink, W.J. (1991). Fluid replacement during prolonged exercise: Effects of water, saline or no fluid. *Medicine and Science in Sports and Exercise,* **23,** 811-817.
6. Bazzano, L.A., Hu, T., Reynolds, K., Yao, L., Bunol, C., Liu, Y., Chen, C.S., Klag, M.J., Whelton, P.K., & He, J. (2014). Effects of low-carbohydrate and low-fat diets:

A randomized trial. *Annals of Internal Medicine,* **161,** 309-318.

7. Beaton, L.J., Allan, D.A., Tarnopolsky, M.A., Tiidus, P.M., & Phillips, S.M. (2002). Contraction-induced muscle damage is unaffected by vitamin E supplementation. *Medicine and Science in Sports and Exercise,* **34,** 798-805.
8. Biolo, G., Tipton, K.D., Klein, S., & Wolfe, R.R. (1997). An abundant supply of amino acids enhances the metabolic effect of exercise on muscle protein. *American Journal of Physiology,* **273,** E122-E129.
9. Burke, L.M., Hawley, J.A., Wong, S.H.S., & Jeukendrup, A.E. (2011). Carbohydrates for training and competition. *Journal of Sports Sciences,* **29**(Suppl 1), S17-S27.
10. Chacko, E. (2016). A time for exercise: The exercise window. *The Journal of Applied Physiology,* **122,** 206-209.
11. Coombes, J.S., & Hamilton, K.L. (2000). The effectiveness of commercially available sports drinks. *Sports Medicine,* **29,** 181-209.
12. Costill, D.L., Bowers, R., Branam, G., & Sparks, K. (1971). Muscle glycogen utilization during prolonged exercise on successive days. *Journal of Applied Physiology,* **31,** 834-838.
13. Costill, D.L., & Miller, J.M. (1980). Nutrition for endurance sport: Carbohydrate and fluid balance. *International Journal of Sports Medicine,* **1,** 2-14.
14. Costill, D.L., & Saltin, B. (1974). Factors limiting gastric emptying during rest and exercise. *Journal of Applied Physiology,* **37,** 679-683.
15. Drenowatz, C., Hand, G.A., Sagner, M., Shook, R.P., Burgess, S., & Blair, S.N. (2015). The prospective association between different types of exercise and body composition. *Medicine and Science in Sports and Exercise,* **47,** 2535-2541.
16. *Dietary Guidelines for Americans, 2015-2020.* Key Elements of Healthy Eating Patterns. Available: https://health.gov/dietaryguidelines/2015/guidelines/chapter-1/key-recommendations [August 22, 2017].
17. Dougherty, K.A., Baker, L.B., Chow, M., & Kenney, W.L. (2006). Two percent dehydration impairs and six percent carbohydrate drink improves boys basketball skills. *Medicine and Science in Sports and Exercise,* **38,** 1650-1658.
18. Fairchild, T.J., Fletcher, S., Steele, P., Goodman, C., Dawson, B., & Fournier, P.A. (2002). Rapid carbohydrate loading after a short bout of near maximal-intensity exercise. *Medicine and Science in Sports and Exercise,* **34,** 980-986.
19. Frizzell, R.T., Lang, G.H., Lowance, D.C., & Lathan, S.R. (1986). Hyponatremia and ultramarathon running. *Journal of the American Medical Association,* **255,** 772-774.
20. Gollnick, P.D., Piehl, K., & Saltin, B. (1974). Selective glycogen depletion pattern in human muscle fibres after exercise of varying intensity and at varying pedaling rates. *Journal of Physiology,* **241,** 45-57.
21. Ivy, J.L., Katz, A.L., Cutler, C.L., Sherman, W.M., & Coyle, E.F. (1988). Muscle glycogen synthesis after exercise: Effect of time of carbohydrate ingestion. *Journal of Applied Physiology,* **64,** 1480-1485.
22. Ivy, J.L., Lee, M.C., Brozinick, J.T., Jr., & Reed, M.J. (1988). Muscle glycogen storage after different amounts of carbohydrate ingestion. *Journal of Applied Physiology,* **65,** 2018-2023.
23. Jeukendrup, A., & Gleeson, M. (2010). *Sport nutrition: An introduction to energy production and performance* (2nd ed.). Champaign, IL: Human Kinetics.
24. Nelson, J.D., Poussier, P., Marliss, E.B., Albisser, A.M., & Zinman, B. (1982). Metabolic response of normal man and insulin-infused diabetics to postprandial exercise. *American Journal of Physiology,* **242,** E309-E316.
25. Ornish, D. (2015). The myth of high-protein diets. *The New York Times.* March 23, A21.
26. Phillips, S.M. (2013). Protein consumption and resistance exercise: Maximizing anabolic potential. Gatorade Sports Science Exchange #107. Barrington, IL: Gatorade Sports Science Institute.
27. Rodriguez, N.R., DiMarco, N.M., & Langley, S. (2009). Position of the American Dietetic Association, Dietitians of Canada, and the American College of Sports Medicine: Nutrition and athletic performance. *Journal of the American Dietetic Association,* **109,** 509-527.
28. Rowlands, D.S., Nelson, A.R., Phillips, S.M., Faulkner, J.A., Clarke, J., Burd, N.A., Moore, D., & Stellingwerf, T. (2015). Protein-leucine fed dose effects on muscle protein synthesis after endurance exercise. *Medicine and Science in Sports and Exercise,* **47,** 547-555.
29. Schabort, E.J., Bosch, A.N., Weltan, S.M., & Noakes, T.D. (1999). The effect of a preexercise meal on time to fatigue during prolonged cycling exercise. *Medicine and Science in Sports and Exercise,* **31,** 464-471.
30. Smith, J.W., Pascoe, D.D., Passe, D.H., Ruby, B.C., Stewart, L.K., Baker, L.B., & Zachwieja, J.J. (2013). Curvilinear dose-response relationship of carbohydrate (0-120 g·h(-1)) and performance. *Medicine and Science in Sports and Exercise,* **45**(2), 336-341.
31. Sundgot-Borgen, J. (1999). Eating disorders among male and female elite athletes. *British Journal of Sports Medicine,* **33**(6), 434.
32. Tang, J.E., Moore, D.R., Kujbida, G.W., Tarnopolsky, M.A., & Phillips, S.M. (2009). Ingestion of whey hydrolysate, casein, or soy protein isolate: Effects on mixed muscle protein synthesis at rest and following resistance exercise in young men. *Journal of Applied Physiology,* **107,** 987-992.
33. U.S. Department of Agriculture, Agricultural Research Service. Nutrient intakes from food: Mean amounts consumed per individual by gender and age, what we eat in America. NHANES 2009-2014. Available: www.ars.usda.gov/ARSUserFiles/80400530/pdf/1314/Table_1_NIN_GEN_13.pdf.
34. U.S. News & World Report Health. Best Diets of 2014. Available: http://health.usnews.com/best-diet [August 15, 2014].

35. Vannice, G., & Rasmussen, H. (2014). Position of the Academy of Nutrition and Dietetics: Dietary fatty acids for healthy adults. *Journal of the Academy of Nutrition and Dietetics,* **114,** 136-153.

36. Wilmore, J.H., Brown, C.H., & Davis, J.A. (1977). Body physique and composition of the female distance runner. *Annals of the New York Academy of Sciences,* **301,** 764-776.

37. Wilmore, J.H., Morton, A.R., Gilbey, H.J., & Wood, R.J. (1998). Role of taste preference on fluid intake during and after 90 min of running at 60% of $\dot{V}O_{2max}$ in the heat. *Medicine and Science in Sports and Exercise,* **30,** 587-595.

Chapter 16

1. Alvois, L., Robinson, N., Saudan, D., Baume, N., Mangin, P., & Saugy, M. (2006). Central nervous system stimulants and sport practice. *British Journal of Sports Medicine,* **40**(Suppl 1), i16-i20.

2. American College of Sports Medicine consensus statement. (2000). The physiological and health effects of oral creatine supplementation. *Medicine and Science in Sports and Exercise,* **32,** 706-717.

3. American College of Sports Medicine position stand. (1996). The use of blood doping as an ergogenic aid. *Medicine and Science in Sports and Exercise,* **28**(6), i-xii.

4. Ariel, G., & Saville, W. (1972). Anabolic steroids: The physiological effects of placebos. *Medicine and Science in Sports and Exercise,* **4,** 124-126.

5. Bailey, S.J., Winyard, P., Vanhatalo, A., Blackwell, J.R., DiMenna, F.J., Wilkerson, D.P., Tarr, J., Benjamin, N., & Jones, A.M. (2009). Dietary nitrate supplementation reduces the O_2 cost of low-intensity exercise and enhances tolerance to high-intensity exercise in humans. *Journal of Applied Physiology,* **107**(4), 1144-1155.

6. Bellinger, P.M., Howe, S.T., Shing, C.M., & Fell, J.W. (2012). The effect of combined beta-alanine and sodium bicarbonate supplementation on cycling performance. *Medicine and Science in Sports and Exercise,* **44**(8), 1545-1551.

7. Bemben, M.G., & Lamont, H.S. (2005). Creatine supplementation and exercise performance: Recent findings. *Sports Medicine,* **35**(2), 107-125.

8. Berning, J.M., Adams, K.J., & Stamford, B.A. (2004). Anabolic steroid usage in athletics: Facts, fiction, and public relations. *Journal of Strength and Conditioning Research,* **18,** 908-917.

9. Bescos, R., Sureda, A., Tar, J.A., & Pons, A. (2012). The effect of nitric-oxide-related supplements on human performance. *Sports Medicine,* **42**(2), 99-117.

10. Bhasin, S., Storer, T.W., Berman, N., Callegari, C., Clevenger, B., Phillips, J., Bunnell, T.J., Tricker, R., Shirazi, A., & Casaburi, R. (1996). The effects of supraphysiologic doses of testosterone on muscle size and strength in normal men. *New England Journal of Medicine,* **335,** 1-7.

11. Black, C.D., Waddell, D.E., & Gonglach, A.R. (2015). Caffeine's ergogenic effects on cycling: Neuromuscular and perceptual factors. *Medicine and Science in Sport and Exercise,* **47,** 1145-1158.

12. Bowtell, J.L., Sumners, D.P., Dyer, A., Fox, P., & Mileva, K.N. (2011). Montmorency cherry juice reduces muscle damage caused by intensive strength exercise. *Medicine and Science in Sports and Exercise,* **43**(8), 1544-1551.

13. Bronson, F.H., & Matherne, C.M. (1997). Exposure to anabolic-androgenic steroids shortens life span of male mice. *Medicine and Science in Sports and Exercise,* **29,** 615-619.

14. Buell, J.L., Franks, R., Ransone, J., Powers, M.E., Laquale, K.M., & Carlson-Phillips, A. (2013). National Athletic Trainers' Association position statement: Evaluation of dietary supplements for performance nutrition. *Journal of Athletic Training,* **48**(1), 124-136.

15. Buick, F.J., Gledhill, N., Froese, A.B., Spriet, L., & Meyers, E.C. (1980). Effect of induced erythrocythemia on aerobic work capacity. *Journal of Applied Physiology,* **48,** 636-642.

16. Calfee, R., & Fadale, P. (2006). Popular ergogenic drugs and supplements in young athletes. *Pediatrics,* **117,** e577-e589.

17. Chung, W., Shaw, G., Anderson, M.E., Pyne, D.B., Saunders, P.U., Bishop, D.J., Burke, L.M. (2012). Effect of 10 week beta-alanine supplementation on competitive and training performance in elite swimmers. *Nutrients,* **4,** 1441-1453.

18. Connolly, D.A., McHugh, M.P., Padilla-Zakour, O.I., Carlson, L., & Sayers, S.P. (2006). Efficacy of a tart cherry juice blend in preventing the symptoms of muscle damage. *British Journal of Sports Medicine,* **40**(8), 679-683.

19. Costill, D.L., Dalsky, G.P., & Fink, W.J. (1978). Effects of caffeine ingestion on metabolism and exercise performance. *Medicine and Science in Sports,* **10,** 155-158.

20. Costill, D.L., Verstappen, F., Kuipers, H., Janssen, E., & Fink, W. (1984). Acid-base balance during repeated bouts of exercise: Influence of HCO-3. *International Journal of Sports Medicine,* **5,** 228-231.

21. Devries, M.C., & Phillips, S.M. (2014). Creatine supplementation during resistance training in older adults: A meta-analysis. *Medicine and Science in Sports and Exercise,* **46,** 1194-1203.

22. Eichner, E.R. (1989). Ergolytic drugs. *Sports Science Exchange,* **2**(15), 1-4.

23. Ekblom, B., & Berglund, B. (1991). Effect of erythropoietin administration on maximal aerobic power. *Scandinavian Journal of Medicine and Science in Sports,* **1,** 88-93.

24. Ekblom, B., Goldbarg, A.N., & Gullbring, B. (1972). Response to exercise after blood loss and reinfusion. *Journal of Applied Physiology,* **33,** 175-180.

25. Evans, N.A. (2004). Current concepts in anabolic-androgenic steroids. *American Journal of Sports Medicine,* **32,** 534-542.

26. Forbes, G.B. (1985). The effect of anabolic steroids on lean body mass: The dose response curve. *Metabolism,* **34,** 571-573.

27. Geyer, H., Parr, M.K., Koehler, K., Mareck, V., Schanzer, W., & Thevis, M. (2008). Nutritional supplements cross-contaminated and faked with doping substances. *Journal of Mass Spectrometry,* **43**(7), 892-902.

28. Gledhill, N. (1985). The influence of altered blood volume and oxygen transport capacity on aerobic performance. *Exercise and Sport Sciences Reviews,* **13,** 75-93.

29. Goforth, H.W., Jr., Campbell, N.L., Hodgdon, J.A., & Sucec, A.A. (1982). Hematologic parameters of trained distance runners following induced erythrocythemia [abstract]. *Medicine and Science in Sports and Exercise,* **14,** 174.

30. Graham, T.E. (2001). Caffeine and exercise: Metabolism, endurance and performance. *Sports Medicine,* **31,** 785-807.

31. Hartgens, F., & Kuipers, H. (2004). Effects of androgenic-anabolic steroids in athletes. *Sports Medicine,* **34,** 513-554.

32. Hervey, G.R., Knibbs, A.V., Burkinshaw, L., Morgan, D.B., Jones, P.R.M., Chettle, D.R., & Vartsky, D. (1981). Effects of methandienone on the performance and body composition of men undergoing athletic training. *Clinical Science,* **60,** 457-461.

33. Hill, C.A., Harris, R.C., Kim, H.J., Harris, B.D., Sale, C., Boobis, L.H., Kim, C.K., & Wise, J.A. (2007). Influence of beta-alanine supplementation on skeletal muscle carnosine concentrations and high-intensity cycling capacity. *Amino Acids,* **32**(2), 225-233.

34. Hurst, P., Foad, A., Coleman, D., & Beedie, C. (2017). Athletes intending to use sports supplements are more likely to respond to a placebo. *Medicine and Science in Sports and Exercise,* **49,** 1877-1883.

35. Ivy, J.L., Costill, D.L., Fink, W.J., & Lower, R.W. (1979). Influence of caffeine and carbohydrate feedings on endurance performance. *Medicine and Science in Sports and Exercise,* **11,** 6-11.

36. Juhn, M.S. (2003). Popular sports supplements and ergogenic aids. *Sports Medicine,* **33,** 921-939.

37. Kuehl, K.S., Perrier, E.T., Elliot, D.L., & Chesnutt, J.C. (2010). Efficacy of tart cherry juice in reducing muscle pain during running: A randomized controlled trial. *Journal of the International Society of Sports Nutrition,* **7**(7), 17.

38. Linderman, J., & Fahey, T.D. (1991). Sodium bicarbonate ingestion and exercise performance: An update. *Sports Medicine,* **11,** 71-77.

39. Magkos, F., & Kavouras, S.A. (2004). Caffeine and ephedrine: Physiological, metabolic and performance-enhancing effects. *Sports Medicine,* **34,** 871-889.

40. Nissen, S.L., & Sharp, R.L. (2003). Effect of dietary supplements on lean mass and strength gains with resistance exercise: A meta-analysis. *Journal of Applied Physiology,* **94,** 651-659.

41. Pärssinen, M., & Seppälä, T. (2002). Steroid use and long-term health risks in former athletes. *Sports Medicine,* **32,** 83-94.

42. Rawdon, T., Sharp, R.L., Shelley, M., & Thomas, J.R. (2012). Meta-analysis of the placebo effect in nutritional supplement studies of muscular performance. *Kinesiology Review,* **1,** 137-148.

43. Rosenfeld, C. (2005). The use of ergogenic agents in high school athletes. *Journal of School Nursing,* **21**(6), 333-339.

44. Roth, D.A., & Brooks, G.A. (1990). Lactate transport is mediated by a membrane-bound carrier in rat skeletal muscle sarcolemmal vesicles. *Archives of Biochemistry and Biophysics,* **279,** 377-385.

45. Rudman, D., Feller, A.G., Nagraj, H.S., Gergans, G.A., Lalitha, P.Y., Goldberg, A.F., Schlenker, R.A., Cohn, L., Rudman, I.W., & Mattson, D.E. (1990). Effects of human growth hormone in men over 60 years old. *New England Journal of Medicine,* **323,** 1-6.

46. Sale, C., Saunders, B., Hudson, S., Wise, J.A., Harris, R.C., & Sunderland, C.D. (2011). Effect of B-alanine plus sodium bicarbonate on high-intensity cycling performance. *Medicine and Science in Sports and Exercise,* **43,** 1972-1978.

47. Smith-Rockwell, M., Nickols-Richardson, S.M., & Thye, F.W. (2001). Nutrition knowledge, opinions, and practices of coaches and athletic trainers at a division 1 university. *International Journal of Sports Nutrition and Exercise Metabolism,* **11,** 174-185.

48. Spriet, L.L. (1991). Blood doping and oxygen transport. In D.R. Lamb & M.H. Williams (Eds.), *Ergogenics: Enhancement of performance in exercise and sport* (pp. 213-242). Dubuque, IA: Brown & Benchmark.

49. Spriet, L.L., & Gibala, M.J. (2004). Nutritional strategies to influence adaptations to training. *Journal of Sports Sciences,* **22,** 127-141.

50. Tamaki, T., Uchiyama, S., Uchiyama, Y., Akatsuka, A., Roy, R.R., & Edgerton, V.R. (2001). Anabolic steroids increase exercise tolerance. *American Journal of Physiology: Endocrinology and Metabolism,* **280,** E973-E981.

51. Tokish, J.M., Kocher, M.S., & Hawkins, R.J. (2004). Ergogenic aids: A review of basic science, performance, side effects, and status in sports. *American Journal of Sports Medicine,* **32,** 1543-1553.

52. Vanhatalo, A., Bailey, S.J., Blackwell, J.R., DiMenna, F.J., Pavey, T.G., Wilkerson, D.P., Benjamin, N., Winyard, P., & Jones, A.M. (2010). Acute and chronic effects of dietary nitrate supplementation on blood pressure and the physiological responses to moderate-intensity and incremental exercise. *American Journal of Physiology: Regulatory, Integrative, and Comparative Physiology,* **299,** R1121-R1131.

53. Vernec, A., Pipe, A., & Slack, A. (2017). A painful dilemma? Analgesic use in sport and the role of anti-doping. *British Journal of Sports Medicine,* **51**(17), 1243-1244.

54. West, D.W., & Phillips, S.M. (2010). Anabolic processes in human skeletal muscle: Restoring the identities of growth hormone and testosterone. *Sports Medicine,* **35**(3), 97-104.

55. Wilkinson, D.J., Hossain, T., Hill, D.S., Phillips, B.E., Crossland, H., Williams, J., Loughnar, P., Churchward-Venne, T.A., Breen, L., Phillips, S.M., Etheridge, T., Rathmacher, J.A., Smith, K., Szewczyk, N.J., & Atherton, P.J. (2013). The effects of leucine and its metabolite β-hydroxy-β-methylbutyrate on human skeletal muscle protein metabolism. *Journal of Physiology,* **591**(11), 2911-2923.

56. Williams, M.H. (Ed.). (1983). *Ergogenic aids in sport.* Champaign, IL: Human Kinetics.

57. Williams, M.H., Wesseldine, S., Somma, T., & Schuster, R. (1981). The effect of induced erythrocythemia upon 5-mile treadmill run time. *Medicine and Science in Sports and Exercise,* **13,** 169-175.

58. Wylie, L.J., Kelly, J., Bailey, S.J., Blackwell, J.R., Skiba, P.F., Winyard, P.G., Jeukendrup, A.E., Vanhatalo, A., & Jones, A.M. (2013). Beetroot juice and exercise: Pharmacodynamic and dose-response relationships. *Journal of Applied Physiology,* **115,** 325-336.

59. Yarasheski, K.E. (1994). Growth hormone effects on metabolism, body composition, muscle mass, and strength. *Exercise and Sport Sciences Reviews,* **22,** 285-312.

60. Yesalis, C.E. (Ed.). (2000). *Anabolic steroids in sport and exercise* (2nd ed.). Champaign, IL: Human Kinetics.

Chapter 17

1. Bar-Or, O. (1983). *Pediatric sports medicine for the practitioner: From physiologic principles to clinical applications.* New York: Springer-Verlag.

2. Bloemers, F., Collard, D., Paw, M.C., Van Mechelen, W., Twisk, J., & Verhagen, E. (2012). Physical inactivity is a risk factor for physical activity-related injuries in children. *British Journal of Sports Medicine,* **46,** 669-674.

3. Cairney, J., Veldhuizen, S., Kwan, M., Hay, J., & Faught, B. (2014). Biological age and sex-related declines in physical activity during adolescence. *Medicine and Science in Sports and Exercise,* **46,** 730-735.

4. Clarke, H.H. (1971). *Physical and motor tests in the Medford boys' growth study.* Englewood Cliffs, NJ: Prentice Hall.

5. Corder, K., Winpenny, E., Love, R., Brown, H.E., White, M., & van Sluijs, E. (2017). Change in physical activity from adolescence to early adulthood: A systematic review and meta-analysis of longitudinal cohort studies. *British Journal of Sports Medicine* Jul 24 [Epub ahead of print]. doi: 10.1136/bjsports-2016-097330.

6. Daniels, J., Oldridge, N., Nagle, F., & White, B. (1978). Differences and changes in $\dot{V}O_2$ among young runners 10 to 18 years of age. *Medicine and Science in Sports and Exercise,* **10,** 200-203.

7. Diamond, A.B. (2015). The cognitive benefits of exercise in youth. *Current Sports Medicine Reports,* **14,** 320-326.

8. Fakhouri, T.H.I., Hughes, J.P., Burt, V.L., Song, M., Fulton, J.E., & Ogden, C.L. (2014). *Physical activity in U.S. youth aged 12-15 years, 2012.* NCHS Data Brief No. 141. Hyattsville, MD: National Center for Health Statistics.

9. Froberg, K., & Lammert, O. (1996). Development of muscle strength during childhood. In O. Bar-Or (Ed.), *The child and adolescent athlete* (p. 28). London: Blackwell.

10. Gunter, K., Baxer-Jones, A.D., Mirwald, R.L., Almstedt, H., Fuller, A., Durski, S., & Snow, C. (2008). Jump starting skeletal health: A 4-year longitudinal study assessing the effects of jumping on skeletal development in pre and circum pubertal children. *Bone,* **4,** 710-718.

11. Kaczor, J.J., Ziolkowski, W., Popinigis, J., & Tarnopolsky, M.A. (2005). Anaerobic and aerobic enzyme activities in human skeletal muscle from children and adults. *Pediatric Research,* **57**(3), 331-335.

12. Katzmarzyk, P.T., Barreira, T.V., Broyles, S.T., et al. (2015). Physical activity, sedentary time, and obesity in an international sample of children. *Medicine and Science in Sports and Exercise,* **47,** 2062-2069.

13. Krahenbuhl, G.S., Morgan, D.W., & Pangrazi, R.P. (1989). Longitudinal changes in distance-running performance of young males. *International Journal of Sports Medicine,* **10,** 92-96.

14. Mahon, A.D., & Vaccaro, P. (1989). Ventilatory threshold and $\dot{V}O_{2max}$ changes in children following endurance training. *Medicine and Science in Sports and Exercise,* **21,** 425-431.

15. Malina, R.M. (1989). Growth and maturation: Normal variation and effect of training. In C.V. Gisolfi & D.R. Lamb (Eds.), *Perspectives in exercise science and sports medicine: Youth, exercise and sport* (pp. 223-265). Carmel, IN: Benchmark Press.

16. Malina, R.M., Bouchard, C., & Bar-Or, O. (2004). *Growth, maturation, and physical activity* (2nd ed.). Champaign, IL: Human Kinetics.

17. Metcalf, B., Henley, W., & Wilkin, T. (2013). Effectiveness of intervention on physical activity of children: Systematic review and meta-analysis of controlled trials with objectively measured outcomes (EarlyBird 54). *BMJ,* **47,** 226.

18. Metcalf, B., Hosking, J., Jeffery, A.N., Henley, W., & Wilkin, T. (2013). Exploring the adolescent fall in physical activity: A 10-yr cohort study (EarlyBird 41). *Medicine and Science in Sports and Exercise,* **47,** 2084-2092.

19. Mostafavifar, A.M., Best, T.M., & Myer, G.D. (2013). Early sport specialization: Does it lead to long-term problems? *British Journal of Sports Medicine,* **47,** 1060-1061.

20. Nes, B.M., Osthus, I.B., Welde, B., Aspenes, A.T., & Wisloff, U. (2013). Peak oxygen uptake and physical activity in 13- to 18-year-olds: The Young-HUNT Study. *Medicine and Science in Sports and Exercise,* **45,** 304-313.

21. Pahkala, K., Hernelahti, M., Heinonen, O.J., Raittinen, P., Hakanen, M., Lagstrom, H., Viikari, J.S.A., Ronnemaa, T., Raitakari, O.T., & Simell, O. (2013). Body mass index, fitness and physical activity from childhood through adolescence. *British Journal of Sports Medicine,* **47,** 71-77.

22. Pitukcheewanont, P., Punyasavatsut, N., & Feuille, M. (2010). Physical activity and bone health in children and adolescents. *Pediatric Endocrinology Reviews,* **7,** 275-282.

23. Ramsay, J.A., Blimkie, C.J.R., Smith, K., Garner, S., MacDougall, J.D., & Sale, D.G. (1990). Strength training effects in prepubescent boys. *Medicine and Science in Sports and Exercise,* **22,** 605-614.

24. Robinson, S. (1938). Experimental studies of physical fitness in relation to age. *Arbeitsphysiologie,* **10,** 251-323.

25. Rowland, T.W. (1989). Oxygen uptake and endurance fitness in children: A developmental perspective. *Pediatric Exercise Science,* **1,** 313-328.

26. Rowland, T.W. (2007). Evolution of maximal oxygen uptake in children. *Medicine and Sport Science,* **50,** 200-209.

27. Santos, A.M.C., Welsman, J.R., De Ste Croix, M.B.A., & Armstrong, N. (2002). Age- and sex-related differences in optimal peak power. *Pediatric Exercise Science,* **14,** 202-212.

28. Sjödin, B., & Svedenhag, J. (1992). Oxygen uptake during running as related to body mass in circumpubertal boys: A longitudinal study. *European Journal of Applied Physiology,* **65,** 150-157.

29. Turley, K.R., & Wilmore, J.H. (1997). Cardiovascular responses to treadmill and cycle ergometer exercise in children and adults. *Journal of Applied Physiology,* **83,** 948-957.

Chapter 18

1. Buskirk, E.R., & Hodgson, J.L. (1987). Age and aerobic power: The rate of change in men and women. *Federation Proceedings,* **46,** 1824-1829.

2. Carter, H.N., Chen, C.C.W., & Hood, D.A. (2014). Mitochondria, muscle health, and exercise with advancing age. *Physiology,* **30,** 208-223.

3. Clarke, P.M., Walter, S.J., Hayen, A., Mallon, W.J., Heijmans, J., & Studdert, D.M. (2015). Survival of the fittest: Retrospective cohort study of the longevity of Olympic medalists in the modern era. *British Journal of Sports Medicine,* **49,** 898-902.

4. Connelly, D.M., Rice, C.L., Roos, M.R., & Vandervoort, A.A. (1999). Motor unit firing rates and contractile properties in tibialis anterior of young and old men. *Journal of Applied Physiology,* **87,** 843-852.

5. Deschenes, M.R., Roby, M.A., & Glass, E.K. (2011). Aging influences adaptations of the neuromuscular junction to endurance training. *Neuroscience,* **190,** 56-66.

6. Dill, D.B., Robinson, S., & Ross, J.C. (1967). A longitudinal study of 16 champion runners. *Journal of Sports Medicine and Physical Fitness,* **7,** 4-27.

7. DeGroot, D.W., Havenith, G., & Kenney, W.L. (2006). Responses to mild cold stress are predicted by different individual characteristics in young and older subjects. *Journal of Applied Physiology,* **101,** 1607-1615.

8. Fitzgerald, M.D., Tanaka, H., Tran, Z.V., & Seals, D.R. (1997). Age-related declines in maximal aerobic capacity in regularly exercising vs. sedentary women: A meta-analysis. *Journal of Applied Physiology,* **83,** 160-165.

9. Frontera, W.R., Meredith, C.N., O'Reilly, K.P., Knuttgen, W.G., & Evans, W.J. (1988). Strength conditioning in older men: Skeletal muscle hypertrophy and improved function. *Journal of Applied Physiology,* **64,** 1038-1044.

10. Glenn, J.M., Gray, M., & Binns, A. (2015). The effects of loaded and unloaded high-velocity resistance training on functional fitness among community-dwelling older adults. *Age and Ageing,* **44,** 926-931.

11. Goodrick, C.L. (1980). Effects of long-term voluntary wheel exercise on male and female Wistar rats: 1. Longevity, body weight and metabolic rate. *Gerontology,* **26,** 22-33.

12. Hagerman, F.C., Walsh, S.J., Staron, R.S., Hikida, R.S., Gilders, R.M., Murray, T.F., Toma, K., & Ragg, K.E. (2000). Effects of high-intensity resistance training on untrained older men. I. Strength, cardiovascular, and metabolic responses. *Journals of Gerontology Series A: Biological Sciences and Medical Sciences,* **55,** B336-B346.

13. Häkkinen, K., Pakarinen, A., Kraemer, W.J., Häkkinen, A., Valkeinen, H., & Alen, M. (2001). Selective muscle hypertrophy, changes in EMG and force, and serum hormones during strength training in older women. *Journal of Applied Physiology,* **91,** 569-580.

14. Hameed, M., Harridge, S.D.R., & Goldspink, G. (2002). Sarcopenia and hypertrophy: A role for insulin-like growth factor-1 and aged muscle? *Exercise and Sport Sciences Reviews,* **30,** 15-19.

15. Hawkins, S.A., Marcell, T.J., Jaque, S.V., & Wiswell, R.A. (2001). A longitudinal assessment of change in $\dot{V}O_{2max}$ and maximal heart rate in master athletes. *Medicine and Science in Sports and Exercise,* **33,** 1744-1750.

16. Hikida, R.S., Staron, R.S., Hagerman, F.C., Walsh, S., Kaiser, E., Shell, S., & Hervey, S. (2000). Effects of high-intensity resistance training on untrained older men. II. Muscle fiber characteristics and nucleo-cytoplasmic relationships. *Journals of Gerontology Series A: Biological Sciences and Medical Sciences,* **55,** B347-B354.

17. Holloszy, J.O. (1997). Mortality rate and longevity of food-restricted exercising male rats: A reevaluation. *Journal of Applied Physiology,* **82,** 399-403.

18. Hunter, S.K., Pereira, H.M., & Keenan, K.G. (2016). The aging neuromuscular system and motor performance. *Journal of Applied Physiology,* **121,** 982-995.

19. Jackson, A.S., Beard, E.F., Wier, L.T., Ross, R.M., Stuteville, J.E., & Blair, S.N. (1995). Changes in aerobic power of men, ages 25-70 yr. *Medicine and Science in Sports and Exercise,* **27,** 113-120.

20. Jackson, A.S., Wier, L.T., Ayers, G.W., Beard, E.F., Stuteville, J.E., & Blair, S.N. (1996). Changes in aerobic power of women, ages 20-64 yr. *Medicine and Science in Sports and Exercise,* **28,** 884-891.

21. Janssen, I., Heymsfield, S.B., Wang, Z., & Ross, R. (2000). Skeletal muscle mass and distribution in 468 men and women aged 18-88 yr. *Journal of Applied Physiology,* **89,** 81-88.

22. Johnson, M.A., Polgar, J., Weihtmann, D., & Appleton, D. (1973). Data on the distribution of fiber types in thirty-six human muscles: An autopsy study. *Journal of Neurological Science,* **1,** 111-129.

23. Kenney, W.L. (1997). Thermoregulation at rest and during exercise in healthy older adults. *Exercise and Sport Sciences Reviews,* **25,** 41-77.

24. Kohrt, W.M., Malley, M.T., Coggan, A.R., Spina, R.J., Ogawa, T., Ehsani, A.A., Bourey, R.E., Martin, W.H., III, & Holloszy, J.O. (1991). Effects of gender, age, and fitness level on response of $\dot{V}O_{2max}$ to training in 60-71 yr olds. *Journal of Applied Physiology,* **71,** 2004-2011.

25. Kohrt, W.M., Malley, M.T., Dalsky, G.P., & Holloszy, J.O. (1992). Body composition of healthy sedentary and trained, young and older men and women. *Medicine and Science in Sports and Exercise,* **24,** 832-837.

26. Lepers, R., Stapley, P.J., & Cattagni, T. (2016). Centenarian athletes: Examples of ultimate human performance? *Age and Ageing,* **45,** 732-736.

27. Lexell, J., Taylor, C.C., & Sjostrom, M. (1988). What is the cause of the aging atrophy? Total number, size, and proportion of different fiber types studied in whole vastus lateralis muscle from 15- to 83-year-old men. *Journal of Neurological Science,* **84,** 275-294.

28. Marcell, T.J., Hawkins, S.A., Tarpenning, K.M., Hyslop, D.M., & Wiswell, R.A. (2003). Longitudinal analysis of lactate threshold in male and female masters athletes. *Medicine and Science in Sports and Exercise,* **35**(5), 810-817.

29. Meredith, C.N., Frontera, W.R., Fisher, E.C., Hughes, V.A., Herland, J.C., Edwards, J., & Evans, W.J. (1989). Peripheral effects of endurance training in young and old subjects. *Journal of Applied Physiology,* **66,** 2844-2849.

30. Pahor, M., Guralnik, J.M., Ambrosius, W.T., et al. (2014). Effect of structured physical activity on prevention of major mobility disability in older adults: The LIFE study randomized clinical trial. *JAMA,* **311,** 2387-2396.

31. Proctor, D.N., Shen, P.H., Dietz, N.M., Eickhoff, T.J., Lawler, L.A., Ebersold, E.J., Loeffler, D.L., & Joyner, M.J. (1998). Reduced leg blood flow during dynamic exercise in older endurance-trained men. *Journal of Applied Physiology,* **85,** 68-75.

32. Robinson, S. (1938). Experimental studies of physical fitness in relation to age. *Arbeitsphysiologie,* **10,** 251-323.

33. Shibata, S., Hastings, J.L., Prasad, A., Fu, Q., Palmer, M.D., & Levine, B.D. (2008). "Dynamic" starling mechanisms: Effects of ageing and physical fitness on ventricular-arterial coupling. *Journal of Physiology,* **586**(7), 1951-1962.

34. Spirduso, W.W. (2005). *Physical dimensions of aging* (2nd ed.). Champaign, IL: Human Kinetics.

35. Tanaka, H., Monahan, K.D., & Seals, D.R. (2001). Age-predicted maximal heart rate revisited. *Journal of the American College of Cardiology,* **37,** 153-156.

36. Trappe, S.W., Costill, D.L., Fink, W.J., & Pearson, D.R. (1995). Skeletal muscle characteristics among distance runners: A 20-yr follow-up study. *Journal of Applied Physiology,* **78,** 823-829.

37. Trappe, S.W., Costill, D.L., Goodpaster, B.H., & Pearson, D.R. (1996). Calf muscle strength in former elite distance runners. *Scandinavian Journal of Medicine and Science in Sports,* **6,** 205-210.

38. Trappe, S.W., Costill, D.L., Vukovich, M.D., Jones, J., & Melham, T. (1996). Aging among elite distance runners: A 22-yr longitudinal study. *Journal of Applied Physiology,* **80,** 285-290.

39. Trappe, S., Hayes, E., Galpin, A., Kaminsky, L., Jemiolo, B., Fink, W., Trappe, T., Jansson, A., Gustafsson, T., & Tesch, P. (2013). New records in aerobic power among octogenarian lifelong endurance athletes. *Journal of Applied Physiology,* **114,** 3-10.

40. Wiswell, R.A., Jaque, S.V., Marcell, T.J., Hawkins, S.A., Tarpenning, K.M., Constantino, N., & Hyslop, D.M. (2000). Maximal aerobic power, lactate threshold, and running performance in master athletes. *Medicine and Science in Sports and Exercise,* **32,** 1165-1170.

41. Zhu, W., Wadley, V.G., Howard, V.J., Hutto, B., Blair, S.N., & Hooker, S.P. (2017). Objectively measured physical activity and cognitive function in older adults. *Medicine and Science in Sports and Exercise,* **49,** 47-53.

Chapter 19

1. American College of Obstetricians and Gynecologists. (2015). Physical activity and exercise during pregnancy and the postpartum period. *ACOG Committee on Obstetric Practice, Committee Opinion,* **650.** Available: www.acog.org/Resources-And-Publications/Committee-Opinions/Committee-on-Obstetric-Practice/Physical-Activity-and-Exercise-During-Pregnancy-and-the-Postpartum-Period.

2. American College of Sports Medicine. (2007). The female athlete triad. *Medicine and Science in Sports and Exercise,* **39,** 1867-1882.

3. Åstrand, P.-O., Rodahl, K., Dahl, H.A., & Strømme, S.B. (2003). *Textbook of work physiology: Physiological bases of exercise* (4th ed.). Champaign, IL: Human Kinetics.

4. Barakat, R., Perales, M., Cordero, Y., Bacchi, M., & Mottola, M.F. (2017). Influence of land or water exercise in pregnancy on outcomes: A cross-sectional study. *Medicine and Science in Sports and Exercise,* **49,** 1397-1403.

5. Blaize, A.N., Pearson, K.J., & Newcomer, S.C. (2015). Impact of maternal exercise during pregnancy on offspring chronic disease susceptibility. *Medicine and Science in Sports and Exercise,* **43,** 198-203.

6. Caudwell, P., Gibbons, C., Finlayson, G., Näslund, E., & Blundell, J. (2014). Exercise and weight loss: No sex differences in body weight response to exercise. *Exercise and Sport Sciences Reviews,* **42,** 92-101.

7. Costill, D.L., Fink, W.J., Flynn, M., & Kirwan, J. (1987). Muscle fiber composition and enzyme activities in elite female distance runners. *International Journal of Sports Medicine,* **8**(Suppl 2), 103-106.

8. Cureton, K., Bishop, P., Hutchinson, P., Newland, H., Vickery, S., & Zwiren, L. (1986). Sex differences in maximal oxygen uptake: Effect of equating haemoglobin concentration. *European Journal of Applied Physiology,* **54,** 656-660.

9. Cureton, K.J., & Sparling, P.B. (1980). Distance running performance and metabolic responses to running in men and women with excess weight experimentally equated. *Medicine and Science in Sports and Exercise,* **12,** 288-294.

10. Deaner, R.O., Carter, R.E., Joyner, M.J., & Hunter, S.K. (2015). Men are more likely than women to slow in the marathon. *Medicine and Science in Sports and Exercise,* **42,** 607-616.

11. Evenson, K.R., & Wen, F. (2010). National trends in self-reported physical activity and sedentary behaviors among pregnant women: NHANES 1999-2006. *Preventive Medicine,* **50,** 123-128.

12. Fink, W.J., Costill, D.L., & Pollock, M.L. (1977). Submaximal and maximal working capacity of elite distance runners: Part II. Muscle fiber composition and enzyme activities. *Annals of the New York Academy of Sciences,* **301,** 323-327.

13. Hermansen, L., & Andersen, K.L. (1965). Aerobic work capacity in young Norwegian men and women. *Journal of Applied Physiology,* **20,** 425-431.

14. Hunter, S.K., & Stevens, A.A. (2013). Sex differences in marathon running with advanced age: Physiology or participation? *Medicine and Science in Sports and Exercise,* **45,** 148-156.

15. Kardel, K.R. (2005). Effects of intense training during and after pregnancy in top-level athletes. *Scandinavian Journal of Medicine and Science in Sports,* **15,** 79-86.

16. Karkazis, K., Jordan-Young, R., Davis, G., & Camporesi, S. (2012). Out of bounds? A critique of the new policies on hyperandrogenism in elite female athletes. *American Journal of Bioethics,* **12,** 3-16.

17. Mudd, L.M., Owe, K.M., Mottola, M.F., & Pivarnik, J.M. (2013). Health benefits of physical activity during pregnancy: An international perspective. *Medicine and Science in Sports and Exercise,* **45,** 268-277.

18. Parks, R.B., Hetzel, S.J., & Brooks, M.A. (2017). Iron deficiency and anemia among collegiate athletes: A retrospective chart review. *Medicine and Science in Sports and Exercise,* **49,** 1711-1715.

19. Peter, L., Rüst, C.A., Knechtle, B., Rosemann, T., & Lepers, R. (2014). Sex differences in 24-hour ultra-marathon performance—A retrospective data analysis from 1977 to 2012. *Clinics (Sao Paulo),* **69,** 38-46.

20. Redman, L.M., & Loucks, A.B. (2005). Menstrual disorders in athletes. *Sports Medicine,* **35,** 747-755.

21. Saltin, B., & Åstrand, P.-O. (1967). Maximal oxygen uptake in athletes. *Journal of Applied Physiology,* **23,** 353-358.

22. Saltin, B., Henriksson, J., Nygaard, E., & Andersen, P. (1977). Fiber types and metabolic potentials of skeletal muscles in sedentary man and endurance runners. *Annals of the New York Academy of Sciences,* **301,** 3-29.

23. Salvesen, K.Å., Hem, E., & Sundgot-Borgen, J. (2012). Fetal wellbeing may be compromised during strenuous exercise among pregnant elite athletes. *British Journal of Sports Medicine,* **46,** 279-283.

24. Schantz, P., Randall-Fox, E., Hutchison, W., Tyden, A., & Åstrand, P.-O. (1983). Muscle fibre type distribution, muscle cross-sectional area and maximal voluntary strength in humans. *Acta Physiologica Scandinavica,* **117,** 219-226.

25. Torgrimson, B.N., & Minson, C.T. (1985). Sex and gender: What is the difference? *Journal of Applied Physiology,* **99,** 785-787.

26. Williams, N.I., McConnell, H.J., Gardner, J.K., Frye, B.R., Richard, E.L., Snook, M.L., Dougherty, K.L., Parrott, T.S., Albert, A., & Schukert, M. (2004). Exercise-associated menstrual disturbances: Dependence on daily energy deficit, not body composition or body weight changes. *Medicine and Science in Sports and Exercise,* **36**(5), S280.

27. Wilmore, J.H., Stanforth, P.R., Gagnon, J., Rice, T., Mandel, S., Leon, A.S., Rao, D.C., Skinner, J.S., & Bouchard, C. (2001). Cardiac output and stroke volume changes with endurance training: The HERITAGE Family Study. *Medicine and Science in Sports and Exercise,* **33,** 99-106.

Chapter 20

1. American College of Sports Medicine. (2018). *ACSM's guidelines for exercise testing and prescription* (10th ed.). Philadelphia: Wolters Kluwer.

2. Bergouignan, A., Rudwill, F., Simon, C., & Blanc, S. (2011). Physical inactivity as the culprit of metabolic inflexibility: Evidence from bed-rest studies. *Journal of Applied Physiology,* **111,** 1201-1210.

3. Booth, F.W., Gordon, S.E., Carlson, C.J., & Hamilton, M.T. (2000). Waging war on modern chronic disease: Primary prevention through exercise biology. *Journal of Applied Physiology,* **88,** 774-787.

4. Borg, G.A.V. (1998). *Borg's perceived exertion and pain scales.* Champaign, IL: Human Kinetics.

5. Byrne, N.M., Hills, A.P., Hunter, G.R., Weinsier, R.L., & Schutz, Y. (2005). Metabolic equivalent: One size does not fit all. *Journal of Applied Physiology,* **99,** 1112-1119.

6. Gibala, M.J., & McGee, S. (2008). Metabolic adaptations to short-term high-intensity interval training: A little pain for a lot of gain? *Exercise and Sports Science Reviews,* **36,** 58-63.

7. Haskell, W.L., Lee, I.M., Pate, R.R., et al. (2007). Physical activity and public health: Updated recommendations for adults from the American College of Sports Medicine and the American Heart Association. *Medicine and Science in Sports and Exercise,* **39,** 1423-1424.

8. Katzmarzyk, P.T., & Pate, R.R. (2017). Physical activity and mortality: The potential impact of sitting. *Trans-*

lational Journal of the American College of Sports Medicine, **2,** 32-33.

9. Lauer, M., Sivarajan Froelicher, E., Williams, M., & Kligfield, P. (2005). Exercise testing in asymptomatic adults. *Circulation,* **112,** 771-776.
10. Manini, T.M., Everhart, J.E., Patel, K.V., Schoeller, D.A., Colbert, L.H., Visser, M., Tylavsky, F., Bauer, D.C., Goodpaster, B.H., & Harris, T.B. (2006). Daily activity energy expenditure and mortality among older adults. *Journal of the American Medical Association,* **296,** 171-179.
11. Matthews, C.E., Moore, S.C., Sampson, J., Blair, A., Xiao, Q., Keadle, S.K., Hollenbeck, A., & Park, Y. (2015). Mortality benefits for replacing sitting time with different physical activities. *Medicine and Science in Sports and Exercise,* **47,** 1833-1840.
12. Murray, A.D., Daines, L., Archibald, D., Hawkes, R.A., Schiporst, C., Kelly, P., Grant, L., & Mutrie, N. (2017). The relationships between golf and health: A scoping review. *British Journal of Sports Medicine,* **51,** 12-19.
13. Persinger, R., Foster, C., Gibson, M., Fater, D.C.W., & Porcari, J.P. (2004). Consistency of the talk test for exercise prescription. *Medicine and Science in Sports and Exercise,* **36,** 1632-1636.
14. Pollock, M.L., Franklin, B.A., Balady, G.J., Chaitman, B.L., Fleg, J.L., Fletcher, B., Limacher, M., Piña, I.L., Stein, R.A., Williams, M., & Bazzarre, T. (2000). Resistance exercise in individuals with and without cardiovascular disease: Benefits, rationale, safety and prescription. *Circulation,* **101,** 828-833.
15. President's Council on Fitness, Sports & Nutrition. (2017). *Facts and statistics: Physical activity.* Available: www.hhs.gov/fitness/resource-center/facts-and-statistics/index.html.
16. Swain, D.P., & Leutholtz, B.C. (1997). Heart rate reserve is equivalent to %$\dot{V}O_2$ reserve, not to %$\dot{V}O_2$max. *Medicine and Science in Sports and Exercise,* **29,** 410-414.
17. Tarumi, T., & Zhang, R. (2015). The role of exercise-induced cardiovascular adaptation in brain health. *Exercise and Sport Science Reviews,* **43,** 181-189.
18. Williams, P.T. (2014). Reduced risk of brain cancer mortality from walking and running. *Medicine and Science in Sports and Exercise,* **46,** 927-932.

Chapter 21

1. American College of Sports Medicine and American Heart Association. (2007). Exercise and acute cardiovascular events: Placing the risks into perspective. Joint position statement. *Medicine and Science in Sports and Exercise,* **39,** 886-897.
2. American College of Sports Medicine. (2014). *ACSM's guidelines for exercise testing and prescription* (9th ed.). Philadelphia: Lippincott Williams & Wilkins.
3. American Heart Association. (2010). Heart disease and stroke statistics—2010 update. *Circulation,* **121,** e46-e215.
4. American Heart Association. (2017). Heart disease and stroke statistics—2017 update. *Circulation,* **135,** e1-e457. Available: http://circ.ahajournals.org/content/circulationaha/early/2017/01/25/CIR.0000000000000485.full.pdf.
5. Berenson, G.S., Srinivasan, S.R., Bao, W., Newman, W.P., Tracy, R.E., & Wattigney, W.A. (1998). Association between multiple cardiovascular risk factors and atherosclerosis in children and young adults. The Bogalusa Heart Study. *New England Journal of Medicine,* **338,** 1650-1656.
6. Blair, S.N., Goodyear, N.N., Gibbons, L.W., & Cooper, K.H. (1984). Physical fitness and incidence of hypertension in healthy normotensive men and women. *Journal of the American Medical Association,* **252,** 487-490.
7. Blair, S.N., & Jackson, A.S. (2001). Guest editorial: Physical fitness and activity as separate heart disease risk factors: A meta-analysis. *Medicine and Science in Sports and Exercise,* **33,** 762-764.
8. Blair, S.N., Kohl, H.W., Paffenbarger, R.S., Clark, D.G., Cooper, K.H., & Gibbons, L.W. (1989). Physical fitness and all-cause mortality: A prospective study of healthy men and women. *Journal of the American Medical Association,* **262,** 2395-2401.
9. Blumenthal, J.A., Babyak, M.A., Doraiswamy, P.M., Watkins, L., Hoffman, B.M., et al. (2007). Exercise and pharmacotherapy in the treatment of major depressive disorder. *Psychosomatic Medicine,* **69,** 587-596.
10. Carnethon, M.R., Gulati, M., & Greenland, P. (2005). Prevalence and cardiovascular disease correlates of low cardiorespiratory fitness in adolescents and adults. *Journal of the American Medical Association,* **294,** 2981-2988.
11. Chow, L., Eberly, L.E., Austin, E., Carnethon, M., Bouchard, C., Sternfeld, B., Zhu, N.A., Sidney, S., & Schreiner, P. (2015). Fitness change effects on midlife metabolic outcomes. *Medicine and Science in Sports and Exercise,* **47,** 967-973.
12. Conroy, M.B., Cook, N.R., Manson, J.E., Buring, J.E., & Lee, I-M. (2005). Past physical activity, current physical activity, and risk of coronary heart disease. *Medicine and Science in Sports and Exercise,* **37,** 1251-1256.
13. Dunlop, D., Song, J., Arnston, E., Semanik, P., Lee, J., et al. (2014). Sedentary time in U.S. older adults associated with disability in activities of daily living independent of physical activity. *Journal of Physical Activity and Health* February 5 [Epub ahead of print]. doi:10.1123/jpah.2013-0311.
14. Eckel, R.H., et al. (2013). 2013 ACC/AHA guideline on lifestyle management to reduce cardiovascular risk: A report of the American College of Cardiology/American Heart Association Task Force on Practice Guidelines. *Circulation* November 12 [Epub ahead of print]. Available: http://circ.ahajournals.org [September 20, 2014]. doi:10.1161/01.cir.0000437740.48606.d1.
15. Enos, W.F., Holmes, R.H., & Beyer, J. (1953). Coronary disease among United States soldiers killed in action in

Korea. *Journal of the American Medical Association,* **152,** 1090-1093.

16. Ford, E.S., & Caspersen, C.J. (2012). Sedentary behaviour and cardiovascular disease: A review of prospective studies. *International Journal of Epidemiology,* **41,** 1338-1353.
17. Gill, J.M.R. (2007). Physical activity, cardiorespiratory fitness and insulin resistance: A short update. *Current Opinion in Lipidology,* **18,** 47-52.
18. Glass, T.W., & Maher, C.G. (2014). Physical activity reduces cigarette cravings. *British Journal of Sports Medicine,* **48,** 1263-1264.
19. Gleeson, M., Bishop, N.C., Stensel, D.J., Lindley, M.R., Mastana, S.S., et al. (2011). The anti-inflammatory effects of exercise: Mechanisms and implications for the prevention and treatment of disease. *Nature Reviews Immunology,* **11,** 607-615.
20. Goff, D.C., et al. (2013). 2013 ACC/AHA guideline on the assessment of cardiovascular risk: A report of the American College of Cardiology/American Heart Association Task Force on Practice Guidelines. *Circulation* November 12 [Epub ahead of print]. Available: http://circ.ahajournals.org [September 20, 2014]. doi:10.1161/01.cir.0000437741.48606.98.
21. *Guidelines for cardiac rehabilitation and secondary prevention programs* (5th ed. with web resource). (2013). AACVPR. Champaign, IL: Human Kinetics.
22. Healy, G.N., Matthews, C.E., Dunstan, D.W., Winkler, E.A., & Owen, N. (2011). Sedentary time and cardio-metabolic biomarkers in US adults: NHANES 2003-06. *European Heart Journal,* **32,** 590-597.
23. Hu, F.B., Stampfer, M.J., Colditz, G.A., Ascherio, A., Rexrode, K.M., Willett, W.C., & Manson, J.E. (2000). Physical activity and risk of stroke in women. *Journal of the American Medical Association,* **283,** 2961-2967.
24. James, P.A., et al. (2014). 2014 evidence-based guideline for the management of high blood pressure in adults: Report from the panel members appointed to the Eighth Joint National Committee (JCN8). *Journal of the American Medical Association,* **311**(5), 507-520.
25. Kannel, W.B., & Dawber, T.R. (1972). Atherosclerosis as a pediatric problem. *Journal of Pediatrics,* **80,** 544-554.
26. Levine, J.A., Vander Weg, M.W., Hill, J.O., & Klesges, R.C. (2006). Non-exercise activity thermogenesis: The crouching tiger hidden dragon of societal weight gain. *Arteriosclerosis, Thrombosis, and Vascular Biology,* **26,** 729-736.
27. Lawler, P.R., Filion, K.B., & Eisenberg, M.J. (2011). Efficacy of exercise-based cardiac rehabilitation post-myocardial infarction: A systematic review and meta-analysis of randomized controlled trials. *American Heart Journal,* **162,** 571-584.
28. Lee, C.D., & Blair, S.N. (2002). Cardiorespiratory fitness and stroke mortality in men. *Medicine and Science in Sports and Exercise,* **34,** 592-595.
29. Libby, P., Ridker, P., & Hansson, G.K. (2009). Inflammation in atherosclerosis: From pathophysiology to practice. *Journal of the American College of Cardiology,* **54,** 2129-2138.
30. McNamara, J.J., Molot, M.A., Stremple, J.F., & Cutting, R.T. (1971). Coronary artery disease in combat casualties in Vietnam. *Journal of the American Medical Association,* **216,** 1185-1187.
31. Montoye, H.J., Metzner, H.L., Keller, J.B., Johnson, B.C., & Epstein, F.H. (1972). Habitual physical activity and blood pressure. *Medicine and Science in Sports and Exercise,* **4,** 175-181.
32. Morris, J.N., Heady, J.A., Raffle, P.A.B., Roberts, C.G., & Parks, J.W. (1953). Coronary heart-disease and physical activity of work. *Lancet,* **265,** 1053-1057.
33. Myers, J., Kaykha, A., George, S., et al. (2004). Fitness versus physical activity patterns in predicting mortality in men. *American Journal of Medicine,* **117,** 912-918.
34. O'Keefe, J.H., Patil, H.R., Lavie, C.J., Magalski, A., Vogel, R.A., et al. (2007). Potential adverse cardiovascular effects from excessive endurance exercise. *Mayo Clinic Proceedings,* **87,** 587-595.
35. Paffenbarger, R.S., Hyde, R.T., Wing, A.L., & Hsieh, C-C. (1986). Physical activity, all-cause mortality, and longevity of college alumni. *New England Journal of Medicine,* **314,** 605-613.
36. Roberts, W.O., Schwartz, R.S., Kraus, S.M., Schwartz, J.G., Peichel, G., Garberich, R.F., Lesser, J.R., Oesterle, S.N., Wickstrom, K.K., Knickelbine, T., & Harris, K.M. (2017). Long-term marathon running is associated with low coronary plaque formation in women. *Medicine and Science in Sports and Exercise,* **49,** 641-645.
37. Siscovick, D.S., Weiss, N.S., Fletcher, R.H., & Lasky, T. (1984). The incidence of primary cardiac arrest during vigorous exercise. *New England Journal of Medicine,* **311,** 874-877.
38. Stone, N.J., et al. (2013). 2013 ACC/AHA guideline on the treatment of blood cholesterol to reduce atherosclerotic cardiovascular risk in adults: A report of the American College of Cardiology/American Heart Association Task Force on Practice Guidelines. *Circulation* November 12 [Epub ahead of print]. Available: http://circ.ahajournals.org. doi:10.1161/01.cir.0000437738.63853.7a.
39. Tanasescu, M., Leitzmann, M.F., Rimm, E.B., Willett, W.C., Stampfer, M.J., & Hu, F.B. (2002). Exercise type and intensity in relation to coronary heart disease in men. *Journal of the American Medical Association,* **288,** 1994-2000.
40. Walther, C., Gielen, S., & Hambrecht, R. (2004). The effect of exercise training on endothelial function in cardiovascular disease in humans. *Exercise and Sport Sciences Reviews,* **32,** 129-134.
41. Williams, M.A., Haskell, W.L., Ades, P.A., Amsterdam, E.A., Bittner, V., Franklin, B.A., Gulanick, M., Laing, S.T., & Stewart, K.J. (2007). Resistance exercise in individuals with and without cardiovascular disease: 2007 update. *Circulation,* **116,** 572-584.

42. Williams, P.T. (2013). Dose-response relationship of physical activity to premature and total all-cause and cardiovascular disease mortality in walkers. *PlOS ONE,* **8**(11), e78777. doi:10.1371/journal.pone.0078777.

43. World Health Organization. (2017). *Fact sheet.* Available: www.who.int/mediacentre/factsheets/fs317/en.

Chapter 22

1. Achten, J., Gleeson, M., & Jeukendrup, A.E. (2002). Determination of the exercise intensity that elicits maximal fat oxidation. *Medicine and Science in Sports and Exercise,* **34,** 92-97.
2. American Diabetes Association. (2017, September 20). *Statistics about diabetes: Overall numbers, diabetes and prediabetes.* Available: www.diabetes.org/diabetes-basics/statistics.
3. Bassuk, S.S., & Manson, J.E. (2005). Epidemiological evidence for the role of physical activity in reducing risk of type 2 diabetes and cardiovascular disease. *Journal of Applied Physiology,* **99,** 1193-1204.
4. Bouchard, C. (1991). Heredity and the path to overweight and obesity. *Medicine and Science in Sports and Exercise,* **23,** 285-291.
5. Bouchard, C., Tremblay, A., Després, J.-P., Nadeau, A., Lupien, P.J., Theriault, G., Dussault, J., Moorjani, S., Pinault, S., & Fournier, G. (1990). The response to long-term overfeeding in identical twins. *New England Journal of Medicine,* **322,** 1477-1482.
6. Bray, G.A. (1985). Obesity: Definition, diagnosis and disadvantages. *Medical Journal of Australia,* **142,** S2-S8.
7. Centers for Disease Control and Prevention. (2011, March 3). *The health effects of overweight and obesity.* Available: www.cdc.gov/healthyweight/effects/index.html.
8. Colberg, S.R., Sigal, R.J., Yardley, J.E., et al. (2016). Physical activity/exercise and diabetes: A position statement of the American Diabetes Association. *Diabetes Care,* **39,** 2065-2079.
9. Dandona, P., Aljada, A., Chaudhuri, A., Mohanty, P., & Garg, R. (2005). Metabolic syndrome: A comprehensive perspective based on interactions between obesity, diabetes, and inflammation. *Circulation,* **111,** 1448-1454.
10. Davidson, L.E., Hudson, R., Kilpatrick, K., Kuk, J.L., McMillan, K., Janiszewski, P.M., Lee, S., Lam, M., & Ross, R. (2009). Effects of exercise modality on insulin resistance and functional limitation in older adults: A randomized controlled trial. *Archives of Internal Medicine,* **169,** 122-131.
11. Donnelly, J.E., Blair, S.N., Jakicic, J.M., et al. (2011). Appropriate physical activity strategies for weight loss and prevention of weight gain for adults. ACSM position stand. *Journal of Medicine and Science in Sports and Exercise,* **41,** 459-471.
12. Flegal, K.M., Carroll, M.D., Kuczmarski, R.J., & Johnson, C.L. (1998). Overweight and obesity in the United States: Prevalence and trends, 1960-1994. *International Journal of Obesity,* **22,** 39-47.
13. Flegal, K.M., Carroll, M.D., Ogden, C.L., & Johnson, C.L. (2002). Prevalence and trends in obesity among US adults, 1999-2000. *Journal of the American Medical Association,* **288,** 1723-1727.
14. Gwinup, G., Chelvam, R., & Steinberg, T. (1971). Thickness of subcutaneous fat and activity of underlying muscles. *Annals of Internal Medicine,* **74,** 408-411.
15. Hall, K.D., Sacks, G., Chandramohan, D., Chow, C.C., Wang, Y.C., et al. (2011). Quantification of the effect of energy imbalance on bodyweight. *Lancet,* **378,** 826-837.
16. Hill, J.O., Wyatt, H.R., & Peters, J.C. (2012). Energy balance and obesity. *Circulation,* **126,** 126-132.
17. Holloszy, J.O. (2005). Exercise-induced increase in muscle insulin sensitivity. *Journal of Applied Physiology,* **99,** 338-343.
18. Jensen, M.D., Ryan, D.H., Apovian, C.M., et al. (2014). AHA/ACC/TOS prevention guideline: 2013 AHA/ACC/TOS guideline for the management of overweight and obesity in adults: A report of the American College of Cardiology/American Heart Association Task Force on Practice Guidelines and The Obesity Society. *Circulation,* **129**(suppl), S102-S138.
19. Katch, F.I., Clarkson, P.M., Kroll, W., McBride, T., & Wilcox, A. (1984). Effects of sit up exercise training on adipose cell size and adiposity. *Research Quarterly for Exercise and Sport,* **55,** 242-247.
20. King, N.A., Caudwell, P.P., Hopkins, M., Stubbs, J.R., Naslund, E., et al. (2009). Dual-process action of exercise on appetite control: Increase in orexigenic drive but improvement in meal-induced satiety. *American Journal of Clinical Nutrition,* **90**(4), 921-927.
21. Knowler, W.C., Barrett-Connor, E., Fowler, S.E., Hamman, R.F., Lachin, J.M., Walker, E.A., & Nathan, D.M. (2002). Reduction in the incidence of type 2 diabetes with lifestyle intervention or metformin. *New England Journal of Medicine,* **346,** 393-403.
22. Ludwig, D.S., & Ebbeling, C.B. (2001). Type 2 diabetes mellitus in children. *Journal of the American Medical Association,* **286,** 1426-1430.
23. Malin, S.K., & Braun, B. (2016). Impact of metformin on exercise-induced metabolic adaptations to lower type 2 diabetes risk. *Exercise and Sport Science Reviews,* **44,** 4-11.
24. Mayer, J., Roy, P., & Mitra, K.P. (1956). Relation between caloric intake, body weight, and physical work: Studies in an industrial male population in West Bengal. *American Journal of Clinical Nutrition,* **4**(2), 169-175.
25. Moore, J.X., Chaudhary, N., & Akinyemiju, T. (2017). Metabolic syndrome prevalence by race/ethnicity and sex in the United States, National Health and Nutrition Examination Survey, 1988-2012. *Prevention of Chronic Diseases,* **14,** 160287.
26. Morton, G.J., Cummings, D.E., Baskin, D.G., Barsh, G.S., & Schwartz, M.W. (2006). Central nervous system control of food intake and body weight. *Nature,* **443,** 289-295.

27. National Institutes of Health. (2000). *The practical guide: Identification, evaluation, and treatment of overweight and obesity in adults* (NIH Publication No. 00-4084). Washington, DC: U.S. Department of Health and Human Services.

28. Ogden, C.L., Carroll, M.D., Fryar, C.D., & Flegal, K.M. (2015). Prevalence of obesity among adults and youth: United States, 2011-2014. *NCHS Data Brief No. 219, November 2015. Centers for Disease Control and Prevention.* Available: www.cdc.gov/nchs/products/databriefs/db219.htm.

29. Organisation for Economic Co-operation and Development (OECD). (2014). *Obesity update.* Available: www.oecd.org/els/health-systems/Obesity-Update-2014.pdf.

30. Ortega, F.B., Ruiz, J.R., Labayen, I., Lavie, C.J., & Blair, S.N. (2018). The fat but fit paradox: What we know and don't know about it. *British Journal of Sports Medicine,* **52**(3), 151-153.

31. Poehlman, E.T. (1989). A review: Exercise and its influence on resting energy metabolism in man. *Medicine and Science in Sports and Exercise,* **21,** 515-525.

32. Seidell, J.C., Deurenberg, P., & Hautvast, J.G.A.J. (1987). Obesity and fat distribution in relation to health—Current insights and recommendations. *World Review of Nutrition and Dietetics,* **50,** 57-91.

33. Singh, G.K., Siahpush, M., Hiatt, R.A., & Timsina, L.R. (2011). Dramatic increases in obesity and overweight prevalence and body mass index among ethnic-immigrant and social class groups in the United States, 1976-2008. *Journal of Community Health,* **36,** 94-110.

34. Slentz, C.A., Aiken, L.B., Houmard, J.A., Bales, C.W., Johnson, J.L., Tanner, C.J., Duscha, B.D., & Kraus, W.E. (2005). Inactivity, exercise and visceral fat. STRRIDE: A randomized, controlled study of exercise intensity and amount. *Journal of Applied Physiology,* **99,** 1613-1618.

35. Tuomilehto, J., Lindstrom, J., Eriksson, J.G., Valle, T.T., Hamalainen, H., et al. for the Finnish Diabetes Prevention Study Group. (2001). Prevention of type 2 diabetes mellitus by changes in lifestyle among subjects with impaired glucose tolerance. *New England Journal of Medicine,* **344,** 1343-1350.

36. Whitaker, K.M., Pereira, M.A., Jacobs, D.R., Jr., Sidney, S., & Odegaard, A.O. (2017). Sedentary behavior, physical activity, and abdominal adipose tissue deposition. *Medicine and Science in Sports and Exercise,* **49,** 450-458.

37. Wilmore, J.H. (1996). Increasing physical activity: Alterations in body mass and composition. *American Journal of Clinical Nutrition,* **63,** 456S-460S.

38. Wilmore, J.H., Stanforth, P.R., Hudspeth, L.A., Gagnon, J., Daw, E.W., Leon, A.S., Rao, D.C., Skinner, J.S., & Bouchard, C. (1998). Alterations in resting metabolic rate as a consequence of 20-wk of endurance training: The HERITAGE Family Study. *American Journal of Clinical Nutrition,* **68,** 66-71.

39. World Health Organization. (2016). *Overweight/obesity, 2014: Prevalence of obesity, ages 20+, age standardized: both sexes.* Geneva: Author.

40. Yaspelkis, B.B. (2006). Resistance training improves insulin signaling and action in skeletal muscle. *Exercise and Sport Sciences Reviews,* **34,** 42-46.

41. Schulz, L.O., Bennett, P.H., Ravussin, E., Kidd, J.R., Kidd, K.K., Esparza, J., & Valencia, M.E. (2006). Effects of traditional and western environments on prevalence of type 2 diabetes in Pima Indians in Mexico and the U.S. *Diabetes Care,* **29,** 1866-1871.

Index

Note: The italicized *f* and *t* following page numbers refer to figures and tables, respectively.

U

V

W

Y

About the Authors

W. Larry Kenney, PhD, is the Marie Underhill Noll Chair in Human Performance and a professor of physiology and kinesiology at Pennsylvania State University at University Park. He received his PhD in physiology from Penn State in 1983. Working at Noll Laboratory, Kenney is researching the effects of aging and disease states such as hypertension on the control of blood flow to human skin and has been continuously funded by NIH since 1983. He also studies the effects of heat, cold, and dehydration on various aspects of health, exercise, and athletic performance as well as the biophysics of heat exchange between humans and the environment. He is the author of more than 200 papers, books, book chapters, and other publications.

Kenney was president of the American College of Sports Medicine from 2003 to 2004. He is a fellow of the American College of Sports Medicine and is active in the American Physiological Society.

For his service to the university and his field, Kenney was awarded Penn State University's Faculty Scholar Medal, the Evan G. and Helen G. Pattishall Distinguished Research Achievement Award, and the Pauline Schmitt Russell Distinguished Research Career Award. He was awarded the American College of Sports Medicine's New Investigator Award in 1987 and the Citation Award in 2008.

Kenney has been a member of the editorial and advisory boards for several journals, including *Medicine and Science in Sports and Exercise, Current Sports Medicine Reports* (inaugural board member), *Exercise and Sport Sciences Reviews, Journal of Applied Physiology, Human Performance, Fitness Management,* and *ACSM's Health & Fitness Journal* (inaugural board member). He is also an active grant reviewer for the National Institutes of Health and many other organizations. He and his wife, Patti, have three children, all of whom were Division 1 college athletes.

David L. Costill, PhD, is the Emeritus John and Janice Fisher Chair in Exercise Science at Ball State University in Muncie, Indiana. He established the Ball State University Human Performance Laboratory in 1966 and served as its director for more than 32 years.

Costill has written and coauthored more than 430 publications over the course of his career, including six books and articles in both peer-reviewed and lay publications. He was the original editor in chief of the *International Journal of Sports Medicine* for 12 years. Between 1971 and 1998, he averaged 25 U.S. and international lecture trips each year. He was president of the ACSM from 1976 to 1977, a member of its board of trustees for 12 years, and a recipient of ACSM Citation and Honor Awards. He has received numerous other honors, including an honorary doctoral degree from the Stockholm School of Physical Education, the Professional Achievement Award from Ohio State University, the President's Award at Ball State University, and the Distinguished Alumni Award from Cuyahoga Falls Public Schools. Many of his former students are now leaders in the fields of exercise physiology, medicine, and science.

Costill received his PhD in physical education and physiology from Ohio State University in 1965. He and his wife of 58 years, Judy, have two daughters. Now retired, Dr. Costill is a private pilot, auto and experimental airplane builder, competitive masters swimmer, and former marathon runner.

Jack H. Wilmore, PhD, retired in 2003 from Texas A&M University as a distinguished professor in the department of health and kinesiology. From 1985 to 1997, Wilmore was chair of the department of kinesiology and health education and the Margie Gurley Seay endowed centennial professor at the University of Texas at Austin. Before that, he served on the faculties at the University of Arizona, the University of California, and Ithaca College. Wilmore earned his PhD in physical education from the University of Oregon in 1966.

Wilmore published 53 chapters, more than 320 peer-reviewed research papers, and 15 books on exercise physiology. He was one of five principal investigators for the HERITAGE Family Study, a large multicenter clinical trial investigating the possible genetic basis for the variability in the responses of physiological measures and risk factors for cardiovascular disease and type 2 diabetes to endurance exercise training. Wilmore's research interests included determining the role of exercise in the prevention and control of both obesity and coronary heart disease, determining the mechanisms accounting for alterations in physiological function with training and detraining, and factors limiting the performance of elite athletes.

A former president of the American College of Sports Medicine, Wilmore received the American College of Sports Medicine's Honor Award in 2006. In addition to serving as chair for many ACSM organizational committees, Wilmore was on the United States Olympic Committee's Sports Medicine Council and chaired their Research Committee. He was a member of the American Physiological Society and a fellow and former president of the American Academy of Kinesiology and Physical Education. Wilmore consulted for several professional sport teams, the California Highway Patrol, the President's Council on Physical Fitness and Sports, NASA, and the U.S. Air Force. He also served on editorial boards of several journals.

Wilmore passed away during the preparation of the sixth edition of this text.